Postharvest Physiology and Hypobaric Storage of Fresh Produce

Postharvest Physiology and Hypobaric Storage of Fresh Produce

---

Stanley P. Burg

Consultant
Miami, Florida
USA

CABI Publishing

CABI Publishing is a division of CAB International

CABI Publishing
CAB International
Wallingford
Oxfordshire OX10 8DE
UK

Tel: +44 (0)1491 832111
Fax: +44 (0)1491 833508
E-mail: cabi@cabi.org
Web site: www.cabi-publishing.org

CABI Publishing
875 Massachusetts Avenue
7th Floor
Cambridge, MA 02139
USA

Tel: +1 617 395 4056
Fax: +1 617 354 6875
E-mail: cabi-nao@cabi.org

A catalogue record for this book is available from the British Library, London, UK.

Library of Congress Cataloging-in-Publication Data

Postharvest physiology and hypobaric storage of fresh produce / edited by Stanley P. Burg.
p. cm.
Includes bibliographical references.
ISBN 0-85199-801-1 (alk. paper)
1. Fruit--Postharvest physiology. 2. Vegetables--Postharvest physiology. 3. Fruit--Storage. 4. Vegetables--Storage. I. Burg, Stanley P., 1933- II. Title.
SB360.P668 2004
631.5′68--dc22

2003018164

ISBN 0 85199 801 1

Typeset by AMA DataSet Ltd, UK.
Printed and bound in the UK by Biddles Ltd, King's Lynn.

Contents

Preface

Postharvest Physiology and Hypobaric Storage of Fresh Produce is dedicated to preserving the existing theoretical knowledge, applied research and technology relating to hypobaric storage in order to promote an understanding and appreciation of the method. The author's views regarding the postharvest behaviour of commodities at atmospheric pressure were developed in an attempt to explain the documented advantages of LP vs. CA storage and do not always agree with currently accepted concepts in postharvest physiology. Scientific readers may respond to some of this background information with scepticism and disapprobation, but if at the end of the day researchers are motivated to test these concepts, *Postharvest Physiology and Hypobaric Storage of Fresh Produce* will have been well worth the time and effort expended in writing.

'Controlled atmosphere storage has been the subject of an enormous number of biochemical, physiological and technological studies, yet it is still not known precisely how it works' (Thompson, 1998). Lipton (1977b) commented 'It seems after 50 years of work on CA storage, we ought to at least try to see whether there is a common basis for our observations'. LP theory has been developed to a higher degree than the present empirical understanding of CA, but this knowledge is difficult for postharvest physiologists to evaluate unless they possess an intimate familiarity with thermodynamics, mass transport, refrigeration, vacuum technology and the physical laws applicable to an environment resembling that in which earth satellites orbit. It is not possible to plan an LP experiment and interpret its result without considering the manner in which mass and heat are transferred in the medium vacuum range and the influence this has on water loss and the gaseous gradients that arise within a commodity's intercellular spaces. The physical laws governing these processes will be reviewed to assist the reader in understanding the mechanisms that give rise to the unusual results attainable with LP described in the following Table. Postharvest physiologists can easily comprehend the biological factors, but may find the physical computations distracting. Therefore, relevant equations are presented in the Appendix, and computations are included as Examples at the end of each chapter, which the reader can ignore if he so chooses because the text summarizes each example's conclusions and interprets its significance.

The low-pressure storage method originally was called LPS or LP. Later, Tolle (1969, 1972) suggested the term 'hypobaric'; Rynearson proposed the trade name 'Dormavac™' (dormant in a vacuum) to describe Grumman's intermodal hypobaric container and the original 'wet' LP method; more recently the service mark 'VacuFresh℠' became synonymous with the 'dry' LP method, and 'TransVac' with a new 'square' LP container design. Frequently used abbreviations throughout the text include RH (relative humidity),

1

Introduction

1.1 Causes of Postharvest Losses

The vast distances and extended distribution times that separate major population centres from their sources of perishable commerce increase the likelihood of spoilage and make the costs associated with transportation, insurance and commodity losses important factors determining the grower's profit margin and the price of fresh foods. This situation has stimulated an ongoing effort to develop improved means of distributing, transporting and storing highly perishable foods.

Horticultural commodities have been genetically engineered or selectively bred to be deficient in ethylene sensitivity or production, and inhibitors of ethylene action, such as 1-methylcyclopropene (1-MCP), are used to delay ethylene responses (Theologis, 1992; Theologis *et al.*, 1993a; Giovannoni, 2001). But the expectation that these techniques will solve the problem of spoilage is unrealistic, since the major causes of postharvest fruit and vegetable losses are desiccation, decay and flavour changes, in that order. Amongst various fruits included in a 1978 US survey, non-climacteric strawberries and bell peppers experienced the largest postharvest losses, 22.9% and 10.6% respectively, mainly because of decay (Burton, 1982; Coursey, 1983; Kader, 1983), independent of ethylene action.

Climacteric fruits that have been rendered insensitive to ethylene or incapable of producing the gas cannot be easily introduced into the marketplace because prior to retail sale they must be ethylene-fumigated and brought to a complete stage of ripeness. Traditionally, the introduction of a step in the distribution cycle during which fruits are artificially ripened has been resisted[1] due to its cost, inconvenience and the difficulty of handling ripe fruit at the wholesale and retail levels. Climacteric fruits that have been 1-MCP treated cannot be ripened on a predictable schedule in advance of retail sale unless they first recover from their induced ethylene-insensitivity in a timely and reproducible manner.

If the hypobaric (LP) system's only advantage were its ability to prevent ethylene action, it would be as restricted as 1-MCP treatment and the engineering of ethylene non-producing or insensitive commodities in coping with the major causes of spoilage. Instead, LP provides a comprehensive solution, protecting commodities from both ethylene-induced and ethylene-independent deterioration, including desiccation, decay and insect infestation, after which ripening proceeds naturally.

1.2 Current Status of LP

During the 35 years that have elapsed since the author filed the first hypobaric (LP) patent application (Burg, 1967), substantial sums and considerable effort were invested by the United Fruit Company, Fruehauf Corp., Grumman Corp., Gelco Internacional s.a., Dormavac Corp., Welfit Oddy Pty. (South Africa), Armour & Co. and others attempting to develop and market the LP process for storing and transporting horticultural commodities and meat. For their part, the Grumman Corp. and Armour Research Center of Armour & Co. earned the 1979 IFT Food Technology Industrial Achievement Award for the most significant advance in the application of food technology to food production (Mermelstein, 1979). To account for the commercial failure of Grumman's hypobaric intermodal container ('Dormavac'), the scientific literature conjectures that LP has implicit technical deficiencies and insoluble economic problems (Lougheed *et al.*, 1977; Abeles *et al.*, 1992; Chapter 13), is wasteful of energy (Giovannoni, 2001), does not satisfactorily prevent fruit senescence (Theologis, 1992; Theologis *et al.*, 1993b), causes severe desiccation due to a reduced partial pressure of water vapour (Hughes *et al.*, 1981) and is unable to prevent ethylene action (Stenvers and Bruinsma, 1975; Purvis, 1981; Goldschmidt *et al.*, 1993). Contrary to this pessimism, the outlook for LP remains more promising now than at any prior time, due to advances in hypobaric technology and a better understanding of the effects that a low pressure has on stomatal opening, gas and vapour mass transport, heat exchange, disease control and insect mortality. There are no technical difficulties, VacuFresh^{SM} and standard refrigerated containers consume the same amount of energy, LP is remarkably effective in preventing senescence, the long life-expectancy of VacuFresh^{SM} containers makes their daily cost competitive with controlled atmosphere (CA) and a new patented square-container design should further improve the economics of hypobaric storage.

It has long been known that LP accelerates gaseous diffusion, but only recently was it demonstrated that LP opens stomates (in darkness) by removing CO_2 from within and around a stored commodity (4.15). The combined effects of enhanced diffusion and stomatal opening can improve gas exchange up to 20,000-fold at a pressure of 1.33 kPa (10 mm Hg), allowing commodities to be stored at remarkably low O_2 tensions without inducing fermentation or low-[O_2] damage. Thirty-five years ago, to avoid low-[O_2] damage, the earliest hypobaric storage trials were carried out at 25–42 kPa (150–250 mm Hg = 3.9–6.6% [O_2]). Later, the pressure was tentatively lowered to 5.33–10.66 kPa (40–80 mm Hg = 0.8–1.9% [O_2]). Eventually, it was found that better results occurred at 1.33–2.67 kPa (10–20 mm Hg = 0.15–0.3% [O_2]), and it now appears that still lower pressures are feasible and may be beneficial. There is no evidence that the low [O_2] limit causing physiological damage has yet been reached. The optimal O_2 tension for LP storage sometimes may be lower by 100-fold compared to the [O_2] concentration that causes fermentation and damage at atmospheric pressure.

Laboratory LP tests carried out more than 20 years ago at universities in various countries occasionally resulted in commodity desiccation. Sometimes this was caused by an improper equipment design that ventilated the storage area with dry air, resulting when humidification was erroneously carried out upstream instead of downstream of the pressure regulator (Hughes *et al.*, 1981). This caused the RH to decrease as a direct function of the expansion that occurs as the humidified atmospheric air enters the vacuum tank (9.03). In other studies, desiccation was caused by the introduction of dry air into the storage area due to equipment leakage (9.06). The lower the pressure, the drier the in-leaking air, and the greater its volume relative to the controlled leak of humidified air. When these mistakes are avoided, water loss is minimal at moderate LP pressures, but below 5.33–8.0 kPa (40–60 mm Hg) commodities sometimes become predisposed to desiccate because the transpirational conductance of air-filled lenticles, the pedicel-end stem scar and stomates decreases as the pressure is reduced (7.05). At a low enough pressure, this pathway for water escape is likely to provide less transpirational resistance than the parallel route through the liquid/solid cuticle, especially if the commodity possesses stomates that have opened. A better understanding of the factors controlling heat transfer and water loss in a hypobaric environment had to be developed before lower pressures could be successfully investigated. After that knowledge was acquired, it became possible without ancillary humidification to avoid water loss at extremely low pressures by simple, inexpensive and reliable means.

1.3 Comparison of LP and CA

Postharvest physiologists have long believed that the best atmosphere for preserving horticultural commodities contains supplementary CO_2 and a reduced tension of O_2, but the reality is that CA atmospheres have limitations that more than 70 years of research had been unable to circumvent or eliminate. Indirect effects on disease resistance attributable to delayed ripening and senescence are cited as evidence that CA prevents decay (Thompson, 1998), but a more discriminating reviewer concluded that 'no atmosphere in which commodities can be stored without injury gives any appreciable protection against fungal attack. The same can be said of rotting by bacteria . . .' (Burton, 1982). 'CA has been used to control insects' but 'the levels which are necessary to control insects within or on the fruit may be phytotoxic to the fruit itself' (Thompson, 1998). Unaware of the unique benefits conferred by heat and mass transport properties at a low pressure, stomatal opening in darkness, and the removal of CO_2, NH_3 and ethylene from around and within stored commodities, some postharvest physiologists have conjectured that LP is simply CA storage bereft of any benefit from added CO_2 (Kader and Morris, 1974; Stenvers and Bruinsma, 1975), and others have added CO_2 during LP storage (Spalding and Reeder, 1976b; Haard and Lee, 1982).

1.4 History of CA and LP Storage

For perspective in evaluating the pace of LP progress, it is instructive to compare the histories of LP and controlled atmosphere storage (Dalrymple, 1967, 1969; Dilley, 1990; Thompson, 1998). The 'discovery' of CA has been attributed to the French chemist Jacques Etienne Berard, who in an 1821 publication reported that the storage life of fruits was increased when they were kept in an atmosphere devoid of O_2. The first recorded attempt to use CA commercially in the USA was an 1865 apple storage performed in northern Ohio by Benjamin Nyce, which apparently provided the example he needed to obtain US patents that discouraged commercial CA development by others. Thirty-five years later, modified atmosphere (MA) fruit shipments were conducted in California, patents issued and again commercial development did not ensue, at least in part because of a high and unpredictable risk of spoilage due to the absence of any basic scientific understanding of the process.

A 1905 USDA publication reviewed early scientific results, and research on CO_2 was again summarized in 1915, but it was not until Franklin Kidd and Cyril West's classic studies in the interval between 1925 and 1927 (Kidd and West, 1925, 1927a,b, 1928) that the scientific basis for CA became sufficiently understood to permit commercial development to proceed in England, the USA and Canada. Laboratories in these countries and elsewhere began compiling results describing the best gas composition and temperature for storing various plant commodities.

The technology needed to produce efficiently and reliably a controlled atmosphere developed more slowly because it required a greater degree of innovation and a larger expenditure of money than that needed for applied descriptive commodity research. Gas composition initially was maintained by ventilation, then caustic soda or other means were used to scrub CO_2, and later ethylene scrubbers were introduced. In the early 1960s, a catalytic O_2 burner, an open-flame burner and the Tectrol process were developed to reduce rapidly the O_2 and enrich the CO_2 content of air by burning propane or natural gas. The Oxy-drain N_2 generator was developed in the 1980s to convert ammonia to N_2 and water vapour, and the use of liquid N_2 supplied from bulk tanks became popular in Nitrol, Oxytrol, TransFresh, Techtrol, Freship and other CA trailers and intermodal containers.

After 1986, the atmosphere in modern transportation vehicles and warehouses began to be regulated by microprocessors that, in response to signals from O_2 and CO_2 sensors, controlled pressure-swing adsorption or hollow-fibre membrane air separators capable of splitting air into N_2 and O_2 on demand (Thompson, 1998). Numerous companies became involved in the commercial application of CA technology in intermodal containers, including CA (Container) Systems, UK; Fresh Box Container, Germany; Finsam International, Norway; Industrial Research, The Netherlands; AgriTech Corporation, USA; Synergen, UK; TransFresh Corporation, USA; TransFresh Pacific, New Zealand; TEM, USA; and Franz Welz, Austria.

More than 100 years of research were required to establish a meagre scientific basis for CA, define the optimal conditions for each commodity, and develop efficient technology, and yet during the past 40 years innumerable attempts to develop a commercially successful CA intermodal container failed. Confidence in CA nevertheless has remained high and CA research continues worldwide.

LP has only existed for approximately 40 years. Commercial development initially was hindered by a lack of basic information and has been slowed by patents. Hypobaric theory was not understood until recently, and LP technology is difficult, more costly and time consuming than CA technology to develop. Inexplicably, because the first attempt to develop a commercial hypobaric intermodal container was unsuccessful, the scientific community seems to have concluded that LP failed to live up to expectations and did not demand further research (Lougheed *et al.*, 1977).

1.5 LP Research and Development

The research and development (R&D) costs associated with the evolution of an LP intermodal container are far greater than for a CA container. Both systems are comprised of a module to control gas composition or pressure, an intermodal container and a refrigerator, but the CA control module is installed in a 'ready-made' refrigerated intermodal container. Therefore, only the CA control module needs to be specially engineered and tested. In contrast, all components of the LP system are unique and need to be independently designed, fabricated and tested. The LP control module and a refrigerator capable of controlling temperature in an evacuated space must be fitted into an ISO-certified intermodal vacuum container. This greatly increases the R&D costs for LP and extends the development time.

1.6 Is LP More Complicated than CA?

A postharvest physiologist complained to the author that 'LP is too complicated to possibly work', whereas paradoxically those researchers who have actually used the method unanimously agree that operationally and mechanically it is far simpler than CA. LP and CA intermodal containers use a compressor/vacuum pump of similar size, but they differ in almost all other respects. N_2-rich gas for use in CA containers is generated by removing some of the O_2 from compressed air using membrane gas separation or pressure swing adsorption (PSA) with molecular sieves, or by an initial N_2 fill of the container using bottled gas, followed by chemical CO_2 absorption, with the respiration of the produce, leakage and time deciding the $[O_2]$ level (Garrett, 1996).

The following description of a PSA intermodal container CA system is condensed and edited from Thompson (1998). Compressed air is cooled and moisture removed and ducted into a reservoir before O_2 is separated from CO_2 and N_2 with a molecular sieve. N_2 and CO_2 are returned to container, but if less CO_2 is required the flow of gas is briefly diverted outside the container, and if more is needed, bottled CO_2 is vented into the container. The O_2 is ducted outside the container, and when additional O_2 is required, a control valve allows ambient air to enter for a short period. Once this air has been mixed with container air, the O_2 level is measured and the valve operated again if required. Air is routed from activated alumina drier beds either directly to the zeolite-containing N_2 and CO_2 beds, or to an ethylene bed that remains pressurized for several hours and then depressurizes, after which the O_2 flow is routed through the ethylene bed for 20 minutes to scrub it, and the process is repeated. The N_2 (and CO_2) retained in the main sieve beds are returned to the alumina drier beds in order to recharge them and remove moisture from them. Air from an instrument air buffer is used to draw reservoir water into an atomizing injection system located upstream of the evaporator discharge in order to replace water vapour condensed or frozen on the coil or removed by the N_2 generator and drier beds. The RH control valve is operated for a short period, and once the additional water spray has been mixed with the container air, the humidity level is measured and the valve operated again if required. [O_2], [CO_2] and RH are continuously monitored and controlled in response to measurements made in a supply of gas taken from the container by a small pump.

The degree of control over the composition of the CA atmosphere is affected by the gas-tightness of the container. To obtain maximum benefit the container leak rate should be no higher than 3 m^3 per hour at 1.3 cm (0.5 inch) water gauge (wg), but leakage occurs around the reefer insert, through the container 'skin' and at the door. Many containers exceed this figure, especially when they have been racked, and others never achieve it when new, even though leakage through the rear doors has been effectively controlled by the use of curtains and gaskets (Garrett, 1996). Air invariably enters or exits during transit when the container is subjected to altitude changes, and leakage limits the lower [O_2] limit that can be controlled to approximately 4% (Table 14.1). Nearly 1 day may be required to establish an operating condition in a CA container unless it initially is flushed with bottled N_2, and subsequently the container must be ventilated before it is safe to enter.

An LP intermodal container is structurally complex, but mechanically simple. The LP environment is established within a few hours; the container can be vented in minutes, and it is safe to enter immediately thereafter. The pressure and humidity are controlled by a mechanical pressure regulator and a flow controller when horticultural commodities are shipped in VacuFresh^{SM}, but if meat is transported, the pressure and flow controllers are not used and the vacuum pump directly evacuates the storage space. The container's leak rate, 0.014 m^3 per hour at one atmosphere pressure differential [390 inch water column (wc)], is so low that the [O_2] can be reduced to 0.000068%. A thermostat and vacuum meter are the only sensors, and the latter displays and records but does not control the pressure. Water and bottled gas are not used, there are no water dryers or beds to 'scrub' CO_2 and ethylene, the refrigeration system does not condense water and it lacks a defrost cycle and evaporator fans. Automatically, the one-pass flow-through system purifies the incoming air by expansion and changes it continuously without imposing a heat load, circulates the air pneumatically without warming it or consuming additional power, and controls the [O_2] level ± 0.014% at any value between 0 and 20%, while removing all ethylene and CO_2 both from the container air and within the commodity. The jacketed refrigeration system and flow controller keep the humidity steady at or above 95% and the wall temperature between −17.1°C and 15.6°C with ± 0.2°C end-to-end variation in the container at ambient temperatures ranging from −17.1°C to 49°C.

1.7 LP Patent Protection and Trade Secrets

The first hypobaric LP patent claimed 'the method of preserving unripe fruit in a continuous stream of water-saturated air at a controlled temperature and a regulated pressure in the range 100–400 mm Hg'. Two examples of a reduction to practice were given; bananas stored at 150 mm Hg and mature-green tomatoes at 130 mm Hg. This was the sum total of LP knowledge extant when a patent application was filed in 1963. The information became public knowledge when it was published in a 1966 scientific journal[2] (Burg and Burg, 1966b) and also when the patent was issued in 1967 (Burg, 1967).[3] The 100–400 mm Hg (13.33–53.33 kPa) pressure range supplied an O_2 partial pressure higher than that which causes low-O_2 damage at atmospheric pressure, but after the patent application was filed the inventor soon learned that pressures of 10–80 mm Hg (1.33–10.66 kPa) gave far better storage results with most plant commodities. This improvement could not be patented because, while it was *new* and *useful*, it did not satisfy the third and most important patent requirement, that it must be *non-obvious*.

Anyone skilled in the art of plant physiology should have realized that because O_2 diffuses into plant tissues at a rate inversely proportional to pressure, plant commodities could be expected to tolerate a lower applied-O_2 partial pressure when the total pressure is reduced. LP patent coverage was effectively negated by this improvement, because an unauthorized use of the LP method at a pressure lower than 100 mm Hg (13.33 kPa) would not have infringed the original patent claims. Experimental results at very low pressures had to be preserved as trade secrets,[4] which could only be revealed to individuals who agreed to maintain the information in confidence. These results could not be made available to the scientific community at large because it would not have been possible to obtain the funding required to develop LP without the economic incentive created by patent and trade-secret exclusivity and the 'due diligence' evidenced by confidential disclosure agreements and other legal precautions and devices.

A patent is, *inter alia*, a 17-year monopoly granted to an inventor by the federal government in exchange for a complete and thorough description of his or her invention. The specification

> must include a written description of the invention and of the manner and process of making and using it, and is required to be in such full, clear, concise, and exact terms as to enable any person skilled in the technological area to which the invention pertains, or with which it is most nearly concerned, to make and use the same.

The complete disclosure of information in a patent is intended to benefit society by serving as an aid for further advancement in the art by others, but in this respect the results with LP have been disappointing. After the initial patent was filed, the legal devices used to protect the technology and improvements prevented the dissemination of new information, which presumably would have dissuaded the majority of independent researchers from continuing to investigate pressures too high to provide substantial benefit, under conditions that often could not control water loss. Amongst more than 100 independent applied LP research papers published during the decade after the storage method was first disclosed, two-thirds described less than optimal results using the relatively ineffectual 100–400 mm Hg (13.33–53.33 kPa) pressure range claimed in the initial patent and publication. This misguided independent research often was uncomplimentary to LP, and these studies provided little useful information about the scientific basis for LP storage. Because small-scale laboratory experiments of this type present vastly different technical problems than those encountered in commercial LP transportation vehicles and warehouses, they also failed to develop new hypobaric technology.

Ostensibly, this situation should have been rectified by the issuance of a 1976 continuation-in-part (CIP) patent (Burg, 1976a), which extended coverage to a lower pressure range by the legal device of appending the primary 'non-obvious' CIP claims[5] to the condition 'maintaining the pressure of said air between 4 mm Hg and 100 mm Hg absolute'. Although more than 50 examples of improved storage in this lower pressure range were included, nevertheless the higher pressure range continued to be studied by the majority of independent investigators. Later, analytic studies were published describing the mechanisms of LP heat transfer, gas exchange and water loss (Burg and Kosson, 1982, 1983; Burg, 1990), disclosing that animal matter is best preserved at 0.6 kPa (4.4 mm Hg), while many types of plant matter prefer 1.33–2.66 kPa (10–20 mm Hg). Only a few subsequent independent studies examined the storage of horticultural commodities at 1.33–2.66 kPa, and none tested meat storage or the validity of the LP heat and mass transport concepts. New LP patents based on these theories were issued in 1987 (Burg, 1987a,b), describing a method and apparatus for storing plant and animal matter *without supplementary humidification*, which prevents water loss even when plant commodities are flushed with < 1% RH rarified air changes.[6] There has been no independent investigation or verification of the unusual concepts set forth in these patents. The 1987 'dry' patents made it possible for the first time to engineer a reliable, energy-efficient, inexpensive hypobaric control module.

A Cleveland patent attorney (now deceased), who assisted with writing and processing many of the hypobaric patents, was fond of reminding the author that it is possible to discern when an invention is about to succeed because at that moment the inventor will feel that the situation is hopeless. During the past decade, apparently because LP was on the brink of success, a '*Bleak House*'-style legal controversy[7] between a licensee, the inventor and owner of the LP patents (the author) and a prospective manufacturer, brought commercial development of hypobaric storage to a standstill, profiting lawyers but no one else. Even the IRS became involved, and rightly so.[8] After several years of mediation proved unsuccessful, the patent licence was terminated and immediately all existing hypobaric patents were abandoned or dedicated to the public to make the license cancellation irrevocable and incontestable, but a controversy regarding the right to use 'trade secrets' followed and prevented commercial production and marketing of a newly designed LP intermodal hypobaric container. It is only recently that the legal controversy has been resolved by a lawsuit and out-of-court settlement, allowing the commercial development of LP again to proceed.

Notes

1. Imported bananas are an exception. They are shipped green and ripened with exogenous ethylene in rooms equipped with special refrigeration and air-distribution systems capable of handling the high climacteric respiratory heat load at the preferred ripening temperature.
2. The right to have an invention patented is lost if a description is published or otherwise disclosed publicly anywhere in the world before a patent application is filed.
3. The US patent office keeps all information contained in patent applications in the strictest confidence until the patent is issued and becomes public knowledge.
4. Trade secret law is common law that has evolved to protect information that cannot be patented, or which for business reasons should not be patented. A trade secret is anything useful or advantageous in business activity but not generally known or ascertainable by others. These secrets have no time limit; they do not require some degree or quantity of innovation or novelty. Even a very minor improvement can be a trade secret so long as it is not generally known and attempts have been made to retain it as a secret.
5. The main claims of the CIP were (i) a method and apparatus for humidifying the incoming expanded dry air by contacting it with *heated* water to overcome the effect of evaporative cooling and (ii) correlated and interdependent conditions of pressure, temperature, relative humidity, air exchange and air circulation for each type of stored commodity.
6. The air change rate is adjusted so that the chamber air is saturated by somewhat less

moisture than that evaporated by a full load of the particular stored commodity when it transfers all of its respiratory heat by evaporative cooling. The container walls are cooled by a jacketed refrigeration system that eliminates cold spots, and uncontrolled leakage of expanded dry air is so small relative to the container's large volume that it does not significantly lower the dew point. Usually the commodity is protected by a water-retentive plastic liner and sometimes by radiation reflecting boxes and wraps.

7. Referring to the Charles Dickens' novel *Bleak House*, in which is depicted a legal controversy that continues for generations, ending only when the lawyers and courts have exhausted the entire estate in dispute.

8. The licensee claimed that sums paid to develop hypobaric storage were a 'business expense', but the IRS concluded that these should have been treated for tax purposes as an investment. They could not be expensed and taken as a tax deduction because they were not directly related to the licensee's business. Either the licensee owed a tax and penalties related to an improper 'deduction', or the licensor (the author) might have received these expenditures as payment for the purchase of his patents, and he in turn used the money to procure a share in a 'partnership' with the licensee. The latter scenario envisioned that the development funds were spent by the partnership, and the licensor owes a capital gains tax larger than his net worth. To reduce his potential tax indebtedness, the licensee argued that 'shop rights' and a presumed obligation entitled him to own the LP patents because they had been developed while the author was engaged as his consultant. Although the licensor had received no money, taken no tax deductions, developed and disclosed one LP patent (Burg, 1987a) years before he first met the licensee, and formulated the other patent (Burg, 1987b) independent of any consulting activity with the licensee, this tax case dragged on for more than 5 years. The licensor was only exonerated by the fortuitous discovery of a 'smoking gun' in an old filing cabinet. Handwritten notes composed by the licensee's partner and signed by the relevant parties described the exact business relationship. The lesson to be garnered from these events is that to avoid a capital gains tax, a patent holder must not take an equity position in a partnership or a corporation in exchange for a patent assignment.

2

Origins of the LP Concept

The idea that plant commodities might be preserved in a continuously refreshed partial vacuum originated from a study on gas exchange in fruits (Burg and Burg, 1965b), in which it was demonstrated that a fruit's internal concentrations of CO_2 (ICC) and ethylene (IEC) vary directly as a function of atmospheric pressure. The laboratory system used in these experiments was later adapted as an assay to determine which developmental processes in plants might be controlled by ethylene, and eventually evolved into the environmental control apparatus disclosed in the first LP storage patent (Burg, 1967, 1975).

2.1 The LP Apparatus

The 'wet' LP method (Fig. 2.1) requires a *vacuum tank* storing the commodity, a

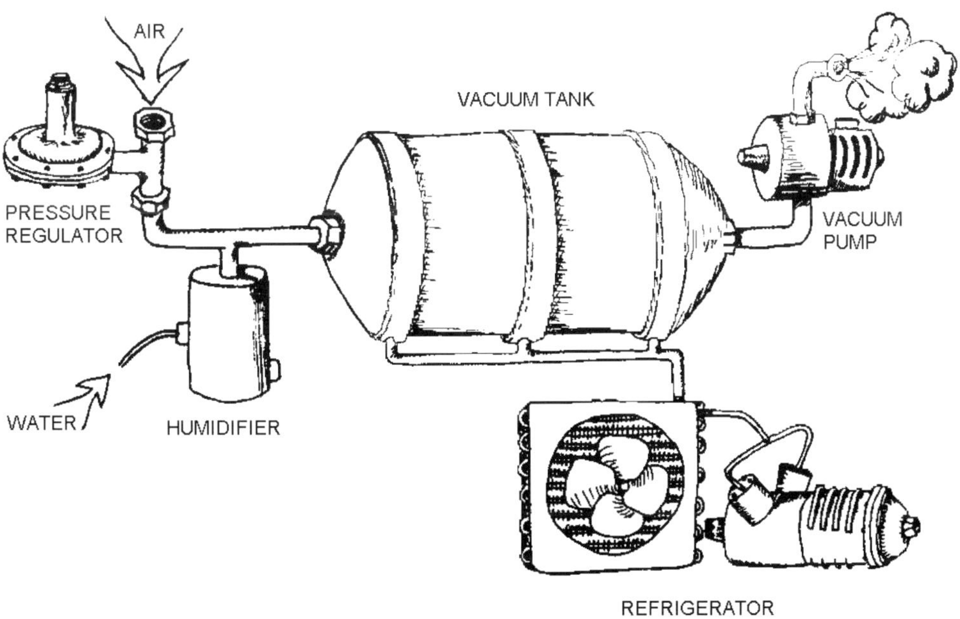

Fig. 2.1. Schematic diagram of hypobaric system.

vacuum pump continuously withdrawing air from the tank, a *pressure regulator* leaking air into the tank at a rate sufficient to maintain the set pressure, a *humidifier* injecting enough moisture into the incoming low-pressure air to saturate the atmosphere in the vacuum tank, and a *refrigerator* controlling the cargo temperature. The 'dry' hypobaric method uses the same equipment, except that it replaces the humidifier with a device which controls the pumping speed.

2.2 Evidence that Gas Exchange in Fruits is Limited by an Air Phase

As rapidly as ethylene and CO_2 are produced by a commodity, they diffuse outwardly into the atmosphere where their concentration is lower, and as quickly as respiratory O_2 is consumed, it diffuses inwardly from the atmosphere where its concentration is higher. The large gradients of these gases that develop across a fruit's surface (Tables 2.1, 2.2, 4.5, 4.6, 5.10 and 5.11; Figs 2.2 and 2.3) prove that the 'skin' is the major barrier to gas exchange. When an apple is peeled, its internal CO_2 (ICC) and ethylene (IEC) concentrations decrease by 80%, even though the ethylene production rate remains unchanged and CO_2 production slightly increases (Burg and Burg, 1965b). Removing a small patch of skin has a similar, but smaller, effect.

In the 1960s, the extent to which the pedicel-end scar, lenticles, stomates and liquid/solid phases of a fruit's peel participated in the gas exchange process was unknown and it was uncertain whether diffusion through the pulp was hindered by occlusion of the intercellular spaces with liquid in the form of Jamin's beads, as had been claimed (Haberlandt, 1927). If the pulp's intercellular spaces were occluded with water, this would markedly retard diffusion and gas exchange because gas diffusion coefficients in water are 10^4-fold lower than in air (Tables 15.4 and 15.5). The influence of atmospheric pressure on a fruit's gas exchange was studied to determine whether the process is regulated by an air phase of restricted surface area or a liquid/solid phase. If gas exchange was limited by the resistance of an air phase, at equilibrium the P/Q ratio [where P = internal gas partial pressure and Q = gas production rate] should decrease in direct proportion to the absolute pressure because lowering the pressure reduces the density of the air and hence the concentration of molecules available to cause collisions that slow diffusion. A low pressure cannot enhance gas diffusion or change the P/Q ratio if the major barrier to gas exchange is a liquid or solid. These phases are incompressible throughout the 0 to –1 atmospheric-pressure range, and therefore the number of molecules available to cause collisions and slow diffusion does not decrease in liquids and solids when the pressure is lowered. The effect of pressure on the P/Q ratio in apples was tested using an apparatus that allowed gas exchange and production to be studied simultaneously, while the pressure and composition of the gas mixture surrounding the fruit was varied (Burg and Burg, 1965b). The O_2 partial pressure was kept high enough to avoid inhibiting ethylene or CO_2 production (Burg and Thimann, 1959), and before the measurements were taken, the tissue was kept at the low pressure for 2 h to ensure the establishment of a new equilibrium. These experiments (Fig. 2.2), and others using a less invasive method (Fig. 2.3), demonstrated that the P/Q ratio for both ethylene and CO_2 is inversely related to pressure. The conclusion that atmospheric pressure influences the internal gas concentration through an effect on gas diffusion was confirmed by

Table 2.1. Ethylene concentration (IEC) at various depths in McIntosh apples (Burg and Burg, 1965b).

	Ethylene concentration (μl/l)		
Experiment	Under peel	In centre of pulp	In ovary
1	364	365	375
2	500	507	513
3	373	369	400
Average	412	414	429

studies in which the binary gas-diffusion coefficient was increased by substituting a light gas, helium, for the atmospheric N_2 in air (Burg and Burg, 1965b). At 22°C, the binary diffusion coefficient of an [80%] helium + [20%] O_2 mixture, calculated

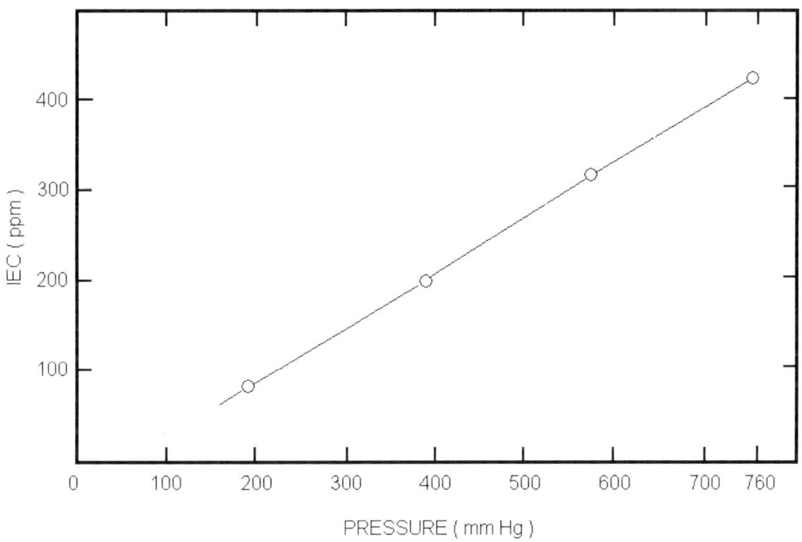

Fig. 2.2. Internal ethylene concentration (IEC) within a McIntosh apple at different hypobaric pressures. The rate of ethylene production at each hypobaric pressure was the same once equilibrium had been established (Burg and Thimann, 1959).

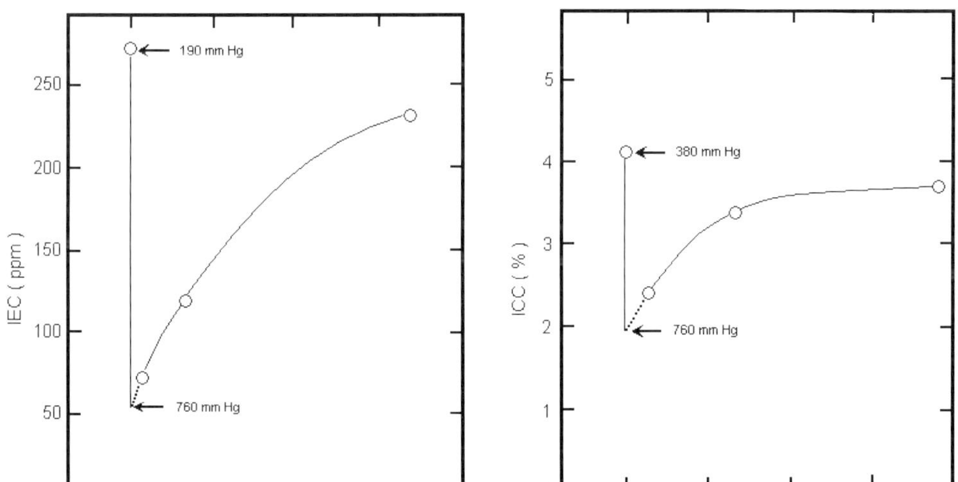

Fig. 2.3. Effect of atmospheric pressure on the ethylene (*left*) and CO_2 (*right*) content of McIntosh apples. The initial reading was taken at 760 mm Hg and then the pressure was reduced to the indicated value for 120 min. The rate of ethylene or CO_2 production was not altered by the treatment. An arrow indicates the time at which the vacuum was broken, and the subsequent curve shows the recovery of internal ethylene or CO_2 at atmospheric pressure. The intercellular partial pressure of CO_2 or ethylene at the initial reduced pressure was calculated by extrapolating the curve to the moment when the vacuum was released (Burg and Burg, 1965b).

Table 2.2. Analyses of the internal atmosphere in various fruits.

Commodity		Temp (°C)	Position under skin (mm)	% [CO_2]	% [O_2]	Reference
Apple						
Lord Derby		12	7	1.2	19.3	1
(32 mm radius)		12	13	1.4	18.7	1
		12	25	1.8	18.3	1
Boskoop	ripe	20	3		18.0	1
(40 mm radius)	ripe	20	6		15.5	1
	ripe	20	18		12.0	1
	ripe	20	24		6.9	1
	ripe	20	30		6.1	1
Summer Pippin		12	Periphery	1.9–3.5	15.5–19.2	2
Cox Orange Pippin		12	Periphery	3.9–5.9	13.6–16.1	2
King Edward VII		12	Periphery	2.5–5.1	13.3–17.9	2
Granny Smith		21.1	Centre	3.2–4.5	11.6–16.7	3
Avocado						
Hass	unripe	20	Centre	1.0–3.0	15.0–19.0	4
	climacteric	20	Centre	5.0–10.0	5.0–10.0	4
	senescent	20	Centre	15.0	0.5–2.0	4
Choquette	ripe	20	Under skin	11.8		5
	ripe	20	Near seed	14.9		5
Banana						
Gros Michel	green	29.4	Under skin	2.0	17.0	6
	green	29.4	Centre	2.2–2.5	15.5	6
	ripe	29.4	Under skin	12.5	15.5	6
	ripe	29.4	Centre	9.0–14.5	3.5–11.0	6
	ripe	20	2	3.5		1
	ripe	20	5	1.1		1
	ripe	20	8	0.7		1
	ripe	20	18	0.2		1
	over-ripe	29.4	Centre	10.2–18.2	1.2–4.2	1
Valery	green	20	Centre	7.5	11.2	6
	ripe	20	Centre	6.0	12.5	6
Cantaloupe	unripe	20	Centre	4.6	17.3	7
	ripe	20	Centre	10.4	14.1	7
	over-ripe	20	Centre	9.8	4.2	7
Carrot		11	Centre	12.1	5.2	8
		24	Centre	28.6	0.9	8
Cucumber			Centre	2.0		9
Orange, Valencia		21	Centre	2.0	18.8	10
Papaya	green	26.7	Centre	1.5–3.9	17.0–17.5	11,12
	ripe	26.7	Centre	7.1	14.2	11,12
	over-ripe	26.7	Centre	13.5	3.5	11,12
Pear, Conference	green	12	Periphery	1.8	18.9	13
	ripe	12	Periphery	5.4	15.4	13
Bartlett		–1	Centre	6.0	19.0	14
		16	Centre	24.0–33.0	8.0–10.0	14
Tomato						
OH7814; VC24320	green	20	Centre	3.0–4.4	16.7–17.0	15,16
	ripe	20	Centre	6.7	14.7	15,16
Castlemart	green	20	Centre	5.0		17
	ripe	20	Centre	12.0		17
Sonato	green	20	Centre	3.0		18
	ripe	20	Centre	10.0–12.5		18

using equations 15.30–15.34, is 2.68-fold higher than the coefficient in air at the same temperature, and apples lost 45% of their internal ethylene when they were placed in this mixture. In 94% [helium] + 6% [O_2], ethylene diffusion is accelerated 3.72-fold and an apple's IEC decreased by 63%, even though the ethylene production rate was unchanged. The proposed inverse relationship between internal gas concentration and the rate of gaseous diffusion also is consistent with data comparing the internal concentrations of CO_2 and ethylene which accumulate in different types of fruits (Table 2.2). If a liquid or solid phase were limiting gas exchange, the rate would depend on $M^{-0.6}$ (Burg, 1990) or $M^{-0.5}$ (equation 15.36), where M is the molecular weight of the gas, and it also would be proportional to each gas's Henry's Law partition coefficient between air and the liquid (Table 15.1) or solid phase. If the liquid phase was water at pH 5.0 (Tables 3.1 and 3.2), the ICC/IEC ratio for equal rates of production would be 9.0, whereas the value based on the binary diffusion coefficients of CO_2 and ethylene in an air phase (Table 15.4) is 0.925. The extent to which the ICC/IEC ratio is less than 0.925 for equivalent production rates (Table 2.3) is an indication of the amount of gas exchange which is occurring in parallel through the cuticle where the resistance to gas mass transfer is much higher than it is through the air-phase pathway, and much less for CO_2 than ethylene (3.20). Additional evidence that gas exchange in fruits is limited by an air phase was provided by experiments showing that the air-filled lenticles of apples have a dimension and frequency consistent with gas transport in an air phase through the thickness of the apple's peel in accord with Fick's diffusion law (Clements, 1935; Burg and Burg, 1965b).

Table 2.3. Relationship between the μl/l internal CO_2 (ICC) and μl/l ethylene (IEC) for gas production rates of 1 μl/kg·h (Burg and Burg, 1965b).

Variety	ICC (μl/l)	IEC (μl/l)	Ratio ICC/IEC
Apple (McIntosh)	3.5	4.7	0.74
Avocado (Choquette)	1.4	1.8	0.78
Mango (Haden)	1.2	1.7	0.71
Passion fruit	6.6	9.0	0.73
Orange (Valencia)	3.7	4.0	0.93
Tomato (Homestead No. 24)	2.5	3.8	0.66
Average			0.76 ± 0.09

2.3 Testing for Ethylene-controlled Developmental Processes

Lowering the IEC by hypobaric treatment may be expected to prevent, delay the onset or slow the advance of any plant developmental process that is initiated or modulated by endogenously produced ethylene. To ascertain whether ethylene is involved in a process, ideally the pressure in a laboratory apparatus (Fig. 2.4) is set at 23.33 kPa (175 mm Hg) and instead of air, pure [O_2] is continuously flowed into the humidified test chamber to provide a normal, atmospheric equivalent of [O_2]. Sometimes 99.8% [O_2] + 0.2% [CO_2] is flowed to provide the equivalent of 0.04% [CO_2] at atmospheric pressure, but this fails to consider that a commodity's ICC is lowered by approximately 87% (Table 3.15) due to enhanced diffusion at 23.33 kPa (175 mm Hg). To restore the normal ICC of a typical harvested fruit, a mixture containing approximately 85% [O_2] + 15% [CO_2] should be flowed. For a typical leafy commodity, a mixture containing approximately 96% [O_2] + 4% [CO_2] would be appropriate.

Table 2.2. *Continued opposite.*

References: (1) Burton, 1982; (2) Kidd and West, 1949b; (3) Trout *et al.*, 1942; (4) Ben-Yehoshua *et al.*, 1963; (5) Burg and Burg, 1965b; (6) Wardlaw and Leonard, 1940; (7) Lyons *et al.*, 1962; (8) Magness, 1920; (9) Wardlaw and Leonard, 1936b; (10) Ben-Yehoshua *et al.*, 1985; (11) Wardlaw and Pratt, 1936a; (12) Jones and Kubpta, 1940; (13) Kidd and West, 1949a; (14) Magness and Ballard, 1926; (15) Lyons, 1964; (16) Knee, 1995; (17) Saltveit, 1993; (18) Sawamura *et al.*, 1978.

Often CO_2 is not added and the pressure is set as low as 16 kPa (120 mm Hg), or in some instances high [CO_2] is provided to inhibit ethylene action (4.12). At a pressure of 23.33 kPa (175 mm Hg), the IEC will be decreased by approximately 77% if the ethylene production rate remains unchanged (Table 3.15). Ethylene effects that have been revealed by this assay include geotropic sensitivity of dark-grown seedlings (Kang and Burg, 1972a), phytochrome-induced anthocyanin synthesis (Kang and Burg, 1973), plumular hook formation and opening (Apelbaum and Burg, 1972; Kang and Burg, 1972b; Fig. 2.5), phytochrome-mediated carotenoid synthesis (Kang and Burg, 1972c), auxin-induced growth inhibition (Apelbaum and Burg, 1971b; Fig. 2.6) and leaf epinasty (Kang and Burg, 1974a; Burg and Kang, 1993), cell division and plumular expansion (Apelbaum and Burg, 1972), flower fading (Mayak and Dilley, 1976; Table 2.4), abscission (Cooper and Horanic, 1973; Table 2.5; Lipe and Morgan, 1972a,b, 1973), induction of peroxidase (Imaseki, 1970), auto-inhibition of ethylene production (Cooper and Horanic, 1973; Saltveit and Dilley, 1978b), leaf senescence and de-greening (Hodges and Coleman, 1984), sex expression (Byers *et al.*, 1972), banana fruit ripening (Burg and Burg, 1966c), seed germination (Ruddnicki *et al.*, 1978) and arginine decarboxylase activity (Apelbaum *et al.*, 1985). Another effect, possibly due to ethylene removal, is the growth of turnip stimulated by LP at an O_2 partial pressure of 21.33 kPa (160 mm Hg) (Mansell *et al.*, 1968). When ethylene is

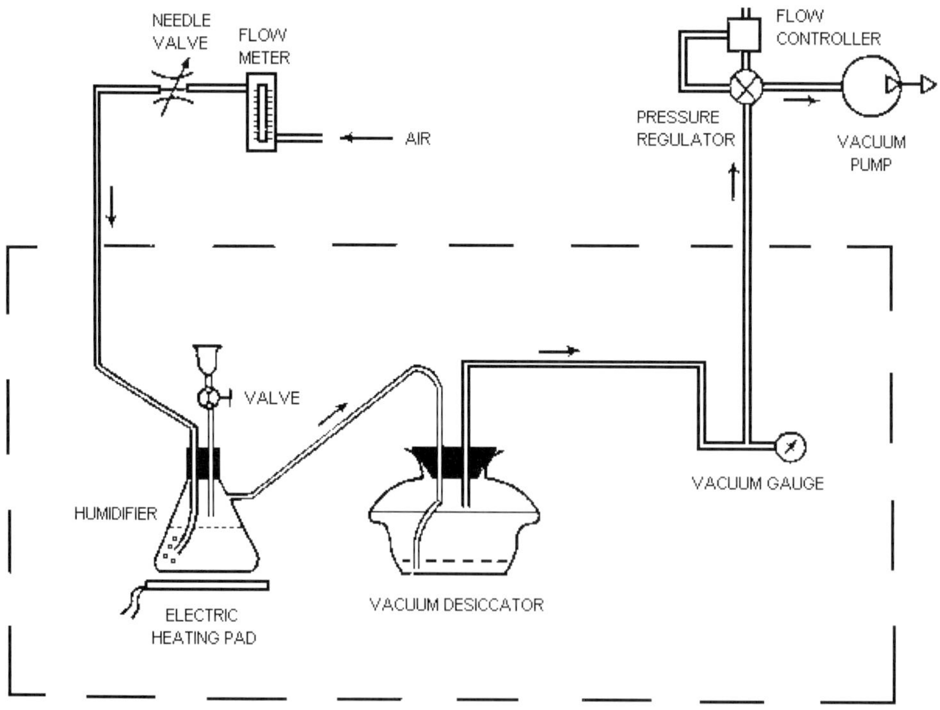

Fig. 2.4. Laboratory hypobaric apparatus. The commodity is stored in the vacuum desiccator. Distilled water is added periodically to the humidifier through the thistle tube to replace evaporated water. A heating pad compensates for evaporative cooling and elevates the water temperature sufficiently to keep the humidity in the vacuum desiccator close to saturation. To test for ethylene effects, the pressure is set at between 16 and 23.33 kPa (120–175 mm Hg), and pure O_2 or a 99.8% [O_2] + 0.2% [CO_2] mixture is flowed through the apparatus (Jamieson, 1979b, 1980a,c).

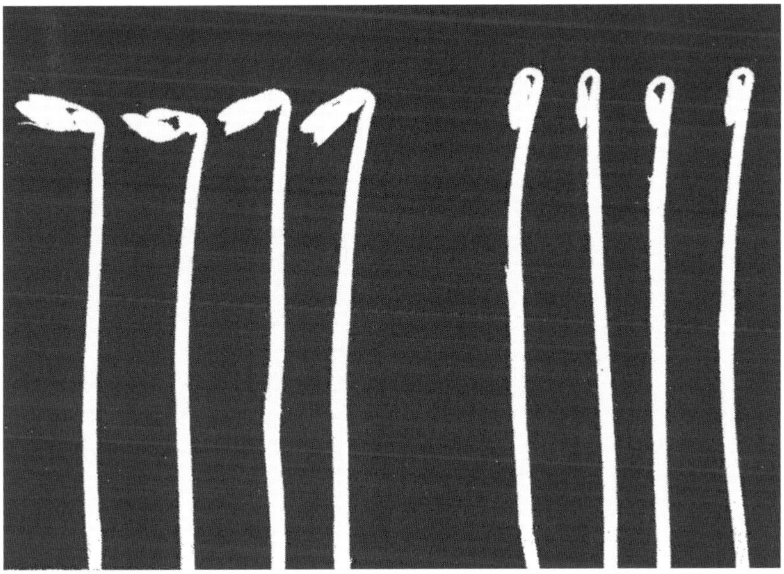

Fig. 2.5. Control of leaf expansion and hook opening by endogenous ethylene in plants grown under hypobaric conditions (*left*) compared to control plants (*right*). Three-day-old etiolated seedlings that had developed in air were grown for four additional days either at a pressure of 16 kPa (120 mm Hg) flowing water-saturated 100% [O_2], or under normal atmospheric conditions (Apelbaum and Burg, 1972).

included along with the hypobaric treatment, the effect of a reduced pressure is negated (Cooper and Horanic, 1973; Mayak and Dilley, 1976; Table 2.5). The LP method has also been used to prove that endogenous ethylene is not involved in processes such as apical dominance (Apelbaum and Burg, 1972), phototropic auxin transport (Kang and Burg, 1974b), stomatal movements, increases in leaf ABA and depletion of GA in turgid or stressed leaves (Aharoni, 1978).

2.4 Involvement of Other Gases and Volatile Substances

The hypobaric assay is not specific for the involvement of ethylene in developmental processes because the method also decreases the internal concentrations of CO_2, NH_3, ethanol and other metabolic gases and vapours. Lowering the ICC causes stomatal opening, even in darkness, and CO_2 and ammonia removal can result in pH shifts and metabolic alterations.

2.5 Converting the LP Test Apparatus into a Storage System

All that is required to convert the LP test apparatus into a storage system is to flow air through the chamber instead of 100% [O_2] or an [O_2] + [CO_2] mixture. The pressure usually is set in the range 1.33–10.67 kPa (10–80 mm Hg), as that is optimal for the storage of most plant commodities.

Table 2.4. Effect of hypobaric ventilation on carnation senescence.

Treatment	Flower longevity (days)
1 atm	5.0
0.2 atm	9.3
0.2 atm + C_2H_4	1.3

A 99.85% [O_2] + 0.15% [CO_2] ± 2.36 μl/l ethylene mixture was flowed at a pressure of 0.2 atm (see a critique of this methodology in 2.3) (Mayak and Dilley, 1976).

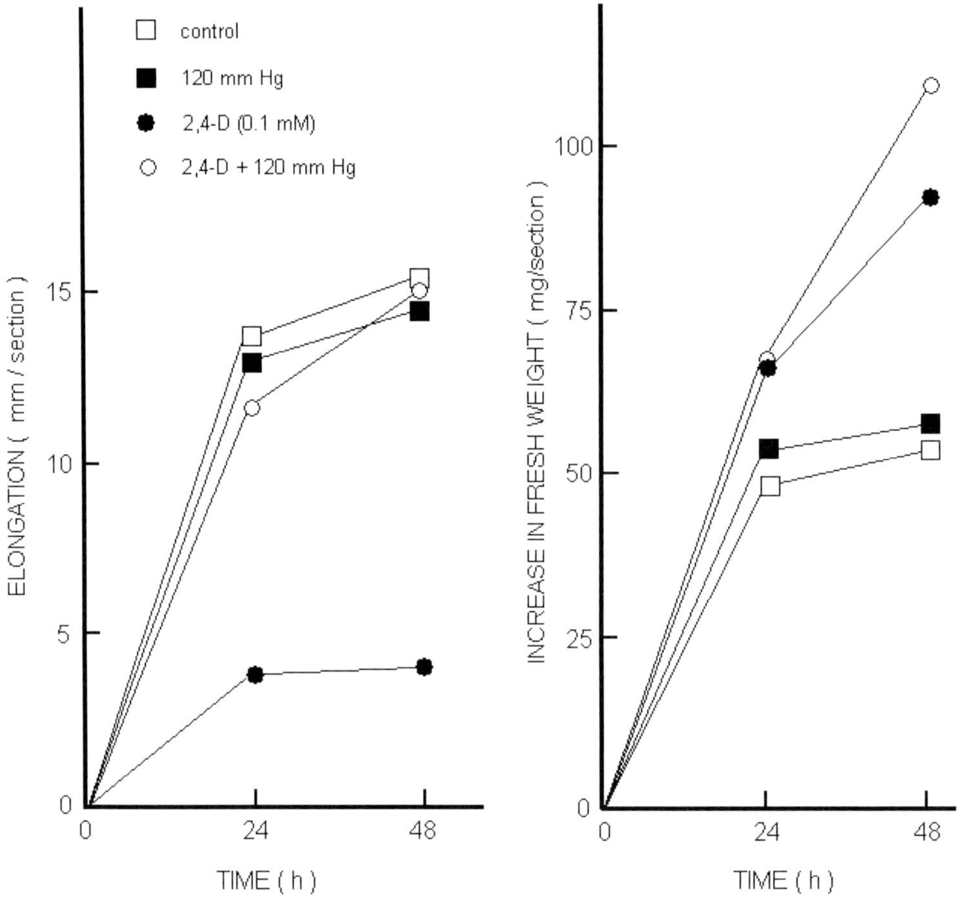

Fig. 2.6. A hypobaric pressure reverses 2,4-D-induced ethylene production's effect on growth. The 5 mm sub-hook region of 5-day-old etiolated pea seedlings was demarcated with two ink spots, and each day the marked zone was excised from ten randomly harvested replicate plants in order to measure its increase in length (left) and fresh weight (right). Treated plants were sprayed with 0.1 mM 2,4-D to induce ethylene production, and then grown either at atmospheric pressure, or at a pressure of 16 kPa (120 mm Hg) flowing water-saturated 100% [O_2] (Apelbaum and Burg, 1971b).

2.6 Who Invented Hypobaric Storage?

W.E. Tolle reviewed the discovery of hypobaric storage (Tolle, 1969, 1972).

> The (LP) system is being gradually evolved. Department of Agriculture scientists laid the groundwork 40 years ago through their studies on the removal of gases from the interior of fruits. They followed with other work on storing fruit in various controlled atmospheres. Workman *et al.* (1957) noted a trivial reduction in the respiration of tomato fruits when the surrounding storage pressure was increased by 120 cm of water pressure (88 mm Hg @ 20°C), and Hummel and Stoddard (1957) reported increases of 20 to 92% in the storage life of several kinds of produce kept in a home-type refrigerator with a compartment maintained at a pressure between 658 and 709 mm Hg. Then about 10 years ago, still other experiments were labeled hypobaric because they concerned the behavior of produce at pressures less than normal. Burg and Burg (1966c) reported similar success with additional kinds of fruits and vegetables stored at pressures from 125 to 360 mm of Hg and in air.

Table 2.5. Effect of a hypobaric pressure on the abscission of calamondin and Persian lime fruits from 0.6 m-high trees sprayed with 20 μg/ml of cycloheximide (CHI) to induce ethylene production.

Variety	Pressure (atm)	Chemical treatment	% fruit drop
Calamondin	1.0	None	0
	1.0	CHI	25
	0.2	None	0
	0.2	CHI + C_2H_4	25
	1.0	Control	0
Persian lime	0.2	CHI	100

The result at a low pressure is after 1 week in 0.11 m^3 bell jars engineered to maintain a leak-proof vacuum of 20 kPa (150 mm Hg), flowing 99.7% [O_2] + 0.3% [CO_2] at a rate of 0.06 m^3 per hour. When the pressure of this mixture is reduced to 20 kPa (150 mm Hg), the O_2 and CO_2 partial pressures approximate that in air at atmospheric pressure (see a critique of this methodology in 2.3). Ethylene was added to the O_2 + CO_2 mixture to provide 5 μl/l at the hypobaric storage pressure (Cooper and Horanic, 1973).

Influenced by Tolle's historical account, Salunkhe and Wu (1975) and Lougheed *et al.* (1974) attributed the development of hypobaric storage to Hummel and Stoddard.

Tolle's description of the 'invention' of hypobaric storage reveals a basic misunderstanding of the method. Varying the pressure by 88 mm Hg above (Workman and Pratt, 1957), or by approximately the same amount below, the atmospheric pressure (Hummel and Stoddard, 1957) should have the same influence on storage as climbing or descending a 1.2 km hill. Typically, the atmospheric pressure is approximately 646 mm Hg at Utah State University where Salunkhe and Wu worked (Logan, Utah), and 760 mm Hg in Miami (Florida), where the author is located. The 114 mm Hg difference 'at atmospheric pressure' between the two locations is similar to the maximum pressure reduction in Hummel and Stoddard's refrigeration compartment, but nobody would claim that under otherwise identical conditions, plant commodities are preserved for 20–92% longer in Logan, Utah, compared to Miami, Florida. In response to questions by the patent examiner regarding the first hypobaric patent application (Burg, 1967), the author had occasion to review the claims of Hummel and Stoddard. It was unclear what they meant by 'storage at a vacuum between 658 and 709 mm Hg'. Did this refer to an absolute pressure between 51 and 102 mm Hg, or a reduction in pressure by that amount? If the Hummel and Stoddard claim disclosed storage at an absolute pressure between 51 and 102 mm Hg, there would have been no innovation in this author's LP patent application, and the patent would have been denied. The question was decided by noting that if Hummel and Stoddard's apparatus reduced the pressure by much more than 51–102 mm Hg below the prevailing atmospheric pressure, surely the refrigerator compartment and/or refrigerator would have imploded.

3

Gas and Vapour Mass Transport

Mass transfer processes profoundly influence postharvest storage life by regulating water loss and the cellular concentrations of O_2, CO_2, ethylene, NH_3 and various volatile substances. The effect that a hypobaric condition has on mass transfer through air phases, and the benefit this provides, distinguish LP from all other storage methods.

Opinions differ widely concerning the relative contributions of the various mechanisms that have been proposed to account for mass transport through the surface ('skin') of harvested fruits, although usually it is tacitly assumed that water vapour and gases move by the same pathway (Ben-Yehoshua *et al.*, 1985; Ben-Yehoshua, 1986). Stomata and lenticles (Burg and Burg, 1965b), the pedicel stem scar (Brooks, 1937; Burg and Burg, 1965b; Cameron and Yang, 1982) and spaces between the epidermal cells (Fockens and Meffert, 1972) have been suggested as air-filled gas and water vapour transport routes. Alternatively, it has been proposed that the stomates and lenticles are non-functional, closed or occluded in mature fruits, causing gases and water to move mainly through the cuticle (Albrigo, 1972), constrained by the properties of the hydrated cuticular polymer (Schönherr, 1976b; Schönherr and Schmidt, 1979) and the resistance of the air-filled spaces separating the cuticular wax platelets (Chambers and Possingham, 1963). Others have suggested that CO_2 moves through the cuticle and O_2 through the lenticles (Marcellin, 1974), or that almost all gas exchange through the fruit skin occurs by diffusion in air-filled pores, while liquid water moves through an aqueous phase in the cuticle (Burg and Kosson, 1983; Ben-Yehoshua *et al.*, 1985). There is general agreement that in flowers and leafy commodities gases and water vapour move through open stomates and, when these close, through the cuticle at a reduced rate.

3.1 Resistance Network

The gaseous diffusion pathway between a commodity's cytoplasm and the atmosphere can be analysed using the equations governing electric circuitry by assigning an 'effective' resistance value (r) to each barrier that a gas or vapour encounters en route. Concentration gradient ($c_o - c_i$) is substituted for $EMF(V)$, and gas or vapour flux (J_v) in place of electric current (I). Then, according to Fick's law, and by analogy to Ohm's law ($V = IR$):

$$(c_o - c_i) = J_v\, r \qquad (3.1)$$

where J_v is the volume flux density across the particular barrier ($cm^3/cm^2 \cdot s$ = cm/s), r is the gas or vapour's resistance to diffusive transport through the barrier [r = atm·s/cm; usually written s/cm], and c_o and c_i are the concentrations (mol/cm^3) outside (o) and inside (i) the barrier. Just as an electric wire's resistance [R (ohms) = $\rho L/A_{cs}$] depends on the wire's resistivity (ρ, ohm-metres), length (L, cm) and cross-sectional area (A_{cs}, cm^2), likewise the resistance value assigned to each diffusion barrier is directly proportional to its thickness (Δx, cm), and inversely related to the surface area (A, cm^2) through which diffusion is occurring, and to the diffusion coefficient (D, cm^2/s) of the particular gas or vapour through the media comprising the barrier. The units used to indicate a barrier's conductance, abbreviated cm/s, refer to the cm^3 of gas transferred each second per cm^2 of surface when there is a 1 atm pressure gradient forcing molecules across the barrier. The barrier's resistance, abbreviated s/cm, is the reciprocal of its conductance, and like conductance is defined per cm^2 of surface area.

At equilibrium, the rate at which each gas or vapour is produced or consumed equals its flux ($A_{com} J_v$) across a commodity's surface area. The same flux must occur through the intercellular system, across the total cellular surface area, and also through the surface of the box in which the commodity is transported or stored. To accurately describe the concentration gradients created when an identical flux passes through the various barriers of the resistance network, the resistance value for each barrier is referenced to the commodity's surface area (A_{com}). This correction adjusts for the fact that when a barrier's surface area is greater, a smaller concentration gradient drives the same flux across it. The overall effect is to create an imaginary network in which each barrier is assigned the same surface area, and the resistance values are adjusted to reflect the relative ease of transport through each step in the network. The 'effective' resistance value (r_{com}) to be used for each barrier in the resistance network formulation is:

$$r_{com} = (A_{com}/A^{\circ})\, r^{\circ} \tag{3.2}$$

where the total surface area of a particular barrier is A°, and the barrier's measured resistance is r° per cm^2 of its surface area. In mesophytic leaves, $\Sigma A_{cell}/A_{com}$ varies between 10 and 40 in different species (Turrell, 1936, 1942); a value of 20 has been used to compute the cytosol, plasmalemma and cell wall resistance values for leaves indicated in Tables 3.3 and 3.4 (Colman and Espie, 1985; Noble, 1991). The $\Sigma A_{cell}/A_{com}$ ratio for an apple is much larger because in a bulky fruit the surface-to-volume ratio is smaller. Typical conditions for a McIntosh apple are: fruit volume = 230 cm^3, porosity = 30% (Clements, 1935), cell volume = 60 to 72×10^{-4} mm^3 per cell (Bollard, 1970), from which $A_{com}/\Sigma A_{cell}$ is computed to be approx. 228. An area correction also is applied to the box resistance referencing it to the commodity's total surface area (equation 3.18). The liquid phase resistance values in Tables 3.3 and 3.4, calculated at 20°C, will be reduced by approximately 50% at 0°C due to the larger air/water partition coefficient at a lower temperature (Tables 15.1 and 15.2). These adjustments allow the laws of parallel and series circuits to be used in summating resistances in the network:

$$r_1 + r_2 + r_3 = r_{1,2,3} \quad \text{(series resistances)} \tag{3.3}$$

$$1/r_1 + 1/r_2 + 1/r_3 = 1/r_{1,2,3} \quad \text{(parallel resistances)}$$

The resistances of boxes and wraps (r_b), stagnant air layers (r_a), the commodity skin ($r_{p,c}$), intercellular air space (r_{ias}), cell walls (r_{cw}), plasmalemma (r_{pl}), cytosol (r_{cyt}) and membranes of cellular inclusions (r_{incl}) act in series. The resistances of lenticles, stomates and the pedicel-end scar act in parallel and together constitute the surface's pore resistance (r_P), which acts in parallel with the skin's cuticular resistance (r_c) and in series with the other resistances.

To simplify the calculation of gas and vapour movement through a resistance network containing both air and liquid phases, the amount of a gas or vapour in each liquid phase is expressed as the equilibrium concentration in an air phase that gives rise to the relevant liquid phase concentration. Gases and vapours then will diffuse through the liquid and air phases toward regions of lower concentration regardless of the actual concentrations in the liquid phases and the partition coefficients that are involved in transport. The liquid phase concentration is multiplied by the partition coefficient (K), where K is the concentration of all forms of the gas or vapour in the liquid phase capable of penetrating the barrier, divided by the equilibrium gas or vapour concentration in an adjacent air phase. Barriers in the resistance network will be examined to determine their magnitude and the extent to which they are diminished by a hypobaric condition. Water movement and loss are considered separately in chapter 6.

3.2 Effect of pH and Temperature on Gas Exchange through Bio-membranes

The exchange of gases and vapours through the plasmalemma occurs by facilitated diffusion, active transport or partitioning into the membrane's lipid phase, passively diffusing through it, and then partitioning into the aqueous cytosol or apoplast. The concentration gradient that drives passive diffusion within the plasmalemma is expressed as a function (k) of the concentration gradient ($c_o - c_i$) in the aqueous phase immediately adjacent to the membrane, where c_o and c_i are the concentrations in the aqueous phase just outside (o) one side of the membrane or adjacent to the other 'inner' side (i), the partition coefficient (k) determines the ratio of the solute concentration within the membrane in equilibrium with c_o and c_i, and the concentration gradient in the membrane is $[k\ (c_o - c_i)]$. Usually k is expressed as the equilibrium ratio of the particular solute molecule's concentration in a lipid phase such as olive oil, compared to its concentration in an adjacent aqueous phase. Depending upon the lipid solvent selected, the oil/water partition coefficient for CO_2 varies from 0.44 to 1.61, and for water from 0.21 to 0.84 (Gimmler *et al.*, 1990). The coefficient for O_2 is 4.4 (Noble, 1991), and this causes the O_2 passive permeability to exceed the CO_2 or water passive permeability. Because ethylene has a lower solubility than CO_2 in water and a greater solubility in organic solvents, it should have a larger oil/water partition coefficient and greater permeability coefficient. Ammonia is soluble in alcohol, ether and organic solvents as well as in water, and penetrates the plasmalemma extremely rapidly.

A plant cell's permeability (P, cm/s) to a penetrating molecule typically is determined by measuring the rate at which a solute enters the cell from an external aqueous solution in which it is maintained at a constant concentration:

$$P = J\,(c_o - c_i) = D\,(c_o - c_i)/\Delta x = 1/r \quad (3.4)$$

where J is the flux into the cell (mol/cm^2·s), Δx is the thickness (cm) of the diffusion barrier, D is the penetrating molecule's apparent diffusion rate (cm^2/s) through the cell wall and plasmalemma, c_o and c_i are the aqueous concentrations (mol/cm^3) of the penetrating molecule outside (o) and inside (i) the cell, respectively, and the cell's resistance (r, s/cm) is the reciprocal of the permeability coefficient (P). When cellular permeability coefficients are used it is not necessary to make adjustments to account for the penetrating molecule's oil/water partition coefficient or its diffusion coefficient in the lipid and aqueous phases.

Many studies have shown that living cells are highly permeable to O_2, CO_2 and NH_3 (Harvey, 1910, 1922, 1928a; Jacobs, 1920, 1922; Curtis and Clark, 1950; Davson, 1960; Keele and Neil, 1961; Kurkdjian *et al.*, 1978; Roberts *et al.*, 1980, 1982). CO_2 passage through the vacuolar and plasma membranes of plant cells may be facilitated by the aquaporin protein-lined pores that are used for water conductance (Nakhoul *et al.*, 1998; 6.7). The permeability coefficient of CO_2 entering plant cells is at least 2×10^{-2} cm/s and may be as high as 10^{-1} cm/s (Noble, 1991); the calculations in Tables 3.3 and 3.4 assume a value of 5×10^{-2} cm/s. In *Chara* and *Nitella* cells the permeability coefficient for NH_3 uptake varies from 6.4 to 18×10^{-4} cm/s (Barr *et al.*, 1974; Walker *et al.*, 1979a,b; Ritchie, 1987); a value of 10^{-3} cm/s has been used for the NH_3 calculations. NH_3 uptake in these algal cells and also in the aquatic liverwort *Riccia fluitans* (Felle, 1980 – referred to in

Kleiner, 1981) is by a specific NH_4^+ uniport that follows Michaelis-Menton kinetics, is inhibited by $CH_3NH_3^+$ and cyanide and develops large concentration gradients. There also is evidence for an inward facing NH_4^+ uniport in *Avena* coleoptiles (Higinbotham *et al.*, 1964 – referred to in Kleiner, 1981) and in *Brassica napus* leaves (Nielsen and Schjoerring, 1998). In *B. napus*, the passive efflux of NH_3 occurs much more rapidly than active NH_4^+ uptake, suggesting that passive NH_3 permeability may be higher than the value assumed in Tables 3.3 and 3.4, which is for NH_4^+ uptake.

Gas solubility increases when the temperature is lowered. The temperature coefficients (Q_{10}) of the Henry constant are approximately 1.33, 1.21 and 1.35 for CO_2, O_2 and ethylene, respectively (Table 15.1). The expression for the temperature dependence of the NH_3 Henry constant is (Farquhar *et al.*, 1980):

$$\log_{10} H = 1477.7/T - 1.6937 \quad (3.5)$$

where T is the absolute temperature (K) and H is the ratio between the dissolved and gaseous concentrations of NH_3. Temperature also influences the constants regulating the formation of various ionic forms of $[NH_3]$ and $[CO_2]$. The 'overall' first and second ionization constants for $[CO_2]$ are:

First:

$$CO_2 + H_2O \rightleftharpoons H_2CO_3 \rightleftharpoons H^+ + HCO_3 \quad (3.6)$$

$$K_1 = (H^+)(HCO_3^-)/(CO_2 + H_2CO_3)$$

$$pK_1 = 6.33 - 0.5\ \omega^{1/2} \text{ at } 25°C$$

Second:

$$HCO_3^- \rightleftharpoons H^+ + CO_3^{2-} \quad (3.7)$$

$$K_2 = (H^+)\ (CO_3^{2-}/(HCO_3^-)$$

$$pK_2 = 10.22 - 1.1\ \omega^{1/2} \text{ at } 25°C$$

where ω is the solution's ionic strength. Table 3.1 illustrates the variation of pK_1 with temperature. For NH_4^+ the temperature dependence of the pK_a is (Farquhar *et al.*, 1980):

$$[H^+]\ [NH_3]_{solution} = K_a\ [NH_4^+]_{solution} \quad (3.8)$$

$$pK_a = 0.09018 + 2729.92/T \quad (3.9)$$

where T is the absolute temperature (K).

The interface between the apoplastic solution and intercellular air is the 'port of entry or exit' for gases produced or consumed by plant cells. The apoplast is defined as the tissue volume that is not separated from the environment by a membrane. It is comprised of interconnected cell walls, intercellular spaces and non-living vascular tissue, together forming a continuous non-cytoplasmic space filled with both air and the 'apparent free space' (Table 3.9; Cosgrove and Cleland, 1983) into and from which substances in solution move by free diffusion (Canny, 1995). Positive and negative ions are present in equal concentrations in an electrically neutral water free-space, while in 'Donnan free-space' the negatively charged -COO^- groups of pectin and other cell wall macromolecules cause there to be more positive than negative diffusible ions in solution (Briggs and Robertson, 1957). Gases in the intercellular air space are in equilibrium with the apoplastic solution in accord with Henry's law (Table 15.1).

The distribution of NH_4^+, HCO_3^- and CO_3^{2-} between the apoplast, cytosol and vacuole depends in part on the pH that exists in each compartment. Measurements of the hydrogen ion concentration in plant cells have been made *in vivo* by introducing 5,5-dimethyloxazolidine-2,4-dione (DMO), methylamine, benzylamine, nicotine or 9-iminoacridine into the cytoplasm and vacuole, by a ^{31}P nuclear magnetic resonance technique (Roberts *et al.*, 1980) and using microelectrodes (Smith and Raven, 1979; Kurkdjian and Guern, 1989). *In vitro* measurements usually are accomplished by expressing juice or preparing a homogenate from a tissue sample and determining the pH of the sap with a pH meter. The measurement is indicative of the vacuolar pH, since

Table 3.1. Values of pK_1 in water at various temperatures (Edsall and Wyman, 1958).

Temperature (°C)	0	5	10	15	20
pK_1 (in water)	6.578	6.517	6.464	6.419	6.381

typically > 90% of the liquid originates from that portion of the cell (Smith and Raven, 1979; Kurkdjian and Guern, 1989). Cytoplasmic pH normally is maintained at a relatively constant value in the range between 6.8 and 7.5, whereas vacuolar pH is more variable, generally in the 5.5–6.5 range (Wray *et al.*, 1985; Kurkdjian and Guern, 1989), but often < 4 in fruits and sometimes as low as 1 or 2.5 (Coombe, 1976; Kurkdjian and Guern, 1989).[1] The apoplastic pH typically is 5.0–6.5 (Grignon and Sentenac, 1991; Pearson *et al.*, 1998); for example, in *Brassica napus* at 25°C it is 5.7 ± 0.10 (Husted and Schjoerring, 1995). Tonoplasts contain proton-translocating ATPases and pyrophosphatases (Rea and Poole, 1993), which pump H^+ from the cytosol into the vacuole, causing the vacuole to be more acidic than the cytoplasm. This creates a trans-membrane potential with the vacuole 20–30 mV more positive than the cytosol (Hopkins, 1995; Willmer and Fricker, 1996), which drives anions such as Cl^-, malate$^-$ and HCO_3^- into the vacuole. The electrochemical proton gradient can also move K^+ and Ca^{2+} into the vacuole by an antiport mechanism, and there is evidence that sugars and certain amino acids enter the vacuole by means of similar antiport transporters (Bush, 1993; Hopkins, 1995). An outward-facing proton-translocating ATPase keeps the apoplasm more acidic than the cytosol.

The total $[CO_2]$ in solution depends mainly on the HCO_3^- concentration because the pH in the various cellular compartments, always < 8, causes the HCO_3^-/CO_3^{2-} ratio to exceed 100 and the dissolved H_2CO_3 concentration to be only about 1/400 that of dissolved CO_2 at equilibrium (Noble, 1991; Table 3.2). The plasmalemma's permeability to CO_2 efflux is not significantly influenced by the cytosolic HCO_3^- concentration, since the resistance to passive diffusion across the plasmalemma is > 10^4 higher for HCO_3^- than for CO_2.

Ammonia is able to move rapidly by passive diffusion from the cytoplasm through the plasmalemma to the apoplast against the $[NH_3 + NH_4^+]$ concentration gradient (Kleiner, 1981). At all physiologically relevant pH values, the total ammonia in solution depends on the concentrations of both NH_4^+ and dissolved gaseous NH_3 (Table 3.2), but the rate of efflux of ammonia from the tonoplast to the apoplast is determined almost exclusively by the $[NH_3]$ concentration gradient because the membranes of plants (and animals – Keele and Neil, 1961) are relatively impermeable to NH_4^+ (Grignon and Sentenac, 1991) and highly permeable to NH_3 (Jacobs, 1922; Curtis and Clark, 1950; Davson, 1960; Kleiner, 1981). The ΔpH across the plasmalemma creates an inward directed NH_4^+ gradient between the apoplast and tonoplast (Nielsen and Schjoerring, 1998), and the pH gradient across the vacuolar membrane causes NH_4^+ to be more concentrated in the vacuole than in the tonoplast, while $[NH_3]$ is highest in the cytoplasm. NH_4^+ is retrieved from the apoplast to the tonoplast by a less rapid transporter with channel-like properties (Nielsen and Schjoerring, 1998), and possibly a similar transporter moves NH_4^+ from the vacuole to the tonoplast. These mechanisms cause the apoplasm of oilseed rape leaves to return to and maintain its initial NH_4^+ concentration of 0.8 mM within 1.5 min after the tissue is infiltrated with an NH_4^+-free solution (Nielsen and Schjoerring, 1998).

Table 3.2. Influence of pH on the CO_2 partition coefficient, K_{CO_2} (Noble, 1991), and the NH_4^+/NH_3 ratio in the liquid phase (equations 3.8 and 3.9), at 20°C. According to Henry's Law (Table 15.1), the CO_2 absorption coefficient (equilibrium ratio of the moles/cm^3 of dissolved gaseous CO_2 vs. moles/cm^3 CO_2 in an air phase) is 0.88 at 20°C (Table 15.2).

Parameter (pH)	4	5	6	7	8
K_{CO_2}	0.91	0.96	1.48	6.6	58
NH_4^+/NH_3 (×10^{-3})	251	25.1	2.51	0.251	0.025

3.3 Cytosol Resistance

The initial diffusion route of metabolic CO_2 and the final pathway for respiratory O_2 is along a concentration gradient in solution within the cytosol of individual cells. When gas consumption or production occurs uniformly throughout a spherical cell (Goddard, 1947; Burton, 1982):

$$p_o - p_x = \frac{\dot{Q}_{O_2}\,(R^2 - x^2)}{6\,D} \qquad (3.10)$$

where for O_2 consumption p_X is the partial pressure of O_2 (atm) at distance x (cm) from the centre of the cell, p_o is the partial pressure of O_2 (atm) at the periphery; $\dot{Q}_{O_2}$ is the O_2 consumption rate (cm^3/min per cm^3 tissue); D is the diffusion coefficient for O_2 in the cell sap (1.4×10^{-5} cm^3/cm^2·min·atm; Goddard, 1947); and R is the radius of the spherical cell (cm). The O_2 gradient from the periphery to the centre (c) of a 5×10^{-3} cm radius apple cell is $p_o - p_c = 0.29_{\dot{Q}_{O_2}}$; and for a pre-climacteric apple cell at 0°C, respiring at a rate of 4×10^{-5} cm^3 O_2/min per cm^3 tissue, would only be equivalent to 1.16×10^{-5} atm of equilibrated partial pressure in an air phase. If the cell periphery contained dissolved O_2 equilibrated with 20% [O_2] in an air phase, the centre would be in equilibrium with 19.9988% [O_2] and the surface-to-centre cellular [O_2] gradient would be only 0.0012%. The binary diffusion coefficients of ethylene, CO_2, NH_3 and O_2 in water are not very different (Table 15.5), and therefore these gases diffuse through the cytosol with similar ease.

A mesophyll leaf cell's transport resistance to photosynthetic CO_2 uptake is lessened by the short diffusion pathway that arises due to the location of the chloroplasts at the cell periphery, and the effective cytosol resistance for respiratory O_2 uptake is diminished since mitochondria also are located peripherally. A representative value for the effective photosynthetic CO_2 cytosol resistance is 0.1 s/cm (Noble, 1991; Table 3.3), and the O_2 cytosol resistance, corrected for the Henry's law coefficient and binary diffusion coefficient in water, should be approximately 32-fold larger. The cytosol's resistance to ammonia transport will be insignificant due to the gas's high solubility in water (Table 15.2), and if ACC oxidase is located in the apoplasm or at the cell wall/plasmalemma complex, as has been proposed (5.3), the cytosol's resistance to ethylene transport will be extremely low.

3.4 Permeability of the Cell Wall

Water moves through the cell wall in numerous tortuous water-filled interstices approximately 10 nm in diameter, occupying slightly less than half of the cell wall volume (Noble, 1991). Gases and vapours must diffuse through this same water when they cross the cell wall, and since the binary diffusion coefficients of various gases in water are not very different (Table 15.5), they will pass through the wall with nearly equal ease. The CO_2 permeability coefficient is not changed by enzymatic digestion of the cell wall, indicating that the wall's CO_2 transport resistance is much smaller than the resistance of the plasmalemma (Gimmler *et al.*, 1990). A theoretical calculation indicates that the CO_2 conductance of a leaf's plasmalemma is approximately threefold larger than that of its cell walls (Noble, 1991).

3.5 Resistance of the Mitochondria

The mitochondria and chloroplasts are similar in size, both are surrounded by two membranes and each has an extensive internal membrane system. The inner mitochondrial membrane is in-folded, forming internal 'cristae', which are freely permeable to water and gases. Measurements made with leaves indicate that the resistance to CO_2 diffusion into chloroplasts and across their stroma is < 2 s/cm, and a theoretical calculation suggests a value of 1 s/cm (Noble, 1991; Table 3.3). The total surface area of a leaf mesophyll cell is similar to that of the chloroplasts it contains, but the mitochondrial surface

Table 3.3. Representative cellular resistance values at 20°C for gaseous diffusion in leaves and McIntosh apples. Values for CO_2 diffusing into leaves are based on a $\Sigma A_{cell}/A_{com}$ ratio = 20 at pH = 6 (Noble, 1991). CO_2 values for McIntosh apples are computed from leaf values based on a ratio $\Sigma A_{cell}/A_{com}$ = 228. The cell wall and cytosol resistance to O_2, NH_3 and ethylene are derived from the CO_2 value, corrected to account for different Henry's Law and binary diffusion coefficients in water. The plasmalemma resistance to O_2 is based on the CO_2 resistance, adjusted to reflect differences in the O_2 and CO_2 oil/water partition coefficients. The oil/water partition coefficient for ethylene is not known, but presumably is higher than for CO_2, since ethylene is less soluble than CO_2 in water and more soluble in oil. The plasmalemma resistance to NH_3 is based on the average measured value in *Nitella* and *Chara* (Barr *et al.*, 1974; Walker *et al.*, 1979a,b; Ritchie, 1987).

	Resistance (s/cm)							
	Leaf				Apple			
Barrier	CO_2	O_2	C_2H_4	NH_3	CO_2	O_2	C_2H_4	NH_3
Cell wall	0.3	6.7	2.7	~0	0.03	0.7	0.3	~0
Plasmalemma	1.0	2.7–10	< 1.0	20	0.09	0.2–0.9	< 0.09	2
Cytosol	0.1	3.2	0.9	~0	0.01	0.2	0.09	~0
Chloroplast	1.0	–	–	–	–	–	–	–

area is considerably less, increasing their effective resistance to O_2 diffusion.

3.6 Total Gas Transport Resistance

Table 3.3 contains representative values for the O_2, NH_3, CO_2, ethylene and ethanol transport resistances of the cell wall, cytoplasm and plasmalemma of an apple and a 'typical' leaf. Table 3.4 summarizes the total gas transport resistance to these gases and ethanol[2] when apples and 'leaves' are stored in cartons.

3.7 Kinetics of Gas Equilibration between Cells, the Intercellular System and Atmosphere

The half-time (τ_L) for gas diffusion from a production or consumption site in the cell to water bathing the cell surface is given by (Noble, 1991):

$$\tau_L = (V_{cell}/A_{cell})\, r_{pl,cw,liq} \ln 2 \qquad (3.11)$$

where $r_{pl,cw,liq}$ is the combined resistance of the plasmalemma (pl), cell wall (cw) and liquid phase (liq); V_{cell} is the cell volume and A_{cell} the cell surface area. In a spherical cell, V_{cell}/A_{cell} = radius/3, and if the radius is 50 μm, τ_L is approximately 0.15 s. The half-time (τ_a) for diffusion between air in the intercellular spaces and the atmosphere (a) is:

$$\tau_a = (V_A/A_t)\, r_{a,c,s} \ln 2 \qquad (3.12)$$

where V_A and A_t are the tissue's total air volume and surface area, respectively. The half-time ($\tau_{a,L}$) for diffusion between cells and the atmosphere is:

$$\tau_{a,L} = \frac{r_{a,c,s}\,(V_A + bV_L)\ln 2}{A_t} \qquad (3.13)$$

where b is the Bunsen absorption coefficient (Table 15.2) and V_L is the tissue's total liquid volume. Equations 3.12 and 3.13 describe the half-time for entry and exit of applied gas, and also for equilibration when the source of gas is an externally applied concentration or a constant production rate by the tissue. The adequacy of these equations was tested with McIntosh apples whose measured skin resistance was 18,200 s/cm, calculated from the IEC, ethylene production rate and the fruit's surface area (Burg and Burg, 1965b; equation 3.14). After the apples were briefly evacuated, upon return to atmospheric pressure their intercellular ethylene half-equilibrated in 90 min (Burg and Thimann, 1959; Fig. 5.18), in close agreement with a 102-min

Table 3.4. Comparison of the CO_2, O_2, ethylene, NH_3 and ethanol cellular and air phase resistances in leaves and McIntosh apples at 20°C. CO_2 data for leaves according to Noble (1991). The box resistance values are for 18.2 kg of 'leaves' contained in a carton used to store carnations (Fig. 3.15; example 9), and for 18.2 kg of apples in a carton with the same permeability properties as a box used to ship 4.5 kg of papayas (example 10).

	Resistance (s/cm)	
Barrier	Leaf	Apple
Total cellular resistance ($r_{cw, pl, cyt}$)		
CO_2	1.4	0.1
O_2	12.6–19.9	1.1–2.0
C_2H_4	< 4.6	< 0.5
NH_3	20	2.2
C_2H_5OH	53,370	4,680
Stomates, lenticels, pedicel stem end and barrier air layer ($r_{a,p}$)	(open stomates)	(no stomates)
CO_2	4.8	17,000
O_2	3.7	13,633
C_2H_4	4.4	18,369
NH_3	3.3	11,848
C_2H_5OH	6.7	23,800
Intercellular air spaces (r_{ias})		
CO_2	0.3	116
O_2	0.2	90
C_2H_4	0.3	125
NH_3	0.2	81
C_2H_5OH	0.4	162
Box (r_b)		
CO_2	14,604	2,338
O_2	11,322	1,813
C_2H_4	15,268	2,444
NH_3	10,641	1,630
C_2H_5OH	19,759	3,164
Total gas phase ($r_{a,p,ias,b}$)		
CO_2	14,609	19,454
O_2	11,326	15,536
C_2H_4	15,273	20,938
NH_3	10,645	13,559
C_2H_5OH	19,766	27,125

value based on equation 3.13 and data for V_A/V_L and b indicated in Tables 3.8 and 15.2. Equilibration half-times, determined by loading various fruits with ethylene (or ethane) and measuring the time course for the applied gas to escape, are indicated in Table 3.5. As the half-time ($\tau_{a,L}$, equation 3.13) is directly proportional to the tissue's air phase resistances, it is decreased by a hypobaric pressure to the extent indicated in Table 3.15. The half-time for equilibration between the intercellular spaces and atmosphere surrounding an apple should be lowered from 90 min at atmospheric pressure to 1.1 min at 1.33 kPa (10 mm Hg).

Tomato seedlings and excised tomato hypocotyl segments emit a pulse of ethylene during the first 2–4 min after their orientation is shifted from vertical to horizontal (Harrison and Pickard, 1984, 1986; Fig. 3.1). Harrison and Pickard suggested that 'deformation of the tonoplasts settling against and past the central vacuoles of some population of cells following horizontal reorientation might transiently increase the capability of EFE located in the tonoplast to attack

Table 3.5. Measured half-times for ethylene equilibration in various fruits at room temperature and atmospheric pressure.

Variety	Half-time (min)	Reference
McIntosh apple	90	Burg and Thimann, 1959
Satsuma mandarin	20	Kitagawa *et al.*, 1977
Valencia orange	10.9	Kitagawa *et al.*, 1977
Trovita orange	10.9	Kitagawa *et al.*, 1977
Natsudaidai	18.8	Kitagawa *et al.*, 1977
Hassaku	20	Kitagawa *et al.*, 1977
Eureka lemon	23	Kitagawa *et al.*, 1977
Navel orange	26	Kitagawa *et al.*, 1977
Tomato*	10	Cameron and Yang, 1982

*Half-time for equilibration with ethane.

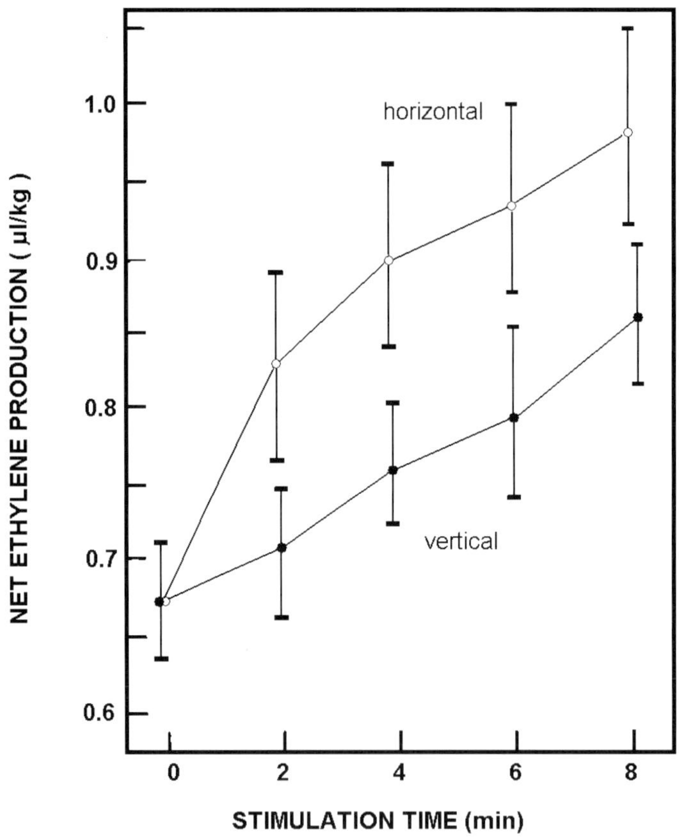

Fig. 3.1. Immediate, transient increase in ethylene production by excised tomato hypocotyls turned horizontal on the bench top (Harrison and Pickard, 1984).

vacuolar ACC', but this is an unlikely explanation. The speed and transient nature of the response indicate that it must be caused by escape of pre-formed intercellular ethylene, rather than accelerated ethylene production. An increase in ethylene emission followed by re-equilibration to the original emanation rate requires essentially ten 'half-times', and since this cycle occurred within 4 min, the half-time for the process could not be much longer than 24 s, which is close to the half-time for diffusive escape of gases from the intercellular spaces of a 1 mm-diameter petiole (equation 3.12). Harrison and Pickard estimated that the root mean square displacement of a diffusing ethylene molecule is roughly 67 mm during the 120-s interval between their measurements. It is not obvious why they concluded from this that there could be no significant boundary layer restricting heat and mass transfer around intact tomato plants or excised hypocotyls, for even at a wind speed as low as 0.1 m/s, a 1-cm-long leaf has a 1.26-mm-thick boundary air layer (Noble, 1991).[3] Example 18 indicates that changing the orientation of a hypocotyl segment from vertical to horizontal should increase the natural convective coefficient for respiratory heat transfer to the surrounding air through the boundary air layer by as much as sevenfold, and changing a leaf's orientation from parallel to vertical will nearly double its convective film coefficient. As the mass flux relation given by Fick's first law is similar to the heat flux expression given by the Fourier law,[4] if free convective heat transfer is enhanced, mass transfer through the boundary air layer also must be accelerated (Özisik, 1985). In leaves with open stomates, the gas mass transport resistance of the boundary air layer is significant compared to the overall 'skin' resistance (Noble, 1991), and if that also is true of tomato hypocotyl segments, increased mass transport through the boundary air layer would give rise to a 'burst' of ethylene emanation during the first 2–4 min after segments are shifted from a vertical to horizontal orientation.

3.8 Measuring the Air Phase Resistance Coefficient

At steady-state, a commodity's ethylene or CO_2 production and O_2 consumption ($\dot{m}$ = mol/kg·s) equal the rate at which each gas passes through the commodity's surface. For CO_2 or ethylene:

$$\dot{m} = \frac{A}{V}\left[\frac{(p_i - p_a)}{r_{p,c,a}}\right]\left[\frac{1}{\rho(R_D T)}\right] \quad (3.14)$$

where p_i and p_a are the partial pressures (atm) of CO_2 or ethylene adjacent to the inner (i) and atmospheric (a) surface of the commodity, R_D is the universal gas constant, T the prevailing temperature (K), ρ the commodity's density (gm/cm^3), V its volume (cm^3) and A its surface area (cm^2). For O_2 transport, the relationship is:

$$\dot{m}\, R_D T \rho V = A\,(p_a - p_i)/(r_{p,c,a}) \quad (3.15)$$

where $\dot{m}$ is the O_2 consumption rate. For CO_2 or ethylene production $p_a <<< p_i$ and $r_{p,c,a}$ can be computed from the ratio between internal gas concentration and production rate. Most air phase resistance values have been evaluated in this way, but for O_2 both p_a and p_i must be considered (equation 3.15). Skin resistance values also can be computed by equilibrating a tissue with ethane (or some other gas) and then measuring the gas's rate of escape as a function of time (Cameron and Yang, 1982). If the volume of the jar into which the ethane escapes is much larger than $V_A + bV_L$ (equation 3.13):

$$\ln\,(1 - C^t_{out}/C^\infty) = \ A\, t/r_{p,c,a} V_{in} \quad (3.16)$$

where $V_{in} = V_A + bV_L$, C^t_{out} is the concentration of ethane in the jar at time t, C^∞ is the concentration in the jar after equilibrium is established (i.e. at time infinity) and A is the commodity's surface area. The total mass of ethane is constant in the closed system at all times, and therefore when $V_{out} >> V_{in}$:

$$V_{in}C^t_{in} + V^t_{out} = C^\infty_{out}\,(V_{in} + V_{out}) = C^\infty_{out} V_{out} \quad (3.17)$$

$C^o_{out} = 0$ at time zero and equation 3.17 simplifies to:

$$V_{in} = C^\infty_{out}\,(V_{out}/C^o_{in}) \quad (3.18)$$

and V_{in} can be estimated from the C^∞_{out} (the concentration of ethane at equilibrium for each fruit at $t = \infty$) and the ratio of V_{out} (the volume of the jar) to C^o_{in} (the internal concentration of ethane in the fruit at $t = 0$, which is equal to the concentration of the loading gas). For a cv. Ace tomato fruit the V_{in}/A ratio was determined to be 0.135 by this method (Cameron and Yang, 1982; Fig. 3.12, *left*). This same V_{in}/A ratio would arise from the relation $V_{in} = V_A + bV_L$ (equation 3.13) if the radius of the fruit was 4.65 cm ($b = 0.04$ for ethane – Table 15.1). The first order ethane efflux rate constant, k, can be directly determined as the negative slope of a plot of ln $(1 - C^t_{out}/C^\infty)$:

$$k = A/(r_{p,c,a}V_{in}) \quad (3.19)$$

from which $r_{p,c,a}$ can be determined. Skin resistance values for mature-green and red cv. Ace tomatoes determined by the ethane efflux and CO_2 steady-state methods (equation 3.15) are in close agreement (Cameron and Yang, 1982).

A plot of ln $(1 - C^t_{out}/C^\infty)$ vs. time (Fig. 3.12) for cv. Ace tomato fruit gave a straight line with an intercept near zero, indicating that diffusion is monophasic and the movement of gases through the cellular liquid phase, to and through the intercellular system, is not limiting relative to gas exchange through the surface of the fruit. This result is consistent with the resistance calculations indicated in Tables 3.3 and 3.4 for an apple, which show that the sum of the cellular liquid phase and intercellular air phase resistances is insignificant compared to the skin resistance.

3.9 O_2, CO_2 and Ethylene Exchange between Cell Fluid and Intercellular Air

The equations and calculations in Sections 3.3–3.8 indicate that even if ethylene formation, CO_2 production and O_2 consumption were uniformly distributed throughout each cell, significant gas gradients would not arise across the cytosolic fluid, and gases in the intercellular spaces and apoplast should be in equilibrium according to Henry's law, independent of the total pressure. As CO_2 production and O_2 consumption occur in peripherally situated mitochondria, and the ACC oxidase isozyme responsible for autocatalytic ethylene production is localized in the apoplasm or plasmalemma of tomato (Bouzayen *et al.*, 1990; Rombaldi *et al.*, 1993; Fig. 5.5) and apple (Latché *et al.*, 1993) cells,[5] the diffusion pathway between cells and intercellular air is shortened, further reducing the magnitude of the gas gradients which arise across the cell fluid.

If the cytosolic O_2, CO_2 or ethylene concentration was disproportionately high relative to Henry's law, this would be revealed when intercellular gas is assayed in samples withdrawn from a bulky fruit with a syringe (equation 3.12), and the result compared with a vacuum extraction measurement (equation 3.13). Because the aqueous phase comprises 60–99% of the total volume of various commodities (Table 3.8), vacuum extraction would recover much more (dissolved) gas from bulky fruits than can be accounted for by assuming that the cytosolic fluid had equilibrated with intercellular gas in agreement with Henry's law and equation 3.12. Instead, when bulky fruits are vacuum extracted, Henry's law accounts for the result.

LP would not improve the low $[O_2]$ tolerance of stored horticultural commodities if cellular $[O_2]$ were limited by transport through a cellular aqueous phase where diffusion is pressure-insensitive. The ability of commodities stored in LP to tolerate remarkably low $[O_2]$ without damage or fermentation (Table 4.7) indicates that the cellular $[O_2]$ concentration depends on the resistance to O_2 exchange through air spaces in the 'skin' and intercellular system where diffusion is pressure-sensitive.

Under optimal conditions, purified apple ACC oxidase has an apparent K_m for $[O_2]$ of 0.3% in the gas phase (Kuai and Dilley, 1992). The K_m is 0.2% $[O_2]$ for ethylene production by 1-mm-thick well-blotted apple discs pre-incubated in a humid atmosphere containing N_2 to increase their [ACC] before exposing them to various $[O_2]$ concentrations (Burg, 1973a). The similar K_{m,O_2} values in a well-aerated enzyme preparation and in thin slices devoid of a skin resistance prove that the cellular diffusional resistance to O_2 mass transport between the intercellular system and the location of EFE (= ACO) is extremely small. As the binary diffusion coefficients of ethylene and

O_2 are not very different (Tables 15.4, gas, and 15.5, water), ethylene formed at its production site as well as O_2 consumed in close proximity to it must be in equilibrium with gas in the intercellular spaces.

3.10 Dependence of Ethylene Action on the IEC

Although there are no obvious barriers that could elevate the cytosolic ethylene concentration to a value significantly above that which equilibrates with the IEC in accord with Henry's law, nevertheless it has been suggested that

> the ethylene produced in ripening tissue may well exert its physiological effect mainly during its passage, within the cell, from the site of biosynthesis to the intercellular space. The concentration of ethylene in the intercellular space is thus an indicator rather than an effectuator of ripening, and its reduction by low atmospheric pressure or absorption is not effective in regulating senescence. (Stenvers and Bruinsma, 1975; see also Stenvers, 1975, 1977)

This concept was proposed to explain why the ripening of Jupiter tomatoes was not delayed during an LP storage performed at a pressure of 25.3 kPa (190 mm Hg) with 20% [O_2] present in the gas phase. Because the ripening time was increased by 50–100% if at that same pressure only 4–5% [O_2] was present, it was concluded that LP affects ripening by reducing the [O_2] rather than by decreasing the IEC. The action of ethylene in seedling hook formation (Schierle and Schwark, 1988; Schwark and Schierle, 1992; Schwark and Bopp, 1993), banana ripening (Whitehead and Bossè, 1991) and citrus de-greening (Purvis and Barmore, 1981; Goldschmidt *et al.*, 1993) has been interpreted in a similar manner.

Tomato fruit ripening

When in the presence of 20% [O_2], the hypobaric assay fails to delay a physiological response that normally would be considered evidence that ethylene is not involved (2.3), but since tomato ripening surely is ethylene-dependent, another explanation for the result is required. Stenvers and Bruinsma's suggestion that ethylene acts within the cell before it emerges into the intercellular spaces is not compatible with a vast literature showing that enhanced ethylene emanation typically is detectable before ripening commences (5.5; 5.9; Figs 5.11, 5.12 and 5.13). The correlation between an increase in the IEC and the onset of ripening indicates that the IEC is in equilibrium with the 'fast' ethylene receptor responsible for most and possibly all actions of the gas (5.11), and argues against the notion that without its emanation becoming evident, endogenously produced ethylene induces ripening. Stenvers and Bruinsma's concept also cannot be reconciled with experiments in which the hypobaric assay has been used with 20% [O_2] present to demonstrate ethylene involvement in abscission (Table 2.5), sex expression, leaf epinasty, auxin-induced growth inhibition (Fig. 2.6), seed germination, hook opening (Fig. 2.5), flower fading (Table 2.4) and numerous other effects for which there is overwhelming evidence for endogenous ethylene's regulatory role (2.3). The same assay delayed banana ripening by 86 ± 15% with 21% [O_2] present (Burg and Burg, 1966c). There is no obvious reason why ethylene produced in tomato cells, but not in other cells, should 'exert its physiological effect mainly during its passage, within the cell, from the site of biosynthesis to the intercellular space', and therefore, before accepting Stenvers and Bruinsma's premise, another explanation for the anomalous tomato ripening result should be sought.

Stenvers and Bruinsma's low-pressure tomato experiment was not performed in a flow-through LP apparatus. Instead, the fruits were sealed in desiccators, gas mixtures were introduced from cylinders and the pressure was adjusted by suction. Dry lime was included to absorb CO_2, potassium permanganate to remove ethylene, and each day the sealed desiccators were vented and flushed, and the pressure was reset. The difficulty of maintaining the IEC of ethylene-sensitive fruits below a physiologically active level by means of permanganate is well documented (Knee, 1990; Sfakiotakis *et al.*, 1993; Knee *et al.*, 2000). When an ethylene scrubber (ethysorb) was used to reduce the ethylene in the atmosphere surrounding apple fruits stored in NA, CA or ultra low CA (ULO), after 2 months the IEC varied between 1.5 and 4.3 μl/l in ULO, 3.6–8.2 μl/l in CA and 68.1–73.8 μl/l in NA (Sfakiotakis *et al.*, 1993). With permanganate included in the tomato experiment, the accumulated ethylene in the desiccators usually was less than 1 μl/l and never exceeded 2 μl/l, and if permanganate was omitted the ethylene concentration surpassed 5 μl/l within a few days, which was judged to be the threshold needed to hasten tomato fruit ripening. Yet, with 20% [O_2] present, regardless of whether the pressure was 101.3 or 25.3 kPa (760 or 190 mm Hg), ripening occurred at the same rate with and without added permanganate. This behaviour would be expected if sufficient ethylene to trigger ripening accumulated in all of the desiccators during the initial days of the experiment regardless of whether or not permanganate was included. Then LP would cause the fruit's IEC to approach the active ethylene concentration that arose within the desiccators, ripening would be accelerated and this would offset any advantage gained by using LP to eliminate the ethylene gradient across the commodity's 'skin'. Both in NA and LP, the tomatoes advanced to a table-ripe stage in 7–9 days at 19°C when 20% [O_2] was present. This is similar to the ripening time of mature-green tomatoes treated with 1–1000 ppm ethylene at 20°C (Table 3.6; Fig. 3.2). An ethylene build-up during the first days of the experiment is

Table 3.6. Effect of ethylene continuously supplied in a flow-through humidified system on the ripening of mature-green Hybrid 'Early Bush' tomatoes at 20°C (Morris *et al.*, 1981).

Ethylene conc. (μl/l)	Elapsed time (days): Mature-green to breaker	Breaker to table ripe	Total
0	15.4	9.5	24.9
0.3	9.6	6.3	15.9
1	7.3	6.0	13.3
3	6.3	5.9	12.2
10	4.4	5.5	9.9
30	4.6	6.3	10.9
100	3.8	5.1	8.9

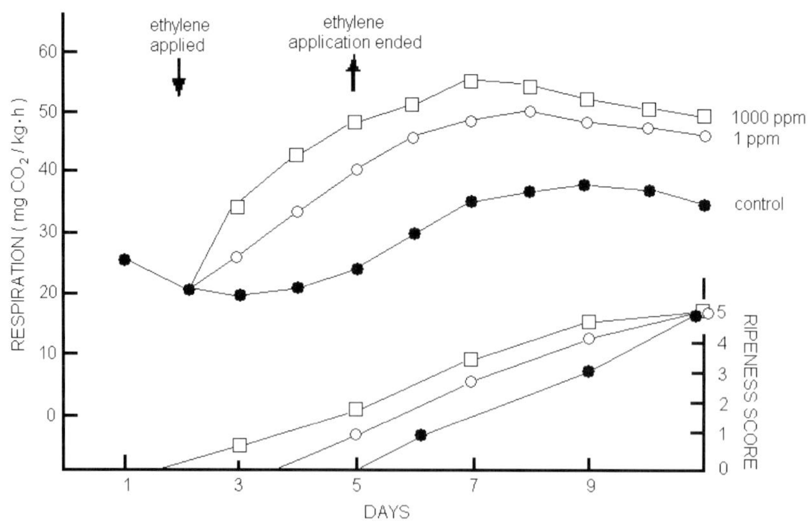

Fig. 3.2. Effect of various concentrations of ethylene applied in a flowing air stream at 20°C on the ripening and respiration of mature-green tomatoes. Ripeness scored on a scale from 0 (mature-green) to 5 (fully ripe). Arrows indicate the time when an ethylene treatment was started or terminated (Takata, 1975).

likely, since 18 h after they are harvested, mature-green tomato fruits may produce ethylene at a rate of 1 μl/kg·h, and then during a few ensuing days the rate progressively decreases to a pre-climacteric minimum of 0.01–0.02 μl/kg·h unless ripening has been stimulated (Lyons and Pratt, 1964).

There are conflicting estimates of the ethylene concentration required to initiate tomato fruit ripening. The 5 μl/l threshold cited by Stenvers and Bruinsma is misleading because it is the minimum concentration in a flowing air stream which hastens the ripening of mature-green cv. 'Sonato' fruits attached to the vine, whereas < 0.2 μl/l suffices with the same cultivar after harvest (Sawamura *et al.*, 1978). Ripening of Floradade tomatoes is promoted by < 0.005 μl/l (Wills *et al.*, 2001), and 0.3 μl/l applied ethylene was highly stimulatory when individual mature-green Hybrid 'Early Bush' tomato fruits were tested in pint jars, each with its own supply of flowing humidified air (Table 3.6; Morris *et al.*, 1981). In another study with mature-green harvested tomatoes, 1 μl/l ethylene in flowing air induced a large respiratory rise within a few hours and was nearly as effective as 10–1000 μl/l (Fig. 3.2). During a 5% $[O_2]$ + 5% $[CO_2]$ CA storage, exogenous ethylene concentrations of 1 and 10 μl/l were equally effective in promoting colouring of mature-green cv. Sonatine tomatoes, and the ethylene had to be scrubbed to < 0.1 μl/l to avoid this occurrence (Geeson *et al.*, 1986). But no concentration of applied ethylene affected tomato ripening when fruits were kept under restricted ventilation, so that the 'endogenous ethylene took effect' (Heinze and Craft, 1953). This same result occurred when the ethylene sensitivity of 'Walter' tomatoes was determined by treating fruits with ethylene in gas-tight chambers with provision for continuous CO_2 absorption, using a static setup similar to that employed by Stenvers and Bruinsma in their low-pressure study (Jahn, 1975). The atmospheres were monitored with a gas chromatograph, and levels of O_2 and ethylene adjusted as needed, or the chambers were opened and aired daily. No consistent effect of 100 μl/l or even 1000 μl/l ethylene applied for 3 days could be discerned. The accumulation of ethylene in the chambers containing control fruits made it impossible to demonstrate a response to applied ethylene, just as it obscured an effect from LP in Stenvers and Bruinsma's experiment. The complicating factor of ethylene accumulation is eliminated in a flow-through LP system.

The interpretation of LP tomato experiments carried out in < 12.5% $[O_2]$ is complicated by the inability of applied ethylene to initiate tomato ripening at so low an $[O_2]$ level. Decreasing $[O_2]$ to 2.5–5% by supplying gas mixtures at atmospheric pressure, or by flowing water-saturated air at a pressure of 12.67–25.33 kPa (95–190 mm Hg), is equally effective in inhibiting the initiation of ripening in mature-green tomatoes and further ripening in fruits harvested at the breaker or light pink stages (Kader and Morris, 1974, 1975a; Stenvers and Bruinsma, 1975). This explains why, in Stenvers and Bruinsma's LP storage experiment, ripening was delayed at a pressure of 25.3 kPa (190 mm Hg) when the $[O_2]$ was 4–5%. This result cannot indicate whether or not a hypobaric condition is able to delay tomato ripening by lowering the IEC because there was insufficient $[O_2]$ to support ethylene action. Ripening was progressively slowed when the pressure was reduced and the $[O_2]$ allowed to vary inversely according to the total pressure, but when the O_2 partial pressure was kept constant at 38 mm Hg (= 5% $[O_2]$), lowering the pressure did not influence tomato ripening (Tolle, 1969). Because ethylene cannot stimulate tomato fruit ripening when the O_2 partial pressure is 38 mm Hg, this result, like that of Stenvers and Bruinsma, does not indicate that LP retards tomato fruit ripening only by lowering the $[O_2]$, as others have claimed (Kader and Morris, 1974, 1975a; Leshuk and Salveit, 1990; Abeles *et al.*, 1992).

Citrus de-greening

Purvis and Barmore (1981) found that a pressure of 0.2 atm (150 mm Hg = 3.6%

$[O_2]$) greatly reduced chlorophyll degradation in the peel of citrus fruit, but the significance of this study had been questioned because the $[O_2]$ may not have been maintained high enough to allow de-greening to proceed (Denny, 1924b; Goldschmidt *et al.*, 1993). Norbornadiene (NBD) and Ag^+ inhibited chlorophyll loss from Shamouti oranges by 55–60% during 4–6 days at atmospheric pressure (Goldschmidt *et al.*, 1993), and flowing pure $[O_2]$ at a pressure of 20 kPa (150 mm Hg = 16.8% $[O_2]$) slowed de-greening by 50% during 15 days (Fig. 3.3, *left*; Apelbaum *et al.*, 1976),[6] but because the hypobaric effect did not set in until the 8th day, Goldschmidt *et al.* (1993) and Purvis (1981) concluded that hypobaric conditions do not prevent ethylene action by lowering the IEC; that 'low levels of ethylene may be bound within the tissue and not show up as ethylene evolution', and 'ethylene binding has not been taken into consideration in connection with studies of the resistance of fruits to gas diffusion and exchange'. This explanation fails to consider an 80% decrease in the ICC caused by a 1/5 atm hypobaric pressure, and CO_2's inhibitory effect on ethylene-induced and natural de-greening of leaves and citrus fruits (Denny, 1924b; Apelbaum *et al.*, 1976; Sisler, 1980a; Philosoph-Hadas *et al.*, 1994). Chlorophyll breakdown is inhibited when > 0.5–1.5% applied CO_2 summates with the 1–3.5% ICC present in harvested Shamouti oranges at atmospheric pressure (Fig. 3.3, *right*; Ben-Yehoshua, 1969; Sawamura, 1981; Ben-Yehoshua *et al.*, 1985), and the opposite result would be expected when a 1/5 atm hypobaric pressure lowers the orange's ICC to between 0.2 and 0.7%. Tobacco leaf de-greening is inhibited by 3% $[CO_2]$, 1–3% $[CO_2]$ markedly reduces the ability of 1000 μl/l ethylene to cause tobacco leaf yellowing (Sisler, 1980a) and 11% $[CO_2]$ is more effective than AVG or Ag^+ in delaying chlorophyll loss in harvested watercress leaves (Philosoph-Hadas *et al.*, 1994). It has been suggested that CO_2 acts on senescence by an additional mechanism that is not necessarily related

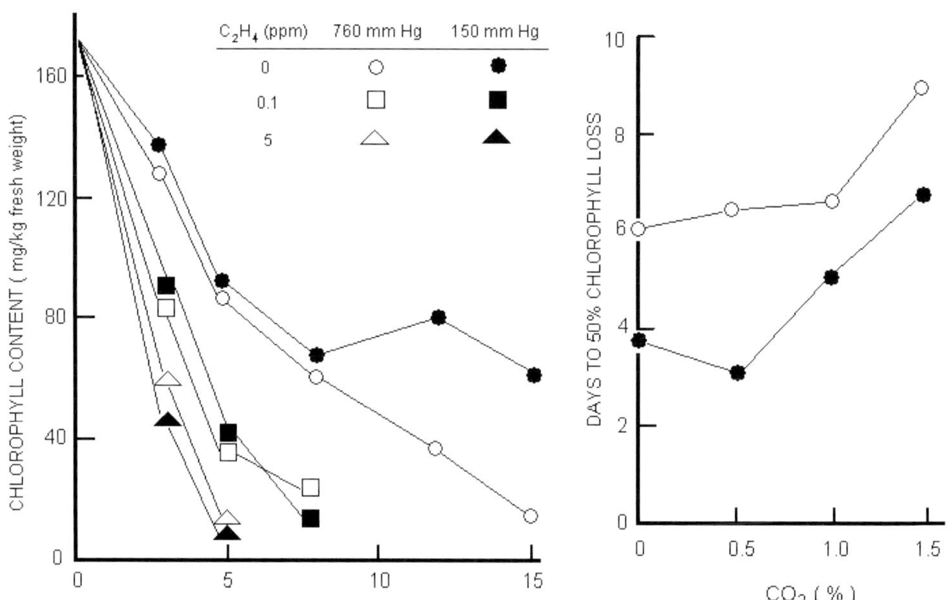

Fig. 3.3. (*left*) Kinetics of rind chlorophyll destruction in detached Shamouti oranges at 24°C in air at atmospheric pressure or in LP at 20 kPa (150 mm Hg) flowing 100% $[O_2]$. (*right*) Effect of $[CO_2]$ on the rate of rind chlorophyll destruction at atmospheric pressure in the presence or absence of 0.1 μl/l ethylene gas (Apelbaum *et al.*, 1976).

to its anti-ethylene activity (Aharoni *et al.*, 1979a; Philosoph-Hadas *et al.*, 1986, 1994).

Although ethylene increases chlorophyll catabolism in leaves and fruits (Poenicke *et al.*, 1977; Aharoni and Lieberman, 1979; Purvis and Barmore, 1981; Abeles *et al.*, 1989; Serek *et al.*, 1996), inhibitors of ethylene synthesis and action often fail to retard chlorophyll breakdown (Aharoni and Lieberman, 1979; Gepstein and Thimann, 1981; Halder-Doll and Bangerth, 1987; Tjosvold *et al.*, 1994) because chlorophyll degrades by both ethylene-dependent and -independent actions (Oeller *et al.*, 1991; Klee, 1993; Murray *et al.*, 1993a,b; Table 5.7). Leaf senescence is delayed in EFE antisense tomato plants, but nevertheless leaf yellowing occurs independent of ethylene (Murray *et al.*, 1993b), and in studies with an *Arabidopsis* mutant that had a completely inhibited response to ethylene, dark-induced leaf yellowing did not require the action of ethylene (Zacarias and Reid, 1990). Chlorophyllase expression at the transcript level is low and constitutive during Valencia orange fruit development, and does not increase significantly during natural colour break and ripening, but ethylene application induces chlorophyllase at any stage of development, and this increase is inhibited by gibberellin-A_3 (Jacob-Wilk *et al.*, 1999). Apparently chlorophyllase is not the regulator of chlorophyll breakdown during natural fruit ripening, and gibberellins play an important role.

Because the 'fast' ethylene binding sites that control ethylene action are dissociable and equilibrate with the IEC within 2–15 min (Fig. 5.20, *left*) many ethylene responses, including de-greening of melons and citrus fruits, are rapidly reversible when ethylene is removed. De-greening of antisense ACO Charentais melons (*Cucumis melo* var. Cantalupensis cv. Védrantais) by 1–5 μl/l applied ethylene stops when the ethylene treatment is terminated (Flores *et al.*, 2001a), and ethylene-induced chlorophyll breakdown in calamondin and Robertson tangerine fruits also is not self-sustaining (Purvis and Barmore, 1981), ceasing after fruits are returned to air following a 24-h treatment during which de-greening was promoted by 7.7 μl/l of exogenous ethylene. A rapid recovery in air would not occur if de-greening was caused by bound ethylene that was unable to equilibrate with the IEC.

LP promotes ethylene synthesis by citrus fruits (Cooper and Horanic, 1973; Brisker, 1980) and etiolated pea epicotyl sections (Saltveit and Dilley, 1978b), and simultaneous application of propylene prevents this effect in pea tissue, indicating that the stimulation results because auto-inhibition is relieved when LP lowers the IEC. This could not occur if endogenous ethylene acts before it leaves cells. Ethylene production is auto-inhibited in citrus peel discs by > 0.1 μl/l applied ethylene and the inhibition is relieved after the tissue is transferred to ethylene-free air. Following a 24-h exposure of citrus leaf discs to 12 μl/l ethylene, the rate of ethylene production is stimulated within 60 min after the applied ethylene is removed, increasing by 500% within 2 h (Riov and Yang, 1982b). Ag^+ promotes ethylene production by preventing endogenous ethylene from auto-inhibiting ethylene biosynthesis (Aharoni and Lieberman, 1979; Mattoo and Lieberman, 1982; Riov and Yang, 1982a,b; Kao and Yang, 1983; Goren *et al.*, 1984; Philosoph-Hadas *et al.*, 1985b, 1994; Mullins *et al.*, 1999) and norbornadiene (Goldschmidt *et al.*, 1993) has this same effect in citrus fruits.

Banana ripening

Whitehead and Bossè (1991) measured ^{14}C-ethylene binding in bananas and concluded that saturation of the ethylene receptor is 'permanent' during ripening. A more likely explanation is that the large increase in the ripening fruit's ethylene production rate diluted the applied ^{14}C-ethylene and increased the fruit's IEC, displacing ^{14}C-ethylene from its binding sites. The bananas were exposed to 100 μl/l non-labelled ethylene for 0–24 h, and then colour development, ethylene production and the binding of 0.1 μl/l ^{14}C-ethylene was measured after the fruit was transferred to air. ^{14}C-binding remained very low throughout the entire ripening period in fruit pre-treated with ethylene for 24 h. In untreated fruit, binding decreased rapidly during the first stages of ripening, and was approximately 93% lower in 4-day-old fruit compared to freshly harvested bananas. The ethylene production rate of fruits treated with ethylene for 24 h had increased at least 100-fold by the time the first binding measurements were made (Burg and Burg, 1965c), and surely at that time accumulated ethylene and the high IEC displaced all ^{14}C-ethylene from the binding sites, creating the impression that ^{14}C-ethylene was unable to bind because the sites were 'permanently saturated'. When ethylene production subsided during the post-climacteric period, the rate still was at least tenfold higher than the initial value before the ethylene treatment, and because post-climacteric gas exchange in bananas is severely hindered by water-logging, the IEC is disproportionately elevated at that time (Fig. 3.9, *left*).

Seedling hook formation

Several studies purport to show that the seedling hook forms because asymmetric ethylene production in its concave and convex sides creates a corresponding inner–outer unequal distribution of ethylene 'irreversibly' bound to a slowly dissociating heat-labile receptor (Schierle and Schwark, 1988; Schwark and Schierle, 1992; Schwark and Bopp, 1993).[7] This implies that ethylene production selectively inhibits growth in the concave side, independent of the IEC, without the gas spreading laterally through the intercellular system and causing an equal growth inhibition in the convex side (Schwark and Bopp, 1993). The heat-extractable ethylene concentration in the plumule, hook and subhook regions of etiolated pea seedlings was 1.96, 3.03 and 0.86 μl/kg, respectively (Schierle *et al.*, 1989), and if all of this ethylene had been freely present in this tissue's 2.8% intercellular air space (Table 3.9), it would have created an IEC of 75.4, 116.5 and 33.1 μl/l, respectively, in these tissues. The extractable concentration of bound and heat-labile ethylene in bean hooks varied between the outer/central/inner tissue in the ratio 2.0/1.0/2.6 (Schwark and Bopp, 1993), and assuming that the bean hook and pea epicotyl have the same porosity, if the heat-labile ethylene recovered from the bean hook was 'intercellular', the IEC would have been 51.5 μl/l in the convex side, 67.3 μl/l in the concave side and 25.8 μl/l in the centre. These calculations indicate that there is no correlation between heat-labile bound ethylene and the IEC, for 0.01–0.03 μl/l applied ethylene tightens bean hooks (Kang and Ray, 1969a) and < 0.16 μl/l suffices with pea hooks (Goeschl and Pratt, 1968), indicating that the IEC in pea and bean hooks could not normally be much larger than these amounts.

The inner half of the bean hook elbow produced 30% more ethylene compared to the outer half (Schierle and Schwark, 1988), and contained 30% more extractable 'bound, heat-labile' ethylene on a fresh weight basis, but since bean hook tissue grows mainly by elongation, and the average cell length is 26% greater in the outer peripheral tissue compared to the inner tissue (Schwark and Schierle, 1992), on a per-cell basis the amount of extractable 'bound' ethylene was the same in both sides. A similar situation exists in etiolated pea tissue, where on a per-cell basis the amount of heat-extractable, bound ethylene is essentially the same in the subhook and hook.[8] The etiolated pea subhook produces no ethylene (Goeschl *et al.*, 1967) and responds to as little as 0.01 μl/l applied gas (Table 5.2), but it has not been affected by its large cellular concentration of bound, heat-labile ethylene. If this slow-dissociating complex causes no physiological effect in the pea subhook, there is no reason to suppose that the same cellular concentration of the slow-dissociating binding component is physiologically active in the hook region of either peas or beans. Bound, heat-labile ethylene also exerts no physiological effect in bean cotyledons. It may represent 'compartmentalized' ethylene in that tissue (Jerie *et al.*, 1979; Bengochea *et al.*, 1980).

Hook closure is initiated within 2 h after ethylene is applied to etiolated bean seedlings, and the effect is fully reversible within 2 h after the ethylene treatment is terminated (Kang and Ray, 1969a). This proves that the receptor involved in the hook's forming and opening responses rapidly associates and dissociates when it equilibrates with applied ethylene, unlike the slow-dissociating heat-labile 'receptor'. That an asymmetric response to ethylene, rather than an ethylene gradient, is the underlying cause of the hook configuration is proved by reformation of an apical hook in light grown pea seedlings exposed to 0.4–1000 μl/l ethylene (Burg and Burg, 1968; Goeschl and Pratt, 1968), and by etiolated pea-seedling hook tightening in response to this same range of ethylene concentrations (Goeschl and Pratt, 1968). When a high concentration of exogenous gas is applied, all cells are exposed to the same ethylene dosage, their ethylene production rate becomes irrelevant and only an asymmetric response, rather than an ethylene gradient, can account for the differential growth that occurs. Exposing pea seedlings to pure O_2 at 1/5 atmospheric pressure ($[O_2] = 20\%$) causes hook opening (Apelbaum and Burg, 1972; Fig. 2.5) and would not do so if ethylene acted to maintain the hook configuration independently of the IEC. Thus, there is no evidence to indicate that the IEC remains asymmetrically distributed in the hook region due to unequal rates of ethylene production in the inner and outer surfaces, or that ethylene action is independent of the IEC and occurs before the gas emerges into the intercellular system.

3.11 NH_3 Exchange between the Apoplast and Intercellular Air

An NH_3 compensation point, influenced by the concentration of NH_4^+ and the pH in a plant's apoplast, determines whether NH_3 will be taken up or released at a given atmospheric NH_3 partial pressure (Farquhar *et al.*, 1980; Lemon and van Houtte, 1980; Sutton *et al.*, 1994; Husted and Schjoerring, 1995; Mattsson *et al.*, 1997). The NH_3 compensation point is equal to the partial pressure difference driving NH_3 from the intercellular system when a plant is ventilated with NH_3-free air (Table 3.7). In the intercellular spaces, the $[NH_3]$ concentration (p_{NH_3}, mol fraction) which is in equilibrium with the apoplastic NH_3/NH_4^+ solution, depends on the solution's temperature (K), and its H^+ and NH_4^+ concentrations (M), according to the expression (Farquhar *et al.*, 1980) (see bottom of page); where R (0.0831 litre·bar/K) is the gas constant. Equation 3.20 predicts that the NH_3 partial pressure in equilibrium with $[NH_4^+]_{soln}$ at a fixed pH will have a strong temperature dependence (Table 3.7; Fig. 3.4).

The ammonia partial pressure in the intercellular spaces of a cold-stored commodity is likely to decrease below the atmospheric level, which varies from 0.1 to 20 nbar (Sutton *et al.*, 1995) and typically is close to 5 nbar (Furutani *et al.*, 1987). NH_3 fumigation experiments have demonstrated that a commodity that is a source of NH_3 may become a strong sink for atmospheric NH_3 when the temperature is lowered (Fig. 3.4; Husted and Schjoerring, 1996). The efflux of gaseous non-elemental N from ten different plant species increased when the temperature was raised from 28 to 35°C (Stutte and Weiland, 1978) and the ammonia compensation point decreased when the temperature was lowered (Table 3.7). Equation 3.20 indicates that if the apoplastic $[H^+]$ and $[NH_4^+]_{soln}$ of *B. napus* were the same at 0°C and 25°C, the apoplastic NH_3 partial pressure would decrease by 21.4-fold when the temperature was lowered from 25 to 0°C. At 25°C the ammonia compensation point was 7.7 nbar and at 0°C it would decrease to 0.35 nbar.

The rate at which NH_3 is lost from a horticultural commodity depends not only

$$P_{NH_3} = RT(10^{1.60352-4207.62/T})\left[\frac{[NH_4^+]soln}{H^+}\right] \tag{3.20}$$

Table 3.7. Ammonia compensation points for various plants determined by gas exchange (Raven *et al.*, 1992).

Plant	Temperature (°C)	Ammonia compensation point (nbar)	Reference
Luzula sylvatica (Huds.)	25	0.67–1.45	1
Phaseolus vulgaris L.	26.6	2.5	2
	33.4	5.5	2
Zea mays	25–26	2.1–5.9	2
Amaranthus edulis	32	4.7	2
Eucalyptus pauciflora	32	3.0	2
Arrhenatherum elatius	25	0.07	3
Triticum aestivum			
pre-senescent	25	16.3	4
senescent	25	28.8–50	5
Brassica napus	25	7.7	6
Hordeum vulgare	10	0.9	7
	40	8.0	7

References: (1) Hill *et al.*, 2001; (2) Farquhar *et al.*, 1980; (3) Hanstein and Felle, 1999; (4) Parton *et al.*, 1998; (5) Morgan and Parton, 1989; (6) Husted and Schjoerring, 1996; (7) Husted *et al.*, 1997.

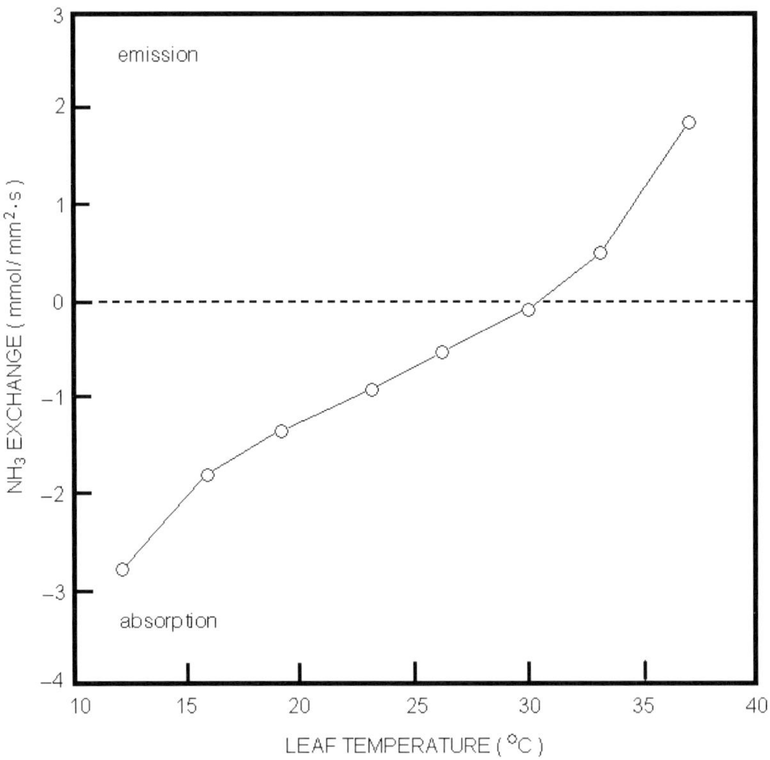

Fig. 3.4. Effect of leaf temperature on NH_3 exchange in vegetative *Brassica napus* plants exposed to 15 nbar NH_3 in air (Husted and Schjoerring, 1996).

on the difference between p_{NH_3} and the NH_3 partial pressure in the surrounding atmosphere, but also on the resistance of the gas exchange pathway (Husted and Schjoerring, 1996). Consequently, the rate of NH_3 escape from plant tissues is proportional to the atmospheric pressure (Farquhar *et al.*, 1980), and will be enhanced by hypobaric storage. NH_3 efflux should decrease and ammonia accumulate when stomates close during NA or CA storage, and efflux must increase and tissue ammonia decrease when the pressure is lowered, not only because diffusion is increased, but also because the stomates remain open in LP (4.15). At 35°C the rate at which *Phaseolus vulgaris* and *Brassica napus* L. plants evolve NH_3 per nbar NH_3 gradient is 74-fold larger in high vs. low nitrogen plants. Yet in both types of plants the conductance of their open-leaf stomates is adequate to account for the flux (Farquhar *et al.*, 1980; Husted and Schjoerring, 1996), at least in part because stomates in plants grown with abundant N open more widely when water is adequate and close more promptly and completely when water becomes deficient, compared to stomates in low-N plants (Pleasants, 1930; Desai, 1937; Radin *et al.*, 1982).

3.12 Jamin's Chains: the Effect of Liquid Water on Intercellular Gas Transport

With few exceptions, every cell in a plant communicates with an air space, which connects with others to form a network extending axially from the leaves through the stem to the root, and radially to the epidermal surface. Even the intercellular spaces of a dense tissue such as a potato are continuous (DeVaux, 1891), connected by long 3 μm diameter passages largely devoid of water (Haberlandt, 1927; Wolley, 1962) through which air can be forced by a modest pressure. The relative volume of intercellular air (V_{air}/V_{total} = porosity), measured by weighing plants before and after forcing water or other liquids into their intercellular system, varies from 0.01 to 0.35 in various fruits, leaves and storage organs (Table 3.8). The air is located along with apoplastic solution in the intercellular space (Table 3.9).

Liquid water in the intercellular system enormously increases the resistance to gas mass transport because the diffusion coefficient for gases is 10^4 less in an aqueous phase than it is in air (Tables 15.4 and 15.5). DeVaux (1891) reported that wetting a potato's surface markedly reduced the rate of O_2 transport into the tissue, and often since that time it has been found that water alters O_2 diffusion and depresses respiration even when tissues are shaken in liquid so rapidly that there is no possibility that the external water could be O_2 deficient (Ohmura and Howell, 1960; Ducet and Rosenberg, 1962). O_2 transport from the atmosphere through the shoot to the root stops when a stem is severed because the cut surface becomes water-occluded and essentially impermeable to gases (Greenwood, 1967). Experiments with pulp sections prepared from bulky commodities often are misleading because cell sap released at a cut surface is drawn into the intercellular system by capillary action (Burg and Burg, 1965b; Sorokin, 1966). The intercellular spaces at the cut surface of a potato disc become water-injected to a depth of several microns, creating a 'skin' resistance to ^{37}Ar diffusive flow equivalent to 3 cm of bulk tissue (Wolley, 1962). The O_2 partial pressure required to half-inhibit the respiration of potato discs (Thimann *et al.*, 1954), and the ethylene production and respiration of apple slices (Burg 1967, unpublished data), are much lower if the discs have been blotted to withdraw water that entered the intercellular spaces by capillary action. Air-filled channels would need to have lengths 21-fold and 65.4-fold longer than the thickness of apple and avocado discs, respectively, to account for the diffusive flow of applied CO_2 through the slices (Burg and Burg, 1965b), whereas the tortuosity expected in tissue comprised of densely packed cells is only approximately $Z = \pi/2 = 1.57$, since a gas molecule must diffuse around half the circumference of each cell in traversing the cell's diameter. Even though a mature avocado's CO_2 production is not significantly changed when the fruit is peeled, its ICC increases because water released from damaged cells immediately occludes the intercellular spaces at the avocado's surface, and eventually the fruit forms a new periderm with restricted ability

Table 3.8. Intercellular fraction (V_{air}/V_{total} = porosity) of various fruits, leaves and storage organs. The extreme limits for 156 types ranged from 0.035 to 0.71, although most were in the 0.20–0.30 range (Spector, 1956).

Commodity	V_{air}/V_{total}	Reference
Apple, Glockenapfel	0.210	1
Golden Delicious	0.210	1
Cox Orange	0.160	1
Alexander Lucas	0.010	1
McIntosh	0.30–0.35	2
Red Delicious	0.250	3
Avocado	0.054	4
Banana, 'Manzana,' ripe	0.157	4
'Pinlo,' ripe	0.147	4
Carrot	0.080	5
Guava	0.170	4
Aubergine	0.15–0.30	6
Leaves (various types)	0.17–0.30	7
Mango, 'Hilacha,' ripe	0.132	4
'La India,' ripe	0.052	4
Papaya, ripe	0.120	4
unripe	0.106	4
Parsnip	0.350	5
Passion fruit, ripe	0.409	4
unripe	0.298	4
Pear	0.010	1, 6
Pineapple, ripe	0.133	4
'Los Andes'	0.105	4
Plantain, ripe	0.159	4
unripe	0.159	4
Potato, various	0.0062–0.0134	8
Lord Derby	0.360	9
Sapote	0.140	4
Soursop	0.190	4
Sweet potato	0.180	5
Tomato	0.050	10

References: (1) Henze, 1969; (2) Smith, 1947; (3) Marcellin, 1963; (4) Pantastico, 1975; (5) Theologis and Laties, 1982a; (6) Culpepper *et al.*, 1936; (7) Turrell, 1936; (8) Burton and Sprague, 1950; (9) Smith, 1929; (10) Knee, 1995.

to exchange gas (Ben-Yehoshua *et al.*, 1963; Burg and Burg, 1965b).

Haberlandt (1927) suggested that the intercellular air system in bulky organs may be discontinuous, in the form of alternating columns of liquid and gas ('Jamin's chains') within microconduits. This arrangement would markedly increase the intercellular-gas mass-transport resistance, and because diffusion through the liquid water is not pressure-dependent, LP would be prevented from significantly increasing intercellular gas conductance. If instead the liquid only increased the thickness of the apoplastic solution without occluding the conduits, LP would be able to increase diffusion in the intercellular system, and the superficial water would not substantially increase the total gas phase resistance since the intercellular and 'skin' resistances act in series, and the intercellular resistance typically is much smaller than the 'skin' resistance (Table 3.4). Burton (1982) concluded that although most plant tissues, including bulky organs, are quite well aerated, 'in senescent material the intercellular spaces may become injected with liquid because of leakage of cell contents into them'. Ben-Yehoshua *et al.*

Table 3.9. Volume of air and water in the free space (FS) of various plant tissues (Cosgrove and Cleland, 1983). In pea internode tissue, the free space solution has an osmotic pressure of 2.9 bars in apical regions and 1.8 bars in basal regions. The volume of apoplastic air in leaves of oilseed rape (*Brassica napus* L.) ranges from 0.2 to 0.25 ml per gram fresh weight, and the apoplastic water from 0.06 to 0.10 ml per gram fresh weight (Nielsen and Schjoerring, 1998).

Plant material	Region	% of tissue volume: Air space	FS water
Pea	Apical	2.8	4.8
	Basal	3.2	4.0
Soybean	Apical	3.6	3.6
	Basal	5.2	4.6
Cucumber	Apical	1.1	4.1
	Basal	1.1	4.6

(1963) suggested that during an avocado's post-climacteric period, leaking cell-exudate may clog the fruit's intercellular air spaces and increase its gas mass-transport resistance (Fig. 3.9, *lower right*). The experimental evidence that gave rise to these views will be reviewed because, to the extent that gas exchange becomes limited by diffusion through intercellular liquid, this could significantly influence the commodity's response to a hypobaric pressure.

3.13 Cell Leakage during Senescence and Ripening

Membrane properties such as fluidity affect solute permeability and membrane enzyme activity (Mazliak, 1983; Shinitzki, 1984), and changes in the structure and function of membranes are a universal feature of plant-cell senescence (Woolhouse, 1981). Many studies suggest that membrane compositional changes occur in senescing leaves, ripening fruit discs, fading flower parts and ethylene-treated tissues, increasing their 'apparent free space' (AFS) and tendency to leak ions, amino nitrogen, sugars, acids and other solutes (Glasziou *et al.*, 1960; Ben-Yehoshua, 1961 – referred to in Laties, 1964; Sacher, 1962, 1966, 1973; Ben-Yehoshua, 1964; Eilam, 1965; Abrams and Pratt, 1966; Brady *et al.*, 1970; Brady, 1970 – referred to in Palmer, 1971; Hansen and Kende, 1975; Mayak *et al.*, 1977; Parups, 1977; Suttle and Kende, 1980a; Thompson *et al.*, 1982; Borochov and Woodson, 1989; Borochov *et al.*, 1990; Serek *et al.*, 1995c). Leakiness of petal-cell membranes, and in particular the tonoplast surrounding the vacuole, is a crucial event in flower ageing and fading (Bruinsma, 1983). Membrane fluidity begins to decrease immediately after petunia flowers are harvested (Borochov *et al.*, 1997), and rapidly declines after the bloom begins to evolve ethylene or the gas is applied (Serek *et al.*, 1995c). Changes in this flower's membrane viscosity are associated with a decrease in microsomal membrane phospholipid, and accompanied by a transient increase in the content of diacetylglycerol, a product of phospholipid metabolism (Borochov *et al.*, 1997; Fig. 5.45). Leakage increases immediately after *Tradescantia* flowers are exposed to ethylene (Suttle and Kende, 1980a), while in roses and carnations the same change occurs at a later stage of senescence when wilting commences (Thompson *et al.*, 1982; Borochov and Woodson, 1989). Changes in lipid composition also occur during fruit ripening.[9]

Increases in AFS have been measured during fruit ripening and leaf and flower senescence by incubating tissue slices or leaf and flower parts in 1-^{14}C-mannitol and determining the extent to which the applied solution has been diluted by free-space water; also, by incubating slices in radioactive mannitol, sucrose, fructose, glycerol or KCl, and measuring the amount of isotope taken up after various intervals of time (Eilam, 1965; Sacher, 1966; Brady *et al.*, 1970). As senescence progressed, the AFS in *P. vulgaris* cotyledons changed from an initial value of 15%, eventually reaching 75%. AFS rose from 10% to reach 50% in the senescing *Arum maculatum* spadix, and increased to 70% in ageing bean endocarp tissue (Glasziou *et al.*, 1960; Eilam, 1965). The AFS of banana pulp discs began to rise 3–4 days in advance of the respiratory climacteric in fruits that had not been treated with ethylene (Sacher, 1966), and after the onset of the climacteric in ethylene-treated fruits (Brady *et al.*, 1970). The rise in banana AFS was associated with a marked enhancement of permeability, measured as leakage or uptake of low molecular weight solutes from pulp slices floated on water or a buffer solution (Sacher, 1962; Baur and Workman, 1964). Avocados behaved in the same manner (Sacher, 1962). Loss of dry matter from banana and avocado slices soaked in water (Ben-Yehoshua, 1961 – referred to in Laties, 1964; Sacher, 1962; Ben-Yehoshua, 1964), and K^+ leakage from banana pulp discs (Baur and Workman, 1964) increased dramatically when ripening progressed, and hexose phosphates, PGA, PEP, AMP, ADP and ATP were more rapidly taken up and leaked as avocado tissue slices advanced from pre-climacteric to climacteric (Laties, 1964). Exosmosis from Cox's Orange Pippin or Golden Delicious apple discs floated on water became more extensive during the course of fruit development until at maturity tissue slices floated on water leaked 90% of their electrolytes in 2 h (Simon, 1977). During a water soak, 90% of the K^+ was lost from tomato tissue in 6 h (Vickery and Bruinsma, 1973 – referred to in Simon, 1977), and 50% of the soluble amino acid escaped from banana slices in 1 h (Brady *et al.*, 1970 – cited in Simon, 1977). The cells in ripening bananas became totally permeable to solutes by simple diffusion (Sacher, 1966), prompting Sacher (1973) to conclude that the climacteric is a direct result of changes in permeability and protoplasmic compartmentalization.

Leakage induces alterations in metabolism, making it difficult to determine what the normal state of a tissue might have been before it was exposed to liquid water during on exosmosis study (Laties, 1964). The rate at which cells leak solute into water depends not only on membrane permeability, but also on the solute concentration gradient across their membranes and the rates of active uptake and efflux of the ion or solute which is escaping. Initially, the K^+ efflux from beet discs floated on water is 4.3-fold greater than the influx, and then within one day these trends reverse as the tissue ages (Van Steveninck, 1962). Sugar efflux from banana discs into water appears to increase when slices are prepared at later stages of ripening, but this is misleading because the sugar content increases as ripening progresses. The slices compensate for the exosmosis of sugar by hydrolysing reserve starch to sustain their initial sugar concentration, and the leakage rate relative to their average sugar concentration remains remarkably constant until the climacteric maximum (Fig. 3.5).

The concepts of nearly total passive permeability and AFS in climacteric banana discs (Sacher, 1973) are difficult to reconcile with the observation that pulp slices cut from both mature-green and eating-ripe banana fruit appear to have normal osmotic properties when they are exposed to solutions of varying tonicity (Burg, 1968b). Aged bean endocarp cells with a measured AFS of 80% also can be plasmolysed and deplasmolysed (Laties, 1964), albeit compared to non-aged tissue there is a marked increase in the speed with which plasmolysis and deplasmolysis occur (Sacher, 1959). As early as 1916, Brooks cautioned that leakage from plant cells occurs when tissues that are not normally exposed to liquid water are placed in contact with it. The propensity to leak may reflect a tissue's ability to reinstate its original apoplastic solution after it has been diluted by applied water during a free space or leakage measurement. For example, *Brassica napus* L. leaves that initially contained 0.8 mM NH_4^+ in their apoplastic solution, recreated a 0.8 mM concentration within 1.5 min after they were infiltrated with an NH_4^+-free solution (Nielsen and Schjoerring, 1998), while at the same time a much slower active uptake of NH_4^+ from the apoplast to the cytoplasm continued.

Leakage measurements carried out by floating fruit tissue on water or a dilute buffer solution almost always indicate that the permeability has increased during ripening, but measurements made by floating tissue on a solution possessing a high tonicity similar to that which must be present in the climacteric apoplastic solution give the opposite result (examples 19 and 20). Wachter (1905) and Iljin (1927) long ago

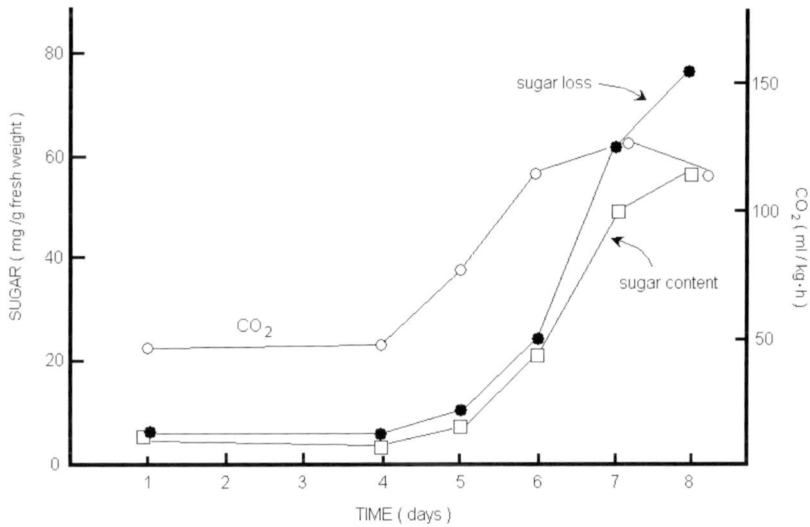

Fig. 3.5. Relationship between sugar leakage and content in tissue cut from non-ethylene-treated bananas at various stages of ripening. CO_2 production by whole fruits was measured immediately before slices were cut. Sugar loss is the total outflow of reducing sugar plus sucrose during a 32-min water soak of slices cut from bananas at the indicated ripening stage (days). The slices leaked almost all of their sugar, but hydrolysed sufficient starch to replace the sugar they had lost. When the duration of the water soak was extended to 120 min, the slices leaked more sugar than they initially contained (Burg *et al.*, 1964).

reported that strips of onion bulb scales and beets give off sugars to distilled water, but less readily to 0.1–0.4 M solutions of KCl, NaCl or KNO_3, and Stiles and Jorgensen (1917a,b) found that 0.2 N NaCl, 0.2 N $CaCl_2$, or a combination of the two, prevented exosmosis from potato slices measured as a change in the conductivity of the external soaking solution. During a 5-min water soak, the fresh weight of ripe apple slices increases by 2.6%, after which water outflow, induced as a consequence of, or in parallel with, solute leakage, causes a progressively intensifying weight loss that reaches nearly 10% within 60 additional min (Burg and Burg, 1960; Fig. 3.6, *upper left*). A massive exosmosis of sugar and acid, which accompanies the cellular water loss, is prevented by increasing the tonicity of the soaking solution with glycerol, mannitol, KCl, sucrose or alanine (Fig. 2.6; Burg and Burg, 1960; Burg *et al.*, 1964). A small but significant water uptake occurs when the applied tonicity is slightly lower than that which initially must have existed in the tissue, and at a higher tonicity the cells behave as typical semi-permeable osmometers (Fig. 3.6, *upper left*). The kinetics of reducing sugar leakage from apples indicate a rapid emptying of free space within 10 min, which in water, but not in 0.6 M glycerol, is followed by a major cellular compartment emptying in 3 h (Fig. 3.6, *upper right*). Slices cut from grape, tomato, plum, banana and melon fruits display a similar magnitude and the same kinetics of electrolyte leakage as apple slices, and in each instance exosmosis is prevented by the addition of 1 M glycerol or 0.5 M KCl to the soaking solution (Simon, 1977; Fig. 3.7). Similar kinetics were reported for leakage into water from *Ipomoea tricolour* Cav. flower rib tissue (Hansen and Kende, 1975), banana slices (Brady *et al.*, 1970), post-climacteric pear discs, potato slices and etiolated pea epicotyl sections (Burg and Burg, 1960; Burg *et al.*, 1964); pre-climacteric honeydew melon tissue (Sacher, 1973), beet root, carrot root, lettuce leaf and cucumber fruit. Potato tuber slices leaked when they were floated on water; and leakage from cucumber, beet root and honeydew melon decreased when the tonicity of the soaking solution was increased.

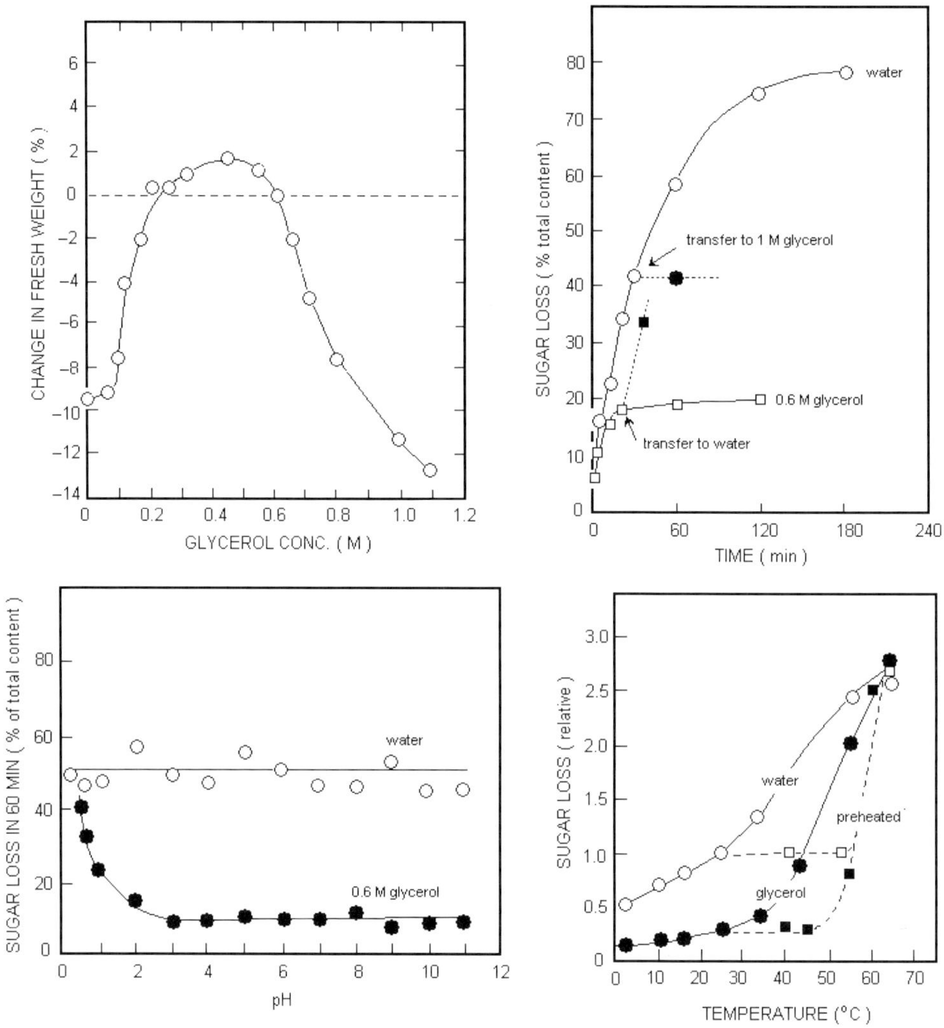

Fig. 3.6. *(upper left)* Changes in the fresh weight of McIntosh apple cylinders after 60 min treatment with water and various concentrations of glycerol (Burg and Burg, 1960). *(upper right)* Leakage of reducing sugar from McIntosh apple tissue slices into water or 0.6 M glycerol. Broken curves indicate sugar loss in sections transferred from 0.6 M glycerol to water or from water to 1 M glycerol at the indicated times. *(lower left)* Effect of pH on reducing sugar leakage from McIntosh apple tissue slices during incubation in water or glycerol (Burg *et al.*, 1964). *(lower right)* Effect of temperature on reducing sugar leakage from McIntosh apple tissue during a 60-min incubation in water or 0.6 M glycerol. The broken curves labelled 'preheated' indicate the amount of leakage that occurs when tissue, which has been incubated in air for 60 min at the specified temperature, is transferred to water or 0.6 M glycerol for 60 min at 24°C (Burg *et al.*, 1964).

Leakage induced by water results not only from the continual emptying of the apoplastic solute and its refilling from the cytosol, but also from membrane 'stretching' and cell 'bursting' caused by osmotic stress. The osmotic pressure of banana pulp increases from 7.15 atm in green fruit, to 16.91 atm in slightly yellow fruit, 25.48 atm in fruits with a green tip and 29.06 atm when the banana is slightly speckled (Stratton and von Loesecke, 1931; von Loesecke, 1950; 7.11). During ripening, the resultant rise in suction pressure (Fukushima *et al.*, 1980) coupled with the degradation of cell wall

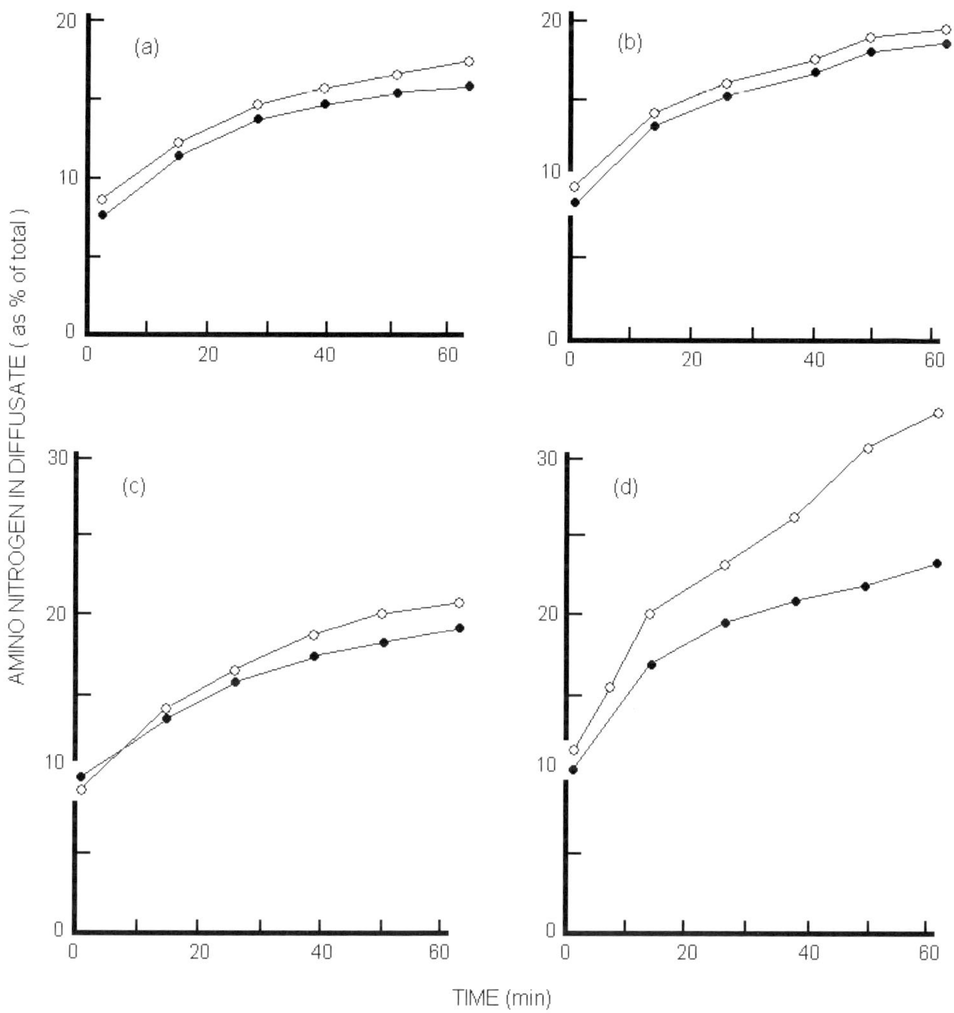

Fig. 3.7. Effect of solution tonicity on amino acid leakage from pulp tissue disks of Dwarf Cavendish bananas after treatment with 10 μl/l ethylene for 0 (a), 24 (b), 48 (c) and 72 h (d). Leakage into water (○) and 0.6 M mannitol (●). The ethanol-soluble carbohydrate contents of the pulp were: (a) 3.08 ± 0.12, (b) 3.48 ± 0.08, (c) 10.72 ± 1.48 and (d) 21.96 + 3.04% of dry matter (Brady *et al.*, 1970).

polymers and the middle lamella, may predispose cells to stretch-induced membrane damage when they are exposed to water during a leakage or free-space measurement. Starch hydrolysis elevates a ripening apple's solute concentration and increases the tonicity of its apoplastic solution, rendering the fruit susceptible to osmotic damage when it is exposed to water. Initially, apple cells exclude Evan's Blue, but during a water soak the number of cells able to be coloured by the dye progressively increases, indicating that they have 'burst' (Simon, 1977). While this may account for the eventual result, it cannot explain the behaviour during the initial hour, for regardless of whether apple tissue is floated on water or 0.6 M glycerol, after 1 h the CO_2 production and O_2 consumption of the slices is unchanged from the original rate in air, indicating that few if any cells have burst (Burg and Thimann, 1959). The ethylene production rate remains high when slices are incubated in 0.6–1.0 M concentrations of glycerol (Burg and Thimann, 1959; Burg *et al.*, 1964) or sorbitol (Mattoo and Lieberman, 1977), but within 1 h decreases by 60% if the tissue is exposed to distilled water or a dilute buffer solution. Leakage immediately ceases (Fig. 3.6, *upper right*) and ethylene production recovers to nearly its initial rate (Fig. 5.4, *lower*) if at any time during a 1 h water soak the tissue is transferred to a 0.6 M glycerol or 0.5 M KCl solution (Burg and Thimann, 1959; Burg *et al.*, 1964; Simon, 1977), but within 3 h in water or buffer the damage becomes irreversible, and microscopic examination at that time reveals the presence of many swollen cells and an appreciable number that have burst open (Mattoo and Lieberman, 1977). The initial behaviour may involve reversible membrane stretching and associated changes in the activity of membrane-bound EFE or ACS, and loss of ACC without cell bursting (Burg and Burg, 1960). By directly disturbing membrane integrity, a high temperature (Fig. 3.6, *lower right*) and pH lower than 3 (Fig. 3.6, *lower left*) negate the ability of a high tonicity to prevent apple tissue exosmosis, and cause the slices to leak as if they were placed in water (Stiles and Jorgensen, 1917b). The same high temperature that causes leakage when apple slices are floated on a high-tonicity solution (Fig. 3.6, *lower right*) inhibits ethylene production when the tissue is incubated in air (Fig. 5.3). Leakage also is promoted when a chilling temperature disturbs membrane integrity (8.10). Electrolyte leakage from *P. vulgaris* leaves into water is stimulated after either excised leaves or intact plants are chilled from 25 to 5°C, and the effect is reversed if the leaves are returned to 25°C within 4 h (Wright, 1974).

Prevention of leakage by high-tonicity solutions does not negate evidence indicating that alterations in membrane structure, fluidity and permeability occur prior to and during ripening and senescence. It does, however, raise serious doubts about the extent to which these changes cause clogging of the intercellular spaces, increase the AFS and lead to a loss of compartmentalization (Brady, 1987). Brady *et al.* (1970) found that 1 μl/l applied ethylene induced a respiratory response in bananas within 8 h, and caused the respiration rate to reach a half-maximal climacteric value within 24 h, but did not increase amino acid leakage or apparent free space to mannitol before 32 h. From this behaviour he surmised that the respiratory increase in bananas is not dependent on changes in tissue permeability, and concluded that the leakage measurement is an indication of the 'propensity to leak' at a particular stage of senescence or development, rather than evidence that massive leakage had occurred into the intercellular system. The porosity of plantains, papayas and passion fruits remains constant during ripening (Table 3.8). If a significant quantity of solute leaked into the intercellular spaces, the porosity would decrease.

3.14 Measuring the Intercellular Gas Concentrations

The concentration of a metabolic gas present in the intercellular air spaces can be calculated if the tissue's surface area, skin resistance and rate of gas consumption or production are known (equations 3.14 and 3.15). Because of stomatal action's relevance to photosynthesis and transpiration, a large database describing the CO_2 and water vapour conductance of the cuticle and stomates of leaves has been accumulated, and while ethylene gas transport resistance values have not been measured in leaves, they can be reliably calculated from this information based on differences in the binary diffusion coefficients of CO_2, ethylene and water vapour in air (Table 15.4; Farquhar and Sharkey, 1982; Sharkey *et al.*, 1982). Example 23 indicates that gas exchange occurs so freely through open leaf stomates that in daylight the highest recorded ethylene production rate by a leaf, 10 μl/kg·h (Ben-Yehoshua and Aloni, 1974b; Gepstein and Thimann, 1981; Kao and Yang, 1983), could not raise the IEC to a threshold for ethylene action, 0.001–0.01 μl/l (Table 5.2). Example 24 demonstrates that if a tobacco leaf were to close its stomates so tightly at night that gas

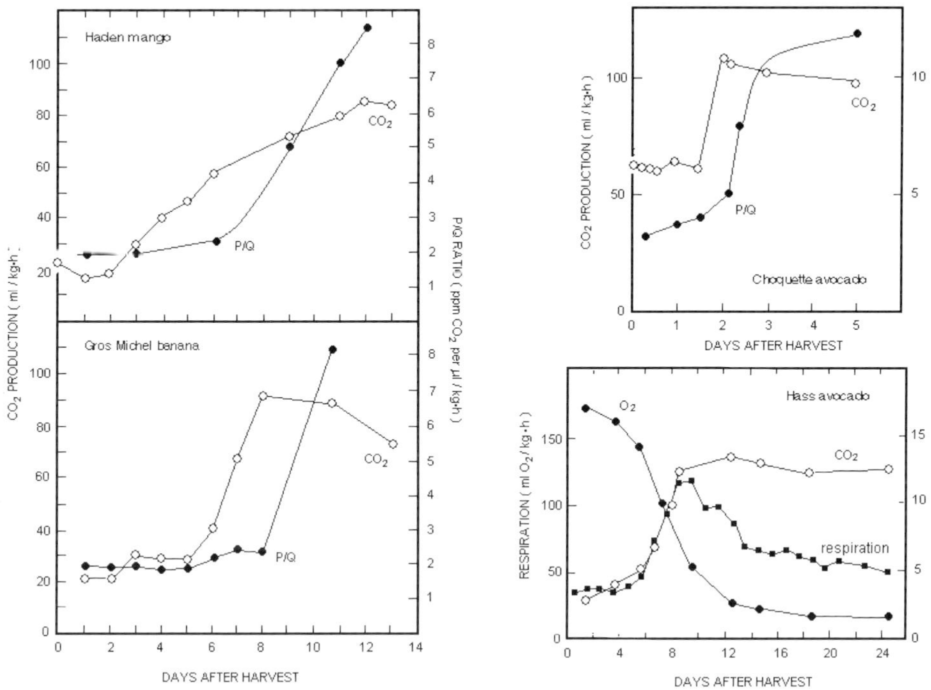

Fig. 3.9. (*upper right*) Changes in resistance to CO_2 exchange during the ripening of Choquette avocados. Abscissa – days after harvest (Burg and Burg, 1965b). (*lower right*) Changes in internal [CO_2], internal [O_2] and respiration during the ripening of Hass avocados at 20°C (Ben-Yehoshua *et al.*, 1963). (*left*) Changes in the resistance to gaseous permeability during the ripening of harvested Gros Michel bananas and Haden mangoes at 24°C. The ratio between the internal CO_2 concentration (P) and CO_2 production rate (Q) expressed as ppm per µl/kg·h, is a measure of the resistance to gas exchange. Ripening is indicated by the CO_2 respiratory climacteric (Burg, 1964, unpublished).

and also in apples (Reid *et al.*, 1973), the ratio between the ICC and CO_2 production rate, and IEC and ethylene production rate remains essentially constant, indicating that the resistance to gas mass transport through the peel and intercellular system has not changed. It is only during the late climacteric and post-climacteric periods that this ratio increases in mangoes and bananas by up to fourfold (Wardlaw and Leonard, 1940; Banks, 1983). The ratio between a Conference pear's ICC and the rate at which CO_2 is produced is the same in hard green and overripe fruits, and only when the pear turns from overripe to mealy does the ICC begin to increase more rapidly than the internal [O_2] decreases (Kidd and West, 1949a; Burton, 1982). In these examples, the exchange of CO_2, O_2 and ethylene is equally affected during ripening, indicating that intercellular gas transport is occurring at a rate limited by diffusion in air. During the post-climacteric period the air spaces act as though they have a lesser effective cross-sectional air-filled surface area and/or a greater tortuosity, or else the skin's resistance to gas exchange has increased.

A tomato fruit's resistance to gas exchange does not increase during ripening. As this fruit advances from a mature-green to a table-ripe stage, its RQ rises from 1.0 to 1.15, the internal O_2 decreases by 2% from 16.7 to 14.7%, and the CO_2 increases by 2.3%, rising from 4.4 to 6.7% (Pratt and Workman, 1962; Lyons and Pratt, 1964; Table 2.2). During the ripening of cantaloupes, the internal [O_2] draw-down and [CO_2] build-up are balanced and the ratio between the ICC and CO_2 production rate remains constant, indicating that the resistance to gas exchange has not increased (Lyons *et al.*, 1962; Table 2.2). As papaya fruits progress from a green to an overripe stage, intercellular [O_2] decreases from 17.5 to 3.5%, while the [CO_2] increases to nearly the same extent, from 1.5 to 13.5% (Jones and Kubpta, 1940; Table 2.2).

Hass avocado fruits behave differently (Fig. 3.9, *lower right*; Ben-Yehoshua *et al.*, 1963). During their climacteric rise in respiration, the ICC increases by 9.4% and the internal [O_2] declines by 9.6%, but during the post-climacteric period when the respiration rate decreases by 55% and the internal [O_2] by 5.6%, the ICC only increases by approximately 0.2%. This indicates that during the post-climacteric period gas exchange was limited by diffusion through water. A similar pattern has sometimes been observed during the banana post-climacteric period. At the climacteric peak, the intercellular [O_2] is drawn down to approximately 10% and the ICC is elevated to the same extent (Table 2.2; Wardlaw and Leonard, 1940; Fig. 3.9, *left*), but when the respiration rate subsides during the post-climacteric period, the internal [O_2] decreases to approximately 2.2%, while for several days the ICC hardly changes, and then it rises progressively, reaching 18% in completely senescent fruits (Wardlaw and Leonard, 1940). In overripe bananas, equivalence between the increased ICC and decreased internal [O_2] may be due to a high RQ caused by low intercellular [O_2], coupled with a small increase in the resistance to CO_2 gas exchange and a much larger increase in the resistance to O_2 diffusion.

Ethylene gradients in horizontal stems

Side-to-side radial diffusion of ethylene through the intercellular system occurs in gravicurving stems and flower spikes after auxin redistributes and stimulates ethylene production in their lower half (Denny, 1936; Burg and Burg, 1966b; Abeles and Gahagan, 1968b; Wright *et al.*, 1978; Wheeler and Salisbury, 1981; Harrison and Kaufman, 1982; Clifford *et al.*, 1983; Harrison and Pickard, 1984, 1986; Clifford *et al.*, 1985; Kaufman *et al.*, 1985; Harrison and Pickard, 1986; Wheeler *et al.*, 1986; Balatti and Willemöes, 1989; Takahashi *et al.*, 1991; Woltering, 1991; Philosoph-Hadas *et al.*, 1996). The magnitude of the *in vivo* ethylene gradient that develops will depend on the tissue's porosity and diameter, the tortuosity of the intercellular system, radial distribution of ethylene production, and the rate of ethylene synthesis at each location. Because the radial path length in stems and flower spikes is one or two orders of magnitude less than the radius of a mango, apple or banana, the gradients that develop in stems and flower spikes should be much smaller than those that have been measured in fruits (Table 2.2). Vacuum extraction measurements of the IEC in tissue halves excised from horizontally positioned tomato and cocklebur stems (Wheeler *et al.*, 1986), snapdragon and *Kniphofia* (Torch-lily) flower spikes (Woltering, 1991; Philosoph-Hadas *et al.*, 1996, 2000), dandelion peduncles (Clifford *et al.*, 1983, 1985) and tomato and cocklebur stems (Wheeler and Salisbury, 1981; Wheeler *et al.*, 1986) suggest that a large lower/upper IEC gradient has developed, with a highly active ethylene concentration in the lower, and sometimes in the upper side. Because the lower half of horizontally positioned *Kniphofia* flower stalks produced ethylene 16.8-fold faster than the upper half, and the IEC determined by vacuum extraction was only 7.6-fold higher in the lower half, it was concluded that the ethylene gradient was reduced by 55% due to radial diffusion from the lower to upper half through the intercellular system (Woltering, 1991). This is misleading because ethylene produced during a vacuum extraction while tissue is submerged under saturated ammonium sulphate caused the measured IEC to bear no relation to the IEC that existed prior to the measurement. The IEC values measured in *Kniphofia* flower stalks would easily have risen during the 3 min that ethylene accumulated in the intercellular spaces while the gas was unable to escape because outward diffusion was prevented due to a low diffusion rate through aqueous ammonium sulphate (example 22). Wheeler *et al.* (1986) measured the IEC in the upper and lower halves of horizontally positioned tomato and cocklebur stems after extracting gas samples with a vacuum of 150 mm Hg applied for 5 min, or 550 mm Hg for 10 min. Essentially no ethylene was recovered from tissue cut from vertical stems or the upper half of horizontal stems, but a substantial IEC was measured in pieces from the lower half excised from horizontally positioned stems, and the IEC was twice as large when it was measured by a 10-min rather than a 5-min vacuum extraction.[12] These vacuum-extraction measurements with *Kniphofia* flower stalks and tomato and cocklebur stems do not indicate whether a significant gradient or active ethylene concentration existed in the tissue during the gravitropic response because the IEC in each half was created by the tissue's ethylene production after the halves were separated and immersed under saturated ammonium sulphate. Inhibitors of ethylene action and production interfered with the gravitropic response of tomato and cocklebur stems, and their effect was reversed and a normal gravitropic response reinstated by a low-applied ethylene concentration (Wheeler *et al.*, 1986). As applied ethylene distributes uniformly within a tissue, this result proves that an ethylene gradient was not required for ethylene's effect on gravitropism.

3.17 Axial Diffusion of Intercellular Gases

Changes in the intercellular gas content within specific tissues, brought about by altered CO_2 and ethylene production or O_2 consumption, conceivably could influence remote parts of a plant (Zimmerman *et al.*, 1931; Greenwood, 1967; Jackson and Campbell, 1975; Zeroni *et al.*, 1977). Either phloem mass transport, dissolution in transpirational water, or diffusion through the continuous intercellular system or specialized aerenchyma tissue might account for the observed transport velocity, but only diffusion through an air phase is compatible with the transport capacity. The quantity of gaseous ethylene moving in the transpiration stream, computed from Henry's Law and the transpiration rate per dm^2 of leaf surface, could account for only a small part of the transport of applied ethylene from the root to the leaves of tomato plants (Jackson and Campbell, 1975).

Changes in the ethylene production rate of a tissue usually do not exert an effect in nearby tissues. The IEC drops off rapidly when ethylene diffuses through *V. faba* petioles (Zeroni *et al.*, 1977), and it must decline more than 140-fold when the gas moves from a cotton stem through the petiole to a leaf, as cotton plants shed their leaves if exposed to more than 0.08–1 μl/l ethylene (Beyer and Morgan, 1971), whereas a single leaf excluded from this treatment and kept in ethylene-free air does not abscise when the rest of the plant is fumigated with 14 μl/l ethylene (Beyer, 1975c). These results indicate that stems and petioles are well protected from changes in the ethylene content of their attached leaves, and such plants do not function as a unit with respect to shifts in IEC. In agreement with this conclusion, it has been found that ACC rather than ethylene is the 'mobile wilting factor' that spreads throughout a flower after pollination induces ethylene production in the stigma (Nadeeau *et al.*, 1993; Woltering, 1993). A theoretical analysis (example 3), predicts that during axial diffusion of ethylene through etiolated pea epicotyl tissue, the density (ρ) of the diffusing gas should decrease by 50% in a 1.9 mm length if the tortuosity (μ) of the intercellular system is 10, or in 19 mm if $\mu = 1$. The predicted rapid IEC dropoff has been confirmed by experiments showing that there is a high endogenous ethylene emanation rate from intact pea plumules, but none from the next 3 cm of epicotyl (Goeschl *et al.*, 1967). The same IEC decrease in tomato stems requires 1–2 cm (Jackson and Campbell, 1975), and in *V. faba* stems, where a central lacuna lowers the value of μ, it occurs in 4–5 cm (Zeroni *et al.*, 1977).

A hypobaric condition cannot cause ethylene to diffuse a farther or lesser distance along the longitudinal axis of a plant tissue before it escapes radially because diffusion through an air phase limits both the radial and axial diffusive movement of gases through stems and petioles, and LP should promote both equally.

3.18 Stomates and Lenticles

Stomates, lenticles and the pedicel-end stem scar are the only air-filled breaks in the continuity of the epidermis and cuticle. The rate at which gases and vapours move through open stomates depends on their binary diffusion coefficient in air (Sharkey *et al.*, 1982), and the same must be true when gases or vapours diffuse through air-filled lenticles and the pedicel-end stem scar. A complex regulatory mechanism modulates the stomatal aperture in response to changes in turgor and water potential, abscisic acid, light and CO_2, keeping the ICC high enough to support photosynthesis during daylight hours, and minimizing the loss of transpired water at night. When the stomates open in the light, the water potential of C-3 plants decreases by about 5 bars, while photosynthesis draws down the ICC to 10–100 μl/l; at night, when the stomates close, dark respiration elevates the ICC to 0.12–0.8% and turgor increases (Raschke, 1975; Noble, 1991; Willmer and Fricker, 1996).

The transpirational resistance of various leaves with open stomates varies from 0.5 to 12 s/cm depending on stomatal size and frequency (Noble, 1991), and their cuticular transpirational resistance is much higher, ranging from 20 to 400 s/cm in different species (Holmgren *et al.*, 1965; Cowan and Milthorpe, 1968; Noble, 1991). Depending on the air velocity and irrespective of whether water vapour moves through the stomatal or cuticular route, it encounters an additional resistance of 0.13–2.1 s/cm in passing through the boundary air layer associated with the leaf's surface (Slatyer, 1967; Noble, 1991). Since conductance in air is 1.7-fold higher for water vapour compared to ethylene, a leaf's stomatal resistance to ethylene transport is in the range 1.1–24 s/cm. The leaf's stomatal resistance to CO_2 and O_2 is 7% and 28% lower, respectively, compared to ethylene.

If the stomates of a typical leaf closed hermetically, forcing all gas transport to occur through the cuticle, the IEC would rise by 30,000-fold above the value with open stomates (example 24). This would have an effect on ethylene action equivalent to that caused by a 3,000,000% increase in the rate of ethylene synthesis. Instead, air conductance through 'closed' leaf stomates, measured with a porometer, tends always to be at least 2–10% of the value at full stomatal opening (Figs 3.8, 4.11 and 4.12; Table 4.9). 'In plants well supplied with water, complete stomatal closure may be rare even in darkness – a hydroactive response may be necessary to shut the stomata tightly' (Raschke, 1975). The ICC in a 'typical' leaf rises to 0.12% at night (Raschke, 1975), and for that to occur, the leaf's resistance to gas exchange must increase by approximately 1380-fold when its stomates close (example 25). Leaves synthesizing ethylene at a rate of ≤ 0.1 μl/kg·h typically do not respond to their IEC, but when the rate increases to 1–3 μl/kg·h, they are likely to abscise, senesce, become epinastic and turn yellow (5.28; Table 5.2). When a 'typical' leaf's stomates are 'closed', an ethylene production rate of ≤ 0.1 μl/kg·h should sustain an IEC of ≤ 3.8 nl/l, which is below the 10 to 20 nl/l threshold for ethylene action in leaves, but if the rate increases to 1–3 μl/kg·h, their IEC rises to an active level of 33.8–100 nl/l (example 25).

With open stomates, the transpirational resistance of a full-sized Valencia orange fruit in the field is approximately 13.5 s/cm (Moreshet and Green, 1980), and the stomatal resistance to ethylene transport, corrected to account for the different binary diffusion coefficients of ethylene and water vapour in air, is 23 s/cm. When a harvested Valencia orange's stomates close, the resistance to ethylene exchange increases to 6886 s/cm (Ben-Yehoshua *et al.*, 1979), and since a fruit's cuticular resistance to ethylene transport is at least tenfold higher (Table 3.12), this implies – and electron scanning micrographs confirm (Fig. 3.10) – that the stomates remain cracked open in harvested oranges and are responsible for most of the ethylene exchange. LP should accelerate gas and vapour diffusive exchange through the air phase of cracked-open stomates.

Because the residual stomatal transpirational resistance (r_s) of 'closed' stomates acts in parallel with the cuticular transpirational resistance (r_c), the total transpirational resistance of the stomates (s) and cuticle (c) is:

$$r_{c,s} = [(r_s)^{-1} + (r_c)^{-1}]^{-1} \quad (3.21)$$

Equation 3.21 indicates that regardless of how tightly the stomates close, transpiration cannot be decreased to less than the rate at which water is conducted through the cuticle because at complete closure $r_{c,s} \cong r_c$. In different leaves, the cuticular transpirational resistance ranges from 20 to 200 s/cm (Holmgren *et al.*, 1965; Raschke, 1975; Noble, 1991), and therefore when the stomates close sufficiently to increase their transpirational resistance to approx. 40–80 s/cm, there is a progressively diminishing advantage to tighter closure since it could not increase the transpirational resistance to more than 20–100 s/cm. Because the diffusive resistance is 1.7-fold higher for ethylene than for water vapour, when gas exchange occurs through closed-leaf stomates having a residual transpirational resistance of 40–80 s/cm, the stomatal resistance to ethylene is 67–134 s/cm. At night, the resistance to ethylene exchange will be determined solely by the residual stomatal resistance, since it acts in parallel with an ethylene cuticular resistance of approximately 200,000 s/cm (3.20). A 90–95% stomatal 'closure' would increase the diffusive resistance to ethylene by 10–20-fold in a 'typical' mesophytic leaf, from 6.7 s/cm with fully open stomates, to approximately 67–134 s/cm when they close.

Stomates close when a commodity is stored or transported in darkness at atmospheric pressure, but they open in LP (4.15; Kirk and Andersen, 1986; Kirk *et al.*, 1986; Veierskov and Kirk, 1986). This response, combined with improved gaseous diffusion at a low pressure, enormously improves intercellular ventilation. Reducing the pressure to 1.33 kPa (10 mm Hg) would by

Fig. 3.10. Mature Hamlin orange surface (500×) showing the heavy wax platelets and a partially occluded stomate (Pantastico, 1975).

itself increase the conductivity of cracked-open stomates by 80-fold (Table 3.15), but coupled with stomatal opening it improves a Valencia orange's gas exchange by 24,000-fold compared to the rate at atmospheric pressure.[13] In LP, the combined stomatal and intercellular resistance to gas mass transport typically is less than the cell's liquid phase resistance, and the box resistance is likely to be the major barrier limiting diffusion (Table 3.4).

The anatomical feature that distinguishes water and gas mass transport in fruits from that in leaves is not the nature of the cuticle. In various fruits the cuticular transpirational resistance has nearly the same range of values, 30–380 s/cm (Wardlaw and Leonard, 1940; Fockens and Meffert, 1972; Johnson, 1976; Sastry *et al.*, 1978; Moreshet and Green, 1980; Jones, 1981; Burton, 1982; George *et al.*, 1982; Jones and Higgs, 1982), as it does in leaves, 20–400 s/cm (Holmgren *et al.*, 1965), and the cuticular resistances to CO_2 determined in various fruits, 5600–16,800 s/cm (Tables 3.11 and 3.12), encompass the average cuticular resistance measured in grape leaves, 13,500 s/cm (Boyer *et al.*, 1997). The main difference between fruits and leaves is the presence in leaves of a much larger number of air-filled stomatal pores, which, when open, provide an alternate low-resistance pathway for water and gas conductance.

Table 3.11. Measured skin resistance ($r_{p,c,a}$) of apples, tomatoes and Valencia oranges to water vapour, CO_2, O_2 and ethylene. The stomates were closed in Valencia oranges (Ben-Yehoshua *et al.*, 1985) and Golden Delicious apples. McIntosh apples and tomatoes do not have functional stomates (Clements, 1935; Clendenning, 1941).

	Resistance (s/cm)				
Fruit	H_2O	CO_2	O_2	C_2H_4	Reference
Valencia orange	106	5,738	6,025	6,886	1
Apple					
McIntosh	–	13,657	–	18,339	2
Granny Smith	398	–	–	–	3
Golden Delicious	55–249	–	–	–	3, 4, 5
Tomato	–	7,400	–	–	2

References: (1) Ben-Yehoshua *et al.*, 1985; (2) Burg and Burg, 1965b; (3) Horrocks, 1964; (4) Jones, 1981; (5) Fockens and Meffert, 1972.

Table 3.12. Relationship between the resistance to ethane and CO_2 mass transport in red tomato fruits before and after sealing the stem scar with lanolin. D_{water} is the binary diffusion coefficient for ethane or CO_2 in liquid water; K_H the Henry's law coefficient for ethane or CO_2 in liquid water; D_{air} the binary diffusion coefficient of ethane or CO_2 in air. Temperature = 25°C. A similar result was obtained for ethane with mature-green tomatoes. The data indicate that 36% of a harvested tomato's CO_2 exchange occurs through its cuticle exclusive of the pedicel-end stem scar, whereas 97% of a tomato's ethane exchange takes place through its stem scar and only 3% through the skin. From Cameron and Yang, 1982, and Burg and Burg, 1965b. The cuticular resistance of a green pepper, measured with the pedicel-end scar sealed with lanolin, is r_c = 10,300 s/cm (Burg and Burg, 1965c).

Parameter	CO_2	C_2H_6	Ratio CO_2/C_2H_6
r_c (s/cm × 10^{-3}) skin – (+lanolin)	19.6	280.0	0.07
$r_{p,c}$ (s/cm × 10^{-3}) whole fruit – (–lanolin)	7.4	7.8	0.95
D_{water} (cm²/s × 10^5)	2.05	1.51	0.74
K_H (mol/dm³ per atm × 10^3) @ 25°C	33.9	1.95	0.057
K_H/D_{water} (cm²·mol per atm × 10^2) @ 25°C	16.54	1.29	0.078
D_{air} (cm²/s) @ 25°C	0.165	0.141	0.85

Functional stomates are present in green stems and petioles; the sepals and petals of flowers; foliage plants and cuttings; the inner and outer surfaces of pods; young quince, apple, grape,[14] banana and tomato fruits; mature limes, oranges, grapefruit, papayas, mangoes, bananas, stone fruits, grapes and soft fruits such as currants and gooseberry; celery, cabbage, asparagus, cucumber, lettuce, Brussels sprouts, leeks, scallions, cauliflower and broccoli (Martin and Juniper, 1970; Pantastico, 1975; Burton, 1982; Kanellis and Roubelakis-Angelakis, 1993; Willmer and Fricker, 1996).

During a Valencia orange's development, the stomatal density reaches a maximum of 160 per mm^2 when the fruit is small, and then the expanding surface area progressively spreads the stomatal spacing until in large fruits the density is reduced to 20–50 per mm^2, compared to 360–500 per mm^2 in a Valencia orange leaf (Moreshet and Green, 1980). In mature Navel oranges, the stomatal frequency is only 13.9 per mm^2 (Turrell and Klotz, 1940), and the older stomates may form a cuticle across the inner tangential wall of the cell surfaces in the substomatal cavity (Scott and Baker, 1947). The peel of mature-green bananas contains 4.8 stomata per mm^2 compared to 1700 per mm^2 in banana leaves (Wardlaw and Leonard, 1936a), and in the fruit they may become partially dysfunctional during final growth and ripening (Johnson and Brun, 1966).

Often the periderm becomes active as a fruit develops, displacing the stomates outwardly, eliminating them or replacing them with lenticles that provide a gas phase continuum from the intercellular system through the cuticle to the atmosphere. Lenticles replace stomates early in the development of pome fruits, often forming in breaks caused by the complete removal of hairs or by skin expansion (Clements, 1935). The stomatal frequency in Golden Delicious apples decreases from 25 per mm^2 in young developing fruits, to < 1 per mm^2 in mature fruits (Blanke and Bonn, 1985), and mature McIntosh apples have 5 lenticles per cm^2 of peel and apparently no functional stomates (Clements, 1935). During an avocado's development, the stem end elongates, decreasing the number of stomates per unit area at the proximal end (Hass, 1936), and eventually many if not all of the stomates are replaced by lenticles (Biale and Young, 1971). In thick-skinned avocado varieties, lenticles are produced where the death of the stomates is followed by cork formation; in thin-skinned varieties such as Fuerte, there is little lenticle formation in the region of dead stomates (Hass, 1936).

Even when a mature fruit's stomates persist, in many instances they are partially or wholly occluded with wax. Stomates are lacking or non-functional in McIntosh apples (Clements, 1935) and mature cranberries (Forsyth *et al.*, 1972), and completely sealed over in mature-green and red tomatoes (Clendenning, 1941; Cameron and Yang, 1982); blueberries have neither functional stomates nor lenticles (Eames and MacDaniels, 1947). Banana stomates remain closed under most environmental conditions from harvest through the stage at which the peel starts to show a blush of yellow colour (Leonard, 1941), even through they can be opened by immersing epidermal sections in water (von Loesecke, 1950), or in response to low intensity light at 90–100% relative humidity (Johnson and Brun, 1966). Between 12 and 60% of a mature orange's stomates are occluded by natural wax (Albrigo *et al.*, 1982; Fig. 3.10), but nevertheless prior to harvest they open and close in response to light and are effective in conducting water and CO_2 (Moreshet and Green, 1980). The transpirational resistance of oranges ranges from approximately 13 s/cm when the stomates are open to 106 s/cm if they are closed (Table 3.11), and based on Graham's diffusion law ($D \sim M^{1/2}$) the resistances to O_2, CO_2 and ethylene should be < 21 s/cm with open stomates. Instead, in harvested oranges the resistance to gas exchange is approximately 6000 s/cm, indicating that the stomates have shut. A small portion of the initial fully open-pore area would account for the residual gas exchange capacity (Ben-Yehoshua *et al.*, 1985). Scanning electron micrographs confirm that most of the stomates in harvested Valencia oranges have closed, except for an occasional few that are slightly cracked open.

Apparently, the stomatal resistance of petioles and stems has not been measured, not even for water vapour. Anatomically, petioles resemble stems rather than blades, and the stomatal density in young stems and petioles is 1/10–1/20 that in leaf blades of the same species (Zimmerman *et al.*, 1931; Fisher, 1980). Young stems of *Phoradendron flavescens* have 28 stomates per mm^3; old stems have fewer than 0.5 functional stomates per mm^3. Since the stomatal density in stems and petioles is similar to the stomatal or lenticular density in fruits such as apples and citrus, presumably the skin resistance of stems, petioles and these fruits is approximately the same and 1000-fold greater than that of a typical leaf with open stomates. To convert these estimates to a ratio comparing the μl/l (IEC) per μl/kg·h production rate in stems, a correction must be applied that takes into consideration the vastly different surface-to-volume ratios of stems and fruits. Example 5 indicates that the μl/l per μl/kg·h ratio of a stem or petiole should be approx. 0.16–0.32. In agreement with this prediction, a 0.13 ratio was calculated for pea subapical stem sections by measuring the biological response observed when differing ethylene production rates were induced by applied IAA (Burg and Burg, 1962b) or physical stress (Goeschl *et al.*, 1966), comparing that result to the experimentally determined responses elicited by a range of applied ethylene concentrations. The ratio in bean hypocotyl hooks was 0.42 for both ethylene and CO_2 (Kang and Ray, 1969a). Because the bean and pea tissues were etiolated and incubated in darkness, these experiments are indicative of the behaviour of stems with closed stomates.

By increasing the tightness of stomatal closure at night or reducing the degree of opening during the day, a treatment or developmental process might induce a response to ethylene without changing the ethylene production rate. Stomatal responses become more rapid as leaves mature and larger apertures are achieved, and then as the leaf continues to age the responses become more sluggish and the maximum aperture is smaller (Fig. 3.11; Ludlow and Wilson, 1971; Vaclavik, 1973; Davis and McCree, 1978; Wilmer *et al.*, 1988). The stomatal conductance of tobacco leaves decreases by 90–95% as senescence advances (Ozuna *et al.*, 1985). Factors within the senescing leaf must limit the aperture, since stomates in senescing tobacco epidermal strips are capable of opening fully.

The maintenance of open stomates in LP and their closure in CA and NP is a significant difference with broad implications because of the importance of gas exchange and internal gas composition during storage. Any commodity possessing stomates can be expected to be better preserved in LP than in CA, MA or NP. Lowering the pressure to 2.67 kPa (20 mm Hg) should increase the CO_2 conductance of a Valencia orange by 10,000-fold (example 4) and cause the gas phase resistance to be close to the liquid phase resistance of the cells (Table 3.4). This would explain why at 10°C Valencia oranges benefit from storage at 2.67 kPa (20 mm Hg = 0.1% $[O_2]$), while at atmospheric pressure they are damaged if $[O_2]$ is lowered to < 5% (Thompson, 1998).

3.19 Pedicel-end Stem Scar

A mature tomato has neither functional lenticles nor stomates (Fig. 3.13). Instead, its calyx is the major avenue for gas exchange (Brooks, 1937; Clendenning, 1941; Burg and Burg, 1965b). This was deduced from the observation that harvested tomatoes ripen more slowly if wax is applied over their pedicel-end stem scar (Wardlaw and McGuire, 1939 – referred to in Clendenning, 1941). It was inferred that covering the stem scar with wax delayed ripening by impeding gas exchange sufficiently to create an internal modified atmosphere enriched in CO_2 and depleted in O_2. Tomato ripening is delayed if the $[O_2]$ falls below 12.5% (Kader and Morris, 1974, 1975a; 3.10).

Gas exchange through the pedicel-end stem scar has been studied before and after sealing one or both of the ends of a fruit with a gas-tight lanolin paste, using the ethane

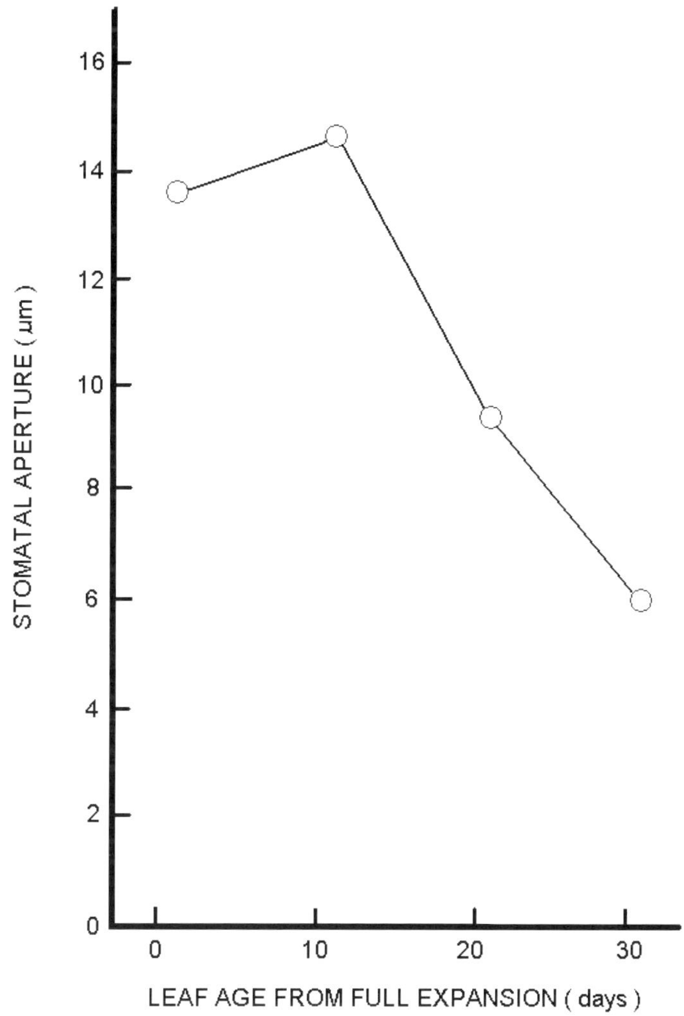

Fig. 3.11. Changing stomatal response in epidermal strips of *Commelina communis* in normal air, as a function of leaf age (Willmer, 1988).

efflux method (Cameron and Yang, 1982; equations 3.16–3.19) or by measuring changes in CO_2 emanation and the ICC (Burg and Burg, 1965b). If a major route for gas exchange has been shut off by this procedure, CO_2 evolution should immediately decline and then it and the ICC should progressively increase to a higher steady-state value. Applying lanolin paste to the floral end of peppers and tomatoes did not produce a measurable response, but sealing the pedicel-end scar of peppers immediately lowered the CO_2 flux by 50% and increased the resistance to CO_2 exchange 2.3-fold. Within 6 h, the initial 2.1% ICC had risen to 4.0%, and CO_2 production was slightly inhibited (Burg and Burg, 1965b; Fig. 3.12, *right*). Sealing the stem scar of a red tomato fruit, which initially contained 6.8% [CO_2], increased the CO_2 mass transport resistance by 2.6-fold, and within 6 h the ICC had risen to 13.3%, while the respiration rate decreased by 30% (Burg and Burg, 1965b). Lanolin paste applied to the stem scar of detached grapefruits and cantaloupes caused an immediate 10% decline in CO_2

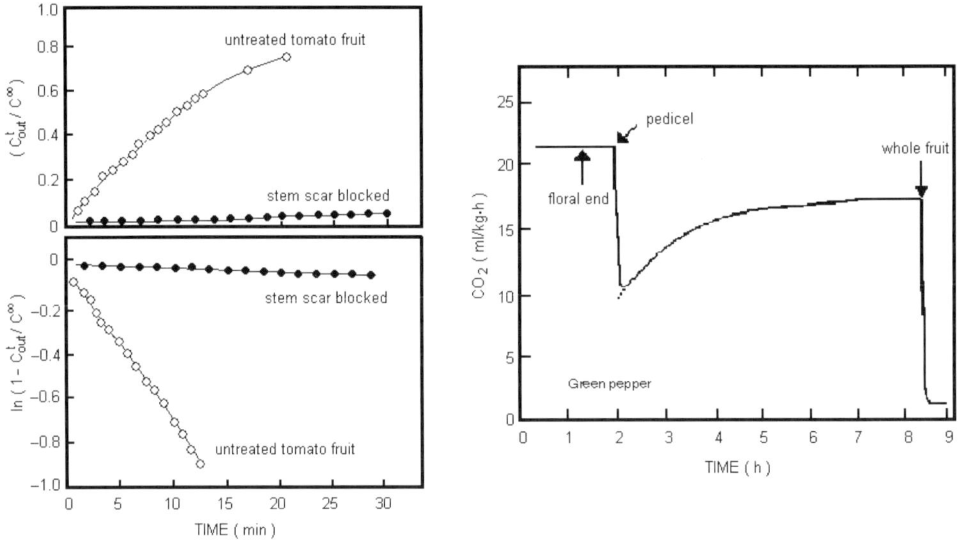

Fig. 3.12. (*left*) Plots of ethane efflux versus time for treated (stem-scar blocked) and untreated tomato fruits (Cameron and Yang, 1982). (*right*) Changes in the apparent rate of respiration of a green pepper upon application of lanolin to the floral end, pedicel and entire surface of the fruit (Burg and Burg, 1965b).

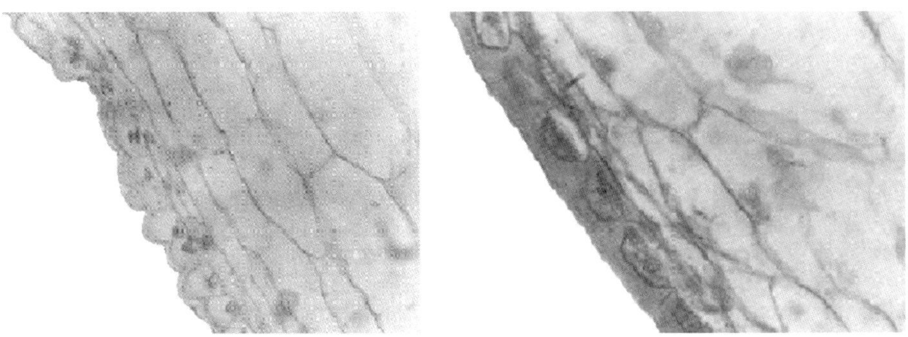

Fig. 3.13. Surface of the tomato fruit. (*left*) Cross-section of the tomato pericarp at the immature-green stage. (*right*) At the mature stage, extensive cutin deposition on the outer tangential and radial walls has occurred, sealing the stomates (450×). Adapted from Baker (1975).

evolution, which possibly is significant, but no change occurred with lemons, McIntosh apples, oranges, avocados, limes, pumpkins, bananas, plums or acorn squash (Burg and Burg, 1965b).[15] Measurements made with the ethane efflux method confirmed that the stem scar is the major pathway of gas exchange in both mature-green and red tomato fruits (Fig. 3.12, *left*).

CO_2 evolution, the ICC, ethylene emanation and the IEC do not change significantly when tomatoes are harvested with an attached stem (Lyons and Pratt, 1964; Sawamura *et al.*, 1978; De Vries *et al.*, 1995; Knee, 1995; Fig. 3.14), and these fruits ripen at the same rate (De Vries *et al.*, 1995) or more rapidly (Sawamura *et al.*, 1978) than tomatoes left attached to the vine. The peak climacteric ICC is lower on the vine, and the climacteric IEC rises to the same maximum during ripening on or off the vine (Sawamura *et al.*, 1978; Saltveit, 1993). All of these parameters indicate that the attached stem differs from applied lanolin paste in that it does not prevent gas exchange through the calyx. This conclusion was confirmed by De Vries *et al.* (1995), using two highly sensitive laser-based detection

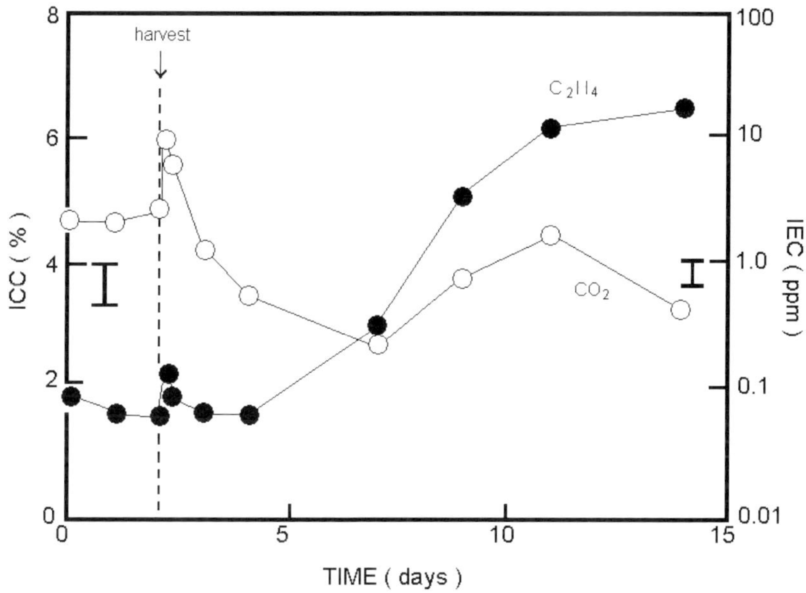

Fig. 3.14. Internal [CO_2] and ethylene concentrations in mature-green tomato fruits shaded by aluminium foil for 2 days prior to harvest and during ripening at 24°C. Vertical broken line indicates the time of harvest. Vertical bars are SE (n = 3) related to adjacent axes. Samples were withdrawn from the fruits with a syringe through a sealed collecting tube (Knee, 1995).

systems, a photothermal deflection instrument and photoacoustic setup, capable of measuring ethylene concentrations of 1 nl/l and 6 pl/l, respectively. Ethylene emission through the stem end of cherry tomatoes was the same regardless of whether the coronet (pedicel + sepals) was removed or left intact, and accounted for 85–90% of the ethylene exchange (De Vries *et al.*, 1995).

Several anomalies in tomato fruit ripening, tentatively explained as resulting because detachment decreases the IEC by enhancing gas exchange through the stem scar, need to be re-examined, since this explanation apparently is incorrect. Detached mature-green transgenic tomato fruits expressing ACC deaminase ripen more slowly than control fruit, but transgenic fruits attached to the vine ripen more rapidly, and almost as quickly as control fruits. Lycopene synthesis is strongly retarded in transgenic tomatoes with a reduced capacity for ethylene synthesis (Oeller *et al.*, 1991; Klee, 1993; Murray *et al.*, 1993b), and while ACO-antisense tomato fruits harvested at the mature-green stage showed the expected reduction in lycopene synthesis compared to control fruit, they and control fruits accumulated lycopene at the same rate when they were attached to the vine (Picton *et al.*, 1993). It was suggested that prior to harvest the 'residual' ethylene in transgenic and ACO-antisense fruits may have been above an active threshold, and that after harvest the IEC rapidly declined due to escape through the stem scar (Oeller *et al.*, 1991; Klee, 1993; Picton *et al.*, 1993; Theologis *et al.*, 1993b). Measurements made with transgenic tomato fruits 0–15 and 15–30 min after removal from the vine revealed that the rate of ethylene emanation was 32.4 ± 12.3% lower during the second interval. Because the half-time for outward diffusion of ethane from normal tomato fruits is 10 min (Cameron and Yang, 1982; Fig. 3.12, *left*; Table 3.5), it was concluded that improved gas exchange through the stem scar of the harvested tomato caused a progressive decline in the IEC and ethylene evolution (Klee, 1993). This explanation is not likely to be valid unless there is a significant difference in control, transgenic and ACO-antisense fruit stem-end anatomy. The IEC of tomato leaves (Bradford and Dilley, 1978; Bradford and Yang, 1980; Else *et al.*, 1993), mango fruits (Figs 5.16 and 5.17) and oranges (Table 5.3; Sawamura, 1981) decreases when they are harvested, not by improved gas exchange, but because they receive ACC from the parent plant, and rapidly exhaust their supply of translocated ACC when they are detached. More than 95% of the intercellular ethylene escapes from a harvested mango with a half-time of 9 min (Fig. 5.17). If transgenic and ACO-antisense tomatoes import ACC from the vine this would cause their IEC to decrease after harvest.

De Vries *et al.* (1995) criticized the experiments of Cameron and Yang (1982) and Burg and Burg (1965b), pointing out that sealing the stem scar of a tomato 'completely disturbs gas exchange', while their laser method of detection allows ethylene exchange to proceed normally. They state that Burg and Burg found that only 60% and Cameron and Yang 97% of the ethylene exchange occurred through the tomato stem scar, whereas the correct value is 85–90%. This is misleading, because Burg and Burg determined CO_2 transport, Cameron and Yang studied ethane escape, and De Vries *et al.* measured ethylene exchange. The variant results obtained by these different protocols are accounted for by the large differences in the cuticle's resistance to CO_2, ethylene and ethane mass transport (Table 3.12).

3.20 Cuticular Gas Exchange

CO_2 and water vapour are transported across a leaf's cuticle despite a complete sealing of the stomata, indicating that gas exchange occurs no matter how tightly the stomates close (Dugger, 1952; Boyer *et al.*, 1997). The mechanism by which gases transit the cuticle has not been extensively studied, and it has been uncertain whether or not the process is pressure-sensitive. A reduction in atmospheric pressure increases foliar transpiration (Sampson and Allen, 1909) and foliar water losses are unusually high in mountainous areas where the atmospheric pressure is low (Hirsch, 1954 – referred to in Raschke, 1960),[16] but these effects probably are due to enhanced diffusion through open leaf stomates.

Typically, the plant epidermis consists of a single layer of cells, approximately 10 μ in diameter, with no discernible intercellular air spaces separating their radial walls. When a gas produced or consumed by cells interior to the epidermis enters into or escapes from a commodity, unless it diffuses through air-filled stomates, lenticles or the pedicel-end stem scar, it necessarily must pass through liquid contained in the epidermal cells, or through the liquid-filled radial walls separating them, as well as through the cuticle. LP would enhance cuticular gas exchange if the process was limited by the diffusional resistance of the air-filled spaces that separate the cuticular wax platelets, as suggested by Chambers and Possingham (1963), but pressure would have little influence if this resistance is small compared to the combined liquid phase resistances of the epidermal cells and cuticle.

Cuticular transpiration accounts for 1.7–28.6% of the water loss from leaves of various species when their stomates are open (Holmgren *et al.*, 1965). The cuticle's CO_2 conductance has been calculated from such transpirational data by assuming that CO_2 and water vapour move along identical paths through both the cuticle and stomates (Gaastra, 1959; Von Caemmerer and Farquhar, 1981), but this fails to consider the possibility that the diffusive properties of water and CO_2 through the cuticle might differ (Kirrschbaum and Pearcy, 1988). Boyer *et al.* (1997) measured water vapour and CO_2 conductance through grape-leaf cuticles (*Vitis vinifera* L.) by sealing their lower surface to force all gas and vapour exchange to occur via the upper stoma-free side. They found the cuticular resistance to be 7.7–40-fold (average = 17.5-fold) higher for CO_2 compared to water vapour, and attributed this difference to a longer diffusional path for CO_2. Water vapour originating in the epidermal cells only needs to move through the cuticle to reach the atmosphere, whereas CO_2 must move through both the epidermal cell layer and cuticle to exchange between the atmosphere and intercellular spaces. The cuticular resistance (r_c) to CO_2 ranges from 1500 to 13,000 s/cm in leaves (Holmgren *et al.*, 1965; Noble, 1974; Boyer *et al.*, 1997), and comparable values have been measured in harvested fruits (Burg and Burg, 1965b; Burton, 1982; Cameron and Yang, 1982; Ben-Yehoshua *et al.*, 1985).

Cameron and Yang (1982) duplicated the experiment of Burg and Burg (1965b) in which the tomato cuticle's CO_2 permeability was measured after sealing the pedicel-end stem scar, but instead of studying CO_2 flux, they equilibrated fruits with ethane and calculated its permeability based on the kinetics of escape (equations 3.16–3.19). A comparison of data from these studies shows that the resistances to ethane and CO_2 mass transport were similar before the pedicel-end stem scar was sealed [ethane = 7400–7800 s/cm; CO_2 = 7400 s/cm (Table 3.12)].[17,18] This indicates that initially gas exchange was limited by the pedicel-end stem scar's air phase in which the binary diffusion coefficients of the two gases are nearly identical (Table 15.4). After the stem scar was sealed, if the cuticular resistance is determined by an aqueous phase, it should be inversely proportional to K_H/D_{water} (Tables 15.1 and 15.5) and therefore 12.9-fold larger for ethane compared to CO_2. In close agreement, cuticular resistance is 14.2-fold larger for ethane compared to CO_2 (Table 3.12). The total cellular resistance to gas exchange ($r_{cw,pl,cyt}$, Table 3.4) is insignificant compared to the tomato's cuticular resistance (r_c, Table 3.11). Equation 15.28 indicates that the 'effective thickness' of the tomato cuticle's aqueous resistance is 2240 μ, which is approximately 350-fold greater than the diameter of a tomato epidermal cell (Fig. 3.13). Therefore, the tortuosity and limited cross-sectional area of the water channels in the cuticle (7.3; Schönherr, 1976b, 1982; Schönherr and Schmidt, 1979) and cell wall (Noble, 1991; 3.4), rather than the thickness of the liquid layer in the epidermal cells, must be responsible for the cuticular resistance. An insignificant amount of respiratory CO_2 is released from tomato and apple fruits when CO_2 dissolved in transpired water vaporizes at the surface of these fruits (example 21). This proves that cuticular CO_2 transport occurs by diffusion through the water-filled channels rather than by passive carriage in transpired water. Measurements of O_2 and CO_2 diffusion through animal and insect tissue support this conclusion. The K_H/D_{water} value is 32-fold greater for CO_2 than for O_2 (Table 15.5), and CO_2 diffuses approximately 35 times more rapidly than O_2 through the insect cuticle (Wigglesworth, 1931, 1950; Miller, 1964) and 20–30-fold more rapidly through animal tissues (Prosser *et al.*, 1952). Consequently, an insect's intake of O_2 occurs through the air-filled spiracles, whereas the elimination of CO_2 takes place to a considerable extent either through the general integument or, in heavily sclerotized insects, through the membraneous regions (Imms, 1949; Wigglesworth, 1950).[19]

If the tomato stem prevented gas exchange through the calyx, the internal respiratory O_2 draw-down in an attached mature tomato respiring with an R.Q = 1 would be 32-fold greater than the ICC build-up. As mature-green tomatoes lacking an attached stem contain 3–5% [CO_2] (Sawamura *et al.*, 1978; Saltveit, 1993; Knee, 1995; Table 2.2), their interior would become 'anaerobic' while they were attached to the tree. This does not occur, proving that the stem does not block gas mass transport through the calyx.

The concept that liquid water determines the cuticular resistance to gas mass transport is consistent with measurements comparing the solubilities of ethane and CO_2 in different solvent systems (Gallardo *et al.*, 1987; Robinson *et al.*, 1987; Table 3.13). Amongst the substances tested, except in water, CO_2 is never more than 2.16-fold more soluble than ethane. Therefore it would seem that only diffusion through an aqueous phase could explain the vastly different cuticular resistances to CO_2 and ethane.

During and after passage through the cuticle, gases encounter two additional resistances in series, the air spaces between the wax platelets and the stagnant air layer at

Table 3.13. Solubility of CO_2 and C_2H_6 in various solvents (Gallardo *et al.*, 1987; Robinson *et al.*, 1987).

		Liquid mole fraction per MPa		
Solvent	Temperature (°K)	CO_2	C_2H_6	CO_2/C_2H_6
N-Dodecane	373.2	0.072	0.14	0.51
Cyclohexane	373.2	0.044	0.078	0.56
Trans-decalin	373.2	0.045	0.097	0.46
Benzene	373.2	0.044	0.067	0.66
Naphthalene	373.2	0.034	0.052	0.65
Phenanthrene	383.2	0.025	0.036	0.69
Pyrene	433.2	0.054	0.025	2.16
Cyclohexanone	298.2	0.213	0.144	1.48
Water	293.2	0.0069	0.00038	18.16

the commodity's surface. The stagnant air layer's resistance to fixed gases and water vapour, at most only a few s/cm, is insignificant compared to the cuticle's resistance (Slatyer, 1967; Table 3.4). The transport resistance of the air spaces separating the wax platelets also contributes little to the overall cuticular gas-transport resistance, and therefore the liquid phase resistance accounts for > 98% of the cuticular resistance to the diffusive transport of O_2, CO_2 and ethylene. Cuticular gas transport is not significantly affected by hypobaric pressure since it is not limited by an air phase.

Even though the CO_2 and ethylene binary diffusion coefficients in air are nearly identical (Table 15.4), a fruit's skin resistance ($r_{p,c}$) to ethylene mass transport typically is 10–50% greater than its CO_2 skin resistance (Table 2.2). If all mass transport occurred through air-filled pores, skin-resistance values would be inversely related to each gas's binary diffusion coefficient in air, in the ratio water vapour (0.62) < O_2 (0.78) < CO_2 (1.00) < ethylene (1.08). Instead, in Valencia oranges, the skin resistance values are in the ratio water vapour (0.018) < CO_2 (1.0) < O_2 (1.05) < ethylene (1.20) (Ben-Yehoshua *et al.*, 1985; Table 3.11). This occurs because these gases are exchanging by parallel pathways through the stomates, lenticles and cuticle, and CO_2 transport through the cuticle is favoured due to the gas's water solubility. These results give qualitative credence to Marcellin's suggestion (1974) that CO_2 is transported through the cuticle and O_2 through the stomates or lenticles of fruits.

3.21 Effect of LP on NH_3 Efflux

Ammonia is the most abundant nitrogen species in the atmosphere after N_2 and N_2O (Söderlund and Svensson, 1976; Raven *et al.*, 1992). NH_3 evolution has been detected from many types of plants when ammonia-free air is passed over them (Stutte and Weiland, 1978; Stutte *et al.*, 1979; Meyer – referred to in Farquhar *et al.*, 1980) and also when the atmospheric content is below the ammonia compensation point (Morgan and Parton, 1989; Husted and Schjoerring, 1995; Hill *et al.*, 2001; equation 3.20). No flux of ammonia into and out of healthy leaves of *Zea mays* occurred when the atmospheric partial pressure of ammonia was 5 ± 3 nbar (0.0038 mm Hg), but in agreement with studies indicating that senescing leaves degrade protein, amides and amino acids and increase in ammonia production (4.20), an efflux of 0.6 nmol/m^2·s was measured from senescing *Z. mays* leaves under the same conditions (Farquhar *et al.*, 1979). NH_3 formation also is stimulated when horticultural commodities are exposed to high [CO_2], during the 'bluing' of roses and certain other flowers, and at high temperatures.

LP promotes NH_3 efflux by opening stomates in darkness, reducing the atmospheric NH_3 partial pressure in the air changes (example 1, chapter 4), lowering the diffusive resistance of the air-filled gas phase barriers (including boxes and wraps), and continuously flushing metabolic NH_3 from the storage area. By preventing senescence and decreasing the ICC and the loss of respiratory carbohydrates, LP also discourages NH_3 production.

The transpirational resistance of mesophytic leaves with open stomates is in the range 0.5–2.5 s/cm (Noble, 1991), and their combined stomatal and barrier air-layer resistance to NH_3 mass transport should be only 10% higher (Table 15.4). Ammonia's extreme water solubility (Tables 15.1 and 15.2) should cause this gas's cuticular resistance (r_{c, NH_3}) to approach 25 s/cm. Even though this is approximately 800-fold lower than the CO_2 cuticular resistance of leaves and fruits, it still is sufficient to explain why ammonia absorption from the atmosphere into *Brassica napus* plants depends on open stomates, and does not occur to any extent through the cuticle in darkness (Husted and Schjoerring, 1996). In contrast, example 16 indicates that in a bulky horticultural commodity, the cuticle should be the favoured pathway for NH_3 escape. At atmospheric pressure, the high 'pore' resistance of bulky fruits should cause them to exchange ammonia more easily through their cuticle than via stomates, lenticles or the pedicel-end scar, and any effect of an LP pressure reduction on ammonia exchange would mainly result from decreasing the diffusive resistances of the intercellular spaces and boxes.

3.22 Changes in Gas Exchange during Storage

Due to continuous cuticular wax deposition, the gas phase and transpirational resistances increase during the storage of harvested leaves and fruits and also as they age or develop on the tree (Trout *et al.*, 1942; Martin and Juniper, 1970; Hulme and Rhodes, 1971; Albrigo, 1973; Baker, 1975; Schönherr, 1976b; Jones, 1981). At atmospheric pressure, the newly deposited wax increases the [CO_2] and [O_2] gradients between the storage atmosphere and an apple's centre, causing a significant amount of CO_2 to be transported directly through the cuticle (example 6). In freshly harvested McIntosh apples, almost all gas transit occurs through lenticles and there is little difference between the skin resistances to O_2, ethylene and CO_2 (Burg and Burg, 1965b). After McIntosh apples were stored at 1°C for 3 months and then transferred to 20°C for several days, the average skin resistance values for ethylene and CO_2 were 24,996 and 10,464 s/cm, respectively (Burg, 1960, unpublished).[20] At that time, 87% of the ethylene and 95% of the O_2, but only 38% of the CO_2, exchanged through the lenticles. The remainder moved directly through the cuticle. The resistance to gaseous diffusion also increases during the storage of orange fruits, at least in part due to peel drying (Ben-Yehoshua, 1969). Initially, the resistance to O_2 and CO_2 mass transport is the same in oranges, but after 60 days of storage at 20°C and 60–80% RH, the resistance to O_2 mass transport has increased 5.5-fold and the resistance to CO_2 mass transport only fourfold. In LP, wax deposition and peel drying during storage should have little influence on the composition of the atmosphere present in the fruit's intercellular spaces because a low pressure levels the gas gradients between the atmosphere and centre of the fruit.

3.23 Resistance of Boxes and Wraps

Commodities typically are transported in boxes containing protective packaging and/or a water-retentive liner. The carton and packing materials influence storage life because they account for a significant part of the total resistance to gas and vapour mass transport between the commodity's centre and the atmosphere (Table 3.4). Inside the packaging, a significant respiratory O_2 draw-down and build-up of metabolic CO_2 and ethylene may occur. For example, during NA storage at 1°C, the [O_2]

inside commercial cartons containing 18 heads of lettuce in a folded-over polyethylene liner declined to 5%, and 12% [CO_2] accumulated (Aharoni *et al.*, 1971).

A box's transport resistance can be determined by injecting a gas and measuring its rate of escape (Fig. 3.14; equation 3.12; example 9), and it also has been computed from cool down measurements during LP storage (Burg and Kosson, 1983; examples 10 and 11). The effective box resistance (r_b) to be used in a resistance network calculation is:

$$r_b = r_{box}\ (\Sigma A_{com}/A_{box}) \qquad (3.22)$$

where A_{box} is the box surface area, ΣA_{com} the total surface area of the commodity in the box and r_{box} is the measured box resistance. The barrier air-layer resistances at the surfaces of the box and commodity can be disregarded, as they are insignificant compared to the box's total air-phase resistance. Equation 3.22 indicates that denser commodity packing in a storage carton increases the effective box resistance by elevating the $A_{box}/\Sigma A_{com}$ ratio. Since the box resistance acts in series with the commodity's intercellular and skin resistances, the total gas phase resistance is the sum of these resistances. When flowers, leafy vegetables, cuttings or potted plants are stored at atmospheric pressure, the box's resistance to O_2, CO_2 or ethylene transport constitutes almost the entire resistance to gas exchange (Table 3.4). With apples and other bulky fruits, r_b represents a smaller, but still significant part of the total resistance.

LP reduces a box's resistance to gas mass transport to such an extent that it does not significantly influence the gaseous composition within the stored commodity (example 14). The box's water-vapour transport resistance also is lowered in LP, but even at 1.33 kPa (10 mm Hg) the carton may limit evaporative cooling sufficiently to elevate the commodity and air temperatures by several degrees above the air temperature adjacent to the box surface. This temperature difference creates the vapour pressure gradient needed to transfer moisture released when evaporative cooling removes the commodity's respiratory heat (example 15). The increased vapour pressure in the box lowers the O_2 partial pressure (Table 3.14), and if the storage pressure is low enough, air-change rate slow enough and temperature rise large enough, respiratory O_2 draw-down can significantly lower the [O_2] available to the commodity (Burg, 1987a; example 15). To prevent this condition from developing and creating the possibility of low [O_2] damage, the storage pressure in an LP intermodal container sometimes is elevated by approximately 0.66 kPa (5 mm Hg) above the pressure which otherwise would be optimal.

Table 3.14. Effect of commodity temperature on the water vapour and O_2 partial pressures inside a box stored at 0°C and total pressure of 1.33 kPa (10 mm Hg).

Commodity Temperature (°C)	Conditions inside box: Vapour pressure (mm Hg)	Conditions inside box: O_2 partial pressure (atm)
0	4.579	0.00149
1	4.926	0.00140
2	5.294	0.00129
3	5.685	0.00119
4	6.101	0.00107

3.24 Effect of LP on Mass Transport through Air Phases

Table 3.4 summarizes the effective air and cellular resistance values for a typical leafy commodity and for McIntosh apples stored in cartons. These data will be used to estimate LP's influence on gas and vapour exchange and the intercellular gas and vapour concentrations. The 'skin' resistance values for a typical leaf presented in Table 3.4 assume that the stomates are open. The effect that LP has on gas and vapour mass transport through an air phase is also discussed in Sections 15.15 and 15.16.

LP increases diffusive gas and vapour transport through stomates, lenticles, the pedicel-end stem scar, intercellular air spaces, boxes, wraps, barrier air layers and the inter-platelet air spaces in the cuticle's wax layer. Internal CO_2, NH_3 and ethylene are not continuously 'evacuated' from the tissue, and LP does not deplete tissue of volatile substances, as some researchers have assumed (Wu *et al.*, 1972; Abeles *et al.*, 1992).[21] As rapidly as they are consumed or produced, gases and vapours continue to enter or escape, but in LP the same rate of inward or outward diffusion is driven by a much lower equilibrium concentration gradient between the atmosphere and interior air phases of the commodity.

The partial pressures of O_2, CO_2, ethylene, NH_3 and biological vapours such as ethanol, acetaldehyde and farnescene, which are present in the storage atmosphere and commodity's intercellular spaces, are small enough relative to the storage pressure that it is not necessary to consider the log mean partial-pressure correction (15.16) in computing the effect of pressure on the mass transfer resistance to these gases and vapours. At high LP pressures, the gas and vapour mass transport resistance does not decrease in direct proportion to the pressure reduction because that portion of the gas and vapour exchange which occurs through the cuticle is pressure-insensitive. The correction for CO_2 is particularly large, since the cuticular conductance is much higher for CO_2 than it is for O_2 and ethylene. At very low LP pressures, the mole fraction of water vapour present in the storage atmosphere and intercellular spaces increases to such an extent that gaseous diffusion is enhanced (15.26), causing the gas and vapour mass-transport resistance to decrease more than otherwise would be expected. The combined effect of pressure, cuticular resistance and water vapour on a 'typical' commodity's skin resistances to CO_2, O_2 and ethylene mass transport is summarized in Table 3.15. The assumed $r_{p,a}$ values are ratiometrically related to each gas's binary diffusion coefficient in air, and the r_c values depend on each gas's K_H/D ratio, where K_H is the gas's Henry's Law coefficient, and D the gas's binary diffusion coefficient in water. LP lowers the diffusive resistance of wraps, liners and boxes to CO_2, O_2, NH_3, ethylene and vapours in direct proportion to the pressure reduction, except that with water vapour a root mean-square correction must be applied at very low pressures (equation 15.26).

3.25 Effect of LP on Vapour Mass Transport

Volatiles that escape through the plasmalemma dissolve in the apoplastic liquid water film at the air–water interface, where they are present in a constant boiling aqueous mixture. Even though the cellular hydrostatic pressure is higher (15.5)[22] than the vapour pressure of many of the low-boiling volatile substances present in plants, and the hypobaric pressure at the air–water interface in the cuticle and intercellular spaces is lower, the volatiles cannot boil from the dilute solution unless the pressure is reduced to the vapour pressure of water at the prevailing temperature. With horticultural commodities the LP pressure never is that low, and the intercellular concentration of a volatile substance is determined by its mole fraction in the interface liquid and the vapour's Henry's Law coefficient, independent of the hypobaric pressure.

LP does not affect gas and vapour mass transport through liquid phases, and therefore lowering the atmospheric pressure would not significantly change the steady-state cellular concentrations of O_2, CO_2, NH_3, ethylene, ethanol and other vapours if their cellular liquid-phase transport resistance was large compared to the summed resistance values of the air-filled barriers in the pulp, skin, boxes and wraps. The data summarized in Table 3.4 indicate that especially in a bulky fruit the combined gas mass-transport resistance of the air-filled barriers is much greater than the cellular liquid-phase resistance to fixed gases. The cellular resistance to ethanol mass transport is large enough to prevent LP from promoting this vapour's escape from leaves, flowers and leafy vegetables, and should restrict ethanol escape from bulky commodities such as corn when they are stored in LP (Spalding *et al.*, 1978; examples 12 and 13).

Table 3.15. Effect of pressure on a fruit's mass transport resistance to CO_2, O_2 and ethylene.

Pressure (mm Hg)	r_{pca} (s/cm)		
	CO_2	O_2	C_2H_4
760	5368.0 (0)	5440.0 (0)	7361.0 (0)
380	3051.0 (+13.7)	2735.0 (+0.6)	3749.0 (+1.9)
150	1395.0 (+31.7)	1083.0 (+0.9)	1497.0 (+3.0)
80	713.2 (+26.2)	572.6 (0)	793.1 (+2.4)
40	355.5 (+25.8)	280.9 (–3.0)	393.5 (+1.6)
20	175.8 (+24.4)	137.8 (–3.7)	191.3 (–1.2)
15	129.7 (+22.4)	107.4 (–5.6)	140.8 (–3.1)
10	84.3 (+19.3)	71.6 (–8.9)	91.3 (–5.8)

Assumed conditions for a 'typical' fruit stored at 0°C and atmospheric pressure are: $r_{p,a}$ = 7070 s/cm for CO_2, 5500 s/cm for O_2, 7640 s/cm for ethylene, r_c = 22,270 s/cm for CO_2, 500,000 s/cm for O_2, 201,660 s/cm for ethylene; $P_{wv,o}$ = 4.579 mm Hg; $P_{wv,i}$ = 5.579 mm Hg, where the subscripts refer to pores (p), cuticle (c), barrier air layer (a), water vapour (wv) inside the commodity (i) and outside the commodity (o). Percentages indicated in parentheses are the difference between the computed value and that which $r_{p,c,a}$ would have if it was inversely related to pressure. The 'skin' resistance of a fruit possessing functional stomates decreases by many additional orders of magnitude at LP pressures that cause the stomates to open. A Valencia orange's 'skin' resistance is lowered by approximately 300-fold when its stomates open (3.18; Table 3.11).

In plant tissues permeability through the plasmalemma limits the rate of escape of most volatile substances because they have a larger molecular diameter than ethanol, and often a lower oil/water partition coefficient. Consequently, LP will have little if any influence on their retention. LP did not enhance the loss of ethylene dibromide from EDB-fumigated grapefruits and lemons during 24–48 h, and only slightly increased the escape of this vapour from lemons during 48–72 h (Table 3.16).

The volatile unsaturated sesquiterpene hydrocarbon α-farnescene is deposited as a component of the cuticular wax of apples and other fruits, and its accumulated oxidation product 6-methyl-5-heptene-2-one may be responsible for the appearance of superficial scald symptoms (Wang, 1990; Wang and Dilley, 1997). The diffusive escape of farnescene and its volatile oxidation products from the wax does not involve penetration of the plasmalemma, and therefore LP might accelerate the loss of these hydrocarbons through the air-filled wax platelets of the cuticle. This has been suggested as an explanation to account for LP's ability to prevent scald symptoms from developing in apples (Bangerth, 1973; Dilley, 1977a; Wang *et al.*, 2000a,b).

Table 3.16. Ethylene dibromide (EDB) content of lemons and grapefruit fumigated and then stored at 20°C in NA or in LP at 80 or 40 mm Hg (Burg, 1975, unpublished).

Time (hours)	EDB (μl/l)		
	760 mm Hg	80 mm Hg	40 mm Hg
Lemons			
0	39.62	–	–
2	38.22	35.36	36.72
4	30.04	32.03	36.62
12	28.27	29.15	26.81
24	21.41	18.54	20.91
48	15.44	13.04	14.21
72	12.69	8.70	8.01
Grapefruit			
0	9.3 ± 2.3	–	–
5	7.9 ± 0.9	8.6 ± 0.4	–
22	4.1 ± 0.6	4.5 ± 0.8	–

3.26 LP Does Not Remove Flavour and Aroma Volatiles from Fruits

Apples stored in LP for a long duration of time at 6.67–10 kPa (50–75 mm Hg) fail to develop normal flavour and aroma when they are removed from storage and ripened in air (Bangerth, 1973, 1984; Bangerth and Streif, 1987). The lower the pressure, the sooner this response sets in. At first it was suspected that a low pressure might 'evacuate' or 'out-gas' flavour and aroma components, but subsequently when it was found that this same symptom arises during prolonged CA storage, and more rapidly at a lower O_2 partial pressure, it became obvious that the effect is caused by O_2 depletion rather than enhanced diffusive escape of volatiles at a reduced pressure (Fidler and North, 1962, 1969a,b; Guadagni *et al.*, 1971; Patterson *et al.*, 1974; Lidster *et al.*, 1983; Bangerth, 1984; Wang, 1990). Both in LP and CA, progressively as the storage interval is increased, apples lose their ability to respond to ethylene supplied during storage or after fruits are transferred to air (Bangerth, 1984; Bangerth and Streif, 1987). After a prolonged storage of Golden Delicious apples, the fruit is capable of recovering and producing aroma volatiles when it is removed from LP and transferred to air for 14 days at 20°C, but the fruit eventually loses this ability if the storage duration is extended further. The decreased synthesis of volatile substances associated with normal apple ripening (Guadagni *et al.*, 1971; Patterson *et al.*, 1974; Hatfield and Patterson, 1974; Knee and Hatfield, 1981a) seems to result from a progressive loss of ethylene sensitivity during storage (Bangerth, 1975, 1984; Bangerth and Streif, 1987). Fruits that had been stored in LP for short periods retained the capacity to generate ACC synthase, but those stored for longer periods in LP (or CA) eventually lost this capacity (Bufler and Bangerth, 1983; Bangerth, 1984). Not all fruits behave in this manner. The ethylene sensitivity of bananas is not decreased by 4 months storage in LP at 14.4°C and a pressure of 5.33–6.67 kPa (40–50 mm Hg = 0.8–1% [O_2]). When the banana fruits are transferred to air they ripen with normal colour, texture, sweetness, flavour and aroma (Liu, 1976; Apelbaum *et al.*, 1977a; Bangerth, 1984), whereas at atmospheric pressure they lose their ability to produce ethylene or to develop acceptable flavour and sweetness after 11 days storage at 18°C in the presence of 1% [O_2] (Shatat *et al.*, 1978; Wang, 1990).

Tomato fruits which did not colour during 100 days at 12.8°C and a pressure of 13.6 kPa (102 mm Hg), failed to produce significant quantities of flavour and aroma volatiles when they were transferred to atmospheric air and ripened (Wu *et al.*, 1972). At 62.8 and 37.1 kPa (471 and 278 mm Hg), the fruits remained green for two and three times longer than air controls, respectively, but eventually they ripened in storage with less than a normal content of volatiles. It was concluded that LP 'evacuates flavours in the gaseous state'. The interpretation of this experiment is complicated by the unusual design of the hypobaric apparatus, which caused the storage to be a hybrid between an uncontrolled modified atmosphere (MA) and LP (example 8). In other studies, when mature-green tomatoes were stored in a proper LP apparatus at 12.8–18.3°C and pressures between 10.13 and 24 kPa (76–180 mm Hg), if they were transferred to air before ripening commenced, normal flavour subsequently developed and there was no discernible difference between the result obtained in LP and CA at a comparable O_2 partial pressure (Burg and Burg, 1966c; Tolle, 1969, 1972; Dilley, 1972; Bangerth, 1973, 1974; Awad *et al.*, 1974; Kader and Morris, 1974; Morris and Kader, 1975; Stenvers and Bruinsma, 1975). There also was no difference between CA and LP when breaker and light-pink tomatoes ripened slowly during storage (Kader and Morris, 1974; Morris and Kader, 1975). In those instances when LP and CA reduce the flavour and aroma volatiles of tomato fruits, they do so by influencing volatile production, not volatile removal (Shatat *et al.*, 1978).

3.27 Examples

1. Transport of respiratory CO_2 from the centre to the surface of a spherical fruit is given by (Burg, 1990):

$$\mathrm{d}m/\mathrm{d}t = \dot{\mathrm{m}} = \left[\frac{a_{\mathrm{ias}} A_{\mathrm{p}} D_{\mathrm{A,CO_2}}}{Z_{\mathrm{ias}}}\right]\left[\frac{\mathrm{d}\rho_{CO_2}}{\mathrm{d}x}\right] \quad (3.23)$$

where ρ_{CO_2} is the density of CO_2 (g/cm^3) in the intercellular air space at a distance x (cm) from the centre of the fruit; a_{ias} is the fractional volume of the air spaces in the tissue ($a_{ias} < 1$), Z_{ias} is the tortuosity of the intercellular air spaces ($Z_{ias} > 1$) and $A_p = 4\pi x^2$. Respiratory production of CO_2 is given by:

$$dm/dx = \frac{d}{dx}\left[\frac{a_{ias}A_p D_{A,CO_2}}{Z_{ias}}\right]\left[\frac{d\rho_{CO_2}}{dx}\right] = \frac{M_{CO_2}\dot{Q}_R \rho_F dV_F}{RTdx} \quad (3.24)$$

where $\dot{Q}_R$ is the fruit's rate of production of respiratory CO_2 (cm^3/g·s), M_{CO_2} is the molecular weight of CO_2, D_{A,CO_2} is the binary diffusion coefficient of CO_2 in air (cm^2/s), ρ_F and V_F are the density and volume of the fruit, respectively ($V_F = 4/3\ \pi x^3$). Equation 3.24 can be rearranged into the form:

$$\frac{d^2\rho_{CO_2}}{dx^2} = \frac{\dot{Q}_R \rho_F Z_{ias} M_{CO_2}}{RTD_{A,CO_2}} \quad (3.25)$$

The boundary conditions are $\rho_{CO_2} = \rho_{CO_2,x=0}$ at $x = 0$; and $d\rho_{CO_2}/dx = 0$ at $x = r$, where r is the radius of the fruit. The solution is:

$$p_{CO_2,x=0} + \Delta p_{CO_2} = (2xr - x^2)\left[\frac{\dot{Q}_R \rho_F Z_{ias}}{2D_{A,CO_2} a_{ias}}\right] + p_{CO_2,x} + \Delta p_{CO_2} \quad (3.26)$$

where p_{CO_2}, $x = 0$ is the partial pressure of CO_2 at the centre of the fruit which decreases to zero at the inner surface of the skin when $\Delta p_{CO_2} = 0$; $p_{CO_2,x}$ is the partial pressure of CO_2 at x; and Δp_{CO_2} is the amount by which partial pressure of CO_2 at the inner surface of the skin exceeds the atmospheric CO_2 partial pressure due to the skin's resistance to CO_2 mass transport. The same equation can be used for ethylene by substituting the ethylene production rate and physical properties in place of the equivalent terms for CO_2. When equation 3.26 and published values for a_{ias} (Clements, 1935; Henze, 1969; Pantastico, 1975; Table 3.8) are used to compute the diffusive flow of ethylene through the intercellular spaces of an intact McIntosh apple and the flow of CO_2 through a nearly spherical Choquette avocado, the measured gas gradients and gas production rates are found to be consistent with gas transport through a continuous intercellular system having a tortuosity of $Z_{ias} = 1$ in the apple and 2.2 in a pre-climacteric avocado (Burg and Burg, 1965b). This indicates that the intercellular system is air-filled and continuous in both fruits, and therefore its resistance must be proportional to pressure. The data indicate that for a cell at the centre of the McIntosh apple, the mass transport resistance of the intercellular system to ethylene, defined relative to the surface area of the fruit, is only 1.09% as large as the skin resistance, and for a typical apple fruit, $r_{ias} = 136$ s/cm. For a cell at the centre of a pre-climacteric Choquette avocado, the resistance of the intercellular system to CO_2 mass transport is 1380 s/cm (Burg, 1990).

2. Radial gas movement through the intercellular system of a spherical fruit can be analysed by the method used in chapter 7 to evaluate cell-to-cell radial liquid water movement (7.6; equations 7.9 and 7.10; Fig. 7.9). Diffusive conductivity of CO_2 in air per metre (K_D) is substituted for hydraulic conductivity, and CO_2 partial pressure (MPa) for water potential:

$$K_D = \frac{D_{A,CO_2} a_{ias}}{Z\,x} \quad (3.27)$$

where D_{A,CO_2} is the CO_2 binary diffusion coefficient in air, a_{ias} is the cross-sectional surface area available for transport in the intercellular system per m^2 of surface and $Zx = 1.57$ m per radial m. For a pre-climacteric McIntosh apple stored at 0°C, typical conditions are:

- respiratory $CO_2 = 3.93 \times 10^{-7}$ m^3 (gas)/m^3·s
- CO_2 concentration in pulp, average = 5.5%
- CO_2 diffusive conductivity, $K_{D,CO_2} = 2.6 \times 10^{-5}$ m^3 (gas)/s·m·MPa (per m^2)
- CO_2 binary diffusion coefficient in air @ 0°C = 1.38×10^{-4} m^2/s per MPa.

The calculated CO_2 gradient from the centre to just beneath the surface of a McIntosh apple is only 0.037%, which is 1/149 the measured CO_2 gradient from just beneath the surface to the ambient atmosphere. Since the apple's measured skin resistance to CO_2 is r_{p,c,a,CO_2} = 14,285 s/cm, the resistance from the centre to the surface of the intercellular system computed relative to the surface area of the fruit, A_{com}, is only r_{ias,CO_2} = 14,285/149 = 96 s/cm. This agrees closely with the value calculated in example 1.

Typical conditions for a McIntosh apple producing 2.44×10^{-14} m³/s·g ethylene at 20°C, are:

radius, $b = 0.038$ m
weight = 169.5 g
volume = 2.30×10^{-4} m³
density of fruit, $\rho = 7.39 \times 10^5$ g/m³
porosity (a_{ias}) = 0.3
tortuosity, $Z = 1.57$
C_2H_4 binary diffusion coefficient in air @ 20°C = 1.44×10^{-4} m²/s (per MPa)
$\Delta p_{C_2H_4} = 412 \times 10^{-6}$ bar
r_{p,c,a,C_2H_4} = 19,180 s/cm.

The calculation indicates that the gradient from the centre of the pulp to the peel of a McIntosh apple should be only 3.38×10^{-6} bar. This is a much smaller value than the 416×10^{-6} bar gradient measured between the inner surface of the skin and the atmosphere (Table 2.1). The skin resistance to ethylene (r_{p,c,a,C_2H_4}) computed relative to the surface area of the fruit (A_{com}) is 123-fold larger than the intercellular system's centre-to-surface resistance to ethylene (r_{ias,C_2H_4}). Therefore, the resistance of the intercellular system to ethylene is r_{ias,C_2H_4} = 19,180/123 = 156 s/cm for a cell at the centre of the fruit. In accord with this prediction, the measured gradient from the centre of the pulp to the surface is very small (Table 2.1), proving that the intercellular spaces are air-filled and not occluded with water.

Typical conditions for a ripe Choquette avocado at 20°C are:

radius, $b = 0.052$ m
weight = 470 g
volume = 5.89×10^{-4} m³
porosity (a_{ias}) = 0.054 (Pantastico, 1975)
respiratory $CO_2 = 2.44 \times 10^{-5}$ m³ (gas)/m³·s
CO_2 diffusive conductivity, K_{D,CO_2} = 0.054×10^{-4} m³ (gas)/s·m/MPa (per m²)
tortuosity, $Z = 1.57$
CO_2 binary diffusion coefficient in air @ 20°C = 1.56×10^{-4} m²/s (per MPa)
r_{p,c,a,CO_2} = 4500 s/cm.

The computation is complicated by the presence of a large seed occupying the space interior to T_3 (Fig. 6.9). If it is assumed that CO_2 is produced at the same rate in the seed and pulp, and that the porosity of the seed and pulp are the same (which obviously is not true), the CO_2 concentration should be 1.9% higher at the surface of the seed compared to just beneath the peel, but instead it is 3.1% higher (Table 2.2). The partial pressure of CO_2 at P_3 would be higher than the computed value if the seed produces less CO_2 than the pulp per m³, or if the tortuosity of the pulp is larger than the assumed value of $Z = 1.57$. The measured 11.8% gradient across the skin of a ripe avocado is 3.8-fold larger than the measured gradient across the pulp. Therefore, the resistance of the intercellular system to CO_2 is r_{ias,CO_2} = 4500/3.8 = 1184 s/cm relative to the surface area of the fruit (A_P). This agrees closely with the value computed in example 1.

3. (Kossons, 1962, unpublished). When axial diffusion occurs through tortuous air passages having a fractional length (l) greater than the axial stem length (Z) traversed by l, then $l = \mu Z$, where μ is the tortuosity ($\mu > 1$). The volume of air in these passages is:

$$V_{air} = A_Z L = A_l\, l = (V_a/V_t)\, \pi R^2 L \quad (3.28)$$

where V_a is the cross-sectional area perpendicular to Z, A_l the cross-sectional area perpendicular to l, R is the stem's radius and L is the stem's length. The surface areas relative to Z and l are:

$$A_Z = \pi r^2 (V_a/V_t) \quad (3.29)$$

$$A_l = A_Z (L/l) = \pi R^2 (V_a/V_t)/\mu \quad (3.30)$$

Starting at a point in the stem where continuous ethylene production keeps the intercellular density (ρ) constant, when ethylene diffuses away axially through adjacent tissues in which no ethylene is produced, the rate of axial transport at steady-state is:

$$\mathrm{d}m/\mathrm{d}t = \dot{m} = -D_{\mathrm{C_2H_4}}A_l(\mathrm{d}\rho/\mathrm{d}l) = -\frac{D_{\mathrm{C_2H_4}}A_l(\mathrm{d}\rho/\mathrm{d}Z)}{\mu} \quad (3.31)$$

The radial loss of ethylene in length dZ is:

$$\mathrm{d}m/\mathrm{d}Z = \frac{\mathrm{d}}{\mathrm{d}Z}\left[\frac{D_{\mathrm{C_2H_4}}A_l(\mathrm{d}\rho/\mathrm{d}Z)}{\mu}\right] = \frac{\rho A_l}{r_{\mathrm{s,c,a}}} \quad (3.32)$$

Rearranging equation 3.32 and combining it with equation 3.31:

$$\frac{\mathrm{d}^2\rho}{\mathrm{d}Z^2} = \frac{\mu A_t \rho}{r_{\mathrm{s,c,a}}D_{\mathrm{C_2H_4}}A_l} = \frac{2\mu^2\rho}{r_{\mathrm{s,c,a}}D_{\mathrm{C_2H_4}}R(V_a/V_t)} \quad (3.33)$$

The boundary conditions are: $\rho = \rho_O$ @ $Z = 0$; dρ/dZ = 0 @ $Z = L$. The solution is:

$$\frac{\rho}{\rho_o} = \frac{\cosh[x(L-Z)]}{\cosh(xL)} \quad (3.34)$$

where:

$$x^2 = \frac{2\mu^2}{r_{\mathrm{s,c,a}}D_{\mathrm{C_2H_4}}R(V_a/V_t)} \quad (3.35)$$

In etiolated pea epicotyls, the values to be used with equation 3.35 are R = 0.05 cm, V_a/V_t = 0.03 (Table 3.10) and $r_{\mathrm{s,c,a}} \cong$ 4500 s/cm (Goeschl *et al.*, 1967; Burg *et al.*, 1971).

4. At atmospheric pressure, when the stomates of a Valencia orange are closed, the transpirational resistance is 110 s/cm and the resistance to CO_2 mass transport is 5700 s/cm. With open stomates, the transpirational resistance is 13.5 s/cm (Moreshet and Green, 1980), and since the stomatal resistance to water vapour and CO_2 mass transport depends on each gas's binary diffusion coefficient in air (Table 15.4), the resistance to CO_2 exchange is 21.5 s/cm. With open stomates, enhanced diffusion at a pressure of 2.67 kPa (20 mm Hg) would reduce the resistance to CO_2 mass transport to 0.57 s/cm. The change in CO_2 conductance induced by lowering the pressure to 2.67 kPa (20 mm Hg) should be 10,000-fold (5700/0.57) and the overall gas-phase resistance at a low pressure might be close to the liquid-phase resistance of the cells (Table 3.4).

5. The surface-to-volume ratio of cylinders such as stems or petioles is S/V = 2/radius. For a radius of 2 mm, which might be typical of a petiole or primary stem, S/V = 10 cm^2/cm^3. This is one-fifth the S/V ratio of a 'typical' leaf, and 6.3-fold greater than the S/V ratio of an apple. The smaller S/V ratio combined with a higher skin resistance might increase the μl/l per μl/kg·h ratio of a stem or petiole by 5000-fold compared to that of a leaf with open stomates. Due to a higher S/V ratio, the μl/l per μl/kg·h ratio of the stem or petiole should be only 1/6.3 that of an apple, or approximately 0.16–0.32 μl/l per μl/kg·h.

6. The resistance to CO_2 and O_2 transport initially is the same and low in Granny Smith apples (Trout *et al.*, 1942), but during 3 months' storage at 1.1°C, both values increase 3.2-fold, and by 7 months the resistance to O_2 and CO_2 had increased 8.3- and 5-fold, respectively. When these fruits are transferred to 21.1°C, the mass transport resistance continues to increase, two to threefold more rapidly for O_2 than CO_2. The different effects which wax deposition has on O_2 and CO_2 exchange are explained by the relative ease with which CO_2 penetrates through the cuticle when the lenticles become occluded, and by the cuticle's impermeability to O_2.

7. The concentration of binding sites (assuming one molecule of ethylene per binding site) varies between 1.9×10^{-9} and 6.8×10^{-9} mol/kg fresh weight in various leafy tissues, flower petals and cotyledons (Sisler, 1991), and in pea epicotyl tissue the binding site is half saturated when the IEC is approximately 0.15 μl/l (Table 5.6). Assuming an average receptor concentration of 4.4×10^{-9} mol/kg fresh weight, the binding site would be half saturated with 2.2×10^{-9} mol/kg fresh weight (0.024 μl/kg fresh weight) of ethylene. As the air space in pea tissue is only 2.8% on a fresh-weight basis (Table 3.9), 28 cc of air containing 0.004 μl of ethylene are present in each kg of tissue (assuming a tissue density of unity), and the quantity of intercellular ethylene is only 1/6 the amount bound to the receptor. In tomato fruits, the apparent concentration of the ethylene receptor is lower (7×10^{-11} mol/kg), the air space is larger (5%, Table 3.8), and the receptor is half saturated at 0.3 μl/l. The amount of ethylene in the intercellular air will be 16.7 times larger than the quantity bound to the receptor. Apple fruits contain 30% air (Table 3.8), the receptor concentration in apple pulp is 3.2×10^{-11} mol/kg, and the receptor is half saturated at 0.097 μl/l. As binding of ^{14}C-ethylene was tenfold greater in whole apples compared to minced pulp (Blankenship and Sisler, 1993), the amount of ethylene in their intercellular air will be 75.8 times larger than the quantity bound to the receptor.

8. The design of this LP apparatus is described by Wu *et al.* (1972). O_2 and N_2 supplied from separate pressurized cylinders were mixed in a 20:80 ratio in a humidifier at room temperature and an unspecified pressure. Airflow through the humidifier was regulated at 30 ml per min in order to maintain the humidity of the resultant mixture at 95%. The mixture was then flowed through individual regulators set to maintain the pressure at 62.8, 37.1 and 13.6 kPa (471, 278 and 102 mm Hg), respectively, in three 19-l vacuum chambers controlled at 12.8°C. The flow also passed through a throttling valve into a fourth temperature regulated 19-l chamber open to the atmosphere at its outlet (atmospheric pressure = 86.1 kPa = 646 mm Hg). When this mixture passed through the pressure regulators, the humidity was decreased in proportion to the expansion, and increased due to the temperature reduction in each chamber. The final humidity entering each chamber was different and at 13.6 kPa (102 mm Hg) was quite low. The flow into each chamber is not specified, but if the total was distributed equally, the amount passing through each chamber would be only equivalent to 0.044 chamber volumes per hour. Assuming that fruit occupied 50% of the chamber's space and respired with an RQ = 1, at such a low air change rate the respiration of green tomatoes at 12.8°C would create an 8% [O_2] + 12% [CO_2] mixture in the 86.1 kPa (646 mm Hg) atmospheric control chamber, and a 2.6% [O_2] + 17.4% [CO_2] mixture in the 62.6 kPa (471 mm Hg) chamber; all O_2 would be consumed at 37.1 and 13.6 kPa (278 and 102 mm Hg), respectively. O_2 draw-down would decrease the respiration rate, but nevertheless the [O_2] in each chamber will be far lower than the indicated storage pressure would suggest, a substantial amount of CO_2 will be present in each chamber and, due to the low air-change rate, the humidity in each chamber will be saturated by water evaporated from the fruits. This explains why the 62.6 kPa (471 mm Hg) chamber doubled the storage life of the fruit, whereas in a well-ventilated system that pressure would have no effect on tomato storage-life.

9. The ethylene transport resistance through a 100 (L) × 50 (W) × 20 (H) cm box containing 18.2 kg of carnations was determined in Bogota, Colombia (barometric pressure = 560 mm Hg), by injecting ethylene and measuring its diffusive escape at 13°C (Fig. 3.15). The box contained newspaper to absorb moisture, and a 1 mil (0.00254 cm) thick 150 × 185 cm polyethylene liner perforated with 80 holes of 1 mm diameter. The box's initial endogenous ethylene concentration was measured and then gas was injected at the centre and both ends. Air samples (1 cc) were withdrawn from the centre at intervals and the ethylene measured by gas chromatography. Initially, the ethylene concentration increased at the sampling point due to diffusive spread of the gas within the box from the points of introduction to the point of sampling. After 15 min, the ethylene concentration stabilized and thereafter decreased with the kinetics of a half-life curve. Based on the box's surface area and volume, and a half-time for ethylene escape of 40 min, equation 3.12 indicates that the box's ethylene mass transfer resistance (r_{box}, equation 3.22) is 3854 s/cm at the prevailing barometric pressure. Corrected for temperature (equation 15.35) and pressure (equation 6.56), the value of r_{box} for ethylene transport at 0°C is 5649 s/cm at 101.3 kPa (760 mm Hg). The box's surface area (A_{box}, equation 3.22) is 15,000 cm^2. The surface area of the leaves, flowers and stems contained in the box is not known, but if the box was filled with 18.2 kg of 'typical' leaves having a surface/volume ratio of 1.4 and density of 0.6 g/cm^3, then $A_{box}/\Sigma A_{com} \cong 0.37$ and r_{b,C_2H_4} = 15,268 s/cm at a pressure of 760 mm Hg. The box mass-transfer resistance values (r_b) to be used in the resistance network calculations for CO_2, O_2, ethanol and water vapour, corrected to take into account each gas's binary diffusion coefficient in air (Table 15.4), are 14,604, 11,322, 19,759 and 9161 s/cm, respectively. At a pressure of 1.33 kPa (10 mm Hg) the values must be divided by 76 to account for increased diffusivity at the lower pressure, and reduced by an additional 12% because diffusivity is increased by the presence of a substantial partial pressure of water vapour in the mixture (equation 15.41). A log mean partial pressure correction further reduces the box's transpirational resistance (15.16; equation 6.8). If it is assumed that the density of carnation stalks is approximately 1 g/cc, the box contained 63,030 cc of free air, and if the average porosity of the flower stalks is 0.1, the value of $V_A + bV_L$ in equation 3.13 (b = 0.1; Table 15.2) would be 2850 cc. This only amounts to 4.5% of the total air within the box, and therefore the flowers should not significantly influence the result.

10. A value for r_{box} was determined from simultaneous dew-point measurements

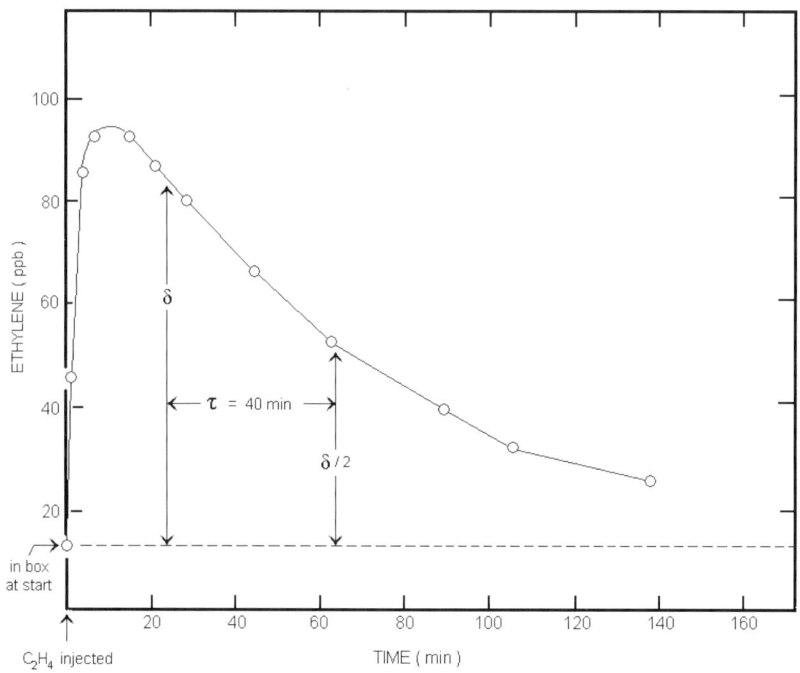

Fig. 3.15. Outward diffusion of ethylene at 13°C and a barometric pressure of 560 mm Hg from a 102 × 51 × 15 cm box containing 18.2 kg of carnation flowers, layers of newspaper and a 1 mil polyethylene liner (150 × 185 cm) perforated with eighty 1-mm-diameter holes. Ethylene was injected one-third of the distance from each end of the box and subsequent air samples were withdrawn from the centre with a syringe and analysed by gas chromatography. The ethylene concentration in the box decreased with a 40-min half-time (τ). Equation 3.12 can be used to compute the box's ethylene mass-transfer resistance. Example 9 indicates that the error introduced by the presence of flowers in the box is insignificant (Burg, 1976, unpublished).

inside and outside papaya boxes during their cool down under hypobaric conditions. The rate of water loss through the surface of a carton can be estimated from the cooling rate because almost all heat transfer at 2.67 kPa (20 mm Hg) results from evaporative cooling (6.22). The box's resistance to water-vapour mass transport was determined from measurements made with a 35.6 cm (L) × 25.4 cm (W) × 15.2 cm (H) papaya carton containing 4.5 kg of fruit. The carton was filled with shredded newspaper to protect the fruit, and sealed at all edges with sealing tape to conform with quarantine regulations. Based on the cooling rate and continuous dewpoint measurements within and outside the box, its effective resistance to water-vapour transport was computed using equation 6.56 to be r_{box} = 10.9 s/cm at 2.67 kPa (20 mm Hg), which is equivalent to a value of 660 s/cm at atmospheric pressure (Burg and Kosson, 1983).

11. Apples have an average load density of 350 kg/m^3 in palletized boxes (Mercantila, 1989b). To hold 18.2 kg of apples, a carton must have a volume of approximately 5.2×10^4 cm^3. Such a box might have dimensions 30 cm (H) × 30 cm (W) × 58 cm (L), and a surface area of 8760 cm^2. The surface area of 18.2 kg of apples is approximately 19,546 cm^2, and therefore $A_{box}/\sum A_{com} \cong 0.45$. The resistance values (r_b) for each gas or vapour indicated in Table 3.4 (apples) are computed assuming that the box containing 18.2 kg of apples has the same permeability characteristics as the papaya box described in example 10. The value for water vapour (660 s/cm) is divided by 0.45 (equation 3.22), and the result is corrected to account for each gas or vapour's binary diffusion coefficient in air (Table 15.4).

12. The cellular resistance to ethanol mass transport, computed relative to the surface area of an apple, is r_{cell} = 4680 s/cm (Table 3.4). At atmospheric pressure, the total air-phase resistance to ethanol transport, exclusive of the box, is $r_{p,a,ias}$ = 23,962 s/cm, and the total resistance of the liquid and air phases to ethanol transport is 28,642 s/cm. At the optimal apple-storage pressure (6.67 kPa = 50 mm Hg), the air-phase resistance to ethanol transport is decreased to 1576 s/cm, the total air and liquid phase resistance is 6256 s/cm, and LP might increase the rate at which alcohol escapes by 74%, whereas under the same conditions LP would enhance diffusive escape of fixed gases by approximately 94%.

13. Sweetcorn was stored in NA, CA and LP at 1.7°C and 98–100% RH, either in 2 or 21% [O_2], or at 6.67 kPa (50 mm Hg = 1.2% [O_2]). After 3 weeks in NA, the corn's ethanol content had increased from an initial value of 25 mg/100 g tissue to 151 mg ethanol/100 g, indicating that this commodity has a propensity to accumulate alcohol even in 21% [O_2]. In 2% [O_2] at atmospheric pressure, the ethanol content increased to 364 mg/100 g in 3 weeks; in LP at 6.67 kPa (50 mm Hg = 1.2% [O_2]), 293 mg/100 g accumulated during the same interval (Spalding *et al.*, 1978). The result indicates that to a considerable extent the escape of ethanol from the tissue was limited by the cellular resistance rather than by diffusion through an air phase.

14. The optimal LP storage condition in laboratory iceberg lettuce trials ostensibly was a pressure of 1.33 kPa (10 mm Hg = 0.149% [O_2]) at 0°C. When lettuce was stored in an intermodal container at this pressure and temperature it developed reddish-brown heart leaves and spots on the inner surfaces of the midribs of the young leaves. This disorder is typical of low O_2 damage in CA at [O_2] concentrations lower than 1% (Ryall and Lipton, 1972). Did the barrier created by the boxes and packaging used in the container trial lower the O_2 partial pressure within the boxes sufficiently to cause low [O_2] damage? Iceberg lettuce has a 175 kg/m^3 packing density in palletized boxes (Mercantila, 1989b) and a maximum O_2 consumption rate at 0°C of 8.65 cm^3/kg·h in 21% [O_2] at atmospheric pressure (Hardenburg *et al.*, 1986). A 60 cm (L) × 45 cm (W) × 38 cm (H) box (surface area = 13,380 cm^2) can accommodate 18.2 kg of lettuce. Assuming that this carton has similar properties to the papaya box described in example 10, at atmospheric pressure its conductance to water vapour is 1/660 cm^3/s per cm^2 of surface when the pressure gradient across the surface is 1 atm. The box's O_2 conductance at 1.33 kPa (10 mm Hg), corrected for the pressure reduction and binary diffusion coefficients in air, will be increased 61.5-fold at the low pressure. If the O_2 consumption rate at 1.33 kPa (10 mm Hg) is reduced by 80% (Fig. 4.2), only a 0.000009 atm O_2 gradient (= 0.00019%[O_2]) across the box's surface is needed to transfer O_2 as rapidly as it is consumed by respiration. This gradient is insignificant and should not influence the gas composition in the commodity's intercellular spaces.

15. At atmospheric pressure, the respiration rate of head lettuce may reach 17 mg/kg·h at 0°C (Hardenburg *et al.*, 1986). When 177,300 kg (39,000 lb) of head lettuce is loaded into a 12.2 m (40 ft) hypobaric intermodal container operated at 0°C and a pressure of 1.33 kPa (10 mm Hg), if LP reduces the respiration rate by 80% (Fig. 4.2) and the RQ remains at unity, the lettuce will consume O_2 and produce CO_2 at a rate of 0.031 m³/h. It will lose 262.8 g/h of water in order to dispel all respiratory heat, and this amount of water will saturate air flowing at a rate of 54.4 m³/h (32 cfm). As the saturated mixture in the container has a vapour pressure of 0.61 kPa (4.59 mm Hg) and a total pressure of 1.33 kPa (10 mm Hg), only 0.39 m³/h of atmospheric air (STP) will enter, supplying 0.082 m³/h of O_2 (STP). The [O_2] in the container will be drawn-down to 0.078% from a starting value of 0.149%, and the [CO_2] will increase to 0.071%. If the box temperature is slightly elevated (Table 3.14), this will further decrease the intercellular [O_2] concentration by increasing the water-vapour pressure within the box and commodity. To some extent, lowering the intercellular O_2 concentration is self-compensating, since it reduces the O_2 consumption rate, but to avoid low [O_2] damage it is prudent to elevate the storage pressure by 0.67 kPa (5 mm Hg) in order to approximately double the O_2 supply flowing through the container.

16. The absorption coefficients of NH_3 and CO_2 in water at 0°C are 1299 and 1.725, respectively, where absorption coefficient = (moles/litre in solution)/(moles/litre in gas phase). According to equation 15.36, the diffusion coefficient in water is 1.6-fold greater for NH_3 than for CO_2. The combined effects of diffusivity and solubility should cause the cuticular resistance to be 1200-fold lower for NH_3 than for CO_2, i.e. approximately 17 s/cm. The binary diffusion coefficients of NH_3 and CO_2 in air at 0°C are 0.198 cm²/s (Özisik, 1985) and 0.134 cm²/s, respectively (Table 15.4). Based on the CO_2 stomatal resistance value indicated in Table 3.4, the NH_3 value should be approximately 3.2 s/cm for a 'typical' leaf.

17. Conditions for an apple at 20°C:

porosity = 0.3 (Clements, 1935)
weight = 169 g
surface area = 181.5 cm²
volume = 230 cm³
volume of air in the apple = 69 cm³
volume of liquid in the apple = 161 cm³
skin resistance to ethylene = 18,339 s/cm (Table 2.3)
ethylene production rate = 100 μl/kg·h
IEC = 474.3 μl/l
bunsen coefficient (b) = 0.126 for ethylene, 0.878 for CO_2 (Table 15.2).

Initially, the apple contained 33 μl of ethylene in the air space, and 9.6 μl in the cell sap. During 5 min under saturated ammonium sulphate, the apple will produce 1.51 μl of ethylene. This will increase the total amount of ethylene by only 3.5%.

18. The natural convective film coefficient of horizontal and vertical tomato hypocotyl cylinders will be analysed to determine the extent to which reorienting a cylinder of uniform wall temperature from vertical to horizontal alters its heat and mass transfer. In the experiments of Harrison and Pickard (1984; Fig. 3.1), fifty 20-mm-long sections were aligned around the inside perimeter of a 15-mm-diameter vial (perimeter = 47.1 mm) forming a 'cylindrical annulus'. The conditions were:

temperature = 24 ± 0.5°C (297 K)
β = 1/297 K
hypocotyl section length (L) = 0.02 m
hypocotyl section diameter (D) = 0.001 m

$g = 9.8\ \mathrm{m \cdot s^{-2}}$
viscosity (v) = $1.6 \times 10^{-5}\ \mathrm{m^2/s}$ @ 24°C
Prandtl number, air = 0.71
Thermal conductivity (k) =
0.026 W/m·K @ 24°C.

When the vertical tomato hypocotyl section transfers its respiratory heat across a 1°C temperature gradient (ΔT) between its surface and the surrounding air, the average Grashof number (Gr_L) based on the section's length (L), is (Kreith and Bohn, 1997):

$$Gr_L = g\beta\,(\Delta T)\,L^3/v^2 = 1028 \quad (3.36)$$

When $10 < Gr_L Pr < 10^8$, and the heat transfer fluid is air (Pr = 0.71), the equation for the average Nusselt number (Nu_L) to be used with a vertical cylinder is (Cebeci, 1974; Kreith and Bohn, 1997):

$$Nu_L = 0.68\ Pr^{1/2}\ [(Gr_L)^{1/4}/(0.952 + Pr)^{1/4}] = hL/k \quad (3.37)$$

According to equation 3.37, $Nu_L = 2.86$ and for the vertical cylinder, when $k = 0.026$ W/m·K, the average natural convective heat transfer coefficient is 2.48 W/m²·°C (0.44 BTU/·h·ft²·°F). The average Rayleigh number (6.12) for a horizontal tomato hypocotyl cylinder, based on its diameter (D), is:

$$Ra_D = g\,\beta\,(\Delta T)\,D^3\,Pr/v^2 = 0.092 \quad (3.38)$$

When $10^{-4} < Ra_D < 1012$, the equation for the average heat transfer coefficient of a single horizontal wire or tube suspended in air is (Özisik, 1985):

$$Nu_D^{1/2} = 0.60 + 0.323\ Ra_D^{1/6} \quad (3.39)$$

If $Ra_D = 0.092$, then $Nu_D = 0.667$, $k = 0.026$ W/m·K, and the average natural convective heat transfer coefficient is $h_D = 17.4\ \mathrm{W/m \cdot °C^{-1}}$ (3.06 BTU/h·ft²·°F). The computations indicate that the coefficient for free convection is significantly higher in the horizontally oriented hypocotyls.

For laminar flow, the mean heat transfer coefficients for free convection from plates to air at atmospheric pressure and moderate temperatures are (Özisik, 1985):

Vertical plate: $h_m = 1.42\ (\Delta T/L)^{1/4}$
Horizontal plate (upper surface):
$h_m = 1.32\ (\Delta T/L)^{1/4}$
Horizontal plate (lower surface):
$h_m = 0.59\ (\Delta T/L)^{1/4}$

As the stomatal frequency is 130 per mm² on the lower epidermis and 120 per mm² on the upper epidermis of tomato leaves (Willmer and Fricker, 1996), for equivalent ΔT values the average mean heat transfer coefficient for the two surfaces of a vertical leaf will be approximately 1.5-fold greater than the average of the mean heat transfer coefficients for the upper and lower surfaces of a leaf in its normal near-horizontal orientation. The mass flux relation given by Fick's first law is similar to the heat flux expression given by the Fourier law, and the mass diffusivity (D) and the heat diffusivity (α) have the same units, m²/s (Özisik, 1985). Therefore, the computation suggests that mass transfer through the boundary air layer is enhanced when hypocotyl sections or plants are switched from vertical to horizontal. If the gas mass-transport resistance of the boundary air layer is significant compared to the overall 'skin' resistance, as it is in leaves with open stomates (Noble, 1991), this could account for the apparent 'burst' in ethylene emanation during the first 2–4 min after hypocotyl sections or seedlings are reoriented from vertical to horizontal.

19. The linear slope of the amino nitrogen leakage curve between 15 and 60 min is only slightly lower when 0.4 or 0.6 M mannitol is added to the liquid on which pre-climacteric or early climacteric Dwarf Cavendish banana slices are floated (Fig. 3.7). At the climacteric peak, the leakage rate in 0.6 M mannitol remains close to the pre-climacteric value, but leakage is > 60% higher if the apoplastic solution of climacteric slices is diluted by floating them on pure water (Fig. 3.7d). This creates the illusion that the cells might leak solution into their intercellular space.

20. The escape of apoplastic amino nitrogen from banana slices can be determined by extrapolating the 15–60-min linear portion of the amino nitrogen leakage curves in water and 0.4 or 0.6 M mannitol back to their intersect close to zero time, before the second phase of leakage commences from within the membrane-bound cell (Fig. 3.7). The amino nitrogen composition of the apoplastic solution should remain relatively constant during banana ripening, since the total amino nitrogen does not change (Brady *et al.*, 1970). As the apoplast empties within a few minutes after fruit slices are floated on solutions (Burg *et al.*, 1964; Brady *et al.*, 1970; Nielsen and Schjoerring, 1998), the loss of amino nitrogen during the initial minutes of an incubation is ratiometrically related to the apoplastic volume. Corrected for a pulp density of 0.76 g/cm^3 (Burg, 1967, unpublished) this technique indicates that the apoplastic solution occupies 8.5% of the pre-climacteric banana's volume, and that it only increased to 11.5% at the climacteric peak. As intercellular air comprises 15% of a ripe banana's volume (Table 3.8), the porosity of the pre-climacteric banana could not be much higher than 3 + 15 = 18%, and the decrease during ripening would not be expected to substantially increase the intercellular resistance.

21. Mature-green tomatoes producing respiratory CO_2 at a rate of 19 cm^3/kg·h at 20°C (Hardenburg *et al.*, 1986) would experience a water loss of 0.0739% per hour if all of their respiratory heat were transferred by evaporative cooling. At 20°C, the ICC in a mature-green cv. Sonata tomato is approximately 3% (Table 2.2) and the CO_2 absorption coefficient is 0.878 cm^3 of gas per cm^3 of water (Table 15.2). The water loss could only carry with it 0.103% of the respired CO_2. Apples respiring at a rate of 2 cm^3/kg·h at 0°C (Hardenburg *et al.*, 1986) would lose water at a rate of 0.00778% per hour if all of their respiratory heat were transferred by evaporating water. At 0°C, the apple's ICC is approximately 1% (Table 4.5) and the CO_2 absorption coefficient is 1.713 cm^3 of gas per cm^3 of water (Table 15.2). The water loss could only passively carry 0.067% of the respired CO_2.

22. Vertical *Kniphofia* stems, elongated at a rate of 1.41 mm/h, produced 0.17 nl/g·h of ethylene, and contained 0.2 μl/l of 'vacuum extractable' ethylene. During a 7-h gravistimulation, the ethylene production rate and IEC in the stem's upper half increased to 0.28 nl/g·h and 0.97 μl/l, respectively, while in the lower half the production rate rose to 4.71 nl/g·h and the IEC reached 7.4 μl/l (Woltering, 1991). Application of 1 or 10 μl/l ethylene retarded the elongation of vertical flower stalks to the same extent, approximately 33%, indicating that ≤ 1 μl/l was a saturating dose for growth inhibition. Ostensibly, the IEC was high enough during gravistimulation to maximally inhibit elongation of both the upper and lower halves, and yet during 6 h of gravistimulation, the average growth rate increased to 1.95 mm/h in the upper half and decreased to 0.2 mm/h in the lower half, while the flower spikes bent by 90°. This result suggests that the vacuum-extraction measurements grossly overestimated the IEC in the tissue. The upper half produced 0.014 μl/kg and the lower half 0.24 μl/kg of ethylene during 3 min, while the *Kniphofia* flower stalk parts were submerged under saturated ammonium sulphate during the vacuum-extraction assay. If the porosity of *Kniphofia* stem is in the range encompassed by cucumber, pea and soybean stems (Table 3.9; 1.1–5.2%), ethylene produced during 3 min would give rise to an IEC of 0.27–1.27 μl/l in the upper half, and 4.7–21.8 μl/l in the lower half, easily accounting for all of the recovered ethylene. The same error accounts for the 'apparent' IEC gradient measured by a 5- or 10-min vacuum extraction of the upper and lower halves of gravistimulated tomato and cocklebur stems (Wheeler *et al.*, 1986).

23. When a leaf blade's stomates are fully open, its transpirational resistance to water vapour varies from 0.2 to 17.8 s/cm (Raschke, 1960; Holmgren *et al.*, 1965; Noble, 1991), and in mesophytes from 0.5 to 5.0 s/cm (Noble, 1991). Depending on the wind velocity and leaf size and shape, the resistance of the barrier-air layer ranges from 0.06 to 1.8 s/cm, so in a 'typical' mesophytic leaf at full stomatal opening, the transpirational resistance, which is the sum of the barrier-air layer and stomatal resistances, is 0.56–6.8 s/cm. Stomatal resistance is inversely related to each gas's binary diffusion coefficient in air, and therefore is 1.7-fold higher for ethylene compared to water vapour (Table 15.4). For computational purposes, an average value of 6.3 s/cm will be considered representative of a 'typical' mesophytic leaf's resistance to ethylene transport when the stomates are open. A leaf's surface (S) to volume (V) ratio = 2/thickness, and therefore $S/V = 50\ cm^2/cm^3$ for a typical leaf of 400 μm thickness (Noble, 1991). The 'typical' leaf's internal air content is approximately 30% (Noble, 1991; Table 3.8), and based on the porosity of the tissue, its density would be approximately 0.7 g/cm^3. With open stomates, the calculated ratio between the IEC and ethylene production rate is only 2.45×10^{-5} μl/l per μl/kg·h. The minimum applied ethylene concentration required to produce a threshold response on leaf movement (Morgan and Baur, 1970), epinasty (Abeles *et al.*, 1992; Leather *et al.*, 1972), petiolar auxin transport (Beyer and Morgan, 1971), abscission (Abeles *et al.*, 1992), leaf de-greening (Sisler, 1980a), ethylene-induced leaf respiration (Sisler, 1979) and numerous other actions of ethylene in plant tissues typically are in the range 0.01–0.05 μl/l (Table 5.2; Abeles, 1973). Ethylene would have to be produced at a rate close to 1000 μl/kg·h to elevate the IEC by 0.02 μl/l/with fully open stomates, but the measured ethylene production rates of intact and excised leaves range only between 0.1 and 10 μl/kg·h (Baur and Morgan, 1969; Ben-Yehoshua and Eaks, 1970; Morgan and Baur, 1970; Beyer and Morgan, 1971; Sisler and Pian, 1973; Aharoni, 1978; Aharoni and Lieberman, 1979; Aharoni *et al.*, 1979b; Henskens *et al.*, 1993; Kefeli *et al.*, 1993), depending on type, maturity and stage of senescence. Therefore, leaves cannot accumulate a significant, active IEC when their stomates are open in the light.

24. The dark CO_2 production rate of a mature tobacco leaf is 250 μl/g·h at 20°C (Sisler and Pian, 1973; Aharoni and Lieberman, 1979; Aharoni *et al.*, 1979a). The average CO_2 cuticular resistance of grape leaves measured through their upper astomatous surface is 13,500 s/cm (Boyer *et al.*, 1997), and assuming that the cuticular resistance of tobacco leaves has this same order of magnitude, if tobacco stomates closed tightly at night, the internal CO_2 concentration (ICC) would increase to 1.3%. The average CO_2 cuticular resistance of grape leaves is similar to that of tomato fruits, and 13.8-fold lower than a tomato fruit's cuticular resistance to ethylene (3.20). For computational purposes, by analogy it can be assumed that a 'typical' mesophytic leaf's cuticular resistance to ethylene is $13.8 \times 13{,}500 = 186{,}000$ s/cm, and accordingly the stomatal mechanism has the potential to cause a 186,000/6.3 = 30,000-fold diurnal shift in a leaf's IEC assuming that the ethylene production rate remains constant and the stomates close tightly. With tightly closed stomata, the μl/l per μl/kg·h ratio is 0.735, and an ethylene production rate of only 0.027 μl/kg·h would give rise to an active IEC of 0.02 μl/l.

25. At 20°C, the average dark respiration rate of a tobacco leaf is 250 μl/g·h, and for the ICC to increase to 0.12% at night, stomatal closure would have to raise the combined CO_2 resistance of the cuticle and stomates to 8700 s/cm. As the binary diffusion coefficients for CO_2 and ethylene in air are almost identical (Table 15.4), the IEC would increase by 8700/6.3 = 1380-fold at night, and the ratio between the IEC and ethylene production rate would be 0.0338 μl/l per μl/kg·h.

26. Conditions for a potato at 20°C:

weight = 120 g potato
volume = 144 cm^3
porosity (V_{air}/V_{total}) = 0.01 (Table 3.8)
respiration rate = 12 μl O_2/g·h (Hardenburg *et al.*, 1986).

Respiratory O_2 consumption will decrease the O_2 in the intercellular spaces by 3.3% per min when the potato is placed under saturated ammonium sulphate to vacuum-extract gases. This explains why during a 5-s vacuum extraction the composition of the extracted gas changes by a detectable amount (Burton and Sprague, 1950; Burton, 1982).

27. Conditions for a 'typical' leaf:

porosity = 0.3% (Noble, 1991)
density = 0.7 g/cm^3
skin resistance to ethylene (open stomates) = 6.3 s/cm
leaf thickness = 400 μm (Noble, 1991)

surface/volume ratio (based on thickness) = 50:1 cm^2/cm^3
volume of air = 0.43 cm^3/g

A 'typical' leaf with open stomates, producing 1 µl/kg·h of ethylene, has an equilibrated IEC of 0.024×10^{-3} µl/l above the background level of ethylene in the ambient atmosphere. During a 5-min submergence under saturated ammonium sulphate, a 0.7 g leaf will produce 5.83×10^{-8} cm^3 of ethylene, which will elevate the IEC in the leaf's 0.3 cm^3 air volume by 0.194 µl/l. This is an 8083-fold increase.

Notes

1. The pH of lemon juice is 2.5, due mainly to its high citric acid content. This is the lowest pH which does not cause significant leakage of apple cell membranes (Fig. 3.6, *lower left*).
2. These values are calculated based on the following assumptions and measurements. In various leaves, the total cellular CO_2 resistance ($r_{cw,pl,cyt}$) varies from 0.4 to 4 s/cm (Noble, 1991); the 1.4 s/cm value in Table 3.4 assumes that $k = 1$ and the cellular CO_2 permeability coefficient is 5×10^{-2} cm/s. Plant cells are nearly equally permeable to small gas molecules such as O_2, ethylene and CO_2 (Davson, 1960; Noble, 1991), and therefore the cellular resistance values for O_2 and ethylene in Table 3.3 have been calculated by multiplying the cellular CO_2 resistance by the ratio ($K_{H,O_2}/K_{H,CO_2}$) or ($K_{H,C_2H_4}/K_{HCO_2}$), where K_{H,CO_2}, K_{H,O_2} and K_{H,C_2H_4} are the Henry's Law coefficients (Table 15.1) for CO_2, O_2 and ethylene, respectively. In *Chara* and *Nitella* cells, the ethanol permeability coefficient is between 10^{-3} and 10^{-4} cm/s (Collander, 1949; Noble, 1991); a value of 4.5×10^{-4} cm/s has been used to calculate the cellular resistance to ethanol (Table 3.4), adjusted to take into account the partition coefficient (K) computed from the Henry's Law coefficient and binary diffusion coefficient in water (Table 15.5).
3. The thickness (mm) of a leaf's boundary air layer (δ^{bl}, mm) is given by the relationship $\delta^{bl} = 4\ (L/v)^{1/2}$, where L is the mean leaf length, m, in the wind direction and v is the wind velocity, m/s (Noble, 1991).
4. Mass diffusivity D and heat diffusivity α have the same units, m^2/s.
5. Grape cells (*Vitis vinifera* L. cv. Gamay) grown *in vitro* generate ethylene mostly from intracellular ACC oxidase (Ayub *et al.*, 1993), avocados also may produce intercellular ethylene (Christoffersen *et al.*, 1993) and a small amount of intracellular EFE enzyme is present in tomato fruits (Bouzayen *et al.*, 1990). Intracellular tomato EFE may be responsible for the system–1 basal rate of ethylene production (5.06).
6. A hypobaric pressure of 23.6 kPa (177 mm Hg), with the [O_2] and [CO_2] maintained at normal ambient partial pressure, also reduced the chlorophyll loss by half in leaves of *Poa pratensis*, stimulated by infection with *Bipolaris sorokiniana* to produce ethylene and fade (Hodges and Coleman, 1984). Abscission of attached Valencia oranges, Persian lime and calamondin fruits, and of Valencia orange fruit and leaf explants containing an attached stem piece, was prevented at a pressure of 20 kPa (150 mm Hg) flowing 99.7% [O_2] + 0.3% [CO_2] to approximate normal atmospheric [O_2] and [CO_2] conditions (Cooper and Horanic, 1973; Table 2.5). A pressure of 0.2 atm (150 mm Hg = 3.6% [O_2]) markedly slowed chlorophyll degradation in the peel of citrus fruit (Purvis and Barmore, 1981), but the significance of this study has been questioned because [O_2] may not have been maintained high enough to allow de-greening to proceed (Goldschmidt *et al.*, 1993).
7. Ethylene derived from the various zones of bean and pea hook and subapical tissue was extracted and measured (Schwark and Bopp, 1993), using the vacuum assay of Beyer and Morgan (1970a), modified by the use of cold temperatures and the addition of $CoCl_2$ to prevent ethylene production during the assay, and by a 20-min heat treatment at 70°C (Bengochea *et al.*, 1980) to release 'bound' and dissolved ethylene before a low pressure of 14.7 kPa (110 mm Hg) was instated to collect the gas sample (Schierle and Schwark, 1988).
8. The pea subhook's diameter is at least 1.5-fold larger than the hook elbow's diameter, and the cell length in the subhook is greater (Goeschl and Pratt, 1968). As the subhook region grows exclusively by cell enlargement (Apelbaum and Burg, 1972), it must contain at least 2.25-fold fewer cells per kg fresh weight, and therefore on a per-cell basis, the amount of heat-extractable (bound) ethylene that was recovered is essentially the same in the subhook and hook.
9. In apples, linoleic acid-rich galactolipids (Hulme and Rhodes, 1971) and plastid membrane lipids (Galliard, 1968) decrease, phospholipids and fatty acid acyl groups remain constant or increase slightly during ripening (Knee, 1993), and phosphotidylcholine increases (Bartley, 1984). Phospholipids (Hobson and Grierson,

1993) and membrane lipids (Kalra and Brooks, 1973) decline in ripening tomatoes, and the temperature at which a tomato fruit experiences chilling injury is lowered during ripening (Hardenburg *et al.*, 1986), indicating that the unsaturation and fluidity of membrane lipids has increased (7.10), but the plasma membrane remains intact in fully ripe fruits (Brady, 1987). The total content of lipids does not change appreciably during banana ripening, although unsaturated acids, especially palmitoleic, decrease by 20% in the pulp (Palmer, 1971). It has also been reported that unsaturation of membrane lipids (Wade and Bishop, 1978) and the fluidity of banana liposomes is enhanced as ripening progresses, possibly due to an increase in the proportion of linolenic and less linoleic acid in phospholipids (Seymour, 1993). Formation of lipid peroxidation products from polyunsaturated fatty acids occurs in an avocado's peel prior to the onset of colour changes (Seymour and Tucker, 1993), the more unsaturated fatty acids (palmitoleate and linolenate) increase in ripe mangoes (Lizada, 1993), and the ratio of unsaturated to saturated fatty acids doubles during the ripening of Honeydew muskmelons (Seymour and McGlasson, 1993).

10. The porosity of apples and leaves is approximately $V_{air}/V_{total} = 0.3$ (Table 3.8). At 20°C, $b = 0.126$ for ethylene and 0.878 for CO_2 (Table 15.2).

11. The free air space in etiolated pea subapical tissue is only 2.6% (Table 3.9), the hook produces 6 µl/kg·h of ethylene (Goeschl *et al.*, 1967), and based on the dose–response curve for a response to applied ethylene, the estimated IEC in the etiolated pea hook is 0.16 µl/l (Goeschl and Pratt, 1968).

12. An error in this publication makes it impossible to quantitatively evaluate the likelihood that the IEC was created by ethylene produced during the period that the tissue was submerged. The measured IEC in the lower half of gravicurving tomato stems is stated to be 1 mm^3/kg dry weight, and since a tomato stem's dry weight is approximately 10% of its fresh weight (Curtis and Clark, 1950), the 'IEC' measured in the lower half was 100 µl/kg on a fresh-weight basis. Presumably, a tomato stem's porosity is in the range 1.1–5.2% (Table 3.9), and therefore the lower half must have contained 1923–9090 µl/l of ethylene, which obviously is incorrect. Perhaps mm^3 is a misprint and the measured IEC was 1.9–9.1 µl/l of ethylene, which is close to the values measured in the lower half of horizontal *Kniphofia* flower stalks (Woltering, 1991). If so, the approximately 10 µl/kg·h ethylene production rate of horizontally positioned tomato plants (Harrison and Pickard, 1986) would have produced the lower half's measured IEC during submergence for 7.2 s if the tissue's porosity is 1.1%, and 34 s if it is 5.2%.

13. The effect may be somewhat less than this during refrigerated cold storage because stomatal apertures decrease by as much as 80% when the temperature is decreased from 20 to 5°C (Wilson, 1948; Willmer and Fricker, 1996).

14. Reports that mature grapes lack functional stomates (Grncarevic and Radler, 1967; Possingham *et al.*, 1967) are incorrect (Kanellis and Roubelakis-Angelakis, 1993).

15. It has been reported that the calyx accounts for the exchange of 42% of the ethylene, 24% of the CO_2 and only 2% of the water vapour transport in Golden Delicious apples; 94%, 81% and 67% of the exchange of ethylene, CO_2 and water vapour, respectively, in tomatoes (Cameron, 1982), and that the point of stem attachment is the path of least resistance for ethylene diffusion from the cavity of Valencia oranges (Barmore and Biggs, 1972).

16. The atmospheric pressure decreases by approximately 3% per thousand feet (305 km) of altitude.

17. In a study with cv. OH 7814 tomato fruits (Knee, 1995), the skin-resistance value for CO_2 varied between 8700 and 14,200 s/cm; for ethylene between 4700 and 12,800 s/cm; and for O_2 from 10,400 to 18,600 s/cm.

18. Cameron and Yang calculated that the stem scar's ethane resistance was 220 and 300 s/cm in green and red tomatoes, respectively, and the whole fruit's resistance values were 7400 and 7800 s/cm, respectively. This is not possible, since the stem scar's resistance acts in parallel with the cuticular resistance. The stem scar's resistance must be larger than the whole fruit's resistance. Based on data presented in Cameron and Yang (1982), the stem scar's ethane resistance values are 7223 and 7551 s/cm in green and red fruits, respectively.

19. In most terrestrial insects, almost all O_2 is taken up by the tracheae, but in *Dixippus* 25% of the CO_2 escapes through the skin, and in the larvae of *Dytiscus* and *Eristalis* somewhat less than 25% (Buddenbrock and Rohr, 1922). In frogs, 74–86% of the respiratory CO_2 escapes cutaneously and 49–67% of the O_2 is taken in by pulmonary respiration. Similar results have been reported for eels and other vertebrates (Dolk and Postma, 1927; Krogh, 1941).

20. The skin resistance of Empire apples, measured with the ethane efflux method (Cameron and Yang, 1982) or from steady-state

CO_2 measurements (Blanpied, 1990a), ranges from 11,000 to 37,000 s/cm.

21. Lowering the pressure may decrease the rate at which a particular gas or vapour is produced.

22. The turgor pressure of a typical plant cell is 0.6–1 MPa, and LP only lowers the hydrostatic pressure by approximately 0.1 MPa.

4

Oxygen, Carbon Dioxide, Ammonia and Cyanide

A unique advantage of LP that is lacking in all other storage methods is an ability to level the O_2, CO_2 and ethylene gradients that develop between the centre and surface of horticultural commodities due to respiration, ethylene production and the gas mass transport resistances of the skin and intercellular system. LP not only increases gas diffusivity into and from the commodity, but in addition continuously replenishes O_2 consumed by respiration and flushes away released metabolic CO_2, NH_3 and ethylene. The storage area is ventilated with air changes in which the atmospheric concentrations of all incoming volatile contaminants have been decreased as much as 132-fold by expansion in the pressure regulator (example 1). These features, and LP's ability to open stomates in darkness (4.15), combine to make LP the only method able to create an almost ethylene-, NH_3- and CO_2-free atmosphere both inside and around the stored commodity. They also allow LP to reliably regulate the O_2 partial pressure ± 0.008% at a low enough value to directly kill insects and prevent bacterial and mould growth without injuring the commodity (Burg, 2000).[1] At atmospheric pressure, the same O_2 level would cause low O_2 damage.

The storage results obtained with CA and LP differ even when both processes are adjusted to provide the same O_2 partial pressure. In LP, the reduction in apple firmness is slower; chlorophyll breakdown and acid catabolism is reduced; sugar content is maintained at a higher value; respiration, ethylene synthesis and ethylene content are remarkably lower; and physiological disorders are reduced or eliminated (Bangerth, 1973; Bangerth, 1974; Dilley, 1978). During Empire and Ida Red apple storages at 0°C in CA at 0.9% $[O_2]$ and in LP at 5.1 kPa (38 mm Hg = 0.9% $[O_2]$), less CO_2 and ethylene were produced in LP, and applied ethylene induced post-storage softening in CA fruits, but not in LP apples until much later (Irwin and Dilley, 1980). Chapter 10 and the maximum storage life table in the Preface contain numerous examples in which LP provides a greater storage benefit than CA at the same $[O_2]$ concentration, contradicting the idea that LP acts exclusively by lowering the O_2 tension (Kader and Morris, 1974; Stenvers and Bruinsma, 1975, 1977). In this chapter, LP's beneficial effects on stored commodities are examined to determine the extent to which they can be explained as physiological responses to unusually low intercellular concentrations of O_2, CO_2 and NH_3 or a reduced pressure. Ethylene is considered separately in chapter 5.

4.1 Measuring Respiration and Ethylene Production under Hypobaric Conditions

LP's effect on respiration and ethylene production usually has been evaluated after a commodity stored at a low pressure was returned to atmospheric conditions. This simplifies the assay, but does not provide accurate information about the rate during storage. A comparison of measurements made with freshly harvested carnations, which had been kept in LP for 5 weeks at 6.67 kPa (50 mm Hg), reveals that during storage the respiration rate decreased by 50–60%, and it returned to the initial at-harvest value within 1 day after the flowers were transferred to atmospheric air (Carpenter and Dilley, 1975; Fig. 10.5).

Low-pressure air samples can be collected for gas analysis by trapping aliquots in a sampling implement (Onoda *et al.*, 1989b), or freezing ethylene and CO_2 in a liquid–air cold trap inserted between the vacuum chamber and pump (Bangerth, 1973). After the aliquot is brought to atmospheric pressure in a gas-tight syringe, or the frozen sample is thawed, it can be analysed by gas chromatography. CO_2 production in an open flow-through system operating at a reduced pressure also has been monitored automatically every hour by passing effluent gas through an infrared gas analyser, and O_2 consumption has been evaluated in a low-pressure circulating closed-loop system flowing gas through a paramagnetic O_2 analyser (Figs 4.4 and 4.5; Tucker and Laties, 1985). Figure 4.1 illustrates an apparatus that has been used to determine O_2 consumption, CO_2 evolution and ethylene production in hypobaric intermodal containers (Burg, 1979a). The levels of these gases are assayed in the vacuum pump exhaust where, depending on the storage pressure, they are concentrated up to 76-fold, allowing rapid measurements to be made using an insensitive, inexpensive method. Analyses of O_2 consumption and CO_2 production made with this apparatus have been used to evaluate a stored commodity's response to the particular combination of pressure, temperature and flow initially selected during an LP intermodal container shipment, and also to determine the flow rate needed to saturate the storage atmosphere when the 'dry' hypobaric method is used (example 2).

The various methods of measuring respiration in LP give similar results.[2] CO_2 evolution by Golden Delicious and Boskoop apples (Bangerth, 1973), and by tomatoes, peppers and cucumbers (Bangerth, 1974) is

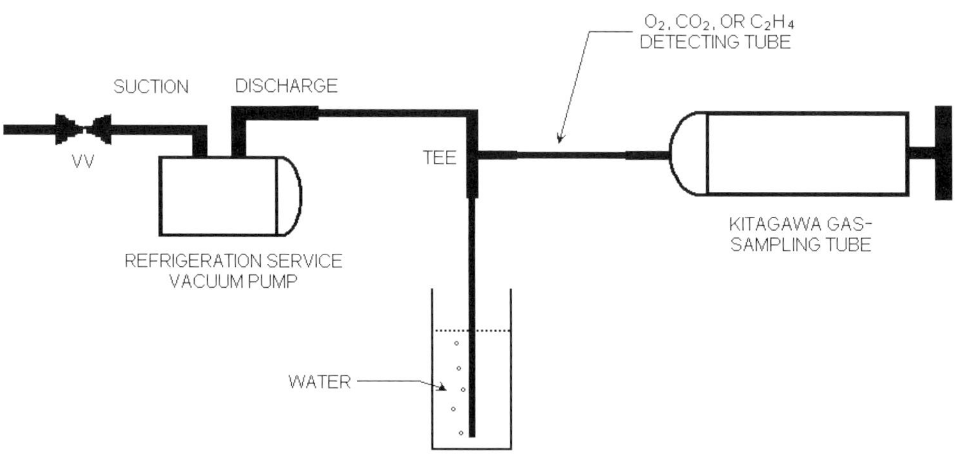

Fig. 4.1. Gas sampling system used with intermodal hypobaric containers to rapidly determine a commodity's CO_2 and ethylene production or O_2 consumption. A refrigeration service vacuum pump (or equivalent) with at least 2 m^3/h (1.2 cfm) capacity is attached to a flare-fitting located on the container's vent valve (VV, Fig. 13.17). If the refrigeration pump has a gas-ballast valve, it must be closed to prevent gas-ballast air from diluting the exhausted container air. First, the refrigeration pump is operated for 10 min with the container's vent valve open in order to equilibrate the pump oil with the atmosphere in the container. Then analyses are made by withdrawing vacuum-pump exhaust samples with a Kitagawa gas-sampling pump through a 2–24% [O_2] tube, a 1–20% [CO_2] tube and/or a 0.1–20 μl/l ethylene tube, depending on which gas or gases are to be measured.

67–75% inhibited at 10.13 kPa (76 mm Hg = 1.7–2% [O_2], depending on the storage temperature); the respiration of Cox apples is 50% inhibited at 2°C and a pressure of 9.33–10.67 kPa (70–80 mm Hg = 1.8–2.1% [O_2]; Sharples, 1974; Bubb, 1975); and at 0°C and 2 kPa (15 mm Hg = 0.28% [O_2]) O_2 uptake by mushrooms is 67% inhibited (Alloca, 1980b; Burg, 1980a). Results with bananas, Ida Red apples, Golden Delicious apples, maize, mushrooms and carnations are presented in Tables 4.1, 4.2 and 4.3, and in Figs 4.2 and 10.1. The respiration rate progressively decreases by approximately 90% when the pressure is lowered from atmospheric to 1.33 kPa (10 mmHg = 0.15% [O_2] at 0°C), and the relationship between pressure and the extent of the respiratory inhibition is remarkably similar for different commodities. This indicates that diffusion of O_2 through the skin and intercellular system is so rapid at a low pressure that there is little difference between the applied and internal [O_2] regardless of the tissue's structure.

Table 4.1. Respiration (μl/g·h CO_2) of Valery bananas stored at 13.3°C in NA or in LP at different pressures. The climacteric rise in respiration begins in NA between the 6th and 13th day; the respiration rate continuously declines in LP (Burg, 1969).

	Pressure, kPa (mm Hg)			
Days in storage	6.67 (50)	10.67 (80)	16.0 (120)	101.5 (760)
2	2.5	2.0	2.5	5.7
6	1.7	1.3	2.5	4.2
13	1.2	1.5	2.4	7.3
27	1.1	1.1	1.5	8.2

Table 4.2. Effect of [O_2] concentration on CO_2 production by White Sims carnations at 2.2°C during CA storage in flowing O_2 +N_2 (Uota and Garazsi, 1967) or in LP storage (Fig. 4.2).

	Relative CO_2 production	
% [O_2]	CA	LP
20.9	100	100
2.0	70	–
1.0	61	–
0.69	–	30
0.5	55	–
0.28	–	22
0.25	53	–
0.15	–	10

4.2 Effect of O_2 on Respiration

Cellular O_2 availability is determined by the intercellular O_2 concentration, which depends not only on the atmospheric O_2 concentration, but also on the commodity's rate of O_2 consumption at the prevailing temperature, the gas exchange resistance of the commodity's surface, the tissue's porosity (Table 3.8), its thickness and intercellular resistance to gas diffusion (Solomos, 1985), the extent to which the respiratory rate declines during storage (Table 4.4), and the storage box's resistance to gas transport. Internal gaseous

Table 4.3. Effect of LP on the respiration and ethylene production of Ida Red apples at 25°C (Dilley and Irwin, 2000, personal communication). The O_2 concentration has been corrected to take into account 3.2 kPa (24 mm Hg) of water vapour present at the storage temperature. The RQ remains constant at pressures providing 20.9–0.66% [O_2].

Pressure		Relative rate of consumption or production		
kPa (mm Hg)	% [O_2]	C_2H_4	CO_2	O_2
101.5 (760)	20.9	1.00	1.00	1.00
45.9 (344)	8.8	–	1.01	0.96
23.3 (175)	4.2	0.55	0.71	0.66
12.5 (94)	1.9	0.48	0.56	0.40
7.3 (55)	0.9	0.43	0.56	–
6.4 (48)	0.7	0.08	0.39	0.34

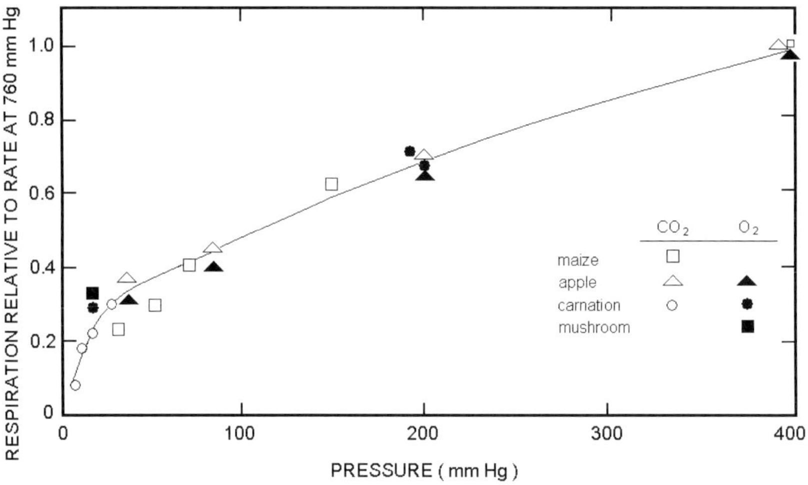

Fig. 4.2. Relation between respiration rate and pressure in various commodities (Burg and Kosson, 1983).

compositions measured in different commodities under various environmental conditions are summarized in Tables 2.2, 4.5 and 4.6.

An influence of air-filled diffusional barriers on the respiratory O_2 isotherm is readily discerned when the relationship between respiration rate and applied $[O_2]$ is compared in intact apples and cylinders or slices prepared from them. A cylinder's diffusional resistance is diminished because the intercellular path length has been shortened and the major diffusional barrier (the 'skin') removed, and a slice's gas conductance is further increased by an additional decrease in the intercellular path length. An intact apple's respiration is half inhibited when it is exposed to approximately 4% $[O_2]$ (example 3; Fig. 4.3, *left*), but the $[O_2]$ must be lowered to 1.2% to achieve the same effect with well-blotted 1-cm-diameter cylinders incubated in moist air (Fig. 4.3, *right*; Burg and Thimann, 1959). The apparent K_{m,O_2} for respiration is less than 0.2% $[O_2]$ in 1-mm-thick well-blotted apple discs incubated in moist air (Banks *et al.*, 1984); and *in vitro* the K_{m,O_2} for O_2 uptake by plant cytochrome oxidase[3] is approximately 0.002–0.05% $[O_2]$ at 25°C, and 0.001–0.02% $[O_2]$ at 0°C.

The respiration rate of Navel oranges, Valencia oranges and lemons progressively doubles as the $[O_2]$ is raised from 21 to 100% (Biale, 1960), but abnormally high $[O_2]$ does not increase the respiration of Bramley's Seedling apples (Fig. 4.3, *left*), or CO_2 production by avocados, bananas, sausage fruits and storage organs such as carrot, rutabaga, red sweet potato, parsnip, turnip, red beet, daicon radish, horse-radish, Jerusalem artichoke and jicama (Biale, 1946; Theologis and Laties, 1982a). At less than a normal atmospheric $[O_2]$ level, the respiratory O_2 isotherms of potatoes (Mapson and Burton, 1962), grapes, peaches, pears and plums (Claypool and Allen, 1948), citrus fruit (Biale, 1960), apples (Fig. 4.3, *left*; Fidler and North, 1967), avocados (Fig. 4.4, *left*) and pineapples (Dull *et al.*, 1967) are biphasic. In phase 1, the respiration rate slowly and progressively decreases when less than 21% $[O_2]$ is present, and in phase 2, depending on the type of commodity, the respiration rate declines very rapidly when the $[O_2]$ is reduced to below 1–7%. At still lower $[O_2]$, an inversion point ('extinction point' = EP) is reached and fermentation increases the RQ (Fig. 4.3, *left*). Usually at atmospheric pressure, the $[O_2]$ must be decreased to between 1 and 7% to half inhibit respiration (Beevers, 1961; Ducet and Rosenberg, 1962; Biale and Young, 1971; Dull, 1971; Burton, 1982; Mattoo *et al.*, 1988), and because the equilibrated gas

Table 4.4. Decline in respiration of asparagus and iceberg lettuce (Platenius, 1942), head lettuce and cut daffodils (Hardenburg *et al.*, 1986), White Rose potatoes (Lutz and Hardenburg, 1968), Majestic potatoes (Burton, 1982) and cut Linda carnations (Uota and Garazsi, 1967) during storage at various temperatures in air or 1% [O_2].

Commmodity	Temp (°C)	Days in storage	Air or 1% [O_2]	Respiration (mg CO_2/kg·h)
Asparagus	0	1	air	60
(Martha Washington)	0	2	air	46
	0	3	air	39
	0	4	air	36
	0	8	air	32
	10	0	air	196
	10	1	air	108
	10	2	air	83
	10	3	air	70
Head lettuce	0	1	air	17
	0	5	air	9
Iceberg lettuce	10	0.5	air	25
	10	2.5	air	16
	10	4.5	air	15
	10	10	air	15
Potato (White Rose)	20	2	air	16
	20	6	air	11
	20	10	air	8
Potato (Majestic)	20	0	air	12
	20	30	air	7.4
	20	60	air	5.7
Daffodil	0	1	air	37.5
	0	2	air	32.5
	0	3	air	25.0
	0	8	air	17.5
	0	20	air	12.5
Linda carnation	2.2	1	air	42.0
	2.2	2	air	38.0
	2.2	4	air	35.5
	2.2	10	air	32.0
	2.2	24	air	26.0
	2.2	1	1% O_2	30.0
	2.2	2	1% O_2	28.5
	2.2	4	1% O_2	24.0
	2.2	10	1% O_2	19.0
	2.2	24	1% O_2	15.0

phase K_{m,O_2} for cytochrome oxidase[3] may be as low as 0.001% [O_2], Mapson and Burton (1962) concluded that cytochrome oxidase is the only oxidase with sufficient O_2 affinity to be operative to any considerable extent at O_2 partial pressures less than 0.01 atm (= 1% [O_2]), and that decreasing the [O_2] from 100 to 1% should have little if any effect on O_2 uptake by this enzyme. They attributed the inhibition of respiration that occurs at higher O_2 partial pressures to the operation of undefined oxidase(s) such as polyphenol or ascorbic acid oxidase, which have a K_{m,O_2} value approximately 300-fold higher than that of cytochrome oxidase (Burton, 1982).[3] Robinson *et al.* (1975) also concluded that the respiratory decrease that they measured at 3% [O_2] with 30 fresh vegetables and soft fruits was too large to result from a suppression of cytochrome oxidase activity.

Table 4.5. CO_2 and O_2 gradients between the external atmosphere and intercellular spaces of pre-climacteric apples stored at 4 and 22°C (Henze, 1969).

Apple variety	Temp (°C)	Int/ext gradient in fruit +% [CO_2]	−% [O_2]
Golden Delicious	22	3.5	4.0
	4	1.0	1.1
Cox Orange	22	4.3	4.9
	4	1.5	1.9
Alexander	22	6.2	5.7
	4	2.0	1.8

Table 4.6. Composition of gases in the intercellular spaces of apples measured by vacuum extraction (Magness, 1920) and a sample tube (Trout *et al.*, 1942), in Hass avocados (Ben-Yehoshua *et al.*, 1963; Biale, 1964), in the centre of a watermelon (Wardlaw and Leonard, 1936a) and in a 105 g Arran Consul potato (Burton, 1950), as influenced by temperature.

Commodity	Temperature (°C)	% [CO_2]	% [O_2]
Apple (vacuum extract)	2	6.7	14.2
	6	8.4	12.9
	11	12.2	10.7
	20	17.2	5.5
	30	21.4	3.2
Apple (sample tube)	7	2.2	16.9
	21	3.7	13.1
	31	6.2	9.1
Potato	5	–	19.4
	10	–	15.6
	15	–	16.9
	20	–	14.2
	25	–	11.5
Watermelon	4	1.8	–
	12	2.2	–
	26	6.8	–
Hass avocado			
unripe	15	0.9	18.6
unripe	20	2.4	17.2
ripe	20	12.5	1.9

Various mathematical models have been devised describing the influence of permeability barriers on the apparent respiratory rate as a function of external [O_2] (Ducet and Rosenberg, 1962). Based on a model that envisioned cytochrome oxidase as the sole terminal oxidase, Chevillotte (1973) predicted that diffusional barriers might cause individual cells in potato tubers to become anaerobic, creating the illusion that a second oxidase limited O_2 consumption. This might account for the shift in the respiratory O_2 isotherm from biphasic in an intact apple (Fig. 4.3, left), to monophasic in apple cylinders (Fig. 4.3, right), but this explanation has been shown to be incorrect. Instead, relatively high [O_2] affects the rate of respiratory metabolism by a process that requires more than 1.5 h to become effective and does not alter cytochrome oxidase activity. Using hypobaric storage and computer modelling techniques, Tucker and Laties (1985) showed that irrespective of whether the total pressure was 101 kPa (760 mm Hg) or diffusion was enhanced by reducing the total pressure to 25 kPa (188 mm Hg), the respiratory O_2 isotherm of Hass avocado fruits was biphasic when measurements were made in a large-volume sealed vessel where the [O_2] changed slowly, and also in a flow-through system at different steady-state [O_2] partial pressures (Fig. 4.5, 'slow'), and it became monophasic when respiration was measured in a small volume sealed vessel under conditions that allowed the ambient [O_2] to change from 21 to 5 kPa (21–5% [O_2]) in less than 1.5 h (Figs 4.4 and 4.5, 'quick'). This explains why the O_2 isotherm in the apple cylinder experiment (Fig. 4.3, *right*) was monophasic, for O_2 consumption was determined during a 1-h period after gas mixtures were vacuum-impregnated into the tissue.

Chevillotte's suggestion (1973) that diffusional barriers create the illusion of a second oxidase was disproved by experiments with potato tubers in which the respiratory O_2 isotherm was plotted as a function of the measured intercellular [O_2] concentration. A quantitative effect of the diffusional barrier was apparent, but it was not sufficient to account for the biphasic nature of the isotherm (Mapson and Burton, 1962). Additional evidence against Chevillotte's theory was provided by experiments demonstrating that the biphasic nature of the respiratory O_2 isotherm is not altered in an open flow-through ('slow') system, when diffusion is accelerated by lowering the

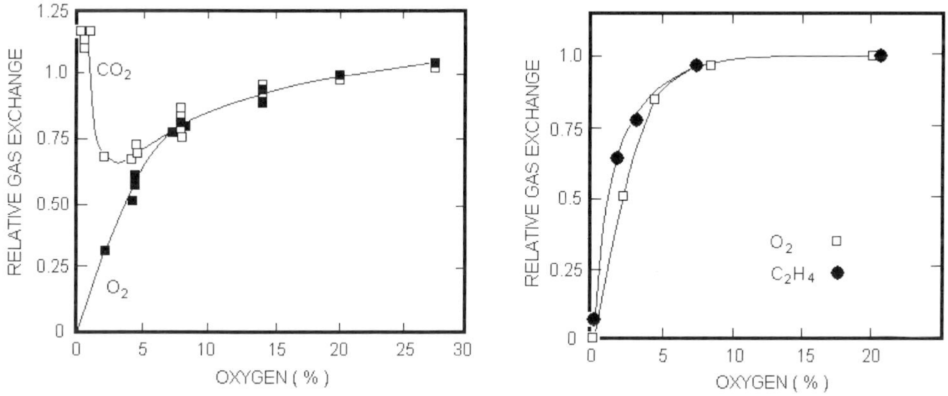

Fig. 4.3. (*left*) Influence of $[O_2]$ concentration on the O_2 uptake and CO_2 evolution of Bramley's Seedling apples at 23°C. The extinction point (EP) below which CO_2 production by fermentation is not completely suppressed is 9% $[O_2]$ (Street, 1963). (*right*) Ethylene production and O_2 consumption by McIntosh apple tissue cylinders at 21°C as a function of $[O_2]$ concentration. Measurements were made over a 1-h period after the gas mixture had been vacuum-impregnated into 3.8 cm long × 1 cm diameter apple plugs which had been briefly rinsed in 0.55 M glycerol solution and blotted dry. All gas mixtures contained only O_2 and N_2; control tissue was treated with 80% $[N_2]$ + 20% $[O_2]$ (Burg and Thiman, 1959). Control rates of ethylene production, CO_2 evolution and O_2 consumption are represented as 1.0 in 20% O_2.

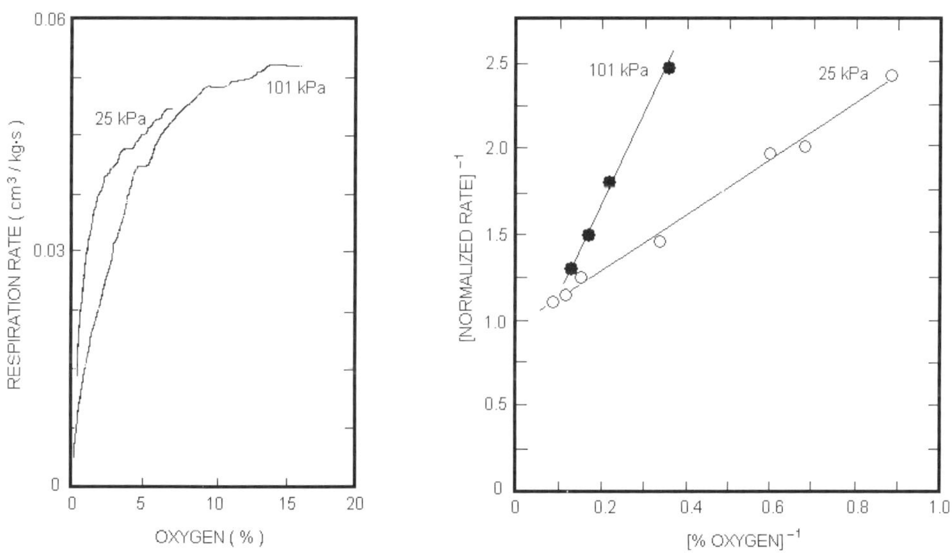

Fig. 4.4. (*left*) Respiratory isotherms for the 'quick' depletion of O_2 at 25°C, measured for individual ethylene-treated climacteric Hass avocado fruits kept at either 101 or 25 kPa (760 or 188 mm Hg) total pressure in a circulating closed system (Tucker and Laties, 1985). (*right*) Double reciprocal plots of respiratory O_2 isotherms shown in figure at left. The K_{m,O_2} is 4.2-fold lower at 25 kPa compared to the value at 101 kPa.

pressure (Tucker and Laties, 1985). The possibility that the cyanide-insensitive 'alternative' oxidase might become engaged when the O_2 partial pressure is lowered was dismissed by keeping Hass avocado fruits at a pressure of 25 kPa (188 mm Hg) in a monophasic 'quick' depletion setup, and treating them either with cyanide to inhibit

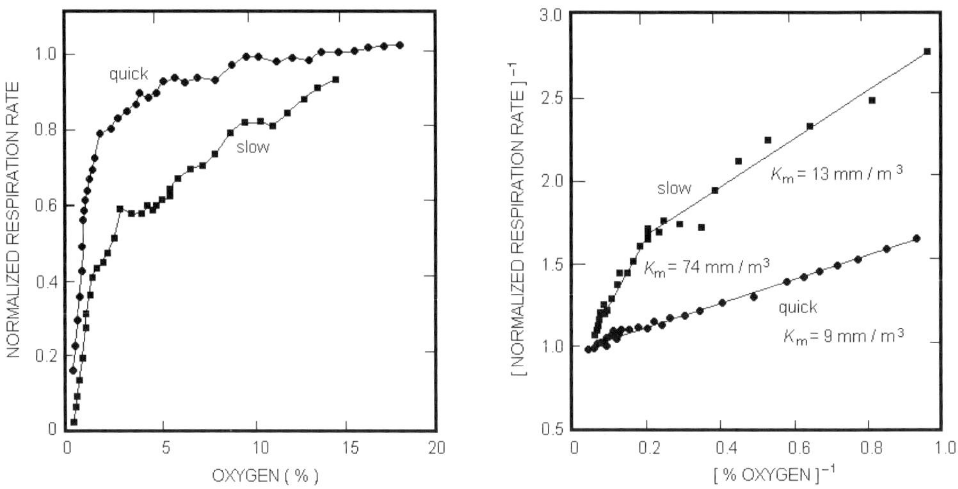

Fig. 4.5. (*left*) Effect of $[O_2]$ concentration on O_2 consumption by pre-climacteric Hass avocado fruits at 25°C in closed systems of different volumes. In the small volume 'quick' system, the O_2 concentration decreased from 21 to 5 kPa in less than 1.5 h. A flow-through system (not shown) and the large volume 'slow' system gave identical results. (*right*) Double-reciprocal plots of respiratory O_2 isotherms shown at left. K_m values for O_2 are calculated from the negative reciprocal of the abscissa intercept and converted from partial pressure of O_2 in the gas phase to the equilibrium concentration in water at 25°C (Tucker and Laties, 1985).

cytochrome oxidase and engage the cyanide-resistant alternative oxidase, or with ethylene to induce a respiratory climacteric rise. Consistent with studies showing that at the surface of yeast cells the K_{m,O_2} values for the cyanide-insensitive oxidase and cytochrome oxidase are 11.5 and 3.8 mmol/m^3, respectively (Henry *et al.*, 1977; Solomos, 1977, 1983), the measured apparent K_{m,O_2} shifted from 3 mmol/m^3 in ethylene-treated control avocados to > 10 mmol/m^3 in cyanide-treated fruits. This experiment proved that the cyanide-insensitive pathway is not normally engaged in ripening avocados and therefore does not participate in the high $[O_2]$ affinity response. It was concluded that

> The biphasic nature of the respiratory O_2 isotherm under steady-state conditions, or conditions of slow depletion of O_2, is neither diffusion-related nor due to the operation of two terminal oxidases, but rather is ascribable to cytochrome oxidase-mediated respiration in conjunction with regulation of an early stage in respiratory metabolism by O_2 at relatively high concentrations.

Blackman (1954) had proposed a similar explanation for oxygen's dual role in controlling respiration. He suggested that the terminal oxidase of apples is not inhibited unless the external $[O_2]$ is reduced below 5%, and envisioned that a gradual decrease in respiration at more than 5% $[O_2]$ is caused by substrate mobilization regulated by relatively high $[O_2]$ concentrations independent of the terminal oxidase. Blackman theorized that a slow change in the level of one or more endogenous substrates or regulators could explain why up to 24 h were required to establish a lower, constant respiration rate when apples were removed from normal air and stored in 5% $[O_2]$.

Tucker and Laties (1985) developed a mathematical computer model describing an avocado's respiration as a function of $[O_2]$ availability and the permeability of the air-filled resistance barriers, and found that the best computer fit to data obtained with the 'quick' measuring system occurred when it was assumed that the terminal oxidase had a K_{m,O_2} value of 3 mmol/m^3. The accuracy of the model's assumptions was verified when different respiration rates were induced by applying ethylene, and also by lowering the total pressure from 101 to 25 kPa (Fig. 4.4) in order to enhance diffusion fourfold. The apparent K_{m,O_2} at atmospheric pressure, 9–13 mmol/m^3, decreased by 4.2-fold when the pressure was reduced to 25 kPa (188 mm Hg; Fig. 4.4, right; Fig. 4.5, right), and at the lower pressure the K_{m,O_2} was similar to the value measured at the surface of yeast cells, 3.8 mmol/m^3 (Henry *et al.*, 1977). The optimal hypobaric storage pressure for most commodities is between 1.33 and 2.67 kPa, where the effect on diffusion is 10–20-fold greater than it is at 25 kPa. Surely at these lower LP pressures the resistance of the air-filled barriers to gaseous diffusion is insignificant in all horticultural commodities.

Tucker and Laties (1985) suggested that during the enlargement of fruits and storage organs, and also when fruits ripen, the high [O_2]-affinity negative-feedback respiratory response might allow cells to adapt to a modest drop in internal [O_2] by lowering their respiration rate. This would prevent O_2 depletion below a level that sustains terminal oxidase activity, and should decrease the critical external [O_2] at which fermentation and an increase in CO_2 evolution begin. In part, this theory was based on the behaviour of avocados and bananas, both of which undergo large changes in internal O_2 as they ripen (Table 2.2). It is less applicable to a hypobaric condition, since LP renders the air-filled diffusional barriers insignificant, keeping the intercellular and external [O_2] the same. At equivalent external [O_2] concentrations, LP would be expected to have a smaller inhibitory effect than CA on O_2 consumption, since in CA, but not in LP, diffusional barriers lower the internal [O_2] below the applied [O_2] concentration. This difference is illustrated in Fig. 4.5 (left), where it can be seen that an avocado's O_2 consumption is only inhibited by 10% at an external [O_2] partial pressure of 5 kPa and a total pressure of 25 kPa (188 mm Hg), compared to a 48% inhibition at atmospheric pressure and the same O_2 partial pressure. Conversely, when the external [O_2] is low, the O_2 gradient between the atmosphere and a commodity's interior may decrease the internal [O_2] in CA, but not in LP, to a level that induces fermentation and elevates the RQ in cells located close to the commodity's centre. This may explain why at 0°C and atmospheric pressure, CO_2 production by corn is only depressed by 12.9% in 3% [O_2] (Robinson *et al.*, 1975), while in LP at the same temperature and a storage pressure of 15.33 kPa (115 mm Hg = 3% [O_2]; Fig. 4.2) the inhibition is 50%. In 25 fruits and vegetables, the average inhibition of CO_2 production caused by 3% [O_2] applied at an optimal storage temperature was only 25% at atmospheric pressure, whereas in LP the inhibition typically is 50% at an equivalent [O_2] concentration (Fig. 4.2). CO_2 production by carnations is inhibited by 65% at a pressure of 5.7 kPa (43 mm Hg = 1% [O_2]), but only by 29% in 1% [O_2] at the same temperature and atmospheric pressure (Table 4.2). When both CO_2 and O_2 were analysed in the vacuum pump exhaust during a full-load LP carnation-storage test at 0°C (Fig. 4.2; Burg, 1979c), the RQ did not increase when respiration was inhibited by 75% at a pressure of 2 kPa (15 mm Hg) or 90% at 1.33 kPa (10 mm Hg), but CO_2 production was stimulated if all O_2 was removed by lowering the pressure to 0.6 kPa (4.6 mm Hg), the vapour pressure of water at the storage temperature, 0°C. The RQ is not increased in LP when an apple's respiration is slowed by 65% at a storage pressure of 6.4 kPa (48 mm Hg = 0.67% [O_2] at 25°C; Table 4.3), but at atmospheric pressure that same [O_2] concentration is below the RQ 'breakpoint' of apples (Gran and Beaudry, 1993; Beaudry and Gran, 1993). LP's ability to inhibit respiration by up to 90% without elevating the RQ is an important advantage contributing to a longer storage life.

Decreasing the storage temperature is one of the most effective ways to preserve commodities because it greatly reduces respiration and metabolism. A temperature reduction from 25 to 0°C decreases CO_2 production by 85–95% (Hardenburg *et al.*, 1986), and a further reduction in respiration, if it is not accompanied by fermentation or low O_2 damage, would be expected to provide an additional storage benefit. Independent of the $[O_2]$ supply, and regardless of the storage temperature, unless a respiratory increase associated with ripening, flower fading, senescence or renewed growth intervenes, the CO_2 production and O_2 consumption rates typically decline for days, weeks or even months after commodities are harvested (Table 4.4). During a 20°C potato storage at atmospheric pressure, the initial O_2 consumption rates of 7 ml/kg·h in air, and 4 ml/kg·h in 5% $[O_2]$, progressively declined for 60 days, at which time they had diminished to 2.3 and 1.7 ml/kg·h, respectively.[4] The CO_2 production rate in air decreased to a lesser extent, from 6 to 2.7 ml/kg·h (Burton, 1982), indicating that the RQ had risen from 0.86 to 1.12 and the respiratory substrates had changed significantly. At the beginning of the potato storage, the respiratory O_2 isotherm was biphasic, but after 1.5–2.5 months it had become monophasic, and O_2 partial pressures ranging from 0.05 to 1 atm had no effect on O_2 uptake (Burton, 1982). The external $[O_2]$ did not change during the entire storage interval, and yet during 60 days, potato tubers had either completely responded to or lost their low O_2-affinity negative-feedback system.

CO_2 production by bananas decreased more rapidly at 6.7–16.0 kPa (50–120 mm Hg) than is typical for bananas kept at atmospheric pressure (Table 4.1; Burg, 1969). The respiration of cut tulip flowers progressively decreased by nearly 80% during 18 days transit from Holland to the USA in a VacuFresh[SM] LP intermodal container, operated at 2°C and a pressure of 2.67 kPa (20 mm Hg). An important factor contributing to the unusually long storage life in LP is the extreme respiratory inhibition that develops from the combined effect of up to an 85–95% decrease due to temperature reduction, as much as an additional 90% retardation due to low $[O_2]$, and a 60–80% diminution due to a prolonged postharvest down-drift.

Decreases in ethylene evolution and respiratory O_2 consumption or CO_2 production caused by lowering the temperature are not compensated by an equivalent temperature-dependent increase in the diffusive resistance to gas exchange because the temperature coefficients (Q_{10}) for respiration and ethylene production are in the range 2–2.5 (Burg and Thimann, 1959),[5] and the Q_{10} for gaseous diffusion in air is only 1.12–1.13 (equation 15.34 and Table 15.4). At atmospheric pressure, the O_2 and CO_2 gradients decrease threefold in apples when the temperature is lowered from 22 to 4°C (Table 4.5), which is the amount to be expected based on the Q_{10} values for respiration and diffusion in an air phase. Consequently, at a lower temperature, a larger reduction in applied $[O_2]$ is required to elicit an RQ breakpoint and cause low $[O_2]$ damage. In addition, the solubility of gases in a liquid phase increases at lower temperatures (Tables 15.1 and 15.2), decreasing the resistance of the cellular liquid phases to gas mass transport. In onion roots, at 10°C the RQ begins to increase at 10% $[O_2]$, at 20°C in 15% $[O_2]$, at 30°C in 25% $[O_2]$ (Beevers, 1961); and with blueberries the RQ breakpoint is 1.8% $[O_2]$ at 5°C, 2.0% $[O_2]$ at 10°C, 2.5% $[O_2]$ at 15°C, 3% $[O_2]$ at 20°C and 4% $[O_2]$ at 25°C (Beaudry and Gran, 1993).

When low $[O_2]$ delays the senescence of fruits with a large food reserve, the beneficial storage result cannot be attributed to a sparing effect on respiratory substrates (Solomos, 1983), but limiting carbohydrate loss may prolong the storage life of leafy vegetables and flowers because they rapidly respire a significant portion of their available carbohydrate. Typically, 30–60% of daily photoassimilate is lost as respiratory CO_2 by plants in the field (Crocker, 1948; Lambers, 1985), and exclusive of translocation, during a single night at 10°C, dark respiration consumes 10% of the starch and sugar present in the leaves of bean plants (Hewitt and Curtis, 1948), and at 20°C it consumes 29.2% of the carbohydrate in the leaflets of potato plants (Bushnell, 1925). The measured respiration rate of the intact flower and stem of fresh-cut White Sims carnation flowers theoretically is sufficient to deplete all carbohydrate by the time fading begins unless other respiratory substrates are consumed (Nichols, 1973, 1975), and by the first signs of wilting the reducing sugar content of the flower petals has decreased 50% and only a trace of sucrose remains.[6] Narcissus blooms also lose half their sugar by the time of incipient wilting. Sucrose absorption from a preservative solution causes a manifold increase in the reducing sugar and sucrose content of carnation flowers, extends their vase-life by 50% and prevents the O_2 consumption of excised petal discs from decreasing. During petal expansion in the daylily, a large amount of carbohydrate, nearly 90% in the form of fructan oligosaccharides and sucrose, accumulates in the vacuoles of the developing flower buds (Bieleski, 1993; Avigad and Dey, 1997). About 24 h prior to flower opening, a rapid and complete degradation of the fructan to fructose and glucose is catalysed by a sudden gene-controlled enhancement of the activity of fructan exohydrolase (FEH), a vacuolar enzyme with a pH 5.0 optimum, causing a three- to fivefold increase in cell-sap osmolarity in conjunction with petal opening. Flower senescence follows when the sugar pool in the petals is completely depleted. A similar process, which was attributed to an increased invertase activity, occurs during carnation flower-petal growth. The large respiratory inhibition caused by LP may in part explain why the method is able to preserve carnations and other flowers for a much longer time than NA or CA, even when the flowers stored in NA and CA have been pre-treated with STS or 1-MCP (Dilley *et al.*, 1975; Goszczynska and Rudnicki, 1982; Staby *et al.*, 1984).

4.3 Low [O_2] and High [CO_2] Injury

The conductance of the gaseous diffusion pathways through the skin and intercellular spaces of different types of horticultural commodities varies widely because of differences in their porosity (Table 3.8), tortuosity and diameter or thickness, and in the frequency and aperture of their stomates and lenticles (3.18). Different respiration rates and these anatomical features create a large disparity in the low [O_2] and high [CO_2] tolerance of individual commodities by influencing the magnitude of the surface-to-centre respiratory O_2 and CO_2 gradients that develop (Chevillotte, 1973). Example 4 indicates that when an applied O_2 concentration of < 1% injures apples, the disorder actually is caused by the presence of < 0.19% O_2 within the fruit's intercellular spaces. LP improves the tolerance to low [O_2] by decreasing the O_2 gradient in direct proportion to the effect that a low pressure has on respiration (Fig. 4.2), diffusional gas transport (Table 3.15; 15.15 and 15.16) and stomatal opening (4.15). Accelerated diffusion lowers the O_2 and CO_2 gradients by approximately 76-fold at 1.33 kPa (10 mm Hg) compared to atmospheric pressure. If the respiratory inhibition caused by LP is also considered, the gradients may be reduced by up to tenfold more (example 1), and the LP-induced opening of a Valencia orange's stomates decreases the gradient by an additional factor of 100. By eliminating the O_2 gradient, LP allows the surface and centre of a commodity to be exposed to the same low, optimal storage O_2 partial pressure, whereas CA is limited to higher applied [O_2] at the surface in order to avoid low [O_2] damage in the commodity's centre. Table 4.7 compares the experimentally determined safe limits of applied [O_2] and [CO_2] for horticultural commodities stored in CA and LP. The optimal [O_2] for LP storage typically is at least threefold lower than that which causes low [O_2] damage at atmospheric pressure, and usually the difference is much greater. Notable examples are green snap beans, which are best preserved in LP at an O_2 partial pressure equivalent to 0.07% [O_2], whereas in CA they eventually are injured by < 5% [O_2]; for asparagus, the comparison is 0.42% [O_2] (LP) vs. < 10% [O_2] (CA); cauliflower 0.15% [O_2] (LP) vs. < 2% [O_2] (CA); avocados 0.3% [O_2] (LP) vs. < 2% [O_2] (CA); mushrooms and papaya 0.15% [O_2] (LP) vs. < 1% [O_2] (CA); mangoes 0.1% [O_2] (LP) vs. < 3–5% [O_2] (CA); apples 0.15% [O_2] (LP) vs. < 1% [O_2] (CA); carambola and Valencia orange fruits 0.25% [O_2] (LP) vs. < 5% [O_2] (CA). It is uncertain that the lowest and most effective

LP pressure has been determined for any commodity, and there is no evidence that a CO_2 'inversion' point has been reached under hypobaric conditions, even at < 0.1% [O_2]. The hypobaric studies summarized in Table 4.7 and example 5 add credence to the view that anatomical differences in air-filled gas-transport pathways are responsible for variations in the tolerance of horticultural commodities toward low [O_2] or high [CO_2] at atmospheric pressure. When a low pressure eliminates the O_2 gradient between the atmosphere and intercellular spaces, with few exceptions different types of commodities tend to have nearly identical low [O_2] tolerances, close to or below 0.15% [O_2].

The cause of low [O_2] injury is uncertain. It is judged to have resulted whenever a physiologically aberrant response results at and below a particular applied [O_2] concentration, and its occurrence is anticipated if

Table 4.7. The highest O_2 and lowest CO_2 partial pressures (atm) that do not cause physiological damage to various horticultural commodities at atmospheric pressure, and the optimal O_2 partial pressure under hypobaric conditions.

Commodity	Temp. (°C)	CA pO_2–injurious–pCO_2		LP optimal pO_2	References
Apple					
McIntosh & Newton	3	< 0.015	0.05–0.15	0.0014	4, 13
All other varieties	0	< 0.01	0.05	0.0014	4, 13
Artichoke	0	< 0.02	> 0.03	–	5, 11
Asparagus	0	< 0.10–0.15	> 0.10	0.0042	2, 5, 8, 11
Avocado	10	< 0.01–0.02	> 0.10–0.15	0.003	1, 2, 5, 11
Banana	13.3	< 0.01	> 0.05–0.10	0.0076	1, 2, 5, 6, 8, 11
Bean (green)	7.2	–	> 0.20	0.0007	2, 3, 8, 11
Broccoli	0	< 0.005	> 0.15	0.0015	2, 3, 11
Cabbage	0	< 0.02	> 0.10	–	3
Cantaloupe	7	< 0.01	> 0.20	0.0015	5, 11
Carnation, bloom	0–2	< 0.005	> 0.04	0.0015–0.0029	2, 12
Cauliflower	0	< 0.01–0.02	> 0.05–0.15	0.0015	2, 3, 5, 11
Cherry, sweet	2	< 0.01	> 0.30	0.0042	2, 11
Chrysanthemum					
bloom	0	–	–	0.0015	2
cutting	0–2	–	–	0.015	10
Corn	0	< 0.02	> 0.10	0.012	2, 8, 11
Cucumber	12	< 0.01	> 0.03–0.05	0.0025	3, 11, 14
Lettuce	0	< 0.01	> 0.01–0.02	0.0015	2, 3, 11
Lime	10	< 0.05	> 0.10	0.039	1, 2, 11
Mango	12.8	< 0.05	> 0.06–0.10	0.001	1, 5, 7, 11, 14
Melon, honeydew	10	–	–	0.003	9
Mushroom	0	< 0.01	> 0.15–0.20	0.0015–0.0029	2, 5, 8, 11
Orange, Valencia	4.4	< 0.05–0.10	> 0.05	0.0025	11, 14
Papaya	10	< 0.01–0.02	> 0.05–0.10	0.0015	1, 2, 5, 8, 11
Pears	–1	< 0.02	> 0.02–0.05	0.003	4, 5, 9, 11
Pepper (green)	12	< 0.01–0.02	> 0.02–0.08	0.019	2, 3, 8, 11
Pineapple	10	< 0.02–0.03	> 0.10	0.003	2, 8, 11
Rose (bloom)	0	–	–	0.0015	2
Strawberry	0	< 0.005–0.02	> 0.20–0.25	0.0015	2, 5, 11
Tomato	13	< 0.02–0.03	> 0.02	0.019	2, 5, 11

References: (1) Hatton and Spalding, 1990; (2) Burg, 1990; (3) Leshuk and Saltveit, 1990; (4) Blanpied, 1990a; (5) Dilley, 1978; (6) Apelbaum *et al.*, 1977a; (7) Bender *et al.*, 2000b; (8) Hatton *et al.*, 1975; (9) Jamieson, 1984; (10) Burg, 1973b; (11) Thompson, 1998; (12) Hanan, 1966; (13) Dilley, 1972, unpublished; (14) Davenport and Burg, 2002, unpublished.

the O_2 tension is low enough to cause an increase in a commodity's RQ. At 0°C, the RQ 'breakpoint' of apples determined by a modified atmosphere packaging (MAP) technique is between 0.7 and 2.0% [O_2] (Beaudry and Gran, 1993; Gran and Beaudry, 1993), the exact value depending upon the variety of apple.[7] Below this range of [O_2] concentrations, a substantial increase in RQ occurs, accompanied by low [O_2] damage and an accumulation of ethanol (Table 4.7). Less than 1% [O_2] causes ethanol and acetaldehyde to increase in apples, bananas, oranges, avocados, artichokes, peppers and other fruits and vegetables (Thomas, 1925, 1929; Fidler, 1931, 1951; Ulrich, 1975; Karaoulanis, 1983; Johnson *et al.*, 1993), and lactate and pyruvate to accumulate in intact sweet potato roots (Solomos, 1983). Although low O_2 damage seems to be associated with alcohol formation (Fidler, 1951), at least in apples, ethanol does not cause the off-flavour or visual symptoms of injury that develop (Blanpied, 1990a).

It is not clear in any study of CO_2-induced injury whether a measured chemical change is causally related to the symptoms that develop. Excess [CO_2] eventually impairs the flavour of most fruits and vegetables, prevents colour changes in fruits, kills growing tissues such as asparagus and potato sprouts, promotes decomposition of exposed tissues of vegetables that are high in moisture content, causes internal discolouration and breakdown, and induces external discolouration (Crocker, 1948; Wilkinson, 1970; Burton, 1982). Succinate accumulation to an apparently 'toxic' level is correlated with CO_2-induced core injuries such as 'brownheart' and 'core flush' in overstored apples (Hulme, 1956; Shipway *et al.*, 1973) and pears (Williams and Patterson, 1964), and with various abnormalities in a wide range of commodities (Chevillotte, 1973; Lipton, 1977a; Burton, 1982). Other factors are likely to be involved. A loss of ascorbic acid occurs in apple tissues that first show signs of CO_2 injury (Wilkinson, 1970). Strawberries, bananas, oranges and apples accumulate acetaldehyde and ethanol in the presence of > 15% [CO_2], and the aldehyde to ethanol ratio, incidence of injury and development of off-flavour and -colour increases (Thomas, 1925, 1929; Fidler, 1951; Ulrich, 1975). A 'scald-like' controlled-atmosphere disorder of Empire apples seems to be a chilling injury induced by CO_2 (Burmeister and Dilley, 1995; Wang *et al.*, 2000a). This disorder only occurs on fruits in CA, especially at 1.5% [O_2] + 2% [CO_2], and at 1°C is much more prevalent than at 3°C. The mechanism responsible for the disorder may be a free-radical-catalysed oxidation of susceptible amino-acid residues in proteins essential for cell function, or in other macromolecules such as membrane lipids containing oxidizable reactive groups. High [CO_2] might also damage tissues by inducing NH_3 formation from amides and amino acids (Fife and Framptom, 1935). The cellular buffering capacity is insufficient to stabilize the cytoplasmic pH against > 1% applied CO_2, and in response to CO_2-induced 'acidosis' the plant cell, like the tubular epithelium of the mammalian kidney, produces NH_3 from glutamine (4.19). In both mammals and plants, ammonia is toxic (4.17), and while in the kidney it is excreted in the urine, in horticultural commodities it is retained unless it is released to the atmosphere (3.11) or recycled into amino acids and amides (4.18).

LP eliminates the possibility of CO_2 injury by inhibiting CO_2 production and levelling the CO_2 gradient between the intercellular system and ambient air. This causes the commodity's ICC to approach the unusually low CO_2 partial pressure present in the incoming low-pressure air changes after they have expanded at the pressure regulator (example 1).

4.4 Effect of O_2 on Ethylene Production

The observation that ethylene production is inhibited by O_2 deprivation (Gane, 1934) was an important step in unravelling the mystery of how ethylene is synthesized. Twenty-five years later, it was demonstrated that anoxia caused a substrate to accumulate that was rapidly converted to ethylene when O_2 was again admitted (Burg and Thimann, 1959), but 20 more years were to elapse before Adams and Yang (1979) identified 1-amino cyclopropane carboxylic acid (ACC) as the ethylene precursor that forms from methionine (Lieberman *et al.*, 1966) in a N_2 atmosphere.

The exogenous [O_2] concentration required to half-inhibit ethylene production often is in the 5–7% range in intact fruits (Abeles *et al.*, 1992). To delay the climacteric rise in ethylene evolution in apple cv. Gala fruits and cut carnation flowers, the [O_2] must be reduced below 6% (Solomos, 2000). The accumulation of ACS and ACO mRNA in carnations was delayed by 6–7 days when the [O_2] was reduced to 4.5%, and during 17 days there was no accumulation of either enzyme at 1.5% [O_2]. K_{m,O_2} values of 1.0–1.4% [O_2] for ethylene production have been measured in protoplasts of senescing *Ipomoea tricolour* flowers (Konze *et al.*, 1980), 2.2% in banana slices (Banks, 1985) and 0.5% in banana fruits (Brown, 1981 – referred to in Mattoo *et al.*, 1988), but these determinations are complicated by an increase in ACC oxidase's O_2 affinity when the ACC concentration is elevated (Yip *et al.*, 1988; Kuai and Dilley, 1992; Smith and John, 1993b), and by a decrease in the apparent O_2 affinity of tissues when a liquid phase infiltrates their intercellular spaces and obstructs the gaseous diffusion pathway between EFE and the O_2 source. Fluid drawn by capillary action into the intercellular spaces of McIntosh apple slices floated with continuous shaking on a 0.55 M glycerol solution increased their apparent K_{m,O_2} for ethylene production to 13% [O_2] (Burg, 1973a). The apparent K_{m,O_2} for ethylene production by apple slices floated on ACC solutions was 16% [O_2], and even more ethylene was produced in 100% [O_2] than in air (Lieberman *et al.*, 1966). When tissue was carefully blotted to withdraw intercellular fluid artificially introduced by cutting and then incubated in a humid atmosphere without a liquid phase, the K_{m,O_2} shifted to 1.5% [O_2] in 1-cm-diameter McIntosh apple plugs (Fig. 4.3, *right*) and to 0.4% [O_2] with 1-mm-thick slices (Burg, 1973a). After the ACC content of 1-mm-thick apple discs was increased to a supra-optimal level by a 2-h pre-incubation in a humid atmosphere containing 100% [N_2], the apparent K_{m,O_2} for ethylene production was 20% [O_2] if the slices were floated on a solution, and 100% [O_2] caused a further stimulation (Burg *et al.*, 1971). But when the ACC-enriched discs were carefully blotted and incubated in humid air, the K_{m,O_2} decreased to 0.2% [O_2], which is similar to the affinity measured with purified-apple ACC oxidase (K_{m,O_2} = [0.3%]) under optimal conditions with a saturating ACC concentration (Kuai and Dilley, 1992). These studies clearly demonstrate that the diffusion pathway's resistance determines the intercellular [O_2] concentration.

Interpreting LP's effect on ethylene production is complicated by the likelihood that ethylene synthesis will be repressed not only by the decreased O_2 partial pressure, but also by a simultaneous reduction in internal [CO_2] (4.10; Fig. 4.6). At 25°C, a pressure of 25.33 kPa (190 mm Hg = 4.7% [O_2]) had little effect on ethylene production by intact McIntosh apples (Burg and Thimann, 1959), and 17.33 kPa (130 mm Hg = 2.9% [O_2]) was half-inhibitory to Ida Red apples (Table 4.3). During LP storage at 1.33–2.67 kPa (10–20 mm Hg), only 0.15–0.28% [O_2] is present and this should inhibit ethylene production by 50% independent of any effect caused by an ICC reduction.

4.5 Effect of O_2 on Ethylene Action

The concept that low [O_2] may interfere with ethylene action has provoked considerable disagreement (Abeles and Gahagan, 1968a; Streif and Bangerth, 1976; Beyer, 1979; Abeles *et al.*, 1992). Experiments with etiolated pea epicotyl sections (Burg and Burg, 1967c) indicated that the effect of [O_2] resembles a 'coupling activation' in which a decreased response to ethylene at a reduced [O_2] concentration, higher than that needed to affect respiration, could be overcome by increasing the ethylene concentration.[8] This concept is supported by the binding of

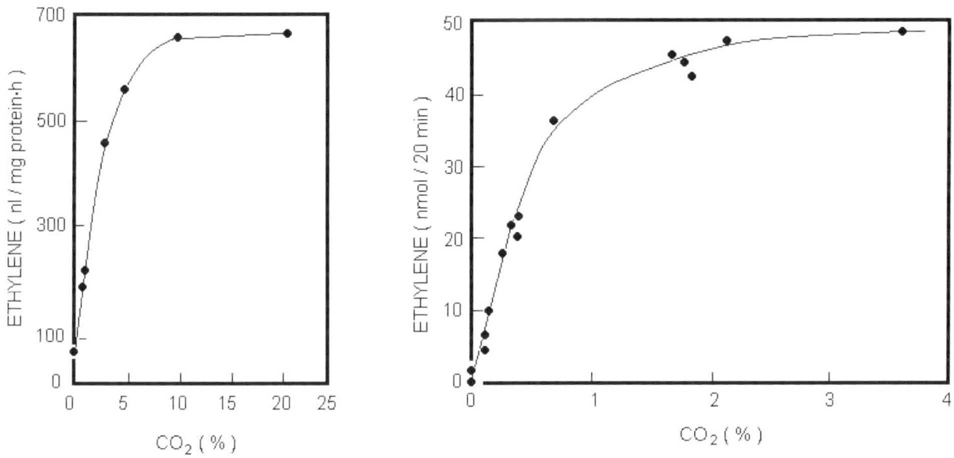

Fig. 4.6. Dependence of ACC oxidase activity on [CO_2] concentration. (*left*) Melon EFE (Smith and John, 1993b); (*right*) apple EFE (Yang *et al.*, 1993).

O_2 to iron-containing cytochromes and copper-containing oxidases, by carbon monoxide's high affinity for both the copper-containing ethylene receptor (5.11) and cytochrome oxidase, and by carbon monoxide's ability to inhibit the binding of O_2 to cytochrome oxidase (Burg and Burg, 1967c).

A 'coupling activation' between ethylene, O_2 and the ethylene receptor cannot be demonstrated or disproved experimentally without ascertaining the response at various O_2 partial pressures in combination with a range of ethylene concentrations lower than a few μl/l. In apples, the ripening effect of ethylene diminishes with a reduction in [O_2] until at or below 0.3% [O_2] applied ethylene does not induce a respiratory climacteric (Kidd and West, 1934). The demonstration (Bangerth, 1988) that apples stored in LP at 3°C respond to 500 μl/l applied ethylene at a storage pressure of 13.33 kPa (100 mm Hg = 2.6% [O_2]) is not evidence against a 'coupling activation', since that theory does not envision that O_2 can affect the activity of so large an ethylene concentration. A report that avocado fruits soften very slowly in 2% [O_2] when they are exposed to 130 μl/l propylene, and hardly at all in 1% [O_2] (Metzidakis and Sfakiotakis, 1993), is consistent with a coupling activation, since 130 μl/l propylene is equivalent in potency to 1 μl/l ethylene (Burg and Burg, 1967c).[9] At 20°C, more than 12.5% [O_2] must be present for 500–5000 μl/l propylene or 10 μl/l ethylene to initiate ripening in mature-green tomatoes and accelerate ripening in fruits harvested at the breaker or light-pink stages (Kader and Morris, 1975a).

Possibly the binding of ethylene to its receptor occurs in the total absence of O_2, after which the ethylene–receptor complex interacts with a rate-limiting reactant that imparts specificity to the ethylene–receptor complex when it elicits a biological response. O_2 may be essential for this step or later steps of transduction and response (Goren and Sisler, 1986). This type of interaction between [O_2] and ethylene is not likely to be accurately described by Michaelis-Menton kinetics.

4.6 Effect of O_2 on Ascorbic Acid Oxidation

'Vitamin C is the only essential component of the diet which is supplied in significant amount, relative to requirement, by food crops' (Burton, 1982), and therefore a storage method's ability to maintain vitamin C has important dietary consequences. Results with NA, CA and MA have been highly variable depending on the commodity and conditions imposed, although with few exceptions high $[CO_2]$ tends to decrease (4.14), low $[O_2]$ elevates or does not alter and ethylene (5.33) has no effect on the ascorbic acid level. All tests performed with LP indicate that a hypobaric condition maintains vitamin C at least as well as, and usually significantly better than any other storage method, presumably because it simultaneously lowers $[O_2]$ and eliminates CO_2 from within the commodity's interior.

Vitamin C can be oxidized to dehydroascorbic acid (DHA), which is irreversibly converted to 2,3-dioxo-L-gulonic acid (Mapson, 1970b) (see equation 4.1 at bottom of page). DHA is not completely stable and is irreversibly converted into 2,3-dioxo-L-gulonate by opening of the lactone ring, and this compound is further metabolized. In plant tissue, a variable proportion of the total ascorbic acid is found as DHA in the reversibly oxidized state. The oxidation of L-ascorbate to DHA does not cause any loss in bio-potency, because DHA is readily reduced *in vivo* to ascorbic acid.

In the presence of O_2, L-ascorbic acid may be oxidatively destroyed by ascorbic acid oxidase, phenolase, cytochrome oxidase, peroxidase and also non-enzymatically in reactions catalysed by copper and iron salts. Only with ascorbic acid oxidase is there a direct reaction between enzyme, substrate and molecular O_2. The other enzymes oxidize the vitamin indirectly. Phenolase catalyses the oxidation of mono- and dihydroxy phenols to quinones, which react directly with ascorbic acid; cytochrome oxidase oxidizes cytochrome c to a form that reacts with ascorbic acid; and peroxidase, in combination with phenolic compounds, utilizes hydrogen peroxide to effect the oxidation (Mapson, 1970b). Except for cytochrome oxidase, these enzymatic activities have a relatively low O_2 affinity, and therefore the oxidation of L-ascorbic acid is likely to be slowed when a commodity is stored at a moderately reduced O_2 partial pressure, and markedly inhibited at a very low $[O_2]$. The situation is complicated by the fact that there are several oxidative steps in the synthesis of L-ascorbate from glucose via L-galactose and L-galactono-1,4-lactone (Smirnoff *et al.*, 2001), making it difficult to predict whether vitamin C production or destruction would be more affected by a diminution in $[O_2]$.

Low $[O_2]$ decreases vitamin C loss in asparagus (Thornton, 1937b), mangoes (Hulme, 1971), Chinese cabbage (Wang, 1983) and apples (Ulrich, 1975), and CA storage has been reported to improve vitamin C retention in spinach (McGill *et al.*, 1966; Izumi *et al.*, 1999), apples (Velkov, 1974), cauliflower (Adamicki and Elkner, 1985), lettuce (Adamicki, 1989) and a few other commodities (Gerber and Bussmann, 1958). In 1.3 or 2.2% $[O_2]$, but not in 5.4–20.8% $[O_2]$, vitamin C is preserved in pineapples, even in the presence of 11.2% $[CO_2]$ (Haruenkit and Thompson, 1996). These results are the basis for the claim that 'reduction in loss of ascorbic acid in CA storage has been reported in many fruits and vegetables' (Wang, 1990), but the preponderance of data indicate that CA has the opposite effect (Burton, 1982; Thompson, 1998).

LP maintains the vitamin C content of tomato and pear fruits (Streif, 1974a; Borecka and Parynow, 1985a), radish and parsley (Fig. 10.4; Bangerth, 1974; Streif, 1974a), apples, cress and peppers (Bangerth, 1973, 1977), sweet cherries (Dilley, 1977a) and limes (Spalding and Reeder, 1976a). In LP, as the pressure is lowered from 10.67 to 2.67 kPa (80–20 mm Hg), the loss of ascorbic acid from asparagus decreases progressively (Table 4.8; Dilley, 1977a). Typically, at the same O_2, partial pressure CA is far less effective than LP in preserving vitamin C (Bangerth, 1973), and consequently fruits and vegetables stored in LP seem always to have a higher ascorbic acid content than

$$\text{L-ascorbate} \underset{O_2}{\overset{NADH_2}{\rightleftharpoons}} \text{dehydro-L-ascorbate} \longrightarrow \text{2,3-dioxo-L-gulonate} \quad (4.1)$$

those kept in CA or NA (Agar *et al.*, 1997; Bangerth, 1977; Izumi *et al.*, 1999). The interpretation of this comparison between LP and CA is complicated by effects on vitamin C retention caused by the removal of CO_2 in LP, and the addition of this gas in CA (4.14).

4.7 Effect of Hypoxia on Ethylene Production, Fruit Softening, ACC Accumulation and Ethylene-induced Enzyme Activity

Low O_2 suppresses or promotes various enzymatic activities independent of its effect on respiration, and this may have a beneficial or detrimental effect when commodities are stored in CA at ultra-low O_2, or in LP at a very low pressure. Hypoxia induces the synthesis of new proteins irrespective of a fruit's developmental stage, and in ripening fruits suppresses *de novo* synthesis of certain enzymes (Kanellis *et al.*, 1993). The onset of climacteric ethylene production was delayed and its intensity reduced in propylene-treated Hass avocados kept in 5% [O_2], progressively more as the [O_2] was lowered to 1%, but even at 1% the IEC began to rise within 5 days and quickly reached a highly active concentration of 17.6 μl/l (Metzidakis and Sfakiotakis, 1993). EFE activity and the IEC increased simultaneously, ACC rose to the highest level in 1 or 2% [O_2], less in air, least in 5% [O_2], and only 1% [O_2] limited the development of EFE activity. Softening and the creation of Cx activity (cellulase = endo-β-1-4-glucanase), cellulase mRNA and polygalacturonase (PG) activity slowed when the [O_2] was lowered from 5 to 1% (Metzidakis and Sfakiotakis, 1993), and synthesis of new alcohol dehydrogenase isozymes was promoted by 2.5–5.5% [O_2] (Kanellis *et al.*, 1991). The activity, immunoreactive protein and abundance of cellulase mRNA decreased (Kanellis *et al.*, 1989a,b) when ripening Hass avocados were transferred to 2.5% [O_2] for 6 days, and the profile of avocado total proteins was changed. Some were suppressed, others enhanced and new polypeptides were induced. Sucrose synthase was expressed at the mRNA level in 100% [N_2].

4.8 Advantages of a CO_2-free Atmosphere

Plant commodities develop intercellular CO_2 concentrations sufficiently elevated above the atmospheric level to result in significant metabolic effects. Since Kidd and West's pioneering work on CA storage (Kidd and West, 1927a, 1928, 1938a), it has been axiomatic that moderately applied [CO_2] generally benefits storage life. Because it is impractical to elevate the [CO_2] in LP, it has been natural to assume that the method lacks the advantages that CO_2 affords to CA (Abeles, 1973; Lougheed *et al.*, 1977; Abeles *et al.*, 1992), and occasionally attempts have been made to add CO_2 under hypobaric conditions to overcome this inferred deficiency (Spalding and Reeder, 1976b; Haard and Lee, 1982). The effects of abnormally low [CO_2] could not be critically tested until LP became available, because prior to that time there was no practical way to lower continuously the

Table 4.8. Effect of hypobaric storage at 0°C on the retention of ascorbic acid in asparagus (Dilley, 1977a).

Pressure kPa (mm Hg)	pO_2 (atm)	Ascorbic acid (mg/100 g fresh weight) at indicated days in storage: 0 days	8 days	14 days	22 days	29 days
101.3 (760)	0.209	65.8	40.4	26.3	21.8	18.0
10.67 (80)	0.021	–	33.7	32.3	20.1	18.6
5.33 (40)	0.0098	–	41.6	35.5	22.3	16.3
2.67 (20)	0.0042	–	52.0	45.2	31.1	21.2

atmospheric content and remove CO_2 from within the intercellular spaces. It is now apparent that contrary to the predictions of classical theory, LP's ability to decrease drastically both the ambient and intercellular [CO_2] is an important advantage, providing benefits that cannot be duplicated by elevating [CO_2]. Some of the most notable documented LP effects resulting from unusually low [CO_2] are stomatal opening in the dark (4.15); depressed growth of aerobic bacteria, fungi and yeasts (7.02, 7.03 and 7.04); retention of ascorbic acid (4.14); inactivation of ethylene-forming enzyme (4.10); and prevention of succinate formation (4.13).

4.9 Effect of CO_2 on Respiration

Prior to the turn of the 19th century, Mangin (1896) had already discovered that high [CO_2] inhibits fruit respiration. This result was confirmed by Kidd (1915) and Willaman and Beaumont (1927), not only for fruits, but also for germinating seeds. Subsequently, it has been found that while applied CO_2 often inhibits O_2 consumption, at other times, depending on the tissue and [CO_2] concentration, respiration may be accelerated or unchanged. During a 20–92-h period at 25°C, the O_2 uptake of potato, onion, tulip, strawberry, asparagus, lima bean, banana and carrot tissue is inhibited by as much as 20% in a 15–20% [CO_2] + 20% [O_2] mixture, and in asparagus and strawberry the inhibition increases to 37% when the [CO_2] is elevated to 65%. Short-term exposure to 10–30% [CO_2] reduces respiration in ripening bananas, pink tomatoes and cucumbers, but it has no effect on guavas, onions and oranges; 5–10% [CO_2] promotes a respiratory climacteric in lemons (Young and Biale, 1962); high [CO_2] stimulates respiration in lettuces, aubergines and cucumbers (Kubo *et al.*, 1989); and O_2 uptake increases when potato, onion, tulip or carrot tissues are continuously exposed to more than 20–30% [CO_2] (Thornton, 1933a, 1935, 1937a; Pal and Buescher, 1993). The respiration of potatoes rises by as much as sixfold after they are exposed to 10–30% [CO_2] in air (Perez-Trejo *et al.*, 1981). Four hours after they are treated with 56–60% CO_2, their respiration decreases by more than 90%, but then progressively the O_2 consumption accelerates, exceeding 200–300% of the control value within 60 h, and up to 400% in 8 days, while simultaneously the RQ falls. These results suggested the possibility that in lemon and potato tissue, and possibly also in onion, tulip and carrot tissue, high [CO_2] might engage and enhance the cyanide-insensitive alternative pathway of respiration just as cyanide, azide or ethylene do within 1–7 days when they enhance potato respiration by four- to sixfold[10] and decrease the RQ (Solomos and Laties, 1974). Initially, CO_2-induced potato respiration is CN sensitive and little affected by low [O_2], in marked contrast to the CO_2-synergized ethylene response, but a 72-h exposure to 10% [CO_2] ultimately yields CN-resistant respiration together with a considerable rise in the residual respiration that is CN-sensitive (Laties, 1982).

4.10 Effect of CO_2 on Ethylene Production

Depending on its concentration, CO_2 may stimulate or inhibit ethylene production (Abeles *et al.*, 1992; Mathooko, 1996). High to very high [CO_2], in the 20–80% range, inhibits ethylene synthesis by ACC oxidase (Burg and Thimann, 1959; Chaves and Thomas, 1984), and relatively high [CO_2] also inhibits autocatalytic ethylene production induced by ACC synthase, with 'competitive' kinetics that depend on the ratio between CO_2 and ethylene (Bufler, 1984; Chaves-Franco and Kader, 1993; Mathooko *et al.*, 1995). Low to moderate [CO_2], in the concentration range that arises naturally in the intercellular spaces of horticultural commodities at atmospheric pressure, stimulates ACC-dependent ethylene production in fruits (Ketsa and Herner, 1987; Chaves-Franco and Kader, 1993), seeds (Esashi *et al.*, 1986; Ziska and Bunce, 1993), maize

(Dunlap, 1988), and in oat (Preger and Gepstein, 1984) and tobacco (Philosoph-Hadas *et al.*, 1985b) leaves. CO_2 removal reduces ethylene production in rice leaves (Kao and Yang, 1982), sweet potatoes (Imaseki *et al.*, 1968), oat leaves, and pear and apple slices (Tan and Thimann, 1989). The dose–response curve for this effect is biphasic with an initial rapid rise in ethylene production up to 1% [CO_2], a levelling off, and then a second rapid rise up to 15% [CO_2] (Philosoph-Hadas *et al.*, 1986).

Both *in vivo* and *in vitro*, low [CO_2] stimulates ACC oxidase up to 40-fold compared to EFE activity in atmospheric air (Kao and Yang, 1982; McRae *et al.*, 1983; Dilley *et al.*, 1993; Fernández-Maculet *et al.*, 1993; Poneleit and Dilley, 1993; Smith and John, 1993a,b; Yang *et al.*, 1993). The responsible mechanism is similar to that for CO_2 activation of RuBP carboxylase ('Rubisco'), involving carbamylation of a lysine residue (Ververidis and Dilley, 1994). CO_2 increases the V_{max} of EFE by tenfold, the apparent K_m for ACC several-fold and the apparent K_m for O_2. The CO_2 concentration required for half-maximal activity is 0.68% (Dilley *et al.*, 1993; Poneleit and Dilley, 1993; Fig. 4.6). NA and CA elevate the ICC sufficiently to support ethylene production at a maximum rate, whereas by creating a nearly CO_2-free atmosphere within a commodity, LP should markedly inhibit its ethylene production. Example 6 indicates that the inhibition of ethylene production caused by LP is more likely to result from a decreased ICC than a reduction in intercellular [O_2].

During CA, MA and even NA storage, the ICC is high enough to stimulate both the development and activity of EFE. ACC-dependent ethylene production by tobacco leaf disks is promoted within 1–2 days by 5–15% [CO_2] (Aharoni *et al.*, 1979a,b; Philosoph-Hadas *et al.*, 1986), and the extent of the stimulation depends on both the [CO_2] concentration and the duration of exposure. This indicates that during a prolonged exposure to CO_2, in addition to enhancing EFE activity, moderate [CO_2] induces the synthesis and accumulation of EFE. Simultaneous application of ethylene with 2% [CO_2] increased the CO_2-enhanced development of EFE, but because CO_2 is an inhibitor of ethylene action, atmospheres with up to 5% [CO_2] are able to limit the ethylene-induced development of ACS and ACO activities (Bufler, 1984, 1986; Blanke, 1991) and the relevant mRNA (Gorney and Kader, 1996).

High to very high [CO_2], in the > 20% range, inhibits ethylene synthesis in whole fruits and tissues slices prepared from them (Burg, 1958; Burg and Thimann, 1959; Chaves and Thomas, 1984; Levin *et al.*, 1993), but the response is not particularly useful for CA or MA storage because [CO_2] concentrations this high damage most horticultural commodities (Table 4.7). Elevated [CO_2] competitively inhibits ethylene-promoted synthesis of ACC synthase in apples (Bufler, 1984), Bartlett pears (Chaves-Franco and Kader, 1993) and tomatoes (Mathooko *et al.*, 1995), but even 20% [CO_2] cannot completely suppress ACC synthase activity present in fruits before they are transferred to the elevated [CO_2]. In apple slices, the CO_2 inhibition affects the ACC-oxidase step, preventing utilization of exogenous ACC (Chaves and Thomas, 1984), but the inhibition is not permanent, and gradual recovery to a normal rate of ethylene production follows transfer to low [CO_2]. High [CO_2] and 1-MCP inhibited ethylene production, ACS-synthase activity, and *PP-ACS1* mRNA accumulation in peach fruits, and although CO_2 had little effect on ACO activity, it and 1-MCP inhibited the accumulation of *PP-ACO1* and *PP-ACO2* mRNA to the same extent (Mathooko *et al.*, 2001; 5.14). High CO_2 blocked wound-induced ethylene production and ACC-synthase activity, and reduced the abundance of *PP-ACS1* mRNA, whereas 1-MCP had the opposite effect, and although CO_2 and 1-MCP had no effect on wound-induced ACO activity, they both inhibited accumulation of its mRNA. These studies indicate that after wounding and during ripening of peach fruits, high CO_2 does not regulate ACC-synthase activity and expression of the *PP-ACS1* gene by antagonizing ethylene action.

4.11 CO_2 as an Activator, Regulator and Inhibitor of Enzymatic Reactions

CO_2 not only serves as a substrate (HCO_3^-) in the PEP carboxylation reaction, but in addition, the gas is a potent activator, regulator and inhibitor of numerous enzymatic processes in plants, microorganisms and animals (Lorimer, 1983). The different storage responses that result in LP, NA, MA and CA are partly a reflection of LP's ability to eliminate intercellular [CO_2], while NA, MA and CA maintain or increase it, producing opposite effects on CO_2-responsive processes.

Low [CO_2] activates Rubisco by carbamylation of a specific lysine residue (O'Leary *et al.*, 1979; Lorimer, 1981). The modified lysine group, stabilized by Mg^{2+}, facilitates the shaping of the active site and permits the rebinding of RuBP, allowing the molecule to participate as a substrate for carboxylation with CO_2 or oxidation with O_2 (Lorimer, 1981; Woodrow and Berry, 1988). Dissociation of RuBP bound to non-activated carboxylase is a slow process, mediated directly or indirectly by an 'activase' enzyme (Mott and Berry, 1986; Portis, 1992; Woodrow and Mott, 1992; Woodrow *et al.*, 1996). The kinetics of this activation are very sensitive to leaf intercellular [CO_2].

By catalysing the degradation of the superoxide free radical to water and hydrogen peroxide, superoxide dismutase (SOD) plays an important role in the cellular defence against free-radical oxidation. Although free radicals are a normal by-product of metabolism, their generation results in oxidative damage to macromolecules such as lipids, proteins, pigments and nucleic acids. O_2^- disproportionates spontaneously, but the rate constant decreases markedly with increasing pH owing to the involvement of H^+ in the reaction:

$$O_2^- + 2H^+ \rightarrow H_2O_2 \qquad (4.2)$$

SOD disproportionates O_2^- to produce hydrogen peroxide, and then hydroxyl radicals (OH*) are produced in a reaction catalysed by transition metal ions, peroxide and O_2^-. Three types of superoxide dismutase (SOD) present in essentially all plant and animal cells differ in their prosthetic metal ligands, Fe, Mn and Cu,Zn (Bowyer and Leegood, 1997). At physiological pH, bicarbonate is required for the peroxidase function of Cu,Zn-superoxide dismutase (Goss *et al.*, 1999; Sankarapandi and Zweier, 1999; Zhang *et al.*, 2000, 2002; Liochev and Fridovich, 2002). HCO_3^- competes with other anions for the anion-binding site (Arg^{141}) of Cu,Zn–SOD, but does not bind directly to the copper. Instead, HCO_3^- bound to Arg^{141} anchors the neutral H_2O_2 molecule at the active copper site, enabling its redox cleavage. Thus Cu,Zn–SOD acquires peroxidase activity at physiological pH only in the presence of HCO_3^- or structurally similar anions (Sankarapandi and Zweier, 1999). The Cu/Zn form has been isolated from pre- and post-climacteric apple, banana, avocado and tomato fruits (Baker, 1976). Both the IEC and SOD activity peaked simultaneously in Fuji and Golden Delicious apples when they were ripened on the tree. During cold storage as the IEC increased, the SOD activity decreased in Golden Delicious apples and increased in Fuji apples (Massia, 1998).

CO_2 is required for the assembly of the binuclear metal centre of phosphotriesterase (Hong *et al.*, 1995), for the ATP-driven carboxylation of the biotin moiety by biotin carboxylase (Harwood, 1997), and for *in vitro* assembly of the urease nickel metallocentre (Park and Hausinger, 1995). Elevated CO_2 represses carbonic anhydrase activity in unicellular green algae such as *Chlorella pyrenoidosa* (Graham and Reed, 1971) and in leaves of C3 plants, including cotton (Chang, 1975) and bean (Porter and Grodzinski, 1984). Bicarbonate ion is a general base in the mechanism of peptide hydrolysis by di-zinc leucine aminopeptidase (Strater *et al.*, 1999), and has a role in iron transport across membranes (referred to in Sanders *et al.*, 1999), and possibly also in the active site of L-lactate monoxygenase (Sanders *et al.*, 1999).

Cyclic 3′,5′ AMP (cAMP) produced by adenylyl cyclase (AC) is a second messenger activating protein kinase A (PKA) in animal cells. All components of this signal transduction system have been identified in plants (Biermann *et al.*, 1990; Bolwell, 1992, 1995; Gatehouse, 1997; Ehsan *et al.*, 1998; Kawai *et al.*, 1998; Volotovski *et al.*, 1998; Lengeler *et al.*, 2000) and cAMP has been found to act as a second messenger in pollen tube growth and reorientation (Moutinho *et al.*, 2001) and in gating a plant cation channel (Leng *et al.*, 1999). Adenylyl cyclase genes also have been isolated from the cyanobacteria *Anabaena cylindrica* (Katayama *et al.*, 1995; Katayama and Ohmori, 1997) and *Spirulina platensis* (Yashiro *et al.*, 1996; Kasahara *et al.*, 1997), where they act as a second messenger for light signal transduction (Ohmori *et al.*, 1998 – referred to in Moutinho *et al.*, 2001), and cAMP signalling is essential for normal growth, morphology and virulence in plant and animal fungal pathogens (Smith, 1999).[11] Soluble adenylyl cyclase (sAC) is a novel AC form, structurally, molecularly and biochemically distinct from transmembrane adenylyl cyclases (tmACs) in that it possesses no transmembrane domains and is insensitive to classical modulators of tmACs. This indicates that it is an independently regulated cAMP signalling system (Wuttke *et al.*, 2001). In both mammalian tissue and *S. platensis*, bicarbonate enhances the activity of sAC (Chen *et al.*, 2000; Zippin *et al.*, 2001; Wuttke *et al.*, 2001; Sun *et al.*, 2003; Litvin *et al.*, 2003). The half maximally effective bicarbonate concentration (18.8 ± 1.6 mM) is in the physiological range,[12] and the reaction is pH-independent both *in vivo* and *in vitro*. Bicarbonate's stimulatory effect is mimicked by bisulfite ion, but not by dissimilar ions such as chloride, sulphate or phosphate, indicating that bicarbonate, rather than CO_2, directly binds and activates sAC. A pollen-specific clone codes for a soluble signalling protein (PSiP) with AC activity, and inhibition of PSiP by antisense assays disturbs apical pollen tube growth (Moutinho *et al.*, 2001), proving that sAC is active in plants. Immunolocalization studies reveal that sAC is abundantly expressed in tissues that respond to bicarbonate or carbon dioxide levels. This suggests that sAC may be the bicarbonate sensor in metabolism, and that because CO_2 is in equilibrium with bicarbonate, sAC and the cAMP-signalling pathway also may indirectly monitor the *in vivo* [CO_2] concentration. LP should limit sAC activity by lowering the cellular ICC. If both sAC and tmAC regulate fungal growth in the same manner,[11] the effect of HCO^- on sAC activity might explain why the inhibition of microbial growth at a low pressure is greater than can be accounted for in response to reduced [O_2] (Tables 7.4 and 7.5; Fig. 11.7, *right*).

4.12 Effect of CO_2 on Ethylene Action and Metabolism

CO_2 is a potent inhibitor of most or all actions of ethylene. The gas antagonizes ethylene-induced abscission (Abeles and Gahagan, 1968a; Lipe and Morgan, 1972b; Abeles, 1973; Wittenbach and Bukovac, 1973; Fig. 4.7, *lower right*), auto-inhibition (Saltveit and Dilley, 1978a; Aharoni and Lieberman, 1979; Aharoni *et al.*, 1979b), synthesis of ACC synthase (Bufler, 1984; Chaves-Franco and Kader, 1993; Mathooko *et al.*, 1995), flower fading (Thornton, 1930; Smith and Parker, 1966; Smith *et al.*, 1966; Nichols, 1968; Uota, 1969a; Kende and Baumgartner, 1974; Akamine, 1976a; Mayak and Dilley, 1976; Mayak *et al.*, 1977), epinasty (Denny, 1935; Leather *et al.*, 1972), chlorophyll loss (Mack, 1927; Apelbaum *et al.*, 1976; Aharoni and Lieberman, 1979; Sisler, 1980a; Gepstein and Thimann, 1981; Fig. 4.8), hook opening (Kang *et al.*, 1967; Kang and Ray, 1969a; Kang and Burg, 1972b), toughening of asparagus spears (Lipton, 1965), stimulation of respiration (Aharoni *et al.*, 1979a; Aharoni and Lieberman, 1979; Sisler, 1980a), ethylene-induced induction of EFE (Wright, 1980), growth inhibition and radial expansion of etiolated pea epicotyls (Burg and Burg, 1965a, 1967c; Fig. 4.7, *upper left* and *right*) and pea root segments (Chadwick and Burg, 1967; Fig. 4.7, *lower left*), ethylene-inhibition of cultured radish roots (Radin and Loomis, 1969), internode elongation in etiolated pea plants (Goeschl

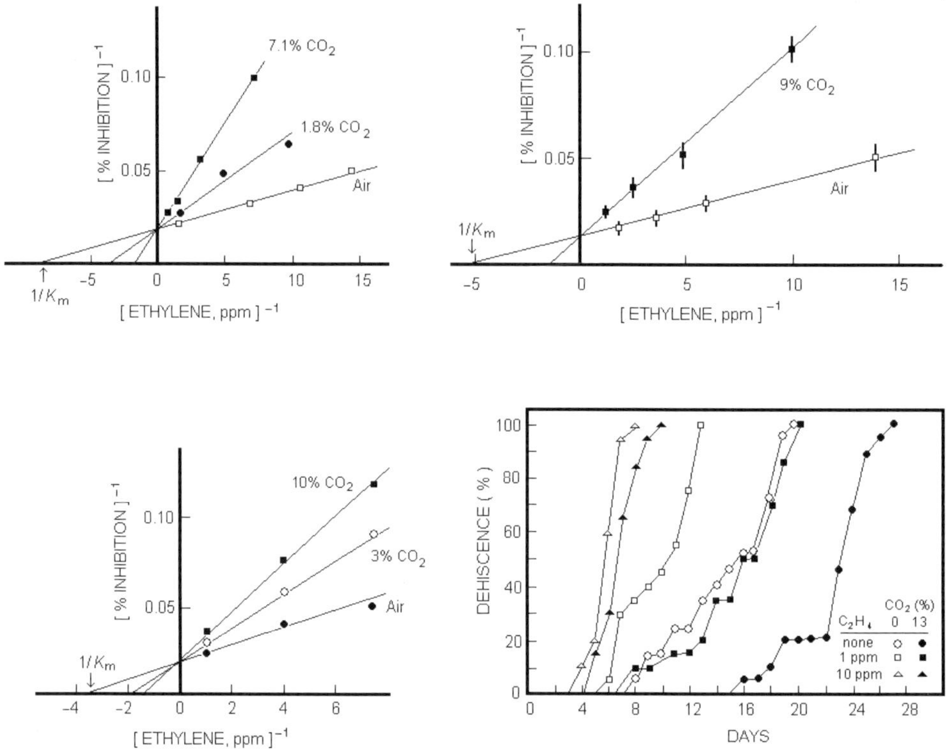

Fig. 4.7. Lineweaver-Burk plots (1/*V* vs. 1/*A*) relating the growth (*V*) of etiolated pea tissue to the ethylene concentration (*A*) at various levels of CO_2. Each line representing a CO_2 concentration intersects the X-axis at $-1/K_p$, where, i is the inhibitor (CO_2) concentration, K_i is the dissociation constant of the CO_2-receptor complex, K_m is the Michaelis-Menton constant for ethylene binding in the absence of inhibitor; and for a purely competitive inhibition $K_p = K_m(1 + i / K_i)$. (*upper left*) Etiolated pea epicotyl sections. The K_i for [CO_2] is 1.55% (Burg and Burg, 1967c). (*upper right*) Second internode of intact etiolated pea seedlings. The apparent K_i is 3.2% [CO_2] (Goeschl and Pratt, 1968). (*lower left*) Pea root segments. The apparent K_i is 5.5% [CO_2] (Chadwick and Burg, 1967). (*lower right*) Competitive effects of CO_2 and ethylene on dehiscence of detached cotton fruits. Lots of 20 Tumcot SP37 fruits were collected 3–5 weeks after bloom and treated in the dark with the indicated levels of CO_2, ethylene, or both (Lipe and Morgan, 1972a).

and Pratt, 1968; Fig. 4.7, *upper right*), fruit ripening (Kidd and West, 1934), carotene biogenesis (Kang and Burg, 1972c), plumular expansion (Goeschl and Pratt, 1968) and also numerous other responses referenced in Abeles (1973), including intumescence formation, loosening of walnut hulls, isocoumarin formation, potato-tuber sprouting and induction of glucanase, phenylalanine ammonia lyase and peroxidase. EFE activity in peel tisssue of AVG-treated apples was strongly promoted by 0.05 µl/l ethylene after a lag of 4 days, but 7% [CO_2] and the same low-applied ethylene concentration immediately promoted this tissue's EFE for several days, and then strongly inhibited the increase that begins after 4 days (Bufler *et al.*, 1980). This result suggests that CO_2 initially antagonized ethylene's ability to auto-inhibit system 1 ethylene production, and after 4 days, inhibited ethylene's ability to promote system 2 autocatalytic ethylene production (5.06, 5.07, 5.08). It is not known whether these effects of CO_2 on ethylene action are caused by CO_2 gas or HCO_3^-.

Binding of ^{14}C-ethylene in plant tissues has been shown to meet all of the requirements of a physiological receptor (Trewavas and Jones, 1981; Sisler, 1991). The criterion

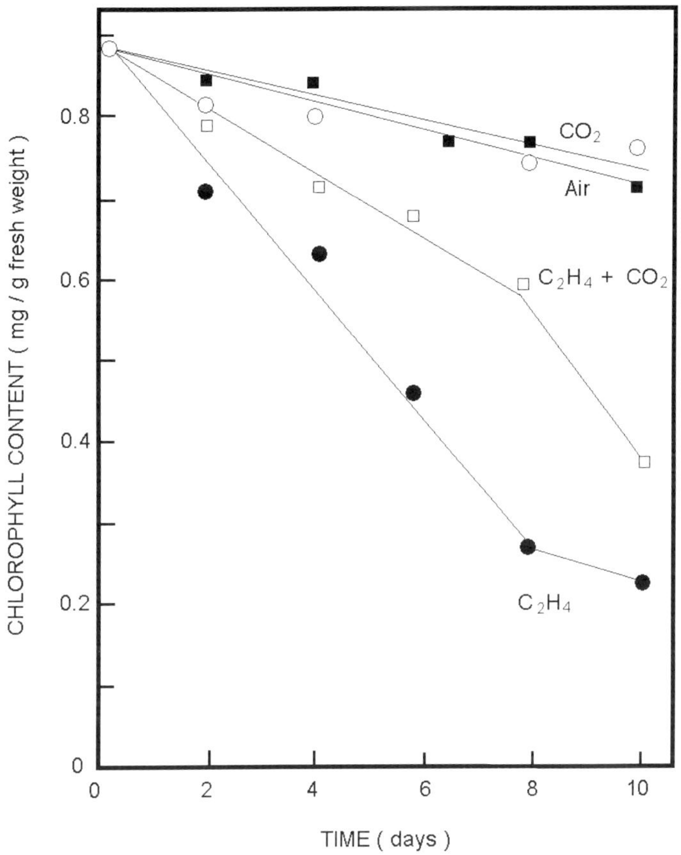

Fig. 4.8. Non-competitive effect of 1% [CO_2] on the loss of chlorophyll from tobacco leaves treated with 1000 μl/l ethylene under laboratory light at 22°C (Sisler, 1980a).

that the binding site's dissociation constant (K_d) is close to the biological concentration that produces a physiological response (K_m) has been satisfied by comparing the ^{14}C-ethylene concentration needed to occupy half of the binding sites vs. values for the applied ethylene concentration, which elicits a half-maximum physiological response (Table 5.6). A plot of the biological response vs. ethylene concentration produces a section of a rectangular hyperbola, which can only result when a process depends on a simple dissociation, in which $XY \leftrightarrow X + Y$, where X is the receptor concentration, and the ethylene concentration [Y] is held constant. A dissociation of this type is the basis of the Michaelis-Menton theory (Dixon and Webb, 1959), and the K_d value for ^{14}C-ethylene binding can only equal the K_m value determined as the biological activity of ethylene if the assumption is valid that equilibrium is maintained between the receptor, ethylene and the receptor–ethylene complex, and that this is the basis for the biological activity. If the ethylene-binding-site measurements are specific to the ethylene receptor, it follows that the Michaelis-Menton theory can be used to evaluate the physiological response, and the kinetics of the response will be determined by the interaction of ethylene with the receptor, and not by rate-limiting reactions occurring downstream in the ethylene-transduction pathway. Because CO_2 interferes with so many diverse actions of ethylene, the inhibitory effect that this gas has on ethylene action must be exerted close to or on the ethylene receptor, before the ethylene-transduction pathway branches out into alternative responses (Fig. 5.24). However, CO_2 may exert additional effects that directly inhibit or promote the phenotype that is being elicited by ethylene.

An irreversible inhibition is progressive and ultimately complete, provided that the amount of inhibitor is sufficient to combine with all of the protein receptor or enzyme. Typically, the CO_2 inhibition of ethylene action is reversible, and quickly reaches an equilibrium value that depends on the inhibitor concentration. Reversible inhibitors may act either on the apparent K_m or maximum velocity (V_{max}) of a reaction. A reversible 'competitive' inhibitor increases the apparent K_m because the inhibitor and substrate tend to drive one another off the binding site, whereas inhibitors which act exclusively by reducing the maximum velocity (V_{max}) are non-competitive. The effect of a competitive inhibitor is abolished by a high substrate (ethylene) concentration so that V_{max} is not affected, whereas the action of a non-competitive inhibitor on velocity remains at a sufficiently high substrate (ethylene) concentration. A competitive type of inhibition is produced by substances structurally related to ethylene, which are unable to induce a physiological response, but combine reversibly with the receptor at the same site as ethylene, whereas substances that combine at another site, sufficiently far removed from the ethylene-binding site to have no influence on ethylene binding, produce the non-competitive type. In the competitive type, the inhibitor (CO_2/HCO_3^-), when it is present in excess, is completely combined with the receptor, and then small amounts of substrate (ethylene) cannot bind. It is also possible that an inhibitor (or activator), while not combining with the binding site, may attach to another site sufficiently close to the binding site, that it reduces (or increases) the receptor's affinity. The activation of RuBP carboxylase and superoxide dismutase by CO_2/HCO_3^- (4.11) exemplifies this type of response. Thus inhibitors can act in more than one way, giving mixed types, and the fact that an inhibitor reduces V_{max} does not exclude an action on K_m, and vice versa. CO_2 might simultaneously change the K_m and V_{max} not only by combining with the receptor, but also by altering the cytoplasmic pH, and at the same time CO_2 might directly influence a physiological response by a direct action downstream of the receptor.

The CO_2 reversal of ethylene's growth inhibitory action on pea root segments (Fig. 4.7, *lower left*; Chadwick and Burg, 1967) and intact (Fig. 4.7, *upper right*; Goeschl and Pratt, 1968) and excised (Fig. 4.7, *upper left*; Burg and Burg, 1967c) etiolated pea epicotyls has competitive kinetics. Likewise, 13% [CO_2] prevents 1 μl/l ethylene from causing abscission of okra (Lipe and Morgan, 1972b) and cotton (Fig. 4.7, *lower right*)] fruits, but does not reverse the effect of 10 μl/l ethylene. Abscission of *Cassia*, bean, *Coleus* and cotton petiole explants is prevented by 5–10% applied [CO_2], and higher ethylene concentrations, varying from 0.1 to 10 μl/l, are required to overcome the inhibitory effect of larger [CO_2] concentrations, ranging from 1 to 5% (Abeles and Gahagan, 1968a). The kinetics also are 'competitive' when CO_2 inhibits ACC synthase-dependent autocatalytic ethylene production in apples, pears and tomatoes (Bufler, 1984; Chaves-Franco and Kader, 1993; Mathooko *et al.*, 1995), and ethylene-induced flower fading (Smith *et al.*, 1966), bean-hook tightening (Kang and Ray, 1969a) and de-greening of oranges (Apelbaum *et al.*, 1976). In mature-green Shamouti oranges, the rate of chlorophyll loss induced by application of 0.1 μl/l ethylene is nearly as rapid as that caused by 5 μl/l ethylene, and is not affected by 0.5% [CO_2], possibly because this fruit's ICC is close to 1% at harvest (Apelbaum *et al.*, 1976), but 1.5% [CO_2] causes a substantial reversal, and 5% almost completely counteracts the ethylene effect (Fig. 3.3, *right*). Other actions of CO_2 and the 'toxic' effect of > 10% [CO_2] on growth, make it impossible to determine whether the apparent K_m increases without limit when [CO_2] is increased, whether K_i remains constant in the presence of an excess of ethylene, and if the maximum effect (V_{max}) of excess ethylene is truly independent of the CO_2 concentration. All of these are a prerequisite for a purely competitive inhibition.

In tobacco leaves, the kinetics of CO_2–ethylene antagonism are non-competitive (Sisler, 1980a, 1991; Fig. 4.8). During 8 days, tobacco leaves lost only half as much chlorophyll in the presence of 1000 μl/l ethylene if 1% [CO_2] was present, and 3% [CO_2] was slightly more effective. The CO_2 effect could not have resulted from competitive displacement of ethylene from its receptor site because the ethylene concentration was 2–3 orders of magnitude higher than that needed to saturate the ethylene response. The data do not, however, indicate whether or not CO_2 lowered the receptor's apparent affinity for ethylene, as that can only be evaluated by testing CO_2's ability to reverse the effect of much lower ethylene concentrations.

Sisler (1979) measured the ability of CO_2 to displace bound ethylene from tobacco leaves exposed to 37 nl/l of ^{14}C-ethylene, either in the absence of CO_2 or with 2–10% [CO_2] present.[13] The ^{14}C-ethylene concentration was sufficient to occupy only 13% of the total available ethylene-binding sites, and during the ^{14}C-ethylene treatment the leaves produced enough unlabelled ethylene to elevate the ethylene concentration to 200 nl/l in the gas phase of the desiccator, which should have displaced most of the bound ^{14}C-ethylene. This does not take into consideration the additional displacement that would be expected because the stomates almost certainly had closed, elevating the tissue's IEC above the ethylene concentration in the desiccator. Of the remaining bound ^{14}C-ethylene, it was calculated that 34–41% was displaced by 2–10% [CO_2], indicating that CO_2 increased the apparent K_m of the receptor for ethylene. The failure of 10% [CO_2] to displace a much higher per cent of the ^{14}C-ethylene than that freed by 2% [CO_2] is compatible with a model in which CO_2 (or HCO_3^-) attaches to the receptor at a site sufficiently close to the ethylene-binding site to decrease the apparent K_m. Once the CO_2/HCO_3^- binding site was saturated, additional CO_2 would have no additional effect on the K_m. This does not consider the complication that 5–10% [CO_2] causes severe wilting and perhaps death of tobacco leaves after about 4 days, whereas no obvious damage results with 3% [CO_2] during 10 days (Sisler, 1980a).

A K_i of 1.15×10^{-3} M CO_2 was reported for the inhibition of ^{14}C-ethylene binding in membranes from *Phaseolus vulgaris*, but no inhibition was observed in Triton X-100 extracts of the membranes (Hall *et al.*, 1987) or in mung bean sprouts exposed to 0.2 μCi of ^{14}C-ethylene per litre in the presence of 5% [CO_2] (Sisler and Wood, 1988). It was not obvious why CO_2 should displace ethylene from tobacco leaves and *P. vulgaris* membranes, but not from mung bean sprouts. In etiolated pea seedlings, 7% [CO_2] had no effect on incorporation of radioactive carbon from ^{14}C-ethylene into tissue components (Beyer, 1975a), but while some of the tissue counts undoubtedly were due to bound ethylene, most were a result of ethylene metabolism to ethylene oxide and other end products (Hall, 1991). This caused the specific activity (dpm/mg dry weight) to increase as a direct function of the ^{14}C-ethylene concentration between 0.3 and 140 μl/l, whereas the amount of ethylene bound to the ethylene receptor would be expected to saturate at < 10 μl/l.

Reversal of the ethylene-induced stimulation of respiration and de-greening was half-maximal at 1% [CO_2] in the tobacco leaf studies. A similar result was obtained with oat leaves (Gepstein and Thimann, 1981), where 1% [CO_2] half-reversed the effect of 100 μl/l ethylene and 10% [CO_2] gave a complete reversal. The half-maximal [CO_2] required to antagonize these effects in the leaf test system is approximately the same as that required for competitive or partially competitive ethylene inhibition in other tissues (Fig. 4.7). Thus, in some experiments CO_2 seems to increase the K_m, in other studies the gas affects V_{max}, and sometimes both seem to be influenced. The kinetic model that most closely conforms to this behaviour is a 'partially competitive or mixed type' of inhibition (Dixon and Webb, 1959) in which CO_2 attaches to the receptor at a different site than that to which ethylene binds, and thereby affects both the receptor's affinity for ethylene and the maximum velocity of the reaction. In addition,

CO_2 may directly affect the de-greening process (Oeller *et al.*, 1991; Klee, 1993; Murray *et al.*, 1993b,c). It is also possible that CO_2 protects CTR1 (5.12) from being down-regulated when ethylene binds to the ethylene receptor. Either scenario would give rise to an interaction between ethylene and CO_2 that, depending on the physiological response used to monitor ethylene action, might have competitive or non-competitive kinetics, without CO_2 actually competing with ethylene for attachment to the same site in the ethylene receptor.

The interpretation of carbon dioxide's influence on ethylene action is complicated by the effect that CO_2 has on ethylene production (4.10), ethylene oxidation (Beyer, 1975a) and other aspects of metabolism (4.11). [CO_2] enhances ACC-dependent ethylene production and induces the synthesis and accumulation of EFE in leaves, and between 0.3 and 1% [CO_2] stimulates ethylene production in seeds (Esashi *et al.*, 1985). This may explain why both ethylene and CO_2 promote seed germination (Toole *et al.*, 1964; Esashi *et al.*, 1985); why 5–20% [CO_2] mimics ethylene's ability to stimulate the growth of underwater stems (Suge, 1971; Suge and Kusanagi, 1975; Ishizawa and Esashi, 1984; Métraux and Kende, 1984); and why 5–10% [CO_2] promotes an ethylene-like respiratory climacteric in lemons (Young and Biale, 1962). An ethylene phenotype also would be produced if CO_2 inactivated the ETR1 ethylene receptor to render it unable to carry out its normal function (5.13).

It has been suggested that CO_2/ethylene interaction may exert a natural regulatory role in plant tissues (Kang and Ray, 1969a; Bufler *et al.*, 1980; Esashi, 1991; Burg and Kang, 1993). The CO_2 concentrations that influence ethylene action are well within the range of the ICCs that normally arise in fruits, and also in vegetative tissues when their stomates close at night.

Ethylene metabolism is inhibited by CO_2 (Beyer, 1975a; Sanders *et al.*, 1986; Evans *et al.*, 1987; Hall, 1991). In pea tissue, 5% [CO_2] inhibited the oxidation of ^{14}C-ethylene to CO_2 by approximately 75% (Beyer, 1975a), but when this effect was reinvestigated using improved techniques, it was found that the inhibition caused by 7% [CO_2] was much less, about 25%, and about the same as that caused by Ag^{2+} (Sanders *et al.*, 1986; Hall, 1991). This effect of CO_2 may complicate the interpretation of ^{14}C-ethylene-binding experiments in which CO_2 is added to determine whether it is able to displace ethylene from its receptor site(s).

Ethylene metabolism is not related to the mode of ethylene action, and in most plant tissues it does not have a significant effect on their IEC (Hall, 1991). The major product of ethylene metabolism by developing cotyledons of *Vicia faba* is ethylene oxide (Jerie and Hall, 1978), which in cell-free preparations is produced by a monooxygenase (Smith *et al.*, 1985):

$$C_2H_4 + O_2 + NADPH + H^+ \rightarrow C_2H_4O + H_2O + NADP^+ \qquad (4.3)$$

Aseptically grown peas metabolize highly purified ^{14}C-ethylene to tissue components (Beyer, 1975a), identified as ethylene glycol and its glucose conjugate β-(2-hydroxy-ethyl)D-glucoside, and also to CO_2. *In vivo*, in addition to ethylene oxide, a small amount of CO_2 and minor quantities of ethylene glycol, glycerol, oxalate, glycolic acid and ethanolamine are produced from ^{14}C-ethylene (Hall, 1991). These products may be formed from ethylene oxide by the action of hydoxylases capable of converting epoxides to glycols. The K_m (220 μl/l) of the pea-system enzyme that converts ethylene to CO_2 is 300–400 times higher than the ethylene concentration that induces a half-maximal physiological response in pea tissue (Tables 5.2 and 5.6).

4.13 Effect of CO_2 on Succinic Acid Accumulation

Normally, succinate is a minor organic acid, but it accumulates in commodities when they are stored with [CO_2] (Ransom, 1953; Frenkel and Patterson, 1969; Ulrich, 1975), and is toxic to fruits (Hulme, 1956). As little as 1 mM applied succinate causes the respiration of apple-peel sections to cease, and the tissue to become brown and die. CO_2 causes succinate to accumulate when it injures apples (Ransom, 1953; Hulme, 1956; Shipway *et al.*, 1973), pears (Williams and Patterson, 1964), sweet cherries (Singh *et al.*, 1970), spinach (Murata and Ueda, 1967), citrus (Davis *et al.*, 1973), peas (Wager, 1973), grapes (Flany, 1967 – referred to in Ulrich, 1970), apricots and peaches (Wankier *et al.*, 1970). More than 0.03% [CO_2] increases the succinate content of certain varieties of pears (Frenkel and Patterson, 1969; Fig. 4.9).

Succinate accumulation may be caused by an inhibitory effect of CO_2 on succinic dehydrogenase (Ransom *et al.*, 1957, 1960; Bendall *et al.*, 1960; Miller and Hsu, 1965; Frenkel and Patterson, 1969; Knee, 1973), an enzyme that is potently inhibited by oxaloacetate (Bonner, 1965; Ulrich, 1970; Brownleader *et al.*, 1997). Propionobacteria form succinate after CO_2 and pyruvate combine to form oxaloacetate (Wood and Werkman, 1940). High CO_2 also causes non-specific structural or conformational changes in apple mitochondria that suppress fumarate, pyruvate, 2-oxoglutarate, citrate and NADH oxidations, while stimulating malate oxidation (Bendall *et al.*, 1960; Shipway *et al.*, 1973); these changes also cause pyruvate, 2-oxoglutarate, glyoxylate, malate and oxaloacetate, but not succinate, to accumulate in bananas (McGlasson and Wills, 1972). By removing CO_2 from around and within a commodity, LP should prevent succinate accumulation.

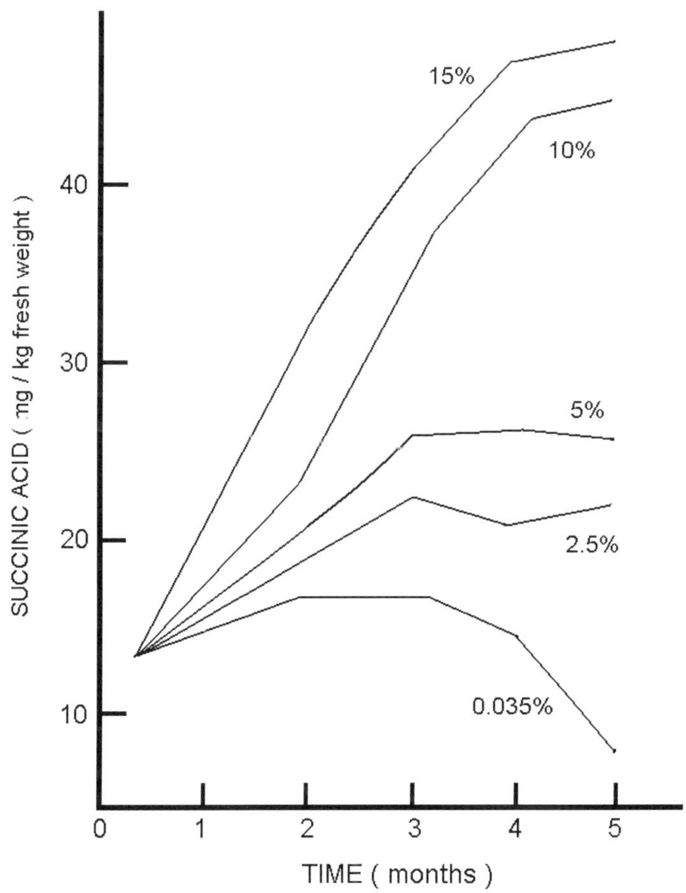

Fig. 4.9. Changes in the succinic acid content of 'Williams' pears in atmospheres containing 0.035% to 15% [CO_2] at 0°C (Williams and Patterson, 1964).

4.14 Effect of CO_2 on Ascorbic Acid Content

Applied or accumulated CO_2 hastens the loss of ascorbic acid in asparagus (Thornton, 1937a; Fig. 4.10), bananas (Thornton, 1943), mangoes (Hulme, 1971), okra (Ogata *et al.*, 1975), strawberries (Thornton, 1937b) and other berry fruits (Agar *et al.*, 1994, 1997), sweet peppers (Wang, 1977; Leshuk and Saltveit, 1990), Chinese cabbage (Wang, 1983), Satsuma mandarin (Thompson, 1998), green pod beans and other plant commodities (Thornton, 1937b, 1946; Bangerth, 1977). In green bananas, 3–5% [CO_2] causes a 10–20% reduction in vitamin C in 4 days, 8% [CO_2] a 30–50% reduction, 24% [CO_2] a 42–80% loss, but CO_2 had little effect in the yellow-ripe stage, possibly because these fruits already have developed a high ICC in the 12.5–14.5% range (Table 2.2). At any state of ripeness where CO_2 caused a reduction, the ascorbic acid returned to normal some time after bananas were removed from CO_2. Broccoli and apples are exceptions. A short-duration high [CO_2] treatment seems to increase ascorbic acid in broccoli (Wang, 1979), while 0–60% [CO_2] in the presence of 20% [O_2] did not change the ascorbic acid content of apples during 10 days (Thornton, 1937b). When high [CO_2] and low [O_2] treatments are combined, the results are highly variable[14] and all that can be concluded is that low [O_2] usually is helpful in preserving ascorbic acid, high [CO_2] often diminishes the effectiveness of low [O_2] and rarely does the combination promote ascorbic acid retention. By removing naturally occurring CO_2 from within a plant commodity's intercellular spaces and simultaneously lowering the O_2 partial pressure, LP seems always to elevate the ascorbic acid content. Hypobaric storage has been found to maintain the ascorbic acid content of apples (10.1), asparagus (Table 4.8), currants (10.12), cress (10.38), parsley (10.46 and Fig. 10.4), radishes (10.49) and spinach (10.50).

It has been suggested that 'there is a causal relation between the rise in pH caused by CO_2 (4.16) and the fall in ascorbic acid, since the latter is protected by acids and is readily oxidized in basic solutions'

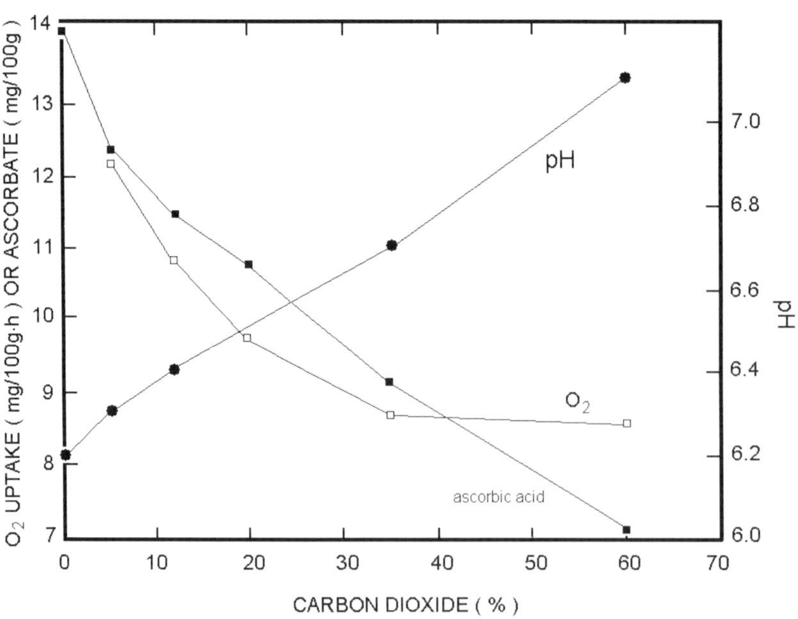

Fig. 4.10. The effect of CO_2 on the ascorbic acid content, respiration rate and pH of asparagus tissue during storage for 24 h at 22°C (Thornton, 1937a).

(Crocker, 1948). The rate of irreversible formation of 2,3 dioxo-L-gulonic acid from ascorbic acid (equation 4.1) is largely determined by pH, occurring rapidly at neutral pH, instantaneously at alkaline pH and much more slowly at pH 3–4 (Burton, 1982). There are exceptions. The pH of snap beans is increased by 40% [CO_2] (Buescher and Adams, 1983), but 0–60% [CO_2] does not enhance their ascorbic acid loss (Thornton, 1937a); the pH of lettuce (measured with NMR) is not elevated in the presence of 15% [CO_2] (Siriphanich and Kader, 1986), but in 5% [CO_2] the ascorbic-acid content of lettuce is reduced (Bangerth, 1977).

The loss of ascorbate in bananas is accompanied by the complete disappearance of dehydroascorbate. However, the dehydroascorbic acid content in the flesh of stored Satsuma mandarins rose even though ascorbic acid decreased in the flesh and peel (Thompson, 1998).

4.15 Effect of CO_2 on Stomatal Action

Many postharvest physiologists assume that the stomates present on fruits are non-functional (Smock and Neubert, 1950; Bean, 1962). To the contrary, Johnson and Brun (1966) used a chlorine-bleaching technique (Janes, 1962) to demonstrate that banana stomates (*Musa acuminata* L. var. Hort. Valery) are capable of opening and closing for many weeks after fruits are harvested. Portions of banana leaves and green fruits with open stomates bleach to a pale yellow after exposure to chlorine gas for a few seconds, while areas with closed stomates remain green after a 1-h exposure. Stomates on intact hanging fruit held in a greenhouse (29–30°C, 20–30% RH) closed after 30 min in darkness and reopened during 2 h of light. They closed within 1 h in darkness on fruits freshly severed from the stem, and reopened during 2 h of sunlight. They opened in dim light at 13°C and 90–100% RH on fruits received after a 10-day commercial shipment from the tropics, and closed in 2 h of darkness. When imported fruits were transferred to 16°C, 90–100% RH, their stomates were open in dim light, and closed in 3 h of darkness, and they displayed the same stomatal response 24 h after ripening was initiated with ethylene gas.

Stomates are sensitive to CO_2 to the same degree in the dark and light, and the responsible mechanism apparently is identical in both cases (Raschke *et al.*, 1970). Guard cells are able to discriminate small changes in external CO_2 in the 0.01–0.036% range (Willmer and Fricker, 1996), and at atmospheric pressure, high ambient [CO_2] closes leaf stomates in the light, while in darkness CO_2-free air opens them (Zelitch, 1969; Willmer and Fricker, 1996; Murray, 1997; Fig. 4.11). The same positive correlation between stomatal aperture and intercellular CO_2 levels occurs within CAM tissues, suggesting that CAM stomates respond to the changing [CO_2] in leaves brought about by the carboxylation and decarboxylation phases of CAM. The half-time for stomatal opening or closing in response to a change in [CO_2] may be as short as 10–20 min (Fig. 3.8; Raschke, 1975), and 290 μl/l CO_2 can initiate closing within 3–16 s (Raschke, 1972). Stomates respond to the [CO_2] concentration in the substomatal chamber (Raschke, 1975; Willmer and Fricker, 1996), and at atmospheric pressure, once stomates are induced to close by high [CO_2], they are not easily forced to open by CO_2-free air because the guard cells remain in equilibrium with the high [CO_2] trapped within the substomatal cavity (Willmer and Fricker, 1996). When a horticultural commodity with closed stomates is brought to a low atmospheric pressure, the air in the substomatal cavity expands and its CO_2 partial pressure is immediately lowered in direct relation to the pressure decrease. [CO_2] in the substomatal cavity is further reduced because respiratory CO_2 production is inhibited due to the lowered O_2 partial pressure, and in addition the CO_2 content of the ambient atmosphere is decreased proportionate to the expansion in the LP pressure regulator, while the rate of CO_2 exchange between the substomatal cavity and ambient atmosphere is increased by enhanced diffusion in a partial vacuum. If the LP pressure is low enough, the [CO_2] in the substomatal cavity should decrease sufficiently to open the stomates. This expectation has been confirmed by measurements indicating low diffusion-resistance values in oat-leaf segments (*Avena sativa* L. – Fig. 4.12, left; Table 4.9; Veierskov and Kirk, 1986) floated on water at pressures ranging from 1.6 to 77 kPa (12–578 mm Hg), in *Hibiscus rosa-sinensis* L. cv. Moesiana cuttings (Kirk *et al.*, 1986; Fig. 4.12, right) stored at 15°C and a pressure of 2–2.67 kPa

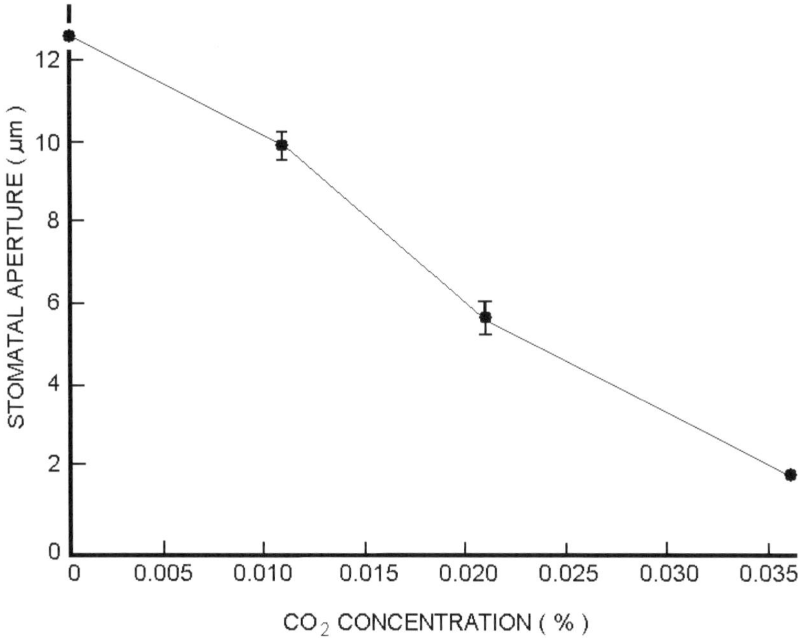

Fig. 4.11. Effect of CO_2 on stomatal opening in epidermal strips of *Commelina communis* (Willmer, 1988).

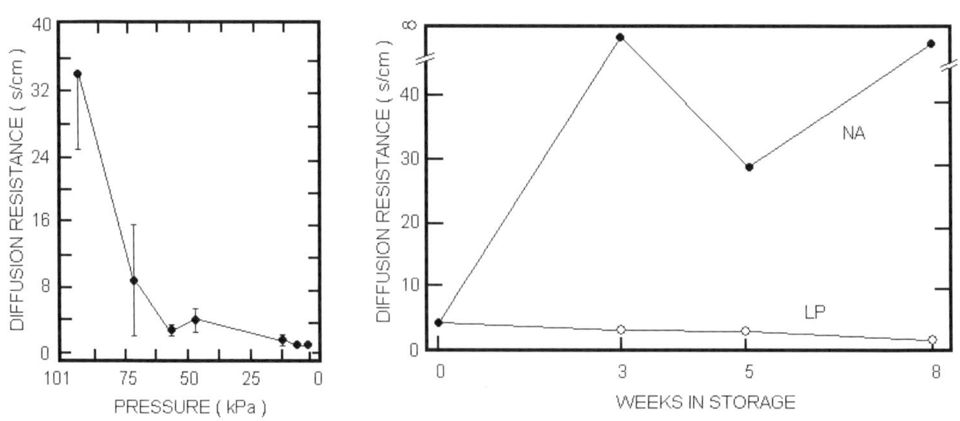

Fig. 4.12. Pressure dependence of stomatal opening. (*left*) Oat leaf segments were floated on water in darkness for 24 h at the indicated pressure before diffusion resistance was measured. Data are means of at least ten measurements (Veierskov and Kirk, 1986). (*right*) Stomatal opening measured as diffusion resistance after *Hibiscus* cuttings were removed from storage at 15°C, either in LP at a pressure of 2.0–2.67 kPa (15–20 mm Hg), or in NA. Each symbol (○ or ●) represents the mean of two replications with four measurements each. The '0-week storage' value was obtained from intact plants before cuttings were excised (Kirk *et al.*, 1986).

Table 4.9. The time course for stomatal opening in 3-cm oat leaf segments kept under hypobaric conditions at 12°C and a pressure of 1.6 kPa (12 mm Hg). Data are means of six independent measurements (Veierskov and Kirk, 1986).

Time (h)	Diffusion resistance (s/cm)	Transmission (μg H_2O/cm·s)
0	32.2	0.5 ± 0.1
7	3.6	2.2 ± 0.9
16	1.3	4.3 ± 0.8
24	1.1	5.9 ± 1.0

(15–20 mm Hg), and in cuttings of 13 ornamental species during 5 weeks when they were stored at 15°C and a pressure of 2 kPa (15 mm Hg; Kirk and Andersen, 1986). Stomatal opening during LP storage has also been confirmed by scanning electron micrographs of *Hibiscus* cuttings (Fig. 4.13, top) and by microscopic examination of acrylic templates prepared from *Caladium bicolor* and *Colocasia esculentum* (Taro) plants that had been kept in darkness for 5 days at 13.3°C and a pressure 2.67 kPa (20 mm Hg), Valencia orange fruits stored in darkness for 16 days at 4 or 10°C and a pressure of 2.67 kPa (20 mm Hg), and cucumbers, green beans and papaya fruits kept in darkness at a pressure of 2.67 kPa (20 mm Hg) for 2 days at 10°C (Davenport and Burg, 2003, unpublished). In all instances, stomates were closed at atmospheric pressure and open in LP. The time course for the opening of oat stomates at 12°C and a pressure of 1.6 kPa (12 mm Hg) is indicated in Table 4.9. *Hibiscus* stomates closed within 2 h after cuttings were placed into 15°C NA storage and remained shut during eight subsequent weeks. In LP at a pressure of 2 kPa (15 mm Hg), they remained open throughout 8 weeks (Kirk *et al.*, 1986; Fig. 4.12, *right*). An elevated ICC should cause stomates to close during CA or MA storage.

Both CO_2-free air (Willmer *et al.*, 1988) and hypobaric conditions (Veierskov and Kirk, 1986) cause stomates to lose their ability to close in response to ABA. In CO_2-free air, a 10^{-5} M ABA concentration in the transpiration stream has virtually no effect on *Xanthium strumarium* stomata if the rate of CO_2 production is not too high (Raschke and Pierce, 1973), and at 1.6 or 6 kPa (12 or 45 mm Hg) a concentration of 5×10^{-6} M ABA was unable to induce stomatal closure in *Avena* leaf sections incubated in darkness (Veierskov and Kirk, 1986). Kinetin (5×10^{-5} M) and hexanol (2.5×10^{-2} M) caused the stomates to open at atmospheric pressure, but had no effect at a low pressure. The effect that LP has on the ABA sensitivity of stomates is a further indication that hypobaric conditions open stomates by lowering the intercellular [CO_2].

Following LP storage, open stomates do not respond to a low humidity or high [CO_2] and need many hours to regain normal function (Kirk *et al.*, 1986). After *Hibiscus* cuttings were removed from LP storage and placed in water culture in the light, stomatal closure proceeded in 1–2 h due to an excessive water loss of up to 50% of the initial fresh weight in 1 h, and closure could not be hastened by ABA, indicating that it was a passive process. When *Hibiscus* cuttings were transferred from LP to NA in darkness under conditions where water loss was minimized, stomatal closure proceeded at a slower rate than is normal for fresh cuttings in the light, and 5% [CO_2] provided during venting of the LP chambers did not cause the stomates to close, and instead seemed to open them wider (Kirk *et al.*, 1986). Even though potted plants of *C. bicolor* and *C. esculentum* were well watered, severe wilting occurred when they were removed from LP storage and immediately placed in the light (Davenport and Burg 2003, unpublished). These results suggest that during LP storage the mechanisms controlling stomatal action may be altered, and that to prevent excess water loss after stomata-bearing commodities are removed from LP storage, it may be prudent to keep them protected in their water-retentive packaging for at least 6 h to induce stomatal closure before the packaging is removed.

Stålfelt (referred to in Zelitch, 1969 and Hanebuth and Raschke, 1972) described a slow, 7-h dark stomatal opening by leaf disks of *V. faba* when they were exposed

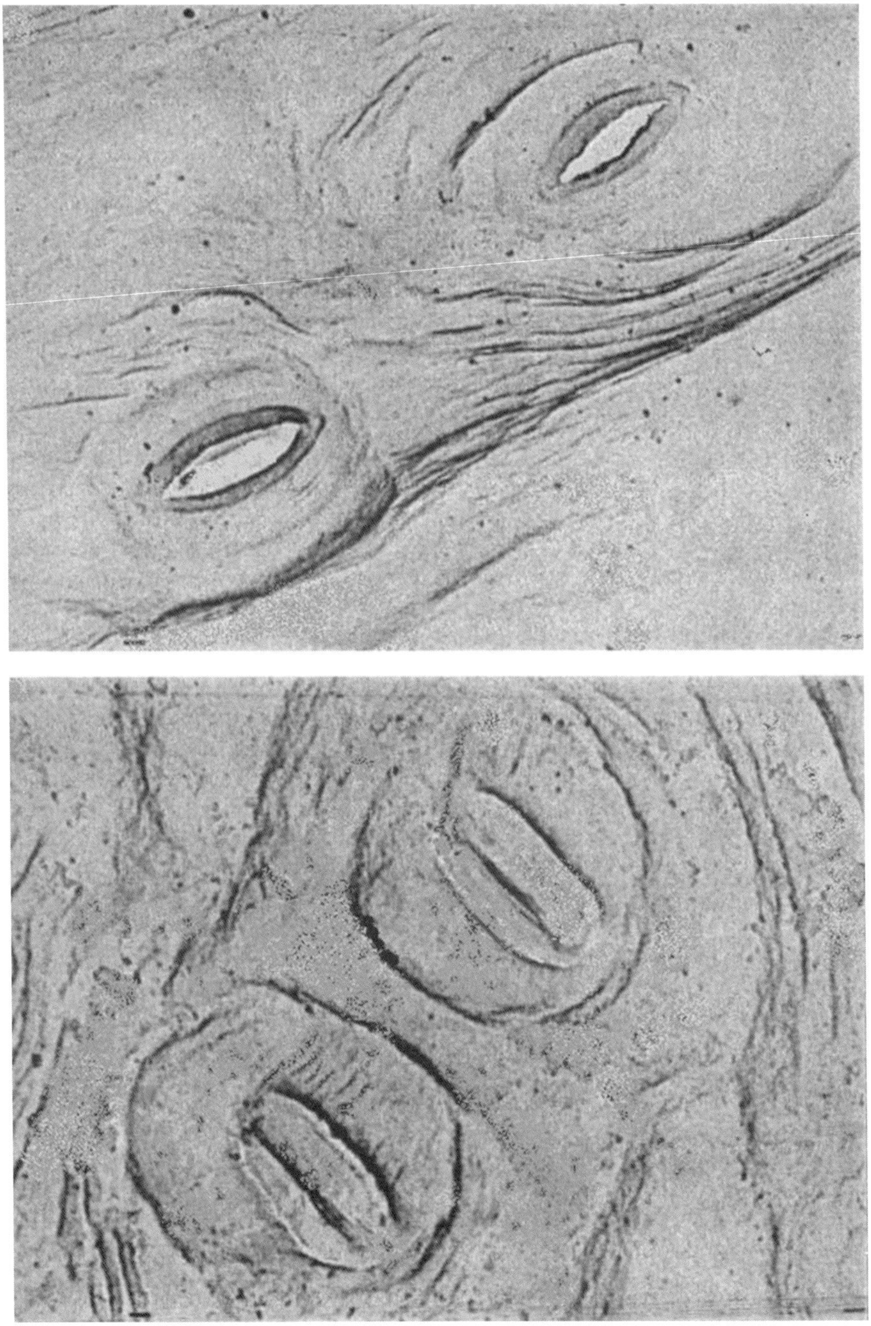

Fig. 4.13. Scanning electron micrographs of *Hibiscus* cuttings stored for 1 week at 15°C. (top) LP at a pressure of 2 kPa (15 mm Hg). (*bottom*) Atmospheric pressure. Colours have been inverted to improve clarity (Kirk *et al.*, 1986).

to CO_2-free air, which has kinetics similar to the opening response in oat segments induced by hypobaric conditions. The diffusion resistance of oat-leaf stomates decreased by 90% within 7 h at 12°C and a pressure of 1.6 kPa (12 mm Hg), and the stomates were completely open within 16–24 h at that pressure (Table 4.9). Because they were partially open after 24 h at 77 kPa (578 mm Hg = 266 µl/l [CO_2]), it was concluded that stomatal opening in LP cannot be explained by the air being CO_2-free (Fig. 4.12, left; Veierskov and Kirk, 1986), but air does not need to be free of all CO_2 to cause stomatal opening (Fig. 4.11). In leaf epidermal strips, the stomatal opening response in dim light is linearly related to the ambient [CO_2] from 360 to 0 µl/l (Willmer and Fricker, 1996), and in darkness, *Pelargonium zonale* stomates close in air, and open when the ambient [CO_2] is decreased to 200 µl/l (Longuet, referred to in Zelitch, 1969), a concentration equivalent to that in air at a pressure of 58 kPa (434 mm Hg). In adequate light, *Zea mays* leaves photosynthesize at a near-maximum rate in air containing 350 µl/l [CO_2] and open their stomates when their ICC is approximately 100 µl/l (Bowyer and Leegood, 1997). A computation based on data from Hanebuth and Raschke (1972) on the relationship between stomatal resistance, ambient [CO_2], and intercellular [CO_2] reveals that at room temperature stomates in *Z. mays* begin to open in darkness when the ambient [CO_2] is less than 60 µl/l and their ICC is 130 µl/l. At 12°C, these stomates should begin to open when the ambient atmospheric [CO_2] is somewhat higher than 112 µl/l because the respiratory production of CO_2 is 2.5-fold lower at 12°C than it is at room temperature.

While there is circumstantial evidence suggesting that LP opens stomates by reducing the ICC, this has not been convincingly demonstrated. To critically evaluate the individual roles of CO_2, O_2 and a low pressure in LP-induced stomatal opening, the response should be studied at 1/5 atm pressure, flowing either air, 100% [O_2] or 85% [O_2] + 15% [CO_2] in order to supply the fruit with a reduced or normal atmospheric [O_2] concentration at a low pressure, without CO_2, or with sufficient CO_2 added to create the 2% [ICC] typical of many fruits at atmospheric pressure (see explanation in 2.3). It is difficult to visualize stomates on fruits by means of acrylic or silicone templates (Johnson and Brun, 1966; Davenport and Burg, 2003, unpublished), and the methods commonly used to measure diffusion resistance in leaves are limited to flat surfaces and, therefore, are inappropriate for most fruits. The ethane efflux method (Cameron and Yang, 1982; Fig. 3.12, *left*) is ideally suited to evaluate quantitatively LP's effect on stomatal opening in bulky fruits. Ethane can be rapidly introduced into a fruit stored in LP by isolating the low-pressure chamber, injecting ethane into it and venting the chamber, forcing ethane into the fruit's intercellular spaces. Immediately after the fruit is removed from the chamber, an analysis of ethane efflux can be initiated. Since a Valencia orange's skin resistance to ethylene mass transport is 6885 s/cm with closed stomates (Table 3.11; Ben-Yehoshua *et al.*, 1979), and only 23 s/cm when they open, the half-time for equilibration of this fruit with ethylene – 10.9 min with closed stomates (Table 3.5) – should decrease to a few seconds when the stomates open (equation 3.13) (3.18; Moreshet and Green, 1980). The changed ethane equilibration time caused by stomatal opening should be equally dramatic.

4.16 Effect of CO_2 on Cellular pH

Horticultural commodities are exposed to a sufficiently high intercellular [CO_2] both naturally and artificially during storage to influence significantly their cellular pH (Crocker, 1948; Brown, 1985). The buffering capacity of the plant cell's cytoplasm cannot compensate for protons released when tissue is exposed to > 1% [CO_2] (Smith and Raven, 1979; Kurkdjian and Guern, 1989), but typically the ICC exceeds this value in bulky tissues (Tables 2.2, 4.5 and 4.6). During storage in CA or MA, the ICC is further elevated, while in LP essentially all CO_2 is removed from within the intercellular system. Storage behaviour should be influenced by the changes induced in cells as they regulate their cytoplasmic pH to compensate for elevated or lowered [CO_2]. These effects undoubtedly contribute to the different results obtained with LP vs. NA, MA and CA.

There are few natural buffers that are operative in the pH 6.0–8.0 physiological range other than bicarbonate (pK_a = 6.38), phosphate compounds (pK_a = 6.8–7.0) and certain amino acids (Raven and Smith, 1976; Smith and Raven, 1979; Kurkdjian and Guern, 1989). Therefore, other mechanisms must be used to stabilize the pH when the ICC varies. The vacuolar and cytoplasmic compartments have similar buffering capacity, but because the vacuolar compartment typically is tenfold larger in volume, if protons were transferred across the tonoplast from the cytoplasm to the vacuole in response to high [CO_2], the resultant change in vacuolar pH would be one-tenth as large and opposite in direction to the cytoplasmic increase. Instead, in response to perturbing treatments, usually the pH of the vacuole and cytoplasm change in the same direction (Kurkdjian and Guern, 1989).

Cultured *Acer pseudoplatanus* cells and *Riccia fluitans* rhizoid cells experience a transient cytoplasmic acidification followed by a partial recovery when they are exposed to 5% [CO_2], or when an acid such as propionate is added in the external medium (Kurkdjian *et al.*, 1978). When the acid is washed out or CO_2 removed, *A. pseudoplatanus* cells react immediately by transient cytoplasmic alkalization (Kurkdjian *et al.*, 1978; Guern *et al.*, 1986; Fig. 4.14). Acidification of the cytoplasm activates an H^+-pumping ATPase located at the plasmalemma, but this cannot account for most of the protons excreted during the recovery phase, and although the pump can remove protons from root hairs or cells in tissue culture, it cannot stabilize the pH in harvested commodities exposed to high [CO_2] because they lack free access to a large extracellular compartment (Kurkdjian and Guern, 1989). A 'pH overshoot' also occurs in acid-loaded animal cells when the acid is

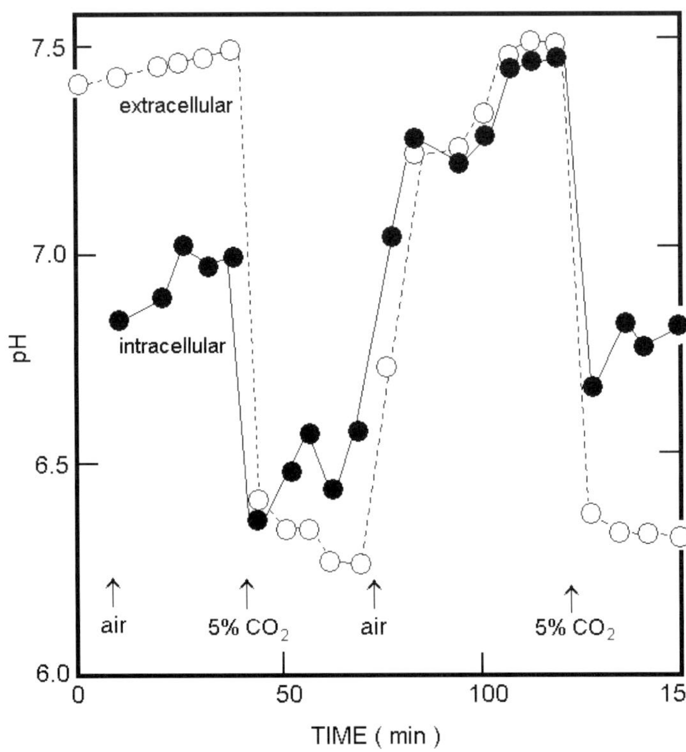

Fig. 4.14. Influence of CO_2 on the extracellular and intracellular pH of *Acer pseudoplatanus* cells cultivated in liquid medium (Kurkdjian *et al.*, 1978).

removed (Roios and Boron, 1981), and in CO_2-treated animal cells after the gas is removed (Thomas, 1974; Boron and De Weer, 1976a,b). From these similarities, Kurkdjian *et al.* (1978) concluded that plant and animal cells respond to CO_2 in the same manner. NH_4^+ has the opposite effect on the pH of cultured *A. pseudoplatanus* cells (Kurkdjian *et al.*, 1978). After 20 mM NH_4^+ is added, the intracellular pH rises from 6.5 to 6.8 and then progressively falls to 5.5 during 4 h, while the extracellular pH falls from 7.0 to 5.5 (Fig. 4.14). This influence of NH_4^+ on pH is similar to that which the base has on animal cells (Thomas, 1974; Boron and De Weer, 1976b).

Most determinations of the effect that high [CO_2] has on plant-tissue cellular pH have been made with expressed cell sap, and therefore are indicative of the vacuolar pH (3.2). When extracted juice is equilibrated with CO_2, the pH decreases as expected, and then after the gas is removed and the excess dissolved CO_2 escapes from the liquid, the pH of the sap returns to its initial value (Fife and Framptom, 1935).[15] The pH of juice extracted from CO_2-treated plants does not change on standing, and is acidified by added CO_2 in the expected manner, indicating that the excess dissolved CO_2 that was present while the tissue was exposed to high [CO_2] escaped during preparation of the cell-sap sample. Juice expressed from CO_2-treated beet leaves had a pH of 7.45; if it was saturated with CO_2, the pH decreased to 5.93; and when the dissolved CO_2 was removed, the pH returned to its original value of 7.45 (Fife and Framptom, 1935). If an intact tissue is treated *in vivo* with high [CO_2] and afterwards juice is expressed or the tissue is homogenized, dissolved CO_2 escapes from the fluid and the pH increases during preparation of the cell-sap sample. This causes the vacuolar pH within an intact tissue continuously exposed to high [CO_2] to be lower than the pH of a cell-sap sample subsequently prepared from the tissue. The pH of the sample prepared from CO_2-treated tissue would be higher than that prepared from control tissue kept in air if a CO_2 treatment decreased the cell's titratable acidity exclusive of dissolved CO_2, but this does not indicate whether the vacuolar pH was elevated or decreased while the high dissolved CO_2 concentration was present within the treated tissue. *In vivo*, high [CO_2] might decrease the pH of cells below the control value in aerated tissue, and yet when a cell-sap sample is prepared from the CO_2-treated tissue, its pH might be elevated compared to a sample derived from control tissue. In a sealed system, limiting the O_2 supply to cell-suspension cultures of *Nicotiana tabacum* increased CO_2 production and accumulation, causing a strong cytoplasmic pH decrease and a lesser decline in the vacuole. When the cultures were gassed with flowing N_2, flushing away the accumulated CO_2, the pH had already returned close to the initial value during the time required to make the first ^{31}P-NMR measurement (Wray *et al.*, 1985).

After plant tissues are exposed to 5–10% [CO_2], the alkalinity of expressed juice usually is increased above that of cell sap derived from control tissue which has not been exposed to the gas.[16] CO_2 causes the pH to be elevated in sap expressed from green bananas (Thornton, 1933b, 1946), beetroots, green lima beans, tulip bulbs, strawberries (Staadelbacher and Shaw, 1969), apricots (Do and Salunkhe, 1975), peaches and oranges (Crocker, 1948), apples (Magness and Diehl, 1924 – referred to in Fife and Framptom, 1935; Crocker, 1948); asparagus (Figs 4.10 and 4.15; Thornton, 1946; Lougheed and Franklin, 1965; Wang *et al.*, 1971), broccoli (Lebermann *et al.*, 1968), spinach (Burgheimer *et al.*, 1967), snap beans (Groeschal *et al.*, 1966; Buescher and Adams, 1983), lettuce (Siriphanich and Kader, 1986; Fig. 4.16, *left*), roses (Fig. 4.15) and other flowers (Thornton, 1934a), sugar beet leaf blades, petioles and roots (Fife and Framptom, 1935; Figs 4.15 and 4.16, *right*), potato tubers, carrot root and onions (Fig. 4.15), *Pisum sativum* (Thornton, 1946), and whole plants of *Aspargus officinalis*, *Atriplex bracyteosa*, *Beta vulgaris*, *Chenopodium murale*, *Erodium cicutarium*, *Lycopersicon esculentum*, *Matthila incana*, *Nicotiana tabacum*, *Oxalis martiana*, *Solanum nigrum* and *Stellaria media* (Fife

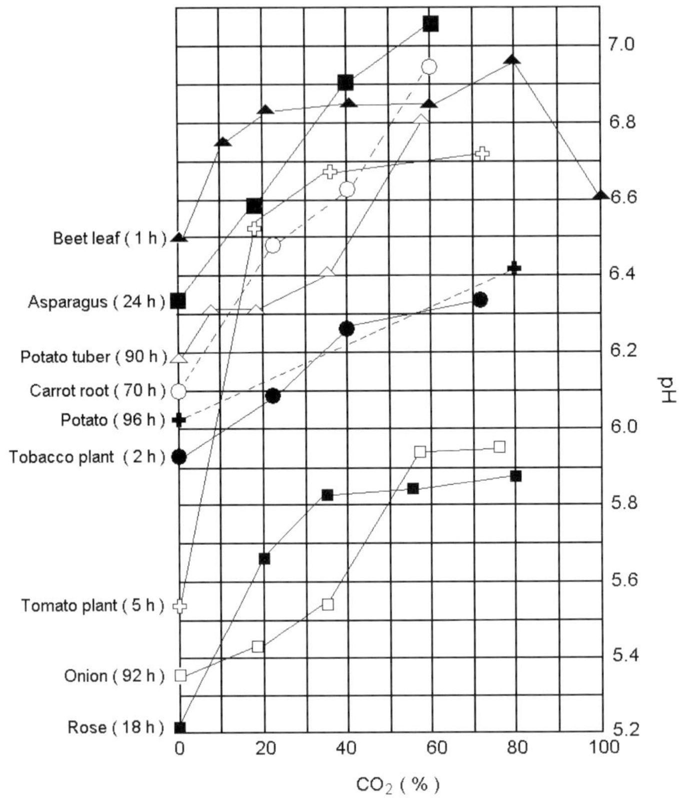

Fig. 4.15. Effect of CO_2 on the pH of various plant tissues kept at 25°C for the indicated time (from Curtis and Clark, 1950; data of Thornton (1934a) and Fife and Framptom (1935)).

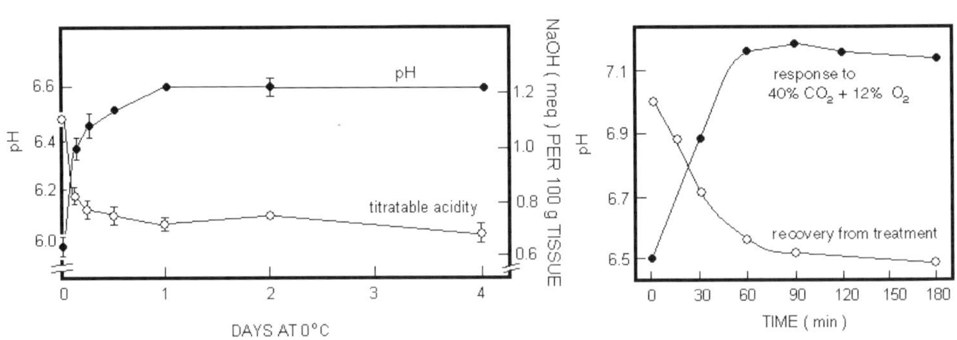

Fig. 4.16. (*left*) Titratable acidity and pH of homogenates prepared immediately after removing lettuce from treatment with an air + 15% [CO_2] mixture for the indicated duration of time at 0°C (Siriphanich and Kader, 1986). (*right*) Rate at which sugar beet leaves change pH in response to, and in recovery from, treatment with a 40% [CO_2] + 12.5% [O_2] mixture. Recovery is the rate at which treated plants return to normal when they were transferred to air after a 1-h exposure to the gas mixture (Fife and Framptom, 1935).

and Framptom, 1935). The pH of expressed sap is increased after tissue is treated with high [CO_2] and low [O_2] together (Lebermann *et al.*, 1968; Buescher and Adams, 1977), but not if a plant is subjected to elevated [CO_2] in the absence of all O_2 (Crocker, 1948), indicating that a metabolic process is involved in the pH response to

asparagus is harvested (Lea, 1997), and this amide could provide a source of ammonia to account for the rise in pH that occurs when asparagus is exposed to high [CO_2] (Figs 4.10 and 4.15).

NH_3 also is readily formed from glycine (Bowyer and Leegood, 1997). About one-third of the soluble protein in mitochondria isolated from leaves is comprised of glycine decarboxylase, an enzyme that transfers a methylene (C1) group to polyglutamine tetrahydrofolate, generating NH_3, CO_2 and NADH. Serine hydroxylmethyltransferase then transfers this C1 group to glycine to form serine. NH_3 also is produced from carbamyl phosphate (Lea, 1997) (see equation 4.16 at bottom of page).

4.20 Ammonia Production during Flower Fading and Leaf Senescence

Carbohydrates are a plant cell's first choice as an energy source, but before carbohydrates become severely depleted, proteins are metabolized and ammonia accumulates (Myer and Anderson, 1952). After leaves are excised, their CO_2 evolution and carbohydrate content rapidly decrease. Carbohydrate loss fully accounts for the CO_2 produced during the first 24 h after bean leaves are harvested, but thereafter the evolved CO_2 exceeds the equivalent of the carbohydrate lost and an increasing fraction of respiration is at the expense of non-carbohydrate substrates (Yemm, 1935). Harvesting barley or tobacco leaves induces a rapid and immediate decrease in protein content (Fig. 4.17), both in the light and darkness, even though in the light the excised leaves carry out photosynthesis and increase in carbohydrate content and dry weight. This proves that excision is responsible for a basic alteration in protein metabolism and that the hydrolysis of protein is not due solely to depletion of the leaf's carbohydrate reserves. The protein loss is accompanied by an increase in glutamine, asparagine and amino nitrogen, especially during the first 48 h (Bonner, 1950). This must be due to secondary synthesis of these amides since they are not contained in protein in appreciable amounts. After 48 h, NH_3 accumulates at the expense of amino acids and amides (Yemm, 1935; Bonner, 1950; Thimann *et al.*, 1977; Fig. 4.17).

Proteolytic activity also increases in senescing tissue, liberating NH_3 (Dalling *et al.*, 1976) and causing an enhanced emission of the gas (Farquhar *et al.*, 1979; Lemon and van Houtte, 1980; Morgan and Parton, 1989; Schjoerring *et al.*, 1993, 1998). No flux of NH_3 into and out of healthy leaves of *Z. mays* occurred when the atmospheric partial pressure of ammonia was 5 ± 3 nbar (0.0038 mm Hg), but an efflux of 0.6 nmol/m^2·s was measured from senescing *Z. mays* loaves at the same atmospheric condition (Farquhar *et al.*, 1979), and the NH_3 compensation point and rate of ammonia volatilization increased as spring wheat approached maturity (Morgan and Parton, 1989). During anthesis the apoplastic NH_4^+ in *B. napus* leaves did not change substantially, ranging from 0.47 to 1.36 mM, but the bulk tissue NH_4^+ increased from 7.54 to 11.98 mM (Husted and Schjoerring, 1996). In both plant and animal cells, continued protein metabolism results in the release of glutamate, which is oxidized by glutamate dehydrogenase (GDH) to 2-oxoglutarate

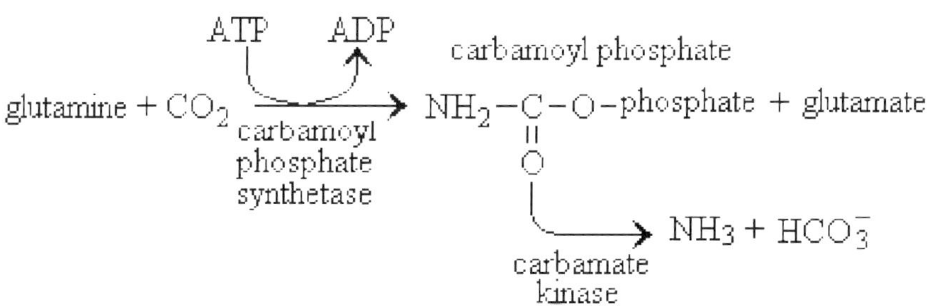

(4.16)

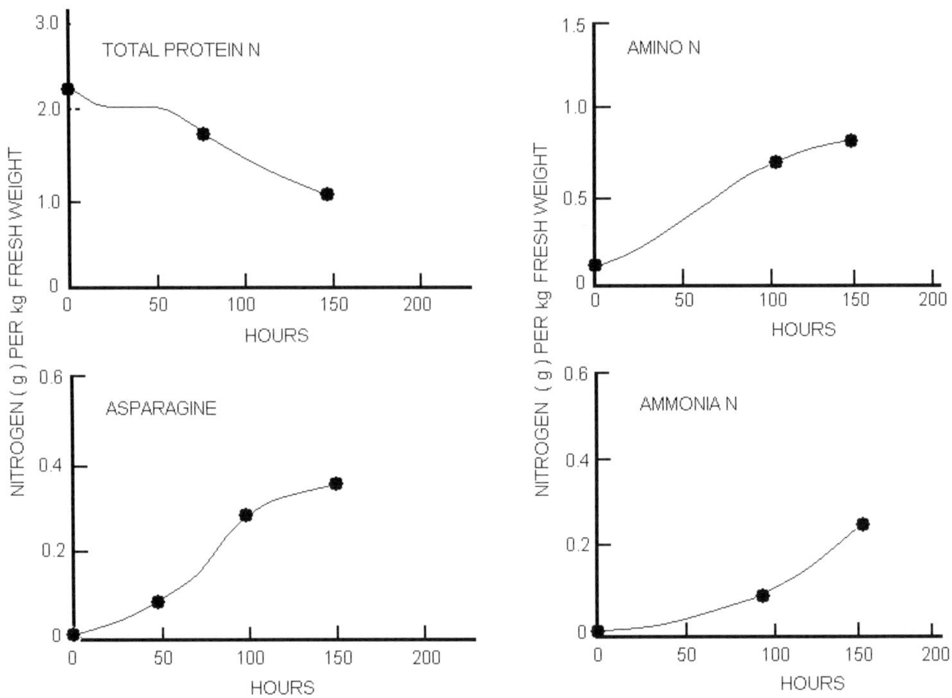

Fig. 4.17. Changes in nitrogenous constituents of excised tobacco leaves cultured in water in darkness (Vickery *et al.*, 1937, referred to in Bonner, 1950).

with the concomitant release of NH_3 (equations 4.12 and 4.13). The oxoglutarate preferentially enters the TCA cycle while the NH_3 accumulates. GDH, which is present in all higher plants and especially in senescing leaves (Srivastava and Singh, 1987), becomes de-repressed in sucrose-starved cells (Robinson *et al.*, 1992), and as senescence progresses, the activity of glutamine synthetase and glutamate synthase decline (Keys *et al.*, 1978; Simpson *et al.*, 1983), but glycine decarboxylase activity remains high and continuously releases NH_4^+ (Peoples *et al.*, 1980). During the natural senescence of tomato leaves, the expression of the glutamine synthase genes is repressed and total GS activity decreased by 96% from the value present during the period of rapid leaf growth (Pérez-Rodríguez and Valpuesta, 1996). This explains why ammonia released during senescence is not reincorporated into organic form.

Ammonia formed from amides and amino acids leads to the increase in pH that causes bluing during rose storage (Mastalerz, 1969). Proteolysis proceeds early and at a fairly constant rate during the senescence of roses, independent of the level of readily available respiratory substrates such as glucose (Weinstein, 1957). Free ammonia nitrogen increases 23-fold, blue portions of petals contain 230% as much ammonia as non-blue areas, and bluing occurs when the petal tissue's pH shifts from 4.0 to 4.5 (Asen *et al.*, 1971). The increase in ammonia coincides with the respiratory loss of various free amino acids during the later stages of senescence. Sucrose in a preservative solution exerts a sparing action on the breakdown of proteinaceous substances in carnations (Rogers, 1973), and holding roses in a solution containing 8-hydroxyquinoline and sucrose delays the pH increase and bluing. 'Forever Yours' roses, which can only be kept for 7–10 days in NA, had excellent vase life after 8 weeks LP-storage at 0°C and a pressure of 6.67 kPa (50 mm Hg), but some bluing typical of stored roses had developed by that time (Dilley, 1977a). This is not surprising because during 8 weeks at 0°C and a pressure of 6.67 kPa (50 mm Hg), a rose's respiration rate would consume 50% more sugar than flowers – such as carnations – initially contain at harvest. At a lower pressure, 1.33–2 kPa (10–15 mm Hg), bluing of Madam Delbar and Visa red roses has not been observed during commercial LP rose storages for periods as long as 6–8 weeks at 2°C. At so low a storage pressure, carbohydrate consumption is inhibited by 80–90% (Fig. 4.2), and it would take nearly four times longer to deplete the carbohydrate reserve compared to storage at a pressure of 6.67 kPa (50 mm Hg). LP also should accelerate the escape of NH_3 from flowers by opening their stomates, increasing gas diffusive transport and lowering the partial pressure of NH_3 in the storage atmosphere.

The decline in protein content, which occurs during the early stages of floral senescence (Borochov *et al.*, 1990), is accelerated when applied ethylene hastens fading, and this response is prevented by a previous treatment with 1-MCP (Serek *et al.*, 1995c). Morning glory flower-rib segments, floated on a citrate-phosphate buffer, roll up in response to applied or endogenous ethylene, and the pH of the incubation medium increases by 0.1–0.4 units during a 5-h incubation (Hansen and Kende, 1975). The flower-rib sections change colour from blue to purple, indicating that their vacuole has increased in pH, presumably due to NH_3 formation. Bluing of the corolla of petunia flowers follows a peak in ethylene production 40 h in advance of flower wilting (Whitehead *et al.*, 1982).

Since NH_3 is highly toxic, and numerous enzymatic reactions and equilibria are influenced by small shifts in pH (Kurkdjian and Guern, 1989; d'Auzac *et al.*, 1993), if LP and CA have opposite effects on cellular pH and ammonia production, that can be expected to give rise to significant differences in the storage result obtained with these systems.

4.21 Cyanide

Cyanide can be produced from the more than 60 cyanogenic glycosides and cyanolipids present in at least 2600 cyanogenic taxa of higher plants (Wink, 1997). An active β-glucosidase of broad specificity is able to hydrolyse cyanoglycosides to yield 2-hydroxynitrile, which is cleaved by a hydroxyl-nitrile lyase to the corresponding aldehyde or ketone, releasing HCN (equation 4.17; see bottom of page). After cyanolipids are hydrolysed by an esterase to yield 2-hydroxynitrile, they also can produce cyanide (equation 4.18; see bottom of page).

Cyanogenic glycosides arise in the cytoplasm and are stored in the central vacuole, preferentially in epidermal cells, while the corresponding β-glucosidase and hydroxynitrile lyase are localized in the adjacent mesophyll cells, safely away from their

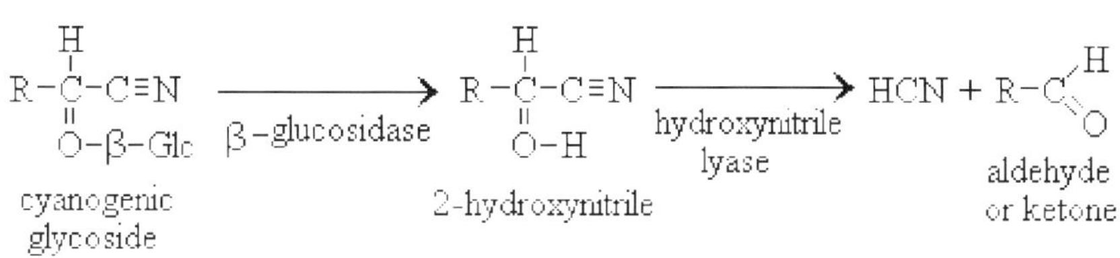

(4.17)

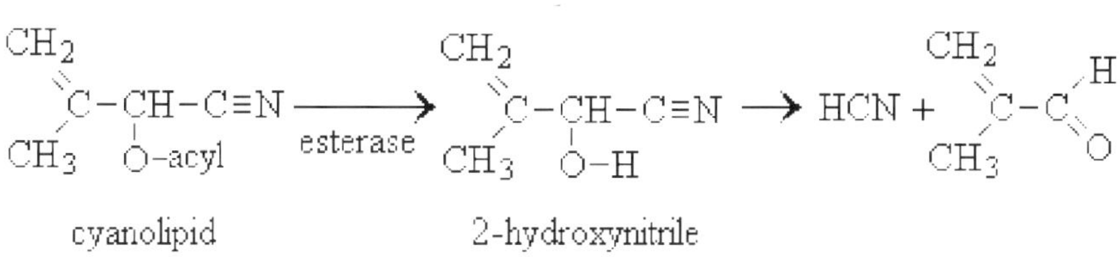

(4.18)

substrates (Wink, 1997). When epidermal tissue is damaged, the enzymes and substrates come into contact, and HCN forms. Cyanide inhibits numerous metallo enzymes, such as cytochrome *c* oxidase (K_i = 10–20 μM), Rubisco (K_i = 6 μM) and catalase (K_i = 5 μM; Theologis and Laties, 1978), and it has been suggested that cyanide may have a role in cell death and in the necrotic lesions that arise during the hypersensitive response to phytotoxic herbicides (Grossmann, 1996). To prevent autotoxicity, plants combine HCN with L-cysteine to produce 3-cyanoalanine in a reaction catalysed by β-cyanoalanine synthase, a pyridoxal-containing enzyme widely distributed in higher plant mitochondria (Akapyan *et al.*, 1975; Miller and Conn, 1980), and especially abundant in cyanogenic species. Cyanoalanine is hydrolysed by β-cyanoalanine hydrolase to L-asparagine (equation 4.19; see bottom of page).

HCN also forms spontaneously by the decarboxylation of cyanoformate, produced as a by-product of ethylene synthesis (Peiser *et al.*, 1984), and for those plant tissues that do not accumulate cyanogenic glycosides, but produce ethylene at a high rate, this can be their major source of HCN (Tucker and Laties, 1985).

Slices cut from ripening apples and auxin-treated mung-bean hypocotyl segments developed steady-state cyanide concentration too low to inhibit respiration, even though they produced ethylene and cyanide at a high rate (Table 4.11). At 25°C, the HCN concentration in apple slices producing ethylene at a rate of 168 μl/kg·h would increase by 7 μM per hour if the tissue was not capable of detoxifying HCN and no HCN gas escaped into the surrounding air (Yip and Yang, 1988). Without an adequate detoxification mechanism or gas exchange rate, the tissue would quickly become cyanogenic, since the K_i for HCN inhibition of cytochrome *c* oxidase and other metal-containing enzymes is in the range 5–20 μM. Apple slices pre-treated with 10 mM ACC evolved 18 nmol of ethylene from 2 g fresh weight of tissue during 2 h, and must have produced an equal amount of HCN (Table 4.11). During that interval less than 0.1 nmol of gaseous cyanide was evolved from the tissue, and yet at the end of the incubation the tissue's HCN concentration was only 0.2 nmol/g. This indicates that approximately 97% of the cyanide that formed must have been recycled into L-asparagine (equation 4.19) or γ-glutamyl-β-cyanoalanine. A similar result was obtained with etiolated mung-bean hypocotyl sections, but both tissues became cyanogenic when they were treated with 10 mM ACC + 5 mM 2-aminoxyacetic acid (AOA) because the pyridoxal phosphate reaction that catalyses

Table 4.11. Effect of AOA and ACC treatment on ethylene production and HCN accumulation in cv. Golden Delicious apple slices (1 cm diameter × 0.5 cm long) during a 2-h incubation ± pretreatment with AOA and ACC (Yip and Yang, 1988).

Treatment	Ethylene produced (nmol/2 g)	HCN recovered (nmol/g)	
		In gas phase	In tissue
Control	7	ND*	0.2
AOA (5 mM)	0.2	ND	0.1
ACC (10 mM)	18	ND	0.2
AOA + ACC	14	1.6	1.7

*ND = not detectable (< 0.1 nmol HCN).

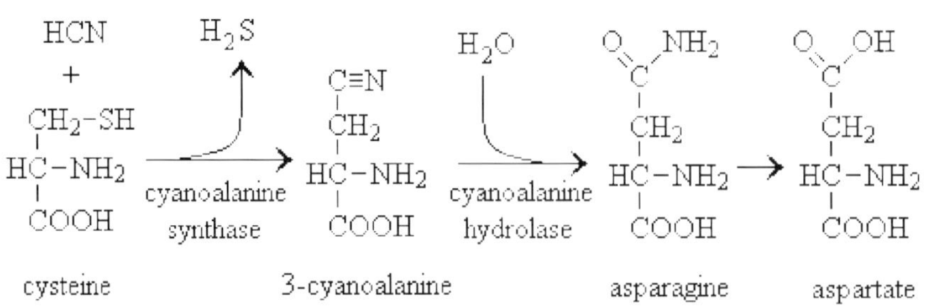

(4.19)

the formation of 3-cyanoloalanine from L-cysteine and HCN is inhibited by AOA. The conversion of SAM to ACC also is inhibited by AOA (Fig. 5.1), but ethylene and cyanide formed directly from the applied ACC, causing the HCN concentration to increase to 1.7 μM in apple slices, and to 8.1 μM in mung bean hypocotyl segments (Table 4.11).

The ease with which HCN gas escapes from apple tissue sections can be estimated from the behaviour of apple slices treated with ACC + AOA. Cyanide is infinitely soluble in water, and in plant tissues exists predominantly as dissolved gaseous HCN, since its pK_A is 9.3 and the pH of the cytoplasm, vacuoles and apoplast typically is much lower. After apple slices were treated with AOA and ACC, 7 mmol of HCN (and ethylene) were produced by 2 g of tissue in 2 h, 1.6 nmol of HCN escaped in the gas phase, 1.7 nmol remained in the tissue and 39% of the HCN was recycled. Example 7 indicates that, although the tissue had been blotted, the intercellular system still was occluded with water from cells damaged by cutting and by solution remaining from pre-incubation in an ACC + AOA solution. The example indicates that cyanide should escape from a fruit with an intact peel about as readily as it escaped from these apple-tissue sections, and that the removal of cyanide by gas exchange could become a significant factor when the diffusional resistance to cyanide is decreased up to 76-fold by LP storage at a pressure of 1.33 kPa (10 mm Hg), and even more when LP opens stomates.

Cyanide could not be detected (< 0.1 nmol/g) in pre-climacteric, climacteric and post-climacteric avocados, unripe apples or etiolated mung bean hypocotyls, and the concentration in ripe apples producing ethylene at rates as high as 190 μl/kg·h was only 0.2 μM (Yip and Yang, 1988). Although ethylene production by apple and avocado fruits increased by several hundredfold during ripening, whereas *β*-cyanolaninine synthase activity only increased one- or twofold, at all times these tissues possessed ample capacity to detoxify HCN formed during ethylene biosynthesis. The conclusion that the cyanide level in these ripening tissues does not increase sufficiently to inhibit cytochrome oxidase and engage the cyanide-insensitive system of respiration is consistent with studies indicating that the respiration in pre-climacteric and climacteric avocado fruit is mediated by the cytochrome respiratory pathway, and that there is no engagement of the cyanide-resistant pathway prior to or during ripening (Tucker and Laties, 1985).

4.22 Examples

1. In LP at 0°C, 94% RH and a pressure of 1.33 kPa (10 mm Hg), the storage atmosphere contains 4.3 mm Hg of water vapour and 5.7 mm Hg of air. The incoming air expands 755.7/5.7 = 132-fold at the pressure regulator, decreasing the incoming atmospheric CO_2 partial pressure ratiometrically. Respiration is inhibited by 90% (Fig. 4.2), diffusion is enhanced approximately 76-fold (Table 3.15), and the combination of these effects decreases the O_2 and CO_2 gradients by 760-fold. A commodity that developed $[CO_2]$ and $[O_2]$ gradients of 2% across its skin at 0°C and atmospheric pressure theoretically will contain 0.0029% CO_2 and 0.146% O_2 when it is stored in LP flowing 0°C saturated air at a pressure of 1.33 kPa (10 mm Hg). At the same temperature and a pressure of 10 kPa (75 mm Hg), expansion decreases the incoming $[CO_2]$ by 10.8-fold, respiration is 60% inhibited (Fig. 4.2) and diffusion is enhanced tenfold; a commodity developing $[CO_2]$ and $[O_2]$ gradients of 2% across its skin at 0°C and atmospheric pressure theoretically will contain 0.082% $[CO_2]$ and 1.86% $[O_2]$. These theoretical limits may not be attained because the gas mixture in the storage space will be augmented to a slight extent by respiratory CO_2 and depleted in O_2, depending on the air-change rate, density of the load, and the stored commodity's respiration rate at the prevailing pressure and temperature.

2. The CO_2 production or O_2 consumption rate in a 6.1 m (20 ft) VacuFresh^{SM} container can be computed from the expression below (see equation 4.20 at bottom of page), where % change = total % [CO_2] in the sample or the decrease in % [O_2] below 20.9%, measured with the apparatus depicted in Fig. 4.1; pump capacity = 77,482,800 cc/h @ 60 Hz; 64,569,000 cc/h @ 50 Hz; compression ratio = (tank pressure)/(atmospheric pressure); % flow = (23–100%) selected with the flow controller; cargo weight = kg; rate = cc/h per kg of plant matter. RQ = (% [CO_2])/(20.9–% [O_2]). If the RQ is close to unity, the stored commodity is considered to be 'healthy'. The required flow rate to saturate the container air is given by (see equation 4.21 at bottom of page), where O_2 consumption = cc/h per full load; (O_2 consumption)/200 = kcal/h of respiratory heat per full load; latent heat of vapourization of water = 592 kcal/kg water; saturated condition = kg/m^3 of water at the storage temperature; required flow rate = m^3/h. This calculation converts the full load's O_2 consumption rate to heat evolution, and then determines how much water that amount of heat will evaporate and the flow rate that will be saturated by the resultant moisture.

3. At 22°C, the O_2 gradient across a pre-climacteric apple's skin is approximately 5% (Table 4.5). Since the O_2 gradient is proportional to the respiration rate, when the O_2 uptake is half-inhibited in 4% [O_2] (Fig. 4.3, *left*), the internal O_2 concentration must be 4–5/2 = 1.5%. When respiration is 75% inhibited at 1.5% [O_2], the internal O_2 tension is only 1.5–5/4 = 0.25%. This is low enough to cause the RQ to rise abruptly.

4. At 4°C, the CO_2 and O_2 gradients in Cox Orange Pippin apples decreased twofold when fruit was transferred from air to a CA atmosphere containing 5% [O_2] + 5% [CO_2] (Hulme, 1975). This indicates that respiration declined by 50%. The respiration rate is linearly related to the O_2 partial pressure below 5% [O_2] (Fig. 4.3. *left*), and therefore in a 1% [O_2] atmosphere, the internal O_2 should decrease to 0.19% and the CO_2 gradient increase to 0.81%. In accord with this prediction, Starking Delicious apples stored in 1% [O_2] at 0.5°C contained 1.1–1.3% [CO_2] even though CO_2 was scrubbed from the air with NaOH (Sfakiotakis *et al.*, 1993). These data suggest that when < 1% applied [O_2] injures apples, the disorder actually is caused by the presence of < 0.19% [O_2] in the fruit's intercellular spaces.

5. At 0°C, the RQ breakpoint for Ida Red apples is 1.1% [O_2] at atmospheric pressure (Gran and Beaudry, 1993), but at 25°C it should be close to 3–4% [O_2] (Fig. 4.3, left). Diffusion through the air phases limiting gas exchange is promoted 16-fold at a storage pressure of 6.4 kPa (48 mm Hg) and a temperature of 25°C, and therefore the O_2 tension at which the 'RQ breakpoint' occurs will be lowered (Andrich *et al.*, 1998). As predicted, the RQ of Ida Red apples remains close to unity in LP (Table 4.3), even though the storage atmosphere contains only the equivalent of 0.66% [O_2] at 6.4 kPa (48 mm Hg).

6. An apple typically contains approximately 5% [CO_2] at atmospheric pressure and 25°C. Two-fifths as much CO_2 will be present at 16.0 kPa (120 mm Hg) because the fruit's CO_2 production is inhibited by 60% at the lower pressure (Table 4.3; Fig. 4.2). In combination with a 6.33-fold enhancement of gas exchange at the lower pressure, the respiratory inhibition should decrease the ICC to 0.3% and reduce ethylene production by nearly 50% (Fig. 4.6). Enhanced gas exchange and a 50% inhibition of O_2 uptake at 16.0 kPa (120 mm Hg) (Table 4.3; Fig. 4.2) lower the O_2 gradient from 5% at atmospheric pressure to 0.39% at 16 kPa, and this in turn reduces the 2.65% [O_2] content of the incoming air to 2.26% in the intercellular spaces. The resultant intercellular

$$\text{rate} = \frac{[(\%\ \text{change})(\text{pump capacity})(\text{compression ratio})(\%\ \text{flow})]}{(\text{cargo weight})} \quad (4.20)$$

$$\text{rate} = \left[\frac{O_2\ \text{consumption}}{200}\right]\left[\frac{1}{592\ (\text{satd. density})}\right] \quad (4.21)$$

O_2 tension should not cause a pronounced inhibition of ethylene production, since in thin discs cut from apples the apparent K_m for ethylene synthesis is 0.2% [O_2] (Burg, 1973a). These data suggest that the removal of CO_2 from an apple's intercellular spaces may be responsible for most of the inhibition of ethylene production that LP causes in apples at a storage pressure of 16.0 kPa (Table 4.3)

7. Henry's Law applies in a very dilute solution of cyanide:

$$K_H = C_a/p_g \quad (4.22)$$

where p_g is the partial pressure of cyanide (atm) in the gas phase, C_a is the concentration of cyanide (mol/m^3) in the aqueous solution and K_H is the Henry's law constant for cyanide [M/atm = (Mol$_{aq}$/dm$^3{}_{aq}$ · atm]. At 25°C, K_H = 9.6 M/atm (Hine and Weimar, 1965; Edwards *et al.*, 1978). During a 2-h incubation, apple slices pre-treated with AOA and ACC (Yip and Yang, 1988; Table 4.11) evolved 1.6 nmol of cyanide per gram of tissue in 2 h (5.43×10^{-9} cm^3/s), and at the end of the incubation contained 1.7 nmol of HCN per gram fresh weight. As the density of apple tissue is 7.39×10^5 g/m^3, therefore C_a = 2.3×10^{-3} mol/m^3 (2.3 μM). Accordingly, at the end of the incubation the HCN vapour pressure (p_g) in the intercellular gas phase should be 2.4×10^{-7} atm. A disc measuring 0.5 cm long × 1 cm diameter weighs 0.31 g and has a surface area of 3.14 cm^2. If the HCN tissue concentration had been constant throughout the incubation, the 'effective' resistance (equation 3.14) of the tissue was 455 s/cm referenced to its 3.14 cm^2 surface area. The surface-to-volume ratio of the apple discs was 7.98 cm^2/cm^3, whereas the *S*/*V* ratio of a 7-cm-diameter apple is approximately 0.86 cm^2/cm^3. Therefore the intercellular resistance of the apple referenced to its surface area should be 9.3-fold higher than the resistance of the apple disc referenced to its surface area. Table 3.4 indicates that the intercellular resistance from the centre to the surface of an intact apple is only approximately 100 s/cm, whereas the present calculation would yield a value of 4231 s/cm referenced to the surface area of a 7-cm-diameter apple. This indicates that, although the tissue had been blotted, the intercellular system still was extensively occluded by water from cells damaged by cutting, and with water remaining from the pre-incubation in ACC and AOA solutions. Table 4.11 indicates that cyanide should escape from a fruit with an intact peel about as readily as it was lost from the apple tissue sections used in this study.

Notes

1. Throughout a 31-day rose test, a VacuFresh^{SM} container was operated at 1.12 ± 0.07°C and a pressure of 1.48 ± 0.039 kPa (11.13 ± 0.29 mm Hg = 0.17 ± 0.008% [O_2]).
2. An exception is a report that respiratory inhibitions as large as 95–98% have been measured with a gas-tight sampling device during LP storage of tomatoes and melons (Onoda *et al.*, 1989b).
3. The equilibrium concentration of O_2 in air-saturated water is 256 μM at 25°C and 453 μM at 0°C. The K_m for plant cytochrome oxidase has been reported as 0.5 μM (James, 1953; Thimann *et al.*, 1954; Solomos, 1983; Hopkins, 1995), but this may be an overestimate due to diffusion barriers, and the actual value may be as low as 0.025–0.07 μM (Stritmatter and Ball, 1954; Ducet and Rosenberg, 1962; Burton, 1974). The affinity of polyphenol oxidase (K_m = 1.5 to 3×10^{-4} M) and ascorbic acid oxidase (K_m = 1.6×10^{-4} M) for O_2 is considerably less, so that only cytochrome oxidase contributes significantly to O_2 uptake at < 1% [O_2] (Thimann *et al.*, 1954; Mapson and Burton, 1962; Burton, 1982).
4. During the first 2 weeks of a tomato storage in 6% [CO_2] + 6% [O_2] at 12°C, the specific activity of two citric acid cycle enzymes, citrate synthetase and malate dehydrogenase, fell substantially, in conjunction with changes in organic acid metabolism (Jeffery *et al.*, 1984).
5. Except at low temperatures approaching 0°C, where the Q_{10} for respiration may approach 10 (Hardenburg *et al.*, 1986).
6. In another study with White Sims carnations, carbohydrate loss did not appear to be a significant factor in either NA or CA storage during 30 days at 2.2°C. Post-storage vase life was not significantly prolonged when the respiration rate was lowered by 45% in 0.5% flowing [O_2] (Uota and Garazsi, 1967).

7. In LP, apples are not injured at 0.3% [O_2], but sometimes they are stored at a pressure that provides the equivalent of 1% [O_2] because both in LP and CA the fruit becomes dormant more quickly when it is stored at a lower O_2 tension (3.26).
8. The accuracy of the O_2/ethylene interaction data for etiolated pea tissue has been disputed (Beyer, 1979).
9. If O_2 increases ethylene's action by a 'coupling activation', a propylene concentration much higher than 130 µl/l would speed avocado ripening in 1–2% [O_2], but not in 21% [O_2].
10. Ethylene chlorohydrin also induces a fourfold increase in potato respiration, as does ethylene bromide, hydrogen sulphide, acetaldehyde, HCl, HCN, ethyl mercaptan, ethylene, carbon monoxide and alkyl, alkylene, and alkylidene halides (Miller, 1933, 1934, 1935; Reid and Pratt, 1970, 1972; Solomos and Laties, 1974, 1975, 1976a; Chi and Frenkel, 1977).
11. Differentiation of the conidia of fungal phytopathogens into an appressorium capable of rupturing the plant cell requires host signals and contact with a hard surface (Lee and Dean, 1993; Smith, 1999). cAMP signal transduction regulates appressorium formation by *Magnaporthe grisea* (Lee and Dean, 1993; Xu and Hamer, 1996; Choi and Dean, 1997; Adachi and Hamer, 1998; Xue *et al.*, 2002), *Colletotrichum gloeosporioides* (Y.-K. Kim *et al.*, 2000), *Cochliobolus heterostrophus* (Lev *et al.*, 1999), and barley powdery mildew (Kinane *et al.*, 2000), and treatments that interfere with this signal transduction system reduce growth, conidiation and conidial germination. cAMP-dependent protein kinase A is required for the accumulation of sufficient glycerol during apopresoria formation to provide the turgor needed to breach the plant cuticle (Thines *et al.*, 2000). Also, cAMP-dependent protein kinase regulates the production of asexual spores in *Aspergillus nidulans* (Shimizu and Keller, 2001) and growth of *A. niger* (Oliver *et al.*, 2002). Signalling through adenylic cyclase is essential for morphogenetic switching from a budding yeast form to a polarized hyphal form, and for growth and virulence in *Saccharomyces cerevisiae* (Lengeler *et al.* 2000) and *Candida albicans* (Rocha *et al.*, 2001). Related signalling cascades play an analogous role in regulating mating and virulence in the plant fungal pathogen *Ustilago maydis* (Lengeler *et al.* 2000).
12. At pH = 7.0, an 18 mM bicarbonate concentration would arise in a solution equilibrated with an air phase containing 6.7% [CO_2] (Table 3.2).
13. The ^{14}C-ethylene was released from mercuric perchlorate. Release of ethylene from perchlorate complexes leads to the formation of other substances, some of which are metabolized by, or incorporated into, higher plants (Hall, 1991).
14. Asparagus was tested at 10–24°C using [O_2] levels between 1.2 and 10.2%, in some instances with 5.4–18.6% [CO_2] added. All combinations except the treatment using 10% [O_2] + 10% [CO_2] at 24°C reduced ascorbic acid loss compared to the air control. However, 2–3 days' CA storage, with increased [CO_2] at temperatures of 20–23°C, was less effective in reducing ascorbic acid loss than low [O_2] alone (Thornton, 1946). Added CO_2 helped retard ascorbic acid loss in spears stored at 10°C for 7 days. In other studies (Carolus *et al.*, 1953), asparagus packaged in low-permeability films developed symptoms of low [O_2] damage but had a higher ascorbic acid content than spears packed in high-permeability films or stored in air. However, when okra was stored at 1°C for up to 10 days in air or in 3% [O_2] + 3, 10 or 20% [CO_2], or at 12°C for up to 20 days in air and also in 3% [O_2] + 3% [CO_2], no CA combination provided any benefit in ascorbic acid retention, and at 12°C CA storage resulted in lowered ascorbic acid retention (Ogata *et al.*, 1975). At 1°C, CA storage in 4% [O_2] + 9.2% [CO_2] slightly increased the ascorbic acid loss in spinach, but at 7°C it reduced the loss by 50% (McGill *et al.*, 1966; Burgheimer *et al.*, 1967). During up to 3 weeks' storage at 5°C in air and various CA combinations containing 8.6–12.5% [O_2] + 8.9–42% [CO_2], the ascorbic acid content was improved by CA, especially by 11% [O_2] + 40% [CO_2] (Murata and Ueda, 1967). At 1°C if more than 2% [CO_2] is present, Empire apples develop a chilling-related scald-like injury during CA storage at 1.5% [O_2]. The injury is associated with a large reduction in ascorbic acid in the scalded peel tissue (Burmeister and Dilley, 1995).
15. The pH of juice expressed from normal beet leaves had a pH of 6.55. After saturation with CO_2, the pH decreased to 5.75, and when the gas was removed, the pH returned to 6.55 (Fife and Framptom, 1935).
16. Mango fruit may be an exception (Lakshminarayana and Subramanyan, 1970).
17. Treatments with < 10% [CO_2] caused no major changes in the acid content of *kalanchoe* leaves during 12 h, carrot roots in 3 days and oat coleoptiles in one day; but 20–90% [CO_2] caused succinate and aminobutyrate to accumulate, while malate, aspartate and alanine were depleted (Ransom, 1953). CA storage has been reported to improve organic and amino acid retention, and in particular malate retention in apples and pears, and it inhibits the decrease in titratable acidity that occurs in air (Ulrich, 1970; Do and Salunkhe, 1975; Wang, 1990), but in other studies with pears, high CO_2 caused malate to decrease and citrate to

increase (Williams and Patterson, 1964); and in apples malate, aspartate, glutamate and alanine became depleted (Ransom *et al.*, 1957). CO_2 caused succinate to accumulate, and malic and citric acids to decrease in apricots and peaches (Do and Salunkhe, 1975), and sweet cherries stored in 10.5% [CO_2] had less tyrosine and more α-amino butyrate and malic acid than those kept in air (Singh *et al.*, 1970). The pH increase caused by high CO_2 was attributed to a loss of malic acid in snap beans (Buescher and Adams, 1977).

18. High-nitrogen plants emit larger amounts of NH_3 and possess higher tissue and xylem sap ammonium concentrations (Parton *et al.*, 1988; Mattsson and Schjoerring, 1996; Schjoerring *et al.*, 1998).

19. Fungal infection has been linked to an increase in phenylalanine ammonium-lyase activity (Habereder *et al.*, 1989; Hahlbrock and Scheel, 1989; Walter, 1989). This may explain why the apoplast of rust-infected barley plants increases in pH (Tetlow and Farrar, 1993).

20. The absorption coefficients of NH_3 and CO_2 in water at 0°C are 1299 and 1.713, respectively (Table 15.2), where absorption coefficient = (moles/litre in solution)/(moles/litre in gas phase).

21. In vesicle and mitochondrial preparations from lemons, NADP-dependent isocitrate dehydrogenase catalyses the carboxylation of 2-oxoglutarate to yield isocitrate (Ulrich, 1970).

5

Ethylene

Ethylene-induced ripening, flower fading, senescence, chlorophyll loss, abscission, physiological disorders, epinasty, and various tropistic and tortional responses contribute to the deterioration of horticultural commodities during transport and storage. Sometimes the ethylene arises naturally during a developmental process, but it also may originate as an atmospheric contaminant from motor exhaust and other industrial sources, or in response to mechanical damage, microbial infection, chilling injury, desiccation or reorientation in the gravitational field. The atmospheric content of ethylene varies from 1–5 nl/l in unpolluted rural areas to 30–700 nl/l in major cities (Abeles, 1973); 17–35 nl/l arises in supermarket stores, 60 nl/l at wholesale markets and distribution centres (Wills *et al.*, 2000); levels as high as 13 μl/l have been detected at flower wholesale and distribution centres, and 0.1–2.0 μl/l in greenhouse packing facilities (Skog *et al.*, 2001). LP protects commodities from atmospheric ethylene by decreasing the concentration within the vacuum storage chamber in proportion to the expansion that occurs as air enters through the pressure regulator (chapter 4, example 1). In addition, a hypobaric pressure decreases a horticultural commodity's IEC by opening stomates (4.15) and enhancing diffusive gas exchange through air phases (3.24), and also by lowering the ICC and internal [O_2] sufficiently to inhibit ethylene production (4.4 and 4.10).

5.1 Biosynthetic Pathway

Three key enzymes participate in ethylene biosynthesis (Fig. 5.1): methionine adenosyl transferase produces *S*-adenosyl methionine (SAM) from methionine and ATP, *S*-adenosylmethionine methylthioadenosine lyase (ACC synthase = ACS) converts SAM to amino-cyclopropane-1-carboxylic acid (ACC) and ethylene-forming enzyme (EFE = ACC oxidase = ACO) oxidizes ACC to ethylene, releasing cyanide. The 'Yang cycle' (Fig. 5.2) recycles 5′-methylthioadenosine (MTA) to methionine, allowing high rates of ethylene production to be sustained in the presence of a low intracellular methionine concentration. The overall result is that the ribose moiety of ATP gives rise to the 4-carbon skeleton of methionine from which ethylene is derived (Yang and Hoffman, 1984).

ACC synthase (ACS) is specific for SAM and utilizes pyridoxal-5-phosphate as a cofactor. Most enzymes that are activated by pyridoxal phosphate have a lysine residue in their active site, and Lys-278 of a tomato ACC synthase isoenzyme, which is conserved in all known ACC synthases, is the site of the pyridoxal-phosphate attachment (Yip *et al.*, 1990). Like other pyridoxal-5-phosphate-linked enzymes, ACS is inhibited by the vinylglycine analogues aminovinylglycine [AVG = NH_2–CH_2–CH_2–O–CH:CH–CH(NH_2)–COOH] and

Exposing intact *Ixora coccinea* plants or petiole explants to chilling (3 days at 3, 7 or 9°C) induced a significant accumulation of ACC in the abscission zone and an increase in ethylene production during the initial 4–5 h after transfer to 20°C, which resulted in 20–80% abscission of mature, non-senescent leaves beginning after 2–3 days (Michaeli *et al.*, 1999, 2000). Application of AVG or 1-MCP prior to chilling inhibited chilling-induced abscission, proving that endogenous ethylene was essential for the process to proceed. Exposing intact plants to either ACC or 0.5–10 μl/l exogenous ethylene for 1–3 days only enhanced abscission if the plants had been pre-exposed to a chilling temperature, suggesting that a low temperature increased the sensitivity of the abscission zone to ethylene. Chilling may also enhance the ethylene sensitivity of Anjou pears (Gerasopoulos and Richardson, 1996).

5.3 Cellular Localization of EFE

When climacteric McIntosh apple-tissue plugs or slices are soaked in water, they leak solutes, decrease in fresh weight and rapidly lose their ability to produce ethylene (Figs 3.6 and 5.4). These effects progressively diminish as the soaking solution's tonicity is increased (Fig. 5.4, *upper right*), until at a 0.5 N soaking-solution concentration, the cells take up water and increase in fresh weight, leakage ceases and ethylene production is not inhibited. At a higher tonicity, the cells behave as typical semi-permeable osmometers, losing water (Fig. 3.6, *upper left*), but ethylene production is not inhibited (Fig. 5.4, *upper right*), and solute leakage does not occur (Burg *et al.*, 1964) unless the temperature is increased to > 35°C (Fig. 5.3). These correlations suggested that ethylene production might be associated with an osmotically active body or a membrane (Burg and Burg, 1960; Mattoo and Lieberman, 1977; Apelbaum *et al.*, 1981), for leakage is caused by membrane changes, and apple cells leak and lose their ability to produce ethylene when osmotic swelling stretches their membranes (Burg and Burg, 1960; Burg *et al.*, 1964). For a significant period of time after water soaking has induced solute leakage and inhibited ethylene production, if tissue is transferred to a high-tonicity glycerol solution, leakage stops (Fig. 5.4, *lower left*) and even a 5-min glycerol soak causes a complete recovery of ethylene production (Fig. 5.4, *lower right*). Arrhenius plots of the logarithm of the ethylene production rate vs. the reciprocal of the absolute temperature show discontinuities in both apple and tomato fruit tissue (Mattoo *et al.*, 1977) that typically are correlated with a physical change of membrane lipid from liquid crystalline to a gel state, and are characteristic of membrane-bound enzymes and transport proteins.

Many experiments support the view that ethylene production requires membrane integrity (Burg and Burg, 1960; Burg, 1962a; Odawara *et al.*, 1977; Anderson *et al.*, 1979; Lieberman, 1979; Yang and Hoffman, 1984; Bouzayen *et al.*, 1990). Apple slices gradually lose their ability to synthesize ethylene if they are incubated with a mixture of enzymes that digest cell walls (Mattoo and Lieberman, 1977), and when the resultant protoplasts are grown in suspension culture, as they deposit a new cell wall their ability to synthesize ethylene is regained, suggesting that EFE is localized in a cell wall-membrane complex. After improved methodology allowed protoplasts to be prepared that formed ethylene from methionine, synthesis of the gas found to be inhibited by the non-ionic detergent Triton X-100, and by lysis caused by osmotic shock (Anderson *et al.*, 1979). *Vicia faba* protoplasts (Mayne and Kende, 1986) and intact vacuoles isolated from *Petunia* (Erdmann *et al.*, 1989) produce ethylene, and the vacuole fraction isolated from pea leaf protoplasts accounted for 80% of that tissue's ethylene production (Guy and Kende, 1984). The substrate specificity for ethylene production, sensitivity to inhibitors and requirement for membrane integrity is the same for these vacuoles and intact tissues, and ethylene production ceases if the vacuoles are lysed (Mattoo and White, 1991). The ability to form ethylene

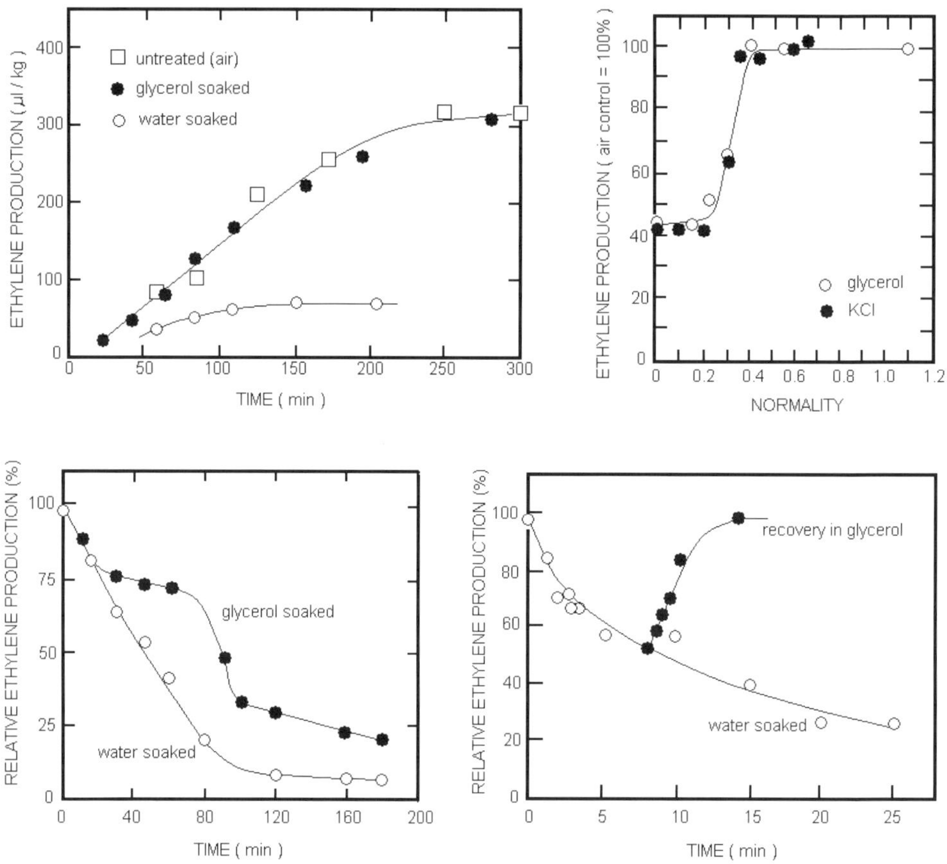

Fig. 5.4. (*upper left*) Effect of a 60-min soak in water or 0.55 M glycerol on the subsequent ethylene production by 4 cm long × 1 cm diameter McIntosh apple tissue plugs, compared to unsoaked tissue plugs (Burg and Burg, 1960). (*upper right*) Relative effectiveness of various concentrations of glycerol and KCl in maintaining ethylene production by 4 cm long × 1 cm diameter McIntosh apple tissue plugs. The solutions were soaked into the tissue for 60 min and ethylene production determined during 90 subsequent minutes (Burg and Burg, 1960). (*lower left*) Ability of glycerol to reinstate ethylene production after it has been inhibited by water soaking. 'Time' is the total duration of the initial water soak before the 4 cm long × 1 cm diameter tissue plugs were removed from water and either their ethylene production was measured during 90 subsequent minutes, or they were soaked in 1 M glycerol for 30 min before ethylene production was determined (Burg and Burg, 1960). (*lower right*) Inhibition of ethylene production caused by incubating 1 cm diameter × 1 mm thick McIntosh apple tissue slices in water, and the kinetics of recovery after transfer to 1 M glycerol. After 8 min in water, the discs were transferred to 1 M glycerol for the indicated time intervals before ethylene production rate was determined (Burg *et al.*, 1964).

from ACC is maximal in microsomal membranes and lacking in cytosol fractions isolated from flowers at the peak of their climacteric (Mayak *et al.*, 1982).

Plant cells contain two sites at which ACC can be converted to ethylene, an extracellular site directly accessible to ACC from the apoplasm, which is suppressed by plasmolysis[2] and digestion of the cell wall, and an intracellular site which is not affected by these treatments (Bouzayen *et al.*, 1990). Ethylene production, but not respiration, is inhibited within 30 min after the protein-binding reagent 2,4,6-trinitrobenzenesulphonic acid (TNBS) is applied to fruit discs, apparently without the substance penetrating the cells (Arteca, 1981). TNBS caused a 68% inhibition of

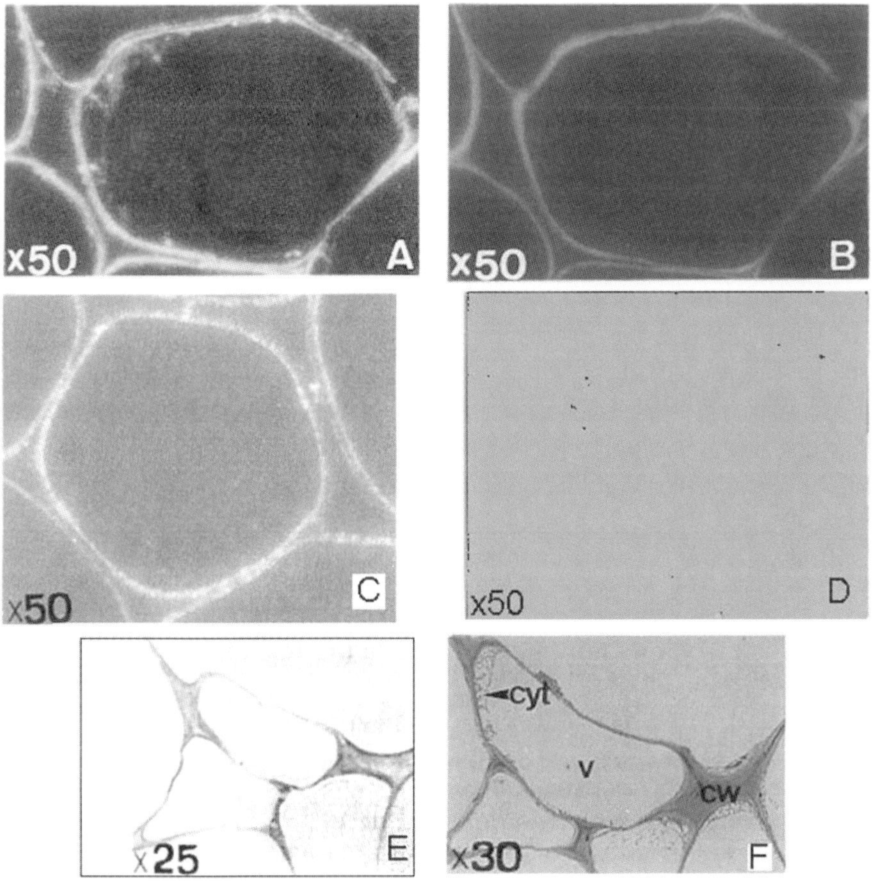

Fig. 5.5. Immunocytolocalization of ACC oxidase in tomato fruit. Labelling with: A: antisynthetic peptide antibodies (PAb) + fluoresceine isothiocyanate (FITC); B: calcofluor white, same cells as in A; C: anti-recombinant EFE antibodies (Rab) + immunogold (IG) of normal tomatoes (epipolarized light); D: Rab + IG of antisense tomatoes (epipolarized light); E: Rab + IG of normal tomatoes; F: toluidine blue, same cells as in E (Rombaldi *et al.*, 1993).

ethylene production by climacteric apple fruit disks incubated in an isotonic solution, 0.6 M sorbitol. An equivalent inhibition resulted from plasmolysis in 2.5 M sorbitol in the presence or absence of TNBS (Latché *et al.*, 1993), and that portion of ethylene production, which remained unaffected by these treatments in discs excised at an advanced stage of ripening, roughly corresponded to the pre-climacteric level of ethylene production. Plasmolysis and TNBS had little or no inhibitory effect on the pre-climacteric ethylene production by melon and apple fruit discs, and ACC oxidase was not released from these discs into a 0.6 M sorbitol bathing medium, but both plasmolysis and TNBS strongly inhibited ethylene production in discs prepared from ripening fruits, and ACO was released from them into the bathing medium. Immunofluorescence labelling corroborated the conclusion that in ripening tomato and apple fruits the ACC oxidase isozyme responsible for autocatalytic ethylene production is mainly localized in the apoplasm or at the cell wall (Latché *et al.*, 1993; Rombaldi *et al.*, 1993; John *et al.*, 2000; Fig. 5.5), with a barely visible secondary site in the cytoplasm. Light microscopy immunocytolocalization has revealed that a major site of ACC oxidase protein in grape (*Vitis vinifera* L. cv. Gamay), a non-climacteric fruit, is localized in the cytoplasm, with a minor site at the cell wall, and that grape cells grown *in vitro* generate ethylene mostly from intracellular ACC oxidase (Ayub *et al.*, 1993). Avocado (Christoffersen *et al.*, 1993) and tomato fruits (Bouzayen *et al.*, 1990) also produce intracellular ethylene.

There also is evidence that EFE is not membrane-bound. The predicted structure of any one of the three polypeptides encoded by three *pTOM13*-homologous genes that encode the complete tomato EFE suggest that the enzyme is located in the cytosol (Hamilton *et al.*, 1993). EFE proteins (EFE1, EFE2, EFE3) encoded by the *ETH* multi-gene family in tomato plants show a strong homology in amino-acid sequence; the three genes are expressed at the RNA level, and hence are functional. Sequence analysis indicates that none of the three proteins have an obvious transit peptide or hydrophobic membrane-spanning domain, but they may have an amphipatic α-helix that would be long enough to span a membrane lipid bilayer (Bouzayen *et al.*, 1993).

5.4 Effect of Ethylene on Respiration

Applied ethylene promotes the respiration of climacteric fruits (Theologis and Laties, 1978; Fig. 5.6, *right*; Fig. 5.8, *left*), potatoes (Burton, 1952; Reid and Pratt, 1970; Theologis and Laties, 1982a; Fig. 5.6, *left*), cocklebur seeds (Esashi *et al.*, 1987), leaves (Sisler and Pian, 1973), flowers (Yip and Hew, 1989) and roots (Solomos, 1983). Within 1 h after 100 μl/l of ethylene was applied to Kent mangoes, their respiration rose substantially; during the next 90 min it fell back to slightly below the initial rate, and then increased in a climacteric pattern (Fig. 5.6, *right*). Even a 15-min treatment with 100 μl/l of ethylene was sufficient to induce a self-perpetuating rise in this

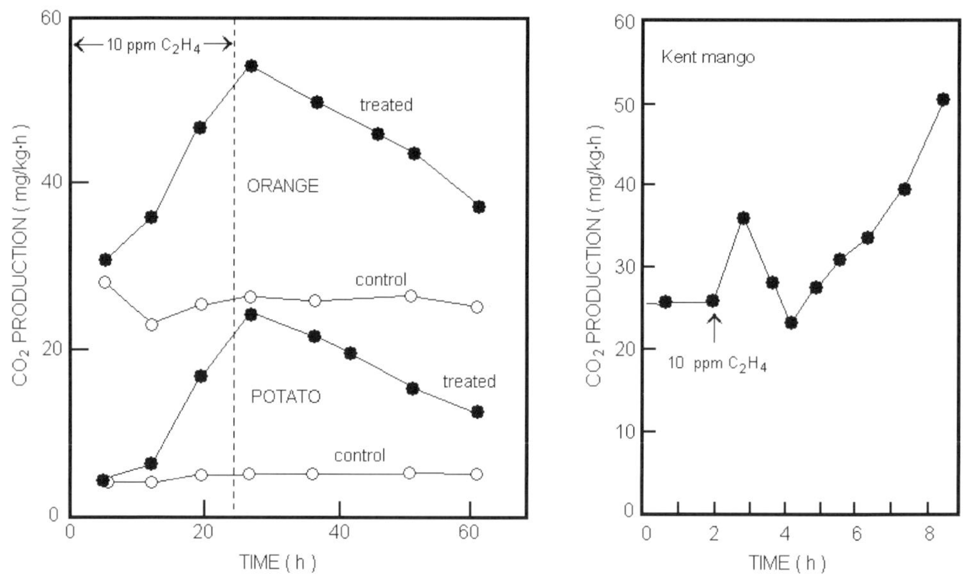

Fig. 5.6. (*left*) Stimulation of respiration by 10 ppm ethylene added during 24 h to a flowing stream of air at 20°C. Mature-green sweet orange (*Citrus sinensis*) and mature potato tuber (*Solanum tuberosum* cv. White Rose) (Reid and Pratt, 1970). (*right*) Stimulation of Kent mango respiration by 10 ppm ethylene applied 3 days after harvest. Respiration was promoted to the same extent during 6 h regardless of whether the ethylene was applied for only the initial 15 min or continuously. With both treatments, the fruit softened and coloured within 24 h. Five additional days were required for control mangoes to reach the same respiration rate and stage of ripening (Burg and Burg, 1962b).

mango's CO_2 production that persisted after the applied ethylene was removed. Application of 2 µl/l ethylene induced the beginning of a respiratory climacteric in 'Dwarf Cavendish' bananas within 3–5 h (Fig. 5.13, *right*), and in a pattern resembling that observed with Kent mangoes, the respiration of Gros Michel and Cavendish bananas doubled within an hour after 1000 µl/l ethylene was applied, then fell back below the initial rate during an additional hour (Regeimbal *et al.*, 1927), before the climacteric respiratory rise set in.

Ethylene accelerates the respiration of citrus fruits (Denny, 1924a,b; Biale, 1960; Reid and Pratt, 1970; Rakitin, 1988; Fig. 5.6, *left*; Fig. 5.9), and possibly all other non-climacteric fruits with the exception of cherries (Li *et al.*, 1994). The respiration of mature Washington navel oranges, Valencia oranges and lemons is stimulated by up to 300% within 3 h during a continuous exposure to 100 µl/l ethylene in a flow-through system, and even 100 nl/l gives a slight effect (Fig. 5.9; Denny, 1924b; Chase and Church, 1927; Biale, 1960; Maier *et al.*, 1973).

The mechanism by which ethylene induces a respiratory rise has been extensively studied, but remains uncertain. HCN induces a cyanide-insensitive respiratory 'climacteric' in avocado fruits resembling that evoked by ethylene, while simultaneously inhibiting cytochrome oxidase and stimulating ripening (Solomos and Laties, 1976a; Tucker and Laties, 1984). Cyanide has several of the chemical characteristics required of an ethylene analogue (Sisler, 1977), and 2000 µl/l HCN inhibits *in vitro* ^{14}C-ethylene binding by 85% (Sisler, 1982b), but initiation of ripening by cyanide binding to the ethylene-receptor site cannot be proved for avocado fruit because cyanide evokes ethylene synthesis, and therefore may elicit ripening by initiating ethylene production (Tucker and Laties, 1984). The amount of cyanide produced as a by-product of ethylene production is not sufficient to cause such an effect (Yip and Yang, 1988; 4.21), and inhibitor studies (Theologis and Laties, 1978) and measurements of the K_{m,O_2} for respiratory O_2 uptake by ethylene- or cyanide-treated fruits indicate that in the absence of applied cyanide, the cyanide-insensitive pathway is not normally engaged in ageing sweet potato slices, or during the respiratory climacteric of senescing ivy leaves and ripening banana and avocado slices (Theologis and Laties, 1978; Solomos, 1983; Tucker and Laties, 1984; Seymour and Tucker, 1993). Ethylene induces cyanide-resistant respiration during lemon ripening (Rakitin, 1988), and it has been suggested that the cyanide insensitive pathway also may participate in the respiratory climacteric of apples (Duque and Arrabara, 1999).

Electron flow through the cyanide-insensitive pathway does not produce an electrochemical gradient and no ATP is generated. The potential energy is lost as heat (Solomos, 1983; Brownleader *et al.*, 1997) and since glycolysis then operates at an increased rate, a large amount of heat is produced. As the alternative pathway provides little usable energy and an abundance of heat to promote water loss (6.1), it is incompatible with a long storage life.

5.5 Climacteric and Non-climacteric Fruits

The respiratory activity of all fruits, expressed on a fresh-weight, dry-weight or unit-protein basis, decreases after fruit set and throughout the stages of cell division, enlargement and maturation, and then when ripening commences, the respiratory rate abruptly rises in fruits of the 'climacteric' type[3] in association with changes in texture, aroma, flavour and composition (Kidd and West, 1925; Fig. 5.6, *right*; Figs 5.7 and 5.8). In contrast, the respiration of commercially mature oranges, grapefruits and lemons steadily declines after harvest (Fig. 5.9, control; Biale, 1954; Bain, 1958; Trout *et al.*, 1960) and no rapid chemical or physical changes occur as these fruits continue ripening (Ting and Attaway, 1971;

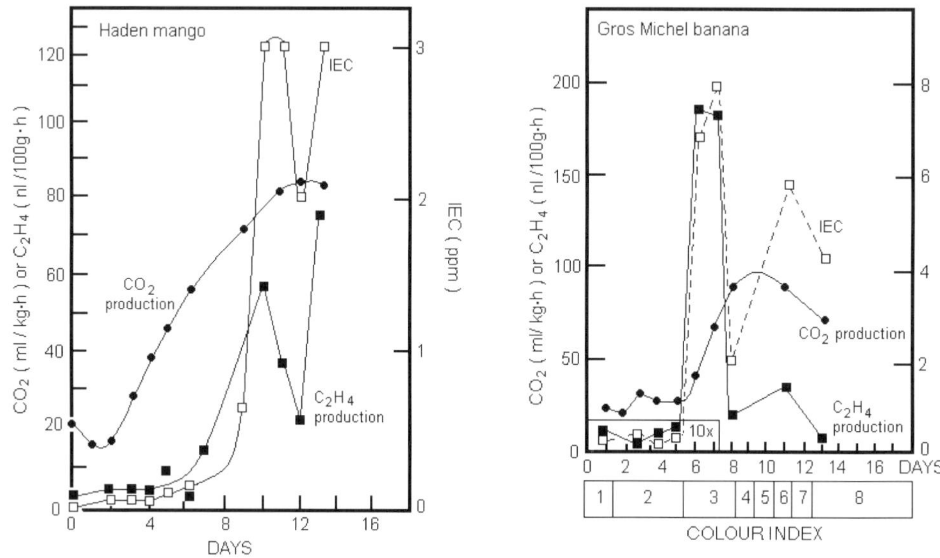

Fig. 5.7. CO_2 production, ethylene production and internal ethylene concentration during the ripening of harvested climacteric fruits at 24°C (Burg and Burg, 1962b). (*left*) Haden mangoes. Kent mangoes behaved in the same manner, reaching a maximum climacteric IEC of 2.7 ppm. (*right*) Gros Michel bananas. Values in the 'box' at the lower left are multiplied by 10. The banana colour index is: 1 = green; 2 = trace yellow; 3 = half yellow/half green; 4 = trace green; 5 = yellow with green tips; 6 = all yellow; 7 = speckled with brown; 8 = overripe.

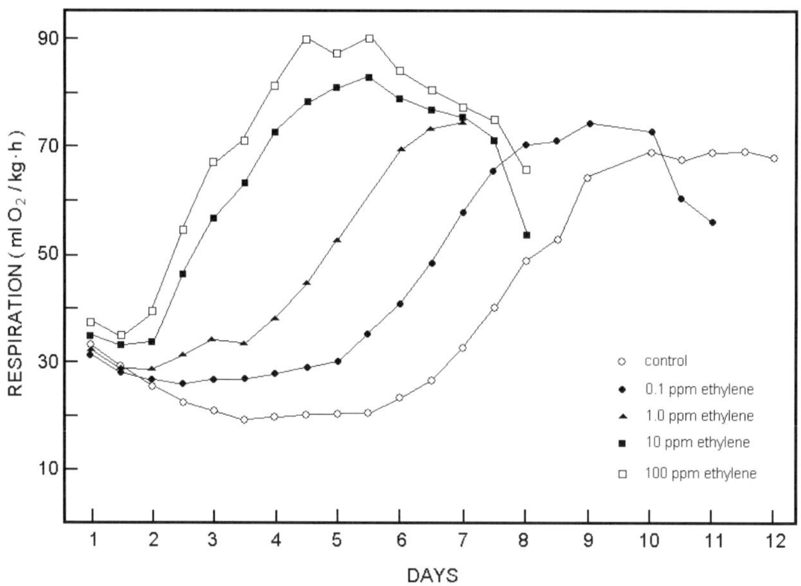

Fig. 5.8. Effect of a continuous exposure to various ethylene concentrations on O_2 uptake by avocado, a climacteric fruit (Biale, 1960).

Baldwin, 1993). For this reason, Biale (1960) challenged the concept that the climacteric pattern is 'universal' and instead proposed that fruits should be classified either as climacteric or non-climacteric based on their postharvest respiratory

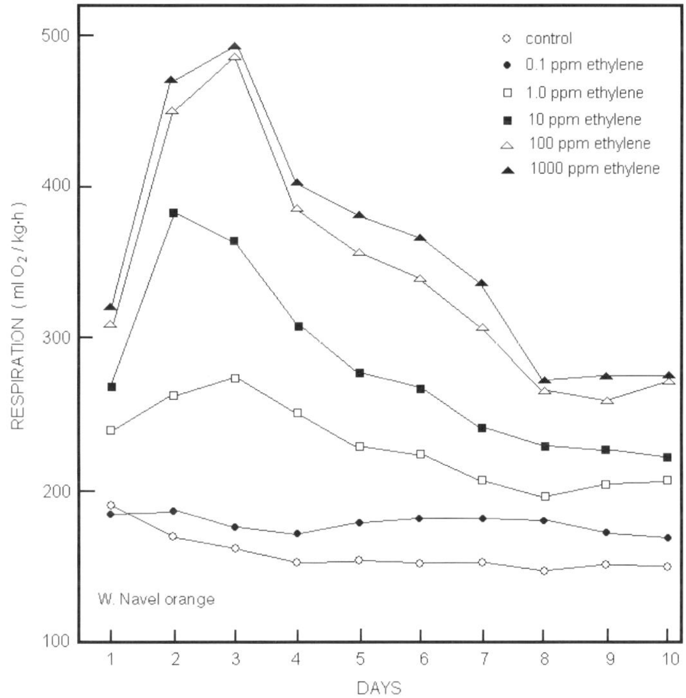

Fig. 5.9. Effect of a continuous exposure to various ethylene concentrations on the O_2 uptake of Washington navel orange, a non-climacteric fruit. A similar result was obtained with lemons and Valencia oranges (Biale, 1960).

Table 5.1. A comparison of the pre-climacteric internal ethylene concentration (IEC) and the minimum effective 'threshold' applied ethylene concentration needed to stimulate ripening in various harvested climacteric fruits.

Fruit	Pre-climacteric IEC (μl/l)	Min. effective applied concentration (μl/l)	Reference
Mango	0.05	0.04–0.40	1
Banana	0.1–0.2	0.1	1, 2, 3
Avocado	0.04	0.1	4, 5
Apple	0.05	0.2	4, 6, 7
Cantaloupe	0.05	0.2	5, 8
Tomato	0.08	0.005–0.100	9, 10, 11

(1) Burg and Burg, 1962b; (2) Palmer, 1971; (3) Burg and Burg, 1965c; (4) Burg and Burg, 1962a; (5) Biale, 1960; (6) Reid *et al.*, 1973; (7) Bufler, 1986; (8) Lyons *et al.*, 1962; (9) Morris *et al.*, 1981; (10) Wills *et al.*, 2001; (11) Lyons and Pratt, 1964.

behaviour.[3,4] Initially, it was thought that citrus fruits do not naturally display a postharvest climacteric respiratory rise because they produce no ethylene (Biale *et al.*, 1954), but this explanation became untenable when improved methods of detection (Burg and Stolwijk, 1959; Burg, 1962b) revealed that they synthesize ethylene at rates comparable to those measured during the pre-climacteric phase of climacteric fruits (Burg and Burg, 1962b, 1965a). The minimum pre-climacteric IEC of climacteric fruits typically is in the 40–100 nl/l range (Table 5.1), and 1 to several days after harvest, the IEC in mature citrus fruits is a trace to 50 nl/l in Shamouti oranges (Apelbaum *et al.*, 1976), 30–50 nl/l in Hamlin oranges (Purvis, 1981), 30–80 nl/l in *Wase* Satsuma mandarin, Yuzu and Dadai fruits (Table 5.3; Sawamura, 1981) and 20–100 nl/l in Persian limes (Noichinda, 1997; Fig. 5.28).

After climacteric fruits are harvested, their ethylene production increases, elevating the IEC prior to the onset of a respiratory upsurge. The 0.2 ± 0.15 µl/l intercellular ethylene concentration present in var. 'Dwarf Cavendish' bananas immediately after a light full three-quarters or a heavier grade is harvested, does not differ significantly from the preharvest level of 0.19 ± 0.05 µl/l (Burg and Burg, 1965c; Fig. 5.13, *left*). The IEC begins to increase 4.1 ± 2.2 h in advance of the respiratory upswing, reaches 1 µl/l before CO_2 evolution rises perceptibly and averages 0.55 ± 0.22 µl/l during the intervening period (Burg and Burg, 1962b). The minimum applied-ethylene concentration, 0.2 µl/l, which consistently accelerates ripening of this fruit, is the same as the IEC present immediately after harvest. The mean IEC in late-season Choquette avocados was 0.15 ± 0.15 µl/l measured while they were attached to the tree (Burg, 1967, unpublished). After harvest, the IEC decreased and then a surge in ethylene production caused the internal ethylene to surpass 0.1 µl/l within 9 h, at least 12–24 h prior to onset of the respiratory climacteric rise (Burg and Burg, 1962a). Before the first increase in CO_2 production could be detected, the IEC reached 0.5–1.0 µl/l, after which it rapidly increased to 300–700 µl/l. Prior to the onset of the respiratory climacteric, the IEC in cv. VC 243–20 tomatoes rose to 0.8 µl/l from a pre-climacteric minimum of < 0.08 µl/l, eventually reaching 28 µl/l (Workman and Pratt, 1957; Lyons and Pratt, 1964; Bruinsma, 1983). At the onset of the respiratory climacteric, the IEC in harvested cv. Castelmart and cv. OH7814 mature-green tomato fruits increased to 0.11 µl/l and 0.1 µl/l, respectively (Saltveit, 1993), suggesting that these low concentrations are adequate to initiate tomato ripening. A pre-climacteric minimum IEC of 20 nl/l in apples increased to 100 nl/l prior to the respiratory rise, and eventually reached 100 µl/l (Reid *et al.*, 1973; Chu, 1984; Watkins *et al.*, 1989). Cantaloupes (Lyons *et al.*, 1962) and passion fruit (Akamine, 1957) behave in this same manner. There can be no doubt that ethylene initiates the ripening of climacteric fruits, since LP and inhibitors of ethylene synthesis and action delay ripening (Saltveit *et al.*, 1978; Takata, 1981; Autio and Bramlage, 1982; Edwards *et al.*, 1983; Hobson *et al.*, 1984; Atta-Aly *et al.*, 1987; Lincoln *et al.*, 1987; Tucker and Brady, 1987; Blankenship and Sisler, 1988; Sisler and Lallu, 1994; Sisler and Serek, 1997; Fan *et al.*, 1999; Harris *et al.*, 2000; Joyce *et al.*, 2000; Kim *et al.*, 2000; Fan *et al.*, 2001; Hofman *et al.*, 2001), and before the respiration rate begins to rise in harvested banana (Fig. 5.7, *right*; Burg and Burg, 1965c), avocado (Burg and Burg, 1962a), mango (Fig. 5.7, *left*; Burg and Burg, 1962b) and cantaloupe fruits (Lyons *et al.*, 1962), their IEC surpasses a concentration which, applied to the same fruit at an earlier time, would hasten the onset of its respiratory climacteric (Table 5.1).

The ICC of attached cv. Sonato tomato fruits decreased from 7 to 4% between the 40th and 45th day after pollination and then rose in a typical climacteric pattern to 8.5% by the 52nd day, after which it steadily declined (Sawamura *et al.*, 1978), but in other studies it has been reported that CO_2 production by attached tomatoes and seven varieties of melon fruits did not appear to rise while their ethylene production increased during ripening on the vine (Miccolis and Saltveit, 1991; Saltveit, 1993). Possibly respiratory CO_2 evolution 'on-the-vine' was masked by photosynthetic CO_2 uptake (Knee, 1995) and dissolution of CO_2 in transpirational water that condensed in the sealed system used to measure tomato respiration (Andres, 1995), but this seemingly cannot explain the result with seven varieties of melons, in which respiration was measured immediately after detaching fruits.[5] In other studies, the ICC remained higher in vine-ripened tomatoes compared to harvested fruits throughout breaker, pink and red stages of maturity (Hong and Lere, 2000), and an increase in respiration associated with ripening could only be demonstrated in saskatoon (*Amelanchier alnifolia* Nutt.) fruits which remained on the vine (Rogiers and Knowles, 1999).

When cycloheximide inhibits the ripening of thick banana slices (McGlasson *et al.*, 1971) and Bartlett pear fruits (Frenkel *et al.*, 1968), a normal respiratory climacteric nevertheless develops independent of other aspects of ripening, indicating that the respiratory rise can be disassociated from the ripening syndrome. The number of fruit types considered 'non-climacteric' has steadily decreased,[4] and the importance of the 'climacteric' vs. 'non-climacteric' distinction has occasionally been questioned (Rhodes, 1970).

5.6 Systems 1 and 2 Ethylene Production

The dissimilar respiratory patterns in climacteric and non-climacteric fruits are caused by the participation of two separate ethylene-producing systems ('1' and '2') that operate with different means of ethylene regulation (McMurchie *et al.*, 1972). System 1 is responsible for the basal low ethylene production rate and for stress-induced ethylene (Oetiker and Yang, 1995); system 2 mediates the autocatalytic rise in ethylene production that in mature climacteric fruits, vegetative tissues and ethylene-sensitive flowers causes or hastens ripening, senescence, abscission and flower fading. System 1 is auto-inhibited and system 2 is auto-stimulated by applied ethylene (or propylene).[6] Exogenous ethylene (or propylene) down-regulates system 1 ethylene production in non-climacteric fruits at all stages of maturity, while simultaneously up-regulating respiration. System 1 ethylene is produced in the cytoplasm, system 2 in the apoplasm/cell-wall area (5.3).

The maximum rate of CO_2 production or O_2 consumption, which develops after a stimulatory ethylene concentration is supplied to a climacteric fruit, is independent of the applied ethylene concentration or duration of exposure since autocatalytic ethylene production is induced, creating a supra-optimal IEC (Fig. 5.8; Fig. 5.13, *right*). Because ethylene does not induce autocatalytic ethylene production in non-climacteric fruits, the magnitude of their ethylene-stimulated increase in CO_2 production or O_2 consumption depends on the applied ethylene concentration in the range between 0.1 and 10–100 μl/l (Fig. 5.9; Biale, 1964). When ethylene is continuously applied to a non-climacteric fruit, the respiration rate reaches a maximum and then slowly declines (Fig. 5.9), and after an orange's or potato's CO_2 production has increased in response to a 1-h treatment with 10 μl/l ethylene, it rapidly decreases when the ethylene application is terminated (Fig. 5.6, *left*). Several days' exposure to > 100 μl/l ethylene causes the respiration rate of immature kiwi fruit, picked nine weeks after anthesis, to rise to a plateau that is sustained only as long as the applied ethylene is present (Pratt and Reid, 1974).

5.7 Auto-inhibition of System 1 Ethylene Production

Auto-inhibition of ethylene production has been detected in climacteric fruits such as green and ripening tomato fruits (Nakatsuka *et al.*, 1998; Atta-Aly *et al.*, 2000a,b), green bananas (Vendrell and McGlasson, 1971; McMurchie *et al.*, 1972), pre-climacteric avocados (Zauberman and Fuchs, 1973) and stage C figs (Zeroni, 1972; Zeroni *et al.*, 1976); also in non-climacteric fruits such as citrus (Riov and Yang, 1982a; Zacarias *et al.*, 1993; Mullins *et al.*, 1999) and strawberry (Atta-Aly *et al.*, 2000b); in pollinated orchids (Bui and O'Neil, 1998), winter squash (Nakajima *et al.*, 1990), tobacco-leaf discs (Aharoni and Lieberman, 1979; Aharoni *et al.*, 1979a,b; Philosoph-Hadas *et al.*, 1985a), leaves of rice (Kao and Yang, 1982) and the tomato cv. 'Never ripe' (Lund *et al.*, 1998), flowers (Anderson *et al.*, 1996 – referred to in Sisler and Serek, 1997), *Arabidopsis thaliana* (Harpham *et al.*, 1996), mung bean seedlings (Kim *et al.*, 1997; Yoon *et al.*, 1997), *Rumex palustris*

(Bangga *et al.*, 1996), and transgenic petunia flowers (Wilkinson *et al.*, 1997). System 1 ethylene production is auto-inhibited (down-regulated) within 15–30 min after ethylene or propylene[6] is applied to subapical etiolated pea sections (Saltveit and Dilley, 1978a,b); immature sycamore figs (Fig. 5.10) and leaf discs and intact leaves of citrus (Riov and Yang, 1982b).

Inhibitors of ethylene action promote ethylene synthesis when the production rate is limited by auto-inhibition of system 1 ethylene production. Ag^+ stimulates ethylene formation in immature tomato fruits (Atta-Aly *et al.*, 1987, 2000a), various leaves (Aharoni and Lieberman, 1979; Riov and Yang, 1982a; Kao and Yang, 1983; Goren *et al.*, 1984; Philosoph-Hadas *et al.*, 1985b, 1994; Mullins *et al.*, 1999) and cut tuberose (*Polianthes tuberose* L.) florets (Waithaka *et al.*, 2001); norbornadiene (NBD) promotes ethylene production in Shamouti oranges (Goldschmidt *et al.*, 1993) and Hass avocado discs at the same time that it inhibits the avocado's climacteric ethylene biosynthesis (Starrett and Laties, 1991a); a 1/5 atm

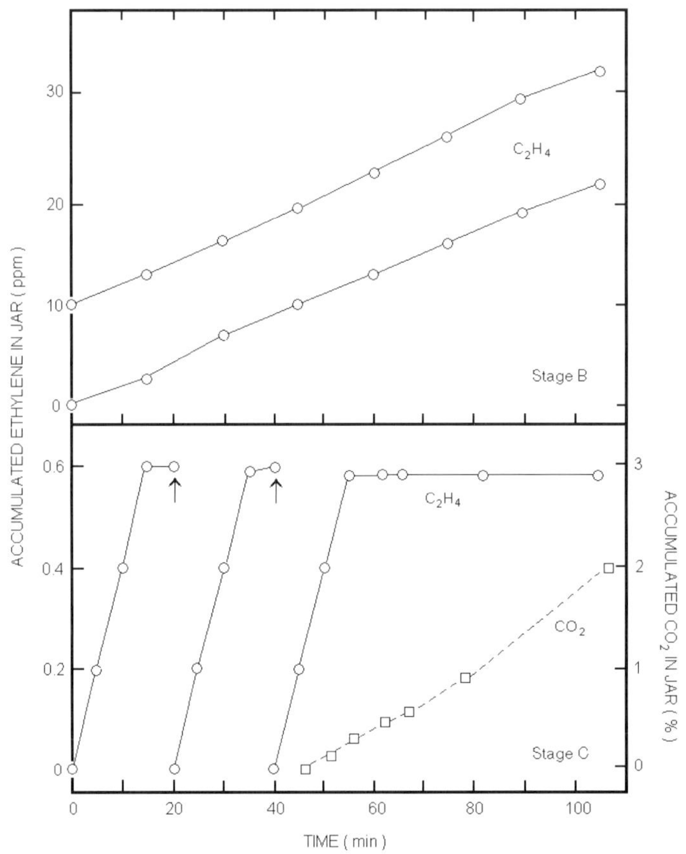

Fig. 5.10. Auto-inhibition of ethylene production in stage B and C figs. (*lower*) Ethylene production by stage C figs enclosed in a 45 ml sealed jar continues for 15 min until the ethylene concentration in the jar reaches 0.6 ppm. Then ethylene production stops even though the $[O_2]$ level is still close to 20% and only 2% $[CO_2]$ accumulates during a 60-min period. When the jar is aerated (arrow), the production begins again. The result is the same during a second and third aeration followed by re-enclosures of the same figs. (*upper*) Stage B figs produce ethylene at a linear rate regardless of whether or not 10 ppm is initially added to the jar, and they continued producing ethylene at the same rate even after more than 30 ppm has accumulated in the jar (Zeroni, 1972; Zeroni *et al.*, 1976).

hypobaric pressure comprised of 100% [O_2] promotes wound-ethylene synthesis in etiolated pea-epicotyl segments (Saltveit and Dilley, 1978a) and ethylene production by citrus fruits (Cooper and Horanic, 1973; Brisker, 1980 – cited in Goldschmidt *et al.*, 1993); 1-MCP treatment enhanced xylanase-induced ethylene production in tomato and *Capsicum annuum* leaves (Anderson *et al.*, 1996 – referred to in Sisler and Serek, 1997) and in grapefruits infected with *Penicillium digitatum* (Mullins *et al.*, 2000).

Ethylene auto-inhibits its own production by decreasing the ACC level (Riov and Yang, 1982a; Atta-Aly *et al.*, 2000a,b), by feedback repression of ACC synthase formation (Yoshi and Imaseki, 1982; Liu *et al.*, 1985b; Philosoph-Hadas *et al.*, 1985a; Hyodo and Fujinami, 1989; Nakajima *et al.*, 1990) and by reducing EFE activity (Ketsa and Herner, 1989; Sawamura and Miyazaki, 1989) or rapidly inducing malonyl ACC transferase, increasing conjugation of ACC to *N*-malonyl ACC (Liu *et al.*, 1985a,b; Philosoph-Hadas *et al.*, 1985a). AVG prevents ethylene production in mature grapefruit flavedo discs by inhibiting SAM to ACC conversion, relieving auto-inhibition and causing a superinduction of ACC synthase activity and cDNA encoding ACS (Mullins *et al.*, 1999). Treating grapefruit flavedo discs with AVG + ethylene suppressed the accumulation of ACS mRNA, indicating that auto-inhibition acts by preventing *de novo* ACS formation. The auto-inhibition of wound-induced ethylene production in ruby red grapefruits results from a suppression of ACC synthesis and a 13-fold promotion of ACC malonylation (Liu *et al.*, 1985b; Mullins *et al.*, 1999), and in green tomato fruits, auto-inhibition causes a rise in ACC-malonyl transferase within 1 h (Liu *et al.*, 1985a).

Total auto-inhibition in young sycamore figs results when the ethylene concentration reaches 0.6 μl/l in the surrounding air (Fig. 5.10), indicating that the ethylene receptor's K_m for auto-inhibition is close to 0.1–0.2 μl/l; in citrus peel the apparent K_m for auto-inhibition is higher, about 1 μl/l (Riov and Yang, 1982a); in transgenic petunia flowers the threshold for auto-inhibition is close to 1 μl/l (Wilkinson *et al.*, 1997).

5.8 Auto-stimulation of System 2 Ethylene Production

Ethylene auto-stimulates (up-regulates) system 2 ethylene synthesis by a slow reaction mechanism (Riov and Yang, 1982b), and antagonists of ethylene action delay the ripening of climacteric fruits and fading of ethylene-sensitive flowers by preventing autocatalytic ethylene production (Potter and Griffiths, 1947; Young and Biale, 1962; Imaseki *et al.*, 1968; Saltveit *et al.*, 1978; Veen, 1979; Mattoo and Lieberman, 1982; Hobson *et al.*, 1984; Sisler and Lallu, 1994; Atta-Aly *et al.*, 2000a). Pre-climacteric avocado, banana and tomato fruits are capable of converting methionine to SAM at an appreciable rate, but because their ability to convert SAM to ACC is restricted, their ethylene production rate and ACC content (< 0.1 nM) are very low. Applied ACC only slightly stimulates ethylene production by pre-climacteric apple and cantaloupe fruit tissue, but their ethylene production increases many hundredfold at the climacteric maximum, indicating that pre-climacteric fruit not only have a limited capacity to convert SAM to ACC, but also to convert ACC to ethylene (Yang and Hoffman, 1984). A Golden Delicious apple's pre-climacteric IEC (0.05 μl/l), its ability to convert ACC to ethylene, and the ACC content of its peel hardly increase during maturation on the tree (Bufler, 1986), and although 100 μl/l applied ethylene increased EFE activity in peel tissue of immature fruit, it did not stimulate ethylene production during 24 h. The peel's ability to produce EFE in response to applied ethylene increased exponentially as fruits approached full maturity, and pre-climacteric apples, which had been treated with AVG prior to harvest, responded to applied ethylene by increasing in CO_2 production and EFE activity within a few days. The threshold and K_m for these ethylene responses was < 0.05 and approximately 1.0 μl/l, respectively. Ethylene does not induce immature apples to ripen or overmature apples to produce autocatalytic ethylene (Brady, 1987).

When autocatalytic ethylene production commences, ACC synthase activity develops in climacteric fruits, and in ethylene-sensitive flowers EFE increases and the ACC content rises dramatically (Kende and Boller, 1981). The transition from auto-inhibition to autocatalysis takes place in mature-green tomatoes at the onset of ripening (Atta-Aly *et al.*, 2000a,b; Barry *et al.*, 2000), and subsequently the ACC declines due to rapid consumption as ethylene production increases to a climacteric maximum. The ACC content again increases when ethylene production slows in overripe fruit due to their reduced ability to convert ACC to ethylene (Yang and Hoffman, 1984).[7]

5.9 Significance of the Pre-climacteric IEC

Researchers tend to focus on the relatively large amounts of autocatalytic system 2 ethylene produced during the ripening of climacteric fruits, rather than the low but potentially active IEC created by pre-climacteric system 1 ethylene production. This same predilection fostered a controversy during the early 1960s concerning whether ethylene was the cause of ripening (Burg and Burg, 1962b) or a by-product of the ripening process (Biale *et al.*, 1954). In many fruits, the climacteric increase in respiration seemed to precede any rise in ethylene production, and some climacteric fruits ostensibly evolved little or no ethylene.[8] The advent of gas chromatography provided the means needed to measure minute amounts of ethylene (Burg and Stolwijk, 1959; Huelin and Kennett, 1959; Meigh, 1959) and settled the matter in favour of the ripening-hormone concept. Extremely small yet physiologically active ethylene concentrations were detected prior to the onset of the climacteric (Burg and Burg, 1962a,b, 1965c), and while in avocados (Burg and Burg, 1962a) and bananas (Fig. 5.7, *right*; Burg and Burg, 1965c) the IEC increases before respiration begins to rise, this does not occur in mangoes to a significant extent (Fig. 5.7, *left*; Burg and Burg, 1962b), and the ripening of this fruit (and perhaps other fruits) seems to be initiated in response to its pre-climacteric IEC. The first detectable indication of tomato ripening is enhanced ethylene synthesis 1–2 days before any visible sign of colour change, but even before this occurs the fruit has become more sensitive to ethylene by developmental changes thought to be ethylene-independent (Hobson and Grierson, 1993). The fruit's ethylene sensitivity may increase sufficiently to induce it to respond to its low 'system 1' IEC (Burg, 1962a) and produce autocatalytic system 2 ethylene, which functions to drive the ripening process toward rapid completion.

While many ripening-related processes are suppressed in fruits with inhibited ethylene production, some aspects of ripening remain unaffected. Cloning studies and the characterization of ripening-related tomato mRNA's reveal that enhanced expression of genes for ACC synthase and EFE occurs early in ripening, and that the regulation of *ACS* gene expression essential for autocatalytic ethylene production (Knapp *et al.*, 1989; Terai, 1993) can be detected before the IEC begins to increase (Lincoln *et al.*, 1987). ACS activity and ACC accumulation are augmented during the pre-climacteric period in both control and *ACO*-antisense tomato fruit (Picton *et al.*, 1993) and also in melon fruits (Guis *et al.*, 1997). In wild-type melons the ACC accumulation is transient, while in *ACO*-antisense melons it is sustained and associated with an increase in ACS activity. An apple's capacity to convert ACC to ethylene also rises prior to the onset of climacteric ethylene production (Lau *et al.*, 1986; Mansour *et al.*, 1986), and ACC accumulates in non-ripening *nor* tomato mutants, which are unable to produce autocatalytic ethylene (Terai, 1993). *ACO* gene expression occurs both in low ethylene-producing fruit such as *ACS*-antisense tomatoes (Oeller *et al.*, 1991) and in tomatoes overexpressing the *ACC* deaminase gene (Klee, 1993). It has been suggested that prior to the onset of autocatalytic ethylene production, the mRNAs required for ACC synthesis and the conversion of ACC to ethylene are developmentally controlled independent of the pre-climacteric IEC and ethylene action (5.15; Sawamura and Miyazaki, 1989; Hobson and Grierson, 1993; Lelièvre *et al.*, 1997a). Alternatively, the developmental change ('ageing'), which initiates autocatalytic ethylene production and ripening, might consist of an increase in the sensitivity of system 2 ethylene production to be up-regulated by the system 1 pre-climacteric IEC.

Many studies suggest that the pre-climacteric IEC plays an essential role in initiating the ripening of climacteric fruits. At 20°C, the ripening time of tomato (cv. Floradade), kiwi fruit (cv. Hayward), mango (cv. Kensington Pride), peach (cvs. Fragar, Coronet, Loring and Floridagold) and custard apple (*Anonona reticulata* cv. Pinks Mammoth) fruits increased linearly with a logarithmic decrease in the applied ethylene concentration over the range < 0.005, 0.01, 1.0 to 10 μl/l (Wills *et al.*, 2001), and the green life of mature Williams and Lady's Finger bananas, assessed as the time to reach the respiratory climacteric, was quadratically related to a logarithmic decrease in the ethylene concentration in air containing 0.001, 0.01, 0.1 and 1 μl/l ethylene (Wills *et al.*, 1999). A plot of the time for detached mangoes and avocados (Fig. 5.11), tomatoes (Morris *et al.*, 1981), cantaloupes (McGlasson and Pratt, 1964a) and bananas (Biale *et al.*, 1954; Palmer, 1971) to ripen, and oranges (Biale, 1960; Jahn *et al.*, 1973; Apelbaum *et al.*, 1976) and lemons (Denny, 1924b) to degreen, is linearly related to the log of the applied ethylene concentration between the threshold for biological activity and the minimum dose which causes a maximum response. An extrapolation of the dose–response curve to the minimum applied ethylene concentration that causes no observable response, indicates that the threshold closely approximates the pre-climacteric fruit's minimum IEC shortly after it was detached from the tree or vine (Table 5.1). The lowest applied-ethylene concentration that accelerates the ripening of harvested bananas, avocados, mangoes, lemons, tomatoes, cantaloupes and oranges, < 0.1–0.2 μl/l (Table 5.2), is close to the 0.04–0.2 μl/l postharvest minimum IEC present within these fruits and also in mature pre-climacteric apples, pears, pineapples, plums, squash and honeydew melons (Table 5.1; Burg and Burg, 1962b, 1965a,c; Lyons *et al.*, 1962; Lyons and Pratt, 1964; Ben-Yehoshua and Eaks, 1970; Rhodes, 1970; Reid *et al.*, 1973; Sawamura, 1981; Chu, 1984). To obtain a benefit from 'scrubbing' ethylene during a Cox Orange Pippin apple CA storage, the fruit's IEC had to be kept below 0.1 μl/l (Stow *et al.*, 2000). It is not surprising that an ethylene concentration close or equal to the minimum pre-climacteric IEC must be applied prior to the onset of natural autocatalytic ethylene production in order to accelerate a climacteric fruit's ripening. A lesser amount of ethylene could not raise the IEC sufficiently to cause a discernible ripening response because ethylene's action is a log function of concentration (Fig. 5.11; Fig. 5.31, *lower left*; Fig. 5.32 *left*), and applied gas summates with the IEC initially present within the fruit (Burg and Burg, 1962b; Pratt and Goeschl, 1968).

The *LE-ACS6* gene responsible for the low rates of system 1 ethylene production is negatively regulated in immature and pre-climacteric tomato fruits (Nakatsuka *et al.*, 1998) and consequently, at these developmental stages, a tomato's ethylene production is promoted by antagonists of ethylene action (Atta-Aly *et al.*, 1987). This indicates that before ripening commences, tomatoes have sensed their pre-climacteric IEC and responded to it by auto-inhibiting ethylene production, and this also explains why a brief ethylene exposure induces ACO activity (Liu *et al.*, 1985c; Bufler, 1986; Starrett and Laties, 1991b) and gene expression (Knapp *et al.*, 1989; Dong *et al.*, 1992; Terai, 1993; Kneissl and Deikman, 1996) in immature tomato fruits and in the non-ripening *rin* tomato mutant, but not ACS activity and ethylene biosynthesis. It also accounts for a transient ACC accumulation

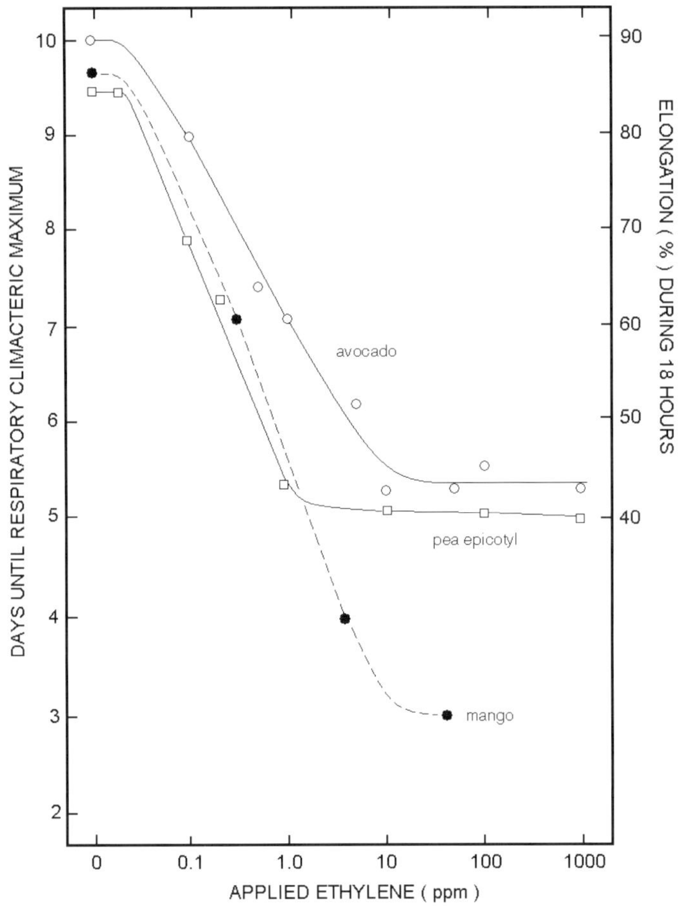

Fig. 5.11. Effect of various applied ethylene concentrations on the elongation of 1 cm etiolated pea epicotyl segments during an 18-h incubation in the dark (Burg and Burg, 1966a), and the ripening of Fuerte avocados (adapted from Biale, 1960) and Kent mangoes (adapted from Burg and Burg, 1962b). Ethylene was continuously applied to the fruits after harvest. Ripening time is judged as the number of days to reach the respiratory climacteric peak.

in wild-type melon fruit, which in *ACO*-antisense fruit[9] is sustained and associated with an increase in ACS activity (Guis *et al.*, 1997). Persistent high-ACS activity in the absence of ethylene indicates that pre-climacteric ethylene production is auto-inhibited. In mature-green tomatoes, the pre-climacteric IEC is 0.08 μl/l (Lyons and Pratt, 1964; Sawamura *et al.*, 1978), and since the K_m is similar for auto-inhibition and most other actions of ethylene (Table 5.2), the tomato has sensed and responded to an IEC level which is able to induce other responses. A pre-climacteric tomato might become developmentally sensitized to respond to an even lower IEC, for as little as 0.005 μl/l applied ethylene normalizes the diageotropic tomato plant (Zobel, 1973, 1974) and promotes ripening in mature-green cv. Floradade tomatoes (Wills *et al.*, 2001); < 0.001 μl/l causes African marigold plants to become epinastic (Table 5.2; Fig. 5.36, *lower*); 0.001 μl/l slightly enhances maize-root growth (Fig. 5.32, *lower left*); and 0.002 μl/l causes orchid sepal wilt (Davidson, 1949).

The ripening phenotype is altered in all transgenic or antisense[9] plants that have a

Table 5.2. Threshold and half-maximal applied-ethylene concentration required to induce ripening and other biological responses. The apparent threshold may be influenced by the tissue's IEC.

	Ethylene conc. (μl/l)		
Response	Threshold	K_m	Reference
Ripening (after harvest)			
banana	0.001–0.100	0.2	Biale *et al.*, 1954; Palmer, 1971; Wills *et al.*, 1999
mango	0.005–0.100	0.6	Burg and Burg, 1962b; Wills *et al.*, 2001
avocado	0.1	0.5	Biale, 1960; Biale and Young, 1971
orange	0.1	1.0	Biale, 1960; Apelbaum *et al.*, 1976
lemon	0.2	0.4	Denny, 1924b
cantaloupe	0.12	1.0	McGlasson and Pratt, 1964a
tomato	0.005–0.100	0.3	Morris *et al.*, 1981; Wills *et al.*, 2001
'honeydew' melon	< 0.5	2.5	Pratt and Goeschl, 1968
Growth (inhibition)			
etiolated pea epicotyl	0.01	0.1	Burg and Burg, 1966a, 1967c
Arabidopsis hypocotyls	0.01	0.1	Bleecker *et al.*, 1988
pea root	0.02	0.1	Chadwick and Burg, 1967, 1970
maize root	0.05	0.2	Whalen and Feldman, 1988
pea bud	0.02	0.1	Burg and Burg, 1968; Burg, 1968a
Cell division (inhibition)			
etiolated pea	0.05	0.3	Apelbaum and Burg, 1972
Plumular expansion (inhibition)			
etiolated pea	0.009	0.04	Goeschl and Pratt, 1968
red-light-treated pea	0.05	0.32	Goeschl and Pratt, 1968
etiolated bean	< 0.01	0.1	Kang and Ray, 1969a
red-light-treated bean	< 0.01	0.1	Kang and Ray, 1969a
Epinasty			
tomato	0.004	0.05–0.10	Hansen, 1943; Abeles, 1973
African marigold	–	0.001	Crocker, 1948
potato	–	0.003	Crocker, 1948
Horizontal mutation			
etiolated pea	0.01	0.1	Goeschl and Pratt, 1968
Lateral IAA transport			
etiolated pea epicotyl	0.01	0.1	Burg and Kang, 1993
Geotropism (inhibition)			
etiolated pea epicotyl	0.01	0.1	Kang and Burg, 1972a
Abscission			
bean explant	0.01	0.1	Abeles and Gahagan, 1968a
Anthocyanin synthesis			
etiolated cabbage	0.01	0.1	Kang and Burg, 1973
Simulation of wound-induced ethylene production	–	0.3	Hoffman and Yang, 1982
Autoinhibition	–	0.1–0.2	Zeroni *et al.*, 1976
Isocoumarin formation	0.1	< 0.5	LaFuente *et al.*, 1996

reduced capacity for ethylene synthesis, and it can be at least partially restored by applying ethylene or propylene[6] (Hamilton *et al.*, 1990; Klee *et al.*, 1991; Oeller *et al.*, 1991; Klee, 1993; Ayub *et al.*, 1996). The rate of ethylene synthesis by antisense *ACC*-oxidase tomato fruits is 97% inhibited compared to wild-type fruits at their climacteric peak (Hamilton *et al.*, 1990, 1991); the maximum rate is reduced by 99.5% in antisense *ACO* melon fruits (Ayub *et al.*, 1996); and ethylene production by various lines of Royal Gala apples transformed by ripening-related antisense *ACS*

Table 5.3. Postharvest changes in the IEC of mature Wase and Yuzu citrus fruits (Sawamura, 1981).

Days after harvest	Internal ethylene (μl/l)	
	Wase	Yuzu
Attached fruits	96.3	99.7
1	2.1	0.04
2	0.03	0.04
3	0.14	0.11
4	0.20	0.09
5	0.13	0.14

genes was down-regulated by 11–97% after 69 days storage at 4°C, with the majority of fruit falling in the 60–80% range (Hrazdina *et al.*, 2001). Colour change in *ACO* antisense tomatoes is initiated at about the normal time, but the extent of reddening is reduced and, after storage for several weeks at room temperature, these fruits are more resistant than controls to overripening and shrivelling (Theologis *et al.*, 1993a). *LE-ACS2* and *LE-ACS4* genes for ACC synthase are expressed during tomato-fruit ripening (Olson *et al.*, 1991; Rottmann *et al.*, 1991), and manifestation of antisense RNA derived from the cDNA of the *LE-ACS2* gene results in an almost complete inhibition of mRNA accumulation for both of these ripening-induced *ACC* synthase genes (Oeller *et al.*, 1991), limiting ethylene production to less than 0.5% of the maximum rate during control-fruit ripening (Fig. 5.12). Normal fruits began to produce autocatalytic ethylene 50 days after pollination, and were fully ripe 10 days later, whereas the *ACS*-antisense fruits were pale orange after 90–150 days, and never softened, turned red or developed aroma. The antisense phenotype was reversed by treatment with ethylene or propylene, and the treated fruits were indistinguishable from naturally ripened tomatoes (Theologis *et al.*, 1993a; Luo *et al.*, 2000).

Wild-type tomato fruits typically develop an IEC of 28–45 μl/l during ripening (Lyons and Pratt, 1964; Sawamura *et al.*, 1978) and therefore a 99.5% inhibition in an antisense fruit's peak climacteric ethylene production should decrease their IEC to 0.14–0.23 μl/l. Thus, while the IEC in *ACS* and *ACO*-antisense fruits may be well below the 1–10 μl/l level that maximally accelerates most ethylene responses, it nevertheless may exceed both a normal tomato's preclimacteric IEC and the minimum applied ethylene concentration that significantly accelerates tomato ripening. This could allow various ethylene-sensitive developmental changes to proceed slowly or even at a moderate rate in antisense fruits, and might account for the gradual but never complete coloration of these fruits by ethylene-sensitive lycopene accumulation (Table 5.7). Several unexpected results suggest that low ethylene-producing antisense fruits may develop an active IEC. That lycopene synthesis is an ethylene-dependent process in tomatoes is indicated by significantly reduced ethylene production, softening and lycopene formation in excised pericarp tissue from breaker fruit soaked in silver nitrate (Saltveit *et al.*, 1978); by an inhibition of ethylene production and ripening when the pericarp tissue is treated with AVG; and by stimulated lycopene synthesis and softening if ethylene is applied (Edwards *et al.*, 1983). Ethylene-insensitive 'Never ripe' (*Nr*) tomatoes (Table 5.5) fail to produce lycopene (Tigchelaar *et al.*, 1978); lycopene synthesis is strongly retarded in harvested transgenic tomatoes, which have a reduced capacity for ethylene synthesis (Oeller *et al.*, 1991; Klee, 1993; Murray *et al.*, 1993c), and *ACO*-antisense tomato fruits detached at the mature-green stage show the expected reduction in lycopene synthesis compared to control fruit. But *ACO*-antisense tomatoes that remained attached to the vine accumulated lycopene at the same rate as control fruits (Picton *et al.*, 1993). It was speculated that 'residual' ethylene present in the attached *ACO*-antisense fruits might have been above an active threshold (Oeller *et al.*, 1991; Klee, 1993; Picton *et al.*, 1993; Theologis *et al.*, 1993b). Tomato-fruit softening by endo-polygalacturonase (PG) is another example of behaviour suggesting that antisense tomatoes with a reduced capacity to produce ethylene may develop an 'active' IEC. PG

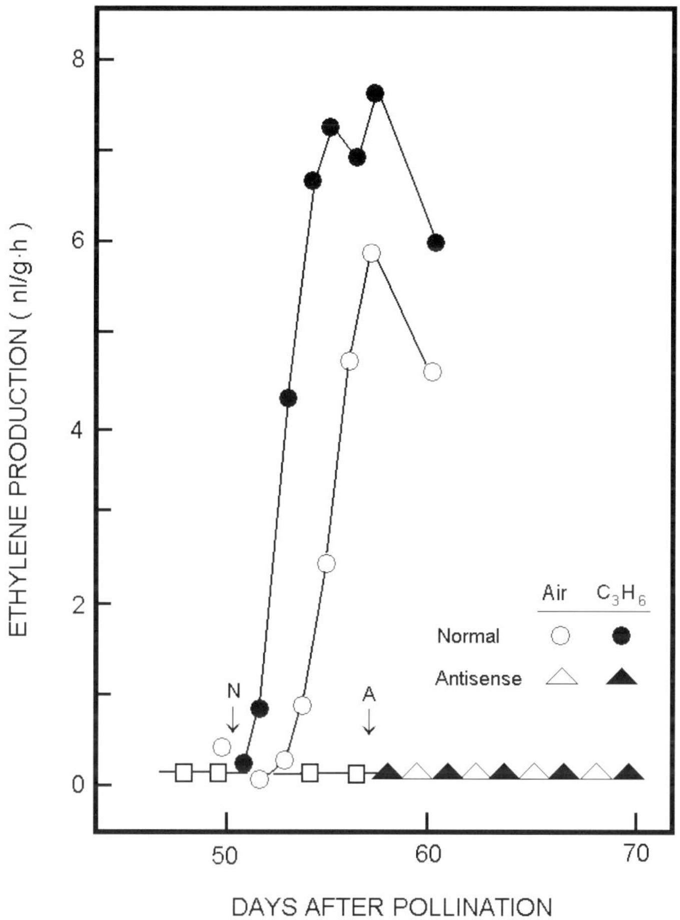

Fig. 5.12. Inhibition of ethylene production in detached tomato fruit by antisense ACC synthase RNA. Arrows indicate the time when propylene was applied to normal (N) or antisense (A) fruits. Open boxes (□) depict the behaviour of antisense tomatoes prior to the time they were treated with propylene or left in air (Oeller *et al.*, 1991).

transcript expression is inhibited by silver (Davies *et al.*, 1988) and increases both during normal ripening and after ethylene is applied to mature-green tomatoes (Abdel-Rahman, 1977; Tigchelaar and McGlasson, 1977; Sawamura *et al.*, 1978; Slater *et al.*, 1985; DellaPenna *et al.*, 1986; Lincoln *et al.*, 1987; Maunders *et al.*, 1987; Theologis *et al.*, 1993b; Yen *et al.*, 1995). Antisense ACC-synthase RNA prevents autocatalytic ethylene production (Fig. 5.12), keeps propylene from promoting ethylene biosynthesis and strongly retards lycopene synthesis, but nevertheless transgenic tomato fruits with 3–10% 'residual' ethylene production soften at the same rate as control fruit (Klee, 1993; Murray *et al.*, 1993b; Picton *et al.*, 1993). PG transcript expression is not altered in either *ACS* (Oeller *et al.*, 1991) or *ACO* (Picton *et al.*, 1993) antisense fruits, and the interval between anthesis and the onset of chlorophyll loss is not delayed. Sitrit and Bennett suggested that PG-transcript accumulation might by 'hyper-sensitive' to ethylene and responsive to even the low levels of ethylene present in antisense fruits (Lelièvre *et al.*, 1997a). In another study, it was found that after harvest, mature-green transgenic tomato fruits expressing ACC deaminase ripened more slowly than control fruit, but transgenic fruits attached to the vine ripened more rapidly, and almost as quickly as control fruits. It was suggested that this might be caused by the higher IEC of transgenic fruits on the vine. In all of these examples, it was speculated that the result was explained by the rapid escape of ethylene through the tomato's stem scar after harvest, but Fig. 3.14 indicates that this does not occur. The IEC in tomato fruits remains remarkably constant throughout their development on the vine and after they are harvested, until ripening commences (Lyons and Pratt, 1964; Sawamura *et al.*, 1978).

5.10 Effect of Harvest

Postharvest physiology is concerned with the events that are set in motion after harvest separates a commodity from its water supply and terminates the import into it of dry matter dissolved in the xylem and phloem saps. In response to detachment, leaves hydrolyse protein, produce ammonia, senesce prematurely, lose their chlorophyll and turn yellow (Fig. 4.17), even when they are illuminated in water culture and photosynthesis increases or maintains their carbohydrate and dry-matter content (4.20). Harvest (or pollination) causes flowers to increase in ethylene sensitivity and rapidly fade (Durkin and Kuc, 1966; Fig. 5.14), and harvested mango (Burg, 1964; Lackshminarayana, 1973), avocado (Burg, 1964; Adato and Gazit, 1977), cantaloupe (Pratt, 1971), peach (Looney *et al.*, 1974), tomato (Sawamura *et al.*, 1978),[10] apple (Kidd and West, 1938b; Wilkinson, 1963; Hulme *et al.*, 1968), banana (Burg and Burg, 1965c), citrus (Ben-Yehoshua and Eaks, 1970) and plum (McGlasson *et al.*, 2000) fruits ripen more swiftly than comparable fruits left on the tree or vine. Fuerte avocados, which mature in California as early as October or November, may be left on the tree until the following June or even August, remaining in a pre-climacteric state with the only apparent change being a slight increase in size and fat content (Biale and Young, 1971). Softening will not take place while this fruit is attached to the tree by a healthy stem (Biale *et al.*, 1954), but immediately after harvest the processes leading to senescence are set into action and the typical climacteric course of respiration ensues. The ethylene production of kiwi fruit does not increase as long as they are attached to the vine (Sfakiotakis *et al.*, 2000). Isolating apples from leaves by bark girdling plus defoliation of spur leaves hastens ripening by about 1 month, judged as an increase in the fruit's IEC, suggesting that a 'tree factor' transported from the leaves through the phloem to the apple inhibits the conversion of ACC to ethylene (McGlasson *et al.*, 2000; Sfakiotakis and Dilley, 1973).[11] It has been proposed that attached flowers are protected by a natural anti-senescence factor supplied from the roots (Rogers, 1973); that anti-ripening 'tree factors' delay ripening of attached fruits (Burg, 1964, 1965a; Meigh *et al.*, 1967; Gazit and Blumenfeld, 1970; Mapson, 1970a; Frenkel and Dyck, 1973; Sawamura *et al.*, 1978; McGlasson, 1985); and ethylene may induce fruit ripening by causing the destruction of ripening inhibitors (Peacock, 1972).

The applied ethylene concentration required to induce ripening decreases as fruits develop on the tree or vine, and rapidly after they are detached, sensitizing mature, harvested fruits to respond to an exogenous concentration equal to or lower than their pre-climacteric minimum IEC (Fig. 5.11; Table 5.1). After harvest, the sensitivity to applied ethylene increases in avocados (Gazit and Blumenfeld, 1970), lemons (Biale, 1960), apples (Knee *et al.*, 1987) and pears (Hansen and Blanpied, 1968). Tomatoes detached as early as 17 days after anthesis require 12–15 days of continuous treatment with 1000 μl/l ethylene to develop red colour (Lyons and Pratt, 1964); the minimum applied ethylene concentration that hastens ripening of mature tomatoes attached to the vine is 4.9 μl/l (Knegt *et al.*, 1974; Sawamura *et al.*, 1978), but < 0.1 μl/l suffices after they are detached (Tables 5.1 and 5.2). A high–exogenous ethylene concentration is required to produce a rapid climacteric rise in early-season attached Fuerte avocados, but as little as 0.1 μl/l ethylene stimulates ripening of late-season harvested fruits (Biale, 1960). 'Honeydew' melons detached 28 days after anthesis do not naturally ripen, but their respiration can be stimulated by applying > 2.5 μl/l of ethylene, whereas < 0.5 μl/l produces the same effect in melons harvested 43 days after anthesis (Pratt and Goeschl, 1968). Mature Haden mangoes attached to the tree are stimulated to ripen by applying 1000 μl/l ethylene (Fig. 5.16), but not by an endogenous IEC of 11.4 μl/l, while after harvest 0.04–0.4 μl/l suffices (Fig. 5.11).

The developmental change in ethylene sensitivity set in motion by harvest is illustrated by the response of detached Dwarf Cavendish bananas to applied ethylene (Fig. 5.13, *right*). The fruit responds sluggishly to 0.5 μl/l ethylene applied immediately after detachment, and more swiftly when the ethylene treatment is deferred until the natural climacteric approaches. Application of 2 μl/l ethylene induced a respiratory climacteric within 3–5 h regardless of the time that had elapsed between harvest and the treatment, and a comparable time to respond to 1 μl/l ethylene has been reported for Gros Michel bananas (Burg and Burg, 1962b). Left on the tree, the Dwarf Cavendish bananas did not ripen for 40–50 days.

The nature of the substances transported in xylem and phloem has been extensively investigated. While most of the water transported in xylem is lost by foliar transpiration, a few per cent is diverted to actively growing fruits or flowers. The clear xylem solution has a pH between 5 and 6, and contains 0.1–0.4% solids of which one-third may be ash (N, P, K, Ca, Mg), including iron and copper. Sugars, organic acids and α-keto acids frequently are present, and coumarin, alkaloids and some enzyme activity has been detected (Bollard, 1960). Nitrate reduction occurs in the roots and therefore NO_3^- represents only a small fraction of the total nitrogen present in the xylem sap, which apart from a small portion of NH_4^+, is transported mainly as amide or amino nitrogen. Usually aspartate, asparagine and/or glutamine predominate in the xylem sap, accompanied by small amounts of alanine, methionine, valine, leucine, serine, threonine and γ-amino butyric acid, as well as a peptide that yields cysteine. ACC is transported from root to shoot in the xylem of tomato plants (Bradford and Yang, 1980; Else *et al.*, 1993), and IAA moves through the

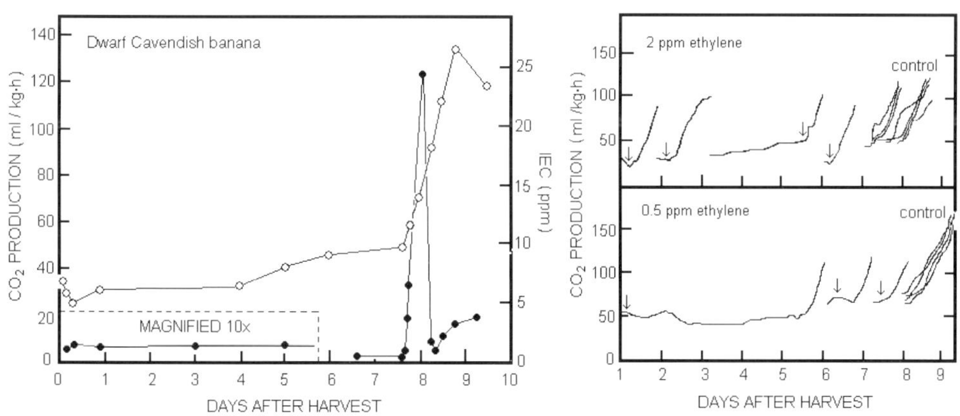

Fig. 5.13. (*left*) Changes in the internal ethylene concentration (IEC) and CO_2 production during normal ripening after Dwarf Cavendish bananas are harvested. (*right*) Induction of the respiratory climacteric in harvested Dwarf Cavendish bananas at 24°C by continuous application of 0.5 or 2 ppm ethylene after the time indicated by the arrow. All fruits were from the same hand and ripened at the same rate in air. The (air) control respiratory drift and climacteric rise in respiration can be discerned by 'connecting' the measurements made with individual fruits that were not exposed to ethylene (Burg and Burg, 1965c).

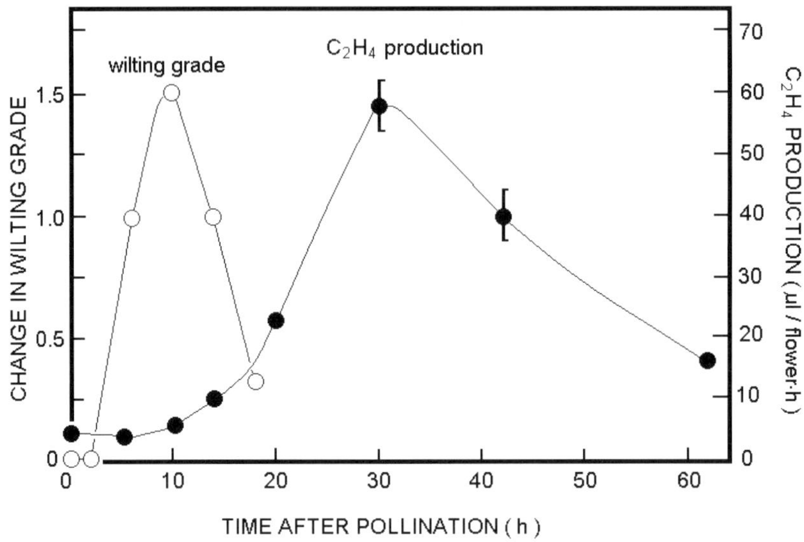

Fig. 5.14. Time course of changes in ethylene production and sensitivity to ethylene following pollination of *Phaleonopsis* flowers. Changes in ethylene sensitivity are represented by the increase in 'wilting grades' at different periods following pollination, measured 12 h after a 4-h exposure to 4 µl/l ethylene (Halevy *et al.*, 1996).

xylem of many varieties of plants. In some species, citrulline, allantoic acid, allantoin and γ-methylene glutamine are major organic nitrogenous constituents, and nicotine, nornicotine, anabasine, scopolamine and other pyridine alkaloids are present. Up to 20% of the phosphorus in a tomato plant's xylem sap and 6% in barley plants is contained in phosphoryl choline (Bollard, 1960).

Sieve tubes represent a highly specialized cell system through which a concentration gradient of foods is transformed to a hydrostatic gradient of solution capable of moving rapidly from source to sink. Foods move from regions of synthesis to utilization, while minor substances are carried along and accumulate in the sinks to concentrations far above those that exist along the channels of movement (Crafts, 1961). Mature leaves and actively synthesizing tissues that produce more assimilate than is required for their own metabolism and growth are sources and net exporters of photo-assimilate, while developing flowers and fruits function as sinks because they consume more photo-assimilate than they produce. Partitioning of phloem-assimilate between competing sinks depends on the nature of the vascular connections between them, sink strength and the proximity of the sink to the source, with a marked bias in favour of translocation toward the closest sink, for example from a cluster of leaves to a subtending fruit (Figs 5.15 and 5.19). Fruit cells absorb the total solution without separation of solutes and solvent because they require more water for their growth than that obtained from the phloem-assimilate stream (Crafts and Lorenz, 1944a).

Phloem exudate's chemical composition is highly variable, depending on the species, age and other factors, but typically at least 90% of the translocated material consists of carbohydrate. Sucrose predominates, but sometimes raffinose, stachyose, verbascose and mannitol are present, while hexoses almost always are absent. Sieve-tube exudates usually contain between 10 and 25% (w/v) sugar, and 0.03–0.4% (w/v) amino acids and amides (Zimmerman, 1960; Crafts, 1961). Amino acids and amides are translocated in the phloem out of ageing leaves and flowers, and aspartic acid, glutamic acid, serine, threonine, alanine, valine, leucine and/or isoleucine, phenylalanine, asparagine, glutamine and possibly α-amino-butyric acid occur in the stylet sap of aphids that have been feeding on trees (Zimmerman, 1960). ACC (Amrhein *et al.*, 1982) and low concentrations of IAA (Huber *et al.*, 1937), cytokinin and GA have been detected, and hormones have been implicated in redirecting assimilates to new sinks (Hopkins, 1995). ABA influences phloem translocation and there is a correlation between ABA concentration and the growth rate of developing fruits. Large amounts of K^+ and lesser quantities of PO_4^{-3}, Mg^{2+}, Ca^{2+}, SO_4^{-2} and Cl^- are present in phloem sap, while NO_3^- is conspicuously absent. Malic acid (2.0–3.2 mg/l) and protein (1.45–2.20 mg/l), including P-protein (= phloem protein)[12] are present in *Ricinus* phloem exudate (Hall and Baker, 1972), and the transport of vitamins produced in tomato leaves, including thiamine, pantothenic acid, pyridoxine and probably also nicotinic acid, has been demonstrated by girdling experiments (Bonner, 1944).

The amount of material translocated through the phloem into fruits, fleshy roots and tubers is incredibly large in relation to the cross-sectional area of the conducting elements and the openings in the sieve plates. The average hourly increase in dry weight of a pumpkin fruit was 0.61 g over a 33-day growing period (Crafts and Lorenz, 1944b), the daily increase in the dry weight of sausage tree fruits (*Kigelia africana*) ranged from 8.7 to 32.6 g throughout a 3-month growing period (Clements, 1940), and the dry weight of the yam tuberous root increased by 45 g during 4 weeks (Mason and Lewin, 1926). Estimates of the velocity of movement through the phloem vary between 21 and 100 cm/h, with occasional estimates up to 300 cm/h (Mason and Lewin, 1926; Huber, 1941; Biddulph and Markle, 1944; Crafts and Lorenz, 1944b; Swanson and El-Shishiny, 1958; Zimmerman, 1960).

In many plants, especially annuals, vegetative growth and food accumulation proceed over a part of the season and then, as the fruits develop, so much of the food that has accumulated in the leaves, stems and roots is completely withdrawn and translocated into the fruits that these other organs die. A striking and rather abrupt change in the direction of solute movement is exhibited by many kinds of flowers, where in the more ephemeral types a marked import of foods and salts into the corolla occurs over a 12–24-h period, followed by an equally rapid export of 40–80% of the organic and inorganic material as the flower fades (Curtis and Clark, 1950). The rate of chlorophyll loss and carotenoid gain in satsuma-mandarin (*Citrus unshiu* cv. Okitsu) epicarp is positively correlated with its sucrose content and negatively with its nitrogen content, and therefore chlorophyll disappearance and carotinoid accumulation are inhibited when leaf removal blocks natural sucrose build-up and nitrogen reduction in the fruit peel (Iglesias *et al.*, 2001). The sucrose-induced colour change is unaffected by ethylene and delayed by GA.

Both steam and bark girdling have been used to study the influence of phloem and xylem transport on fruit ripening and ethylene sensitivity. Bark girdling ('ringing') a stem, so that all tissues external to the xylem are removed, does not prevent the movement of water to organs attached above the ring, whereas cutting through the stem's xylem causes almost immediate wilting of leaves attached to the stem upstream of the ring. When the ring wound is protected by melted paraffin, there is no visible plugging by gums or tyloses, and dye solutions introduced through the xylem readily move past a ring that prevents phloem-solute transport (Curtis and Clark, 1950). When the xylem is removed, the phloem must be kept moist and the tissues above supplied with water in order for phloem transport to continue to function.

Deleano (1911) may have been the first to 'scald' petioles in order to interfere with phloem transport. By this method, he reduced the loss of starch from grape leaves by two-thirds. Phloem transport of organic materials through the stem over long distances is dependent upon living cells, and killing by heat (or anaesthetics) produces effects equivalent to those of a ring girdle (Curtis, 1929). When $C^{14}O_2$ was applied to limited areas of soybean leaves in order to study translocation of labelled sugar, steam ringing the petiole prevented export, while steaming below a leaf node did not prevent acropetal transport to younger leaves (Aronoff, 1955).

The possibility that 'tree factors' delay ripening of Choquette avocados was investigated by interrupting phloem transport to attached fruits. The stalk supporting each avocado was heated for 20 min with steam applied according to the description in Fig. 5.15, either (i) 5 cm from the fruit (Fig. 5.15 – *below*); (ii) 30 cm from the fruit with a cluster of leaves intervening between the fruit and steam girdle (Fig. 5.15 – *above*); or (iii) the same as (ii), but with the leaves cut away (Fig. 5.15 – *above, trimmed*). If ripening is prevented by one or more substances conducted to the fruit through the phloem, a steam girdle at (i) ought to induce the climacteric by interrupting phloem transport; a transport block at (ii) would not be as effective in stimulating ripening, since sap is preferentially supplied to the fruit from the adjacent leaves; the response in (iii) should resemble that in (i) because the leaves have been removed. The result (Fig. 5.15) supports the contention that phloem continuity between the fruit and shoot system is required to keep the fruit in a pre-climacteric state.

The mean ethylene content in 47 unripe, mature Haden mangoes was 1.6 µl/l while the fruits were attached to the tree, with values ranging as high as 11.4 µl/l (Burg, 1963, unpublished). Left on the tree, none of the fruits reached a stage of half-ripe and turning (stage 1.5, Figs 5.16 and 5.19) for at least 30 additional days even though, after the mangoes were harvested, the threshold for ethylene activity was close to 0.04 µl/l, and 1.6 µl/l induced rapid softening (Burg and Burg, 1962b). Attached mangoes became half-ripe and turning within 2.2 ± 0.3 days when they were confined in a polyethylene bag through which 1000 µl/l ethylene was continuously flowed (Fig. 5.16); and harvested, non-gassed fruits required only 4.5 days to reach the same stage of ripeness. Regardless of their initial IEC, within 1 h after Haden mangoes were removed from the tree, their internal ethylene concentration decreased to 0.06 ± 0.06 µl/l (Figs 5.16 and 5.17; Burg and Burg, 1962b). The IEC then

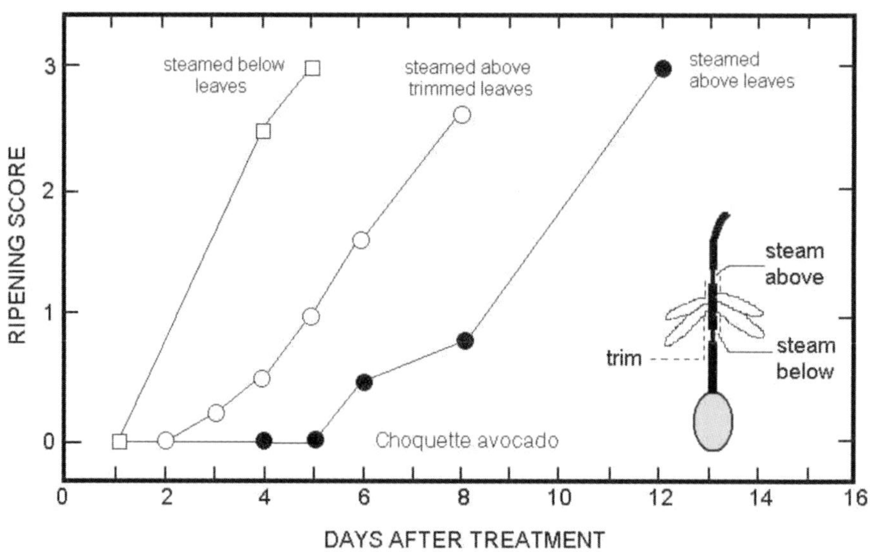

Fig. 5.15. Ripening of Choquette avocados on the tree after the stalk supporting the fruit was steam girdled for 20 min. Ripening was scored empirically on the basis of coloration and softening using as categories: 0 = hard green; 1 = slight softening; 2 = soft/ripe; 3 = overripe. Controls on the tree ripened to stage 1.5 in approximately 30 days. Results obtained with 238 fruits (Burg, 1964, unpublished).

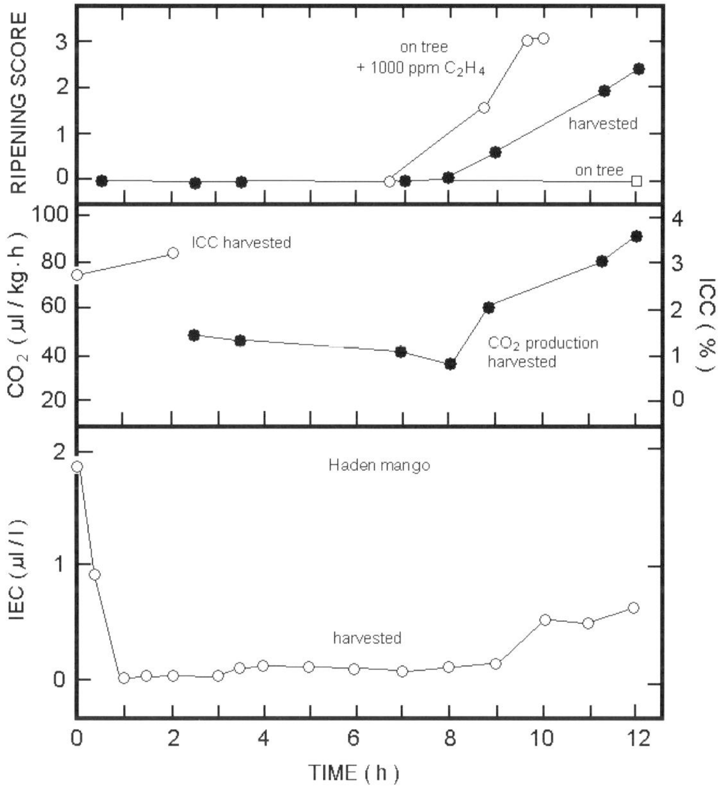

Fig. 5.16. Ripening, internal ethylene (IEC), internal CO_2 (ICC) and CO_2 production of harvested Haden mangoes at 21.1°C compared to ripening on the tree with and without 1000 ppm applied ethylene. Ripening was scored empirically on the basis of coloration and softening using as categories: 0 = hard green; 1 = slight softening; 2 = soft/ripe; 3 = overripe (Burg, 1963, unpublished).

increased to 0.15–0.2 µl/l before ripening commenced, and peaked during the climacteric at between 0.6 µl/l (Fig. 5.16) and 3 µl/l (Burg and Burg, 1962b; Fig. 5.7, *left*). This behaviour proves that the threshold for ethylene action is much higher in attached vs. harvested mangoes. Prior to detachment, the IEC sometimes reached 11.4 µl/l, and yet these mangoes did not ripen for at least 30 days, whereas 0.04–0.4 µl/l applied ethylene stimulated harvested fruits to ripen, and their IEC at the climacteric maximum did not exceed 0.6–3 µl/l.

That the postharvest decrease in the mango's IEC is not caused by enhanced gas exchange is readily apparent, since the ICC did not change or slightly increased, while the IEC rapidly decreased by > 95% (Fig. 5.16). The harvested mango's total gas and liquid phase resistance to ethylene mass transport is approximately 4000 s/cm (Burg and Burg, 1965b), and if ethylene production ceased immediately after harvest, depending on the porosity of the fruit ($V_{air}/V_{total} = 0.052–0.132$ for different varieties of ripe mangoes – Table 3.8), according to equation 3.13 ($b = 0.126$ @ 20°C – Table 15.2), half of the gas initially present would escape in 7.6–10.5 min. The measured half-time for the loss of ethylene after harvest was 9 min (Fig. 5.17), indicating that ethylene production by the fruit, or the supply of ethylene to it, must have immediately ceased after harvest.

Example 1 indicates that in order to supply an attached mango containing 1.9 µl/l ethylene with sufficient ethylene dissolved in transpirational water to replace that escaping to the atmosphere through the fruit's skin, the water would need to have

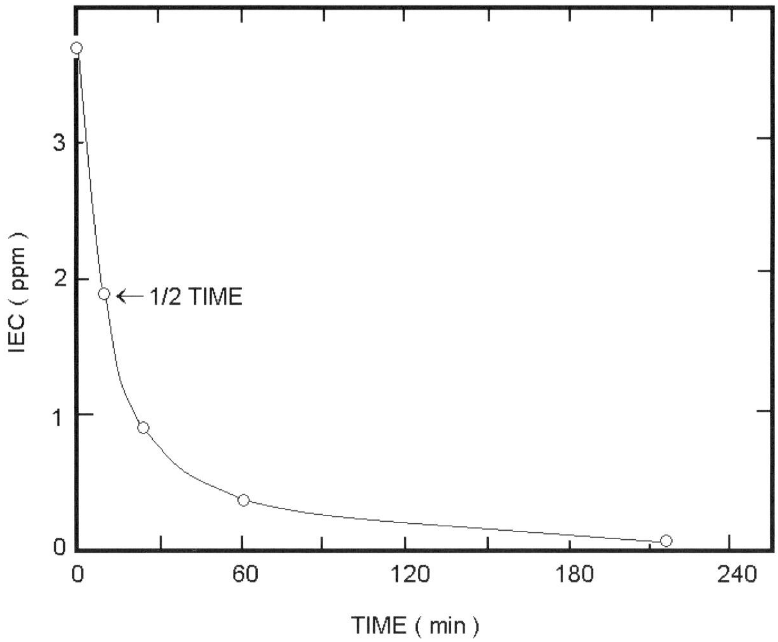

Fig. 5.17. Time course for the loss of ethylene from Haden mangoes after harvest (Burg, 1963, unpublished).

been equilibrated with at least 38,400 µl/l ethylene in a gas phase. This obviously has not occurred, but the same ethylene production and evolution rate would result if the transpirational water entering the fruit contained 160 µM ACC. Tomato plants transport sufficient ACC from their roots in xylem sap to account for a foliar ethylene production rate of 25 ml/g·h (Bradford and Dilley, 1978; Bradford and Yang, 1980; Else *et al.*, 1993). Prior to harvest, ACC may have been supplied to the mango in this manner.

After fruits are induced to accumulate ACC by placing them in a N_2 atmosphere, upon subsequent return to air they temporarily produce ethylene at an elevated rate (Fig. 5.18). There is a close similarity between the kinetics of the decrease in ethylene evolution after mangoes are harvested, and in the loss of ACC accumulated during a period of anaerobiosis and consumed when apples (Fig. 5.18), cherry tomatoes (De Vries *et al.*, 1995) and subapial sections of etiolated pea (Saltveit and Dilley, 1978a) are returned to air.

Labelling and heat girdling experiments show that ACC transport occurs through phloem as well as xylem (Amrhein *et al.*, 1982), and since the attached mango fruit contains constitutive EFE, if it lacked the ability to synthesize ACC but continuously received ACC in the xylem and/or phloem sap, its supply of ACC would be quickly depleted after harvest. To investigate this possibility, the stalk supporting attached mangoes was either steamed or bark girdled below the nearest cluster of leaves (Fig. 5.19, *below*), or the stem was split at that same location leaving the bark 'intact' while removing the wood, or wood and bark girdles were applied simultaneously 5 cm apart. Ninety minutes later, the IEC was measured. The IEC only diminished when both xylem and phloem transports to the fruit were simultaneously obstructed, suggesting that ACC is transported to the leaves and fruit in the xylem sap, and that ACC received by the leaves or produced in them, is 'source to sink' translocated to the closest fruit via the phloem. Apparently steaming not only interfered with phloem transport,

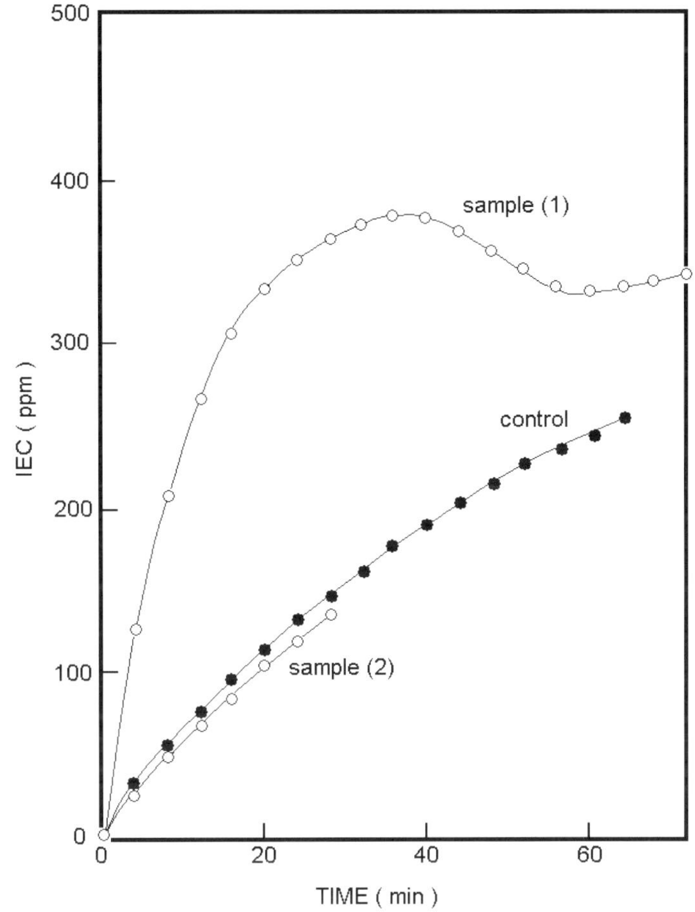

Fig. 5.18. Effect of N_2 treatment on ethylene production by whole McIntosh apples. After apples were evacuated and returned to atmospheric pressure, the IEC increased along the curve shown as 'control'. When the fruits were re-evacuated at the last point indicated on the control curve, and again returned to atmospheric pressure, the IEC increased along an identical control curve. Other apples were vacuum-impregnated with N_2 and left for 4 h. The fruits were then evacuated and placed in air, after which the IEC increased along the curve marked 'sample 1'. When the tissue was re-evacuated at the last point indicated on the sample 1 curve, the IEC now increased at the same rate as the control (sample 2). Each curve is an average compiled from the behaviour of three apples (Burg and Thimann, 1959).

but in addition caused production of 'toxic' substances and/or embolism (plugging) of xylem vessels, as has often been reported, for after bark or wood girdling there was only a slight, inconsistent decrease, but after steam girdling the IEC rapidly declined by 84%. A combination of bark and wood girdling caused a 90% reduction, and harvesting the fruit lowered the IEC by 95%. The loss of ethylene obviously was not caused by 'escape through the stem scar', since the stem was left attached to harvested fruits, a substantial length of stem remained intact between the fruit and the closest wood or bark girdle and the ICC did not decrease when the fruit was detached.

The pre-harvest ethylene level in 'Carabao' mangoes also is higher than the IEC in harvested fruits (Cua and Lizada, 1990), and an immediate effect of harvest on the IEC has been reported for three varieties of citrus (Sawamura, 1981).[13] Removing citrus fruits from the tree at the time when they were developing colour resulted in a

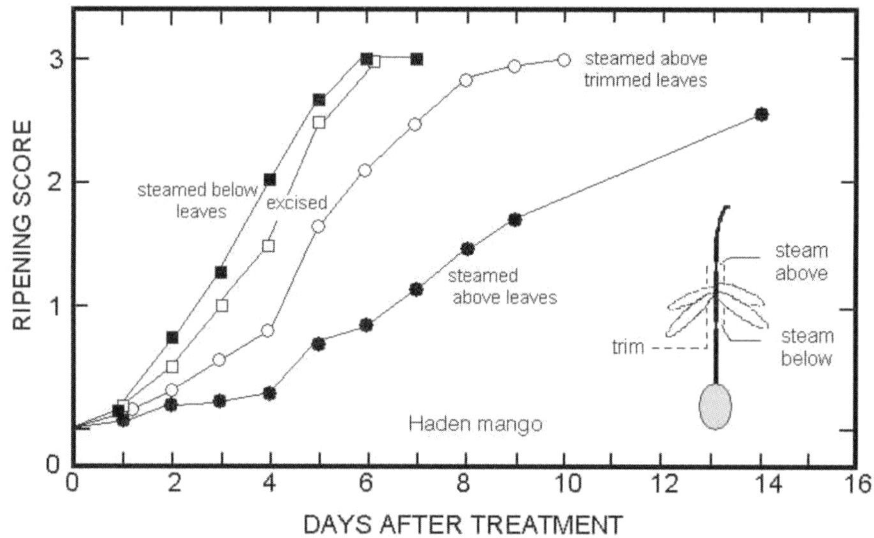

Fig. 5.19. Ripening of fully mature Haden mangoes on the tree when the stalk supporting the fruit was steam girdled, and after harvest at 21.1°C. Steam generated in a portable vaporizer was piped through flexible rubber tubing to a grooved rubber cup, which fitted snugly around the stem and confined the treatment to a 2.5 cm length. The steam was applied for 20 min either 5 cm from the fruit, or 30 cm from the fruit with a cluster of leaves intervening between fruit and steam girdle. In some instances the leaf cluster was removed by trimming, as indicated. Ripening was scored empirically on the basis of coloration and softening using as categories: 0 = hard green; 1 = slight softening; 2 = soft/ripe; 3 = overripe. Untreated control fruit on the tree ripened to stage 1.5 in approximately 30 days. Results obtained with 121 fruits (Burg, 1963, unpublished).

rapid reduction in their IEC (Table 5.3), whereas the opposite might have been expected, since citrus fruits immediately close their stomates after harvest (Ben-Yehoshua *et al.*, 1985). Sawamura (1981b) suggested that the materials involved in ethylene biosynthesis might be translocated continuously from the leaves to citrus fruits, and this explanation is consistent with studies on the promotion of orange-fruit abscission by cycloheximide (Cooper *et al.*, 1969). The IEC in attached Valencia orange fruits was 0.02 μl/l, but after leaves were sprayed with 7.1–355 μM cycloheximide (CHI) to induce ethylene production, not only did they become epinastic, but the IEC in the fruits rose to 0.22–0.98 μl/l even though they had been protected from the CHI treatment. This indicates that ACC produced by the leaves in response to CHI stress was transported to the fruit. Harvested tomatoes display a sharp peak in ethylene production at the transition from breaker to turning, and then ethylene production rapidly decreases as ripening progresses (Klee *et al.*, 1991). In contrast, attached fruits continue to produce ethylene at a high rate through the red-ripe stage, possibly because they import ACC from the vine. ACS is up-regulated in carnation styles 1 h after pollination (Jones *et al.*, 2000), and subsequently translocated to the ovary (Jones and Woodson, 1999); pollination induces ACC production in the stigma of *Phaleonopsis* orchids and the ACC is transported to the petals where EFE converts it to ethylene, causing petal fading (Nadeeau *et al.*, 1993); and when emasculation induces ethylene production in the rostellum of *Cymbidium* and *Vanda* orchids, the ACC is transported from the rostellum to other parts of the flower. Lips excised from *Cymbidium* flowers 15 h after emasculation initially evolved ethylene at a rate of 2 nl/l·h, but ceased evolving the gas within 13 min, suggesting that they did not produce ACC but rather imported it from elsewhere (Burg and Dijkman, 1967; Woltering, 1990, 1993).

The effect of a 20-min steam girdle on Haden mango ripening was investigated using the same methodology described for avocados. The time to reach ripening stage 1.5 was 3.5 days after the fruit was harvested; 4 days when steamed below the leaves; 5 days when steamed above trimmed leaves; 8.5 days when steamed above intact leaves; and 30 days for control fruit left on the tree (Fig. 5.19). This result is similar to that obtained with avocados (Fig. 5.15). An identical study was carried out with 239 Palmer mangoes, except that the duration of the steaming was decreased to 5 min. The time to reach ripeness stage 1.5 was 7 days after harvest; 4.5 days after fruits were steam girdled below the leaves; 7 days after fruits were steam girdled above trimmed leaves; 15 days for fruits steam girdled above intact leaves; and approximately 42 days for fruits left untreated on the tree.

It has been suggested that harvesting or steam girdling a fruit might produce sufficient water stress to cause ripening by inducing ethylene production (Abeles *et al.*, 1992). In the mango studies, leaves located downstream of a steam girdle often wilted within 1–2 weeks, and soon thereafter the adjacent fruit ripened prematurely and abscised. Although this indicates that steaming sometimes caused an eventual breakdown of the xylem water-transport system, water stress cannot explain why trimming away the adjacent leaves accelerates the ripening rate when a steam girdle is applied distal to the leaf cluster, and why ripening is delayed in fruits attached to stems that had been steam girdled upstream of the leaf cluster. There is no reason to suppose that steam would preferentially interfere with water conduction to the fruit if applied above vs. below the leaf cluster, and certainly there can be no selectivity in this respect when comparing the *above* with the *above-trim* condition. Unripe fruits typically possess a lower osmotic concentration than the adjacent leaves, so that when water conduction is prevented the leaves remain fresh and the fruit soon withers (Bartlemew, 1926). Allowing the leaves to remain intact below a stream girdle would be expected to increase the water stress on the adjacent fruit, and yet it delays ripening.

Mangoes that were exposed to 1000 μl/l applied ethylene while attached to the tree ripened somewhat more quickly than harvested fruits kept in air (Fig. 5.16). This indicates that prior to harvest a sufficient concentration of the ethylene-receptor molecule was present to rapidly initiate ripening after it combined with ethylene, and that the ethylene transduction pathway was not rate-limiting when the gas was supplied in excess. As attached mango fruits did not ripen prematurely even though their IEC sometimes exceeded 11 μl/l, the ethylene receptor's apparent K_{m,C_2H_4} must have been in the range 11–1000 μl/l prior to harvest, whereas after harvest it was able to bind ethylene with an apparent K_{m,C_2H_4} of 0.04–0.4 μl/l (Burg and Burg, 1962b; Fig. 5.11).

How harvest increases the ethylene sensitivity of a mango by at least 30-fold, and disposes it and other types of fruits to ripen in response to their low postharvest IEC, remains a mystery. All that can be deduced with a modicum of certainty is that the preharvest sensitivity to ethylene is influenced by a substance or substances transported to the fruit. As the growth of many types of fruits ceases and they stop increasing in dry matter one to several weeks in advance of the climacteric rise in respiration (Archbold, 1932; Barnell, 1940; Roux, 1940; Krotkov, 1941; Kidd and West, 1945; Pratt, 1971), ripening on the tree sometimes may resemble that after harvest in that it only occurs when phloem transport to the fruit has ceased. If a fruit has reached a stage of maturity where phloem transport has ceased to deliver further dry matter, and ripening is about to be initiated, the fruit may not ripen more quickly when it is harvested. This might explain the divergent results that have been obtained when the effect of harvest on tomato ripening has been studied (Sawamura *et al.*, 1978; Saltveit, 1993; De Vries *et al.*, 1995).

During the last weeks on the tree a fruit may increase in weight by 10–20%. Growers would like to harvest more mature, larger fruits to increase the productivity of their farms and orchards, and also because such fruits command a higher sales price since they ripen with better texture, aroma, flavour, colour and eating quality. Because of the high cost of air transportation, many fruits must be exported by sea container, and therefore they need to survive the somewhat unpredictable arrival and departure schedules of container ships, a long transport time and the interval required to accumulate a load for shipment and distribute it after arrival. The longer a fruit remains on the tree, the sooner ripening occurs after harvest. To extend storage life so that the fruit can survive normal sea transport, it must be harvested before it is horticulturally mature. LP overcomes this limitation, allowing fully mature fruits of maximum weight to be successfully shipped for long periods in sea containers.

5.11 Ethylene Receptor

The prediction that the ethylene receptor contains a metal was based on experiments in which the biological activity of ethylene and its physiologically active analogues was shown to be related to their ability to bind silver ion on a gas chromatographic column (Burg and Burg, 1967c).[14] Thirty-two years later, the ethylene receptor's metal ligand was identified as copper (Hirayama *et al.*, 1999; Rodriguez *et al.*, 1999). The amounts of copper required are so minute that tomato plants grown in its absence and exhibiting extreme copper-deficiency symptoms nevertheless develop strong leaf epinasty in response to applied ethylene.[14]

Binding of ^{14}C-ethylene to the ethylene receptor present in plant tissues meets the criteria for involvement of a receptor protein (Trewavas and Jones, 1981; Sisler, 1991). The binding is reversible, the site is saturable and specific, the dissociation constant is at or lower than the physiologically active ethylene concentration, the number of binding sites is small, and chemical modification leads to changes in binding affinity and biological activity (Sisler and Goren, 1981). When plants are exposed to very low ^{14}C-ethylene concentrations, the gas dissolves in the tissue fluids according to Henry's Law, and some binds to the ethylene receptor. The amount of dissolved ^{14}C-ethylene is the same regardless of whether or not sufficient unlabelled ethylene to saturate the receptor is applied before the low ^{14}C-ethylene concentration is introduced. The binding site will be occupied by unlabelled ethylene if a saturating concentration of unlabelled ethylene is present, but by ^{14}C-ethylene if unlabelled ethylene has not been added. The difference between the two conditions is a measure of the binding.

^{14}C-ethylene binding has been detected in leaves, sprouts, flowers and fruits that respond to ethylene, and also in tissues which do not respond, such as certain seeds and the cotyledons of *Phaseolus vulgaris*. Tobacco leaves (Fig. 5.20, *left*), mung bean sprouts (Fig. 5.20, *right*), pea epicotyls and other ethylene-responsive tissues contain a fast-binding component with a 2–15-min half-life and a slow-binding component with a half-life of many hours (Sanders *et al.*, 1991), whereas cotyledons of *P. vulgaris*, seeds of mung bean and other tissues that do not respond to ethylene, seem to possess only the slow half-life component (Sisler, 1991). The fast-binding site can be saturated with ethylene rather quickly, while the slow-binding site often requires at least 20 h (Moshkov *et al.*, 1993). When tissue is disrupted, the short half-life component disappears, while longer dissociative components persist. The partially purified 'slow'-binding component prepared from *P. vulgaris* cotyledons has a 3-h half-life for association and 20 h for dissociation, and bound ^{14}C-ethylene cannot be easily displaced from this site by unlabelled ethylene, nor does a vacuum increase ethylene's rate of dissociation and release (Bengochea *et al.*, 1980; Sisler, 1991), but ethylene is freed from the *Phaseolus* complex by a 20-min heat treatment at 70°C (Bengochea *et al.*, 1980). Ethylene has no obvious physiological effect in cotyledons of *P. vulgaris*, and therefore the function of the 'slow' receptor is unknown. It may represent a mechanism for 'compartmenting' ethylene to control its level in plants (Jerie and Hall, 1978).

Only the fast-associative and -dissociative receptor can account for rapid physiological responses to applied ethylene. Inhibition of stem and root elongation (Fig. 5.21), auto-inhibition of ethylene production (5.07), inhibition of auxin gravitransport (Fig. 5.37, *right*) and prevention of spontaneous pea-section curvatures

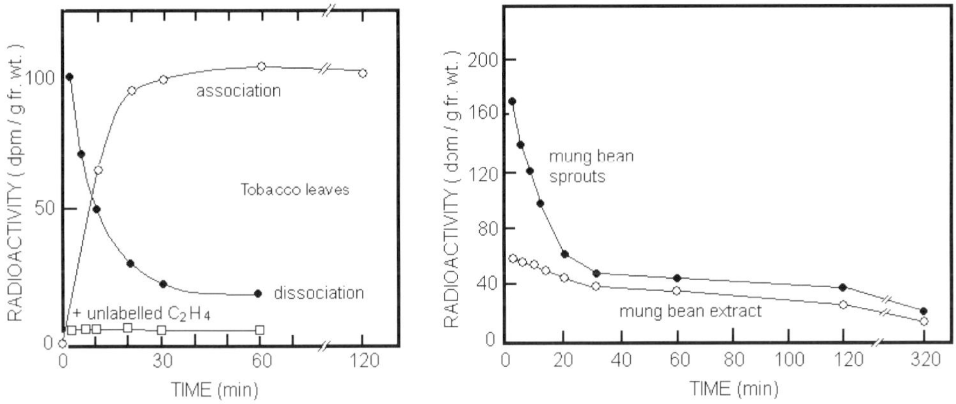

Fig. 5.20. (*left*) Association and dissociation of ^{14}C-ethylene in tobacco leaves (Sisler, 1991). (*right*) Diffusion of bound ethylene from mung bean sprouts and a mung bean extract prepared from an equivalent amount of sprouts (Sisler, 1991).

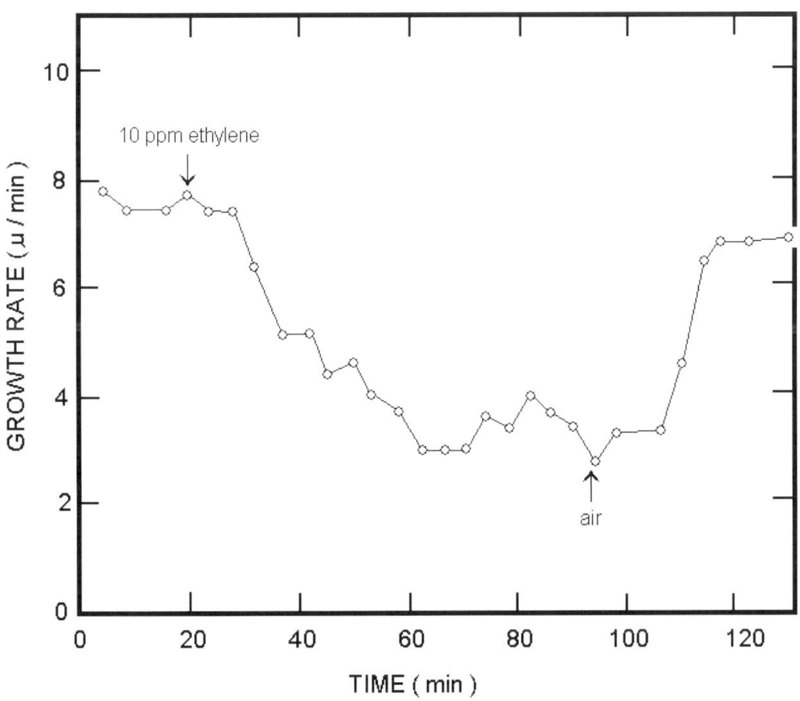

Fig. 5.21. Growth rate of intact etiolated pea plants following application of ethylene, and the subsequent removal of the gas 72 min later (Warner and Leopold, 1971). Similar kinetics for ethylene-induced growth inhibition have been measured in pea and cucumber shoots and in pea roots (Burg, 1973a).

(Fig. 5.40) are induced within 15–30 min after ethylene is applied, and complete recovery occurs in the same time after the gas is removed, indicating that ethylene rapidly associates and dissociates from its receptor and that products of ethylene action do not persist for very long. Other ethylene responses compatible with the kinetic characteristics of the fast, but not the slow binding site, include the lowering of the auxin content of *V. faba* and pea stems within 1 h (Laan, 1934), a shift in microtubule orientation within 30 min (Roberts *et al.*, 1985), induction of epinasty (Leather *et al.*, 1972) and enzyme (Liu *et al.*, 1985a) and messenger-RNA synthesis (Lincoln *et al.*, 1987) within 1 h, abscission within 60–90 min (Evensen *et al.*, 1993), post-harvest development of a respiratory climacteric in banana fruits (Fig. 5.13, *right*) and mangoes (Fig. 5.6, *right*) within a few hours, and increased PO_4^{-3} incorporation into membrane polypeptides of etiolated pea-epicotyl tips within 1 h (Novikova *et al.*, 1993). In LP, the IEC approaches zero, and the slow ethylene receptor causing a physiological action will completely dissociate.

The ethylene receptor also has been studied in mutants of *A. thaliana*, *Lycopersicon esculentum*, *Pisum sativa*, *Nicotiana plumbaginifolia* and *Medicago truncatula*, which are deficient in ethylene perception (Tables 5.4 and 5.5). Three classes of *Arabidopsis* mutants with impaired responses to ethylene have been identified using the triple-response phenotype as a morphological marker (Hirayama and Alonoso, 2000; 5.24; Figs 5.33 and 5.35):

- Ethylene insensitive mutants: *etr1* (Bleecker *et al.*, 1988), *etr2* (Sakai *et al.*, 1988), *ein2* (Guzman and Ecker, 1990), *ein3*, *ein4* and *ein6* (Roman *et al.*, 1995).
- Constitutive ethylene-response mutants: *eto1*, *eto2*, *eto3* and *ctr1* (Guzman and Ecker, 1990; Kieber *et al.*, 1993).
- Tissue-specific ethylene-response mutants: *hls1* (Lehman *et al.*, 1996) and *eir1* (Roman *et al.*, 1995).

In *Arabidopsis*, ethylene is sensed by receptor ETR1 and its related proteins, localized in the plasma membrane. ^{14}C-ethylene binding is reduced by 82% in the *etr1* mutant of *A. thaliana* that does not respond to ethylene with respect to organ elongation, chlorophyll content, peroxidase activity, ethylene biosynthesis and seed germination (Bleecker *et al.*, 1988), whereas leaves and fruits of the *rin* and *nor* non-ripening tomato mutants (Table 5.5) do not display reduced ethylene-binding ability because the ethylene receptor is not the site of these mutations (Sisler, 1991). That *ETR1* encodes a receptor for ethylene also is indicated by the conferring of saturable ethylene-binding activity to yeast after heterologous expression of the ETR1 protein (Schaller and Bleecker, 1995). The binding is prevented by competitors of ethylene action, and mutant alleles of *ETR1* encode proteins that fail to display detectable ethylene-binding activity when expressed in yeast.

The functional ETR1 protein is a disulphide-linked dimer (Schaller *et al.*, 1995), with an N-terminal hydrophobic sensor domain adjacent to the histidine kinase domain, and a putative receiver domain in the C-terminal region (Chang, 1996). The ethylene binding N-terminal domain consists of three putative membrane-spanning subdomains that are modelled as alpha helices (Rodriguez *et al.*, 1999). ETR1 protein forms a membrane-associated disulphide-linked histidine kinase similar to sensor proteins of the two component regulatory systems in prokaryotes and eukaryotes (Chang, 1996; Gamble *et al.*, 1996). The C-terminus has homology to the two-component signal-transducing mechanism used by bacteria to respond to environmental stimuli (Chang *et al.*, 1993; Chang and Stewart, 1998), which contains a sensor protein comprised of an 'input' domain that perceives the stimulus, and a 'transmitter'. A second component protein, the 'response regulator', consists of a module that receives the signal from the transmitter via phosphorylation, and an 'output' domain that triggers the response (Bowler and Chua, 1994). ETR1 has a 'transmitter'-homologous domain and a 'receiver'-homologous domain downstream

Table 5.4. Ethylene-related *Arabidopsis thaliana* mutants (Smalle and van der Straeten, 2000).

Locus	Name	Phenotype and comments	Reference
etr1	Ethylene resistant	Ethylene insensitive; delay in bolting time; increase in rosette size; *ETR1* is homologous to 2-component regulators	5
ers	Ethylene response sensor	Ethylene insensitive; *ERS* is homologous to *ETR1*; mutation was induced by reverse genetics	7
ein2	Ethylene insensitive	Ethylene insensitive; delay in bolting; increased rosette size	3
ein3	Ethylene insensitive	Ethylene insensitive	2
ein6	Ethylene insensitive	Ethylene insensitive	1
ein7	Ethylene insensitive	Ethylene insensitive	1
ain1	ACC insensitive	Ethylene insensitive; increase in rosette size	8
eti	Ethylene insensitive	Ethylene insensitive	9
eto1	Ethylene over-producer	Constitutive ethylene response in etiolated seedlings, due to higher ethylene biosynthesis level	3
eto2	Ethylene over-producer	Constitutive ethylene response in etiolated seedings, due to higher ethylene biosynthesis level	2
eto3	Ethylene over-producer	Constitutive ethylene response in etiolated seedings, due to higher ethylene biosynthesis level	2
ctr1	Constitutive triple response	Constitutive ethylene responses at all developmental stages tested, not due to higher ethylene biosynthesis; phenocopied by ethylene treatment; *CTR1* is homologous to Raf kinases	2
hls1	Hookless	No differential growth in apical hook of etiolated seedlings; phenocopied by treatments with auxins or auxin transport inhibitors; *HLS1* is homologous to *N*-acetyltransferases	3, 10
eir1	Ethylene-insensitive root	Root is ethylene insensitive and agravitropic	1
aux1	Auxin insensitive	Root is agravitropic and insensitive to auxin and ethylene; apical hook slightly ethylene insensitive; negatively gravitropic hypocotyls. *AUX1* is homologous to amino acid permeases	1,11 4
aux2	Reduced auxin sensitivity	Normally gravitropic roots, hypocotyls negatively gravitropic	6
axr1	Auxin resistant	Root is agravitropic and insensitive to ethylene, auxin and cytokinin; the shoot is short and bushy; etiolated seedlings have a short hypocotyl and are defective in apical hook formation; *AXR1* is homologous to ubiquitin-activating enzyme E1	12
axr2	Auxin resistant	Agravitropic root lacking root hairs, and insensitive to auxin, ethylene and ABA; shoot agravitropic and dwarfed	13
axr3	Auxin resistant	Increased apical dominance, increased root proliferation, seems to have an increased auxin response; root is insensitive to auxin, ethylene and cytokinin	14
rhd6	Root hair defective	Altered root hair initiation; phenotype can be rescued with exogenous ACC or auxin	

References: (1) Roman *et al.*, 1995; (2) Kieber and Ecker, 1993; (3) Guzman and Ecker, 1990; (4) Mirza *et al.*, 1984; (5) Bleecker *et al.*, 1988; Chang *et al.*, 1993; (6) Estelle and Somerville, 1987; Lincoln *et al.*, 1990; Leyser *et al.*, 1993; (7) Hua *et al.*, 1995; (8) Van der Straeten *et al.*, 1993; (9) Harpham *et al.*, 1991; (10) Lehman *et al.*, 1996; (11) Maher and Martingale, 1980; Pickett *et al.*, 1990; (12) Wilson *et al.*, 1990; (13) Leyser *et al.*, 1996; (14) Masucci and Schiefelbein, 1994.

Table 5.5. Ethylene-related mutants of *Lycopersicon esculentum*, *Medicago truncatula*, *Nicotiana plumbaginifolia* and *Pisum sativum* (Smalle and van der Straeten, 2000).

Locus	Name	Phenotype and comments	References
Lycopersicon esculentum			
Nr	Never ripe	*Nr* is homologous to *ETR1* and *ERS*, and structurally similar to *ERS*. The fruit slowly assume an orange colour. Less softening, polygalacturonase and lycopene synthesis compared to wild type. Has low, residual sensitivity to 1 μl/l ethylene	1, 3
dgt	Diageotropic	Horizontal shoot and root growth, hyponastic leaves, no lateral roots, marked insensitivity to exogenous auxin with respect to ethylene synthesis and hypocotyl elongation; reduced levels of auxin-binding proteins; phenotype is partially rescued by ethylene treatment. Sluggish response to gravity	2
epi	Epinastic	Epinastic, swollen stems and petioles, excessive root branching, over-production of ethylene in the apical region. Normal auxin and ethylene sensitivities; inhibitors of ethylene synthesis or action fail to revert phenotype to wild-type	4
rin	Ripening inhibitor	A transcription factor. Only very low levels of ethylene are produced; fruit grow normally and then slowly turn pale yellow; hardly soften and contain very low polygalacturonase activity; do not ripen when exposed to exogenous ethylene, but the respiration rate rises. High O_2 causes a slight pink colour to develop	3, 5
nor	Non-ripening	A transcription factor. Ethylene production is very low. Final fruit colour is deep yellow; when ripe contain < 1% of wild type poly-galacturonase; do not ripen when exposed to endogenous ethylene, but their respiration rate rises. More extreme in its action than *rin*	3, 5
Medicago truncatula			
Sik1	Sickle	Ethylene insensitive; hyper-infectable by rhizobial symbiont	6
Nicotiana plumbaginifolia			
aux1	Auxin sensitive	Mild leaf epinasty, short primary root, increased root branching, no root hairs; increased sensitivity to ethylene and auxin	7
Pisum sativum			
agt	Ageotropum	Agravitropic roots, shoots are only gravitropic in the dark; two- to fivefold less ethylene production; possibly a lesion in auxin transport or sensitivity	8
ert	Erectoides	Reduced internode and petiole length, GA insensitive. Higher ethylene biosynthesis	9, 10

References: (1) Hobson, 1967; Lanahan *et al.*, 1994; Wilkinson *et al.*, 1995; Yen *et al.*, 1995; (2) Zobel, 1973; Bradford and Yang, 1980; Kelley and Bradford, 1986; Lomax *et al.*, 1993; (3) Hobson and Grierson, 1993; (4) Fujino *et al.*, 1988; Fujino *et al.*, 1989; Ursin and Bradford, 1989; (5) Brady, 1987; (6) Penmetsa and Cook, 1997; (7) de Souza and King, 1991; (8) Olsen and Iversen, 1980; Takahashi *et al.*, 1991; (9) Reid, 1986; (10) Ross and Reid, 1986.

of an N-terminal region of 313 amino acids whose sequence is novel. Based on the homology of the C-terminus of ETR1 with the prokaryotic proteins, it is likely that perception is signalled via phosphorylation reactions. This two-component ethylene receptor, once bound with ethylene, initiates a signal incorporating either the dephosphorylation or phosphorylation of several gene products downstream of the membrane-bound ETR protein (Chang *et al.*, 1993), and once the signal is perceived, an ethylene response can be triggered. The prokaryotic transmitters are phosphorylated on a histidine residue, and the conservation of this residue in ETR1 suggests that histidine also may be phosphorylated in ETR1. *Arabidopsis ETR1* is one of a five-member gene family (Hall *et al.*, 1999), and in addition several *ETR1* homologues exist (Chang *et al.*, 1993). *ETR2* and *EIN4* (Sakai *et al.*, 1988; Hua *et al.*, 1998) have structures similar to *ETR1*, and *ERS1* and *ERS2* encode receptor proteins lacking the carboxy-terminal response-regulator domain (Hua *et al.*, 1995, 1998).

Ethylene receptors are negative regulators of the ethylene-signalling pathway with redundant or partially redundant functions. Binding of ethylene to the receptors inhibits their biochemical activity (Hua and Meyerowitz, 1998; Fig. 5.24), and in *Arabidopsis* the loss of ability to bind ethylene by any of the five receptor proteins results in dominant insensitivity to ethylene, while double, triple or quadruple loss-of-function mutations in which *ETR1*, *ETR2*, *EIN4* and *ERS2* were knocked out showed a constitutive ethylene-response phenotype. Transgenic *Arabidopsis* plants expressing *ers1* and *ers2* genes (Hua *et al.*, 1995, 1998), and *etr2* or *ein4* mutants, are unable to bind ethylene (Sakai *et al.*, 1988; Roman *et al.*, 1995), but the loss of function of any one of these receptors has no effect on ethylene sensitivity, indicating that functional redundancy exists.

Tomato contains a family of ethylene receptors, designated LeETR1, LeETR2, NR, LeETR4 and LeETR5, with homology to the *A. thaliana* ETR1 ethylene receptor (Zhou *et al.*, 1996; Lashbrook *et al.*, 1998; Tieman *et al.*, 2000). Tomato fruit containing the 'never ripe' mutation (*Nr*) fail to ripen because they show no accumulation of NR mRNA; they cannot bind ethylene, and therefore are insensitive to the gas (Payton *et al.*, 1996). *Nr* results from a dominant mutation in the trans-membrane domain of the NR protein, which is an ERS-like gene product capable of binding ethylene when expressed in a transgenic yeast strain (Wilkinson *et al.*, 1995). *Nr* mutants synthesize reduced amounts of ethylene compared to wild-type fruit and retain residual ethylene responsiveness (Lanahan *et al.*, 1994), which may account for the induction of some ripening-related pathways in *Nr* fruits (Yen *et al.*, 1995). *NR* transcripts rapidly increase developmentally from basal levels in mature-green fruit to maximal levels at the breaker stage when autocatalytic ethylene production is initiated, and expression of the *NR* gene is up-regulated by ethylene in wild-type mature-green fruits (Wilkinson *et al.*, 1995; Lashbrook *et al.*, 1998), but not in *Nr* mutant tomatoes. The level of NR mRNA is extremely low in immature green fruit, and is not elevated by a 1-day exposure to ethylene or a 4-day exposure to propylene, but it begins to increase in mature-green tomatoes when they first display a slight rise in ethylene production. The low concentration of NR mRNA in immature fruits is consistent with reports that 12–15 days of continuous treatment with 1000 μl/l ethylene are required to develop red colour in young tomato fruits harvested 17 days after anthesis (Lyons and Pratt, 1964), whereas as little as 0.1 μl/l rapidly ripens mature-green fruits. 1-MCP inhibits the accumulation of NR mRNA associated with ripening (Nakatsuka *et al.*, 1998), and antisense inhibition of the *Nr* gene results in normal ripening, indicating that *NR* is not required for ripening to proceed and ethylene acts to inhibit the function of the NR protein (Hackett *et al.*, 2000).

Expression of both *NR* and *LeETR4* is elevated during tomato ripening, and transgenic lines with reduced NR mRNA levels exhibit elevated levels of LeETR4 mRNA due to a feedback mechanism resulting in compensation for the missing NR by increased *LeETR4* expression (Giovannoni, 2001). Repression of *NR* had no obvious effect on ethylene signalling other than to elevate expression of *LeETR4*, but reduced *LeETR4* gene expression in transgenic tomato plants affected multiple developmental processes and resulted in symptoms of extreme ethylene sensitivity, indicating that *LeETR4* is a negative regulator of ethylene responses. Repression of *LeETR4* did not elicit any alteration of *NR* expression, but resulted in extreme leaf epinasty, premature floral senescence and accelerated ripening

(Tieman *et al.*, 2000), while over-expression of *NR* in lines with lowered *LeETR4* gene expression eliminated the ethylene-sensitive phenotype, indicating that despite marked differences in structure, these ethylene receptors are functionally redundant (Tieman *et al.*, 2000; Giovannoni, 2001).

The ethylene receptor's metal ligand (Burg and Burg, 1967c) has been identified as copper. An isolated and partially purified glycoprotein-binding component from *P. vulgaris* contains Cu (Smith *et al.*, 1988), and a copper ion in the N-terminal hydrophobic domain of ETR1 is required for ethylene binding. The *etr1-1* mutation in subdomain 2 prevents ethylene binding (Schaller and Bleecker, 1995) by abolishing the capacity of the receptor to coordinate copper (Rodriguez *et al.*, 1999). Cloning and characterization of the *ETR1* gene from *Arabidopsis* provided an opportunity to directly investigate the role of Cu(I) in ethylene sensing (Chang *et al.*, 1993). The sequences that are necessary and sufficient for the protein to bind ethylene are contained in 165 amino acids, and for high-binding activity a copper ion is required in association with the ethylene-binding domain (Rodriguez *et al.*, 1999). A mis-sense mutation in the domain that renders the plant insensitive to ethylene eliminates both ethylene binding and the interaction of copper with the receptor. This ethylene/receptor complex resembles the long half-life receptor revealed by the *in vivo* ethylene-binding assay in that it has an 11-h apparent half-life (Schaller and Bleecker, 1995). The active site consists of five different, associated polypeptide strands, each containing Cu(I). A genetically coded deficiency in any one of the polypeptide chains renders the entire site inactive even though each individual polypeptide chain seems to promote a different ethylene response by negative regulation (Theologis, 1995). An X-ray diffraction and spectroscopy study of a series of Cu(I) mono-olefin complexes has demonstrated that Cu(I)-olefin coordination chemistry is consistent with a role of copper at the ethylene receptor site of plants (Thompson *et al.*, 1983). Cu(I) complexes with imidazole-like ligands and these could participate at the active site of Cu(I)-olefin protein and provide a long half-life for the complex.

Trans-cyclooctene (TCO), a potent competitive inhibitor of *in vitro* and *in vivo* ethylene binding (Sisler, 1991; Schaller and Bleecker, 1995), prevents ethylene from inducing a triple-response phenotype in etiolated wild-type *Arabidopsis*, but TCO induces the triple response in *ran1* (responsive to antagonist1) mutant *Arabidopsis* seedlings (Hirayama *et al.*, 1999). *RAN1*, which acts at or upstream of the ethylene receptor, encodes a protein with significant amino-acid sequence similar to human Menkes/Wilson disease proteins and yeast Ccc2p, both of which are copper-transporting P-type ATPases. These proteins are localized in a post-Golgi compartment where they function to transport copper ions, delivering them to secreted or membrane-bound proteins that require copper for functionality. That RAN1 functions as a copper transporter was indicated by the demonstration that expression of *RAN1* complemented the defects of a ccc2Δ mutant, and also by experiments in which $CuSO_4$ (10–15 μM) had no effect on the germination and seedling growth of wild-type *Arabidopsis*, but prevented *ran1* seedlings from expressing a triple-response phenotype in response to TCO. The suggested role of RAN1 is to deliver copper ions to create functional ethylene receptors (Fig. 5.22).

In vitro studies indicate that dimerized ethylene receptors must contain at least two molecules of copper to assume proper structure and function (Rodriguez *et al.*, 1999). Since mutations in both alleles (*ran1–1* and *ran1–2*) of *ran1* are localized in the functional domains of copper transporter, it is possible that a reduction in copper-transport activity of RAN1 protein creates a state of suboptimal copper: apoprotein stoichiometry that results in altered protein confirmation and reduced ligand specificity, allowing TCO to act as an agonist (rather than antagonist) that de-represses the entire ethylene response pathway. In *RAN1*-co-suppressed plants (Fig. 5.22, *ran1* – no copper) the metal-deficient receptors become non-functional,

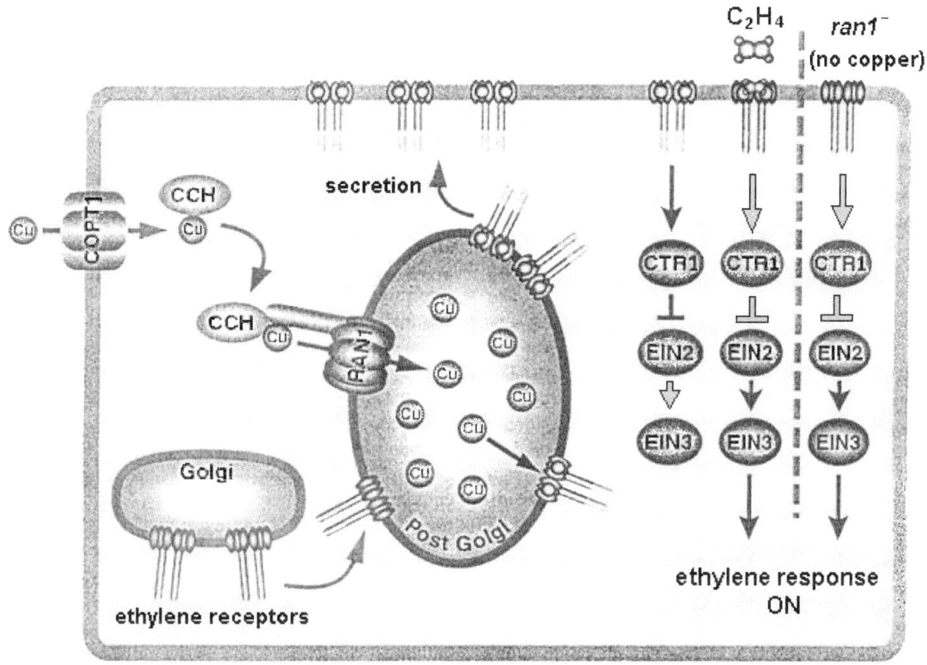

Fig. 5.22. A model for the function of RAN1 in the *Arabidopsis* ethylene-signalling pathway. RAN1 is presumed to be localized in the membrane of a post-Golgi compartment. Copper ions brought into the cell by a copper-trafficking protein are received from a putative copper chaperon (Atxi-like protein), and transported by RAN1 into a post-Golgi compartment, delivering the metal to membrane-targeted ethylene-receptor apoproteins. After the incorporation of copper ions, the receptors are able to coordinate ethylene. In the absence of ethylene, the receptors are active and negatively regulate downstream signalling components, preventing ethylene phenotypes. Ethylene inactivates the receptors upon binding, presumably by causing a reduction in histidine kinase/phosphatase activity. This depresses downstream signalling pathway components EIN2 and EIN3, and gives rise to ethylene phenotypes. Due to reduced copper transport in *RAN1*-co-suppressed plants (*ran1* – no copper) the metal-deficient receptors are non-functional, resulting in a constitutively activated signalling pathway and plants that show constitutive ethylene phenotypes. Black and grey shapes indicate the active and inactive states of ethylene signalling-pathway components, respectively (Hirayama *et al.*, 1999).

resulting in ethylene phenotypes and a constitutively activated signalling pathway and ethylene phenotypes. In contrast to *ran-1* and *ran-2* mutants, which are morphologically indistinguishable from wild-type *Arabidopsis* plants, a *ran-3* allele results in a rosette-lethal phenotype that is not suppressed in ethylene-insensitive mutants, suggesting that *ran-3* also affects a non-ethylene-dependent pathway regulating cell expansion (Woeste and Kieber, 2000a). The similar phenotype of *ran1–3* and loss-of-function ethylene-receptor mutants is consistent with the proposed role of *RAN1* in the formation of functional ethylene receptors, and indicates that copper not only is required for ethylene binding, but also for the signalling function of the ethylene receptors.

Silver thiosulphate (STS) inhibits flower fading and many other actions of ethylene (Beyer, 1979; Veen, 1983) and prevents binding of ^{14}C-ethylene to its receptor site both *in vitro* and *in vivo* (Sisler, 1982b; Goren *et al.*, 1984). Because the binding of ethylene and its physiologically active analogues on a chromatography column containing Ag(I) is correlated with their biological activity (Burg and Burg, 1967c), it has been suggested that Ag(I) may replace the ethylene receptor's metal (Beyer,

1976), or compete with copper for an active site on a secondary proteinaceous subunit of the receptor (Veen, 1983). Silver ion [Ag(I)] exhibits coordination chemistry similar to that of copper ion [Cu(I)], and therefore might be incorporated into ethylene receptor(s), where it may inhibit ethylene binding or proper folding of a functional receptor, resulting in constitutively active receptor proteins that are unable to be inhibited by ethylene binding. Ag(I) also might interfere with copper binding to the receptor, and fail to induce the changes in the receptor that are needed to elicit downstream signalling (Schaller and Bleecker, 1995).

To explain the attachment of ethylene to its metal-containing receptor, and the competitive inhibition of ethylene action by 1-MCP and other cyclopropenes, Sisler proposed a model based on the *trans* effect in inorganic chemistry (Sisler, 1991; Sisler and Serek, 1997; Fig. 5.23). The *trans* effect may be due to the ability of a coordinated ligand to accept electron density by back-donation from the central metal, a mechanism known as π-acceptance. Ligands may bind by donating electrons to a metal and back-accepting electrons into vacant orbitals, causing a change in the charge on the metal, which leads to a substitution of the ligand in the *trans* position. As compounds such as ethylene, carbon monoxide, isocyanides and phosphorous trifluoride, which give an 'ethylene-like response', are all high in the platinum (II) *trans* effect series, Sisler proposed that π-acceptance was the characteristic required for ethylene binding. According to this concept, ethylene induces the withdrawal of electrons from Cu(I) in the ethylene receptor, promoting a ligand-substitution process that results in a response. Because 1-methylcyclopropene (1-MCP) is a highly strained molecule capable of binding to metal, it may attach to the receptor so strongly that it remains bound to Cu(I) and prevents the formation of an active complex. Ethylene (E) must leave the receptor for the final active complex to form. The proposed L_1, L_2, L_3, L_4 and L_5 ligands are unknown. One or more of the hypothetical ligands may be localized on the *ETR1* gene product (Schaller and Bleecker, 1995).

5.12 Ethylene Transduction Pathway

The presence in the ethylene-signalling pathway of a downstream component, CTR1, which encodes a putative serine/threonine kinase related to the MAPKK kinase, suggests that ethylene responses are controlled by a phosphorylation cascade (Kieber *et al.*, 1993; Bowler and Chua,

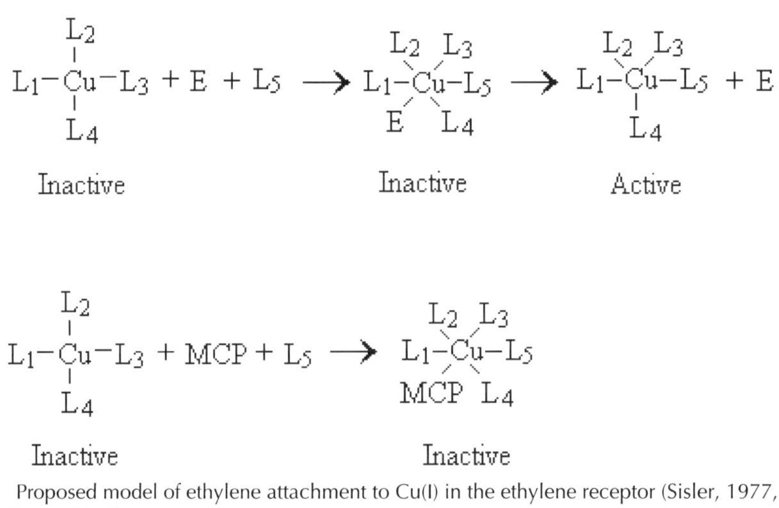

Fig. 5.23. Proposed model of ethylene attachment to Cu(I) in the ethylene receptor (Sisler, 1977, 1991; Sisler and Goren, 1981).

1994). A tomato homologue of the *Arabidopsis CTR1* gene is up-regulated by ethylene during fruit ripening and in response to applied gas (Giovannoni, 1997; Giovannoni *et al.*, 2000), and applied ethylene increases MAP kinase in cytosolic fractions of *A. thaliana* (Smith *et al.*, 2000). Additional evidence that ethylene signal transduction involves protein phosphorylation comes from the use of inhibitors (Raz and Fluhr, 1993), the identification of a LAMMER family kinase that is regulated by ethylene (Sessa *et al.*, 1996) and a sixfold increase in $^{32}PO_4^{-3}$ incorporation into membrane polypeptides of etiolated pea-epicotyl tips after a 1-h treatment with 1 μl/l ethylene (Novikova *et al.*, 1993). A ctr2 mutation which results in constitutive activation of ethylene responses (Woeste and Kieber, 2000b). The *ctr2* mutation disrupts both ethylene signalling and a second, non-ethylene related pathway that is involved in cell expansion, resulting in a rosette-lethal phenotype.

EIN2 acts downstream of CTR1 (Roman *et al.*, 1995; Chang and Shockey, 1999; Fig. 5.24), and further downstream EIN3, a nuclear DNA-binding protein of the AP2/EREBP family (Chang and Shockey, 1999), and its related proteins EIL1, EIL2 and EIL3, are responsible for the ethylene-mediated transcriptional activation of ethylene-inducible genes (Chao *et al.*, 1997; Solano *et al.*, 1998). Because of the complete ethylene insensitivity of the *ein2 Arabidopsis* mutant, EIN2 is believed to be a pivotal signal transducer in the ethylene-transduction pathway. EIN2 is required for the transduction of the ethylene signal from the Raf kinase CTR1 to the transcription factor EIN3, and therefore the functional loss of the EIN2 protein completely blocks the ethylene response. A model for ethylene transduction, derived from genetic studies with *Arabidopsis* (Fig. 5.24), suggests that physiological selectivity occurs at the level of EIN 2,3,5,6,7 and AUX1, EIR1 and HLS1.

The *EIN2* gene encodes a unique membrane-anchored protein (Alonso *et al.*, 1999), whose N-terminal half has a significant amino-acid sequence similarity to the N-ramp protein family that transports divalent cations such as Mn^{2+}, Zn^{2+} and Cu^{2+} (Supek *et al.*, 1996). Applied ethylene induces a large increase in the level of mung bean root ACO mRNA, neomycin almost completely blocks ethylene-induced *VR-ACO1* transcript accumulation, the addition of Ca^{2+} restores *VR-ACO1* gene activation, and there is evidence that Ca^{2+} acts as a second messenger in ethylene-mediated pathogenesis responses (Raz and Fluhr, 1992). The metal ion-transporting activity of EIN2 protein has been analysed extensively, but no transporter activity has been detected.

A tobacco cDNA encoding an EIN3 homologue (TEIL = tobacco EIN3-like) has a DNA-binding domain localized in the N-terminal half, which shows 92% identity in amino-acid sequence with the corresponding region of EIN3. Like EIN3 overexpressing plants, transgenic *Arabidopsis* seedlings overexpressing TEIL cDNA exhibited constitutive triple-response mutants. TEIL's function as a transcription activator may be regulated by modulation of DNA-binding activity through ethylene signalling (Kosugi and Ohasgi, 2000).

5.13 Regulation of Ethylene Sensitivity

Ethylene responses are controlled both by the rate of ethylene biosynthesis and the sensitivity to ethylene perception (Wilkinson *et al.*, 1995). The dissociation constant (K_d) for ^{14}C-ethylene binding ranges from 0.09 to 0.63 μl/l in leaves of tobacco, bean, citrus, *Ligustrum*, carnation, cucumber and tomato, and in carnation and morning-glory petals, mung bean sprouts, bean cotyledons, pea epicotyl and apple pulp (Sisler, 1991), while the apparent K_m for ripening and other actions of ethylene varies over that same range (Table 5.2). The close agreement between the K_d for ^{14}C-ethylene binding and each tissue's apparent Michaelis-Menton constant (K_m) for ethylene action has been cited as proof that the K_m is the dissociation constant for the ethylene–receptor complex. This

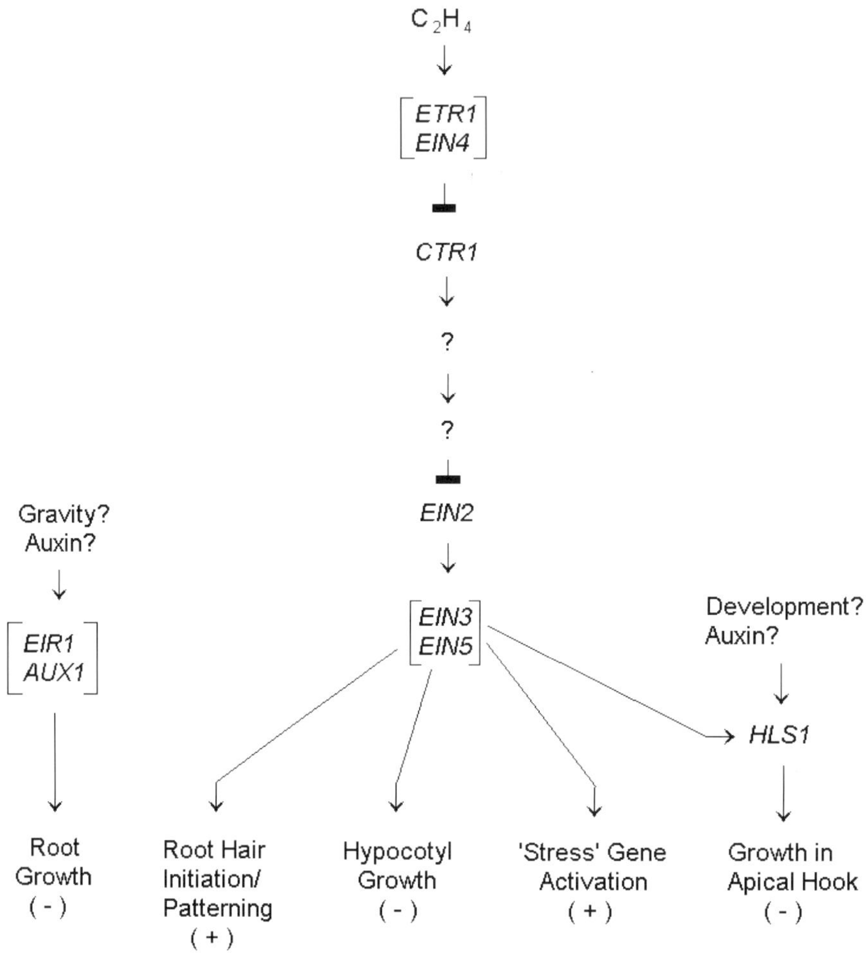

Fig. 5.24. Model for the genetic pathway of ethylene signal transduction. *CTR1* is shown acting after *ETR1* and *EIN4* because *ctr1-1* masks the phenotype of *etr1-3* and *ein4*. The *etr1* and *ein4* mutations are assumed to be dominant negative; therefore, these genes would negatively regulate *CTR1*, which negatively regulates the ethylene-response pathway (Kieber *et al.*, 1993). Negative control points are indicated by a bar. As *ein2, ein3, ein5, ein6, ein7, hls1, eir1* and *aux1* are all epistatic to *ctr1*, they probably act after *CTR1*. The *EIR1/AUX1* effect on root growth is distinct from the *EIN3/EIN5* root response and therefore is shown in a separate pathway affecting root growth. *EIN2* is required for the *EIN3/EIN5* and *EIR1/AUX1* responses and is shown acting before these genes. The gravity and/or auxin responses of *hls1, eir1* and *aux1* are defective. Brackets indicate uncertain gene order. Responses that are regulated positively or negatively by ethylene are indicated (+) or (–), respectively (Roman *et al.*, 1995).

equality depends on the truth of the assumption implicit in equation 5.2 (example 2), that equilibrium is maintained between the receptor–metal complex (RM), ethylene (E) and the receptor–metal–ethylene complex (RME), and that the action of ethylene depends on the RME concentration. If this were valid, a lower apparent K_m for ethylene action after harvest would be an indication that the dissociation constant (K_d) of the RME complex has decreased. Alternatively, if ethylene action depends on the RME concentration, and the RM concentration increased after harvest, this might give the illusion that the affinity for ethylene was amplified, because a lower ethylene concentration would now produce the same amount of RME (example 2).

Because ethylene is a negative regulator, lowering the ethylene-receptor concentration might be expected to have the same effect as applying ethylene, and when the receptor concentration approaches zero, ethylene phenotypes should arise. This expectation has been confirmed by the constitutive ethylene-response phenotype of quadruple *Arabidopsis* mutants in which *ETR1, ETR2, EIN4* and *ERS2* are knocked out (Hua and Meyerrowitz, 1998); by the ethylene phenotype expressed in *RAN1*-co-suppressed plants (Fig. 5.22, *ran1* – no copper) in which the metal-deficient receptors become non-functional; in *ran1*-mutant seedlings treated with *trans*-cyclooctene (Hirayama *et al.*, 1999); and by extreme leaf epinasty, premature floral senescence and accelerated ripening in transgenic tomato plants with reduced *LeETR4* gene expression (Tieman *et al.*, 2000).

If increased sensitivity to ethylene was brought about by a decrease in the receptor concentration (example 2, less *r*), the apparent K_m for ethylene action might increase whenever the receptor concentration became elevated. Instead, in ^{14}C-ethylene-binding experiments, the apparent K_m does not appear to be influenced by the receptor concentration. The K_m for ethylene action in apple and tomato fruits that have an apparent ethylene-receptor concentration in the range $3.2–7.0 \times 10^{-11}$ M is similar to the K_m in leaves and floral parts in which the apparent receptor concentration is 100-fold higher ($1.9–6.9 \times 10^{-9}$ M; Sisler, 1991; Table 5.6). Although apples pass through a maturation phase during which their ethylene sensitivity increases and the fruits become predisposed to ripen (Hartmann *et al.*, 1987), the total number of ^{14}C-ethylene-binding sites in apple pulp decreases or does not change (Blankenship and Sisler, 1989), while the K_d for ^{14}C-ethylene binding increases from 0.1 to 0.57 μl/l (Blankenship and Sisler, 1993). Ethylene production and the apple's 'IEC' should not have influenced the result by displacing ^{14}C-ethylene during the assay, since the concentration of ethylene-binding sites was determined in minced pulp, but 'labile' ethylene-binding sites might have eluded detection as binding of ^{14}C-ethylene was tenfold greater in whole apples compared to minced pulp (Blankenship and Sisler, 1993). There was little change in the ethylene receptor's ability to bind ^{14}C-ethylene in tobacco leaves while senescence advanced during 96 h (Sisler, 1991), and the number of binding sites declined as the leaves aged for 2–6 days in light or darkness (Goren *et al.*, 1984). The decrease in ^{14}C-ethylene binding during the later stages of senescence has been ascribed to a decrease in the affinity of the membrane-associated ethylene-binding sites caused by changes in the physical properties of cell membranes (Brown *et al.*, 1986). A small decline in a morning glory flower's binding capability was observed during the period spanning 24 h before full bloom until 24 h later, when the flowers had in-rolled and senesced

Table 5.6. The measured dissociation constant (K_d) for various ^{14}C-ethylene binding sites, and the apparent Michaelis-Menton constant (K_m) for ethylene action in the same plant tissues.

Tissue	K_d (μl/l)	K_m (μl/l)	Response	References
Tobacco leaves	0.27	0.30	Respiratory stimulation	1
Bean leaves	0.14	0.10	Abscission	2
Etiolated pea epicotyl	0.12	0.10	Growth inhibition	3
Tomato leaf	0.30	0.10	Epinasty	4
Carnation petal	0.07–0.10	0.125	'Sleepy' blooms	5

References: (1) Sisler, 1979; (2) Abeles and Gahagan, 1968a; Goren and Sisler, 1986; (3) Burg and Burg, 1967c; Smith *et al.*, 1987; (4) Abeles, 1973; Sisler, 1982a; (5) Uota, 1970; Brown *et al.*, 1986; Sisler *et al.*, 1986.

(Blankenship and Sisler, 1989). During carnation-petal senescence, the K_d for ethylene binding changed from 0.07 to 0.22 µl/l (3.7×10^{-9} to 1.2×10^{-8} M) and the number of binding sites decreased from approximately 1.1×10^4 per cell in petals of young fully expanded flowers to 0.6×10^4 per cell in in-rolled petals of senescent flowers (Brown *et al.*, 1986).[15] As there is evidence that the receptor protein for ethylene is lipophilic and associated with microsomal membranes (Dodds and Hall, 1980; Sisler, 1980b; Evans *et al.*, 1982a,b), it was suggested that the decline in the number of binding sites with advancing age might be caused by the extensive breakdown of membranes that accompanies senescence, and that the decreased affinity for ethylene in the older tissue possibly resulted from an alteration in the association of the binding site and membrane.

In contrast to the ^{14}C-ethylene-binding results, mRNA measurements indicate that the ethylene-receptor level increases in fruits, flowers and vegetative plants during ripening, flower fading and in response to ethylene. In mango fruits, the level of METR1 mRNA, a cDNA homologue of the ethylene-receptor gene *ETR1*, increases in the mesocarp during ripening, and transiently during wounding (Gutierrez Martínez *et al.*, 2001). In wild-type and the 'Never ripe' (*Nr*) tomato fruits, NR and LeETR4 mRNA increase in anticipation of ripening and as the process commences (Lelièvre *et al.*, 1997a; Giovannoni, 2001). Exposing miniature roses (*Rosa hybrida*) to a low ethylene concentration hastens senescence and induces the expression of genes coding for RhETR, a putative ethylene receptor (Müller *et al.*, 2000a). Both miniature potted-rose cv. 'Bronze', which has a short flower life, and long-lasting cv. 'Vanilla', exhibit a climacteric increase in ethylene production when the flowers senesce, and their postharvest characteristics can be partly explained by differences in their sensitivity to ethylene. Amongst four ethylene-receptor gene fragments isolated from rose plants, *RhETR2* (88% identity to *Arabidopsis ETR1*) was barely regulated during flower development, *RhETR1* (72% amino-acid identity to *A. thaliana ERS1*) was distinctly higher in cv. 'Bronze' than in cv. 'Vanilla' and *RhETR3* (67% amino-acid identity to *A. thaliana ETR2*) increased in senescing flowers of 'Bronze', while in 'Vanilla' this gene was constitutively expressed at a low level (Müller *et al.*, 2001). These experiments contradict the ^{14}C-ethylene binding results referred to in the preceding paragraph, and raise serious concerns about the equivalence of these methods for evaluating ethylene sensitivity.

There are exceptions. Geranium PhETR1 and PhETR2 cDNAs share a 79% amino acid sequence identity with the ethylene receptor ETR1 isolated from *A. thaliana*, and the corresponding mRNAs are expressed at moderate levels in geranium leaves, pedicles, sepals, pistils and petals, and at a very low level in roots (Dervinis *et al.*, 2000). Message levels of PhETR1 and PhETR2 in pistils and receptacles are unaffected by self-pollination or application of 1 µl/l ethylene, but the ethylene treatment induces petal abscission. As *PhETR1* and *PhETR2* expression occurs long before florets are receptive to pollination, and their transcript levels remain constant throughout floral development, their expression is not indicative of a geranium floret's ethylene sensitivity.

Ethylene responsiveness may also depend on a basic PR-1 promoter (Eya *et al.*, 1993), on the availability of EIN2 and downstream transduction pathway elements, and on modulation of the copper supply to ethylene receptors (Hirayama and Alonso, 2000; example 3). Copper also serves as an essential cofactor for enzymes such as cytochrome *c* oxidase, copper-zinc superoxide dismutase, lysyl oxidase, ascorbate oxidase, diamine oxidase, phenol oxidase and dopamine β-hydroxylase, but since copper ion is highly toxic, its intracellular concentration has to be tightly regulated. Extra copper ions are sequestered in non-reactive forms, and a plasma membrane-localized copper transporter imports copper ions into the cytoplasm, where they are immediately bound by copper-trafficking proteins which differ in their intracellular targets. Because ethylene causes leaf senescence and the copper level decreases by half in *Arabidopsis* leaves a few days after the onset of senescence (Himelblau *et al.*, 1998), senescing leaves might become more sensitive to ethylene if a copper deficiency inactivated the ethylene receptor's normal function in the same manner that ethylene does when it binds to its receptor (Hirayama and Alonoso, 2000). However, *ran1* mutants that presumably provide a reduced copper supply to their ethylene receptors did not show enhanced ethylene sensitivity compared to wild-type plants (Hirayama *et al.*, 1999), and copper-deficient tomato plants that exhibit deficiency symptoms such as dwarfing of young leaves, chlorosis and shortened internodes (Twigg and Link, 1951) are not epinastic, and retain an ability to respond epinastically to applied gas.[14]

Changes in membrane fluidity caused by compositional changes and superoxide radical production may alter the K_m of membrane-associated ethylene-binding sites during flower senescence (Thompson *et al.*, 1982; Mayak *et al.*, 1983; Brown *et al.*, 1986). The increase in ethylene sensitivity of *Petunia hybrida* flowers after pollination may be caused by accumulation of short-chain saturated fatty acids ranging in length from C_6 to C_8 (Whitehead and Halevy, 1989). Treating bananas with 100 μM octanoic acid seemed to increase slightly ^{14}C-ethylene binding in fruits pretreated with ethylene for 6 h, but it hardly affected their subsequent ripening regardless of whether they were exposed to ethylene for 0, 6 or 24 h (Whitehead and Bossè, 1991).

Because ethylene oxide seemed to 'synergize' with ethylene in causing cotton-leaf abscission, rice growth stimulation and the triple response in peas (Beyer, 1980), it was suggested that ethylene oxide's production during ethylene metabolism might modulate a tissue's ethylene sensitivity (Beyer, 1985; Sanders *et al.*, 1986). The opening of rose and carnation blooms is delayed by 0.25% ethylene oxide (Lieberman *et al.*, 1964), but the effect is due to phytotoxicity, and does not extend their longevity (Asen and Lieberman, 1963). The concentration of ethylene oxide, which partially prevents the ethylene-induced triple response of etiolated pea seedlings and delays tomato ripening, also is toxic (Lieberman *et al.*, 1964; Veen, 1987), and > 0.5% ethylene oxide induced a sharp, transitory rise in ethylene production by tomatoes, almost certainly due to tissue damage (Lieberman and Mapson, 1962; Lieberman *et al.*, 1964; Ben-Yehoshua *et al.*, 1966).

5.14 Initiation of Ripening in Climacteric Fruits

The mechanisms initiating climacteric fruit ripening have been most extensively studied in tomatoes. Ripening-related biochemical changes, which cause alterations in a tomato's colour, texture, flavour, taste, aroma, physiological response and enzyme activity, are summarized in Tables 5.7 and 5.8. Initially, ACC synthase, EFE, endo-polygalacturonase (PG) and phytoene synthase increase, followed by enhanced malic enzyme activity, and elevated concentrations of fructose-2,6,-biphosphate and cell-wall hydrolases. The climacteric respiratory rise that occurs in response to autocatalytic ethylene production is associated with enhanced ATP levels and a surge in protein synthesis. Although pectic polymers are extensively modified by PG during ripening, this is not the sole cause of softening, and antisense suppression of *PG* expression does not maintain a tomato fruit's firmness (Sheehy *et al.*, 1988; Smith *et al.*, 1990). Transgenic tomatoes expressing 1-aminocyclopropane-1-carboxylic acid (ACC) deaminase remained significantly firmer than control fruits during ripening, even though the same PG level developed in the control and transgenic fruits (Klee, 1993), and when a *PG* gene was introduced into the tomato-ripening mutant *rin*, which normally fails to express *PG* or soften, *PG* was expressed but the fruit remained firm. An endo-β-1-4-glucanase induced during tomato fruit ripening may contribute to softening by degrading cell-wall hemicelluloses (Lashbrook and Bennett, 1993).

Table 5.7. Ethylene-dependent and -independent developmental changes during ripening of climacteric fruits (adapted from Lelièvre *et al.*, 1997a).

Process	Experimental evidence	References
Ethylene-dependent		
Softening	Softening prevented in *ACO*-antisense melons and inhibited in *Nr* tomato mutants. Treatment of immature pear fruits with propylene or > 0.1 μl/l ethylene selectively induces softening, but not other aspects of ripening	5, 10, 15
PG transcript expression	Ethylene is required for *PG* transcript expression in tomatoes, avocados and cantaloupes. Ethylene antagonists prevent *PG* transcript expression in tomatoes	12, 13, 16
Lycopene synthesis	Strongly retarded in transgenic tomatoes. The *Nr* tomato mutant fails to synthesize lycopene	2, 5, 9, 17
Chlorophyll loss	Totally prevented in the rind of transgenic melons	3, 4, 8
pTOM4, pTOM36 and pTOM75 cDNA clones	Accumulate at higher levels in antisense than in wild-type tomato fruits, indicating negative regulation by ethylene	1
pTOM5, pTOM99 and pTOM137 cDNA clones	Levels of mRNA homologous to pTOM5 (phytoene synthase), pTOM99 (E8 homologue) and pTOM 137 were significantly lower in *ACO*-antisense tomatoes. Ethylene treatment partially restores these genes	1
E4, E8, J49 gene induction	Induced by ethylene in tomatoes	6, 7
Peduncular cell separation	Does not occur in antisense *ACO* cantaloupes without applied ethylene	3
Ethylene-independent		
Chlorophyll loss	Interval between anthesis and onset of chlorophyll loss is not delayed in transgenic tomatoes with reduced ethylene	2, 17
Carotinoid synthesis	Carotinoid accumulation in melon flesh is similar in transgenic and wild-type fruits, and phytoene synthase expression starts well before autocatalytic ethylene production	10, 11, 17
PG transcript expression	Expression is unaffected in both *ACS*- and *ACO*-antisense tomatoes and occurs in antisense *ACO* cantaloupes. In the ripening-blocked *rin* mutant, exogenous ethylene does not restore the expression of PG mRNA	1, 2, 12, 14
pTOM31, pTOM96 and pTOM129 cDNA clones	These cDNA clones show similar expression in wild-type and *ACO*-antisense tomatoes	1
E8 and *J49* gene expression	E8 is expressed in *ACS-antisense* tomatoes. J49 showed higher expression in antisense than in wild-type fruits, and this expression was repressed by ethylene	13
Softening and membrane disintegration	Continue in antisense *ACO* cantaloupes after applied ethylene is removed	3

References: (1) Picton *et al.*, 1993; (2) Oeller *et al.*, 1991; (3) Flores *et al.*, 2001a; (4) Guis *et al.*, 1997; (5) Tigchelaar *et al.*, 1978; (6) Lincoln *et al.*, 1987; Lincoln and Fischer, 1988; (7) Yen *et al.*, 1995; (8) Flores *et al.*, 2001b; (9) Klee, 1993; (10) Ayub *et al.*, 1996; Guis *et al.*, 1997; (11) Karvouni *et al.*, 1995; (12) Botondi *et al.*, 2000; (13) Theologis *et al.*, 1993b; (14) Knapp *et al.*, 1989; (15) Wang *et al.*, 1972; Gerasopoulos and Richardson, 1996; (16) Buse and Laties, 1993; Lashbrook *et al.*, 1994; (17) Murray *et al.*, 1993a,b.

Ripening is a genetically programmed event involving the regulated expression of specific genes (Grierson, 1987). Two cooperative promoter elements are required for ethylene-responsive transcription of fruit-ripening genes, and cDNAs encoding DNA-binding proteins that interact with ethylene-responsive promoter elements have been detected and isolated from tomato fruits and carnation flowers (Deikman, 1997). Both ethylene-dependent and -independent pathways of gene expression

Table 5.8. Ripening-related changes in tomato fruit (Hobson and Grierson, 1993).

Change	Mechanism
Colour	Chlorophyll breakdown; disintegration of the light-harvesting complexes and dissolution of the chloroplast lamellae. β-carotene and lycopene accumulation in the plastids as they are converted to chromoplasts
Texture	Reduction in galactan, araban and polyuronide content of cell walls; solubilization of calcium–pectin complexes; particularly the solubilization and partial depolymerization of polyuronides; loss of electron density in the middle lamella and cell-wall erosion observed from light- and electron-microscopy studies
Flavour, taste and aroma	Decrease in malate and an increase in citrate; depolymerization and degradation of starch to sugars; destruction of alkaloids such as α-tomatine; reduction in polyphenol and polyamine content; increase in the complexity of the volatile fraction
Physiological response	Decrease in cytoplasmic volume; increase in hydraulic conductivity; redistribution of K^+ between cell compartments; decrease in phospholipid content
Enzyme activity	Increase in invertase, malic enzyme, β-1-4-glucanase, endopolygalacturonase, phosphofructokinase, ACC synthase, EFE and many other enzymes

have been distinguished during ripening based on the behaviour of transgenic and antisense plants[9] and the molecular analysis of naturally occurring mutant lines such as *Nr* (Lelièvre *et al.*, 1997a). Cloning and partial characterization of members of the *ACS* and *ACO* multi-gene families indicate that family members are selectively regulated in mature fruits during the transition from system 1 auto-inhibited ethylene production to system 2 autocatalytic ethylene synthesis.

ACO gene family

Accumulation of ACC oxidase gene transcripts is enhanced when ethylene production increases and also after ethylene is applied to tomato (Barry *et al.*, 1996), apple (Ross *et al.*, 1992), melon (Lasserre *et al.*, 1996), banana (Huang *et al.*, 1997), kiwi fruit (Whittaker *et al.*, 1997), pear (Lelièvre *et al.*, 1997b), peach (Mathooko *et al.*, 2001) and Japanese apricot (Mita *et al.*, 1999). The tomato *ACO* gene family is comprised of at least three transcriptionally active members. *LE-ACO1*, the main gene for ACC oxidase, is expressed at the onset of ripening and LE-ACO1 continues to accumulate along with LE-ACO4 by a process that is largely prevented by 1-MCP. *LE-ACO3* transcripts only accumulate transiently at the breaker stage and then decline sharply as ethylene production rises (Barry *et al.*, 1996), indicating that *LE-ACO3* is down-regulated by ethylene (Lelièvre *et al.*, 1997a). Of the three *ACO* genes expressed in melon fruit, only *LE-ACO1* is stimulated by ethylene (Lasserre *et al.*, 1996). In banana fruits (cv. Grand Nain; AAA group, Cavendish subgroup), expression of *MA-AC01* was detectable at the pre-climacteric stage, increased when ripening commenced, and remained high throughout the later ripening stage, despite a rapid reduction in ACC-oxidase activity (Liu *et al.*, 1999).

ACS gene family

Nine *ACS* genes have been identified in tomato, melon (Yamamotto *et al.*, 1995) and winter squash fruits (Nakajima *et al.*, 1990; Nakagawa *et al.*, 1991). Among the eight-member (at least) *ACS* gene family expressed in tomato fruits (Barry *et al.*, 2000), the *LE-ACS6* gene is responsible for system 1 ethylene production, and is negatively regulated in pre-climacteric fruit. *LE-ACS6* is expressed in immature and mature-green tomatoes, but not in ripening fruits unless they are treated with 1-MCP, indicating that expression of this gene is regulated by a negative-feedback mechanism. After signals for LE-ACS6 were detected in immature green fruit, they were eliminated by exposing the fruit to propylene, proving that LE-ACS6 is auto-inhibited in the presence of sufficient ethylene. During the natural onset of tomato ripening, the abundance of LE-ACS6 mRNA gradually decreases to an undetectable level at the turning stage. *LE-ACS1* and *LE-ACS3* genes also are responsible for pre-climacteric system 1 ethylene production. Their transcripts accumulate constitutively throughout fruit development and ripening, and the abundance of their mRNAs is not influenced by treatment with either 1-MCP or propylene, indicating that expression of these genes is independent of ethylene action (Nakatsuka *et al.*, 1998). *LE-ACS2* and *LE-ACS4* are the dominant genes responsible for system 2 ethylene production. Increased ethylene synthesis during the transition from system 1 to system 2 ethylene production is caused by accumulation of transcripts for *LE-ACS2*, *LE-ACS4* and *LE-ACO1* (Rottmann *et al.*, 1991; Lincoln *et al.*, 1993; Barry *et al.*, 1996, 2000) Expression of these genes is prevented by 1-MCP via a positive-feedback mechanism (Nakatsuka *et al.*, 1997), and their transcripts, absent in mature-green fruits, are inducible by treatment with ethylene (Olson *et al.*, 1991; Lincoln *et al.*, 1993). Expression of *LE-ACS2* first occurred in mature-green fruit when ethylene production increased above the basal pre-climacteric system 1 level, and propylene did not induce accumulation of *LE-ACS2* or *LE-ACS4* transcripts in immature green fruit within 4 days, but did in 8 days of treatment, indicating a possible lack of a rapid autocatalytic system for ethylene biosynthesis in young fruit. This finding is in agreement with the observation that tomatoes harvested as early as 17 days after anthesis require 12–15 days of continuous treatment with 1000 μl/l ethylene to develop red colour (Lyons and Pratt, 1964). During the ripening of peach fruits, the abundance of PP-ACS1 mRNA increased in parallel with ethylene production and ACC-synthase activity (Mathooko *et al.*, 2001). In banana fruits (cv. Grand Nain; AAA group, Cavendish subgroup), *MA-ACS1* is inducible by ethylene, and MA-ACS1 mRNA increased along with a sharp rise in ethylene production at the onset of ripening, and then decreased rapidly.

NR-gene expression

Expression of the *NR* gene was extremely low in immature and mature-green tomato fruits, but increased greatly at the turning stage (Wilkinson *et al.*, 1995; Lashbrook *et al.*, 1998; Nakatsuka *et al.*, 1998). 1-MCP transiently inhibited the accumulation of NR mRNA, but recovery occurred after 2–4 days, simultaneously inducing the re-accumulation of LE-ACS2, LE-ACS4, LE-ACO1 and LE-ACO4 mRNAs. This indicates that NR mRNA is positively regulated by ethylene in a developmentally specific manner, and suggests that the synthesis of the NR receptor protein during ripening may lead to the recovery of gene transcripts that are regulated by a positive-feedback mechanism.

Figure 5.25 presents a model summarizing the cascade of ethylene-dependent gene expression and ethylene-independent developmental changes that may occur during the transition to autocatalytic ethylene production in a climacteric fruit. A tomato fruit's pre-climacteric system 1 ethylene production is mediated by the *LE-ACS1*, *LE-ACS3* and *LE-ACS6* genes, together with *LE-ACO1* and *LE-ACO4*, and ethylene production shifts to system 2 at the climacteric stage with a burst in the accumulation of LE-ACS2, LE-ACS4, LE-ACO1 and LE-ACO4 mRNAs as a result of positive-feedback regulation. The shift from system 1 to system 2 ethylene synthesis during the transition from the mature-green to the turning stage may be controlled by the accumulation of NR ethylene-receptor protein that occurs at that time (Nakatsuka *et al.*, 1998).

In ripening antisense fruits of tomato, but not in their leaves, ethylene is 'overproduced' when applied gas activates transcription of a negative regulator of

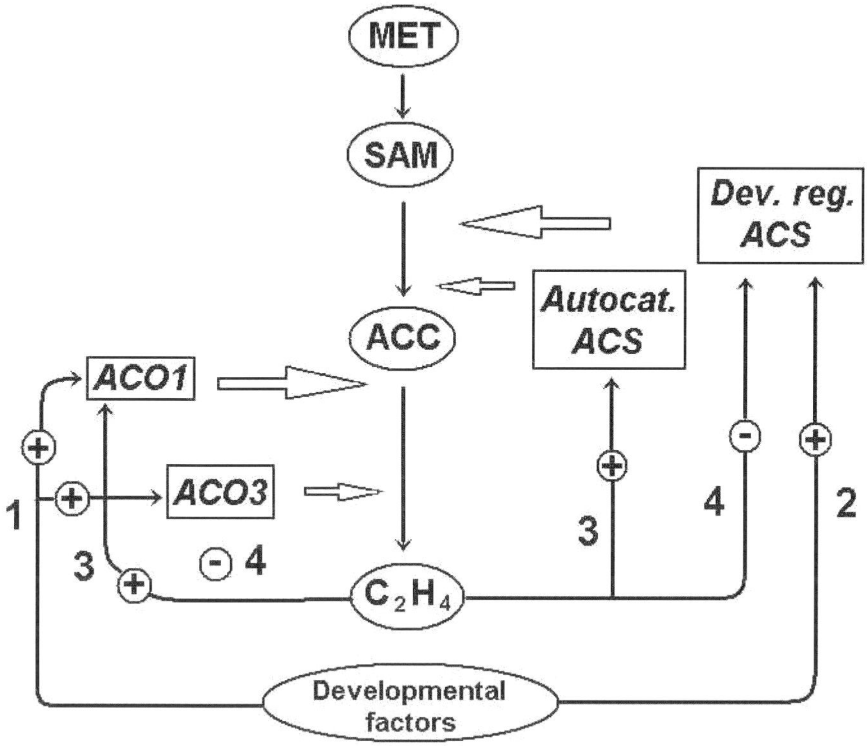

Fig. 5.25. A model for the cascade of *ACS* and *ACO* gene expression during the transition to autocatalytic ethylene production in a climacteric fruit. A rise in ethylene production is initiated by (1) a developmentally regulated increase in *ACO1* and *ACO3* gene expression, followed by (2) a developmentally regulated increase in ACS (Dev. regl. *ACS*). (3) This induces 'autocatalytic' *ACS* gene expression and stimulates *ACO1*, and finally (4) down-regulates Dev. regl., *ACS* and *ACO3* (Lelièvre *et al.*, 1997a).

ethylene synthesis, the '*E8*' gene (Kneissl and Deikman, 1986; Lincoln *et al.*, 1987; Peñarrubia *et al.*, 1993; Deikman, 1997). The E8 protein and its mRNA (Yen *et al.*, 1995) are induced both in *ACS*-antisense tomato fruits (Theologis *et al.*, 1993b), and also after a relatively high ethylene concentration is applied to normal fruits (Lincoln *et al.*, 1987), suggesting that expression of this gene may be controlled by both ethylene-dependent and -independent developmental processes (Table 5.7). The amino-acid sequence of *E8* indicates that it is related to a large family of iron (II) dioxygenases (Deikmann and Fisher, 1988). Because *E8* gene expression requires a functional NR receptor, it does not occur in *Nr* fruit. Mutation of the *EPI* (Epinastic) allele in tomato causes over-expression of *LE-ACS2* and *LE-ACO1* and overproduction of ethylene, but significantly suppresses the expression of *LE-ACS4* and *LE-ACO3* (Ying *et al.*, 1999).

5.15 Ethylene-independent Developmental Regulation of Ripening in Climacteric Fruits

Senescence and ripening have many common features, but ripening includes processes that are not part of the senescence syndrome. Non-ripening tomato mutants, which do not exhibit an ethylene or CO_2 climacteric, lose chloroplast components and cell-wall hemicelluloses, eventually senesce, and become more susceptible to disease and death. Normal fruit undergo these same changes at an earlier time when they ripen, but in addition accumulate pigment and soften due to cell-wall processes not usually associated with senescence. Ripening hastens the onset of senescence and the probability of injury and death, but even when ethylene action is averted and ripening prevented, ethylene-independent developmental changes associated with senescence inevitably continue after harvest (Brady, 1987). Table 5.7 summarizes the ethylene-dependent and -independent[16] ripening processes which have been revealed by studies with transgenic and mutant fruits, and Fig. 5.26 presents a model scheme for their role in the ripening of climacteric and non-climacteric fruits.

Amongst 16 cDNAs corresponding to mRNAs whose abundance in Charentais melons (*Cucumis melo* var. reticulates – a climacteric fruit) is ripening-regulated, one was encoded ACC oxidase, three were associated with pathogen responses, two were involved with sulphur amino-acid biosynthesis, and two had significant homology to a seed-storage protein or a yeast-secretory protein (Hadfield *et al.*, 2000). The remaining eight cDNA sequences did not reveal significant similarities to previously characterized proteins. The majority of the 16 ripening-regulated cDNAs corresponded to mRNAs that were fruit-specific, although they were expressed at low levels in vegetative tissues. Three distinct patterns of mRNA accumulation were identified in experiments with transgenic antisense ACC-oxidase melon fruit: (i) one group of cDNAs corresponded to mRNAs whose abundance was reduced in transgenic fruits but inducible by ethylene treatment, indicating that these genes were directly regulated by ethylene; (ii) a second group was not significantly altered in transgenic fruits or induced by applied ethylene, indicating that these genes are regulated independent

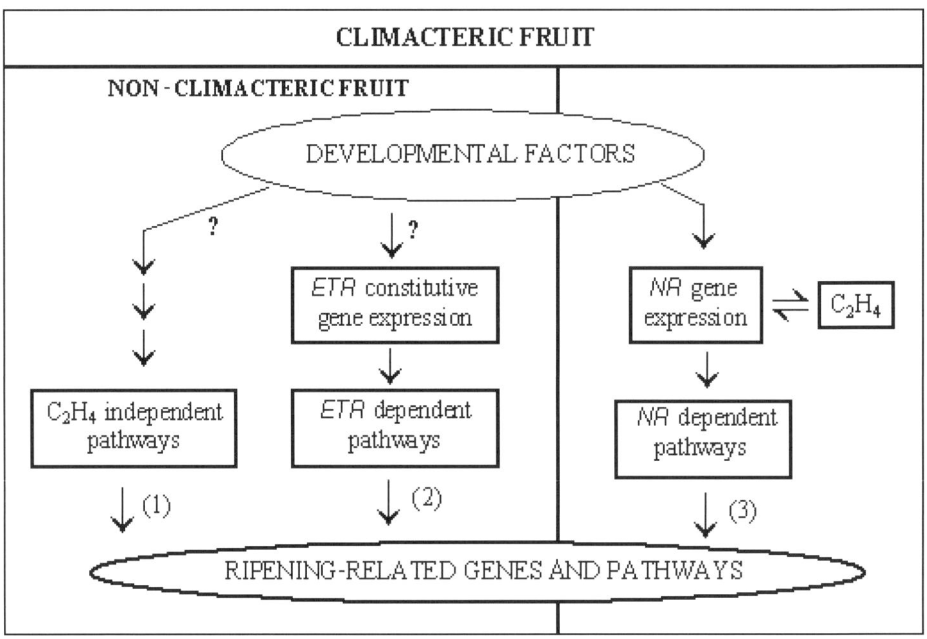

Fig. 5.26. A model for ethylene-dependent and -independent ripening pathways in climacteric and non-climacteric fruits. (1) Ethylene-independent events; (2) events regulated through the *ETR* transduction pathway; (3) events regulated through the *NR* transduction pathway. Ethylene-dependent events are ascribed to the *ETR* pathway when they occur in the absence of *NR* gene expression (in non-climacteric fruit or immature fruit). Events considered to proceed through the *NR* pathway are those that are activated when *NR* is repressed (ripening phase of climacteric fruits) and that cannot be stimulated by exogenous ethylene in *Nr* tomato mutants (Lelièvre *et al.*, 1997a).

of ethylene; and (iii) the mRNA levels corresponding to the third and largest group of cDNAs were reduced in transgenic fruits and remained low after ethylene was applied, indicating that they also were regulated independently of ethylene.

Ripening begins at the same time in wild-type and ethylene-inhibited tomato fruits, suggesting that the initiation of ripening associated with ethylene biosynthesis can occur as an ethylene-independent developmental phenomenon. An increase in ACC synthase activity at the early stages of ripening is induced by an unknown developmental signalling system (Barry *et al.*, 2000) and several members of the *ACO* and *ACS* gene families, the NR ethylene receptor and E8 are induced during ripening by developmental or non-ethylene-mediated regulation of a subset of ripening genes present in climacteric fruits (Giovannoni, 2001). In kiwi fruit, much of the softening process occurs by an ethylene-independent process prior to the respiratory climacteric and autocatalytic production of ethylene (Wang *et al.*, 2000b). Of three polygalacturonase cDNA clones isolated from kiwi fruit, expression of *CkPGA* and *CkPGB* was only detected in fruits producing endogenous ethylene, in flower buds, petals at anthesis and senescent petals. *CkPGC* expression occurred in softening fruit and reached maximum levels 50-fold higher than the quantities of *CkPGA* and *CkPGB* produced as fruit passed through the climacteric. Expression of *CkPGC* was also readily detected during fruit development, in fruit harvested prior to the onset of softening, in root tips and in senescent flower petals. Experiments with *ACO* antisense cantaloupes indicate that their aroma synthesis is controlled by ethylene (Bauchot *et al.*, 2000), whilst amongst physiological disorders, chilling injury is ethylene-dependent and water soaking ethylene-independent (Lelièvre *et al.*, 2000). The activities of some cell-wall degrading enzymes (pectin methylesterase and exo-polygalacturonase) are not regulated by ethylene, but galactanase, α-arabinosidase, β-galactosidase and endo-polygalacturonase were higher in wild-type fruit, indicating a regulatory role for ethylene (Pech *et al.*, 2000).

The identification of ethylene-independent changes in Sonatine tomatoes was simplified by storing fruits for 12 weeks at 12°C in a 6% $[O_2]$ + 6% $[CO_2]$ CA atmosphere to slow ripening (Jeffery *et al.*, 1984). When these fruits were first removed from the vine, they produced approximately 0.2 μl/kg of 'wound' ethylene during a 1-h measurement, but within 1 week in storage, when the next measurement was made, the rate had fallen to an 'undetectable' level. The specific activity of citrate synthase and malate dehydrogenase decreased during the first 2 weeks of storage in association with changes in organic-acid concentration, but ethylene production only began to increase again between weeks 4 and 6. Regardless of whether 27 μl/l exogenous ethylene was continuously present or absent, fruits kept in NA at 22°C showed similar quantitative changes in the specific activity of citrate synthase and malate dehydrogenase in association with shifts in organic-acid concentrations, before any detectable increase in invertase and polygalacturonase. Changes in polygalacturonase, invertase and pigmentation occurred 2–3 days earlier in fruit exposed to ethylene. It was concluded that immediately after tomatoes are harvested at the mature-green stage, and also during ripening on the vine, the breakdown of starch to equal quantities of glucose and fructose, and alterations in the specific activities of citrate synthase, malate dehydrogenase and malic enzyme occur independently of ethylene action, whereas chlorophyll loss, the continued formation of lycopene and increases in polygalacturonase and invertase are enhanced by ethylene. In a similar study, increases in ethylene evolution, red colour development and loss of fruit firmness were delayed when mature-green tomato fruits were stored in 3% $[O_2]$ at 20°C (Kim *et al.*, 1999). The ripening-related decrease in cell-wall galactosyl residues and increase in soluble galactose were suppressed for 9 days, but between 9 and 13 days these processes commenced, either because relatively low baseline levels of ethylene were required and sufficient, or else the cell wall-related changes were ethylene-independent.

The retarding effect that LP has on the developmental regulation of banana ripening is illustrated in Fig. 5.27 and Table 5.9. Within 30 days, control Dwarf Cavendish fruits ripened in NA at 14°C (Table 5.9). The time required for autocatalytic ethylene production and the respiratory climacteric to begin after fruits were removed from 30 days' LP storage and transferred to atmospheric air was inversely related to the storage pressure (Fig. 5.27, left), indicating that the developmental change that sensitizes bananas to respond to their pre-climacteric IEC (Fig. 5.13, right) was progressively delayed as a function of the degree to which the pressure was lowered. Freshly harvested control fruits and bananas that had been stored in LP for 30 days developed the same maximum rates of ethylene and respiratory CO_2 production when they were transferred to air and ripened. Bananas kept at 10.7–20 kPa (80–150 mm Hg) did not begin to ripen during 120 days storage at 14°C, but experienced a slight reduction in firmness and an increase in glucose content during the first 30 days of storage by an ethylene-independent process (Table 5.9). Subsequently, the firmness and glucose content hardly changed during 90 additional days. When these fruits were transferred to 20°C air after 120 days' LP storage, they and freshly harvested control fruits developed the same glucose content, reduction in firmness and loss of tannins as they ripened. All fruit stored in LP had good texture, aroma and taste after ripening. Figure 5.27 (*right*) illustrates the ripening patterns of Cavendish bananas after 12 (A), 11 (B), 10 (C), 9 (D) and 7 (E) weeks storage in LP at 14°C and a pressure of 6.67 kPa (50 mm Hg). In agreement with Fig. 5.27 (*left*), ripening did not occur during storage, but it is evident that the sensitivity of the fruits to their endogenous ethylene progressively increased as the storage period was extended. The bananas did not respond to their pre-climacteric IEC and increase in respiration rate during LP storage because at a low pressure their endogenous ethylene was eliminated, but after 11 weeks' storage they responded as soon as they were transferred to air and allowed to reinstate a normal IEC. The time for Hass avocados to ripen after removal from LP also is inversely related to the storage pressure (Cicale and Jamieson, 1978; 10.3).

Table 5.9. Changes in glucose content and fruit firmness during storage and ripening of Dwarf Cavendish bananas at 14°C and various pressures. Fruit became yellow-green in 30 days at atmospheric pressure, began to colour after 60 days at 33.3 kPa (250 mm Hg), but remained dark green during 120 days at 20 or 10.67 kPa (150 or 80 mm Hg). Fruit stored at 760, 250, 150 and 80 mm Hg was transferred to 20°C atmospheric air after 30, 60, 90 and 120 days, respectively, and allowed to ripen. During ripening, all fruits developed a glucose content of 12.3 mg/g, a firmness of 1.6 kg, and their tannin content decreased by the same extent (Apelbaum *et al.*, 1977a).

Characteristics	Days in storage at 14°C	Pressure (mm Hg) 760	250	150	80
Glucose (mg/g)	0	3.0	3.0	3.0	3.0
	30	12.3	4.5	5.4	6.2
	60	–	7.2	5.9	6.9
	90	–	–	7.6	7.2
	120	–	–	7.3	6.9
Firmness (kg)	0	6.0	6.0	6.0	6.0
	30	2.4	5.5	5.5	5.6
	60	–	3.8	5.0	4.9
	90	–	–	4.9	5.0
	120	–	–	5.0	4.9

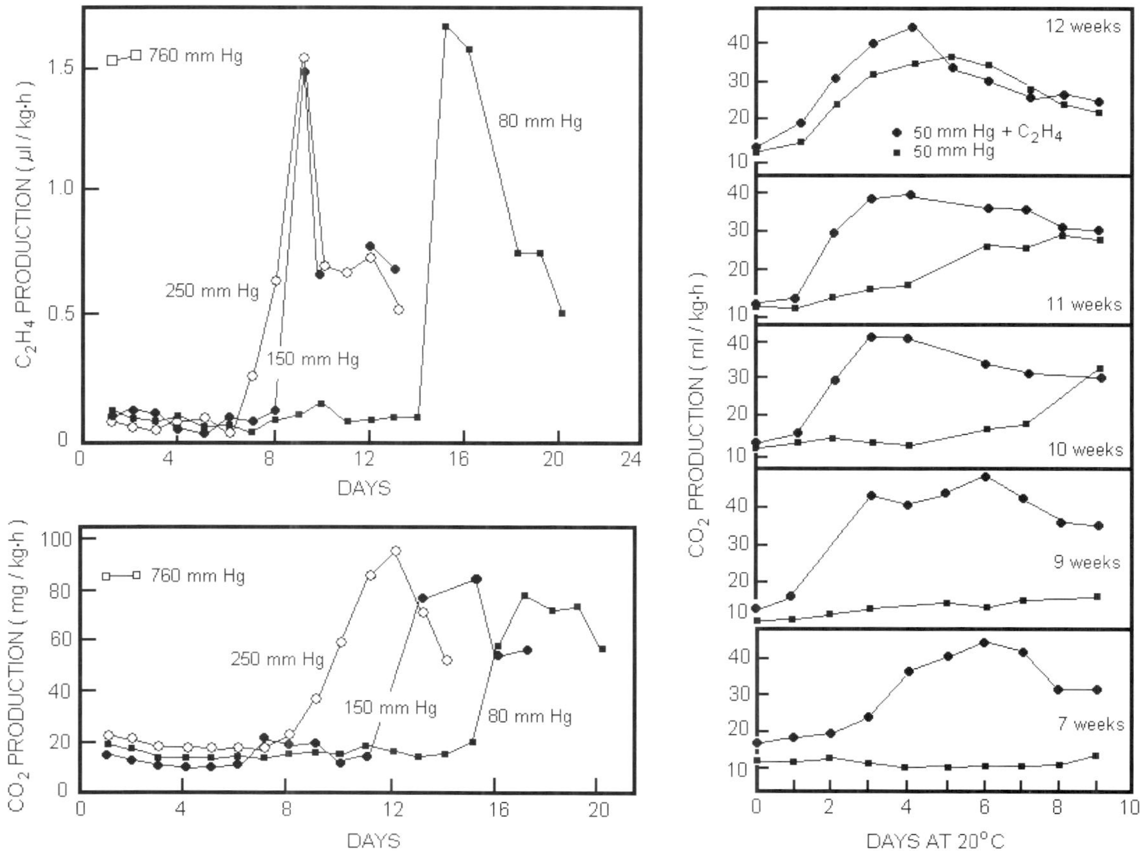

Fig. 5.27. Ethylene (*upper left*) and CO_2 (*lower left*) production by 'Dwarf Cavendish' [*Musa* (AAA Group)] bananas after removal from 30 days' storage at 14°C and various pressures. Measurements were taken beginning 24 h after fruits were transferred from storage to 20°C air at ambient pressure (Apelbaum *et al.*, 1977a). (*right*) Respiration of Cavendish bananas after 12 (A), 11 (B), 10 (C), 9 (D) and 7 (E) weeks' storage in LP at a pressure of 6.67 kPa (50 mm Hg), 14°C. During the 9-day shelf-life period at 20°C, fruits were either untreated or treated with 50 µl/l ethylene, as indicated (Bangerth, 1984).

5.16 Ripening in Non-climacteric Fruits

The natural role of endogenous ethylene in the ripening of non-climacteric fruits remains controversial, but there is evidence that at certain stages of citrus-fruit development ethylene may control or hasten respiration and abscission. Aharoni (1968) measured typical climacteric patterns of ethylene production and respiration in young, immature Marsh grapefruits, Washington navel oranges, Shamouti oranges and Valencia oranges. The rise in respiration and ethylene production coincided with a young fruit's colour change and the abscission of its stem ends, but as the fruits neared horticultural maturity, their respiration rate gradually declined and eventually they produced very little ethylene. Aharoni suggested that citrus fruits may be harvested for commercial consumption in a 'post-climacteric' state, and Trout *et al.* (1938, 1960) reported that a respiratory rise occurs in oranges if they are picked prior to the commercial ripening stage.

Endogenous ethylene may influence de-greening of citrus fruit at certain developmental stages.[17] Shamouti oranges are degreened by 0.1 μl/l applied ethylene nearly as rapidly as by 5 μl/l (Apelbaum *et al.*, 1976). The time required for exogenous ethylene to colour lemons is 5–8 days using 5–1000 μl/l, 6–9 days with 0.5 μl/l and 0.2 μl/l is close to the threshold for a meaningful effect during 14 days at 26°C (Denny, 1924b). When data for applied ethylene's effect on Hamlin orange de-greening is plotted on a log scale, the curves approach a straight line up to 10 μl/l and the minimum active concentration, 0.03–0.05 μl/l, is close to the fruit's IEC immediately after harvest (Jahn *et al.*, 1973). A role for endogenous ethylene in the natural de-greening of Shamouti oranges (Goldschmidt *et al.*, 1993) and calamondins (Purvis, 1979, 1981) is suggested by the effect that ethylene antagonists have on the process. Ethylene production by harvested mature-green Shamouti oranges increased from an initial value of 0.007–0.01 μl/kg·h to 0.072–0.1 μl/kg·h within 2–4 days after harvest, the IEC rose to approximately 0.2 μl/l within 3–10 days (Apelbaum *et al.*, 1976), and within 8 days they lost more than 80% of their chlorophyll. Norbornadiene (NBD) and silver nitrate inhibited this loss of chlorophyll by a maximum of 60 and 55%, respectively, indicating that although endogenous ethylene was involved in the de-greening process, chlorophyll loss continued to some extent when ethylene action was prevented. This same ambiguity is apparent in Table 5.7, where de-greening is listed as having the characteristics of both an ethylene-dependent and -independent process.

A somewhat different pattern of respiration and de-greening can be deduced from a series of internal ethylene and CO_2 measurements made with attached Myers lemons, Key limes, Robinson tangerines, Persian limes and Marsh grapefruit (Tables 5.10 and 5.11). In agreement with Aharoni (1968), high ICC values accompanied by a potentially active IEC were measured in young fruits and, as they matured, their IEC and ICC declined, but both the IEC and ICC tended to increase again in full-sized fruits as they changed colour on the tree. A similar correlation between the IEC and fruit colouring was reported for *Citrus unshiu* Marcov. Var. *Praecox* Tanaka cv. Okitsu-wase (*Wase* Satsuma mandarin), *Citrus junos* Tanaka (Yuzu) and *Citrus aurintium* Linn. var. *cyathifera* Y. Tanaka (Daidai). Young fruits contained 2.3–3.2% [CO_2] and extraordinarily high levels of ethylene. The concentrations of both gases continuously declined for several months until eventually only 0.1 μl/l ethylene and 0.9–1.5% [CO_2] remained. Then, during three subsequent months the IEC again increased and the fruit coloured (Sawamura, 1981). The IEC in attached and harvested Persian limes and in harvested Key limes increased to a stimulatory level of 0.55–2.5 μl/l before colouring occurred (Tables 5.10 and 5.11). After mature-green Persian limes were harvested, their IEC and ICC decreased for several days, and then both increased in a 'climacteric' pattern as the fruits coloured. In another study, the IEC in harvested mature-green Persian limes increased to 0.6 μl/l, and ethylene production and respiration rose as

Table 5.10. Internal ethylene and CO_2 measurements made at various stages of citrus fruit development while the fruit was attached to the tree. Fruits were momentarily immersed under water while internal air samples were withdrawn with a gas-tight syringe for analysis by GC (Burg, 1965, unpublished data).

	Grapefruit		Persian lime		Lemon		Key lime*		Tangerine	
Size/ colour	% CO_2	µl/l C_2H_4	% CO_2	µl/l C_2H_4	% CO_2	µl/l C_2H_4	% CO_2	µl/l C_2H_4	% CO_2	µl/l C_2H_4
1/8	11.2	–	10.0	–	10.8	1.44	–	–	9.4	1.54
1/4	6.8	–	10.2	0.34	4.4	0.22	2.70	0.31	6.4	1.22
3/8	3.0	0.67	–	–	–	–	–	–	3.8	0.92
1/2	2.0	0.34	8.2	0.29	3.2	0.13	–	–	2.3	0.48
3/4	1.9	0.07	4.7	0.22	–	–	–	–	–	–
Full-green	6.4	0.70	4.95	0.27	2.7	0.02	2.90	0.20	2.4	0.40
Turning	–	–	6.8	0.57	3.2	0.17	3.95	0.55	0.6	0.73
Ripe	5.2	0.40	9.8	0.78	4.7	0.10	4.05	0.90	0.9	1.45
Over-ripe	2.2	0.17	12.5	2.78	4.0	2.01	–	2.30	5.0	1.05

*Within 2 days after key limes were harvested, the internal ethylene increased to 2.5 µl/l in fruits that initially were full-green, and to 12.4 µl/l in fruits that initially were ripe or turning.

Table 5.11. Changes in the internal atmosphere of harvested full-sized mature–green Persian limes during ripening at 20°C. Fruits were momentarily immersed under water while internal air samples were withdrawn with a gas-tight syringe for analysis by GC (Burg, 1965, unpublished).

Days after harvest	% CO_2	µl/l ethylene	colour
0	3.5	0.33	green
1	2.4	0.10	green
4	1.4	0.02	green
8	2.8	1.44	yellow–green
13	3.0	0.38	yellow

the fruits coloured, abscised their buttons and began to emit aroma (Fig. 5.28; Noichinda, 1997). The ratio between the IEC and ethylene production rate remained constant throughout these events, indicating that the increased IEC was not caused by a change in the fruits' resistance to gas exchange. The limes developed a maximum IEC and ethylene production rate quantitatively similar to values measured in harvested mangoes at their climacteric peak (Fig. 5.7, *left*).

Applied ethylene has effects on citrus fruit other than causing de-greening and increasing their respiration and aroma production. A 3-h ethylene treatment of lemons, grapefruit and navel oranges accelerates limonate A-ring lactone metabolism (Maier *et al.*, 1973), both during the exposure and after it is terminated, and the resulting substantial loss of limonate A-ring lactone reduces the bitterness of juice. Ethylene induces abscission of citrus fruits in explant tests, and applied to whole trees under a canopy it causes mature fruits to dehisce. The primary mode of action of cycloheximide, a chemical that has been tested to loosen citrus fruits in order to improve the efficiency of mechanical shakers, is through stress-induced ethylene production (Ting and Attaway, 1971). Ethylene increases a lemon's volatile production (Winston, 1955 – referred to in Ting and Attaway, 1971), and it releases free amino acids, causes sugar accumulation, and promotes phenylalanine lyase activity in citrus fruits. Studies of ethylene-regulated genes in citrus have revealed the presence of 12 cDNAs corresponding to ethylene-inducible mRNAs, of which three showed increased expression as the fruit coloured (Alonso *et al.*, 1995a,b; Canel *et al.*, 1995). One gene encodes a putative vacuolar processing protease (Alonso and Granell, 1995); another encodes a putative non-photosynthetic ferredoxin (Alonso *et al.*, 1995a).

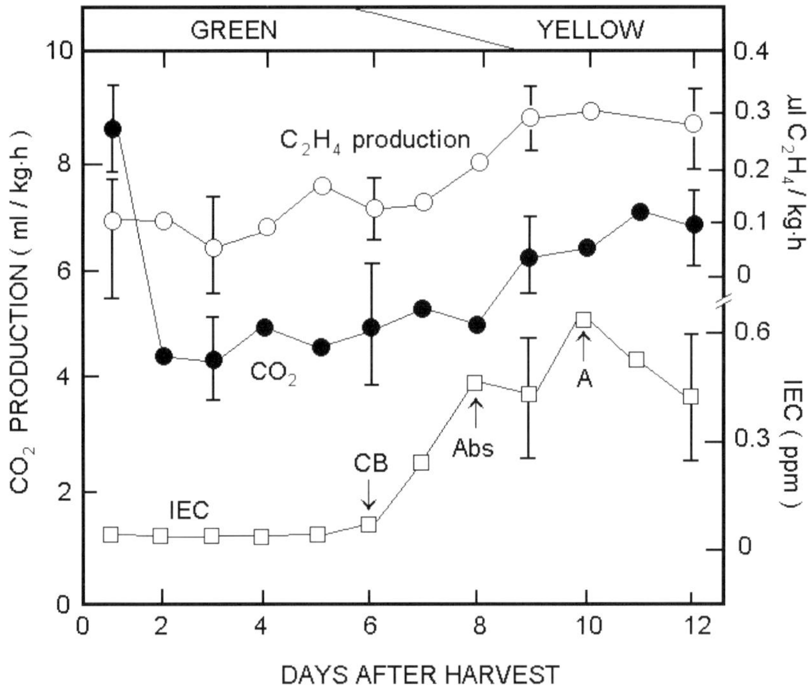

Fig. 5.28. Respiration, ethylene production, internal ethylene concentration and peel colour changes in lime fruits at ambient temperature. Means ± SD bars (n = 4). CB = colour break, Abs = abscission, A = aroma (Noichinda, 1997).

5.17 Circadian Rhythms of Ethylene Production

Ostensibly, the stomatal system is poorly suited to accommodate a gaseous hormone such as ethylene because stomatal opening and closing should cause large daily IEC fluctuations. The same rate of ethylene production would cause the IEC to be at least 20-fold higher at night, even though the stomates remained slightly cracked open when they closed (3.18). Have plants developed a compensatory response to dampen day/night IEC cycles in order to reconcile the ethylene growth regulatory and stomatal systems? The most important factors controlling stomatal opening, water status and [CO_2] would seem to be logical candidates to provide 'feedback' control of the IEC.

In leaves (Jordan *et al.*, 1972; McMichael *et al.*, 1972; El-Beltagy and Hall, 1974; Wright, 1974, 1977, 1980; Aharoni *et al.*, 1977; Aharoni, 1978; Apelbaum and Yang, 1981), avocados (Adato and Gazit, 1974) and various other climacteric fruits (Littmann, 1971), water stress promotes both ACC synthase and oxidase activity, stimulating ACC to ethylene conversion, and when leaves are rehydrated they return to their original ethylene production rate within a few hours if the stress was not too severe or prolonged (Jordan *et al.*, 1972; Ben-Yehoshua and Aloni, 1974a,b; Wright, 1974). Ethylene production by cotton petioles decreases during a normal nightly increase in water potential caused by stomatal closure, and increases during a normal daily decrease (Jordan *et al.*, 1972). A light-dependent diurnal ethylene-production cycle in attached cotton fruits (Lipe and Morgan, 1972a; Fig. 5.29, *lower right*) might reflect a water potential cycle, since the turgor of bolls fluctuates in parallel with leaf water potential as water moves from bolls to leaves along a potential gradient. The water potential of lemons also is determined by leaf evaporation

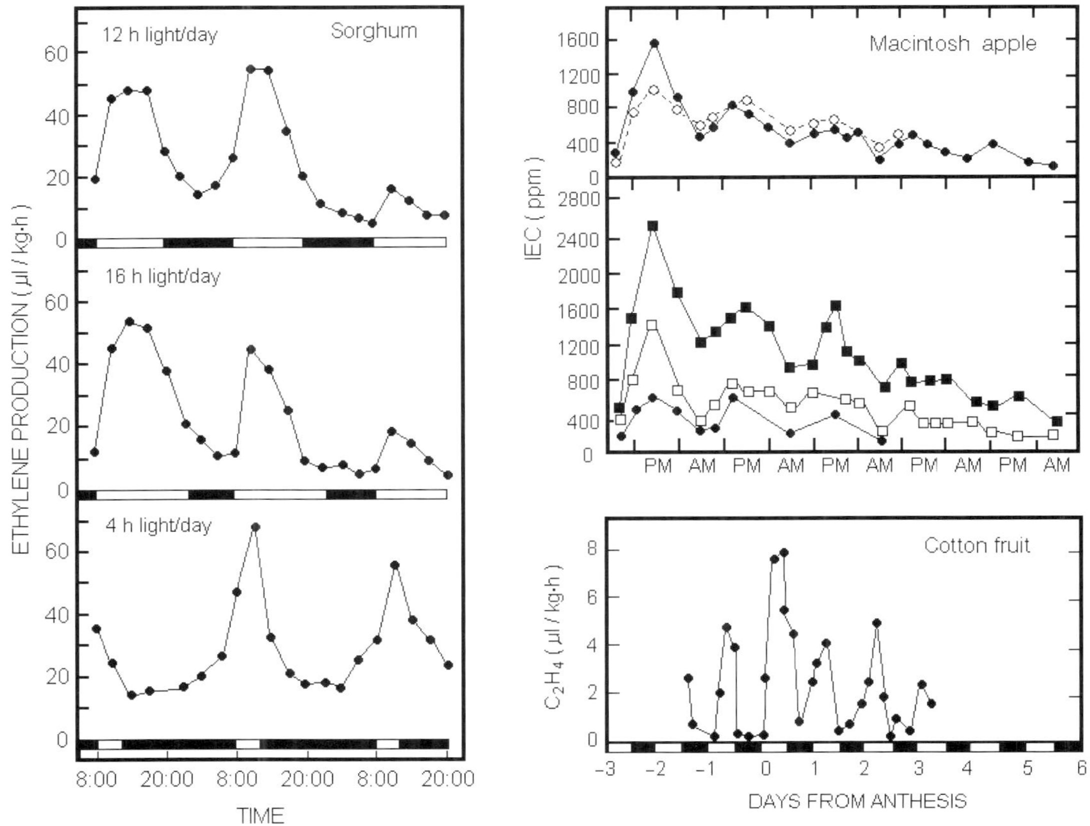

Fig. 5.29. (*upper right*) Diurnal IEC cycle in McIntosh apples removed from 1.1°C cold storage, transferred to the laboratory bench, and exposed to daylight and night darkness at room temperature. Each of the five curves depicts the behaviour of an individual fruit. Ethylene production cycled in an identical pattern. The pulp temperature re-equilibrated within the first 10 h, after which CO_2 was produced at a constant rate (Burg, 1958). (*lower right*) Diurnal pattern of ethylene production by an intact young cotton fruit during a period when young fruit abscission was rarely occurring (Lipe and Morgan, 1973). (*left*) Diurnal ethylene production rates by sorghum plants (Finlayson *et al.*, 1998).

(Bartlemew, 1926). Ethylene emanation in wheat leaves is stimulated within 1 h by a –8 to –9 bar decrease in water potential, reaches a maximum at –11 bars, and may depend on Ψ_{LEAF} throughout the normal diurnal range of –4 to –7 bars (Wright, 1977). Water stress also induces ABA accumulation (Wright, 1969; Wright and Hiron, 1969; Loveys and Kriederman, 1974) and decreases guard cell turgor, causing stomates to close by hydroactive and hydropassive feedback (Raschke, 1975), and if the stress is enduring and severe enough, the stomates remain closed (Jordan *et al.*, 1972; Hsiao, 1973). About –10 bars causes ABA to begin to accumulate in wheat leaves (Wright, 1977) and between –7 to –18 bars in a variety of other leaves (Hsiao, 1973). When stomates open, ethylene production should increase before the water potential decreases sufficiently to induce ABA formation, and at night ethylene production should recover when the leaf rehydrates. However, environmental and climactic variance makes water potential an unreliable regulator of day/night IEC oscillations in stomate-bearing organs. IEC fluctuations in the leaf sheath and panicle of rice plants, and in *V. faba*, *Kalanchoe daigremontiana*, *Bryophyllum*, *Caltha polypetala* and tomato leaves, are not related to stomatal aperture or water stress (El-Beltagy and Hall, 1974; El-Beltagy *et al.*, 1976b; Kapuya and Hall, 1977; Michiyama and Saka, 1988), and low water potential did not promote ethylene production by *intact* plants of *P. vulgaris* L., *Gossypium hirsutum* L. and *Rosa hybrida* L., cv. Bluesette (Morgan *et al.*, 1990).

The ICC increase caused by stomatal closure should inhibit ethylene action at night (4.12), but promote ethylene production (4.10), further elevating the IEC. When stomates close, the IEC might reach a level which auto-inhibits ethylene synthesis, but this would have the disadvantage of limiting the internal ethylene to a level that is partially active for most developmental processes (Table 5.2).

In many plants a circadian rhythm of ethylene production dampens the day/night IEC fluctuations caused by stomatal action. Diurnal IEC and ethylene-production cycles have been repeatedly recorded since they were first described for ripening McIntosh apples in 1958 (Fig. 5.29, *upper right*; Burg, 1958). This circadian rhythm certainly was not caused by a cycling water potential, stomatal action or auto-inhibition, as this fruit lacks stomates (Clements, 1935), steadily loses water after harvest and had an IEC vastly in excess of that needed to saturate an auto-inhibition response. Ethylene production cycles have since been measured in cotton petioles (Jordan *et al.*, 1972), cypress branches (Blake *et al.*, 1980), orchids (Gregg, 1973, 1982), cotton fruits (Lipe and Morgan, 1972a, 1973; Fig. 5.29, *lower right*), cotton and bean seedlings (Gregg, 1982; Rikin *et al.*, 1984; Morgan *et al.*, 1990), in the leaf sheath and panicle of rice plants (Lee *et al.*, 1981), sorghum (Finlayson *et al.*, 1998, 1999, 2000; Fig. 5.29, *left*), etiolated barley, rye and wheat seedlings (Ievinish and Kreicbergs, 1992), *Stellaria longipes* (Emery *et al.*, 1994; Kathiresan *et al.*, 1996), *Cupressus arizonica* (Blake *et al.*, 1980), *Acetabularia mediterranea* (Driessche *et al.*, 1988) and *Chenopodium rubrum* (Macháchová *et al.*, 1997). The ethylene production rate always is maximal during the chronological midday and lowest at night, tending to equalize stomate-induced IEC variations, but the opposite is true of CAM plants such as *Kalanchoe diagramontiana* and *Bryophyllum* since their stomates open to a varying extent in darkness.

Circadian rhythms in stomatal movement also occur, and have been repeatedly documented (Stålfelt, 1963; Willmer and Fricker, 1996), since they were first described by Darwin (1898) and Lloyd (1908). In continuous light, the stomata of many and probably all species open and close with a periodicity of about 24 h (Mansfield and Heath, 1963; Martin and Meidner, 1971; Hennessey *et al.*, 1993; Willmer and Fricker, 1996), followed by a damping or a gradual decrease in the amplitude of the response with time. Stomatal circadian rhythms in continuous darkness have also been observed in many species, although they tend to dampen-out after a few cycles (Willmer and Fricker, 1996).

The circadian system can be divided into three conceptual parts: input pathways that entrain the clock, the central oscillator (clock) and output pathways to generate overt rhythms. Although it is likely that the circadian rhythms of ethylene production and stomatal opening utilize different clocks, the naturally occurring input factors that entrain the clock, and the output pathways that generate overt rhythms, should keep these cycles reasonably well synchronized. The most potent entraining stimuli in plants are light and temperature pulses or temperature steps (McClung, 2001). The periodicity of ethylene production in cotton seedlings continues in constant light, but stops in darkness (Rikin *et al.*, 1984), and in cotton fruits is dampened by a steady temperature and eliminated if both light and temperature are kept constant, with light the predominant influence (Lipe and Morgan, 1972a). When cotton seedlings grown under a photoperiod of 12 h darkness and 12 h light were transferred to an inverted photoperiod, the ethylene production rate continued to oscillate according to the chronological time, irrespective of the immediate light conditions. In sorghum, a diurnal rhythm can be produced with either light or temperature cycles, but both light and temperature cycles are required for circadian entrainment, and the temperature signal overrides the light signal in the production of circadian rhythms (Finlayson *et al.*, 1998). The cycle in etiolated barley, wheat and rye seedlings is induced following germination rather than by external factors, and is endogenous and self-sustaining in constant light and darkness (Ievinish and Kreicbergs, 1992). In *Stellaria longipes*, the oscillations show damping under constant light, darkness and temperature, and light/dark cycles have a greater entraining effect than temperature cycles (Kathiresan *et al.*, 1996). The cycle in *Chenopodium rubrum* was entrained by a 12-h photoperiod and no regular fluctuations in ethylene production occurred in either constant light or constant darkness (Macháchová *et al.*, 1997). Underlying rhythms in mRNA abundance for the *SbACO2* gene encoding ACC oxidase, and in ACC-oxidase activity, seem to modulate the circadian ethylene-production cycle in sorghum (Finlayson *et al.*, 1999, 2000). The ability to convert [3,4-^{14}C] methionine to ethylene is correlated with the ethylene circadian cycle in cotton cotyledons, while conversion of ACC to ethylene is independent of the dark/light cycle (Rikin *et al.*, 1984). The abundance of mRNA for and activity of ACC oxidase fluctuated diurnally in *Stellaria longipes*, with a maximum in the middle of the light phase and a minimum by the middle of the dark phase (Kathiresan *et al.*, 1996), while in *C. rubrum* ACC-oxidase activity did not vary diurnally, rhythmic ethylene production seemed to result from variations in ACC-synthase activity (Macháchová *et al.*, 1997) and the photoperiodic regime affected both ACC and malonyl-ACC levels as well as the conversion of ACC to ethylene. The circadian ethylene-production rhythm in *Acetabularia* was correlated with the ability of cells to convert added ACC to ethylene, and the cap-promoting effect of ethephon seemed to cycle, with maximum sensitivity displayed when ACC was applied during the day, with minimum effectiveness at night (Driessche *et al.*, 1988).

Not only is ethylene production controlled by a biological clock, but in addition the gas affects several rhythmic nutations. Circumnutation, nyctinastic sleep movements and leaf in-rolling of tomato and sunflower plants are reversibly prevented by 2 μl/l applied ethylene (Crocker *et al.*, 1932). The gas acts as an 'anaesthetic', producing growth rigour in *Coleus*, *Ricinus* and *Datura stramonium* plants (Doubt, 1917);[18] 5–8 μl/l of ethylene causes the leaflets of upright *Mimosa pudica* plants to fold, lose their sensitivity to touch and droop as they normally do at night or after stimulation; and gassed *M. pudica* plants recover completely when ethylene is removed (Doubt, 1917; Crocker, 1948). *M. pudica* also displays a gravitational response when the petiole bearing the primary pulvinus is displaced from its normal orientation (Fig. 5.30). This movement seems to be identical to the gravitropic bending of a typical stem or root in that it is prevented by the

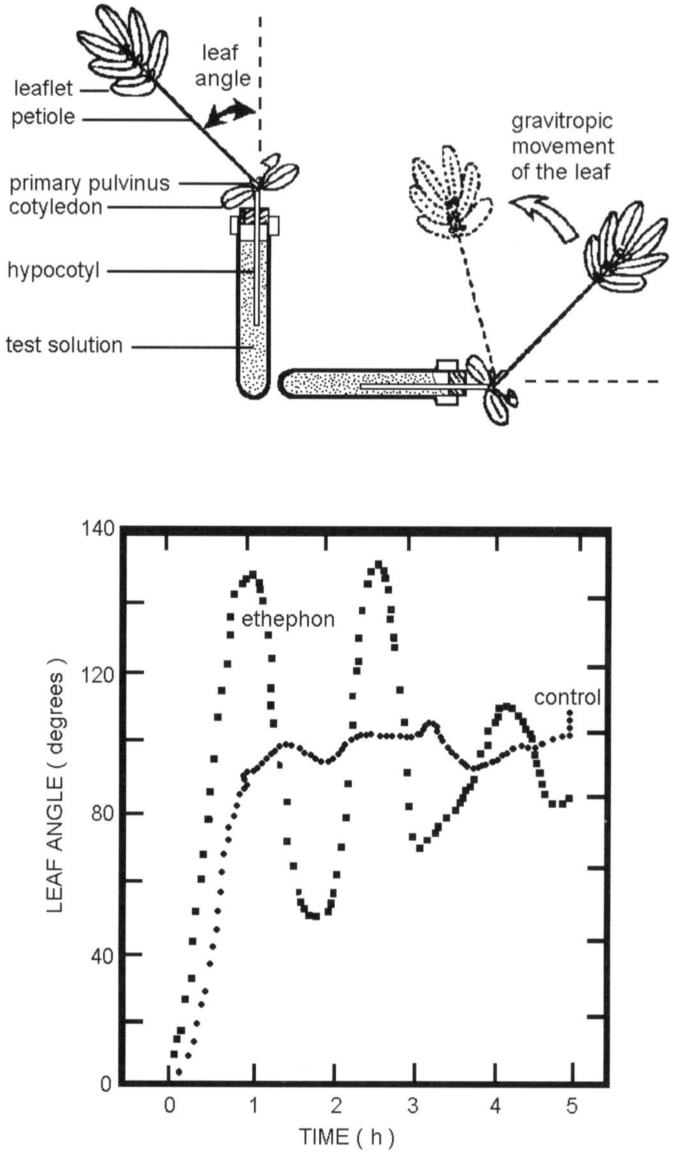

Fig. 5.30. (*upper*) Experimental procedure for the observation of the gravitropic movement of the primary pulvinus of *Mimosa pudica* seedlings. (*lower*) Oscillatory time course of the gravitropic movement following treatment with water for 2 h after excision, or with 10 μM ethephon from 2 to 5 h after excision. Plants were placed horizontally 5 h after excision (Roblin and Pérault, 1985).

auxin-transport inhibitor, triiodobenzoic acid (Roblin, 1976), has a 15-min latency period (Roblin and Pérault, 1985) and apparently perceives gravity with a statolith apparatus (Fleurat-Lessard, 1981). Gravi-stimulated *M. pudica* petioles experience small angular oscillations after they bend, which have been attributed to the antagonistic reaction of the flexor and extensor halves of the pulvini (Aimi, 1963; Roblin, 1979) caused by an endogenous cycle controlled by a biological clock. Neither the latency period for the gravitational response nor the 80 ± 14-min period

Fig. 5.34. Effect of a 5-day ethylene exposure on the orientation of *Vicia faba* seedling roots. (*left*) Lateral roots – air control; (*middle*) lateral roots – 10 μl/l ethylene (Abeles *et al.*, 1992); (*right*) primary roots – 100 μl/l ethylene, except air control (Harvey and Rose, 1915).

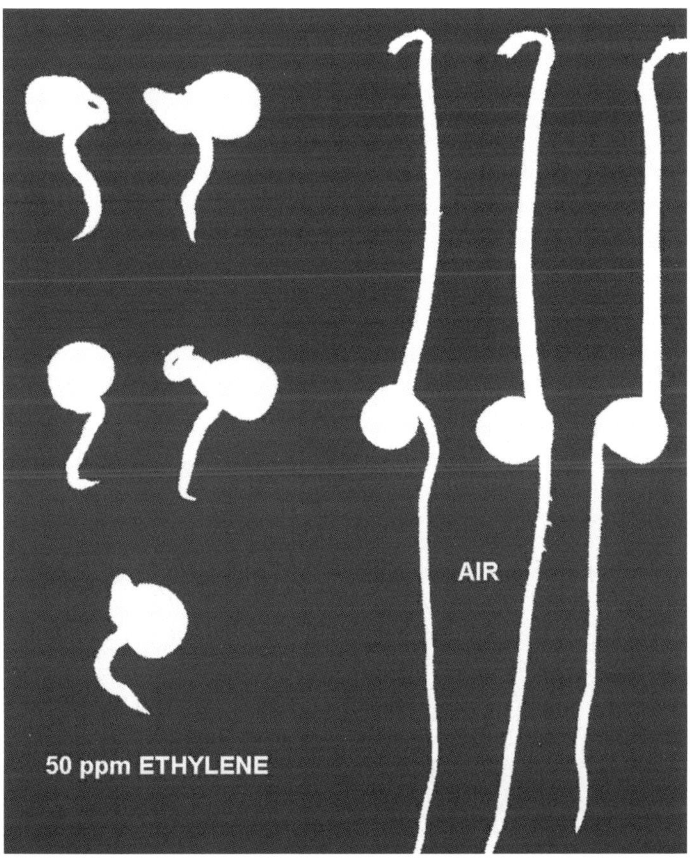

Fig. 5.35. Effect of ethylene on pea seed germination and growth in darkness. Seeds were imbibed for 4 h, sown in pots containing vermiculite and immediately placed for 4 days in a desiccator ± 50 μl/l ethylene. (*right*) Etiolated control plants grown in air have root and shoot systems of equal size. (*left*) Seedlings treated with 50 μl/l ethylene have diminished, swollen shoot systems that are nutating horizontally and swollen roots bending plageotropically (Apelbaum and Burg, 1972).

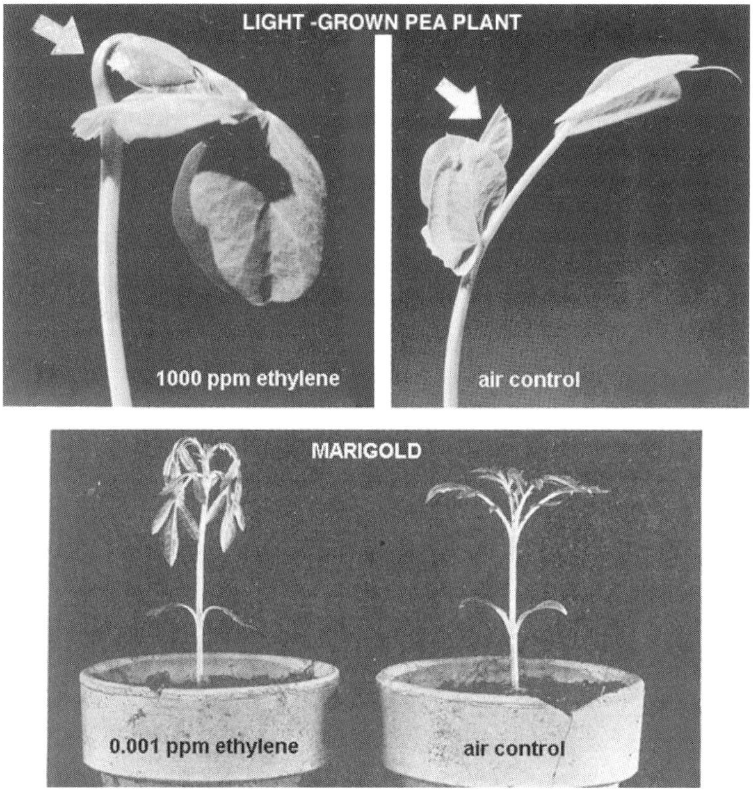

Fig. 5.36. (*upper right*) Normal appearance of expanding bud (arrow) of the shoot of a light-grown pea plant. (*upper left*) After 36-h treatment with 1000 μl/l ethylene, an emerging bud (arrow) has resumed the epinastic hook configuration characteristic of an etiolated pea plant, and the developed terminal leaves have become epinastic (Burg and Burg, 1968). (*lower*) Effect of ethylene on marigold plants. (*lower right*) Control plant in air. (*lower left*) Epinasty induced by a 20-h exposure to 1 nl/l ethylene. This is one of the most sensitive ethylene bioassays (Crocker, 1948).

ethylene in normal shoot gravitropism, and striking tropistic responses to the gas by specific tissues under certain environmental or experimental conditions. Applied ethylene has little or no apparent influence on gravitropic curving or auxin gravitransport in light-grown tomato stems, etiolated black bean hypocotyls, maize and *Avena* coleoptiles, and other tissues that do not respond tropistically to the gas (Burg and Burg, 1966a; Kang and Burg, 1974a,b; Kaufman *et al.*, 1985; Harrison and Pickard, 1986). None of the *Arabidopsis* and tomato mutants that are specifically ethylene-insensitive show defects in shoot gravitropism (Bleecker *et al.*, 1988; Guzman and Ecker, 1990; Van der Straeten *et al.*, 1993; Smalle and van der Straeten, 2000), and inhibitors of ethylene action do not substantially alter the gravitational response of horizontally positioned oat segments (Kaufman *et al.*, 1985) and tomato hypocotyls (Harrison and Pickard, 1986).

Modern reviews have questioned whether applied ethylene affects gravitropic curvatures except by inhibiting growth (Hensel and Iversen, 1980; Bucher and Pilet, 1982; Jackson, 1991; Suttle, 1991; Abeles *et al.*, 1992; Raskin, 1992), but there are innumerable examples of gravity-sensitive ethylene responses in roots and shoots that cannot be explained in this manner. Horizontally positioned pea seedling roots curve downward by > 40° during 1.5 h in air, and even though ethylene only reduces their growth rate by 50–60% (Fig. 5.32, *upper left*), they remain absolutely straight during 3 h of gravistimulation in an ethylene atmosphere (Chadwick and Burg, 1970). Horizontally oriented pea subapical sections cease curving in ethylene within 30 min (Fig. 5.37, *right*), but their elongation is not affected for at least 150 min (Fig. 5.31, *upper right*). Ethylene increases the liminal angle of lateral roots of maize, *Avena* and *P. sativum* (Zimmerman and Hitchcock, 1933; Borgström, 1939b; Lyon, 1972; Goeschl and Kays, 1975); induces curvatures in vertically positioned pea root segments (Bucher and Pilet, 1982); creates characteristically curved root tips in barley by reorienting the direction of root growth from near-vertical towards horizontal (Crosset and Campbell, 1975); and causes bean roots to grow in an undulating pattern (Goeschl and Kays, 1975). In ethylene, the emerging root tips of germinating maize occasionally bend back toward the seed (Harvey and Rose, 1915; Curtis, 1968; Mueller and Brown, 1982b), and vertical maize roots curve away from the plumb-line in both darkness and light (Bucher and Pilet, 1982).[25] Pre-treating maize roots with 0.1 mM AVG eliminates their geotropic sensitivity (Mulkey *et al.*, 1982); the *dgt* tomato mutant, which is not truly diagravitropic, but rather has a sluggish response to gravity, is normalized by a low applied-ethylene concentration (Zobel, 1974; Lomax *et al.*, 1993; Muday *et al.*, 1995); ethylene promotes upward auxin migration in lateral pea roots (Lyon, 1972); and the gas produces turning and coiling of primary roots of *V. faba* (Fig. 5.34, *right*), radish, mustard (Harvey and Rose, 1915), *Ipomoea pentaphylla* (Vyas *et al.*, 1973), *Vigna radiata* (Curtis, 1987) and tomato (Gold and Facciolli, 1972; Woods *et al.*, 1984). Ethylene enhances (Feldman, 1984; Woods *et al.*, 1984) and ethylene action inhibitors prevent (Zacarias and Reid, 1972) circumnutational movements in the tomato root tip. The enhancement of circumnutation by ethylene results in overcoiling of tomato seedling roots, whereas roots of the *dgt* tomato mutant, which are ageotropic, do not form helical spirals in ethylene (Zobel, 1973; Jackson, 1979; Woods *et al.*, 1984; Kelley and Bradford, 1986). Darwin (1981) described experiments in which the tips of secondary radicles of *V. faba* and *P. sativum* bent upwards by 90° when they were stimulated by contact, and Fig. 5.34 (*middle* and *right*) and Fig. 5.35 depict a similar curvature response induced by ethylene applied to lateral roots of *V. faba* (Abeles *et al.*, 1992) and primary roots of *P. sativum*, respectively (Goeschl and Kays, 1975). The similarity between the responses to ethylene application and contact arises because physical impedance elevates the rate of ethylene production by *V. faba* and maize roots three- to sixfold within a few hours (Kays *et al.*, 1974; Sarquis *et al.*, 1991; Morgan *et al.*, 1993), causing them eventually to produce ethylene at a rate one order of magnitude higher than that which causes a supraoptimal ethylene response in pea roots (Chadwick and Burg, 1970). Contact also stimulates tomato roots to produce ethylene (Zacarias and Reid, 1972), and exogenous ethylene causes roots of *Chrysanthemum*, tomato and white clover to grow upward (Doubt, 1917; Borgström, 1939b; Goodlass and Smith, 1979). Ethylene produced in roots subjected to submergence causes them to grow upward rather than downwards (Bond, 1952; Sena Gomes and Kozlowski, 1980; Ellmore, 1981), and high auxin concentrations that induce ethylene production cause the roots of *Allium cepa*, *A. fistulosum*, *A. ramosum*, *A. hymenorrhizum*, *A. sphaerocephalum* and *A. tuberosum* to grow upward in tissue culture (Johni *et al.*, 1965). The diageotropica (*dgt*) tomato mutant, which is characterized by horizontal stem growth (Zobel, 1973; Muday *et al.*, 1995), agravitropic roots (Muday *et al.*, 1995), reduced auxin sensitivity (Zobel, 1973; Jackson, 1979) and a decreased inhibition of root growth in response to ethylene (Muday *et al.*, 1995), resumes near-normal vertical growth when it is treated with 5 ml/l ethylene (Zobel, 1973). The gas alters the direction of stem growth from vertical to horizontal in cotton plants (Hall *et al.*, 1957), and from prostrate to erect in strawberry clover stolons (Hansen and Bendixen, 1974). Light treatments that change lateral-branch orientation from diatropic (horizontal) to erect increase the groundnut plant's rate of ethylene production, and applied ethylene has the same effect as light on the plant's branch orientation (Ziv *et al.*, 1976). Bermuda grass (*Cynodon dactylon* L.) stolons grow horizontally when they are attached to the plant, but when they are excised and kept horizontal, their apical growth reorients to upward (Montaldi, 1969). After excised stolons that had been adapted in a vertical orientation in darkness were reoriented horizontally, their ethylene production began to increase within 3 h, and by 15 h had risen 3.3-fold (Balatti and Willemöes, 1989). There was a positive correlation between the increased ethylene production rate and the upward curvature that resulted. Incubation of stolons with $AgNO_3$ did not inhibit ethylene production, but slowed curvature development, while ACC, ethylene and ethephon accelerated the rate of curvature.

The auxin- and ethylene-insensitive *Arabidopsis* mutants *axr1*, *axr2*, *axr3* and *aux1* all have a defect in root gravitropism (Smalle and van der Straeten, 2000), and the *AUX1* gene is expressed in root-apical tissues that are regulating root gravitropic curvatures (Bennett *et al.*, 1996).[26] The *eir1* *Arabidopsis* mutant has agravitropic and ethylene-insensitive roots that display normal auxin responses. A possible function of the *EIR1/AUX1* pathway may be the regulation of auxin in response to gravity, and the ethylene insensitivity of *aux1* and *eir1* roots possibly reflects a role of ethylene in regulating gravity responses and auxin transport (Roman *et al.*, 1995). The *ain1-1* *Arabidopsis* mutant, isolated by screening for ACC resistance, has a normal ability to bind ethylene, but produces less of the gas. Elongation of its hypocotyl and root is not affected by applied gas, 10 μl/l ethylene fails to accelerate the senescence of its leaves, and it displays a reduced gravitropic responsiveness (Van der Straeten *et al.*, 1993). Both the root and shoot are agravitropic in the tomato *dgt* mutant, which has a reduced sensitivity to both auxin and ethylene (Kressin Muday *et al.*, 1995).

Plagiotropic organs usually have dorsal and ventral surfaces that are morphologically distinct and therefore potentially able to respond differentially to applied ethylene, even though the gas distributes uniformly within tissues, but the tropistic effect that ethylene causes in radially symmetric organs such as roots requires a different explanation (Molisch, 1884; Zimmerman and Hitchcock, 1933; Chadwick and Burg, 1970; Lee *et al.*, 1990; Jackson, 1991; Suttle, 1991). Gravitropic curvatures are a differential growth response, which according to the Cholodny–Went hypothesis (Cholodny, 1927; Went, 1927) results from a gravity-imposed auxin asymmetry. Auxin migration to the lower surface of both roots and shoots under the influence of gravity has been confirmed experimentally, and Ca^{2+}, which is essential for the gravitropic response, is redistributed to the upper side of gravistimulated sunflower hypocotyls and maize coleoptiles, and to the lower side of roots (Goswami and Audus, 1976; Slocum and Roux, 1983; Moore and Evans, 1986). At least four processes contribute to the overall gravitropic response: sensing the signal, transduction of the signal, the motor response and adaptation to the gravity stimulus (Hild and Hertel, 1972; Feldman, 1985; Bjorkman, 1988). The sensing step requires sedimentation of amyloplasts (statoliths), the transduction step for ortho-gravitropism involves a disturbance in cytoplasmic calcium concentration, and the motor step results from auxin redistribution. Comparative studies of the kinetics of curvature and timing of the development of hormone asymmetry indicate that lateral auxin redistribution cannot by itself account for the intricacies of the gravitropic response (Evans, 1991). Time-dependent gravity-induced changes in the sensitivity of the gravity receptor and the response to auxin play an important role as an organ 'adapts' to the gravitational stimulus, after which it reacts to a change in gravitational orientation, and not to a static position (Bennet-Clark *et al.*, 1959; Rufelt, 1961; Palmer and Phillips, 1963; Burström, 1971; Larsen, 1973). A plagiotropic attitude arises when a root or leaf continues growing in the direction it achieved if a single geo-induction failed to induce a curvature to the plumb line, and afterwards another stimulus can cause a further reaction (Palmer and Phillips, 1963; Burström, 1971).

An ethylene gradient is not likely to play a role in the gravity response of roots or stems as they have extensive intercellular spaces in their cortex that preclude them from sustaining a substantial upper/lower IEC ratio in response to asymmetric production of the gas (Harrison and Pickard, 1986; Suttle, 1991). Nevertheless, vacuum-extraction measurements, made with upper and lower tissue halves prepared from vertically and horizontally oriented shoots (Wheeler *et al.*, 1986), floral stems (Woltering, 1991; Philosoph-Hadas *et al.*, 2000) and roots (Mulkey *et al.*, 1982), ostensibly have shown that they develop a large lower/upper IEC gradient, which seems to interfere with their gravitropic response by an effect that is partially reversed by inhibitors of ethylene action and production (Wheeler and Salisbury, 1981; Kelley and Bradford, 1986; Wheeler *et al.*, 1986; Balatti and Willemöes, 1989). This is misleading because essentially all of the ethylene recovered during the vacuum-extraction assay was produced while the tissues were submerged under saturated ammonium sulphate during the measurement (3.15). Although the vacuum extraction assay confirms that a radial gradient of ethylene production had developed in the horizontally positioned organs, with the highest rate in the lower side, the methodology cannot assess the extent to which radial diffusion through the intercellular system levelled the IEC gradient within the intact organs before they were split for analysis. Treating tomato and cocklebur plants with AVG to inhibit their ethylene production slowed stem bending by 50%, and applying 0.1 μl/l ethylene largely reversed the inhibition, proving that it was not caused by an ethylene gradient (Wheeler *et al.*, 1986). Even if a significant ethylene gradient cannot be sustained in a horizontal root or shoot, that does not preclude the possibility that a gross increase in the IEC might affect the organ's ability to perceive gravity, transduce and respond to the gravitational stimulus, transport auxin laterally and adapt to the gravitational stimulus.

5.23 Effect of Ethylene on Auxin Gravitransport

Within 45 min after *V. faba* epicotyls are positioned horizontally, they develop a lower/upper diffusible auxin ratio of 1.66 in air, and throughout a 90-min transport period that ratio is sustained. The ratio is inverted when they nutate horizontally in response to a continuous ethylene treatment. Laan (1934) favoured the view that the auxin asymmetry and plagiotropic growth of the ethylene-treated seedlings resulted independent of gravity, but his data do not exclude the possibility that gravity caused auxin to move upward in the presence of ethylene to the same extent that it normally moves downward in air (Thimann, 1952). To determine which interpretation is correct, ethylene's effect on gravitropic auxin transport was measured in horizontally positioned etiolated pea subapical sections and epicotyl explants. A lower/upper specific activity ratio of 2.5 developed in the etiolated pea subapical sections within 135 min after a donor agar block containing 1 μM ^{14}C-IAA was applied to the apical end (Burg and Burg, 1966a). Throughout the transport period, 100 μl/l applied ethylene did not alter uptake from the donor block or polar transport of radioactivity to the section's base (Burg and Burg, 1967a), but it decreased the lower/upper specific activity ratio to 1.2. When pea subapical sections were floated without shaking on a solution containing 3.6 μM 5-^{3}H-IAA, within 30 min gravitropic transport caused an identical downward auxin movement and upward curvature in horizontally positioned sections regardless of whether or not ethylene was applied (Fig. 5.37, *right*). Downward gravitropic auxin transport continued unabated in control sections and curvature increased at a linear rate, but after 30 min ethylene-treated sections abruptly ceased transporting auxin downward and stopped curving even though the gas did not cause their elongation rate to decline for at least 150 min (Fig. 5.31, *upper right*).[27] This result confirms a report that the 'geo-sensitivity' of etiolated pea tissue persists in

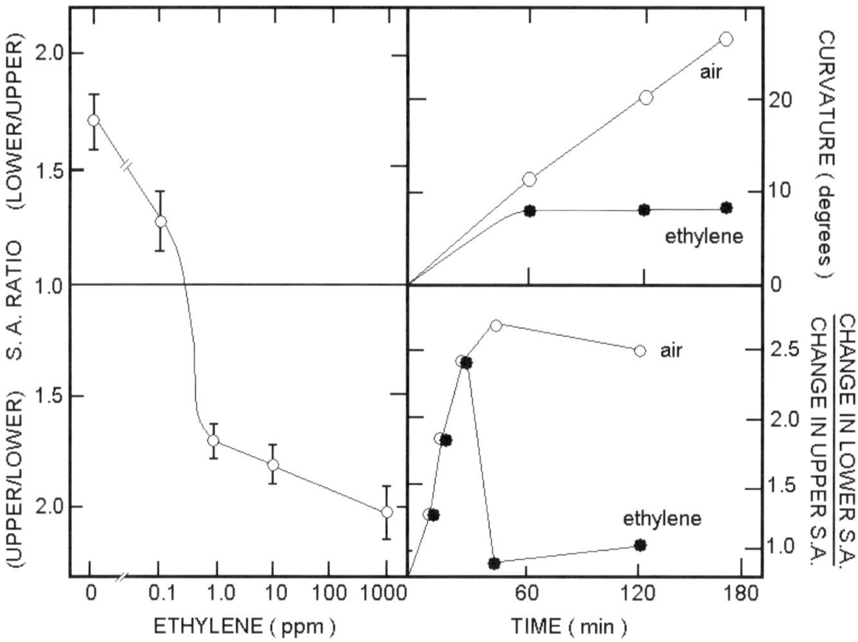

Fig. 5.37. (*left*) Effect of various concentrations of ethylene on gravitransport from a droplet of 3.6 μM 5-^{3}H-IAA in lanolin paste applied to the apical end of horizontally positioned etiolated pea explants containing a hook elbow. The plumular leaves were removed and the basal end of the explant inserted in 1% agar contained in a Petri dish. Transport period = 4 h in darkness. The hook elbow was aligned at a right angle to the gravitational field to eliminate any effect of tissue asymmetry. (*upper right*) Effect of 10 μl/l ethylene on the curvature of 1 cm etiolated pea subapical segments floated without shaking on a solution containing 1 μM IAA. (*lower right*) Effect of 10 μl/l ethylene on the change in (upper S.A.)/(lower S.A.) 1 cm subapical etiolated pea segments during each time interval while they are floated without shaking on a solution containing 1 μC/ml of 3.6 μM 5-^{3}H-IAA (Burg and Kang, 1993).

ethylene for a short time, and then rapidly decreases within 1 h (Guttenberg, 1910). A different result was obtained when ethylene's effect on gravitropic transport was studied with etiolated pea explants whose plumular leaves had been excised. The intact hook elbow was oriented at a right angle to the gravitational field to ensure that its asymmetry did not influence the result (Burg and Kang, 1993). A few μl/l of ethylene reversed the direction of lateral IAA gravitropic movement as labelled auxin passed through the horizontally positioned U-shaped hook elbow into the subtending subapical zone. Ethylene caused 5-^{3}H-IAA to move upward as intensely as it normally moves downward in air, with a compensation point at 0.2 μl/l ethylene (Fig. 5.37, *left*). In 4 h, the epicotyls bent upward by 51 ± 3.6° in air, and downward by 13 ± 4.5° in 100 μl/l ethylene. The ethylene-induced inversion of gravitransport was not accompanied by a decline in the velocity or capacity of polar auxin transport or an alteration in IAA uptake or destruction (Burg and Burg, 1967a). It was concluded that ethylene reverses the direction of IAA gravitransport in the hook region, and prevents lateral auxin transport in the subapex. Studies with the *Ageotropum* mutant confirm this interpretation. During 5 h, horizontally positioned *Alaska* (wild-type) pea epicotyl subapical and *Ageotropum* mutant segments, supplied with ^{3}H-IAA through their cut apex, developed 64/36 and 46/53 lower/upper specific-activity ratios, respectively, indicating that auxin moved downward in wild-type and slightly to the upper side in *Ageotropum* (Takahashi *et al.*, 1991). The wild-type epicotyl segments curved upward by 38°, while *Ageotropum* failed to curve. Ethylene did not alter polar auxin transport in *Ageotropum* during the lateral-transport studies. When 5-day-old *intact* etiolated *Ageotropum* plants were placed horizontally, during 5 h they did not bend in air, but in the presence of 2 μl/l applied ethylene they curved downward by 25°, indicating that ethylene caused upward IAA gravitransport in *Ageotropum*, just as it does in wild-type seedlings.

The pattern of expression of two primary auxin-regulated genes, *AtAux2–11* and *SAUR-AC1*, is only altered in tissues of the *hls1 Arabidopsis* mutant (5.07) that form the apical hook, and not in other parts of the seedling. It has been suggested that *HLS1* may play a unique role in auxin metabolism or transport in the hook region. Possibly *HLS1* acetylates the amino terminus of a group of proteins involved in auxin transport (Kieber, 1997), or regulates auxin function in the apical hook by acetylating an IAA-related metabolite, thereby altering its activity (Lehman *et al.*, 1996). The pattern of *HLS1* expression might explain why ethylene inverts gravitropic auxin transport in the hook elbow and shank of the pea epicotyl, but only inhibits it in the subapex.

A disagreement about whether ethylene affects auxin gravitransport during the epinastic response of leaf petioles (Palmer, 1976; Abeles *et al.*, 1992) arose in part because of technical difficulties associated with lateral auxin transport measurements. In published IAA gravitransport studies, often an auxin concentration high enough to stimulate ethylene production had to be applied in order to produce a measurable transported count using the IAA-specific activity available to the researcher. The lower/upper specific activity ratio of 2.5 that developed when pea subapical segments were supplied with 1 μM ^{14}C-IAA was reduced to 1.4 when they transported 10 μM ^{14}C-IAA, a concentration high enough to stimulate ethylene production (Burg and Burg, 1966a). In addition, auxin concentrations above a physiological level 'saturate' the gravitropic transport system, masking the development of a lateral gradient in horizontally positioned organs (Gillespie and Thimann, 1961, 1963). When very high labelled IAA concentrations were used to study transport, ethylene caused only a slight accumulation of radioactivity in the upper half of the leaf axis of tomato plants (Lyon, 1970), pepper plants (Lyon, 1970) and pea roots (Lyon, 1972), and no accumulation in the upper half of *Coleus* plants (Palmer, 1976).[28]

The lateral IAA transport data presented in Table 5.12 show that tomato petiole explants incubated upright in air or inverted in 10 μl/l ethylene developed an adaxial/abaxial specific-activity ratio close to unity when they were supplied with a physiological IAA concentration, and the ratio changed to approximately 2.1 when they were incubated upright in ethylene or inverted in air. This indicates that ethylene caused upward gravitransport regardless of the explants' orientation, offsetting the 'autonomic tendency' of auxin to move into the adaxial surface of inverted explants (De Vries, 1872), and causing auxin to move into the adaxial surface of upright explants. The lateral transport system overloads when a very high ^{3}H-IAA concentration is transported, creating an adaxial/abaxial specific activity ratio close to unity, regardless of whether ethylene is applied or the explants are upright or inverted. In other studies, during an 18-h period after 6 μM 2,4-D-^{14}C in lanolin paste was applied to the acropetal ends of intact tomato or *Coleus* petioles, the specific activity ratio that developed between the upper and lower halves was 0.62 in air-control tomato plants vs. 1.96 in plants treated with 0.5 μl/l ethylene, and 0.73 in air-control *Coleus* plants vs. 1.35 in plants treated with 0.1 μM ethephon (Parups, 1973). Pepper plants behaved in a similar manner.

Ethylene may invert the direction of gravitropic transport by affecting adaptation to a gravitational stimulus (Kang and Burg, 1974a; Kang, 1979; Lee *et al.*, 1990; Burg and Kang, 1993). During the first 15 min after vertically grown *Avena* (Brauner and Zipperer, 1961) or maize (Filner *et al.*, 1970) coleoptiles are positioned horizontally, auxin is transported to the *upper* half of the organ, causing a downward curvature

Table 5.12. Effect of 10 μl/l ethylene on polar and gravitropic transport of 5-^{3}H-IAA, and the epinastic curvature of tomato petiole explants, during an 18-h incubation. The bases of the explants were inserted into 1% agar contained in an upright or inverted Petri dish (adapted from Burg and Kang, 1993).

Measured property IAA concentration (μM – in lanolin)	Upright		Inverted	
	Air	Ethylene	Air	Ethylene
Degrees Δ curvature				
0	2 ± 3	12 ± 6	14 ± 11	24 ± 23
5.7	0 ± 2	33 ± 12	24 ± 15	33 ± 18
57	9 ± 6	73 ± 12	41 ± 15	45 ± 21
5700	40 ± 20	66 ± 27	52 ± 20	65 ± 24
Adaxial/abaxial SA ratio				
3.4	1.05 ± 0.17	1.85 ± 0.12	2.14 ± 0.42	1.10 ± 0.17
57	0.89 ± 0.08	2.04 ± 0.11	2.24 ± 0.37	1.01 ± 0.14
57,000	0.86 ± 0.16	0.85 ± 0.16	0.89 ± 0.07	0.89 ± 0.07
^{3}H-IAA (μM) in basal 5 mm				
3.4	0.023 ± 0.007	0.046 ± 0.012	0.023 ± 0.007	0.050 ± 0.006
57	0.220 ± 0.046	0.32 ± 0.055	0.18 ± 0.044	0.28 ± 0.099
57,000	29.6 ± 9.1	43.5 ± 11.2	32.5 ± 0.8	44.8 ± 16.6

(Fig. 5.38; Hild and Hertel, 1972). It is only after 20 min have elapsed that auxin distributes downward and promotes an eventual upward curvature. Within 10–15 min after tomato plants are oriented horizontally, their hypocotyls bend downward by 7–8°, and then an upward bending begins to develop (Harrison and Pickard, 1986). *Kniphofia* flower stalks bend downward by 8° during the first hour of gravistimulation, before they curve upward (Woltering, 1991). A similar result has been noted with other seedlings (Pickard, 1885). The existence of a negative curvature during the initial 0.5–4 min also can be deduced from studies of the presentation time for the geotropic response of *Artemesia* roots (Larsen, 1956). These and other observations led to the suggestion that gravistimulation has a biphasic dose–response curve (dose = intensity of stimulus × time) and that the effective stimulus decreases and may even reverse direction if it is greater than the optimal dose. This theory proposes that very strong and effective gravistimulation leads to auxin transport in the 'wrong' direction, until the adaptation process counteracts the overstimulation, bringing the 'effective stimulation' to a point on the dose–response curve that promotes auxin gravitransport in a downward direction. Greater *g*-forces have been found to increase the intensity and extend the duration of the negative geotropic response (Hild and Hertel, 1972), and after a continuous, long exposure to gravity, adaptation may eventually offset the gravitational stimulus. Then there will be no difference in the activity of the auxin permeases on opposite sides, and lateral auxin transport should cease. In lateral organs, the adaptation level needed to counteract constant geo-stimulation might even be intrinsic.

^{3}H-IAA initially migrates to the upper side of the primary roots of 2-day-old gravistimulated maize seedlings, and the roots curve upward. Then auxin moves into the lower side and they curve downward (Young *et al.*, 1990). Studies in which the roots were pre-adapted to gravity or pre-stimulated and rotated indicated that the initial upward curving and auxin migration resulted from overstimulation of the gravity-sensing mechanism and that subsequently as the roots adapted to gravity, downward auxin transport and curving set in. The growth inhibition caused by 1–100 μl/l ethylene delayed the gravitropic root curvature, but the final extent of curvature was not altered by 1 μl/l ethylene, and 10–100 μl/l prolonged the duration of the period during which curvature development occurred (Lee *et al.*,

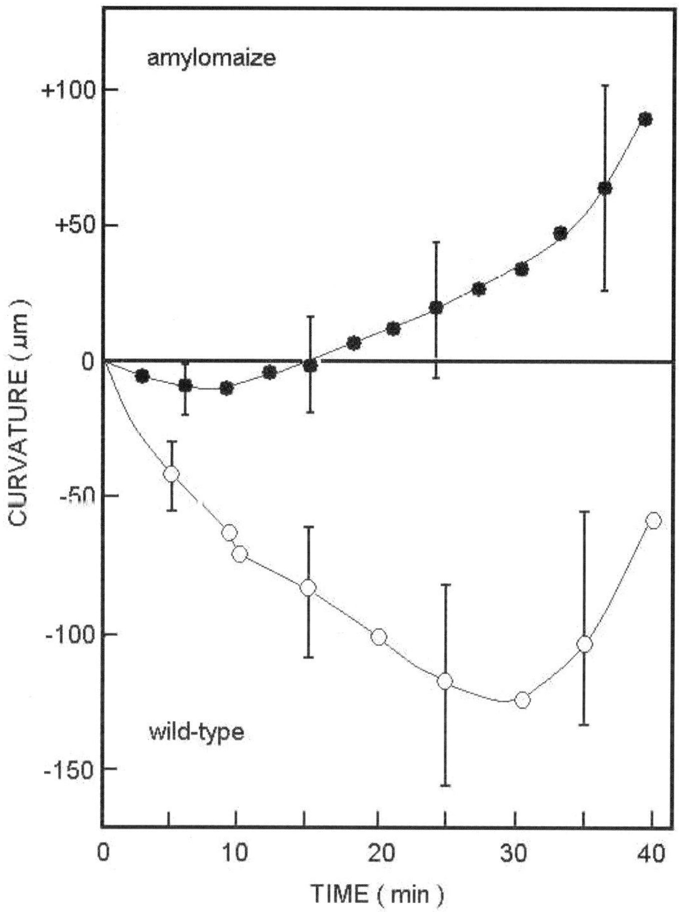

Fig. 5.38. Curvature of maize wild-type (*lower*) and amylomaize (*upper*) coleoptiles. After 60 min in the vertical position, excised 3 cm-long coleoptiles were placed horizontally at $t = 0$ min. Upward or downward displacement of the tip (curvature) was measured with 5–6 µm accuracy during 40 subsequent minutes. The amylomaize mutant has smaller starch statoliths and a weaker transport and curvature georesponsiveness, and therefore it is less likely to be 'overstimulated' by a 1 *g* gravitational force. The downward curvature changes direction when the adaptation process counteracts the gravity 'overstimulation' (Hild and Hertel, 1972).

1990). During 45 min, 0.4 µM ^{3}H-IAA auxin was transported upward (Lower/Upper (L/U) = 0.92) in the presence of 100 µl/l of ethylene, and downward in control roots (L/U = 1.33; Table 5.13). Within 6 h, horizontally positioned roots bent downward by 60° in air and 105° in 100 µl/l ethylene. The exaggerated downward curvature induced by ethylene was prevented by silver nitrate. Progressively as the duration of the prestimulation period was increased to 300 min, control roots adapted, the L/U ratio approached unity and ethylene-treated roots now transported IAA downward with a L/U ratio = 1.35. It was concluded that ethylene interferes with the auxin lateral-transport system's ability to adapt to gravity stimulation, that an 'overstimulation' caused upward auxin transport initially and then, in the presence of exogenous ethylene, a prolongation of the period required for complete adaptation to the gravity stimulus extended the duration of downward transport and curvature development.

Table 5.13. Asymmetric movement of label from ^{3}H-IAA donor agar block across the elongation zone of gravistimulated maize roots in the presence and absence of 100 μl/l applied ethylene. Roots were oriented horizontally for the indicated times ± ethylene, and then a donor block containing 15,000 cpm ^{3}H-IAA was applied either on the top or bottom side of the elongation zone about 3 mm from the root tip. After a 45-min transport period, the donor was removed and a 0.5 cm semi-cylindrical section was excised from the side of the root immediately across from the donor. Ratio = ratio of movement of label from top-to-bottom (T → B) ÷ bottom-to-top (B → T) (Lee *et al.*, 1990).

Treatment	Pre-stimulation (min)	Radioactivity (cpm)		
		T → B	B → T	Ratio
Control	45	330 ± 27	248 ± 15	1.33
	150	174 ± 19	151 ± 16	1.15
	300	128 ± 18	132 ± 11	0.97
Ethylene	45	257 ± 25	280 ± 20	0.92
	150	172 ± 5	164 ± 14	1.05
	300	149 ± 10	110 ± 7	1.35

5.24 The 'Triple Response' of Seedlings

Ethylene causes *P. sativum* (Figs 5.33 and 5.35), *Vicia sativa*, *V. faba*, *Ervum lens*, *Lathyrus odorata*, *Tropaeolum* (Neljubow, 1901, 1910), potato sprouts (Singer, 1903; Mingo-Castel *et al.*, 1964), *Arabidopsis* (Smalle and van der Straeten, 2000), tomato (Yen *et al.*, 1995), and certain other seedlings that contain a reflexed apical hook, to undergo a 'triple response' consisting of a slowed elongation rate, stem swelling and a horizontal nutation. Placed upright in ethylene, the etiolated pea epicotyl assumes a horizontal position (actually diagravitropism – at a slight angle below horizontal) and continues to grow in that direction (Fig. 5.33). If it is horizontally oriented in ethylene, it continues growing in approximately that orientation. Aligned in an oblique position in ethylene, either above or below the horizontal, it assumes a nearly horizontal position. Regardless of its initial spatial orientation, it always bends in the direction of the closed side of the hook (Crocker *et al.*, 1932), and eventually the hook is held uppermost (Neljubow, 1901, 1910). This ethylene response has been used to select genetic mutants of *A. thaliana* and tomato that are deficient in ethylene sensitivity. These mutants have proved extremely useful in the study of fruit ripening, flower fading and vegetative growth.

The 'triple response' was first observed in etiolated seedlings grown in heated laboratories during winter months. The cause of this 'winter disease' was traced to ethylene released from leaking pipes used to distribute illuminating gas for heating purposes (Neljubow, 1901). The response remained a curiosity until Goeschl *et al.* (1966) provided evidence that it may be the mechanism by which seedlings avoid obstacles encountered during emergence through the soil. Ethylene production is stimulated when a pea epicotyl's elongation is restricted by pressure, and the resultant 'stress' ethylene causes the stem to thicken (strengthen) and the seedling's elongation rate to slow as it nutates diagravitropically to avoid the obstacle. A similar behaviour has been reported for seedling shoots of maize (Sarquis *et al.*, 1991). Darwin (1981) and Haberlandt (1877 – referred to in Darwin, 1981) had described the underground movements of hypocotyls bearing U-shaped hooks, and concluded that they were a mechanism for avoiding obstacles and overcoming the pressure exerted by the soil. Borgström (1939a) found that etiolated plants such as pea and seedlings of *Sinapsis alba* are extremely reactive to contact stimulus, and that this causes them to swell and grow transversely in order to avoid an obstacle. The ability of germinating wild-type and mutant *Arabidopsis* seedlings to grow through sand is directly proportional to their ethylene sensitivity (Harpham *et al.*, 1991).

The longer transport path through the etiolated pea hook's convex side would be expected to create an inner/outer IAA gradient (inner = side facing the hook) in the subapex because as IAA moves basipetally through this tissue, the auxin concentration declines by 50% in each 2.5 mm interval (Varder and Kaldewey, 1972). After ^{3}H-IAA in lanolin paste was applied to the cut stump of an etiolated pea explant's debladed hook, regardless of whether the explant was upright or inverted, the inner/outer specific-activity ratio was 1.53 and the side-to-side ratio 1.03 (Kang and Burg, 1974a; Burg and Kang, 1993). ^{3}H-IAA also is preferentially transported from the apex into the 'inner' concave side of the hooks of *Phaseolus* (Schwark and Schierle, 1992). The inner peripheral hook tissue of *Phaseolus* (cortex plus epidermis) contained 3.7–10-fold more radioactivity than the outer hook tissue, and after a 5-h treatment with 10 μl/l ethylene, during a subsequent 3-h transport period with ethylene present, the amount of ^{3}H-IAA transported to the inner side of *Phaseolus* hooks nearly doubled, even though ethylene inhibited polar transport. Ethylene synthesis is asymmetrically distributed in *Phaseolus* hooks (Schierle and Schwark, 1988), the higher rate occurring in the 'inner' concave side where the native auxin concentration presumably is higher.

The 'natural' auxin gradient across the pea hook causes 'back-bending' of the subapex away from the hook (Fig. 2.5, *right*), orienting it in the same direction it will assume when ethylene causes it to nutate horizontally.[29] Weisner (1878) and Richter (1906, 1910) suggested that the horizontal nutation results when ethylene prevents a geotropic correction, causing the normal backbending to intensify, and Richter concluded that since the tissue had lost its ability to sense gravity, its final orientation must be independent of gravity. A comparison of Figs 5.33 (*left*) and 5.37 (*left*) indicates that progressively as the ethylene concentration is increased to 0.1 μl/l, auxin gravitropic transport is inhibited, allowing the auxin asymmetry in the hook to increase the backbending of the hook; and when the ethylene concentration is increased further, the direction of auxin gravitransport in the hook becomes inverted, amplifying the intrinsic subapical auxin asymmetry, promoting growth in the upper side and forcing the epicotyl toward and even slightly below a horizontal orientation with the hook uppermost. The final sustained plagiotropic orientation must involve an adaptation to the gravitational stimulus.

A back-and-forth 'bobbing' nutation with a 2–3-h period and an amplitude of approximately 15°, which rhythmically corrects the subapical backbending of etiolated pea plants (Fig. 5.39; Galston *et al.*, 1961), must be dampened by ethylene when the gas induces a horizontal nutation. Circumnutations also cease when ethylene is applied to tomato and sunflower shoots (Crocker *et al.*, 1932). A clue to the manner in which ethylene may affect the bobbing nutation is provided by the effect that the gas has on a spontaneous curvature that develops in etiolated pea subapical segments when they are incubated on solutions. During the initial 3-h interval after etiolated pea (or sunflower) subapices are excised and floated in darkness on a 1 μM IAA solution, the sections curve by 20° even though gravity has been compensated by back-and-forth shaking of the incubation flasks. Then the sections progressively straighten (Fig. 5.40). The 'bobbing' nutation and the curving and straightening of the excised segments have similar kinetics, and a spot of lanolin containing lamp black applied to the inner side of the subapex before the sections were cut revealed that the same side of the segment curves and straightens during pea-section assay and the 'bobbing' nutation. This also is the side that overgrows during the ethylene-induced horizontal nutation. Applied ethylene does not inhibit elongation of pea sections for 2.5 h (Fig. 5.31, *upper right*), and had no effect on their fresh weight increase, polar auxin transport, and IAA conjugation, decarboxylation and uptake (Burg and Burg, 1966a, 1967a; Eisinger and Burg, 1971), but immediately prevented the spontaneous curvature from developing, and promoted rapid straightening if applied

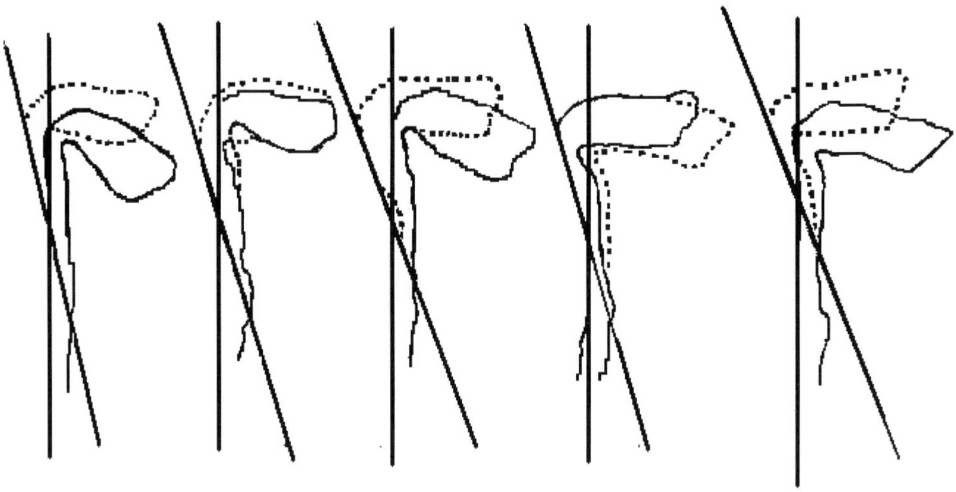

Fig. 5.39. The 'bobbing' nutation of the etiolated pea plumule (dotted outline) has a 2–3-h period (Galston *et al.*, 1961).

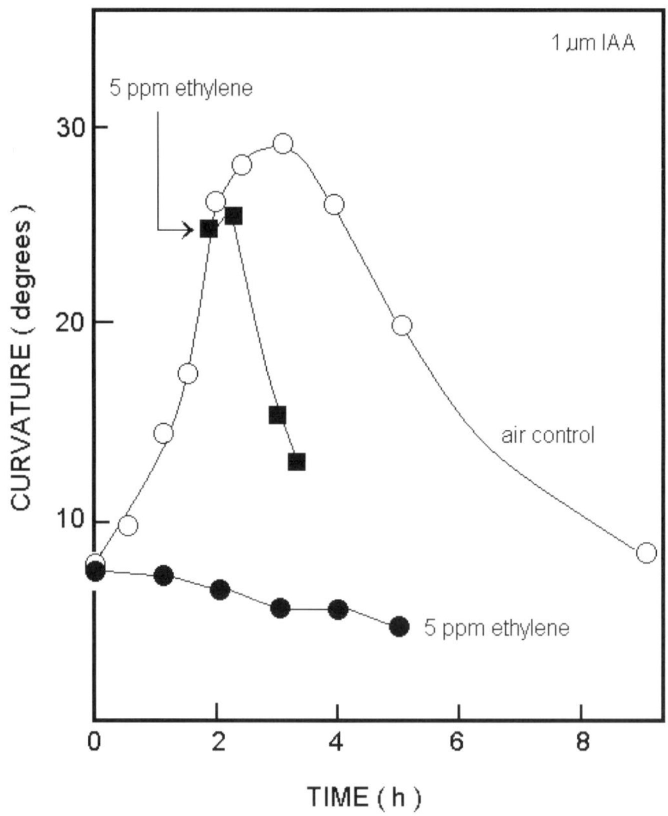

Fig. 5.40. Effect of 5 µl/l ethylene on the development of a spontaneous curvature in 1 cm etiolated pea subapical stem sections floated on a 1 µM IAA solution in darkness with continuous shaking to compensate for gravity. The sections curved in the direction of the subapical side that originally faced the closed hook. Application of 5 µl/l ethylene at the time indicated by the arrow caused the growing sections to straighten (Burg and Burg, 1966a).

after curving had progressed. The subapical sections behave as though a previous auxin asymmetry rapidly 'cross-transfers' when ethylene is applied, as Borgström (1939b) proposed in his theory of transverse reactions in plants.

5.25 Ethylene-induced Epinasty

Of 202 different species and varieties of plants tested, 89 showed ethylene-induced petiole epinasty and 113 did not (Crocker *et al.*, 1932; Fig. 5.36). One of the most sensitive responses to ethylene is the extreme epinasty that 1 nl/l of the gas induces in African marigold plants within 20 h (Crocker, 1948; Fig. 5.36, *lower*). Leaf petioles develop epinastic curvatures when cells in the adaxial surface resume elongation, or younger adaxial cells in the petiolar mid-region increase their rate of elongation (Crocker *et al.*, 1932; Palmer, 1972; Abeles *et al.*, 1992). The adaxial surface overgrows because cells in the abaxial surface either are not stimulated to expand or their growth is slightly inhibited by ethylene. The epinastic response is reversible, and after the stimulus has been removed, plants usually regain their original orientation. Because the application of 1–10 μl/l ethylene was only partially effective in increasing epinastic curvatures in tomato and marigold plants rotated on a horizontal clinostat, and exogenous ethylene induced epinastic petiole curvatures in upright plants and explants, but not when they were inverted (Table 5.14), Crocker *et al.* concluded that

> Ethylene-induced epinasty is tied up with the orientation of the petioles to the direction of the pull of gravity. Ethylene was most effective in inducing epinasty when plants were upright, about 0.4 [times] as effective when plants were rotated on a horizontal clinostat, and only slightly, or not at all effective, when the plants were inverted or the upper sides of the petioles faced the earth.

Borgström (1939b) reported similar observations. Any explanation of the epinastic response to ethylene must be able to account for its gravity dependence.

Many studies have shown that the epinasty induced by high-auxin concentrations or clinostat rotation results at least in part from auxin-induced ethylene production, and that ethylene's ability to cause petiole epinasty is diminished if insufficient IAA is supplied to auxin-depleted petiole explants (Tables 5.12 and 5.15), debladed plants (Lieke and von Guttenberg, 1961; Lyon, 1963a,b; Leather *et al.*, 1972; Palmer, 1972, 1973; Burg and Kang, 1993) or plants lacking auxin because they are zinc-deficient (Skoog, 1940).[14] Studies utilizing inhibitors of ethylene action or production have implicated ethylene in the epinastic

Table 5.14. Effect of horizontal clinostat rotation, inversion and 10 μl/l ethylene on the change in petiolar angle (epinasty) in tomato plants. The terminal apex was removed to prevent torsional bending of the main stem (Crocker *et al.*, 1932).

	Δ petiolar angle (degrees)	
Treatment	Air	C_2H_4
Horizontal clinostat (24 h)	30	60
Upright plants (22 h)	6	56
Inverted plants (22 h)	11	14

Table 5.15. Effect of 10 μl/l ethylene and a hypobaric treatment enriched in CO_2 on the epinastic curvature of tomato petiole explants exposed to various concentrations of IAA for 18 h. IAA dissolved in lanolin was applied to the apical petiolar stump. The hypobaric condition was a flowing, saturated mixture containing 11.2% O_2 + 2% [CO_2] at a pressure of 16.3 kPa (122 mm Hg). Curvature = change in degrees (Burg and Kang, 1993).

IAA concentration	Curvature (degrees)		
(μM – in lanolin)	Air	Ethylene	Hypobaric
0.0	2 ± 3	12 ± 6	2 ± 3
5.7	0 ± 3	33 ± 12	5 ± 4
57	9 ± 6	73 ± 12	7 ± 6
570	24 ± 12	–	11 ± 7
5,700	40 ± 20	66 ± 27	15 ± 10
57,000	108 ± 27	–	29 ± 8

response to high auxin concentrations and clinostat rotation (Amrhein and Schneebeck, 1980; Salisbury and Wheeler, 1981; Ursin and Bradford, 1989; Cardinale *et al.*, 1995) and heat-treated tomato plants (40°C for 4 h), which became temporarily unable to produce ethylene (5.2), do not develop petiole curvatures when they are exposed to excess auxin (Stewart and Freebairn, 1969), but the same plants became epinastic if ethylene is applied. Both clinostat-induced ethylene production and a 3–4-h exposure of upright tomato plants to 10 μl/l ethylene caused essentially the same epinastic petiole curvature response, and 10% [CO_2] prevented epinastic development both during clinostat rotation and when upright petiolar explants were exposed to ethylene (Leather *et al.*, 1972). CO_2 also prevents ethylene-induced epinasty in potato plants (Denny, 1935). When auxin-treated tomato petiole explants are kept at a hypobaric pressure in the presence of ample [O_2] and 2% [CO_2], epinastic curvatures are inhibited (Burg and Kang, 1993; Table 5.15).

Epinastic curving of petiole explants in response to 10 μl/l exogenous ethylene increases in intensity as the IAA concentration in agar blocks applied to the cut petiolar surface of VFN8 tomato petioles is raised from 0 to 100 μM (Ursin and Bradford, 1989; Tables 5.12 and 5.15). It is only at extremely high auxin concentrations that a further increase in auxin has no additional effect on curvature development (Kazemi and Kefford, 1974). The epinastic response of petiole explants prepared from the *dgt* tomato mutant also increased when the IAA concentration was raised from 0 to 100 μM, but approximately a tenfold higher auxin concentration was required to obtain the same curvature in *dgt* compared to VFN8 (Ursin and Bradford, 1989). This result is consistent with data showing that the IAA sensitivity of *dgt* is reduced by about tenfold, and that the mutant's leaves produce one-sixth as much ethylene as wild-type leaves in response to the same auxin concentration (Zobel, 1973; Kelley and Bradford, 1986). Epinasty was not induced in *dgt* plants when the leaves were sprayed with 10 μM IAA, but 100 μM IAA caused a strong effect, and ethrel spray [10^{-4} or 10^{-5} M] induced curvatures to the same extent in normal and *dgt* petioles.

Ursin and Bradford (1989) concluded that there is no auxin requirement for epinasty, that ethylene acts directly on petiolar cells to induce differential expansion (as previously suggested by Osborne (1982)), and that the differential growth responsible for epinasty is unlikely to result from ethylene-induced lateral auxin redistribution. This proposal was based on the finding that AOA (Amrhein and Schneebeck, 1980), AVG and STS prevent wild-type VFN8 tomato petiole explants from curving epinastically in response to 100 μM IAA, and that 1 μl/l ethylene reversed the curvature inhibition that AVG caused in explants treated with 50 μM 2,4-D (Ursin and Bradford, 1989). This explanation does not account for the gravity-dependence of the epinastic response (Crocker *et al.*, 1932; Tables 5.11 and 5.14). In agreement with other studies, Ursin and Bradford found that the curvature induced by ethylene depends on the applied auxin concentration, and therefore it is not obvious why they excluded the possibility of an intermediate gravity-dependent step in which ethylene causes auxin to distribute into the adaxial surface.

Many studies have shown that high exogenous auxin concentrations have a differential effect on the elongation of the adaxial and abaxial petiolar surfaces, and that this causes an epinastic curvature to develop independently of ethylene action (Lyon, 1963a, 1970; Palmer, 1964, 1972; Leather *et al.*, 1972). Leather *et al.* applied agar droplets containing 1 mM IAA to the cut petiole surface of tomato explants and measured curvature development in the presence of 10 μl/l ethylene. When the surface was notched in a manner that allowed IAA to be supplied exclusively through the upper or lower halves, it preferentially stimulated growth of the upper half. Auxin induces epinasty in tomato-leaf midribs (Kazemi and Kefford, 1974), *Chrysanthemum* (Furutani *et al.*, 1987) and poinsettia (*Euphorbia pulcherrima* Willd. ex Klotzch) petioles (Reid *et al.*,

1981) treated with inhibitors of ethylene synthesis and action, and the constitutive epinastic phenotype of the ethylene-overproducing *epi* tomato mutant (Fujino *et al.*, 1988; Table 5.5) is intensified by applying auxin, independent of ethylene (Ursin and Bradford, 1989). Ethylene production by *epi* was 96% inhibited by AVG, but this only partially relieved the epinastic curvature caused by 100 μM 2,4-D, or 1–100 μM IAA. Transgenic petunia and *Arabidopsis* plants that overproduce both ethylene and auxin also display severe laminar leaf epinasty of this type (Klee *et al.*, 1987b). Applied ethylene causes 'curled' leaf growth in sunflower (Palmer, 1985), fuchsia and young rose leaves (Crocker *et al.*, 1932), due to differential expansion of the adaxial and abaxial laminar surfaces during the growth phase. This phenotype develops with the same intensity in auxin-overproducing transgenic *Arabidopsis* plants regardless of whether they have high or low ethylene production rates, or ethylene-insensitivity conferred by *ein1-1* or *ein2-1*. It has been suggested that in *Arabidopsis* the curling is caused by a differential growth response of the adaxial and abaxial leaf surfaces to excess auxin, independent of ethylene (Romano *et al.*, 1993).[30]

A differential growth response to ethylene or auxin by cells in the adaxial and abaxial sides of the petiole, and an auxin requirement for ethylene's effect on growth, cannot explain the gravity-dependence of the epinastic response depicted in Tables 5.12 and 5.14. The data in Table 5.12 indicate that the gravity-dependence of the response arises because ethylene inverts the direction of auxin gravitransport in the petiole, while at the same time it promotes the accumulation of auxin at the petiole base. When explants were supplied with a physiological range of IAA concentrations, applied ethylene induced marked epinasty in upright petioles, but had no significant effect on curvature development in inverted explants, in agreement with Crocker *et al.* (1932) and Borgström (1939b). Exogenous ethylene did not increase the curvature of inverted explants because it caused auxin to be transported upward into the abaxial petiolar surface, offsetting the gas's tendency to cause preferential growth of the adaxial surface. The adaxial/abaxial IAA ratio in air was unity in upright petioles and they did not curve significantly. When they were inverted or exposed to ethylene, the adaxial/abaxial IAA ratio doubled and the resultant auxin asymmetry caused a curvature to develop. The extent of the curvature depended on the amount of auxin transported into the tissue, which was a function of the applied IAA concentration, and increased 1.5–2-fold when ethylene was applied. Upright petioles in air and inverted petioles in ethylene had adaxial/abaxial auxin ratios close to unity, but even when there was not a marked difference between the IAA concentrations that accumulated at the base of the respective petioles, inverted ethylene-treated petioles developed substantial curvatures and upright petioles remained essentially straight in air. This indicates that in addition to its influence on gravitransport and IAA accumulation, ethylene had a direct, differential effect on the growth of the adaxial and abaxial surfaces.

When the petiole explants were supplied with a high IAA concentration (5.7–57 μM IAA), the adaxial/abaxial IAA ratio was close to unity regardless of whether they were upright, inverted, kept in air or treated with ethylene, and the curvature that developed depended on the auxin concentration that accumulated at the petiole base. As the high applied IAA concentration must have induced sufficient ethylene production to expose upright, inverted and ethylene-treated petioles to an active IEC, these data suggest that IAA had a direct, differential effect on the growth of the abaxial and adaxial surfaces.

5.26 Effect of Ethylene on Correlative Inhibition

Auxin-induced ethylene production prevents the release of buds from correlative inhibition when etiolated pea nodal explants are floated on IAA-containing solutions (Burg, 1968a). In this assay, at all applied IAA concentrations, there is a close correlation between the intensity and duration of ethylene production and the bud growth inhibition that results. Cytokinins promote bud expansion in intact seedlings and nodal explants, overcoming applied auxin's inhibitory effect (Wickson and Thimann, 1958; Sachs and Thimann, 1964). Likewise, kinetin negates ethylene's ability to prevent bud outgrowth in the pea nodal-explant assay (Fig. 5.41) and in *Avena* stem segments (Harrison and Kaufman, 1982). Ethylene produced when nodal sections are floated on auxin-containing solutions, and by decapitated or intact plants treated with auxin, makes it difficult to experimentally evaluate auxin's role in correlative inhibition (Burg and Burg, 1968a). Russel and Thimann (1990)[31] suggested that ethylene might be the sought after second messenger in the auxin control of apical dominance.

Ethylene is produced mainly in the apical hook and buds of etiolated pea plants (Goeschl *et al.*, 1967; Burg, 1968a). The gas's production by nodal sections was decreased by 24% in plants decapitated one day before the measurement was made, and elevated if IAA had been applied to the cut stump (Burg and Burg, 1968; Burg *et al.*, 1972), but no reduction in ethylene production following decapitation could be detected in nodal sections possessing buds and lacking leaf scales (Burg, 1968a). Flowing water-saturated 100% [O_2] at 20°C and a pressure of 16 kPa (120 mm Hg = 13.5% [O_2]) did not affect stem growth (Apelbaum and Burg, 1971b), but caused the apical hook to open (Apelbaum and Burg, 1972; Kang and Burg, 1972b; Fig. 2.5) and reversed the inhibitory effect that 2,4-D-induced ethylene production had on the elongation of intact pea seedlings (Fig. 2.6). The IEC in the hook region of etiolated pea seedlings is sufficient to prevent hook opening until ethylene production is

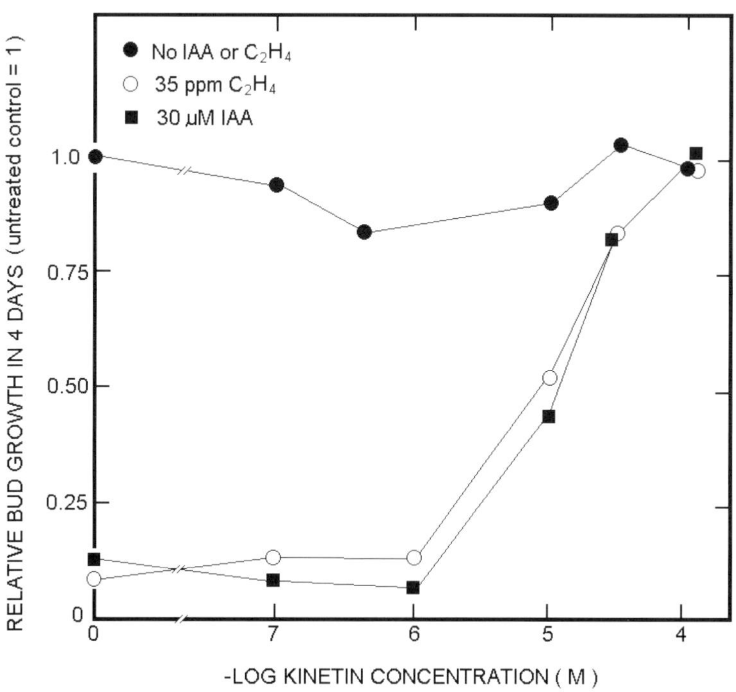

Fig. 5.41. Kinetin reversal of IAA- and ethylene-induced bud inhibition in *Pisum sativum* nodal sections. Control buds with no added IAA, ethylene or kinetin measured 2.46 mm in length after 4 days (Burg, 1968a).

depressed by red light, or CO_2 is applied to interfere with ethylene action (Kang and Ray, 1969a), and if endogenous ethylene participated in the normal control of apical dominance, the same hypobaric treatment that promoted hook opening and prevented 2,4-D-induced ethylene production from affecting elongation, should have caused the lateral buds to grow out, but it did not (Apelbaum and Burg, 1972). This indicated that although under experimental conditions the application of sufficient auxin to induce ethylene production may account for applied auxin's inhibitory action on bud outgrowth, ethylene might not normally be involved in correlative inhibition (Burg, 1968a).

After plants are released from apical dominance by removing the stem apex, the mitotic activity (Apelbaum and Burg, 1972), outgrowth (Burg *et al.*, 1972) and cellular differentiation (Apelbaum *et al.*, 1972) of buds is repressed if either ethylene or enough auxin to induce ethylene production is applied, but after an ethylene treatment is terminated, buds often are released from apical dominance. Inversion of *Pharbitis nil* shoots increased ethylene production in the apex and initiated the release of lateral buds from apical dominance (Prasad and Cline, 1985), and buds in potato plants (Catchpole and Hillman, 1976), rose plants (Zieslin *et al.*, 1972), mesquite seedlings (Morgan *et al.*, 1969) and geraniums (Fonteno, 1992) are released from correlative inhibition by an ethephon treatment or a transient exposure to ethylene. Lateral buds do not develop in control petunia plants or during a continuous 7-day treatment with 100 μl/l ethylene, but axillary buds in the subapical zone are released from correlative inhibition if the gas is applied for just 2 h and then removed (Table 5.16; Ramos and Burg, 1972 – cited in Burg, 1973a). When petunia plants are transferred to air after an 8–12-h ethylene exposure, essentially all buds are released from apical dominance by an effect quantitatively almost as great as that caused by excising the apex. Polar transport inhibitors overcome apical dominance (Yeang and Hillman, 1984), and it has been suggested that applied ethylene may release buds from correlative inhibition by inhibiting auxin polar transport and lowering the tissue's auxin content (Burg, 1973a).

Table 5.16. Ethylene-induced release of apical dominance in 5-week-old *Petunia* × hybrid Grandiflora calypso seedlings. Ethylene (100 μl/l) was applied for between 2 and 24 h during the first day, or else continuously for a 7-day period. Bud growth was appraised continuously throughout the same period (Ramos and Burg, 1972 – from Burg, 1973a).

Duration of C_2H_4 treatment	% buds released from apical dominance at indicated time (days) 2	4	5	7
2 h	0	0	9	18
4 h	0	10	13	23
6 h	0	15	30	36
8 h	4	27	33	37
12 h	7	36	36	43
24 h	3	43	46	46
7 days	0	0	0	0
Control – no C_2H_4	0	0	0	0

The roles of IAA and auxin-induced ethylene in the control of apical dominance are unresolved and remain controversial (Cline, 1991). Decapitation reduces ethylene production in *P. vulgaris* (Abeles and Rubenstein, 1964), and during bud break of cherry vegetative and flower buds, their IEC seems to decrease (Blanpied, 1972a). IAA's ability to prevent bud outgrowth when applied to the cut stump of decapitated *V. faba* and *P. vulgaris* plants is negated by AVG or norbornadiene, and also if the decapitated IAA-treated plants are kept at 33.8 kPa (253 mm Hg) or 67.7 KPa (507 mm Hg) (Russell and Thimann, 1990; Fig. 5.42). Although these results confirm that applied auxin is able to act by inducing ethylene production, they do not indicate whether or not ethylene normally is a 'second messenger' controlling the apical dominance of intact plants. To the contrary, AVG applied to the lateral buds of *Phaseolus* did not release them from correlative inhibition (Yeang and Hillman, 1981a, 1984), and based on the effects of ethylene, kinetin, AVG, ACC and IAA on the growth of tiller

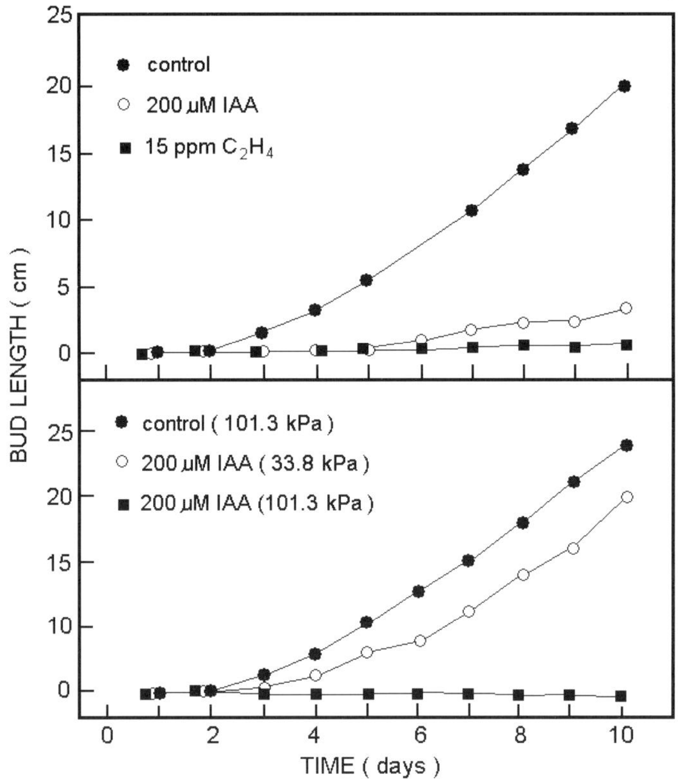

Fig. 5.42. (*upper*) Effect of 15 μl/l ethylene and a daily application of 200 μM IAA on the growth of lateral buds of decapitated *Vicia faba* plants, in continuous light at 25 ± 1°C. Measurements are an average for 25 plants in each group. (*lower*) Effects of 1/3 atmosphere pressure (33.8 kPa) and 200 μM IAA on elongation of the lateral buds of decapitated *V. faba* plants. Average of 12 plants in each group (Russell and Thimann, 1990).

buds in isolated *Avena* stem segments, Harrison and Kaufman (1982) concluded that IAA-induced ethylene production most probably is not involved in maintaining apical dominance in these shoots. An atmosphere containing 5% [CO_2] released some lateral buds in intact pea plants (Thimann, 1972), possibly by inhibiting ethylene action, and ethylene production by the nodal region decreased by two-thirds following decapitation and increased greatly if IAA was added to the cut stump (Blake *et al.*, 1983). Applied ethylene and ACC reduced bud expansion, and inhibitors of ethylene synthesis (canaline) and action (silver nitrate) enhanced bud growth. It was suggested that the 'second messenger' in apical dominance might be ACC. However, the behaviour of plants with transgene-mediated auxin (Romano *et al.*, 1991) and ethylene (Klee *et al.*, 1991) deficiencies, and mutants insensitive to auxin (Lincoln *et al.*, 1990) or ethylene (Bleecker *et al.*, 1988; Guzman and Ecker, 1990), suggests that auxin mediates apical dominance independent of ethylene action. Auxin-deficient transgenic plants display a loss in apical dominance (Romano *et al.*, 1991) and auxin overproduction has the opposite effect (Klee *et al.*, 1987a; Romano *et al.*, 1993). To investigate the roles of auxin and ethylene in apical dominance, transgenic tobacco and *Arabidopsis* plants expressing the auxin-overproducing tryptophan monooxygenase transgene were crossed with plants expressing an ethylene synthesis inhibiting ACC deaminase transgene to produce plants with elevated auxin and normal levels of ethylene

(Romano *et al.*, 1993). Transgenic auxin-overproducing *Arabidopsis* plants were also crossed with the ethylene-insensitive *ein1* and *ein2* mutants to produce high auxin/ethylene-insensitive plants. Analysis of these plants indicated that apical dominance was controlled mainly by auxin to cytokinin ratios rather than by ethylene. However, these experiments do not consider possible auxin and ethylene interactions resulting from the regulation of sensitivity and transport rather than biosynthesis (Smalle and van der Straeten, 2000). Plants carrying a mutation in the *HLS1* gene located on the intersection of the auxin and ethylene pathways display a loss of apical dominance (Lehman *et al.*, 1996), so the control of apical dominance by an auxin–ethylene interaction remains a possibility.

5.27 Effect of Ethylene on Microfibrillar and Microtubular Orientation

Cells normally are constrained from 'swelling' by the orientation of cellulose microfibrils in their walls, which are deposited like the hoops around a barrel, coiled around the cell in a manner that permits elongation but restricts radial growth (Frey-Wyssling, 1959; Green, 1963). In polarized light, the major extinction position (m.e.p.) corresponding to the mean direction of cellulose crystallites (microfibrils) is transverse and fairly uniform in elongated pea subapical cells, but in ethylene-treated swollen cells the m.e.p. is directed along the cell axis, indicating the presence of longitudinally oriented microfibrils (Burg and Burg, 1968; Eisinger and Burg, 1971; Ridge, 1973). Electron micrographs confirm that ethylene causes newly deposited microfibrils to reorient in a longitudinal direction both in pea seedlings (Apelbaum and Burg, 1971a) and maize root cortex cells (Gunning and Hardham, 1982; Mueller and Brown, 1982a,b), and auxin concentrations that induce ethylene production and cause swelling have the same effect as ethylene on the optical birefringence pattern and microfibrillar orientation (Veen, 1970a,b). Cell width begins to increase within 2.5–3 h after ethylene is applied to etiolated pea plants (Eisinger, 1983; Eisinger *et al.*, 1983; Fig. 5.31, *upper right*) and while cortical cells are constrained in their radial expansion by surrounding cells, epidermal cells in both the stem subapex and pea root often bulge out from their exterior surface into hair-like structures (Borgström, 1939b; Apelbaum and Burg, 1971a; Fig. 5.43). Ethylene also causes root hair development in tulips (Munk and Rooy, 1971), cucumber (Pierik *et al.*, 2000), *Arabidopsis* (Cao *et al.*, 1999), *Coleus*, *Cosmos*, *Heliotrope*, African marigold and tomato plants (Zimmerman and Hitchcock, 1933; Crocker, 1948; Abeles *et al.*, 1992). The phenotype of the root hairless *rhd6* mutation of *Arabidopsis* can be reverted by auxin or ACC treatment (Masucci and Schiefelbein, 1994), and the *ctr1 Arabidopsis* mutant (Table 5.4), as well as wild-type *Arabidopsis* plants treated with ethylene or ACC, develop root hairs from epidermal cells that do not normally form them (Tanimoto *et al.*, 1995). Ectopic root hairs also develop in the *transparent testa glabra* (*ttg*) and *glabra 2* (*gl2*) glabrous mutants (Galaway *et al.*, 1994; Masucci and Schiefelbein, 1996). The *TTG* and *GL2* genes control cell patterning in the root epidermis by negatively regulating the auxin and ethylene pathways, inhibiting root-hair differentiation in specific cells of the developing root epidermis.

The orientation of microfibrils in the cell wall is determined by cortical microtubules that are aligned parallel to the newly deposited microfibrils (Gunning and Hardham, 1982; Shibaoka, 1994). The cortical microtubule array in the cytoplasm occurs in close proximity and parallel to the plane of the plasma membrane to which it is linked through numerous cross bridges. The microtubules tend to be arranged as a single, multi-start, tightly compressed helix, orientated at right angles to the cell axis and direction of elongation (Gunning and Hardham, 1982; Roberts *et al.*, 1985; Fosket and Morejohn, 1992). Rapid effects of ethylene on microtubular orientation have been detected, but there is no information to suggest how they might be brought about.

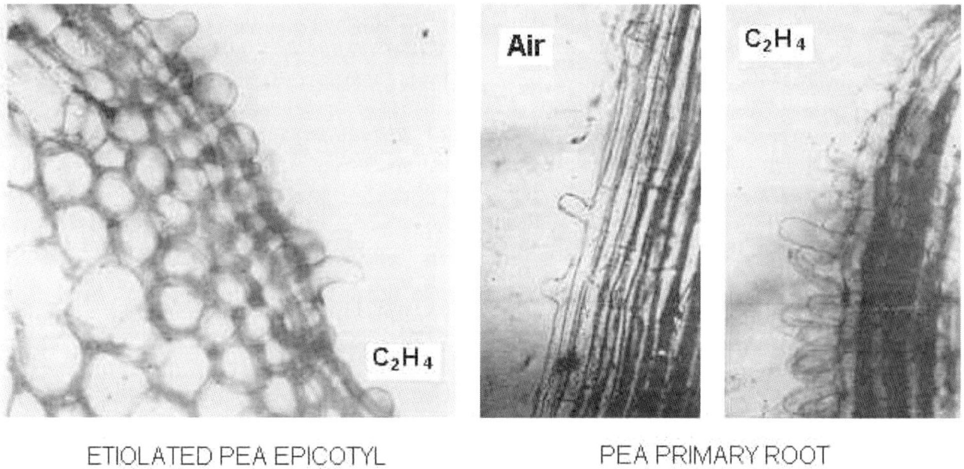

Fig. 5.43. (*left*) Cross section (×105) through the subapical zone of an etiolated pea seedling treated with ethylene for 96 h. Ethylene causes the epidermal and cortical cells to continue expanding and eventually to develop hair-like structures, which are absent in control tissue (Apelbaum and Burg, 1971a). (*middle*) Longitudinal section (×100) of a 2-day-old primary pea root in air. Root hairs, rare at this time, are shown emerging from elongated epidermal cells. (*right*) Proliferation of root hairs emerging from swollen epidermal cells after 2-day-old pea roots were treated with ethylene for 2 additional days (Chadwick and Burg, 1967).

Normally, the polylamellate walls of pea subapical cortical cells contain predominantly transversely oriented microfibril layers, and transverse microtubules are six times more frequent than those which are longitudinally aligned, but within 30 min after ethylene is applied to pea or mung bean segments the arrangement of the microtubule changes (Roberts *et al.*, 1985). After ethylene treatment, longitudinal deposition of microfibrils predominates in cortical and epidermal pea cells and longitudinal and oblique arrays of microtubules become eight times more frequent than transverse microtubules (Steen and Chadwick, 1981; Lang *et al.*, 1982). Both the microtubules that constitute the mitotic spindle, and those in the cytoplasm that control the direction of cellulose microfibrillar deposition and the transfer of vesicles from the Golgi, are relatively unstable, and can disassemble and reassemble according to cell requirements, but microtubules were not seen to be depolymerized during a 15–60-min ethylene treatment (Roberts *et al.*, 1985), indicating that the entire microtubule array reoriented from transverse to oblique to longitudinal without intervening phases of depolymerization. The transverse and longitudinal microtubules constitute part of flat-pitched and steeply pitched helices, respectively, and apparently they are capable of sliding along one another and with respect to the plasma membrane. Ethylene also reorients microtubules in maize roots (Baluska *et al.*, 1993).

Within 30 min after ethylene is applied to pea seedlings, at the same time that ethylene inverts the direction of gravity-directed lateral auxin transport (Fig. 5.37, *left*) and prevents the spontaneous curving of pea segments (Fig. 5.40), the arrangement of the microtubules changes from transverse to longitudinal, reorienting subsequently deposited microfibrils in a direction that causes cellular swelling to begin after 2.5 h (Roberts *et al.*, 1985; Fig. 5.31, *upper right*). The timing of these events, the close association of the microtubules with the plasma membrane where auxin transport occurs, the indirect association of the microtubules with the ER and the pattern of microtubular reorientation during the graviresponse of both shoots (Nick *et al.*, 1990) and roots (Blancaflor and Hasenstein, 1995) suggest the possibility that the direction of auxin transport in the gravitational field might be influenced by an interaction between the microtubule skeleton and auxin-transport system. An EM examination of cortical cells from 5-day-old dark-grown soybean hypocotyls after 5–180 min of gravistimulation revealed no significant orientational changes in the cytoplasmic microtubules or cell-wall microfibrils, even though ethylene synthesis was initiated during this interval, but there were slightly more microtubules associated with the plasma membrane in cells on the lower side of gravistimulated stems, with predominantly helical or lateral orientation. Maize coleoptiles treated with ethyl-*N*-phenylcarbamate and propizamide displayed both microtubular depolymerization and inhibited gravitropic bending (Nick *et al.*, 1991). In vertically growing roots, microtubules in the elongation zone are transverse to the longitudinal axis of the root, but when a root is gravistimulated in a horizontal orientation and a downward curvature begins to develop, microtubules of the outer cortical cells on the slower growing concave side reorient longitudinally (Blancaflor and Hasenstein, 1993). However, an intact microtubular cytoskeleton is not required for a normal gravitropic curvature by maize roots (Baluska *et al.*, 1996). Although the microtubule inhibitors taxol and oryzalin changed the orientation of the microtubule cytoskeleton and caused radial cellular swelling of maize roots, these substances did not alter the gravitropic response (Hasenstein *et al.*, 1999).

The microtubules control the orientation of various cellular components (Staehelin and Moore, 1995; Dey *et al.*, 1997). Protein translation from mRNA occurs at the rough ER, and the final protein product enters the luminal space of the ER, where it is enclosed in membrane-bounded vesicles that migrate from the lumen, through the cytosol to the Golgi body. There glycoproteins are enzymatically modified within the Golgi's lumen and then they leave the cisternae in vesicles that bud off and migrate through the cytoplasm, directed by the microtubules, to fuse with either the plasmalemma or the tonoplast. In this manner, non-cellulosic cell-wall polysaccharides (hemicellulose and pectin) and hydroxy-proline-rich protein (extensin) are transferred to the plasma membrane to be incorporated into the growing cell-wall architecture (Staehelin and Moore, 1995; Dey *et al.*, 1997). When the microtubular arrangement is altered by ethylene, this might influence the incorporation of these substances into the wall, and both increases and decreases in hydroxy-proline-rich wall-associated protein have been reported after ethylene treatment (Eisinger and Burg, 1971; Ridge and Osborne, 1973; Eisinger, 1983).

5.28 Ethylene-induced Leaf Senescence

Leaf RNA, chlorophyll and protein degradation are accelerated by applied ethylene and ACC, delayed by inhibitors of ethylene synthesis and action, and often as leaves age and yellow, ethylene production and respiration increase (Mack, 1927; Yemm, 1935; Abeles *et al.*, 1967; Thimann *et al.*, 1977; Aharoni and Lieberman, 1979; Aharoni *et al.*, 1979b; Morgan and Durham, 1980; Sisler, 1980a; Gepstein and Thimann, 1981; Jana and Choudhuri, 1982; Kao and Yang, 1983; Solomos, 1983; Alejar *et al.*, 1988; Abeles *et al.*, 1989; Aharoni, 1989; Makhlouf *et al.*, 1989; Fig. 5.44, *left*). Cytokinin and auxin levels decrease during leaf 'ageing', and application of these hormones as well as gibberellins retard the ability of ethylene to promote senescence, even though they stimulate ethylene production (Burg and Burg, 1966a; Burg, 1968a; Gepstein and Thimann, 1981; Kao and Yang, 1983; Fig. 5.44, *left*).

During leaf ontogeny, white clover (*Trifolium repens*) produces significant ethylene at its apex, from newly initiated leaves, and in senescent leaf tissue (McManus *et al.*, 2000). Three distinct *ACO* genes have been identified; *TRACO1* and *TRACO2* are expressed in the apex, *TRACO2* also is expressed in developing leaves and especially in mature-green leaves, while *TRACO3* is expressed in senescent leaf tissue. Three isoforms of ACC oxidase were distinguished, MG1 in

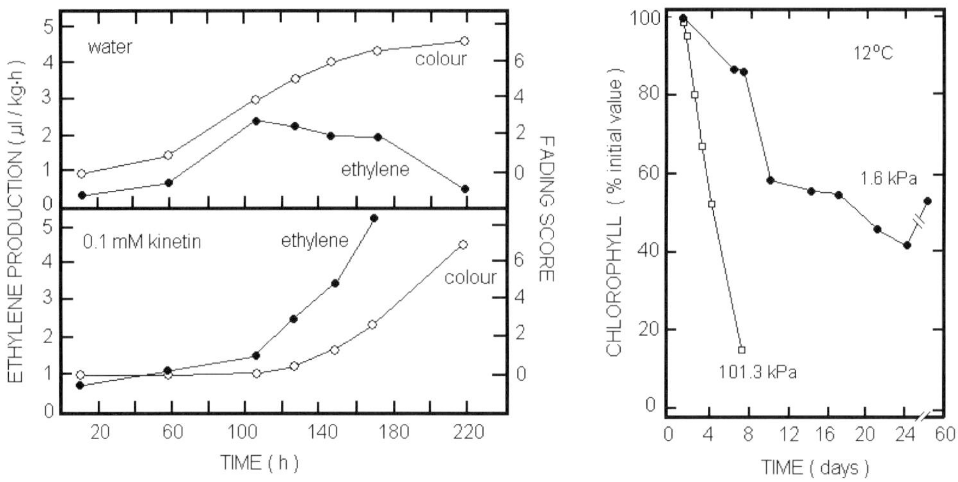

Fig. 5.44. (*left*) Ethylene production and fading of 9-day-old *Avena sativum* leaves in darkness at 21.1°C. *Upper*, floated on water; *lower*, floated on 0.1 mM kinetin. Fading (colour score): green = 0; trace yellow = 1; 1/8 yellow = 2; quarter yellow = 3; half yellow = 4; three-quarters yellow = 5; all yellow = 6; brown = 7 (Burg, 1965, unpublished). (*right*) Fading of 7-day-old light-grown *Avena sativum* subapical segments. Fifteen segments were incubated in 100 ml of test solution in the light at 12°C and atmospheric pressure (101.3 kPa) or at a pressure of 1.6 kPa. Data are means of two experiments (Veierskov and Kirk, 1986).

mature-green tissue, and both SEI and SEII in senescent tissue.

Endogenous ethylene hastens chlorophyll loss, but not the proteolysis associated with the senescence of watercress (*Nasturtium officinale* R. Br.) leaves (Philosoph-Hadas *et al.*, 1994). Wound-ethylene production is stimulated when the leaves are detached, and the harvested leaves rapidly lose chlorophyll and protein, while free amino acids increase. After wound-ethylene production slows, chlorophyll loss, but not proteolysis, can be stimulated by applying ethylene. Both chlorophyll and protein loss are retarded by 11% [CO_2], AVG and Ag^+ affected chlorophyll loss much more than proteolysis. This indicates that protein loss is not ethylene-mediated. LP prevents both chlorophyll and protein loss in leaf segments of oat (Veierskov and Kirk, 1986) and barley (Bangerth, 1974).

Although endogenously produced ethylene accelerates leaf senescence, the gas is not essential for the process to proceed (McAfee and Morgan, 1971; Mayak and Halevy, 1972). Leaves of an ethylene-insensitive *Arabidopsis* mutant (Bleecker *et al.*, 1988; Bleecker and Patterson, 1997) and tomato plants transformed with an antisense ethylene-forming enzyme gene (Gray *et al.*, 1993; Murray *et al.*, 1993c) still senesce, albeit more slowly than control leaves or plants. Dark-induced leaf yellowing did not require ethylene action in an *Arabidopsis* mutant that had a completely inhibited response to ethylene (Zacarias and Reid, 1990). Applied ethylene stimulates leaf yellowing in roses and many other plants (Serek *et al.*, 1996), and leaf yellowing is delayed in transgenic rose plants whose ethylene biosynthesis has been reduced by antisense expression or co-suppression of ACC synthase (Picton *et al.*, 1993), but STS treatment of rose plants increases leaf yellowing (Tjosvold *et al.*, 1994). A peroxygenase or peroxygenase-like peroxidase catalyses the oxidative loss of chlorophyll *a* in radish cotyledons (Adachi *et al.*, 2000) and most of the chlorophyll lost from spinach leaves also is degraded by a peroxidase pathway, where the porphyrin ring is opened and the resulting compound is colourless (Yamauchi and Watada, 1991). Peroxidase activity in spinach stored at 25°C increased by approximately 30% in 1 day and by 2.5-fold within 3–4 days regardless of whether or not ethylene was applied, and chlorophyllase played a minor role in de-greening both in the presence and absence of applied ethylene.

Excised leaves senesce more rapidly in darkness than in light, in part because photosynthesis slows ethylene production by lowering the ICC (4.10; Gepstein and Thimann, 1980; De Laat *et al.*, 1981; Grodzinski *et al.*, 1981). Ethylene synthesis by excised rice (Gepstein and Thimann, 1981) and oat (Kao and Yang, 1983) leaves is 300% and 200% more intense in darkness than in light, respectively. The light inhibition of EFE activity is completely overcome by 1% [CO_2], and the stimulation of EFE activity by CO_2 is rapidly reversible (Kao and Yang, 1982; Philosoph-Hadas *et al.*, 1986). Ethylene production is half-maximally promoted by 0.06% and 0.18% [CO_2] in rice and tobacco leaf tissue, respectively, and the effect of 0.5% [CO_2] can be detected within 15 min (Dhawan *et al.*, 1981; Fig. 4.6). While enhanced ethylene production may be a contributing factor promoting leaf senescence in darkness, an alternative explanation, that stomatal closure in darkness increases the IEC, seems not to have been considered. Example 23 in chapter 3 indicates that the IEC cannot reach an active level when a typical leaf's stomates are open, and Kuraishi (1976; Kuraishi and Ishikawa, 1977) provided evidence that there is a relationship between stomatal aperture and leaf senescence. When leaf discs cut from plants that bear stomata on one side only were floated with their stomatal surface downward in contact with a kinetin solution, the hormone did not delay senescence, but if the stomates faced upward and exchanged gas freely, applied kinetin was effective. Leaf senescence is hastened by darkness, ABA and other treatments or substances that induce stomatal closure, while light, and substances and treatments that promote stomatal opening always delay the process (Thimann and Satler, 1979a,b). Senescence should be accelerated when the stomates close in NA, MA or CA, and LP would be expected to have the opposite effect when it causes stomates to open. ABA hastens senescence by causing stomatal closure and promoting ethylene production in leaves (Aharoni, 1989), petiole explants (Abeles, 1967), flowers (Koning, 1986) and fruits (Plich, 1987).

Oat leaves bear stomates with equal frequency on both surfaces (Willmer and Fricker, 1996). Their stomates remain open and chlorophyll loss is slowed in CO_2-free air (Satler and Thimann, 1983) and also when excised oat-leaf segments are floated on solutions containing kinetin (5×10^{-5} M) and hexanol (2.5×10^{-2} M). ABA solutions cause stomatal closure and accelerate senescence in the light, but have only a slight effect on fading when the stomates close naturally in darkness (Thimann and Satler, 1979a). At 12°C and a pressure of 1.6 kPa (12 mm Hg; Table 4.9), or at 25°C and a pressure of < 60 kPa (< 450 mm Hg; Fig. 4.12, *left*), LP opens oat leaf stomates; protein degradation and chlorophyll loss are retarded (Fig. 5.44, *right*); and ABA, kinetin and hexanol do not affect stomatal opening or chlorophyll loss (Veierskov and Kirk, 1986). Although the fading of *Avena* leaves is half-maximally accelerated by 0.2 μl/l applied ethylene (Burg, 1968b), Veierskov and Kirk concluded that ethylene removal was not likely to be involved in the retardation of senescence caused by LP because oat leaves did not produce a sufficient amount of the gas to influence their fading. However, in other studies, oat leaves have been found to produce a significant amount of ethylene (Denny, 1937; Crocker, 1948; Burg and Burg, 1968; Gepstein and Thimann, 1981). Initially, when freshly cut *Avena* leaves are floated on water or 0.1 mM kinetin in darkness, they produce ethylene at a low rate, and then as senescence proceeds, the rate increases to > 2–4 μl/kg·h (Fig. 5.44, *left*).

RNA, protein and chlorophyll in barley-leaf segments degrade more slowly in LP when pure water-saturated O_2 is flowed at a pressure 20 kPa (150 mm Hg = 17.2% [O_2]) than in atmospheric air, and LP's effect on barley leaf senescence is reversed when ethylene is added to the flowing O_2 (Bangerth, 1974). Senescence was further retarded, and applied ethylene no longer reversed LP's effect on senescence, when air instead of pure O_2 was flowed at a pressure of 10 kPa (75 mm Hg), decreasing the [O_2] to 1.8%. Hypobaric storage at 15°C and a pressure of 2 kPa (15 mm Hg) has an unusual senescence-retarding effect on *Hibiscus* cuttings (Kirk *et al.*, 1986). At that pressure and temperature only 0.27 kPa (2 mm Hg) of air is present along with 1.7 kPa (13 mm Hg) of water vapour, and the [O_2] concentration is so low (0.06%) that respiration should be inhibited by approximately 95% (Fig. 4.2). Respiration immediately returned to the at-harvest value when, after 8 weeks' storage in LP, cuttings were removed. During that same storage period in NA, the respiratory rate increased by 50% due to a 'climacteric' rise associated with leaf senescence (Solomos, 1983). Although the improved storage of *Hibiscus* cuttings in LP must be due in part to very low [O_2], an equivalent O_2 partial pressure did not cause the same result at atmospheric pressure. The measured ethylene production rate of *Hibiscus* leaves was only 0.0005 nl/cm^2 of leaf surface per 24 h regardless of whether the tissue was freshly harvested or excised from cuttings stored for 8 weeks in NA or LP (Kirk *et al.*, 1986).[32] This rate was judged to be too low to implicate ethylene in the accelerated senescence that occurred when stomates closed at atmospheric pressure, or in the delayed senescence associated with LP-induced stomatal opening, but the same assay procedure (Veierskov and Kirk, 1986) failed to detect ethylene production by *Avena* leaves, which are known to produce an active amount of the gas (Fig. 5.44, *left*; Burg and Burg, 1968; Gepstein and Thimann, 1980, 1981), and at least some types of *Hibiscus* leaves produce detectable, significant ethylene.[32] *Hibiscus* abscission-zone explants produce ethylene at a rate that increases prior to and during senescence and peaks after senescence is completed (Wilkins and Swanson, 1975).

There are many examples of delayed senescence and de-greening of leaves stored in LP (10.59; 10.62; 10.66; Table 10.11; Fig. 10.7, *left*; Burg, 1973b; Eisenberg, 1977; Eisenberg *et al.*, 1977a; Jensen and Rasmussen, 1978). Chlorophyll was better maintained at a pressure of 10 kPa (75 mm Hg) than it was in NA during a 12-day storage of cress (*Lepidium sativum* L.) at 3°C (Bangerth, 1973, 1974). 1-MCP retards leaf yellowing of *Epipremnum pinnatum* (L.) Engl. cuttings stored in NA (Müller *et al.*, 1997); they degreen at 15°C during 6 weeks' storage in NA, and in LP at a pressure of 2–2.7 kPa (15–20 mm Hg) their stomates open and they remain green for at least 8 weeks (Kirk *et al.*, 1986). Endogenous ethylene within *Poa pratensis* leaves infected with *Bipolaris sorokiniana* begins to increase 12 h after inoculation, reaches a maximum at 48 h, and then progressively decreases during an additional 48 h (Hodges and Coleman, 1984). Necrotic lesions surrounded by chlorotic halos developed on infected leaves between 24 and 48 h, and mid-vein chlorosis and complete chlorosis of all tissues not directly affected by the lesions occurred between 72 and 96 h. A hypobaric pressure of 23.6 kPa (177 mm Hg), with [O_2] and [CO_2] maintained at normal ambient partial pressures, reduced the chlorophyll loss by half compared to that in leaves kept at atmospheric pressure.

5.29 Role of Ethylene in Abscission

The abscission process in fruits, floral organs and leaves has many similarities to senescence. It is induced by the same low applied-ethylene levels that are active in promoting senescence; both processes are delayed or prevented by a hypobaric pressure in the presence of 21% [O_2] (Lipe and Morgan, 1972b; Cooper and Horanic, 1973; Table 2.5) and by inhibitors of ethylene action or production (Lipe and Morgan, 1972a; Iannetta *et al.*, 2000); and senescing petiole explants produce sufficient ethylene to elevate their IEC to a level that initiates abscission (Abeles and Rubenstein, 1964; Abeles and Gahagan, 1968a; Abeles *et al.*, 1992). Auxin supplied to freshly excised leaf explants induces ethylene production and prevents abscission, but if the treatment is delayed until the explants have 'aged' for several days, abscission is promoted when ethylene production is induced (Rubenstein and Abeles, 1965; Hallaway and Osborne, 1969).[33] The ability of applied ethylene to induce abscission in cotton seedling leaves of different physiological age depends on the IAA content in their abscission zones (Suttle and Hulstrand, 1991); and it has been suggested that an ethylene-induced decline in leaf blade auxin levels might explain why the gas hastens 'ageing' and abscission (Hall, 1952). A decrease in auxin content occurs when ethylene induces abscission in cotton (Suttle and Hulstrand, 1991), bean and *Prunus serrulata* leaves (Rubenstein and Abeles, 1965).

Ethylene's involvement in the abscission process is apparent in CO_2 inhibition studies (4.12; Fig. 4.7, *lower right*) and LP experiments. Attached cotton fruits and detached cotton, pecan and hibiscus fruits dehisce more slowly in the presence of 21% $[O_2]$ if the pressure is lowered to 26.7 kPa (200 mm Hg) (Lipe and Morgan, 1972b; Gregg, 1982) and after the ethylene production of Valencia orange leaves and fruits, calamondins, and Persian limes has been elevated by a cycloheximide treatment, their dehiscence is delayed by lowering the pressure to 20 kPa (150 mm Hg), flowing 99.7% $[O_2]$ + 0.6% $[CO_2]$ (Cooper and Horanic, 1973) (Table 2.5). This mixture creates a normal atmospheric $O_2 + CO_2$ content at the reduced pressure, and does not affect the dehiscence of abscisic-acid treated citrus fruits (Cooper and Horanic, 1973; Rasmussen, 1975).

Ethylene antagonists rarely completely block abscission (Sexton, 1997), and exogenous ethylene has no effect on abscission in as many as 60% of the species that have been investigated, while in many others the correlation between ethylene production and abscission is poor (Brown, 1997). Ethylene has no direct effect on the abscission of cereal grain fruits (Sargent *et al.*, 1984), oil palm tepals (Chan *et al.*, 1972; Henderson and Osborne, 1994), and tepals of tulips (Sexton *et al.*, 2000) and many other monocotyledonous plants (Reid and Wu, 1992). Neither STS nor AVG prevent abscission of tulip tepals, and tulips produce ethylene at a constant, very low rate (approximately 0.1 nl/g·h) throughout the period from harvest to abscission (Sexton *et al.*, 2000).

5.30 Role of Ethylene in Flower Fading

The senescence of flowers, like that of leaves from which they are developmentally related, is controlled by 'ageing' and ethylene production (Abeles *et al.*, 1992). As little as 2 nl/l applied ethylene causes sepal wilt of orchids (Davidson, 1949), 30–125 nl/l causes 'sleep' of carnations (Uota, 1970), morning glory rib segments are stimulated to roll up by 10–100 nl/l (Hansen and Kende, 1975); and 60–200 nl/l causes roses to drop their petals (Kaltaler and Boodley, 1970). Applied ethylene affects numerous other flower types,[34] and endogenously produced gas hastens floral senescence and induces a respiratory climacteric in various flowers (Wilkins, 1965).[35]

A flower's ethylene sensitivity often depends on its stage of development. Mature carnation blooms respond to as little as 30–125 nl/l ethylene, immature closed blossoms whose petals show through the calyx are not affected by a 24-h exposure to 1000 nl/l (Crocker and Knight, 1908; Camprubi and Nichols, 1978; Woodson and Lawton, 1988), and during the ageing of cut carnation blooms, their ethylene sensitivity increases (Whitehead *et al.*, 1984). The K_m for rapid ethylene-induced abscission of *Pelargonium hortorum* blooms is 0.7 μl/l, but 0.1 μl/l elicits the same response in more mature 'post-receptive' blooms (Evensen *et al.*, 1993), and petals on freshly opened *P. domesticum* flowers do not abscise after exposure to 100 μl/l, while petals from unpollinated 2–3-day-old flowers abscise after a 1-h exposure to as little as 0.5 μl/l (Evensen, 1991). Morning glory, *Hibiscus*, *Digitalis*, *Petunia* and rose blossoms also display an 'ageing' requirement for the development of ethylene sensitivity (Suttle and Kende, 1980b; Lovell *et al.*, 1987;

Lukaszewska and Gorin, 1988). Ethylene does not affect petals taken from *Tradescantia* flower buds 1–2 days before opening, but they increase in ethylene sensitivity as they near the day of opening, and the gas hastens pigment efflux from petals excised from open flowers (Suttle and Kende, 1980a). Unpollinated *Cyclamen* flowers are ethylene insensitive and eventually lose turgor and wilt, whereas pollinated blooms become ethylene sensitive and abscise their turgid corolla (Halevy *et al.*, 1984). Following pollination of *Phaleonopsis* hybrid cv. Herbert Hager cut orchid flowers, the first event detected is an increased ethylene sensitivity, measured as the time required for 0.5 μl/l ethylene to cause 50% wilting of the flowers after a 12-h exposure (Halevy *et al.*, 1996; Fig. 5.14). Ethylene sensitivity begins to increase about 4 h after pollination and peaks 6 h later. An increase in ethylene production could not be detected for 12–14 h, and the accelerated wilting is completed in 2 days. Treating these flowers with STS or 1-MCP completely inhibited the pollination-induced increase in ethylene production, proving that it resulted in response to endogenous ethylene (Porat *et al.*, 1995a). Ethylene sensitivity was increased by applying short-chain saturated fatty acids 7–10 carbons in length on to the stigmas of *Phaleonopsis* blooms, and within 6 h after pollination these same fatty acids, and especially octanoic acid (C_8), increased in the flower's perianth and column (Halevy *et al.*, 1996).

Although it has been suggested that possibly all floral crops produce ethylene (Staby, 1976a),[6] there are exceptions and floral senescence often proceeds without ethylene involvement. The senescence of cv. Sandra carnation flowers is characterized by a loss of petal colour and a gradual necrosis of the petals; they never show normal corolla in-rolling and produce no ethylene (Reid and Wu, 1992). When the petals of 'Queen Fabila' *Brodiaea* (*Triteleia laxa* Benth = *B. laxa*) flowers start to senesce, their stigma becomes receptive and pollination induces a burst of ethylene synthesis, but unpollinated flowers of *Brodiaea* produce no measurable ethylene during their entire lives (Bufler *et al.*, 1980), and consequently STS does not affect their vase-life (Han *et al.*, 1991). Treatment with 0.03 μl/l ethylene induced senescence of open *Brodiaea* flowers, promoted ovary growth and completely inhibited petal growth and bud opening, and these effects of applied ethylene predictably were counteracted by STS. *Cyclamen* flowers (*Triteleia laxa*) only produce ethylene after pollination, but unlike *Broidiaea* flowers they are insensitive to applied ethylene (Halevy *et al.*, 1984). The dark respiration of petal slices of *Saxifraga cernua* progressively increases as the petals develop and fully expand, and then decreases during senescence, indicating that the flowers do not undergo an ethylene-induced climacteric rise as they fade (Collier and Cummins, 1991).

A survey of ethylene's role in the petal senescence of 93 floral species representing 23 families indicates that except for Campanulaceae, Caryophyllaceae, Malvaceae and many Orchidaceae, most of those flowers in which wilting was the primary senescence symptom were insensitive to ethylene.[37] Petal abscission is the initial symptom of senescence in Geraniaceae, Labiatae, Ranunculaceae, Rosaceae and Scrophulariaceae, and these flowers are mostly ethylene sensitive (Woltering, 1984, 1987; Woltering and van Doorn, 1988; Table 5.17). Flowers of many species do not abscise a turgid corolla, but instead wilt before dropping desiccated flower parts by a process that does not appear to be active abscission. Although endogenous ethylene production may regulate abscission in blooms of *Digitalis*, petunia, *Pelargonium*, apple and cherry, in other flowers there often is no clear correlation between ethylene evolution and abscission in the absence of pollination-induced ethylene production (Sexton *et al.*, 1985). Senescence of the petals of many cut flowers, including commercially important varieties such as gladiolus, iris, narcissus and tulip, seems not to be related to ethylene (Reid and Wu, 1992). China asters, forget-me-nots, *Lobelia*, sweet peas, *Viola*, *Acacia*, calla lily, cyclamen, *Calendula* and tulips also have been reported to be resistant to applied gas (Abeles, 1973; Abeles *et al.*,

Table 5.17. Ethylene's role in the senescence of flowers from different families. Sensitivity from 0 = insensitive to 4 = sensitive (Reid and Wu, 1992 – from data by Woltering and van Doorn, 1988).

Family	Symptoms	Sensitivity
Monocotyledonae		
Amaryllidaceae	Wilting	0–2
Iridaceae	Wilting	0–1
Liliaceae	Wilting	0–1
Orchidaceae	Wilting	3–4
Dicotyledonae		
Campanulaceae	Wilting	3–4
Caryophyllaceae	Wilting	4
Compositae	Wilting	0–1
Dipsaceae	Wilting	2–4
Euphorbiaceae	Wilting	4
Geraniaceae	Abscission	4
Labiatae	Abscission	4
Malvaceae	Wilting	4
Primulaceae	Abscission	3–4
Ranunculaceae	Abscission	3–4
Rosaceae	Abscission	3–4
Scrophulariaceae	Abscission	2–4
Umbelliferae	Wilting	0

1992). *Dendrobium* and *Oncidium* orchids are almost insensitive (Goh *et al.*, 1985); and although cut *Ranunculus asiaticus* flowers of the Aviv cultivar group produce ethylene in a climacteric-like pattern, they are not sensitive to applied ethylene, and their flower quality is not improved by treatments with AOA or STS (Kenza *et al.*, 2000).

Flower fading, like fruit ripening, is a genetically programmed event involving the regulated expression of specific genes. cDNAs that are up-regulated during carnation-petal senescence have been cloned, and these genes are transcriptionally activated by ethylene (Lawton *et al.*, 1990). A 126 bp sequence from the 5′ flanking region of the *GST1* gene which encodes glutathione-*S*-transferase is necessary for ethylene regulation in carnation petals, and a protein that interacts with this sequence was identified that has DNA-binding activity in pre-senescent, senescent and ethylene-treated petals (Itzhaki *et al.*, 1994; Maxson and Woodson, 2000).

Changes in membrane composition and permeability are two of the first symptoms of floral fading. In senescing *Ipomoea tricolour* flowers, phospholipid breakdown precedes in-rolling and a change in ethylene production, and increases dramatically when autocatalytic ethylene production sets in (Beutelmann and Kende, 1977; Fig. 5.45, *left*). During petunia flower senescence, microsomal-membrane phospholipid declines and diacylglycerol increases in advance of changes in the ethylene production rate, and then both decline precipitously after autocatalytic ethylene production begins (Borochov *et al.*, 1997; Fig. 5.45, *right*). In both *Ipomoea* and petunia, developmental changes in membrane structure and function precede autocatalytic ethylene. Treatment of *Tradescantia* blooms with ethylene accelerates the onset of membrane leakiness and phospholipid deterioration in their petals (Suttle and Kende, 1980a). An increase in the lipid microviscosity of microsomal membranes from senescing cut 'White Sims' carnation flowers is initiated within 3–4 days of cutting flowers, rises with advancing senescence and coincides temporally with the in-rolling that denotes the climacteric-like rise in ethylene production (Thompson *et al.*, 1981). AVG treatment of young cut carnation flowers prevents petal in-rolling and delays the rise in membrane viscosity until 9 days after cutting. The microviscosity of microsomal membranes rose sharply within 24 h after freshly harvested or AVG-treated flowers were exposed to 1 μl/l ethylene, in-rolling of the petals was clearly evident at that time and STS delayed the rise in microsomal-membrane viscosity both in control flowers and those exposed to exogenous ethylene. Membrane rigidification was accompanied by an increased sterol: phospholipid ratio, reflecting the selective loss of membrane phospholipid that accompanies senescence (Thompson *et al.*, 1982).

Often floral fading can be delayed and vase life improved by treating blooms with inhibitors of ethylene synthesis or action such as rhizobitoxin and sodium benzoate (Baker *et al.*, 1977), silver thiosulfate

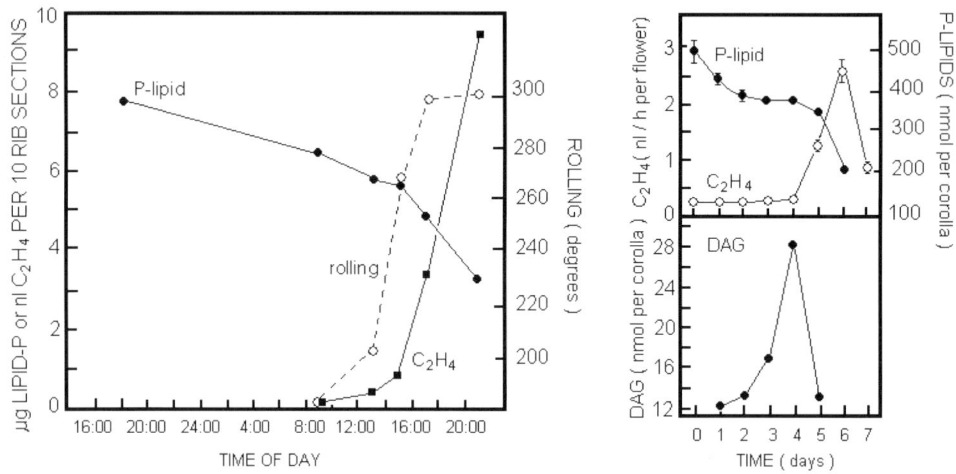

Fig. 5.45. (*left*) Time course of change in phospholipids, wilting (rolling) and ethylene production of petal sections of *Ipomoea tricolor Cav.* (Beutelmann and Kende, 1977). (*right*) Time course during petunia flower senescence of changes in ethylene production rate, microsomal membrane phospholipid content (P-lipid) and plasma membrane diacylglycerol content (DAG) (Borochov *et al.*, 1997).

(STS), amino ethoxyacetate, amino vinyl glycine (AVG), Co^{2+}, 1-methylcyclopropene (1-MCP), 2,5-norbornadiene (Wang and Woodson, 1989) and other cyclo-olefins. Flowers exported from Holland and Colombia have for many years been pulse-treated with STS to delay senescence, and 1-MCP, diazocyclopentadiene (DACP) and STS have a similar effectiveness (Sisler *et al.*, 1993; Serek *et al.*, 1994b, 1995a, 1996; Porat *et al.*, 1995b; Wawrzynczak and Goszczynska, 1999). STS significantly improves the marketing of carnations, *Aconitum*, *Delphinium*, *Euphorbia*, *Lathyrus*, *Alstroemeria*, *Antirrhinum*, *Campanula*, *Freesia*, *Phlox* and *Physostegia* (Hardenburg *et al.*, 1986), and commercial suppliers feel that STS is helpful with cut tulips (Sexton *et al.*, 2000). Because STS is a potential environmental hazard, many countries are currently working to prohibit its use, and 1-methylcyclopropene (1-MCP) has been introduced as a substitute. Results with STS and 1-MCP are summarized in Table 5.18 and footnote 38. Floral senescence also is delayed (but not prevented) in ethylene-insensitive transgenic *CaMV35S-etr1-1* petunias (Clark *et al.*, 2000).

The effectiveness of STS and 1-MCP usually has been studied at ambient temperature, measuring vase life in water ± ethylene. This does not indicate whether these substances improve long-term dry storage at a reduced temperature. 1-MCP's protective effect on *Pelargonium pelatum* petal abscission is transient, indicating that the inhibition is able to dissociate from the ethylene receptor and be lost to the atmosphere. Although, silver ions inhibit tomato ripening (Atta-Aly *et al.*, 1987; Tucker and Brady, 1987; Davies *et al.*, 1988), eventually silver-treated fruit ripen either by sequestering the inhibitor or by synthesis of new binding sites (Davies *et al.*, 1990). Because most flowers treated with inhibitors of ethylene action deteriorate for other reasons before they again become ethylene-sensitive, it often has not been possible to determine the length of the protection period provided by 1-MCP and STS. Banana and tomato fruits treated with 1-MCP at 24–25°C remain insensitive to ethylene for 11–13 days after which they ripen normally (Sisler *et al.*, 1996b; Sisler and Serek, 1997; Macnish *et al.*, 2000b; Roh *et al.*, 2001; Sisler and Serek, 2001), and carnations are protected for 12–15 days (Sisler and Serek, 1997), but *Grevillea* flowers are only protected for two days and cut flowers of *Chamelaucium uncinatum* (cvs. Lollypop, Alba and Mid

Table 5.18. Effect of 1-methylcyclopropene (1-MCP) and ethylene on the vase life of cut flowers (in water), and the shelf-life of potted flowering plants (at room temperature). Similar effects of 1-MCP on floral abscission and fading have been reported for *Chamelaucium uncinatum* (Serek *et al.*, 1995b), *Phlox paniculata* (Porat *et al.*, 1995b) and *Dianthus barbados* (Sisler and Serek, 1997).

	Shelf-life (days)				
Commodity	Control	1-MCP	1-MCP + C_2H_4	C_2H_4	Reference
Cut flowers					
Petunia hybrida L. cv. Pink Cascade*	5.4	6.2	5.3	3.0–3.2	1
Alstroemeria			6.8	5.0	2
Antirrhinum majus			6.5	3.0	2
Consolida ambigua			5.8	1.5	2
Dianthus caryophyllus	4.0	7.0	5.3	2.3	2
Matthiola incana			5.0	2.3	2
Penstemon hartwegii × *P. cobaea*			5.3	1.8	2
Potted plants in flower					
Begonia × *eliator hybrida*	26.2	26.1	7.3	2.3	3
Rosa hybrida 'Victory Parade'	21.2	29.5	9.3	3.3	3, 4
Kalanchoe blossfeldiana, 'Tropicana'	35.2	35.5	13.3	2.0	3
Campanula carpatica 'Blue Clips'			9.0	3.2	5

*STS extended the vase life of *Petunia hybrida* L. cv. Pink Cascade to 9.1 days from 5.1 days (Borochov *et al.*, 1997).
References: (1) Serek *et al.*, 1995c; (2) Serek *et al.*, 1995a; (3) Serek *et al.*, 1994d; (4) Serek *et al.*, 1994b; (5) Sisler *et al.*, 1996b.

Pink) for 2–4 days (Macnish *et al.*, 2000b). STS (0.5 mmol Ag^+ per litre) applied at 2 or 20°C protected *C. uncinatum* from the effects of ethylene for at least 10 days (Macnish *et al.*, 2000b). These times are short compared to the storage durations attainable with hypobaric storage. Tomatoes (Burg, 1978, unpublished) and bananas (Apelbaum *et al.*, 1977c) ripen normally after 8 and 17 weeks in LP, respectively, and the vase life of carnations is increased after 9 weeks of LP storage (Dilley and Carpenter, 1973; Dilley *et al.*, 1975).

At low storage temperatures, higher concentrations of 1-MCP and DACP are required to protect carnation flowers (Sisler *et al.*, 1996a), *Grevillea* inflorescences (Macnish *et al.*, 2000b), *Kalanchoe* plants (Reid *et al.*, 1996 – referred to in Sisler and Serek, 1997) and banana fruits (Macnish *et al.*, 2000b) against exogenous ethylene. DACP is ineffective during storage in darkness, and *Penstemon* flowers (Serek *et al.*, 1995a), banana fruits and *Grevillea* inflorescences treated with 1-MCP at 2°C were not protected from the effects of ethylene by the same exposure to 1-MCP that is effective when applied at 20°C. Based on a determination of applied ethylene's ability to cause abscission of detached 1-MCP treated flowers of *Pelargonium peltatum* ('Pink Blizzard' = ivy geranium), the half-life of 1-MCP activity was 2, 3 and 6 days after 1-MCP treatment at 25, 20.7 and 12°C, respectively, and there was no evidence of a residual effect after 4–5 days at 25 or 20.7°C (Cameron and Reid, 2001).

STS and 1-MCP protect many flowers from the action of exogenous ethylene, but usually do not appreciably extend storage or vase life in the absence of applied ethylene (Table 5.18). Although STS and 1-MCP improve (> 50%) the vase life of *Lisianthus* [*Eustoma grandiflorum* (Raf.) Shinn] flowers exposed to 1 μl/l ethylene for 2 days, these substances did not significantly affect the vase life of control flowers (Cho *et al.*, 2001). In tests with 14 native Australian cut flowers, the vase lives of *Ceratopetalum gummiferum*, *Chamelaucium uncinatum*, *Grevillea* cvs. Kay Williams and Misty Pink, *Leptospermum petersonii*, *Telopea* cv. Shady Lady and *Verticordia nitens* were reduced by applied ethylene, and 1-MCP generally protected these flowers from the applied gas, but usually it did not extend the storage life of flowers kept in ethylene-free air (Macnish *et al.*, 2000c). Amongst 21 types of flowers tested, with the exception of poinsettia 'Freedom' and Easter lily 'Nellie White', 1-MCP protected all types from the effect of 2–10 µl/l exogenous ethylene, but in only nine varieties was normal vase life increased (Skog *et al.*, 2001). Pre-treatment of cut spikes of tuberose (*Polianthes tuberose* L.) with STS eliminated the effects of applied ethylene, but had no effect on the vase life of cold-stored flowers and substantially increased ethylene production by their florets (Waithaka *et al.*, 2001). STS failed to prevent the abscission of bracts of *Bougainvillea* flowers in the absence of applied ethylene (Gago *et al.*, 2001), and 1-MCP had relatively little effect (applied alone) on petal fresh weight, electrolyte leakage or membrane lipid fluidity in *Petunia*, although it prevented a response to applied ethylene (Serek *et al.*, 1995c). Pre-treatment with STS only increased the vase life of cut florets of *Lilium* Oriental Hybrid 'Casa Blanka' by 1.4 days from 6 to 7.4 days (KiCheol and JungNam, 1999). Post-production life of miniature potted rose plants is terminated by flower senescence, yellowing and loss of buds, and senescence and abscission of leaves. Applied ethylene enhances these symptoms, and 1-MCP increases display life and reduces bud, flower and leaf abscission caused by exogenous ethylene (Serek *et al.*, 1994b,d), but does not stop applied ethylene from causing leaf yellowing (Sisler and Serek, 1997), or prevent a rose plant's leaves from yellowing when they are not exposed to ethylene (Serek *et al.*, 1996). 1-MCP also fails to significantly increase the shelf life of flowers on rose plants held in ethylene-free air (Table 5.18). It was concluded that ethylene is not the primary agent causing leaf yellowing in miniature roses (Serek *et al.*, 1996). Even though a miniature potted rose (*Rosa hybrida* cv. Bronze) produces ethylene, and 1-MCP does not increase its normal display life, surprisingly when it was 'transport-stressed', regardless of whether or not ethylene production increased, 1-MCP partially prevented the decrease in vase life which otherwise would have resulted (Müller *et al.*, 2000b). Although 1-MCP prevents the response of *Kalanchoe* flowers to applied ethylene, it has no effect on their normal vase life and did not inhibit ethylene action when applied to *Kalanchoe* plants whose ethylene production already was high (Serek *et al.*, 1994d; Reid, 1996 – referred to in Sisler and Serek, 1997; Serek and Reid, 2000; Table 5.18).

1-MCP, STS and other ethylene antagonists do not inhibit floral senescence of cut rose flowers (Veen, 1983), tulips (Nichols and Kofranek, 1982; Veen, 1983; Sexton *et al.*, 2000), iris (Reid and Wu, 1992), *Sandersonia* (Eason and Vre, 1995), narcissus (Reid and Wu, 1992; Kenza *et al.*, 2000), gladiolus (Serek *et al.*, 1994a), *Montbretia* (McKenzie and Lovell, 1992), *Hemerocallis* (Stead, 1992) and freesia (Spikman, 1989); ethephon had little influence and STS shortened the vase life of bird-of-paradise (*Strelitzia reginae*) flowers (Finger *et al.*, 2000); and STS did not prevent ethylene-induced stamen damage in selected *Myrtaceae* (Sun *et al.*, 2001).

Although laboratory experiments place great emphasis on ethylene's role in floral senescence, often water loss and wilting, depletion of respirable carbohydrates, disease, insects, chilling injury, mechanical damage, water spotting, tissue browning, petal 'tip burn' and colour changes caused by ammonia production are more important than ethylene for flower preservation during transit and commercial storage. Bud-cut flowers may lose their ability to open, or blooms may 'blow open' and rapidly fade during vase life (Rogers, 1973). Several studies indicate that developmental processes occur independently of ethylene action (Fig. 5.45), and along with the loss of respiratory substrates, protein breakdown and ammonia formation, ultimately lead to flower senescence. AVG completely suppresses ethylene production by *Tradescantia* flowers and extends their vase life (Suttle and Kende, 1980b), but it only partially reduces the loss of membrane semi-permeability and the decline in phospholipids. Membrane degeneration starts in the petal cells of morning glory and petunia long before ethylene production increases (Fig. 5.45). When AOA suppresses a carnation's ethylene synthesis, and STS or 1-MCP prevent ethylene action, in-rolling of the petals is delayed and instead the flowers ultimately fade by desiccation and petal necrosis (Bufler *et al.*, 1980).

A hypobaric condition preserves flowers regardless of whether they fade in response to ethylene or due to any other cause. LP maintains flowers in an at-harvest condition for a much longer time than is possible using STS or 1-MCP (Staby *et al.*, 1984) because in addition to preventing ethylene production, removing endogenous ethylene and purifying the storage air, a hypobaric condition slows developmental processes, opens stomates, reduces respiration, slows water loss, kills insects, inhibits microbial growth and limits accumulation of the ammonia that causes bluing of red roses and certain other flowers. Young flower buds of Easter lily, gladiolus, snapdragon and chrysanthemum, which do not open after NA or CA storage, open normally after LP storage (Burg, 1969, 1970, 1973b; Dilley, 1972, 1977a; Carpenter and Dilley, 1975; Patterson, 1975b; Staby *et al.*, 1976; Marousky, 1977b; Hardenburg *et al.*, 1986; Grumman, unpublished), and the hypobaric method prolongs the storage life of flowers reputed to be insensitive to applied ethylene (Woltering, 1987; Woltering and van Doorn, 1988), including Compositae such as chrysanthemum and aster, Liliaceae such as Easter lily and tulips, and Iridaceae such as freesia and gladiolus. *Protea*, *Fynbos* and *Leucodendron* blooms show massive blackening of flowers and leaves within a few hours after they are removed from NA or CA storage (Akamine, 1976a), but retain their at-harvest appearance and shelf life after removal from 31 days' LP storage (Spearpoint, 2000a,b,c; 10.80). After 40 days in LP, red roses still have their initial vase life. Successful long-term LP storage has been reported for more than 35 other varieties of flowers (10.53–10.86). LP intermodal containers have been used for nearly a decade to accumulate roses and other types of flowers at Miami International Airport during 4–6 weeks in advance of major holidays in order to benefit from holiday price appreciation. Although many of these blooms had been treated with STS, without hypobaric storage they could not be stored for 4–6 weeks without an excessive loss of vase life.

5.31 Induction of Physiological Disorders by Ethylene

Numerous disorders of fruits and vegetables are induced or enhanced by exogenous ethylene, including softening and reduced rind thickness of watermelons (Risse and Hatton, 1982; Elkashif and Huber, 1988), 'peel puffing' in Satsuma mandarin (*Citrus unshiu*; Maotani *et al.*, 1983), russet spotting in lettuce (Hardenburg *et al.*, 1986; Leshuk and Saltveit, 1990), *Botrytis cinerea* growth in strawberries (Barkai-Golan, 1990), chlorophyll loss in broccoli (Leshuk and Saltveit, 1990), deterioration of cabbage (Leshuk and Saltveit, 1990), toughening of green asparagus spears (Haard *et al.*, 1974) and etiolated asparagus plants (Chang and Han, 1975), bitterness in carrots (Carlton *et al.*, 1961), discoloration of Hubbard squash (Yeager *et al.*, 1945) and formation of toxic pisatin in the pods of green peas (Chalutz and Stahman, 1969), and phenolic substances in sweet potato roots (Buescher *et al.*, 1975).

Ethylene also causes disorders of flowers and foliage plants (Rogers, 1973; Staby *et al.*, 1978a; Hardenburg *et al.*, 1986), including drying and bleaching of *Cattleya* sepals (Davidson, 1949), sleep of carnation flowers (Uota, 1970), increased branching and defoliation of roses, decreased keeping-quality of cut roses, shattering of snapdragon florets, shelling of *Calceolarias* and snapdragons, early flowering and reflexed, twisted leaves in bulbous iris, and yellowing or abscission of container-grown plants. In tulips, > 0.05–0.1 μl/l applied ethylene causes bud necrosis (Munk, 1971, 1972, 1973a; Kamerbeek and deMunk, 1976), > 0.3 μl/l induces flower-bud blasting (Munk, 1973b; Kamerbeek and deMunk,

1976) and the gas causes premature development of axillary buds (Munk, 1973b), retardation of root development, formation of root hairs (Munk and Rooy, 1971), reduced development of leaf width in forced tulips (Munk, 1973a), suppression of wax deposition on the foliar leaves, formation of more than one tip on the basal foliar leaves, reduced starch level in the stamens (Munk, 1972), faint yellowing coloration of the outer cell layers of the bulb scales (Munk, 1973a), and increased fresh-weight loss and accelerated respiration of bulbs during storage (Munk, 1972).

5.32 Effect of Ethylene on Stomatal Opening

A connection between stomatal functioning and ethylene action has been sought after because water stress and CO_2 increase ethylene production, CO_2 inhibits ethylene action and stomatal responses are sensitive to water stress and CO_2. It has been reported that ethylene and ethephon close stomates (Browning, 1974; Vitagliano and Hoad, 1978; Erkan and Bangerth, 1980; Kays and Pallas, 1980; Bradford and Hsiao, 1982; Pallas and Kays, 1982; Kirkham, 1983; Madhavan *et al.*, 1983; Squier *et al.*, 1985; Tissera and Ayres, 1986; Taylor and Gunderson, 1986, 1988), but in other studies no effect could be detected (Pallaghy *et al.*, 1970; Pallaghy and Raschke, 1972; El-Beltagy and Hall, 1974; Mayak *et al.*, 1977; Aharoni, 1978; Bradford, 1983; Madhavan *et al.*, 1983; Taylor and Gunderson, 1986; Horton, 1991; Pallaghy and Raschke, 1972), and sometimes ethephon opens stomates (Frommhold, 1982; Levitt *et al.*, 1987; Cliquet and Morot-Gaudry, 1989). Usually, when ethylene reduces the stomatal aperture it causes a parallel reduction in photosynthesis, and it is not clear whether ethylene directly caused the closure, which inhibited photosynthetic CO_2 uptake, or instead ethylene inhibited photosynthesis, which, in turn, increased the ICC and closed the stomates (Willmer and Fricker, 1996). Ethylene-induced stomatal closure also might result from ethylene-enhanced ABA levels (Mayak and Halevy, 1972; Vitagliano and Hoad, 1978).

5.33 Effect of Ethylene on Ascorbic Acid Content

At atmospheric pressure, ethylene had no effect on the ascorbic acid content of tomatoes, bananas, celery (Thornton, 1940) or grapefruit (Maier *et al.*, 1973). Applying ethylene to apples and red currants while they were being stored in LP did not affect their vitamin C content (Bangerth, 1973).

5.34 Examples

1. During a typical Florida day during the season when mature Haden mangoes are available on the tree, the daytime temperature may reach 30°C and the relative humidity 70%. This would create a vapour pressure gradient of 9.5 mm Hg between the fruit and atmosphere if the fruit remained at the ambient temperature. A typical mango may weigh 400 g, and have a surface area of 320 cm^2, a transpirational resistance of 100 s/cm and a skin resistance to ethylene mass transport of 5000 s/cm. Equation 3.14 predicts that such a fruit will lose water vapour at the rate of 0.04 cm^3/s, in 1 h transpire 0.12 cm^3 of liquid water and with an IEC of 1.9 µl/l will evolve ethylene at a rate of 4.39×10^{-4} cm^3/h. The ethylene absorption coefficient in water is 0.095 cm^3 (gas) per cm^3 (water) at 30°C (Table 15.2), and to supply a 400 g fruit with 4.39×10^{-4} cm^3/h of ethylene gas dissolved in a flow of 0.12 cm^3 of liquid water per hour, the water would have to be in equilibrium with an ethylene concentration of 38,400 µl/l. Alternatively, the ethylene could be derived from 160 µM ACC delivered to the fruit in the flowing transpired water. As some mangoes contain 3.7 µl/l ethylene on the tree (Fig. 5.17), and even 11.4 µl/l (5.10), they would require a correspondingly higher concentration of ACC in the xylem sap to support their ethylene production rate.

2. Substrate concentration is one of the most important factors determining the velocity of an enzymatic reaction. Usually, when initial velocity (v) is plotted against substrate concentration (s), a section of a rectangular hyperbola is obtained. Such a result is obtained whenever a process depends upon a simple dissociation of the type:

$$XY \leftrightarrow X + Y \quad (5.1)$$

in which the concentration of [Y] is kept constant. The ethylene dose–response curve for a variety of ethylene actions has this shape, and therefore can be treated by the same kinetic method used for enzyme action. The simplest model for ethylene action that conforms to a dissociation of this type consists of a receptor protein (R) containing a metal (M = copper), which binds ethylene (E) according to the expression:

$$RM + E \leftrightarrow RME \quad (5.2)$$

When the concentration of free ethylene (the IEC) remains constant, the action of ethylene depends on the steady-state concentration of RME. If r is the total concentration of the receptor protein, e the concentration of ethylene (E) in the IEC and p is the concentration of RME, at steady-state the concentration of RM is $(r - p)$ and the equilibrium constant (K_e) for the dissociation of RME into RM + E is given by:

$$K_e = e(r - p)/p \quad (5.3)$$

In this model, the action of ethylene (v = velocity) depends on the concentration of p:

$$v = kp \quad (5.4)$$

where k, the rate constant for the ethylene action, would take into account any other substrates that might be involved. Combining equations 5.3 and 5.4:

$$v = kr/[1 + (K_e/e)] \quad (5.5)$$

RM becomes saturated with ethylene when e is large in comparison with K_e. Then the maximum velocity (V) equals kr and equation 5.5 can be written as:

$$v = V/[1 + (K_e/e)] \quad (5.6)$$

according to which v will be equal to $V/2$ when $e = K_e$. Under these conditions, the value of e that is found experimentally to give a half-maximal velocity or 'activity' (which corresponds to the Michaelis-Menton constant = K_m) is equal to K_e.

3. In enzyme kinetics, the combination of a metal-ion activator with components of a system can affect the velocity of the reaction. A plot of reaction velocity against concentration of added metal ion commonly gives a rectangular hyperbola, resembling the normal substrate concentration curve of an enzymatic reaction, and such data can be treated by any of the usual graphical methods. However, the 'K_m' for the metal so obtained rarely is the dissociation constant of the enzyme–metal complex. Instead, the constant has a number of different meanings, depending upon the mechanism of action and the conditions. Assuming that the activating metal ion (M) acts by combining with the receptor independent of the substrate (ethylene), and that ethylene cannot bind to the receptor protein when it lacks the metal 'activator', the following reactions must be considered:

$$R + M \leftrightarrow RM \quad (5.7)$$

$$RM + E \leftrightarrow RME \quad (5.8)$$

If the action of ethylene depends on the concentration of RME:

$$v = kP_e \quad (5.9)$$

where k, the rate constant, takes into consideration the concentration of any reactant that may be consumed, and P_e is the concentration of RME. Assuming equilibrium between the receptor (R), activator (metal = M), and substrate (ethylene = E):

$$(r - P_r - P_e)m = P_r K_M \quad (5.10)$$

$$eP_r = P_e K_E \quad (5.11)$$

where the dissociation constants of reactions 5.10 and 5.11 are K_M and K_E, respectively, r is the total concentration of receptor protein, m the steady-state concentration of free metal, P_e is the concentration of RME and e, the ethylene concentration, is independent of the equilibrium reactions. Eliminating P_e by combining equations 5.9–5.11:

$$v = (kr)/(1 + K_E/e + K_E K_M/em) \quad (5.12)$$

When m is present in excess and much larger than K_m, equation 5.12 reduces to the form of equation 5.6, and a half-maximal effect should be produced when $e = K_E$, whereas if $m = K_M$ and $e = K_E$, the ethylene effect would be 1/3 of maximum. Ostensibly, when $m = 0$, the velocity would be zero, but this interpretation is confounded by the finding, based on genetic data, that the receptor–metal complex is required to prevent the development of ethylene phenotypes (5.11). Therefore, lowering the Cu concentration should promote instead of inhibit the ethylene response. This can be formulated as follows:

$$R + M \leftrightarrow RM \quad (5.13)$$

$$RM + E \leftrightarrow RME \quad (5.14)$$

where R is the receptor protein, M is the metal 'activator' (Cu) and E (ethylene) negatively regulates the activity of RM by binding to form RME. The velocity (v) of the 'reaction' regulated by RM is:

$$v = kP_a \quad (5.15)$$

where P_a is the concentration of RM and k is a rate constant that includes any 'substrates' or transducers involved in the reaction that is catalysed by RM. The mass action equations are:

$$(r - P_a - P_e)a = K_1 P_a \quad (5.16)$$

$$eP_a = K_2 P_e \quad (5.17)$$

where r is the *total* concentration of receptor protein in all forms, a is the steady-state concentration of free copper ion (M), P_e is the concentration of RME, e is the ethylene concentration and K_1 and K_2 are the dissociation constants for reactions 5.16 and 5.17, respectively. Eliminating P_e and substituting into equation 5.15:

$$v = kr/[1 + (e/K_2) + (K_1/a)] \quad (5.18)$$

If the binding of the metal to the receptor is essentially irreversible, $K_1/a \rightarrow 0$ and the velocity will be half-maximal when $e = K_2$. In this instance, when applied ethylene causes a half-maximal response (= 50% inhibition), K_2 is the Michaelis-Menton constant for the binding of ethylene to the receptor, and to the extent that the metal can dissociate from the complex, or if the metal's concentration is rate-limiting, K_1/a will increase and v will decrease. When $a \rightarrow 0$, K_1/a becomes infinite and $v \rightarrow 0$.

Notes

1. The promotive effect of Co^{2+} and Ni^{2+} on leaf expansion (Miller, 1951), stem elongation (Thimann, 1972), chlorophyll retention in senescing leaves (Yang and Hoffman, 1984), and hook opening (Kang *et al.*, 1967), has been attributed to the ability of these metals to inhibit ethylene biosynthesis.
2. Plasmolysis in 1.1 N glycerol or KCl did not inhibit ethylene production in climacteric apple-tissue slices (Burg *et al.*, 1964), and ACC oxidase activity was better maintained in mango slices that were osmotically dehydrated (Tovar *et al.*, 2001).
3. A partial list of climacteric fruits includes (Biale *et al.*, 1954; Wills *et al.*, 1989; Abeles *et al.*, 1992): apple, apricot (Chahine *et al.*, 2000), avocado, banana, blueberry, cantaloupe, cherimoya (Martinez *et al.*, 1993), egg plant, *Feijoa*, fig, guava, kiwi fruit (Bonghi *et al.*, 2001), mango, melons, olive, papaya, passion fruit, peach, pear, persimmon, vanilla beans, cucumber, (Sexton *et al.*, 1993), date (Serrano *et al.*, 2001), walnut (Zhang *et al.*, 2001), Chinese bayberry (Hu *et al.*, 2001), sapota, nectarine and tomato.
4. A partial list of non-climacteric fruits includes strawberry (Atta-Aly *et al.*, 2000b), raspberry (Perkins-Veazie *et al.*, 1982; Sexton *et al.*, 1993), blackberry (Perkins-Veazie *et al.*, 2000), citrus, bell peppers, lychee, carambola and grape. Chinese jujube, pineapple, snap bean, cherry, raspberry, tamarillo and watermelon (Wills *et al.*, 1989; Abeles *et al.*, 1992) are considered to be non-climacteric. However, applied ethylene accelerates the ripening of carambola (Mercantila, 1989b), jujubes (R.B. Harvey, 1928b) and raspberry (Sexton *et al.*, 1993) and there have been conflicting reports about whether or not cherries exhibit a respiratory climacteric and are ripened by endogenous ethylene (Blanpied, 1972a; Hartmann *et al.*, 1987; Li *et al.*, 1994).

Late-picked pineapples display the beginning of a climacteric rise, and the respiration of early-picked but not late-picked fruits is stimulated by ethylene (Hulme *et al.*, 1968; Dull, 1971). A ripening-induced cDNA encoding ACC and another encoding ACO have been isolated from pineapple, and ACC expression increased 16-fold in ripe fruit compared to green fruit, while ACO was mainly induced in wounded pineapple tissue (Cazzonelli *et al.*, 2000). Ethylene-enhanced ripening has been reported in peppers (Sims *et al.*, 1970) and other members of that genus, including ornamental pot plants, pimento, paprika and chilli peppers (Abeles *et al.*, 1992), and applied ethylene induces a ripening-related cellulase in bell peppers (Ferrarese *et al.*, 1995; Harpster *et al.*, 1997). Domestication may have led to the selection of non-climacteric bell-pepper cultivars resembling the *rin* tomato mutation, which do not abscise, redden or display increased respiration (Saltveit, 1977). The ethylene production and respiration of strawberries doubles during the period when changes in colour, texture and flavour occur (Abeles and Takeda, 1990), and ethylene removal delays the loss of firmness in detached berries (Wills and Kim, 1995). It has been reported that applied ethylene does (Atta-Aly *et al.*, 2000b) and does not (Tian *et al.*, 2000) stimulate the respiration of strawberry fruits and, depending on tissue maturity, 1-MCP may or may not affect the respiratory rise induced by ethylene. The respiration rate increases in blackberries during the mottled (part-black) and black-ripeness stages, and ethylene production and ACC increase in the dull black (overripe) stage (Perkins-Veazie *et al.*, 2000). Santa Rosa plums exhibit a climacteric ripening pattern, including increased respiration and ethylene production, whereas Golden Japan plum cultivars do not, possibly because polyamines such as putrescine increase in this cultivar and may inhibit ethylene production during ripening (Zuzunaga *et al.*, 2001).

5. Respiration data for mature-green tomatoes (Pratt and Workman, 1962; Lyons and Pratt, 1964; Sawamura *et al.*, 1978; Knee, 1995) and 'mature slip' cantaloupes (Lyons *et al.* 1962) reveal that after detachment their CO_2 production progressively declines to a pre-climacteric minimum and then increases to a climacteric maximum, and in passing from the minimum to maximum the rate increases by approximately 110% in melons and 60% in tomatoes. If the postharvest respiration drift of these fruits is extrapolated back to zero time when harvest occurred and the CO_2 production had not yet begun to decline, and that rate is compared to the postharvest climacteric maximum, in tomatoes there is no difference, and with cantaloupes the climacteric rate is higher by approximately 30%. Thus, most or all of the postharvest climacteric rise in CO_2 production might be a recovery to the respiration rate that existed before these fruits were harvested. The endogenous ethylene that accumulated as they ripened on the vine may have been incapable of elevating the respiration rate above that value (Miccolis and Saltveit, 1991).

6. The biopotency of 1400 μl/l of propylene is equivalent to that of 10.8 μl/l of ethylene (Burg and Burg, 1967c).

7. In some fruits, the post-climacteric decline in ethylene production may be influenced by low internal $[O_2]$ caused by the leakage of cell fluids into the intercellular spaces (3.12 and 3.13), and/or by a decrease in EFE activity due to an altered membrane confirmation or fluidity induced by compositional changes in membrane lipids.

8. At their climacteric maximum, Kent and Haden mangoes contain only 2.7–3 μl/l of ethylene and produce the gas at a rate of 0.45–0.55 μl/·kg·h (Fig. 5.7, *left*).

9. Antisense genes have a coding sequence downstream from a suitable gene promoter, joined in the inverse orientation of the normal endogenous gene. Transferred and expressed, the antisense gene is transcribed by RNA polymerase 2 to produce an antisense RNA, copied from the 'wrong' DNA strand. When the target gene and antisense gene are both transcriptionally active in a cell, the antisense RNA and complementary mRNA may interact to form an RNA–RNA helix that is rapidly degraded. The net result is that the expression of the corresponding endogenous gene is greatly reduced (Hobson and Grierson, 1993).

10. Harvest did not accelerate the ripening of cherry tomatoes (De Vries *et al.*, 1995) or mature-green cv. Castelemart tomato fruits that had reached 80% of their full size and were about 6 days from showing a colour break (Saltveit, 1993).

11. Knee *et al.* (1987) claim that if individual apples are separated to prevent cross-diffusion of ethylene, detachment may have little effect, and that apples on the tree seem to be as sensitive to ethylene as detached fruits.

12. P-protein tends to accumulate in the vicinity of the sieve plates in such large quantities that it interferes with phloem transport and is visible as 'slime' under the light microscope. Initially, 'slime plugs' were thought to be a carbohydrate. Fischer (1885) suggested that slime accumulation at sieve plates was an artefact resulting from cutting the stem while the sieve tube contents were in a fluid condition. He concluded that slime plugs arise when protein particles are filtered from the

moving sap. The 'slime' was identified as a protein by Cronshaw (1975).

13. To determine the IEC in apples for the purpose of estimating the most appropriate harvest date (Chu, 1984; Watkins *et al.*, 1989), often their internal atmosphere is not sampled with a syringe until many hours after they are picked (Graell *et al.*, 1993). Since harvest can cause a rapid decrease in the IEC of fruits and flower parts, it might be prudent always to extract these intercellular samples from apples immediately after picking the fruit. D.R. Dilley (2000, personal communication) has confirmed that the IEC of apples tends to decrease after harvest.

14. A mineral-deficiency experiment with hydroponically grown tomato plants suggested that zinc might be the metal required for an ethylene response. During a 4 h treatment with 2.5 μl/l ethylene, the initial 47.5 ± 11.4° petiolar angle of 20-day-old tomato plants grown in complete nutrient media increased to 108.1 ± 14.1° and in 24 h to 141.7 ± 5.1°. There was no significant difference (± 20°) in the initial petiole angle or the curvature that a 24 h ethylene fumigation caused in control plants vs. 23–30-day-old plants grown without iron, copper or manganese, all of which displayed acute deficiency symptoms typical of the metal that had been excluded. Because 32-day-old zinc-deficient plants with an initial petiole angle of 52.8 ± 8.1° only developed a 63.1 ± 3.4° angle during a 5 h exposure to 2.5 μl/l ethylene, and 77.0 ± 13.1° during a 24 h treatment, it appeared that the ethylene receptor might contain zinc (Burg and Burg, 1967c), but there is another, more plausible explanation. The ethylene-induced epinastic response of tomato plants requires auxin to support tropistic growth (5.25), and while stem length is reduced to the same extent, by approximately 60% in zinc-, copper- or manganese-deficient tomato plants, their auxin contents differ markedly (Skoog, 1940). Sections cut from zinc-deficient plants contain no detectable auxin, whereas copper- and manganese-deficient plants contain 47 and 97% as much auxin, respectively, compared to control plants. The auxin requirement for an epinastic response and the effect which zinc deficiency has on auxin content, rather than a zinc-deficient ethylene receptor, accounts for the reduced ethylene-induced epinasty in zinc-deficient tomato plants.

15. Ethylene produced by fading carnation flowers displaced ^{14}C-ethylene during the binding assay, influencing the result. To avoid this complication, the measurements were run in duplicate with AOA-treated tissue, but another technical difficulty may have masked the extent to which the receptor's K_d increased as senescence advanced. If the receptor site concentration is not at least ten times lower than the true dissociation constant, the apparent dissociation constant obtained from a Scratchard plot will vary as a linear function of the receptor concentration (Chang *et al.*, 1975). In the carnation study there were 1.84×10^{-9} and 0.59×10^{-9} moles of binding site per kg tissue in the young and senescing petals, respectively. The concentration of receptor sites in the senescing petals was 20-fold lower than the K_d determined from Scratchard plots, and in the young petals twofold lower. Therefore, the true K_d in the young petals is less than the 'measured' value, and the decreased affinity for ethylene that developed during senescence is more than had been estimated. In a *Phaseolus* system (Bengochea *et al.*, 1980), where the concentration of binding sites varied from 3×10^{-9} to 4×10^{-8} M, and the apparent K_d was 6.4×10^{-10} M, the real K_d, determined by diluting the concentration of binding sites, was found to be 0.88×10^{-10} M, i.e. 7.3-fold lower.

16. See Section 5.9 for a discussion of the possible involvement of the pre-climacteric IEC in some of the responses considered to be ethylene-independent.

17. In the early 1900s, it had been the practice of growers to pick lemons according to size, and to colour green fruit by placing them in storerooms heated by kerosene stoves. The prevailing view that yellowing and aroma production were brought about by the temperature and humidity conditions was proved incorrect; instead, it was found that the effect was caused by an active combustion product (Sievers and True, 1912), later identified as ethylene gas (Chase and Denny, 1924; Denny, 1924b). F.E. Denny obtained a patent on the de-greening process (Denny, 1923), and his method of colouring grapefruits and oranges is still used commercially (Ting and Attaway, 1971; Hardenburg *et al.*, 1986; Baldwin, 1993).

18. It is not clear in the description of this response whether the 'rigour' was an inability to recover from ethylene-induced epinasty, or an effect on circumnutational and/or nyctinastic movement.

19. Stress ethylene is induced by cutting, bruising, radiation, virus infection, pressure, insect infestation, freezing, chilling, occasional high temperatures, drought, flooding or fungal exhudates; or high concentrations of calcium chloride, cycloheximide, bisulphite, methyl bromide, salicylate, trichloroacetic acid, herbicides (endothal, paraquat, glyphosphate), defoliants (monosodium cyanamide, potassium cyanate, sodium chloroacetate, sodium chlorate, ammonium thiocyanate), pesticides (methomyl,

fenitrothion), SO_2, ozone, NH_3, various ions (cadmium, copper, iron, zinc, lithium, silver) and pollutants (Yang and Hoffman, 1984; Abou Hadid *et al.*, 1986; Abeles *et al.*, 1992).

20. Horizontal orientation stimulates ethylene production in tomato seedlings and their excised hypocotyls (Denny, 1936; Harrison and Pickard, 1984, 1986; Wheeler *et al.*, 1986), *Coleus blumei* plants (Abeles and Gahagan, 1968b), *Avena sativa* seedlings (Harrison and Kaufman, 1982; Kaufman *et al.*, 1985), Alaska pea shoots (Takahashi *et al.*, 1991), Bermuda grass stolons (Balatti and Willemöes, 1989), *Echinochloa* grass nodes (Wright *et al.*, 1978), dandelion peduncles (Clifford *et al.*, 1983; Clifford and Oxdale, 1989) and snapdragon flower spikes (Philosoph-Hadas *et al.*, 1996). Ethylene production also is stimulated when *Pharbitis nil* (Prasad and Cline, 1985, 1987; Prasad *et al.*, 1989) and tomato plants (Harrison and Pickard, 1986) are inverted. After etiolated bean seedlings (Abeles and Rubenstein, 1964), light-grown tomato plants (Wheeler *et al.*, 1986), *Kniphofia* flower stalks (Woltering, 1991) and snapdragon floral spikes (Philosoph-Hadas *et al.*, 1996) are placed in a horizontal position, sections cut from the lower side produce considerably more ethylene than those cut from the upper side, and ACC accumulates in the lower half.

21. Andreae *et al.* (1968) claimed that the growth inhibition could not involve IAA-induced ethylene production because the auxin effect occurs rapidly, whereas ethylene production is not induced for 3 h, and the inhibition by auxin is reversible, while that by 100 μl/l of applied ethylene is not. To the contrary, ethylene production is induced by low concentrations of applied auxin within 1 h; the gas inhibits the elongation of pea (Chadwick and Burg, 1970; Burg, 1973a; Rauser and Horton, 1975), maize (Jackson *et al.*, 1981; Whalen and Feldman, 1988), radish (Jackson, 1983) and barley (Hall *et al.*, 1977) roots within 15–30 min; and after ethylene is removed, normal root elongation resumes within 20 min in maize (Jackson *et al.*, 1981; Whalen and Feldman, 1988) and radish (Jackson, 1983), 60 min in peas (Rauser and Horton, 1975) and 2 h in barley (Smith and Robertson, 1971).

22. NAA's ability to restore the ethylene response could not be duplicated with IAA, possibly because the intracellular level of applied IAA rapidly decreases due to conjugation and decarboxylation.

23. Eliasson *et al.* (1989) and Bertell *et al.* (1990) measured ethylene production and growth after a 48-h incubation, long after auxin-induced ethylene production and its inhibitory effect on growth had subsided. They concluded that low auxin concentrations (0.01–1 μM), which partially inhibit root elongation in intact seedlings, have little or no stimulatory effect on ethylene production. When they increased the intensity and extended the duration of ethylene production by applying ACC or benzylaminopurine, or by exposing the roots to light, fourfold more ethylene was produced during 24 h. This caused the inhibition of root growth to increase dramatically (Bertell *et al.*, 1990). Applied ACC caused the roots to swell and become diagravitropic, and the ACC effect, but not the long-term growth inhibition caused by IAA in the absence of ACC, was reversed by silver ions and cobalt. Bucher and Pilet (1983) suggested that the stimulation of ethylene production that auxin causes in maize roots during a 6-h incubation is not sufficient to account for the growth inhibition that resulted, but seemed not to have considered the possibility that it accounted for part of the growth inhibition.

24. Applied ethylene does not inhibit polar auxin movement through petiole or stem sections excised from control plants (Guttenberg and Steinmetz, 1947; Abeles, 1966; Burg and Burg, 1967a; Osborne and Mullins, 1969) because, during the standard transport assay, polar IAA movement ceases in about 3.5 h, before ethylene significantly affects transport.

25. Bucher and Pilet (1982) reported that application of 500 nl/l ethylene inhibited the elongation of vertical maize roots by 26% during 6 h, and the roots developed strong curvatures in response to the exogenous ethylene, but 500 nl/l of applied ethylene had little effect on the growth and geotropic curving of the roots when they were oriented horizontally for 6 h (Bucher and Pilet, 1982). Eliasson *et al.* (1989) suggested that this behaviour does not support the idea that ethylene influences the gravitational response of roots, but there is another plausible explanation. In air, horizontal maize roots elongated by 26% less than vertical roots, and since 500 nl/l of applied ethylene inhibits the elongation of vertical maize roots by 26% (Whalen and Feldman, 1988; Lee *et al.*, 1990; Fig. 5.32, *lower left*), the growth inhibition in horizontal roots may have resulted from auxin-induced production of sufficient ethylene to raise the IEC to 500 nl/l (Filner *et al.*, 1970; Lee *et al.*, 1990). The growth and gravitational curving of horizontal roots might not be significantly influenced by 500 nl/l of applied ethylene because the inhibition of maize-root growth by ethylene is a log function of the applied concentration (Fig. 5.32, *lower left*), and applied ethylene summates with the root's IEC.

26. Except for *axr2*, shoot gravitropism is not altered in these mutants.

27. Ethylene does not alter phototropic lateral auxin transport in pea epicotyls (Kang and Burg, 1974b).
28. In the *Coleus* experiment, auxin was transported at a velocity of 66 mm/h from the leaf surface to the tissue halves, which were analysed. Obviously, the transport was taking place in the vascular system as polar transport moved IAA through stem and petiole tissue at 3–6 mm/h, and non-polarly in the phloem at rates as high as 100–200 mm/h (Goldschmidt *et al.*, 1974).
29. This may explain why etiolated pea plants with their hook facing towards unilaterally applied light remain vertical after 24 h, while in the same interval, plants illuminated from the opposite side develop 45° curvatures in the direction of the light (Weisner, 1878). When the hook faces toward the light, phototropic lateral auxin transport to the non-illuminated side is counteracted by the natural tendency for auxin to accumulate in the hook's inner side. With the hook facing away from the light, phototropic auxin transport and the hook's geometry move auxin to the same side, causing the hook to bend in the direction it assumes during an ethylene-induced horizontal nutation. Ethylene does not alter phototropic lateral auxin transport in pea epicotyls, and they bend strongly toward unilateral illumination when their hook is oriented at right angles to the light source (Kang and Burg, 1974b).
30. Other experiments employing a genetic approach have given conflicting results relating to the roles of ethylene and auxin in epinastic growth. A recessive nuclear mutant of *Arabidopsis*, *rooty (rty)*, which displays extreme proliferation of roots, inhibition of shoot growth, epinasty and other alterations suggestive of responses to auxin or ethylene, was found to contain 2–17 times as much auxin compared to wild-type plants (King *et al.*, 1995). Dose–response assays measuring root and shoot elongation in the presence of exogenous IAA indicated that *rty* and wild-type had equal sensitivities to auxin. Mutations conferring resistance to auxin (*axr1-3*) or ethylene (*etr1-1*) were combined with *rty* in an effort to distinguish between auxin and ethylene effects. Resistance to auxin promoted root elongation and leaf expansion, and reduced but did not eliminate exfoliation of the hypocotyls' tissue, root proliferation and epinasty. Inhibition of hypocotyl elongation caused by *rty* was not normalized by expression of *AXR1-3*. Insensitivity to ethylene decreased the epinastic response and inhibition of hypocotyl elongation, moderately promoted leaf expansion, but had no effect on proliferation of lateral roots.
31. This research has often been misrepresented as claiming that ethylene is the second messenger in apical dominance (Yeang and Hillman, 1984; Romano *et al.*, 1993; Smalle and van der Straeten, 2000).
32. Ethylene production was measured by incubating 1–2 g leaf samples in 100 ml serum bottles for 24 h, analysing the final ethylene concentration in 1 ml air samples. During a 350-min interval, 57.2 g of fresh-cut *Hibiscus rosa-sinensis* leaves incubated in a 500 cc chamber produced ethylene at a rate of 0.04 nl/g·h (S.P. Burg, 1967, unpublished). If 1–2 g of these same leaves were incubated in a 100 ml bottle for 24 h and the ethylene production rate remained constant, they would only increase the ethylene concentration in the bottle by 9.6–19.2 nl/l. This would be difficult to detect in 1 ml air samples against the background ethylene level of 30 to several hundred nl/l typical of urban areas (Abeles, 1973).
33. Auxin stimulates banana ripening (Mitchell and Marth, 1944; Freiberg, 1955) and inhibits the ripening of Bartlett pears (Frenkel and Dyck, 1973).
34. Applied ethylene causes fading of carnations, narcissus, rose, larkspur, *Calceolaria*, snapdragon, petunia, daffodil and orchids such as *Cattleya*, *Dendrobium*, *Phaleonopsis* and *Vanda* (Lumsden *et al.*, 1940; Fisher, 1950; Kaltaler and Boodley, 1970; Mayak and Halevy, 1972; Burg, 1973b; Goh *et al.*, 1985; Serek *et al.*, 1995c).
35. An elevated rate of ethylene synthesis causes fading or ageing of carnations, orchids, *Ageratum*, larkspur, zinnias, morning glory, petunia, roses, snapdragon, *Calceolaria* and many other floral types (Ferguson, 1942; Fisher, 1950; Kende and Baumgartner, 1974; Nichols, 1966).
36. Flowers of hollyhock, marigold, *Ranunculus*, *Primula*, *Ulex*, *Taraxacum*, *Diplotaxis*, *Sarothamnus*, *Endymion*, *Stellaria*, *Chrysanthemum*, calla lily, *Dahlia*, *Viola*, *Cerise*, *Dianthus* (carnation), *Gladiolus*, *Peony*, iris, *Begonia*, *Aster*, *Fuchsia*, *Antirrhinum*, *Verbena*, *Prunus*, *Iberis*, *Buddleia*, geranium, gardenia, larkspur, lily-of-the-valley, lilac, petunia, tulip, *Calceolaria*, rose, *Convolvulus sepium*, *Lilium candidum*, most types of orchids, morning glory, *Vaccinium angustifolium*, strawberry and pompom chrysanthemum are known to produce ethylene (Fisher, 1950; Phan, 1963, 1965, 1970; Hall and Forsyth, 1967; Mayak and Halevy, 1972; Abeles *et al.*, 1992; Michael *et al.*, 1993; Nadeeau *et al.*, 1993; Woltering, 1993).
37. The claims and family distinctions classified in Table 5.17 are to some extent contradicted by the commercial awareness that ethylene accumulation in closely packed boxes caused by in-transit

stress is a significant problem during rose distribution (Durkin, 1992), and by studies showing that roses produce ethylene and increase in respiration when they fade (Liao *et al.*, 2001; Yan and Fan, 2001).[21] Easter lilies also evolve ethylene if they are stressed (Prince *et al.*, 1987); 1–2 μl/l of the gas in a greenhouse causes lily-bud abortion and blasting within a few days (Hitchcock *et al.*, 1932; Rhoads *et al.*, 1973; Weiler, 1993). Exogenous ethylene induces numerous physiological disorders in tulips and other Liliaceae (Kamerbeek and deMunk, 1976); ethylene production by *Lilium* Oriental Hybrid 'Casa Blanca' florets increased steeply from the 5th day after flowering in association with a loss in vase life (Son and Suh, 2000); and illuminating gas injures hyacinth blooms (Hitchcock *et al.*, 1932). Pollination induces a rise in ethylene production in *Dendrobium* 'Pompadour' flowers, exogenous ethylene and pollination induce senescence, and AOA and NBD delay premature senescence of pollinated *Dendrobium* blooms (Ketsa and Rugkong, 2000).

38. STS and 1-MCP reduce the amount of natural and ethylene-induced floral-organ abscission during the distribution of *Chamelaucium uncinatum* Schauer (Geraldton waxflower) (Serek *et al.*, 1995b; Taylor *et al.*, 2001). Although the quality of *Begonia* plants was not affected by 3 days in darkness, the presence of as little as 0.1 μl/l of ethylene gas for 24–72 h in combination with 3 days' darkness caused flower and bud drop (Høyer, 1985). An STS spray was effective in preventing this response (Moe and Fjeld, 1985). Exogenous ethylene causes petal shatter in geraniums (Armitage *et al.*, 1980; Fonteno, 1992), and application of STS to flowers showing colour can reduce the incidence of this disorder, although at the same time it significantly increases the susceptibility of some cultivars to *Pythium* root rot. STS prevents exogenous ethylene from inhibiting *Kalanchoe* flower opening and prevents the gas from causing premature flower fading and desiccation (Love, 1980). Both exogenous and self-generated ethylene can cause shattering or premature dropping of snapdragon florets (Haney, 1952), and STS (Nowak, 1981), 1-MCP (Sankhla *et al.*, 2001) and inhibitors of ethylene production (Wang *et al.*, 1977) improve this flower's vase life. Foliar application of STS significantly reduced flower and bud abscission for as long as 4 weeks when *Schlumbergera truncata* plants were stressed by exposure to ethylene and dark storage (Cameron and Reid, 1981). STS is used to prevent abscission in cut flowers such as snapdragons, sweet peas and lily-of-the-Nile, and in potted plants such as Christmas cactus, *Impatiens*, *Pelargonium*, *Calceolaria*, *Bougainvillea* and *Schefflara* (Veen, 1987). STS also prevents abscission of lily flower buds, and is useful in preventing shedding of leaves of potted plants when they are subjected to stress conditions during transport and storage. 1-MCP or DACP delayed and reduced the climacteric rise in ethylene production by carnation flowers (Sisler *et al.*, 1993, 1996a), and completely inhibited the pollination-induced ethylene-production increase in *Phaleonopsis* flowers (Porat *et al.*, 1995a). A 6-h pre-treatment with a low 1-MCP concentration, in the nl/l range, eliminates the abscission or wilting of *Alstroemeria*, *Begonia*, carnation, *Malthiola*, *Phaleonopsis*, rose, *Petunia* and *Phlox* (Porat *et al.*, 1995a,b; Serek *et al.*, 1994c,d, 1995c) caused by applied ethylene. 1-MCP prevented applied ethylene from decreasing the vase life of *Boronia heterophylla* flowers (Macnish *et al.*, 1999); protected *Alloxylon pinnatum*, *Telopea speciosissima* and *Grevillea* cv. Sandra Gordon cut flowers against the effects of applied ethylene and extended their vase life (Macnish *et al.*, 2000a); and prevented abscission and extended longevity of cut flowers of *Grevillea* cv. Sylvia (Macnish *et al.*, 2000c). Pulsing *Oncidium* cv. Grower Ramsey with 1-MCP extended vase life and improved orchid flower quality (RueySong, 1999). Veen (1983) summarized some additional responses to STS: *Bougainvillea glabra* Chois bracteole drop decreased; reduced percentage of dropped flowers on *Calceolaria herbeohybrida* Voss plants following 4 days in the dark or 2 days exposure to 1 μl/l ethylene; prevented flower drop of cut *Delphinium* flower spikes after a simulated shipment period; increased vase life of *Dendrobium* orchid cut flower; doubled shelf life of carnations (*Dianthus*); prevented flower-bud drop and lengthened vase life of *Lathyrus odouratus*; *Lilium*' Enchantment' vase life was improved and the effect persisted during long-term dry storage; doubled vase life of *Matthiola incana* L. (stock); inhibited petal drop normally associated with postharvest handling of *Pelargonium hortorum*.

39. Unexpectedly, in young, pre-climacteric cantaloupes, wound-induced ethylene production was inhibited by STS (Hoffman *et al.*, 1982). Usually, stress ethylene is promoted by inhibitors of ethylene action.

6

Heat Transfer and Water Loss

Exclusive of mechanical injury and improper harvest time, water loss usually is the most important factor shortening storage life and reducing the quality of horticultural crops (Van den Berg and Lentz, 1978; Burton, 1982; Kader, 1983; Ben-Yehoshua, 1986; Hardenburg *et al.*, 1986). Most fruits and vegetables lose their freshness when they transpire more than 3–10% of their weight (Robinson *et al.*, 1975; Burton, 1982), the exact value depending on the type of commodity. The degree of weight loss that can be tolerated before the product becomes unsaleable ranges from as low as 5% for apples (Pieniazek, 1942) and oranges (Kaufman *et al.*, 1956) to as high as 37% for green beans (Hruschka, 1977), but the point at which shrivel is visible usually is about half of this total figure (Grierson, 1987). Not only does the weight loss cause shrivelling, wilting, shrinkage and loss of firmness, but in addition the resultant water stress induces hormonal changes such as enhanced ethylene production and increased abscissic acid (ABA) levels, which hasten senescence, membrane disintegration and the leakage of cellular contents (Boyer, 1976).

It has long been a goal of plant biologists and postharvest physiologists to be able to prevent or delay fruit ripening and senescence in a reversible manner by controlling ethylene action or production (Theologis *et al.*, 1993a). The belief that this will reduce distribution losses has prompted attempts to prevent ethylene synthesis, or action and lessen sensitivity using inhibitors, genetic breeding and molecular engineering (Gray *et al.*, 1993; Murray *et al.*, 1993c), or a combination of CA and ethylene scrubbers (Sisler *et al.*, 1993), but none of these methods address the problem of water loss. Jacketed refrigeration systems (Jorgensen, 1968; Van den Berg and Lentz, 1978), HumiFresh high-humidity storages (Krahn and Darby, 1971; Meredith, 1973) and humidified intermodal containers often are preferred to CA because preventing water loss has proved to be more important and cost-effective than controlling the gaseous environment (Ben-Yehoshua, 1986, 1989).

Water loss is a complex phenomenon resulting from mechanical, biological and physical interactions. In an air-conditioned storage, the process begins when water moves from within individual cells to an exposed interface in the cuticle where 'latent' heat causes the liquid to change state and evaporate. Usually the process ends when the vaporized water condenses on the refrigeration system's evaporator coil. Biologists traditionally have formulated the problem in terms of a vapour-pressure gradient between the commodity and the air passing over it (Burton, 1982; Ben-Yehoshua, 1986). The larger the gradient, the more rapidly water will escape. To diminish the loss, either the relative humidity of air leaving the evaporator coil must be increased, or the water conductance of the stored commodity decreased by waxing its surface or protecting it with a water-retentive wrap.

Thermodynamics examines water loss in a different, more comprehensive manner, which provides a better understanding of the mechanisms involved. Evaporation can be analysed as an interaction between three interdependent variables: the availability of latent energy at the evaporating surface, the vapour-pressure gradient that develops at equilibrium and the resistances in the water-vapour pathway (Raschke, 1960; Slatyer, 1967). When there is a temperature difference in a system, heat flows from regions of high to low temperature, and once the temperature distribution is known, the rates of heat transfer can be determined from the laws relating heat flux to the temperature gradient. In systems containing biological material, the combined effects of four heat transfer modes, conduction, convection, radiation and evaporation (or condensation), modulate the vapour-pressure and temperature gradients that develop, and heat transferred by evaporation determines the rate of water loss. 'Even if the air is saturated, a commodity can still transpire,' (Raschke, 1960), provided that it is kept warmer than the air by metabolic processes or radiation. In water-saturated air, respiratory heat and radiation increase a fruit's temperature, elevating the water activity at the surface. This causes moisture to evaporate, and then the water vapour condenses in the ambient air forming a fog as it cools. Apples (Lentz and Rooke, 1964) and bananas (Wardlaw and Leonard, 1940) lose water at a significant rate even when the relative humidity of the surrounding air is 100%.

6.1 Dependence of Water Loss on Respiratory Heat

Water loss results in evaporative cooling, which would lower a stored commodity's temperature if the latent energy used to change the state of water from liquid to vapour was not replaced from a heat source. When lettuce (specific heat = 0.87 kcal/kg·°C) is vacuum pre-cooled, each kg of water that is vaporized generates 595.4 kcal of evaporative cooling at 0°C, and a 1% evaporative water loss decreases the product's temperature by $(595.4 \times 0.01)/0.87 = 6.84$°C. Yet when Golden Delicious, Jonathan, McIntosh or Red Delicious apples experience a 6.7% average weight loss during a 12-week storage in 95% RH air at 0°C (Hruschka, 1977), which could result in a 46°C decrease in pulp temperature, the fruit remains at 0°C. What supply of heat offsets the effect of evaporative cooling and keeps the apple's temperature constant? Solar and infrared radiation are the main heat sources available to the aerial portion of field-grown plants (Noble, 1991; example 9), but this situation changes dramatically when commodities are harvested and stored isothermally in a darkened storage room.

According to the first law of thermodynamics, the quantity of water that can be evaporated from an adiabatic system depends on the amount of heat added. Accordingly, commodity water loss in a refrigerated room depends on respiratory heat, sometimes augmented or reduced by additional heat transferred to or from the stored product by convection and radiation. Respiratory heat is immediately available for this purpose since it is generated within each cell, and therefore does not have to be acquired from the environment. When a commodity remains at a constant temperature, if the heat necessary to evaporate the transpirational water is less than the respiratory heat, the commodity must transfer heat to its environment, but if the heat used to transpire water exceeds the respiratory heat, the commodity is acquiring heat from its environment (Gac, 1956). A commodity stored in a refrigerated space cannot remain at a constant temperature and lose more water than its respiratory heat is capable of vaporizing unless the commodity is colder than its environment and receiving heat from it.

A close correlation between water loss and respiratory heat[1] was demonstrated for bananas more than 50 years ago, when it was recorded that a plot of transpiration rate vs.

time resembles the climacteric curve for respiration (Fig. 6.1) (Leonard, 1941). A similar result was described for papayas and mangoes (Wardlaw and Leonard, 1936a, 1940), and the same behaviour would be expected when any climacteric fruit ripens, stored flowers begin to fade or a leafy vegetable becomes senescent, since typically at that time the respiration rate rises. At atmospheric pressure an apple's respiratory heat generation and weight loss decrease in unison when the ambient [O_2] is lowered and/or the applied [CO_2] is increased (Roberts *et al.*, 1965; Fidler and North, 1967; Toledo *et al.*, 1969; Haffner, 1992; Gran and Beaudry, 1993; Andrich *et al.*, 1998). Only a slowing of the evaporative cooling rate caused by less respiratory heat generation at a lower O_2 or higher CO_2 tension could account for this correlation, since the weight savings are of an order of magnitude greater than that which would result from the reduced rate of respiratory CO_2 loss.[1] Figure 6.2 and Table 6.1 indicate that during NA and LP storage the weight loss from Golden Delicious and Boskoop apples is proportional to their respiratory heat production. According to Table 6.2, the weight loss from pre-cooled bananas stored in LP is only slightly greater than the amount required to transfer all respiratory heat (Burg, 1969, 1970, 1971), and a similar behaviour has been reported for limes (example 13) and roses (chapter 9, example 6; Burg, 1992) stored in LP.

Refrigerated fruits can easily lose all of their respiratory heat by evaporative cooling when the ambient humidity is close to saturation. When an apple equilibrates with a 0°C storage air temperature, its transpirational resistance will cause it to transfer 100% of its respiratory heat to the air by evaporative cooling if the storage RH is < 95% (example 3; Fig. 6.3). At 20°C and

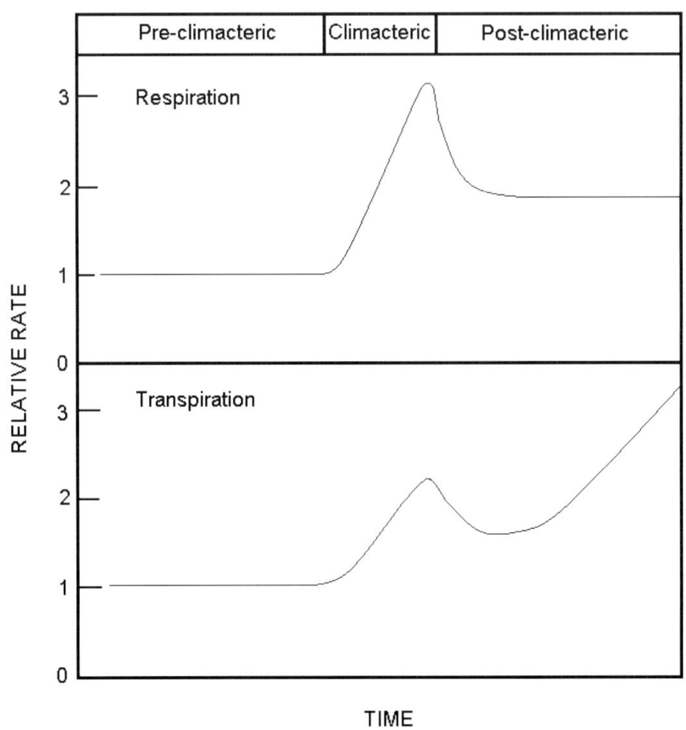

Fig. 6.1. Generalized diagram to illustrate the relationship between transpiration and respiration during banana ripening. Actual values for respiration and transpiration vary widely according to conditions, and therefore their trends are indicated in arbitrary units (Simmonds, 1959; Wardlaw, 1961).

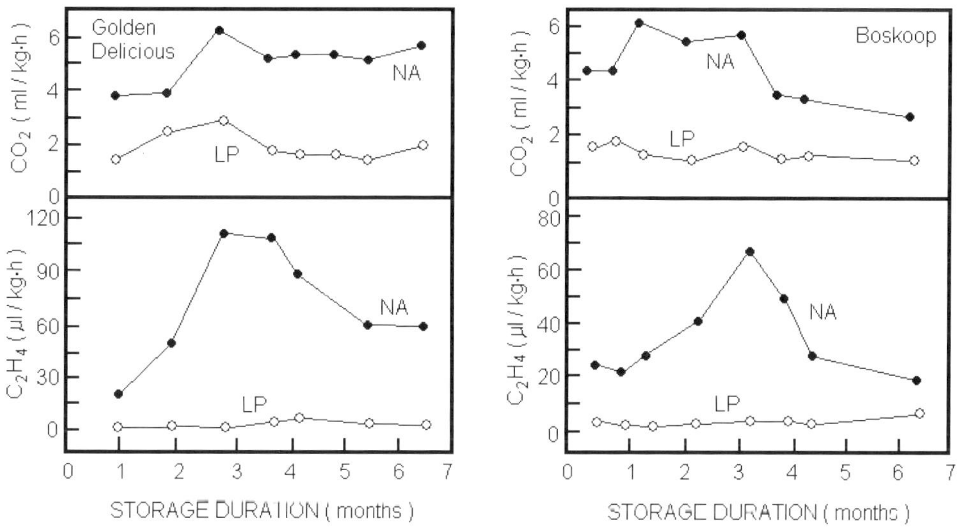

Fig. 6.2. CO_2 and ethylene production at 3°C by Golden Delicious and Boskoop apples during storage in NA or in LP at a pressure of 10 kPa (75 mm Hg) (Bangerth, 1973).

Table 6.1. Water loss from Golden Delicious and Boskoop apples during 2 and 5 months' storage at 2°C in LP at a pressure of 10 kPa (75 mm Hg) or in NA (Bangerth, 1973).

Variety	Storage duration (months)	% water loss NA	% water loss LP	% water loss to dispel resp. heat NA	% water loss to dispel resp. heat LP
Golden	2	2.9	1.0	2.9	1.1
Delicious	5	6.3	3.5	7.6	3.2
Boskoop	2	2.1	0.8	3.2	0.8
	5	5.1	1.2	6.8	1.6

93% RH, 100% of an orange's respiratory heat will be eliminated by evaporative cooling when the fruit's temperature equals the storage air's dry-bulb temperature (example 2). If the RH is less than the indicated value, these fruits will develop the vapour-pressure gradient required to transfer their respiratory heat by evaporative cooling when they are colder than the storage air's dry-bulb temperature and will gain heat by convection and radiation from the warmer air and adjacent surfaces. The extra heat will be dispelled by additional evaporative cooling, creating a greater weight loss than is needed to transfer the respiratory heat.

At atmospheric pressure, the combined heat transfer capacity of radiation and convection is so much greater than the evaporative cooling generated by the same temperature gradient that a fruit's temperature will stabilize close to the storage air temperature even when the humidity is significantly lower than 90%. The heat taken in by convection and radiation across a small dry-bulb temperature gradient will offset heat lost due to evaporation occurring over a much larger wet-bulb temperature gradient. In equally sized apple and orange fruits, for a 1°C temperature difference the heat transfer coefficients for free convection (equations 6.21 and 6.22) and radiation (equation 6.33) are 1.82 W/m^2·°C (0.32 BTU/ft^2·h·°F) and 5.35 W/m^2·°C (0.81 BTU/ft^2·h·°F), respectively, whereas the heat transfer coefficient for evaporative cooling (equation 6.39) is much lower, 0.18 W/m^2·°C (0.032 BTU/ft^2·h·°F) for an apple at 0°C and 0.28 W/m^2·°C (0.049 BTU/ft^2·h·°F) for an orange at 20°C. Consequently, apples and oranges cannot lose respiratory heat by convection and radiation unless the storage air's relative humidity is elevated very close to saturation. For this reason the RH required to dispel part of the respiratory heat by heat transfer means other than

Table 6.2. Weight loss from Valery bananas stored in LP for 27 days at various pressures and 13.3°C, 95% RH.

Pressure, kPa (mm Hg)	Respiration rate µl CO_2/g·h	% Weight loss in 27 days	
		Measured	Theoretical
5.33 (40)	–	1.29	–
6.67 (50)	1.54	1.80	1.06
8.00 (60)	–	1.55	–
10.67 (80)	1.46	1.30	1.01
13.33 (100)	–	1.42	–
16.00 (120)	2.24	1.56 ± 0.65	1.45
20.00 (150)	–	2.02	–
101.30 (760)	7.19	6.56 ± 0.89	4.95

Data are average values from 23 experiments. Hands of bananas weighing approximately 1.5 kg (3.3 lb) were stored in 10-l glass vacuum desiccators at 13.3°C and at the indicated pressures, flowing one humidified air change per hour. Respiration varied during the course of the experiment (Table 4.1); an average value for the 27-day period is used for the computation. The theoretical weight loss is the amount of water that needed to be evaporated to transfer all of the respiratory heat plus the weight of carbon lost due to respiration. Controls at atmospheric pressure (1013 mbar = 760 mm Hg) were stored in 'Chiquita' perforated polyethylene bags to minimize weight loss. They had ripened by 27 days. Fruits in LP were stored 'naked', except at 5.33 kPa (40 mm Hg) the fruit was protected by a polyethylene film wrap (Burg, 1969, 1970, 1971).

evaporative cooling can only be provided by high-humidity HumiFresh or jacketed refrigeration systems. Otherwise, unless an additional transpirational resistance such as surface waxing or a water-retentive plastic barrier (6.4 and Tables 6.3 and 6.4) is installed in series with the fruit to artificially restrict evaporation, all respiratory heat will be transferred by evaporative cooling before a positive temperature gradient develops between the commodity's surface and the surrounding air.

Calculations by the author based on: (i) weight loss and respiration data for 55 types of tropical and subtropical fruits and vegetables stored at an optimal temperature in an air-conditioned room at 85–95% RH (Pantastico *et al.*, 1975a); (ii) data tabulating the percentage daily weight loss per mbar water vapour pressure deficit (Burton, 1982)[2] for 23 varieties of fruits and 11 varieties of leafy vegetables stored at various humidities and temperatures; and (iii) data for 32 commodities stored in 3% [O_2] either at 10°C and a 3–5 mbar water vapour pressure deficit (wvpd), or 15°C and a 6–9 mbar wvpd (Robinson *et al.*, 1975), indicate that with few exceptions[3] out of 121 examples, evaporative cooling (rate of water loss × latent heat of vaporization of water) consumed at least as much heat as respiration generated. Usually the weight loss was considerably greater than this amount, suggesting that the commodity evaporated sufficient water to cool below the prevailing storage temperature and acquire heat from its surroundings by convection and radiation, after which evaporative cooling eliminated this extra heat in addition to the respiratory heat. This cannot always be concluded with certainty, since often it is not specified whether or not the stored commodity was adequately pre-cooled before its initial weight was recorded. When cool down occurs during storage, evaporation removes part of the excess respiratory heat produced while the commodity is hot, as well as some portion of the sensible heat that must be transferred to lower the commodity's temperature. If the evaporative weight loss associated with cool down is included as part of the published total loss during storage (Fig. 6.4), this would create the false impression that the commodity evaporated water rapidly enough to decrease below the storage temperature and acquire heat from its environment. Examples 22–24 and Fig. 6.3 indicate that in those instances where there are explicit data indicating that cool down is not a complicating factor, either the entire weight loss that occurs during NA storage of cabbage, cauliflower, asparagus, snap beans and apples is caused by respiratory heat, or sometimes heat has been acquired from the environment and the weight loss is further elevated. At 8 kPa (60 mm Hg) respiration is inhibited by 60% (Fig. 4.2) and the weight loss during asparagus storage is entirely due to respiratory heat, while during the storage

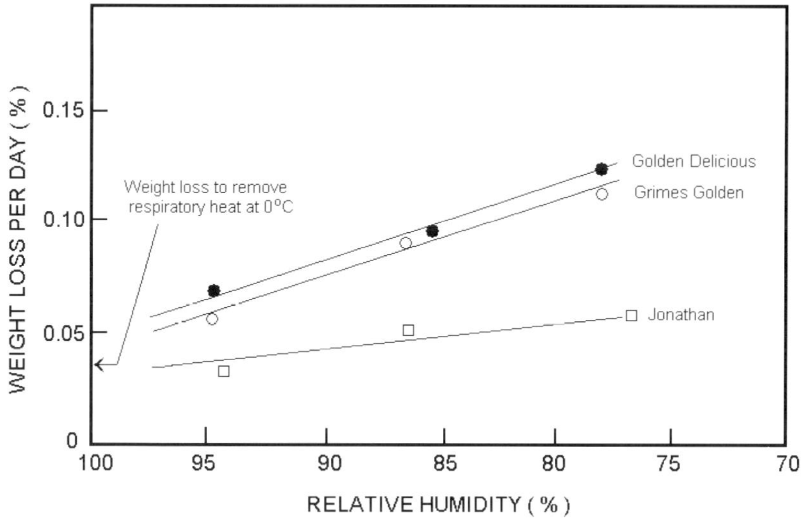

Fig. 6.3. Effect of relative humidity on Golden Delicious apple weight loss at 0°C (adapted from Wells, 1962). Arrow indicates weight loss required to remove all respiratory heat by evaporative cooling (Tolle, 1962). When the weight loss is greater than this amount, the apples are colder than their environment and receive heat from it.

Table 6.3. Transpirational resistance of apples with and without wax. Condition: air = 50% RH at 25°C (Horrocks, 1964).

		Resistance (s/cm)	
Variety	Treatment	Wax intact	Wax removed*
Granny Smith	Whole fruit	398	–
	Isolated peel	662	9.4
Golden Delicious	Whole fruit	249	–
	Isolated peel	311	10

*Corrected for stagnant air layer.

Table 6.4. Gas and water vapour resistance of Valencia orange fruits, HDPE film, and both combined in seal-packages (Ben-Yehoshua *et al.*, 1985).

	Resistance (s/cm)		% Increase in resistance after sealing the fruit in HDPE
Vapour	Fruit	HDPE film	
H_2O	110	1500	1400
O_2	6000	14,000	230
CO_2	5700	4100	72
C_2H_4	6900	1700	25

of snap beans it exceeds that needed to remove respiratory heat.

Tropical commodities such as avocados, papayas and mangoes have a higher respiration rate than apples or oranges at each fruit's optimal storage temperature, and may not lose all of their respiratory heat by evaporative cooling when the storage humidity is very high. A pre-climacteric Choquette avocado's transpirational resistance has not been measured, but if it is the same as that of an apple (50 s/cm), at 13°C an individual fruit would only lose 74% of its respiratory heat by evaporative cooling when the RH of the storage air was 90%, and at that condition the fruit and air would have identical temperatures (Table 6.5). The remaining respiratory heat would warm the fruit above 13°C, and while most of this residual heat would be removed by convection and radiation, an additional part would be cooled by evaporation. Thus water evaporation from the pre-climacteric avocado would still depend mainly on respiratory heat.

The average inhibition of respiratory heat production is 50% in CA (Robinson *et al.*, 1975) and 90% in LP (Fig. 4.2) at the optimal storage condition for various commodities. Therefore the propensity to

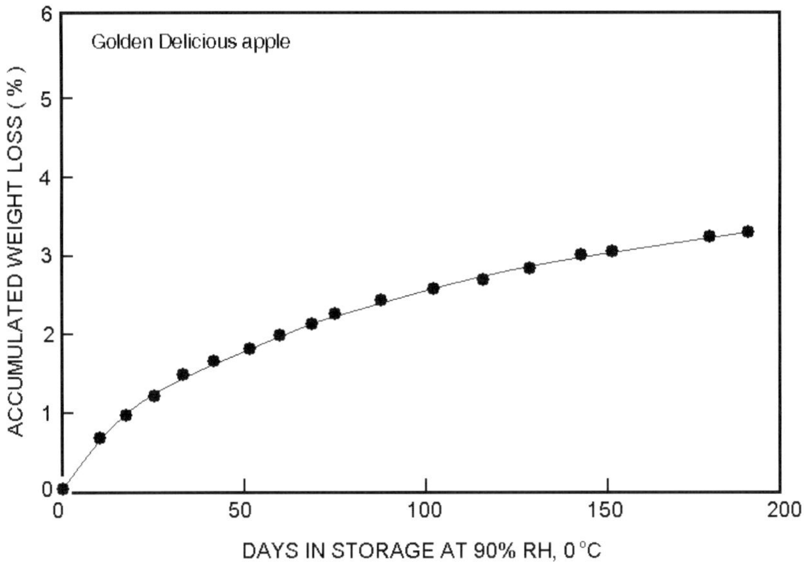

Fig. 6.4. Cumulative weight loss of a 30-lb bin sample of Golden Delicious apples stored in a controlled atmosphere room at 90% RH, 0°C (Waelti *et al.*, 1992). The weight loss during the first 2 weeks was rapid because the apples were cooling.

Table 6.5. Vapour pressure deficit and temperature rise needed to transfer all respiratory heat produced by pre-climacteric Choquette avocados stored at 13°C either in LP at 2.67 or 10.67 kPa (20 or 80 mm Hg) or in NA.

	kPa (mm Hg) storage pressure			
Property	101.3 (760)	10.67 (80)	2.67 (20)	2.67 (20)
Respiratory heat, W/kg (BTU/ton·day)	0.051 (3854)	0.026 (1542)	0.013 (771)	0.013 (771)
Transpirational resistance (s/cm):				
Lenticles	2741	252	32.6	32.6
Cuticle	50	50	50	150
Total	49.1	41.7	19.7	26.8
To transfer all respiratory heat:				
v.p. gradient (mm Hg)	1.30	0.45	0.137	0.187
$T_F - T_{DP}$ (°C)	0.93	0.32	0.098	0.133

Conditions: Fruit weight = 470 g; fruit surface area = 340 cm^2, measured lenticular resistance to CO_2 transport = 4500 s/cm @ 1 atm (Burg and Burg, 1962a); calculated lenticular transpiration resistance = 2741 s/cm @ 1 atm (Table 15.4); measured pre-climacteric respiratory rate = 60 ml/g·h @ 24°C and 1 atm (Burg and Burg, 1962a); computed respiratory heat = 0.14 W/kg (10,560 BTU/ton·day) @ 24°C and 1 atm (Hardenburg *et al.*, 1986) and 0.051 W/kg (3854 BTU/ton·day) @ 13°C and 1 atm (Ryall and Lipton, 1972);[8] respiratory heat in LP is computed from Fig. 4.2; assumed relative humidity = 95% in LP and 90% in NA; LP dew point = 12.2°C; NA dew point = 12.4°C; assumed cuticular transpiration resistance = 50 or 150 s/cm. T_F and T_{DP} are the fruit and air dew point temperatures, respectively.

lose water due to latent heat provided by respiration is in the order 10 (NA):5 (CA):1 (LP). The amount of respiratory heat that has to be transferred in LP is indicated in Table 6.6, but these are maximum values and do not consider the decrease in heat evolution that occurs during storage (Tables 4.1 and 4.4). Until ripening, fading or senescence commences, the respiration rate of leaves (Blackman, 1953), fruits (Kidd and West,

Table 6.6. Respiratory heat production at optimal storage pressure and temperature (Burg and Kosson, 1983). An asterisk (*) indicates that storage may be improved at a pressure lower than that indicated.

Commodity	Pressure (mm Hg)	Temperature (°C)	W/kg (×10^2) (BTU/ton·day)
Apple	50	0	0.34 (252)
Asparagus	10–20	0	2.95 (2200)
Avocado	20	5–13	1.3–3.21 (770–2400)
Banana*	40	13.3	2.22 (1656)
Beans (green)	10	7.2	1.72 (1280)
Broccoli	10	0	0.59 (440)
Carnation (bloom)	10	0	2.61 (1950)
Cauliflower	10	0	0.52 (390)
Cherry (sweet)*	20	0	0.20 (150)
Chrysanthemum (bloom)	10	0	0.62 (462)
Cucumber*	80	7.2	1.41 (1050)
Grape	10	0	0.07 (53)
Lime	150	8.9	1.05 (780)
Maize	50	0	3.60 (2685)
Mango	15–20	13	1.47 (1100)
Melon (cantaloupe)	15	7.2	0.34 (255)
Melon (honeydew)	15	7.2	0.17 (128)
Mushroom	10	0	2.36 (1760)
Papaya	20	10	0.84 (625)
Pepper (sweet)*	80	10	2.48 (1848)
Pineapple	15	8.3	0.52 (385)
Rose (bloom)	10	0	3.18 (2376)
Strawberry	10	0	0.97 (726)

1945), flowers (Siegelman *et al.*, 1958; Coorts *et al.*, 1965) and vegetables (Hardenburg *et al.*, 1986) declines during NA, CA and LP storage, providing less heat to evaporate water. This is reflected in a progressive decrease in the rate of water loss during the CA storage of apples (Fig. 6.4; example 4), and also when cabbage or cauliflower is stored in NA (example 22). After the initial apple cool down was completed, the average rate of weight loss during 135 subsequent days was 43% of that needed to transfer all of the respiratory heat produced by the same type of apple when it is stored in NA. This is consistent with the respiratory inhibition of apples measured during a comparable CA storage (Table 4.3).

The relationship between respiration rate and water loss is not significantly altered when the temperature is shifted throughout the physiological range, 0 to 20°C, because the temperature coefficients (Q_{10}) for water vapour pressure and respiratory heat production are both approximately 2.0. The summary equation commonly used to represent respiration is:

$$C_6H_{12}O_6 + 6\ O_2 = 6\ CO_2 + 6\ H_2O_{(liq)} + 673\ \text{kcal} \tag{6.1}$$

but if all of the released energy is used as latent heat to vaporize water, the equation can be rewritten in the form:

$$C_6H_{12}O_6 + 6\ O_2 + 64\ H_2O_{(liq)} = 6\ CO_2 + 70\ H_2O_{(vap)} \tag{6.2}$$

Respiration generates only about 9% of the liquid water needed to remove respiratory heat by evaporative cooling. The remainder must be drawn from the cellular reserve, but there is a limit to how much can be lost without adverse consequences. Water loss does not normally limit LP storage life due to the marked reduction in respiratory heat generation under hypobaric conditions and the elimination of access to other heat sources (6.18).

Equation 6.2 indicates that 1152 ml (64 mol) of water are evaporated for each mole of sugar consumed, which is equivalent to losing 0.86 M solution. Since the osmotic concentration in plant cells typically is lower than 0.86 M, stored commodities tend to increase in percentage water content and decrease in solute concentration unless they hydrolyse reserve starch rapidly enough to produce more sugar than respiration consumes.

6.2 The Cuticle

The surfaces of the aerial parts of vascular plants are covered by a waxy layer of cutin secreted over the epidermis, closely following the contour of the cells. The design is similar in leaves and fruits, except that in some mature fruits the cuticle is thicker (Fig. 6.5). The cuticular layer covers all above-ground exterior plant/air interfaces including leaf mesophyll cell walls, guard cells, sub-stomatal cavities, flower parts, fruits, seed coats, and possibly the inner surfaces of epidermal cells where they abut on air spaces in the underlying tissue, and even internal parts such as the juice sacs of citrus (Martin and Juniper, 1970; Kolattukudy, 1980; Burton, 1982).

The cuticle has a structural matrix composed of large polymers of hydroxylated C_{16} and C_{18} fatty acids, oligosaccharides, phenolic compounds and polyuronic acids interspersed amongst cellulose microfibrils and proteins, in combination with varying amounts of soluble and insoluble waxes imbedded in the polymer matrix and deposited superficially over its surface (Martin and Juniper, 1970; Kolattukudy, 1980, 1981). The paraffins, aliphatic acids, aliphatic alcohols and their monoesters that are extractable by dissolution in organic solvents are referred to collectively as 'waxes', and the term cuticular membrane (CM) is used to describe cuticles from which the waxes (soluble cuticular lipids = SCL) have been removed to study the structure of the remaining cutin–polymer matrix (PM). The cuticular mass varies from < 20 to 600 μg per cm^2 in leaves, and up to 1.5 mg per cm^2 in fruits. Morphologically, two

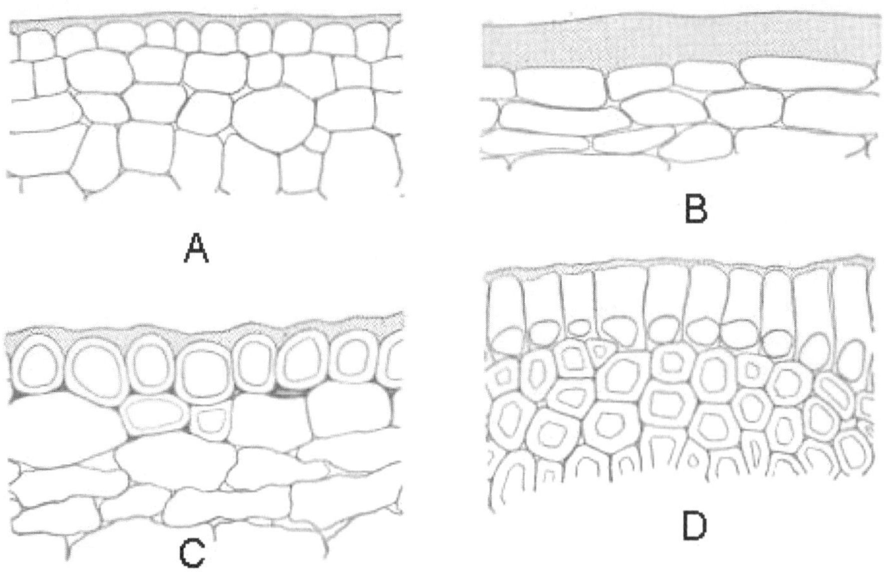

Fig. 6.5. Sections of the epidermis and adjacent tissues showing various degrees of thickness and extent of the cuticle between cells. A, *Citrus sinensis* (orange), fruit; B, *Malus pumila* (apple var. Ben Davis), fruit; C, *Dracaena goldieana*, stem; D, *Dasylirion serratifolia*, leaf (Eames and MacDaniels, 1947).

cuticular zones can be distinguished based upon the staining pattern and appearance in the electron microscope (Hollaway, 1982), a lower zone impregnating the outer cellulose cell wall layers and a wax layer of varying thickness deposited on the outer surface of the cuticle. The 'bloom' of grapes and plums is the result of deposits of epicuticular wax platelets, and the surface of the orange is covered unevenly with such platelets (Figs 3.10 and 6.6) with gaps over epidermal cells that have most recently divided (Scott and Baker, 1947; Bain, 1958). Some commodities have a peripheral periderm consisting of several cell layers that ultimately lose their living contents and become corky as a result of suberin and wax deposition on their walls. Potato tubers and avocados have typical periderms with visible lenticles (Burton, 1982).

The driving force for transpiration is the vapour-pressure gradient across the continuous non-porous layer of SCL that separates liquid water and vapour at the interface in cuticular membranes (Schönherr and Schmidt, 1979). Many studies have shown that the cuticular transpirational resistance is due mainly to the soft cuticular waxes (Albrigo, 1972; Schönherr, 1976a,b, 1982; Sastry *et al.*, 1978; Schönherr and Schmidt, 1979), which often form overlapping hydrophobic platelets separated by air-filled pores and microcapillaries. Disruption of the surface wax on grape berries greatly increases water loss (Possingham *et al.*, 1967), and while the stomatal transpirational resistances of leaves of glaucous and non-glaucous sibling lines of *Brassica oleracea* are not significantly different, the cuticular transpirational resistance of glaucous siblings is on the average 40% higher than their non-glaucous counterparts (Denna, 1970). During wound healing of potato tubers, if the tissue is treated with 4 mM trichloroacetic acid to preferentially inhibit wax synthesis without altering the deposition of the major aliphatic components of the suberin polymer, the development of diffusion resistance to water vapour is severely impaired (Soliday *et al.*, 1979). The transpirational resistance is the same measured in intact apples and in isolated cuticles obtained by peeling the fruit and enzymatically removing the underlying cellular matter from the skin, but if the cuticular wax is removed from the isolated cuticle by dissolution in hot chloroform, the permeability to water vapour increases 30–60-fold (Table 6.3). Extracting the wax from enzymatically isolated cuticular

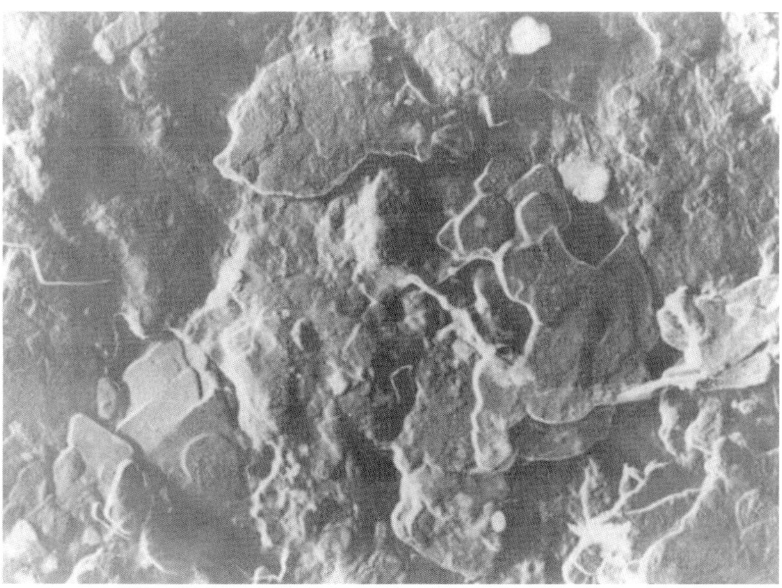

Fig. 6.6. The wax plates on the surface of a Cox's Orange Pippin apple (approx. 9000×) (Baker, 1975).

membranes of citrus leaves, pear leaves and onion bulb scales increases the liquid water permeability 300–500-fold (Schönherr, 1976b).

6.3 Dependence of Transpirational Resistance on Relative Humidity

Polymers may be classified according to the effect that humidity has on their permeability (Schönherr, 1982). 'Solubility membranes' such as polyethylene plastic films have permeability coefficients that are independent of water vapour pressure because these materials are hydrophobic polymers (Schönherr and Schmidt, 1979). Exposed to water on one side, all hydrogen bonds must be broken before a water molecule can enter this type of membrane. Its water solubility is very low, and fluxes are proportional to the vapour-pressure gradient. The liquid/vapour-phase transition occurs before water molecules enter into the membrane; water vapour permeates as individual molecules; and water clusters do not form regardless of the vapour pressure.

The cuticle and periderm of plants contain polar substituents and differ from solubility membranes in that Henry's Law (Table 15.1) is not obeyed and water solubility is not proportional to water vapour pressure. Progressively as the vapour pressure is elevated, the number of water molecules surrounding the polar groups multiplies, and provided that the polar groups are sufficiently close to each other, water sorption leads to the formation of water clusters or to continuous water-filled channels, causing the water permeability coefficient to increase as a function of the water content of the polymer. Membranes of this type are termed 'porous', and have permeability coefficients that depend on vapour pressure. Isolated cuticular membranes swell in the presence of water, and this causes their transpirational resistance to decrease (Fig. 6.7). Water continuity exists across the entire membrane when it is exposed to water on one side and vapour on the other. The liquid/vapour transition occurs at the membrane/vapour interface and the water vapour concentration gradient is not the force driving water across the membrane (Schönherr, 1982).

An increase in ambient RH decreases the transpirational resistance in many and possibly in all horticultural commodities, including apples (Fig. 6.7, *upper left*), plantains[4] (George *et al.*, 1982), various leaves and isolated cuticular membranes prepared from them (Meidner, 1986; Santrucek and Slavik, 1990; van Gardingen and Grace, 1992; Hoad *et al.*, 1996, 1997; Boyer *et al.*, 1997; Schreiber *et al.*, 2001; Fig. 6.7, *upper right* and *lower*), aubergine fruit membranes (Fig. 6.7, *middle left – Solanum melongena*), non-isolated cuticles of onion bulb scales (Schönherr and Merida, 1981), and potato and birch periderm layers (Fig. 6.7, *middle right*). The response to RH is self-regulating and advantageous, slowing water loss when the humidity is low, and facilitating the removal of respiratory heat from harvested crops by evaporative cooling when the humidity is high. This mechanism allows plants more easily to evaporate water from their leaves and other surfaces on humid days. It may have evolved to assist in regulating field temperature by improving a plant's ability to dissipate heat received as solar and infrared radiation (Gates, 1968; Noble, 1991; example 9).

Within the lipid matrix (PM) of the cuticular membrane, approximately 10^{10} water-filled pores per cm^2 with a mean equivalent pore radius of 0.45 nm (about 3 water molecules wide), cross the membrane in tortuous paths (Schönherr, 1976b, 1982; Schönherr and Schmidt, 1979). These pores, which only develop on hydration of polar groups of the membrane matrix (Schönherr and Bukovac, 1973), swell and shrink in response to RH and water, causing the water permeability to increase when the RH is raised (Chamel *et al.*, 1991). Methylating wax-free isolated polymer matrix eliminates the response to high humidity, indicating that non-esterified polymer matrix contributes to the humidity effect (Schreiber *et al.*, 2001). These and other polar groups sorb water, which in turn increases the water permeability of polar

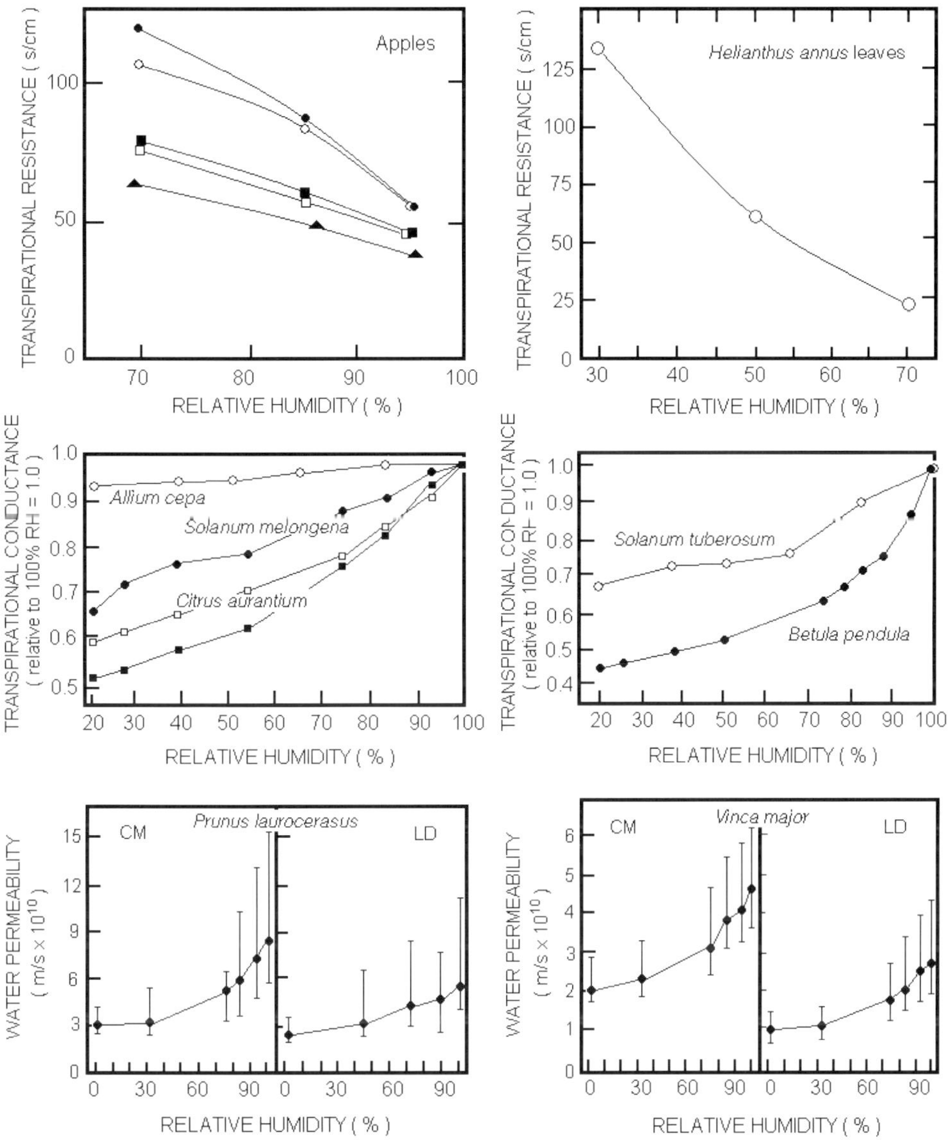

Fig. 6.7. (*upper left*) Effect of relative humidity (RH) on the transpirational resistance (s/cm) of apples; □ Jonathan; ○ Laxton's Superb; ● Lombartscalville; ■ Golden Delicious; ▲ Belle de Boskoop. Transpirational resistance has been computed by dividing the product of the diffusional resistance factor (μ) and thickness of the skin (*s*) by the binary diffusion coefficient of water vapour in air at 0°C (adapted from Fockens and Meffert, 1972).[5] (*upper right*) Effect of relative humidity on the cuticular resistance to water vapour (r_c) in *Helianthus annus* leaves (Moreshet, 1970). (*middle left*) Effect of relative humidity of the vapour phase on transpirational conductance of cuticular membranes. Values are averages of 5–30 membranes. Results of two separate experiments are indicated for *Citrus aurantium* (Schönherr, 1982). (*middle right*) Effect of water activity of the vapour phase on the transpirational conductance of potato (*Solanum tuberosum*) tuber and birch (*Betula pendula*) trunk periderm. Each result is the mean value for six membranes (Schönherr, 1982). (*lower*) Water permeances (*P*, m/s) of isolated cuticular membranes (CM) and intact leaf discs (LD) of *Prunus laurocerasus* and *Vinca major* at different external air humidities (Schreiber *et al.*, 2001).[6]

domains in the cuticle. Although the cuticle's resistance to water-vapour transport is almost entirely due to the deposited wax, the RH dependence of liquid water permeability is equally large in intact membranes and those that have had the cuticular waxes extracted (Schönherr and Schmidt, 1979). The diffusion of a very lipophilic substance, octadecanoic acid, in the recrystallized cuticular wax from 24 different species, is highly correlated with the water permeability of the same species' cuticular membranes (Schreiber and Riederer, 1996b), and co-permeability experiments have shown that the cuticular permeability to benzoic acid, salicylic acid, 2,4-D and water also are correlated (Niederl *et al.*, 1998). This indicates that water and organic compounds penetrate plant cuticles via the same lipophilic path composed of cuticular waxes and cutin monomer. Apparently there are two parallel paths of water diffusion across plant cuticles, a humidity-sensitive path along hydrated polar groups (–OH and –COOH) that form pores, and a dominating and humidity-independent, non-polar path comprised of the lipophilic wax components of the cuticle (wax and cutin):

$$P_{Pores} = (D_{Pores}K_{Pores})/\Delta x_{Pores} \quad (6.3)$$

$$P_{Lipids} = (D_{Lipids}K_{Lipids})/\Delta x_{Lipids} \quad (6.4)$$

where permeance (P) is given by:

$$P = F/A\ (C_{Don} - C_{Rec}) \quad (6.5)$$

and D (m^2/s) is the diffusion coefficient in each of the paths, K represents the dimensionless partition coefficient between water and the respective phase in which it diffuses, Δx (m) is the path length for diffusion across the cuticle, $(C_{Don} - C_{Rec})$ is the driving force of transpiration (mol/m^3) and F the water flow (mol/s) across the cuticle.[6] The rates of water flow, F_{Pores} and F_{Lipids}, across these parallel paths are:

$$F_{Pores} = P_{Pores}A_{Pores}(C_{Don} - C_{Rec}) \quad (6.6)$$

$$F_{Lipids} = P_{Lipids}A_{Lipids}(C_{Don} - C_{Rec}) \quad (6.7)$$

where A_{Pores} and A_{Lipids} are the areas of the diffusion paths.

The parallel pathway theory is consistent with the increase in cuticular transpiration caused by wax extraction, temperature elevation (Schönherr *et al.*, 1979), and the action of plasticizers. Lipophilic plasticizers increase the fluidity of the transport-limiting barrier of cuticles by sorbing amorphous cuticular waxes (Schreiber and Riederer, 1996a), temperature elevation has the same effect, and plasticizers (Riederer and Schönherr, 1990) and temperature (Schönherr and Schmidt, 1979; Baur *et al.*, 1997) increase the cuticular permeability to organic lipophilic compounds and water, but not to calcium (Schönherr, 2000), proving that water is able to diffuse along the lipophilic path, whereas calcium can only diffuse in an aqueous environment. Extracting waxes, increasing the temperature or treating with plasticizers enhances cuticular permeability by 10–1000-fold, whereas increasing the humidity leads only to a two- to threefold increase in water permeability across the polar pores.

6.4 Water-retentive Packaging

The calculations summarized in Section 6.1, indicating that naked fruits lose all or most of their respiratory heat by evaporative cooling even at relative humidities as high as 93–95%, do not consider the reduction in water loss that results when a commodity is stored or transported in cartons containing water-retentive plastic wraps or liners that create a near-saturated microclimate around the stored commodity. The effectiveness of water-retentive wraps is based in large measure on the low transpirational resistance and high skin resistance to gas exchange of horticultural commodities (Table 6.4). If the product of the wrap's resistance to water vapour and gas transport × the wrap's surface area is a multiple of the commodity's transpirational resistance × the total surface area of the commodity enclosed by the wrap, but significantly lower than the commodity's resistance to gas mass transport × the total commodity surface area within the wrap, the humidity can be elevated without significantly changing the gaseous composition in the microclimate. At atmospheric pressure, a harvested fruit's cuticular transpirational resistance (30–380 s/cm) is much lower than its resistance to water vapour transport through air-filled lenticles, the pedicel-end stem scar or 'closed' stomates (2750–11,000 s/cm) (Burton, 1982; Burg and Kosson, 1983; Ben-Yehoshua *et al.*, 1985). When fruits are protected by a plastic liner that is relatively impervious to gases and water vapour, the transport resistances of the plastic and the fruit's surface summate because they act in series. If the plastic's permeability and extent of 'perforation' are properly selected relative to the commodity weight, respiration rate and the surface areas of the carton and contained fruit, an amount of packaging resistance can be added that substantially elevates the water-vapour-transport resistance above 30–380 s/cm without causing a significant percentage increase in the gas-transport resistance above 2750–11,000 s/cm. This effect is illustrated for seal-packaged Valencia oranges in Table 6.4. Although the gaseous permeability of various plastics usually is in the order $CO_2 > O_2 >$ water vapour (Zagory, 1992; Table 6.4), sealing the fruit individually in high-density polyethylene film (HDPE) reduces water loss by up to tenfold without significantly changing the fruit's endogenous O_2, CO_2 or ethylene content (Ben-Yehoshua *et al.*, 1979, 1985).

Plastic liners and seal packages force fruits to increase sufficiently in temperature to transfer an additional amount of respiratory heat by convection, in lieu of a lesser amount transferred by evaporative cooling (Ben-Yehoshua, 1978). Because the plastic 'insulates' the fruit and interferes with convective cooling, it causes the commodity temperature rise to be higher than otherwise would be required. The wrap may impede cool down to such an extent that sometimes the commodity must be precooled before the water-retentive plastic is applied. An additional disadvantage is that seal packaging and water-retentive liners create a saturated microenvironment within the package or box, with the fruit slightly warmer than the inner surface of the plastic. Evaporated water tends to condense on the colder inner surface of the plastic and drip back on to the fruit, increasing the frequency of decay (7.8; Ben-Yehoshua, 1978).

6.5 Dependence of Transpirational Resistance on Pressure

LP increases a commodity's transpirational conductance by enhancing the diffusion of water vapour through lenticles, stomates, the pedicel stem-end scar, boxes and wraps, and by opening stomates. Cuticular water conductance also would be pressure-dependent if it was determined by the barrier-air layer's thickness and the diffusive resistance created by the tortuosity and small cross-sectional profile of the air spaces separating the overlapping wax platelets, as proposed by Possingham *et al.* (1967). This was the prevailing view (Burg and Kosson, 1983; Kirk *et al.*, 1986; Burg and Kang, 1993) until proponents of the parallel pathway theory of cuticular transpiration demonstrated that cuticular water conductance is limited by the rate at which liquid water diffuses through water-filled pores and non-polar lipophilic wax components in the cuticle (6.3). As neither of these paths contain an air phase, cuticular transport should not be influenced by pressure. Apparently, the pressure-dependent gas-phase resistances of the wax platelets and barrier air layer, which act in series downstream of the liquid-water/vapour transition in the cuticular matrix, do not comprise a significant part of the overall cuticular transpirational resistance.

Pressure-dependent water transport through lenticles, the pedicel-end stem scar and stomates occurs in parallel with the pressure-independent cuticular pathway (equation 3.3, parallel). Therefore LP will not have a substantial effect on transpirational resistance if at atmospheric pressure the resistance of a commodity's pressure-dependent pathways is much larger than the cuticular resistance. McIntosh apples are representative of this condition since they lack stomates (Clements, 1935) and at 95% RH the transpirational resistances of their lenticles and cuticle are 8928 s/cm (example 10) and 50 s/cm (Fig. 6.7, *upper left*), respectively. Lowering the pressure to 6.65 kPa (50 mm Hg) at 2°C and 95% RH should only reduce the apples' transpirational resistance to 45.7 s/cm (example 10; Fig. 6.8). At atmospheric pressure, the transpirational resistances of the closed ('cracked open') stomates and cuticle of a harvested orange are 3385 s/cm (Ben-Yehoshua *et al.*, 1985) and 106.1 s/cm, respectively. If the stomates remain 'closed' when the pressure is lowered at 10°C and 95% RH, a 38-fold pressure reduction to 2.67 kPa (20 mm Hg) would only lower the transpirational resistance by 3.1-fold, to 33.9 s/cm.

Pressure has a larger effect on water evaporation if the resistance of the pressure-dependent air-phase pathway is much smaller than the pressure-independent cuticular resistance, or when this condition arises at a reduced pressure. The transpirational resistance of an orange with open stomates should decrease 67.3-fold, from 14.8 s/cm at atmospheric pressure, to 0.22 s/cm when the pressure is lowered to 2.67 kPa (20 mm Hg) at 95% RH, 10°C (Fig. 6.8; example 25). When LP causes the closed stomates of a harvested orange to open at 2.67 kPa (20 mm Hg), their conductance should increase by 482-fold and reduce the transpirational resistance from 106.1 s/cm at atmospheric pressure to 0.21 s/cm at the low pressure. A significant part of this effect is caused by a progressively increasing

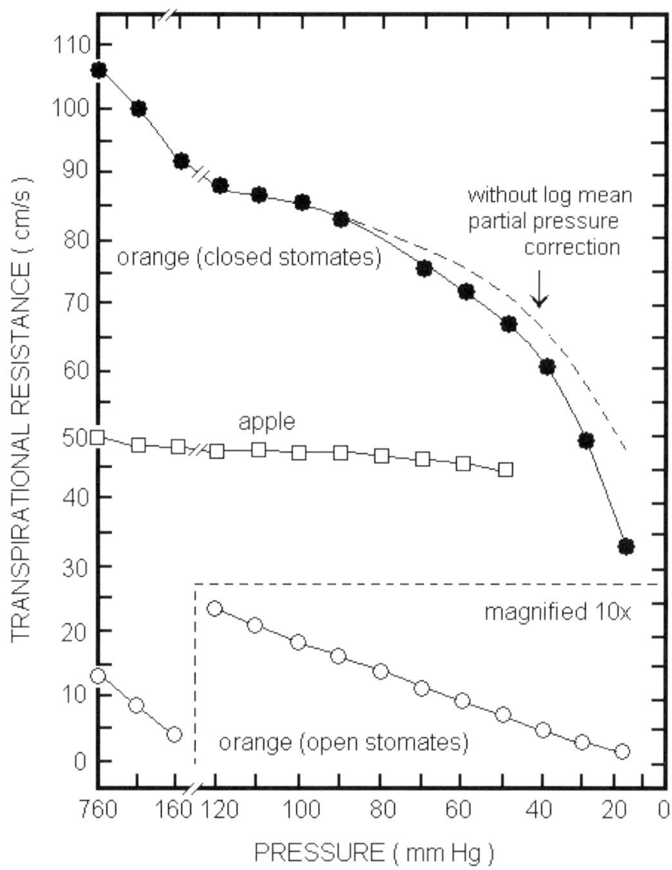

Fig. 6.8. Analysis of the effect of pressure on the transpirational resistance of Valencia oranges with open (○) and closed (●) stomates, and McIntosh apples (□). Conditions are described in examples 25 and 10.

proportion of water vapour present in the hypobaric atmosphere when the pressure is decreased below 10.67 kPa (equation 6.8; Fig. 6.8 – 'without log mean water-vapour partial-pressure correction'). A typical leaf has a low water vapour transport resistance through open stomates and the barrier air layer (1–6 s/cm; Noble, 1991), and a much higher cuticular transpirational resistance, 25–200 s/cm. When a storage pressure of 1.33 kPa (10 mm Hg) causes the stomates of a harvested leafy commodity to open at 0°C and 95% RH (4.15), the transpirational resistance should decrease by 136.5-fold,[7] from 25–200 s/cm at atmospheric pressure, to 0.04–0.007 s/cm at the low pressure.

Table 6.5 presents an analysis of evaporative heat transfer from an individual Choquette avocado stored at 13°C in NA or in a laboratory LP apparatus (Fig. 2.4) at pressures of 1.33 or 6.66 kPa (20 or 80 mm Hg). An avocado was selected for this example because it has a high respiration rate and therefore a demanding heat-transfer requirement. Reducing the pressure decreases both the avocado's respiratory heat load (Fig. 4.2) and the transpirational resistance of its lenticles, until below approximately 10.67 kPa (80 mm Hg) the mole fraction of water vapour present in the storage mixture increases to such an extent that the log mean partial-pressure correction (equations 15.18, 15.19 and 6.8) begins to cause an additional decrease in the lenticular transpirational resistance. The resistance to water vapour flux through air-filled pores (A) is given by (Burg and Kosson, 1983):

$$r_A = r_{A,R}\frac{\ln\left[\dfrac{p_R - p_{v,o}}{p_R - p_{v,i}}\right]}{\ln\left[\dfrac{p - p_{v,o}}{p - p_{v,i}}\right]} \tag{6.8}$$

where p_R is the reference pressure (atmospheric pressure), $r_{A,R}$ the pore resistance (s/cm) measured at atmospheric pressure and 273°K for the particular vapour pressure values $p_{v,i}$ and $p_{v,o}$ inside (i) and outside (o) the surface, respectively, and r_A is the pore resistance at pressure p. Regardless of whether it is assumed that the cuticular resistance at atmospheric pressure is 50 or 150 s/cm (Table 6.5), when the pressure is lowered to 2.67 kPa (20 mm Hg), the avocado can transfer all of its respiratory heat by evaporative cooling when its temperature is < 0.2°C above the dew point temperature, and well below the storage air's dry-bulb temperature. This is misleading because it does not take into account the transfer of heat into or from the fruit by radiation and convection across the temperature difference between the fruit and air. This heat raises (or lowers) the fruit temperature, increases (or decreases) the vapour-pressure gradient, and elevates (or lowers) the amount of water that will be evaporated. The thermodynamic solution to this problem is deferred to 6.23.

The avocado example in Table 6.5 does not consider the effect that a storage box has on transpirational resistance and the commodity's temperature rise. When roses are stored in a standard commercial flower box at 0°C and a pressure of 1.33 kPa (10 mm Hg), even though LP reduces their respiration rate by 90%, if the low pressure did not also increase water conductance through the box and protective wraps, the flowers would steadily warm. This does not occur because the box resistance (r_{box}) depends on the total pressure and water vapour partial pressure according to equation 6.56. At a pressure of 1.33 kPa (10 mm Hg) only a 0.36 mm Hg vapour-pressure gradient is required to transfer all respiratory heat through the standard box and wraps used with roses, and the commodity temperature rise needed to develop this vapour-pressure gradient is only 1°C at a storage temperature of 0°C (example 7). The increased water vapour concentration in the box lowers the O_2 partial pressure by an insignificant amount, from 1.13 mm Hg in the absence of flowers, to 1.06 mm Hg when they are present. In agreement with this calculation, an average temperature rise of several °C has been measured inside rose boxes stored at 1°C in commercial LP intermodal containers fully loaded at a storage pressure of 2.0 kPa (15 mm Hg). When tomatoes are stored at 15°C and a pressure of 10.67 kPa (80 mm Hg), only a 0.6 mm Hg

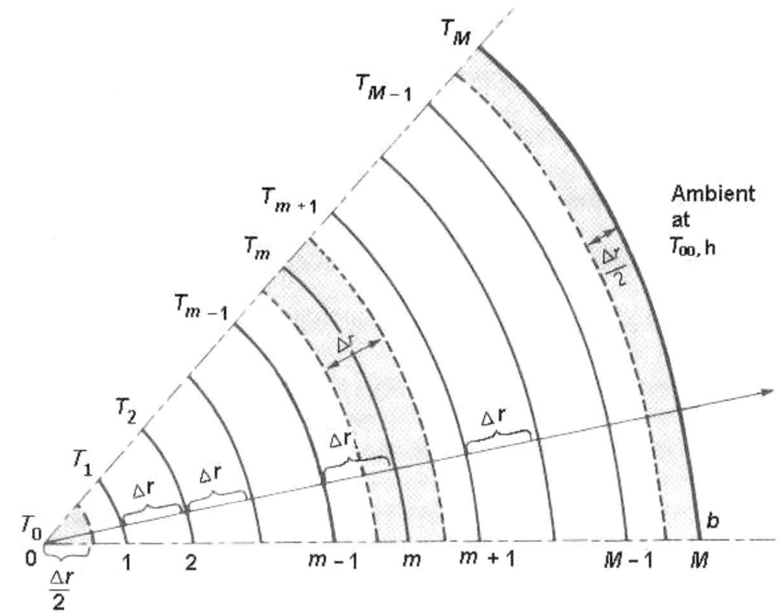

Fig. 6.9. Nomenclature for finite-difference formulation in cylindrical or spherical symmetry (Özisik, 1985).

vapour-pressure gradient is required to transfer water vapour through a typical storage box's surface at the same rate that it is evaporated in response to the fruit's respiratory heat (example 6). This elevates the fruit's temperature by only 0.8°C.

6.6 Radial Transport of Respiratory Heat

At atmospheric pressure, more than 85–90% of a stored horticultural commodity's transpiration occurs through its cuticle (Sastry *et al.*, 1978; example 8). Almost all of this water evaporates at the air–water interface in the cuticular membrane, and therefore respiratory heat has to be transported from the commodity's interior to its surface before it can be removed by evaporative cooling. The finite-difference method can be used to compute the temperature gradient required to transfer the respiratory heat from the centre to the surface of a commodity that has three-dimensional geometry. The partial differential equation of heat conduction is approximated by a set of coupled algebraic equations that can be solved by the Gaussian elimination method (Özisik, 1985). Radial heat conduction in a sphere such as an apple of radius $r = b$, metres (m), in which energy is generated at a rate of $g(r)$ (W/m^3) can be analysed by dividing the region $0 \le r \le b$ into ten spherical subregions each of thickness $\Delta r = b/10$ (Özisik, 1985; Fig. 6.9). The temperature at the centre node ($m = 0$) is described by the relationship:

$$6(T_1 - T_O) + (\Delta r)^2\, g_O/k = 0 \qquad (6.9)$$

and at internal nodes $m = 1$–9 by:

$$(1 - 1/m)T_{m-1} - 2T_m + (1 + 1/m)T_{m+1} + (\Delta r)^2 g_O/k = 0 \qquad (6.10)$$

where g_O is the energy generation term (W/m^3), k is the thermal conductivity (W/m·k) and T the temperature (K). This system provides 10 simultaneous algebraic equations for the unknown node temperatures T_m where $m = 0$–9. For an apple stored in air at 0°C, assuming that all respiratory heat is transferred by surface evaporation, typical conditions are:

$b = 0.0380$ m
$\Delta r = 0.0038$ m

weight = 169 g
volume = 230 cm^3
surface area = 181.5 cm^2
porosity (a_{ias}) = 0.3 (Clements, 1935; Table 3.8)
respiratory heat, g = 7.41 W/m^3 (Tolle, 1962)
thermal conductivity,[11] k = 0.273 W/m·K

where the thermal conductivity has been computed as an averaged value for air and human muscle[9] assuming that the apple fruit is 70% 'muscle' and 30% air (a_{ias} = 0.3). The energy generation term is evaluated as:

$$(\Delta r)^2\, g/k = (0.0038)^2(7.41)/0.273 = 3.919 \times 10^{-4}\ \text{K}$$

The system of equations is written in a banded matrix and then the matrix system is transformed to an upper diagonal form (for details of the method, see Özisik, 1985) (see equation at bottom of page). Once the upper diagonal form is obtained, the last equation in the system immediately gives T_9; after T_9 is known, the next-to-last equation gives T_8, and the process is repeated until T_1 is calculated from the first equation. Only a 0.011°C temperature gradient is needed to continuously conduct all respiratory heat from the centre of the pulp to the peel of a 0°C apple respiring at atmospheric pressure. This is more than an order of magnitude smaller than the temperature gradient needed to transfer the same amount of heat by convection between the apple surface and ambient air at atmospheric pressure. For a cylindrical fruit such as a banana, the temperature at the centre node $m = 0$ is described by the relationship (Özisik, 1985):

$$4(T_1 - T_O) + (\Delta r)^2 g_O/k = 0 \qquad (6.11)$$

and at internal nodes m = 1–9 by:

$$(1\text{–}0.5\ m)T_{m-1} - 2T_{m+}(1 + 0.5\ m)T_{m+1} + (\Delta r)^2 g_O/k = 0 \qquad (6.12)$$

The conditions for a banana ripened at 29.2° C are:

respiration rate at climacteric peak = 273 mg CO_2/kg·h (Wardlaw and Leonard, 1940)
pre-climacteric respiration rate = 45 mg CO_2/kg·h (Wardlaw and Leonard, 1940)
respiratory heat production (g) = 610 W/m^3
weight of banana = 169 g
density of banana = 0.76 g/cm^3
surface to volume ratio of banana = 0.91 cm^2/cm^3
surface area = 202 cm^2
b = 0.022 m
Δr = 0.0022 m
porosity of banana (a_{ias}) = 0.15 (Pantastico, 1975)
thermal conductivity (k)[10] = 0.352 W/m·K
transpirational resistance @ 85–90% RH (green banana) = 27.7 s/cm (Burton, 1982)

The energy generation term is evaluated as:

$$(\Delta r)^2\, g/k = (0.0022)^2(610)/0.352 = 8.387 \times 10^{-3}\,^{\circ}\text{K}$$

$$\begin{bmatrix} -6 & 6 & 0 & 0 & 0 & 0 & 0 & 0 & 0 & 0 \\ 0 & -2 & 2 & 0 & 0 & 0 & 0 & 0 & 0 & 0 \\ 0 & 0 & -1.5 & 1.5 & 0 & 0 & 0 & 0 & 0 & 0 \\ 0 & 0 & 0 & -1.333 & 1.333 & 0 & 0 & 0 & 0 & 0 \\ 0 & 0 & 0 & 0 & -1.25 & 1.25 & 0 & 0 & 0 & 0 \\ 0 & 0 & 0 & 0 & 0 & -1.2 & 1.2 & 0 & 0 & 0 \\ 0 & 0 & 0 & 0 & 0 & 0 & -1.167 & 1.167 & 0 & 0 \\ 0 & 0 & 0 & 0 & 0 & 0 & 0 & -1.143 & 1.143 & 0 \\ 0 & 0 & 0 & 0 & 0 & 0 & 0 & 0 & -1.125 & 1.125 \\ 0 & 0 & 0 & 0 & 0 & 0 & 0 & 0 & 0 & -1.111 \end{bmatrix} \begin{bmatrix} T_0 \\ T_1 \\ T_2 \\ T_3 \\ T_4 \\ T_5 \\ T_6 \\ T_7 \\ T_8 \\ T_9 \end{bmatrix} = \begin{bmatrix} -3.919\times10^{-4} \\ -3.919\times10^{-4} \\ -3.919\times10^{-4} \\ -3.919\times10^{-4} \\ -3.919\times10^{-4} \\ -3.919\times10^{-4} \\ -3.919\times10^{-4} \\ -3.919\times10^{-4} \\ -3.919\times10^{-4} \\ -14.17\times10^{-4} \end{bmatrix}$$

At the climacteric peak, the temperature gradient necessary to conduct all respiratory heat from the centre of the pulp to the peel of a banana, ripening at an air temperature of 29.2°C (84.6°F), is only 0.068°C (0.12°F).

The steady-state temperature of bananas has been measured by inserting a mercury thermometer calibrated to within 0.11°C (0.2°F) accuracy into the pulp (Wardlaw and Leonard, 1940). Prior to the climacteric peak, bananas were 0.1–0.2°C colder than 65% RH flowing air, and 0.2°C warmer than 95% RH air (Fig. 6.10), while at the climacteric peak the pulp temperature was 0.6°C warmer than the air due to the large amount of respiratory heat produced. These temperature shifts are not caused by the centre-to-surface transfer of respiratory heat. Instead, they are adjustments that balance heat input or loss by convection and radiation, and loss by evaporation vs. heat input from respiration (example 5). The centre-to-surface temperature gradient can only be smaller than the overall temperature shifts measured in this experiment and could easily be close to the 0.068°C (0.12°F) value computed by the finite-difference method.

6.7 Cellular Water Conductance

Only a small percentage of the liquid water lost from a stored commodity evaporates from the cells in which it is produced. The rest is transported to the fruit's surface before it is vaporized. The conductance of water molecules through a plant cell's vacuolar and plasma membranes is facilitated by aquaporin protein-lined micropores in which the water permeability coefficient (P_W) is increased approximately tenfold compared to passage through a lecithin bilayer[10] (Macey, 1984). P_W ranges from 0.1 to 3×10^{-4} cm/s in different types of plant cells (Stadelmann, 1969; Gimmler *et al.*, 1990; Maurel, 1997), and is always higher for osmotic or hydraulic flow than

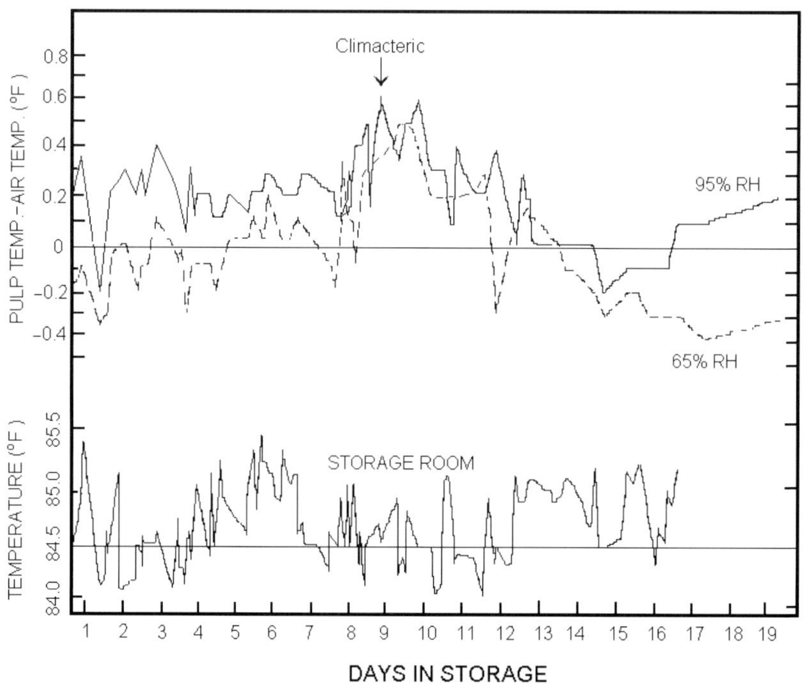

Fig. 6.10. Difference between pulp and air temperatures during ripening of two three-quarters-full Gros Michel bananas held at the same room temperature but at 95 and 65% relative humidity (Wardlaw and Leonard, 1940).

for diffusion.[11] This indicates that water molecules pass through the pores in single file, moving in groups or clusters. The polar groups of the aquaporin protein that lines the pores recreate the polar environment of the external solution, allowing the water molecules to remain close together in the liquid state as they move through the pores.[12] Plant cells are able to alter their membrane permeability to water and transcellular water flow by modulating aquaporin abundance and activity.[13]

6.8 Apoplastic Water Transport

Although liquid water moves through both the apoplastic and symplastic transcellular routes at all times (Chrispeels and Maurel, 1994), usually the preferred pathway is cell-to-cell when osmotic driving forces predominate (Steudel and Boyer, 1985; Maurel, 1997). The solute contribution to the driving force in the apoplast is small, the tensions normally present in the wall discourage liquid water films from forming at the surface of the intercellular system, and although the liquid-filled cell wall conducts water freely (Noble, 1991), it none the less affords a significant transport resistance because the mean fraction of total cross-sectional area available for apoplastic transport is only a few per cent of that available for cell-to-cell transport (Molz and Boyer, 1978).

6.9 Symplastic Cell-to-cell Water Transport

The symplasm is a cytoplasmic continuum created by protoplasmic threads, the plasmodesmata, that pass through pores in the cell wall, into an adjacent cell's protoplast. Even though plasmodesmata may only occupy about 0.2% of the cellular surface area, and aqueous channels only comprise approximately 10% of the plasmodesmata's area, diffusion through the plasmodesmata per unit area of the cells is many hundredfold greater than across the plasmalemma (Noble, 1991).

Mass transfer by molecular diffusion in a fluid at rest, resulting from concentration gradients, is analogous to heat diffusion resulting from temperature gradients, and therefore radial cell-to-cell water mass transport through the symplasm can be analysed by the same finite difference method used to evaluate radial heat transfer (Özisik, 1985; Fig. 6.9), substituting pressure for temperature, hydraulic conductance per unit length (K_h) for thermal conductivity and mass flow in place of thermal flow. For low concentrations of mass in the fluid and low mass-transfer rates, the mass-transfer equations and coefficients can be obtained by analogy from the corresponding heat-transfer equations.

The plasma and vacuolar membranes are considered to be the primary impediment to flow through the symplastic route, and therefore the presence of aquaporin proteins in the tonoplast, as well as in the plasma membrane, reduces the resistance of this pathway by increasing the effective cellular cross-sectional area through which water flows freely (Maurel, 1997). To obtain K_h per metre for a banana, the hydraulic conductivity of a typical higher plant cell, $L_P = 7 \times 10^{-7}$ m/s·MPa per m^2 of cell membrane (Steudel and Boyer, 1985; Noble, 1991), is divided by $(2 \cdot t)/(0.85\ \Delta x)$, where the factor 2 corrects for the condition that water must pass through the membranes twice in traversing each cell, $t = 1$ m of tissue thickness, Δx is the mean cell diameter (6.7×10^{-5} m for a banana – von Loesecke, 1950) and 0.85 corrects for the porosity of the tissue (Pantastico, 1975). For a banana, $K_h = 1.99 \times 10^{-11}$ m^4/s·MPa. The ΔP across the banana peel[14] is likely to be somewhat larger than the 0.1297 MPa value indicated in Table 6.7 because cells close to the surface are smaller than the mean value assumed for the calculation (von Loesecke, 1950).

The volume of a mature apple cell is $60–72 \times 10^{-4}$ mm^3 (Bollard, 1970); the mean cell diameter is $\Delta x = 2.26 \times 10^{-7}$ m; the apple's porosity = 0.3; and $K_h = 5.68 \times 10^{-11}$ m^4/s·MPa. In normal air at 0°C, if an apple's respiratory heat is eliminated by surface evaporation and no heat is gained from the environment, symplastic water

Table 6.7. Computed water-pressure gradient (relative to a pressure of 0 MPa at the centre) in a mature-green banana stored in air at 13.3°C assuming that all respiratory heat is transferred by surface evaporation and no heat is gained from the environment. Liquid water is lost from the banana at a rate of 1.9×10^{-8} m^3/s per m^3 of tissue.

P_m	Finite difference, MPa	
P_0 (centre)	0	pulp
P_1	–0.0012	
P_2	–0.0048	
P_3	–0.0109	
P_4	–0.0193	
P_5	–0.0392	
P_6	–0.0525	
P_7	–0.0682	peel[14]
P_8	–0.0863	
P_9	–0.1068	
Surface	–0.1297	

transport would create a 0.039 MPa water potential gradient. This is in close agreement with a value of approximately 0.2 bar (0.02 MPa) computed for potato tubers by analogy to leaves (Burton, 1982).

The magnitude of the vapour-pressure gradient driving water from a commodity stored at a high RH depends in part on the saturation vapour pressure of the commodity's water at the air/water interface in the cuticular membrane. Intact leaves may develop negative tensions in excess of 3 MPa at midday, which lowers the RH of water vapour at their air/water interface by 2.3–4.0% (equation 15.11; Slatyer, 1967). The much smaller negative tensions computed for radial water transport in bananas and apples would only lower the RH at the air/water interface by 0.1% in bananas and 0.01% in apples, and therefore will not significantly influence their rate of water loss.

6.10 Water Vapour Transport through the Intercellular System

When a 5–10°C temperature gradient is imposed across an apple or tomato, water moves from the warm to the cold side and causes the warm side to desiccate and the cold side to expand and split within a few days (Curtis, 1937; Lessler, 1947).[15] The direction of the water flux can be repeatedly reversed by inverting the temperature gradient, and the movement occurs even if the fruits are covered with vaseline or paraffin wax to prevent transpiration. A 10°C temperature rise increases the volume of water in the warm side by 0.2% due to the cubical expansion coefficient of water, and the osmotic pressure[16] in the warm side by 3.7%. In a typical plant cell, the increased water volume will elevate the turgor pressure by approximately 0.02 MPa, and the osmotic pressure will increase by 0.025 MPa, including a correction for a 0.015 MPa decrease caused by water expansion. The net result should be a slight decrease in water potential in the warm side relative to the cold side, which would tend to draw liquid water toward the warm side. Instead, water moves in the opposite direction because water vapour condenses on the cold surfaces after diffusing from the warm side where the vapour pressure is higher, through the intercellular spaces, to the cold side where the vapour pressure is lower (Curtis, 1937). This unusual situation resembles the condition that arises on a small scale in stored fruits and vegetables due to the conduction of respiratory heat along a temperature gradient from the centre to surface. Therefore, it is reasonable to inquire whether the respiratory temperature gradient induces a significant movement of water vapour through the intercellular system of bulky fruits.

Transport through the intercellular system can be analysed by the method used to evaluate cell-to-cell symplastic movement of liquid water, substituting diffusive conductance per metre in an air phase ($K_{D,WV}$) for K_h, and water vapour pressure for hydrostatic pressure:

$$K_{D,wv} = a_{ias} D_{wv,a} M_W / Zx\, RT \qquad (6.13)$$

where $D_{wv,a}$ is the binary diffusion coefficient of water vapour in air at the prevailing temperature, M_W is the molecular weight of water, a_{ias} is the cross-sectional surface area available for transport through the intercellular system per m^2 of surface and

$Zx \geq 1.57$ m per radial metre.[17] For a pre-climacteric apple, the calculation indicates that at 0°C a 2.09 MPa water potential gradient is required to diffuse all of the water through the intercellular system that must be evaporated to remove the respiratory heat. Since apoplastic transport of the same amount of water requires a 0.039 MPa gradient, only 1.9% of the radial transport of water will occur through the intercellular system. The 0.011°C temperature gradient created by the conduction of all respiratory heat from the pulp to the surface of an apple would give rise to a 0.081% centre-to-surface RH gradient, which would only transfer 2.1% of the required moisture to the surface. These calculations are in agreement with data showing that, depending on the ambient humidity and duration of storage, only 0.3–5.3% of an apple's transpiration occurs through its lenticles (example 8).

While the respiratory temperature gradient does not significantly promote centre-to-surface water transport through the intercellular system, the calculations do not necessarily indicate how much water vapour normally is transported through the intercellular system. This can be discerned by comparing the water-potential gradients required to transport the same amount of water through the intercellular spaces and cell-to-cell through the symplastic system. To transport all of the water that must be evaporated to remove an apple's respiratory heat at 0°C, a 0.039 MPa gradient is required for cell-to-cell symplastic transport and a 2.09 MPa gradient for diffusion through the intercellular system. Therefore, only 1.8% of the radial transport of water will occur through the intercellular system. In a green banana stored at 13.3°C, a 0.13 MPa gradient is required for cell-to-cell symplastic transport, and a 5.53 MPa gradient for diffusive movement through the intercellular system; only 2.4% of the water will be transported radially by diffusion through the gas phase of the intercellular system. The amounts moved through the intercellular system are barely adequate to provide the vapour transpired at atmospheric pressure through banana stomates and apple lenticles.

In LP, the amount of water vapour moving through the intercellular system may be increased by enhanced diffusion, condensation and capillary action. Condensation occurs whenever water vapour contacts a surface that is colder than the vapour-saturation temperature, and if the surface is hydrophilic so that it is wetted by the condensate, the water forms a smooth film that flows down the surface under the action of gravity. The thermal resistance to heat flow of the liquid water film determines the convective coefficient for film condensation, which is enormous if only water vapour is present, varying from 4000 to 11,000 W/m^2·°C on vertical surfaces, and from 9000 to 25,000 W/m^2·°C on horizontal tubes, compared to 5 W/m^2 for free convection on a 0.25 m vertical plate in air with a 25°C temperature gradient (Özisik, 1985). But if even a small amount of a non-condensable gas such as air is mixed with the vapour, the coefficient for film condensation decreases because the incoming vapour must diffuse through this air before it contacts the surface (15.27). The resistance to this diffusion process decreases the partial pressure of the condensing vapour at the surface, lowering the heat-transfer coefficient for film condensation by half if 0.5% air is present, by a factor of 5 in 5% air and by much more at atmospheric pressure (Özisik, 1985). Water-vapour diffusion is enhanced up to 76-fold when most of the air is evacuated from the intercellular spaces in LP, and therefore the convective coefficient for film condensation is much larger under hypobaric conditions than it is at atmospheric pressure.

Capillary condensation can occur at a higher temperature than that needed to change the state of water above a plane surface.[18] When apples previously equilibrated at 20°C and 40% RH are cooled with rapidly flowing 6°C, 85% RH air, although the fruit's temperature starts higher than the cooling air's 3.7°C dew point, nevertheless the apple's weight initially increases (Fig. 6.11). Surface tension causes the saturation vapour pressure e over the concave liquid surfaces in the interstices between the hydrophilic cellulose microfibrils of the cell

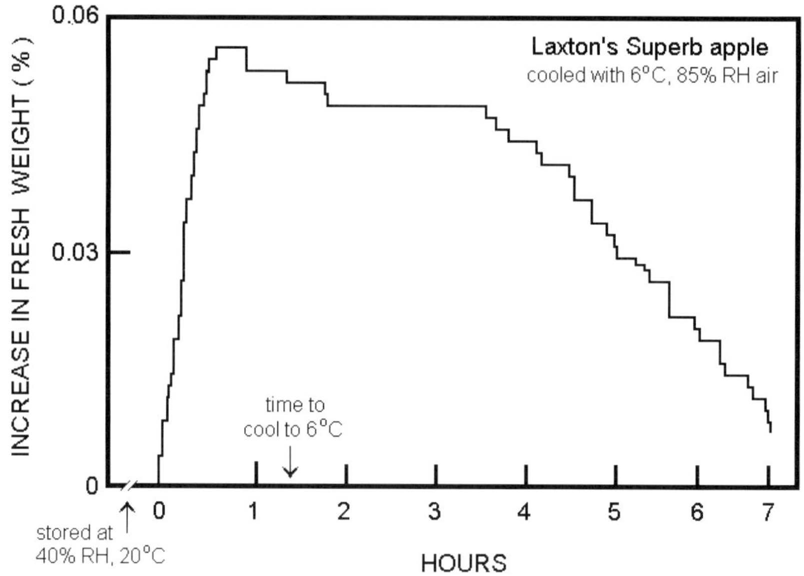

Fig. 6.11. Water uptake by a 6 kg lot of Laxton's Superb apples during cool down with 6°C, 85% relative humidity air. Apples were stored at 20°C, 40% relative humidity, prior to cooling (Fockens and Meffert, 1972).

walls to be higher than the saturation pressure e^o over a large plane surface, in accord with the Kelvin equation:[18]

$$\ln e/e^o = -2\sigma V/r\mathrm{RT} \qquad (6.14)$$

where σ is the surface tension, V the molar volume and r the radius of curvature. For water at 20°C, equation 6.14 predicts[19] that the ratio e/e^o is 1.001 at $r = 10^{-4}$ cm, 1.011 at $r = 10^{-5}$ cm, 1.114 at $r = 10^{-6}$ cm and 2.95 at $r = 10^{-7}$ cm. The radius of the interstices between the hydrophilic cellulose microfibrils typically is between 2×10^{-7} and 1.5×10^{-6} cm (Noble, 1991), and therefore vapour will readily condense at the entrance to the smallest capillaries at the temperature prevailing during the apple cool down depicted in Fig. 6.11.

At a hypobaric pressure, heat and liquid water may not have to be transferred cell-to-cell from the centre to the surface in order to dispel respiratory heat. Respiration is reduced to such an extent in LP that all of a commodity's respiratory heat might be transferred directly from respiring cells into the evacuated intercellular system by evaporative cooling. The vapour would rapidly diffuse to the cuticular surface through the rarified intercellular atmosphere, and if the commodity possessed stomates that opened in LP, their transpirational resistance might decline sufficiently (6.5; example 25) to permit the escape of most or all of the evaporated vapour into the surrounding air across the prevailing vapour-pressure gradient. If the commodity's transpirational resistance through pressure-dependent pathways did not decrease sufficiently at a low pressure to accommodate the required vapour flow, liquid water might reform on the inner surface of epidermal cells by capillary condensation made efficient by an unusually high convective film coefficient of condensation. The latent heat of vaporization acquired from the respiring cells that originally evaporated the water would be transferred in vapour that diffused through the intercellular system, and then released when the water condensed in the wall capillaries of the epidermal cell. This heat would be conducted through the epidermal cells to the air–water interface in the exterior cuticular membrane and removed by evaporative cooling, while the negative tension generated by transpiration drew condensed water into the epidermal cells, replacing the evaporated water. In this manner an efficient evaporation/condensation cycle might transport respiratory heat and water through the intercellular system from an interior cell, to the surface, through the cuticle, to the atmosphere.

6.11 Skin Shrivel and 'Peel Puffing'

The likelihood that skin shrivel will develop during storage depends mainly on the ambient temperature and humidity, the commodity's transpirational resistance and respiration rate, and the presence or absence of water-retentive wraps. Pears stored without polyethylene box liners at 0°C and 95% RH develop severe skin shrivel after 8 weeks, apples slight shrivel by 12 weeks (Hruschka, 1977), and the water loss from both fruits exceeds that needed to remove all respiratory heat by evaporative cooling. These commodities must cool below the temperature of the circulating refrigerated air and acquire heat from it.

Skin shrivel in bananas results during storage and ripening when selective changes in the osmotic content of specific tissues cause water to move from the peel to the pulp. The pulp-to-peel ratio of three-quarters-full bananas remains constant for 14 days at 11.7°C, and then during 7 additional days increases from 1.48 to 1.65. The ratio in heavy three-quarter-full fruit is constant for 9 days and then rises from 1.64 to 1.82 in 7 additional days (Table 6.8). These results indicate that the water-potential gradient, which initially moved water from the pulp to peel, reversed in direction. Eventually the peel began to draw down its water reserve when it eliminated respiratory heat by evaporative cooling without a commensurate transport of water from the pulp. During the last 7 days of storage, for each mole of water evaporated from the peel, 1.7 moles of water were transferred from the peel to the pulp. Since the osmotic concentration in both green banana pulp and peel is approximately 0.3 M, when sugar and water are lost in accord with equations 6.1 and 6.2, unless starch is hydrolysed, the cellular osmotic concentration should be lowered and the water content increased in both peel and pulp. Instead, starch decreases and the sugar content increases in the pulp and peel of green bananas during transport and distribution at 11.7°C (Stratton and von Loesecke, 1931; von Loesecke, 1950), and water increases in the pulp and decreases in the peel of stored green bananas (Simmonds, 1959). Measurements made at the time of discharge from a ship indicate that by then the osmotic concentration in the pulp of green bananas, 0.32 M, is slightly higher than in their peel, 0.28 M. The 0.04 M difference in osmotic concentration ($\Delta\Pi$ = 0.09 MPa) has the same magnitude as the

Table 6.8. Variation in the pulp-to-peel ratio of bananas stored at 11.7°C and ripened at 20°C (Wardlaw *et al.*, 1939; Barnell, 1941; von Loesecke, 1950).

Three-quarter-full fruit		Heavy three-quarter-full fruit		Three-quarter-full fruit		Heavy three-quarter-full fruit	
Days in storage	Pulp-to-peel ratio	Days in storage	Pulp-to-peel ratio	Days in ripening room	Pulp-to-peel ratio	Days in ripening room	Pulp-to-peel ratio
0	1.48	0	1.61	0	1.51	0	1.63
3	1.51	2	1.64	2	1.53	2	1.60
6	1.48	4	1.63	4	1.69	4	1.70
10	1.48	6	1.60	6	1.87	7	1.97
14	1.49	9	1.64	8	2.04	9	2.16
17	1.58	11	1.65	10	2.28	–	–
19	1.63	14	1.77	12	2.42	–	–
21	1.65	16	1.82	–	–	–	–

computed centre-to-surface ΔP needed to move all of the water to the surface that must be evaporated to remove respiratory heat. Extensive starch hydrolysis elevates the osmotic content of the pulp to 1.24–1.30 M as bananas ripen during a 10-day period, while the peel only increases to 0.55–0.66 M (Fig. 6.12). The large osmotic pressure gradient that develops between these tissues draws water from the peel to the pulp, progressively changing the pulp-to-peel ratio from 1.51–1.71 in green fruits to 2.16–2.79 in ripe fruits (Table 6.8), while the water content of the pulp increases from 64 ± 4% to 74 ± 3% (Palmer, 1971). At the same time that water moves inward to the pulp, the rates of transpiration and respiration double and the pulp continues transferring all of its respiratory heat to the surface by conduction. The peel is forced to draw down its water reserve in order to release the extra respiratory heat by evaporative cooling, and this increases the pulp-to-peel ratio. It is not known whether hypobaric conditions alter this behaviour, but LP provides a longer storage life for high-quality bananas than any other method thus far developed. After 4 months at 10.67 (80 mm Hg), the fruit retains excellent firmness and there is no visible evidence of peel shrivel (Apelbaum *et al.*, 1977a).

'Peel puffing' in citrus fruit is characterized by an increase in the ratio of peel to fruit weight and a decrease in peel specific gravity. Apparently an excessive pulp-to-peel migration of water occurs similar to the peel-to-pulp movement in bananas, except in the opposite direction. This disorder may be similar in morphology to the increased albedo formation seen when citrus are grown in hot, dry climates. Peel puffing in Satsuma mandarin (*Citrus unshiu*) is promoted by ethephon, and reduced by treating the fruits with AVG, or by storing them in the presence of $KMnO_4$, indicating ethylene involvement (Maotani *et al.*, 1983; Abeles *et al.*, 1992). Tahiti limes kept in CA maintain an acceptable green colour, but their juice content rapidly diminishes and within a few weeks they develop thick rinds resembling peel puffing (Spalding and Reeder, 1974). Limes stored

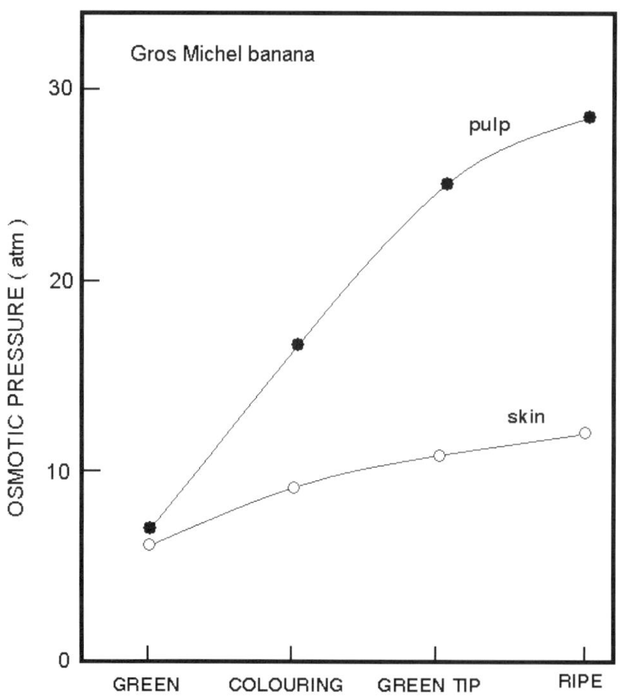

Fig. 6.12. Osmotic pressure changes in the skin and pulp of ripening Gros Michel bananas (Stratton and von Loesecke, 1931).

in LP at a pressure of 20–22.67 kPa (150–170 mm Hg = 3.8–4.3% [O_2]) remain green for a longer time, retain their normal juice content and their peels do not thicken[20] (Spalding and Reeder, 1974, 1976a; Burg, 1990). Peel puffing in limes during CA storage might be promoted by ethylene production and accumulation, and the disorder may be prevented in LP at 20–27 kPa (150–170 mm Hg) because the IEC is lowered by enhanced gaseous diffusion and presumably also by stomatal opening.

6.12 Dimensionless Groups used for Convective Heat Transfer Calculations

After a commodity's temperature stabilizes during storage or distribution, the heat lost by evaporative cooling equals the sum of the respiratory heat ± heat transferred by convection and radiation between the commodity and its surroundings. The vapour -pressure and dry-bulb temperature gradients, which arise when the energy balance requirement is satisfied, determine the relative contributions of each heat-transfer mode, and evaporative cooling accounts for the rate of water loss. The thermodynamic methods used to compute the rates of heat transfer by convection, radiation and evaporation will be reviewed, and the influence that a hypobaric pressure has on each mode scrutinized to ascertain how LP prevents commodity overheating and limits water loss.

Heat transfer by convection occurs when warm or cold air flows over a surface. The air velocity decreases near the surface due to viscous forces created by shear stress between the surface and flowing air. The velocity of the air layer immediately adjacent to the surface is zero, and in this region heat is transferred exclusively by conduction at a rate limited by the thermal conductivity and thickness of the stagnant boundary-air layer. This layer's thickness is designated as the distance from the surface at which 99% of the free-stream velocity is reached, and the region beyond that point is called the undisturbed free stream. The temperature gradient at the surface is determined by the ability of air farther from the surface to transport energy to the surface. Higher air velocities are able to produce larger temperature gradients and heat-transfer rates. The situation is similar in free (natural) and forced convection, the main difference being that in forced convection the velocity approaches the free-stream value imposed on the system by an external pump or fan of known capacity, whereas free convective flow is induced by buoyancy forces created by changes in the density of the air as it is warmed or cooled in passing over a surface. The air velocity in free convection depends on the temperature difference between the air and surface, the density change per unit temperature difference caused by the air's coefficient of thermal expansion, and the force field, which is simply the gravitational force (g). The natural convective velocity reaches its maximum a short distance from the surface and then approaches zero farther away because the buoyant force decreases as the air density approaches that of the surrounding fluid. Convective heat transfer ($\dot{Q}_c$,W) to or from an object is given by:

$$\dot{Q}_c = h_m A(\Delta T) \tag{6.15}$$

where A is the object's surface area (m^2), ΔT is the temperature difference (°C) between the object and the ambient air, and the mean convective heat-transfer coefficient, h_m (W/m^2·°C or BTU/h·ft·°F), depends on the density, viscosity, temperature gradient, velocity and thermal properties of air. The convective film coefficient is reduced in LP because the air density (ρ, kg/m) varies directly as a function of the storage pressure. The most important unknown in convective heat transfer is the value of the heat-transfer coefficient, h_m.

The solution of heat-transfer problems involving convection is simplified by the use of dimensional analysis to determine the appropriate value of the convective heat-transfer coefficient to be used in the thermodynamic calculation. Empirically derived correlations based on dimensionless groups (Table 6.9) provide reasonably accurate

Table 6.9. Dimensionless groups in the engineering system of dimensions.

Grashof number (Gr_r)	$L^3\rho^2\beta g\Delta T/\mu^2$
Nusselt number (Nu)	HL/k
Prandtl number (Pr)	$c_P\mu/k$
Reynolds number (Re)	$Lv\rho/\mu$
Rayleigh number (Ra)	Gr·Pr

Symbols: k is the thermal conductivity (W/m·K); g the local acceleration due to gravity (9.8 m/s^2); μ the dynamic viscosity (kg/m·s); c_P the specific heat at constant pressure (kJ/kg·K); ρ is the density (kg/m^3); β is the linear coefficient of expansion (K), ΔT is the temperature difference between the surface and fluid flowing over it (K); v is the velocity of flow (m/s); h is the convective film coefficient (W/m·°C); and L is a characteristic dimension (m) depending on the geometry of the system (adapted from Hoerl *et al.*, 1984).

solutions to convective heat transfer, mass transfer and flow problems. The appropriate dimensionless form of the mean heat-transfer coefficient (h_m) is the Nusselt number ($Nu_m = h_m L/k$), where L is a characteristic dimension (m), k the thermal conductivity (W/m·K) and the mean Nusselt number (Nu_m) is the ratio between heat transfer by convection vs. conduction across a fluid of thickness L. When the Nusselt number equals unity, heat transfer is by pure conduction and there is no convection; if the Nu_m is large the convective film coefficient is enhanced. Depending on the geometry of the system, when air flows over an object or through a circular duct, usually the characteristic dimension is either the diameter (D) or length (L) of the heated or cooled surface, but in an enclosed space or a duct that is not circular, the characteristic length is the hydraulic diameter (D_h), defined as four times the cross-sectional area ($4A_{CS}$) divided by the wetted perimeter (P). For a circular tube, D_h becomes the tube diameter (D) because $A = (\pi/4)D^2$, $P = \pi D$, and therefore $4A/P = D$.

The Prandtl number (Pr) represents the relative importance of momentum and energy transport by the diffusion process. It is defined as the ratio of the kinematic viscosity of the fluid (also called molecular diffusivity of momentum) to the thermal diffusivity of the liquid (also called molecular diffusivity of heat), and relates the temperature distribution to the velocity distribution:

$$Pr = c_P\mu/k \quad (6.16)$$

where c_P is the heat capacity at constant pressure (W/s·kg·K), k is the thermal conductivity (W/m·K) and μ is the dynamic viscosity (kg/m·s). The kinematic viscosity (ν, m^2/s) is defined as:

$$\nu = \mu/\rho \quad (6.17)$$

where ρ is the density (kg/m^3) of the air. Thermal diffusivity (α, m^2/s) is given by:

$$\alpha = k/\rho c_p \quad (6.18)$$

and therefore $Pr = \nu/\alpha$ = momentum diffusivity/thermal diffusivity. The relative thickness of the velocity and thermal boundary-air layers depends on the Prandtl number, and for flow in a pipe or over a flat plate, the velocity and temperature profiles are similar for a fluid having a Prandtl number of unity. When $Pr < 1$, the temperature gradient near the surface is less steep than the velocity gradient, and if $Pr > 1$, the temperature gradient is steeper than the velocity gradient. For air in the physiological temperature range, $Pr_{air} = 0.71$. Since both α and ν are inversely related to ρ, the Prandtl number is independent of pressure.

The Reynolds number (Re) is the ratio of inertia to viscous force. Viscous forces predominate for small Reynolds numbers and inertial forces for large Reynolds numbers:

$$Re = Lv\rho/\mu \quad (6.19)$$

where L is a characteristic dimension determined by the geometry of the system. For flow over a flat plate, L is its length; for flow over a sphere and cylinder, or through a circular duct, L is the diameter; and if the duct cross-section for flow is not circular, L is the hydraulic diameter (D_h). Re is used as the criterion to determine the change from laminar to turbulent flow. At higher Reynolds numbers, the inertia forces amplify small disturbances in the fluid, causing a moving stream of air to undergo a transition from ‘smooth’ laminar flow to turbulent flow. The turbulence decreases the thickness of the boundary-air layer, the high velocity increases the temperature gradient, and both effects combine to improve convective heat transfer. In forced convection, the transition from laminar to turbulent flow begins when the Reynolds number exceeds approximately 3000, the exact value depending in part on the system's geometry. Since the Reynolds number is directly proportional to the density of air, it depends on the air pressure.

The Grashof number (Gr) represents the ratio of buoyancy force to viscous force acting on the air:

$$\mathrm{Gr} = L^3\rho^2\beta g\Delta T/\mu^2 \quad (6.20)$$

The buoyancy for natural convection arises from the change in density (ρ), represented by the ($\rho\beta\Delta T$) product in the numerator of the Grashof number, and for a perfect gas in the absence of mass transfer, $\beta = 1/T$. The Grashof number in free (natural) convection plays the same role as the Reynolds number in forced convection. Because Gr is proportional to ρ^2, it varies with the square of the pressure.

The parameter Gr/Re^2 is a measure of the relative importance of free convection in relation to forced convection. When $\mathrm{Gr/Re}^2 \cong 1$, free and forced convection are of equal magnitude, both must be considered, and the Nusselt number depends on Re, Gr and Pr. If $\mathrm{Gr/Re}^2 > 10$, flow is primarily by forced convection (Noble, 1991) and only Re and Pr need be considered in calculating h_m (equation 6.41). If the ratio is much less than unity, free convection becomes dominant and the Nusselt number depends only on Gr and Pr (equations 6.21–6.26).

When heat transfer begins as soon as a fluid enters a duct, the velocity and temperature profiles start developing simultaneously. The hydrodynamic entrance length is the length required from the duct inlet to achieve a maximum velocity of 99% of the corresponding fully developed value. The thermal entrance length is the length required from the beginning of the heat-transfer section to achieve a local Nusselt number equal to 1.05 times the corresponding fully developed value. For laminar flow through ducts of the shape created by the box-stacking arrangements in VacuFresh[SM] and Grumman/Dormavac hypobaric containers (Figs 6.13, 6.14 and 6.15), the thermal entrance length (L_t) for a constant wall heat flux, and the hydrodynamic entrance length (L_h) are $L_h/D_h = 0.011$ Re and $L_t/D_h = 0.012$ Re·Pr, respectively (Özisik, 1985). The Nusselt number and friction factor for laminar flow in ducts of various cross-sections has been determined in the region at a distance from the duct entrance where velocity and temperature profiles are fully developed (see tables in Özisik, 1985 and Kreith and Bohn, 1997). The value of the Nusselt number in the region of hydrodynamically and thermally developed laminar flow varies for different duct shapes, and also depends on whether the surface temperature and surface heat flux are uniform along the duct length, and the duct is straight or curved. Example 14 considers the thermal (L_t) entrance length required to achieve a constant-wall heat flux, and the hydrodynamic (L_h) entrance length in a loaded LP intermodal container.

Values for viscosity (15.22), thermal conductivity (15.23), specific heat (15.24), Pr and Re numbers (15.25) and diffusion coefficients (15.26) to be used in LP heat and mass transfer calculations must be adjusted to reflect the large proportion of water vapour present (example 18; Table 15.6).

In the application of any empirical equation for forced convection, the predicted heat transfer coefficients are not exact, and the results obtained by various experimenters, even under carefully controlled conditions, differ appreciably. In turbulent flow, the accuracy of a heat-transfer coefficient predicted from any available equation or graph is no better than ± 20%, and with laminar flow, the accuracy may be in the order of ± 30% (Kreith and Bohn, 1997).

6.13 Heat Transfer Coefficient for Free Convection at Atmospheric Pressure

In the range $1 < \mathrm{Ra_D} < 10^5$, the average Nusselt number ($\mathrm{Nu_m}$) for free convection on a single isothermal sphere such as an apple or orange suspended in air, is given by (Özisik, 1985):

$$Nu_m = hD/k = 2 + 0.43\ Ra_D^{1/4} \quad (6.21)$$

where:

$$Ra_D = Gr_D \cdot Pr = [g\beta\ D^3\rho^2(\Delta T)/\mu^2][C_P\mu/k] \quad (6.22)$$

and D is the diameter (m) of the fruit, μ is the dynamic viscosity (kg/m·s), ρ is the density (kg/m^3) and $C_P\mu/k$ is the Prandtl number (Pr = 0.71 for air). At very low temperature differences, $Ra_D \rightarrow 0$, and equation 6.21 reduces to $Nu_D \rightarrow 2$. At this limiting value for free convection, when a Choquette avocado with a diameter of 0.053 m is stored at 13°C (Table 6.5), the mean film coefficient for heat transfer is h_m = 0.96 W/m^2·°C (0.17 BTU/h·ft^2·°F) at atmospheric pressure. If ΔT = 1°C, $Ra_D = 2.09 \times 10^{-4}$, Nu_m = 7.17 and the heat-transfer coefficient for free convection is h_m = 3.38 W/m^2·°C (0.6 BTU/h·ft^2·°F).

In the range $10^{-4} < Ra_D < 10^{12}$, the average Nusselt number for free convection on a single isothermal horizontal cylinder such as a banana suspended in air, is given by (Özisik, 1985):

$$Nu_D^{1/2} = 0.60 + 0.323\ Ra_D^{1/6} \quad (6.23)$$

For a 0.03 m diameter horizontally positioned banana stored at 13.3°C, at a very low temperature difference $Nu_m \rightarrow 0.6$ and $h_m \rightarrow 5.7$ W/m^2·°C (0.09 BTU/h·ft^2·°F). When ΔT = 1°C, Ra_D = 3650, Nu_D = 3.49 and the heat transfer coefficient for natural convection is h_m = 2.91 W/m^2·°C (0.51 BTU/h·ft^2·°F). For Pr > 0.5 and Grash of numbers ranging from 10^3 to 10^9, the average heat-transfer coefficient for free convection from a single horizontal tube or narrow cylinder can also be computed from the relationship (Kreith and Bohn, 1997):

$$Nu_m = 0.53\ (Gr_D Pr)^{1/4} \quad (6.24)$$

According to equation 6.24, when ΔT = 1°C, Nu = 3.77 and h_m = 3.15 W/m^2·°C (0.55 BTU/h·ft^2·°F) for the horizontal banana, in close agreement with equation 6.23.

In the range $10 < Gr_L Pr < 10^8$, the average Nusselt number for free convection on a single isothermal vertical banana (cylinder) is given by (Kreith and Bohn, 1997):

$$Nu_L = 0.68\ Pr^{1/2}\ [(Gr_L)^{1/4}/(0.952 + Pr)^{1/4}] = hL/k \quad (6.25)$$

When a 0.18 m-long vertical banana stored at 13.3°C transfers its respiratory heat at atmospheric pressure across a 1°C temperature gradient (ΔT) between its surface and the surrounding air, the average Grashof number (Gr_L) based on the fruit's length (L) is 1.68×10^5, Nu_m = 10.19 and the heat-transfer coefficient for free convection is h_m = 1.42 W/m^2·°C (0.25 BTU/h·ft^2·°F).

Values for h_m also can be based on data correlations for natural convection from short vertical plates to air (McAdams, 1954), in the form:

$$Nu_L = c(Gr_L\ Pr)^n \quad (6.26)$$

where Nu_L is computed across a fluid of thickness L, and constants c and n depend on the value of Nu_L as shown in Table 6.10. This relationship is useful in computing the heat-transfer coefficient from stacks of cartons, where the characteristic dimension (L) is the height of the stack.

6.14 Radiative Couplings

Radiant heat transfer through a non-participating medium separating the surfaces of objects occurs independently of the pressure. A portion of the total radiation incident on a surface is absorbed by the material, a part is reflected from the surface and the remainder is transmitted through

Table 6.10. Values of c and n for equation 6.26.

Nu_m	c	n
1–10	1.44	0.120
10–10^2	1.37	0.411
10^2–10^3	1.20	0.170
10^3–10^4	1.04	0.190
10^4–10^9	0.59	0.250
10^9–10^{12}	0.13	0.333

the body. The *absorptivity* (α) of a surface is the fraction of the total irradiation absorbed by the body; the *reflectivity* (ρ) is the fraction reflected; and the *transmissivity* (τ) is the fraction transmitted. If an energy balance is made on a surface involving radiation as the only heat-transfer mode:

$$\alpha G + \rho G + \tau G = G \tag{6.27}$$

where G is the rate at which irradiation is incident on the surface. According to equation 6.27, the sum of absorptivity, reflectivity, and transmissivity must equal unity:

$$\alpha + \rho + \tau = 1 \tag{6.28}$$

Bodies such as boxes and plant and animal commodities are opaque and do not transmit incident irradiation. For opaque bodies, $\tau = 0$ and equation 6.28 reduces to:

$$\alpha + \rho = 1 \tag{6.29}$$

If the surface of an opaque body also is a perfect reflector from which all irradiation is reflected, $\rho = 1$ and the transmissivity as well as the absorptivity is zero. In the physiological temperature range, the polished 'shiny' aluminized side of Mylar film has a reflectivity of $\rho \cong 0.96$ and an emissivity of only $\varepsilon \cong 0.04$ (Table 6.11). Therefore Mylar serves as a highly effective radiation shield.

The emissivity (ε) of a surface is the ratio of the energy emitted by a real surface to that of an equally sized and shaped blackbody emitting at the same temperature. Since a blackbody emits the maximum possible radiation at a given temperature, the emissivity of a surface is always between zero and unity. Greybodies are surfaces with monochromatic emissivities and absorptivities whose values are independent of wavelength.[21] The greybody assumption that the spectral hemispherical emissivity (ε_λ), spectral hemispherical absorptivity and spectral hemispherical reflectivity (ρ_λ) are uniform over the entire wavelength spectrum, simplifies the analysis of radiative heat transfer since then, according to the Kirchoff law, $\alpha = \varepsilon$.

Table 6.11. Emissivity of opaque surfaces in the physiological temperature range (Knudsen *et al.*, 1984; Özisik, 1985; Kreith and Bohn, 1997).

Material	Emissivity
Aluminium	
Bright foil	0.04
Oxidized	0.20
Painted (white)	0.95
Cardboard or paper	0.90
Water or ice	0.97

In an enclosure containing only two zones with surface areas A_1 and A_2, between which heat exchange takes place because they are at different temperatures, the net radiation heat exchange from zone 1 to zone 2 (Q_{1-2}) equals the radiation energy leaving A_1 that strikes A_2 minus the radiation energy leaving A_2 that strikes A_1. If zone 1 has an emissivity ε_1 and is maintained at a uniform temperature T_1, and zone 2 has an emissivity ε_2 and is maintained at a uniform temperature T_2, radiative energy transfer (Q_{1-2} = watts) is described by the relationship (Özisik, 1985):

$$Q_{1-2} = \sigma (T_1^4 - T_2^4)/[(1 - \varepsilon_1)/(A_1\varepsilon_1) + 1/(A_1F_{1-2}) + (1 - \varepsilon_2)/(A_2\varepsilon_2)] \tag{6.30}$$

where F_{1-2} is the greybody view factor[22] and σ is the Stefan-Boltzman constant (5.6697×10^{-8} W/m^2·K^4). In a fully loaded intermodal container, the external box surface area facing the container wall (A_1) is essentially the same as the wall surface area (A_2). Therefore in a Grumman/Dormavac hypobaric container, radiation occurs between two large parallel plates, and in a VacuFresh[SM] container between two concentric cylinders. Since in both cases the graybody view factor for external boxes $\cong 1$, equation 6.30 reduces to the form:

$$Q_{1-2} = A\sigma (T_1^4 - T_2^4)/[(1/\varepsilon_1) + (1/\varepsilon_2) - 1] \tag{6.31}$$

where $A_1 = A_2 = A$. Because the aluminium interior of a Grumman/Dormavac container is painted white, its emissivity is close to unity and nearly equal to that of the external commodity's cardboard boxes (Table 6.11). When $\varepsilon_1 = \varepsilon_2$ and the view factor between parallel surfaces is $F_{1-2} = 1$, equation 6.31 reduces to the form:

$$Q_{1-2} = A\sigma (T_1^4 - T_2^4) \tag{6.32}$$

If a small opaque object such as a single fruit or piece of meat, with surface area A_2 and emissivity ε_2, is maintained at a uniform temperature T_2 inside a large opaque cavity such as an intermodal container, with a wall surface of area A_1, emissivity ε_1 and temperature T_1, since $A_1/A_2 \rightarrow 0$, and in both cases the greybody view factor is approximately 1, equation 6.32 reduces to the form (Özisik, 1985):

$$Q_{1-2} = A_1\varepsilon_1\sigma\,(T_1^4 - T_2^4) \qquad (6.33)$$

To include radiation in a thermal network involving convection, it is convenient to define a radiation heat-transfer coefficient, h_R:

$$Q_{1-2} = h_R\, A\Delta T \qquad (6.34)$$

and for a small temperature difference, the value of h_R to be used with equation 6.34 is $h_R = 5.35$ W/m^2·°C (0.94 BTU/ft^2·h·°F). The combined heat-transfer coefficient for radiation and convection is $h_R + h_c$.

In the nodal heat transfer model illustrated for a Grumman/Dormavac container in Fig. 6.17, the heat transferred by radiation (R) from the box (B) to the wall (W) is:

$$\dot{Q}_{R,BW} = R_2\sigma(T_B^4 - T_W^4) \qquad (6.35)$$

where $R_2 = F_{BW}\,A_{BF}$. For exterior boxes with a direct line-of-sight view of the container wall, the greybody view factor $F_{B,W}$ is set = 1. To conform to the same nodal model, the heat transferred by radiation from the fruit (F) to the box wall (B) is expressed as:

$$\dot{Q}_{R,FB} = R_1\sigma(T_F^4 - T_B^4) \qquad (6.36)$$

where the radiation coupling is:

$$R_1 = F_{FB}\,A_{FB} \qquad (6.37)$$

To account for the restrictive radiative path when intervening layers of plastic and paper are present in the box,[23] F_{FB} is set = 0.1. The interior of a VacuFresh^{SM} container is not painted and therefore the emissivity of its aluminium walls is approximately 0.2 (Table 6.11). As the emissivity of exterior boxes is close to unity, in the VacuFresh^{SM} container equation 6.31 reduces to the form:

$$Q_{1-2} = A\sigma(T_1^4 - T_2^4)/5 \qquad (6.38)$$

6.15 Evaporative Couplings

Evaporative heat transfer from a commodity to ambient air ($\dot{Q}_v$, kcal/s) is given by:

$$\dot{Q}_v = m_v H_v = \left[\frac{A(\Delta p_v)H_v}{r_{v,c,p}}\right]\left[\frac{M_v}{R_u T}\right] \qquad (6.39)$$

where m_v (kg) is the weight of water vapour (v) evaporated, H_v (kcal/kg) the latent heat of water evaporation [595.4 kcal/kg @ 0°C; 590.2 kcal/kg @ 10°C; 584.9 kcal/kg @ 20°C], A (m^2) the surface area of the commodity, M_v the molecular weight of water (18), T the temperature (K), $r_{v,c,p}$ the combined transpirational resistance of the cuticle and pores (s/m), Δp_v the water-vapour pressure gradient between the commodity and surrounding air (atm), and the gas constant $R_u = 0.08295$ m^3·atm/kg·mol·K. For a commodity enclosed within a carton, the heat transferred by vapour flow through the box walls to the atmosphere is:

$$\dot{Q}_v = m_v H_v = \left[\frac{A(\Delta p_v)H_v}{r_{v,\,box}}\right]\left[\frac{M_v}{R_u T}\right] \qquad (6.40)$$

where $r_{v,box}$ (s/m) is the box's resistance to water vapour transfer (equations 3.22 and 6.56).

6.16 Heat Transfer in CA Storage at Atmospheric Pressure

Apples are usually packed in pallet bins for CA storage, and free (natural) convection and evaporative cooling transfer heat between the fruits and the air in the bins. The main function of the forced airflow in the storage room is to continuously flush away water vapour evaporated from the fruit, as well as the air that has been warmed or cooled by the fruit and escaped from the boxes by natural convection. Some heat is transferred between the boxes and surrounding air by forced convection, but the amount is relatively small, in part due to the insulating properties of the wooden pallet boxes. The fruit's temperature is not the same as the room air temperature, and the dry-bulb and dewpoint temperatures are not identical inside and outside the bins. While these temperature and humidity differences are often too small to be accurately measured, they nevertheless profoundly influence weight loss.

Radiation does not play an important role in heat transfer to or from the majority of boxes in a CA storage since they do not have a view of the room's walls, ceiling or floor. Golden Delicious apples stored in a 2–3% $[O_2]$ + 1–2% $[CO_2]$ CA atmosphere at 0°C and 88% RH lost 0.97% of their weight between the 20th and 60th day (Fig. 6.4). This fruit consumes 0.028 mmol/kg·h of O_2 (Gran and Beaudry, 1993), generating enough respiratory heat to evaporate 0.5% of its fresh weight in 40 days. As cool down had been completed prior to the 20th day, the data show that an additional 0.47% of the fruit's water was evaporated to balance convective heating during the 20–60-day storage period. Convection provided 2.9 cal/kg·h and respiration 3.1 cal/kg·h of the latent heat used to evaporate water (example 3).

Measurements made during a remotely instrumented CA pear storage (Tables 6.12 and 6.13) indicate that the air inside the bins had the highest average humidity, vapour pressure and density, and the coldest average temperature; air outside the stack had the lowest average humidity, vapour pressure and density, and highest average temperature; air in the runner spaces had intermediate values for these parameters. Since the air in the bins had the coldest average temperature, it was cooled by the pears, and the air density was higher inside the pallet bins than elsewhere in the storage room, producing a natural convective downdraught in the bins. In the Golden Delicious CA apple-storage example, since the (refrigerated) air is warming the fruits, the air

Table 6.13. Computed values for air density, vapour pressure and dew point during the CA storage described in Table 6.12. The density is calculated from the expression $D = 1.2929 (273.13/T)[(B–0.3738e)/760]$, where T is the absolute temperature, B the barometric pressure in mm Hg and e the vapour pressure of the moisture in the air (mm Hg). The barometric pressure is assumed to be 760 mm Hg.

CA storage	Vapour pressure (mm Hg)	Dew point (°C)	Density (kg/m^3)
Centre of bins	4.3455	–0.68	1.2896
In runner spaces	4.2829	–0.92	1.2893
Outside the stacks	4.2625	–0.98	1.2884

Table 6.12. Average and range of temperature and relative humidity during CA storage of pears (SD = standard deviation).

	Temperature, °C		Relative humidity, %	
Location	Average (SD)	Range	Average (SD)	Range
Centre of bins				
2	0.13 (0.05)	–0.04–0.44	93.1 (1.3)	87–99
4	0.10 (0.06)	0.01–0.41	94.3 (0.8)	90–97
6	0.07 (0.04)	–0.05–0.23	94.9 (1.0)	91–99
average	**0.10**		**94.2**	
In runner space				
3	0.19 (0.17)	–0.70–0.73	91.6 (1.8)	85–95
5	0.14 (0.11)	–0.18–0.44	92.6 (1.2)	90–95
9	0.22 (0.06)	0.07–0.52	92.7 (1.2)	88–97
average	**0.18**		**92.3**	
Outside the stacks				
7	0.41 (0.85)	–2.04–3.84	91.7 (2.2)	84–86
1	0.29 (0.43)	–1.80–1.85	89.6 (2.6)	80–95
10	0.40 (0.09)	–0.03–0.75	90.5 (1.7)	84–93
average	**0.37**		**90.6**	

Source: from Waelti *et al.*, 1992.

density in the bins should decrease, creating a downdraught.

6.17 Controlling Water Loss at Atmospheric Pressure

Initially it was thought that a nearly saturated humidity was detrimental for fruit and vegetable storage because it encouraged decay, but recommended RH levels have steadily increased, and now close to 100% RH is advocated with many commodities to decrease water loss, improve quality and control decay (Hardenburg, 1973; Van den Berg and Lentz, 1978). This changed attitude has come about in part due to the development of jacketed refrigeration (Van den Berg and Lentz, 1978; Van den Berg, 1981) and HumiFresh (Meredith, 1973; 12.5) improvements in humidity maintenance. In combination with modulating temperature control, these systems allow a high humidity to be maintained without the temperature cycles typical of conventional 'ON–OFF' forced-air refrigeration. Fruit has a much higher specific heat (kcal/kg·°C = BTU/lb·°F) than air, and therefore when 'ON–OFF' refrigeration control causes the air temperature to cycle by several °C, the fruit temperature hardly changes and instead remains at an intermediate value between the maximum and minimum air temperatures. If the storage humidity is very high, during the OFF refrigeration cycle, water will condense on the colder surface of the stored commodity and stimulate mould growth (7.8).

The air-conditioning systems used in conventional storage rooms typically cannot raise the relative humidity above 92% when the refrigeration compressor is operating because, in order to provide sufficient refrigerant velocity to ensure proper oil return through the evaporator coil to the compressor, and also for proper thermal expansion valve operation, the equipment must be sized so that the refrigerant emerging from the coil is at least 2.8°C colder than that entering. This temperature difference reduces the RH of the discharge air by causing water vapour present in the circulating air to condense on the evaporator coil. The temperature difference will be greater than 2.8°C and the discharge RH less than 92% if the evaporator coil has less than the maximum permissible surface area relative to the compressor's capacity. Humidity can be increased to approximately 95% in an air-conditioned refrigerated storage by continuously adding moisture into the air discharged from the evaporator coil, and by turning the evaporator fans off except when cooling is required, but when the outside ambient temperature is higher than the storage temperature, air circulating within the room is warmed and the RH lowered by heat conducted through the enclosure's insulation. During a CA pear storage (Table 6.12) in which the evaporator fans were only operated to circulate air during the cooling cycle plus 10 min each hour, the average room air temperature was 0.37°C higher than the 0°C set-point, decreasing the humidity by 3%. An LP intermodal container's jacketed refrigeration system prevents this type of humidity decrease by capturing and removing heat transmitted through the insulation before it enters the storage area.

Moisture is not removed from the box runners when the evaporator fans in a CA room are intermittently cycled off. Instead it accumulates in the storage bins elevating the RH until evaporation practically ceases. Essentially all respiratory heat is retained and progressively warms the fruit. During a 2–3% [O_2] CA apple storage, the fruit temperature should increase by approximately 0.076°C per day when respiratory heat accumulates and is not removed. In close agreement, in a study of intermittent fan cycling during CA apple storage, when the evaporator fans were turned off, the measured rate of temperature rise in apple bins was 0.074°C per day during a 9-day period (Hellickson and Koca, 1992). The transient buildup of respiratory heat when the fans are not operating does not result in eventual weight savings because the fruit remains at a constant average temperature throughout the entire storage period. The respiratory heat retained when the fans are off is lost by extra evaporative cooling during periods when the fans are operating. The main advantage is an energy cost savings gained by decreasing the fan operating time. A slight reduction in water loss might result because convective heat uptake essentially ceases while the fans are off.

The hollow-fibre membrane air-separator systems used to generate nitrogen in CA intermodal containers and storages lower the humidity because water vapour, O_2 and CO_2 have a fast permeation rate through the membrane and therefore are removed by it, while N_2 moves through more slowly and is enriched in the output stream. Air must be dried before it is passed through the carbon molecular sieve (CMS) used to split N_2 from O_2 in a pressure swing adsorption air separator (PSA), because the sieve's O_2 adsorption capacity is reduced when the material adsorbs and entrains water. Usually a refrigerated dryer is used, causing the N_2 produced by the PSA to be bone dry, with at least a –40°C dew point. Both systems withdraw moisture from the storage area.

6.18 LP Heat Sources

The jacketed refrigeration system used to cool LP containers removes heat transmitted through the insulation before it enters the container. Since the tank is vacuum-tight there is no infiltration heat except that which enters in the controlled air changes. The kilograms of air that have to be passed through a container to effect an air change under hypobaric conditions are much less than those needed at atmospheric pressure, in proportion to the pressure reduction, and therefore the amount of sensible heat admitted per air change is correspondingly lower. The incoming air decreases in RH when it expands because the same amount of water vapour now is present in a much larger volume. Even when the expanded air is cooled to 0°C, it does not reach its dew point, and therefore no latent heat needs to be eliminated. Assuming an ambient condition of 80% RH at 38°C, only 23.2 W (79 BTU/h) must be removed to change the air 3.3 times each hour when a 12.2 m Grumman/Dormavac hypobaric container is operated at an interior temperature of 10°C and a pressure of 2.67 kPa (20 mm Hg), compared to 5862 W (20,000 BTU/h = 1.6 tons of refrigeration) at atmospheric pressure for the same rate of air change! At atmospheric pressure, nearly half the available capacity of a typical intermodal-container refrigeration system would be required to remove this amount of heat. A standard refrigerated intermodal container is limited to a lower air-change rate because the remaining refrigeration capacity would be insufficient to remove respiratory heat, transmitted heat, fan heat and other heat loads. In contrast, the air-change rate in LP containers can be adjusted to any value needed to prevent O_2 drawdown or CO_2 and ethylene accumulation, without burdening the refrigeration system. In LP the RH, interior pressure and wall temperature are precisely controlled and kept constant to prevent an evaporation/condensation cycle from developing between the commodity and tank wall due to enhanced diffusion at a low pressure. Essentially no heat is produced by the pneumatic air horn used to circulate air in a VacuFresh^{SM} container. Therefore, after cool down is completed, the only heat that has to be transferred from the product to the cold plate or via air changes is the small amount produced by respiration.

When meat is stored in LP at –1°C and a pressure of 0.57 kPa (4.3 mm Hg), cold steam in the chamber is changed approximately once each hour to prevent odour buildup and to flush away any in-leaking residual O_2. This does not introduce any heat into the vacuum tank.

6.19 Humidity Control in LP

Until the advent of the ‘dry’ hypobaric method, it had been axiomatic that the LP air changes must be humidified to prevent weight loss. To accomplish this in a laboratory apparatus, usually the expanded, dry, low-pressure air is bubbled through water before it is admitted into the LP chamber (Burg, 1967; Fig. 2.4). This same procedure created difficulties in the first commercial-scale LP prototype intermodal container (13.1) because evaporative cooling caused ice to form in the humidifier due to the container's large airflow rate relative to a small volume of humidifying water. A patent entitled 'Low pressure storage of metabolically active material with open-cycle refrigeration' (Burg and Hentschel, 1974) describes the cooling of a hypobaric intermodal container by means of the evaporative effect generated in the humidifier, but this proved impractical because the method could not simultaneously bring the container humidity close enough to saturation. An electric heater was installed in the humidifier to provide the latent energy needed for humidification, and the original LP patent claims had to be modified to include a step in which the incoming air was contacted with a heated water supply (Burg, 1976b). The Grumman/Dormavac container's humidification system (Fig. 13.8, *right*) contained a 3 kW modulating heater for this purpose.

Maintenance of a nearly saturated humidity in Grumman/Dormavac and VacuFresh^{SM} intermodal containers is made possible by the jacketed refrigeration system, precise pressure and temperature regulation, elimination of infiltration and fan heat, removal of respiratory heat by evaporative cooling without a temperature change and careful design of the tank structure to eliminate end-to-end and localized temperature gradients. The entire surface of the aluminium LP vacuum tank serves as a heat exchanger cooled by refrigerated glycol. Use of a secondary coolant eliminates moisture condensation on a cold evaporator coil and the need for defrost cycles.[25] A proportional integral derivative thermostat (PID – 12.14) keeps the temperature uniform ± 0.2°C throughout the entire interior wall surface of a VacuFresh^{SM} container and eliminates ON/OFF refrigeration cycles. LP does not use devices that remove water from the storage atmosphere, such as the atmosphere generators used in CA containers, and consequently the LP system is able to maintain the highest possible humidity that can be sustained in a dynamic system by mechanical equipment, while preventing condensation/evaporation cycles from drying or wetting the stored commodity.

The VacuFresh^{SM} system resolves the apparent incongruity that in a Grumman/Dormavac container humidity is added to saturate the atmosphere and prevent water loss, and yet the commodity can only remain at a constant temperature if it loses sufficient water to transfer its respiratory heat by evaporative cooling. In VacuFresh^{SM} the air-change rate is slowed until commodity water, evaporated responsively to respiratory heat, is more than sufficient to saturate the incoming dry, rarified air changes (Table 6.14; Burg, 1987b).[26] The supplementary heat and water consumed in the Grumman/Dormavac humidification step are replaced by respiratory heat and product water in the 'dry' hypobaric method. The VacuFresh^{SM} sytstem is similar to intermittent fan cycling in a CA room, except that the flow is modulated to a steady, low, optimal rate, whereas CA uses a timed ON–OFF cycle. The 'wet' and 'dry' hypobaric processes cause the same commodity weight loss (Burg, 1992), but the vaporized water containing the heat of respiration is evacuated from the VacuFresh^{SM} tank by the vacuum pump, whereas almost all of this water condenses on the wall panels of a Grumman/Dormavac container, transferring heat into the secondary coolant flowing through tubes welded to the walls (Fig. 13.7). Consequently, in VacuFresh^{SM}, the steady-state commodity temperature is closer to the thermostat set point, the refrigeration requirement is reduced, water does not condense on the ceiling and fall back on to the boxes, and they maintain their initial strength and do not become soggy. In VacuFresh^{SM} the optimal air-change rate for each commodity always is sufficient to flush away metabolic gases and prevent respiratory O_2 drawdown from significantly lowering the O_2 partial pressure, because the set pumping speed is determined by the rate at which the commodity produces respiratory heat, and this depends on the CO_2 production and O_2 consumption rates.

To avoid desiccating that portion of the cargo in a VacuFresh^{SM} container that is first contacted by the incoming dry air,[26] it is admitted into the tank through the collar of a pneumatic air mover, pressurizing the air-horn jets to induce circulation of up to 40 volumes of chamber air previously saturated with evaporated commodity water, per each volume of expanded incoming air. The RH of the air discharged from the air mover always is within a few percentage points of the steady-state humidity in the storage area, and due to rapid diffusive mixing in the under-shelf discharge duct (Fig. 13.12, *middle right* and *lower middle*), it closely approaches the chamber-air humidity before it is released at the door end of the container. Substituting a simple flow controller in place of Grumman/Dormavac's complex and unreliable humidification system (13.6), eliminates the requirement for stacking 'chimneys' to facilitate the flow of water vapour from the boxes, and simplifies the equipment design, reducing its size, weight and cost.

A variable low-pressure system (VLPS), in which the pressure is cycled between 13.33 and 40.0 kPa (100 and 300 mm Hg) without humidifying the incoming air (Onoda *et al.*, 1989a), is a modification of the same humidification principle used in the dry hypobaric method. VLPS does not require a moisture supplement because the inlet air to the chamber is saturated with water vapour produced by evaporation from the stored commodity. When cabbage and turnips were stored using VLPS, it created less weight loss and resulted in better commodity appearance compared to the 'wet' LP system operated at a constant pressure. An oscillating pressure storage system has also been used to store apples (Gimenes and Sommer, 1988).

6.20 LP Stacking Arrangements

Depending on the storage pressure, diffusion is accelerated by one to two orders of magnitude in LP, but this cannot by itself remove evaporated water from the centre of a tightly stacked commodity load at a rate sufficient to prevent respiratory heat from warming the interior boxes. To prevent a temperature rise, air must be circulated through the load by either natural convection or forced flow. Natural convective airflow is facilitated in a Grumman/Dormavac intermodal container by a stacking arrangement that exposes every carton to an unobstructed vertical chimney. The boxes preferably contain ventilating holes in their ends, oriented toward the chimneys, so that buoyancy caused by vapour released from the commodity and boxes (equations 6.42–6.44), and by the warming of the air as it passes over the boxes (example 17), can create a natural convective updraft in the vertical channels. The main purpose of this airflow is not so much to support convective cooling, but rather to move water vapour released from the boxes into and through the chimneys, to the container walls where condensation occurs. The stacking arrangement determines: (i) the surface area of each box that is exposed to a ventilating channel; (ii) the percentage of boxes (exterior) that have a line-of-sight view of the tank wall allowing them to be effectively cooled or warmed by radiation (exterior boxes); (iii) the percentage of boxes (interior) that lack a radiant coupling with the tank wall and consequently must transfer heat primarily by evaporation and to a much lesser extent by convection (interior boxes); and (iv) the stability of the load when the boxes are exposed to in-transit shock and vibration. For stationary flower storage in a Grumman/Dormavac container, rows of boxes, 11 high, are separated by 8-cm-wide vertical chimneys (Fig. 6.13), positioning 84% of the boxes with a view of the container wall ('external boxes'). This stacking arrangement is intrinsically unstable and can only be used for transportation if vertical braces are inserted between rows to prevent in-transit shock and vibration from obstructing the ventilating chimneys by causing fore or aft displacement of the boxes. The braces must be installed in a manner that does not interfere with vertical air movement.

Theoretically, Grumman/Dormavac containers can be pallet loaded, but normally

Table 6.14. Optimal pressure, temperature and flow settings for full loads of various types of commodities in a 6.1 m VacuFresh^{SM} container.

Commodity	Pressure (mbar)	Temp (°C)	Flow Set (No.) 60 Hz (50 Hz)	Air Horn (mbar) 60 Hz (50 Hz)
Alstroemeria (cut flower)	20	4	–	–
Anthurium (cut flower)	60	13	–	–
Apple	65	0	3 (3)	255 (212)
Asparagus	25	0	5 (5)	145 (121)
Avocado				
Lula, Booth 1 and 8, Taylor	30	4.4	3 (4)	95 (110)
Fuerte, Hass and Booth 7	30	7.2	3 (4)	103 (137)
Fuchs, Pollock, Waldin	30	13	4 (5)	88 (102)
Banana*	60	14	3 (3)	194 (162)
Bean, snap (green)	20	7.2	4 (5)	58 (67)
Blueberry*	100	0	3 (3)	415 (346)
Broccoli	20	0	3 (3)	61 (51)
Carnation (cut flower)	20	0	3 (3)	61 (51)
Carnation (cutting)	60	1	2 (2)	439 (439)
Cauliflower	20	0	3 (3)	61 (51)
Cherry (sweet)	25	0	3 (3)	84 (70)
Chrysanthemum (cut flower)	20	0	3 (3)	61 (51)
Chrysanthemum (cutting)	60	1	5 (6)	411 (415)
Chrysanthemum (potted plant)	80	0	–	–
Cucumber*	110	10	5 (6)	489 (489)
Grape (Vinifera)	20	–1	3 (3)	63 (53)
Grape (American)	20	–0.5	3 (3)	63 (53)
Foliage cuttings	45	12.8	–	–
Lettuce	35	0	3 (3)	128 (107)
Lily, Easter (potted plant)	80	0	–	–
Lime	205	9	3 (3)	584 (584)
Maize	70	0	2 (2)	449 (449)
Mango	30	13	3 (4)	88 (102)
Melon (cantaloupe)	27	5	3 (3)	81 (68)
Melon (honeydew)	27	7	3 (3)	75 (63)
Mushroom	20	0	3 (3)	61 (51)
Onion (green)	20	0	3 (3)	61 (51)
Papaya	30	10	3 (3)	78 (65)
Pepper, bell (green and red)*	110	7	3 (3)	442 (369)
Pear	65	0	3 (3)	260 (217)
Pineapple*	110	8.3	3 (3)	438 (354)
Rose (cut flower)	20	0	2 (2)	228 (190)
Strawberry	20	0	3 (3)	63 (53)
Tomato (mature-green)*	110	12.8	2 (4)	421 (468)
Tomato (breaker)*	110	10	3 (3)	432 (360)

*The optimal pressure may be lower than the indicated value.
Flow control switch settings: number (% maximum flow) = 2 (100%), 1 (79%), 8 (69%), 7 (63%), 6 (57%), 5 (44%), 4 (36%), 3 (27%). Air horn (mbar) = air mover jet pressure upstream of the air horn collar at the indicated tank pressure, temperature and flow, limited to 379 mbar above the tank pressure by a pressure relief valve set at a 37.4 kPa (5.5 psig) cracking pressure.

this is impractical because the usable interior width of the container, 198 cm (6.5 ft), requires a non-standard pallet size. It is not possible to pallet load a cylindrical VacuFresh^{SM} container, but there are economic advantages to hand-loading. The cost is minimal, since only 4–8 man-hours are required per load, and this fee is more than

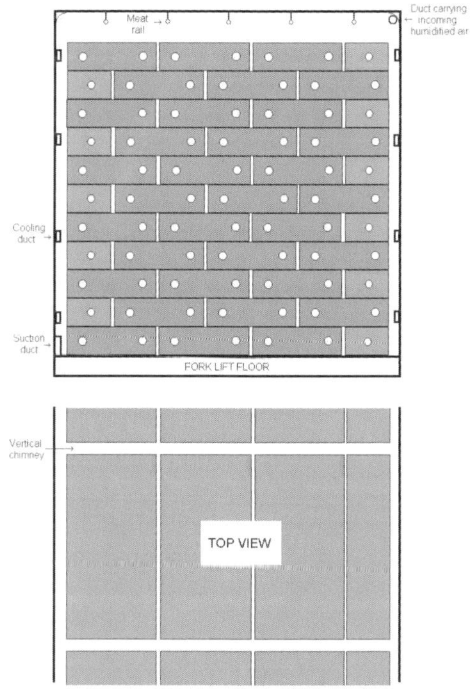

Fig. 6.13. Longitudinal cross-section (*upper*) and top view (*lower*) of a carnation and rose box stacking pattern in a Grumman/Dormavac 12.2 m container. Large boxes = 102 × 51 × 15 cm (18.2 kg); small boxes = 51 × 51 × 15 cm (9.1 kg); full load = 7007 kg. Large box ends are perforated with two, and small box ends with one, 5-cm diameter hole(s). Vertical chimneys between longitudinal rows of boxes are 8 cm wide.

offset during both the forward and return trips by substituting cargo for the space and weight required for pallets, and by eliminating the expense of maintaining or replacing pallets that are damaged or not returned. An even greater cost savings results from the avoidance of structural injury to the container interior due to forklift traffic, which is a major cause of damage to conventional intermodal containers.

International Standards Organization (ISO) regulations require intermodal containers to have a width of 2.43 m (8 ft), a height of either 2.43 (8 ft) or 2.61 m (8.5 ft) and a length of either 6.1 or 12.2 m (20 or 40 ft). It might seem obvious that the available storage space inside the cross-sectional area encompassed by a 2.43 m square Grumman/Dormavac container is greater than the space inside a cylindrical VacuFresh^{SM} tank designed to fit within the same 2.43 m square envelope, but the opposite is true. The thick wall structure required to sustain the vacuum forces in the square Grumman/Dormavac vacuum tank reduces the cross-sectional interior area so that it is slightly less than the space available above the floor and shelves in a thin-walled cylindrical VacuFresh^{SM} container (Fig. 6.14). To maximize the usable cross-sectional area in VacuFresh^{SM}, the outer diameter of the stiffening rings extends to within 1.25 cm of the insulation's external surface at the horizontal centreline, and the insulation is applied in a pattern that creates nearly an external 2.43 × 2.61 m rectangle. Grumman/Dormavac containers also have less available length than VacuFresh^{SM} because their equipment section requires a 94-cm depth to house a special built-in diesel-motor generator capable of satisfying a high-power requirement, and additional usable length was sacrificed to house manifolding for a parallel cooling system. VacuFresh^{SM} has the same power requirement as a conventional refrigerated container, which is satisfied by a 12.5–15 kW clip-on or under-chassis standard auxiliary motor-generator set. The VacuFresh^{SM} 'series' cooling system requires no manifolding, allowing the equipment to be packaged in a depth of 40 cm. A nearly square 'Transvac' 'barrel' design for the next generation of 12.2 m (40 ft) containers will simplify the stacking arrangement and provide a 20% increase in available space with almost no increase in cost and weight (13.24).

VacuFresh^{SM} stagger-stack box-loading arrangements, illustrated for carnation and papaya boxes in Fig. 6.15, are modelled after intrinsically stable patterns that have been used for more than five decades transporting perishable foods in over-the-road trailers (Ashby, 1970). Continuous evacuation causes air to flow vertically downward toward a longitudinal suction duct located at the floor mid-line (Fig. 13.12, *lower middle*), while at the same time the air horn (13.09, Air mover; Fig. 13.12, *lower middle*;

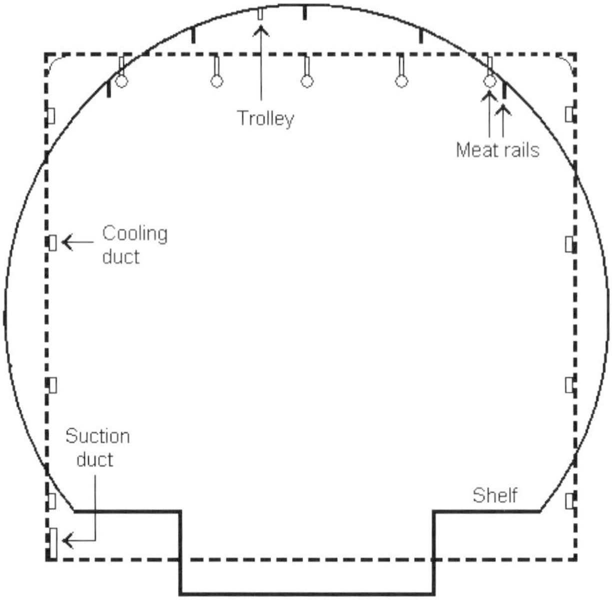

Fig. 6.14. Comparison of the available above-floor usable cross-sectional stacking area in VacuFresh^{SM} cylindrical and Grumman/Dormavac square intermodal hypobaric containers. VacuFresh^{SM} tank shown as solid line (—); Grumman/Dormavac tank shown as dotted line (----). The cylindrical tank has slightly more usable space.

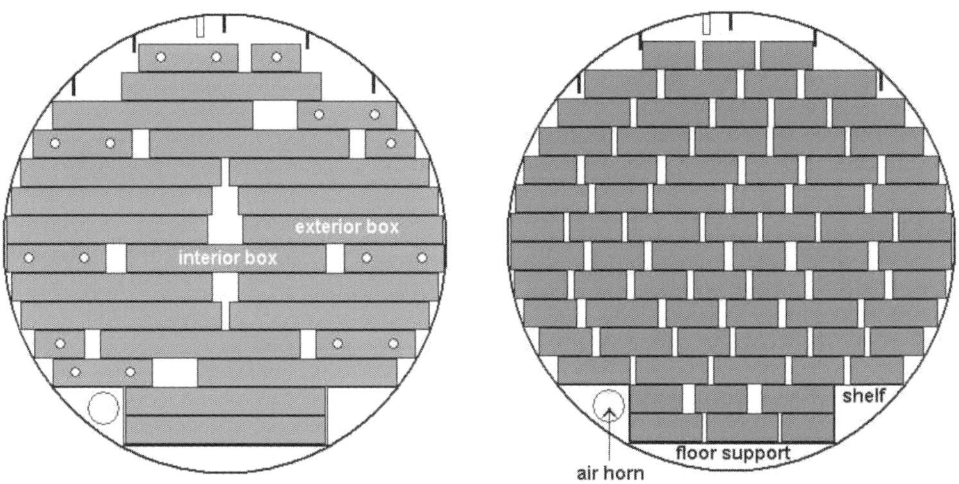

Fig. 6.15. (*left*) Stacking pattern for a 3273 kg (7200 lb) carnation load in a 6.1 m VacuFresh^{SM} container. Large boxes (18.2 kg) = 101 × 50 × 16 cm; small boxes (9.1 kg) = 101 × 25 × 16 cm. Interior boxes = 13.3%; exterior boxes = 86.7%. (*right*) Stacking pattern for a 4954 kg papaya load in a 6.1 m VacuFresh^{SM} container. Box size (4.56 kg) = 36 × 25 × 16 cm. Interior boxes = 47.9%; exterior boxes = 52.1%.

Figs 13.11 and 13.17) induces up to a 40-fold greater end-to-end forced-air circulation through the longitudinal ventilating ducts formed by the stacking arrangement. Openings at the ends of each 'inner' carnation storage carton are oriented toward a longitudinal ventilating channel, and the boxes are distributed so that the maximum possible percentage are exterior with a radiant coupling to the container's wall. The openings in exterior boxes do not have to be directed toward a longitudinal ventilating channel because these boxes have a strong radiant coupling with the container wall, which suffices to remove respiratory heat and prevent their temperature from rising.

6.21 Forced Convection in LP

Due to the low specific heat of rarified air, a full load of flowers stored in a Grumman/ Dormavac LP container at 1.33 kPa (10 mm Hg) would overheat if the flowers' respiratory heat had to be removed by forced convective flow through the vertical chimneys separating box rows. In order for carnation flowers to remain at a constant temperature after they have cooled, 262 cal/ min (62.5 BTU/h) of respiratory heat must be transferred into the air moving through each chimney, but the Grumman/Dormavac vacuum pump only induces 190 g (0.42 lb) per hour of the 1.33 kPa (10 mm Hg) air/water-vapour mixture to flow in each vertical duct. Saturated 1.33 kPa air increases in temperature by 1°C/g when it absorbs 33 cal of heat (0.33 BTU/lb·°F). With forced convection as the only heat-transfer mode and the vacuum pump as the only inducer of airflow, the air would warm by 250°C (450°F) in transferring the respiratory heat load even though the respiration of carnations is inhibited by 90% at 1.33 kPa (Fig. 4.2; example 11).

If the only air movement in a Grumman/ Dormavac container was flow induced by vacuum-pump suction, the dew point would progressively increase in a downward direction as air passed through the boxes toward the floor-level suction duct. One might expect that the upper layers of the stack would be cooler and experience noticeably more weight loss than the lower boxes, but observations made during a 21-day papaya shipment and during flower storages indicate that this does not occur. Since the net flow must be vertically downward, it would appear that natural convection causes a significant amount of recirculation within the stack, with downward flow in the narrow gap regions near the box surfaces, and upward flow in the centre of the chimneys (Burg and Kosson, 1982, 1983; Table 6.15). When both natural and forced convection can occur simultaneously, a decision as to which predominates can be made by comparing the Reynolds number (Re) for forced flow to the Grashof number (Gr) describing the buoyancy needed for natural convection (example 16). Forced convection dominates if the Re^2/Gr ratio is ≥ 10, natural convection prevails if it is ≤ 1, and in-between both natural and forced convection must be considered (Noble, 1991). In a Grumman container, and when hanging meat is stored in a VacuFresh^{SM} container, natural convection predominates. Forced convection between the carton and container air must be considered when plant commodities are stored in a VacuFresh^{SM} container, but free convection predominates inside the cartons.

The heat transfer coefficient for forced convection in a VacuFresh^{SM} container is difficult to compute because of the complex geometry. The net flow induced by the vacuum pump varies from 0.8 to 3.6 air changes per hour, depending upon the selected air-change setting and electrical phase (50 or 60 Hz). The flow is directed uniformly downward through tightly stacked boxes

Table 6.15. Analytic results with papaya in a Grumman/Dormavac container.

	$T_{DP,C}$ = 11.11 °C		$T_{DP,C}$ = 8.89 °C	
Computed values	Interior box	Exterior box	Interior box	Exterior box
Papaya temperature, T_F	11.18	11.17	9.06	9.44
Box air temperature, T_A	11.18	11.16	9.11	9.85
Box temperature, T_B	11.18	11.13	9.28	10.59
Box dew point, $T_{DP,A}$	11.13	11.12	8.92	8.89
21-day weight loss, %	0.63	0.51	1.19	4.00

Conditions: 2.67 kPa (20 mm Hg) pressure; container air temperature (T_C) = container wall temperature (T_W) = 13°C. Container dew point temperature ($T_{DP,C}$) = 11.11°C (100% RH) or 8.89°C (86% RH); assumed papaya transpirational skin resistance = 7 s/cm at 2.67 kPa (20 mm Hg); r_{box} = 660 s/cm for water vapour at atmospheric pressure (Burg and Kosson, 1982).

lacking vertical chimneys (Fig. 6.15), toward equally spaced holes drilled beneath the full length of the longitudinal suction duct located at the floor centre-line (Fig. 13.12, *lower middle*). Since the downward flow is laminar, the coefficient for forced convective heat transfer will not be very different from that computed for a Grumman/Dormavac container (example 14). Due to the air's low specific heat at the pressures used to store most commodities, the temperature difference will be levelled by convective heat transfer between the air and the first box that it encounters.

A pneumatic air mover (13.09, Air mover; Fig. 13.11; Fig. 13.12, *lower middle*) improves forced convection in a VacuFresh^{SM} container by increasing the volume of pre-cooled heat-carrying air passing over each kilogram of cargo.[27] One rarified volume of incoming make-up air induces the flow of up to 40 additional volumes of saturated low-pressure air through the air horn.[28] The discharge is circulated longitudinally through the cargo and continuously recooled in the under-shelf discharge duct and by the container walls and fore and aft dish-heads. With the vacuum pump operating at full capacity, the air-change rate in a VacuFresh^{SM} container fully loaded with carnations is only 12% higher per weight of flowers than it is in a Grumman/Dormavac container, but the quantity of air recycled through the load by the air horn is 45-fold greater than the forced flow in a Grumman/Dormavac container. At 0°C and a pressure of 2 kPa (15 mm Hg), interior carnation boxes (Fig. 6.15, *left*) adjacent to each longitudinal duct, and the two exterior boxes at the forward end, produce a total of 34 cal/min (8.1 BTU/h) of respiratory heat, of which theoretically 18.9 cal/min (4.5 BTU/h) could be removed by forced convection if the temperature difference was 1.2°C (2.2°F). Convective cooling will, however, be limited by the heat-carrying capacity of the circulating air if the temperature difference between the carton and air is less than 1.2°C (example 15). At higher storage pressures and pumping speeds, heat transfer by forced convection between the box and air increases because the convective coefficient and heat-carrying capacity of the air are larger than the values computed for carnations at 1.33 kPa (10 mm Hg). Convective heat transfer between the commodity and air within each box, and between the air in the box and the box wall, occurs by free convection in both Grumman/Dormavac and VacuFresh^{SM} LP containers.

6.22 Effect of LP on Free Convective Heat Transfer

An understanding of the influence that atmospheric pressure and dew point temperature have on heat transfer and water loss is a prerequisite for successful hypobaric storage. The design of the LP structure and associated equipment, and the recommended loading and operational procedures are based in part on thermodynamic computations, which reveal that certain requirements must be met in order to reduce weight loss to the theoretical minimum determined by respiratory heat production at the lowest pressure that does not cause low-O_2 injury. Although laboratory tests had proved the benefits of hypobaric storage, it remained to be demonstrated that the same performance would result in a commercial application many times greater in size. Could an entire commodity load be cooled in a reasonable period? Can the commodity dispel its respiratory heat and remain cold without drying excessively? Is commodity water loss so accelerated at a low pressure that it limits LP storage life? There was a need for a better understanding of internal modes of heat transfer because laboratory tests indicated excellent heat-transfer cooling capability under hypobaric conditions, whereas a theoretical analysis indicated a very low convective heat-transfer film coefficient. A series of tests conducted in a hypobaric intermodal container and in small bell jars resulted in the following conclusions (Alloca, 1980a):

- Radiation, depending upon the stacking arrangement, can be significant. The heat-source/heat-sink surface-area ratio is nearly unity in a laboratory test, resulting in rapid cooling by radiant heat transfer from the commodity to the laboratory vessel's colder walls. Exterior boxes in a fully loaded container are cooled in part by radiation, but interior boxes are not, since they lack a view of the container wall.

- Evaporative cooling due to commodity moisture loss is the predominant heat-transfer mode in an LP intermodal container.
- Convective cooling is ineffective at a low pressure because the air's low density reduces its specific heat and the convective heat-transfer coefficient.

In CA or a conventional air-conditioned storage, a significant portion of the heat transfer may occur by forced convection, but LP would require an unfeasible air velocity to function by that same heat-transfer mode. At velocities (v) typical of an air-conditioned storage, the Reynolds number is in the range $10^3 < \mathrm{Re} < 3 \times 10^5$, where according to data correlations for flow over a flat plate, the average film coefficient (h_m, W/m$^2\cdot$°C) for forced convection is (Knudsen *et al.*, 1984):

$$\mathrm{Nu}_m = 0.678\,(\mathrm{Re}_L)^{1/2}\,(\mathrm{Pr})^{1/3} = h_m L/k \quad (6.41)$$

Since the Reynolds number is pressure-dependent and the Prandtl number pressure-independent, h_m is proportional to the square root of the pressure. The average forced-convection heat-transfer coefficient at a storage pressure of 1.33 kPa (10 mm Hg) is reduced to approximately 11% of the value at 1 atm, and to return h_m to its original atmospheric pressure capability, the velocity (v) would need to be increased 76-fold. The evaporator fans in a conventional air-conditioned intermodal container typically provide approximately 57 m^3 per minute (2000 cfm) of air flow, and therefore in LP the unfeasible rate of 4332 m^3 per minute (152,000 cfm) would be required to match the performance at a pressure of 1 atm! Unlike the coefficient for forced convection, which is independent of the temperature gradient (equation 6.41), the free-convective film coefficient depends on the temperature difference between air and the surface being cooled as well as pressure (6.13). This makes free convection an inefficient heat-transfer mode when the temperature difference is small and the pressure low. For these reasons, LP relies on evaporative cooling and a reduction in respiratory heat output to prevent heat build-up in boxes containing horticultural commodities.

The extent to which natural convection is curtailed by the restricted heat-carrying capacity of the low-pressure air/water-vapour mixture can be discerned by determining the Nusselt number for forced convective flow that is the same as the Nusselt number for free convection. From this relationship the Reynolds number, total free convective flow and heat capacity of the convective flow can be computed. At 2.67 kPa (20 mm Hg), free convection is limited by the heat-transfer coefficient rather than by the heat-carrying capacity of the mixture (example 18). The situation improves at higher pressures because the natural convective film coefficient is approximately proportional to $p^{1/2}$ (equations 6.21–6.26), whereas the heat-carrying capacity of the mixture is proportional to the air density (ρ), which varies as a direct function of the pressure (p). Natural convection is likely to be limited to some extent by the restricted heat-carrying capacity of the mixture at a pressure of 1.33 kPa (10 mm Hg), and the effect is dramatic when meat is stored at 0.57 kPa (4.3 mm Hg; chapter 11, example 2).

The buoyancy force (β) for free convection arises from the change in density of air as it passes over a surface. The density change is represented by the ($\rho\beta\Delta T$) product in the numerator of the Grashof number (Table 6.9), and for a perfect gas, in the absence of evaporative cooling, $\beta = 1/T$. Because in biological systems the commodity transfers heat by evaporative cooling, there is an additional buoyancy term due to the lower molecular weight of the evaporated water ($M_v = 18$) compared with that of air ($M_A = 28.9$). At atmospheric pressure, this effect is small enough to be disregarded, but it can be significant under hypobaric conditions because the evaporating water represents a much larger fraction of the gas-vapour mixture present inside the vacuum chamber. The expression for the buoyancy term β can be written (Burg and Kosson, 1983):

$$\beta = (1/T)\,(1 + \alpha_1) \qquad (6.42)$$

$$\alpha_1 = (M_A - M_V)\,(c_P\rho K_{ge}/h_m)\,(dp_V/dT)\,[(T_F - T_{DP,A})/(T_F - T_A)] \qquad (6.43)$$

where K_{ge} is the mass transfer coefficient (m/s), ρ the air density (kg/m) at atmospheric pressure and temperature T_A, dp_v/dT is the change (atm/K) in the commodity's water vapour pressure as a function of temperature, T_F is the commodity (F = 'fruit') temperature (K), T_A is the dry-bulb temperature of the air (A) surrounding the commodity and $T_{DP,A}$ is the dew point temperature of the air/water-vapour mixture around the commodity. In this expression, the $[(T_F - T_{DP,A})/(T_F - T_A)]$ ratio accounts for the fact that the β term is multiplied by the dry-bulb temperature difference in the expression for the Grashof number (Table 6.9), while evaporation depends on the dew point difference between the air and the fruit. When the commodity is being heated by the air ($T_F < T_A$) and at the same time losing heat by evaporating water ($T_F > T_{DP,A}$), the sensible and evaporation buoyancy effects are in opposite directions, and to prevent computational difficulties, it is necessary to set $\beta = /\beta/$. The buoyancy augmentation term for natural convection from external surfaces of a box (B) to the container air/water vapour mixture (a) is given by:

$$\alpha_2 = (M_A - M_v)\,(c_P\,\rho/r_{box}\,h_m)\,(dp_v/dT)\,[(T_{DP,A} - T_{DP,a})/(T_B - T_a)] \qquad (6.44)$$

where r_{box} (equation 3.22; example 1) represents the resistance to water-vapour flow of the box, $T_{DP,a}$ is the dew point temperature of air/water-vapour mixture in the storage area, and T_B and T_a are the dry-bulb temperatures of the air in the box (B) and storage area (a), respectively.

Evaporative cooling is enhanced and becomes the predominant heat transfer mode under hypobaric conditions because lowering the pressure increases the transpirational conductance of boxes, wraps and the air-filled pores in the commodity's skin, while simultaneously it lowers the convective film coefficient, air density and specific heat of the air. The importance of evaporative cooling in an LP system is illustrated by examples 11 and 12. The low air density created by a hypobaric condition causes each calorie of heat transferred into the rarified air by convection to elevate the air temperature by an amount proportional to the extent of the pressure reduction. For convection to remove the respiratory heat produced at a low pressure by a full load of carnations, the temperature of the flowing air would need to increase by 250.5°C in a 12.2 m (40 ft) Grumman/Dormavac container and by 155.5°C in a 6.1 m (20 ft) VacuFresh^{SM} container. When the same amount of heat is transferred by evaporation, the temperature of the surrounding air is not elevated.

6.23 Thermal Balance under Hypobaric Conditions

If the commodity is to remain at a constant temperature after cool down has been completed, the respiratory heat added to or diminished by any heat taken in from or lost to the environment must equal the net heat loss from the commodity. This equivalence provides a basis for computing the commodity's steady-state temperature and weight loss during LP storage, and the rates of heat transfer by convection, radiation and evaporation. The result of a calculation for a 'naked' Choquette avocado stored in a laboratory apparatus (Fig. 2.4) under the conditions described in Table 6.5, is presented in Fig. 6.16 for assumed avocado transpirational resistance values of 50 and 150 s/cm at atmospheric pressure.[29] The computed commodity storage-temperature values are in agreement with measurements made in a laboratory LP set-up with Keitt mangoes shielded from radiation with Mylar film (9.9) and kept at 98% RH (dew point = 12.8°C),[30] flowing one saturated 13°C air change per hour at a pressure of 1.33 kPa (20 mm Hg). Within the accuracy

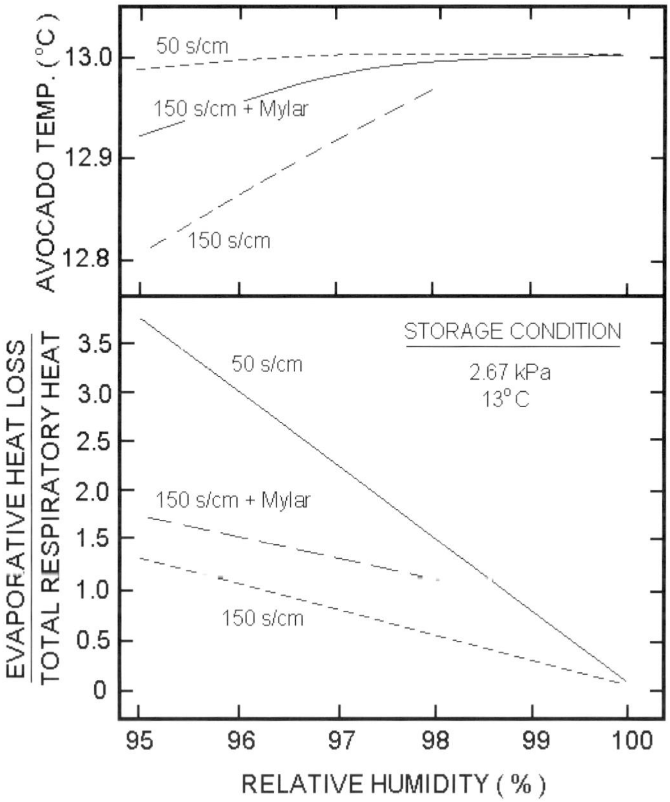

Fig. 6.16. Analytic study of Choquette avocado weight loss and equilibrium temperature as a function of relative humidity and Mylar radiation shielding. Conditions: fruit weight = 470 g; fruit surface area = 340 cm^2; measured lenticular resistance to CO_2 transport = 4500 s/cm @ 1 atm (Burg and Burg, 1962a); calculated lenticular transpiration resistance = 2741 s/cm @ 1 atm; assumed transpirational resistance = 50 or 150 s/cm @ 1 atm; calculated transpirational resistance = 19.7 or 26.8 s/cm @ 2.67 kPa (Table 6.5); measured pre-climacteric respiratory rate = 60 μl/g·h @ 24°C and 1 atm (Burg and Burg, 1962a); computed respiratory heat = 0.013 W/kg (771 BTU/ton·day) @ 13°C and a storage pressure of 2.67 kPa (20 mm Hg). The calculated effect of shielding the fruit from radiation with a single layer of Mylar assumes that the emissivity of the Mylar is 0.1. The ratio of evaporative loss/total respiratory heat compares the total heat transferred by evaporative cooling to the total respiratory heat produced. A ratio of unity indicates that only respiratory heat was transferred by evaporative cooling.

of the thermistors that were used (± 0.07°C), no difference between the mango surface temperature and air temperature could be detected.

The avocado analytic example described in Table 6.5 considers weight loss and commodity temperature when an isolated fruit is suspended in a laboratory chamber. The situation is strikingly different in a fully loaded LP intermodal container, where the box resistance, stacking configuration and internal geometry are of paramount importance in determining weight loss and the commodity temperature. The measured transpiration resistance (r_{box}) of standard boxes used to ship 10 kg of fruit (papaya) is 660 s/cm at atmospheric pressure, and while this is only equivalent to 10.9 s/cm at a storage pressure of 2.67 kPa (20 mm Hg) (Burg and Kosson, 1983; equation 6.56; example 7), in a resistance network a correction must be applied that references the box resistance to the surface area of the contained fruits (equation 3.22). This elevates the 'effective' box resistance to r_b = 85 s/cm, acting in series with an avocado transpirational resistance likely to be in the range 19.7–26.8 s/cm (Table 6.5). Thus, the box affords 76–81% of the resistance to evaporative cooling when avocados packed in cartons are loaded into a hypobaric container.

Figure 6.17 illustrates a schematic model that has been used to analyse the thermal balance in a Grumman/Dormavac intermodal hypobaric container. The stacking arrangement determines the portion of each box's surface that is exposed to radiant transfer from the intermodal container's walls, and since adjacent boxes are essentially isothermal, convective heat transfer is limited to that portion of each box that abuts on a 'ventilating' chimney, or is exposed to air at the container-wall surface. Heat transfer occurs from a fruit (F) contained in a box (B) to the container air (a) and container wall (W). The model assumes uniform temperatures for the commodity (T_F), air in the box (T_A), dew point in the box ($T_{DP,A}$), box wall (T_B), container air (T_a), container dew point ($T_{DP,a}$) and container wall (T_W). Other symbols used are for respiratory heat production ($\dot{Q}_F$), convective heat transfer from the commodity to the box air ($\dot{Q}_{C,FA}$), convective heat transfer from the box air to the box wall ($\dot{Q}_{C,AB}$), convective heat transfer from the box wall to the container air ($\dot{Q}_{C,Ba}$), radiation from the commodity to the box wall ($\dot{Q}_{R,FB}$), radiation from the box wall to the container wall ($\dot{Q}_{R,BW}$), evaporative heat flux from the commodity to the box air ($\dot{Q}_{V,FA}$) and evaporative flux from the box air to the container air ($\dot{Q}_{V,Aa}$). All convective couplings are assumed to be natural convection,[31] and the radiation coupling from the box to the wall is set = 0 for interior boxes. The corresponding nodal model (Fig. 6.17, *right*) involves the determination of four nodal temperatures (T_F, T_A, $T_{DP,A}$, T_B), given a specified respiratory heat source ($\dot{Q}_F$) applied to node F, with three fixed boundary nodal temperatures (T_a, $T_{DP,a}$, T_V). Example 1 describes the seven temperature-dependent heat-transfer couplings, including three natural convective ($\dot{Q}_{C,FA}$, $\dot{Q}_{C,AB}$, $\dot{Q}_{C,Ba}$), two radiative ($\dot{Q}_{R,FB}$, $\dot{Q}_{R,BW}$) and two evaporative ($\dot{Q}_{V,FA}$, $\dot{Q}_{V,Aa}$). Analytic results obtained with the nodal model are presented in Fig. 6.18 and Table 6.15 for pre-cooled papaya fruit (Burg and Kosson, 1983).

The result in Table 6.15 for an r_{box} value of 10.9 s/cm at 2.67 kPa (20 mm Hg), and an assumed papaya transpirational resistance of 6.9 s/cm, at that pressure, indicates that at 100% RH ($T_{DP,a}$ = 11.11°C) the papaya temperature is very close to the container-air temperature (T_a), and the weight loss in three weeks is only 0.51% for exterior boxes, and 0.63% for interior boxes. At 86% RH ($T_{DP,a}$ = 8.89°C) the exterior box weight loss increases to 4.0%, while interior boxes lose 1.19%. Figure 6.18 shows the heat loss by mode for exterior (*left*) and interior (*middle*) boxes, as a function of dew point temperature. Exterior boxes receive more heat from their surroundings due to radiation from the wall panel, and therefore lose more weight than interior boxes. At 100% RH ($T_{DP,a}$ = 11.11°C), the papaya are marginally warmer than their surroundings, with

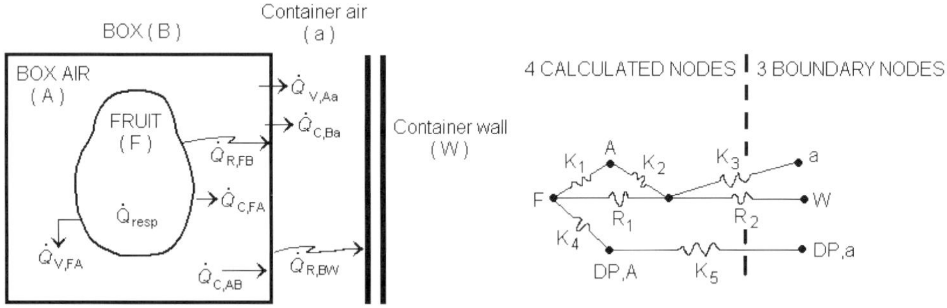

Fig. 6.17. (*left*) Schematic model for heat transfer. (*right*) Nodal model for LP heat transfer. K_1, K_2, K_3, K_4, K_5, R_1 and R_2 are the rate constants for convection (K) and radiation (R), respectively (Burg and Kosson, 1982, 1983).

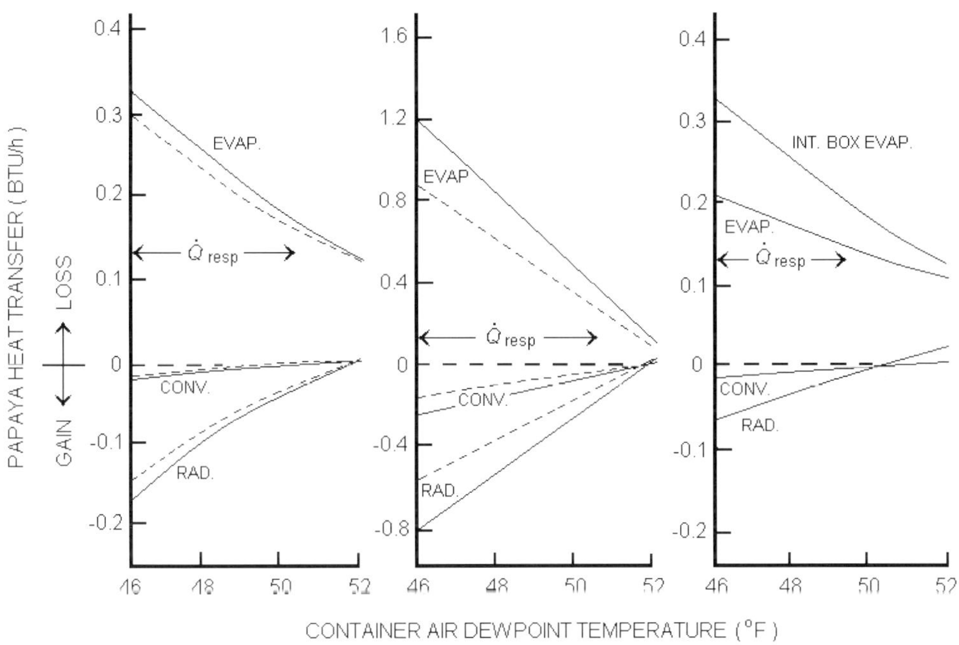

Fig. 6.18. Papaya heat transfer by mode in a Grumman/Dormavac container. Each box contains 4.55 kg (10 lb) of fruit; r_{box} = 10.9 s/cm at 20 mm Hg. (*left*) Interior box: r_{skin} at 2.67 kPa (20 mm Hg) = 6.9 s/cm (—) or 21 s/cm (----); $T_a = T_W$ = 11.11°C. (*middle*) Exterior box: same conditions as *left*. (*right*) Exterior box (—); interior box (----); $T_{DP,F,a} = T_W$; T_a = 11.11°C (Burg and Kosson, 1983).

all three modes of heat transfer in the same direction, and evaporation as the dominant mode of heat loss. At lower dew point temperatures, radiation and convection provide heat gains and the evaporative weight loss increases substantially. When the RH is close to saturation, because exterior boxes radiate heat to the container walls, they are kept colder than interior boxes (Fig. 6.18, *right*), and the colder boxes provide less heat to the papayas, reducing their weight loss. This theoretical analysis of papaya storage temperatures and weight loss has been confirmed during Grumman/Dormavac full-load commercial shipments of fruits ranging from green to half-coloured. However, the excessive respiratory heat load of papaya initially displaying more than a 50% colour break caused these fruits to overheat during shipment (example 19). In part, this may have resulted because the cartons used to ship the papaya did not have vent holes and were sealed with tape to satisfy quarantine regulations.

The weight loss problem is more severe with leafy commodities such as cut flowers than it is with fruits because the surface/volume ratio of flowers is approximately 40 cm^2/cm^3, whereas the ratio for a papaya is only 0.67 cm^2/cm^3. To prevent excessive weight loss from flowers, especially from exterior cartons, it is essential that the boxes not be allowed to decrease below the container air and wall temperature. This is accomplished by adding water-retentive plastic wraps or liners as an additional resistance in series with the skin and box resistances. The effect is similar to that shown in Fig. 6.18 (*left* and *middle*) for the two assumed values of cuticular transpirational resistance (400 and 1200 s/cm @ 1 atm = 6.9 and 21 s/cm @ 2.67 kPa). By selecting an appropriate total resistance of the box + plastic wrap, it is always possible to prevent weight loss in excess of that required to dispel respiratory heat, without causing the commodity to overheat.

The calculations summarized in Fig. 6.18 and Table 6.15 do not consider the influence of evaporated commodity water on the dew point of the container air ($T_{DP,a}$). In 21 days the respiratory heat from a full load of papayas would evaporate 226.5 kg of water into 86% RH incoming air changes, but during that time only 106.9 kg of water would be required to raise the air's humidity from 86 to 100%, so in fact the humidity in the container would be close to saturation and the rate of weight loss less than the computed values. In addition, the calculations were based on the erroneous assumption that the transpirational resistance of papayas at atmospheric pressure is likely to be between 400 and 1200 s/cm, and that it varies with pressure according to equation 6.8. Instead, the cuticular transpirational resistance is independent of pressure (6.5), and at atmospheric pressure is much smaller than the assumed values. These errors are offsetting, and the computed skin-resistance values of 6.9 and 21 s/cm at 2.67 kPa (20 mm Hg) that were used in the analysis were not very different than the 19.7–26.9 s/cm values computed for a Choquette avocado at the same pressure and temperature (Table 6.5), based on correct assumptions. The avocado surface possesses lenticles, whereas a papaya's surface has functional stomates, which presumably open at a low storage pressure, further reducing the transpirational resistance of the cuticle and stomates. This should not have a major influence on the result, however, because the overall resistance to water mass transport in a fully loaded hypobaric intermodal container is mainly determined by box properties rather than by the papaya's transpirational resistance.

6.24 Cool down under Hypobaric Conditions

Vacuum cooling in an LP intermodal container is not recommended even though at atmospheric pressure commercial vacuum coolers are used to remove field heat from many of the commodities that can be stored in LP.[32] If a container is loaded with a warm commodity whose vapour pressure exceeds the optimal LP storage pressure, during pump down product water will 'flash' and begin to boil when the tank pressure decreases to the commodity's vapour pressure. The resultant 'agitated' evaporative cooling progressively lowers the commodity's vapour pressure so that the tank can be pumped down further, causing additional 'vacuum cooling'. Until the commodity's temperature has been reduced sufficiently to lower its vapour pressure below the set storage pressure, the LP container cannot be evacuated to the optimal storage pressure. Only water vapour is present in the intermodal container's atmosphere after the commodity's water 'flashes' and boiling begins. Because O_2 is lacking until the commodity's vapour pressure decreases below the set pressure and the vacuum breaker opens, a prolonged exposure to a slow rate of vacuum cooling can eventually cause 'anaerobic' physiological damage to certain commodities. Fruits such as plums (Anon., 1920) and avocados (Pesis *et al.*, 1993) can withstand anaerobiosis for a considerable period of time without being killed or developing an alcoholic unpleasant flavour. Tomatoes and bananas do not develop poor flavour when they are stored in 100% [N_2] for 4 days at 15.5°C. An anaerobic condition had little or no effect on the flavour of lettuce stored at 1.1°C, but off-flavours were detected in peaches after 4 days in pure N_2 (Parsons *et al.*, 1964). The duration of a low [O_2] exposure that plant commodities can tolerate is far less when they are hot than after they have been cooled. Fairtime peaches, which were not injured by 0.21% [O_2] for up to 6 days at 21°C, could withstand 19 days at 0°C and the same low O_2 partial pressure. At atmospheric pressure, carnations were not damaged by 0.4–0.5% [O_2] for 6–9 weeks at 0.6°C, or during 3 weeks at 7.2–10°C, but they were injured in a matter of days at 15°C (Hanan, 1966). At 11.1–15.6°C, carnations can withstand 2–3 h of anaerobiosis in a hypobaric environment at their boiling point, but they are damaged in 5 h (Dressler, 1979a; Burg, 1980, unpublished), and depending on their initial temperature, it may take longer than 5 h to bring a full carnation load below 11.1°C in an LP intermodal container (example 26). Injury to *Chrysanthemum* cuttings did not occur when boxes containing a normal commodity density were cooled to 0°C at a hypobaric pressure of 8 kPa (60 mm Hg), but cuttings were damaged during cool down if the boxes were filled with twice the normal weight of cuttings. Meat is not damaged by anaerobic exposure, and can be vacuum cooled in an LP container regardless of its starting temperature, although preferably it should be pre-cooled to 0°C in the slaughterhouse. Commercial vacuum coolers do not harm commodities during a typical cooling cycle because they are equipped with sufficient refrigeration and/or vacuum capacity to reduce the temperature of full loads to the desired end-point in less than a few hours, often in less than 15 min, whereas a 6.1 m VacuFresh^{SM} intermodal container's refrigeration and vacuum systems can only decrease the temperature of a full load by at most 1°C per hour. Consequently, if certain plant commodities are loaded hot enough, and if the storage pressure is set low enough, anaerobic damage may occur during cool down.

A small weight of commodity can be safely vacuum cooled in an LP intermodal container provided that the cargo is capable of losing water at a sufficient rate. The container pump down time (t, min) is given by:

$$t = (V/S)\ \ln(101.3/p) \qquad (6.45)$$

where V is the container volume (m^3), S the effective pumping speed (m^3/min) and p the pressure (kPa). Table 6.16 indicates the pump down schedule for a 6.1 m VacuFresh^{SM} container ($V = 21.1$ m^3) and the theoretical cool down to be expected when the weight of a sample to be vacuum cooled is small enough to ensure that the cooling rate is not limited by refrigeration capacity.

Preferably, plant commodities should be pre-cooled before they are loaded into a hypobaric container, but if this is not practical, field heat can be removed in the container without 'vacuum cooling' provided that certain precautions are observed (example 20). A commodity can be slowly and safely cooled in LP after the tank is evacuated to the set pressure if the commodity's vapour pressure is significantly lower than the optimal storage pressure at the time of loading (Table 6.17). This condition invariably is satisfied if the commodity's optimal storage pressure is higher than approximately 6.67 kPa (50 mm Hg). To remove field heat from a commodity whose vapour pressure initially is higher than the optimal storage pressure, the following procedures can be used:

- Initially set the tank pressure at least 25 mbar (19 mm Hg) higher than the vapour pressure of the water in the commodity at the starting commodity temperature.
- As the commodity temperature and vapour pressure decrease, the tank pressure can be reduced at convenient intervals by a corresponding amount.
- When the commodity temperature decreases to within 5°C of the desired storage temperature, set the pressure at the optimal value for storage.

Temperature measurements made while cooling a full load of carnations in a 6.1 m VacuFresh^{SM} container (Fig. 6.19, *left*) verify the accuracy and utility of the

Table 6.16. Pump down from atmospheric pressure (1013 mbar) and the vacuum cooling profile expected in a 6.1 m VacuFresh^{SM} container operated at 60 Hz. Initial commodity temperature = 25°C. Pump down and cool down will occur at 5/6 rate @ 50 Hz.

Time (min)	Pressure (mbar)	Commodity temp. (°C)
0	1013	25
8.6	600	25
26.5	200	25
37.8	100	25
52.8	40	25
57.4	30	24
64.1	20	22
68.7	15	13
75.4	10	7
83.8	6	0

Table 6.17. Vapour pressure of water.

Temperature °C (°F)	Vapour pressure (mbar)
0 (32.0)	6.1
1 (33.8)	6.6
2 (35.6)	7.1
3 (37.4)	7.6
4 (39.2)	8.1
5 (41.0)	8.7
6 (42.8)	9.3
7 (44.6)	10.0
8 (46.4)	10.7
9 (48.2)	11.5
10 (50.0)	12.3
11 (51.8)	13.1
12 (53.6)	14.0
13 (55.4)	14.0
14 (57.2)	16.0
15 (59.0)	17.0
16 (60.8)	18.2
18 (64.4)	20.6
20 (68.0)	28.4
25 (77.0)	31.7
30 (86.0)	42.4
35 (95.0)	56.2

thermodynamic assumptions and heat-transfer analyses. A theoretical cool down computed for these flowers, based on the average air temperature and wall temperature during each 10-h interval of the actual cool down, is in close agreement with the experimentally determined result (example 21; Fig. 6.19, *right*). Initially, the rate of cooling was limited by heat transfer rather than refrigeration capacity, and the vacuum pump was operated at 57% of full capability, exhausting 85% of the water evaporated from the carnations. The rest of the evaporated water condensed on the container wall panels, and possibly promoted radiant heat transfer to the external boxes by increasing the wall's emissivity (Table 6.11). The theoretical calculation indicates that between the 18th and 24th hour the cool down should slow dramatically, in agreement with measurements recorded during that interval. At that time the pumping speed was increased to 100% in order to hasten evaporative cooling by causing nearly twice as many dry-air changes to be

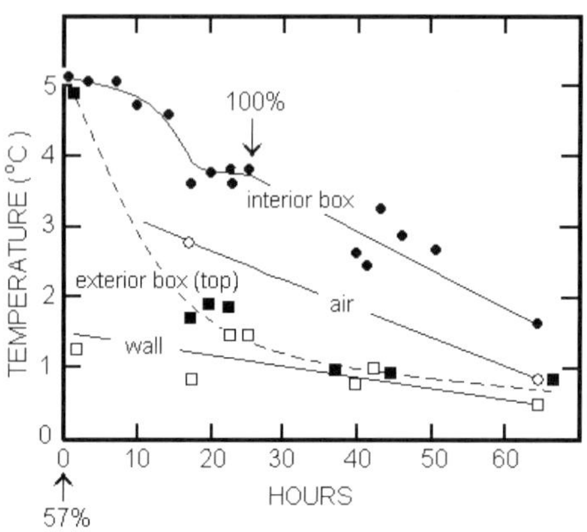

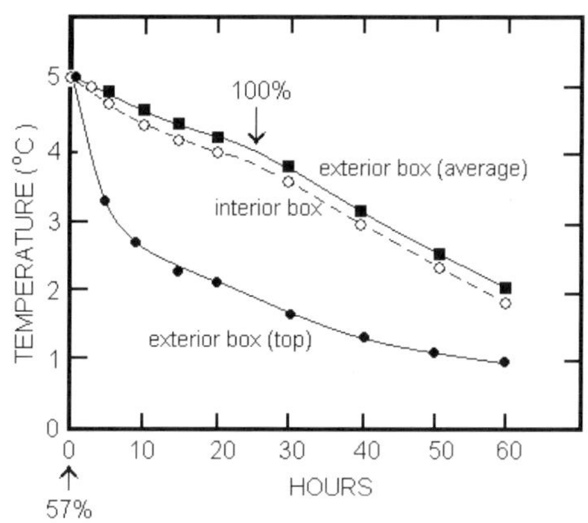

Fig. 6.19. (*left*) Cool down of 3273 kg (7200 lb) of carnation flowers in a 6.1 m VacuFresh^{SM} container operated at 0°C and a pressure of 2 kPa (15 mm Hg).[33] Flowers stacked according to Fig. 6.15, *left*. Interior box (●) located in the centre of the load; exterior box (■) located at the mid-point on top of the load; wall temperature (□) measured at the mid-point along one side; average air temperature (○) measured at front, rear and middle in the space between the load and wall. The vacuum pump's capacity, initially set at 57%, was increased to 100% at 24 h. (*right*) The theoretical cool down of 3273 kg (7200 lb) of carnation flowers in a VacuFresh^{SM} container assumes that the fruit, container air and container air dew point temperatures are the same, and the temperature of the container walls equals the dew point temperature of the container air. Symbols used for the theoretical cool down are: exterior box on top of the load (●); average of all exterior boxes (■); interior box (○). The total surface area available for each mode of heat transfer by the entire load was: external box evaporation = 284.5 m^2; external box convection = 84.1 m^2; external box radiation = 46.6 m^2; internal box evaporation = 35.9 m^2; internal box convection = 7.1 m^2.

drawn through the container each hour. The cooling rate still was limited by heat transfer rather than refrigeration capacity, but all evaporated water now was pumped out of the container and none condensed on the wall panels, causing them to dry and perhaps decrease in emissivity. The strong radiant coupling between the container wall and the surface of the top exterior boxes caused them to cool much faster than internal boxes lacking a view of the container wall. The surface area available for radiant transfer is much less for the 'average' external box than it is for an external box on top of the load, so the 'average' external box should cool more slowly.

The more rapidly boxes cool, the sooner their respiration rate declines. Since a large portion of the total heat that must be removed during a cool down arises from the extra respiratory heat produced when the commodity is warmer, and as evaporation is the dominant heat transfer mode, speeding the cool down significantly reduces the overall weight loss.

6.25 Examples

1. The heat transferred by evaporation from the papayas (F) to the air in a box (A) is given by:

$$\dot{Q}_{V,FA} - m_v h_{fg} - K_4(T_F - T_{DP,A}) \quad (6.46)$$

where h_{fg} is the latent heat of vaporization of water (kcal/kg), m_v is the weight of water evaporated (kg), T_F is the papaya temperature and $T_{DP,A}$ is the dew point temperature of the air (A) inside the box. The conductance (K_4) is:

$$K_4 = A_F\, K_{ge}(dp_v/dT)h_{fg} \quad (6.47)$$

where K_{ge} is the mass-transfer coefficient defined in equation 6.43, A_F is papaya fruits' surface area and dp_v/dT is the change in water vapour pressure per °C change in fruit temperature. The heat transferred by vapour flow through the box walls is given by:

$$\dot{Q}_{V,AC} = m_v h_{fg} = K_5(T_{DP,A} - T_{DP,a}) \quad (6.48)$$

where $T_{DP,a}$ is the dew point temperature of air in the container. The conductance (K_5) is:

$$K_5 = A_B\,(m_v/R_\mu T\, r_{box})\,(dp_V/dT) \quad (6.49)$$

where A_B is the total box surface area including surfaces in contact with adjacent boxes, R_μ is the universal gas constant, T is the temperature of the vapour and r_{box} is the box's transpirational resistance. Free convection from the commodity to the air is given by:

$$\dot{Q}_{C,FA} = K_1(T_F - T_A) \quad (6.50)$$

$$K_1 = h_{F,A}\, A_F \quad (6.51)$$

where $h_{F,A}$ is the fruit's natural convective-film coefficient and T_A is the temperature of the air inside the box. Values for $h_{F,A}$ are based on data correlations for natural convection from short vertical plates to air in the form:

$$hL/k = cX^n \quad (6.52)$$

where:

$$X = Gr\,Pr = (L^3\rho^2 g\, \beta\Delta T/\mu^2)(c_P\, \mu^2/k) \quad (6.53)$$

and Gr and Pr are the Grashof and Prandtl numbers, respectively; L is the characteristic 'height' of the fruit (m); g the gravitational constant (9.8 m/s^2); ΔT is the temperature difference (K); and ρ (kg/m^3), μ (kg/m·s), c_P (W·s/kg·K) and k (W/m·K) are the density, dynamic viscosity, heat capacity at constant pressure and thermal conductivity of the air/water-vapour mixture at the storage pressure, respectively. Constants c and n in equation 6.52 depend on the value of X as shown in Table 6.10. Natural convection from the air in the box to the box is given by:

$$\dot{Q}_{C,AB} = K_2(T_A - T_B) = h_{A,B}(A_{BX} + A_{BE})(T_A - T_B) \quad (6.54)$$

where T_B is the temperature of the box, A_{BX} is the portion of the surface area of a box that is exposed to a ventilating air passage (chimney) and A_{BE} is the surface area of an exterior box that faces the container wall. Portions of the box surface that contact adjacent boxes in the stack are considered adiabatic and are excluded from the heat transfer analysis. The convective film coefficient $h_{A,B}$ for the inner surface of the box is assumed to be equal to that of the external surface of the fruit, $h_{F,A}$, in part because of the strong coupling between the two natural convective flows, but also because the internal packing and wrapping geometry is rather complicated for analysis. Natural convection from external surfaces of the box to the container air is given by:

$$\dot{Q}_{C,BC} = K_3(T_B - T_A) = h_{B,a}(A_{BX} + A_{BE})(T_B - T_a) \quad (6.55)$$

where $h_{B,a}$ is the convective film coefficient for exterior box surfaces. Values for $h_{B,a}$ are determined using the same equation as for $h_{F,A}$, except that the characteristic length is taken as the height of the stack and the buoyancy augmentation term depends on the resistance to vapour flow of the box (r_{box}), expressed as a 'skin' resistance (equation 6.44).

Grumman/Dormavac container conditions for Fig. 6.18 and Table 6.15 (Burg and Kosson, 1983) are:

Pressure (p_a) = 2.67 kPa (20 mm Hg)
Air temperature (T_a) = 11.11°C
Dew point temperature ($T_{DP,a}$) = 7.78 or 11.11°C
Wall temperature (T_W) = 11.11°C

Box property determined from simultaneous internal and external measurements of dew point during cool down:

Resistance to water vapour mass transport (r_{box}) = 10.9 s/cm @ 20 mm Hg = 660 s/cm @ 760 mm Hg

Papaya properties:

Respiration at 20 mm Hg ($\dot{Q}_P$) = 8.4×10^{-3} W/kg (625 BTU/ton·day)
Density (ρ) = 1.11 g/cm^2
Surface to volume ratio (A_F/V_F) = 0.67 cm^2/cm^3
Cuticular transpirational resistance @ 2.67 kPa (20 mm Hg) = 6.9 or 20 s/cm (see note 33)
Characteristic length for natural convection (x_F) = 15.2 cm

Papaya are assumed to be packed 4.54 kg (10 lb) in $35.6 \times 25.4 \times 15.2$ cm boxes (surface area = 0.36 m^2). With the chimney stack arrangement (10 boxes high, 6 boxes wide and 32 boxes long), there are 1920 boxes, of which half are interior boxes surrounded by other boxes or interior gaps. The stacking arrangement causes the interior boxes to have the equivalent of 48.3 cm of side length and a 5.1×25.4 cm strip on top and bottom surfaces (total area = 0.07 m^2) exposed to the container air. The remaining surfaces (0.29 m^2) are in contact with other interior boxes. The average exterior box has an additional 0.072 m^2 facing one or more walls.

2. A harvested Valencia orange produces 12 ml CO_2/kg·h. A 200 g fruit (surface area = 177 cm^2) must evaporate 2.08×10^{-6} g of water per second to transfer the respiratory heat. The transpirational resistance of the harvested fruit, measured at 20°C and 70% RH, is 106 s/cm (Ben-Yehoshua *et al.*, 1985), and the vapour-pressure gradient between the fruit and air required to evaporate water at the required rate is 1.29 mm Hg. The orange would lose water at this rate if the RH is 93%, the fruit's temperature 20°C and the transpirational resistance the same at 70 and 93% RH. As the transpirational resistance typically decreases when the humidity is raised from 70 to 93% (6.3; Fig. 6.7), sometimes up to twofold, all of the orange's respiratory heat will be removed by evaporative cooling when the RH is significantly higher than 93%.

3. The average transpirational resistance of six varieties of apples decreased from 100 to 50 s/cm when the RH was increased from 70 to 95% (Fig. 6.7, *upper left*). At 0°C the average heat evolution by Golden Delicious, Jonathan, McIntosh and Red Delicious apples stored in air at 0°C, 95% RH and atmospheric pressure, is 9.4×10^{-3} kcal/kg·h (Tolle, 1962). This amount of heat can vaporize 2.1% of the fruit's weight in 8 weeks. Initially, weight is lost at an abnormally rapid rate for several weeks while the fruit is cooling (Hruschka, 1977) because (i) warm fruit have a high rate of respiratory heat production; (ii) evaporative cooling is an important heat-transfer mode, removing sensible heat during the cool down process; and (iii) wicking of the 'dry' wood and cardboard packing materials draws down the humidity. To simplify the thermal analysis, only events occurring after the cool down has been completed will be considered. The weight loss from the 4th to 12th week was 2.9%, so not only did respiratory heat vaporize 2.1% of the fruits' weight during that interval, but in addition the refrigeration system warmed the fruit by convection, providing enough heat to evaporate an additional 0.8% of the fruit's weight. In a study of weight loss as a function of humidity (Fig. 6.3), all of the respiratory heat produced at 0°C by Golden Delicious apples was removed by evaporative cooling when the RH exceeded 95%. These apples would lose 3.3% of their weight during 8 weeks' storage at 0°C, 95% RH.

4. During the first two weeks of a CA apple storage (Fig. 6.4), moisture absorbed from the fruit saturates the dry wooden bins, and evaporative cooling assists in lowering the fruit's temperature, both together causing an unusually rapid weight loss. Subsequently, although the humidity and fruit temperature in the storage room were kept constant, the rate at which weight was lost steadily declined for 26 weeks. As the resultant 3% weight loss in 200 days could not significantly lower the fruit's vapour pressure, either the transpirational resistance must have increased due to the deposition of cuticular wax (3.22), or the respiration rate of the fruit decreased (Table 4.4), or both occurred simultaneously.

5. At 29.4°C, 70% RH, the transpirational resistance of Gros Michel bananas is 27.7 s/cm (Wardlaw and Leonard, 1940; Burton, 1982), the pre-climacteric respiration rate is 45 mg/kg·h, and evaporative cooling will remove all pre-climacteric respiratory heat when the vapour-pressure gradient between the fruit and air is 2.0 mm Hg. The measured pulp temperature of pre-climacteric fruits was approximately 0.1°C lower than the storage air temperature, and the vapour-pressure gradient between the fruit and air was 9.4 mm Hg. Therefore, these bananas must lose all of their respiratory heat by evaporative cooling, acquire heat by convection and radiation, and re-emit this extra heat by evaporating additional water. At the same temperature and 95% RH, a pre-climacteric fruit was approximately 0.1°C hotter than the air, and the vapour-pressure gradient between the fruit and air, approximately 1.92 mm Hg, was able to transfer 96% of the respiratory heat by evaporative cooling. At 12.8–15.5°C, the respiration rate of bananas is only 60% as high as it is at 29.4°C (Wardlaw and Leonard, 1940; Leonard and Wardlaw, 1941; Pantastico *et al.*, 1975a; Hardenburg *et al.*, 1986) and therefore if their transpirational resistance remained at 27.7 s/cm, only a 1.22 mm Hg vapour-pressure gradient would be required to transfer all of their respiratory heat. A calculation based on data for green Lacatan, Latundan and Cavendish bananas stored for 3–4 weeks in 12.8–15.5°C (55–60°F) air at 85–90% RH (Pantastico *et al.*, 1975a) indicates that during the pre-climacteric phase, evaporative cooling removed 80–97% of their respiratory heat, and if the temperature of these pre-climacteric fruits was the same as the air temperature, the vapour-pressure gradient was in the range 1.2–2.1 mm Hg. The average respiration rate of Gros Michel bananas was 117 mg/kg·h when they ripened during 13 days at 29.4°C (Wardlaw and Leonard, 1940), and this amount of heat could result in a 1.2% weight loss per day. The fruit lost 1.8% of its weight each day at 70% RH, and 0.8% per day at 100% RH.

6. Tomatoes must evaporate 4.48×10^{-2} cm^3/s of water vapour per 4.54 kg (10 lb) fresh weight to dispel the respiratory heat they produce at 15°C and a storage pressure of 10.67 kPa (80 mm Hg; Fig. 4.2; Hardenburg *et al.*, 1986). The tomato's pedicel-end stem scar's CO_2 mass-transfer resistance is 11,887 s/cm at atmospheric pressure (3.20), and the stem scar's transpirational resistance is 7338 s/cm at 101.3 kPa (760 mm Hg) and 130 s/cm at 10.67 kPa (80 mm Hg), adjusted to reflect the difference between the binary diffusion coefficients of CO_2 and H_2O in air (Equation 6.8; Table 15.4). If the tomato's cuticle's transpirational resistance was 50 s/cm, the total transpirational resistance of the cuticle and stem scar would be 36.1 s/cm, conducting water vapour in parallel at a storage pressure of 10.67 kPa (80 mm Hg). The surface area of a 6-cm-diameter tomato is 113 cm^2, and its weight approximately 90 g. If all heat transfer occurred by water evaporation, the tomato would require a 0.22 mm Hg vapour-pressure gradient to dispel its respiratory heat by evaporative cooling (equation 6.39). At 95% RH, this gradient would develop at a 14.5°C fruit temperature. When 4.54 kg (10 lb) of tomatoes are stored in the same box used for papaya transport (example 10, chapter 3):

surface area of the box = 3900 cm^2
r_{box} (H_2O) = 660 s/cm at 760 mm Hg;
69.5 s/cm at 10.67 kPa (80 mm Hg)
water vapour evaporated to remove resp. heat = 4.48×10^{-2} cm^3/s per 4.54 kg of fruit

At a pressure of 10.67 kPa (80 mm Hg), a vapour-pressure gradient of only 0.6 mm Hg would be required to transfer water vapour across the surface of the box at the rate needed to transfer all respiratory heat. This will elevate the box temperature by 0.8°C.

7. A 2–3°C temperature rise was measured in interior rose boxes when a fully loaded hypobaric container was operated at 0°C and a pressure of 1.33 kPa (10 mm Hg). The ethylene mass-transfer resistance of the box used to store 18.2 kg (40 lb) of roses is r_{box} = 5649 s/cm at 0°C and atmospheric pressure (example 9, chapter 3). Corrected to take into consideration the difference between the binary diffusion coefficients of ethylene and water vapour (wv) in air, $r_{box,wv}$ = 3389 s/cm at 0°C and atmospheric pressure. The respiratory heat produced by 18.2 kg (40 lb) of roses at 0°C and a pressure of 1.33 kPa (10 mm Hg) can be removed by evaporating cellular water at a rate of 0.31 cm^3/s (Table 6.14; Fig. 4.2). A 53.2 mm Hg vapour-pressure gradient would be required to transfer moisture through the box surface at this rate if LP did not increase water-vapour conductance through the 15,000 cm^2 surface area of the box. The air within the box would need to be saturated at 40.8°C to dispel the respiratory heat to a storage atmosphere saturated at 0°C. Instead, a much lower vapour-pressure gradient is required to transfer the respiratory heat because the box resistance (r_{box}) depends on the storage pressure and water-vapour pressure according to the expression:

$$r_{box} = r_{box,R}\frac{\ln\left[\frac{p_R - p_{v,o}}{p_R - p_{v,i}}\right]}{\ln\left[\frac{p - p_{v,o}}{p - p_{v,i}}\right]} \qquad (6.56)$$

where $r_{box,R}$ is the box resistance measured at a reference pressure p_R for particular vapour pressure values $p_{v,i}$ and $p_{v,o}$ inside (i) and outside (o) the box, respectively. Conditions:

$r_{box\,R}$ = 3389 s/cm at 0°C
p_R = 1 atm
$p_{v,o}$ (saturated at 0°) = 0.00603 atm (4.579 mm Hg)

The rate of water-vapour transfer across the box surface is:

$$0.31\ cm^3/s = 15{,}000\ (p_{v,i} - p_{v,o})/r_{box} \qquad (6.57)$$

Solving by repeated approximations after combining equations 6.56 and 6.57, $p_{v,i}$ = 0.0065 atm (4.94 mm Hg), and at 1.33 kPa (10 mm Hg) only a 0.36 mm Hg vapour-pressure gradient is required to transfer the respiratory heat by evaporative cooling. The transpiration resistance of the box (r_{box}) is lowered from 3389 s/cm at atmospheric pressure, to 23 s/cm at 1.33 kPa (10 mm Hg). The temperature rise needed to provide the required vapour-pressure gradient at 1.33 kPa (10 mm Hg) is approximately 1°C. The elevated water-vapour concentration in the box only lowers the O_2 partial pressure from 1.13 mm Hg in the absence of flowers, to 1.06 mm Hg when they are present. If the transpirational resistance of the box was calculated assuming that it is an inverse function of pressure, and the log mean partial-pressure correction for water vapour was not applied (equation 6.56), the box-resistance value would have been erroneously computed to be 44.6 s/cm, and the required temperature rise would have been nearly twice as large.

8. At atmospheric pressure, a harvested apple's transpirational resistance through air-filled lenticles (approximately

17,000 s/cm; Burg and Kosson, 1983), is so much larger than its cuticular resistance to water-vapour transport (50–398 s/cm; Fockens and Meffert, 1972; Sastry *et al.*, 1978), and a harvested orange's stomatal transpirational resistance (5000 s/cm) is so large compared to its cuticular resistance to water vapour (106 s/cm; Ben-Yehoshua *et al.*, 1985), that all but a few per cent of the water loss from both types of fruit occurs through the cuticle.

9. On a 30°C day (RH = 60–80%) in the field at midday, an attached Persian lime may transpire water through open stomates rapidly enough to lower its temperature to 25°C (Davenport, 2002, unpublished). Immediately after it is cut off from its water supply by harvest and placed in a carton, the lime's stomates close, slowing transpiration. Then, exposed to sunlight in the field, solar radiation transfers sufficient heat to elevate the temperature of boxed limes to 45°C (Davenport and Campbell, 1977b), even though simultaneously they are losing heat by natural convection. Harvested dark-green tomato fruits, when exposed to strong sunlight, reach temperatures as high as 53°C (Harvey, 1924).

10. At atmospheric pressure and 95% RH, an apple's transpirational resistance is approximately 50 s/cm (Fig. 6.7, *upper left*). A McIntosh apple's resistance to CO_2 exchange is 14,285 s/cm (chapter 3, example 2), and since the diffusion coefficient for water vapour in air is 1.6-fold higher than the CO_2 binary coefficient (Table 15.4), the water-vapour transport resistance of the lenticles is 8928 s/cm. When the apple is stored at 2°C, 95% RH, and a storage pressure of 6.65 kPa (50 mm Hg), if it remains at the air dry-bulb temperature, diffusion through the lenticles will increase by 16.8-fold (equation 6.8), reducing their transpirational resistance to 531 s/cm. As the loss of water vapour occurs in parallel through the lenticles and cuticle, the transpirational resistance of the surface will only be lowered from 50 s/cm at atmospheric pressure, to 45.7 s/cm at 6.65 kPa (50 mm Hg).

11. If all respiratory heat produced in a Grumman/Dormavac container by a full carnation load (7007 kg) is transmitted into the 3.6 saturated air changes per hour drawn through the 45.3 m^3 (1600 ft^3) container by the vacuum system, and the respiratory heat is carried out of the container without evaporative cooling or heat exchange with the walls, at a 0°C set temperature and a pressure of 1.33 kPa (10 mm Hg) the air temperature would increase by 250.5°C. If instead the commodity evaporated sufficient water to transfer the latent heat of respiration into the container atmosphere, and the walls provided no cooling, the air temperature would increase by only 9°C.

12. The air changes in a VacuFresh^{SM} container originate from an insulated compartment that is maintained at 21°C (at atmospheric pressure) by an air/glycol–water heat exchanger. The humidity in the compartment is kept at or below 85% by coolant flowing through the heat exchanger, and when this air is drawn into the container, upon expanding to 2 kPa (15 mm Hg), its humidity at 0°C cannot be greater than 6.6%.[26] Without evaporative cooling caused by transpiration, and if no circulation and heat exchange with the walls occurred, when all of the respiratory heat produced by a full carnation load (3410 kg) was transmitted into the 3.6 air changes per hour drawn through the 21.1 m^3 (750 ft^3) container by the 1.27 m^3 per minute (45 cfm) full capacity of the vacuum system, the air temperature would increase by 155.5°C. If instead the flowers evaporated sufficient water to saturate the air, all respiratory heat would be removed, the air temperature would remain at 0°C, and the commodity temperature would decrease by 0.04°C per hour due to evaporative cooling in excess of that needed to remove respiratory heat. At the optimal pumping speed for carnations, the air is changed 1.4 times per hour, all respiratory heat is removed, and the temperature of both the carnations and air remains at 0°C.

13. Persian limes (var. Tahiti) were cooled to 10°C, weighed and then 4.54 kg (10 lb) sample lots were stored in 17-litre vacuum jars at 10°C and a pressure of 21 kPa (155 mm Hg). Incoming atmospheric airflow was measured with a rotameter, controlled with a needle valve and admitted at the discharge side of an internal fan rotating at a speed which discharged 25 chamber volumes of rarified air per volume of incoming expanded atmospheric air. Respiratory heat evolution was determined to be 49.2 kcal/kg·h (1311 BTU/ton/day) based on analyses of the CO_2 content of the vacuum pump exhaust made during a 21-day storage period at 10°C and a pressure of 21 kPa (155 mm Hg). To dispel this amount of heat by evaporative cooling, 1.3% of the initial fresh weight of the limes needs to be vaporized in 3 weeks. The evacuation rate required to exhaust exactly this quantity of moisture as saturated 10°C cold steam is 215 cm^3/min. The evacuation rate was set slightly below this value and the limes lost 1.4% of their initial fresh weight during the 21-day LP storage (Burg, 1987a).

14. The velocity and temperature profiles start developing simultaneously if heat transfer begins as soon as a fluid enters a duct. For laminar flow through ducts of the shape created by the stacking arrangement in VacuFresh[SM] and Grumman/Dormavac containers, the thermal entrance length for a constant-wall heat flux (L_t) and the hydrodynamic entrance length (L_h) are (Özisik, 1985):

$$L_h/D_h = 0.011 \text{ Re } L_t/D_h = 0.012 \text{ Re}\cdot\text{Pr} \qquad (6.58)$$

In the Grumman/Dormavac container, L_h = 6.5 cm, L_t = 7.0 cm and the flow is fully developed before it passes the uppermost row of boxes. Assuming uniform-surface heat flux in a duct approaching the shape of two parallel plates, the Nusselt number (Nu = hD_h/k) for hydrodynamically and thermally developed laminar flow is Nu = 8.235 (Özisik, 1985; Fig. 6.20). At 0°C and a pressure of 1.33 kPa (10 mm Hg), the thermal conductivity (k) of saturated air is 0.021 W/m·K (0.0113 BTU/h·ft·°F), the dynamic viscosity (μ) of air is 0.175×10^{-4} Pa.s and the mean heat-transfer coefficient (h_m) for forced flow through the vertical chimneys of a Grumman/Dormavac container is 0.57 kcal/m^2·h·°C (0.12 BTU/h·ft^2·°F).

15. The pneumatic horn induces air to flow through the longitudinal cooling ducts adjacent to the exposed ends of interior boxes at a velocity (v) of nearly 1.28 m/s (4.2 ft/s) when carnations are loaded according to the pattern illustrated in Fig. 6.15 (*left*), and stored at 0°C and a pressure of 2 kPa (15 mm Hg) with the vacuum pump operating at maximum capacity. The hydraulic diameter of the ducts is 10.2 cm (0.33 ft), the Reynolds number at a pressure of 2.0 kPa (15 mm Hg) is Re = 187 and the flow is laminar. The flow reaches its limiting Nusselt number for thermodynamic development in approximately 70 cm (2.3 ft), and for hydrodynamic development in 1.41 m (4.6 ft). The mean forced-convective heat-transfer coefficient will vary from approximately 1.06 kcal/m^2·h·°C (0.2 BTU/h·ft^2·°F) at the second closest box to the door, to 0.73 kcal/m^2·h·°C (0.15 BTU/h·ft^2·°F) for the boxes at the opposite end, provided that the interior box temperature is constant along the duct from end to end. The heat-transfer coefficient will be slightly higher than this computation indicates if at steady-state the box temperatures vary along the duct, with warmer boxes situated toward the door end. The surface area of an interior carnation box available for forced convective heat transfer is 0.31 m^2 (3.33 ft^2) and each longitudinal row adjacent to a cooling duct contains 5 boxes (total weight = 90.9 kg = 200 lb), including the exterior boxes at both ends. At a pressure of 2 kPa (15 mm Hg), with the vacuum pump operating at full capacity, the air horn circulates 3.7 kg (8.1 lb) per hour of saturated air having a specific heat of 0.29 kcal/kg·°C (0.29 BTU/lb·°F) through each longitudinal duct. A 0.88 kcal/h·m^2·°C (0.18 BTU/h·ft^2·°F) convective film coefficient and 0.56°C (1°F) temperature difference theoretically could transfer 0.756 kcal/h (3 BTU/h) from the boxes adjacent to each longitudinal duct, and that quantity of heat would elevate the temperature of the flowing air by 1.33°C (2.4°F). Forced convective heat transfer will not be limited by the heat-carrying capacity of the air unless the temperature difference between the boxes and air is less than 0.44°C (0.8°F). To limit carnation weight loss, the vacuum pump should be throttled to 27% of the maximum pumping speed. Then

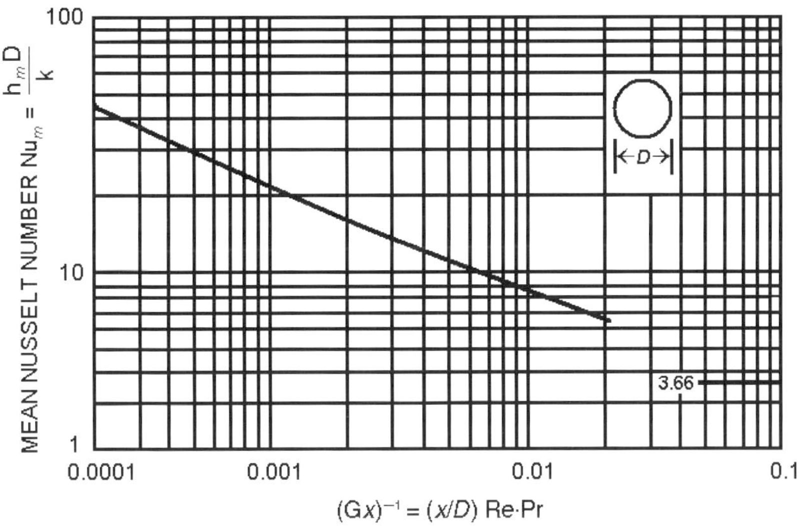

Fig. 6.20. Mean Nusselt number for simultaneously developing flow in air (Pr = 0.7) inside a circular tube subjected to constant wall temperature. Gr = Graetz number; *D* is the diameter of the tube; *x* is the axial distance along the tube measured from the beginning of the 'heated' zone; Re = Reynolds number (Özisik, 1985).

Re = 94.5, and the mean heat-transfer coefficient for forced convection decreases to h_m = 0.63 kcal/m^2·h·°C (0.13 BTU/h·ft^2·°F).

16. In a Grumman/Dormavac hypobaric container, and also when meat is stored in a VacuFresh^{SM} container, invariably Re^2/Gr < 1 and free (natural) convection predominates. When plant commodities are stored in a VacuFresh^{SM} container, the stacking arrangement (Fig. 6.15) causes the characteristic dimension for natural convection to be the box height. Re^2/Gr > 10 and forced convection predominates (6.12).

17. When vertical tiers of carnation flower boxes are loaded in a Grumman/Dormavac container (Fig. 6.13), if the container air is saturated at 0°C and a pressure of 2 kPa (15 mm Hg), and the commodity and box temperatures are close to 1°C, α_2 = 1.18 and the mean natural convective film coefficient for heat transfer from the boxes to the vertical chimneys is h_m = 0.24 kcal/m^2·h·°C (0.05 BTU/h·ft^2·°F). Instead, if both the commodity and box temperatures are 0.1°C, α_2 = 1.96 and the natural convective film coefficient is h_m = 0.15 kcal/m^2·h·°C (0.03 BTU/h·ft^2·°F). These very low natural convective coefficients are approximately one-third the value for forced convection in a VacuFresh^{SM} container at the pumping speed optimal for carnation storage.

18. The natural convective film coefficient for the outer surface of interior boxes (35.6 × 25.4 × 15.2 cm = 14 inches × 10 inches × 6 inches) containing 4.54 kg (10 lb) of papaya fruit stored in a Grumman/Dormavac container at 11.1°C (52°F) and a pressure of 2.67 kPa (20 mm Hg), computed from data correlations from short vertical plates to air (Table 6.10 and equation 6.26), is 0.225 kcal/h·m^2·°C (0.046 BTU/h·ft^2·°F) when the temperature difference between the boxes and container air is 1.83°C (Burg and Kosson, 1983). To simplify the calculation it is assumed that exterior boxes have the same properties and behave in the same manner as interior boxes. The boxes are stacked ten high; the height of each stack is L = 1.52 m (5 ft); the surface area accessible for natural convection is 0.08 m^2/box (0.86 ft^2/box); the properties of a mixture containing partial pressures of 0.0133 atm air and 0.0130 atm water vapour are (15.21–15.25):

viscosity (μ) = 13.1 × 10^{-6} kg/m·s
(8.83 × 10^{-6} lbm/ft·s)

specific heat of air/water-vapour mixture = 0.317 kcal/kg·°C (0.317 BTU/lbm·°F)
thermal conductivity (k) = 0.022 W/m·°C (0.0127 BTU/h·ft·°F)
Prandtl number (Pr) = 0.793

The corresponding Nusselt number ($Nu = hL/k$) for natural convection is Nu = 18.1.

Equation 6.41, which describes forced convective flow over a flat plate at a constant wall temperature in the range $Re \leq 5 \times 10^5$ and $Pr < 1$, indicates that Re = 911 when Nu = 18.1. The Reynolds number can be written $Re = G_{max}(L/\mu) = 911$. Therefore, the mass velocity $G_{max} = 0.0078$ kg/m^2·s (0.0016 lbm/ft^{-2}·s). The container is stacked with 1920 boxes occupying a cross-sectional area of 5.4 m^2 (185 ft^2), situated on a container cross-sectional area of 7 m^2 (240.5 ft^2), leaving a flow area of 1.6 m^2 (55.5 ft^2). The total free convective flow through the stacks is $G_{max}A = 40.2$ g mass per second (0.0888 lb m/s) and the flow can carry 46 kcal/h·°C (101.3 BTU/h·°F). If the flow were to cool or heat the entire 1.83°C (3.29°F) temperature difference, the heat transferred would be 84 kcal/h (333.4 BTU/h). Theoretically, a 1.83°C (3.29°F) temperature difference could only transfer 62.9 kcal/h (249.8 BTU/h) when $h_m = 0.22$ kcal/h·m^2·°C (0.046 BTU/h·ft^2·°F), and therefore the heat-carrying capacity of the rarified air/water-vapour mixture will not limit natural convection.

19. Boxes containing either less than half-coloured papaya, or nearly ripe, more than half-coloured fruits were shipped together in a Grumman/Dormavac container operated at 10°C and a pressure of 2.67 kPa (20 mm Hg). The respiration rate of climacteric, nearly-ripe fruits was so high that during 21 days they 'broke away' and their temperature progressively increased to 22.5°C, which is the boiling point of water at 2.67 kPa (20 mm Hg). Vapour 'boiling' from these fruits continuously flushed air from the boxes in which they were stored, causing the ripening fruits to ferment and experience low [O_2] damage. The vast majority of boxes contained less than half-coloured fruits. These remained at 10°C and were well-preserved.

20. At the time of loading, if the temperature of limes initially is 25°C (77°F), their vapour pressure equals 31.7 mbar (Table 6.17). They can be safely cooled in the container since the vapour pressure of the fruit is much lower than 197 mbar (= 20 kPa = 150 mm Hg), which is the optimal storage pressure for limes (Table 6.6). The vapour pressure of roses loaded at an initial temperature of 20°C (68°F) is 28.4 mbar (Table 6.17), and the optimal storage pressure for roses is 20 mbar in VacuFresh^{SM} (Table 6.6). These roses would not satisfy the cool down prerequisite.

21. To simplify the carnation cool down calculation, the container air is assumed to be saturated and, because flowers have a large surface-to-volume ratio and a low water-vapour mass-transport resistance, the flower temperature, box-air temperature and box-air dew point temperature are considered to be equal. The pumping speed was adjusted to 57% as soon as the set pressure (2 kPa = 15 mm Hg) was reached, and then when commodity cooling slowed between 18 and 24 h, the evacuation rate was increased to 100%, as indicated. For the stacking geometry, the computed average forced-convective heat-transfer coefficient from the box to container air was 0.75 kcal/m^2·h·°C (0.15 BTU/h·ft^2·°F) when the vacuum pump operated at 58% capacity, and 0.9 kcal/m^2·h·°C (0.18 BTU/h·ft^2·°F) when the pump operated at full capacity. The boxes, described in the legend to Fig. 6.15, had a transpirational resistance of 1320 s/cm at atmospheric pressure (Fig. 3.15), which is reduced to 16.5 s/cm at 2 kPa (15 mm Hg; equation 6.56). Initially, the amount of water evaporating from the flowers exceeded that needed to saturate the container air, causing the container walls to be wetted by condensed water, which increased the wall emissivity to approximately 0.97 (Table 6.11, water) and caused radiant heat transfer to obey equation 6.32. The container walls gradually dried after the pumping speed was increased to 100%, the wall emissivity decreased to 0.2 and radiation then was described by equation 6.38. The predicted change in a box's temperature during the first 5-h interval, calculated from the appropriate heat-transfer equations, was used to compute the starting box temperatures for the next 5-h interval, and the process was repeated for each subsequent cooling period to create a theoretical cool down curve (Fig. 6.19, *right*).

22. At 0–1.7°C and 85–95% RH, the average heat production by cauliflower was 0.07–0.086 W/kg (5200–6400 BTU/ton·day). During 7 weeks, the cauliflower lost 30.4% of its weight (Pantastico *et al.*, 1975a), whereas to remove the respiratory heat it would only need to lose 3.86% in each 2-week interval. Other references cite a heat-production rate of 0.048–0.056 W/kg (3600–4200 BTU/ton·day; Johnson and Buford, 1963) and 0.052 W/kg (4000 BTU/ton·day; Robinson *et al.*, 1975) for different cauliflower varieties stored at 0°C, indicating that the commodity would need to lose approximately 2.6% if its weight in each 2-week period to transfer all of its respiratory heat by evaporative cooling. During the first 2 weeks after cauliflower was placed into storage at 0–2°C and 85–87.5% RH it lost 17.02% of its weight, and then during three subsequent 2-week periods the weight loss decreased to 9.07, 8.14 and 8.51%, respectively. The total weight loss was 42.7% in 8 weeks (Burton, 1982), in close agreement with results cited in Pantastico *et al.* (1975a). While the unusually rapid weight loss during the first 2 weeks must have resulted from evaporation during cool down, surely this would have been completed within a few weeks, and the substantial and relatively constant rate of water loss between 2 and 8 weeks could only result if the cauliflower was colder than its environment and acquired heat from it.

At 0–1.7°C, cabbage produces 0.0134–0.019 W/kg (1000–1400 BTU/ton·day) of respiratory heat (Lutz and Hardenburg, 1968), which can be removed by evaporating 0.7–1.0% of the commodity's weight in each 2-week period. Another reference cites heat-production rates of 0.008–0.0268 W/kg (600–2000 BTU/ton·day) for different varieties of cabbage stored at 0°C (Robinson *et al.*, 1975), which could be transferred by an evaporative weight loss of 1.4–2.8% in each 2-week period. During storage at 0–2°C, 85–87.5% RH, in the first 2-week period cabbage lost 7.06% of its weight, and then during three subsequent 2-week periods the weight loss decreased to 4.8, 4.55 and 3.85% (Burton, 1982). The high rate of weight loss from 2 to 8 weeks could only occur if the cabbage was colder than the storage environment and acquired heat from it. In another study, cabbage stored at 0–1.8°C and 92–95% RH produced 0.027–0.032 W/kg (2000–2400 BTU/ton·day) of respiratory heat and lost 13% of its weight in 12 weeks, whereas only a 7.8–10.9% weight loss was needed to dispel all of its respiratory heat by evaporative cooling (Pantastico *et al.*, 1975a).

23. Snap beans (cv. Improved Tendergreen) were stored at 10°C either in NA, or in LP at a pressure of 8 kPa (60 mm Hg) flowing 0.25 air changes per hour of 'saturated' air (McKeown and Lougheed, 1980). The result is indicated in Table 18, where the values in parentheses are the theoretical weight loss to dispel all respiratory heat. The weight decreased at a nearly linear rate during the 14-day period, indicating that it does not include an extra evaporative loss due to a cool down. In NA, respiratory heat was

Table 6.18. Weight loss from cv. Improved Tendergreen snap beans stored at 10°C either in NA or in LP at a pressure of 8 kPa (60 mm Hg) flowing one-quarter saturated air-change per hour (McKeown and Lougheed, 1980).

	% Weight loss	
Days	NA	LP
7	3.5 (4.2)	3.1 (1.7)
14	8.0 (8.4)	6.8 (3.4)

Conditions: Respiration rate @ 10°C = 58 mg/kg·h at atmospheric pressure (Hardenburg *et al.*, 1986); weight loss to remove respiratory heat @ 10°C and atmospheric pressure = 0.6% per day; % inhibition of respiration in LP = 60% (Fig. 4.2). The values in parentheses are the theoretical weight loss to dispel all respiratory heat.

responsible for the entire weight loss, but in LP the snap beans must have acquired additional heat by radiation and convection.

24. Asparagus was stored at 3°C in NA, or in LP at a pressure of 8 kPa (60 mm Hg) flowing 0.25 air changes per hour of 'saturated' air (McKeown and Lougheed, 1981). The weight decreased at approximately a linear rate during the 42-day period, indicating that water loss during storage did not include an extra evaporative loss due to a cool down. The result is indicated in Table 19, where the values in parentheses are the theoretical weight loss to dispel all respiratory heat. In both NA and LP, respiratory heat was responsible for essentially the entire weight loss.

25. At atmospheric pressure, a mature orange's transpirational resistance is 13 s/cm with open stomates (Moreshet and Green, 1980) and 106 s/cm when they close (Ben-Yehoshua *et al.*, 1985). With closed stomates, the resistance to CO_2 exchange is 5738 s/cm (Table 3.11), and since the binary diffusion coefficient for water vapour in air is 1.6-fold higher than for CO_2 (Table 15.4), the water-vapour transport resistance of the closed stomates is 3590 s/cm. The transpirational resistances through the cuticle and open stomates, computed for parallel pathways according to equation 3.3, are 106.1 and 14.8 s/cm, respectively. If a harvested orange remains at the air dry-bulb temperature when its stomates are opened in LP by a pressure of 2.67 kPa (20 mm Hg), at 4 or 10°C, 95% RH, diffusion through the stomates will increase by 54.3-fold at 4°C and 68.1-fold at 10°C (equation 6.8). The transpirational resistance will be reduced to 0.27 s/cm at 4°C and 0.21 s/cm at 10°C. If the stomates did not open in LP, the initial transpirational resistance of 106 s/cm at atmospheric pressure, would only be reduced to 39.3 and 33.9 s/cm at 4 and 10°C, respectively.

Table 6.19. Weight loss during the storage of asparagus at 3°C in NA, or in LP at a pressure of 8 kPa (60 mm Hg) flowing one-quarter saturated-air-change per hour (McKeown and Lougheed, 1980).

	% Weight loss	
Days	NA	LP
21	6.2 (7.8)	4.5 (3.1)
42	14.5 (15.6)	7.7 (6.2)

Conditions: Respiration rate @ 3°C = 36 mg/kg·h at atmospheric pressure (Robinson *et al.*, 1975); weight loss to remove respiratory heat @ 3°C = 0.37% per day at atmospheric pressure. At a pressure of 8 kPa, the respiration should be inhibited by approximately 60% (Table 4.2), decreasing the weight loss needed to remove all respiratory heat to 0.15% per day. The values in parenthesis are the theoretical weight loss to dispel all respiratory heat.

26. During the cool down of a full load of carnation flowers from 15.6 to 1.7°C in a Grumman/Dormavac container, with the pressure regulator set to open at a pressure of 1.33 kPa (10 mm Hg), product water flashed and boiling commenced when the flower temperature was 14°C and the pressure reached 1.6 kPa (12 mm Hg). Immediately, the CO_2 content of the vacuum pump exhaust increased significantly, indicating that fermentation was taking place. During the next 5 h, the water-vapour pressure in the container, computed from dew point measurements, equalled the total pressure; the glycol and flower temperatures were identical, both slowly decreased; and the cool down rate was limited by refrigeration capacity. The vacuum breaker opened after the flower temperature had decreased below 11.7°C and the tank pressure reached 1.33 kPa (10 mm Hg). Thereafter, in the absence of 'boiling', the cool down was much slower and limited by the rate of heat transfer from the flower boxes to the cold plate and air. The carnations were damaged during this cool down, causing their leaves to turn black several days later while they were still in storage.

Notes

1. Heat evolution is calculated by multiplying the respiration rate expressed as mg/kg·h by 2.55 to yield cal/kg·h, by 10.676 to yield J/kg·h, or by 220 to yield BTU/ton·day (Hardenburg *et al.*, 1986). Calculated in this manner, the heat output

in relation to the release of respiratory energy ranged from 95 to 110% in measurements made with apples, carrots, lettuce, oranges, peas, potatoes and strawberries (Green *et al.*, 1941). Water comprises 94.3%, and respiratory substrate approximately 5.7%, of the total weight loss during storage.

2. The author made this calculation using respiration data from Hardenburg *et al.* (1986) and Burton (1982), and weight loss data from Burton (1982).

3. In the tabulation of 55 varieties of fruits and vegetables stored at 85–90% RH, bananas, two types of mango and sweetcorn were the only examples that did not lose essentially all of their respiratory heat by evaporative cooling.

4. At 20°C the transpiration resistance of green plantains is 130 s/cm at 97% RH, and 382 s/cm at 59% RH (George *et al.*, 1982). The average transpirational resistance of bananas as they ripen from green to overripe is approx. 27.7 s/cm at 70% RH, 29.4°C (Wardlaw and Leonard, 1940; Burton, 1982).

5. Fockens and Meffert (1972) assumed that water vapour moved through hypothetical air-filled spaces separating the epidermal cells of apples, and that these spaces expanded and contracted when the cells gained or lost water in response to humidity changes. If this explanation was valid, the rate at which water vapour and gases move through an apple skin would depend on the binary diffusion coefficients of water vapour and gases in air, and the skin resistance to water vapour and the fixed gases would be similar ± 27%. Instead, the peel is somewhat more permeable to CO_2 and 50–350-fold more permeable to water than would be predicted for diffusion through air-filled pores, disproving Fockens and Meffert's explanation.

6. In the experiments depicted in Fig. 6.7 (*lower*), the driving force for transpiration is expressed as a liquid water gradient across the membrane (Schreiber *et al.*, 2001), whereas typically for stomatal and transpirational studies the driving force is expressed as the concentration of water in the vapour phase. At 25°C, permeance values based on a liquid phase driving force can be converted to gas phase permeances by multiplying the former by the liquid/vapour partition coefficient (×43,384). A permeance of 1×10^{-10} m/s based on a liquid phase driving force is equivalent to a resistance of 23 s/m based on a vapour phase driving force.

7. Without a log mean partial pressure correction that takes into consideration the large proportion of water vapour present at 1.33 kPa (10 mm Hg) the transpirational resistance would only decrease by 76-fold (15.16; equation 6.8).

8. The respiration rate is computed from the expression $Q_{10} = (R_2/R_1)\exp[10/(t_2 - t_1)]$, where R_1 and R_2 are the respiration rate at temperatures t_1 and t_2 (°C), respectively (Ryall and Lipton, 1972). The Q_{10} for avocado respiration is assumed to be 2.5 (Hardenburg *et al.*, 1986).

9. No citations relating to the thermal conductivity of plant tissues could be found. The presence of dissolved substances reduces the thermal conductivity of the cellular phase (Liley *et al.*, 1984), and the thermal conductivity of lipid membranes and cell walls is less than that of water. As muscle contains approximately 75% water (Hamm, 1975), which is comparable to the 83.1 and 74.8% water content of apples and bananas, respectively (Lutz and Hardenburg, 1968; Hardenburg *et al.*, 1986), the thermal conductivity of plant cells is likely to have a value similar to that of muscle, 0.41 W/m·K @ 30°C (Özisik, 1985). By analogy to water, which has a thermal conductivity of $k = 0.565$ W/m·K @ 0°C and 0.614 W/m·K @ 30°C, for muscle $k = 0.38$ W/m·K @ 0°C. The thermal conductivity of apples at 0°C is calculated as the averaged value for air and human muscle assuming that the apple fruit is 70% 'muscle' and 30% air ($a_{ias} = 0.3$, Table 3.8); for bananas it is the averaged value at 30°C assuming that the fruit is 85% 'muscle' and 15% air ($a_{ias} = 0.15$, Table 3.8); where for air, $k = 0.0245$ W/m·K @ 0°C and 0.0263 W/m·K @ 30°C.

10. The water is not actively transported; i.e. the channels are not 'pumps'.

11. In isolated *Citrus aurantium* L. leaf cuticular membranes, at pH 6 when flow is driven by a pressure gradient, the hydrodynamic permeability P_f is 1.08×10^{-4} cm/s. When water transport is driven by a concentration gradient, the diffusive permeability P_d is 0.416×10^{-4} cm/s (Schönherr, 1976a).

12. In both *C. aurantium* L. leaves and aubergine fruits, the polar poles swell and shrink in response to water activity, a_w, maintaining liquid continuity down to at least $a_w = 0.22$ (–2180 bar). This suggests a strong interaction of the water molecules with the polar groups of the aquaporin protein, since liquid water normally cavitates at –350 bar (Noble, 1991). The tensile strength of water varies depending on the wall material, the pore diameter and presence of solutes, so presumably the small diameter of the pores and presence of polar groups in the aquaporin protein explain how it is possible that such enormous negative tensions can develop in these water columns, vastly exceeding the tensile strength of liquid water measured in other systems.

13. The *in vivo* phosphorylation rate of the putative aquaporins in spinach leaves depends on the apoplastic water potential (Johansson *et al.*, 1996), suggesting that aquaporins may have a role in regulating cell turgor. Red beet cell vacuoles have a stretch-activated channel that is sensitive to osmotic pressure and modulates cation and osmotically driven fluxes (Alexandro and Lassalles, 1992). The nucleotide sequence of an *Arabidopsis thaliana* transmembrane protein is turgor-responsive (Shagan and Bar-Zvi, 1993). The resistance of *Xenopus* oocytes to liquid water (1000 s/cm), changes to 100 s/cm when aquapores form in the membrane. Changing the pH from 3 to 9 causes the number of pores/cm^2 to increase 3.1-fold, from 5.1×10^{10} to 15.8×10^{10} per cm^2 in isolated citrus leaf cuticular membranes (Chrispeels and Maurel, 1994).
14. The peel comprises 20% of the thickness and 40% of the fresh weight of the mature-green banana (von Loesecke, 1950).
15. Water vapour does not move from the hot side and condense on the cold side of oranges (Curtis, 1937; Lessler, 1947).
16. The assumed properties are a 0.3 osmolar solute concentration, and the volumetric elastic modulus $\varepsilon = 10$ MPa. The cubical expansion coefficient of water is 0.207×10^{-3} at 20°C.
17. The intercellular system's minimum tortuosity, Z, would be 1.57 if water molecules diffuse around half the circumference of each cell in moving a linear distance equal to the cell diameter. Thus $Z_x = 1.57$ m per linear distance of 1 m. The tortuosity might be considerably larger than this minimum estimate.
18. Water droplets condense on the internal walls of epidermal cells in leaves under certain environmental circumstances (Sheriff, 1977).
19. There is some question as to the exactness of the equation, since the vapour pressure lowering is greater than predicted (Kraemer, 1952). Equation 6.13 cannot be applied to aquaporin-lined channels, because they have a mean effective radius of 4.5×10^{-8} cm. The equation has no meaning at the molecular level (Schönherr and Schmidt, 1979).
20. Peel thickening occurs in limes at 10.67 kPa (80 mm Hg = 1.9% [O_2]) and at lower pressures (Spalding and Reeder, 1974, 1976a).
21. Even though real surfaces do not behave in exactly this manner, suitable average values for emissivity and absorptivity can be chosen to make the greybody assumption acceptable for engineering analyses.
22. The view factor represents the fraction of the radiative energy leaving one surface that strikes the other surface directly. The greybody shape factor combines the greybody assumption and view factor, adjusting the Stefan-Boltzmann radiation law to take into account the orientation (view factor) between differently shaped surfaces that are exchanging radiation, and the effect that intervening media that absorb, emit or scatter radiation, have on the fraction of radiation leaving one surface that strikes the other.
23. The emissivities of cardboard, shredded paper and newspaper are approximately equal (Table 6.11). When the emissivities are equal, the ratio of heat transfer rates for parallel-plate systems having no shield and N intervening shields, is $Q_N/Q_O = 1/(N+1)$, where Q_N and Q_O are the heat transfer rates with N shields and no shields, respectively (Özisik, 1985).
24. With certain commodities that are best stored in a Grumman/Dormavac container at 1.33 kPa (10 mm Hg), the pressure in a VacuFresh^{SM} container is elevated by 0.67 kPa (5 mm Hg) to eliminate the possibility of excessive respiratory O_2 draw-down (Table 6.14). The lowest permissible (optimal) pressure has never been established for these commodities. Conceivably the pressure does not have to be elevated above 10 mm Hg when the 'dry' method is used, but without critical data, the pressure has been elevated as a precaution.
25. In intermodal containers equipped with forced-air cooling systems, defrost cycles typically interrupt temperature and humidity control during six 30-min cycles each day.
26. The humidity of the incoming air will be less than 6.6% if the dew point of the ambient (outside) air is lower than 18.5°C.
27. The main function of the air mover is to premix each volume of incoming dry rarified air with up to 40 volumes of air previously saturated with commodity water, so that the incoming air does not desiccate the first commodity that it encounters. At a pressure of 1.33 kPa (10 mm Hg), the RH of the incoming air may be less than 1%, but after mixing with cargo air in the pneumatic air mover's discharge, the RH is increased to approximately 98%.
28. Pumping at a maximum rate with the storage pressure set higher than 6.67–8 kPa (50–60 mm Hg), the induced circulation may be less than 40:1 because part of the incoming air bypasses the pneumatic air mover through a pop-off relief valve set to discharge when the pressure upstream of the air horn jets exceeds the downstream pressure by more than 379 mbar (284 mm Hg) (example 1, chapter 13).
29. Under the assumed conditions of the analytic study, the heat transfer coefficients were 5.35 W/m^2·K (0.91 BTU/h·ft^2·K) for radiant transfer, and

0.91 or 0.67 $W/m^2 \cdot K$ (0.16 or 0.12 $BTU/h \cdot ft^2 \cdot K$) for evaporative cooling at transpirational resistances of 19.7 and 26.8 s/cm, respectively. Depending on the relative humidity, the coefficient for free convection varied between 1.067 and 1.096 $W/m^2 \cdot K$ (0.188–0.193 $BTU/h \cdot ft^2 \cdot K$).

30. Relative humidity was measured with a wet and dry bulb, using Teflon-coated thermistors with ± 0.07°C accuracy (Fig. 9.3). An additional thermistor probe was attached to the fruit surface with insulating tape.

31. The coupling between the box wall and the container air is by forced convection in a VacuFresh^{SM} container.

32. Commodities that can be vacuum cooled include lettuce, spinach, escarole, parsley, Brussels sprouts, endive, cabbage, asparagus, leeks, mushrooms, sweet corn, cauliflower, broccoli, celery, bell peppers, grapes, strawberries, artichokes, green onions, peas and snap beans (Hayes, 1954; Isenberg and Hartman, 1958). Cut flowers, snow peas, cuttings and nursery greens can be vacuum cooled to 0–2°C in 45–60 min from an initial temperature of 10–30°C; strawberries cool from 21.7 to between 1.1 and 3.3°C in 45 min after the pressure is lowered to the flash point; breadfruit cools from 31 to 9.4°C during 90 min after the pressure is reduced to the flash point (Burg, 1995, unpublished); in a 13°C laboratory incubator the cooling half-time for naked hands of bananas was 62 min at atmospheric pressure and 170 min in LP at 80 mm Hg, but when the pressure was decreased to the flash point and continuously lowered, the temperature decreased from 26.7 to 15.6°C in 90 min (Burg, 1969).

33. The calculation in Fig. 6.19 includes a correction for respiratory heat production, assuming the respiratory temperature coefficient $Q_{10} = 2.5$. More than half of the total heat that had to be removed during the carnation cool down was provided by respiration.

7

Postharvest Diseases and Physiological Disorders

Postharvest diseases are an unpredictable cause of commodity loss of quality during distribution, and therefore a storage method's ability to control decay often is more important than its capacity to prevent physiological deterioration and water loss. The bacteria *Erwinia* and 25 species of fungi are responsible for the major diseases of plant commodities (Eckert and Ratnayake, 1983). Fruits are most often rotted by fungi and vegetables by bacteria, because fungal growth is favoured by the low pH, close to 4.0, which is characteristic of fruits, and bacteria develop more readily at the higher pH, near neutrality, typical of vegetables (Wills *et al.*, 1989). Approximately 20 species of bacteria and an equal number of mould and yeast varieties account for the spoilage of stored meat, poultry, fish and shrimp (Mossel *et al.*, 1975).

Preharvest field infections, which occur in surface lesions on stems, leaves and floral parts, often enter a latent or quiescent stage due to the host's resistance to further invasion. Eventually, these infections spread inward where they are difficult to control with postharvest fungicides, as these cannot penetrate in effective concentration to the required depth in the commodity. This type of infection is best prevented by periodic preharvest fungicide applications throughout the growing season, but it also may respond to the 'systemic' gas composition within a commodity stored or transported in LP, CA or MA. The major cause of postharvest decay losses is infections that occur through cut stems or in superficial damage to tissues caused by harvesting, packaging and transportation (Eckert, 1975). These injuries occur at the surface, and while they are sensitive to the composition of the storage atmosphere, the ability to control decay by this means is limited by the commodity's tolerance to low [O_2] or elevated [CO_2] (Table 4.7). At atmospheric pressure, the host usually is damaged if the storage mixture contains less than 1–2% [O_2] or more than 5–10% [CO_2] (Table 4.7).

7.1 Direct Effects of O_2 on Microbial Growth

The [O_2] concentration required to inhibit the growth of obligate aerobic and microaerophilic bacteria and most fungi varies considerably amongst species, but generally is < 1% for a substantial effect (Cochrane, 1958; Figs 7.1, 7.2 and 7.3; Tables 7.1 and 7.2; Couey *et al.*, 1966; Follstad, 1966; Tabak and Cooke, 1968; Wells, 1974; Imoolehin and Grogan, 1980; Barkai-Golan, 1990). CA and MA usually cannot directly inhibit the growth of bacteria and moulds by regulating [O_2] because most horticultural commodities develop off-flavours and are damaged by < 1% [O_2] at atmospheric pressure (Tables 4.7 and 7.2), and existing CA and MA technology cannot accurately

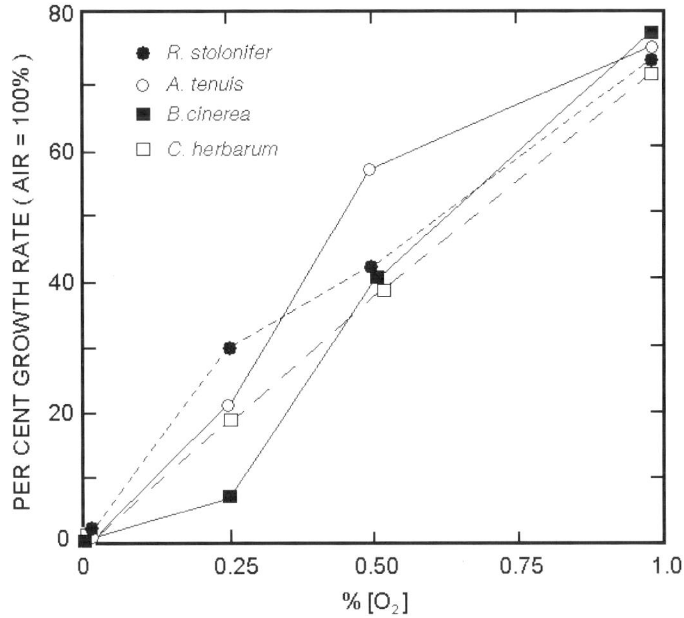

Fig. 7.1. Average growth response of *Rhizopus stolonifer*, *Alternaria tenuis*, *Botrytis cinerea* and *Cladosporium herbarum* cultured on agar media at 15°C in atmospheres containing 1% [O_2] or less (Follstad, 1966).

and reliably reduce [O_2] below 1% (Couey *et al.*, 1966; Couey and Wells, 1970; Wells, 1970). Low [O_2] does not kill bacteria and moulds, and their growth resumes when a higher O_2 tension is reinstated.

Very low O_2 concentrations are required to inhibit the development of aerobic and microaerophilic bacteria. The growth of *Escherichia coli* 0157:H7 on raw salad vegetables, such as shredded lettuce, sliced cucumber and shredded carrot, is not affected by 3% [O_2] (Abdul Raouf *et al.*, 1993). To inhibit bacterial growth on meat (Table 11.3), the [O_2] must be decreased to such an extent that an effect on bacterial proliferation cannot be demonstrated in sealed barrier bags that have been twice evacuated and flushed with N_2 (Huffman, 1974; Huffman *et al.*, 1975). *Pseudomonas fluorescens* and *Ps. putida* grow on the surface of agar plates incubated in N_2-flushed packages in which the residual [O_2] is 0.2–0.5% (Eyles *et al.*, 1993), and [N_2] levels approaching 100% have no inhibitory effect on the typical spoilage microorganisms growing on fish (Brody, 1989b). The growth of cultures of *Ps. fluorescens* and two facultative anaerobes, *Erwinia carotovora* and *E. atroseptica*, is reduced at 1–3% [O_2] and decreases linearly below 1% [O_2] as a log function of the O_2 partial pressure, but 3–12.5% of normal growth still occurs in cultures flushed with N_2 containing a maximum of 0.2% [O_2] (Fig. 7.2). Cultures of *Phytophthora cactorum* flushed with N_2 containing 0.2% residual [O_2] developed at 30% of the normal rate in air (Covey, 1970).

Tabak and Cooke (1968) reviewed literature from 1889 to 1968 describing the varied effects which O_2 and CO_2 have on the growth, sporulation and spore germination of fungi and yeasts. *Ascophanus carneus*, *Phoma* sp., *Rhizoctonia solani*, *Ophiobolus graminis*, *Claviceps purpurea*, *Neurospora sitophila*, *Penicillium roqueforti*, *Geotrichum candidum*, *Penicillium expansum*, *Aspergillus flavus* and *Aspergillus niger* are listed amongst fungi whose growth is inhibited in low [O_2]. In general, these early results agree with more recent studies indicating that the O_2 content of air must be reduced to < 1% and often to < 0.2% before

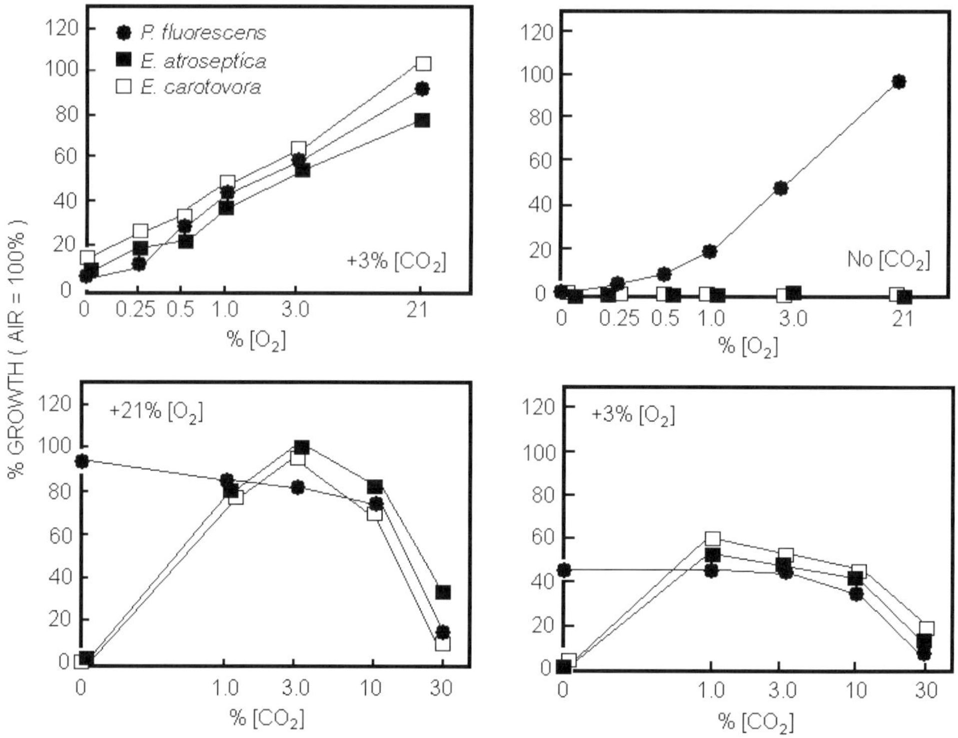

Fig. 7.2. Mean per cent growth, compared to growth in air, of *Pseudomonas fluorescens*, *Erwinia carotovora* and *Erwinia atroseptica*, in (*upper left*) different concentrations of O_2 + 3% [CO_2]; (*upper right*) different concentrations of O_2 + 0% [CO_2]; (*lower left*) different concentrations of CO_2 + 21% [O_2]; (*lower right*) different concentrations of CO_2 + 3% [O_2]. Growth in the dark at 21°C on a buffered asparagine–yeast -extract broth during 16 h for *E. carotovora*, 20 h for *Ps. fluorescens* and 24 h for *E. atroseptica*. High-purity N_2 gas was used to formulate the gas mixtures (Wells, 1974).

growth of most fungi is strongly inhibited (Follstad, 1966; Banks and Annis, 1990; Figs 7.1 and 7.3). Some types grow as well in 0.5–5% [O_2] as in air, and amongst 18 varieties of fungal pathogens tested at 5.5–12.5°C, growth of the majority was not inhibited by 2.3% [O_2] (El-Goorani and Sommer, 1979). Atmospheres containing 3% [O_2], and even less, do not inhibit the growth of *Rhizopus* and *Alternaria* cultures (Parsons *et al.*, 1970), and while the growth of *Rhizopus* sp., *Penicillium* sp., *Phomopsis* sp. and *Sclerotinia* sp. cultures is inhibited by total N_2, they develop at a nearly normal rate in 99% [N_2] + 1% [O_2] (Ryall, 1963). On solid media the colony diameter of *Botrytis cinerea*, *Alternaria tenuis*, *Cladosporium herbarum* and *Rhizopus stolonifer* is decreased by one-third to one-half in 0.5% [O_2], and by even more at 0.25% [O_2] (Follstad, 1966), and the development of *Botrytis* on carnations (Table 7.1) and strawberries (Table 7.2) is inhibited to that same extent at comparable low O_2 partial pressures. Growth of *Gloeosporium album* and *Fusarium oxysporum* in liquid media is inhibited when [O_2] is lowered to 2.5% (Lockhardt, 1969), and the dry weight of *A. tenuis*, *B. cinerea*, *C. herbarum*, *Fusarium roseum* and *R. stolonifer* decreases when the [O_2] in the aerating mixture is $< 4\%$, but spore germination is only inhibited at much lower O_2 partial pressures (Fig. 7.3). The development of moulds usually is reduced substantially at the 0.3% [O_2] concentration present in technical N_2, but to completely inhibit fungal growth it often is necessary to provide even lower [O_2] (Paster, 1990).

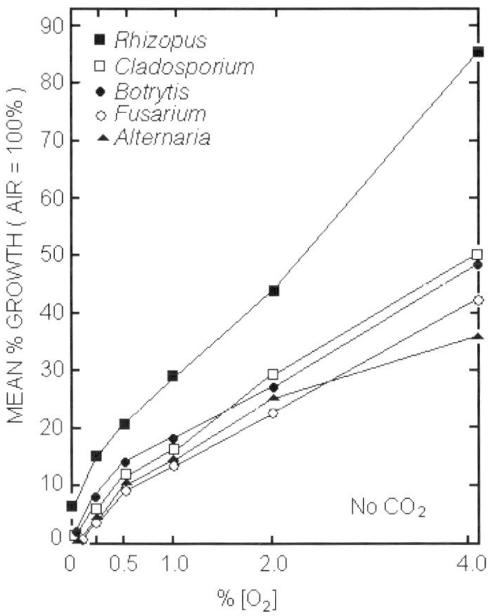

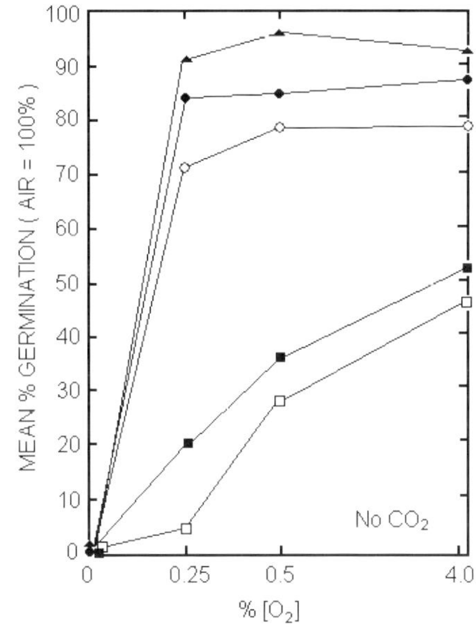

Fig. 7.3. Growth (*left*) and germination (*right*) of *Rhizopus stolonifer*, *Cladosporium herbarum*, *Botrytis cinerea*, *Fusarium roseum* and *Alternaria tenuis* in atmospheres containing different concentrations of O_2 in the absence of CO_2 (Wells and Uota, 1970).

Completely anaerobic conditions prevent spore germination of *Gloeosporium musarum* (= *Colletotrichum musae*), but do not prevent its vegetative growth (Goos and Tschirsch, 1962). Tomatoes (cv. Homestead) inoculated with *G. candidum* and held at 12.8°C decayed less in air than in CA with 3% [O_2], and adding 5% [CO_2] to the CA atmosphere did not alter the total number of decayed fruits (Parsons and Spalding, 1971). Inoculated with *A. tenuis*, *B. cinerea* or *Colletotrichum coccodes*, tomatoes stored no better at 12.8°C in 3% [O_2] + 5% [CO_2] than in air, but in 0.25% [O_2] and 5% [CO_2] all three decays were significantly reduced.

Table 7.1. Effect of low-O_2 atmospheres on decay of carnations caused by *Botrytis* during 30 days' storage at 2.2°C (Uota and Garazsi, 1967).

% [O_2]	% decayed blooms
21 (air)	57.2
2	50.8
1	37.5
0.5	19.0

7.2 Direct Effects of CO_2 on Microbial Growth

[CO_2] ranging from 10 to 100% selectively inhibits the growth of aerobic and

Table 7.2. Decay of strawberries after 5 days' storage at 3°C in the indicated [O_2] concentration plus two additional days at 15°C. Flavour preference was determined after 5 days' storage at 3°C (Couey *et al.*, 1966).

	[O_2] concentration (%)				
Attribute measured	21	1	0.5	0.25	0
Decay of strawberries (%)	10.3	8.3	4.0	4.5	2.5
Flavour preference rating	64.9	61.4	57.3	40.3	26.1
Relative growth of *Botrytis cinerea* in culture (Follstad, 1966)	100.0	83.0	38.0	14.0	0.0

facultative anaerobic bacteria such as *Ps. fluorescens*, *E. atroseptica* and *E. carotovora* (Fig. 7.2, *lower left*), but anaerobic or facultative anaerobic *Streptococci* and microaerophilic *Lactobacilli* are not repressed (Johnson, 1974; Gill and Harrison, 1989). High [CO_2], in the 20–50% range, also inhibits the mycelial growth and sporulation of many species of fungi that cause decay of horticultural commodities (Tabak and Cooke, 1968; Ulrich, 1975; Tian *et al.*, 2001), but often more than 35% [CO_2] is required (Paster, 1990), and at the CO_2 partial pressures tolerated by plant commodities (Table 4.7), the inhibition seldom reaches 50%.

Many fungi require CO_2 to develop and sporulate (Tabak and Cooke, 1968). *Aspergillus oryzae* is not inhibited by 10% [CO_2], both *Mucor* and *Aspergillus* apparently require CO_2 to grow, and spores of *A. niger* do not germinate in the absence of CO_2. Low [CO_2] promotes, and high concentrations inhibit, growth of *P. roqueforti*; *Puccinia* cannot infect cereal if the CO_2 concentration is less than atmospheric; and 100% [CO_2] does not inhibit the growth of yeasts. *Fusarium* tolerates 75% [CO_2], its growth is promoted by 3–10% [CO_2], and the germination of its spores is stimulated by 4–16% [CO_2] (Fig. 7.4; Stover and Frieberg, 1958; Toler *et al.*, 1966). *A. tenuis*, *B. cinerea*, *C. herbarum* and *R. stolonifer* grow reasonably well in the presence of 10–20% [CO_2] (Fig. 7.4, *left*), and *Botryodiploidia theobromae*, a causative agent of banana stem-end rot and anthracnose, is suppressed by 2.3% [O_2] to the same extent in the presence or absence of 5% [CO_2] (El-Goorani and Sommer, 1979). Because very high [CO_2] is required to slow fungal growth, a direct inhibition by MA packaging or applied CO_2 has only been possible with strawberries, cherries, blueberries and a few other horticultural commodities able to withstand more than 10–20% [CO_2] without developing off-flavours, odours and colours (Couey and Wells, 1970; Wells, 1970; Prince, 1989). Fresh red meat tolerates pure CO_2 provided that O_2 is absent (Gill and Harrison, 1989), and therefore it can be successfully packaged in anaerobic barrier bags containing 100% [CO_2].

Although CO_2 cannot be utilized as an exclusive carbon source by heterotrophic bacteria, yeasts and moulds, it nevertheless is assimilated into lactic, fumaric, citric and other acids of the Krebs cycle (Werkman and Wood, 1942; Stephenson, 1950; Thimann, 1955), and used for energy and growth. This may explain why the development of certain aerobic bacteria and moulds is stimulated by low [CO_2] concentrations, and retarded or

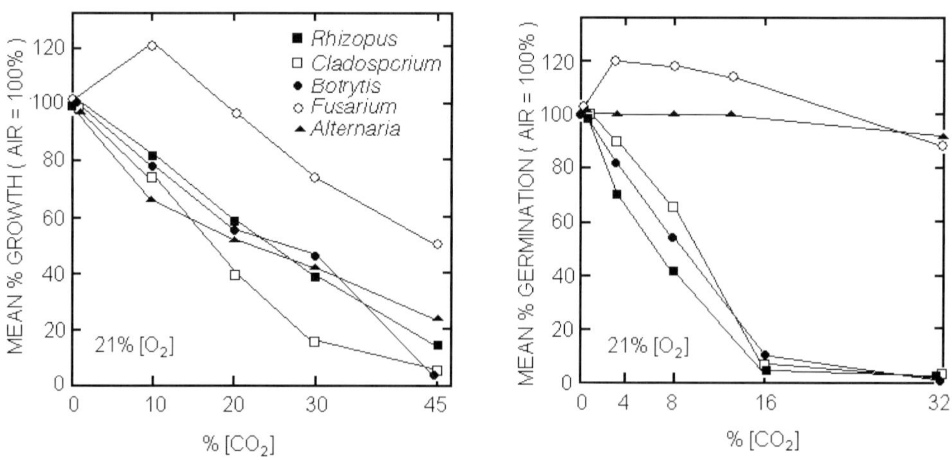

Fig. 7.4. Growth (*left*) and germination (*right*) of *Rhizopus stolonifer*, *Cladosporium herbarum*, *Botrytis cinerea*, *Fusarium roseum* and *Alternaria tenuis* in atmospheres containing different concentrations of CO_2 in the presence of 21% O_2 (Wells and Uota, 1970).

prevented when CO_2 is removed from the atmosphere around them (Rochwell and Highberger, 1927; Valley, 1927; Valley and Rettger, 1927; Rahn, 1941; Krebs, 1943; Tabak and Cooke, 1968). Growth often is reduced if aerobic bacterial cultures are vigorously oxygenated with CO_2-free air, and anaerobic bacterial growth also may be dependent on CO_2, especially amongst coliforms (Thimann, 1955). When *Erwinia*, a facultative anaerobe and *Pseudomonas*, an aerobic bacterium, were incubated at 21°C on a buffered asparagine/yeast-extract broth, there was no growth by six isolates of *E. atroseptica* in CO_2-free air, and trace levels of growth in only one of six isolates of *E. carotovora* (Fig. 7.2, *lower left*; Wells, 1974). This is highly significant because *Erwinia* is the major cause of postharvest bacterial decays.[2] Only four of six isolates of *Ps. fluorescens*[2] were able to grow in CO_2-free air (Fig. 7.2, *lower left*), and when 3% [CO_2] was added to cultures incubated in 1% [O_2], their growth was doubled (Fig. 7.2, *upper*). With no O_2 present they were unable to grow without CO_2, but when 3% [CO_2] was included, their average growth was 3% of that in air. *Bact. lactis aerogenes* (Fig. 11.3) and *Streptococcus haemolyticus* (Table 11.5) also are stimulated to grow by low [CO_2]; less than 0.03% [CO_2] promotes, and relatively high [CO_2] levels inhibit *F. roseum*'s mycelial growth and spore germination (Figs 7.3 and 7.4); and the generation time of *Pseudomonas aeruginosa* is decreased 2–3-fold in gas mixtures containing high [CO_2] (King, 1966 – referred to in Wells, 1974). The primary isolation of *Brucella abortus* requires the presence of 25% [CO_2], and 10% [CO_2] is needed to promote its growth on subsequent transfers (Difco, 1953; Bryan *et al.*, 1962). Primary isolation of *Neisseria meningitidis*, *N. gonorrhoeae* and *N. intracellularis* requires 10% [CO_2]; this same CO_2 concentration improves their subsequent growth, and also the growth of various anaerobic bacteria (McClung and Lindberg, 1957). CO_2 incubators are common in bacteriology laboratories, where they are used to accelerate the growth of CO_2-dependent bacteria. CO_2 at concentrations tolerated by most plant commodities would not be the gas of choice to control decay.

7.3 Combined Direct Effects of CO_2 and O_2 on Microbial Growth

Low [O_2] and moderate [CO_2] do not usually cause additive inhibitions of fungal growth and spore germination. Elevating [CO_2] to between 4 and 10% occasionally increases the inhibition of sporulation caused by low [O_2] (Fig. 7.5; Littlefield *et al.*, 1966), but normally the effects of low [O_2] and moderate [CO_2] tend to be mutually offsetting, so that any benefit which low [O_2] might provide for decay control is likely to be counteracted by adding a tolerable CO_2 concentration (Banks and Annis, 1990). In 2.5–15% [O_2], the growth of *G. album* and *F. oxysporum* is stimulated by 5–10% [CO_2] (Lockhardt, 1967, 1969). Anaerobic growth of *G. candidum* occurs if CO_2 is added, but not in its absence, and at all [O_2] concentrations 3% [CO_2] stimulates this pathogen's development by 33% (Wells and Spalding, 1975). With 1% [O_2] present, germination of *Alternaria alternata*, *B. cinerea*, *C. herbarum* and *F. roseum* spores is stimulated by 4% [CO_2] (Fig. 7.5, *right*), and in 2% [O_2] the development of these organisms is promoted by 4–16% [CO_2] and possibly by even lower concentrations (Fig. 7.5, *left*). *F. oxysporum* growth is reduced by decreasing [O_2], but stimulated when low [O_2] is supplemented with 2.5–15% [CO_2] (Lockhardt, 1968). Growth of *Ps. fluorescens* in 0.25% [O_2] is slightly stimulated by 3% [CO_2] (Fig. 5.2, *upper*), and 5% [CO_2] promotes the development of several species of *Phytophthora* in the presence of 1% [O_2] (Mitchell and Zentmyer, 1971). Proliferation of *Monilinia fructicola* is 50% suppressed in 2.3% [O_2], but the addition of 5% [CO_2] almost completely obviates this effect (El-Goorani and Sommer, 1979). CA and MA are relatively ineffective in controlling decay because of these O_2/CO_2 interactions and the limited tolerance of most plant commodities to moderate or high [CO_2] and low [O_2]. Secondary effects on disease resistance attributable to delayed ripening and senescence (7.5) are cited as evidence that CA prevents decay (Thompson, 1998), but a more discriminating review concludes that at atmospheric pressure,

> no atmosphere in which commodities can be stored without injury gives any appreciable protection against fungal attack. The same can be said of rotting by bacteria, different species of which cover the whole range from obligate aerobes, through facultative anaerobes, to obligate anaerobes
>
> (Burton, 1982)

While it is true that CA atmospheres that are tolerated by horticultural commodities usually have no direct effect on microbial growth, these gas mixtures sometimes indirectly reduce decay by delaying ripening and senescence (7.5, 7.7). Even though the development of *Rhizopus* and *Alternaria* cultures is hardly affected by 3% [O_2] (Figs 7.1, 7.3 and 7.5) and their growth in the presence of low [O_2] is promoted by 3–5% [CO_2], storing mature-green tomatoes in 3% [O_2] ± 3% [CO_2] markedly reduced decay during 6 weeks at 13°C and the effect persisted during a subsequent 1–2 week shelf-life test at 15–21°C (Parson *et al.*, 1970). Adding 5% [CO_2] slightly increased decay in the presence of 3% [O_2].

7.4 Microbial Development in LP

LP pressure regulation is so accurate that pathogen development and spore germination can be directly inhibited by reliably providing 0.1–0.25% ± 0.008% [O_2] or even less, whereas CA cannot control the [O_2] low enough or with sufficient precision to obtain this benefit. The suggestion that O_2 tensions in this range are not useful because they cause low O_2 damage to most types of crops (Lougheed *et al.*, 1978) is true for CA, but fails to consider LP's ability to eliminate the O_2 gradient that develops at atmospheric pressure between the ambient gas mixture and the intercellular spaces of stored horticultural commodities. The LP pressure that causes the greatest prolongation of storage life without physiological damage to the commodity typically provides 0.1–0.25% [O_2] and therefore is highly inhibitory to mould and bacterial growth and development. In addition, LP prevents CO_2 from stimulating pathogen growth and spore germination because it eliminates CO_2 from within the commodity's intercellular spaces by opening stomates and enhancing the diffusive escape of gases, while continuously ventilating the storage area at a rate which prevents CO_2 from accumulating, flowing air changes in which the CO_2 content has been

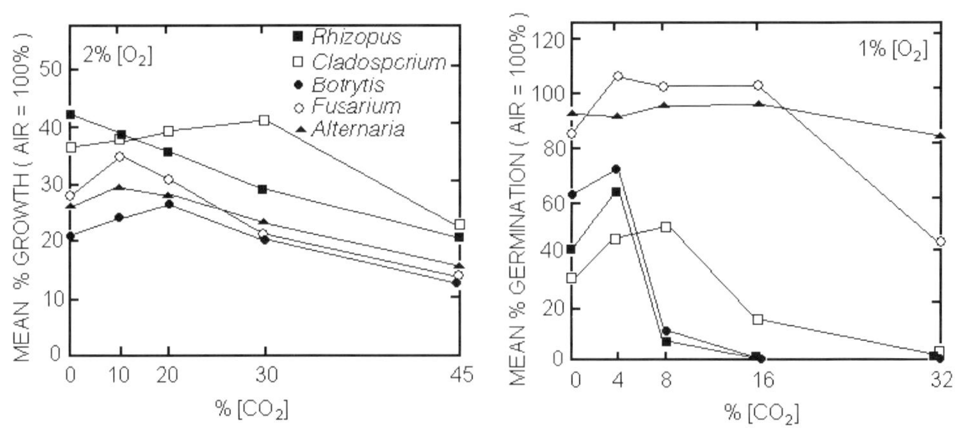

Fig. 7.5. Growth (*left*) and germination (*right*) of *Rhizopus stolonifer, Cladosporium herbarum, Botrytis cinerea, Fusarium roseum* and *Alternaria tenuis* in atmospheres containing different concentrations of CO_2 in the presence of 1 or 2% [O_2] (Wells and Uota, 1970).

decreased by as much as 132-fold (chapter 4, example 1). LP does not kill moulds and bacteria. It prevents them from growing.

Direct inhibition of microbial growth and sporulation at low LP pressures has been confirmed in experiments with pure cultures of pathogenic fungi. On solid media at 23°C, the growth and spore germination of various fungal pathogens exposed to water-saturated air progressively decreases when the storage pressure is reduced below 13.3 kPa (100 mm Hg = 2.2% [O_2]), and at 3.33 kPa (25 mm Hg = 0.11% [O_2]) the growth of *Penicillium digitatum*, *B. cinerea*, *A. alternata* and *G. candidum* var. *citri-aurantii*, and the sporulation of *P. digitatum*, *B. cinerea* and *A. alternata* essentially ceases (Fig. 7.6; Wu and Salunkhe, 1972a; Salunkhe and Wu, 1975; Apelbaum and Barkai-Golan, 1977; Borecka and Olak, 1978). The gaseous composition in these studies, corrected for the presence of 2.8 kPa (21 mm Hg) of water vapour at 23°C, decreased from 2.2% [O_2] + 0.003% [CO_2] at 13.32 kPa (100 mm Hg), to 0.14% [O_2] + 0.00024% [CO_2] at 3.33 kPa (25 mm Hg). Fungal growth resumed at a normal rate after cultures were transferred from hypobaric to atmospheric conditions. During 18 days at a pressure of 2 kPa (15 mm Hg = 0.14% [O_2]), growth of *Colletotrichum gloeosporioides* on 10% vegetable juice agar is severely inhibited at 10°C (Table 7.3), sporulation is prevented, and a fibrillar network that covers the mycelium of cultures grown in air is absent in LP (Chau and Alvarez, 1983). Growth of *C. gloeosporioides* resumed after LP cultures were returned to 24°C air. Mycelial growth of *P. expansum* and *P. patulin*, and patulin production by *P. patulin*, are inhibited at 20 kPa (150 mm Hg; Adams *et al.*, 1976), and a pressure of

Table 7.3. The effects of LP on mycelial growth and sporulation of *Colletotrichum gloeosporioides* on 10% vegetable juice agar during 18 days of incubation (Chau and Alvarez, 1983).

Pressure kPa (mm Hg)	Temp (°C)	Colony radius (cm)	Cultures with spores (%)
2 (15)	10	0.4	0
101.3 (760)	10	1.3	100
101.3 (760)	24	8.0	100

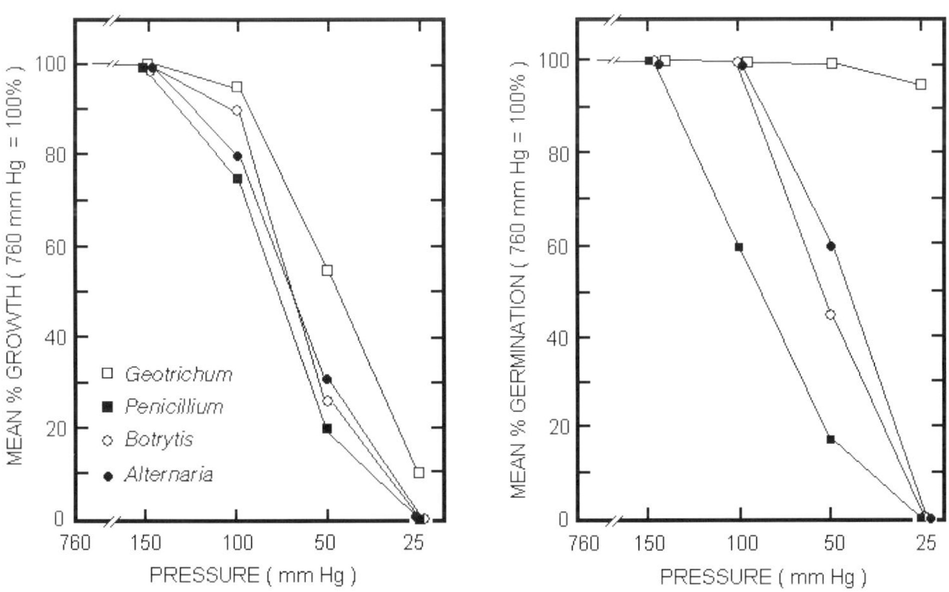

Fig. 7.6. Effect of hypobaric pressure on (*left*) colony growth of *Geotrichum candidum* var. *citri-aurantii*, *Penicillium digitatum*, *Botrytis cinerea* and *Alternaria alternata*; (*right*) spore germination at 23°C (Apelbaum and Barkai-Golan, 1977).

5.33 kPa (40 mm Hg) causes a slight reduction in mycelial growth of *Rhizoctonia* and *Pythium*, but *Botrytis* is unaffected (Eisenberg, 1977).

At the same O_2 partial pressure, the inhibition of mould and bacterial growth is greater in LP than it is in CA at atmospheric pressure. Growth of *P. digitatum*, *B. cinerea*, *A. alternata*, *Diplodia natalensis* and *G. candidum* var. *citri-aurantii* cultures is markedly inhibited at a pressure of 6.67 kPa (50 mm Hg = 0.8% [O_2]) flowing water-saturated air, and much less inhibited when the same O_2 partial pressure and humidity is provided at atmospheric pressure (Table 7.4). The intensity of the inhibition due to pressure reduction is approximately equal to and additive with that caused by lowering the O_2 partial pressure from 0.21 to 0.008 atmospheres while the total pressure is maintained at 1 atm. Apelbaum and Barkai-Golan (1977) suggested that the reduction in atmospheric [CO_2] which occurs at a low pressure, augments the anti-fungal effect attributable to a simultaneous reduction in [O_2]. In other studies, the growth and sporulation of *P. digitatum*, *Rhizopus nigricans*, *A. niger*, *Botrytis alli* and *Alternaria* sp. were inhibited to a greater extent at a pressure of 13.6 kPa (102 mm Hg = 2.3% O_2) than at atmospheric pressure and a comparable O_2 tension (Table 7.5; Wu and Salunkhe, 1972a), and an 'extra' effect of pressure on bacterial growth also was demonstrated when Atlantic cod fish (Haard *et al.*, 1979) were stored either at atmospheric pressure in 21% [O_2], N_2 containing 0.2% [O_2] or in LP at a pressure of 2.23–2.67 kPa (16.7–20 mm Hg = 0.33–0.42% [O_2]). Pseudomonads were the primary bacterial flora that developed on the fish regardless of the storage method. At atmospheric pressure, the total plate count increased at essentially the same rate in 0.2% and 21% [O_2]), but in LP the lag phase for bacterial growth was

Table 7.4. Effect of LP on postharvest pathogen growth at equivalent O_2 partial pressures and different total pressures. Cultures were incubated at 21.1°C for 8 days on potato dextrose agar either at 6.67 kPa (50 mm Hg) ventilated with water-saturated O_2 (pO_2 = 0.008 atm); or at 101.3 kPa (760 mm Hg = atmospheric pressure) ventilated with water-saturated air (pO_2 = 0.21 atm) or with water-saturated air in which the partial pressure of O_2 had been lowered to pO_2 = 0.008 atm. Results are means of nine replicated experiments (Apelbaum and Barkai-Golan, 1977).

	% Growth relative to growth at 101.3 kPa (760 mm Hg), pO_2 = 0.21 atm	
Fungus	6.67 kPa (50 mm Hg) (0.008 atm pO_2)	101.3 kPa (760 mm Hg) (0.008 atm pO_2)
Pencillium digitatum	35	55
Botrytis cinerea	37	60
Alternaria alternata	46	82
Diplodia natalensis	53	83
Geotrichum candidum var. *citri-aurantii*	75	100

Table 7.5. Effect of LP, NA and CA on mycelial spread and the average sporulation rating of various fungi at 90–95% RH. The indicated [O_2] has been corrected to take into account the water vapour pressure in LP, and an atmospheric pressure of 646 mm Hg in NA. Mycelial growth was measured as per cent coverage of the agar in a Petri dish during 8 days for *Penicillium expansum*, 5 days for *Rhizopus nigricans* and 6 days for *Aspergillus niger*, *Botrytis alli* and *Alternaria* sp. Sporulation was rated on the basis of 1 = slight to 10 = very profuse (Wu and Salunkhe, 1972a).

	NA (17.8% [O_2])		CA (2.3% [O_2])		LP (102 mm Hg = 2.3% [O_2])	
Fungus	Growth	Sporulation	Growth	Sporulation	Growth	Sporulation
Penicillium expansum	100	8	95	6	85	3.5
Rhizopus nigricans	100	10	90	9	75	5
Aspergillis niger	100	10	85	8	65	2
Botrytis alli	100	8	90	7	60	4
Alternaria sp.	100	7	90	6	65	4

significantly extended and the log phase abbreviated in the presence of 0.33–0.42% O_2 (Fig. 11.7, *right*). The bacterial growth inhibition that occurred in LP was similar in magnitude to that which the same pressure causes on lamb (11.24), and the flora that grows on lamb at that pressure also is comprised predominantly of pseudomonads. Because vacuum-packaged haddock is spoiled predominantly by group III and IV pseudomonads, while growth of the oxidative pseudomonad groups, I and II, is favoured in air-packaged samples (Licciaedello *et al.*, 1967), Haard and Lee (1972) suggested that species differences may exist between the pseudomonads that grow at a hypobaric and normal atmospheric pressure. They theorized that volatile substances that directly or indirectly influence bacterial proliferation might have been removed as a result of hypobaric conditions, and stressed the lack of odour development in LP-stored cod fish as evidence supporting this explanation. CO_2 is probably the volatile substance whose removal causes an extra anti-microbial effect. At least some strains of pseudomonads require this gas to grow (Wells, 1974), and CO_2 accumulates in vacuum-packaged fish. Studies at atmospheric pressure indicate that the growth of *E. carotovora* and *E. atroseptica* also should be prevented when LP eliminates CO_2 (Fig. 7.2).

There is a close correlation between the behaviour of pure cultures grown at low [O_2] and the incidence of decay in the host when it is stored in the same partial pressure of O_2. At 0°C, the growth and spore germination of *B. cinerea* cultures is prevented by 0.15% [O_2], both at atmospheric pressure (Figs 7.1 and 7.3) and in LP at a pressure of 1.33 kPa (10 mm Hg; Fig. 7.6). At atmospheric pressure, the development of *B. cinerea* in carnation flowers (Table 7.1) and strawberries (Table 7.2) almost ceases when [O_2] is lowered to 0.15%, and this decay also does not develop when strawberries are stored in LP at 0°C and a pressure of 1.33 kPa (10 mm Hg = 0.14% [O_2]; Table 7.6). *C. gloeosporioides*, the causative agent of papaya anthracnose, develops normally at a pressure of 20 kPa (150 mm Hg), but at 10°C and a storage pressure of 2 kPa (15 mm Hg = 0.14% [O_2]), its growth and sporulation are severely inhibited (Chau and Alvarez, 1983). At the same temperature, a pressure of 2 kPa (20 mm Hg) prevents papaya stem-end rot *in vivo*, as well as anthracnose caused by *C. gloeosporioides*, stem-end rots due to *Mycosphaerella* sp., *Ascochyta caricae-papayae* and *Botryodiploidia theobromae*, and peduncle infection caused by *C. gloeosporioides* (Fig. 7.7; Alvarez, 1980). After mature-green papaya fruit inoculated with *C. gloeosporioides* had been stored for 21 days at 10°C and a pressure of 2 kPa

Table 7.6. Low-pressure (LP) inhibition of mould growth on Jerseybelle strawberries stored at 1.1°C. Shelf-life at atmospheric pressure (760 mm Hg) was determined at 1.1°C. The flavour of LP-stored berries was normal (Jamieson, 1980a).

	% infected berries at indicated pressure, kPa (mm Hg)		
Storage time (days)	101.3 (760) = 21% [O_2]	2.67 (20) = 0.4% [O_2]	1.33 (10) = 0.15% [O_2]
9	0	0	–
14	–	1.7	–
14 + 2 (shelf-life)	–	2.4	–
14 + 5 (shelf-life)	31.6	10.5	–
18	18.9	5.0	–
18 + 2 (shelf-life)	–	10.5	–
18 + 5 (shelf-life)	54.2	7.0	–
21	28.8	8.4	5.6
Relative *in vitro* growth of *B. cinerea* (Apelbaum and Barkai-Golan, 1977)	100	10	0

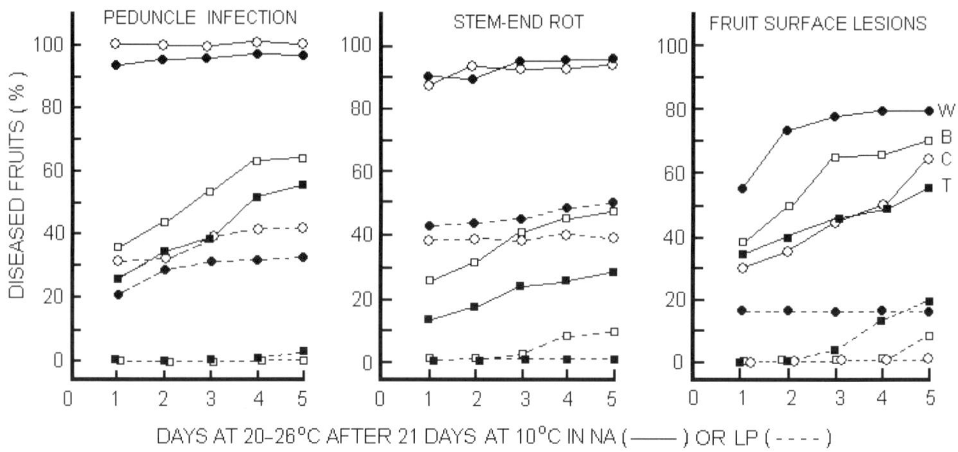

Fig. 7.7. Incidence of decay in papayas shipped from Hawaii to New York City in an intermodal hypobaric container operated at 10°C and a pressure of 2.67 kPa (20 mm Hg), and in a random sampling of the same lot of fruit stored in a refrigerated container at the same temperature and atmospheric pressure. There was no significant advancement in softening or colour development during the 21-day interval from harvest to the first day of the shelf-life test. Hypobaric stored fruits had 63% less peduncle infection, 55% less stem-end rot and 45% fewer fruit surface lesions than papayas stored at atmospheric pressure in a refrigerated container. Treatments included: W = wax (●); B = benomyl plus wax (□); T = thiabendazole (TBZ) plus wax (■); C = untreated control (○) (Alvarez, 1980).

(15 mm Hg), subsequent development of the pathogen was suppressed during five additional days when the fruit was transferred to atmospheric air and ripened at room temperature (Chau and Alvarez, 1983). Following LP storage, fewer than 5% of the conidia germinate and form appressoria within 7 days at 10°C, whereas during NA storage 30–50% germinate, forming numerous appressoria and a mycelial mat with hyphae that cover the fruit's surface. Infection of LP-stored fruit was observed 2 weeks later. Bacterial soft rot develops in asparagus at 0°C and pressures ranging from 101.3 to 10.57 kPa (760 to 80 mm Hg), but not at 2.67–5.33 kPa (20–40 mm Hg). This indicates that at the lower pressures there is insufficient CO_2 and O_2 present to support growth of *E. carotovora*, the causative agent of bacterial soft rot in asparagus (Dilley, 1977a). Direct suppression of mould growth by LP at 10°C and a pressure of 2.67 kPa (20 mm Hg = 0.3% $[O_2]$) has also been demonstrated in carambola (11.8), cucumber (11.41) and mango fruits (11.19).

LP is far more effective than CA in inhibiting the growth and spore germination of pathogens when commodities are stored at the CA or LP condition that is optimal for prolonging storage life without causing physiological damage. *G. candidum* does not proliferate at 23°C and a pressure of 3.33 kPa (25 mm Hg = 0.14% $[O_2]$) (Apelbaum and Barkai-Golan, 1977), but at atmospheric pressure in liquid (Lockhardt, 1967) and on solid media (El-Goorani and Sommer, 1979), regardless of the CO_2 partial pressure, the growth of this pathogen is 50% greater at the low $[O_2]$ levels typical of CA storage, 1–3% $[O_2]$, compared to growth in air, and at all $[O_2]$ concentrations growth is 33% greater in the presence of 3% $[CO_2]$. LP completely prevents spore germination and growth of *B. cinerea* and *A. alternata* at 23°C and a pressure of 3.33 kPa (25 mm Hg; Apelbaum and Barkai-Golan, 1977), whereas the maximum growth inhibition in CA at 1–21% $[O_2]$ in combination with up to 16% $[CO_2]$ is never more than 80% (Wells and Uota, 1970), and while spore germination of *B. cinerea* is prevented by 16% $[CO_2]$ in the presence of 1–21% $[O_2]$ (Apelbaum and Barkai-Golan, 1977), no combination of O_2 and CO_2 effects the germination of *Alternaria* spores at atmospheric pressure.

7.5 Ripening, Ageing and Senescence Influence the Host's Susceptibility to Disease

Physiological changes that occur during ageing, fruit ripening, flower fading and leaf senescence decrease the host's natural resistance to postharvest infection and cause quiescent infections to proliferate. By delaying or postponing these developmental changes in the host, LP, CA and MA sometimes are able to indirectly control disease development. Atmospheres containing approximately 2.5% [O_2], a concentration commonly used in CA storage, are most likely to act upon host resistance rather than directly on the pathogen, since most fungal and bacterial cultures grow profusely at that O_2 partial pressure. Tomato decay by *Rhizopus* sp. and *Alternaria* sp. is markedly reduced by lowering the [O_2] to 3% at atmospheric pressure (Parsons *et al.*, 1970), although both fungi grow well in cultures at even lower O_2 partial pressures (Fig. 5.1). An atmosphere containing 2% [O_2] + 10% [CO_2] has almost no effect on the growth of *C. gloeosporioides* cultures, but reduces anthracnose during avocado storage (Hatton and Reeder, 1972; Spalding and Reeder, 1972, 1975). The susceptibility of apples to *P. expansum* increases with both fruit maturity and ripeness; as oranges mature they become more vulnerable to infection by *Penicillium* moulds; a striking increase in *Alternaria* stem-end rot occurs when stored lemons pass a certain threshold of ripeness; latent infections of *Colletotrichum* on bananas, papaya and mango fruits seldom become a serious problem until the fruit approaches full ripeness; potatoes become more susceptible to *Fusarium* dry rot during storage; and the susceptibility of carrots to *Botrytis* and *Centrospora* increases as the storage duration is extended (Eckert and Ratnayake, 1983).

The enhanced susceptibility to decay that develops during ageing, senescence and ripening may be caused by a decrease in the host's ability to synthesize anti-fungal compounds, by an increase in membrane permeability resulting in the release of nutrients and water into intercellular spaces, and by an increased susceptibility of the plant cell wall to attack by the pathogen's macerating enzymes. Postharvest disease development depends on the pathogen's ability to secrete enzymes that solubilize pectic substances of the middle lamella, leading to a separation of individual cells and softening of the tissue. Cells in the digested tissue increase in permeability and die, releasing their metabolites to be used as substrates for pathogen growth (Eckert and Ratnayake, 1983). Pectin solubilization, cellular dissociation, tissue softening and alterations in membrane permeability occur naturally during fruit ripening and leaf senescence (Sacher, 1962), making the pectic substances more vulnerable to maceration by the pathogen's proteolytic enzymes at that time.

7.6 Ethylene

Ripening and senescence is hastened by ethylene produced either by an invading pathogen (Williamson, 1950; Achilea *et al.*, 1985) or by the host in response to disease, wounding and chilling, and during the normal course of ripening, flower fading and abscission. Therefore it is not surprising that applied gas usually promotes disease development. Ethylene stimulates *Alternaria* rot of tomatoes (Segall, 1967), anthracnose in citrus fruits inoculated with *C. gloeosporioides* (Brown and Barmore, 1977; Brown, 1978), development of *B. cinerea* in inoculated strawberry fruit and *Penicillium italicum* in Valencia oranges (El-Kazzaz *et al.*, 1983), *Diplodia natalensis* stem-end rot in citrus (McCormack, 1972), appressoria formation and germination of spores of the crown rot pathogen (*C. musae*) in bananas (Ranawaka and Wijeratnam, 2001), and decay of plums, celery, vanilla beans, rose and carnation flowers, watermelon fruit, wheat plants and ornamental pot plants of geranium and *Ruscus hypoglossum* (Abeles *et al.*, 1992). Stimulation of infections due to *Diplodia* is brought about in part by ethylene-induced abscission of the fruit button, which provides an avenue for the fungus to enter into the fruit (Brown and Wilson, 1968). The ability of CO_2-enriched atmospheres to decrease rot development in broccoli florets inoculated with *B. cinerea* spores is completely nullified by 10 µl/l ethylene, indicating that CO_2 prevents decay in broccoli by inhibiting ethylene's senescence-promoting action (Aharoni *et al.*, 1985).

Ethylene directly affects the growth of certain pathogens. The gas stimulates spore germination in cultures of *P. digitatum*, *P. italicum* and *Thielaviopsis paradoxa*, but not in *A. alternata*, *B. cinerea*, *B. theobromae*, *C. gloeosporioides*, *M. fructicola*, *P. expansum* and *R. stolonifer*. Ethylene also promotes germ-tube development in most of these moulds, and significantly increases the glucosamine content in *B. cinerea* and *P. italicum*, indicating that the synthesis of fungal chitin has been accelerated and growth stimulated (El-Kazzaz *et al.*, 1983).

7.7 Phytoalexins and Anti-fungal Compounds

Phytoalexins are anti-microbial compounds of low molecular weight that are absent in healthy plants but accumulate in them when the plants are exposed to micro-organisms (Paxton, 1981). Infection of tobacco by tobacco mosaic virus (TMV) is a typical example. After small necrotic lesions form, the plant becomes resistant to subsequent infection by the virus and a variety of bacterial and fungal pathogens (Ross, 1961; Kuc, 1982). Nearly millimolar phytoalexin concentrations are required to inhibit microbial growth, and as these amounts are toxic to higher plant cells, phytoalexins are catabolized or detoxified by the plants themselves, as well as by microbial monoxygenation, reduction and hydration (Darvill and Albershein, 1984). Physiological concentrations of phytoalexins usually do not kill bacteria or fungi, but they accumulate in sufficient quantity at the site of an infection, and quickly enough, to inhibit microbial growth. Pectic oligogalacturonides from plant cell walls, fungal cell-wall glucans and glycoproteins, and microbial enzymes such as a fungal polygalacturonase, are 'biotic' elicitors that signal plants to synthesize phytoalexins. Potatoes synthesize the phytoalexin rishitin in response to *E. carotovora*; apples form benzoic acid at the site of *Nectria galligena* infections; phytoalexins play a role in the latency of *C. musae* on banana fruit (Eckert and Ratnayake, 1983; Barkai-Golan, 1990); and an endopolygalacturonic acid lyase from *E. carotovora* elicits phytoalexin accumulation by releasing plant cell-wall fragments (Davis *et al.*, 1984). 'Abiotic' elicitors include heavy metals, UV light, chloroform, detergents, wounding, freezing and heating (Darvill and Albershein, 1984). Elicitors activate the expression of the genes encoding mRNA for the *de novo* synthesis of enzymes that catalyse phytoalexin production (Ragg *et al.*, 1981). As many as 20 enzymes may be required to synthesize a phytoalexin from shikimate, acetate or mevalonate (Stoessl, 1982).

Isocoumarin (8-hydroxy-3-methyl-6-methoxy-3,4-dihydro-isocoumarin = 6-ME), the 'bitter' principle in carrots, arises in carrot-root tissue exposed to ethylene (Carlton *et al.*, 1961; Chalutz *et al.*, 1969; Sarkar and Phan, 1979; Hoffman *et al.*, 1988; LaFuente *et al.*, 1996), or inoculated with the 'black rot' organism, *Ceratocystis fimbriata* (Condon and Kuc, 1960) or spores of *R. stolonifer*, *B. cinerea* and other storage pathogens (Abeles *et al.*, 1992). In carrot tissue, the threshold applied-ethylene concentration for inducing isocoumarin formation is 0.1 µl/l, and a half-maximal response occurs at < 0.5 µl/l (Table 5.2). Isocoumarin induction is prevented by 1-methylcyclopropene (1-MCP) and norbornadiene (NBD) regardless of the method used to induce its production (Hoffmann and Heale, 1987; Fan *et al.*, 2000), indicating that the synthesis of this compound requires ethylene mediation. Isocoumarin's fungitoxic action may account for the ethylene-induced resistance to *C. fimbriata* infection that results, along with increased peroxidase and polyphenol oxidase activity, when

Table 7.7. Effect of hypobaric pressure on the concentration of antifungal resorcinols in the peel of Tommy Atkins mango fruits, and on disease expression following inoculation with *Alternaria alternata* (Droby *et al.*, 1986).

	Resorcinols (μg/g fresh weight)		% Infected sites	
Days after harvest	101.3 kPa (760 mm Hg)	20 kPa (150 mm Hg)	101.3 kPa (760 mm Hg)	20 kPa (150 mm Hg)
1	195	195	0	0
12	160	–	16	0
17	115	162	49	8

delaying ripening and maintaining the antifungal resorcinol level since the growth of *A. alternata* cultures is not decreased at a pressure of 20.3 kPa (Fig. 7.6). At the optimal pressure for mango storage, 2.0–3.33 kPa (15–25 mm Hg), the incidence of decay is further decreased because the growth and sporulation of *A. alternata* is directly limited by the low [O_2] and [CO_2] present (Figs 7.3 and 7.6).

7.8 Water Condensation

Although it has been claimed that 'The high relative-humidity conditions possible in LP appear ideal for growth of fungi and resultant decay in storage' (Lougheed *et al.*, 1978), to the contrary, a humidity approaching saturation has been found to discourage mould growth on most plant commodities (Van den Berg and Lentz, 1978; Van den Berg, 1981), provided that water does not condense on their surface (Cook and Papendick, 1978; Eckert, 1978a). Leakage occurs when tissues that are not normally exposed to liquid water are placed in contact with it (Brooks, 1916), and all plant parts leach solute upon immersion in water (Helder, 1956). Leakage is enhanced at cut surfaces, and if liquid water condenses on the commodity during storage it can cause an intense, rapid solute exosmosis in the vicinity of wounds (Burg *et al.*, 1964; Figs 3.5–3.7). The escape of these substances increases disease susceptibility by providing a substrate stimulatory to pathogen development, spore germination and germ-tube growth (Eckert and Ratnayake, 1983), and causes infections to originate at and spread from superficial wounds (Lutz and Hardenburg, 1968; Eckert, 1975; Ben-Yehoshua, 1989). Leakage increases as leaves senesce, flowers fade and fruits ripen (2.13; Simon, 1977), and in a few cases the leachate has been found to contain specific substances stimulatory to the germination and growth of spores and conidia. Anthranilic acid, a leachate of banana fruit, is converted to 2,3-dihydroxybenzoic acid by *C. musae*, the organism causing banana anthracnose, and the acid stimulates the pathogen's conidial germination and appressorium formation (Harper and Swinburne, 1979). Leachates of apple fruits contain chlorogenic acid and *p*-coumarylquinic acid, both of which stimulate spore germination and formation of appressoria of the apple-rotting pathogen *Diaporthe perniciosa* (Brown and Swinburne, 1978).

The jacketed refrigeration system used in LP intermodal containers precludes transmitted heat from entering the container and causing the temperature and humidity to fluctuate sufficiently to promote water condensation on the commodity's surface. A proportional-integral-derivative (PID) thermostat controls the algorithm of a hot-gas bypass solenoid valve, eliminating ON–OFF refrigeration cycles that are a major cause of humidity and temperature fluctuations (9.7, 11.13, Figs 11.4 and 11.5). Infiltrating heat is kept to a minimum in LP due to the vacuum tightness of the container and expansion of the incoming air at the pressure regulator (6.18). Defrost cycles, which contribute to surface condensation in NA and CA containers, are eliminated in LP containers by using a secondary refrigerant coolant and substituting a brazed-plate heat exchanger in place of a conventional evaporator coil (13.12). At ambient temperatures ranging from −17.8 to 49°C, the wall temperature remains constant ± 0.2°C throughout the length of the LP container, and water cannot condense on the commodity because respiratory heat keeps its surface warmer than the dew point of the surrounding air.

7.9 Hypochlorous Acid Vapour

In a laboratory LP apparatus, an amount of available chlorine capable of preventing microbial growth can be continuously produced and passed over the stored commodity by adding alkaline sodium hypochlorite or other salts of hypochlorite to the water used to humidify the storage chamber (Burg and Burg, 1976). Air chlorinated in this manner is remarkably efficacious both at atmospheric (examples 12, 13, 19–21) and sub-atmospheric pressures (examples 14–21; Table 7.9) in preventing the growth of a wide variety of moulds and bacteria, both in pure culture and on stored commodities. The anti-microbial effectiveness of chlorine solutions is a function of the concentration of undissociated hypochlorous acid present. The active ingredient of chlorinated air also is hypochlorous acid vapour:

- The partial pressure of chlorine (pCl_2) in an air stream equilibrated with a solution of sodium hypochlorite is given by:

$$pCl_2 = [(H^+)^2(Cl^-)(OCl^-)]/[K_B'K_A'Q_C'] \quad (7.1)$$

  where Q_C' is the Henry's Law constant for chlorine, corrected for salting out; K_A' is the dissociation constant for hypochlorous acid, corrected for salting out; and K_B' is the equilibrium constant for the reaction of water on elementary chlorine, corrected for salting out. Equation 7.1 indicates that air equilibrated with a solution containing 2.63% sodium hypochlorite at pH 8.6 should contain 0.3 μl/l of chlorine gas, but when air passed through such a solution is analysed by bubbling it for 5 min through 10 ml of *o*-tolidine reagent contained in a darkened test tube, it produces a yellow colour as intense as that caused by an air stream containing 18 μl/l authentic chlorine. Thus, the chlorine dissolved in a 2.64% hypochlorite solution at pH 8.6 could only account for about 2% of the total evolved Cl^+ detected by the *o*-tolidine test.
- At an alkaline pH greater than 8.0, the amount of available chlorine measured in the effluent decreases about tenfold for a pH increase of one unit. The hypochlorous acid content also decreases tenfold under the same conditions, for it is related directly to the hydrogen ion concentration according to the expression:

$$HOCl = (H^+)(OCl^-)/K_A' \quad (7.2)$$

  whereas the dissolved chlorine content decreases 100-fold per unit increase in pH since it is proportional to $(H^+)^2$, according to the expression:

$$Cl_2 = (H^+)^2 (Cl^-)(OCl^-)/K_B'K_A' \quad (7.3)$$

  When allowance is made for the effect of salting out on the equilibrium constant for the dissociation of hypochlorous acid, the rate of generation of available chlorine is directly proportional to the calculated hypochlorous acid content between pH 8.5 and 10.4, and throughout the 0.014–0.7 M range of sodium hypochlorite concentrations tested (example 2). If the pH is increased above 8.5, dissolved chlorine contributes progressively less than 2% to the total evolved available chlorine, until eventually at pH 9 or a higher alkalinity, its contribution is insignificant.
- The available Cl^+ concentration in the effluent air is independent of the rate of airflow when air is passed at 0.015–0.15 standard cubic metres per hour through 500 ml of a 2.63% sodium hypochlorite solution (example 1). At 0–20°C, when the solution is violently and continuously agitated by mechanical means at a low flow rate, the concentration of available Cl^+ in the

effluent does not increase, indicating that the air stream is completely equilibrated with the hypochlorite solution. The reaction between hypochlorous acid and chloride ion that is the rate-determining step in the reversible formation of chlorine from hypochlorous acid has so low a rate constant that equilibrium is not attained at 0°C for several hours. Therefore chlorine does not form rapidly enough to account for the detected active chlorine.

- When the effluent vapour is passed over liquid elementary mercury, it produces a brown product on the surface of the mercury. Hypochlorous acid passed over mercury produces Hg(OH)Cl, a brown product, whereas chlorine gas produces a white–grey reaction product, $HgCl_2$.

Although hypochlorous acid is thermodynamically unstable, it is much more stable than any of the other hypohalite acids and, in fact, is the only one that can be distilled and recovered without extensive decomposition. Thus, hypochlorous acid can exist in the vapour phase for a sufficient period of time to control microbial development on materials that it contacts. The method by which the hypochlorite solution is brought into contact with the air is not critical. As rapidly as hypochlorous acid is lost from the solution it is regenerated from the large excess of hypochlorite ion present, causing the solution's alkalinity to increase by 0.1–0.3 pH units during several weeks regardless of the airflow rate. The tendency to change pH is partly overcome by the buffering action of hypochlorous acid, which has a pK of approximately 7.5. This stabilizes the pH between 6 and 9.3, but a much more important consideration is equilibration of the solution with atmospheric CO_2. To preserve electric neutrality in a dilute sodium hypochlorite solution in equilibrium with atmospheric CO_2, the concentrations of the various ionic components in solution must give rise to a total electric charge carried by the cations that is equal and opposite in magnitude to that carried by the anions:

$$(Na^+) + (H^+) = K_1'q'(pCO_2)/(H^+) + 2\,K_1'K_2'q'(pCO_2)/(H^+)^{2+}C_AK_A'/[(H^+) + K_A'] + (Cl^-) + K_W'/(H^+) \qquad (7.4)$$

where K_1' is the dissociation constant for carbonic acid, corrected for salting out according to the expression $pK_1 = pK - 0.5\omega^{1/2}$, where $\omega = ½\Sigma m_1 Z_1^2$, m_1 is the molality of the ion and Z_1 its valence; K_2' is the dissociation constant for bicarbonate corrected for salting out according to the expression $pK_2' = pK_2 - 1.1\omega^{1/2}$; q' is the Henry's Law constant for CO_2 corrected for salting out according to the expression log $(Q/Q') = 0.11\omega$, where $q' = Q'/760$; pCO_2 is the partial pressure of CO_2 in air (mm Hg); C_A is the hypochlorite concentration ($C_A = OCl^{-1} + HOCl$); K_A' is the dissociation constant of hypochlorous acid corrected for salting out; and K_W' is the dissociation constant of water corrected for salting out. This equation predicts, and tests verify, that sodium hypochlorite solutions ranging in concentration from 0.1 to 2.6% (weight/volume) equilibrate with atmospheric CO_2 at pH values in the range 8.5–9, and that when the [CO_2] is increased or the hypochlorite concentration decreased, the equilibrium pH is lower (Table 7.8; examples 2–7).

When a hypochlorite solution is exposed to a [CO_2] concentration substantially greater than that present in atmospheric air, equilibrium occurs at a pH lower than 8.5 and results in an initial rapid evolution of both hypochlorous acid vapour and gaseous chlorine. Subsequently both rates of evolution decline precipitously as the reservoir of hypochlorite ion is depleted through loss of these vapours and instability

Table 7.8. Equilibration pH of a 2.6% sodium hypochlorite solution with various concentrations of CO_2 in air at 25°C.

	Equilibrium pH	
% CO_2	Calculated	Observed
0.035	8.8	9.10
0.7	8.1	8.25
5.5	7.6	7.55

of hypochlorite solutions at pH values below 8.5 (example 7). To sustain the controlled evolution of a predetermined titre of hypochlorous acid vapour for a much longer duration, the solution must be stabilized against high [CO_2] by adding an appropriate amount of the sodium or potassium salt of bicarbonate or carbonate (example 8) in order to increase the cation concentration on the left side of equation 7.4 so that at any pH the solution will equilibrate with a much larger anion concentration, the source of which must be carbonate and bicarbonate ions derived from and in equilibrium with atmospheric CO_2. The potassium salts of carbonate and bicarbonate are approximately two to three times more soluble in water than their respective sodium analogues and, therefore, if the [CO_2] concentration is relatively high, the potassium salts of carbonate and bicarbonate must be used, preferably with potassium rather than sodium hypochlorite. At 0°C the method cannot be used with more than 0.8 M sodium or 1.6 M potassium ion. In the presence of 3% [CO_2] at 0°C or 15% [CO_2] at 25°C, it is possible to maintain the equilibrium pH at 8.5 with the potassium ion.

If the sodium or potassium salt of a very weak acid is substituted for the carbonate–bicarbonate mixture, the cations on the left side of equation 7.4 are supplemented with cations supplied from the salt of the weak acid, while the anions on the right side of the equation are augmented by the dissociated form of the acid expressed by the term $C_CK_C'/(H^+ + K_C')$, where C_C is the total concentration of the weak acid (associated and dissociated forms) and K_C' is the dissociation constant of the weak acid corrected for salting out. When the dissociation constant of the weak acid is small, the dissociated form of the acid contributes little to the total anion content at pH 8.5–10, and the situation approaches that of a pure carbonate–bicarbonate mixture (examples 9–12).

Alkyl amines such as diethanolamine or a mixture containing a large amount of diethanolamine with a small amount of thiourea, provide a ready source of cations, since at an alkaline pH diethanolamine is present almost exclusively in the positively charged form, NH_2^+R. This adds a term $C_B(H^+)/(K_B' + H^+)$ to the left side of equation 7.4, where C_B is the total concentration of amine added and K_B' is the basic dissociation constant for the amine corrected for salting out. Although alkyl amines sometimes are used to stabilize solutions against high [CO_2], they are oxidized by hypochlorite, and in addition have a significant vapour pressure, unpleasant odour and toxic properties that make them an undesirable choice for food treatment, in contrast to bicarbonates and carbonates.

In the presence of sodium or potassium salts of bicarbonate or carbonate, a hypochlorite solution maintains a high pH for a longer period of time than it would in their absence. Enough bicarbonate–carbonate buffer can be added to cause the solution to equilibrate at a higher pH, making it more stable, allowing a lower concentration of hypochlorous acid to be generated while providing a much larger total hypochlorite concentration. This greatly increases the size of the hypochlorite reservoir, which otherwise would be so small when low titres of hypochlorous acid vapour are required that it would not sustain a pH of 8.5 and hence the solution would tend to be unstable as well as low in reservoir capacity. The solution must be adjusted so that the titre of hypochlorous acid vapour in the effluent air is not high enough to damage the stored commodity (example 12).

In LP, the [CO_2] in the air passing through the hypochlorite solution is lower than it is in atmospheric air in proportion to the pressure reduction, so the equilibrium pH of the solution is elevated, causing it to evolve a lower concentration of Cl^+ for a longer duration of time. This makes it possible to use relatively high hypochlorite concentrations to stabilize the hypochlorous acid titre in the rarified air at a safe µl/l value for an extended period of time. At LP pressures ranging from 7.67 to 20 kPa (50–150 mm Hg), the method has proved useful for long-term decay control with tomatoes (example 16), bananas (examples 12 and 17), peppers, cucumbers (example 21), green beans, grapes (example 19), oranges (example 20), grapefruit, limes, avocados (example 14), mangoes, lettuce, strawberries (example 15 and Table 7.9) and pineapples. At a pressure of 13.33 kPa (100 mm Hg), this procedure prevented the growth of pure cultures of *Colletotrichum* sp., *C. gloeosporioides*, *T. paradoxa*, *Verticillium theobromae*, *B. cinerea*, *Sclerotinia sclerotiorum* and *Gloeosporium musarum*. It slightly inhibited the growth of *Pseudomonas marginalis* and caused *F. roseum* to develop abnormally (example 13).

Hypochlorous acid, chlorine and chlorine monoxide are strong oxidizing agents that destroy many materials of construction. Aluminium, glass, polyethylene and polypropylene plastics, and cadmium-plated steel were not affected during a 6-month exposure to 5 μl/l available chlorine in air, but stainless steel and brass were corroded and carbon steel and copper strongly corroded (Burg, 1970). The oxidative reaction not only destroys these materials, but it also depletes the available chlorine in the air. Rubber tubing rapidly reacts with the available chlorine without suffering any visible damage; cellulose fibres do not remove an appreciable amount of Cl^+; but plastic foam removes available chlorine so effectively from an air stream that it is useful as a filter, turning from white to brown as it reacts with chlorinated air. By proper choice of materials, it is possible to minimize loss of available chlorine and deliver a titre of chlorinated air through a laboratory vessel that is sufficient to eliminate microbial decay as a limiting factor during a laboratory storage test.

Cl^+ also can be generated by using an LP intermodal container's pneumatic air mover to pass air over a dry mixture of high-test calcium hypochlorite (HTH) fortified with sodium metaborate or potassium carbonate (Burg, 1971; Burg and Burg, 1976; Table 7.9). A mixture containing three parts potassium carbonate to one part HTH + 2 parts chlorox generates 2.5 μl/l in the air and 1–1.5 μl/l within boxes (Woodruff, 1971). The container's interior aluminium is not reacted by available chlorine, but an adequate Cl^+ titre is difficult to maintain because the vapour is rapidly depleted by reaction with the cardboard storage boxes and packing.

Table 7.9. Effect of HOCl vapour generated in an LP intermodal container on decay of Shasta strawberries. Fungicide was generated in an LP prototype 6.1 m intermodal container from 12 kg of HTH (high test calcium hypochlorite – 70% available chlorine) mixed with 1 gallon of commercial 'Clorox'. The container was equipped with a pneumatic air mover and operated at 3°C and a pressure of 21.33 kPa (160 mm Hg). The fungicide concentration within the container was approximately 20 μl/l and in the boxes 10 μl/l. A second LP intermodal container was operated without fungicide at 3°C and the same pressure. Strawberries (var. Shasta) were placed in both containers and in a commercial storage room at 3°C. After 10 days the berries were removed and scored for rot and mould (Woodruff, 1971).

	% Rot and mould
Initial inspection	1.6
After 10 days' storage	
LP + fungicide	4.2
LP w/o fungicide	35.3
NA control	30.3

7.10 Chilling Injury

Fruits and vegetables that have been injured by chilling are particularly susceptible to decay. *Alternaria* rot is unusually severe on chilled tomatoes, squash, aubergine, peppers and melons (McColloch and Worthington, 1952; McColloch, 1962; McColloch *et al.*, 1966; Hardenburg *et al.*, 1986), and anthracnose on chilled avocados (Spalding and Reeder, 1976b). At a critical 'transition' temperature, chilling changes the state of membranes from liquid to gel or semi-crystalline, disrupting the integrity of the membrane channels, causing metabolic dysfunction, solute leakage, tissue degeneration and a loss of compartmentation (Lyons, 1973; Wright and Simon, 1973; Wright, 1974). Sometimes the symptoms of chilling damage are not expressed during cold storage, but become evident when the commodity is returned to a warmer temperature for several days and its metabolism is accelerated (Hardenburg *et al.*, 1986).

Because fatty acids solidify at a higher temperature if they are saturated, the ratio of unsaturated to saturated fatty acids in membrane lipids influences the transition temperature at which chilling injury occurs (Fig. 7.8). This ratio and the chilling temperature can change depending upon available nutrients, time of year, stage of ripeness and growth conditions (Lyons, 1973). The duration of exposure to a 10–11°C chilling temperature required to injure bananas varies between clones, and even fruit to fruit, from a few hours to up to several weeks (Simmonds, 1959; Palmer, 1971). 'Full' fruits and bananas maturing at higher temperatures (tropical or summer subtropical climates) tend to be more susceptible to chilling than thinner grades and fruits maturing at lower temperatures (subtropical winter fruit). The chilling temperature decreases when mangoes (Bender *et al.*, 2000a) and tomatoes (Cook *et al.*, 1958;

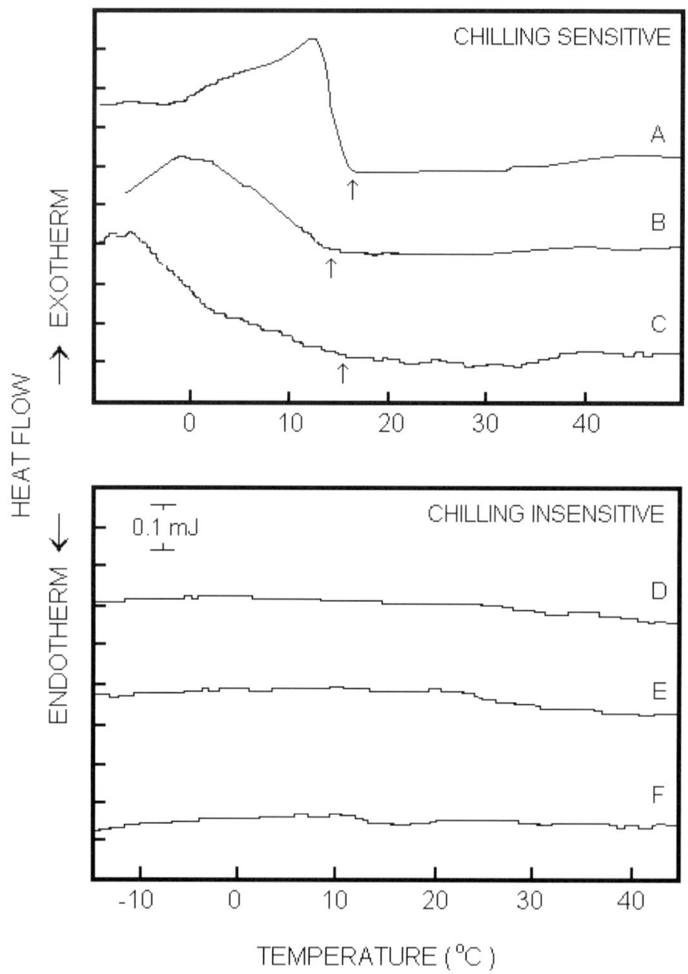

Fig. 7.8. Thermographs prepared by differential scanning calorimetry of polar lipids from mitochondria of chilling-sensitive and -insensitive plants. Each sample was scanned 10°C/min at a sensitivity of 0.2 mcal/s; (*upper*) chilling-sensitive plants with critical temperatures indicated by the thermally induced physical changes around 14–15°C: A, cucumber; B, mung bean; C, sweet potato roots; (*lower*) chilling-insensitive plants with no indication of a thermally induced physical change in the temperature range −15 to 45°C: D, artichoke roots; E, wheat roots; F, barley roots. Sample size was 3.6–11.0 mg of lipid (Raison and Orr, 1986).

Hardenburg *et al.*, 1986) ripen, but bananas suffer chilling damage below 7.2°C irrespective of their stage of ripeness (Poland and Wilson, 1933; Anon., 1961).

There is evidence that ethylene sometimes accentuates chilling injury. Chilling stimulates ethylene production by avocados (Eaks, 1983), cucumbers (Wang and Adams, 1982; Cabrera and Saltveit, 1990), okra (Lipe and Morgan, 1973), papaya (Chan *et al.*, 1985), pears (Wang *et al.*, 1972) and mature-green tomatoes (Autio and Bramlage, 1986), and ethylene absorbents and AVG application reduce chilling injury in stored tomatoes (Hong and Gross, 2000). Antisense ACC oxidase RNA considerably reduces the chilling sensitivity of Charentals cantaloupe melons, allowing them to withstand 3 weeks' storage at 2°C without developing pitting and rind browning when they were subsequently rewarmed (Ben-Amor *et al.*, 1999). Their improved tolerance to chilling is associated with a lower accumulation of ethanol and acetaldehyde, reduced membrane deterioration, a higher capacity to remove active oxygen species and increased activities of catalase, superoxide dismutase and peroxidase. Treating the antisense melons with 10 μl/l ethylene for more than 1 day prior to cold storage resulted in a restoration of chilling sensitivity. Conversely, if 1–10 μl/l ethylene was present during a 6-week storage of cv. Hermosa peach fruits at 0°C, the incidence of woolly breakdown (WB) was reduced during 5 subsequent days at 20°C without significantly affecting fruit softening (Sonego *et al.*, 2000).

Molecular species remodelling, changes in unsaturation, alterations in the proportions of lipid classes and increases in lipid-to-protein ratios may cause a membrane bilayer to remain fluid at a lower temperature. The production of unsaturated species of phosphotidylglycerol in temperature-sensitive plants depends on the specificity of their glycerol-3-phosphate acyltransferases. Plants such as squash, which produce large amounts of the 'saturated' molecular species of phosphatidylglycerol, seem to contain glycerol-3-phosphate acyltransferase enzymes that have a substrate specificity for palmitate in addition to oleate. Transferring the genes for these enzymes into a tobacco plant rendered it more chilling-sensitive, and when the genes for acyltransferases from chill-resistant *Arabidopsis* plants were introduced into the tobacco plant, it became more chilling-resistant (Harwood, 1997). The *fad* mutants of *Arabidopsis* have a reduced amount of polyunsaturated fatty acids and an elevated sensitivity to chilling (Miquel *et al.*, 1993), and when transgenic tobacco plants were engineered to contain increased levels of unsaturated trienoic fatty acid by the introduction of FAD7 cDNA, chilling injury was suppressed (Kodama *et al.*, 1994).

The proportion of unsaturated membrane fatty acids increases and the critical transition temperature decreases when chill-sensitive plants are acclimated to a low temperature (Graham and Patterson, 1982; Mazliak, 1983; Williams *et al.*, 1988; Kodama *et al.*, 1995). *Brassica napus* plants develop an increased ability to desaturate fatty acids in the major-membrane diacylglycerols of their leaves after temperature acclimation at 5°C for several weeks (Williams *et al.*, 1996), and keeping grapefruits for 1–3 weeks at an intermediate temperature before exposing them to 1–5°C counteracts some of the adverse effects of a low storage temperature (Hardenburg *et al.*, 1986). Preconditioning Valery bananas at 17.8°C for 8 days at pressures ranging from 8.0 to 21.3 kPa (60–160 mm Hg) protected them from developing under-peel discoloration (UPD) when the storage temperature subsequently was lowered to 11.7°C for eight additional days, but the LP fruits developed traces of a dull colour typical of slightly chilled bananas after they were gassed with ethylene and ripened for 4 days at 17.8°C (Table 7.10). The same temperature regime at atmospheric pressure had little or no effect on UPD development prior to or during ripening, and these fruits developed poor colour and intense browning. Bananas preconditioned in LP at 14.4°C for 6–12 days at pressures below 21.3 kPa (160 mm Hg) were protected against UPD development when the LP temperature was lowered to 4.4°C for up to 20 additional days, and UPD was not evident after storage at 8.0 kPa (60 mm Hg). Control fruit preconditioned and stored at atmospheric pressure and the same temperatures developed severe UPD (Table 7.11). LP did not diminish UPD development when green Valery bananas were continuously stored at 4.4°C without preconditioning, and only slightly ameliorated chilling injury during storage at 11.7°C. A similar result has been reported for Cavendish bananas stored in LP at 5–7°C (Olorunda *et al.*, 1978). UPD development in ripe Valery bananas was progressively suppressed when the storage pressure was lowered from 21.3 to 8.0 kPa (160–60 mm Hg) at 5–7°C (Burg, 1969), possibly because the low [O_2] present in LP inhibits polyphenol oxidase, the enzyme that converts tyrosine to dopamine (Bonner, 1957).

Table 7.10. Effect of LP on chilling-induced under-peel-discoloration (UPD) of Valery bananas while in storage, and colour development during subsequent ripening. After 16 days' storage, fruit was gassed with ethylene and ripened for 4 days at 17.8°C. UPD was scored on a 0–4 scale when the fruit was still green, before it was gassed with ethylene; 0 = no UPD and a score of 4 = severe UPD. The air control was stored in 'Chiquita' bags (Burg, 1971).

Storage condition	Temp °C 0–8 days	Temp °C 9–16 days	UPD at 16 days	Colour during ripening
Air control	17.8	17.8	1	fair
	17.8	11.7	3	dull-brown
	11.7	17.8	4	dull-brown
	11.7	11.7	5	poor-brown
100 or 160 mm Hg	17.8	17.8	0	excellent
	17.8	11.7	1	dull
	11.7	17.8	3	dull
	11.7	11.7	3	dull
60 mm Hg	17.8	17.8	0	excellent
	17.8	11.7	1	dull-sl. green
	11.7	17.8	3	dull-sl. green
	11.7	11.7	3	dull-sl. green

Table 7.11. Effect of LP on chilling-induced under-peel-discoloration (UPD) of Valery bananas while in storage. UPD was scored on a 0–4 scale at the indicated times; 0 = no UPD and a score of 4 = severe UPD (Burg, 1971).

Storage condition	Temp °C 0–6 days	Temp °C 7–12 days	Temp °C 13–33 days	UPD at 12 days	UPD at 33 days
Air control	14.4	14.4	4.4	–	4*
	14.4	4.4	–	4	–
	4.4	14.4	–	4	–
	4.4	4.4	–	4	–
160 mm Hg	14.4	14.4	4.4	–	3
	14.4	4.4	–	3	–
	4.4	14.4	–	4	–
	4.4	4.4	–	4	–
100 mm Hg	14.4	14.4	4.4	–	2
	14.4	4.4	–	2	–
	4.4	14.4	–	4	–
	4.4	4.4	–	4	–
60 mm Hg	14.4	14.4	4.4	–	1
	14.4	4.4	–	1	–
	4.4	14.4	–	4	–
	4.4	4.4	–	4	–

*Fruit was 'green ripe'.

In apple cultivars, the symptoms of chilling and low [O_2] injury often are indistinguishable, and the symptoms can be alleviated either by raising the temperature or the [O_2] level (Blanpied, 1990a). The temperature must be slightly elevated to prevent low-temperature breakdown of Cox Orange Pippin (Sharples and Johnson, 1982) and Ida Red apples (Johnson and Ertan, 1983) in storage atmospheres containing less than 2% [O_2], and parenthetically it also has been reported that an increased storage temperature is needed to reduce the susceptibility of apples to low [O_2] injury (Lau, 1985; Olsen, 1985). Because of this correlation, low-temperature breakdown has been diagnosed as an expression of chilling injury induced by a low O_2 partial pressure, and it has been suggested that low [O_2] may raise the temperature threshold for chilling injury (Blanpied, 1990a). During CA storage, the percentage of polyunsaturated fatty acids in the pericarp of Fuerte avocados progressively decreases in inverse relation to the O_2 partial pressure (Mazliak, 1965), and possibly a similar response explains why low [O_2] seems to elevate the temperature threshold for chilling injury in some apple varieties during CA storage. The same response might also occur in LP, and therefore in order to avoid chilling injury at very low pressures, the storage temperature sometimes may need to be elevated slightly above the value that otherwise would be suitable.

An [O_2] concentration of 7% is optimal in alleviating chilling injury in bananas, limes and grapefruits at atmospheric pressure, and a relatively high LP pressure has the same effect on these fruits (Pantastico *et al.*, 1967, 1968; Pantastico, 1968). Rind pitting of Marsh grapefruits held at 4.4°C for 7 weeks was decreased from 23.5% in NA, to 4.4% in LP at 29.33 kPa (220 mm Hg = 5.9% [O_2]), but a pressure of 40 kPa (380 mm Hg = 10.3% [O_2]) had no effect on chilling susceptibility (Grierson, 1971). Hypobaric storage at 29.33 kPa (220 mm Hg) completely prevented rind pitting in Persian limes stored for 4 weeks at 4.4°C, while 65.4% of limes stored in NA at the same temperature suffered damage (Pantastico *et al.*, 1968). In other studies with limes, chilling damage decreased at 33.3 kPa (250 mm Hg) and increased when the pressure was lowered to 10.7 kPa (80 mm Hg) (Burg, 1970). These effects of LP on chilling injury probably result from the low O_2 level and nearly saturated humidity present during hypobaric storage. A high humidity ameliorates low-temperature injury in bananas (Wardlaw, 1961; Maltei, 1972; Ulrich, 1975), lychee fruits, cucumbers, peppers, limes and grapefruits (Morris and Platenius, 1938; McColloch, 1962; Pantastico *et al.*, 1975b, Hardenburg *et al.*, 1986). Prevention of desiccation may be an important factor (Pantastico *et al.*, 1975b) and grapefruits, limes, bananas, cucumbers and peppers possess stomates which are capable of opening and promoting postharvest transpirational water loss (4.15). Chill-hardened or chill-resistant plants are able to reduce stomatal apertures and maintain positive leaf turgor during a cold treatment, but the stomates of chill-sensitive species are locked open during chilling periods despite severe wilting (Willmer and Fricker, 1996) especially if they are rapidly cooled (Willis *et al.*, 1989). In ohter studies saturated humidity and pressures ranging from 5.33 to 29.33 kPa (40–220 mm Hg) did not prevent chilling symptoms from developing in mature-green tomatoes, cucumbers, limes, bananas, green peppers and avocados (Morris and Kader, 1975; Burg, 1976b; Lyons, 1973).

CO_2 sometimes alleviates chilling injury or lowers the temperature required to induce chilling damage. High [CO_2] applied before or during low-temperature storage reduces chilling damage in okra (Ilker and Morris, 1975), summer squash (Thompson, 1998), peaches (Kajiura, 1975) and chilli peppers (Kader *et al.*, 1975). Chilling injury also is decreased when grapefruit are conditioned with a short pre-storage exposure to 40% [CO_2] at 21°C, before the fruit is transferred to 4°C (Hatton and Cubbedge, 1982). Both intermittent exposure to 20% [CO_2] and CA storage in 3–9% [CO_2] + 1–3% [O_2] limits chilling injury in avocados kept at temperatures as low as 4°C (Spalding and Reeder, 1975; Marcellin and Chaves, 1983;

Hardenburg *et al.*, 1986; Thompson, 1998), and high [CO_2] in the storage atmosphere lowers the incidence of chilling-induced internal decay in guava, bananas and peaches (Wade, 1975; Thompson, 1998). Peaches suffer low-temperature injury at 1°C when they are stored in 3% [O_2], but injury is almost completely controlled if 3% [CO_2] is included (Kajiura, 1975). Conversely, high [CO_2] accentuates low-temperature injury in cucumbers and tomatoes, and no combination of [O_2] or [CO_2] reduces chilling injury in cucumbers (Eaks, 1956), mangoes (Bender *et al.*, 2000a) or tomatoes (Kader and Morris, 1975b). A 'scald-like' chilling injury of Empire apples, which is particularly severe during CA storage at 1°C, is exacerbated by > 2% [CO_2] (Burmeister and Dilley, 1995).

At atmospheric pressure with 2 or 21% [O_2] present, and in LP at a pressure of 10–20 kPa (75–150 mm Hg = 1.8–3.9% [O_2], extensive chilling damage and anthracnose resulted during the storage of Lula and Waldin avocados at 10 and 7.2°C (Tables 7.12 and 7.13), respectively, and in Fuchs avocados at 4.4°C (Reeder and Hatton, 1971; Spalding and Reeder, 1975; Hatton and Spalding, 1990). Chilling injury and decay development was essentially eliminated when 10% [CO_2] was included in the 2% [O_2] storage, or added during LP storage at a pressure of 12.13 kPa (91 mm Hg = 2% [O_2]) by flushing the storage chamber with a 16.7% [O_2] + 83.3% [CO_2] mixture to provide 2% [O_2] + 10% [CO_2] at the low storage pressure (Tables 7.12 and 7.13). The

Table 7.12. Quality of 'Lula' avocados stored at 10°C and 98–100% relative humidity for 6 weeks in various atmospheres under normal or sub-atmospheric pressure and then softened at 21.1°C in air at 760 mm Hg (Spalding and Reeder, 1976b).

Storage conditions					
mm Hg	% O_2[z]	%CO_2[z]	Decay Index[y]	Chilling index[y]	Acceptable fruit (%)[x]
760	21.0	0	4.0	4.0	0
760	2.0	10	1.8	0.1	70
152	3.9	0	4.0	4.0	0
76	1.8	0	4.0	2.7	0

[x]Acceptable fruit had good appearance, were free of moderate decay and chilling injury and had no off-flavours.
[y]In the rating system for decay and chilling, 0 = none; 1 = trace; 2 = slight; 3 = moderate; 4 = severe. Fruit with 3 or 4 ratings were considered unacceptable.
[z]The % O_2 values at 152 or 76 mm Hg have been corrected to account for the presence of water vapour in the saturated mixture.

Table 7.13. Quality of 'Waldin' avocados stored at 7.2°C for 25 days in various gas mixtures at 760 or 91 mm Hg and then softened at 21.1°C in air at atmospheric pressure (Spalding and Reeder, 1976b).

Storage condition					
mm Hg	% [O_2][y]	% [CO_2][y]	Decay (%)	Chilling (%)	Acceptable fruit (%)[x]
760	21.0	0	76	100	0
760	2.0	0	100	100	0
760	2.0	10.0	4	4	92
91	2.3	0	64	100	0
91	1.8	9.2	0	0	100
Unstored fruit			4	0	96

[x]Acceptable fruit had good appearance, were free of moderate decay and chilling injury and had no off-flavours.
[y]The % O_2 and % CO_2 values at 91 mm Hg have been corrected to account for the presence of water vapour in the saturated mixture.

added CO_2 must reduce decay indirectly by preventing chilling damage, since a 2% $[O_2]$ + 10% $[CO_2]$ mixture has almost no direct effect on the colony growth of the causal agent of avocado anthracnose, *C. gloeosporioides* (Spalding and Reeder, 1975).

LP avoids decay and chilling damage in avocados by taking advantage of this fruit's ability to tolerate extremely low $[O_2]$ under hypobaric conditions (Table 4.7). Although avocados can be stored for long periods in LP at a non-chilling temperature and a pressure of 8.13 kPa (60 mm Hg = 1.3% $[O_2]$; Apelbaum *et al.*, 1977b), the result is much better at 2.67 kPa (20 mm Hg = 0.28% $[O_2]$; Spalding and Reeder, 1980, unpublished results; Burg, 1990). At 10°C, growth and spore germination in *C. gloeosporioides* cultures are not reduced at a pressure of 20 kPa (150 mm Hg = 3.9% $[O_2]$), but they are severely inhibited at 2 kPa (15 mm Hg = 0.14% $[O_2]$; Table 7.3). Therefore, the storage pressure used in the studies depicted in Tables 7.12 and 7.13 had little if any direct effect on *C. gloeosporioides*' development. A 2.67 kPa (20 mm Hg) storage pressure almost completely prevents *C. gloeosporioides*' spore germination and growth both in pure cultures (Table 7.3) and in papayas (Fig. 7.7), and chilling damage will not occur if the temperature is maintained high enough at that pressure.

To slow decay and physiological change, the storage temperature preferably should be set at the lowest value that does not cause chilling damage, but since low-temperature injury depends on both temperature and exposure time, a temperature that is suitable for short-term transport or storage may cause damage when the duration is extended. Normally bananas are transported overseas at 11–13°C with minimum chilling injury, but even at 13°C, chilling occurs if the voyage is longer than 8 days (Simmonds, 1959; Palmer, 1971). It can be anticipated that because LP is used to prolong storage life, instances will arise in which chilling injury will occur unless the LP temperature is slightly elevated above that normally used for NA storage.

Chill-sensitive commodities such as cucumber, green pepper, tomato and banana have been successfully kept at 10.67 kPa (80 mm Hg) for long durations of time (Burg, 1990), but decay eventually limits storage life at that pressure (Bangerth, 1974). In view of the improved decay control and storage life demonstrated with avocados, melons, pineapples, mangoes and papayas when the pressure is lowered from 10.67 to 2.67 kPa (80 to 20 mm Hg), and the superior decay control in strawberries, mushrooms, cherries, asparagus and cut flowers stored at 0°C and 1.33 kPa (10 mm Hg), the optimal pressure/temperature combination for cucumber, green pepper, tomato and banana should be reinvestigated to determine whether these commodities are able to tolerate a pressure of 2.0–2.67 kPa (15–20 mm Hg). The major pathogens of green peppers [*A. tenuis* (alternaria rot), *C. gloeosporioides* (anthracnose) and *E. carotovora* (bacterial soft rot)]; tomato [*A. tenuis* (alternaria rot) and *E. carotovora*]; and cucumbers [*Colletotrichum* sp.] cannot grow at 2.67 kPa (20 mm Hg). In a preliminary 2-week trial, tomatoes, green peppers and cucumbers benefited from LP storage at 13°C and a pressure of 2.67 kPa (20 mm Hg), and moulds that developed prematurely on cucumbers in NA were suppressed in LP (Davenport and Burg, 2003, unpublished data).

7.11 Physiological Disorders other than Chilling Injury

Apple scald may result from injury to epidermal cells caused by 6-methyl-5-hepten-2-one (MHO), a conjugated oxidation product of a volatile terpene constituent of apple wax, α-farnescene (Huelin and Coggiola, 1970; Anet, 1972). The idea that harmful compounds of low volatility accumulate in an apple's cuticle and cause scald is based on the observation that (i) loosely packed or well-ventilated apples develop less scald than tightly packed or poorly ventilated fruit; (ii) if parts of an apple are covered with an impervious material, the area of the skin under the material tends to scald; (iii) the presence of activated carbon near the surface of an apple prevents scald; (iv) removal of the cuticular wax sometimes prevents scald; (v) when the apple skin is punctured, the surrounding skin tends to be resistant to scald; and (vi) there is a general correlation between the cuticular content of α-farnescene and the intensity of scald development (Huelin and Kennett, 1958, 1959; Meigh, 1969).[3] The auto-oxidation of α-farnescene requires O_2, and the incidence of scald decreases when $[O_2]$ is reduced to 1% at atmospheric pressure, but a further reduction in $[O_2]$ induces scald (Chen *et al.*, 1985; Little *et al.*, 1985; Johnson *et al.*, 1990, 1993; Lau and Yastremski, 1993; Sfakiotakis *et al.*, 1993).

Over the course of many seasons, superficial apple scald has never been observed during or following hypobaric storage, although often a high incidence developed on comparable fruits kept in NA or CA (Dilley, 1972, 1977a; Bangerth, 1973). Scald development did not occur in apples stored at 1°C for 8 months in LP at 5 kPa (37.6 mm Hg = 0.9% $[O_2]$) provided that the fruits were placed in LP immediately after harvest or transferred from NA to LP within 1 month, but the disorder developed in five out of six cultivars of apples stored in NA (Wang and Dilley, 2000). After 6 months' storage, the concentration of 6-methyl-5-hepten-2-one (MHO) in the epicuticular wax was 97% lower in Granny Smith apples stored in LP, and 65% lower in Law Rome apples compared to storage at atmospheric pressure.[4] It was suggested that hypobaric storage decreases the rate of formation of MHO by lowering the $[O_2]$, and possibly enhances the diffusive escape of scald-related compound(s) that otherwise partition into and accumulate in the epicuticular wax of fruit stored at atmospheric pressure (Dilley, 1963).

Ethylene promotes scald in apples. Ethylene and ethephon accelerate and AVG inhibits cuticular wax formation and α-farnescene accumulation (Ju and Curry, 2000; Golding *et al.*, 2001; Ju and Bramlage, 2001); ethylene removal delays the production of α-farnescene in the surface wax and lowers the levels of its oxidation products (Knee and Hatfield, 1981b; Blanpied, 1990b; Johnson *et al.*, 1993); and 1-MCP inhibits ethylene production, α-farnescene accumulation and scald development in McIntosh, Delicious, Granny Smith, Empire and Law Rome apples (Rupasinghe *et al.*, 2000; Watkins *et al.*, 2000; Zanella, 2001). At atmospheric pressure, it is difficult to lower the ethylene concentration. During an 8-month CA storage in 1% $[O_2]$, even though ethylene was continuously scrubbed to < 0.3 μl/l in the storage atmosphere, the interior of the fruit nevertheless contained 2.5–8 μl/l ethylene (Sfakiotakis *et al.*, 1993). LP controls scald by slowing ethylene production and removing the gas from within the storage atmosphere and the apple's interior.

Crumbly breakdown occurred as an internal disorder of McIntosh apples stored in LP, especially at 3.5°C, and less so at 0.5°C (Bérard, 1982), but with this exception, internal breakdown had not been observed in any other type of apple stored hypobarically (Dilley, 1977a). LP prevented internal breakdown and controlled low-temperature-induced flesh browning in other apple varieties (Dilley, 1971), and reduced the incidence of Jonathan spot and soft scald (Kim *et al.*, 1969a), bitter pit and core flush (Bangerth, 1973). Even a short exposure of Spartan apples to LP prior to cold storage seemed to reduce the incidence of breakdown (Porritt *et al.*, 1975). If acetaldehyde and acetic acid are causative agents for internal breakdown and soft scald, as has been proposed, their accumulation might be minimized by hypobaric ventilation because they are quite volatile (Dilley, 1977a) and penetrate the plasma membrane with reasonable ease, making it likely that LP will enhance their diffusive escape (3.25).

By reducing the internal $[O_2]$, $[CO_2]$ and ethylene, LP creates the optimal environment for limiting scald and low-temperature breakdown, since these disorders are promoted by ethylene and > 1% $[CO_2]$, and alleviated by 0.8–1.2% $[O_2]$ storage and ethylene removal (Anon., 2001; Zanella, 2001).

7.12 Examples

1. When air was bubbled through 500 ml of a 2.6% sodium hypochlorite solution at a rate of 0.27 standard cubic metres per hour (SCMH), the initial pH of 11.0 equilibrated at 9.2, and 8–10 μl/l of hypochlorous acid vapour was released in the effluent. Similar pH and μl/l Cl^+ values resulted with airflow rates of 0.06 and 0.015 SCMH, and also when the solution was continuously mixed with a magnetic stirrer and air was passed over its surface. Flowing 0.27 SCMH without stirring continuously released 5 μl/l of hypochlorous acid vapour at a pH of 9.2 even though the fluid surface area was only 28 cm^2. When 200 ml of the same solution was placed in a trough, increasing the surface area in contact with the air to 240 cm^2, the pH declined to 9.2 from an initial value of 11 within 18 h at an airflow rate of 0.27 SCMH, and thereafter 8–10 μl/l of hypochlorous acid vapour was continuously released into the air stream.

2. Flowing 0.27 SCMH of atmospheric air through 500 ml of 0.1–2.63% sodium hypochlorite solutions initially released chlorinated air containing between 0.8 and 8 μl/l of Cl^+, and the titre did not decline appreciably for a long duration of time. A 1.4% solution of sodium hypochlorite still emitted 2 μl/l of Cl^+ into the effluent air after 88 days. This entire range of Cl^+ concentrations is sufficient to exert control over microbial development.

3. Bubbling 0.13 SCMH of air containing 0.035% [CO_2] through 500 ml of a 2.63% sodium hypochlorite solution yielded 8 μl/l of hypochlorous acid vapour for several weeks, and within a few days the pH of the solution declined from an initial value of 10.9 to approximately 9.1. Subsequently, the pH increased by a few tenths of a unit during several months while the hypochlorous acid vapour progressively declined to 3 μl/l. When CO_2-free air was bubbled through the pH 9.1 solution, within a few days the pH rose to 10.1 and the Cl^+ content of the effluent air declined to a very low value. The process is completely reversible; the pH fell to 9.1 and the chlorine content of the effluent air increased soon after atmospheric air was again introduced. When CO_2-free air was passed through an identical solution, the pH did not decrease significantly and only 0.25 μl/l of Cl^+ were present in the effluent air. The addition of 5.5% [CO_2] to the air, which was bubbled through the same solution, increased the titre of available chlorine in the effluent to 120 μl/l within a few hours; after 1 day it decreased to 25 μl/l, and within 6 days to 6 μl/l, by which time the pH was 7.5. Clearly, the reaction generating hypochlorous acid is driven not only by the law of mass action, but also by equilibrium with atmospheric CO_2, which prevents the solution's pH from rising when hypochlorous acid is vapourized and lost.

4. When 0.13 SCMH of atmospheric air containing 0.035% [CO_2] was bubbled through 500 ml of a solution containing 0.3% sodium hypochlorite, the pH initially equilibrated at 8.7 and the effluent air contained approximately 2 μl/l of hypochlorous acid vapour. These values remained constant for several weeks, and then by 45 days the pH increased to 8.9 and the hypochlorous acid vapour decreased to 0.7 μl/l. When the same solution initially was exposed to 0.7% [CO_2], the pH stabilized at 7.85 and the effluent air contained 14 μl/l of hypochlorous acid vapour, and then within 5 days the pH increased to 8.22 and the hypochlorous acid vapour titre decreased to 5 μl/l.

5. Equation 7.4 predicts that solutions containing 2.63, 0.53 and 0.11% sodium hypochlorite should equilibrate with atmospheric CO_2 at pH 8.8, 8.65 and 8.45, respectively. Using dilute solutions of commercial sodium hypochlorite bleaching solution, it was found that the stated concentrations had initial pH values of 10.9, 10.5 and 10.05, respectively, and that within a few days after atmospheric air was bubbled through 500 ml of each solution at a rate of 0.22 SCMH, the pH stabilized at 9.05, 8.85 and 8.50, respectively, in excellent agreement with the calculated values (Table 7.8). At slower rates of airflow, equilibrium was reached in progressively longer times. If air containing 0.7% or 5.5% [CO_2] was flowed, the pH equilibrated at 8.25 and 7.55, respectively, in agreement with values of 8.1 and 7.6 predicted by equation 7.4.

6. When 2.6% sodium hypochlorite was added to a carbonate–bicarbonate buffer containing 3 M potassium ion, the solution initially had a pH of 9.6 at 25°C, but within 12 h after 0.22 SCMH of air containing 5.5% [CO_2] was bubbled though 500 ml of the mixture, the pH stabilized at 9.0 and thereafter the effluent air contained 14 μl/l of hypochlorous acid vapour.

7. Adding 2.6% sodium hypochlorite to a carbonate–bicarbonate buffer containing 0.16 M potassium ion resulted in a solution with an initial pH of 9.1 at 25°C. When 0.22 SCMH of 0.7% [CO_2] in air was bubbled through 500 ml of this solution, within 4 h the pH stabilized at 8.85 and thereafter the effluent air stream contained approximately 14 μl/l of hypochlorous acid vapour. When the same solution was ventilated with air containing 0.035% [CO_2], the pH increased to 9.55 and remained steady for 48 days while the hypochlorous acid vapour content of the effluent air changed from an initial value of 3.6 μl/l to a final value of 2.0 μl/l.

8. At 25°C, a 2.6% sodium hypochlorite solution containing 3 M potassium ion equilibrated at pH 9.0 with 5.5% [CO_2], pH 8.7 with 7.5% [CO_2] and pH 8.45 flowing 17% [CO_2]. When 3 M sodium ion was added, a white precipitate, sodium bicarbonate, resulted when 5.5% [CO_2] was flowed, causing the pH to decrease below 9.1.

9. The sodium and potassium salts of metaborate are highly water soluble, relatively stable, known to be safe when applied to food materials, and endowed with fungicidal activity. By titration it was found that their pK_A at infinite dilution is approximately 10.2 at 25°C. These salts should be almost as efficacious as sodium and potassium carbonate–bicarbonate in buffering hypochlorite solutions against various concentrations of CO_2. A solution containing 2.6% sodium hypochlorite and 0.16 M sodium metaborate equilibrated with air containing 0.035% [CO_2] at pH 9.5, and a solution containing an equivalent concentration of sodium bicarbonate equilibrated at pH 9.6. The same solutions ventilated with 0.7% [CO_2] equilibrated at pH 8.7 and 8.85, respectively, and both yielded approximately 12 μl/l of hypochlorous acid vapour. Without the added salts, the solution equilibrated at pH 8.25 and initially yielded 24 μl/l of hypochlorous acid vapour.

10. A 0.5% hypochlorite solution initially stabilized at pH 8.9 and released 3.6 μl/l of hypochlorous acid vapour when it was ventilated with air containing 0.035% [CO_2]. Within 48 days, the pH rose to 9.2 and the hypochlorous acid vapour titre declined to 0.8 μl/l. When, to a 2.6% hypochlorite solution, 0.16 M sodium ion was added as the salt of a carbonate–bicarbonate mixture, initially 3.6 μl/l of hypochlorous acid vapour was released when the solution was ventilated with air containing 0.035% [CO_2]. The solution equilibrated at pH 8.65 and the pH remained constant for 48 days, while the titre of hypochlorite acid vapour only declined to 2 μl/l.

11. A 0.1% sodium hypochlorite solution ventilated with 0.035% [CO_2] initially released 1.5 μl/l of hypochlorite vapour and equilibrated at pH 8.5. Within 45 days, the pH rose to 8.85 and the hypochlorous acid titre declined to 0.12 μl/l. When 0.8 M sodium ion was added as a carbonate–bicarbonate buffer to a solution containing 2.6% sodium hypochlorite, and the mixture was ventilated with 0.035% [CO_2], initially it released 2.3 μl/l of hypochlorite vapour and equilibrated at pH 10.0. After 45 days it still released 1.2 μl/l of vapour at the same pH.

12. Air was flowed at a rate of 0.07 SCMH through 500 ml of 0.5% hypochlorite solution and then over several hands of green Valery bananas maintained at 16°C. The solution generated 3.6 μl/l of hypochlorous acid vapour, which scalded the peel of the bananas within 5–6 days. A 0.1% hypochlorite solution, which initially gave rise to effluent air containing 1.67 μl/l of hypochlorous acid vapour, and after 8 days to only 0.89 μl/l, also scalded the surface of banana fruits. When 0.2% hypochlorite was supplemented with 0.16 M sodium carbonate, the effluent air continuously contained 0.35 μl/l of hypochlorous acid vapour during a 14-day period, the fruit remained green, the surface was not scalded and crown rot did not develop, whereas when air was passed through water not enriched with hypochlorite and then over bananas, 100% of the fruits developed crown rot in 14 days. Fruit treated with hypochlorous acid vapour ripened normally when it was gassed with ethylene.

13. A concentration of 0.5–1 μl/l of Cl^+ was produced in the effluent low-pressure air stream when a 0.25% sodium hypochlorite solution was present in the humidifying flask of a laboratory hypobaric set-up (Fig. 2.4) operating at a pressure of 13.33 kPa (100 mm Hg). This titre of Cl^+ almost completely inhibited the growth of cultures of *Colletotrichum* sp., *C. gloeosporioides*, *T. paradoxa*, *V. theobromeae*, *B. cinerea*, *Sclerotinia sclerotiorum* and *G. musarum* incubated in the LP desiccator; slightly inhibited the growth of *Pseudomonas marginalis*; and caused *F. roseum* to develop abnormally.

14. Avocados stored at 12.8°C and a pressure of 13.33 kPa (100 mm Hg) developed stem-end rot after 30 days, whereas if they were treated with low-pressure air that had passed through 0.1, 0.25, 0.5 or 1.6% sodium hypochlorite solution, they did not develop this disorder during 30 days' storage or subsequently, when they ripened with normal taste and colour.

15. Heavy mould development occurred within 3–8 days when Florida 90 strawberries were stored at 3.3°C in LP at a pressure of 21.33 kPa (160 mm Hg) or in NA, but if sodium hypochlorite solutions ranging from 0.1 to 5.25% were used in place of the humidifying water, no mould development occurred during 21 days. When a box of berries that had just started to develop mould in atmospheric air was transferred to a hypobaric chamber and treated with vapours derived by passing air through 1.5% sodium hypochlorite, the mould was killed. Sequoia strawberries stored at 3.3°C and a pressure of 21.33 kPa (160 mm Hg) developed severe mould in 22 days, but not if the fruits were treated at the same pressure with vapours derived by passing air through a 3.0% sodium hypochlorite solution.

16. Breaker tomatoes (var. Homestead No. 9) were stored at 14.4°C in an LP apparatus operated at a pressure of 10.67 kPa (80 mm Hg). After 4 weeks, 25% of the fruit had severe *Fusarium* rot and an additional 25% showed a trace of mould development. At that time, none of the fruit stored at the same pressure had *Fusarium* rot if they had been treated with air flowed through a 1.5% sodium hypochlorite solution, and a 0.5% sodium hypochlorite solution gave partial protection. When tomatoes that had been exposed to chlorinated air were transferred to atmospheric air at 14.4 or 26.7°C, they ripened normally without developing *Fusarium* rot.

17. Valery bananas stored at 14.4°C and a pressure of 13.33 kPa (100 mm Hg) developed slight crown rot in 15 days and severe crown rot in 27 days, but if 0.1% sodium hypochlorite was present in the humidifying water, mould development was prevented for 42 days and the fruit ripened with good appearance and taste when it was removed and gassed with ethylene. LP fruit stored in air passed through a 0.25% sodium hypochlorite solution did not develop crown rot within 63 days. Fruit that was not treated with HOCl vapour began to develop crown rot after 24 days of storage, and when at that time it was exposed to the vapours generated from a solution containing 0.25% sodium hypochlorite, the crown rot was killed, and it did not reappear even after 103 days of storage.

18. During a 5-week storage at 3.3°C and a pressure of 13.33 kPa (100 mm Hg), iceberg lettuce was not injured by air passed through solutions containing 0.01–0.5% sodium hypochlorite. None of these heads developed bacterial soft rot, whereas 33% of an equivalent sample of lettuce developed this disorder when it was stored in LP at the same pressure without sodium hypochlorite added to the humidifying water.

19. Red Emperor grapes were stored at 1.7–4.4°C in NA or at pressures of 4.0, 8.67, 10.67 and 21.33 kPa (30, 65, 80 and 160 mm Hg) flowing air which had been contacted with either 0, 0.25, 1.5 or 5.25% sodium hypochlorite solution, respectively. After 60 days, none of the grapes exposed to hypochlorous acid vapour had developed mould, whereas there was moderate mould development in all lots lacking HOCl.

20. Temple oranges were stored at 7.2°C either in NA or in flowing air that had been passed through a 2.6% sodium hypochlorite solution. After 37 days, all fruit in NA was heavily contaminated with blue and green mould, whereas no mould developed on fruit treated with HOCl vapour. Florida Valencia oranges were stored at pressures of 6.67, 9.33, 14.67, 21.33 and 101.3 kPa (50, 70, 110, 160 and 760 mm Hg) flowing air that had been bubbled through water or 2.5% sodium hypochlorite solution. In 110 days, 24% of the fruit stored without HOCl vapour in LP or NA had developed mould, whereas after 157 days none of the fruit treated with HOCl vapour had decayed.

21. Cucumbers were stored at 7.2°C in NA or in LP at a pressure of 10.67 kPa (80 mm Hg), flowing either normal air or air which had been passed through 1.1 or 2.5% sodium hypochlorite solution. After 17 days, 50% of the cucumbers in NA which had not been treated with HOCl developed soft spots and decay, whereas all other cucumbers were in excellent condition. By 31 days, 100% of the control cucumbers were rotten and yellowing with mould, and 100% of the cucumbers in LP without HOCl had mould and were beginning to soften. After 43 days, the LP cucumbers treated with HOCl began to develop soft spots, but even after 46 days no mould was visible.

Notes

1. 2.3% [O_2] did not inhibit the growth of *A. alternata*, *B. cinerea*, *C. gloeosporioides*, *Dothiorella gregaria*, *G. candidum*, *P. digitatum*, *P. italicum*, *P. expansum*, *Phomopsis citri*, *Phytophthora cactorum*, *R. stolinifer* and *T. paradoxa*, and caused a 15–50% growth inhibition of *Ascochyta caricae*, *Botryodiplodia theobromae*, *M. fructicola*, *Phytophthora parasitica*, *V. theobromae* and *Whetzelinia sclerotiorum* (El-Goorani and Sommer, 1979).
2. *E. carotovora* causes soft rot in carrots, lettuce, cabbage, celery, cucumber, aubergine, iris, muskmelon, hyacinth, onion, parsnip, pepper, potato, radish, tomato, turnip, potted *Chrysanthemum* and other plants. *E. atroseptica* causes a black rot on the stems and tubers of potatoes and other vegetables; *Ps. fluorescens* has been associated with pink eye disease in potato tubers, and its metabolic and cultural characteristics are closely related to those of *Ps. marginalis*, the bacterium that causes marginal leaf blight in lettuce (Wells, 1974).
3. Spray application of α-farnescene does not induce scald (Meigh, 1969).
4. LP reduced the cuticular concentration of α-farnescene by 99% in Granny Smith apples, but elevated it by 225% in Law Rome apples.

8

Insect Quarantine

The USA and other countries have regulations preventing the importation of horticultural commodities that have not been treated by an approved method ensuring the mortality of specified insect pests (Table 8.1). Fumigation with ethylene dibromide (EDB), carbon disulphide, hydrogen cyanide and methyl bromide satisfied this requirement in the past, but the use of carbon disulphide and hydrogen cyanide has been discontinued because high toxicity and flammability make these substances hazardous to workers and potentially dangerous to the environment, and EDB was banned in the USA and other countries due to suspected carcinogenicity (Wills *et al.*, 1989). The quarantine problem became acute when methyl bromide, the fumigant of choice, was recently scheduled to be discontinued. Public concern about chemical residues, and also the damaging effects that insecticides have on many fruits and vegetables, had already prompted a shift to high- or low-temperature exposures as alternative procedures for killing insects, but temperature extremes tend to be injurious to many horticultural commodities. Significant differences in the gas-exchange systems of insects and plants cause insects to perish at the low hypobaric pressures that are beneficial for the storage of horticultural commodities, but not at the [O_2] and [CO_2] levels commodities are able to tolerate in CA or MA at atmospheric pressure. Preliminary trials indicate that LP intermodal containers may be able to disinfest fruits, vegetables and floral crops residue-free during transit, replacing chemical fumigation, cold temperature and hot water or vapour treatments.

8.1 Effect of High Temperature

Amongst numerous fruit fly species that infest the tropical areas where mangoes are grown, the most economically important are members of the genus *Anastrepha*, including the Caribbean fruit fly (*A. suspensa* Loew), the Mexican fruit fly (*A. ludens* Loew), and *A. striata*, *A. distincta*, *A. fraterculus*, *A. serpentina* and *A. obliqua* (Aluja and Liedo, 1990). A USDA/APHIS-approved 90-min hot-water treatment at 46°C, which was developed to kill Caribbean fruit fly eggs and larvae in mangoes prior to packing (Sharp, 1986), is effective in killing 99.9968% of the eggs and larvae in a host of tropical fruit fly species present in mangoes up to 750 g in weight (Sharp *et al.*, 1988, 1989a,b). Nearly all Hawaiian fresh commodities subject to attacks by the oriental fruit fly (*Dacus dorsalis* Hendel), melon fly (*D. curcurbitae* Coquillet) and Mediterranean fruit fly (*Ceratitis capitata* Wiedermann) must be disinfected for shipment to the US mainland, Japan and elsewhere where these insects do not exist but are able to survive

Table 8.1. Insects and mites distributed by fruits and vegetables (Hill *et al.*, 1989).

Insect type	Common name	Common hosts	Approximate distribution
Cryptophlebia leucotreta Meyr.	False codling moth	Citrus, avocado, guava, stone fruit	Africa
Cydia pomonella (L.)	Codling moth	Apple, pear, walnut, quince, peach, *Prunus* sp.	Worldwide
Cylas formicarius (Fab.)	Sweet potato weevil	Sweet potato	Africa, Asia, N. and S. America, Pacific Islands
Graphognathus leucoloma (Boh.)	White-fringed weevil	Root vegetables	S. Africa, Australia, New Zealand, USA, S. America
Halotydeus destructor (Tucker)	Red-legged earth mite	Leafy vegetables	Australia, New Zealand, Africa
Lobesia botrana (Schiff.)	Vine moth	Grape	Europe, Japan, Africa
Maruca testulalis (Geyer)	Bean pod borer, mung moth	Legumes	Africa, Asia, Australia, S. America, Central America, Pacific Islands
Panonychus ulmi (Koch)	European red mite	Apple and other deciduous fruits	Europe, Africa, Australia, New Zealand, S. America, N. America
Phthorimaea operculella (Zell.)	Potato tuber moth	Potato, tomato, aubergine	Worldwide
Sternochaetus mangiferae (Fab.)	Mango seed weevil	Mango	Africa, Asia, Australia
Fruit flies			
Anastrepha fraterculus (Wied.)	S. American fruit fly	Peach, guava, citrus, *Spondias* sp., *Eugenia* sp.	S. America, West Indies, Central America
A. ludens (Lw.)	Mexican fruit fly	Citrus, other tropical and tropical fruits	Central America, Mexico
Ceratitis capitata (Wied.)	Mediterranean fruit fly	Deciduous and subtropical fruits, especially peach	Southern Europe, Central America, W. Australia, Africa, Hawaii
C. rosa (Karsch)	Natal fruit fly	Many deciduous and subtropical fruits	Africa
Dacus ciliatus (Lw.)	Lesser pumpkin fly	Curcurbits	Africa, India, Pakistan, Bangladesh
D. curcurbitae (Coq.)	Melon fly	Curcurbits, tomato	Asia, Hawaii, Papua New Guinea
D. dorsalis (Hend.)	Oriental fruit fly	Most fleshy fruits and vegetables	Asia, Hawaii
D. tryoni (Frogg.)	Queensland fruit fly	Many deciduous and subtropical fruits	Australia, Pacific Islands
Rhagoletis cerasi (L.)	Cherry fruit fly	Cherry, *Lonicerra* sp.	Europe
R. cingulata (Lw.)	Cherry fruit fly	Wild and cultivated cherry, *Prunus* sp.	North America
R. pomonella (Walsh)	Apple maggot	Apple, blueberry	USA, Canada
Scale insects			
Aonidiella aurantii (Maskell)	Red scale	Citrus	Worldwide
Lepidosaphes beckii (Newm.)	Purple scale	Citrus	Worldwide
Quadraspidiotus perniciosus (Comst.)	San José scale	Deciduous fruits	Worldwide
Mealybugs			
Planococcus citri (Risso)	Citrus mealybug	Citrus	Worldwide
Dysmicoccus brevipes (Ckll)	Pineapple mealybug	Pineapple	Africa, Asia, Australia, Pacific Islands, S. America

in the prevailing climate. Two non-chemical disinfestation methods are currently approved to kill Oriental and Mediterranean fruit flies infesting Hawaiian papayas: (a) double dipping in hot water and (b) vapour heat. In (a) the fruit is immersed in 42°C water for 30 min, and then in 49°C water for 20 min (Hardenburg *et al.*, 1986); for (b) the fruits are conditioned with dry heat at 43.3°C for 6–8 h to improve their high-temperature tolerance, and then they are warmed in a saturated atmosphere for 4 h or more until the fruit temperature reaches 47.2°C, after which they are cooled with circulating air for several hours before packing (Akamine, 1976b). Heat treatment also is effective in controlling diseases such as *Phytophthora citrophthora* on lemons, *Rhizopus* and *Monilinia* on peaches, *Penicillium* and *Diplodia* on oranges, *Colletotrichum gloeosporioides* on mangoes and papayas, and *Gloeosporium* sp. on apples (Eckert, 1967).

At a lethal temperature, an insect's O_2 consumption increases, then an inability to move ('heat stupor') sets in, followed by death. The critical temperature depends on the species, the duration of exposure and other factors, especially humidity and $[O_2]$. The high-temperature mortality of insects may be caused by enzyme inactivation, irreversible protein coagulation, metabolic imbalance, production of toxic products, changes in the physical state of cellular lipids, desiccation or some combination of these factors. An old theory, that the O_2 supply is inadequate at a high temperature, has been disproved (Fraenkel and Herford, 1940), but might still be relevant for insect life stages imbedded within bulky fruits, since at a high temperature the fruit's respiration depletes the intercellular $[O_2]$ to a low level and elevates the $[CO_2]$.

Evaporation from eggs (Fig. 8.1, *left*) and adult insects (Fig. 8.1, *right*) increases enormously at the temperatures required for the USDA/APHIS approved hot water and vapour treatments, possibly due to melting of wax in the chorion layer of eggs and in the adult's cuticle (Chapman, 1998). It is difficult to selectively kill insects at a high temperature without simultaneously damaging the host commodity because the temperature required to kill insects in an economically practical time is close to the commodity's injury threshold (Eckert, 1967). Temperatures higher than 30°C disrupt cellular membranes in apples (Fig. 3.6, *lower right*; Burg *et al.*, 1964) and other plant tissues (3.13; Stiles and Jorgensen, 1917b) and reversibly decrease the Q_{10} for plant respiration. After exposure to 40°C air, apples (Fig. 5.3; Burg and Thimann, 1959), pears (Hansen, 1942; Romani and French, 1977), bananas (Burg, 1969, unpublished), tomatoes (Biggs *et al.*, 1988; Yang *et al.*, 1990) and other fruits (5.2) temporarily stop producing ethylene. Near 50–60°C, respiratory

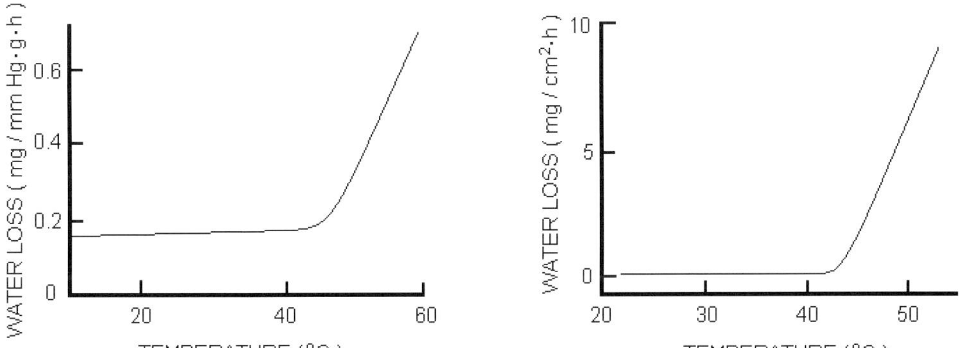

Fig. 8.1. Dependence of insect water loss on temperature. (*left*) Water loss from the egg of *Rhodinius* (Hemiptera) (Loveridge, 1968); (*right*) water loss through the cuticle of an adult *Locusta* (Orthoptera) (Beament, 1946).

enzymes are thermally denatured, membranes are irreversibly damaged and respiration permanently ceases (Hopkins, 1995). Consequently, commercial use of the USDA/APHIS-approved hot-water and vapour treatments often results in injury to the host commodity (Eckert, 1967; Akamine, 1976b; Heather, 1994; McDonald and Milhjer, 1994). This becomes an acute problem when, to avoid the prohibitive cost of air transport, mangoes are harvested at less than a fully mature state in order to slow ripening sufficiently to allow their distribution by low-cost oceanic transport. Hot-water treatment frequently bursts the immature fruit's latex ducts, causing the peel to undergo 'browning', which can lead to a complete loss of shipments (Davenport, 2000, unpublished). Because of the added cost for equipment upkeep and USDA personnel to oversee hot-water treatments, and the inconvenience of having to carry out all subsequent operations in screened-in areas until the cargo is sealed in an insect-proof container for shipment, some South American packers elect to forgo US-mandated disinfestation and ship their mangoes to Europe where quarantine treatments are not required.

8.2 Effect of Low Temperature

Many insects do not tolerate a low temperature. The wax-moth *Galleria* can survive as a larva for 60–120 days at 20°C with only 20% mortality, but 100% are killed in 20–50 days at 10°C and in 1–3 days at 0°C (Chapman, 1998). The lethal effect of a 13.3°C storage temperature on Caribbean fruit fly egg mortality is illustrated in Fig. 8.9.

The Japanese quarantine requirement for Australian oranges is satisfied by a cold treatment at 1.0 ± 0.6°C. The US requirement for Mediterranean fruit fly stipulates 10, 11, 12, 13 and 16 days at or below 0, 0.6, 1.1, 1.7 and 2.2°C, respectively (Hill *et al.*, 1989). Low-temperature insect disinfestation is useful with cold-tolerant vegetables and deciduous fruits, but not for most tropical and subtropical commodities because of their susceptibility to chilling injury. Preconditioning Marsh and Ruby Red grapefruit for 7 days at 16° or 21°C counteracts some of the adverse effects of a subsequent 1°C oceanic shipment for up 21 days (Hatton and Cubbedge, 1982), satisfying the Japanese quarantine requirement, but not without some degree of residual chilling damage.

8.3 Dependence of Insect O_2 Consumption on Temperature, $[O_2]$ and $[CO_2]$

In both plants and insects, the Q_{10} for O_2 consumption is between 2 and 3, but at the same temperature the O_2 consumption rate typically is higher for various types of resting insects and their life stages (Melampy and Willis, 1939; Edwards, 1946b; Wigglesworth, 1972; Navarro and Calderon, 1979; Herreid *et al.*, 1981; Sláma, 1982; Chapman, 1998) than it is for different fruits and vegetables (Hardenburg *et al.*, 1986). In air containing less than 5% $[O_2]$, the respiration rates of insects and plants tend to be proportional to the O_2 tension (Wigglesworth, 1983; 4.2). There are many exceptions, however, and in some insect species and plant tissues the respiration rate varies linearly as a function of $[O_2]$ in a range extending to and above 21% $[O_2]$, while in others the $[O_2]$ must be lowered to < 3.3–4.2% before an inhibition occurs (Hiestand, 1931; Fox *et al.*, 1937; Bishop, 1952; Prosser *et al.*, 1952; Wigglesworth, 1983). Usually the critical O_2 tension (t_C) below which the O_2 consumption of insects is decreased (Table 8.2) is an order of magnitude higher than the optimal $[O_2]$ for LP storage (Table 4.7). In some insects CO_2 acts as a respiratory stimulant and decreases the t_C (Hiestand, 1931). Insects kept at very low $[O_2]$ may lessen their O_2 consumption and become quiescent[1] in an effort to avoid or delay the effects of anoxia, but in the near or total absence of O_2, anaerobic glycolysis invariably causes insects to accumulate lactic acid, elevating the osmotic pressure and

Table 8.2. Critical O_2 tensions (t_c) of various insect life stages; 159 mm Hg = 20.9% O_2 (adapted from Prosser *et al.*, 1952).

Insect	t_c (mm Hg)	Temp. (°C)
Cleon dipterum (mayfly nymph)	32	10
Leptophlebia marginata (mayfly nymph)	53	10
Baetis (mayfly nymph)	240	10
Dixippus morosus (stick insect)	150	–
Calliphora erythrocephala (blowfly)	57	29
Melanoplus differentialis eggs (grasshopper)	80	–
Termopsis nevadensis (termite)	40	40
Chironomus thummi (midge larval fragments)	70	16–23
Chironomus plumosus (midge)	60	17
Limnephilus rhombicus (caddis fly)	25	–
Drosophila (fruit fly)	58–102	–

reducing the pH of their blood and tissue fluids. If this situation persists for long enough, death is likely to ensue.

8.4 Lethal Effects of High [CO_2], Low [O_2] and Desiccation

Effects of low [O_2], high [CO_2] and RH on insect mortality have been extensively studied in relation to grain storage. The most important feature of these atmospheres is their lack of O_2 (Banks and Annis, 1990). Depending on the insect species, relative humidity, life stage and temperature, in order to reliably kill stored-cereal insects and their larvae, the atmosphere must contain less than 0.9–5% O_2 (Oxley and Wickenden, 1963; Jay and Pearman, 1971; Shejbal *et al.*, 1973; Banks and Annis, 1990; Fleurat-Lessard, 1990; Carpenter and Potter, 1994), with a balance of N_2, CO_2 and inert gases, or mixtures of these gases. The atmospheres function mainly by anoxic effect, gases other than O_2 exert a minor influence and death presumably results from the accumulation of lactate and pyurate produced by glycolysis (Price and Walter, 1987; Banks and Annis, 1990). The insect's survival depends on its ability to accumulate and tolerate glycolytic products, reduce its metabolic rate and restrict water loss (Carpenter and Potter, 1994). At 20–30°C, most insect species and developmental stages display > 95% mortality

Table 8.3. Exposure times necessary for 100% mortality (LT_{100}) of some adult stored-cereal insects in the interstitial atmosphere of silos fed with N_2 containing various [O_2] concentrations at 22°C and 70% RH, flowing 50 l/h (Shejbal *et al.*, 1973).

[O_2] concentration (%)	LT_{100} (days)		
	Tribolium confusum	*Tribolium castaneum*	*Sitophilus granarius*
0.1	1	1	2
0.5	2	2	4
0.8	3	3	9
1.0	6	5	> 10

within < 10 days[2] in 0 or 1.0% [O_2] (Abe and Kondoh, 1989; Banks and Annis, 1990; Table 8.3).[3] The time required for low [O_2] to cause insect lethality increases when the temperature is decreased or the humidity increased, and is longer for eggs and pupae than for larvae and adults (Yoshida, 1975). At atmospheric pressure, there often is a correlation between daily percentage weight loss, [O_2], and the exposure time required for 95% mortality (Jay *et al.*, 1971; Yoshida, 1975).

All life stages of most stored-product insects, including eggs, are killed at 26°C by exposure to pure CO_2 for 10–48 h, to 60% CO_2 for 4 days or 35% CO_2 for 10 days (Fleurat-Lessard, 1990). CO_2-anaesthesia causes permanent spiracular opening, reduces membrane permeability, and acts on the insect's central nervous system hormones. The symptoms of CO_2 poisoning initially include a narcotic effect leading to an immobilization. There is no decrease in O_2 consumption, and mortality is independent of water loss. While elevated [CO_2] sometimes enhances the action of low [O_2] (Fig. 8.2), usually more than 35% [CO_2] is required to obtain a meaningful effect (Fleurat-Lessard, 1990), and concentrations that high are injurious to most plant commodities (Girsch, 1978; Calderon and Navarro, 1979; Table 4.7). In < 1% [O_2] there usually is little or no influence on insect mortality from CO_2 or any other gas added to the atmosphere (Banks and Annis, 1990).

A large part of an insect's transpirational water loss occurs through the respiratory surfaces (Mellanby, 1934; Wigglesworth and Gillett, 1936; Bursell, 1957), and since weight loss and mortality increase in parallel when the relative humidity is decreased at a low O_2 partial pressure, presumably the increased mortality at a lower RH is caused by desiccation (Pearman and Jay, 1970; Jay *et al.*, 1971). This is not helpful in CA or LP, since the RH must be kept close to saturation to prevent excess weight loss from the stored commodity. Moreover, insect stages located within the commodity are protected by the high relative humidity present in the host's intercellular spaces.

CA has been used to control insects (Thompson, 1998), but because the gas concentrations that are required cannot be tolerated by fruits and vegetables for very long, the most promising approach has been to expose the commodity to < 1% [O_2] for short periods (Ke and Kader, 1989).[4] Without risk of significant damage to papaya fruits, 0.4% [O_2] can be used for insect control at 20°C for periods of less than 3 days, but after 5 days abnormal ripening occurs and off-flavours develop (Yahia *et al.*, 1989, 1992). A 5-day exposure to 0.2–0.3% [O_2] at 20°C controls fruit flies on Keitt mangoes without damage or an adverse effect on organoleptic quality when fruit is transferred to air and ripened during five additional days at 20°C (Yahia and Hernandez, 1993). The 0.2–0.4% [O_2] range used at atmospheric pressure to kill insects that infest mangoes and papayas encompasses the [O_2] concentration provided by the optimal pressure for the long-term LP storage of these and most other commodities (chapter 10). Therefore, hypobaric storage might have general utility as an in-transit insect-quarantine method.

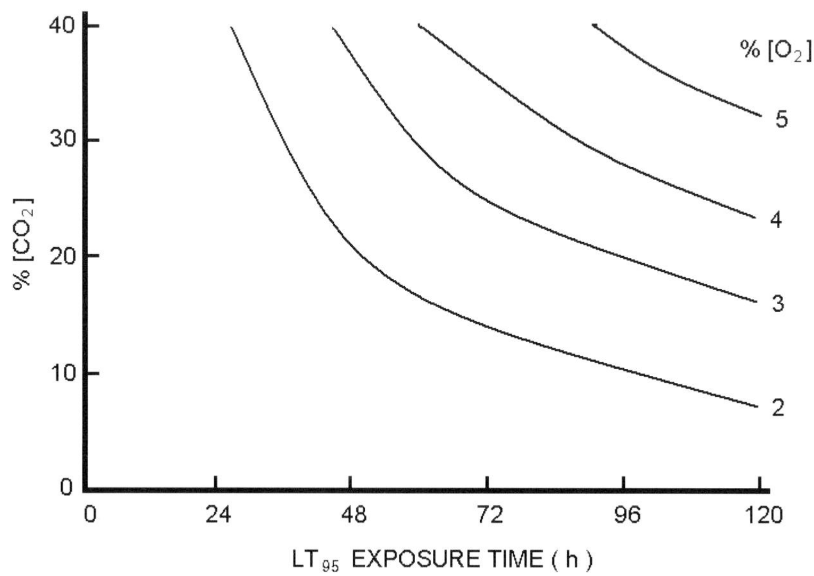

Fig. 8.2. Lethal time (LT_{95}) to kill 95% of *Tribolium castaneum* adults as a function of the [CO_2] and [O_2] content in the atmosphere at 26°C and 57% RH (Calderon and Navarro, 1979).

8.5 The Gas Exchange System of Insects

Even though plant cytochrome oxidase and the terminal oxidase of insects seem to have the same O_2 affinity (Chapman, 1998), structural features may cause insects to be less suited than plants to supply the O_2 needs of their cells at a hypobaric pressure. The eggs of terrestrial insects are more tolerant than other developmental stages to low $[O_2]$ because extensive air spaces usually are present in their chorion adjacent to the oocyte. The cytoplasm inside the eggshell is arranged as a bounding periplasmic layer around a relatively large amount of yoke within which the zygote nucleus is located. Immediately following fertilization, the zygote nucleus divides, but this is not accompanied by cell division, and the daughter nuclei, accompanied by a halo of cytoplasm, migrate to the egg's periphery to form the blastoderm, a layer of cubical cells surrounding the yoke (Fig. 8.3, *upper middle* – b). Subsequently, the blastoderm cells become columnar in the ventral region of the egg, forming the germ band from which the embryo develops (Fig. 8.3, *upper right* – gb), but in most insects some of the daughter nuclei remain behind as vitellophages (Fig. 8.3, *upper middle* – yoke cells, yc), and secondary vitellophages form from the blastoderm cells. The vitellophages break

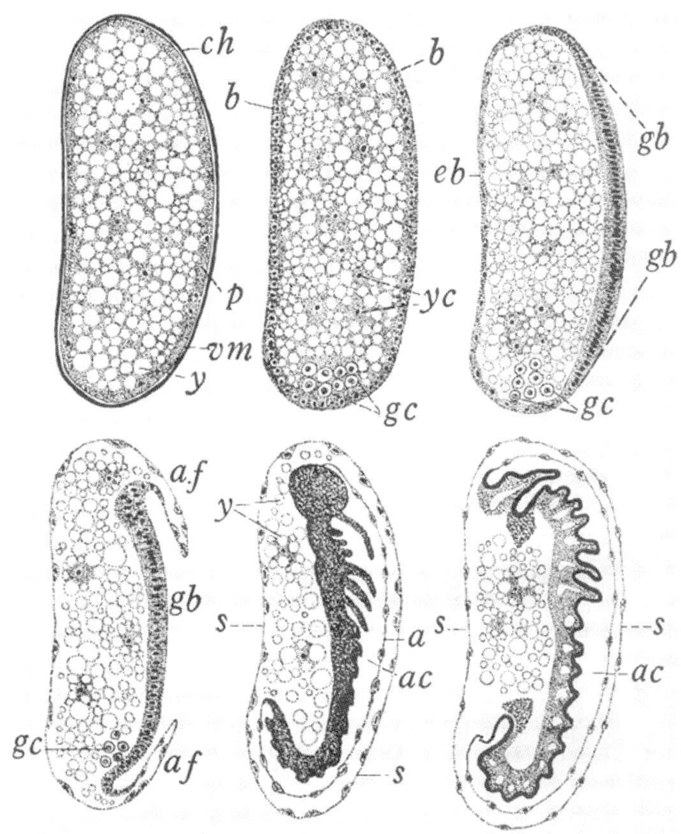

Fig. 8.3. Embryology of an insect. (*upper left*) Cleavage and migration of nuclei to the periplasm (*p*); (*upper middle*) formation of blastoderm (*b*); (*upper right*) development of the germ band (*gb*); (*lower left*) developing amniotic folds (*af*); (*lower middle*) embryo enclosed in amniotic cavity (*ac*); (*lower right*) section through *lower middle*. Other lettering: *eb*, extra embryonic blastoderm; *gc*, germ cells; *s*, serosa; *vm*, vitelline membrane; *y*, yolk; *yc*, yolk cells (vitellophages). The chorion and vitelline membrane are present at all stages, but are only shown in *upper right* (adapted from Imms, 1949).

down the yoke and engulf it in vacuoles; they are involved in the formation of new cytoplasm, are responsible for yoke contractions, and may form part of the mid-gut epithelium. Vitellophages may be particularly susceptible to low [O_2] damage in LP, since they are located throughout a liquid layer of considerable thickness through which O_2 diffusion is not enhanced at a low pressure. In Orthoptera, Lepidoptera and Coleoptera, the yolk may become temporarily subdivided by membranes into large masses of spherules, each containing one or more vitellophages, and these and the germ cells (Fig. 8.3, *upper middle* and *upper right* – gc) that arise early in the development of the egg might easily suffer low [O_2] damage in LP due to the surrounding liquid layer.

Subsequent stages of embryonic development are illustrated in Fig. 8.3 (*lower*). Soon after the germ band is differentiated, amniotic folds appear at its periphery (*lower left* – af) and form the amniotic cavity (ac). This is bounded by the multicellular amnionic membrane (*lower right* – a), which along with the multicellular serosal membrane (s) outside the yoke, constitutes a significant liquid phase resistance to gas exchange, interfering with O_2 transport to respiring cells, making them more susceptible to low [O_2] damage in LP.

The insect's integument is comprised of a cuticle and single layer of epidermal cells covering the whole of the insect's body and lining the tracheae (Figs 8.4 and 8.6). Adjacent epidermal cells are not tightly bound together, but the lateral lymph spaces between them are liquid-filled, and therefore gases that exchange through the integument must pass through a liquid phase in the epidermis, just as they do in plants (Fig. 6.5). The insect's cuticle has a 1–4 μm thick outer epicuticle layer devoid of chitin, an inner endocuticle layer up to 200 μm thick containing both chitin and insoluble protein, and a hard and rigid sclerified exocuticle that protects the insect (Fig. 8.4). The inner 0.5–2.0 μm-thick epicuticle is composed primarily of lipoproteins and covered with a thin outer epicuticle comprised of polymerized lipid ('cuticulin'), over which is secreted an epicuticular wax layer similar in composition to the plant cuticle, containing hydrocarbons, aliphatic alcohols, esters and free fatty acids. In many insects a thin layer of cement, possibly a mucopolysaccharide, is deposited over the cuticle, and in both insects and plants a wax layer waterproofs the surface.

The membranous coverings of many insects and the surfaces of leaves and fruits typically have similar permeability properties. In both insects and plants, the cuticular transpirational resistance often is in the 20–400 s/cm range (3.20; Table 8.4), and the cuticular layer is practically impervious to O_2, but sufficiently permeable to CO_2 to allow this gas's considerable outward diffusion through the insect integument and plant skin (Table 3.12; Imms, 1949). The insect cuticle's high permeability to CO_2 and low permeability to O_2 must be caused by the different water solubilities of these gases

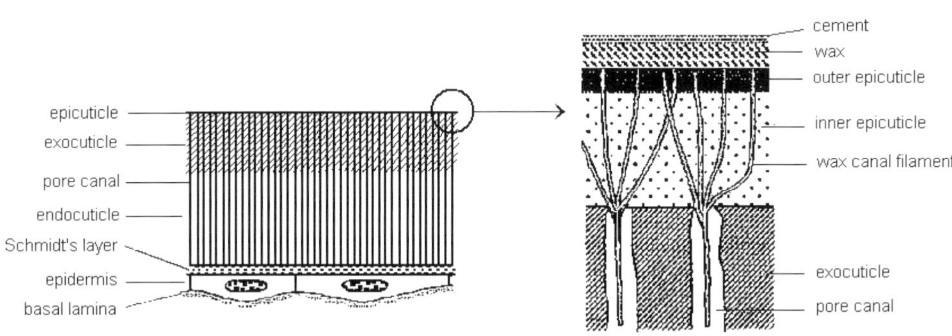

Fig. 8.4. Basic structure of the insect integument: (*left*) section through mature integument; (*right*) section through the epicuticle at greater magnification (Chapman, 1998).

Table 8.4. Cuticular transpirational resistance of various adult insects (computed from data in Chapman, 1998).

Insect	Cuticular resistance (s/cm)
Diceroprocta (desert cicada)	3.4×10^1
Eleodes (beetle)	2.0×10^2
Onymacris (beetle)	4.0×10^3
Trimerotropis (grasshopper)	2.3×10^2
Aeropedellus (grasshopper)	5.0×10^1
Glossina (tsetse fly)	1.1×10^4

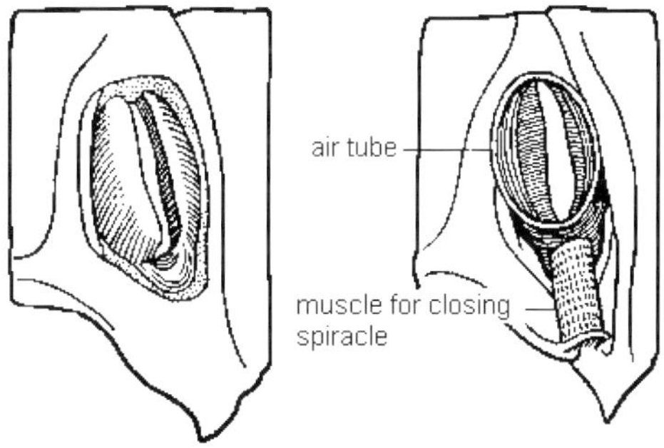

Fig. 8.5. Closed (*left*) and open (*right*) one-muscle metathoracic spiracle of a grasshopper (Buchsbaum, 1948).

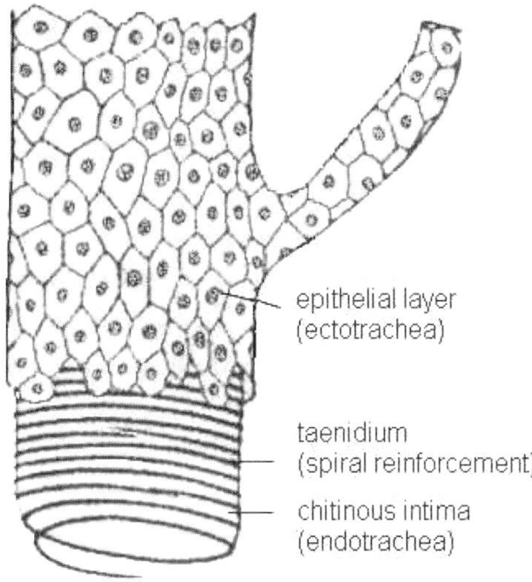

Fig. 8.6. Portion of an insect trachea stained with iron-alum haemotoxylon (highly magnified). Chitinous intima (endotrachea) with striated lining (taenidium) keep the trachea extended. The polygonal cells secrete a cuticular lining, which chemically resembles the epicuticle covering the insect's surface (Imms, 1949).

(Tables 3.12, 3.13, 15.1 and 15.2). It has been suggested that the impermeability of insect cuticles to O_2 arises in the epicuticle, and not from the wax layer that renders the cuticle impermeable to water (Buck, 1962).

The tracheal system in most terrestrial larvae, all pupae and smaller winged adult insects, and the aeropyles and chorionic air spaces in the eggs of most terrestrial insects, serve the same function as the intercellular system of plants, facilitating O_2 and CO_2 exchange to and from centrally located cells. Air enters through the spiracles (Fig. 8.5), and diffuses to every part of the insect body through interconnected, ramifying tracheal tubes, the smallest of which are approximately 2 μm in diameter. The trachea are lined with an intima composed of outer epicuticle beneath which a protein–chitin layer differentiates to form thread-like ridges, the taenidia (Chapman, 1998; Fig. 8.6), which run spirally around the inner circumference of the tracheal lining, keeping the tubes distended to allow free passage of air and prevent the trachea from collapsing when the pressure within the tube is reduced. In larger, more active insects, tracheal diffusion is supplemented by rhythmic muscular pumping of air through the tracheal system, stimulated by either a lack of O_2 or an excess of CO_2. The pulsation rate is highest during activity, is lower in pupae, decreases when the temperature is reduced and almost ceases during hibernation. These respiratory movements are generally absent from Onychophora, Arachnida, Mytiapoda and Chilopoda, from most terrestrial larvae, from all pupae and from many small winged insects (Krogh, 1941). Sometimes the tracheae are expanded to form thin-walled air sacs, which are widely distributed along the main tracheal trunks of many insects and play an important part in ventilation. Gaseous exchange is enhanced by the alternating collapse and expansion of the air sacs due to rhythmic haemolymph pressure changes coupled with the synchronized opening and closing of spiracles. Although ventilation often is continuous, there may be extended periods during which all spiracles are closed and gas exchange and water loss practically cease. Then movement of O_2 into and CO_2 from the trachea occurs in discrete bursts when the spiracles open. The ventilatory movements are stimulated by centres sensitive to CO_2 accumulation and to a lesser extent to O_2 depletion. Discontinuous ventilation of this type is common in adult insects and many pupae when they are inactive, but it may not occur at a low storage pressure because when O_2 is scarce the spiracles should always be open to prevent CO_2 accumulation.

At various points along their length, especially distally, the trachea are subdivided into tracheal capillaries or tracheoles, proximally about 1 μm in diameter, tapering to 0.1 μm or less at their ends (Fig. 8.7). These ramify and pass between the cells of the gut and salivary glands without penetrating their cytoplasm, while in the fat body and rectal papillae they may enter cells, and in flight muscles they form an intracellular network. The tracheae and tracheoles are most abundant in areas of high metabolic activity, reflecting the O_2 demands of different tissues. Tracheoles have a greater gas and water permeability than trachea because the tracheolar walls are thinner and more delicate (Wigglesworth, 1983). A variable amount of fluid often is present in the tracheole endings of both terrestrial and aquatic insects, through which gas diffusion occurs more slowly than in air, limiting the transfer of O_2 to the adjacent tissues and cells (Fig. 8.7). The tracheoles are so small, often with a diameter less than 0.3 μm, that the force of capillarity in the innermost ends of the tubules is approximately 10 atm, and therefore an opposing force of this magnitude is required to prevent fluid from being withdrawn from adjacent tissues and rising in these microcapillaries. In the insect body there is only one tissue fluid, the blood (haemolymph) which occupies a single cavity, the haemocoel, and freely bathes all of the tissues. The osmotic pressure of this fluid may normally exceed 10 atm, providing the force needed to prevent the capillary rise of water in the tracheole endings. During periods of high-energy consumption, water is osmotically withdrawn from the tracheole system as a consequence of lactate produced during activity, and air is drawn further into the tracheoles increasing the surface area available for O_2 diffusion into the tissue. At other times, when activity ceases, the air retreats and fluid again enters the tracheoles. If tracheoles are exposed to salt solution somewhat hypertonic to the insect's blood, more fluid is extracted from the tracheoles and the air extends downward towards the tissues, and if the fluid is replaced by a hypotonic solution, the process is reversed and fluid rises in the tubules (Fig. 8.7). The

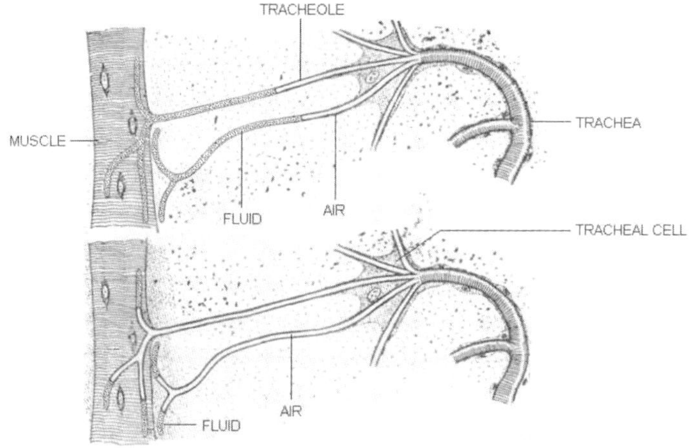

Fig. 8.7. Response of tracheoles to osmotic change during rest (*upper*) and after activity (*lower*). In fatigued muscle (*lower*) air extends far into the tracheoles (Wigglesworth, 1930).

'diameter' of the intercellular spaces in plant tissues is much larger than 1 μm, and so there is far less tendency for capillary filling of their 'tubes' with water (equation 6.14), and except when plant tissues become senescent, fluid is prevented from entering their intercellular system by the water potential of the cells (3.13).

The tracheal system is liquid-filled in a newly hatched insect, rendering it susceptible to injury at a low O_2 tension both at atmospheric pressure and in LP. Later the liquid is replaced by gas in a process caused by pneumatization. In terrestrial insects, the gas which fills the tracheae may enter via the spiracles, but pneumatization also occurs in aquatic insects and in many terrestrial species without access to air, apparently using gas forced out of solution by physical forces. The filling process can begin at any point within a main tracheal trunk and can occur in the near absence of O_2, but is totally inhibited by N_2 or a low temperature near 0°C (Prosser *et al.*, 1952). Often in ichneumonid and braconid (Hymenoptera) larvae, the tracheal system of the first instar is liquid-filled, and even when it becomes gas-filled the spiracles remain closed up to the last instar. In *Sciara* larvae, the filling seems to involve an internal source of gas produced metabolically (Buck and Keisster, 1949), and the tracheoles, even in the terminal regions, never normally contain fluid after initially filling with gas. At the opposite extreme, the entire tracheal system of *Drosophila* larvae is permeable to respiratory gases and water, and various mechanical stimuli may cause fluid uptake by the tubes (Edwards, 1946a).

In most terrestrial insects, the spiracles typically are closed by the activity of one muscle, and opened by the elasticity of the cuticle associated with the spiracle (Fig. 8.5), or sometimes by an elastic filament, or a second muscle. Even when there is an opening muscle, its activity often involves an elastic component so that in the absence of any muscular activity a spiracle opens. Both spiracles and stomates close to protect against water loss, and open to facilitate gas exchange, in insects to provide O_2 for respiration, and in plants to furnish CO_2 for photosynthesis. Insect spiracles normally are kept closed, and opened for the shortest time necessary to satisfy the respiratory O_2 requirement (Hazelhoff, 1927; Wigglesworth, 1935) and in both plants and insects desiccation induces closure. The systems differ in that full spiracular opening is initiated by either high [CO_2] or an O_2 deficiency, whereas low [CO_2] opens and high [CO_2] closes a plant's stomates. Less than 1% [O_2] and, depending on the type of insect and its stage of development, 2–20% [CO_2], cause sustained opening of the spiracles of resting insects (Hazelhoff, 1927; Mellanby, 1934; Bursell, 1957). Both of these conditions lead to a reduction in action potential frequency, which causes relaxation of the opening muscle of 'two muscle' spiracles. The ensuing closure tends to perpetuate itself because CO_2 accumulates and O_2 is consumed by respiration. CO_2 also acts on the closure muscle of 'one muscle spiracles' to cause opening. Unless desiccation intervenes, in LP both the stomates and spiracles should be open because low O_2 takes precedence over CO_2 with respect to opening the spiracles (Bishop, 1952). While the spiracles should be open in NA or CA, darkness and accumulated or applied CO_2 cause the stomates to close, making it more difficult for CA to kill insects by means of low [O_2] without simultaneously injuring the host horticultural commodity.

Tracheal systems and spiracles vary widely, and therefore it might be expected that the sensitivity of insects to low [O_2] would be equally divergent. The spiracles are atrophied or always closed in some insects, and gas exchange occurs through the skin. In others, the openings lack any means to regulate their size; a few types are devoid of all traces of a tracheal system; in some only an anterior or posterior pair of spiracles is functional, while the remainder are closed; and in others the ten normal pairs of spiracles, two per segment, are functional (Imms, 1949). Some gas exchange takes place through the cuticle of most insects, but usually this does not amount to more than a small percentage of the total.

The insect body cavity contains the circulating blood, which is pumped by the insect's open-heart system. All organs and tissues are bathed by this fluid, whose main function is to convey nutrient substances to the tissues and transfer waste products to the excretory organs. The blood does not contain a biochemical carrier of O_2 or CO_2 and therefore it takes up no more of these gases than can be accounted for by simple physical solution.[5] Its part in the respiration of tracheate forms sometimes may be small, but typically there are many cells separated from the nearest tracheal tubes by an appreciable space. Organs in the pupa may be entirely devoid of a tracheal supply, and in certain aquatic insects the tracheal system is completely filled with water. Under these circumstances the blood acts as a carrier of O_2 to the deprived tissue, acquiring dissolved O_2 when it passes over tracheae, which are to some extent O_2-permeable.

In both insects (Herreid, 1980) and plants, the limiting factor in respiratory exchange is the diffusion of O_2, which depends on the O_2 partial pressure difference between the atmosphere and mitochondria and is inversely related to the system's total resistance to O_2 transport. Only a 2–3% $[O_2]$ gradient between the open spiracles and tracheal endings is required at atmospheric pressure to satisfy an insect's O_2 need (Krogh, 1920). At atmospheric pressure, the required gradient between the atmosphere and intercellular spaces of plants often is similar in magnitude, but in LP it is much lower. There are two distinct phases in the transport of gases into and from insects: air-tube diffusion through the tracheal system and tissue diffusion (Weis-Fogh, 1964). Since O_2 diffuses 100,000 times faster through air than it does through water or tissues (Tables 15.4 and 15.5), even though the path for air-tube diffusion through tracheae is much longer, tissue diffusion limits O_2 entry. 'It may take much longer for gas to diffuse from the tracheolar endings to the mitochondria than from spiracles to the tracheolar endings' (Chapman, 1998).[6] If that conclusion is valid, in insects the resistance of the liquid-filled pathways exceeds that of the air-filled pathway, whereas in plants the opposite is true (Tables 3.3 and 3.4). This causes plants to be more tolerant than insects to a hypobaric condition.

8.6 Comparison of the Gas Exchange Systems of Insects and Plants

A hypobaric pressure promotes O_2 and CO_2 diffusion to the same extent through air-filled tracheae, spiracles, the chorion layer and aeropyles of insects, and the stomates, lenticles, pedicel end-scar and intercellular spaces of plant tissues. When the atmospheric pressure is reduced, a horticultural commodity's total O_2 mass-transport resistance approaches a lower limit established by cellular gas permeability (Table 3.3) because each plant cell communicates directly to the atmosphere through interconnected air-filled intercellular spaces, lenticles and stomates. A hypobaric condition will decrease an insect's O_2 mass-transfer resistance to a lesser extent because O_2 must exchange between the tracheal air and a secondary fluid of considerable thickness, the circulating blood, in order to reach larval, pupal and adult tissues that are not directly serviced by tracheoles. In addition, sufficient fluid is present at the tracheolar endings in the egg yolk and in cells of the amnion and serosa membranes to significantly hinder O_2 diffusion into associated cells. Each trachea is lined with a cuticle and surrounded by an epithelial layer of cells (Fig. 8.6), which must impede diffusive gas exchange to approximately the same extent that the epidermal cell layer and cuticle of plants hinders gas transport between the atmosphere and intercellular system. In some insect life stages, all O_2 exchange occurs directly by diffusion through the cuticle, restricted by passage through the liquid in epidermal cells. These structural differences between insects and plants are likely to determine their low $[O_2]$ tolerance. Hypoxia and anoxia cause toxic levels of lactate and pyruvate to accumulate in insects (Price and Walter, 1987), but in plant commodities the RQ remains close to unity at pressures as low as 1.33 kPa (10 mm Hg), indicating that ethanol and acetaldehyde are not being produced at an appreciable rate (4.2).

8.7 Effect of a Low Pressure on Insect Mortality

The killing by vacuum of various developmental stages of stored-product insects was first reported in 1925 (Back and Cotton, 1925), and the sensitivity of numerous types of insects to low pressures was listed at that time. Exposure of *Ephestia elutella* (Hbn.) and *Lasioderma serricorne* (F.) to high vacuum in a chamber containing insects and tobacco together (Bare, 1948) results in mortality within the range cited in Back and Cotton, and favourable results with high vacuum were reported by others (El Nahal, 1953; Thornton and Sullivan, 1964). A sub-atmospheric pressure's effect on water loss from stored-product insects also has been studied (Narayanan and Bhambhani, 1956; Munro, 1959; Sharplin and Bhambhani, 1963; Thornton and Sullivan, 1964). Lowering the relative humidity in the exposure chambers causes desiccation, enhancing insect mortality at a low pressure (Bhambhani, 1956). However, when insects oviposit their eggs in a stored horticultural commodity, the various life stages that develop are protected from desiccation by the saturated atmosphere present in the intercellular spaces of the host.

Mortality of six species of common stored-product insects at 18 or 25°C and a low pressure was examined in steel containers of 20-l capacity filled with a wheat load having a moisture content of 12.5%, which corresponds to a relative humidity in the intergranular air of 60% (Calderon *et al.*, 1966). Initially, the containers were pumped down to approximately 1.33 kPa (10 mm Hg) at 18°C, or 1.6 kPa (12 mm Hg) at 25°C, which is close to the wheat's vapour pressure at each temperature. Subsequently, the pressure rose to a final value of 2.33–2.67 kPa (16–20 mm Hg) depending on the duration of the experiment. When a correction is applied that takes into account the vapour pressure created by a 60% RH, if the insects caused no respiratory O_2 draw-down, the final [O_2] concentration would have been in the 0.16–0.19% range. Within 120 h at 18 or 25°C, the low pressure caused 100% mortality of larval and adult forms of *Ephestia cautella* (Wlk.), *Oryzaephilus surinamensis* (L.), *Tribolium castaneum* (Hbst.), *Callosobruchus maculatus* (F.), *Sitophilus oryzae* (L.) and *Trigoderma granarium* Everts. Adults of *E. cautella*, *O. surinamensis* and *C. maculatus*, and adults and larvae of *T. castaneum* were killed within 7 h. At the same [O_2] concentration and temperature, a nearly identical time is required to kill *T. castaneum* (Hbst.) adults at a low pressure and at atmospheric pressure (Table 8.3). Insect mortality was higher in LP at 25°C than it was at 18°C (Calderon *et al.*, 1966), and the same relationship arises when insects are exposed to low [O_2] at atmospheric pressure and various temperatures (Yoshida, 1975). This may be due to the respiratory Q_{10}, which should cause a lethal dose of lactate to accumulate sooner at a higher temperature. In addition, the water solubility of O_2 decreases at a high temperature (Tables 15.1 and 15.2), lowering the [O_2] concentration available in the cellular liquid phase. These relationships suggest that efforts to kill insects by simultaneously lowering both the temperature and [O_2] are likely to be offsetting to some extent.

If the exposure time in a small sealed chamber is longer than several days, the atmospheric composition changes due to insect respiration, and water lost by the insects increase the humidity and suppress further water loss (Bhambhani, 1956; Calderon *et al.*, 1966). Yet even when a high relative humidity prevents desiccation, and [O_2] and [CO_2] are kept constant, at a reduced pressure low [O_2] causes insect mortality (Navarro, 1975). Studies with *Aedes aegypti* adults and housefly pupae seemed to indicate that 'mortality of mosquitoes at very low pressures stems from at least 3 factors acting independently: a) dehydration; b) lack of O_2; and c) low pressures as such', and it was deduced that 'very low pressures have an adverse effect on respiration *per se*, independent of that of the concomitant low O_2 tension' (Galun and Fraenkel, 1961, referred to in Navarro, 1975). However, no effect of pressure on respiration other than that which could be accounted for by low $[O_2]$ could be detected in studies with *Tenebriodes mauritanicus* (L.) larvae (Dumas, Buckland and Munro, 1969, referred to in Navarro, 1975) or when low $[O_2]$ was tested on *E. cautella* (Wlk.) pupae at both atmospheric and reduced pressure using a special apparatus which kept the relative humidity at 93–99% to eliminate the possible effect of desiccation (Navarro and Donahaye, 1972). Insect mortality and changes in CO_2 production were due exclusively to low $[O_2]$ independent of pressure (Navarro and Calderon, 1979; Fig. 8.8). LP improves the low O_2 tolerance of *E. cautella* pupae far less than might be expected based on an anticipated 12.5–15.8-fold increase in the rate of O_2 diffusion through air at 6.4–8.33 kPa (48–61 mm Hg), suggesting that aqueous phases restrict gas exchange within these pupae. CO_2 production was similar at 760 and 61 mm Hg in the presence of 1% $[O_2]$, but the time to reach 100% mortality increased from 3 days at 760 mm Hg to 5 days at 61 mm Hg, indicating that a low pressure improved gas exchange through an air phase in the pupae. After 6 days in 3% $[O_2]$, mortality was slightly greater at 760 mm vs. 48 mm Hg.

In LP a pressure of 2–2.67 kPa (15–20 mm Hg) at 2°C is useful for long-term storage of lettuce and 100% lethal within 52 h to the green peach aphid (*Myzus persicae*) that infests this commodity (Aharoni *et al.*, 1986). The mortality of the aphid is pressure- and time-dependent (Table 8.5). Death must be caused by low $[O_2]$ rather than desiccation or a reduction in total pressure, since within 24 h there is a significant difference in mortality at 2.67 vs. 6.67 kPa (20 vs. 50 mm Hg), even though

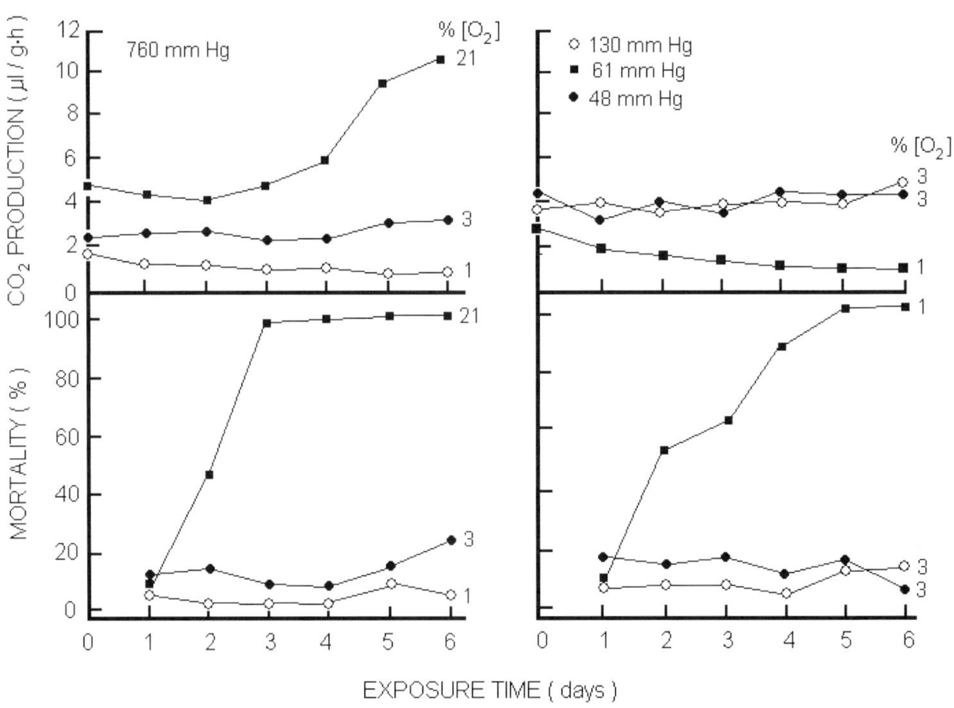

Fig. 8.8. CO_2 production, and mortality of *Ephestia cautella* (Wlk.) pupae exposed to various O_2 tensions at different pressures (Navarro and Calderon, 1979). The pupae were confined in copper mesh and studied during 6 days at 26°C, flowing either 1, 3 or 21% $[O_2]$ at 93% RH and atmospheric pressure, or 1 and 3% O_2 at 99% RH and pressures of 15.33, 8.33 or 6.4 kPa (130, 61 or 48 mm Hg).

out without humidification (Tolle, 1969), while in other studies the RH ostensibly was controlled at 75% (Kim and Oogaki, 1986), or at 90–95% (Salunkhe and Wu, 1973; Apelbaum *et al.*, 1977b). Often the RH is 'estimated' to be close to saturation (Spalding and Reeder, 1976a; McKeown and Lougheed, 1981), but usually it is neither measured nor reported, and rarely does a Materials and Methods section provide information concerning whether or how the humidity was controlled, making it difficult to interpret and reconcile conflicting results.

Horticultural commodities contain approximately 90% water, have a large surface area relative to their volume, produce respiratory heat and in LP may have a low transpirational resistance. Unless the LP air-change rate is excessively high relative to the commodity weight, water evaporated from the commodity will elevate the humidity in the vacuum chamber close to saturation even when the RH of the incoming air changes is relatively low (Burg, 1992). A measurement of the chamber's relative humidity cannot distinguish whether the moisture originated from the commodity or from a humidification system. Therefore, it is difficult to conceive how the humidity could be 'controlled' at 75%, 90% or any other value unless the incoming air stream is humidified after it passes through the pressure regulator and the humidity measurement is made before the air enters the vacuum chamber. This problem was highlighted during the development of Grumman's Dormavac intermodal hypobaric container. Whenever the humidity decreased below approximately 95%, a highly accurate humidity sensor installed in the storage space close to the inlet air supply was supposed to activate the humidification boiler's electric heater (Fig. 13.7, *right*). Instead, the immersion heater rarely (or never) was powered because the enormous weight and surface area of commodity present in the container always elevated the humidity close to saturation. The humidistat had to be replaced by a manual controller that set the heater wattage at a value that continuously evaporated the calculated amount of water needed to saturate the air changes. A similar problem can be anticipated with the humidification systems used in several published laboratory hypobaric storage studies:

> A pump drew gas from the storage chamber and circulated it through a tank which contained baffles and absorbent sponges dipping in water. The water level was maintained by a needle valve from a low-pressure source. When the humidity in the storage chamber reached a pre-set level, the lithium chloride sensor actuated an electric relay. This in turn stopped the pump motor and closed a solenoid valve to stop further water vapour additions. When the vacuum system lowered the humidity in the storage chamber below the pre-set limit, the humidifying cycle was repeated. In a simpler method, which worked about as well, the gas supply was normally bubbled through a gas-washing bottle about 7/8ths filled with clean water. When the humidity in the storage chamber reached a satisfactory level, a solenoid valve bypassed the supply gas around the humidifier until the sensor indicated the cycle was to be repeated.
>
> (Tolle, 1969, 1972)

When Tolle's system was used at 4.4°C to preserve strawberries at a pressure of 25.33 kPa (190 mm Hg), even though the electro-sensors recorded equal humidities independently of air flow, berry weight loss increased at higher flow rates, and when tomatoes were stored without humidification, water droplets condensed as beads inside the lids of the storage chamber. In both instances the stored commodity was saturating the humidity.

9.3 Factors Determining the Steady-state Relative Humidity

Before low-pressure air is admitted into a laboratory LP apparatus, it is often humidified by passage through a water tower located within the same temperature-controlled space in which the vacuum chamber is situated (Figs 2.4 and 9.6). If instead the air was humidified at atmospheric pressure upstream of the pressure regulator, it would expand as it entered the chamber and decrease in relative humidity in proportion to the pressure reduction, negating the effect of humidification. Often it is difficult to discern from a Materials and Methods description whether the air was humidified at atmospheric pressure upstream of the regulator, or downstream at a low pressure. In an experiment with green peppers in which CA and LP storage was compared:

> Gas mixtures containing varying amounts of O_2 and CO_2 were humidified at atmospheric pressure and passed at a rate of 5 litres per hour through 10-litre sealed plastic buckets used for CA storage, creating ½ air change per hour. Humidified air also was passed through 100-litre LP chambers at a rate of 5 l/h.
>
> (Hughes *et al.*, 1981)

This description does not indicate whether the LP flow was humidified at atmospheric pressure or at the LP storage pressure. It also is not discernible whether the humidification was carried out at the storage temperature or ambient temperature, and whether the flow in LP was expressed as 5 l/h of atmospheric pressure air or 5 l/h of reduced pressure air. If 5 l/h of air had been humidified at each storage pressure, the volume rate of rarified LP air flow would have provided only 1/20 of an air change per hour and been grossly inadequate to supply O_2 to the stored commodity as rapidly as it was consumed. If 5 l of air was humidified at atmospheric pressure and flowed through a 100-l LP chamber, the air change rate at the various storage pressures that were tested (5, 10.1 and 20.2 kPa = 38, 76 and 152 mm Hg, respectively) would have ranged from ¼ to 1 per hour, which is adequate and typical for an LP experiment, and similar to the flow passed through the CA chambers. Depending on the storage pressure and assuming that the temperature of the incoming air was the same at atmospheric pressure and the storage pressure, expansion would have reduced the relative humidity of the humidified atmospheric pressure air to between 5 and 20% as it entered the LP storage chambers. The green peppers stored in LP became 'wrinkled and flaccid, showing severe desiccation', and they lost weight 4–5 times faster than peppers stored in CA. The researchers concluded that LP was the least effective method of storing green peppers, and that 'water loss thus appears to be a major problem in hypobaric storage, *but any improvement in humidification is difficult because of the reduced partial pressure of water vapour under these conditions.*' This statement confirms that the incoming air was humidified upstream of the regulator at atmospheric pressure and was dried by expansion as it entered the LP chambers. The researchers had compared storage and weight loss in CA at 100% relative humidity vs. LP at 5–20% relative humidity! 'Work carried out at the National Vegetable Research Station in England also showed desiccation to be the main problem in hypobaric storage' (Tucker, 1979, personal communication – cited by Hughes *et al.*, 1981). Were all of these studies carried out using equipment with this same basic humidification design defect? Others have found that when the incoming air is properly humidified in a laboratory apparatus, LP is extremely effective in preserving green peppers without weight loss (10.47).

Incoming air expands as it rises and passes through humidifying water contained in a tower. Although the expansion is slight at atmospheric pressure, at a low enough chamber pressure it is significantly larger, making it more difficult to humidify the air because additional evaporation is required to compensate for the decreased RH caused by the greater expansion at a lower pressure (example 1). The tubing connecting the humidifier and vacuum chamber must have an adequate diameter to prevent the humidity from being further reduced by expansion as the air flows from the humidifier to the vacuum chamber. The relative humidity can easily be decreased by 6% or more if the diameter of this tubing is too small (example 2).

Unless heat is continuously supplied to the humidifier at a rate sufficient to replace the latent energy consumed by evaporation, the temperature and vapour pressure of the humidification water, and therefore the moisture content of the discharged air, will decrease. Because of the low density and heat capacity of subatmospheric pressure air, it easily and rapidly equilibrates at the temperature of the humidification water as it bubbles through the tower. The humidifier must compensate for both air expansion and the water's tendency to decrease in temperature, or else the air will exit the tower partially humidified at a temperature lower than that existing in the vacuum chamber, and the relative humidity will further decrease when the discharged air warms to the storage temperature. The air-change rate was so large relative to the size of the humidifying tower in a prototype hypobaric intermodal container that without heat input, evaporative cooling generated in the humidifier was sufficient to cool the entire container (13.1; Burg and Hentschel, 1974). In laboratory studies, an excessive pumping speed can lower the relative humidity and increase the weight loss, not only by shortening the contact time between the heated water and incoming air, but also by lowering the water temperature and rapidly flushing away moisture evaporated from the commodity (example 3).

In a laboratory set-up, the humidification water is normally warmed by convective heat transfer from air in the temperature-controlled space within which the apparatus is situated, and by radiation to the humidifier's walls, followed by conduction of the heat through the walls to the water. A thermal analysis reveals that if the air-change rate is not excessive and the tower is properly sized relative to the laboratory chamber's volume, the temperature decrease in the water should be < 0.3°C. If complete equilibration between air and water occurred, the humidity might be as high as 98% when the discharge air rewarmed by 0.3°C. Because equilibration may not be complete, and vacuum leaks can depress the humidity (9.6; Table 9.3), it is prudent to compensate for such effects by using a heating means to keep the water temperature at least several degrees centigrade above the chamber temperature (Apelbaum *et al.*, 1977b; Kim and Oogaki, 1986; Figs 2.4 and 9.6).

Water supplied inside a hermetically sealed, non-ventilated space can raise the humidity to 100%, but in a dynamic system it is not possible to maintain a saturated relative humidity because flow necessarily involves a pressure gradient, and the associated expansion of the flowing air lowers its RH. When unheated water is provided inside a chamber through which air is flowing, heat must be transferred into the water from the surrounding environment at a rate sufficient to continuously replace that which is lost in evaporating water. In laboratory studies with mangoes, avocados, limes and sweetcorn (Spalding and Reeder, 1972, 1974, 1975, 1976a,b, 1977; Spalding, 1977; Spalding *et al.*, 1978), humidity was provided by placing a polyethylene tray filled with water inside each vacuum chamber. One air-change per hour flowed through tubing into the tray of water and was dispersed through an air-diffusion stone. Locating the humidifier inside the vacuum chamber has the advantage that it compensates for chamber leaks (Table 9.3) by humidifying the air changes after they have entered the storage area, whereas an external tower cannot provide this benefit, since it is located upstream of chamber leaks. An interior tray of water has the disadvantage that the heat transfer coefficient for natural convection to the liquid water in the tray from the low-pressure air/water-vapour mixture inside the storage chamber is reduced in proportion to the square root of the pressure. This lowers the humidity by increasing the temperature gradient between the water in the tray and the surrounding low-pressure air. The use of a porous stone to disperse the air changes improves heat transfer into the water by thorough mixing, but significantly increases the upstream pressure, causing a greater expansion when the air is released into the water, making it more difficult to saturate the air. A thermodynamic analysis reveals that without heat conduction through the base of the plastic tray, convection and radiation could not raise the relative humidity of the air change above 85% using the described configuration. Highly efficient heat conduction through the contact resistance between the base of the plastic tray and the floor of the storage chamber[2] would be required to elevate the humidity to the claimed value of 98–100% measured electronically. A thermal resistance develops between different conducting surfaces when they are placed in contact because the materials only touch at peaks in their surfaces, while in this instance the valleys between the mating surfaces are occupied by a vacuum that greatly retards conductive heat transfer. The interface resistance is primarily a function of the pressure holding the two surfaces together, which in this particular LP apparatus was not very great, and also depends on the surface roughness and insulating properties of the vacuum. At the interface, heat is transferred by conduction through the contact points of the solids across the temperature difference which develops through the contact resistance.[1] An additional temperature gradient is required to conduct heat through the plastic tray to the water. It is difficult to calculate the level of humidity that might be sustained by this method because the contact resistance is not known, but certainly this procedure cannot by itself produce a humidity of 98–100%. This raises the question 'to what extent was the humidity elevated by vapour evaporated from the commodity in response to respiratory heat, rather than from the humidification tray in response to heat transferred by convection and radiation?' Because the water in the humidification tray must be slightly colder than the walls of the container, moisture evaporated from the commodity, and any moisture brought into the container in the air changes, will tend to condense on the colder surface of the water, and the released latent heat of condensation will elevate the tray water's temperature. This reduces the temperature gradient between the storage chamber and water layer and causes proportionately less heat to be transferred to the water layer, offsetting the temperature rise caused by condensation.

A VacuFresh^{SM} container is not equipped with a humidification system, nor is any required when full loads are shipped, but the floor must be inundated with water to elevate the humidity whenever the container is used to test a small weight of commodity. When water is provided directly on the floor in an LP intermodal container or laboratory chamber, energy removed from the water by evaporative cooling is replaced by heat conduction and convection into the water from the floor, and by heat transferred to the surface of the water by radiation and convection from the chamber walls and air. During a 1.5 kPa (11.3 mm Hg) LP storage with the wall and floor temperatures controlled at 1.1°C, the average water temperature was −0.6°C when air changes were flowed at the lowest practical rate determined by the requirement that the discharge from the pneumatic air horn must be supersonic for proper operation (13.9). The 1.7°C dew point depression corresponds to a relative humidity of 88.5% (Table 9.2), in close agreement with an RH value of 87% measured during the test (Fig. 9.5).

The surface of the water layer develops a colder temperature than the air and walls of an intermodal container operated with a flooded floor. Moisture evaporated from a small weight of commodity in response to its respiratory heat, and moisture brought into the container in the air changes, condenses on the cold surface. The latent heat of condensation elevates the water's surface temperature, but this reduces the temperature gradient between the tank floor and the surface of the water layer, so that proportionately less heat is transferred to the surface by transfer through the water, offsetting the temperature rise caused by condensation. Likewise, although heat transferred by radiation and convection from the container wall to the water layer elevates the water's surface temperature, it also decreases the amount of heat conducted from the floor through the water layer, resulting in a compensatory decrease in the surface temperature. Consequently, when the floor is flooded with water and only a small weight of commodity is present in the container, moisture brought in by the air changes or evaporated from the commodity will have little, if any, effect on the steady-state humidity. The relative humidity is determined exclusively by the water surface temperature. This situation initially caused serious difficulties in designing small-scale marketing demonstrations for customers who did not wish to risk a full container load of commodity in an initial static trial. At 87% RH, evaporative cooling lowers the commodity temperature below the wall temperature. This predisposes the commodity to acquire heat by radiation, evaporate additional water and desiccate. The problem was solved by installing a Mylar slip-sheet inside each carton to shield the stored commodity from radiation (9.9).

When under-floor water is frozen at −1°C, it humidifies an intermodal container by sublimation. The amount of energy required to accomplish this is the sum of the heat of fusion (79.71 Kcal/kg) and the heat of vaporization (595.9 Kcal/) at −1°C, and therefore 13.4% more heat must be conducted through the ice layer to sublime an amount of water equal to that which would evaporate from the same surface area of liquid water. The thermal conductivity is 3.79-fold higher for ice than for water (Perry and Given, 1984), but there is considerable uncertainty in any calculation based on thermal conductivity because a layer of insulating 'snow' is deposited on the surface of the ice when it forms at a low pressure. Under-floor ice and water seemed to be approximately equally effective in elevating the humidity in tests run in a VacuFresh^{SM} container.

An operator who did not understand the dry VacuFresh^{SM} concept of humidification loaded a VacuFresh^{SM} container with cartons of tulips, and operated the container at 1.5°C and a pressure of 2.66 kPa (20 mm Hg), continuously evacuating the container at a maximum rate with the floor flooded. The temperature of interior flower boxes decreased by 1.7°C below the wall temperature, indicating that the dew point was lower than this value. Exterior boxes that received radiant heat from the walls stabilized at 0.85°C below the wall temperature and were severely desiccated during the 18-day shipment.

9.4 Measuring the Relative Humidity

Properly calibrated, the best electronic humidity-sensing instruments have an accuracy no better than ± 2% in the 95–100% RH range. Although these instruments have been used to measure the humidity in laboratory LP chambers (Tolle, 1969; Spalding and Reeder, 1972; Salunkhe and Wu, 1973), one researcher noted that 'precise measurement of relative humidity in the 95%+ range is not easy nor inexpensive in an experimental system using many chambers, even with the availability of sophisticated electronic instrumentation' (Lougheed *et al.*, 1978). Figure 9.2 illustrates a simple, small, inexpensive wet-and-dry bulb that has been used to accurately measure the dew point in laboratory chambers and in Grumman/Dormavac and VacuFresh^{SM} hypobaric intermodal containers. Shielded from radiation by Mylar film, the device takes advantage of the lack of effective convective heat transfer in a vacuum. The wet-bulb temperature closely tracks the dew point.[2]

9.5 Other Sources of Moisture

The humidification system is not necessarily the major source of moisture in a

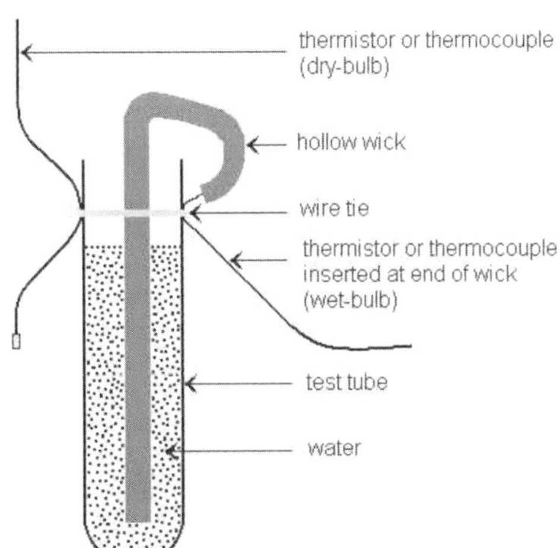

Fig. 9.2. A simple device for measuring the dew point in an LP chamber.

hypobaric storage. Depending on the air-change rate and commodity weight, the amount of water vapour released when all respiratory heat is transferred by evaporative cooling may be larger than the quantity of water brought into a laboratory apparatus by humidified air changes. Strawberries stored at a pressure of 1.33 kPa (10 mm Hg) will lose 0.0028% of their fresh weight each hour when evaporative cooling transfers all of their respiratory heat. If the fruit is loaded at maximum density into a laboratory apparatus (or intermodal container), this rate of water loss would more than saturate one air change every 2 h even if the air change was bone dry when it entered the storage chamber. At higher pressures, the respiration rate is greater, more water is evaporated from the commodity and it would seem that humidification might be unnecessary. How could the humidity decrease under these conditions and cause the commodity to lose more water than that required to transfer its respiratory heat, especially in a laboratory set-up in which the incoming air has already been saturated or nearly so in a humidifier?

9.6 Chamber Leak Rate

Whenever a laboratory apparatus is humidified downstream of the pressure regulator, and the weight loss exceeds that needed to transfer the respiratory heat, air leakage is the likely cause of the extra weight loss. The assembly of a leak-free laboratory vacuum apparatus is far more difficult than most researchers imagine. All components should be rated for use in at least the medium vacuum range, pipe threads are inherently leaky in a vacuum and should be avoided, proper sealants must be used with all fittings and gaskets carefully selected. Usually, it is not evident that a laboratory system is leaking because the pressure regulator automatically controls the storage pressure at the set value:

> The past season we noted a phenomenon which could explain the desiccation of products reported by other researchers. In some of our chambers it was impossible to obtain the high relative humidity achieved in similar chambers, despite equivalent conditions of flow rate, pressure and temperature of the humidifier water. We traced the problem to leaks which allowed non-humidified air at atmospheric pressure to enter chambers, making it difficult to increase the humidity to the desired level. In our case, the pump capacity was sufficiently high that the leaks did not noticeably affect the overall pressure of the system nor that of the chamber.
>
> (Lougheed *et al.*, 1978)

The only reliable way to determine whether a laboratory system is hermetically tight is to pump it down to a substantial vacuum, isolate the system, and measure the rate at which the pressure rises (the 'decay rate'). The mass of air that leaks in each minute depends only on the atmospheric pressure, and not on the downstream pressure, because flow through a leak is critical ('choked') when the ratio between the discharge and inlet pressures is less than 0.55. The mm Hg/min rate at which the pressure rises due to leakage will be constant until the pressure ratio increases above 0.55 atm,[3] and if the system volume is known, the mass of air which enters per unit time can be computed from the rate of pressure rise. The in-leaking air expands as it enters the vacuum chamber, increasing in volume and decreasing in RH in direct proportion to the ratio between the atmospheric pressure and storage pressure. Consequently, at a lower storage pressure, for any given chamber 'decay rate', leakage represents a greater proportion of the air passing through the chamber, and the in-leaking air is drier. The effect that various chamber 'decay rates' have on relative humidity is indicated in Table 9.3 for lots of ten Hass avocados stored in 250-mm-diameter glass vacuum desiccators, assuming that at each storage pressure the commodity evaporated enough water to transfer all of its respiratory heat. The analytic result indicates that the leak rate must be kept below 6 mbar/h in order to avoid excess water loss at a pressure of 2.33 kPa (10 mm Hg); it should not exceed 10 mbar/h at 2 kPa (15 mm Hg) and 15 mbar/h at 2.67 kPa (20 mm Hg), while somewhat more than 50 mbar/h is tolerable at 5.33 kPa (40 mm Hg).

The relationship between leakage and relative humidity will vary depending upon such factors as load density, respiration rate, air-change rate, whether or not the air change has been saturated in a humidifier, the commodity's transpirational resistance at the storage pressure, and the temperature uniformity of the system (9.7). Table 9.3 indicates that the pressure, range from 2.67 to 5.33 kPa (10–40 mm Hg) is particularly susceptible to a weight-loss problem during LP storage at chamber 'decay rates' that might easily develop in a poorly designed and constructed laboratory apparatus. This would explain the result of the avocado experiment referred to in the first paragraph of this chapter (Apelbaum *et al.*, 1977b; Fig. 9.1), in which desiccation occurred at pressures lower than 6.67 kPa (50 mm Hg), as well as the aberrantly high weight loss from Hass avocados (Table 9.1, boldface type) stored for 31 days at 5.33 kPa (40 mm Hg) and 38 days at 2 kPa (15 mm Hg). The sporadic, excessive weight loss from these particular lots of Hass avocados was attributed to undetected leakage (Cicale and Jamieson, 1978).

All structural parts of a VacuFresh^{SM} container are welded and leak-tested, and the door seal has been specially designed (Fig. 13.10) to prevent leakage. The container's leak rate is only approximately 0.4 mbar/h. Attempting to operate an LP container with a leaky door seal would be analogous to trying to make a shipment in a CA container which could not sustain the desired atmosphere because the door no longer fitted or sealed properly. When this happens, as it eventually does in any CA container due to in-transit racking, the container is 'retired' from CA service because it no longer is suitable for a controlled atmosphere application. The door frame of a VacuFresh^{SM} container is much stronger than that of a conventional container, and the 38,590 kg (85,000 lb) force on the door created by the operational vacuum draws the dish-head inward, causing it to conform with the door jamb, preventing it from being displaced or distorted by racking. The VacuFresh^{SM} door seal is inexpensive and easily replaced, allowing it to be changed whenever there is any indication of damage.[4] Grumman's Dormavac containers have a similar door-seal design, and during decades of usage their leakage rate has remained remarkably stable.

The nominal leak rate of a VacuFresh^{SM} container has almost no influence on the operational RH. In a 6.1 m (20 ft) VacuFresh^{SM} container loaded with 8170 kg (18,000 lb) of avocados, operated at a pressure of 2.67 kPa (20 mm Hg), with the flow set to deliver one air change (21,225 cm^3; 750 ft^3) per hour, the water vapour produced when all respiratory heat is transferred by evaporative cooling will exceed that needed to saturate the air change by 23.6%. This extra water will condense within the container, since it cannot be carried out by the air change. At this pumping speed, a 0.4-mbar/h leak rate only comprises 1.5% of the total air flowing through the container, and any effect this has on RH is compensated by less water condensation on the container wall. Because 'uncontrolled' leakage automatically is offset by a decrease in 'controlled (modulated) leakage', the total rate at which ambient air enters is not influenced by leakage as long as the regulator is able to control the pressure. The leak rate would have to exceed 20 mbar/h before it would affect the pressure regulator's ability to maintain a 2.7 kPa (20 mm Hg) pressure. The situation usually is different in a laboratory set-up because typically the air-change rate is adjusted to a specific value, and any 'uncontrolled' leakage adds to the total airflow through the chamber.

9.7 Chamber Temperature Uniformity

If there is a cold spot on the chamber wall of a laboratory hypobaric apparatus, water will evaporate from the commodity, diffuse to and condense on the cold area, and the dew point in the chamber will approach the temperature of the cold spot. Several unusual thermodynamic properties make a hypobaric apparatus unusually susceptible to the establishment of this type of evaporation/condensation cycle between the commodity and chamber wall. In a mixture containing water vapour and a non-condensable gas such as air, when the vapour condenses on a cold surface, the non-condensable gas is left at the surface and the incoming condensable vapour must diffuse through the body of vapour–gas mixture collected in the vicinity of the condensate surface before the vapour reaches the cold surface to condense (Özisik, 1985). The presence of non-condensable gas adjacent to the condensate surface acts as a thermal barrier of resistance to heat transfer, greatly reducing the heat-transfer coefficient for condensation. Lowering the pressure increases the water-vapour diffusion coefficient, decreases the concentration of non-condensable gas and elevates the heat-transfer coefficient for condensation, making it easier for water vapour to condense on the container wall. A low pressure also reduces the commodity's transpirational resistance (6.5) and increases the rate at which transpired water can diffuse to the wall.

Lougheed *et al.* (1978) suggested a method of estimating the humidity in LP laboratory chambers without actually measuring it: 'We seemed to have solved this problem in our preliminary studies by operating at or near saturation as established by noting the formation of condensation on transparent viewing plates in each chamber'. Unfortunately, that can have other implications. Even if the air changes are not close to saturation, water evaporated by the commodity into a fully loaded chamber elevates the humidity close to saturation, and part of this vapour is likely to condense on the chamber walls if they are not completely uniform in temperature. During a 12-day hypobaric storage at 18.3°C and a pressure of 24 kPa (180 mm Hg), only a 2.09% weight loss is required to remove all respiratory heat from tomatoes by evaporative cooling. When this fruit was stored in a laboratory LP apparatus without humidification, in 12 days the weight loss increased progressively from 3.5 to 7.3% as the pressure was lowered from 100.66 to 24 kPa (755 to 180 mm Hg). Yet at all pressures, even though the chambers were not humidified, beads of water condensed on their transparent rims (Tolle, 1969). Often when LP studies are carried out in humidified glass laboratory chambers, a zone of condensation spread over a relatively small area of the chamber wall is evident and indicative of a 'cold spot'. The chamber dew point cannot be significantly higher than the temperature of the cold spot, and an evaporation–condensation cycle between the commodity and the cold spot may be stripping water from the commodity, fuelled by respiratory heat. To prevent this condition from arising, the temperature must be precisely controlled so that it is uniform over the entire surface of the storage chamber.

Laboratory LP chambers usually are incubated in an air-conditioned cold room or growth chamber cooled by a refrigeration compressor controlled by an 'ON–OFF' thermostat with a dead band of at least 2–3°C. The enclosure is likely to have colder and warmer locations, and cold discharge air may preferentially pass over one side of the chamber and create a 'cold spot' when the compressor is operating. The chamber walls and commodity decrease in temperature during the compressor 'ON' cycle, and when the compressor turns 'OFF' in response to the thermostat, or when the evaporator coil defrosts, the wall and commodity temperatures increase. These cycles are illustrated in Fig. 9.3, composed from data obtained during the commercial storage of a full load of roses in a 12.2 m Grumman/Dormavac intermodal hypobaric container operated at a pressure of 2 kPa (15 mm Hg).[6] During midday, solar radiation and a high ambient temperature kept the refrigeration compressor operating almost continuously, causing the wall temperature to rapidly decrease by 1°C while the boxes slowly cooled by as much as 0.35°C in 4.5 h. As the container dew point cannot be higher than the wall temperature, during the midday period evaporative cooling lowered the commodity temperature, transferring all of the respiratory heat and most of the sensible

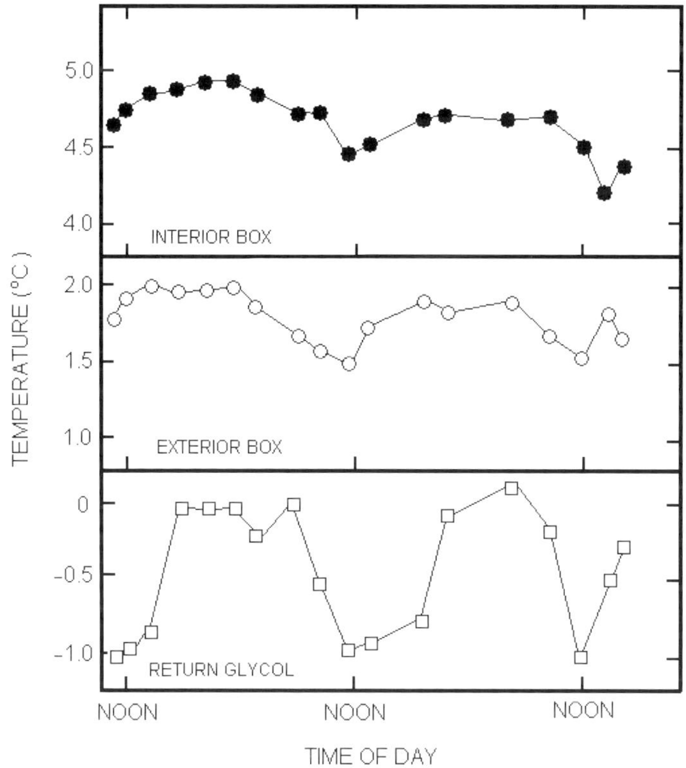

Fig. 9.3. Return glycol, interior box and exterior box temperature fluctuations in a 12.2 m Grumman/Dormavac hypobaric intermodal container loaded with boxes of roses according to the stacking arrangement in Fig. 6.13, operated at a pressure of 2 kPa (15 mm HgA). Interior wall temperature ≅ return glycol temperature. The flowers had almost cooled to their steady-state temperature at the time these measurements were made. The return glycol temperature was controlled with an ON–OFF thermostat with a dead band of 1°C. For other conditions see example 6.

heat that was removed during each cooling cycle (example 4). At night the cycle reversed, the average wall and glycol temperature increased by 1°C, the air dew point tracked the commodity temperature, and the retention of all respiratory heat accounted for the rate at which the commodity temperature increased. The extra water evaporated to remove sensible heat during the cooling cycle was 'recovered' by less water loss during the heating cycle.

The behaviour of roses in a commercial hypobaric intermodal container operated with an ON–OFF thermostat differs in several important respects from that of a naked commodity stored in a laboratory set-up that has a 'cold spot' on the chamber wall. The wall temperature is uniform throughout the intermodal container, and the flowers were protected by water-retentive packaging and boxes that elevated their temperature by 2.5–5.5°C, preventing heat uptake by radiation and convection. Condensation on a cold spot in a laboratory chamber draws down the average 'steady-state' humidity and causes the temperature of a naked commodity to fall below the wall temperature (Fig. 6.16). Then, because the radiant heat source-to-sink ratio is unity in a laboratory chamber (6.22), radiant heat transfer into the commodity fuels an excessive evaporative weight loss.

To avoid evaporation/condensation cycles caused by cold spots, VacuFresh[SM] containers are cooled by a jacketed refrigeration system comprised of cooling surfaces and conduits specially designed to work in combination with a proportional-integral-derivative (PID) modulating thermostat in order to prevent the wall temperature from cycling or varying by more than −0.2 to +0.3°C from the set temperature throughout a −17.2 to 49°C ambient temperature range (13.12; 13.14). Defrost cycles that interrupt the temperature uniformity of conventional forced-air intermodal container cooling systems are not required in hypobaric intermodal containers because glycol-water is utilized as a secondary coolant.

9.8 Elevating the Humidity

There has been an ongoing emphasis seeking ways to elevate the humidity in conventional and CA forced-air-cooled intermodal containers. Modulating hot gas-refrigeration systems are used to lower the refrigerant vs. air temperature difference that causes moisture to condense on the evaporator coil, and in reverse-flow intermodal containers the fan-driven flowing air is returned over the top of the load to capture heat transmitted through the roof before it can circulate throughout the cargo. Reverse-flow intermodal refrigeration systems also provide a high velocity of flow through and around the cartons, improving the convective-film coefficient, speeding cool down with minimum water loss (Fockens and Meffert, 1972), and reducing the commodity steady-state temperature rise. Nevertheless, the humidity cannot be brought close to saturation because the evaporator coil's refrigerant is cold enough to strip water from the circulating air, and on warm days heat transmitted through the container's insulation elevates the air dry-bulb temperature and lowers the humidity. Several other factors influence the humidity that can be sustained. Up to two air changes varying in temperature and humidity are introduced each hour into conventional refrigerated containers and several times each day the compressor cycles OFF to defrost for periods of approximately 30 min. The systems used in CA containers to remove O_2 from the air, extract water vapour (6.17), and CA containers not only leak but also occasionally admit ambient air varying in humidity and temperature in order to provide O_2. In some CA and NA water that condensed on the evaporator coil is atomized or converted to steam and injected into the evaporator discharge air to elevate the humidity in CA and NA systems, but nevertheless the humidity cannot be maintained above 90–95% (Table 14.1). Because of these limitations, commodities often are protected by water-retentive wraps or packaging during transport or storage in NA or CA (6.4), yet the same commodities have been stored 'naked' in almost all laboratory LP studies.

To be useful in an LP intermodal container, a wrap must be beneficial both at atmospheric and low pressures because before and after a commodity is transported or stored in LP, the wrap is required to provide protection during the distribution cycle when the commodity is removed from LP. Lowering the pressure decreases the amount of heat generated by the commodity and the amount of water that needs to be evaporated (6.1), and it also increases the wrap's ability to transmit water vapour (equation 6.56). Nevertheless, standard commercial wraps often are adequate to provide the degree of protection needed in LP containers. Example 5 describes a result obtained in a Grumman/ Dormavac container loaded with Madam Delbar, Scandia and Visa roses packed in commercial boxes with standard perforated polyethylene liners (Fig. 3.15). The flowers are prevented from acquiring heat by radiation and convection because the water mass-transport resistance of the box and wrap elevates the temperature of the roses above the wall temperature (Fig. 9.3).

The small size of a laboratory LP chamber precludes the use of commercial boxes and water-retentive liners in laboratory experiments, but plastic wraps can be used advantageously to elevate the commodity temperature and decrease water loss even though they are less effective in a laboratory set-up than in a commercial box. The wrap is less able to prevent weight loss in a laboratory apparatus because the ratio

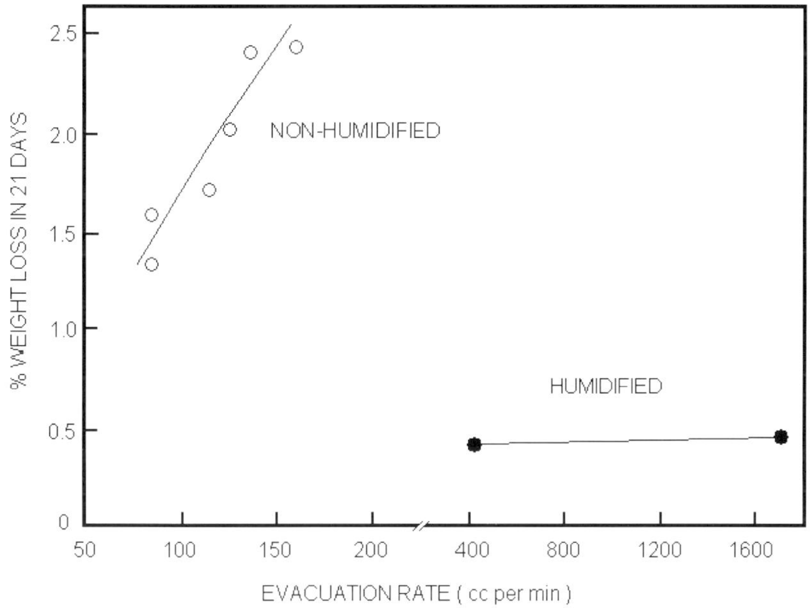

Fig. 9.4. Weight loss during rose storage (3 bunches weighing 850 g per bunch) at 2°C and a pressure of 3.33 kPa (25 mm HgA), in a 15-cm-diameter × 90-cm-long PVC tube, with and without humidification. The weight change includes a 0.2% loss due to carbon respired as CO_2 (Burg, 1992).

between the weight of commodity enclosed vs. the surface area of the 'wrap' is much lower than it is in a commercial box, and the entire surface area of the laboratory wrap transfers heat by radiation and convection, whereas only one end of an exterior box in an LP container transfers heat by radiation, and two ends by convection (Fig. 6.15). In a laboratory rose-storage experiment (example 6; Fig. 9.4) performed at 2°C and a pressure of 3.33 kPa (25 mm Hg), a polyethylene wrap's resistance to water mass transport was only 1.44 s/cm at the storage pressure, whereas at that pressure the same polyethylene material used as a liner in a standard commercial carton has a resistance of 31.4 s/cm due to the box geometry. The flower temperature only needed to be 0.033°C above the wall temperature to transfer all respiratory heat[6] when the laboratory chamber was humidified. If the chamber dew point was significantly decreased by leakage or any other cause, the laboratory wrap used in this study would have been inadequate to prevent desiccation. A Mylar radiation-reflecting wrap would have been a better choice (Fig. 9.5).

9.9 Mylar Radiation Shielding

Mylar radiation shielding's ability to protect commodities from excess weight loss at a low storage pressure takes advantage of the ineffectiveness of convective heat transfer in LP (6.22). When radiant heat transfer is also prevented, the only significant energy source available to evaporate commodity water is respiratory heat. In a test with flowers (example 5; Fig. 9.5), a Mylar wrap caused the temperature of roses and *Protea* blooms to decrease below the wall temperature and approach the container-air dew point, as would be expected in the absence of radiant and convective heat transfer. The flower temperature remained sufficiently above the container-air dew point temperature to create the vapour pressure gradient needed to transfer all respiratory heat by evaporation through the box's water-vapour mass-transport resistance. Because the flowers were unable to receive radiant and convective heat, they lost very little moisture during a 1-month storage even though they were colder than their environment.

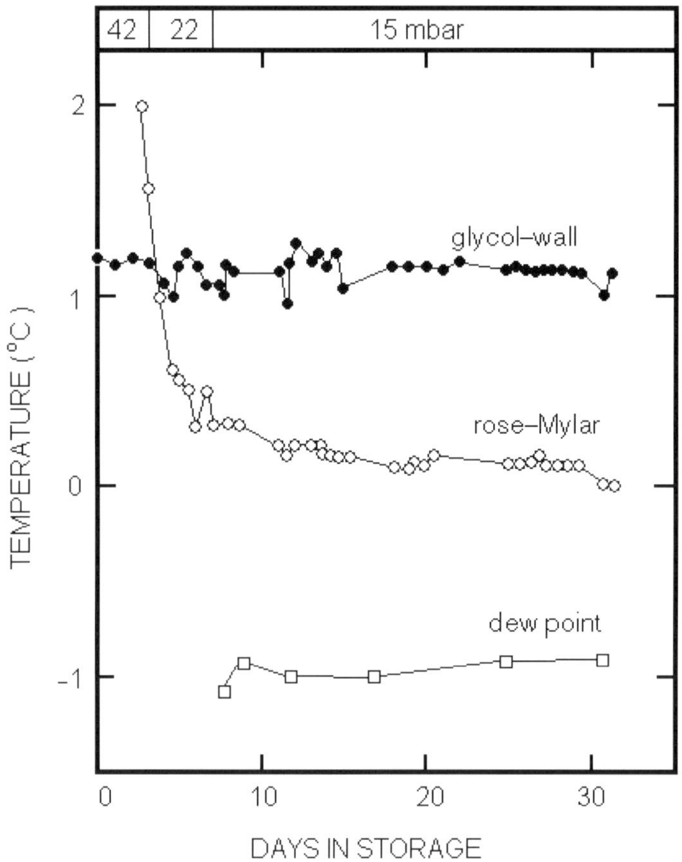

Fig. 9.5. Glycol (wall), interior of box containing roses, and dew point temperatures measured in a 6.1 m VacuFresh[SM] container operated at 1.1°C, flowing 0.81 air changes per hour. The floor was flooded with water. Two boxes, each containing 13.62 kg of roses, were protected by perforated Mylar slip sheets, and four boxes each containing 6.5 kg of *Protea* blooms by perforated water-retentive Mylar or PVC box liners. Initially, the pressure was set at 43 mbar (4.4 kPa = 33 mm Hg) to avoid subjecting the commodity to anaerobic conditions while it was cooling (6.24), but as cooling progressed, the pressure was lowered to 22 mbar (17 mm Hg) and then to 15 mbar (11.5 mm Hg). The RH and air temperature were measured with a data logger, and the rose box and glycol temperature by ± 0.1°C thermistor probes. After 31 days, the rose and *Protea* blooms protected by Mylar had not changed from their initial appearance, and after transfer to water opened normally, with at-harvest vase life. Mylar gave a much better result than PVC with *Protea* blooms.

Although a polished aluminium surface has an emissivity of 0.04 in the physiological temperature range, in commercial practice the emissivity is increased to 0.2 by oxidation (Table 6.11), causing a single layer of Mylar to be 80% effective as a radiation shield. Mylar also is a highly efficient water barrier and is capable of causing an excessive temperature rise when a boxed commodity protected by a Mylar liner is transferred from LP to atmospheric pressure. To avoid this occurrence, the Mylar wrap may be perforated to reduce its resistance to water mass transport without significantly reducing its ability to shield radiation. Multiple layers of Mylar are used in space vehicles to provide essentially 100% radiation shielding (note 23, chapter 6), and a double layer of perforated Mylar can provide 96% radiation shielding for commodities stored in LP. If additional shielding is required, the box surface can be coloured by a thermally reflective coating. Gloss white, tan and aluminium surfaces have reflectivity values varying from 0.5 to 0.83. In studies carried out at the US National Institutes of Health (Kwon and M.B. Burg, 1999, personal communication) a monolayer of tissue-culture cells, suspended on a mesh grid and shielded from radiation by Mylar film, remained viable in LP and did not lose water during an 11-day storage at a constant temperature, very low pressure and precisely controlled, slow rate of air change. This result demonstrates that by using appropriate equipment, commodity weight loss can be kept to an absolute minimum in a laboratory LP set-up.

9.10 Air Change Rate

An excessive air-change rate provides no advantage in LP and is difficult to saturate because it causes a high rate of evaporative cooling in a laboratory humidifier and may result in a pressure drop in the line connecting the humidifier to the vacuum chamber (example 2). Conversely, if the air-change rate is too low, this can cause an excessive $[O_2]$ draw-down and accumulation of CO_2 and ethylene.

9.11 Design of a Laboratory Apparatus

According to Lougheed *et al.* (1978), in LP 'control of airflow rate, relative humidity, and atmospheric pressure is not easy. Once the desired conditions are obtained, it is best to leave the system undisturbed'. This observation refers to the LP apparatus illustrated in Fig. 2.4, which has most often been used in laboratory research. A better design (Fig. 9.6) functions with extreme ease and a minimum of attention. The conditions can be altered easily, reliably, precisely and very rapidly, and the chamber can be quickly repressurized, opened and re-evacuated. Preferably the chamber should be kept in an incubator with a uniform airflow pattern designed to provide an even temperature distribution, and the wet- and dry-bulb temperatures should be measured inside the storage area. To prevent commodity water loss, a Mylar radiation screen installed at the inner surface of the storage chamber is strongly recommended.

9.12 Measuring the Pressure

An unexpected problem was encountered with the popular closed-end style laboratory mercury manometer which often is used to measure absolute pressures lower than 40–50 mm Hg (Fig. 9.7, *left*). These manometers have a restriction in their measuring tube to prevent breakage caused by a rapid rise of mercury when the vacuum is released. New manometers of this type, carefully cleaned and filled according to the manufacturer's instructions with preheated, evacuated, fresh laboratory-grade mercury, fail to operate reliably below about 40 mm Hg. In that pressure range, the flow-restriction invariably 'clogs' and prevents the outflow of the mercury from the centre tube. Two manometers attached to the same vacuum chamber side-by-side gave readings of 40 and 19 mm Hg for many days, even with continual tapping, while both a McLeod gauge and a home-made closed-end U-tube manometer without a restriction (Fig. 9.7, *right*) gave identical, correct readings of 4 mm Hg. Closed-end manometers with a restriction, obtained from a laboratory that had used these instruments to measure pressure during hypobaric studies could not read lower than 30–40 mm Hg. As these published studies were carried out at pressures of 50 mm Hg or higher, the measuring error may not have been very large, but the problem is so serious at 10–20 mm Hg that it interferes with efforts to carry out research in that range and gives the false impression that the vacuum system is not able to lower the pressure further, when in fact it has reduced the pressure to the vapour pressure of water at the storage temperature. Conceivably, this instrumentation error has caused spurious results in some laboratory studies at very low pressures. The problem

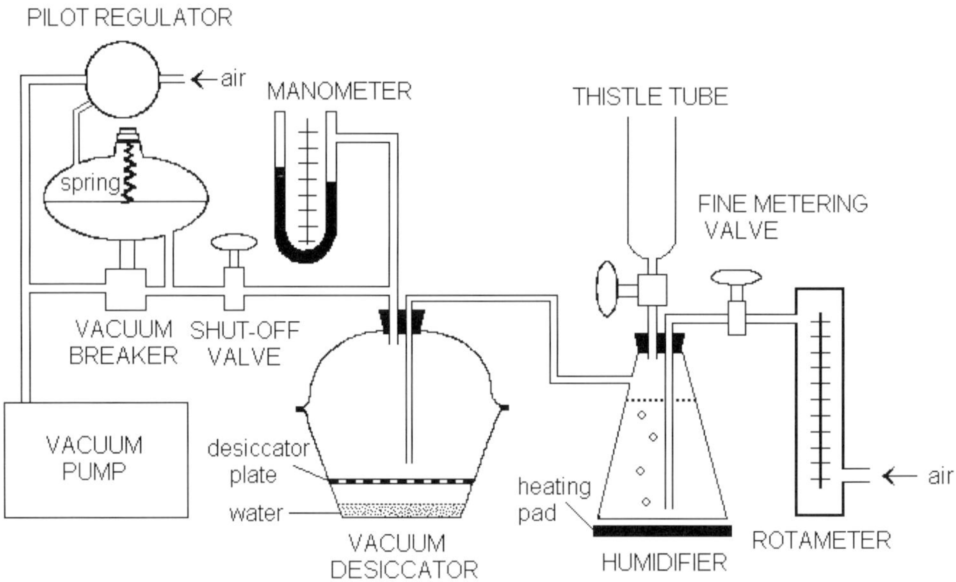

Fig. 9.6. Laboratory apparatus. The vacuum desiccator and humidifier are located in a temperature-controlled room. A null-balance absolute-pressure pilot regulator establishes the reference pressure above the diaphragm of the main regulator, and the pressure in the vacuum desiccator is registered below the diaphragm. The main regulator spring is adjusted to create a downward pressure of approximately 26.7 kPa (200 mm Hg) and the set pressure in the vacuum desiccator is adjusted by varying the reference pressure with the pilot regulator. The main regulator will crack open when the pressure below its diaphragm decreases to 2.7 kPa (20 mm Hg), if the reference pressure above the diaphragm is set at 24 kPa (180 mm Hg). This regulator system controls the absolute pressure ± 0.5 mbar (0.4 mm Hg). Initially, when the chamber is pumping down, the fine metering valve is kept closed until the vacuum in the desiccator reaches the desired level. Then the valve is adjusted to the desired flow rate, which is indicated on the rotameter. This configuration has the advantage that the flow responds instantly and immediately stabilizes whenever the fine metering-valve setting is changed, whereas using the arrangement shown in Fig. 2.4, a change in the metering-valve setting is followed by a long adjustment period before the flow stabilizes at a new value, and it may be necessary to repeat the adjustment many times before the desired flow is achieved. A single main regulator can be used to control numerous desiccators operating at the same pressure by plumbing the chambers in parallel, using individual fine metering valves to set the flow independently in each chamber, at different values if desired. Each desiccator requires a downstream shut-off valve so that it can be disconnected separately. The shut-off valve also serves as a coarse adjustment during pump-down. In the configuration shown in Fig. 2.4, desiccators should not be plumbed in parallel because it is not possible to predict how the flow from a single upstream regulator would distribute between the individual desiccators. Voltage to the humidifier heating pads is adjusted with a rheostat to elevate the temperature of the humidifying water by 2–3°C above the controlled room temperature. Additional water is added to each humidifier through the thistle tube as required.

is eliminated by using a manometer without a restriction (Fig. 9.6, *right*).

9.13 Examples

1. When the water height in a tower is 130 mm and the chamber pressure is 2 kPa (15 mm Hg), the pressure upstream of the water tower is 3.33 kPa (25 mm Hg), and the air expands 1.67-fold as it passes through the tower. If instead the chamber pressure is 20 kPa (150 mm Hg), the pressure upstream of the water tower is 23.33 kPa (175 mm Hg), and the air expands 1.17-fold as it passes through the tower. At atmospheric pressure the expansion would only be 1.013-fold.

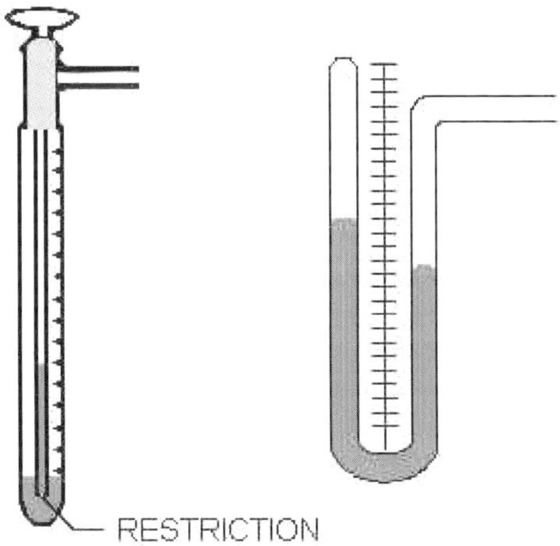

Fig. 9.7. Manometer design. (*left*) Closed-end absolute pressure laboratory manometer with restriction (orifice) at the base of the centre tube to limit mercury flow. (*right*) Absolute pressure U-tube manometer without a flow restriction. The U-tube manometer is accurate in the 0–30 mm Hg range. The closed-end laboratory manometer is unreliable in that range due to clogging of the restriction in the U-tube at pressures lower than 30 mm Hg.

2. Conditions:

storage pressure, p_2 = 3.33 kPa (25 mm Hg)
temperature = 2°C
chamber volume = 10 l (0.353 ft^3)
air-change rate = one per hour = 0.00278 l/s @ 3.33 kPa (25 mm Hg)
copper tubing OD = 0.3175 cm (0.125 in.)
copper tubing wall thickness = 0.0762 cm (0.03 in.)
copper tubing ID (D) = 0.165 cm (0.065 in.)
copper tubing length, L = 91.44 cm (3 ft)
p_1 = pressure above the water in the humidifier

In vacuum technology, the 'throughput' (q_{pV}) is the product of the pressure x the volume of gas that flows through a conducting element, divided by the time (Leybold-Heraeus, 1991):

$$q_{pV} = 33.3 \times 0.00278 = 0.0925 \text{ (mbar/l)}$$

The Reynolds number (Re) is estimated to be 8.4 from the expression:

$$\text{Re} = 15q_{pV}/D \qquad (9.1)$$

Below Re = 2000, the flow is laminar and the conductance (C, l/s) of a pipe flowing low-pressure air nearly saturated with water at 2°C is:

$$C = \phi 135(d^4/L)[(p_1 + p_2)/2p] \qquad (9.2)$$

where $(p_1 + p_2)/2$ is the average pressure of air entering and leaving the pipe, and ϕ corrects for the amount of water vapour present. In this example $\phi \cong 1.15$ (Leybold-Heraeus, 1991) and the result is computed as:

$$C = [1.15 \times 135 \times (0.165)^4/91.44]$$
$$[(p_1 + 33.3)/2] = 0.000629(p_1 + 33.3)$$

The relationship between conductance and throughput is:

$$q_{pV} = C(p_1 - p_2) \qquad (9.3)$$

and the solution is:

$$0.0925 = [0.000629(P_1 + 33.3)][P_1 - 33.3]$$
$$p_1 = 35.44 \text{ mbar}$$

The pressure above the water in the humidifier is 2.14 mbar higher than it is in the vacuum chamber. If air passes through the tube connecting the humidifier and vacuum chamber at a constant temperature, it will expand 1.06-fold and decrease in humidity by 6%.

3. Bananas stored in 10-l desiccators at 15°C and a pressure of 16 kPa (120 mm Hg) were humidified by bubbling the incoming air through 500 ml of 17.8°C 'heated' water contained in a side-arm Erlenmeyer flask (Fig. 2.4). The weight loss in 33 days was 1.4, 1.5 and 3.8%, when the pumping speed was varied to create 0.9, 5.3 and 15 air changes per hour respectively (Burg, 1969). Based on data in Table 4.1, a 1.5% weight loss in 27 days would be expected if all respiratory heat produced at 15°C and a pressure of 16 kPa (120 mm Hg) was transferred by evaporative cooling. Flow had little effect on weight loss in the 0.9–5.3 air-change-per-hour range commonly used in a laboratory apparatus or intermodal container, but at 15 air changes per hour, the weight loss increased significantly.

4. At a pressure of 2 kPa (15 mm Hg), under the conditions of the rose storage (Fig. 9.3) the film coefficient for natural convection is only 0.24 kcal/m^2·h·°C (0.05 BTU/ft^2·h·°F) for a 1°C temperature difference (6.22). Assuming an emissivity of 0.95 for the flower boxes and white painted walls of the container (Table 6.11), the radiant heat-transfer coefficient (equations 6.36 and 6.37) is 3.5 kcal/m^2·h·°C (0.73 BTU/ft^2·°F). The exterior box ends that participate in the radiant process had a surface area of 0.077 m^2 (0.83 ft^2), and both ends of interior and exterior boxes transferred heat by convection. The 18.16 kg (40 lb) of flowers in each box have a specific heat of 0.95 kcal/kg·°C (0.95 BTU/lb·°F) and produce 0.95 kcal (3.76 BTU) per hour of respiratory heat. During a midday cycle, the boxes cooled by 0.3°C in 4.5 h. This required removal of all the respiratory heat plus a total of 2.87 kcal (11.4 BTU) of sensible heat. Convection would require nearly a 31°C temperature difference to exchange this amount of heat and radiation, a 4.3°C temperature difference, whereas evaporative cooling would suffice with the vapour pressure gradient created by a 2.5°C difference between the commodity and air-dew point temperature. The container-air dew point could not be higher than the wall temperature, and the vapour pressure gradient between the box and surrounding air accounts for the entire commodity temperature decrease in both interior and exterior boxes, including removal of both respiratory and sensible heat.

During the nightly warming cycle, the glycol and wall temperature rapidly increased by 1°C and the box temperatures rose by as much as 0.25°C in a few hours. Because the wall temperature always was colder than the box temperature, the warming could not be caused by convection and radiation. The container dew point rapidly increased when the wall temperature rose, preventing evaporative cooling, and if during this interval all respiratory heat was retained, the boxes would warm by 0.21°C in 2 h. Thus respiratory heat accounts for the entire box-temperature rise. Most of the 'extra' water evaporated to remove sensible heat during the cooling cycle was 'recovered' by less water loss during the heating cycle.

5. Conditions: 12.2 m (40 ft) Grumman/Dormavac intermodal container

- pumping speed = 11.13 m^3 (40 ft^3)/min at 2.67 kPa (20 mm Hg)
- load = 7028 kg (15,480 lb) of Visa, Scandia and Madam Delbar roses packed in commercial boxes with perforated polyethylene liners; two vent holes in each end
- wall temperature = 0°C
- container is humidified
- box resistance = 953 s/cm for water vapour at 1 atm (Fig. 3.16).
- box surface area = 10,495 cm^2

The container, loaded according to the stacking diagram illustrated in Fig. 6.13, had average interior and exterior box temperatures of 3.3 ± 0.3°C and 1.1 ± 0.2°C, respectively. A wet/dry-bulb thermometer located in the centre of an air chimney in the middle of the load indicated an air dry-bulb temperature of 2.6°C at 92.5% relative humidity. The CO_2 production of the roses, measured by the method indicated in Fig. 4.1, was 20.5 mg/kg·h, which produces 0.95 kcal (3.76 BTU) per hour of respiratory heat per box. The vapour-pressure gradient computed between an interior box and the air surrounding it was 0.00092 atm, and since at 2.67 kPa (20 mm Hg) the box's resistance to water vapour mass transport is 25.1 s/cm (Fig. 3.15), the amount of water vapour evaporating from the box, across the vapour-pressure gradient existing between the box and air, could remove 1.07 kcal (4.21 BTU) per hour. This accounts for all of the respiratory heat. In 21 days, the flowers would be expected to lose 3.5% of their weight due to water evaporation, and 0.22% from the loss of respiratory carbon. Exterior boxes constituting up to 85% of the load lose a significant amount of respiratory heat by radiation, and therefore less by evaporation, but internal boxes mainly lose heat by evaporation. When the container is filled with carnations and operated under the same conditions, the temperature gradients that arise between the boxes and container wall are approximately half as great because carnations respire at half the rate of roses. Consequently, they lose half as much weight during an identical storage period.

6. One bunch each of Madam Delbar, Scandia and Visa roses (25 flowers/bunch = approximately 850 g) were precooled to 2°C, over-wrapped with perforated 1 mil polyethylene film, weighed and placed end-to-end in a 90 cm long × 15 cm diameter horizontal PVC vacuum chamber (tube), with thermocouples inserted within each bunch, in the chamber air and on the chamber wall (Fig. 9.4; Burg, 1992). Duplicate chambers were operated at various flow rates for 21 days at 2°C and a pressure of 3.33 kPa (25 mm Hg), after which the roses were removed and weighed. The humidified air was flowed at 400–1800 cm^3/min through the chambers, or else non-humidified air was flowed at 80–160 cm^3/min. In none of the chambers did the temperature difference between the flowers, chamber wall and chamber air exceed the inherent variance of the calibrated thermocouples (± 0.1°C). If all respiratory heat was transferred by water evaporation, the expected weight loss was 2.5% in 21 days, plus 0.2% due to carbon converted to CO_2 by respiration. When the flowers were humidified, the weight loss in 21 days was 0.5% independent of the air-change rate, of which 0.2% was due to the conversion of carbon to CO_2 by respiration. This result indicates that the flower temperature remained slightly above the wall temperature, and that most of the respiratory heat was lost by radiation and a small amount by convection. When the experiment was repeated without humidification, at an air-change rate of approximately 180 cm^3/min, the weight loss was accounted for by the amount of water that needed to be evaporated to remove all respiratory heat. This also corresponded to the quantity of water required to saturate the air change at that flow rate. At lower rates of air change the weight loss declined, approaching the value with humidification. This indicates that progressively as the flow was decreased, a temperature gradient developed that allowed part of the respiratory heat to be transferred by radiation and convection.

A simple correlation for free convection from a horizontal isothermal cylinder, such as the wrapped flowers in this example, is given in the form (Özisik, 1985; Table 6.9):

$$\mathrm{Ra_D} = \{[g\beta\,(T_1 - T_2)L^3\rho^2]/u^2\}(\mathrm{Pr}) \qquad (9.4)$$

$$\mathrm{Nu_m} = hL/k = c\mathrm{Ra_D}^n \qquad (9.5)$$

where constant c and exponent n are given in Table 9.4, and $\mathrm{Ra_D}$ is the Raleigh number; $\mathrm{Nu_m}$ the Nusselt number; the gravitational constant g = 9.8 m/s^2 (32.17 ft/s^2); β = 1/275 K; T_1 is the flower temperature, K; T_2 the wall temperature, K; L is the diameter of the cylinder, m (ft); the Prandtl number (Pr) is 0.793; and at 2°C the dynamic viscosity (μ) is 1.257×10^{-5} (N·s)/m^2 (8.418×10^{-6} lbm/ft·s), the thermal conductivity k is 0.0206 W/m·°C (0.0121 BTU/h·ft·°F) and the density (ρ) of the air/water-vapour mixture at 3.33 kPa (25 mm Hg) is 0.0380 kg/m^3 (0.00237 lb/ft^3).

If T_1 = 0.033°C, $\mathrm{Ra_D} = 2.8545 \times 10^4$; $\mathrm{Nu_m}$ = 6.239 and the natural convective-film coefficient h_m is 0.874 W/m^2·°C (0.154 BTU/ft^2·°F·h). For a small temperature difference, the coefficient for radiant transfer (equation 6.34) is 5.35 W/m^2·°C (0.94 BTU/ft^2·h·°F), and assuming an emissivity of ε = 0.8, the combined coefficient for radiation and natural convection is 5.01 W/m^2·°C (0.883 BTU/ft^2·°F·h). With humidification, the 0.033°C temperature difference that developed between the flowers and wall could not be detected by the instrumentation used in this experiment, but nevertheless would allow radiation and convection to transfer an amount of heat equal to 88% (65.5 cal/h = 0.26 BTU/h) of that generated by respiration. The remaining 12% was transferred by

Table 9.4. Constant *c* and exponent *n* for equation 9.5 (Özisik, 1985).

$\mathrm{Ra_D}$	c	n
10^2–10^4	0.850	0.188
10^4–10^7	0.480	0.250
10^7–10^{12}	0.125	0.333

evaporation. The wrap's surface area was 4594.6 cm^2, and if the humidifier saturated the chamber air at 2°C and the commodity was 0.033°C warmer than the wall, the vapour pressure gradient between the commodity and air was 1.65×10^{-5} atm and, according to equation 6.39, the transpirational resistance of the wrap (r_{box}) was only 1.44 s/cm. Based on a typical leaf's surface-to-volume ratio of 50 cm^2/cm^3 (chapter 3 – example 23), the rose/wrap surface area ratio ($\Sigma A_{com}/A_{box}$ – equation 3.22) would be approximately 9.3, and the value of r_b to be used in a resistance-network calculation is $9.3 \times 1.44 = 13.4$ s/cm. When the stomates open in LP, a rose leaf would have a negligible transpirational resistance (6.5), and the water-retentive plastic film's transpirational conductance will determine the commodity temperature, the amount of heat acquired by radiation and convection, and the rate of water loss.

Notes

1. It is assumed that the plastic tray containing water was in contact with the base of the vacuum chamber. If it were elevated, either intentionally or due to warpage of the floor or tray, little if any additional heat would be transferred to the water through the contact resistance.
2. The temperature probes' wires can be brought out from a laboratory vacuum chamber by passing them through a short length of 6 mm OD copper tubing in which they are sealed by means of silicon rubber cement. After the sealant has dried, the tube is crimped to prevent the silicon from being pulled through by the force of the vacuum. The copper tube can be installed in a vacuum chamber either through a one-hole rubber stopper, or in a 'bored-through' swaged fitting. The temperature readings are converted to a humidity value using a wet-bulb relative-humidity table (Table 9.2) or water-vapour-pressure/temperature table. To improve accuracy when thermocouples are used, each should be individually calibrated in an ice bath to reduce its intrinsic error. A ±0.07°C thermistor is recommended for maximum accuracy (±0.5% RH). Thermistors have the additional advantage that their electrical resistance can be measured with a multimeter or ohm-meter, after which the ohm reading is converted to a temperature using a resistance-conversion chart.
3. When the atmospheric pressure is 101.3 kPa (760 mm Hg), the leak rate will be constant until the chamber pressure reaches 55.7 kPa (418 mm Hg).
4. In a low-temperature trial carried out in a −20°C test chamber, the rubber of a VacuFresh^{SM} container's door gasket became too rigid to make an initial seal. The operator entered the test chamber, in a few minutes removed the door seal, brought it outside to warm, reinstalled it and the container immediately drew its vacuum, pumped down and subsequently did not leak.
5. The container was equipped with a 1.0°C dead-band thermostat sensing the return glycol temperature. If the span of the ON–OFF thermostat was any smaller, the refrigeration compressor would short-cycle, lose oil during each start-up and not have sufficient continuous running time to recover the oil. Eventually the compressor will become oil-starved. See Fig. 11.4 for another example of temperature cycling in a hypobaric intermodal container operated with an ON–OFF thermostat.
6. A more complex thermodynamic calculation indicates that all of the respiratory heat could be transferred by a 0.02°C temperature gradient (Burg, 1992).

10

Horticultural Commodity Requirements

During the past 70 years, researchers have determined the optimal CA storage condition for most horticultural commodities, and even today such studies are ongoing. The optimal $CO_2 + O_2$ combination differs widely depending on the type of fruit, vegetable or flower, in-part because at atmospheric pressure structural diversity determines each commodity's resistance to gas exchange and the magnitude of the gaseous gradients that develop across the commodity's surface and through its intercellular system. The best LP condition for different products should be less variable, since stomatal opening and enhanced diffusion at a low pressure cause the gas concentration in intimate contact with each cell to resemble that in the storage atmosphere. Although there are exceptions,[1] and the optimal LP pressure has yet to be determined for tomatoes, bananas, peaches, green peppers, cucumbers, and many other fruits and vegetables, usually when an 'optimal' LP pressure and temperature have been established, two general storage classes have been defined: cold-tolerant commodities at 0°C and a pressure of 1.3–2 kPa (10–15 mm Hg), and tropical commodities at 4.4–16°C and pressures of 2–4 kPa (15–30 mm Hg). When a correction is made for the presence of water vapour in the saturated atmosphere around and within the commodity, in both classes the applied $[O_2]$ is in the range 0.1–0.3%. Occasionally, even lower $[O_2]$ has been found to be satisfactory or preferable: 0.04% $[O_2]$ with oat leaves (Veierskov and Kirk, 1986); 0.06% with green beans (S.P. Burg, 1976, unpublished), *Hibiscus* cuttings and *Epipremnum pinnatum* cuttings (Kirk *et al.*, 1986); and 0.003% has given better results with lettuce than any higher concentration (Grumman Allied Industry, 1979). It is uncertain whether the low $[O_2]$ limit for LP storage has been established for any commodity, or that a CO_2 'inversion' point has yet been reached. Reducing the pressure further is likely to improve storage in those instances where the 'optimal' pressure has been reported to be in the 8–10.67 kPa range (60–80 mm Hg). A pressure of 2.67 kPa (20 mm Hg) at a temperature of 10–13°C is highly effective with numerous cold-intolerant commodities, but a pressure of 2 kPa (15 mm Hg) may give a better result, since that small change lowers the $[O_2]$ by an additional factor of two.

It is difficult to quantify a commodity's storage life in LP because different varieties of the same agricultural commodity, and even the same variety grown in diverse seasons and locations, may not have the same keeping quality. The amount of rainfall during the growing season, stage of maturity at harvest, the cool down procedure, method of packaging and handling, elapsed time after harvest before the commodity is shipped, the pretreatments used to control insects and moulds, and sanitation procedures and care exercised in picking, handling, packing and shipping are all important factors.

There are several ways to compare storage systems. The time required for a commodity to spoil with no remaining shelf life is one criterion, or shelf life after various storage periods can be compared to that which the commodity had at the time of harvest. These methods are useful indicators of the relative advantages of different types of storage, but commercially the 'maximum useful storage life' ends when a product's shelf life diminishes to the minimum that is acceptable within a reasonable safety margin for the particular distribution method which has been used. For each commodity there tends to be a major hazard responsible for quality deterioration and the termination of storage life (Hall *et al.*, 1975; Grierson and Wardowski, 1978; Kader, 1983), but always the result is tempered by the uncertainty of decay development, insect pests and human error.

Results obtained in laboratory studies are not necessarily indicative of the behaviour to be expected under commercial conditions. Laboratory CA studies often are performed by continuously flowing a humidified gas mixture over the commodity, whereas commercial storage systems are sealed and the gaseous composition regulated by CO_2 scrubbing or addition, ethylene removal, N_2 generation and occasional admission of air. It is not unusual to find that a laboratory CA result cannot be duplicated commercially because of the build-up of volatile metabolic by-products in the sealed system. The heat and mass-transfer processes, air circulation and distribution, gas gradients between the commodity and atmosphere, humidification, and temperature uniformity are vastly different in commercial containers filled with a dense stack of cartons vs. a small-scale laboratory apparatus in which the commodity may be stored 'naked'. Typically, the control of pressure, temperature, air change and air circulation is better in a commercial LP intermodal container than it is in a small-scale laboratory apparatus. All of these factors influence the result.

FRUITS

10.1 Apple (*Malus* × *domestica* Borkh.), American Summer Pearmain, Jonathan, Cox Orange, Red King, Golden Delicious, Boskoop, Schweizer Glockenapfel, McIntosh, Red Delicious, Ida Red, Spartan, Staymen, Northern Spy, Red Rome, Lodi, Empire, Ontario

The storage life of American Summer Pearmain, Jonathan, Cox Orange, Red King and Golden Delicious apples improves progressively when the pressure is reduced from atmospheric (101.3 kPa = 760 mm Hg) to 13.3 kPa (100 mm Hg) (Kim *et al.*, 1969a,b; Sohn *et al.*, 1970; Sharples, 1971; Thornton, *et al.*, 1971; Dilley, 1972; Langridge and Sharples, 1972; Kajiura, 1973b; Salunkhe and Wu, 1973, 1975; Bubb, 1975a; Ha and Sohn, 1985; Fig. 10.1). At a pressure of 20 kPa (150 mm Hg), apples store about the same as in an optimal CA atmosphere, but a better result is obtained with Golden Delicious, Boskoop, Schweizer Glockenapfel, Cox Orange, McIntosh, Jonathan, Red Delicious, Ida Red, Spartan, Staymen, Northern Spy, Red Rome, Lodi and Empire apples at lower pressures ranging from 4.67 to 13.33 kPa (35–100 mm Hg) (Dilley, 1972; Bangerth, 1973, 1975, 1977, 1980, 1984; Kajiura, 1973b; Sharples and Langridge, 1973; Sharples, 1974; Dilley *et al.*, 1975; Bérard *et al.*, 1976; Dilley, 1977b; Bérard and Lougheed, 1978, 1982; McKeown *et al.*, 1978; Irwin and Dilley, 1980; Fig. 10.2). All apple cultivars with the exception of McIntosh respond well at 0°C and a pressure of 6.67 kPa (50 mm Hg), typically remaining in good condition for 9–11 months. Following 4 months in LP, McIntosh apples retain flesh firmness even after a week at room temperature (Bérard *et al.*, 1976), and fruit harvested early and stored at 6.67–8 kPa (50–60 mm Hg) maintained their initial firmness for 7 months at 0.5°C. After removal from storage, these apples softened less rapidly than NA controls, but McIntosh apples harvested later and stored in LP at 3.5°C softened more rapidly (Bérard and Lougheed, 1978), and

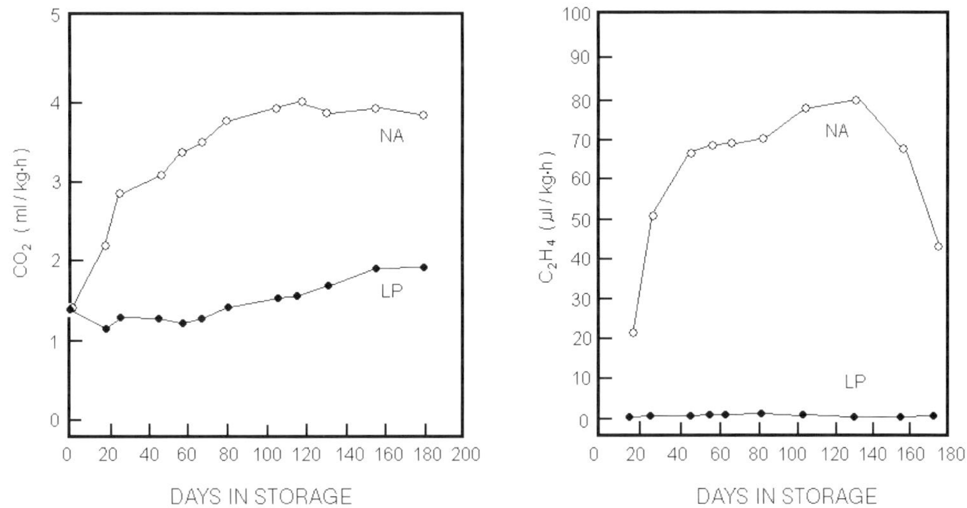

Fig. 10.1. CO_2 (*left*) and ethylene (*right*) production during 3°C storage of Golden Delicious apples at atmospheric pressure (NA) or in LP at 13.3 kPa (100 mm Hg). Ethylene and CO_2 were continuously frozen in a liquid-air trap situated between the storage container and vacuum pump. After thawing, the gases were measured by gas chromatography (Bangerth, 1975).

after 8 months at 0°C, apples kept at a pressure of 10.13 kPa (76 mm Hg) may suffer breakdown (Dilley, 1972). A storage temperature of 2–3°C is required in NA and CA to minimize chilling damage of McIntosh apples, and also Jonathan, Yellow Newton and Grimes Golden apples from certain locales (Hardenburg *et al.*, 1986).

At a comparable O_2 partial pressure, LP is more effective than CA in extending apple storage life (Kim *et al.*, 1969a,b; Dilley, 1972; Bangerth, 1973, 1975; Kajiura, 1973b; Salunkhe and Wu, 1973). The reduction in firmness and chlorophyll breakdown is slower in LP, abscissic acid synthesis is prevented, the ascorbic acid content is increased, acid catabolism is reduced, sugar is maintained at a higher value, electrolyte leakage and physiological disorders such as internal breakdown are reduced or eliminated, shelf life after transfer to ambient conditions is greatly improved, and respiration, ethylene synthesis and ethylene content are remarkably lower (Bangerth, 1973, 1974, 1975, 1977, 1980; Dilley, 1978; Bérard and Lougheed, 1982). At 3°C and a pressure of 13.33 kPa (100 mm Hg), apples evolved less than 0.5 µl/kg·h of ethylene during an LP storage, compared to up to 80 µl/kg·h in NA (Fig. 10.1). After storage at 3°C, during an 8-day shelf-life test at 25°C, the CO_2 production by Boskoop apples was high regardless of whether they had been stored in NA or in 3% [O_2] + 4% [CO_2], and low when they had been stored in LP at a pressure of 10 kPa (75 mm Hg) (Bangerth, 1973). Ripening, senescence, chlorophyll breakdown, ethylene production, internal ethylene and acid catabolism were greatly reduced in LP, internal breakdown and scald were decreased or prevented and firmness was improved compared to CA or NA. After LP storage, ethylene production recovered at ambient temperatures, but shelf life was remarkably improved. Excellent storage with very low weight loss also occurred when McIntosh, Jonathan, Red Delicious and Golden Delicious apples were stored at 0°C and a pressure of 10.13 kPa (76 mm Hg) (Dilley, 1972), and storage life was further extended by decreasing the pressure from 13.33 kPa (100 mm Hg) to between 4.67 and 6.67 kPa (35–50 mm Hg) (Sharples and Langridge, 1973; Sharples, 1974; Bangerth, 1984). During 11 months, fruits at this low-pressure range never produced autocatalytic ethylene or developed a respiratory climacteric (Bangerth, 1984),

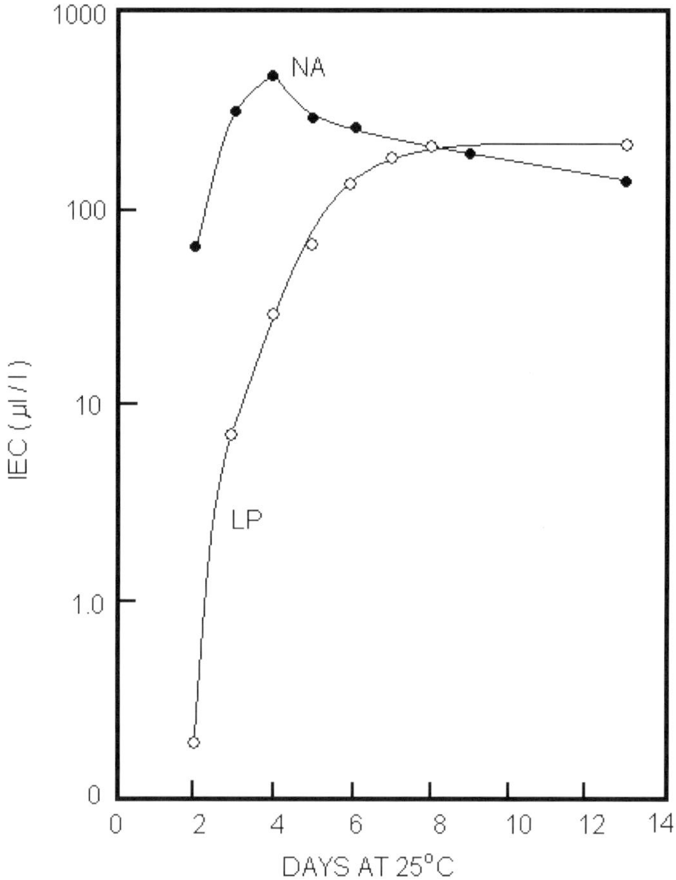

Fig. 10.2. Ethylene synthesis by Empire apples during ripening at 25°C following 239 days' storage at –1.0 to –0.6°C in NA or in LP at a pressure of 5.3 kPa (40 mm Hg). Note the log scale for the IEC (D.R. Dilley, 1979, unpublished data, referred to in Jamieson, 1980c).

and their firmness only decreased slightly. LP also increased the ascorbic acid content (Bangerth, 1977), decreased electrolyte leakage from daminozide-treated McIntosh apples (Bérard and Lougheed, 1982) and prevented abscissic acid synthesis (Bangerth, 1980). Both in LP and CA when the $[O_2]$ is decreased below 2%, apples eventually lose their sensitivity to ethylene and their ability to produce aroma and flavour volatiles (Bangerth, 1973; Shatat *et al.*, 1978; Bangerth and Streif, 1987; 3.26). In LP, weight loss is so low that even sensitive apples such as Golden Delicious do not shrivel.

Ethylene accumulation in a CA storage room hastens softening and ripening, and CO_2 may build up sufficiently to cause or accentuate physiological disorders such as scald, brown core and brown heart. These disorders have never been observed in LP, even when they developed in identical apple samples stored in CA and NA (7.11).

Respiration is promoted, titratable acidity and fruit firmness decreased, tissue permeability increased, and the activities of malic enzyme, acid phosphatase and PAL stimulated when 500 µl/l ethylene is provided in air changes during an LP apple storage at 3°C and a pressure of 13.3 kPa (100 mm Hg), but chlorophyll breakdown is hardly influenced (Bangerth, 1975). This proves that ethylene is able to act at a low temperature and O_2 partial pressure (Fidler,

1950; Forsyth *et al.*, 1969; Blanpied, 1972b; Stoll, 1972), and indicates that the improved apple storage in LP vs. CA and NA is due at least in part to diffusive escape of ethylene from the tissue and the gas's removal from the storage area by continuous air changes.

A full container-load of Empire apples, picked at the pre-climacteric stage and stored in a 12.2 m Grumman/Dormavac intermodal container operated at −1.1 to −0.6°C and a pressure of 5.33 kPa (40 mm Hg), was still in a pre-climacteric condition after 239 days of LP storage. The internal ethylene was very low when the fruits were removed, and then the IEC rose to a normal high level within 1 week at 25°C (Fig. 10.2; Jamieson, 1980c). The ability of these fruits to respond to ethylene was impaired (Dilley, 1972) and the time to ripen extended. After 150 days of storage, the firmness, total soluble solids and taste were evaluated for the LP Empire apples and a comparable lot kept in CA. LP apples were firmer than CA apples (116 vs. 87 kPa), had a higher percentage of total solids (13.3% vs. 12.1%), and a better taste-panel rating (31.3 for LP vs. 41.2 for CA, where 23.5 is a perfect apple and 60 is unacceptable). After 239 days, the taste rating of LP apples was 38.4 vs. 46.5 for CA apples.

The major diseases of stored apples are lenticle rot caused by *Phlyctaena vagabunda* Desm. (= *Gloeosporium album* Osterw.), and blue mould rot due to *Penicillium expansum* (Lk.) Thom. (Wills *et al.*, 1989). At atmospheric pressure the growth and sporulation of *P. expansum* and *P. digitatum* is inhibited at a low $[O_2]$ concentration (Tabak and Cooke, 1968; Table 7.4). LP pressures as high as 13.57 kPa (102 mm Hg = 2.3% $[O_2]$) significantly inhibit growth and sporulation of *P. expansum* (Table 7.5), and at 23°C and a pressure of 3.33 kPa (25 mm Hg = 0.11% $[O_2]$), both the growth and sporulation of *P. digitatum* cease (Fig. 7.6). The vegetative growth and sporulation of *G. musarum* is not prevented by anaerobiosis at atmospheric pressure (Goos and Tschirsch, 1962).

10.2 Apricot (*Prunus armeniaca* L.), var. Large Early Montgament

Apricots normally can be stored for 1–2 weeks in NA at −1 to 0°C (Welby and McGregor, 1997). They tolerate 1.5–2% $[O_2]$ but 0.3% $[O_2]$ induces fermentation and physiological disorders. While 2–3% $[O_2]$ + 2.5–3% $[CO_2]$ improves flavour retention, it sometimes accentuates internal browning (Hardenburg *et al.*, 1986; Thompson, 1998).

LP delayed ripening and softening, slowed colour development and decreased the rates of sugar and acid loss in proportion to the pressure reduction when Large Early Montgament apricots, harvested at a hard-mature stage, were stored at 0°C in NA (646 mm Hg) or in LP at pressures of 62.8, 37.1 or 13.6 kPa (471, 278 or 102 mm Hg). Storage life was limited to 53 days in NA, whereas fruit was still in a marketable condition after 90 days in LP at 13.6 kPa (102 mm Hg). By the end of the LP storage, the fruits had reached a ripe stage, with normal colour, sugar and acid (Wu and Salunkhe, 1972b; Salunkhe and Wu, 1973).

The greatest hazard in handling and shipping apricots is decay caused by brown rot (*Monilinia* sp.) and rhizopus rot (*Rhizopus* sp.) (Hardenburg *et al.*, 1986). Development of *M. fructicola* on potato dextrose agar is half-suppressed in 2.3% $[O_2]$ (El-Goorani and Sommer, 1979), and very low $[O_2]$ prevents essentially all growth and sporulation of *R. stolonifer* (Figs 7.1 and 7.3). Inhibitions of *M. fructicola* and *R. stolonifer* growth in 2.3% $[O_2]$ are reversed by 4–5% $[CO_2]$, (Fig. 7.5; El-Goorani and Sommer, 1979). Very low $[O_2]$ has not been tested on *Monilinia*, and apricots have not been stored at pressures below 13.6 kPa (102 mm Hg) and temperatures higher than 0°C. If this fruit tolerates storage at 2.0–2.67 kPa (15–20 mm Hg), that would prevent decay caused by *Rhizopus* and *Monilinia*.

10.3 Avocado (*Persea americana* Mill.), Choquette, Waldin, Booth 8, Lula, Hass

The storage life of avocados in NA varies depending on each variety's cold tolerance (Hardenburg *et al.*, 1986). CA storage in 3–9% [CO_2] + 1–3% [O_2] decreases chilling damage (Tables 7.11 and 7.12) and extends storage life by allowing avocados to be kept at a lower temperature. Lula avocados have been preserved for 60 days in CA (Hardenburg *et al.*, 1986; Thompson, 1998).

Choquette avocados stored in NA at 14.4°C began to ripen after 8–9 days, were half-ripe and turning by 10 days and fully ripe by 14 days. Fruits kept at the same temperature in LP at pressures of 5.3–13.3 kPa (40–100 mm Hg) were still firm with no external blemishes after 17 days, but softening commenced in 25 days, and by 32 days ripening was well advanced. Within 3 days after they were transferred from LP to 20°C atmospheric air, all fruits developed normal taste without internal blackening or decay (Burg, 1969). Storage life at 12.8°C improved when the pressure was lowered to between 13.3 and 20 kPa (100 and 150 mm Hg), but the result was much better at 5.3–10.7 kPa (40–80 mm Hg). In the lower pressure range, fruit remained firm for 25–36 days, compared to 8–9 days in NA (Burg, 1976a). Ripening proceeded normally at room temperature after fruits were removed from LP, and they had an acceptable shelf life. Ripening and decay development were accelerated during storage at 15.6°C, but LP still had a beneficial effect. Chilling damage occurred at 10°C. In subsequent experiments, the optimal storage pressure for Choquette avocados was found to be 2.7 kPa (20 mm Hg) (S.P. Burg, 1980, unpublished).

The storage life of Waldin avocados at 10°C improved as the pressure was lowered from 13.3–20 kPa (100–150 mm Hg) to 8.0–10.7 kPa (60–80 mm Hg). Fruit remained firm for 30 days at the lower pressures vs. 12–16 days in NA (Burg, 1976a). Similar results were obtained at 12°C, except that all fruits ripened more rapidly at the higher temperature. Chilling damage occurred at 7.6°C in both NA and LP.

Early-season Lula avocados began to soften within 3 weeks when they were stored in NA at 7.2°C; 40% softened in 23 days, 50% in 29 days and 100% in 41 days. Fruits that were still firm after 23 days in NA storage ripened in 3 days at 26.7°C. Regardless of whether the avocados softened during NA storage or after transfer to a higher temperature, during ripening they developed skin darkening (anthracnose) that often penetrated into the pulp. After 88 days in LP, fruits stored at 10.7 or 16 kPa (80 or 120 mm Hg) were beginning to soften. At that time, avocados kept at 8 kPa (60 mm Hg) were still firm and, after transfer to atmospheric air at 26.7°C, they ripened in 3–4 days without surface blackening. In 7.2°C air they began to soften in 8 days and became eating ripe in 14 days (Burg, 1969, 1970). In another study, Lula avocados stored at 8°C ripened in 23–30 days in NA, but remained firm in LP for 75–100 days at pressures ranging from 5.3 to 10.7 kPa (40–80 mm Hg) (Burg, 1969, 1970, 1976a). After 102 days' storage at 8 kPa (60 mm Hg), fruits became eating ripe in 3–4 days at 26.7°C. If instead they were transferred to 8°C air, they did not begin to soften for 8 days, and became eating ripe in 14 days. Pressures ranging from 13.3 to 20 kPa (100–150 mm Hg) were far less effective. Chilling damage resulted at 6°C.

Early-season Booth 8 avocados were tested either in NA or in LP at pressures of 5.3, 8.0, 10.7 or 16 kPa (40, 60, 80 or 120 mm Hg). After transfer to 20°C, fruits which previously had been kept in NA for 30 days at 4.4°C ripened in 2–3 days without decay development, but with poor flavour. Avocados which remained in NA storage for 45 days at 4.4°C were slightly desiccated. Within 60 days, 27% were soft and 67% had extensive peel and flesh blackening indicative of anthracnose. By 75 days, 100% had softened. All fruits stored in LP remained firm for 64 days; in 75 days the percentage softening was 100% at a pressure of 16 kPa (120 mm Hg), 44% at 10.7 kPa (80 mm Hg), 43% at 8 kPa (60 mm Hg) and 0% at 5.3 kPa (40 mm Hg). At that time, fruit transferred from LP to air at 26.7°C ripened within 2–3 days with considerable blackening of the skin and pulp. If instead the fruit was transferred to air at 14.4°C, it ripened in 2–3 days with less blackening of the skin and no darkening of the pulp (Burg, 1969). In another study, depending upon their state of maturity, waxed Booth 8 avocados ripened in 8–22 days during NA storage at 7.8–10°C, while at the same temperature and a pressure of 5.3–10.7 kPa (40–80 mm Hg), they did not soften during 45–50 days, after which they ripened normally without skin darkening when transferred to 23.9°C air, (Burg, 1976a; Dressler, 1978a). Their shelf life was 4–5 days compared 7–10 days at harvest. After 64 days in LP storage, although the fruit was still firm, it no longer ripened with acceptable quality. The optimal storage condition for Booth 8 avocados is 2.7–5.3 kPa (20–40 mm Hg) at 10°C. After 49 days, fruits stored in this pressure range ripened normally during 4 days at 21°C (Dressler, 1978a).

Hass avocados were tested at pressures ranging from 2 to 8 kPa (15–60 mm Hg). Within 30 days at 5°C, the fruit ripened and became soft in NA. At that time, all fruit kept at the same temperature and 2.0–5.3 kPa (15–40 mm Hg) was still hard green, whereas nearly half the fruit kept at 8 kPa (60 mm Hg) had begun to ripen (Cicale and Jamieson, 1978). After 38–45 days of storage, fruit kept at a pressure of 5.3 kPa (40 mm Hg) began to soften, but at 2–2.7 kPa (15–20 mm Hg) the avocados remained firm. The at-harvest 5.4-day ripening time in air at room temperature was retained during 45 days' storage at a pressure of 2.0–2.7 kPa (15–20 mm Hg). At higher pressures, the time to ripen decreased after 31–38 days of LP storage. The maximum percentage of marketable fruit after 31–38 days' storage was obtained at pressures in the range 2.7–5.3 kPa (20–40 mm Hg). In another storage trial, mature Hass avocados were selected for size and uniformity, and lots of 100 fruits were stored at 6°C either in NA or in LP at pressures of 26.7, 13.3, 10.7 or 8.0 kPa (200, 100, 80 or 60 mm Hg). After various storage times, fruits were checked for weight loss, general appearance, skin damage, firmness, ethylene evolution, respiration rate, pulp colour and taste. During 35 days' storage, fruits lost 5.7% of their weight in NA and only 1.2% in LP. After 70 days, fruits stored at 13.3, 10.7 or 8 kPa (100, 80 or 60 mm Hg) lost 1.7, 1.7 and 3.0% of their weight, respectively. Hass avocados stored in NA and in LP at 26.7 kPa (200 mm Hg) ripened after 35 and 50 days, respectively, while those kept at pressures below 13.3 kPa (100 mm Hg) remained unripe for 70 days. The best result was obtained at 8 kPa (60 mm Hg) because at pressures lower than 6.7 kPa (50 mm Hg), substantial desiccation occurred (9.2; Fig. 9.1, right). Softening was markedly retarded by LP storage, and fruits ripened normally several days after being transferred to 14°C air (Apelbaum *et al.*, 1977b). In another experiment with Hass avocados, when fruit was stored at 5°C in NA and at pressures of 2.0, 2.7, 5.3 and 8.0 kPa (15, 20, 40 and 60 mm Hg), in LP, the optimal pressure was in the range 2.0–5.3 kPa (15–40 mm Hg). At 21°C, the ripening time of fruit removed from LP storage varied inversely with the previous storage-pressure regime. Freshly harvested fruit ripened in 5.4 days. After 45 days of LP storage, fruits that had been kept at 2.0 kPa (15 mm Hg) ripened in 6.3 days; those that had been stored at 2.7 kPa (20 mm Hg) in 5.1 days, the ripening time was 4 days after storage at 5.3 kPa (40 mm Hg), 3.1 days for fruits that had been held at a pressure of 8 kPa (60 mm Hg) and 2.1 days after NA storage (Cicale and Jamieson, 1978). Hass avocados were successfully shipped from Mexico to Stockholm in a 12.2 m (40 ft) Grumman/Dormavac LP-intermodal hypobaric container operated at 5°C, 95% RH, and a pressure of 2.7 kPa (20 mm Hg) (Jamieson, 1984).

In tests carried out in Japan, avocados were successfully stored for 40 days with delayed ripening. Deterioration was prevented, especially if the fruits had been hot-water-dipped. The results were less favourable at 10°C, 70–80% RH and a pressure of 20.26 kPa (152 mm Hg) (Gemma *et al.*, 1989).

Lula, Booth 8 and Waldin avocados stored at 4.4, 7.2 or 10°C for 25–42 days either at 10.13, 12.13, 20.26 or 101.33 kPa (76, 91, 152 or 760 mm Hg) may develop an unacceptable incidence of anthracnose when they are softened at 20°C, and in the lower storage-temperature range, they suffer chilling injury (Spalding and Reeder, 1976b; Spalding, 1977; Gemma *et al.*, 1989; Tables 7.11 and 7.12). A 2% $[O_2]$ + 10% $[CO_2]$ mixture delays avocado softening and keeps *C. gloeosporiodes* dormant, preventing the development of anthracnose (Hatton and Reeder, 1972). Continuously flushing Lula avocados with a mixture containing 5 parts CO_2 and 1 part O_2 to provide 10% $[CO_2]$ at a pressure of 12.1 kPa (91 mm Hg), prevented chilling damage and skin darkening during 6 weeks' storage at 10°C, followed by ripening at 20°C (Spalding and Reeder, 1976b; Spalding, 1977; Table 7.11). The addition of 10% $[CO_2]$ also prevented chilling and anthracnose when Waldin avocados were stored for 25 days at 7.2°C and pressures of 12.13 and 101.3 kPa (91 and 760 mm Hg), followed by ripening at 20°C (Fig. 7.12). In the presence of 2% $[O_2]$ + 10% $[CO_2]$, fruits stored equally well in LP and CA (Spalding and Reeder, 1976b; Spalding, 1977), but LP storage improved dramatically when 'Waldin' and 'Lula' avocados were kept at a non-chilling temperature, 10°C, and a much lower pressure, 2.7 kPa (20 mm Hg), without added CO_2. The fruits remained firm during five weeks of storage, and 90–100% were acceptable after they were allowed to ripen at 20°C (Spalding and Reeder, 1980, unpublished).

The major rots of avocados are caused by *Dothiorella gregaria*, *Colletotrichum gloeosporioides*, *Fusarium* sp. and *Alternaria alternata* (Eckert *et al.*, 1975). A storage pressure of 20 kPa (150 mm Hg) hardly affects the *in vitro* growth of *C. gloeosporiodes*, but it suppresses development of this pathogen on inoculated avocado discs by delaying fruit softening and maintaining the concentration of a pre-formed natural antifungal agent (7.7; Prusky *et al.*, 1983). The optimal LP storage pressure for avocados, 2.0–2.7 kPa (15–20 mm Hg), inhibits the growth of *C. gloeosporioides*, *Fusarium* sp. and *A. alternata* cultures (Figs 7.3, 7.5, 7.6, 7.7; Table 7.4). The growth of *D. gregaria* on potato dextrose agar at 5.5 and 12.5°C is not inhibited by 2.3% $[O_2] \pm 5\%$ $[CO_2]$ at atmospheric pressure (El-Goorani and Sommer, 1979), but lower $[O_2]$ has not been tested. HOCl vapour inhibits the growth of *C. gloeosporiodes* and *F. roseum* (example 13, chapter 7), and prevents decay from developing during LP avocado storage (S.P. Burg, 1976, unpublished; example 14, chapter 7).

10.4 Banana (*Musa paradisica* var. Sapientum), Lacatan, Valery, Gros Michel and Dwarf Cavendish

Green bananas can be stored in NA for 2–3 weeks at 13°C (Mercantila Publishers, 1989b). A gas mixture containing 1.5–5% $[O_2]$ + 2–10% $[CO_2]$ increases the storage time to 6–8 weeks (Hardenburg *et al.*, 1986; Thompson, 1998). Usually the CA atmosphere is created in refrigerated break-bulk ships that have specialized equipment installed in their holds, but sometimes CA intermodal containers or MA packaging is used. Less than 1% $[O_2]$ causes low-O_2 damage (Table 4.7).

Gros Michel and Valery bananas, which ripened in 10 days during NA storage at 13.3–14.4°C, remained green for 40 days at 20 kPa (150 mm Hg), but often by 50 days the fingers had developed mould. The LP result was the same at air-change rates varying from 0.8 to 6.8 per hour (Burg and Burg, 1966c, 1969; Burg, 1969; Burg and Kosson, 1983). Valery banana storage life increased progressively when the pressure was lowered from 20 to 5.3 kPa (150 to 40 mm Hg). At 13.3°C, caliper 19 fruits ripened within 30 days when they were stored in 'Banavac' modified-atmosphere polyethylene bags or in CA, but they remained green for more than 105 days at pressures of 6.4, 9.5 or 12.7 kPa (48, 71 or 95 mm Hg). Fruits ripened normally when they were transferred from LP to air and gassed with ethylene (Burg, 1969). Favourable LP results with bananas have been reported in other studies (Tolle, 1972; Poulssen *et al.*, 1982).

A long exposure to a pressure of 6.7 kPa (50 mm Hg) does not lessen a banana's ability to respond to ethylene (Burg, 1969; Bangerth, 1984), and has no deleterious influence on the flavour or aroma that develops during ripening. A loss of 'squeak' when fruits are rubbed together was evident after a successful 45-day shipment of Valery bananas in a 6.1 m (20 ft) intermodal Fruehauf hypobaric container operated at 15°C and a pressure of 20 kPa (150 mm Hg). 'Squeak' is used by jobbers as an indication of 'old' bananas. The result was confirmed in 30-day laboratory experiments at pressures of 5.3, 8.0, 10, 13.3, 16 and 22 kPa (40, 60, 75, 100, 120 and 165 mm Hg). None of these fruits had begun to ripen and no particular quality differences were noted at the various pressures, except that the weight loss, which varied from 1.07 to 3.6%, was greater at elevated pressures because the respiration rate was higher and provided more heat to evaporate water (6.1; Burg, 1969). During a 28-day LP storage, the CO_2 production by bananas progressively declined to less than 20% of the initial value in air (Table 4.1).

Storage life and the incidence of decay in Dwarf Cavendish bananas were inversely related to pressure between 10.7 and 33.3 kPa (80 and 250 mm Hg; Apelbaum *et al.*, 1977a). No injuries attributable to subatmospheric conditions were observed, except that mechanical damage during picking and handling became progressively more pronounced, and tissue around wounds more depressed during storage. All fruit stored in LP had good texture, aroma and taste after ripening. Table 5.9 documents the changes in glucose content and fruit firmness that occurred during storage. LP also maintained the tannin content. When bananas were transferred to air after 120 days of storage at 10.7 or 20 kPa (80 or 150 mm Hg), they developed the same glucose content, decrease in firmness and loss of tannins as freshly harvested control fruits ripened in air. It was concluded that 'Of all the storage methods, the subatmospheric pressure system is the one that provides the longest storage life for high quality banana fruits'. The longer the LP storage period, the faster the fruit ripened when it was transferred to atmospheric pressure (Fig. 5.27, *left*). In another study (Bangerth, 1984), Cavendish bananas were successfully stored for at least 12 weeks at 14°C and a pressure of 6.67 kPa (50 mm Hg), and the time required for the respiratory climacteric to be initiated after the fruit was removed from LP storage progressively decreased as the storage period was extended (Fig. 5.27, *right*). Regardless of the duration of LP storage, when the fruit was removed and gassed with 50 μl/l ethylene, a respiratory climacteric resulted that was identical in rate and timing to that induced in control, non-stored, ethylene-treated fruits. Volatile production by gassed fruits is essentially the same in non-stored control bananas and in LP fruits when they ripened after removal from storage (Shatat, 1977). Fruit continuously exposed to 7.5 μl/l of ethylene during storage became overripe in 4 weeks regardless of whether it was kept in NA or LP (Bangerth, 1984). These experiments demonstrate that, unlike apple fruits (3.26), bananas retain their ethylene sensitivity during low $[O_2]$ storage.

LP is able to delay banana ripening after it has been initiated by applied ethylene. Pre-climacteric Dwarf Cavendish bananas (*M. cavendishii* Lambert), were pretreated with 10 μl/l of ethylene for various intervals at 21°C, and then stored for 28 days at 14°C, either in LP at pressures of 6.7 and 10.4 kPa (50 and 78 mm Hg = 0.86–1.63% $[O_2]$), or in CA with 1% $[O_2]$. If the duration of the ethylene pretreatment was less than the minimum time (TLT = threshold length of time) required to induce a ripening response in 21°C air, the bananas remained green and firm when they were subsequently stored in either LP or CA, and they usually did not start to ripen immediately after they were removed from storage unless they were treated with ethylene (Liu, 1976). Fruits that had been ethylene-pretreated for a few hours longer than the TLT did not remain green and firm in CA, and developed impaired eating quality at the ripe stage due to insipid taste, mealy texture and off-flavour, even though it has been reported that bananas ripen normally at a diminished

O_2 tension (Quazi and Freebairn, 1970). Bananas that had been ethylene-pretreated for 2–5 h longer than the TLT did not ripen during LP storage, and began to ripen without ethylene treatment 1–2 days after removal from LP. Fruit only ripened in LP if it was ethylene-pretreated for 16 h longer than the TLT before it was placed into storage. All fruit that remained green during LP or CA storage had normal eating quality when it subsequently ripened. The improved storage in LP compared to CA was not caused by differences in O_2 partial pressure, as the $[O_2]$ in the LP chamber was 0.86–1.63% and in the CA chamber 1.0%.

Banana fruits kept in a hypobaric atmosphere for 3 h daily, and then returned to atmospheric pressure each day, did not develop a climacteric and remained green during a 15-day experiment (Awad *et al.*, 1975).

The major postharvest storage decays of bananas are caused by *Thielaviopsis paradoxa, Botryodiplodia theobromae, Gloeosporium musarum, Deightoniella torulosa, Phomopsis* sp., *Verticillium theobromae* and *Fusarium* sp. (Wardlaw, 1961; Eckert *et al.*, 1975). The low $[O_2]$ present in 16°C water-saturated air at 8 kPa (60 mm Hg) would only slightly decreases the growth and sporulation of these organisms, but development of *B. theobromae* stem-end rot (Fig. 7.7; El-Goorani and Sommer, 1979), *F. rosea* (Fig. 7.3), *V. theobromae* and *Phomopsis* sp. (note 1, chapter 7; 7.1) should be largely prevented at a pressure of 2.67 kPa (20 mm Hg). At atmospheric pressure and 12.5°C, the growth of *T. paradoxa* on potato dextrose agar is not significantly depressed by 2.3% $[O_2]$ ± 5% $[CO_2]$, but lower $[O_2]$ was not tested (note 1, chapter 7). Apparently, *G. musarum* can grow but not sporulate in N_2 (Goos and Tschirsch, 1962; El-Goorani and Sommer, 1979). Bananas should be evaluated at 2.67 kPa (20 mm Hg) to determine if this pressure provides improved decay control without injuring the fruit. Banana decay is prevented by exposing the fruits to HOCl vapour during LP storage (chapter 7, examples 12 and 17).

It is likely that banana stomates open in LP (4.15), and has been suggested that enhanced diffusion and continuous ventilation in LP removes noxious volatiles that may limit storage life by causing peel pitting (Pantastico *et al.*, 1967) and abnormal ripening (Fuchs and Temkin-Gorodeiski, 1971).

10.5 Blueberry (*Vaccinium corymbosum* L.), var. Jersey

First-harvest blueberries can be stored in NA for 12–15 days at −0.5 to 0°C (Welby and McGregor, 1997). CA retards blueberry breakdown and decay, but causes off-flavours to develop (Hardenburg *et al.*, 1986).

Blueberries spoiled in NA within 4 weeks at 0–2°C, and at pressures ranging from 10.7 to 26.7 kPa (80–200 mm Hg) mould development limited their storage life to 6 weeks in LP (Burg, 1976a). After 44 days of storage at 101.3, 50.7, 21.3, 10.7, 5.3 and 2.7 kPa (760, 380, 160, 80, 40, 20 mm Hg), the maximum percentage of sound var. Jersey blueberries resulted at 2.7–5.3 kPa (20–80 mm Hg), and decay progressively decreased from 87% at 100 kPa (760 mm Hg) to 10.8% at 2.7 kPa (20 mm Hg) (D.R. Dilley, 1989, personal communication).

The major diseases of blueberries are alternaria rot (*Alternaria* sp.), grey mould rot (*Botrytis cinerea*), anthracnose (*Gloeosporium* sp.), mummy berry (*Monilinia vaccinii-corymbosi*), phoma rot (*Phoma vaccinii*) and rhizopus rot (*Rhizopus nigricans*) (Harvey and Pentzer, 1966). Provided that no CO_2 is present, the 0.4% $[O_2]$ contained in 0–2°C water-saturated air at a pressure of 2.67 kPa (20 mm Hg) directly inhibits the growth and sporulation of *A. alternata, A. tenuis, B. cinerea* and *R. nigricans* cultures (7.1; Tables 7.1 and 7.2; Figs 7.1, 7.3, 7.5 and 7.6). A 0.4% $[O_2]$ concentration retards spore germination, but not the growth of *Gloeosporium* sp. (7.1), and less than 2.3% $[O_2]$ slows the growth of *M. fructicola* (note 1, chapter 7) and *Phoma* sp. (7.01). These effects may explain why blueberry decay was progressively reduced when the pressure was lowered to 2.7 kPa (20 mm Hg = 0.4% [O_2]). The result should be even better at 1.3 kPa (10 mm Hg = 0.14% [O_2]).

10.6 Cantaloupe (*Cucumis melo* L.), cv. Earl's Favourite

Cantaloupes can be stored in NA at 2–5°C for 1–2 weeks (Mercantila Publishers, 1989b; Welby and McGregor, 1997). An atmosphere containing 2% [O_2] + 10–20% [CO_2] slightly improves storage at 5°C, and less than 1% [O_2] causes low-O_2 damage (Table 4.7). Melons harvested at the hard-ripe stage (3/4 to full-slip), chill at temperatures lower than 2.2–4.4°C, whereas full-slip cantaloupes are not injured at 0–1.7°C during 5–15 days' storage in NA at 85–90% RH. Cantaloupes should be harvested at a fully mature, pre-climacteric stage for LP storage, when they are likely to possess their maximum insensitivity to chilling damage.

California cantaloupes developed decay, and no advantage was noted from LP vs. NA storage during 18 days at 4.4°C and pressures of 8.0, 10.7, 13.3 and 101.3 kPa (60, 80, 100 and 760 mm Hg) (Burg, 1970). The flavour in cv. Earl's favourite muskmelons stored in 2.5% [O_2] + 1% [CO_2] was better than in fruit stored in LP at 10.67–13.3 kPa (80–100 mm Hg = 2.1–2.6% [O_2]) (Zhao and Murata, 1988). However, storage in LP was significantly improved during a 20-day trial by raising the temperature to 7.2°C, and lowering the pressure to 2 kPa (15 mm Hg; Jamieson, 1984). Melons were successfully shipped from the US to Taiwan in a 12.2 m Grumman/Dormavac intermodal hypobaric container operated at this condition (Jamieson, 1984).

The major diseases of cantaloupes are rots caused by *A. tenuis*, *Penicillium* sp., *Cladosporium cucumerinum*, *Fusarium* sp. and *Rhizopus* sp. (Ryall and Lipton, 1972). Provided that no CO_2 is present, the growth and sporulation of all of these organisms should be suppressed by the 0.22% [O_2] concentration contained in 7.2°C water-saturated air at a pressure of 2 kPa (15 mm Hg) (Figs 7.1, 7.3 and 7.6; Table 7.4).

10.7 Carambola (*Averrhoa carambola* Linn.)

Carambolas can be stored for 3–4 weeks in NA at 9–10°C (Hardenburg *et al.*, 1986). CA storage at 7°C and 85–95% RH with either 2.2% [O_2] + 8.2% [CO_2] or 4.2% [O_2] + 8% [CO_2] resulted in low losses of about 1.2% in 1 month, and fruit retained a bright red colour and good retention of firmness, Brix^2 and acidity compared to fruits stored in air (Rennel and Thompson, 1994; Thompson, 1998).

The storage of Florida carambolas was compared in NA and LP, flowing one air change per hour of 98% RH air at 13°C. The fruit initially was heavily contaminated with mould and had not been chlorine-washed. Within a few weeks, carambolas stored in NA were totally engulfed by a 'white' filamentous mould, and after 1 month they were completely dissolved and unrecognizable as fruits. At that time, carambolas stored at a pressure of 2.0 kPa (15 mm Hg = 0.1% [O_2]) were still in their initial condition. The fungal growth had been remarkably well controlled at a low pressure, except that a few small spots of fungus had just begun to develop (T.L. Davenport and S.P. Burg, 2002, unpublished).

10.8 Cherimoya (*Annona cherimola* Miller)

Cherimoyas can be stored at 8–9°C for 1–2 weeks in NA (Welby and McGregor, 1997), and in CA for 22 days in a 2% [O_2] + 10% [CO_2] mixture (Hatton and Spalding, 1990). Changes in texture, protein content and the activities of polyphenol oxidase and peroxidase have been noted when cherimoyas ripen in LP or after treatment with sulphite (Plata *et al.*, 1987), but apparently no detailed storage studies have been performed in LP to determine the optimal condition or maximum storage time.

10.9 Cherry, Sweet (*Prunus avium* L.) cvs Bing, Lambert, Grosse Schwarze Knorpelkirsche, Emperor Francis, Picota

NA storage of sweet cherries is limited to 14–21 days at −1 to −0.5°C (Welby and McGregor, 1997). Controlled atmospheres containing 20–25% [CO_2] or 0.5–3% [O_2] help to maintain firmness, green stems and bright berry colour at 0–5°C (Borecka, 1986; Hardenburg *et al.*, 1986; Thompson, 1998). During 35 days at −1.1°C, the best CA gas mixture for storing Bing cherries was 0.03% [CO_2] + 0.5–2% [O_2] (Chen *et al.*, 1981).

Sweet cherries kept equally well in high-[CO_2] controlled-atmosphere storage and in LP at pressures ranging from 5.33 to 26.7 kPa (40–200 mm Hg). The major benefits of LP storage were delayed decay development and retention of stem condition, fruit colour and brightness (Patterson and Melsted, 1977). LP fruit could be kept in good condition for 6–10 weeks, depending on its initial quality and condition. Lower pressures and airflows gave the best result, but due to technical difficulties (9.2, 9.6, 9.7) desiccation occurred at 5.33 kPa (40 mm Hg).

'Bing' sweet cherries were well preserved in LP at 0–5°C during 30–60 days at pressures ranging from 10.7–26.7 kPa (80–200 mm Hg) (Tolle, 1972; Wu and Salunkhe, 1972b; Salunkhe and Wu, 1973; Burg, 1976a). Loss of sugar was prevented in proportion to the pressure reduction, red colour was retained and pedicels, which remained green at 13.6 kPa (102 mm Hg) in LP, became brown and mouldy in NA. During a 6–10-week storage at 0°C and pressures ranging from 5.33 to 6.67 kPa (40–50 mm Hg), the stems of Bing cherries remained green and fresh-looking, and ascorbic acid was retained better than in NA (Dilley, 1977a; 4.06, 4.14). The eating quality of the fruits remained good during 8 weeks of LP storage, but subsequently the flavour became bland. Similar results were reported for Picota sweet cherries (Alique *et al.*, 1979).

'Emperor Francis' and 'Grosse Schwarze Knorpelkirsche' cherries stored at 2°C and a pressure of 5 kPa (38 mm Hg) had significantly better taste than cherries stored in CA, which were better than those stored in air (Borecka, 1986; Table 10.1). A pressure of 25 kPa (188 mm Hg) was less effective, and a temperature lower than 2°C seemed to damage these cultivars of cherries. The major decay-causing organisms in this test were identified as *B. cinerea* and *Monilinia* sp., with occasional traces of *Penicillium* sp. and *Rhizopus* sp. After a 42-day storage, the percentage of rots caused by *B. cinerea* was 41.6% in NA, 31.9% in LP at 25 kPa (188 mm Hg) and 10.6% in LP at 5 kPa (38 mm Hg). The percentage of rots caused by *Monilinia* sp. was 4.0% in NA, 0.6% in LP at 25 kPa (188 mm Hg) and 0.7% in LP at 5 kPa (38 mm Hg). These results are consistent with the inhibitory effect on the growth of these fungi caused by the 0.9% [O_2] concentration present at 2°C and a pressure of 5 kPa (38 mm Hg). This [O_2] level is 75–85% effective in preventing the development of *B. cinerea* cultures (Figs 7.6 and 7.7), 80% inhibitory with *P. digitatum* (Fig. 7.6), 74% with *R. nigricans* (Fig. 7.1) and in the absence of CO_2 *M. fructicola* growth is reduced by 55% at 2.3% [O_2] (El-Goorani and Sommer, 1979). An even greater effect on decay would be expected when the pressure is reduced below 5 kPa (38 mm Hg = 0.9% [O_2]). In agreement with this prediction, at 0°C sweet cherries were found to

Table 10.1. Taste of sweet cherries stored for various durations at 2°C in NA, CA (5% [CO_2] + 3% [O_2]) or in LP at either 5 kPa (38 mm Hg) or 25 kPa (188 mm Hg). The taste of the fruit was evaluated by six persons according to a scale from 1 to 5, where 5 was the best taste (Borecka, 1986).

	Taste rating			
Cultivar	NA	CA	5 kPa	25 kPa
Emperor Francis				
15 days (Expt 1)	5.0	–	5.0	5.0
30 days (Expt 1)	3.6	–	5.0	4.1
42 days (Expt 1)	2.0	–	3.5	2.0
30 days (Expt 2)	3.3	3.9	5.0	–
60 days (Expt 2)	3.6	3.6	4.0	–
Gross Schwarze Knorpelkirsche				
30 days	2.9	3.6	5.0	–

store best at 2.67 kPa (20 mm Hg = 0.4% [O_2]; Jamieson, 1984). Three full loads of sweet cherries were successfully shipped from Washington State to Stockholm in a Grumman/Dormavac 12.2 m LP container operated at 0°C, 95% RH and a pressure of 2.7 kPa (20 mm Hg). Still better results might be anticipated at 0°C and 1.33 kPa (10 mm Hg = 0.15% [O_2]), but this pressure has yet to be tested.

Rapid changes in pressure may contribute to cherry splitting and loss of firmness (Borecka, 1986). After LP storage of sweet cherries, the pressure should be released (vented) slowly during 30–120 min (Jamieson, 1984).

10.10 Cherry, Barbados (*Malpighia glabra* L.) and Surinam (*Eugenia uniflora* L.)

Randomized samples containing half-coloured, red and overripe Surinam cherry berries were stored at 13.3°C and 95% RH, either in NA or in LP at a pressure of 4 kPa (30 mm Hg), flowing one nearly saturated air change per hour. After 7 days in NA, 3.8% of the berries were still firm and fully red-ripe, no half-coloured berries remained, and 96.2% were overripe and rotten with a fermented smell. A large amount of juice had escaped from these berries into the storage jar. In LP, 14.7% of the berries were still half-coloured, 45.1% were firm and red, and 40.2% were overripe and soft. There was no odour of fermentation and no juice had escaped from any of the LP berries (T.L. Davenport and S.P. Burg, 2002, unpublished).

Barbados cherries stored in LP at 2°C and a pressure of 20 kPa (150 mm Hg) were well preserved during 13 days, whereas in NA they spoiled in 5 days (Burg and Burg, 1966c). Lower pressures were not tested.

10.11 Cranberry (*Vaccinium macrocarpon* Ait.)

Cranberries can be stored for 2–4 months in NA at 2–4°C and 90–95% RH (Hardenburg *et al.*, 1986). CA does not increase their storage life. Provided that sound fruit were selected for the storage trial, cranberries that had been kept in LP at 11 kPa (80 mm Hg) exhibited depressed rates of respiration and ethylene evolution and subsequently had a longer post-storage shelf life compared to control fruit that had been stored in air (Pelter, 1975).

10.12 Currant (*Ribes vulgare* Lam.), Red, White and Black

Currants can be stored for 1–4 weeks in NA at –0.5 to 0°C and 90–95% RH (Welby and McGregor, 1997). The incidence of fungal rots is reduced by storing this fruit in 12–25% [CO_2]. The optimal CA condition, 20% [CO_2] + 2% [O_2], sometimes provides satisfactory storage for as long as 20 weeks (Thompson, 1998).

Currants stored in LP at 2.2–2.8°C and a pressure of 10 kPa (75 mm Hg) were still marketable at 38 days, with better taste, less decay, longer shelf life, less abscission of single berries, increased ascorbic acid and more sugar compared to NA controls (Table 10.2; Bangerth, 1977). Extensive loss of vitamin C and sugar occurred during NA storage within 24 days, and the fruits were unmarketable at that time. LP storage did not affect the loss of titratable acidity.

Grey mould rot caused by *B. cinerea* is the major decay affecting currants during storage and shelf life (Harvey and Pentzer, 1966). To inhibit the growth of *B. cinerea* cultures, the [O_2] must be decreased below 1%, and [CO_2] should be excluded (Tables 7.1 and 7.2; Figs 7.1 and 7.6). At 2.2–2.8°C, the [O_2] content of water-saturated air is 1.9% at 10 kPa (75 mm Hg), and therefore this pressure would not be particularly effective in preventing decay of currants. Storage of currants should be improved by lowering the pressure to 1.3–2.0 kPa (10–15 mm Hg) and the temperature to –0.5 to 0°C.

Table 10.2. Changes in ascorbic acid and sugar content of white, red and black currants after 24 or 38 days' storage at 3°C in air (NA) or in LP at a pressure of 10 kPa (75 mm Hg). Losses primarily resulted from decay (Bangerth, 1973).

Type	Storage time (days)	Spoiled berries (%)		Ascorbic acid (mg/100 g)		Reducing sugar (mg/g)		Total sugar (mg/g)	
		NA	LP	NA	LP	NA	LP	NA	LP
White	0	–	–	48.4	–	–	–	–	–
	24	19	4	22.8	37.1	–	–	–	–
	38	37	3	16.6	34.4	51.6	68.0	56.0	76.1
Red	0	–	–	55.5	–	–	–	–	–
	24	8	0	44.0	53.0	–	–	–	–
	38	21	0.5	30.5	44.0	54.4	60.6	56.2	62.5
Black	0	–	–	282.0	–	–	–	–	–
	24	11	1	222.0	241.0	–	–	–	–
	38	33	5	204.0	249.0	63.7	71.8	72.5	87.0

10.13 Grape (*Vitis vinifera* L.), American, European and var. Red Emperor

In NA, the storage life of Vinifera grapes at –1 to –0.5°C varies depending on the cultivar. Thompson Seedless grapes will store for 1–2 months, Ribier for 2–4 months, and late-season cultivars such as Emperor, Ohanez and Calmeria for 3–6 months. A high RH is required because appearance is adversely affected when grapes transpire more than 1.2% of their weight. Grey mould (*B. cinerea*), cladosporium rot (*Cladosporium herbarium*) and alternaria rot (*Alternaria* sp. and *Stemphylium* sp.) are the most important diseases of Vinifera grapes, and anthracnose (*Elsinoe ampelina*), blue mould rot (*Penicillium* sp.) and powdery mildew (*Uncinula necator*) occasionally are troublesome (Harvey and Pentzer, 1966; Hardenburg *et al.*, 1986). To reduce spoilage caused by these decay organisms, grapes are fumigated with sulphur dioxide immediately after harvest and at intervals during storage, but this procedure is illegal in many countries (Mercantila Publishers, 1989b), and excessive fumigation can cause pitting and bleaching of the grapes. CA (2–5% [O_2] + 1–8% [CO_2]) is effective in maintaining the shelf life of grapes, but does not control decay (Hardenburg *et al.*, 1986). Infections caused by *B. cinerea* can be prevented by 20–25% [CO_2], and a mixture of 45% [CO_2] + 11.5% [O_2] controls several insect pests that infest grapes (Thompson, 1998)

Red Emperor grapes, which usually can be stored for 3–8 weeks in NA at –0.5 to 0°C (Welby and McGregor, 1997), remained in good condition except for occasional mould development during 90 days in LP at 1.7–4.4°C and pressures of 4.0, 8.7, 10.7 and 21.3 kPa (30, 65, 80 or 160 mm Hg). In NA no berries remained in a saleable condition by that time (Burg, 1971). HOCl vapours controlled decay development during LP storage of this grape variety (Burg, 1970; example 19, chapter 6).

C. herbarum, *B. cinerea*, *Penicillium* sp. and *Alternaria* sp. cannot grow in the 0.15% [O_2] present in 0°C water-saturated air at a pressure of 1.3 kPa (10 mm Hg; Figs 7.1, 7.3, 7.5 and 7.6; Table 7.4; Harvey and Pentzer, 1966) provided that [CO_2] is excluded. 'American' and 'European' grapes were successfully stored for more than 20 days in laboratory studies (Jamieson, 1984) carried out at –1 to 0°C, 95% RH, and a pressure of 1.3 kPa (10 mm Hg).

10.14 Grapefruit (*Citrus* × *paradisi* Macfad.), Florida Ruby Red, Florida Marsh White, Texas Ruby Red

The optimal temperature for grapefruit storage varies depending upon the source of the fruit and time of year. In Florida, where grapefruit harvest begins in September and extends through to June, fruit picked before January must be kept at or above 15.6°C to avoid cold-temperature-induced rind pitting, but fruit can be stored at 10°C after January. California and Arizona grapefruit should be kept at 14.4–15.6°C, and Texas grapefruit at 10°C. Cold-temperature-induced pitting is worse at 4.4°C than at higher or lower temperatures (Hardenburg *et al.*, 1986). CA provides little if any benefit for grapefruit storage (Thompson, 1998).

All LP tests with grapefruit have been carried out using waxed fruit. At atmospheric pressure, waxing improves storage life and the resistance to chilling damage, but it also changes the gaseous permeability of the fruit's surface by plugging stomates. LP-stored grapefruit require a slow pressure release to avoid physical damage due to 'crushing'.

Originally, it was believed that the optimal LP pressure for grapefruit storage might be between 29.3 and 46.7 kPa (220 and 350 mm Hg; Burg, 1976a), and very low pressures were not tested. Early-season Florida fruit picked before January are not prone to decay and can often be successfully stored in LP for 3–4 months at 7.2–10°C and pressures ranging from 29.3 and 46.7 kPa (220–350 mm Hg). The incidence of chilling damage is reduced from 23.5% at atmospheric pressure to 4.4% at 29.3 kPa (220 mm Hg) during 7 weeks' storage of Marsh grapefruits at 4.4°C (Pantastico, 1968; Pantastico *et al.*, 1968), but not at 50.7 kPa (380 mm Hg) (Grierson, 1971). Grapefruit developed peel pitting and lost flavour in NA within 4–6 weeks at 6°C, and had improved flavour retention with no peel pitting when stored at pressures ranging from 33.3 to 53.3 kPa (250–400 mm Hg). LP storage was limited by mould development (Burg, 1976a).

A full-load static test with Florida and Texas grapefruit was carried out in a 12.2 m Fruehauf intermodal LP container. Biphenyl pads were placed in each carton of Texas grapefruit for decay control, and the fruit was pretreated with *o*-phenylphenate before it was stored. Florida fruit was treated with *o*-phenylphenate only. All fruit was waxed and various sizes packed in domestic cardboard containers, 18 kg per carton. Field heat was removed in the LP container and subsequently the fruit was stored at 6.7°C, 98% RH, and a pressure of 46.7 kPa (350 mm Hg). Decay became evident in the load after 30 days, and reached an average value of 30% by 3 months. Middle-sized fruit had the lowest percentage decay, followed by small fruits, while the incidence of decay amongst large fruits was almost total. The remaining sound fruit had good firmness and flavour. A second simulated storage was carried out with Texas fruit under the same conditions. Again decay became evident within 30 days and by 60 days had advanced to the point that the test had to be terminated.

Grapefruit are subject to the same decay organisms as oranges (10.20). The pressure range between 2.0 and 2.67 kPa (15–20 mm Hg) should be tested with grapefruit, since it directly controls the growth of this fruit's major decay organisms. Successful grapefruit storage at 10.7 kPa (80 mm Hg) has been reported by the Israeli Institute for Technology and Storage of Agricultural Products. If grapefruit are able to tolerate very low pressures without experiencing low [O_2] injury, this should effectively control decay during storage.

10.15 Guava (*Psidium guajava* L.)

Ripe guavas can be stored in NA for 2–3 weeks at 5–10°C and 90% RH. Below 5°C the fruit chills (Hardenburg *et al.*, 1986; Welby and McGregor, 1997). The storage life of guavas is only slightly extended by CA (Thompson, 1998).

Fully mature guavas, which ripened in 6 days during NA storage at 13.3°C, did not colour during 30 days in LP at the same temperature and a pressure of 20 kPa (150 mm Hg). After LP storage, the fruit ripened normally when it was transferred to air at atmospheric pressure (Burg and Burg, 1966c). An extension in guava storage life at a subatmospheric pressure also has been reported in other studies (Salunkhe and Wu, 1974). When firm ripe guavas were stored at 13°C and 4 kPa (30 mm Hg) to test their tolerance to a pressure low enough to provide insect control, after 7 days the fruit remained unchanged, it had a fresh fragrance and there was no odour of fermentation or any indication of low- [O_2] damage (T.L. Davenport and S.P. Burg, 2002, unpublished). Longer storage times at a very low pressure have not yet been tested.

Anthracnose caused by *Colletotrichum* sp. is the most common decay of guavas. The fruit also is susceptible to soft watery rot caused by *B. theobromae* and *R. nigricans* (Eckert, 1975). Decay caused by these organisms should be suppressed (Figs 7.1 and 7.3; El-Goorani and Sommer, 1979) if guavas can tolerate the 0.1–0.3% [O_2] concentration present in 5°C water-saturated air at 1.3–2.0 kPa (10–15 mm Hg).

10.16 Honeydew Melon (*Cucumis melo* L.)

Honeydew melons can usually be kept for 2–3 weeks in NA at 7–14°C, 85–90% RH (Mercantila Publishers, 1989b; Welby and McGregor, 1997). Their storage life is slightly extended by 2–5% [O_2], and they are damaged by less than 1% [O_2] (Dilley, 1978; Thompson, 1998). After 20 days' storage at 7.2°C and a pressure of 2 kPa (15 mm Hg), melons were free of decay and had up to 4 days' remaining shelf life depending upon their initial stage of ripeness (Jamieson, 1984). A report from Israel claims that honeydew melons are damaged by a rapid pressure release, but not if the pressure is released over a 2-h period.

The major fungal storage rots of honeydew melons are *A. tenuis*, *Penicillium* sp., *Cladosporium* sp., *Fusarium* sp. and *R. stolonifer*. This fruit also is susceptible to bacterial soft rot caused by *Erwinia* sp. and bacterial spot by *Pseudomonas lachrymans* (Ryall and Lipton, 1972). Provided that CO_2 is absent, the growth of these organisms in culture is directly suppressed in LP when 0.2% [O_2] is present in 7.2°C water-saturated air at a pressure of 2 kPa (15 mm Hg) (Figs 7.1, 7.3, 7.5 and 7.6). There are no data for *P. lachrymans*, but *P. fluorescens* also cannot grow at so low an [O_2] tension (Fig. 7.2).

10.17 Lime Persian [*Citrus aurantifolia* (Christm.) Swingle], var. Tahiti

The storage life of 'Tahiti' limes is limited to 14–35 days in NA, at which time de-greening progresses far enough to make them unmarketable (Lutz and Hardenburg, 1968). CA storage slows lime de-greening (Thompson, 1998), but has not been successful because it increases decay, and causes rind breakdown, peel thickening and juice loss (Spalding and Reeder, 1976a; Spalding, 1977). The sour-rot organism, *Geotrichum candidum*, which grows faster under low [O_2] than in air (Wells and Spalding, 1975), sometimes causes decay in limes stored at 10.13 kPa (76 mm Hg = 1.8% O_2) or in CA at a low-O_2 tension (Spalding and Reeder, 1974).

Limes stored in LP at 7–10°C and a pressure of 10.7–20 kPa (80–150 mm Hg), remained green with good flavour and appearance for 60–90 days, whereas fruits yellowed within 14–35 days in NA (Burg, 1976a). This result was confirmed by an experiment in which limes were well preserved for more than 56 days at 10°C and a pressure of 20 kPa (150 mm Hg) and were unmarketable after 10 days in NA (Burg and Burg, 1966c; Burg, 1987b). Ethylene induces peel 'puffiness' (peel thickening) in Satsuma mandarin (*Citrus unshiu*) (Maotani *et al.*, 1983), and since limes produce ethylene and respond to their IEC (Tables 5.10 and 5.11; Fig. 5.28), the removal of endogenous ethylene by enhanced diffusion may explain why peel thickening does not occur when limes are stored in LP at a pressure of 20 kPa (150 mm Hg). Limes have stomates that presumably open at a low pressure, but remain closed in CA, and this should help to decrease their IEC during LP storage. In another test at 10°C, depending upon whether or not the limes had been dipped in Benlate fungicide, 87–100% of fruit kept at 10.7 kPa (80 mm Hg) was marketable after 71 days (Jamieson, 1980e). The result was less favourable at 2.7 and 5.3 kPa (20 and 40 mm Hg). At pressures lower than 20.3 kPa (152 mm Hg), decay and juice loss increases and the peel thickens (Spalding and Reeder, 1976a; Spalding, 1977).

In a test carried out at the USDA, limes stored at 10 or 15.6°C and a pressure of 22.7 kPa (170 mm Hg) retained their green colour, juice content and flavour for up to 6 weeks, and were acceptable for marketing at that time. Only a low incidence of decay developed, due mainly to green mould rot (*P. digitatum*) and stem-end rot (*Phomopsis citri*). Pressure ranging between 20.3 and 25.3 kPa (152–190 mm Hg = 3.9–4.2% [O_2]) seemed to maintain a high resistance to fungal infection by retarding fruit senescence.

Slight de-greening was evident in limes stored for 23 days in NA at a chilling temperature, 4.4°C. By 40 days, yellowing had progressed noticeably, whereas in LP at pressures of 8.0, 13.3 or 20 kPa (60, 100 or 150 mm Hg) all fruits were still green, without any visible indication of cold damage. After 70 days, the percentage of fruit that was still green, with excellent taste, without any mould development, was 72% at 20 kPa (150 mm Hg), 50% at 13.3 kPa (100 mm Hg) and 24% at 8 kPa (60 mm Hg), compared to 0% in NA. The fruits that had 'spoiled' in LP had superficial browning, indicative of chilling damage (Burg, 1970).

A 1000 kg sample of unwaxed, field-run limes was packed in wooden pallet bins, cooled in a 12.2 m Fruehauf intermodal LP container operated at 20 kPa (150 mm Hg), and stored at that pressure for 8 weeks at 10°C. After removal, the fruit was warmed ('sweated'), dried and passed through a packing line where it was washed, graded, waxed and boxed. Recovery was 98.2%. The fruit was sold through normal distribution channels and accepted as 'fresh'. No claims were received from buyers. The stored fruit retained nearly at-harvest appearance, rind thickness, flavour and content of vitamin C, total solids and juice.

It is important at certain times of the year, especially during hot and rainy weather, to harvest limes at midday when their turgor is low. They should be segregated for 1 or 2 days to permit stylar end-rot development so that susceptible fruit can be culled before hypobaric storage is commenced. This disease, which is caused by excessive turgor in the juice sacs, is aggravated by thermal expansion of the juice, and therefore the fruit should be handled in a shaded area to avoid radiant warming (Pantastico *et al.*, 1975c; Davenport and Campbell, 1977a,b).

A full load of limes was transported from Brazil to Rotterdam in a 12.2 m Grumman/Dormavac LP intermodal container operated at 10°C and a pressure of 10.7 kPa (80 mm Hg; Jamieson, 1984). The shipment was reported to be a success, but no details were made available.

10.18 Mango (*Mangifera indica* L.), cvs Haden, Irwin, Tommy Atkins, Kent, Keitt, Maya, Okrang

Mangoes can be kept in NA for 2–3 weeks at 13°C (Welby and McGregor, 1997). The storage life of Florida mangoes is increased only slightly or not at all by CA (Hatton and Spalding, 1990), and variable results with CA have been obtained with mangoes from other locations (Thompson, 1998). The optimal CA condition, 3–5% [O_2] + 5–10% [CO_2] at 10–15°C, is unsatisfactory because after 2 weeks at 12°C, the fruit has no subsequent shelf life (Bender *et al.*, 2000b). Some varieties have been stored in 2% [O_2], but CO_2 production by mangoes is stimulated and ethanol production elevated in atmospheres containing less than 9.2% [O_2] (Singh *et al.*, 1937). Below 5% [O_2], mangoes usually suffer low-[O_2] injury (Table 4.7), ripening and chlorophyll degradation are prevented, and an abnormal colour develops. Less than 1% [O_2] invariably results in off-flavours (Thompson, 1998), and within 2–3 weeks off-flavours developed in Haden and Tommy Atkins mangoes stored in 2–3% [O_2] at 12–15°C (Bender *et al.*, 2000b). Symptoms of high-[CO_2]

injury include an irreversible inhibition of ripening upon transfer to 20°C air, characterized by abnormal greyish epidermal colour, off-flavours and a lack of normal aroma development.

Haden mangoes ripened during 3–4 days when they were stored at room temperature in NA; in 8 days during LP storage at a pressure of 26.7 kPa (200 mm Hg); and in 13 days at 20 kPa (150 mm Hg) (Burg and Burg, 1966c). After LP storage at 13°C and pressures of 10.1 or 20.3 kPa (76 or 152 Hg), Irwin, Tommy Atkins and Kent mangoes ripened at 21°C with less anthracnose (*C. gloeosporioides*) and stem-end rot (*Diplodia natalensis*) and a higher percentage of acceptable fruits compared to mangoes stored in NA (Spalding, 1977; Spalding and Reeder, 1977, 1980). CA was ineffective with the same fruits primarily because it did not prevent decay development. In LP, the reduction in decay coincided with a retardation of ripening, permitting storage of the fruit for at least 3 weeks at 13°C. Mangoes kept at pressures of 10.1 and 20.3 kPa (76 and 152 mm Hg) required 3–5 days longer to soften after storage, compared to fruits that had been stored in NA. LP had no effect on Keitt mangoes at the pressures studied. Chilling damage did not occur within 4 weeks when Irwin, Kent, Keitt, Haden and Tommy Atkins mangoes were stored at 10°C and a pressure of 10.7 kPa (80 mm Hg) (Spalding, 1977; Spalding and Reeder, 1977).

Mature-green Okrang mangoes were hydrocooled from 32.2°C to 30°, 20° or 15°C, and dipped in 0.7, 0.85 or 1% concentrations of a commercial wax formulation before they were stored at 20°C for 2–5 weeks, or at 13°C for 4 weeks, either at atmospheric pressure or in LP at 8.0, 10.67 or 13.3 kPa (60, 100 or 150 mm Hg). Weight loss, pulp firmness, sugar content, percentage of decay and the degree of skin colour change were measured at the end of the storage period (Ilangantileke *et al.*, 1989; Ilangantileke and Salokhe, 1989). Quality after storage was best in fruits hydrocooled to 15°C, dipped in wax (any concentration) and stored at 13°C either at 8.0 or 13.3 kPa (60 or 100 mm Hg).

In tests run in Israel, the storage life of mangoes at 13°C improved when the LP pressure was reduced below 13.3 kPa (100 mm Hg). Fruit stored in NA began to ripen after 16 days; in LP at pressures of 13.3 and 10 kPa (100 and 75 mm Hg) softening occurred after 25 and 35 days, respectively; and in LP at 6.7 kPa (50 mm Hg) fruit remained firm for 35 days (Apelbaum *et al.*, 1977c). Pressures higher than 33.3 kPa (250 mm Hg) had no effect. Mangoes stored in LP ripened in 3–4 days with normal flavour and aroma after transfer to 25°C air at atmospheric pressure, but green 'Haden' and 'Maya' fruits that had been stored in LP for a prolonged period did not develop proper red or orange colour during shelf life, and instead turned pale yellow. Treatment with ethylene upon removal from storage only slightly improved their colour development. Keitt mangoes did not benefit from LP storage.

In a study performed at the USDA, during 5–6 weeks at 10–13.3°C, a pressure of 3.6 kPa (25 mm Hg = 0.4% [O_2]) was far superior to higher pressures in extending the storage life of mangoes and controlling decay (D.H. Spalding, 1982, personal communication). Off-flavours, chilling or low-[O_2] damage did not occur, clearly demonstrating the effectiveness of a hypobaric condition in promoting O_2 diffusion into fruits. LP may also improve a mango's gas exchange by opening its stomates. In subsequent tests, the pressure was reduced to 2 kPa (15 mm Hg = 0.1% [O_2]) at 10°C with no indication of low-[O_2] injury or off-flavours (T.L. Davenport and S.P. Burg, 2002, unpublished). End-of-season fully mature Keitt mangoes were stored at 13.3°C either in NA or in LP at 1.33 kPa (20 mm Hg). The RH, measured with a wet-and-dry bulb using Teflon-coated thermistors with ± 0.07°C accuracy (Fig. 9.3), was maintained at 98%. Although at receipt the mangoes were heavily infected with anthracnose, after 45 days in LP all fruits were still hard and capable of ripening with little apparent increase in decay, but within 60 days the LP fruit became unmarketable due to anthracnose development. In NA, the fruit ripened within 1 week at 13.3°C

(T.L. Davenport and S.P. Burg, 2002, unpublished).

Decay in mangoes can be controlled for periods in excess of 30 days by hot-water treatment (Heather, 1994), but this has the disadvantage that while under laboratory conditions it does not usually damage the fruit, under commercial conditions it reduces fruit quality (McDonald and Milhjer, 1994). Therefore, a better method of microbial control is highly desirable. A reduction in resorcinol concentration during mango ripening allows previously latent *A. alternata* infections to resume development. By delaying ripening, a hypobaric pressure of 20.3 kPa (152 mm Hg) maintains the level of preformed antifungal resorcinols in the peel of Tommy Atkins mangoes. This suppresses disease development by maintaining latency, and decreases the incidence of infected sites in inoculated fruits (Table 7.7; Droby *et al.*, 1986). At pressures of 2.67–3.33 kPa (20–25 mm Hg), the incidence of decay in mangoes is further decreased (D.H. Spalding, 1982, personal communication) because in the absence of CO_2, the growth and sporulation of *A. alternata* is limited by low $[O_2]$ (Figs 7.3 and 7.5; Table 7.4). Many of the other organisms that infect mangoes, including *C. gloeosporioides* (Table 7.3, Fig. 7.7), *Aspergillis niger* (7.1; Table 7.5), *Fusarium* sp. (Fig. 7.3), *Rhizopus* sp. (Fig. 7.3), *Erwinia* sp. (Fig. 7.2) and at least one strain of *Pseudomonas* (Fig. 7.2; Eckert, 1975) are suppressed by 0.23–0.37% $[O_2]$, the concentration present in water-saturated air at 13.3°C and a pressure of 2.7–3.6 kPa (20–25 mm Hg). This is an important factor contributing to optimal mango storage, that results in the 2.0–2.7 kPa (15–20 mm Hg) range.

10.19 Nectarine (*Prunus persica* L.), var. Nectarina

Nectarines can be kept for 2–4 weeks in NA at –0.5 to 0°C (Welby and McGregor, 1997). At 0–1°C, fruit ripened with normal flavour and aroma during 11–20 days in NA, and within 18–35 days when it was stored in LP at 10.7–16 kPa (80–120 mm Hg; Burg, 1976a). Pressures lower than 10.7 kPa (80 mm Hg) have not been tested, but should be far more effective in delaying ripening and controlling the major decay organisms that cause nectarines to rot in storage (see peaches – 10.22).

10.20 Orange (*Citrus sinensis* Osbeck), Valencia

Provided that stem-end rot and other decays are controlled by a fungicidal treatment, Florida and Texas Valencia oranges can be stored in NA for 8–12 weeks at 0°C and 85–95% RH (Welby and McGregor, 1997). At 4.4°C, Valencia oranges, which could be kept for 72 days in NA, were successfully stored for 157 days in LP at 95% RH and pressures ranging from 6.7 to 14.7 kPa (50–110 mm Hg) (Burg, 1970, 1976a). The pressure was not low enough to offer any advantage with respect to decay control (Kim and Oogaki, 1986), which limited storage in an unpredictable manner. The main advantage was improved flavour and juice retention. Decay was effectively controlled both in NA and LP by HOCl vapour continuously flushed over the oranges during storage (example 20, chapter 7; Burg, 1970).

The principle decays of oranges are caused by stem-end rots (*D. natalensis* and *Phomopsis citri*), alternaria rot (*Alternaria citri*), phytophthora brown rot (*Phytophthora* sp.), green mould (*P. digitatum*), blue mould (*Penicillium italicum*) and sour rot (*G. candidum*) (Eckert, 1978b). The growth and sporulation of *P. digitatum*, *P. expansum*, *A. alternata*, *D. natalensis* and *G. candidum* var. *citri-aurantii* are directly suppressed by a low pressure independent of the $[O_2]$, apparently due to CO_2 removal (Tables 7.4 and 7.5). At atmospheric pressure, growth of *A. tenuis*, *Penicillium* sp. and *Phomopsis* sp. is strongly inhibited at less than 1% $[O_2]$ (Ryall, 1963; Follstad, 1966; Fig. 7.1). Very low pressure has this same effect on *P. digitatum*, *A. alternata* and *G. candidum* var. *citri-aurantii* (Fig. 7.6;

Apelbaum and Barkai-Golan, 1977; Alvarez, 1980). Valencia oranges stored at 4 and 10°C in NA, and in LP at a pressure of 2.67 kPa (20 mm Hg), had indistinguishable flavour, aroma and appearance after 16 days. They remained in that condition during a subsequent 5-day shelf-life test at 4, 10 and 20°C (T.L. Davenport and S.P. Burg, 2003, unpublished). This indicates that, without adverse side effects, they had tolerated an LP pressure low enough to provide direct control over decay development. No suitable CA combination of [O_2] and [CO_2] prevents decay development in oranges (Seberry and Hall, 1970).

10.21 Papaya (*Carica papaya* L.), var. Solo

Papayas can be kept for 1–3 weeks in NA at 7°C (Hardenburg *et al.*, 1986). Their storage life is only slightly extended by CA (Hatton *et al.*, 1975; Hatton and Spalding, 1990; Thompson, 1998). *Mark's Fruit Crop Encyclopedia* (http://www.uga.edu/fruit) estimates storage of 12 days is possible at the optimal CA condition, 12.8°C, in the presence of 1–4% [O_2]. Decay is the primary factor limiting papaya storage. To minimize anthracnose development, heat-treated fruits have been used in most papaya trials.

The major benefits of hypobaric papaya storage at 10°C and a pressure of 2.7 kPa (20 mm Hg) are slower coloration, increased shelf life, improved colour development after storage, reduced decay, and less surface pitting and desiccation. Good results have been obtained with green papaya and also with fruits showing up to a 50% colour break at harvest. Fruit with a colour break of more than 50% should not be shipped in LP as their temperature tends to rise in transit due to the extra heat produced during the respiratory climacteric (Burg, 1978b). When the temperature of these advanced papayas increases to 22°C, their vapour pressure equals the storage pressure (2.67 kPa = 20 mm Hg) and water 'flashes' at the fruit surface. The resultant 'vacuum cooling' prevents a further temperature rise, but the interior of the fruit ferments because only water vapour (and no O_2) is present when this occurs. In the same hypobaric shipment, boxes containing fruits with a colour break of less than 50% were well-preserved and remained close to the 10°C storage temperature, while fruits with a colour break greater than 50% increased in temperature and were spoiled. Exported papaya must have a colour break of at least 5% to exceed the minimum total solids required by law.

In Japan, satisfactory LP storage of papayas has been reported for 40 days with delayed ripening and prevention of deterioration (Gemma *et al.*, 1989), especially with hot-water-dipped fruit. The result was less favourable at 10°C, 70–80% RH, and a pressure of 20.27 kPa (152 mm Hg), compared to lower pressures (Gemma *et al.*, 1989). In laboratory tests performed by Grumman Allied Industries, 'Solo' (Maui strain) papayas were stored at 7.2, 10 and 12.8°C either in NA or in LP at pressures of 6.67, 8.0 and 16.0 kPa (50, 60 and 120 mm Hg). Fruit kept in LP still had 5 days' shelf life after 21 days' storage at 10°C and a pressure of 6.67 kPa (50 mm Hg; Jamieson, 1980e). Chilling damage resulted at 7.2°C, and a storage temperature of 12.8°C hastened colour development, reduced shelf life and enhanced decay. In a subsequent test at 10°C, when papayas that had an at-harvest 5% colour break were stored in NA or in LP at pressures of 1.33, 2.67, 5.33, 8.0, 10.67 and 16 kPa (10, 20, 40, 60, 80 and 120 mm Hg), fruit kept in NA displayed surface wrinkling, black spotting, extensive fungal growth, advanced softening and colour break within 22 days. All fruit kept in LP at 1.33–10.67 kPa (10–80 mm Hg) still retained its at-harvest appearance, but at 16 kPa (120 mm Hg) significant superficial black spotting had developed (Dressler, 1978b; Jamieson, 1980e). After ripening for 5 days in 24.4°C air, papayas that had been stored at 1.33–2.67 kPa (10–20 mm Hg) were in the best condition, firm, with good (yellow) colour. Fruits stored at 5.33–10.67 kPa (40–80 mm Hg) were softer with some black spotting, and those that had been kept in NA or at 16 kPa (120 mm Hg) were unmarketable, soft, with black spotting. By 28 days' storage, the controls were unmarketable, but fruits kept at 2.67 kPa (20 mm Hg) still had their initial appearance, while at all other pressures the papayas had developed some black spotting. Following a 5–6-day shelf-life test in room-temperature air, fruit that had been kept at 2.67 kPa (20 mm Hg) had the best, excellent appearance, with good colour and no desiccation. In another trial, papayas were well preserved when they were stored at 10°C for 18 days at 2 kPa (15 mm Hg) (Chau and Alvarez, 1983).

Papayas are one of the few fruits damaged by rapidly venting an LP container (Chau and Alvarez, 1983). The injury, which appears as a watery internal breakdown, softening and sometimes complete collapse of the fruit, is far worse on softening fruit and is absent if the fruit is green-hard. Half-coloured fruit tolerate evacuation to 2.7 kPa (20 mm Hg) in 60 min, but should not be exposed to a pressure release in less than 30 min. Commercial shipments have been vented in 2 h (Jamieson, 1984).

The major postharvest diseases of papaya are anthracnose caused by *C. gloeosporioides*, peduncle infection due mainly to *Fusarium* sp. and *C. gloeosporiodes*, and stem-end rots by *Mycosphaerella* sp., *Ascochyta caricae-papayae* and *Botryodiplodia theobromae* (Alvarez, 1980). At 10°C, radial growth and sporulation of *Colletotrichum* cultures is retarded at a pressure of 2.0 kPa (15 mm Hg), and the entire infection process is delayed on papaya fruit inoculated with *C. gloeosporioides* (Table 7.3). When these fruits were stored for 21 days at 10°C either in NA or at 2.0 kPa (15 mm Hg), less anthracnose developed in LP, and 1 day after the fruits were removed and transferred to 21°C air at atmospheric pressure, 88% of the LP fruit were marketable vs. 64% in NA; 5 days after removal 26% of the LP fruit were still marketable vs. 0% in NA (Chau and Alvarez, 1983).

Both ripening and disease development were inhibited during 18–21-day shipment of a full load of heat-treated Solo papayas in a 12.2 m Grumman/Dormavac intermodal container operated at 10°C and a pressure of 2.7 kPa (20 mm Hg). The fruit exhibited much less anthracnose caused by *C. gloeosporioides*, peduncle infection caused mainly by *Fusarium* sp. and *C. gloeosporiodes*, and stem-end rots due to *Mycosphaerella* sp., *A. caricae-papayae* and *B. theobromeae*, compared to fruits stored in refrigerated containers operated at the same temperature and atmospheric pressure (Alvarez, 1980; Fig. 7.7). The incidence of stem-end rot at 2.7 kPa (20 mm Hg) was extremely low because the growth and spore germination of *Fusarium*, the causative agent of papaya stem-end rot, is severely inhibited in the absence of CO_2 at the 0.28% [O_2] concentration present in 10°C saturated air at a pressure of 2.67 kPa (20 mm Hg; Wells and Uota, 1970; Fig. 7.3). *A. alternata* and *R. stolonifer* also may contribute to papaya rotting (Eckert *et al.*, 1975), and both organisms cannot grow at the O_2 partial pressure present in 10°C water-saturated air at a pressure of 2.67 kPa (20 mm Hg; Tables 7.2 and 7.4; Figs 7.1, 7.3, 7.5 and 7.6). After 15–19-day winter shipments of papaya from Hawaii to the US, 76% of unwaxed fruits carried in a Grumman/Dormavac LP intermodal container and 63% of NA-stored fruits transported in a standard container were marketable; after summer shipments, 72% of LP fruits and only 15% of NA fruits were marketable. Weight loss was greater in LP than in NA.[3] Three days after removal from LP, papayas had 63% less peduncle rot, 55% less stem-end rot and 45% less anthracnose compared to fruit kept in NA (Fig. 7.7; Alvarez, 1979). Insects infesting papaya fruit are killed at 10°C during 7 days' storage at an LP pressure of 2 kPa (15 mm Hg; A.M. Alvarez, 1980, personal communication to W. Jamieson).

Papaya were successfully shipped from Hawaii to the US West Coast in a 12.2 m Grumman/Dormavac hypobaric container operated at 10°C and a pressure of 20 mm Hg (Grumman Allied Industry, 1978). The entire load, 2100 cases, sold in the marketplace after 18 days of elapsed time, had 5–9 days' shelf life remaining. The test was intentionally extended to simulate a long journey, including storage in Hawaii, a sea journey to Oakland and an over-the-road trip to Los Angeles where the fruit was distributed through normal marketing channels.

10.22 Peach (*Prunus persica* L. Batsch.), cvs White Okubo, Cardinal, Red Haven, Gleason Early Elberta

Peaches and nectarines have similar storage requirements and neither is well adapted to prolonged storage in NA or CA. Most varieties can be stored in NA for up to 2–4 weeks at –0.6 to 0°C, but only for 1 week at 4.4°C (Lutz and Hardenburg, 1968). CA provides a modest increase in storage life. With longer storage times decay becomes a problem, and peaches may develop mealy breakdown (wooliness), internal browning and abnormal peeling, lose their flavour and natural colour, and if picked at an immature stage, become impaired in their ability to ripen. Intermittent warming to 22°C for 1–3 days at intervals of 1–4 weeks during storage at 0°C sometimes is used to reduce the incidence of these 'low-temperature' disorders (Thompson, 1998).

At 20°C, ripening of 'Okubo' peaches, measured as a 'peeling index', was slowed by CA storage in less than 5% [O_2], but low-O_2 damage resulted. A fermented flavour and high ethanol content developed at 0–1% [O_2] and even at 3–5% [O_2]. CO_2 production increased below 3% [O_2], softening was only retarded at 0% [O_2], and brown pit occurred when all O_2 was removed (Kajiura and Iwata, 1971). At 4°C, varying the [O_2] from 3 to 21% did not delay ripening or improve storage, and below 1% a fermented flavour developed. All [O_2] concentrations failed to reduce the incidence of chilling damage, and no benefit from CA was noted. An apparatus that regulated pressure by adjusting the height of water towers situated at the inlet of the storage vessel was used to test the effect of LP. Weight loss in LP was 2.5–3.5 times higher than in NA or CA, indicating that LP chamber leakage resulted in poor humidity control (chapter 9). At atmospheric pressure, 'Okubo' peaches ripened in 2.0–3.2 days when they were stored in 21% [O_2] at 30°C; 4 days in 10% [O_2]; and 7 days in 5.7% [O_2]. In LP they ripened in 5.5 days at a pressure of 48.7 kPa (365 mm Hg = 9.2% [O_2]), and 9 days at 27.3 kPa (205 mm Hg = 4.8% [O_2] (Kajiura, 1973a,b). In another study using the same LP apparatus (Kajiura, 1975), white 'Okubo' peaches were stored at 1°C for 3 weeks in either NA, CA (3% [CO_2] + 3% [O_2] or 0% [CO_2] + 3% [O_2]), and in LP at 14.7 or 86.7 kPa (110 or 650 mm Hg). After removal from storage, the fruits were ripened in 20°C air at atmospheric pressure. Flesh browning, mealy breakdown and abnormal peeling developed in NA, indicating that the fruit had been injured by the low temperature. Ripening after storage was accelerated and low-temperature injuries were not controlled by keeping fruit in 3% [O_2] without CO_2, but when 3% [CO_2] was added, the post-storage ripening rate was not enhanced and low-temperature injuries were almost completely controlled (7.10). After LP storage at 14.7 kPa (110 mm Hg = 3% [O_2]), the ripening rate was slowed and mealy breakdown was reduced, but flesh browning and abnormal peeling were not prevented. LP storage at a pressure of 86.7 kPa (650 mm Hg) had no effect.

Cardinal peaches stored at 0–1°C ripened and softened in 14–21 days in NA, and in 28–35 days in LP at a pressure of 10.7 kPa (80 mm Hg) (Burg, 1976a). At 0°C, Gleason Early Elberta peaches softened in 66 days in NA vs. 93 days in LP at a pressure of 13.6 kPa (102 mm Hg). Pressures of 62.8 and 37.1 kPa (471 and 278 mm Hg) were progressively less effective. Carotenoid steadily increased during ripening in NA, but for 45 days no carotenoid formation occurred at 13.6 kPa (102 mm Hg). LP slowed the loss of titratable acidity and sugars as a function of pressure reduction (Salunkhe and Wu, 1973). Red Haven peaches were successfully stored at 2°C for periods of 4–6 weeks at 10–20 kPa (75–150 mm Hg). No wooliness developed and ripening was normal, but results were quite dependent on the condition of the peaches when they were placed in storage (D.R. Dilley, 1978, personal communication).

M. fructicola, *Botrytis* sp., *Penicillium* sp. and *Alternaria* sp. cause most of the decays of peaches and nectarines stored in air or in CA at 1% [O_2] + 5% [CO_2] (Smith and Anderson, 1975). To directly and substantially inhibit the growth of *Penicillium*, *Botrytis* and *Alternaria*, the [O_2] must be lowered to 0.15–0.25% at atmospheric pressure, and in LP the pressure must be decreased to 1.33–2.0 kPa (10–20 mm Hg) to create this same range of O_2 tensions (Tables 7.1, 7.2 and 7.4; Figs 7.1, 7.3, 7.5 and 7.6). At atmospheric pressure, the growth of *M. fructicola* on potato dextrose agar is 50% suppressed in 2.3% [O_2], but lower [O_2] was not tested (El-Goorani and Sommer, 1979). Pressures below 10 kPa (75 mm Hg) have not been tested with peaches, but they are likely to give better results both for decay control and preservation. The lower pressure range should be tested in combination with a storage temperature between 10 and 16°C to prevent internal browning and other disorders induced by a cold temperature.

10.23 Pear (*Pyrus communis* L.), cvs Bartlett, Clapp, Commice

Depending upon the cultivar, growing condition and locale, more than 3–5% [CO_2] and sometimes less than 1% [CO_2] causes core and flesh browning of pears (Fig. 4.7; Table 4.7) and markedly reduces this fruit's storage life (Hardenburg *et al.*, 1986; Thompson, 1998). LP improves pear storage by removing CO_2 from within and around the fruit. In CA, pears benefit from [O_2] concentrations as low as 0.5–1%, and therefore it is likely that at −1 to 0°C the optimal LP pressure for pear storage is lower than 3.1 kPa (23 mm Hg = 0.5% [O_2]), but pressures less than 5.3 kPa (40 mm Hg) have not yet been tested with this fruit.

At 0°C, the storage life of Bartlett pears was 2.5–3.5 months in NA at an ambient pressure of 86.1 kPa (646 mm Hg); 5 months in LP at 63.8 kPa (471 mm Hg); 7 months at 37.1 kPa (278 mm Hg); and at 13.6 kPa (102 mm Hg) the green colour was maintained for up to 5 months, and the pears were still marketable after 8 months (Salunkhe and Wu, 1973 – see example 8, chapter 3, for a description of technical difficulties inherent in the apparatus used for these studies). Pears ripened normally when they were removed from LP storage at a firm-green stage, whereas comparable pears stored in NA often developed brown core as they softened.

At −1 to +1°C, the storage life of Bartlett, Clapp and Commice pears in LP was improved at a pressure of 20 kPa (150 mm Hg), but much better results were obtained during 4–6 months at 8 kPa (60 mm Hg) (Burg, 1976a; D.R. Dilley, 1976, personal communication). Upon removal from LP storage, fruit ripened normally, with no internal browning and a normal shelf life. A pressure of 5.3 kPa (40 mm Hg) was superior to 8 kPa (60 mm Hg) with Clapp pears. Pears stored at 25.3 and 40 kPa (190 and 285 mm Hg) for 36 days (Kim and Oogaki, 1986), kept better in LP than in CA (Streif, 1974b).

The major diseases of pears are lenticle rot caused by *Phlyctaena vagabunda* Desm. (= *G. album* Osterw.) and blue mould rot due to *P. expansum* (Lk.) Thom. (Wills *et al.*, 1989). The effect of LP on these moulds is discussed in 10.1.

10.24 Pineapple (*Ananas comosus* L. Merr.), var. Smooth Cayenne

Depending on their initial stage of ripeness, pineapples can be stored in NA for 2–3 weeks at 7–10°C (Welby and McGregor, 1997). Chilling injury occurs at temperatures lower than 13°C (Hatton and Spalding, 1990; Welby and McGregor, 1997), rather than below 7°C as formerly believed (Hatton *et al.*, 1975). After removal from storage at < 13°C, internal browning due to chilling typically appears within 2 days at 22°C. The injury is not prevented by storage in 3% [O_2] regardless of whether 5% [CO_2] is included (Hatton and Spalding, 1990), and while 1.4–2.3% [O_2] + 11.2% [CO_2] reduces the severity of the disorder, it does not eliminate the problem. A CA atmosphere containing 2–3% [O_2] sometimes extends shelf life by several days, but often CA provides no significant benefit (Thompson, 1998).

Green pineapples treated with Dowcide A fungicide were stored at 4.4°C in NA or in

LP at pressures of 13.3, 22.7 or 33.3 kPa (100, 170 or 250 mm Hg). In NA the fruits developed loosening of the crown after 10 days. Flesh softening occurred in 17 days both in NA and LP, colour development did not progress during 35 days and the interior of the fruits suffered severe chilling damage (Burg, 1969).

Half-coloured pineapples were treated with Dowcide A and stored at 13.9°C either in NA or in LP at pressures of 13.3, 17.3, 21.3 or 30.7 kPa (100, 130, 160 or 230 mm Hg). In NA, fruit began to soften within 15 days, was marginally acceptable after 19 days and spoiled within 35 days. At that time about 50% of the fruit stored in LP at 13.3 or 17.3 kPa (100 or 130 mm Hg) was still in good condition, but all fruit kept in LP at 21.3 or 30.7 kPa (160 or 230 mm Hg) had internal breakdown. The experiment was terminated after 35 days due to decay (Burg, 1969).

Green and yellow 'sour' pineapples, air-shipped from the Dominican Republic, were treated with Dowcide A and stored at 10°C. After 20 days in LP at a pressure of 16 kPa (120 mm Hg), the green fruit was in perfect condition with its original colour and no exterior or interior mould development. By that time fruit kept in NA had yellowed and softened, but was still without fungal infection. All LP fruit was still firm after 30 days' storage, with excellent colour and no mould development, whereas due to softening, surface and internal browning, and occasional mould on the surface of the fruit and leaves, none of the pineapples stored in NA were in saleable condition at that time (Burg, 1970).

Fully coloured pineapples that had been treated with Dowcide A were stored at 13.9°C in NA or in LP at pressures of 9.3, 20.0 or 33.3 kPa (70, 150 or 250 mm Hg). In NA, fruits began to soften in 9 days and displayed considerable internal breakdown within 16 days. A small percentage of pineapples remained marketable after 19 days' storage in NA, but by then their flavour was only fair, and within 23 days all controls had to be discarded. At that time fruits stored in LP were still in excellent condition, with normal flavour, but within 34 days pineapples stored at 20 or 33.3 kPa (150 or 250 mm Hg) had decayed, and the first signs of mould growth were evident at 9.3 kPa (70 mm Hg), although the flavour of these fruits was still excellent (Burg, 1969). In another laboratory study, an LP pressure of 10.13 kPa (76 mm Hg) extended the storage life of pineapples to 30–40 days (Staby, 1976b).

A partial load of Hawaiian smooth Cayenne pineapples was tested in a 12.2 m hypobaric intermodal container operated at 8.3°C and a pressure of 10 kPa (75 mm Hg). The fruit was harvested in late May and was of poor quality. Sunburn and translucency of the flesh make such fruit unsuitable for export at that time of year. At receipt, the fruit ranged in colour from green to less than one-quarter gold. During 15–19 days' storage, the colour increased one-quarter to one-half units in LP and one-and-a-quarter units in NA. Shell condition increased by one-and-a-quarter units in LP and one-and-a-quarter to 2 units in NA. Crown condition also was improved by LP storage, especially in fruits initially at colour stage 1 or 2. LP decreased the loss of sugar and acid during storage, and improved the Brix:acid ratio.[2] After 19 days' of LP storage, the fruit was transferred to NA for 4–5 additional days. At that time, the taste of LP fruit was rated marginal to outstanding, whereas the taste of control fruit was rated poor to marginal. It was concluded that LP storage enhances the retention of flavour, colour, acid, sugar, shell condition and crown condition, and improves the Brix:acid ratio.

A successful full-load shipment of pineapples was made from Taiwan to Japan in a 12.2 m Grumman/Dormavac LP intermodal container operated at 8.3°C and a pressure of 2 kPa (15 mm Hg) with humidity controlled at 95% (Jamieson, 1984). At this temperature and pressure combination, the 0.19% [O_2] concentration present in water-saturated air prevents the growth of *Penicillium* sp. and *Fusarium* sp., the causal agents of pineapple black spot (Figs 7.3 and 7.6; Table 7.4). The major disease of pineapples is black rot caused by *Ceratocystis paradoxa* (= *T. paradoxa*). At atmospheric pressure the growth of *T. paradoxa* is not inhibited in 2.3% [O_2]

(El-Goorani and Sommer, 1979), but that is generally true of fungi whose development is prevented at 0.19% [O_2] (7.1; Eckert *et al.*, 1975). The optimal LP storage condition for pineapples may be a pressure of 2 kPa (15 mm Hg) to control physiological processes and prevent decay, coupled with a temperature of 13°C to avoid chilling injury.

10.25 Prune (*Prunus domestica*), cv. Richards Early Italian

The Italian prune plum is one of the most important commercial cultivars, but its distribution, storage and marketing is limited by internal browning, softening and decay. At 0°C, the maximum storage life of Richards Early Italian prune plums is less than 3 weeks in NA (Hardenburg *et al.*, 1986; Mercantila Publishers, 1989b). CA (1–2% [O_2] + 2.5% [CO_2]) provides a significant benefit at 2°C (Thompson, 1998), and LP pressures in the range between 5.3 and 13.3 kPa (40 and 100 mm Hg) effectively reduce losses (Patterson and Melsted, 1978). Lower LP pressures might be more effective, since they should control the plum fruit moth *Cydia funebrana*, and even 2.3% [O_2] causes a 50% growth inhibition of *Monilinia* sp. (El-Goorani and Sommer, 1979), a major cause of Italian prune decay.

10.26 Strawberry (*Fragaria* × *ananassa* Duch.), cvs Tioga, Shasta, Florida 90, Donna, Z-5A, Long Island Jerseybelle, Cambridge Favourite

Fresh strawberries cannot be stored in NA for more than 5–7 days at 0°C (Hardenburg *et al.*, 1986; Mercantila Publishers, 1989b). Decay is reduced and storage improved in 0.5–2% [O_2], but off-flavours result (Table 7.2). Better results are obtained with high [CO_2] (Thompson, 1998). At 20°C, strawberries produce approximately 0.1 μl/kg·h of ethylene, and in wholesale markets this gives rise to 30–360 nl/l of ethylene in strawberry punnets. Storage life at atmospheric pressure is extended by reducing the ethylene level around the fruit to 50 nl/l at 20°C, or 5 nl/l at 0°C (Wills and Kim, 1995). LP preserves strawberries by removing ethylene from within and around the berries, and by lowering the [O_2] to 0.15% in order to inhibit mould development.

Storage at and above 3.3 kPa (25 mm Hg)

At 4.4°C, a pressure of 25.3 kPa (190 mm Hg) improved the storage life of several strawberry varieties by 1 week in LP compared to NA; flavour was retained, and less decay occurred. Desiccation increased with higher airflow rates, although measurements indicated no difference in the RH (9.2; Tolle, 1972). Tioga and Florida 90 strawberries sometimes were well preserved at 0–2°C for up to 4–5 weeks in LP at pressures between 10.7 and 26.7 kPa (80 and 200 mm Hg), but storage life was unpredictable due to mould growth (Burg, 1976a). After 15 days at 2.8°C and pressures of 8.0, 13.3, 18.7 or 30.7 kPa (60, 100, 140 or 230 mm Hg), Florida 90 berries had good appearance and adequate flavour, but had developed mould. Berries stored at the same temperature in NA shrivelled within 7 days and developed mould within 15 days. Similar LP results were obtained with Shasta, Donna and Z-5A strawberries (Burg, 1969). Full loads of Tioga and Florida 90 berries were kept in a 6.1 m (20 ft) prototype Fruehauf hypobaric intermodal container for 3 weeks at 1°C and a pressure of 21.33 kPa (160 mm Hg). The berries had 4–5 days' shelf life after removal from storage. An over-the-road shipment of a mixed load of berries from Homestead, Florida, to Salinas, California, gave satisfactory results with Shasta berries, but the softer Florida 90 variety experienced extensive bruising damage.

Cambridge Favourite strawberries were stored at 3.3°C with and without a dehydroacetic-acid sodium-salt monohydrate (fungicide) dip either in NA, in LP at 13.33 or 3.33 kPa (100 or 25 mm Hg) or in CA at equivalent [O_2] levels (Sharples, 1974; Bubb, 1975b). The improved flavour retention of strawberries held at 3.33 or 13.33 kPa (25 or 100 mm Hg) was not duplicated by keeping fruit in CA at comparable low [O_2] concentrations. Flavour remained excellent in LP but was tainted in CA. Undipped samples developed mould during shelf life at 18.5°C, and after 10 days in storage all samples that were undipped developed a high incidence of decay. Weight loss was less than 2% in all treatments after 10 days.

Storage below 3.3 kPa (25 mm Hg)

The most important strawberry postharvest decays are caused by *B. cinerea* (grey mould rot) and *R. nigricans* (rhizopus rot) (Harvey and Pentzer, 1966). Provided CO_2 is absent, the growth of both organisms is almost completely suppressed by the 0.15% [O_2] present in 0°C water-saturated air at a pressure of 1.3 kPa (10 mm Hg) provided that CO_2 is absent (Figs 7.1, 7.3 and 7.6; Tables 7.1, 7.2 and 7.4). Anthracnose (*C. gloeosporiodes*), leather rot (*Phytophthora cactorum*), rhizoctanis rot (*Rhizoctania solani*), sclerotinia rot (*Sclerotinia sclerotiorum*), stem-end rot (*Dendrophoma obscurans*) and tan brown rot (*Discohainesia oenotherae*) also occasionally infect strawberries (Harvey and Pentzer, 1966). The growth of *C. gloeosporiodes* (Table 7.3) and *S. sclerotiorum* (7.1) is inhibited at less than 1% [O_2] and *Phytophthora parasitica* is slightly inhibited in 2.3% [O_2], but lower concentrations were not tested (El-Goorani and Sommer, 1979).

Long Island Jerseybelle strawberries were stored at 1.1°C either in NA or in LP at pressures of 1.33, 2.67, 5.33 and 10.67 kPa (10, 20, 40 and 80 mm Hg). The best LP condition, 1.33 kPa (10 mm Hg), was effective in preventing *Botrytis* development without inducing off-flavours due to fermentation (Table 10.3). Poulssen *et al.* (1982) reached a similar conclusion after storing berries at 5–20°C and pressures ranging from 1.33 to 48 kPa (10–360 mm Hg). To prevent *Botrytis* development and water loss for 3–5-day periods prior to sale, individual trays of strawberries with a plastic overwrap were vacuum-cooled and then stored in LP at 0°C and 95% RH either at 1.3 or 2 kPa (10 or 15 mm Hg). A pressure of 1.33 kPa (10 mm Hg = 0.15% [O_2]) was noticeably better than 2 kPa (15 mm Hg = 0.29% [O_2]) for this purpose (Burg, 1995 unpublished). A partial load of Long Island Jerseybelle strawberries was kept at 1°C in a Grumman/Dormavac 12.2 m hypobaric intermodal container operated at 2.67 kPa (20 mm Hg) and a comparable sample was stored in NA at the same temperature. After 18 days in storage, the LP berries scored better with respect to firmness, colour, stem and leaf appearance, gloss, flavour and shelf life. These fruits developed only 5% decay during storage, and had 7% decay after an additional 5 days' shelf life, compared to 18.9% and 54.2% in NA, respectively, after the same number of days' shelf life (Grumman Allied Industries, 2000, unpublished test data). Pressures lower than 1.33 kPa (10 mm Hg) have not been tested with strawberries.

Table 10.3. Condition of Long Island Jerseybelle strawberries after 21 days' storage at 1.1°C in NA or LP. A taste panel judged the berries kept at a pressure of 1.33 kPa (10 mm Hg) acceptable with no off-odour or fermentative taste (Jamieson and Dressler, 1977).

	Per cent		
Pressure kPa (mm Hg)	Decay	Soft berries	'Brown' berries
101.30 (760)	28.8	78	69
2.67 (20)	8.4	69	48
1.33 (10)	5.6	62	0

Effect of hypochlorous acid vapour

Florida 90 strawberries stored at 3.3°C either at a pressure of 21.33 kPa (160 mm Hg) or in NA, developed heavy mould within 3–8 days, but when sodium hypochlorite solution was added to the humidifying water, no mould development occurred during 21 days. The mould was killed in an infected box of berries when it was transferred to a hypobaric chamber and treated with HOCl vapours derived by passing air through sodium hypochlorite solution (Burg, 1970). Sequoia strawberries stored at 3.3°C and 21.33 kPa (160 mm Hg) developed severe mould in 22 days, but not when the fruits were treated with vapours derived by passing the air changes through a sodium hypochlorite solution (chapter 7 – example 15). HOCl generated from a mixture of HTH (calcium hypochlorite – 70% available chlorine) and commercial 'Clorox' prevented mould development on var. Shasta strawberries during storage in a 6.1 m (20 ft) Fruehauf prototype intermodal hypobaric container operated at 3°C and a pressure of 21.33 kPa (160 mm Hg) (Woodruff, 1971; Table 7.9).

10.27 Tomato (*Lycopersicon esculentum* Mill.), vars Homestead 2, Jupiter, Michigan-Ohio Hybrid, Heines Hochzucht, NM Super II, NM Milo I, Waldin, Ace

Mature-green tomatoes have a storage life of 1–3 weeks in NA at 13–21°C (Hardenburg *et al.*, 1986; Mercantila Publishers, 1989b), but riper tomatoes tolerate lower temperatures. Mature-green and breaker fruit stored at 13°C for longer than 2 weeks fail to develop a deep red colour and may deteriorate due to decay. Tomatoes eventually suffer chilling injury and become susceptible to anthracnose decay if they are exposed in the field or during storage to a temperature of 10 to 16°C. At 13°C, the storage life of mature-green and breaker tomatoes is increased to up to 6 weeks in 3% [O_2] + 0% [CO_2] (Thompson, 1998). CO_2 provides no benefit, 3–5% [CO_2] sometimes is injurious and 1% [O_2] can cause an off-flavour to develop (Hardenburg *et al.*, 1986; Thompson, 1998). The optimum temperature for ripening tomatoes is 18–21°C.

Mature-green tomatoes

Mature-green tomatoes kept in NA at 15°C reached full-red colour in 7 days, with fair flavour. A significant percentage were overripe and inedible by 18 days, and all were discarded after 22 days. Stored in LP at a pressure of 32.8 kPa (246 mm Hg) they reached a full-red stage of ripeness in 22 days, but the locules within the fruit were still green. After several days in air at 15.6°C, these fruits ripened completely with flavour similar to that of the control kept in NA. Tomatoes stored in LP at 16.67 kPa (125 mm Hg) reached a dull-red stage in 33 days, and failed to develop a satisfactory taste and shelf life when they were transferred to air at 15°C. It was concluded that, because tomato fruits do not ripen with acceptable colour and flavour in the presence of less than a normal ambient [O_2] concentration, LP storage should be terminated before the fruit advances to a pink stage (Burg, 1969). During storage of Michigan–Ohio hybrid tomatoes at 12.8°C, a progressive decrease in ripening, chlorophyll degradation, β-carotene synthesis and lycopene formation occurred as the pressure was lowered from 86.1 to 13.6 kPa (646 to 102 mm Hg), and the content of tomato volatiles was reduced if the fruits ripened during storage[4] (Wu *et al.*, 1972; Wu and Salunkhe, 1972b; see example 8 – chapter 3). When var. Santa Cruz Kada tomatoes were stored at 23–27°C the fruits reached a breaker stage during 7 days in NA, in 21 days at 33.3 kPa (250 mm Hg), and under hypobaric conditions, the fruit did not reach full-red colour in 40 days (Awad *et al.*, 1974). A comprehensive laboratory test with Homestead 2 tomatoes included temperatures of 7.2, 10, 12.8 and 15.6°C, and at each temperature, pressures of 8.0, 10.67, 13.33, 16.67 and 101.33 kPa (60, 80, 100, 125 and 760 mm Hg). Fruit ripened during 2 weeks when it was stored in NA at 12.8°C, chilling damage caused black lesions and heavy fungal infection to develop within 4–5 weeks in all tomatoes kept at 7.2 and 10°C, and ripening and decay were accelerated at 15.6°C compared to 12.8°C. Ripening was delayed progressively as the pressure was lowered, and after 8 weeks the fruit colour had hardly changed at pressures of 8–10.67 kPa (60–80 mm Hg). At the optimal condition in this study, 12.8°C and a pressure of 8–10.67 kPa (60–80 mm Hg), decay limited storage life to 7–8 weeks (Burg, 1969). Ace tomatoes were stored with good retention of ripening ability for 8 weeks at 12°C and a pressure of 10.13 kPa (76 mm Hg). Following LP storage fruit ripened with acceptable quality, good colour development, and a proper balance of sugar and acids (Dilley, 1977a). There was a low incidence of internal decay caused by *Fusarium* sp. In another study at the same pressure and temperature, the incidence of decay was severe after 56 days of LP storage (Dilley, 1972). Heines Hochzucht, MM Super II and MM Milo I tomatoes stored at 12°C and a pressure of 10 kPa (75 mm Hg) displayed retarded chlorophyll breakdown and lycopene synthesis, less pectin degradation and loss of acidity, a slower decrease in fruit firmness, lower ethylene production and greater retention of ascorbic acid, compared to control fruits. Tomatoes kept in NA had a maximum storage life of 3–4 weeks; in LP they were stored for 7 weeks without appreciable losses and with normal post-storage ripening (Bangerth, 1974). Favourable results have also been reported with Jupiter, Waldin and Japanese pear tomatoes (Burg and Burg, 1966c; Tolle, 1969, 1972; Kader and Morris, 1974; Salunkhe and Wu, 1974, 1975; Stenvers and Bruinsma, 1975; Staby, 1976b; Stenvers, 1977; Kim and Oogaki, 1986).

To test the effect of intermittent-pressure cycling on mature-green tomatoes they were stored at 13°C and a pressure of 10.67 kPa (80 mm Hg), and each day the LP chambers were vented to atmospheric pressure for periods of 0, 2, 4, 6 or 8 h. Then the fruits were returned to 10.67 kPa (80 mm Hg) for the remainder of a 24-h period. Controls kept at atmospheric pressure in NA at 13°C ripened in 12 days. Interrupting the vacuum for 2–6 h each day had little effect on storage life and ripening during 8 weeks, except that an 8-h cycle increased the number of ripe and turning fruits to 56% from 30 ± 10% without pressure cycling (Burg, 1976a). In another test of intermittent cycling, mature-green to 'turning' Homestead 2 tomatoes were stored at 12.8°C for 2, 3 and 4 weeks either in NA or in LP at pressures of 2.67 and 10.67 kPa (20 and 80 mm Hg). At each LP pressure, half of the sample was returned to atmospheric pressure for 4 h each day. None of the LP fruit advanced significantly in colour during 4 weeks, but all fruits kept at 2.67 kPa (20 mm Hg) developed a high incidence of decay (Shum, 1981).

The storage life of full container-loads of tomatoes kept in LP at pressures of 8.0–10.67 kPa (60–80 mm Hg) often is limited by decay development. Mature-green non-staked Ace tomato fruits, packed in wooden pallet bins and field crates, were loaded at field temperature, 4540 kg (10,000 lb) and 18,160 kg (40,000 lb), respectively, in to 6.1 m (20 ft) and 12.2 m (40 ft) Fruehauf hypobaric intermodal containers. The fruits received no pretreatment for decay, and both containers were operated at 12.8°C, 95% RH, and a pressure of 10.67 kPa (80 mm Hg). After 8 weeks the tomatoes still had not reached a light pink colour, but ripened with normal colour and flavour when they were transferred to atmospheric air. Visually the fruits were rated excellent, but at an eating-ripe stage, nearly 25% were found to be infected internally with *Fusarium* (D.R. Dilley and S.P. Burg, 1977, unpublished data).

Breaker tomatoes

Breaker tomatoes were stored at 12.8°C either in NA or in LP at pressures of 5.33, 8.0, 10.67, 13.33, 15.33 and 17.33 kPa (40, 60, 80, 100, 115 and 130 mm Hg). Controls became eating-ripe in 8 days at 25°C, and fruits stored in NA at 12.8°C developed a red colour in 8 days, became eating-ripe in 11 days, and were still acceptable up to 19 days. Ripening occurred slowly in LP at a pressure of 17.33 kPa (130 mm Hg), and was almost completely prevented at 5.33 kPa (40 mm Hg). Between these extremes there was a progressive effect related to pressure. All fruits removed from LP at 19 days ripened normally, except possibly those stored at 5.33 kPa (40 mm Hg), which seemed to develop less flavour. After 26 days, tomatoes stored at 17.33 kPa (130 mm Hg) had reached an eating-ripe condition; fruit kept at 10.67, 13.33 and 15.33 kPa (80, 100 and 115 mm Hg) had partially ripened, and at 5.33–8 kPa (40–60 mm Hg) the tomatoes had not advanced significantly in colour. A longer storage period was not feasible due to decay, and also because it was found that if ripening proceeded to an advanced stage in LP, flavour and taste were not normal. Fruit that ripened during LP storage tended to develop excess yellow-orange carotenoid pigments (Burg, 1969).

Vine-ripe tomatoes

Vine-ripe tomatoes stored at 0–2°C were well preserved in NA for 8–10 days, but spoiled within 30 days. At the same temperature in LP at a pressure of 13.33 kPa (100 mm Hg) they were still firm, with good flavour and colour after 30–45 days (Burg, 1976a). In a test at 2.8°C, vine-ripe fruits softened in 8 days and spoiled within 15 days in NA, whereas in LP at 12 kPa (90 mm Hg) they remained unchanged for 18 days with excellent flavour and no indication of decay. After 18 days of LP storage, tomatoes transferred to 15.6°C air at atmospheric pressure remained in good condition for 4 days, with normal flavour (Burg, 1969).

Decay control

The major diseases of tomato fruits are alternaria rot (*A. tenuis*), anthracnose (*Colletotrichum coccodes*), bacterial canker (*Corynebacterium michiganense*), bacterial soft rot (*Erwinia carotovora*), bacterial speck (*Pseudomonas tomato*), bacterial spot (*Xanthomonas vesicatoria*), buckeye rot (*Phytophthora* sp.), and fusarium rot (*Fusarium* sp.) (McColloch *et al.*, 1966). Pressures in the 8.0–10.67 kPa (60–80 mm Hg = 1.3–1.9% [O_2]) range do not directly control the growth of any of these organisms, but at 2.0–3.33 kPa (15–25 mm Hg), in the absence of CO_2, the 0.1–0.38% [O_2] concentration present in water-saturated 13°C air should prevent the development of *A. tenuis*, *Fusarium* sp., *E. carotovora* and at least some species of *Colletotrichum*, *Pseudomonas* and *Phytophthora* (7.1; Figs 7.1, 7.2, 7.3 and 7.6; Table 7.4; El-Goorani and Sommer, 1979). At 12.8°C and atmospheric pressure, 0.25% [O_2] has been found to control the development of *G. candidum*, *A. tenuis*, *B. cinerea* and *C. coccodes* after these organisms were inoculated into Homestead tomatoes (Parsons and Spalding, 1971). While most horticultural commodities develop off-flavours (Table 7.2) and are damaged by less than 1% [O_2] at atmospheric pressure (Table 4.7), this does not usually occur in LP. Therefore, the 2.0–2.67 kPa (15–20 mm Hg) pressure range should be tested to determine whether tomatoes can tolerate a low enough [O_2] level to control decay in LP. Temperatures at least as high as 15–17°C should be included in these studies to avoid chilling injury (7.10).[5]

A postharvest chlorine rinse helps to control decay during LP storage. In a test with 'breaker' Homestead No. 2 tomatoes, the incidence of decay which developed during 8 weeks' storage at 13°C and a pressure of 10.67 kPa (80 mm Hg) was reduced from 56 to 29% by a chlorine wash prior to LP storage (Burg, 1976a). Chlorine-rinsed mature-green tomatoes were successfully shipped from the Dominican Republic to New York in a 12.2 m Grumman/Dormavac intermodal LP container operated at a pressure of 10.67 kPa (80 mm Hg), 12.8°C, and 95% RH (Jamieson, 1984).

During 4 weeks' storage at 14.4°C and a pressure of 10.67 kPa (80 mm Hg), 25% of var. Homestead No. 9 breaker tomatoes developed severe fusarium rot and 25% showed a trace of mould development. At that time, no fruits treated with air changes passed through hypochlorite solution had fusarium rot and when these tomatoes were transferred to atmospheric air at 14.4 or 26.7°C, they ripened normally without mould development (Burg, 1970; example 16, chapter 7).

A pressure of 5.06 kPa (38 mm Hg) enhanced ascorbate retention in tomato fruits treated with methylthiophanate (Borecka and Parynow, 1985a) and reduced the active benomyl and methylthiophanate residues during 14 days' storage (Borecka and Parynow, 1985b).

VEGETABLES

10.28 Asparagus (*Asparagus officinalis* L.)

Asparagus can be kept in NA for 3 weeks at 2°C. Chilling damage occurs within 10 days at 0°C, and storage life is markedly reduced above 5°C (Hardenburg *et al.*, 1986; Mercantila Publishers, 1989b). The major causes of spoilage during storage are growth, desiccation, decay and loss of tenderness, flavour and vitamin C. Storage life is improved by 7–10% [CO_2], but [O_2] must be kept above 10% and [CO_2] below 12–15% to avoid injury (Table 4.7). High [CO_2] reduces *Phytophthora* rot and other decays (Thompson, 1998).

Storing asparagus for 2 weeks at 1.5–4.5°C and pressures in the range between 22.67 and 24.0 kPa (170–180 mm Hg = 4.5–4.8% [O_2]) provided no advantage compared to NA (Chu *et al.*, 1976). Asparagus remained firm and green for 42 days when it was stored in 2% [O_2] static CA with slaked lime included to prevent CO_2 accumulation, or in LP at 3°C and a pressure of 7.33–8 kPa (55–60 mm Hg = 1.4–1.5% [O_2]) (McKeown and Lougheed, 1981). Spears stored in NA were senescent by that time. An initial 793 kPa pre-storage asparagus tenderometer reading increased by 83, 76 and 34 kPa during storage in air, CA and LP, respectively. This indicates that LP was unusually effective in preventing toughening of the spears.[6] As this disorder is induced by ethylene, LP may preserve tenderness by preventing ethylene production and removing the gas from within and around the tissue (Haard *et al.*, 1974; Chang and Han, 1975). Weight loss was 10.5, 6.2 and 6.0% in air, LP and CA, respectively. Undesirable and objectionable off-odours and flavours developed in CA, but not at a hypobaric pressure even though the [O_2] was lower in LP. McKeown and Lougheed (1981) postulated that any off-odours and off-flavours produced in LP must have volatized and escaped, but this is an unlikely explanation (see 2.25 and 2.26). The useful storage life seemed to be limited by chilling injury.

In a test at lower pressures, asparagus was stored at 0°C in NA or in LP at 2.67, 5.33 and 10.67 kPa (20, 40 and 80 mm Hg) (Dilley, 1977a,b). Because the humidity was very high, it was not necessary to place the butts on moist absorbent material or in pans of water to prevent loss of moisture and firmness. LP kept spears in marketable condition for at least 4–6 weeks. Ascorbic acid retention improved as the storage pressure was reduced (Table 4.8) and all LP storage pressures provided good retention of colour, tenderness and flavour. Bacterial soft rot affected some spears at pressures of 10.67 and 101.3 kPa (80 and 760 mm Hg), but not at 2.67 or 5.33 kPa (20 or 40 mm Hg). Chilling injury, which results within 3–4 weeks if spears are stored below 2.8°C in NA (Leshuk and Saltveit, 1990), did not occur in LP (Dilley, 1977a). Successful asparagus shipments were made from the Dominican Republic to New York City in a Grumman/Dormavac 12.2 m hypobaric intermodal container operated at 0°C and a pressure of 2.67kPa (20 mm Hg).

The major decays of asparagus are bacterial soft rot (*E. carotovora*), fusarium rot (*Fusarium* sp.) and phytophthora rot (*Phytophthora* sp.) (Ryall and Lipton, 1972; Smith *et al.*, 1982). Provided that CO_2 is absent, the growth of *E. carotovora* (Fig. 7.2) and *Fusarium* sp. (Fig. 7.3) is inhibited by the 0.4% [O_2] concentration present in air at 0–2°C and a pressure of 2.67 kPa (20 mm Hg). Development of *P. parasitica* is slightly inhibited by 2.3% [O_2] (El-Goorani and Sommer, 1979). Asparagus storage should be tested at 1.33 kPa (10 mm Hg = 0.14% [O_2]), as this pressure ought to provide the best decay control.

At atmospheric pressure, low-[O_2] injury occurs if the [O_2] is reduced to less than 10% (Table 4.7), and yet a pressure of 2.67 kPa (20 mm Hg) preserves asparagus for long periods of time without causing low-[O_2] injury, even though it only supplies 0.42% [O_2] at 0°C. This striking difference in susceptibility to low-$[O_2]$ damage at atmospheric vs. subatmospheric pressures may be due not only to enhanced diffusion, but also may result from stomatal opening in LP (4.15). Stomatal opening delays senescence in vegetative tissue (5.28).

10.29 Beans (*Phaseolus vulgaris* L.), Pole, Snap and Bush; cvs Improved Tendergreen, 'McCaslan 42', Sprite

Green beans can be stored in NA for 7–10 days at 4–7°C, but eventually this temperature range causes chilling injury. Snap beans may be chilled in a few days below 3°C. CA (2–3% $[O_2]$ + 5–10% $[CO_2]$) retards yellowing at 7.2°C, and a 24-h exposure to 20–25% $[CO_2]$ prevents discoloration of the cut ends of beans (Hardenburg *et al.*, 1986; Thompson, 1998).

Freshly harvested green pole beans were stored at 7.2°C either in NA wrapped loosely in plastic wrap or in LP at pressures of 8.0, 12.0 or 17.33 kPa (60, 90 or 130 mm Hg) (Burg, 1970). After 16 days, beans in NA had no snap, but their colour was still good, and within 26 days they had shrivelled. Beans stored in LP at 8.0 kPa (60 mm Hg) were still in excellent condition after 26 days; at 12.0 kPa (90 mm Hg) their condition was good; at 17.33 kPa (130 mm Hg) their condition was only fair. At all pressures mould development had commenced. In other LP studies, relatively high pressures ranging between 8.0 and 20 kPa (60–150 mm Hg) extended the storage life of green snap beans compared to NA, but decay always terminated the experiments within 2–3 weeks (Burg, 1976a). Pole beans, which had a storage life of 10–13 days in NA at 8°C, were successfully stored for 30 days in LP at the same temperature and a pressure of 8 kPa (60 mm Hg). Snap beans, which deteriorated in NA after 7 days at 5°C and 10 days at 8°C, remained in marketable condition for 26 days in LP at pressures of 8 kPa (60 mm Hg). The LP storage of both snap and pole beans was limited by mould development. During 2 weeks at 7°C, 'McCaslan 42' pole beans and 'Sprite' bush beans were better preserved at pressures of 10.13 and 20.3 kPa (76 and 152 mm Hg) than in NA (Spalding and Reeder, 1980, unpublished). During a 14-day storage trial at 10°C, weight loss was reduced and the appearance of cv. Improved Tendergreen snap beans improved both at an LP pressure of 7.33–8.0 kPa (55–60 mm Hg) and by static CA storage in 2% $[O_2]$ with slaked lime included to prevent CO_2 build-up (McKeown and Lougheed, 1981). The result with LP was somewhat better than with CA, even though the beans lost twice as much weight in LP. Both in CA and LP, snap beans were acceptable after 14 days, but not in NA.

A pressure of 1.33 kPa (10 mm Hg) is far more effective than 7.33–8.0 kPa (55–60 mm Hg) in preserving beans and preventing decay. 'Naked' and polyethylene-wrapped green snap beans were stored at 4.4 and 7.2°C in NA, and in LP at pressures of 1.33, 2.67, 4.0, 8.0 and 10.67 kPa (10, 20, 30, 60 and 80 mm Hg). At receipt 24 h after harvest, the temperature of the beans was 26.7–29.4°C. They were cooled in the storage chambers and stored for 38 days, after which they were rated for snap, colour, aroma, taste, weight loss, juiciness, tip and surface browning, mould and slime development, surface pitting and desiccation. De-greening, slime and decay development were prevented at 1.33 kPa (10 mm Hg) and to some extent at 2.67 kPa (20 mm Hg), but not at higher pressures. All samples held at 4.4°C, except beans held at 1.33 kPa (10 mm Hg), developed symptoms of chilling damage, evidenced by surface pitting and a glassy appearance. The best condition for preserving the beans and suppressing mould development was 7.2°C at a pressure of 1.33 kPa (10 mm Hg). These beans retained at-harvest flavour and aroma, colour, general appearance, snap and juiciness (Burg, 1978a, 1990).

The major decays of beans are watery soft rot (*Sclerotinia* sp.), cottony leak (*Pythium butleri*), grey mould (*B. cinerea*) and rhizopus rot (*Rhizopus* sp.) (Smith *et al.*, 1982). The growth of these moulds is strongly inhibited in the absence of CO_2 at the $[O_2]$ concentration present in 7.2°C water-saturated air at a pressure of 1.33 kPa

(10 mm Hg) (7.1, 7.4; Figs 7.1, 7.3 and 7.6; Tables 7.1 and 7.2).

Scanning electron microscopy (SEM) was used to study and compare the structural characteristics of black beans (*P. vulgaris* L.) after storage for 2 years either at 23–25°C and 30–50% RH or in LP at 4.5°C, 50–60% RH, and a pressure of 16.67 kPa (125 mm Hg) (Berrios *et al.*, 1998, 1999). In cross-sections of the seed coats, the parenchymal cell layers of AC beans (ambient conditions = AC) were disrupted and frequently exhibited large intercellular spaces between the cell layers, whereas the parenchymal cell layers of LP beans had little disruption of the cell layers. In cotyledonary cross-sections, the cells of LP beans had many large intercellular spaces, a condition characteristic of normal beans. AC beans had fewer and much smaller intercellular spaces typical of hard-to-cook beans. The LP beans exhibited quality factors characteristic of fresh beans, such as shorter cooking time, smaller solids loss, less electrolytes leached and a lower percentage of hard-shell. AC beans had quality factors characteristic of hard-to-cook beans. The germination rate was higher (93%) after LP storage than it was after AC storage (72%), and beans that had been kept in LP absorbed water faster.

10.30 Beet Leaf (*Beta vulgaris* L.), var. Conditiva Alef

Bunched beets can be stored in NA at 0°C for 10 days to 2 weeks (Hardenburg *et al.*, 1986). CA is ineffective with bunched beets, and atmospheres containing more than 5% [CO_2] are injurious (Thompson, 1998).

At 1°C, the chlorophyll content of beet leaves decreased less in LP at a pressure of 7.33–8.0 kPa (55–60 mm Hg) than in NA or in a 2% [O_2] static CA storage with slaked lime present to prevent CO_2 build-up (McKeown and Lougheed, 1981). During 28 days in NA, CA and LP, the content of chlorophylls (a + b) decreased by 0.259, 0.252 and 0.202 mg per gram fresh weight, and the total fresh weight loss was 7.7, 4.3 and 0.7%, respectively. Off-odours and -flavours developed in CA, but were not evident in LP even though the [O_2] level (1.5%) was slightly lower in LP. It was suggested that if off-flavours and -odours developed in LP, they may have volatized and escaped at the low pressure, but this is an unlikely explanation (3.25 and 3.26).

Cercospora leaf spot (*Cercospora beticola* Sacc.) and bacterial soft rot (*E. carotovora*) are the major decays of beet leaves (Ramsey *et al.*, 1967; Ryall and Lipton, 1972). The growth of *E. carotovora* is prevented by the [O_2] present in 0°C water-saturated air at a pressure of 1.33 kPa (10 mm Hg), and also when no [CO_2] is present (Fig. 7.2). Apparently the [O_2] and [CO_2] sensitivity of *C. beticola* has not been determined. Bunched beets should be tested at 1.33–2.67 kPa (10–20 mm Hg), as this pressure range is likely to cause a maximum depression of decay and delay in de-greening.

10.31 Broccoli (*Brassica oleracea* L. – Italica Group)

After 10–14 days in NA at 0°C, broccoli leaves discolour, buds may yellow and drop off, and tissues soften (Hardenburg *et al.*, 1986). A controlled atmosphere containing 5–10% [CO_2] + 1–3% [O_2] at 5–10°C can increase shelf life to 3–4 weeks by retarding mould growth and retaining green colour and tenderness (Hatton *et al.*, 1975; Hardenburg *et al.*, 1986; Thompson, 1998). Off-odours are induced by 15% [CO_2], and 0.1–0.25% [O_2] causes severe injury (Lipton, 1975) and results in off-odours and off-flavours when the broccoli is cooked (Hardenburg *et al.*, 1986). Modified-atmosphere packaging is commonly used for spears prepared for sale in supermarkets (Thompson, 1998). During broccoli senescence, ACC oxidase activity and the rate of ethylene production increase to a climacteric maximum that is hastened by applied ethylene and suppressed by NBD (Kasai *et al.*, 2000). Chlorophyll loss is correlated with the increase in ethylene production (Wattanabe *et al.*, 2001). Yellowing of florets is suppressed by NBD and 1-MCP (Fan and Mattheis, 2000) and accelerated by ethylene, indicating that endogenous ethylene plays a major role in broccoli senescence.

Freshly harvested broccoli was kept in a walk-in refrigerator at 1.1°C for 4 days to simulate the interval between harvest and the time when the commodity might be placed into LP storage in a commercial situation. The broccoli then was stored at 0°C in NA or in LP at pressures of 1.33, 2.67 or 5.33 kPa (10, 20 or 40 mm Hg). After 21 days it was evaluated for appearance and eating quality, and samples were tested for shelf life at 10°C. At the end of the shelf-life test, 60–90% of the buds had turned yellow in broccoli stored in NA or in LP at 2.67–5.33 kPa (20 or 40 mm Hg). Broccoli that had been kept at 1.33 kPa (10 mm Hg = 0.15% [O_2]) had the best appearance, with less than 40% yellowing of the buds. No off-odours or off-flavours were noted in any samples (Shum, 1979; Jamieson, 1984).

The major postharvest decays of broccoli are alternaria leaf spot (*Alternaria brassicae*), bacterial leaf spot (*Pseudomonas maculicola*), bacterial soft rot (*E. carotovora*) and grey mould rot (*B. cinerea*) (Ryall and Lipton, 1972). At a pressure of 1.33 kPa (10 mm Hg), in the absence of CO_2 the low [O_2] present in 0°C water-saturated air markedly inhibits the growth of *E. carotovora*, *B. cinerea* and some species of *Alternaria* and *Pseudomonas* (Figs 7.1, 7.2, 7.3 and 7.6; Tables 7.1 and 7.2).

10.32 Brussels Sprouts (*Brassica oleracea* L. – Gemmifera Group)

Brussels sprouts can be kept in NA for 3–5 weeks at 0°C (Hardenburg *et al.*, 1986). Deterioration is much more rapid at higher temperatures. A controlled atmosphere containing 2.5–5% [O_2] + 5–7.5% [CO_2] improves quality during storage at 5 or 10°C, but not at 0°C (Hardenburg *et al.*, 1986; Thompson, 1998). High [CO_2] reduces decay.

According to Ward (1975), storing Brussels sprouts at 10 kPa (75 mm Hg ≅ 1.8% [O_2]) provides no benefit, which is surprising since a comparable low [O_2] concentration delays yellowing of this commodity at atmospheric pressure (Thompson, 1998).[11]

The major decays of Brussels sprouts are grey mould rot (*B. cinerea*), alternaria leaf spot (*A. brassicae*), bacterial soft rot (*E. carotovora*) and rhizopus rot (*R. stolonifer*) (Ryall and Lipton, 1972). In the absence of CO_2, the low [O_2] present in water-saturated air at a storage pressure of 1.33 kPa (10 mm Hg) prevents the growth of these organisms (Figs 7.1, 7.2, 7.3 and 7.6; Tables 7.1 and 7.2). A pressure of 1.33 kPa (10 mm Hg) should be tested with Brussels sprouts.

10.33 Cabbage (*Brassica oleracea* L. – Capitata Group), cv. Quick Green Storage

Cabbage can be held in NA for many months at 0°C. Storage life is significantly improved by a controlled atmosphere containing 2.5–5% [O_2] + 2.5–5% [CO_2] (Hardenburg *et al.*, 1986; Thompson, 1998).

White cabbage was successfully stored at 0°C and a pressure of 8.0 kPa (60 mm Hg) in mixed loads containing apple fruits (McKeown *et al.*, 1978). LP minimized the effect that ethylene, produced by the apples, had on the senescence and abscission of cabbage leaves. Cabbage held with apples for 69 days at 1 or 5°C and an LP pressure of 7.33–8.0 kPa (55–60 mm Hg), remained as if fresh-harvested. Ten panellists gave the LP cabbage a sensory evaluation of 6.1 after storage, compared to 3.5 for NA and 3.3 for CA, where 10 is a maximum rating and 0 a minimum rating (McKeown and Lougheed, 1981).

Summer cabbage (cv. Quick Green Storage) stored in LP for 42 days at 1°C and a pressure of 7.33–8.0 kPa (55–60 mm Hg = 1.4–1.5% [O_2]) was visually greener than the commodity held in flowing air at the same temperature (McKeown and Lougheed, 1981) and had a better appearance than cabbage kept in a static 2% [O_2] CA with slaked lime included to prevent

CO_2 build-up. Weight loss during storage and from trimming was 16.4, 11.7 and 2.4% in NA, CA and LP, respectively. Objectionable off-odours and -flavours developed in CA, but not in LP even though the [O_2] was lower (1.5%) than in CA. It was suggested that off-flavours and off-odours did not develop in LP because they must have volatized and escaped, but this is an unlikely explanation (3.25 and 3.26). It has been claimed that even better results with cabbage, including less weight loss and improved appearance, can be obtained using a variable low-pressure system (VLPS), cycling the pressure between 13.3 and 40 kPa (100 and 300 mm Hg) without humidification (Onoda *et al.*, 1989a).

The major decays of cabbage are watery soft rot (*S. sclerotiorum*), bacterial soft rot (*E. carotovora*), grey mould rot (*B. cinerea*), alternaria leaf spot (*A. brassicae*), bacterial zonate spot (*Pseudomonas cichorii*) and black rot (*Xanthomonas campestris*) (Ryall and Lipton, 1972). In the absence of CO_2, the reduced concentration of [O_2] present in 0°C water-saturated air at a pressure of 1.33–2 kPa (10–15 mm Hg) strongly inhibits the growth of *E. carotovora*, *S. sclerotiorum*, *B. cinerea*, and some species of *Alternaria* and *Pseudomonas* (7.1; Figs 7.1, 7.2, 7.3 and 7.6; Tables 7.1 and 7.2). No data are available for *X. campestris*. This low-pressure range should be tested with cabbage.

10.34 Carrot (*Daucas carota* L.)

Carrots in mixed loads containing Red Delicious apples and cabbage were stored at 1 or 5°C either in air, in static 2% [O_2] with slaked lime present to prevent CO_2 build-up, or in LP at a pressure of 7.33–8.0 kPa (55–60 mm Hg) (McKeown *et al.*, 1978; McKeown and Lougheed, 1981). The subatmospheric pressure prevented bitter flavour formation in response to ethylene produced by the apples. A sensory evaluation by 10 panellists rated the LP carrots decidedly superior, compared to those kept in CA. The sensory ratings were 2.1 in CA and 6.7 in LP, where 10 is a maximum and 0 a minimum rating. Weight loss during 69 days' storage was 11.2 and 0.3% in CA and LP, respectively.

10.35 Cauliflower (*Brassica oleraceae* L. – Botrytis Group)

Cauliflower is usually harvested with the surrounding leaves serving as protection against damage to the head. For consumer acceptance, the attached leaves should remain green and fresh with no signs of wilting, and the head must remain white or light cream in colour (Mercantila Publishers, 1989b). After 5 weeks at 3°C, leaves were still green and turgescent when cauliflower was stored in LP at a pressure of 10 kPa (75 mm Hg), whereas by that time in NA the leaves had senesced, turned yellow and either had abscised or had to be trimmed (Bangerth, 1973). In another study (Ward, 1975), LP did not improve the storage of cauliflower at a pressure of 10 kPa (75 mm Hg).[11]

Freshly harvested cauliflower was kept in a walk-in refrigerator at 1.1°C for 4 days to simulate the possible delay from harvest to the time the commodity might be placed into LP storage in a commercial situation. Then the cauliflower was kept at 0°C either in NA or in LP at pressures of 1.33, 2.67 or 5.33 kPa (10, 20 or 40 mm Hg). After 21 days it was evaluated for appearance and eating quality, and samples were tested for shelf life at 10°C. By the end of the shelf-life test, the leaves of cauliflower stored in NA were yellow and so dry that they abscised with minimal handling. Although the leaves of LP-stored cauliflower were also quite yellow by this time, they remained attached to the stem. Cauliflower stored in LP had a superior appearance, especially if it had been kept at 1.33 or 2.67 kPa (10 or 20 mm Hg). No off-flavours or -odours were noted in any LP samples (Shum, 1979; Jamieson, 1984).

The major rots of cauliflower are alternaria leaf spot and brown rot (*A. brassicae* and *A. oleracea*), bacterial leaf spot (*P. maculicola*), bacterial soft rot (*E.

carotovora), bacterial zonate spot (*Pseudomonas chichorii*), black rot (*Xanthomonas campestris*), downy mildew (*Peronospora parasitica*), grey mould rot (*B. cinerea*) and *Rhizopus* soft rot (*R. stolonifer*) (Ryall and Lipton, 1972). In the absence of CO_2, the low $[O_2]$ concentration present in 0°C water-saturated air at a pressure of 1.33 kPa (10 mm Hg) prevents the growth of *E. carotovora*, *B. cinerea*, *R. stolonifer* and species of *Alternaria* and *Pseudomonas* (7.1; Figs 7.1, 7.2, 7.3 and 7.6; Tables 7.1 and 7.2). No data are available for *X. campestris* and *P. parasitica*.

10.36 Celery, Transplants (*Apium graveolens* L.), cv. Utah 52–70

A 2-day exposure of celery transplants to temperatures below 10°C may cause bolting in the field after they are planted. To test the effect on bolting of prior storage in LP, celery transplants of cv. 'Utah 52–70' were kept for 2, 4 or 6 weeks at 0°C either in NA or at a pressure of 3.33 kPa (25 mm Hg). Plants stored for 4 weeks in NA developed longer internal seed stalks than those stored in LP. There were no differences in yield between the plants that had been stored for 2 weeks, but after 6 weeks' storage the yield was 4 Mt/ha after NA storage and 25 Mt/ha after LP storage. The increase in yield was due largely to the decrease in number of bolters caused by the LP treatment (Jardine *et al.*, 1984).

10.37 Corn, Sweet (*Zea mays* L.), cvs Iobelle, Florida Sweet, Wintergreen

The quality of green sweetcorn is determined primarily by sugar content. Soon after harvest a rapid conversion of sugar to starch commences[7] even when sweetcorn is hydro-cooled, top-iced and refrigerated. Subsequently, sugar and flavour loss continue, limiting storage to 4–8 days in NA. In CA, sweetcorn is damaged by less than 2% $[O_2]$ or more than 20% $[CO_2]$ (Hardenburg *et al.*, 1986).

LP storage of cv. Wintergreen sweetcorn at 1.7 ± 1°C was studied at pressures ranging from 1.33 to 101.3 kPa (10–760 mm Hg) (Burg, 1976a; Fig. 10.3). To evaluate flavour, after storage the sweetcorn either was cooked for 15 min in boiling water and immediately sampled by a three-member tasting panel, or it was blanched for 17 min in boiling water, cooled in ice water for 20 min, and then shelf-frozen in a polyethylene bag. After 30 days' storage in a freezer, the frozen processed sweetcorn was thawed, boiled for 15 min, and tasted for sweetness and flavour. Samples also were sent to the Birdseye Division of General Foods for evaluation by their trained tasting panel. During a 7-day storage, the sugar content decreased precipitously in NA, and in LP was inversely related to the storage pressure between 5.33 and 2.0 kPa (40 and 15 mm Hg). The shifts in sugar content were accounted for by loss or gain of starch, and there was no significant change in total carbohydrate. The respiration rate decreased progressively as the pressure was lowered to 1.33 kPa (10 mm Hg), without any indication of an 'inversion point' below which CO_2 production increased (Fig. 4.2). After 7 days in LP, sweetcorn stored at 2.67 kPa (20 mm Hg) consistently scored highest for sweetness and flavour. Quality decreased progressively at higher pressures. Sweetcorn kept in NA was judged to be bland and scored lowest for flavour, while that kept at 2.7 kPa (20 mm Hg) in LP was rated significantly better than field-harvest quality by a trained tasting panel at the Birdseye Division of General Foods, primarily because it was sweeter. A low percentage of panellists detected a 'hay-like' flavour in the hypobaric sweetcorn, but still rated it better than field-harvest quality.[8] By the 11th day of storage, the taste pattern had shifted. Sweetcorn that had been kept at 3.33 kPa (25 mm Hg) had deteriorated in flavour, and the best result occurred at 6.67 kPa (50 mm Hg). A tasting panel detected an off-flavour in all sweetcorn stored for 20 days at less than 8.67 kPa (35 mm Hg), which became more intense when the pressure was lowered further. Maximum sweetness with no off-flavour occurred at 6.67 kPa (50 mm Hg). Higher

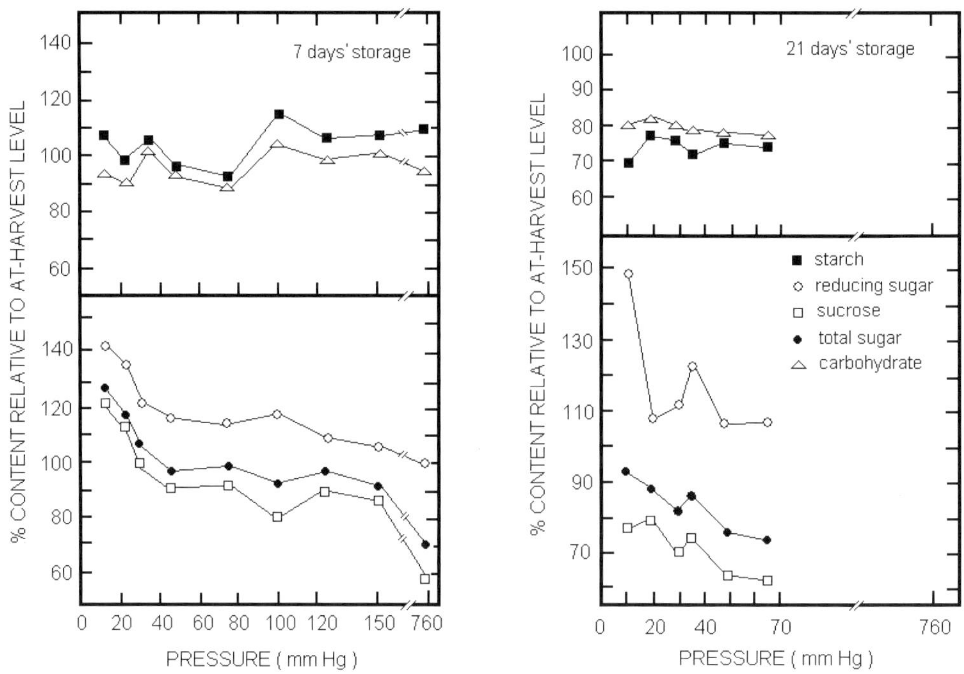

Fig. 10.3. Effect of pressure on changes in the content of various carbohydrates in sweetcorn stored at 1.7°C. (*left*) 7 days' storage, average of three experiments; (*right*) 21 days' storage, one experiment. The sweetcorn used in the 7-day trials initially contained 3 ± 0.5% sucrose, 1.2 ± 0.2% reducing sugar, 8.0 ± 1.8% starch, 12.2 ± 2.2% total carbohydrate and 4.2 ± 0.6% total sugar. In the 21-day trial, the sweetcorn initially contained 3.5% sucrose, 1.3% reducing sugar, 9.4% starch, 19.0% total carbohydrate and 4.8% total sugar (Burg, 1976a).

storage pressures resulted in a greater loss of sweetness.

Sweetcorn cvs 'Iobelle' and 'Florida Sweet' were stored for 3 weeks at 1.7°C either in NA, in CA at 2 or 21% [O_2] + 0 or 25% [CO_2], or in LP at pressures of 6.67 kPa (50 mm Hg) and 10.1 kPa (76 mm Hg) (Spalding *et al.*, 1978). Sucrose remained highest in 2% [O_2] + 0% [CO_2], and in LP at a pressure of 10.1 kPa (76 mm Hg). After 3 weeks the sweetcorn's initial ethanol content of 25 mg/100 g tissue had increased to 151 mg/100 g tissue in NA, 364 mg/100 g tissue in 2% [O_2] + 0% [CO_2], 651 mg/100 g tissue in 2% [O_2] + 25% [CO_2] and 293 mg/100 g tissue in LP at 6.67 kPa (50 mm Hg = 1.25% [O_2]). The higher the [CO_2] concentration, the larger the increase in ethanol. The ethanol content increased less in LP than in CA, even though the [O_2] was significantly lower in LP. Both diffusive loss of ethanol vapour from the sweetcorn at a low pressure (2.25), and the unusually low [CO_2] content around and within the tissue may account for this difference.

Sweetorn stored at 10.1 kPa (76 mm Hg) was not tested for flavour, but in all other storage conditions flavour was unacceptable or barely acceptable after 3 weeks. Sweetcorn stored at 21% [O_2] + 25% [CO_2] had the poorest flavour rating, and that stored at 2% [O_2] + 25% [CO_2] was rated the best, even though it had the highest ethanol content. Sweetcorn stored in LP or at atmospheric pressure in 21 or 2% [O_2] showed slight browning on the cut end of the shank. Browning did not occur in treatments with elevated [CO_2], but both 15 and 25% [CO_2] bleached some of the kernels.

A partial load of sweetcorn was stored successfully in a 6.1 m (20 ft) prototype Fruehauf LP intermodal container operated for 3 weeks at 1°C, 95% RH and a pressure of 10 kPa (75 mm Hg). A Michigan State University tasting panel rated the flavour of the sweetcorn excellent after storage. The sweetcorn was successfully processed and frozen.

10.38 Cress (*Lepidium sativum* R. Br.)

The ascorbic acid content of cress was better maintained in LP at a pressure of 10 kPa (75 mm Hg) than it was in NA during a 12-day storage at 3°C. In LP, there was no appreciable loss of protein and chlorophyll (Bangerth, 1973, 1974). Lower pressures were not tested.

10.39 Cucumber (*Cucumis sativus* L.), var. Poinsett

Cucumbers stored in NA at 10–13°C begin to yellow in about 10 days and can only be kept for 10–14 days (Hardenburg *et al.*, 1986; Thompson, 1998). They are subject to chilling injury if held for longer than 2 days at 10°C. Modified atmospheres containing 5% [O_2] delay yellowing, and at 12.5°C a 5% [O_2] +5 % [CO_2] mixture reduces the virulence of mucor rot (*Mucor mucedo*) and grey mould (*B. cinerea*). Ethylene promotes rapid yellowing of cucumbers (Van Uffelen, 1974).

Non-waxed cucumbers (var. Poinsett), which had a storage life of 10–14 days in NA at 10°C, were preserved for up to 49 days in LP at 9–12°C and a storage pressure of 10.0–10.67 kPa (75–80 mm Hg) (Burg, 1971, 1976a; Bangerth, 1974; Jamieson, 1984). Five days after the fruits stored in LP were transferred to 25°C atmospheric air they were still green and crisp, with a dry weight content of 3.5%. In NA, the fruits ripened progressively during 3 weeks' storage at 10–12.2°C, and by the end of 5 additional days at 25°C their dry weight content had decreased to 2.8% and they were completely yellow, with no crispness left. LP storage was limited by mould development. A chlorine wash prior to storage reduced the incidence of decay. LP decreased cucumber respiration by 67–75% (Bangerth, 1973). Excellent LP results with cucumbers at 10.13 kPa (76 mm Hg) also were reported by Staby (1976b).

Cucumbers stored at a chilling temperature, 7.2°C, in NA or in LP at pressures of 13.33, 16.0 or 21.3 kPa (100, 120 or 160 mm Hg), developed chilling injury within 21 days. This caused them to soften and induced 30% of the fruits to develop decay. No benefit was noted from LP (Burg, 1970). At a lower pressure, 10.67 kPa (80 mm Hg), cucumbers stored for 7 weeks had a reduced incidence of decay and excellent retention of green colour, crispness, firmness and overall appearance, but within 1–2 days after they were transferred to atmospheric conditions at a higher temperature, they developed chilling lesions (Burg, 1971).

The major rots of cucumbers are anthracnose (*Colletotrichum lagenarium*), bacterial soft rot (*E. carotovora*), bacterial spot (*Pseudomonas lachrymans*), black rot (*Mycosphaerella citrullina*), cottony leak (*Pythium aphinidermatum*) and scab (*Cladosporium cucumerinum*) (Ryall and Lipton, 1972). The growth of *E. carotovora*, and species of *Pseudomonas*, *Colletotrichum*, *Pythium*, *Cladosporium* and *Mycosphaerella* (7.4; Figs 7.2, 7.3 and 7.7; Table 7.3; Follstad, 1966) is inhibited by the low [O_2] present in water-saturated 10°C air at a pressure of 2–2.67 kPa (15–20 mm Hg = 0.14–0.28% [O_2]).

Cucumbers tolerate relatively low O_2 concentrations at atmospheric pressure and can be expected to benefit from even lower [O_2] in LP without low-[O_2] injury. At 4°C, cucumbers stored in 0.5% [O_2] at atmospheric pressure were free from decay for 2 weeks, whereas at higher [O_2] up to 32% were unacceptable (Leshuk and Saltveit, 1990). Off-flavours and other symptoms of low-O_2 injury only became evident in 0.5% [O_2] after 3 weeks. This result suggests that the optimal LP storage pressure for cucumbers may be in the 2–2.67 kPa range (15–20 mm Hg = 0.16–0.3% [O_2]) at 10°C.

'Pickling' cucumbers that were consumed by extensive mould development within 2 days at 10 °C flowing 98% RH atmospheric air were completely free of mould when they were stored at the same temperature and RH at a pressure of 2.67 kPa (20 mm Hg = 0.3% [O_2]). Their stomates, viewed with acrylic templates, were found to be open in LP and closed at atmospheric pressure. In LP, the fruits remained in excellent condition with no indication of fermentation or off-odour (T.L. Davenport and S.P. Burg, 2002, unpublished). Temperatures as high as 16°C should be investigated at 2–2.67 kPa (15–20 mm Hg) to determine if this is the best condition for long-term cucumber storage at a non-chilling temperature.

Cucumbers stored in LP at 7.2°C and a pressure of 10.67 kPa (80 mm Hg) were protected from decay development during 43 days by flowing air that had been passed through a 1.1 or 2.5% sodium hypochlorite solution. After 17 days' storage, 50% of the control cucumbers in NA that had not been treated with HOCl had soft spots and decay lesions (Burg, 1970; chapter 7, example 21).

10.40 Kohlrabi (*Brassica caulorapa*)

In NA, the leaves of kohlrabi became senescent and 60% of the petiole stumps abscised within 5 weeks at 2.5–4°C. The leaves were better retained during LP storage at 95% RH and a pressure of 10 kPa (75 mm Hg). Lower pressures have not been tested (Bangerth, 1973). CA had slight or no effect on kohlrabi (Thompson, 1998).

10.41 Leek (*Allium ampeloprasum* L.)

Leeks can be stored in NA at 0°C for 2–3 months, but yellowing and decay develop rapidly at higher temperatures. A CA atmosphere containing 1–3% [O_2] + 5–10% [CO_2] prolongs storage to 4–5 months with some loss in quality (Hardenburg *et al.*, 1986; Thompson, 1998). Ward (1975) found no improvement in leeks stored in LP at 10 kPa (75 mm Hg = 1.9% O_2), which is surprising since that same [O_2] concentration is effective in CA.[11]

10.42 Lettuce (*Latuca sativa* L.), var. Iceberg

Head lettuce can be kept in good condition in NA for 2–3 weeks at 0°C. The storage life at 3°C is only about half as long (Hardenburg *et al.*, 1986). Lettuce is afflicted with numerous disorders at atmospheric pressure, some of which are difficult to distinguish from each other, or from diseases (Ryall and Lipton, 1972). No matter what the cause, the affected cells die in groups, turn some shade of red or brown and collapse. Distortion necrosis (DN), internal rib necrosis (IRN), pink rib (PR), rib discoloration (RD) and russet spotting (RS) start and develop most intensely along some portion of the midrib. Marginal browning (MB) is confined to wrapper leaves, rusty-brown discoloration (RBD) occurs on midribs of leaves or entire leaves and tends to follow veins but is not confined to them, and tipburn (TB), characterized by light tan to dark brown leaf margins, is initiated before harvest by the rupture of latex ducts and the resultant spreading of latex among other cells.

Atmosphere modification may cause or influence the development of the various disorders. The incidence of PR increases when [O_2] is lowered below 2% at temperatures higher than 2.2°C, and is not influenced by ethylene (Lipton, 1971). The intensity of RS is diminished by 1–8% [O_2] and increased by the accumulation of as little as 0.5 μl/l ethylene at temperatures higher than 2°C, but at 0°C even 20 μl/l of ethylene has little or no effect. A 1-week exposure to 1–2% [CO_2] at 3.3°C can cause noticeable brown stain (Lipton, 1975), and longer exposures, higher [CO_2] concentrations, less than 2–3% [O_2] and elevated temperatures, increase the intensity of this characteristic symptom of [CO_2] injury. Severe low-[O_2] injury results in shiny to water-soaked, grey, dead patches on wrapper and cap leaves, in reddish-brown spots on the inner (adaxial) surface of midribs of young head leaves and in reddish-brown heart leaves. Usually low-[O_2] injury on the midribs only affects the inner surface. A flat, sweet flavour, completely unlike that of normal lettuce, accompanies visible low-[O_2] injury, but disappears within 3–4 days after lettuce is transferred to air. The first and sometimes only sign of low-[O_2] injury is discoloration of heartleaves, which can occur at 10°C in less than 0.4% [O_2]. Other types of low-[O_2] injury only develop in the presence of less than 0.25% [O_2]. The severity of low-[O_2] injury is greater at higher temperatures, and the symptoms increase in intensity when the lettuce is transferred to air.

CA has limited benefit with lettuce. A 2–3% [O_2] + 1.5% [CO_2] mixture causes a reduction in decay and PR (Singh *et al.*, 1972), and butt discoloration and RS are reduced, but these are transient effects, which disappear during subsequent shelf life in air. In some studies, low [O_2] either has no effect or slightly aggravates PR (Lipton, 1971), and during short storage periods [CO_2] does not affect PR development (Stewart and Uota, 1971). The optimal [O_2] concentration for CA storage is approximately 2.5% (Singh *et al.*, 1972), which is equivalent to the [O_2] in 2°C water-saturated air at a pressure of 12 kPa (90 mm Hg).

Wrapped and naked lettuce was stored at 2°C either in NA or in LP at pressures of 6.67, 10.67, 13.3, 16.0, 20.0, 24.0 or 26.7 kPa (50, 80, 100, 120, 150, 180 or 200 mm Hg). At 6.67 kPa (50 mm Hg), PR developed within 21 days, while at 10.67 kPa (80 mm Hg) there was little or no evidence of PR. In the 13.3–24 kPa (100–180 mm Hg) pressure range, the major benefits of LP storage were reduced butt browning, improved crispness, and a reduced incidence of brown stain and RS but the diminished butt browning was a cosmetic effect, which rapidly disappeared when the lettuce was returned to atmospheric air (Burg, 1969, 1970, 1976a). Haard and Salunkhe (1975) reported that the storage of lettuce was increased from 14 days in NA, to 40–50 days in LP, but Ward (1975) found no improvement in lettuce kept at 9.33 kPa (70 mm Hg),[11] and LP had no beneficial effect at 3°C and a pressure of 10 kPa (75 mm Hg) (Bangerth, 1973).

Full loads of lettuce were tested in Fruehauf LP intermodal containers operated at 0–1°C and a pressure of 13.3–24 kPa (150–180 mm Hg). Storage was not improved significantly during 4 weeks, and at pressures of 10.67 and 4.0 kPa (80 and 30 mm Hg) the lettuce developed PR and was rated inferior to that stored in NA (Burg, 1970).

To study the effect of lower pressures, lettuce was stored for 33, 47 and 63 days at 0.5°C and 90–95% RH, either in NA or in LP at pressures of 1.33, 2.67, 5.33 or 10.67 kPa (10, 20, 40 or 80 mm Hg) (Jamieson, 1980a,e). NA controls were still marketable after 37 days' storage, but had a significant incidence of PR, decay and black heart, whereas all LP samples were in good condition at that time. After 47–63 days, lettuce held at 1.33 kPa (10 mm Hg) still was in excellent condition. There was a slight decrease in quality at 2.67 kPa (20 mm Hg), but much less so than that which occurred at 5.33 and 10.67 kPa (40 and 80 mm Hg). No control lettuce was marketable after 47 days. Heads that had been stored at 1.33 and 2.67 kPa (10 and 20 mm Hg) were still in excellent condition after 6 days' shelf life at 4.4°C, whereas lettuce that had been stored at 5.33 and 10.67 kPa (40–80 mm Hg) needed trimming.

When very low pressures were tested in 12.2 m intermodal Grumman/Dormavac LP containers operated at 1.7°C and 1.33 or 2.67 kPa (10 or 20 mm Hg = 0.14 or 0.41% [O_2]), within 14–21 days a tan discoloration developed on the inner leaf margins and ribs regardless of whether the heads were packed naked or wrapped (Grumman Allied Industry, 1979). This disorder resembled injury in CA caused by less than 1% [O_2] (Ryall and Lipton, 1972; Singh *et al.*, 1972), but differed in that heartleaves did not turn brown and only leaf margins were affected. Butt discoloration was suppressed, and the lettuce remained free of decay, TB, RS, RD and brown stain for up to 56 days. In the interval between 28 and 56 days, the lettuce developed a low incidence of PR, just as it sometimes does in low [O_2] CA storage (Lipton, 1971).

The cause of the marginal and rib discoloration that developed in the Grumman/Dormavac test at 1.33–2.67 kPa (10–20 mm Hg) was investigated in a series of laboratory studies (Grumman Allied

Industry, 1979; chapter 3, examples 14 and 15). Inexplicably, the disorder was eliminated during 35 days by lowering the pressure to 0.67 kPa (5 mm Hg ≅ 0% [O_2]). The injury was progressively accentuated when the pressure was elevated from 1.33 to 2.67 kPa (10–40 mm Hg = 0.14–0.95% [O_2]), and then decreased in frequency as the pressure was increased from 10.67 to 21.33 kPa (80–160 mm Hg = 2.1–4.3% [O_2]). The best storage condition during 21 days was 2°C at a pressure of 0.67 kPa (5 mm Hg ≅ 0% [O_2]), but in 35 days this gave rise to a high incidence of RS.

Bacterial soft rot caused by *E. carotovora* is the most serious disease of lettuce (Hardenburg *et al.*, 1986). *P. marginalis* causes marginal leaf blight (Wells, 1974), and infections also arise due to *Bremia lactucae* (downy mildew), *B. cinerea* (grey mould rot), *S. sclerotiorum* (watery soft rot), *Stemphylium botryosum* and *Aspergillus tenuis* (Eckert *et al.*, 1975). In the absence of [CO_2], the low [O_2] present in water-saturated air at 0–2°C and a pressure of 0.67–1.33 kPa (5–10 mm Hg) prevents the growth of *E. carotovora*, *Pseudomonas* sp., *B. cinerea*, *Sclerotinia* sp. and *A. tenuis* (7.1; Figs 7.1, 7.2, 7.3 and 7.6; Tables 7.1 and 7.2). No data are available for *S. botryosum* and *B. lactucae*.

In a laboratory experiment, iceberg lettuce stored in LP for 5 weeks at 3.3°C and a pressure of 13.33 kPa (100 mm Hg) did not develop bacterial soft rot when the air changes were passed through chlorinated water, while during that same interval 33% of the heads stored in NA developed this disorder (Burg, 1970; chapter 7, example 18).

10.43 Mushroom [*Agaricus bisporrus* (Lge.)], cvs Golden Light, White Button, Shitake, Enokitake, Matsutake

Freshly picked, rapidly cooled mushrooms can be kept in NA for 5 days at 0°C, 2 days at 4.5°C and only 1 day at 10°C. Loss of quality is caused by brown discoloration, elongation of the stalks and opening of the veils (Hardenburg *et al.*, 1986). The storage of mushrooms is not improved by lowering the [O_2] at atmospheric pressure (Smith, 1967), and for that reason 10–21% [O_2] is recommended (Leshuk and Saltveit, 1990; Thompson, 1998). Whiteness is maintained by storage in 5% [CO_2] (Hardenburg *et al.*, 1986), and 10–15% [CO_2] retards cap and stalk development and inhibits mould growth. Sometimes more than 10% CO_2 causes a pinkish discoloration, and while a combination of low [O_2] and high [CO_2] inhibits cap opening and internal browning, it also causes yellowing of the cap's surface (Thompson, 1998).

Mushrooms were stored at 2.8°C in baskets wrapped with cellophane, either in NA, or in LP at pressures of 5.33, 8.0, 13.33 or 21.33 kPa (40, 60, 100 or 160 mm Hg). After 4 days they were slightly whiter at 5.33 kPa (40 mm Hg) than in NA, while higher pressures gave intermediate results (Burg, 1970). In a more comprehensive laboratory study (Dilley, 1976, 1977a,b), cvs Golden Light and White Button mushrooms were packed in 1 lb commercial paper-pulp trays with a completely sealed resinite shrink-film overwrap. Two 3-mm-diameter holes in the side of each package facilitated vacuum cooling and precluded the possibility that anaerobic conditions might arise and allow botulism to develop during shelf-life testing at a higher temperature (Sugiyama and Yang, 1975). The mushrooms were vacuum cooled to 1.7°C and received for testing 36 h later, but their temperature had risen to 13°C in transit. Storage was studied at 0 or 5°C in NA and in LP at pressures of 1.33, 2.0, 2.67, 5.33, 10.67, 21.33 or 48.0 kPa (10, 15, 20, 40, 80, 160 or 360 mm Hg). The mushrooms were recooled in the NA or LP chambers and evaluated at 1-week storage intervals for weight loss, texture, odour, external browning, cap opening, microbial infection and internal whiteness measured by the reflectance method at 525 nm. A subjective rating of marketability based on overall appearance was made according to a scale of 1–9, where 1 was unmarketable and 9 excellent, and the initial rating was 9. Shelf life was evaluated during 3–4 days after the mushrooms were removed from storage and transferred to 10°C air at atmospheric pressure. Both in NA and in LP at 1.33 kPa (10 mm Hg), the total weight loss during 21 days' storage was approximately 3.8%. Cap opening did not progress irrespective of the pressure, but was delayed during the shelf-life evaluation of mushrooms that had been stored at 1.33 kPa (10 mm Hg). Microbial infection contributed to browning and pitting of the pileus during storage at atmospheric pressure, and an off-odour developed that intensified during shelf-life evaluation. These mushrooms were unfit to eat after 14 days' storage, whereas those kept at 1.33 kPa (10 mm Hg) were free of microbial infection and off-odours for 21 days and were edible at that time. Internal browning developed at atmospheric pressure, but whiteness was retained at 1.33–2 kPa (10–15 mm Hg) even during shelf-life evaluation (Table 10.4). After 21 days' storage, the subjective marketability ratings were 5.7 in NA and 8.0 at 1.33 kPa (10 mm Hg). During 3 days' shelf life, the ratings declined to 1.3 and 6.0, respectively. The decline during shelf-life evaluation at 1.33 kPa (10 mm Hg) was similar to that which fresh mushrooms experience at atmospheric pressure in the same period of time at 10°C. Similar results were obtained with Michigan-grown mushrooms stored in tubs. In all of these studies, spoilage occurred as rapidly or more rapidly at pressures higher than 2.67 kPa (20 mm Hg) than it did at atmospheric pressure, and effective storage at 1.33–2.0 kPa (10–15 mm Hg) required a 0°C temperature. The prerequisites of precise, very low [O_2] and exact temperature are readily satisfied in LP intermodal containers, making this method uniquely suited for long-term mushroom storage or transportation. Mushroom storage has also been studied in Germany at 5–20°C and pressures ranging from 1.33 to 48.0 kPa (10–360 mm Hg) (Poulssen *et al.*, 1982), and favourable results with Shitake, Enokitake and Matsutake cultivars stored in LP have been reported from Japan.

Table 10.4. Effect of hypobaric storage on internal browning of mushrooms (*Agaricus bisporus*) at 0°C. Stipe reflectance in cross-section at 520 nM was 60% at the beginning of storage (Dilley, 1977a).

Pressure (mm Hg)	pO_2 (atm)	% of initial reflectance after indicated days in storage		
		6	13	18
760	0.2090	63	46	34
25	0.0056	82	54	37
15	0.0029	90	81	74

There is a theoretical risk of botulism developing in mushrooms at a storage pressure of 1.33 kPa (10 mm Hg) and in MA-packed fresh vegetables (Church and Parsons, 1995) at temperatures in excess of 10°C (11.12). Between 1971 and 1975, botulism toxins or organisms were identified in marinated mushrooms and in 41 cans of mushrooms from 20 lots packed by seven domestic and two foreign producers, including canned mushrooms from eight packers in 1973 (Sugiyama and Yang, 1975). The responsible organisms are always type A or B, which are able to multiply and produce toxin under the conditions of pH, temperature and [O_2] that prevail in canned mushrooms (11.12). These reports indicate that freshly harvested mushrooms occasionally are contaminated by the botulism organism under natural conditions, and therefore it has been suggested that less than 6% [O_2] in mushroom packages poses a possible risk of botulism (Roy *et al.*, 1995). Although botulism toxigenesis may not arise in MA-packaged vegetables without some sensory indication (Church and Parsons, 1995), nevertheless vacuum-packaged lettuce has been implicated in a botulinum-poisoning outbreak (Betts, 1996).

Mushrooms traditionally were packed for retail display in semi-permeable plastic (0.75 mil PVC film), which restricts O_2 movement and water loss. Not only does this prevent drying, but also the high-respiration rate of mushrooms depletes O_2 and causes CO_2 to accumulate in the packages (Nichols and Hammond, 1973). Within 2 h after they are prepared for retail sale in this manner, 16 out of 24 packages kept at 20°C had only 0–0.9% [O_2] remaining (Sugiyama and Yang, 1975).[9] When mushrooms were inoculated heavily with type A *Cl.*

botulinum and then sealed in retail packages with PVC overwraps, the organism grew, and at 20°C toxin production first appeared in 3 days (Sugiyama and Yang, 1975). Between 3 and 4 days, the mushrooms were botulinic in spite of their fresh, edible appearance, but then they spoiled and for that reason presumably would not be eaten. When mushrooms were inoculated with *Cl. botulinum* type B, they spoiled before they become botulinic, and in some instances the type B organism does not develop at all (Sugiyama and Yang, 1975). The reason why retail packages of mushrooms have never developed botulism is that the product does not start with a large enough natural contamination to develop toxin before spoilage occurs. The natural inoculum on mushrooms never exceeds 100 organisms per 100 g, or about 10 organisms per mushroom, and the smaller the initial inoculum, the longer it takes to become botulinic (Hauschild *et al.*, 1975). To present a health hazard, at least 1000 spores are needed per mushroom (Sugiyama and Yang, 1975). Tests have revealed that the growth of type A *Cl. botulinum* can occur in a significant percentage of inoculated retail mushroom packages when they are perforated with a single 0.3 cm diameter hole, but not with two such holes. As a precaution to eliminate the possibility of botulism during retail display, Campbell Co. perforates the wrap with two 0.6-cm-diameter holes.[10] Such packages are ideally suited for LP storage, and ensure that the mushrooms will be safe during retail display.

A full load of mushrooms was successfully transported for 5 days in a 12.2 m Grumman/Dormavac intermodal hypobaric container operated at –1°C and a pressure of 2.0 kPa (15 mm Hg) (Grumman Allied Industries, 1980, unpublished data; Alloca, 1980a). During transit, individual packages in the container ranged in temperature from –0.6 to 3.3°C, which was safely below the temperature that could allow the development of botulism (11.15). It was recommended that the pressure should be lowered to 1.33 kPa (10 mm Hg) in subsequent tests to decrease the respiration rate and temperature spread by as much as 50% (Alloca, 1980b).

10.44 Onion, Green (*Allium cepa* L.)

Provided that moisture loss is prevented, green onions can be stored in NA for up to 3 weeks at 0°C, and for approximately 1 week at 5°C. Higher temperatures favour more rapid yellowing and decaying of the leaves. CA mixtures containing 2–3% $[O_2]$ + 0–5% $[CO_2]$ have only a slight effect on storage life, but 1% $[O_2]$ + 5% $[CO_2]$ can extend storage life to 6–8 weeks (Hardenburg *et al.*, 1986; Thompson, 1998).

The quality of green onions deteriorated within 6 days during NA storage at 0–3°C, and LP storage at a pressure of 13.33–20 kPa (100–150 mm Hg) had only a slight beneficial effect, but at 6.67–10.67 kPa (50–80 mm Hg) the onions remained saleable for nearly 3 weeks (Burg, 1970, 1976a). The maximum storage time in LP was limited by mould development. At 1°C, the leaves of green onions retained more chlorophyll in LP at a pressure of 7.33–8 kPa (55–60 mm Hg) compared to storage in NA or in 2% $[O_2]$ static CA with slaked lime present to prevent $[CO_2]$ build-up (McKeown and Lougheed, 1981). Off-flavours and -odours developed in CA, but not in LP at the same $[O_2]$ concentration. It was suggested that if off-flavours and -odours developed in LP, they may have been volatized and escaped at the reduced pressure, but this is an unlikely explanation (3.25). Ward (1975) reported that LP did not improve the storage of green onions at 9.33 kPa (70 mm Hg).[11]

Bacterial soft rot caused by *E. carotovora*, watery soft rot due to *S. sclerotiorum* and grey mould rot from *B. cinerea* are the major diseases of leafy vegetables. Provided that CO_2 is absent, the low $[O_2]$ present at 0°C and 1.33 kPa (10 mm Hg = 0.14% $[O_2]$) is effective in controlling the growth of these organisms (7.1; Figs 7.1, 7.2, 7.3, 7.6). This LP pressure/temperature combination should be tested with green onions.

10.45 Onion, Dry (*Allium cepa* L.), cvs Autumn Spice, Rocket, Trapps #6

Bulb onions held in LP for 21 days at 26°C, 58% RH and a pressure of 8 kPa (60 mm Hg) lost 12.2% of their weight, whereas onions cured in simulated field conditions lost only 2%. As the average acceptable weight loss is 3–5% (Hardenburg *et al.*, 1986), this result indicates that the onions in LP were overcured (McKeown and Lougheed, 1981).

10.46 Parsley [*Petroselinum sativum* (Mill.) Nym.]

Parsley can be kept in NA for 2–3 months at 0–5°C. CA storage in 10% [O_2] + 11% [CO_2] resulted in 95.6% saleable leaves after 45 days vs. 43.9% in NA (Hardenburg *et al.*, 1986; Leshuk and Saltveit, 1990; Thompson, 1998). During 56 days at 3°C and 95% RH, LP storage at 10 kPa (75 mm Hg) prevented appreciable loss in protein, ascorbic acid and chlorophyll, whereas during NA storage losses were high after 35 days (Bangerth, 1974; Fig. 10.4).

The major diseases of parsley are bacterial soft rot (*E. carotovora*), brown spot (*Cephalosporium apii*), grey mould rot (*B. cinerea*), late blight (*Septoria apiicola*) and watery soft rot (*Sclerotinia* sp.). No data are available for *C. apii* and *S. apiicola*, but in the absence of [CO_2] *E. carotovora*, *B. cinerea* and *Sclerotinia* sp. cannot grow at the low [O_2] concentration present in 0–5°C water-saturated air at a pressure of 1.33–2.0 kPa (10–15 mm Hg) (7.1; Figs 7.1, 7.2, 7.3 and 7.6; Tables 7.1 and 7.2). LP pressures lower than 10.0 kPa (75 mm Hg) have not been tested with parsley.

10.47 Pepper, Sweet (*Capsicum annum* L.), Neusiedler Ideal, Bellboy

Ripening limits the NA storage of green peppers to 2–3 weeks at 7.2–10°C (Lutz and Hardenburg, 1968). Chilling damage occurs below 7°C (Hardenburg *et al.*, 1986), and sometimes at less than 11.1–12.8°C (Pushka and Srivastava, 1963). CA atmospheres containing 2–5% [O_2] + 2–10% [CO_2] slightly extend storage life. Less than 2% [O_2] causes low-[O_2] damage, and more than 2–10% [CO_2] eventually causes calyx discoloration (Hatton *et al.*, 1975; Dilley, 1978; Leshuk and Saltveit, 1990). The storage of peppers is improved by treatments that restrict water loss, including pre-packaging in perforated polyethylene, MA packaging and waxing (Hughes *et al.*, 1981; Hardenburg *et al.*, 1986).

Non-waxed peppers were stored at 7.2°C either in NA or in LP at pressures of 6.7, 10.7 or 16.0 kPa (50, 80 or 120 mm Hg; Burg, 1970). Peppers began to deteriorate after 16 days in NA and were in poor condition by 21 days. Those kept at a pressure of 12.7 kPa (80 mm Hg) were still in excellent condition after 28 days, while at 6.7 and 16 kPa (50 and 120 mm Hg) they were only in fair condition. Peppers were marketable with excellent taste and quality after 46 days' LP storage at 12.7 kPa (80 mm Hg), except for a trace of mould, which appeared on the stem ends. Storage was limited by decay. A chlorine rinse prior to storage reduced the incidence of decay, and a Benlate dip was effective for up to 7 weeks during LP storage. Chilling damage occurred at 7°C in LP, but only became apparent 1–2 days after peppers were removed (Burg, 1976a).

Storage of Neusiedler Ideal peppers in NA and in LP at a pressure of 10 kPa (75 mm Hg) was compared during a 23-day test at 10–12°C. The peppers remained firmer and greener in LP with slightly higher ascorbic acid content and significantly lower ethylene production (Bangerth, 1973). LP storage was limited by decay. Respiration was reduced by as much as 67–75% at a low pressure (Bangerth, 1974). NA storage was compared to LP at lower pressures, 2.0, 5.3 and 10.7 kPa (20, 40 and 80 mm Hg), in laboratory tests run by Grumman Allied Industries. Some of the peppers stored in NA were pre-treated with chlorine and/or Benlate; the remainder and all peppers stored in LP were untreated and unwaxed. After 50 days, the percentage of saleable peppers was

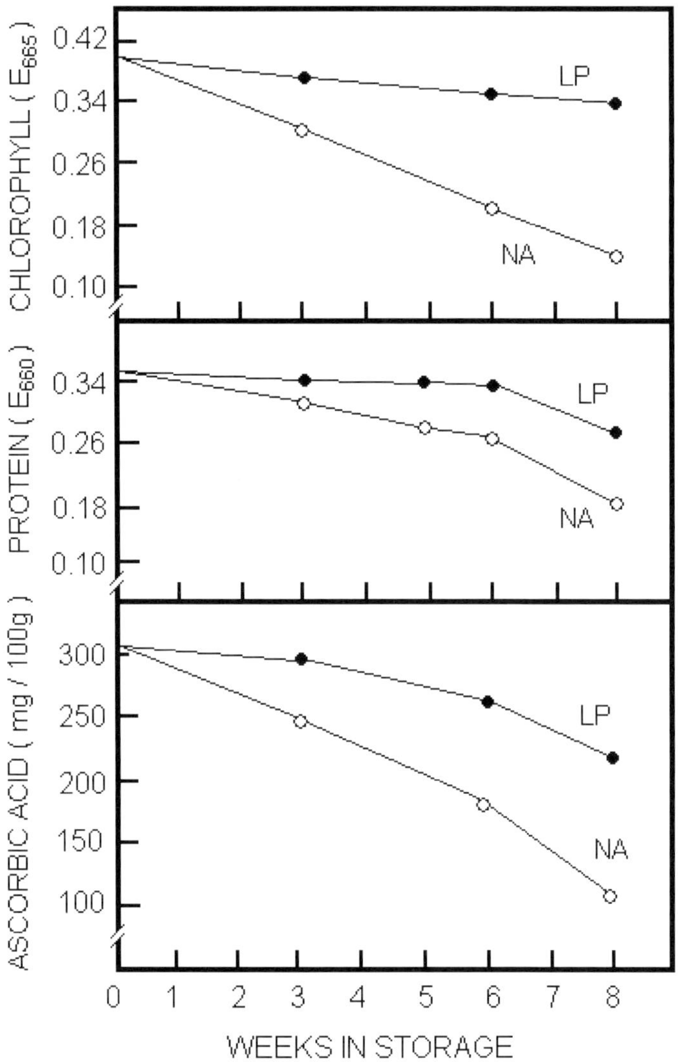

Fig. 10.4. Changes in the protein, chlorophyll and ascorbic acid content of parsley stored at 3°C in LP at a pressure of 10 kPa (75 mm Hg) or in NA (Bangerth, 1974).

80–87% at 5.3–10.7 kPa (40–80 mm Hg), respectively; 47% at 2 kPa (15 mm Hg); and none in NA regardless of the pre-treatment (Jamieson, 1980a). The storage temperature was not specified. Excellent results with peppers stored at 10.13 kPa (76 mm Hg) were reported by Staby (1976b).

Hughes *et al.* (1981) reported that no benefit resulted when pepper cultivar 'Bellboy' was stored at 8.8°C in CA at 2% $[O_2]$ supplemented with 0, 3, 6 or 9% $[CO_2]$, or in LP at 20.3, 10.1 and 5.1 kPa (152, 76 and 38 mm Hg). At atmospheric pressure, film wrapping reduced wastage, mainly due to less water loss, while storage in 2% $[O_2]$ + 6% $[CO_2]$ resulted in a significant increase in decay during subsequent shelf life at 20°C. Weight loss in LP was at least five times higher per day compared to NA, and 7–10 times greater than that needed to remove respiratory heat by evaporative cooling (6.1).[11] Indicating that the humidity was not properly maintained in this LP study.

The main storage diseases of peppers are alternaria rot (*A. tenuis*), anthracnose (*C. gloeosporioides*), bacterial soft rot (*E. carotovora*), *Cladosporium* rot (*C. herbarum*), grey mould rot (*B. cinerea*), bacterial spot (*X. vesicatoria*), phoma rot (*Phoma destructor*), sunken spots (*Vermicularia capsici*) and rhizopus rot (*R. stolonifer*) (McColloch *et al.*, 1966; Ryall and Lipton, 1972; Eckert *et al.*, 1975; Hardenburg *et al.*, 1986). There are no data for *X. vesicatoria* and *V. capsici*, but provided that CO_2 is absent, the growth of the other organisms is suppressed by the [O_2] present in 7–13°C water-saturated air at pressures of 2.0–2.7 kPa (15–20 mm Hg) (7.1; Figs 7.1, 7.2, 7.3, 7.6, 7.7; Tables 7.1 and 7.2). This low-pressure range should be tested at non-chilling temperatures as high as 12–13°C to determine if decay in peppers can be controlled without causing low-[O_2] or low-temperature injury. Peppers tolerated a 2-day exposure to a pressure of 2.67 kPa (20 mm Hg) at 10°C with no adverse effects (T.L. Davenport and S.P. Burg, 2003, unpublished), but longer exposure periods and higher temperatures have not yet been tested.

10.48 Potato (*Solanum tuberosum* L.)

During storage, the appearance, texture and nutritive value of potatoes and potato products are affected by greening and the development of bitter taste (solanine), off-flavour and potentially dangerous solanine glycoalkaloids. According to US standards, greening is defined as damage if more than 5% of the total weight must be removed to eliminate the greened tissue, and as serious damage if the loss is over 10%. According to the same standards for grades, green potatoes are unfit for consumption.

Holding potatoes in LP at 15°C and a pressure of 16.8 kPa (126 mm Hg) inhibited greening in the light, but had no effect on solanine content (Jadhav *et al.*, 1973). Lower pressures were not tested and no benefit was obtained at pressures higher than 33.7 kPa (253 mm Hg). The result was the same at 2.5, 5, 10 or 20°C in storages performed at pressures ranging from 10.1 to 12.67 kPa (76–95 mm Hg), and also at atmospheric pressure in 2–2.5% [O_2] (Kader *et al.*, 1975, unpublished data). This indicates that low [O_2] was responsible for the result. These mixtures and conditions slightly delayed sugar accumulation at 10°C or lower temperatures, but this effect diminished after 4 weeks in storage. Both in LP and CA, a low-[O_2] atmosphere enhanced sprouting at 20°C. Subatmospheric pressures also influence the physiochemical properties of potato starch (Morrow and Lorenz, 1974) and prevent starch-to-sugar conversion during potato storage (Apelbaum, 1979). A high-sugar content is undesirable because it promotes carmelization during the preparation of French-fried potatoes and potato chips.

10.49 Radish (*Raphanus sativa* L.), var. sativus

Topped spring radishes can usually be kept for 3–4 weeks in NA at 0°C, and for a shorter time at 5°C. At temperatures higher than 0°C, CA storage in 1–2% [O_2] improves quality by reducing top growth, root growth and softening (Hardenburg *et al.*, 1986; Thompson, 1998).

An LP pressure of 10.0 kPa (75 mm Hg) decreased protein, ascorbic acid and chlorophyll loss in radishes stored at 2.2–2.8°C and 95% RH (Bangerth, 1974). During 7–28 days at 1°C weight was reduced to the same extent by LP storage at 7.33–8 kPa (55–60 mm Hg = 1.4–1.5% [O_2]), as by static CA storage at 2% [O_2] with slaked lime added to prevent CO_2 accumulation (McKeown and Lougheed, 1981). Off-flavours and -aromas were not evident in LP, but developed in CA. It was suggested that if off-flavours and -odours developed in LP, they may have volatized and escaped at the low pressure, but this is an unlikely explanation (3.25). In contrast to these studies, Ward (1975) reported no advantage from storing radishes in LP at a pressure of 10.0 kPa (75 mm Hg).[11]

The major diseases of radishes (Ryall and Lipton, 1972) are bacterial black spot (*X.

vesicatoria), downy mildew (*Perenospora parasitica*) and rhizoctonia root rot (*R. solani*). The growth of *R. solani* is inhibited by low [O_2] (7.1). No data are available for *X. vesicatoria* and *P. parasitica*. Radishes should be tested at pressures lower than 7.33 kPa (55 mm Hg).

10.50 Spinach (*Spinacia oleracea* L.)

Spinach can only be stored for 10–14 days in NA at 0°C, even with top ice (Lutz and Hardenburg, 1968). A CA atmosphere containing 10% [O_2] + 10–40% [CO_2] is beneficial in preventing yellowing and maintaining quality for up to 30 days (Hardenburg *et al.*, 1986; Thompson, 1998). Spinach stored at a pressure of 10 kPa (75 mm Hg) and 95% RH retained its vitamin C, green colour and total protein for nearly 7 weeks in LP (Bangerth, 1973, 1974).

The main storage diseases of spinach are bacterial soft rot (*E. carotovora*) and downy mildew (*Peronospora effusa*) (Ryall and Lipton, 1972). *E. carotovora* cannot grow in the absence of CO_2 in the low [O_2] present in 0°C water-saturated air at a pressure of 1.33–2.0 kPa (10–15 mm Hg) (Fig. 7.1). No data are available for *P. effusa*. Pressures lower than 10 kPa (75 mm Hg) have not been tested with spinach. The optimal LP storage condition is likely to be 0°C and a pressure of 1.33 kPa (10 mm Hg).

10.51 Squash (*Curcurbita maxima* Duch.), var. Acorn and (*Curcurbita pepo* L. var. Melopepo Alef.), Yellow Crookneck 'Summer' Squash

Acorn-type squashes can be kept for 5–8 weeks in NA at 10°C. They suffer chilling damage at 0–4°C and when they are kept at 15–20°C, their flesh acquires a 'stringiness' and they become yellow. Summer squashes can be held for 1–2 weeks at 5–10°C, but chill below 5°C (Hardenburg *et al.*, 1986). CA slightly extends storage life and reduces the severity of chilling injury in some types of squash. There is considerable disagreement about the best combination of O_2 + CO_2 (Thompson, 1998).

Yellow Crookneck squash storage was studied at 7.2°C either in NA with the commodity loosely wrapped in plastic, or in LP at pressures of 8.67, 10.67 and 20.0 kPa (65, 80 and 150 mm Hg). All squashes remained in excellent condition for 11 days, but began to russet slightly and soften by 19 days. Within 36 days they had developed mould on the stem ends and were in poor condition. No benefit from LP was noted (Burg, 1970). LP had no effect on Yellow Crookneck squash stored in combination with pole and bush beans at 7°C and pressures of 10.13 or 20.3 kPa (76 or 152 mm Hg) (Spalding and Reeder, 1980, unpublished).

In the same period of time, acorn squash stored in LP at 70% RH and pressures of 7.33–8.0 kPa (55–60 mm Hg) lost 10.9% of its weight; in 35–50% RH air it lost 10.0% of its weight; and at 90–95% RH the weight loss was 4.2% in flowing 2% [O_2] (McKeown and Lougheed, 1981). The outer surface of squashes kept in air had become covered with a wax which was soft, 'wet' and slippery, and the rinds were yellow to orange in colour, which is a sign of ageing. The rinds of squash held in LP and CA were firm, green and non-waxy. A considerable number of rots occurred in air, but very few in LP and CA. It was concluded that LP holds promise for holding acorn squash at the low humidities recommended for this commodity (Hardenburg *et al.*, 1986).

Black rot (*Mycosphaerella citrullina*), cottony leak (*Pythium aphinidermatum*) and bacterial soft rot (*E. carotovora*) are major diseases of acorn squash. Summer squashes are attacked by fusarium rot (*Fusarium* sp.) and rhizopus soft rot (*R. stolonifer*), and both types of squash are susceptible to alternaria rot (*A. tenuis*) if they have been weakened by chilling damage (Ryall and Lipton, 1972). *Fusarium* sp., *E. carotovora*, *A. tenuis* and *Mycosphaerella* sp. are inhibited by very low [O_2] (7.4; Figs 7.1, 7.2 and 7.3). No data are available for *Pythium* sp. Very low pressures, which should directly control decay, have not been tested with squash.

10.52 Turnip (*Brassica campestris* L. – Rapifera Group)

Storing turnips by cycling the pressure between 13.33 and 40.0 kPa (100 and 300 mm Hg) with a variable low-pressure system (VLPS) gave less weight loss and retained better appearance compared to LP storage at a constant pressure (Onoda *et al.*, 1989a). CA storage provides only a slight benefit with turnips (Thompson, 1998).

CUT FLOWERS, CUTTINGS AND POTTED PLANTS

Several reviews of LP and CA floral crop storage are available (Uota, 1969b; Burg, 1973b; Hassek, 1975; Staby, 1977; Larson, 1992). CA storage is not recommended for cut flowers because, although increased [CO_2] benefits some flowers, it can also cause injury, and while respiration often is slowed and sometimes deterioration slightly delayed by 0.5–3% [O_2], the display life of flowers stored in low [O_2] usually does not equal or even approach that of freshly cut blooms (Uota, 1969b; Hardenburg *et al.*, 1986).

10.53 Diseases of Cut Flowers

The most prevalent storage disease of cut flowers is botrytis blight (*B. cinerea*; Tammen, 1961; Dimock *et al.*, 1964; Larson, 1992; Rogers, 1992; Whealy, 1992). The growth and sporulation of this organism is prevented by the [O_2] present in water-saturated air at 0–10°C and a pressure of 1.33–2.67 kPa (10–20 mm Hg) (Figs 7.1, 7.3 and 7.6). Both at atmospheric and hypobaric pressures, decay of carnation blooms (Table 7.1) and other commodities (Table 7.2) is suppressed in direct proportion to the reduction in [O_2] below 1%. Low pressure *per se* has an additional direct inhibitory effect on the growth of *B. cinerea* (Table 7.4). To control decay, flowers should be stored in LP at the lowest pressure that does not cause injury, usually 1.33–2.67 kPa (10–20 mm Hg). Even at atmospheric pressure, some flowers tolerate very low [O_2]. At 0–4°C, King Alfred daffodils can be kept for 3 weeks in 100% [N_2] without any loss and even an extension of vase life (Parsons *et al.*, 1967).[12]

Occasionally decay may be caused by *Helminthosporium* sp. and *Alternaria* sp. infections on snapdragon flowers (Rogers, 1992); fusarium bud rot (*Fusarium tricinctum*) of cut carnation flowers (Whealy, 1992); powdery mildew (*Sphaerotheca pannosa* var. *rosae*) of roses (Dimock and Tammen, 1969; Durkin, 1992); bacterial blight (*Erwinia chrysanthemi*) and ray blights of chrysanthemum blooms due to *Stemphylum* sp., *Alternaria* sp. and *Ascochyta chrysanethemi* (Kofranek, 1992); and curvularia blight (*Curvularia trifolii*) of *Gladiolus* (Wilfret, 1992). Less than 1% [O_2] inhibits the growth of *Alternaria* sp., *Fusarium* sp., *Ascochyta* sp. and *Erwinia* sp. (Figs 7.1–7.7; note 2 – chapter 7). No data are available for the O_2 tolerance of the other disease organisms.

10.54 *Alstroemeria*, Cut Flowers (*Alstroemeria* sp.)

Alstroemeria blooms can only be preserved in NA for 2–3 days at 4°C (Hardenburg *et al.*, 1986). In anticipation of holiday-price appreciation, during the past 10 years this flower has been stored in LP for 3–4-week periods in Grumman/Dormavac 12.2 m (40 ft) intermodal hypobaric containers operated at 1–3°C and a pressure of 2 kPa (15 mm Hg). Standard commercial boxes of *Alstroemeria* are pressure-cooled in Colombia, but because they warm during air transport and subsequent drug, USDA and customs inspections at Miami International Airport, they are vacuum-recooled in Miami prior to loading into the intermodal hypobaric containers. Following LP storage, the boxes have been distributed and sold by normal commercial practices.

10.55 *Anthurium*, Cut Flower (*Anthurium* sp.), var. Ozaki

With their stem ends in water, *Anthurium* cut flowers can be stored in NA for 2–4 weeks at 13°C. Chilling occurs below 7°C and causes the leaves to blacken. CA is beneficial at ambient temperatures (Hardenburg *et al.*, 1986).

Variety 'Ozaki' blooms, which initially had 13.4 days' display life, were stored in NA or LP at 15°C wrapped in moist shredded paper. After 15 and 30 days' storage in NA, the vase life decreased to 6.5 and 2 days, respectively. Blooms kept at 2.67 kPa (20 mm Hg) were 92% marketable with 6 days' shelf life remaining after 42 days' storage in LP, and 60% were marketable with 3 days' shelf life remaining after 67 days. Flowers stored at 5.33 kPa (40 mm Hg) also kept well, but their appearance had slightly deteriorated. LP reduced fungal growth, stem browning, abscission, browning of the spadix, purpling of the spathe and black spots on the spadix. A partial load of anthuriums was tested in a 12.2 m (40 ft) Grumman/Dormavac intermodal hypobaric container operated at 17.2°C and a pressure of 4.0 kPa (30 mm Hg). After 21 days' storage, the blooms still had their initial appearance and vase life (Jamieson, 1984).

10.56 Aster, China, Cut Flower (*Callistephus chinensis*)

China asters can be kept for 1–3 weeks in NA at 0–4°C (Hardenburg *et al.*, 1986). Storage of this flower improved progressively as the pressure was lowered from 20.0 to 5.33 kPa (150 to 40 mm Hg). Best results were obtained at 0–3°C and a pressure of 5.33–8.0 kPa (40–60 mm Hg). After 30 days, these blooms still displayed their initial 3–6-day vase life, and their colour and fresh appearance was unchanged (Burg, 1970, 1973b, 1976a). At pressures higher than 8.0–10.67 kPa (60–80 mm Hg), colour fading was evident. Lower pressures have not been tested.

10.57 Azalea, Potted Plant (*Rhododendron* sp.)

Azalea flower-bud dormancy can be broken by storing potted plants in NA at 4°C for 6 weeks. Blooming occurs in an additional 4 weeks, but often the foliage deteriorates during the 6-week storage period even when the plants are continuously illuminated and watered. Potted azalea plants with dormant flower buds were stored at 4°C either in NA or in an LP intermodal container operated at a pressure of 8 kPa (60 mm Hg). Although the plants in LP were kept in complete darkness for 6 weeks, and not watered, they did not lose green colour or develop leaf drop. Dormancy was broken and flowering occurred within 4 weeks after the plants were transferred to daylight at atmospheric conditions (Burg, 1976a).

10.58 Carnation, Cut Flower (*Dianthus caryophyllus* L.), cvs White Sims (White), Linda (Pink), Scania (Red), Laddie and Various Other Colours

Carnations can be kept for 3–4 weeks in NA at −0.5 to 0°C (Hardenburg *et al.*, 1986). Flowing 0.5, 1.0 or 2% [O_2] decreases the respiration rate of carnation blooms by 50–60% and slightly improves their appearance during 30 days' CA storage, but there is no noticeable difference compared to NA controls after 4 days' vase life. Less than 0.5% [O_2] injures blooms and more than 10–15% [CO_2] decreases vase life and causes the petals of coloured varieties to lose their lustre and develop a slightly faded appearance. Quality is not improved by 5% [CO_2], and clean air is as effective as low [O_2] in maintaining shelf life. Carnations that were held in a shipping container (static air) had a much lower quality rating than those kept in continuously flowing streams of air or low [O_2], demonstrating the effect that volatile accumulation has on the display life of carnations (Uota and Garazsi, 1967; Uota, 1969b). Although LP pressures ranging from 3.07 to 4.67 kPa

(23–37 mm Hg = 0.5–0.8% [O_2]) greatly extend carnation storage life, CA atmospheres containing the same amount of [O_2] are not beneficial (Staby *et al.*, 1979, 1984).

When pink carnations with an initial vase life of 14 days in water were stored at 1.7°C either in NA or in LP at pressures of 5.33, 8.0 and 10.67 kPa (40, 60 and 80 mm Hg), storage life was 10 days in NA and 18 days in LP (Burg, 1969). LP extended the storage life of carnations at all stages from tight buds to full open when they were stored at 4°C and a pressure of 10.0 kPa (75 mm Hg; Bangerth, 1973). At 1.7–2.2°C and 95% RH (Burg, 1976a), the storage life of cut carnation blooms improved progressively as the pressure was lowered from 20.0 to 2.0–3.33 kPa (from 150 mm Hg to 15–25 mm Hg). LP storage at 0–4°C and pressures ranging from 1.33 to 10.0 kPa (10–75 mm Hg) eliminated 'sleepiness' and extended the storage life of 'Laddie', 'Scania', 'White Sims' and other cultivars of cut carnations harvested at all stages from tight buds to full open (Dilley, 1972; Bangerth, 1973; Burg, 1973b, 1976a; Carpenter and Dilley, 1975; Dilley *et al.*, 1975; Goszczynska and Rudnicki, 1982; Jamieson, 1984; Staby *et al.*, 1984) with little or no decrease and sometimes even an extension in post-storage vase life (Tables 10.5, 10.6, 10.7 and 10.8).

After 6 weeks' storage at 0°C and a pressure of 6.67 kPa (50 mm Hg), the vase life of White Sims carnations was extended (Dilley and Carpenter, 1973). Carnations do not lose their sensitivity to ethylene during a prolonged stay in LP, and after removal from storage if they are treated with 150 μl/l propylene, they produce ethylene and fade (Carpenter and Dilley, 1975; Dilley *et al.*, 1975). Carnations stored at 0–2°C and a pressure of 5.33 kPa (40 mm Hg) were well preserved for more than 63 days in LP (Burg, 1973b), and at 0°C and a pressure of 6.67 kPa (50 mm Hg), bud and full-stage White Sims carnations were kept for up to 9 weeks without 'sleepiness' and with exceptional post-storage shelf life due to a low ethylene production rate and delayed climacteric rise in respiration (Dilley *et al.*, 1975; Fig. 10.5). When carnations were removed from LP storage after 5 weeks at 0°C and a pressure of 6.67 kPa (50 mm Hg), their rate of ethylene production (0.1 μl/flower/day) was similar to that of fresh-cut carnations.

'Scania 3C' flowers, with either tight buds or coloured petals just visible as a red cross, were preconditioned in a solution containing silver thiosulphate (STS) and sucrose, and their storage in NA and LP compared. After 20 weeks, quality was much better in LP, and all buds developed into fully open acceptable flowers regardless of whether they initially were tight-budded or cracked open. In NA, blooms were acceptable for 14 weeks, but subsequently lost vase

Table 10.6. Mean vase life of Laddie and Scania carnations after storage at 4°C in LP at a pressure of 5.33 kPa (40 mm Hg) or in NA (Bangerth, 1973).

Storage time (days)	Vase life after removal from storage (days)			
	Laddie		Scania	
	NA	LP	NA	LP
0	7.0	–	8.1	–
42	2.2	7.2	3.0	8.5
58	0.0	7.9	0.0	7.9

Table 10.5. Vase life of carnations following storage in a prototype Fruehauf 6.1 m (20 ft) hypobaric container operated at a pressure of 6.67 kPa (50 mm Hg) and temperature of 2°C (Dilley, 1977a,b).

Cultivar	Vase life at 22°C (days) after storage for indicated days at 0°C and 6.67 kPa (50 mm Hg)							
	0	22	34	43	50	57	64	79
White	10.0	11.8	9.5	9.2	8.3	9.2	8.2	8.1
Red	–	12.2	10.6	10.1	10.0	10.0	–	–
Pink	–	13.0	12.2	8.3	10.6	9.7	8.8	10.2

Table 10.7. Vase life of California carnations after storage in a 12.2 m (40 ft) Grumman/Dormavac intermodal hypobaric container operated at a pressure of 2.0 kPa (15 mm Hg) and a temperature of 1.1–1.7°C at 98% RH (Burg, 1979a; Dressler, 1979b).

	Vase life (days) after indicated storage time				
Colour	0	23	29	36	43
Red	9.9	8.9	7.5	9.2	7.5
White	9.6	8.4	7.0	7.5	8.25
Light pink	7.1	8.6	7.9	8.2	5.5
Dark pink	9.2	8.0	6.9	8.8	–
Red and white	10.6	9.5	7.3	5.6	8.0
Salmon	–	9.4	10.5	10.5	7.9
Novelty	9.0	7.3	–	–	–
Orange	–	–	7.9	–	–

Table 10.8. Vase life of Colombian carnations after storage in a 12.2 m (40 ft) Grumman/Dormavac intermodal hypobaric container operated at a pressure of 2 kPa (15 mm Hg) and a temperature of 2.2–2.8°C at 98% RH (Burg, 1976, unpublished data).

	Vase life (days) after indicated storage time			
Colour	0	23	30	37
Red	10.2	7.0	8.5	9.8
White	9.1	7.4	7.1	8.8
Pink	9.2	6.8	6.9	8.8
Sims	9.5	9.0	8.0	9.5

life. Fungicides such as Rovral or Sumilex helped to extend the LP storage life of flowers for as long as 140 days if they were preconditioned with STS and sucrose (Goszczynska and Rudnicki, 1982). When carnation storage at 0°C was compared in NA, CA at 0.5–8.0% [O_2], and in LP at pressures ranging from 1.33 to 4.67 kPa (10–35 mm Hg = 0.15–0.83% [O_2]), blooms could be stored for 6 weeks in NA if they were pre-treated with STS and vapour barriers were utilized, and they lasted 8 weeks in LP without STS. Low [O_2] at atmospheric pressure was not beneficial (Staby *et al.*, 1984).

The results of small-scale laboratory carnation experiments have been confirmed in partial and full-load hypobaric intermodal static trials and shipments. A full load of California carnations packed in boxes lacking polyethylene liners was stored in a 12.2 m (40 ft) Grumman/Dormavac intermodal hypobaric container operated at a pressure of 2.0 kPa (15 mm Hg). The temperature was set at 1.1–1.7°C to improve the storage life of red blooms, which may develop dark spots prematurely at a lower temperature (Lutz and Hardenburg, 1968). After various storage times, samples were removed, their stems were recut under water and vase life was measured at 21.1°C in double-strength 'floralife' preservative (Table 10.7; Burg, 1979c; Dressler, 1979b). The red blooms retained their colour exceedingly well in this test. A partial load of pink and red carnations was tested in Japan for Zen-Noh in a 12.2 m (40 ft) Grumman/Dormavac hypobaric intermodal container operated at 1.7°C, 80–90% RH and a pressure of 3.33 kPa (25 mm Hg). During 44 days storage, the blooms retained their at-harvest appearance and still displayed their initial vase life of 20 days in water (S.P. Burg, 1978, unpublished data). Similar results were obtained with both 50-degree bud-cut and 80-degree medium-open ('3 o'clock') blooms. Storage of these same blooms was limited to about 10 days in NA. STS-treated carnations were flown from Colombia to Miami and a full load was kept in a 12.2 m (40 ft) Grumman/Dormavac LP container operated at 2.2–2.8°C and a pressure of 2.0 kPa (15 mm Hg). After 37 days there was no significant loss of vase life (Table 10.8). Carnations grown in Denver, Colorado, were air-shipped to Long Island and stored for 61 days in a 12.2 m (40 ft)

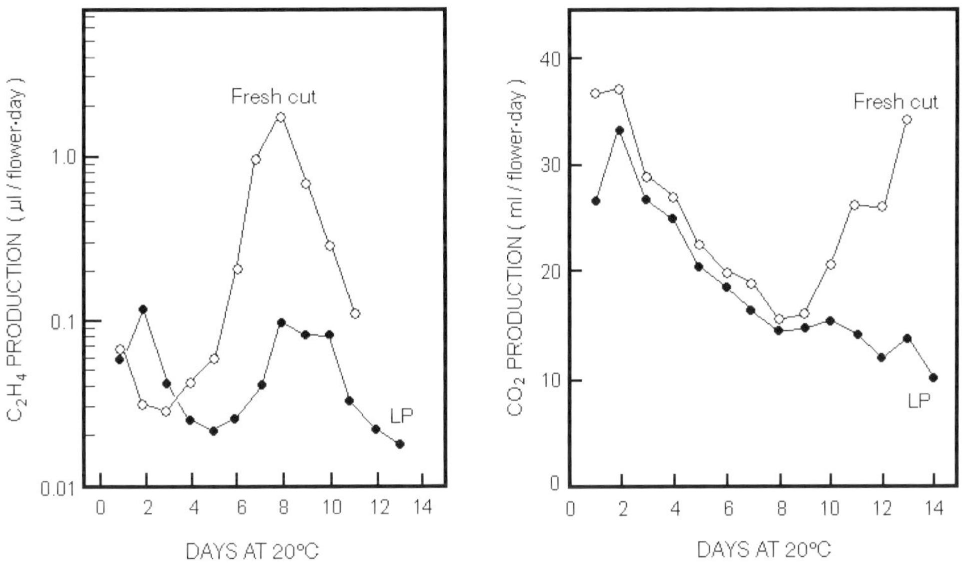

Fig. 10.5. Ethylene (*left*) and CO_2 (*right*) production of bud cut 'White Sim' carnations at 20°C immediately after harvest (fresh cut) or following 5 weeks' storage at 0°C and an LP pressure of 6.67 kPa (50 mm Hg). Ethylene production is expressed on a logarithmic scale (Carpenter and Dilley, 1975).

Grumman/Dormavac container operated at 2.2–2.8°C and a pressure of 2.0 kPa (15 mm Hg). The flowers were distributed to 1000 shareholders at Grumman Corporation's annual meeting (Grumman Allied Industry, 1978). Successful shipments of carnations from California to Chicago and from the Dominican Republic to the USA were made in Grumman/Dormavac 12.2 m (40 ft) LP intermodal containers operated at 0°C, 95% RH and pressures of 1.33–2.0 kPa (10–15 mm Hg) (Jamieson, 1984).

During the past 10 years, nearly 10,000,000 stems of Columbian carnations, often in mixed loads with other flowers, have been stored in LP intermodal containers for periods of 4–6 weeks in anticipation of each major holiday when flowers are in high demand. The blooms are packed with perforated polyethylene slip liners in conventional boxes containing newspaper to absorb liquid water droplets. They are pressure-cooled in Colombia to 0–2°C, warmed by 10–25°C during air shipment and subsequent drug, insect and customs inspections at Miami International Airport, and vacuum cooled in Miami prior to loading into 12.2 m (40 ft) Grumman/Dormavac hypobaric containers operated at 1°C and a pressure of 2.0 kPa (15 mm Hg). Each container holds 270,000 carnation stems, and during a 4–6-week period in advance of a holiday, the price of carnations will generally increase by 7–14 cents (US) per stem. Hypobaric containers also have been used to accumulate carnations for a mail order company that is unable to purchase sufficient inventory from a single grower to satisfy their needs during major holiday sales periods.

For successful LP storage, it is important to load carnation boxes with adequate spacing (Figs 6.13 and 6.15). The normal commercial carnation pack containing cardboard, polyethylene, newspaper and waxed paper, which is designed to assist in conventional NA distribution, restricts heat transfer and the exchange of water vapour and gases. By interfering with evaporative cooling, this box can slow cool down and elevate the steady-state temperature of cartons in the centre of densely packed LP loads. This is avoided by opening the ventilating holes in the box ends, and aligning the boxes so that the holes have access to ventilating ducts or chimneys in the stack (Fig. 6.13).

Cut carnations that had been treated with STS were stored at 1°C either in NA wrapped in polyethylene, or in LP at pressures ranging from 3.33 to 5.33 kPa (25–40 mm Hg). Some of the blooms stored at 5.33 kPa (40 mm Hg) were returned to atmospheric pressure for 4 h each day. After 4 weeks, most of the blooms in NA failed to open properly, and those that did had a relatively short shelf life. The vase life of flowers kept at a pressure of 3.33–4.0 kPa (25–30 mm Hg) was slightly longer compared to those held at 5.33 kPa (40 mm Hg). Cycling the pressure from 5.33 kPa (40 mm Hg) to atmospheric each day did not decrease vase life (Burg, 1976a).

10.59 Chrysanthemum, Rooted and Non-rooted Cuttings (*Chrysanthemum morifolium = Dendranthema morifolium*), cvs Delaware, Neptune, Yellow Delaware, Blue Marble, Golden Anne, Bright Golden Anne, Regal Anne, Yellow Snowdown, Hurricane, Yellow Mandalay, Polaris, Fandango, Super White, Heyday, Rose Chip

Extensive testing of *Chrysanthemum* cuttings has been carried out in 6.1 m (20 ft) and 12.2 m (40 ft) intermodal hypobaric containers and at various university and industrial laboratories throughout the world. After removal from LP storage, cuttings still retain their initial shelf-life and can be shipped by conventional means. Over 200 varieties have been investigated and all respond favourably, both in the rooted and unrooted state.

Storage of non-rooted cuttings of cvs Delaware Neptune and Yellow Delaware is improved at 0–4°C and a pressure of 13.33–20.0 kPa (100–150 mm Hg), but a much better result occurred when the pressure was lowered to 5.33–10.67 kPa (40–80 mm Hg) (Burg, 1969, 1970, 1971, 1976a). Non-rooted hardy cultivars could be stored for 84 days at 0–2°C and a pressure of 8 kPa (60 mm Hg) vs. 21–42 days in NA. Even the tenderest non-rooted varieties, such as Blue Marble, which normally deteriorate during 10–14 days in NA, were well preserved in LP and rooted normally with no subsequent dieback after 6–8 weeks' storage at 0–2°C and a pressure of 8.0 kPa (60 mm Hg) (Burg, 1973b). Hardier varieties such as Neptune, Golden Anne and Bright Golden Anne were successfully preserved for more than 12 weeks in LP. Fresh cuttings and those which had been stored in LP developed the same number of roots in an identical period of time in a rooting bed. Non-rooted cuttings have been stored in LP for 6–12 weeks, rooted during a 2-week period, and then the rooted cuttings were stored in LP for 6–12 additional weeks before they were removed and potted. The plants developed normally and bore the usual number of blooms (Burg, 1976a). In tests carried out in England, storage at 0°C and a pressure of 6.67 kPa (50 mm Hg) markedly improved the quality of Yellow Snowdown, Hurricane, Bright Golden Anne, Polaris and Fandango non-rooted cuttings during a 5–6-week period, and rooted cuttings of Super White, Heyday and Rose Chip during 4 weeks (Butters, 1976, 1977). The Danish low-pressure storage group reported that more than 75% of *Chrysanthemum* cuttings rooted after 10 weeks' LP storage, but details of variety, temperature and pressure were not provided (Danish Research Service for Plant and Soil Science, 1980).

The effect of temperature on the storage of *Chrysanthemum* × *hortorum* L.H. Bailey 'Yellow Mandalay' cuttings was studied in LP at 6.33, 12.67 and 25.33 kPa (47.5, 95 and 190 mm Hg) and in NA. At 2°C, all of the cuttings survived for 5 weeks in NA and LP regardless of the storage pressure, while at 6–10°C cuttings stored at pressures of 6.33–12.67 kPa (47.5–95 mm Hg) survived, but those stored in NA or at 25.33 kPa (190 mm Hg) did not. Rooting was tested after 5 weeks' storage. All cuttings that had been stored at 2°C rooted, but after storage at 6–10°C, satisfactory rooting only occurred if the cuttings had been stored in LP at a pressure of 6.33–12.67 kPa (47.5–95 mm Hg) (Jensen and Rasmussen, 1978).

Chrysanthemum cuttings that have been stored in NA often root more rapidly than fresh-cut or LP-stored cuttings because root initials form during NA storage but not in LP. Premature rooting is followed by poor development of the root system, so that ultimately NA cuttings are less vigorous and develop more slowly than fresh-cut or LP-stored cuttings.

Rooted cuttings are particularly difficult to store in NA because the new growth that forms during storage is chlorotic and disfigured, making the cuttings unmarketable within a few weeks. Rooted cuttings of Neptune, Bright Golden Anne and Regal Anne were stored in LP at 3.3°C and pressures of 8.0, 10.67 and 13.33 kPa (60, 80 and 100 mm Hg). After 44 days at 8.0 or 10.67 kPa (60 or 80 mm Hg), there was no apparent growth or change in appearance. Both in laboratory and intermodal container tests, rooted cuttings of cvs Neptune, Golden Anne and Bright Golden Anne have been preserved at 0–2°C for more than 84 days at 5.33–8.0 kPa (40–60 mm Hg), whereas in NA they can only be kept for 21–42 days. Rooted cuttings of more sensitive varieties, such as Blue Marble, grow excessively and become chlorotic at 0–2°C within 1–2 weeks in NA. They are preserved for 6–12 weeks in LP without yellowing or growth at pressures ranging between 5.33 and 8 kPa (40–60 mm Hg) (Burg, 1973b, 1976a).

The effect of daily cycling the LP storage pressure to atmospheric was studied to determine the feasibility of opening a hypobaric warehouse each day to add and remove inventory. Unrooted Blue Marble cuttings were stored at 2°C, loosely wrapped in polyethylene (PE) film either in NA or in LP at a pressure of 8 kPa (60 mm Hg), flowing saturated air. Within 12 days in NA the cuttings became chlorotic and did not root properly. Each day the LP pressure was cycled to atmospheric for periods of 0, 2, 4, 6, 8, 10, 12 or 14 h while continuing the saturated airflow. Then the pressure was returned to 8 kPa (60 mm Hg) for the remainder of a 24-h period. Cuttings cycled 14 h daily showed slight leaf yellowing after 6 weeks' LP storage, but those cycled for 2–8 h remained green (Burg, 1976a).

A hypobaric warehouse has been used to inventory sensitive-rooted *Chrysanthemum* cuttings such as Blue Marble (Hardenburg *et al.*, 1986; Fig. 11.1). The warehouse is operated on a 16-h night cycle at a pressure of 8 kPa (60 mm Hg), and kept at atmospheric pressure during an 8-h working day to facilitate adding and removing inventory. The cuttings are stored at a density of 160/box with a polyethylene liner, eight boxes strapped together to form an '8-pack'. Attempts to double the number of cuttings per box slowed the cool down, increased weight loss during cooling, and in both LP and NA resulted in injury during cooling (Tepper, 1978). If cuttings are precooled in a single-density pack with the polyethylene liners laid open, they can be stored double density in LP after cool down is completed.

If the $[O_2]$ is too low, eventually a *Chrysanthemum* cutting's ability to root after LP storage seems to decrease even though the cutting's appearance is better preserved than it is at a higher pressure. At 0°C, *Chrysanthemum* cuttings lose chlorophyll less rapidly at 5.33 kPa (40 mm Hg = 1.0% $[O_2]$) compared to 8.0 kPa (60 mm Hg = 1.5% $[O_2]$), and the result is even better at 2.67–3.33 kPa (20–25 mm Hg = 0.4–0.6% $[O_2]$), but rooting is improved if the cuttings are stored at 5.3–8.0 kPa (40–60 mm Hg) rather than 1.3–3.3 kPa (10–25 mm Hg) (Burg, 1973b; Eisenberg, 1977; Eisenberg *et al.*, 1977a; Jensen and Rasmussen, 1978).

10.60 Chrysanthemum, Cut Flower (*Chrysanthemum morifolium* = *Dendranthema morifolium*)

Standard and pompom types of *Chrysanthemum* flowers harvested when they are almost completely open can be stored in NA at 0°C for 3–4 weeks. Botrytis grey rot may develop at temperatures higher than 2–5°C. Stems harvested with medium to large buds in the bud-cut stage can be kept in NA for 2–3 weeks at −0.5 to 0°C, but if the storage is extended beyond that time the buds may not open (Hardenburg *et al.*, 1986).

Fully open *Chrysanthemum* blooms were tested at 4.4°C, stored in LP at pressures of 8.0, 9.33 or 13.33 kPa (60, 70 or 100 mm Hg), and in NA at the same temperature, but with the cut ends in water. The initial vase life of the flowers was 6–9 days in water at 20°C. After 21 days' storage, all blooms kept in LP still had excellent appearance, but flowers held at 8.0–9.33 kPa (60–70 mm Hg) had better colour and clearer bloom centres than those stored at 13.33 kPa (100 mm Hg). At that time flowers stored in NA had a vase life of 2–4 days at 20°C; flowers that had been stored at 13.33 kPa (100 mm Hg) had a 4–6-day vase life; and blooms stored at 8.0–9.33 kPa (60–70 mm Hg) still retained their initial vase life of 6–8 days (Burg, 1973b, 1976a). A similar result was obtained with full-open blooms stored at 0–2°C and pressures ranging from 5.33 to 20 kPa (40–150 mm Hg) (Burg, 1970). Storage at 1.33–6.67 kPa (10–50 mm Hg) prolonged the duration of storage beyond 4 weeks with some loss in vase life (Patterson, 1975a,b; Grumman Allied Industries, 1979, unpublished test data).

After 6 weeks' storage at 0–1.7°C and a pressure of 1.33–3.33 kPa (10–25 mm Hg), bud-cut *Chrysanthemum* blooms stored in boxes with polyethylene liners to prevent wilting of the leaves and petals were in excellent condition, with no loss in their ability to open and little or no loss in vase life. Waxed paper was placed between the flowers and polyethylene liner to prevent water spotting due to condensation on the inner surface of the polyethylene during storage (Grumman Allied Industries, 1979, unpublished test data). Tight bud-cut flowers stored well in LP and performed like freshly harvested flowers when they were opened in floral preservative. LP buds opened more rapidly than fresh-cut buds, but both had approximately the same vase life. The rate of ethylene production was the same in buds regardless of whether they were fresh-cut or had been stored in LP. Staby *et al.* (1976; Staby, 1976b) found that after 6 weeks' storage at 3°C, bud-cut blooms that had been kept at 3.33 kPa (25 mm Hg) had double the vase life of equivalent blooms stored in NA.

In a test performed in Japan for Zen-Noh, a partial load of cut *Chrysanthemum* varieties that could only be stored for 10 days in NA was kept for 30 days in a 12.2 m (40 ft) Grumman/Dormavac LP intermodal container operated at 0–1°C, 95% RH and a pressure of 3.33 kPa (25 mm Hg). The flowers were protected with a polyethylene over-wrap. They retained their fresh appearance during LP storage, and drew water when the stems were recut under water or hardened in hot water overnight. Subsequently, the buds opened properly, but vase life was reduced from the initial value of 20 days (Grumman Allied Industries, 1980, unpublished data).

10.61 Chrysanthemum, Potted Plant (*Chrysanthemum morifolium* = *Dendranthema morifolium*)

Potted chrysanthemums are produced year-round in Florida and California. Growers would like to warehouse production 1 month prior to peak demand times and presently attempt to do so using NA facilities not only for these periods but also throughout the year as the requirement arises. More than 5–7 days of NA storage results in yellowing of the lower leaves, greatly reduces shelf life and prevents flower buds from subsequently opening. Growers have experienced claims for spoilage that more than offset any benefit gained by storage.

Potted plants of cvs Neptune, Golden Anne, Delaware and Bright Golden Anne benefit from LP storage at 0–4.4°C and a pressure of 13.33–20.0 kPa (100–150 mm Hg), but are better preserved at 5.33–10.67 kPa (40–80 mm Hg). They may be kept in that pressure range for more than 28 days in LP, compared to only 7 days in NA (Burg, 1969, 1970, 1973b, 1976a). After 4 weeks in LP storage, even very young buds develop and open fully when plants are transferred to air at atmospheric pressure. Young buds did not develop and open after only 1 week of NA storage. The flowers that developed from the buds of LP-stored plants had a normal shelf life of 3–4 weeks before fading occurred.

A full load of equal numbers of Yellow Anne, Neptune and Delaware potted *Chrysanthemum* plants was shipped from Florida to California in a 6.1 m (20 ft) Fruehauf prototype intermodal LP container operated at 2°C and a pressure of 8.0 kPa (60 mm Hg). Plants initially were selected at four stages of flower opening: tight-budded, slightly open buds, three-quarters open (commercial grade) and full-open. Six plants were packed per carton, and each plant was sleeved with polyethylene. The plants were inspected, scored and photographed just prior to loading. After arrival in Salinas, California, they were examined by State inspectors, and then immediately or 14, 21 and 28 days later, one-quarter of each variety and stage of flower development was removed and tested for shelf life. All plants in each sampling were judged to be in excellent condition at receipt in California and displayed at least 3 weeks' shelf life. Flower opening did not progress during storage, and even the smallest buds opened normally and fully after plants were removed to air. No leaf yellowing was observed, the soil remained moist during storage, the plants did not lose turgidity and no wilting occurred when they were transferred to air at atmospheric pressure and room temperature. Even better results can be expected at 1.22–3.33 kPa (10–25 mm Hg), because this pressure range should protect against disease development. Plastic sleeves are recommended.

10.62 Cuttings (Foliage and Woody)

At 2°C, after 5 weeks' storage, all cuttings of *Ligustrum* and *Hedera* were acceptable and rooted regardless of whether they had been kept in NA or in LP at pressures of 6.33, 12.67 or 25.33 kPa (47.5, 95 or 190 mm Hg). When instead they were stored at 6–10°C, they did not root after NA storage, and did after LP storage at 6.33–12.67 kPa (47.5–95 mm Hg). At 2°C, cuttings of *Rosa* 'Sweet Promise' Sonia were preserved for 5 weeks in NA, but subsequently could not be rooted. Stored at 6°C, they did not survive for 5 weeks in NA, but when they were kept for that same duration of time in LP at either 2 or 6°C and a pressure of 6.33 kPa (47.5 mm Hg), they survived and subsequently rooted. The result was less favourable at higher pressures (Jensen and Rasmussen, 1978). Excellent LP storage results have also been reported for poinsettia, *Hibiscus*, *Scindapsus*, *Chrysanthemum*, *Kalanchoe*, *Hedera*, *Cissus*, *Dieffenbachia* and other cuttings (10.59; Table 10.9). The Danish Low Pressure Storage Group (Danish Research Service for Plant and Soil Science, 1980)[13] reported a rooting percentage of at least 75% after LP storage of *Dieffenbachia* cuttings for 10 weeks, *Hibiscus* (12 weeks), *Hypocyrta* (12 weeks), *Scindapsus* (14 weeks) and *Chrysanthemum* (10 weeks). Tropical foliage cuttings (green *Peperomia*, *Philodendron cordatum*, Marble Queen Pothos, Golden Pothos, *Aphelandra* and grape ivy) were stored for 23 days at 12.8°C either in NA or in LP (Dressler and Jamieson, 1977; Jamieson, 1980c). The loss of viable cuttings was less than 10% when they were kept at 2.67 kPa (20 mm Hg), but some of these cuttings were chlorotic; at 4.0 kPa (30 mm Hg) the cuttings retained excellent appearance with no loss; at 5.33 kPa (40 mm Hg) the loss was less than 5%; and in NA most cuttings were dead or dying by that time. After storage the cuttings were rooted in soil under mist for 4 weeks, along with fresh cuttings of the same varieties. Of those stored at 2.67 kPa (20 mm Hg), 80% rooted and survived; at 4.0 kPa (30 mm Hg) 100% survived, rooted and then grew at the same rate as fresh cuttings; after storage at 5.33 kPa (40 mm Hg), 73% rooted and survived; after NA storage all cuttings died and none rooted. Based on visual appearance and the rooting response, the optimal LP pressure was 4.0 kPa (30 mm Hg).

Hibiscus cuttings stored in NA at 15°C no longer rooted satisfactorily after 3 weeks' storage in flowing saturated air, whereas if they were kept at a pressure of 2 kPa (15 mm Hg = 0.06% [O_2]) they still displayed an at-harvest rooting response. Within 5 weeks in LP they no longer rooted in an acceptable manner (Kirk *et al.*, 1986).

Table 10.9. Storage life of cuttings of various plant genera in LP and NA, after which normal rooting occurred.

Genera	Storage life – days		Temp. (°C)	Pressure (mm Hg)	Reference
	NA	LP			
Aphelandra	< 12	> 23	12.8	30	1
Cissus rhombifolia (grape ivy)	< 12	23	12.8	30	1
Carnation	90–120	> 240	0–2	60	2
Epipremnum pinnatum	42	> 56	15	15–20	3
Euphorbia pulcherrima					4
Rooted cutting	7	21	7.7	35	
Unrooted cutting	3	7	7.7	35	3
Hedera canariensis	< 35	> 35	2–10	47.5	5
Hibiscus rosa-sinensis	< 21	21–35	15	15–20	3
	14	> 42	12	47.5	5
Ligustrum obtusifolium	< 63	63	3.3	35	4
Ligustrum vulgare	21	42	4.4	25	6
	< 35	70	2	47.5–95	5
Peperomia (green)	< 20	> 23	12.8	30	1
Poinsettia (unrooted)	3	7	4.4	35	6
Poinsettia (rooted)	7	35	5.0	35	7
Pothos (Marble Queen)	< 12	> 23	12.8	30	1
Pothos (Golden)	< 12	> 23	12.8	30	1
Philodendron cordatum	< 12	> 23	12.8	30	1
Rhamnus	21	42	4.4	25	6
Rosa, Sweet Promise-Sonia	< 35	> 35	2	47.5–190	5
Viburnum opulus cv. Compactum	21	42	4.4	25	6

References: (1) Dressler and Jamieson, 1977; (2) Burg, 1973b; (3) Kirk *et al.*, 1986; (4) Eisenberg, 1977; (5) Jensen and Rasmussen, 1978; (6) Eisenberg *et al.*, 1977a; (7) Eisenberg *et al.*, 1978.

Apparently the 2 kPa (15 mm Hg) pressure was too low for optimal rooting, for 100% of LP-stored *Hibiscus* cuttings rooted after 6 weeks' storage at 12°C and a pressure of 6.33 kPa (47.5 mm Hg = 1% [O_2]) (Jensen and Rasmussen, 1978), while those stored in NA for 2 weeks at the same temperature did not root satisfactorily.

10.63 Foliage Plants, Potted

Dieffenbachia picta, *Codiaeum variegatum* and *Dracaena* sp. usually can be kept in NA at 16–18°C without significant loss of quality for up to 15, 21 and 15 days, respectively (Hardenburg *et al.*, 1986). Foliage plants do not benefit from CA storage (Poole *et al.*, 1985). They are injured by several days' exposure to more than 1 µl/l ethylene (Coniver, 1992), and in trial shipments have been damaged by ethylene build-up inside sealed see-through plastic packages (Harbaugh *et al.*, 1976).

Potted foliage plants of *Dieffenbachia amoena*, *D. picta*, *C. variegatum* hybrid, *Maranta leuconeura* Massangeana, *Syngonium* hybrid and *Dracaena* sp. were tested at 16.7°C and 95% RH in NA or in LP at pressures of 2.0, 4.0 and 10.67 kPa (15, 30 and 80 mm Hg = 0.014, 0.43 and 1.8% [O_2]). All varieties were well preserved in LP during a 21-day storage period, but not in NA. The optimal LP pressure was 4.0 kPa (30 mm Hg). A partial load of the same varieties was stored in a 12.2 m (40 ft) Grumman/Dormavac intermodal hypobaric container operated at 16.7°C, 95% RH and a pressure of 4.0 kPa (30 mm Hg). Initial shelf life was retained in LP during the 21-day storage. These varieties were successfully shipped in a 12.2 m Grumman/Dormavac LP intermodal container from Hilo, Hawaii, to

Oakland, California, and then over-the-road to Bay City. No loss of leaves or colour was noted during the 12-day journey, and no greenhouse rejuvenation was required prior to distribution (Grumman Allied Industry, 1978; Jamieson, 1980c, 1984).

The major fungal diseases of potted foliage plants are species of *Alternaria, Fusarium, Myrothecium, Phytophthora, Pythium, Rhizoctonia* and *Sclerotinia*. The most important bacterial diseases include *Erwinia, Pseudomonas* and *Xanthomonas* (Coniver, 1992). Growth of *Alternaria, Fusarium, Phytophthora, Pythium, Rhizoctonia, Sclerotinia, Erwinia* and *Pseudomonas* is inhibited by the 0.45% [O_2] present at 16.8°C and a saturated humidity when the LP pressure is 4.0 kPa (30 mm Hg) (7.1 and 7.4; Figs 7.2, 7.3 and 7.6). No information is available about the low [O_2] tolerance of *Xanthomonas* and *Myrothecium*.

10.64 *Freesia*, Cut Flower (*Freesia* sp.)

Freesia cut blooms can be stored for 2 weeks in NA at 0–0.5°C, or 1 week at 2°C (Hardenburg *et al.*, 1986). A sample of flowers wrapped in polyethylene film was included in an LP intermodal test performed in Japan for Zen-Noh in a Grumman/ Dormavac 12.2 m (40 ft) intermodal hypobaric container operated at 2°C, 90% RH and a pressure of 3.33 kPa (25 mm Hg). After 10 days the flowers were in excellent condition, opened normally and retained their initial vase life. Although the flowers were stored in a horizontal orientation, they did not develop gravicurvatures in LP, as they do in air. A longer storage time was not attempted. In a laboratory trial, a higher pressure, 6.67 kPa (50 mm Hg), only slightly benefited *Freesia* flower storage (Bangerth, 1973).

10.65 Fynbos, Cut Flower (*Erica sparsa, Stoebe plumosa, Phylica imberbus*)

Four boxes containing *Erica sparsa* (Heather), *Stoebe plumosa* and *Phylica imberbus* were stored in a 6.1 m (20 ft) VacuFresh^{SM} intermodal container operated at 1°C and a pressure of 14.8 ± 0.5 mbar (1.5 kPa = 11.1 ± 0.4 mm Hg). The container floor was flooded to simulate a full cargo load, but the measured RH was only 87% at a dew point of −1°C (9.3; Fig. 9.5). To protect the flowers from drying, the boxes were lined with slip-sheets of perforated Mylar, non-perforated Mylar, perforated PVC or non-perforated PVC (6.4, 9.9). The blooms looked exceptionally well after 31 days' LP storage, and had retained their initial vase life (Spearpoint, 2000a,b,c).

10.66 Geranium, Cutting (*Pelargonium* × *hortorum* Bailey)

The quality of unrooted geranium cuttings stored at 2–7°C improves when the pressure is lowered from 26.67 to 5.33 kPa (200 to 40 mm Hg). The tenderest geranium cuttings were unsaleable after 2–4 days' storage in NA at 2°C, and the hardiest in 1 week. They lost their leaves when an attempt was made to root them. At the same temperature, all varieties of geranium cuttings were well preserved in LP for at least 3–4 weeks at a pressure of 5.33 kPa (40 mm Hg), and they subsequently rooted normally. Interrupting the vacuum for 8 h each day gave the same result as continuous vacuum during a 3-week LP storage at 5.33 kPa (80 mm Hg), and cuttings from both treatments rooted properly with no leaf yellowing. Loosely fitting polyethylene wraps improved storage life in LP (Burg, 1976a).

The foliage of unrooted geranium cuttings died during a 2-week storage in NA at 4.4°C. At the same temperature and a pressure of 3.33–6.67 kPa (25–50 mm Hg), the cuttings were well preserved for 4–6 weeks in LP, and subsequently rooted normally. The result was less favourable at 13.33 kPa (100 mm Hg). In another study at 4.4°C and a pressure of 4.67 kPa (35 mm Hg), rooted cuttings remained in good condition for 4 weeks and unrooted cuttings for 6 weeks (Staby *et al.*, 1976; Eisenberg, 1977; Eisenberg *et al.*, 1977a,b, 1978; Fig. 10.6).

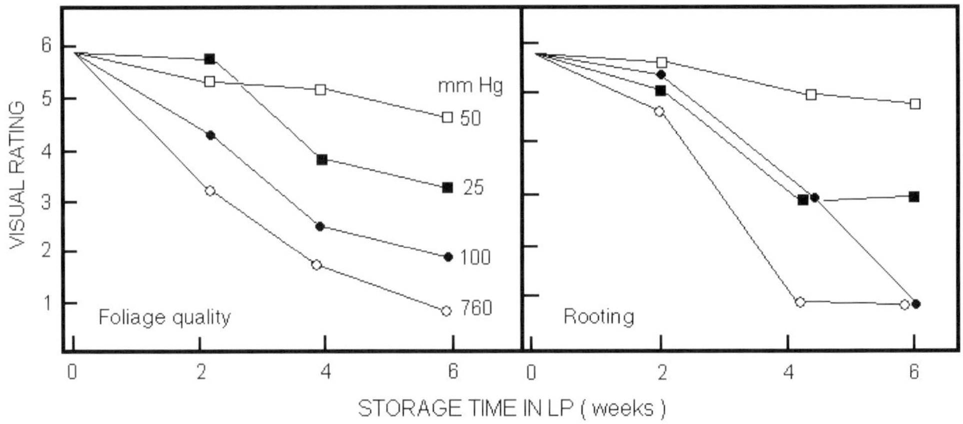

Fig. 10.6. Visual evaluation of the foliage (*left*) and root quality (*right*) of unrooted geranium cuttings after 26 days in the propagation bed following 2, 4 or 6 weeks' storage at 4.4°C in NA or in LP at pressures of 3.33, 6.67 or 13.33 kPa (25, 50 or 100 mm Hg), respectively. Ratings for foliage: 1 = cutting dead; 2 = all leaves have deteriorated (lost turgor, yellowed, developed necrotic areas or abscissed); 3 = more than 50% of leaves have deteriorated; 4 = less than 50% of leaves have deteriorated; 5 = one or two leaves deteriorated; 6 = leaves in good condition, with no loss of turgor. Rating for rooting: 1 = cutting dead; 2 = callus with 1–3 roots; 3 = showing few roots; 4 = light rooting; 5 = medium rooting; 6 = heavy rooting (Eisenberg, 1977).

A pressure of 6.67 kPa (50 mm Hg) gave a favourable result with rooted geranium cuttings at 0°C (Dilley, 1977a).

10.67 Geranium, Potted Plant (*Pelargonium* × *hortorum* Bailey), Red

Potted geranium plants with tight buds or with buds showing a trace of red were stored at 12.8°C either in NA or in LP at pressures of 10.67 and 21.33 kPa (80 and 160 mm Hg). Within 13 days all buds had opened moderately and to the same extent in NA and LP but leaf yellowing, browning and leaf drop occurred only in NA, and these flowers were flaccid and shaded half-blue/half-red. Bluing occurred to a lesser extent in LP at 21.33 kPa (160 mm Hg), and was absent at 10.67 kPa (80 mm Hg), indicating that ammonia formation had been prevented (4.20). Plants stored in LP retained excellent appearance during 13–28 days' storage, but 3 days after they were transferred to air in shaded light at room temperature they developed extensive yellowing and leaf drop in shaded light (Burg, 1970). Lower pressures were not tested.

10.68 Ginger, Red, Cut Flower (*Alpinia purpuata*)

Red ginger flower spikes can be kept in NA for 4–7 days at 13°C (Hardenburg *et al.*, 1986), but during storage their leaves wilt and darken (Akamine, 1976a).

At 15°C, the leaves and flowers of cut floral spikes of plastic-wrapped red ginger turned black within 12 days in NA. In LP at the same temperature and pressures of 6.67–10.67 kPa (50–80 mm Hg), the at-harvest appearance of both leaves and flowers was retained during 28 days' storage. After 44 days in LP, the tips of the leaves began to show darkening, although the flowers were still in excellent condition. The result was somewhat better at 6.67 kPa (50 mm Hg) than it was at 10.67 (80 mm Hg). After 44 days, the vase life of the LP flowers was determined in water ± Bloomlife at ambient temperature. In-rolling and fading began after 5–6 days in water and was completed in 8 days; in Bloomlife the leaves and blooms remained in excellent condition for 16 days (Burg, 1969).

During storage at 11°C, the leaves began to yellow after 5–7 days in NA, and soon thereafter the flowers turned brown and desiccated, even though the spikes were wrapped in polyethylene or wet newspaper. Flowers and leaves remained in good condition during 28–35 days of LP storage at pressures of 6.67 and 8.0 kPa (50 and 60 mm Hg), and the floral spikes subsequently displayed a normal vase life (Burg, 1973b). A pressure of 13.33 kPa (100 mm Hg) provided only slight benefit (Burg, 1976a), and pressures lower than 6.67 kPa (50 mm Hg) have not been tested.

10.69 Ginger, Wax, Cut Flower (*Alpinia speciosa*)

Flower stalks of wax ginger, cut to a 30-cm length containing the bloom and a few appended leaves, were stored at 11.7 or 15°C either in NA or in LP at pressures of 7.33, 10.67, 16.0 or 21.33 kPa (55, 80, 120 or 160 mm Hg). The stalks were stored in NA with their cut ends in water, or wrapped in moist newspaper. In LP they were protected with polyethylene overwraps. Floral wilting and fungal spotting were evident after 9–10 days in NA, and by 12 days the leaves had turned brown, the flowers were black and there was severe fungal contamination. LP storage improved progressively as the pressure was reduced. At 16 or 21.33 kPa (120 or 160 mm Hg), fungal spotting marred the appearance of the leaves within 15 days, and some leaves had become chlorotic. The bloom of stalks stored at 10.67 kPa (80 mm Hg) was still in good condition at that time, with a few spotted leaves, while at 7.33 kPa (55 mm Hg) there was no change from the initial appearance of either the blooms or leaves (Burg, 1969). Lower pressures have not been tested.

10.70 *Gladiolus*, Cut Flower (*Gladiolus* sp.)

Gladiolus can be stored in NA for 5–8 days at 2–5°C. An atmosphere containing 1% $[O_2]$ + 5% $[CO_2]$ slightly extends storage life at 0–4.4°C (Uota, 1969b). The flower stems must be kept upright in NA to prevent geotropic curving.

Gladiolus spikes were stored at 0–3°C either in NA or in LP at pressures ranging from 5.33 to 20.0 kPa (40–150 mm Hg). Storage improved progressively as the pressure was lowered. Although stored in a horizontal position, none of the spikes curved geotropically during 2 weeks in LP, whereas curvature developed within a few days in NA. At 1.7–2°C and a pressure of 8.0 kPa (60 mm Hg), the blooms were well preserved for 28 days, and after removal from LP the buds opened completely and normally. Bud opening was impaired in NA after 1 week of storage (Burg, 1973b, 1976a). In another study, LP gave a favourable result at a pressure of 6.67 kPa (50 mm Hg) and a storage temperature of 0°C (Dilley, 1977a).

The stage of harvest is important for *Gladiolus* bud opening. Flowers should be cut when one or two of the lowest florets show colour but are still at a relatively tight-bud stage (Marousky, 1977b; Hardenburg *et al.*, 1986). It has been reported that when flowers are cut at this stage, there are no differences in the opening of florets on spikes held in CA, in flowing air, and at 6.67 kPa (50 mm Hg), and that the maximum storage duration is 2 weeks (Marousky, 1977a).

When *Gladiolus* was stored in a prototype 12.2 m (40 ft) intermodal hypobaric container operated at 1.7°C, 95% RH and a pressure of 5.33 kPa (40 mm Hg), after 4 weeks all florets opened when spikes were transferred to water, but vase life was somewhat diminished compared to fresh-cut flowers. Pressures lower than 5.33 kPa (40 mm Hg) should be tested.

10.71 *Gypsophila*, Cut Flower (*Gypsophila* sp.)

Gypsophila (Baby's-breath) can be stored at 4°C for up to 3 weeks in water or preservative solution (Hardenburg *et al.*, 1986). It does not keep well in dry storage.

Standard commercial boxes of *Gypsophila* have been routinely dry-stored in LP for holiday-price appreciation during periods of 4–6 weeks. The boxes are cooled in Colombia, S.A. but re-warm during air shipment from Bogotá, Colombia, to Miami, Florida. The boxes are vacuum-cooled in Miami, and loaded into Grumman/Dormavac 12.2 m (40 ft) intermodal hypobaric containers as part of mixed flower loads. The containers are operated at 1–2°C and a pressure of 2.0–2.67 kPa (15–20 mm Hg), using the 'dry' hypobaric method (2.1). After storage the flowers are distributed by normal commercial means.

10.72 *Heliconia*, Cut Flower (*Heliconia humilis, Heliconia latispatha*)

Heliconia humilis (red), *H. latispatha* red and yellow bloom spikes were stored at 10–12.8°C either in NA or at pressures of 8.0, 13.33 and 20.0 kPa (60, 100 and 150 mm Hg). In NA, the leaves developed necrotic spots and flower fading was evident within 10–15 days. Storage life was improved at 13.33–20.0 kPa (100–150 mm Hg), but 8.0 kPa (60 mm Hg) gave a much better result, preserving the blooms for at least 23 days and in the case of *H. latispatha* (yellow) for as long as 41 days. Storage was limited by mould development (Burg, 1969, 1973b, 1976a). Lower pressures have not been tested.

10.73 *Leucadendron*, Cut Flower (*Leucadendron argenteum*), vars Pisa, Jubilee Crown

Four boxes of *Leucadendron* blooms (Silver-tree, a member of the *Protea* family) were stored in a 6.1 m (20 ft) VacuFresh^{SM} container operated at +1°C and a pressure of 14.8 ± 0.5 mbar (11.3 ± 0.4 mm Hg). The container floor was flooded to simulate a full load, but the measured RH in the container was only 87% at a −1°C dew point (9.3; Fig. 9.5). To protect the flowers from drying, the boxes were lined with slip-sheets of perforated Mylar, non-perforated Mylar, perforated PVC or non-perforated PVC (6.4, 9.9). During 31 storage days, the best result was obtained with non-perforated Mylar, and the worst with perforated PVC. The 'Pisa' and 'Jubilee Crown' varieties showed very little degradation during storage, except for the crown tips of 'Jubilee Crown' kept in perforated PVC. After removal and transfer to water, the flowers continued opening, and 10 days later they were still in excellent condition (Spearpoint, 2000a,b,c).

10.74 *Liatris*, Cut Flower (*Liatris* sp.)

The optimal storage temperature for *Liatris* in NA is 0–2°C (Hardenburg *et al.*, 1986). Standard commercial boxes of *Liatris* have been routinely stored in LP for price appreciation during periods of 4–6 weeks prior to major holidays. The boxes are cooled in Colombia, they re-warm during air shipment from Bogotá to Miami, Florida, are vacuum cooled in Miami, and loaded into Grumman/Dormavac 12.2 m (40 ft) intermodal hypobaric containers as part of mixed-flower loads. The equipment is operated at 1–2°C and a pressure of 2.0–2.67 kPa (15–20 mm Hg), using the 'dry' hypobaric method (2.1). After storage the flowers are distributed by normal commercial means.

10.75 Lily, Easter, Cut Flower (*Lilium longiflorum speciosum rubrum*)

Easter lily cut blooms can be stored for 2–3 weeks in NA at 0–2°C (Hardenburg *et al.*, 1986). A sample of cut Easter lily blooms, wrapped in polyethylene, was included in a 10-day test performed in Japan for Zen-Noh in a 12.2 m (40 ft) Grumman/Dormavac LP intermodal container operated at 1.7°C, 90% RH and a pressure of 3.33 kPa (25 mm Hg) (Grumman Allied Industry, 1980, unpublished data). After 10 days in LP, the blooms were in excellent condition, and had retained their initial vase life. Longer storage periods were not attempted.

10.76 Lily, Easter, Potted Plant (*Lilium longiflorum speciosum rubrum*)

Depending upon weather conditions, potted lily plants sometimes come into bloom as much as 3–6 weeks before Easter. When this occurs, they are unsaleable. Storage in NA is limited to 2 weeks at 1.7°C because subsequently the leaves yellow and flowers lose their ability to open.

Plants with one open flower were stored at 4.4°C either in NA or in LP at a pressure of 10.67 kPa (80 mm Hg). At room temperature, control plants opened all buds within 3–4 days and fading was complete in 9–10 days. After 22–26 days in NA, a second blossom had opened and the first showed signs of ageing. At that time, plants in LP still had a fresh first flower, and a second bud that was barely cracked open. Plants removed from LP storage after 23 days displayed 10–15 days floral shelf life (Burg, 1970).

During 6 weeks at 1.7°C and 95% RH in LP, plants stored with their lower bud just cracked open did not develop further at 5.33–8.0 kPa (40–60 mm Hg), leaf yellowing did not occur, there was no loss in the ability of the lower buds to open, and the flowers had a normal shelf life after plants were transferred to atmospheric pressure (Burg, 1973b).

Plants with their most mature flower bud just swelling and showing white were stored at 4.4°C either in NA or in LP at a pressure of 8.0 kPa (60 mm Hg). Initially, at room temperature and atmospheric pressure, all buds opened within 7 days and flower fading was complete in 17 days. After 15 days in LP or NA, all flower buds were still tightly closed. When these plants were transferred to room temperature, buds from both NA and LP plants began to open in 3–4 days, and fading was complete in 15 days. However, the plants stored in NA developed extensive yellowing of their lower leaves, extending half-way up the stem, and eventually all of these leaves died and were shed, whereas the LP plants did not display this symptom (Burg, 1970). In another test, plants with their most mature flower bud just swelling and showing white were stored at 2°C in NA and in LP at pressures of 5.33 and 8.0 kPa (40 and 60 mm Hg). In NA, the leaves became chlorotic within 2 weeks. After 6 weeks in LP the appearance was excellent and the flowers had retained their normal shelf life (Burg, 1976a).

A pressure of 6.67 kPa (50 mm Hg) gave favourable result with Easter lily plants in a laboratory test at 0°C (Dilley, 1977a). This was followed by a trial in which a partial load of Easter lilies was stored at Michigan State University in a 6.1 m (20 ft) prototype Fruehauf intermodal hypobaric container operated at 1.7°C, 95% RH and a pressure of 6.67 kPa (50 mm Hg). During 4 weeks, the lilies retained their at-receipt appearance, and after transfer to air, their flower buds opened properly and they displayed normal shelf life.

10.77 *Narcissus*, Cut Flower (*Narcissus* sp.)

Bangerth (personal communication, 1979) found that a pressure of 6.67 kPa (50 mm Hg) provided only slight benefit for the storage of narcissus blooms.

10.78 Orchid, Cut Flower (*Vanda Agnes Joaquim*)

Vanda Agnes Joaquim blooms can be kept for only 16 days in NA at 10°C. During 29 days' storage in LP at the same temperature and 13.33 kPa (100 mm Hg), 31% of the blooms faded; at 8.67 kPa (65 mm Hg), 13% faded; and at 5.33 kPa (40 mm Hg), none faded. All blooms kept at 5.33 kPa (40 mm Hg) were still in excellent condition after 41 days' storage, and they did not begin to fade for 10–14 days after they were transferred to water at 10°C and atmospheric pressure (Burg, 1969, 1973b, 1976a). Pressures lower than 5.33 kPa (40 mm Hg) were not tested.

Pressures between 16.67 and 25.33 kPa (125 and 190 mm Hg) prevented fading of *Vanda* flowers during simulated transit and subsequent holding conditions (Akamine,

1976a). Ethylene production was suppressed after LP storage.

10.79 Poppy, Cut Flower (*Papaver* sp.)

Cut poppy flowers can be stored in NA for 3–5 days at 4°C (Hardenburg *et al.*, 1986). A sample of poppy cut flowers, wrapped in polyethylene, was included in a 10-day LP intermodal test performed for Zen-Noh in Japan. The 12.2 m (40 ft) Grumman/Dormavac hypobaric intermodal container was operated at 1.7°C, 90% RH and a pressure of 3.33 kPa (25 mm Hg). After 10 days in LP, the flowers were in excellent condition, opened normally and had their initial vase life. Longer times and lower pressures have not been tested.

10.80 *Protea*, Cut Flower (*Protea* sp.), vars Eximia, Cardinal, Sylvia, Pink Ice, Renata, Repens

Protea blooms can be kept refrigerated for long periods of time at 0°C, but massive blackening of both flowers and leaves begins within a few hours after they are removed from storage (Haasbroek *et al.*, 1973; Akamine, 1976a; Akamine *et al.*, 1979; Paull *et al.*, 1980; Ferreira, 1983; Brink and de Swart, 1986; Newmann *et al.*, 1989). The market value of susceptible species such as *Protea nerifolia* and *P. exima*, which have an at-harvest vase life of 3–4 weeks, is severely reduced by 'blackening' within 3–7 days (McConchie *et al.*, 1991). Deterioration is promoted by water loss and high temperatures, and slowed at 2–7°C (De Swardt, 1977; Mulder, 1977; Uys, 1980; Ferreira, 1983). The blackening may occur when water stress leads to a total loss in the membrane integrity of mesophyll cells adjacent to the leaf veins, causing decompartmentation and the subsequent oxidation of phenolic compounds (Mulder, 1977) and/or flavonoids such as proanthocyanidins (Ferreira, 1983). Leucoanthocyanins, which occur in *P. nerifolia* R. Br. leaves (Elsworth and Martin, 1971; Du Plessis, 1978), darken due to auto-oxidation in the presence of O_2 or after enzymatic oxidation (Ribereau-Gayon, 1972). Oxidative enzymes also participate in petal blackening of roses (Zieslin and Halevy, 1969; Moe, 1975). The high phenyl ammonia lyase and polyphenol oxidase activity in *Protea* may be responsible for the pigment oxidation that occurs (Paull *et al.*, 1980).

Water loss from *P. nerifolia* flowers accounts for 25–50% of the transpiration from the leafy stem with an attached flower, and the floral transpiration accentuates the water stress that triggers leaf-blackening (Paull *et al.*, 1980). The leaves do not use much water and they brown more slowly if the inflorescence is removed (Paull *et al.*, 1980; Ferreira, 1983).[14] Browning also is delayed when *Protea* flowers are shipped in plastic bags to limit water loss but inclusion of an ethylene absorbent (Purafil) inside the plastic bags hastens browning (Paull *et al.*, 1980). During vase-life tests, complete browning of *Protea* leaves is induced within 2 h at 40°C even though the humidity is kept at 100% (Ferreira, 1983). Possibly this occurs because high temperatures damage membranes and cause cell leakage (3.13; Fig. 3.6, *lower right*).[15] In addition, *P. nerifolia* leaves display a typical respiratory climacteric pattern during senescence, and elevating the temperature from 25 to 50°C increases the respiration rate from 176.2 to 1877.3 ml O_2/kg·h ($Q_{10} \cong 2.5$), and the extra respiratory heat is removed by evaporative cooling (6.1). Assuming that a 5% weight loss is sufficient to cause water stress (Burton, 1982), if all the respiratory heat produced by dry-packed *Protea* blooms is transferred by evaporative cooling, the leaves will become water-stressed within 33.4 h at 25°C, and in 3 h at 50°C. Approximately 9 days at 2°C would cause the same weight loss, assuming that the Q_{10} is 2.5 in the 2–25°C range. In agreement with this calculation, flowers preconditioned with water and then held in dry packs for 1 week at 2°C displayed 50% leaf blackening by the 4th day of a vase-life test in water (Paull *et al.*, 1980).

Leaf blackening is reduced by postharvest treatments that increase the carbohydrate content of *Protea* leaves, such as illumination, preconditioning flowers with a preservative containing 0.2–1% sucrose and adding sugar during vase-life display (Ireland *et al.*, 1967; Haasbroek *et al.*, 1973; Paull *et al.*, 1980; Brink and de Swart, 1986; Ferreira, 1986; Newmann *et al.*, 1989; McConchie *et al.*, 1991). In *Protea* leaves, as in other leaves, one of the initial events to be expected after harvest is a shift from normal carbohydrate metabolism to the breakdown of protein, amino acids and amides. The ammonia that results is responsible for the bluing of roses and some other flowers (4.20), and conceivably could contribute to leaf blackening in *Protea*, since tissue darkening is a common symptom of ammonia toxicity. LP storage at a pressure of 1.33 kPa (10 mm Hg) should lower the respiration rate of *Protea* flower stalks by 90% (Fig. 4.2), conserve carbohydrate, prevent ammonia formation and reduce water loss to a minimum.

A few boxes of *Protea* blooms were included in a test carried out with roses stored in a 6.1 m (20 ft) VacuFresh^{SM} container operated at 1°C and a pressure of 14.8 ± 0.5 mbar (11.1 ± 0.4 mm Hg). The operating condition was chosen to be optimal for rose storage. Neither the cold sensitivity nor proper LP storage pressure for *Protea* was known, as this flower had never been tested in LP previously. The container floor was flooded to simulate a full load, but the measured RH was only 87% at a –1°C dew point. To protect the flowers from drying, the boxes were lined with either perforated Mylar, non-perforated Mylar, perforated PVC or non-perforated PVC slip-sheets (6.4, 9.9). The measured temperature inside the Mylar-lined rose boxes was 0°C during the LP test (Fig. 9.5). During 31 days *Protea* was well preserved in boxes lined with non-perforated Mylar. 'Renata' (*P. nerifolia* × *P. burchellii*) gave the best result, with no leaf blackening or damage to the flowers. 'Repens' showed signs typical of heat stress (random blackening of bracts), but this also might have resulted from cold damage. The Eximia types of *Protea* ('Eximia', 'Cardinal', 'Sylvia') developed leaf blackening during storage, possibly due to cold damage. 'Pink Ice' showed signs of 'frostbite' at the tips of the primary bracts, but little damage on the leaves. After removal from LP and transfer to water, all flowers continued opening normally, they had their normal 8-day vase life and leaf blackening did not develop or worsen. The flower grower who provided the blooms concluded that if the burn/frostbite problem could be overcome by elevating the temperature to at least 2°C, which is the lowest recommended temperature for *Protea* (Hardenburg *et al.*, 1986), the method had considerable promise for these flowers.

10.81 Rapeseed, Cut Flower (*Brassica napus*)

A sample of rapeseed cut flowers, wrapped in polyethylene, was included in a 10-day LP intermodal test performed for Zen-Noh in Japan. The 12.2 m (40 ft) Grumman/Dormavac LP intermodal container was operated at 1.7°C, 90% RH at a pressure of 3.33 kPa (25 mm Hg). After 10 days in LP, the flowers were in excellent condition, opened normally and had their initial vase life. A longer storage time was not attempted.

10.82 Rose, Cut Flower (*Rosa* sp.), 'Sweetheart', 'Tanbeedee Belinda', 'Forever Yours', 'Bacarra', 'Merko Mercedes', 'Sweet Promise', 'Sonia', 'Visa', 'Madam Delbar', 'Sandia', 'Samantha', 'Royalty', 'Spanish Sun', 'Town Crier'

The vase life of roses is significantly reduced when they are dry-stored for more than 5 days in NA at –0.5 to 0°C. After 7 days, flowers of some cultivars may 'blow open' in as little as 4 h, whilst flowers of other cultivars may fail to open (Durkin, 1992). Dry storage at –0.5 to 0°C in moisture-proof containers can increase storage life to nearly 2 weeks (Hardenburg *et al.*, 1986). CA storage in 0.5–8% [O_2] has not proved beneficial (Staby *et al.*, 1984) and neither STS nor 1-MCP improve rose storage (5.30).

Atmospheres containing 0.5, 1 or 2% [O_2] at 0°C, delayed bud opening of 'Red Delight', 'Bacarra' and 'Pink Sensation' roses, but did not contribute to a satisfactory vase life (Uota, 1969b).

When 'Sweetheart' roses were stored at 1.7°C with their cut ends in water, blooms kept in NA opened in a few days and fading occurred in 14–18 days. At that time, all blooms stored at 5.33 kPa (40 mm Hg) were still tightly budded, and the foliage was in excellent condition. At 10.67 kPa (80 mm Hg), and to a lesser extent at 8.0 kPa (60 mm Hg), the leaves were slightly wilted and the flower buds had 'bent neck'. Blooms stored at 10.67 kPa (80 mm Hg) did not open normally due to bent neck, but blooms stored at 5.33 kPa (40 mm Hg) opened when they were transferred to water in 26°C air, and had a vase life of 3–4 days. After 26 days' storage in LP, even the blooms kept at 5.33 (40 mm Hg) developed 'bent neck' (Burg, 1969).

'Forever Yours' roses were dry-stored in LP at 0–4°C and pressures of 5.33–6.67 kPa (40–50 mm Hg). After 8 weeks in LP at 6.67 kPa (50 mm Hg) these roses had good residual shelf life, but often decay had developed and 'bluing' typical of stored red roses was evident (Dilley, 1972, 1977a; Burg, 1973b, 1976a; Dilley and Carpenter, 1973; Carpenter and Dilley, 1975; Dilley *et al.*, 1975; 4.20).

LP storage at 4°C and a pressure of either 5.33 or 10 kPa (40 or 75 mm Hg) preserved bud-cut 'Bacarra' roses for more than 42 days (Table 10.10). Bluing was evident in NA controls within 28 days, but not in LP. Possibly because of leakage and inadequate humidity control at 5.33 kPa (40 mm Hg) (Table 9.3), the result was better at 10 kPa (75 mm Hg).

Table 10.10. Mean vase life of Baccara roses in water after storage at 4°C in LP at 10.0 and 5.33 kPa (75 and 40 mm Hg) or in NA (Bangerth, 1973).

Storage time (days)	Shelf life (days)		
	NA	75 mm Hg	40 mm Hg
0	7.0	7.0	7.0
14	4.0	7.5	–
28	0.5	6.1	5.1
42	0.0	6.5	1.1

Rosa (Floribunda) 'Tanbeedee Belinda' blooms harvested at two stages of development, were dry-stored at 3°C and 98% RH, either wrapped or unwrapped, for up to 28 days at 6.33 kPa (47.5 mm Hg). Shelf life was determined at 22°C, 60% RH with 15 W/m^2 emitted light during a 14-h day. The average keeping quality of the roses after 3 weeks in LP was judged as good or better than that of roses kept in NA for 1 week. LP prevented bluing of the petals, which is a common problem with this type of rose during extended storage. The flower diameter was 7% larger for open LP flowers compared to NA controls. Bud-cut roses kept 20% better and their flower diameter was 20% smaller compared to flowers harvested with the petals fully developed. The flowers progressively lost vase life between 28 and 42 days of LP storage. LP doubled the dry-storage life of wrapped 'Merko Mercedes' 'Sweet Promise' and 'Sonia' roses, but after 30 days leaf quality deteriorated (Jensen and Rasmussen, 1978; Jensen and Bredmose, 1979; Bredmose, 1979, 1980a,b).

When 'Visa', 'Madam Delbar' and 'Sandia' roses are stored using the 'dry' LP method, after 21 days at 2°C and a pressure of 3.33 kPa (25 mm Hg) the flowers still retain their initial vase life. They open normally, except for 'Visa', which usually does not open, but did after LP storage (Burg, 1992). A sample of red and pink roses was stored in Japan at 1.7°C and 90% RH in a 12.2 m (40 ft) Grumman/Dormavac LP intermodal container operating at 3.33 kPa (25 mm Hg). The test duration was 10 days. At that time the blooms still displayed their initial appearance and shelf life (S.P. Burg, 1980, unpublished data).

An atmosphere containing 0.5–0.8% [O_2] at 0°C provides the equivalent of the [O_2] present in LP at 3.07–4.67 kPa (23–35 mm Hg) and the same temperature. This CA atmosphere did not prolong the storage life of roses, whereas the same [O_2] concentration supplied at a low pressure was highly beneficial (Staby *et al.*, 1979, 1984). After 4 weeks' dry storage at –0.6°C either in LP at a pressure of 1.33 kPa

(10 mm Hg), NA or CA at 0.5–0.8% [O_2], the shelf life of roses packed in boxes was: 'Samantha' = 7.57 days in LP, 0.41 days in NA and 4.31 days in CA; 'Forever Yours' = 7.85 days in LP, 4.62 days in NA and 0.50 days in CA; 'Sonia' = 3.94 days in LP, 2.37 days in NA and 2.54 days in CA (Dressler, 1980).

The incidence of bent neck in roses was reduced by a pre-treatment in LP. When the flowers were scored after 6–9 days of vase life, bent neck had been reduced from 90% without a pre-treatment in LP, to 77% if they were pre-treated for 3 h at 10 kPa (75 mm Hg), and to 55% with a 3.33 kPa (25 mm Hg) pre-treatment (Staby, 1976a, 1977).

During the past 10 years, LP has been used to store many millions of stems of Colombian roses in anticipation of holiday-price appreciation. The flowers are packed with perforated polyethylene sleeves in conventional boxes, cooled to 0°C in Colombia, air-shipped with 'ice'-packs to Miami, vacuum-cooled to 0–2°C at Miami International Airport, loaded into 12.2 m (40 ft) Grumman/Dormavac hypobaric intermodal containers, and stored at 1°C and a pressure 2.0 kPa (15 mm Hg) using the 'dry' hypobaric method. Each container holds approximately 270,000 blooms, and the price per bloom may increase by as much as US$1.00–US$1.50 per stem within a 3–4-week period prior to Mother's Day.

It has been reported that leaf disorders sometimes develop in 'Forever Yours', 'Royalty', 'Spanish Sun' and 'Town Crier' roses during LP storage (Staby *et al.*, 1979, 1984). Leaf disorders have only been observed on a few occasions during commercial LP rose storages. The cause is obscure. In the same commercial storage, roses from one farm may sporadically develop this disorder, whereas the same variety from several other farms does not. It has been speculated that the disorder is caused by fungicide residues, and that the spray regime differs depending on the farm. Leaf disorders also might be related to differences in the packaging used by various farms, which could cause desiccation when leaf stomates are opened by LP storage, especially if they remain 'locked open' for an extended period after the flowers are removed from storage. This sometimes causes wilting of potted plants and cuttings after they are removed from LP storage (4.15). If this is the cause of the disorder, it can be alleviated by leaving the roses in their protective packaging at atmospheric pressure for at least 1 day to allow the stomates to close before the floral spikes are removed from the boxes and exposed to light.

Successful shipments of 'Visa' roses from California to Chicago and from the Dominican Republic to the USA have been made in 12.2 m (40 ft) Grumman/Dormavac LP intermodal containers operated at 0°C, 95% RH and pressures of 1.33–2.0 kPa (10–15 mm Hg; Jamieson, 1984). In a test performed for South African growers, six boxes of 'Grand Gala' roses were stored at 1.1°C and a pressure of 14.8 ± 0.5 mbar (11.1 ± 0.4 mm Hg) in a 6.1 m (20 ft) VacuFresh^{SM} container. The floor was flooded to simulate a full load, but the measured RH in the container was only 87% at a −1°C dew point. To protect the flowers from drying, the boxes were lined with either perforated Mylar, non-perforated Mylar, perforated PVC or non-perforated PVC slip-sheets (6.4, 9.9). The measured temperature inside the Mylar-lined rose boxes was 0°C during the LP test (Fig. 9.5). After 31 days of LP storage, the flowers still had their at-harvest vase life of 8.5 days when they were transferred to water in ambient air (Spearpoint, 2000a,b,c).

10.83 Snapdragon, Cut Flower (*Antirrhinum majus*), 'Promenade Crimson'

Snapdragon spikes can be stored in NA at 4°C for 1–2 weeks, or by the dry-pack method at −0.5 to 0°C for 3–4 weeks (Hardenburg *et al.*, 1986). They must be stored upright because when they are placed in a horizontal orientation they undergo rapid geotropic bending, and within 2 h are stimulated to produce ethylene (Philosoph-Hadas *et al.*, 2000).

Snapdragons were stored at 0°C in NA or in LP at pressures ranging from 5.33 to 20.0 kPa (40–150 mm Hg). Bloom spikes that had 3 weeks' storage life in NA were well preserved during 42 days at pressures of 5.33–8.0 kPa (40–60 mm Hg). Bud opening was normal after 6 weeks in LP, but not after 3–4 weeks in NA. Higher LP pressures were less effective. Flowers stored in a horizontal position at 0–3°C curved geotropically in NA within a few days, but they remained straight during several weeks in LP (Burg, 1973b, 1976a).

The stage of harvest is important for snapdragon bud opening. Flowers should be cut just after the lower five to six flowers have opened (Marousky, 1977b; Hardenburg *et al.*, 1986). Snapdragon var. 'Promenade Crimson' was harvested with only one to three open florets, and stored at 0°C in NA or in LP at 6.67 kPa (50 mm Hg). When the flowers were transferred to air after 8 weeks' LP storage, the florets of the entire spike opened normally, and abscission and dropping of the florets was retarded. The beneficial effect was attributed to a reduced respiration rate and lower IEC. However, a high incidence of decay developed during LP storage of snapdragons harvested at the commercial stage with 30–50% of the florets open (Dilley, 1972, 1977a; Carpenter and Dilley, 1975). Contrary to this result, it has been reported that at 2°C there was no difference in the number of florets that opened on spikes stored in flowing air (NA), in CA or at 6.67 kPa (50 mm Hg in LP). Floral preservatives aided opening on stored spikes, but in no case did the stored spikes have as many florets as freshly harvested spikes (Marousky, 1977a).

It is likely that snapdragon decay can be alleviated by storage at 1.33 kPa (10 mm Hg), but the lower pressure range has not been investigated with this flower.

10.84 Statice, Cut Flower (*Statice ameria*)

Statice can be stored in NA at 2–4°C for 3–4 weeks (Hardenburg *et al.*, 1986). Standard commercial boxes have been routinely stored in LP for holiday-price appreciation during periods of 4–6 weeks. The boxes are cooled in Colombia, air-shipped from Bogotá to Miami, Florida, vacuum-cooled in Miami and loaded into Grumman/Dormavac 12.2 m (40 ft) intermodal hypobaric containers as part of mixed-flower loads. The equipment is operated at 1–2°C and a pressure of 2.0–2.67 kPa (15–20 mm Hg). After storage, the flowers are distributed by normal commercial means.

10.85 Stock, Cut Flower (*Malcomia maritima*), Purple and Rose Red

Stock can be stored for 3–5 days in NA at 4.4°C (Hardenburg *et al.*, 1986). Samples of purple and rose-red stock, wrapped in polyethylene, were included in an LP intermodal container test performed in Japan for Zen-Noh at 1.7°C, 90% RH and a pressure of 3.33 kPa (25 mm Hg). After 10 days, the flowers were in excellent condition, opened normally and had retained their initial vase life (Grumman Allied Industy, 1980, unpublished data).

10.86 Tulip, Cut Flower (*Tulipa* sp.)

The storage life of tulips in NA is 2–3 weeks at –0.5 to 0°C. If stored dry in boxes, tulips should be kept vertical with the blooms upright to prevent bending (Hardenburg *et al.*, 1986).

A full load of cut tulips was shipped in commercial cartons from Holland to Miami, Florida, in a 6.1 m (20 ft) Vacu-Fresh^{SM} intermodal container operated at 1.5°C and a pressure of 2.67 kPa (20 mm Hg). The bunches were plastic sleeved and the boxes contained wood shavings at each end. After the 18-day shipment, the flowers were distributed and sold (Burg, 1997).

According to F. Bangerth (personal communication, 1979) a pressure of

6.67 kPa (50 mm Hg) provides only slight benefit for tulip blooms.

10.87 Seeds [*Apium graveolens* var. Dulce (Celery), *Brassica oleracea* (Cabbage), *Allium cepa* (Onion)]

Storing seeds in a partial vacuum in sealed chambers or packages does not have much effect upon their viability, although experiments with high-moisture feed grains stored at a reduced pressure in static or semi-static systems have been encouraging. Seeds were stored at 26.5 ± 2°C for 14 weeks in the dark at the local atmospheric pressure (98.7 kPa = 740 mm Hg) with just sufficient vacuum to retain lids tightly on the chambers (97.3 kPa = 730 mm Hg) and also at 50.7 kPa (380 mm Hg) or at 10.5 kPa (79 mm Hg). The relative humidity of the flowing air was maintained at 20 ± 5% (ambient, 'dry') or 80 ± 5% ('wet') by means of a humidifier, and the moisture content of the air entering the LP chambers was reduced in proportion to the pressure reduction. The most striking effect observed with a 'wet' condition was the higher germination percentage of seeds that had been held at a reduced pressure, compared with seeds stored at 97.3 kPa (730 mm Hg). No significant differences in seed germination were observed among 'dry' treatments, but those seeds that had been stored at 10.5 kPa (79 mm Hg) under 'wet' or 'dry' conditions germinated the best. The data for celery seeds suggest that reducing the storage pressure increased the percentage germination, but this trend was not evident with cabbage or onion seeds. It was concluded that the observed effect on germination was due to control of RH rather than reduced [O_2] at a low pressure. LP storage of short-lived seeds such as *Apium graveolens* var. dulce (celery), *Brassica oleracea* (cabbage) and *Allium cepa* (onion) reduced the deleterious effects of high relative humidity and temperature upon germination (Lougheed *et al.*, 1976; Elliot and Lougheed, 1977).

10.88 Tissue Cultures (*Nicotiana tabacum* 'Wisconsin 38', *Chrysanthemum* × *morifolium*)

Successful LP storage of *Nicotiana tabacum* 'Wisconsin 38' and *Chrysanthemum* × *morifolium* tissue cultures has been reported (Bridgen *et al.*, 1978; Bridgen and Staby, 1979, 1981, 1983).

10.89 Pressure Cycling (Tomatoes, Bananas, Apples, Cabbage, *Chrysanthemum* Cuttings, Cut Carnation Blooms)

A hypobaric storage can be rapidly pumped down to an optimal pressure, and quickly vented to return it to atmospheric pressure. This allows commodities to be added or removed at frequent intervals, inventory to be accumulated and stored, and small lots removed for sale at convenient times. A container can be opened for inspection at ports of entry, partial unloadings can be made at different locations en-route, and at destination the contents can be kept in the container and sold in small lots to satisfy demand. Some commodities can be returned to atmospheric pressure for 3–8 h each day without a significant diminution in LP storage life (see bananas, tomatoes, chrysanthemum cuttings, carnation flowers, apples, cabbage and turnips). This capability is important for certain types of warehouse applications (chapter 12).

A pressure cycle has occasionally been used to improve the uptake of water by severely wilted cut flowers. In a large-scale study carried out in Japan, cut carnations, roses and *Chrysanthemums* were allowed to wilt until they were unable to rehydrate when their bases were recut under water and the stems were placed in pails filled with water. The buckets then were transferred to a hypobaric intermodal container. During a 2-h pump-down to 1.33 kPa (10 mm Hg), air contained in the vascular system expanded 76-fold and escaped, pushing water from the xylem vessels. Immediately thereafter the intermodal container was rapidly vented in minutes, forcing water from the pails into the xylem vessels before the air pressure could equalize through the stomates and intercellular system. This procedure vacuum-infiltrated the xylem vessels with water and eliminated embolisms, allowing the flowers to draw water, hydrate and open normally (S.P. Burg, 1980, unpublished data).

10.90 Mixed Loads (Cabbages, Carrots, Bananas, Tomatoes, Apples)

Mixing loads of ethylene-producing and ethylene-sensitive commodities must be avoided at atmospheric pressure, but such commodities can be stored together in LP because the hypobaric process lowers a commodity's rate of ethylene production and continuously flushes the gas from within the storage area. Stored at atmospheric pressure in the presence of ethylene-producing apples, cabbage loses its green colour and its leaves abscise, and bitterness develops in carrots (Hardenburg *et al.*, 1986). Cabbage, carrots, and apples were successfully stored together in LP with no adverse effects (McKeown *et al.*, 1978).

The following combinations of breaker tomatoes and Valery bananas were stored in LP and NA at 14.4°C (Burg, 1970): (i) bananas and tomatoes stored separately in NA with ample ventilation; (ii) bananas and tomatoes stored together in NA in unsealed plastic bags; (iii) bananas and tomatoes stored together in LP at a pressure of 10.67 kPa (80 mm Hg) at air-change rates of 10, 4 or 1.3 chamber volumes per hour; and (iv) tomatoes stored alone in LP at a pressure of 10.67 kPa (80 mm Hg). After 7 days, all tomatoes except those in LP developed a pink-red colour indicating that they were producing ethylene. At that time, all bananas were green. In 14 days, all tomatoes except those in LP were red, and during 28 days all bananas remained green except those stored together with tomatoes in unsealed plastic bags. Bananas stored separately in NA broke colour in 19–20 days, whereas bananas stored with tomatoes in unsealed bags yellowed in 14 days. The results show that breaker tomatoes do not cause ripening of bananas when they are stored together in LP, whereas in NA ethylene evolved by the tomatoes initiated banana ripening when the fruits were confined together.

Notes

1. A lime's juice content decreases and its peel thickens when it is stored at a pressure lower than 20 kPa (150 mm Hg); apples become 'dormant' and will not ripen and develop full flavour and aroma if they are stored for an extended period at a pressure lower than 6.7 kPa (50 mm Hg); *Chrysanthemum* and certain other types of cuttings may not root as well if they are stored at a pressure lower than 8 kPa (60 mm Hg).
2. A fruit's percentage of soluble solids is determined primarily by sugar content. Degrees Brix relates the specific gravity of a solution to an equivalent concentration of pure sucrose. The fruit's tartness and flavour depends on the 'Brix-to-acid ratio' (ratio of sugar to acid).
3. The author of this journal article was not aware that the humidity control system in the Grumman/Dormavac intermodal container malfunctioned in all papaya test shipments. The heater in the humidification boiler was programmed to turn ON/OFF in response to a signal generated by a humidity-sensing device located in the storage area, but instead the heater never turned ON because the sensor could not distinguish whether the humidity was elevated by water evaporated from the cargo or by vapour supplied in the incoming air changes. The container was operated at 4.5 air changes per hour, flowing 5 times the air volume that could be saturated by the water that needs to be evaporated to transfer the amount of respiratory heat generated by papayas stored in LP. Evaporative cooling caused the temperature of the papayas to decrease below the wall temperature; the fruit gained heat by radiation and convection, which caused it to lose additional weight. Subsequently, Grumman eliminated the humidity sensor and operated the humidification heater with a pre-selected power setting.
4. Reduced volatile production occurs in both LP and CA when tomatoes ripen in the presence of low [O_2]. This is discussed in 3.26.
5. Pressures between 1.33 and 48.0 kPa (10 and 360 mm Hg) have been tested at 5–20°C (Poulssen *et al.*, 1982), but the results could not be accessed.

A storage test in which mature-green fruit kept at 20 mm Hg developed a high incidence of decay at 12.8°C (Shum, 1981) was performed at a chilling temperature. Decay was less severe at higher pressures, suggesting the possibility that very low [O_2] increases a tomato's susceptibility to chilling. An increased susceptibility to chilling injury at a very low [O_2] has been noted with apples (7.10).

6. Inexplicably, the LP and CA results were combined and it was concluded that the pooled result was not statistically significant?

7. Unless the sweetcorn is a variety containing the 'stay-sweet' gene.

8. According to the Birdseye Division of General Foods, only trained panellists can detect this 'hay-like' flavour.

9. The reported [O_2] content of the packages was 1.0–1.9%, but this probably was an overestimate. The method used to measure O_2 was gas chromatography on a 180 cm 5A Linde molecular sieve column at 70–75°C, using helium carrier gas. Under these conditions, O_2 cannot be separated from argon (Krejci *et al.*, 1959), and since the sensitivity of the thermal conductivity cell used in these studies is nearly the same for O_2 and argon (Burg, 1962b), the reading always measured the sum of the O_2 and argon present. The argon content of the atmosphere is 0.96%, and therefore the O_2 content of the retail packages was 0–0.9% with a mean value of 0.4% in 67% of the packages after 2 h, rather than 1–1.9% as reported.

10. When mushrooms are displayed in supermarkets, the retailer often pastes a price tag over one of these holes. Presumably the remaining hole suffices?

11. The respiration rate of peppers at 8.8°C is approximately 12 mg CO_2/kg·h (Hardenburg *et al.*, 1986). At 10–12°C the respiration of peppers is 75% inhibited at 10 kPa (75 mm Hg) (Bangerth, 1974), and at 8.8°C and a pressure of 5.1 kPa (38 mm Hg) the inhibition is at least 80% (Fig. 4.2). All of the respiratory heat generated during this LP storage trial could have been transferred by a water loss of less than 0.024% of the pepper's weight per day; instead, the peppers lost 0.15–0.22% of their weight each day. The rapid deterioration of peppers stored in LP was caused by the high rate of weight loss created by a faulty humidification procedure, which only provided 5–20% RH in the air changes (9.3). Possibly the same error in equipment design caused the poor LP storage results reported by Ward (1975) for leeks, lettuce, cauliflower, Brussels sprouts and radishes, contrary to favourable results by others testing the same commodities.

12. The nitrogen used in this study may have contained as little as 0.1 mole per cent O_2 (extra dry grade) and as much as 0.5 mole per cent O_2 (technical grade) (Anon., 1966).

13. The temperatures and pressures were not specified.

14. Water stress also contributes to bent neck in roses (Zieslin *et al.*, 1978) and to a failure of floret development in other types of cut flowers (Mayak and Halevy, 1971; Halevy and Mayak, 1974).

15. The luggage compartments of commercial jet aircraft in which *Protea* blooms are transported occasionally reaches 45–54.5°C (Maxie *et al.*, 1974). Another study reports a range from –10 to 55°C (Edmond and Nunes, 1999).

11

Meat Storage

Meat is preserved in VacuFresh^{SM} by setting the thermostat at –1.0 to 0°C and continuously evacuating the storage area without pressure regulation or humidification. Evaporative cooling lowers the meat temperature until it stabilizes near the container dew point, about 0.5–1°C below the thermostat set-point and slightly higher than the meat's freezing point. Cold steam continuously 'boils' from the meat, nearly saturating the storage atmosphere, flushing away the last traces of O_2 and any off-odours. The storage pressure depends on the meat's temperature, which determines the vapour pressure of the cold steam.

Vacuum packaging (VP) and the CAPTECH system are the most important modified-atmosphere methods currently used to preserve fresh meat. Vacuum wrapping evacuates much of the air that is present within a package, and then the respiration of the meat and associated contaminating microbes lower the residual O_2 in the limited 'head space' to less than 1% within the gas-impermeable wrap, replacing it with approximately 20% [CO_2] derived from respiration (Abe and Kondoh, 1989; Garcia-López *et al.*, 1998; Stanbridge and Davies, 1998). The CAPTECH system reduces the [O_2] further by flushing the package with 100% [CO_2] before the gas-impermeable film is sealed (Gill and Harrison, 1989; Stanbridge and Davies, 1998), Shelf life is twice as long in CAPTECH compared to storage in vacuum packages due to the lower residual [O_2] and higher [CO_2] in CAPTECH (Stanbridge and Davies, 1998). Transport of meat, fish and poultry in CA intermodal containers has been unsuccessful because the equipment cannot produce a suitable atmosphere for meat storage. The [CO_2] must be in the range 80–100% to provide effective control of microbial growth, and if all [O_2] is not removed, the high [CO_2] + low [O_2] causes discoloration and undesirable changes in texture, odour and flavour (Haard and Lee, 1982; Haard, 1983; Brody, 1989c; Hotchkiss, 1989).

LP meat storage differs from vacuum (VP) and CO_2-modified atmosphere (MAP) meat packaging in several important respects. All CO_2 is removed in LP, whereas high [CO_2] is present in vacuum and CO_2 packages; and in VP and MAP, but not in LP, water-tight packaging creates a saturated atmosphere where microbial growth is promoted, odours accumulate and the meat soaks in its purge. A major disadvantage of LP is that the protective atmosphere is lost when meat transported in a hypobaric intermodal container is transferred to cold storage at atmospheric pressure.

LP is well suited for carcass meat too large to be vacuum or CO_2 packaged. An important potential market is the shipment of lamb or mutton carcasses from Australia to the Near East for sale in the open-air market where this commodity traditionally is distributed. The animals must be butchered in accord with Muslim tradition, shipped in carcass form and have sufficient shelf-life to survive several days hanging in the open air. Trial hypobaric container shipments of hanging carcass lamb from Australia to the Near East have proved that the method is capable of providing this benefit (11.23).

11.1 Optimal LP Storage Condition for Meat

All meat stored or shipped in Grumman/Dormavac hypobaric intermodal containers during an 8-year period, except for the final two loads, was transported at a sub-optimal pressure, 1.33 kPa (10 mm Hg). This mistake contributed substantially to the demise of the Grumman/Dormavac programme, since meat potentially was a major market segment, and most demonstration shipments carried this commodity. The author's initial laboratory experiments at 0°C indicated that meat storage improved when the pressure was reduced from 10.67 to 1.33 kPa (80–10 mm Hg), but lower pressures could not be tested because of limitations in the laboratory equipment available at that time. Later, Grumman/Dormavac engineers warned that container pressures lower than 1.33 kPa (10 mm Hg) might not be reliably sustained due to door-seal leakage, the porosity of aluminium weld seams and tolerance variations in the evacuation system. Although prototype Grumman/Dormavac containers had leak rates of less than 1.33 kPa (10 mm Hg) per hour, engineers doubted that this performance could be improved upon or sustained in production.[1] It was theoretically obvious that meat should remain in better condition and store for a longer time at pressures lower than 1.33 kPa (< 10 mm Hg), but permission to test the 0.57–1.33 kPa (4.3–10 mm Hg) range was withheld until a series of lamb shipments from Australia to Iran[2] failed, causing cancellation of a large production order. Meanwhile, the author had conceived of the 'dry' hypobaric meat-storage concept and determined that the lowest pressure sustainable in a prototype Grumman/Dormavac container fully loaded with meat was approximately 0.61 kPa (4.6 mm Hg) at −1°C. In container tests at a Grumman facility (Alloca, 1979), the feasibility of achieving pressures close to 0.67 kPa (5 mm Hg) was confirmed, provided that all leaks were located and sealed during container manufacture.[3]

Laboratory tests of the 'dry' meat-storage concept demonstrated that weight loss at −1°C was not excessive at a pressure of 0.57 (4.3 mm Hg), and the meat stored three times longer than identical samples humidified at 1.33 kPa (10 mm Hg). Commercial trials of the 'dry' method were postponed because Grumman's thermodynamics specialists were understandably concerned that without humidification the high water-diffusion rate at a low pressure might desiccate the meat by establishing an evaporation/condensation cycle due to small temperature differences along the length of the container's wall panel. Meanwhile, customers complained that the surface of meat shipped with humidification at 1.33 kPa (10 mm Hg) sometimes 'browned' and often was 'slimy'. The residual O_2 present at this pressure caused the browning (11.5), and the slime, which was not of bacterial origin, resulted because the saturated humidity induced the meat to exude mucopolysaccharides. To alleviate the excess humidity problem, the humidification system was turned off 1 day before a meat shipment reached destination, and then during successive shipments it was shut down for progressively longer times. It soon became apparent that the weight loss was determined by the rate at which the vacuum pump exhausted steam from the container, and would not be excessive during a prolonged trip without humidification. The first test shipment using the 'dry' method was attempted carrying Australian lamb carcasses to the Middle East, and soon thereafter a second shipment was sent. Both 42-day shipments were judged by the Australian Meat and Livestock Commission to be equivalent to 2 days' cold storage (11.23), but this remarkable result came too late. Grumman's management had already decided to terminate the 'Dormavac' programme.

11.2 Partial Pressures of O_2, CO_2 and H_2O Vapour Present in a VacuFresh[SM] Container during Meat Storage

Leak rate and pumping speed determine the steady-state O_2 partial pressure in an evacuated hypobaric container. The 655 linear metres of welded seams in the inner wall (vacuum barrier) of a Grumman/Dormavac container increase the probability of leakage due to the inherent porosity of welds and the likelihood of 'pin-holes' or 'burn-throughs'. The leak rate is dramatically improved in a VacuFresh[SM] container because it has only 38 linear metres of welded seams in the tank.

The gas and vapour composition determined in a 6.1 m VacuFresh[SM] container operated at −1°C and maximum pumping speed, 71.4 m^3/h (42 cfm), with water-ice frozen under the floor grating to simulate the humidity created by a full load, was O_2 = 0.000069 atm (0.052 mm Hg); N_2 = 0.000267 atm (0.197 mm Hg); $CO_2 = 1.18 \times 10^{-7}$ atm (8.98×10^{-5} mm Hg); $(H_2O)_V$ = 0.00533 atm (4.051 mm Hg); total pressure = 4.3 mm Hg; relative humidity = 95.1%; meat temperature = −1.5°C (example 1).

11.3 Heat Transfer and Weight Loss

Juice can be seen 'frothing' at the surface of meat during an initial evacuation in LP (Cicale, 1979a,d), and purge conceivably might be squeezed from meat by the pressure build-up that occurs during venting (repressurization). Measurements during a 3-day test (S.P. Burg, 1990, unpublished) and after much longer periods (Cicale, 1979a) have shown that the same amount of purge is released from beef pieces in NA, LP and vacuum packages (VP) regardless of the LP storage pressure and independently of how rapidly the meat is evacuated or vented.

An evaporation/condensation cycle cannot develop between the meat and the walls of a hypobaric container, since post-rigor meat produces an insignificant amount of respiratory heat to support evaporation (Govindarajan, 1973). The meat temperature always is close to the storage-air dew point temperature, and colder than the constant and uniform temperature of the container's wall (Fig. 11.5). Because LP is a dynamic system, a vapour-pressure gradient between the meat and container atmosphere develops, causing moisture to continuously evaporate from the meat and replace water evacuated from the container. The steady-state RH in the container can never reach saturation, and in a fully loaded container usually is approximately 95% at −1°C. The rate of weight loss depends on the density of the water-vapour mixture present in the container (ρ_V, kg/m^3), the pumping speed (S, m^3/h), and the weight of meat (W, kg):

$$\% \ \Delta \ \text{weight}/h = 100\ (S\rho_V)/W \qquad (11.1)$$

An 8090 kg (20,000 lb) meat load will lose 2.73% of its weight in 30 days when a 6.1 m (20 ft) VacuFresh[SM] container is evacuated at a maximum rate (71.4 m^3/h = 42 cfm). The weight loss will be proportionately smaller at slower pumping speeds, and less than half as large in a 12.2 m (40 ft) VacuFresh[SM] container because it holds more than twice as much meat and is exhausted by the same vacuum system.

Convection and radiation must continuously transfer heat from the container wall to the meat at a rate sufficient to replace the latent energy lost when water evaporates from the meat. If no heat is transferred, each tenth of a per cent weight loss will lower the meat's temperature by 0.8°C, until the freezing point of the cellular solution is reached. Then 10% of the meat's weight will freeze for each additional per cent weight loss.[4] When cellular water rapidly freezes, it forms minute ice crystals that are not very damaging, but if it freezes slowly, as it does when inadequate heat transfer occurs during a prolonged hypobaric storage of meat, large ice crystals and clusters of crystals form, which can cause severe separation and rupture of cells (Potter, 1973; 11.14).

Therefore, it is essential to prevent freezing, except for superficial crust freezing of pure water on the meat's surface. The dew point depression corresponding to 95% RH at 0°C is –0.7°C (Table 9.2). Since the meat closely tracks the dew point, the thermostat should be set at least 0.7°C higher than the meat's freezing point. If the thermostat is set at –0.5°C, the meat temperature will be close to –1.5°C, the optimal temperature for meat preservation without danger of freezing (Gill and Harrison, 1989).

A heat and mass-transfer analysis of the meat-basket loading arrangement illustrated in Fig. 11.1 indicates that although the natural convective heat-transfer coefficient is adequate to replace heat lost from the meat by evaporation at a storage pressure of 0.61 kPa (4.59 mm Hg), the low heat capacity of the water vapour inside the container limits heat transfer from the container wall to interior boxes (example 2). The entire temperature difference between the wall and meat is cooled by convection to exterior boxes, and in addition they have a strong radiant coupling with the wall. The exterior meat re-emits this convective and radiative heat by evaporating the water that keeps the container RH close to 95%. The interior boxes are protected from weight loss, since they stop losing water when their temperature approaches the container dew point created by the exterior boxes. In 34 days, pumping at 24.3 m^3/h (14.3 cfm), theoretically the average weight loss from an 8182 kg (18,000 lb) load is 1.1%, varying from 0% in interior boxes to between 1.1 and 3.72% in exterior boxes, depending upon the weight of meat in each exterior box and the surface area that each exterior box presents for heat transfer by radiation and natural convection (example 2). When hanging carcasses rather than boxed meat are shipped, the weight loss is spread more uniformly throughout the load because each carcass tends to have at least a partial line-of-sight view of the container wall, and direct access to both convective and radiative heat. During 42-day shipments of full loads of hanging-carcass lamb from

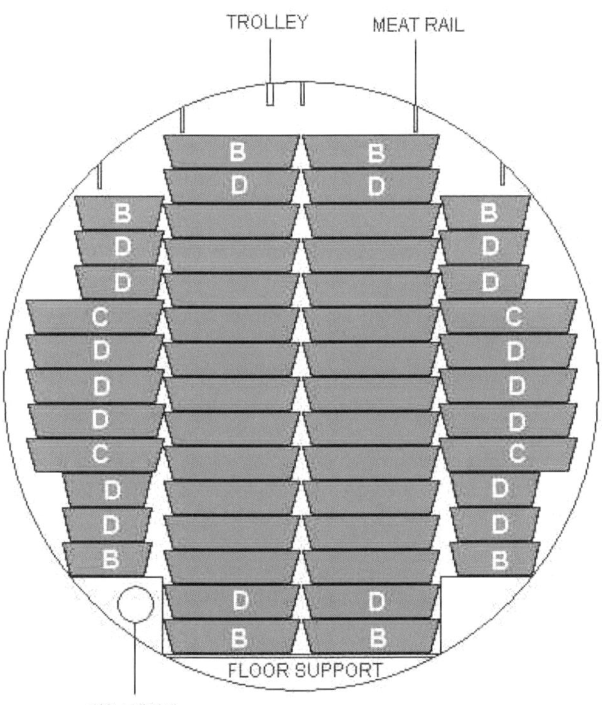

Fig. 11.1. Meat basket stacking arrangement in a VacuFresh[SM] container. Basket size: large baskets (27.3 kg) = 52.1 (W) × 50.1 (L) × 14 cm (H); small baskets (18.2 kg) = 31.8 (W) × 50.1 (L) × 14 cm (H). Total weight = 8182 kg. The B, C, D labels are explained in example 2 and the Table 11.25 spreadsheet.

Australia to the Middle East in 12.2 m (40 ft) Grumman/Dormavac hypobaric intermodal containers, weight loss was rather evenly distributed throughout the load and not excessive when containers were operated at 0.6 kPa (4.5 mm Hg) without humidification (11.23).

Pumping at full speed improves meat storage by lowering the residual $[O_2]$ in an LP container, but increases the weight loss from exterior boxes. It might seem that a relatively easy way to reduce evaporative weight loss and distribute it more uniformly throughout the load would be to add water or freeze ice under the floor grating. The container vapour pressure then should be close to the water's saturation value and the meat might lose little or no weight, since dissolved solutes lower its vapour pressure. The meat would need to be 0.12°C warmer than ice to develop the same vapour pressure (example 3). The expectation that under-floor water or ice will prevent exterior boxes of meat from evaporating water in a fully loaded container fails to consider the complication that the latent energy required for sublimation or evaporation must be provided by heat conducted across a temperature gradient from the container floor through the ice or water to the surface. The temperature of the water or ice is lowered by continuous evaporation or sublimation, decreasing the vapour pressure in the container. Heat radiating from the container walls keeps the exterior boxes slightly warmer than the container dew point and provides these boxes with the latent energy needed to continuously evaporate water. The 'warmer' moisture evaporating from this meat condenses on the surface of the 'colder' under-floor ice or water layer, releasing energy that warms the ice or water and lowers the temperature gradient between its surface and the floor. This reduces evaporation or sublimation by decreasing heat conduction from the floor to the surface of the under-floor ice or water. Will under-floor water or ice prevent the meat from evaporating water, or will the meat prevent sublimation or vaporization of the under-floor water or ice? Under-floor ice did not decrease in amount during a 15-day horse meat shipment in a 6.1 m (20 ft) VacuFresh^{SM} intermodal hypobaric container operated at −1°C, whereas almost half of the 248.2 kg (546.7 lb = 69.6 gal water) of ice would have sublimed if it provided the humidity during the shipment. Evaporation from the meat protected the ice from sublimation, rather than the opposite.

11.4 Packaging Parameters

Title 21 FDA 175–300 of the United States Food and Drug Administration code of food regulations, and similar requirements issued by foreign agencies pertaining to indirect food additives, require meat and other fatty and aqueous foods to be protected from incidental contact with the aluminium walls of VacuFresh^{SM} or Grumman/Dormavac intermodal containers. This is not a significant problem with boxed meat, since packaging satisfying Title 21 FDA 175–300 invariably is used to protect meat from dehydration and exposure to bacteria and other forms of contamination regardless of the method used to store, transport, distribute or display. To satisfy Title 21 FDA 175–300, exterior hanging meat carcasses shipped in LP containers have been protected with a PVC mummy wrap or sterilized cotton stockinette.

Aluminium is widely used for processing and handling meat and the metal is popular for truck and cold-room liners, floors, hanger rails and truck exteriors. Blood is corrosive to many aluminium alloys, but smudging caused by dressed meat rubbing against aluminium linings is minimal with the 5000-series marine-grade alloy used in the VacuFresh^{SM} container shell. The 6000-series aluminium interior of the Grumman/Dormavac container was protected with a paint approved under FDA regulation 153.000 for accidental contact with dry and wet foods. This satisfies Title 21 FDA 175–300, but painting aluminium is costly and impractical. Eventually, the paint chipped and peeled in the high-humidity hypobaric environment. Therefore, the interior of the VacuFresh^{SM} container is not painted. This has the additional advantage of reducing the radiation emissivity of the aluminium wall (Table 6.11).

The most important consideration in designing a package for displaying and preserving meat is the susceptibility of myoglobin to oxidation at a low O_2 partial pressure (Fig. 11.2). A packaging film must have a permeability of at least 5000–9000 ml $O_2/m^2 \cdot 24$ h·atm (resistance to O_2 mass transport = r_{O_2} = 96,000–172,800 s/cm) to prevent myoglobin oxidation and maintain good meat colour during 48 h in atmospheric air (Landrock and Ayres, 1955; Brody, 1989c). PVC plastic with a high enough O_2 permeability to delay pigment browning, and low enough water permeability to prevent desiccation, is routinely used to display meat shelf-packs at atmospheric pressure. PVC and polyethylene wraps are equally effective in providing resistance to water vapour mass transport in a vacuum (Table 11.1), but the PVC wrap used for display at atmospheric pressure is preferred for LP storage because it improves case life and appearance when the meat is removed from LP and returned to atmospheric air.

At atmospheric pressure, the movement of gases and vapours through plastics depends on the diameter, tortuosity and

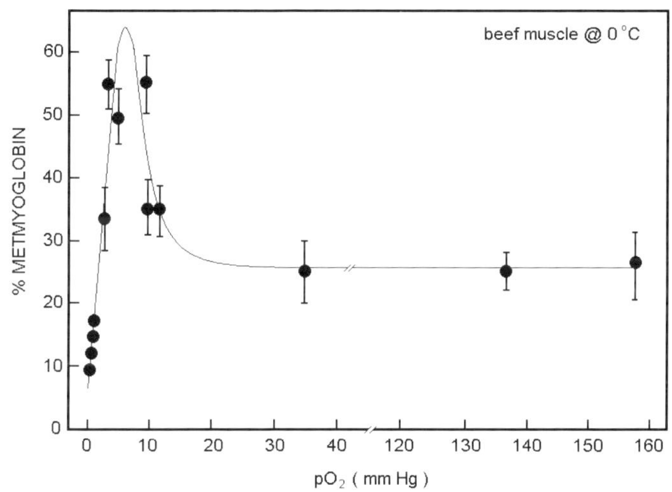

Fig. 11.2. The relationship between metmyoglobin formation at the surface of sterile beef muscle and the partial pressure of O_2 during storage for 12 ± 2 days at 0°C (Ledward, 1970).

Table 11.1. Effect of boxes and 1 mil polyethylene, polyvinyl chloride and Mylar wraps on the cooling half-time and weight loss from a small sample of pork loins stored in a modified 12.2 m (40 ft) Grumman/Dormavac container operated at 0.61 kPa (4.6 mm Hg) with a –0.5°C thermostat set-point. Water-ice was frozen under the floor. All samples had a line-of-sight view of the container wall and experienced a weight loss of approximately 0.2% during cool down, and an additional purge loss of 1.5% during the first few days regardless of whether they were kept at atmospheric or a hypobaric pressure. Each temperature indicated in the table is an average of 87 measurements made with individually calibrated thermocouple probes during the 14–31-day storage period (S.P. Burg, 1998, unpublished data).

Treatment	Cooling half-time (h)	Temp. (°C)	% weight loss per day (14–31 days)
Naked	1.3	–0.83	0.300
Polyethylene (PE) wrap	1.9	–0.72	0.176
Polyvinylchloride (PVC) wrap	2.0	–0.67	0.150
PE wrap, in box	3.8	–0.72	0.115
PE wrap, in box + Mylar liner	4.5	–0.78	0.067

frequency of the pores, thickness of the plastic, the diffusing molecule's molecular diameter, its solubility and diffusivity in the plastic, and in some plastics on the humidity and the diffusing molecule's polarity. At a pressure of 0.57 kPa (4.3 mm Hg), a package's mass-transport resistance usually is determined by the manner in which the plastic film is applied rather than by the nature of the plastic, because mass transport at a low pressure is so rapid through small openings and air spaces between overlapping layers of plastic that unless the plastic is meticulously sealed around a piece of meat, it cannot be applied too snugly for successful storage. Theoretically, if the plastic were applied tightly enough, cold steam boiling from the meat would continuously flush the last vestiges of O_2 from within the package in the same way that boiling water purges air from jars used to preserve food by 'canning'. The protective plastics used for hypobaric meat storage have never fitted as snugly as this, and if they did, the package might 'explode' when the container was pumped down.

To simulate the behaviour of a full load, all tests performed with small samples of meat in a VacuFresh^{SM} container have been carried out with a flooded floor, or with water-ice frozen beneath the floor. Without humidification, a partial load would lose excessive weight unless the pumping speed was slowed in proportion to the small amount of meat stored. This is not possible to accomplish with very small samples because the VacuFresh^{SM} flow-control system (13.18) cannot throttle the vacuum pump below 27% of full capacity. In small-scale tests with under-floor water or ice present, the weight loss is quite sensitive to the type of wrap used to protect the meat, the 'tightness' with which it is applied and the amount of radiation received by the meat (Table 11.1). Preferably, small samples of meat should be protected from radiation by installing a Mylar shield within the boxes or baskets, or a Mylar mummy wrap covering hanging carcasses. Mylar increases the initial cooling half-time by interfering with radiant heat transfer from the warmer meat to the colder container wall (Table 11.1), and after cool down has been completed Mylar decreases the weight loss by depriving the meat of radiant (latent) heat required to evaporate water. The radiant heat transfer that takes place between the cardboard of an exterior box and meat contained in the box occurs between two essentially parallel opaque plates with greybody shape factors $F_1 = F_2 \cong 1$ (note 22 – chapter 6) and nearly identical surface areas ($A = A_1 = A_2$):

$$Q_{1-2} = A\sigma\,(T_1^4 - T_2^4)/[(1/\varepsilon_1) + (1/\varepsilon_2) - 1] \qquad (11.2)$$

where T_1 and ε_1 are the temperature and emissivity of the cardboard, and T_2 and ε_2 the temperature and emissivity of the meat. The radiant transfer of heat from the container wall to the Mylar is described by the relationship (Özisik, 1985):

$$Q_{1,2,3} = [A\sigma\,(T_1^4 - T_2^4)]/[(1/\varepsilon_1 + 1/\varepsilon_2 - 1) + (1/\varepsilon_{3,1} + 1/\varepsilon_{3,2} - 1)] \qquad (11.3)$$

where $\varepsilon_{3,1}$ and $\varepsilon_{3,2}$ are the emissivities of the Mylar at the surface of the cardboard and meat, respectively. The emissivities of a cardboard box and 'oxidized' Mylar are $\varepsilon_1 = 0.95$ and $\varepsilon_2 = 0.20$ (Table 6.11). A single layer of oxidized Mylar will reduce radiant transfer from the cardboard to the meat by 90%, a double layer of Mylar by 96%, and if the exterior of the box is tinted with a heat-reflecting 'pigment', radiation from the wall to the cardboard can be reduced by an additional 45–80%, depending on whether or not the container wall has been painted white.

Convective heat transfer is ineffective at a low pressure, and therefore when meat is shielded from radiation its sensible heat is the only energy source available to evaporate water and elevate the container RH. A small meat sample protected with Mylar and stored in LP without under-floor water or ice will lose sensible heat, progressively decrease in temperature and eventually freeze, but with under-floor water or ice present, the meat's temperature cannot decrease below the container dew point established by the water or ice (Table 6.17; Fig. 9.2). Then the humidity in the container will be derived almost exclusively from the under-floor water or ice because the meat has been prevented from acquiring heat, while the ice or water continuously is warmed by heat transmitted by conduction and convection from the floor, and by radiation and convection from the walls. The combination of a Mylar radiation shield plus water or ice beneath the floor not only reduces the total weight loss, but in addition spreads it more uniformly between exterior and interior boxes. Although the temperature and vapour pressure of a small sample of 'naked' meat also decrease during LP storage (Table 11.1, naked), this increases weight loss because in the absence of a water-retentive wrap such as Mylar, evaporative cooling is promoted by a low water-vapour mass-transport resistance and by increased radiant heat transfer due to the larger temperature gradient that develops between the meat and wall. These effects more than offset the advantage conferred by a lower vapour-pressure gradient.

11.5 Oxidation of Meat Pigments

Myoglobin, the principal haem pigment responsible for fresh red meat's colour, exists in three inter-convertible forms: reduced myoglobin (Mb) is purplish, oxygenated myoglobin (MbO_2) is bright red and oxidized myoglobin (Mmb) is brown. The consumer judges meat freshness in part by the natural bright red colour of the oxygenated form, since he/she is aware that the brown colour of metmyoglobin results from oxidation during prolonged storage (Govindarajan, 1973).

Myoglobin and haemoglobin are oxidized most rapidly at the O_2 partial pressure which half-saturates each pigment (George and Stratmann, 1952a,b). At 30°C, a solution of ox-blood oxymyoglobin is half-saturated at 1–1.3 mm Hg [O_2], and 77.5% is converted to metmyoglobin within 3 h when the [O_2] partial pressure is 2 mm Hg. Higher temperatures promote dissociation of O_2 from oxygenated myoglobin and the conversion of haemoglobin to methaemoglobin (Q_{10} = 1.65), causing the oxidation rate to be maximal both at 30°C and 1.16 mm Hg [O_2] and at 0°C and 0.3 mm Hg [O_2] (George and Stratmann, 1952b). Aqueous solutions of horse heart haemoglobin oxidize most rapidly at 20 mm Hg [O_2] (Brooks, 1935), and at temperatures of 0 and 7°C the rate of muscle myoglobin oxidation is maximal at O_2 partial pressures of 6 ± 3 mm Hg and 10 ± 3 mm Hg, respectively (Fig. 11.2; Ledward, 1970). At higher O_2 partial pressures, oxidation decreases to a 'plateau', while in the absence of O_2, metmyoglobin formation ceases and oxidized pigment may be reduced to reform myoglobin (Saleh and Watts, 1968; Pierson *et al.*, 1970).

The resistance of meat-tissue fluids to gaseous diffusion causes a steep O_2 gradient to develop in an inward direction due to the meat's respiration and bacterial O_2 consumption. Exposed to air, the surface of fresh meat is bright (cherry) red due to the formation of oxymyoglobin (MbO_2), but a short distance below the surface where the O_2 partial pressure is optimal for metmyoglobin (Mmb) formation, a brown layer of varying thickness develops. The depth at which this brown layer occurs depends on the square root of the external [O_2] concentration (Brooks, 1929). At higher ambient O_2 partial pressures, the brown colour occurs deeper in the tissue, at lower [O_2] the layer forms nearer to the surface, and in air at 0°C it is found approximately 5 mm under the surface. The purplish colour characteristic of reduced myoglobin (Mb) develops beneath the metmyoglobin layer. Deterioration in meat colour due to a decreased O_2 partial pressure beneath patches of micro-organisms may cause the consumer to reject meat even before spoilage is fully manifested (Stanbridge and Davies, 1998). LP has no effect on the penetration of O_2 into meat because diffusion through animal muscle is limited by a liquid phase.

An initial [O_2] level higher than 0.15% compromises the colour of beef and lamb, but pork may remain acceptable even at 1% [O_2] (Stanbridge and Davies, 1998). Multiple CO_2 flushes of CAPTECH packages (MAP) reduce the [O_2] to less than 0.05%, and it can be further reduced to below 0.01% if a sachet containing an O_2 scavenger is included. Flushing with technical N_2, vacuum packaging (VP), and sealing in a CAPTECH bag without an O_2 scrubber were equally ineffective in preserving beef jerky, but a combination of CAPTECH plus an oxygen scrubber maintained flavour and prevented discoloration (Abe and Kondoh, 1989). LP intermodal containers lower the [O_2] to 0.0068% when meat is stored.

Browning of polyethylene (PE) or PVC-wrapped ribs and rounds of freshly slaughtered beef increases at –2 to +2°C when the saturated storage pressure is decreased below 10 kPa (75 mm Hg = 15 mm Hg [O_2]) (Fig. 11.2). The 'extra' brown colour tends to form at the outermost surface where it is most evident because it is not obscured by overlying layers of non-oxidized myoglobin. Metmyoglobin formation peaks at 8 mm Hg [O_2] and decreases at lower pressures, but still is significant at 1.33 kPa (10 mm Hg = 0.15% [O_2] = 1.3 mm Hg [O_2]) and only ceases in the absence of all O_2. When rib steaks, prepared from hanging beef quarters previously stored for 21 days at 1.33 kPa (10 mm Hg) in a 12.2 m (40 ft) Grumman/Dormavac intermodal hypobaric container, were allowed to bloom along with similarly cut 5-day-old Cryovac trimmed ribs, after 18 h, the LP sample was dark red, and the VP sample was noticeably browner. Within 48 h, there was a substantial difference in the metmyoglobin content determined by spectro-photometric analysis (Table 11.2; Burg, 1976a; Armour, 1978). A much better result was obtained by lowering the pressure to 0.57 kPa (4.3 mm Hg) at –0.5°C. A beef round was fabricated into top round, bottom round and knuckle, and these pieces were sliced into 1-inch steaks. After 19 days in storage, samples held at 1.33 kPa (10 mm Hg) had excessive browning, while those kept at 0.57 kPa (4.3 mm Hg) remained in excellent, marketable condition without browning or off-odour (Cicale, 1979b,c).

Table 11.2. Metmyoglobin – per cent of total pigment in beef samples (Armour, 1978).

Sample description	Bloom time (hours @ 4.4°C)	Metmyoglobin (%)
Cryovac beef	1	0.5
Cryovac beef	48	22.5
LP beef	1	0.0
LP beef	48	14.0

11.6 Fat Oxidation

Metmyoglobin is able to catalyse lipid oxidation and cause a stale or 'rancid' odour to develop in meat and fish (Brown *et al.*, 1963). Cellular breakdown brings the myoglobin and lipids present in adipose tissue into contact by a diffusion process, and the interaction of fats and haem pigments can lead to 'mutual co-oxidation' (Govindarajan, 1973). Pork is more vulnerable than beef to oxidative rancidity because the triglycerides in beef fat tend to be saturated, while in pork fat the unsaturation level is higher.

Flavour retention is improved and the development of rancidity in 'fatty' fish prevented by vacuum- or 100% [CO_2]-packaging (Shank and Lundquist, 1963). Enzymes in raw meat reduce metmyoglobin under anaerobic conditions (Cutaia and Ordal, 1964; Stewart *et al.*, 1965), and if no additional O_2 enters a vacuum or CO_2 package, theoretically pigment and lipid oxidation should cease. Vacuum packaging is only successful in preventing oxidation of myoglobin, and in turn, lipids, if the commodity contains sufficient enzyme activity to establish anaerobic conditions quickly to completely reduce metmyoglobin (Green, 1969). LP prevents lipid oxidation by rapidly evacuating all O_2, and does not depend on the commodity's ability to remove O_2.

11.7 Meat Odour and Flavour

The interior muscle tissue of animals killed under reasonably hygienic conditions is usually sterile. Microorganisms are restricted to the surface until spoilage counts develop and the bacteria produce sufficient amounts of proteolytic enzymes to enable them to penetrate the muscle. Storage methods that eliminate O_2 are primarily concerned with events that occur at the surface, since even in air, muscle is essentially anaerobic at greater than 5 mm depth. Initially, bacteria preferentially metabolize substrates such as glucose and other simple sugars that are present in dilute concentrations at the surface (Stanbridge and Davies, 1998). Offensive odours are not produced during this phase of bacterial growth. By the time bacteria reach a count of approximately 10^7 per cm^2 (10^8 organisms per gram), these carbohydrate substrates become depleted, and because their outward diffusion rate from within the meat is too slow to meet the increasing demand of the bacteria, protein degradation products and amino acids are attacked, causing ammonia formation, a rise in pH, and the formation of varying amounts and types of malodorous volatiles (Nychas *et al.*, 1998). The fresh meaty smell ($< 10^7$ bacteria per gram) gradually changes to an inoffensive but definitely non-fresh odour, to a dairy/buttery/fatty/cheesy odour (10^8 bacteria per gram), and eventually to a sweet/fruity and finally a putrid odour (10^9 bacteria per gram) (Ingraham and Dainty, 1971). The meat surface develops a 'tacky' feel, and when the bacterial load reaches 10^8 organisms per gram, bacterial slime becomes evident.

The high [CO_2] present in VP and MAP packages causes odoriferous volatile compounds to preferentially form, and because these packages are sealed and essentially vapour-impermeable, the odours build up to an offensive level long before the meat has actually spoiled (Licciaedello *et al.*, 1967; Nychas *et al.*, 1998). When the packages are opened, if the meat or fish has not spoiled the odour eventually dissipates. LP keeps meat and fish 'odour-free' for remarkably long periods of time by continuously ventilating these commodities with cold steam, and removing all CO_2. This attribute of LP meat storage has frequently attracted favourable comment.

The natural fat covering of a hanging carcass protects it from bacterial proliferation and extends its storage life. Microbes growing on fatty surfaces utilize the same soluble carbohydrates and proteins present in muscle tissue, but the nutrients in fatty tissue exist at a lower concentration than in muscle, the rate of diffusion of fresh substrates to the surface is significantly slower, and because the fat covering loses moisture during chilling, consequently it contains less moisture to support the growth of microbes. It is not practical to vacuum- or CO_2-package large carcasses possessing a fat covering,[5] but they can easily be hung in an LP intermodal container. 'Wagyu' beef shipped to Japan must have the fat covering exposed for inspection, and for religious reasons lamb shipped to the near East must be in carcass form for display in the open-air market. In these and other unique circumstances, LP can be helpful in shipping carcass meat. For an equivalent weight of usable meat, the shipping cost can be as much as 30% greater per kg for meat left in carcass form with bones and fat, compared to boxed meat. To eliminate the problem of disposing fat and bone, the European common market restricts or discourages the importation of hanging carcasses, and Japan has in the past imposed a tax on imported bone and fat.

11.8 Microbial Spoilage under Aerobic Conditions

The major spoilage types on chilled fresh meats stored in air are gram-negative aerobic rods and coccobacilli belonging to the genera *Pseudomonas* (*Ps. fragi*, *Ps. fluorescens* and *Ps. lundensis*), *Moraxella*, *Psychrobacter* and *Acinetobacter* (Brody, 1989c; Garcia-López *et al.*, 1998), sometimes accompanied by yeasts (*Candida* sp.) and moulds (Dillon, 1998). In addition, psychrotrophic Enterobacteria grow aerobically on high-pH meat, especially on fatty pork and lamb tissues when the temperature is ≥ 4°C. The predominant aerobic spoilage organisms on eviscerated, chilled poultry are pigmented and non-pigmented pseudomonads, especially *Ps. putrefaciens* (= *Alteromonas putrefaciens*) and, to a lesser extent, *Acinetobacter*, *Psychrobacter* and *Shewanella putrefaciens*, with small populations of *Flavobacterium* and Enterobacteria (Cox *et al.*, 1998; Garcia-López *et al.*, 1998). The primary bacterial floras on fish tend to be *Ps. putrefaciens* and fluorescent pseudomonads, with *Achromobacter* and *Acaligenes* species also represented.

11.9 Microbial Spoilage under Anaerobic Conditions

Slow-growing gram-positive *Lactobacillus, Leuconostoc, Lactococcus, Enterococcus* and *Carnobacterium* sp. predominate in MAP and VP anaerobic packages along with lesser populations of Enterobacteria, *Brochothrix thermospacta*, and several other types (Pierson *et al.*, 1970; Mossel *et al.*, 1975; Beebe *et al.*, 1976; Gill and Harrison, 1989; McMullen and Stiles, 1993; Holzapfel, 1998; Stanbridge and Davies, 1998). Pseudomonads generally do not form a numerically significant proportion of the microbial population that develops because their growth is inhibited both by low $[O_2]$ and high $[CO_2]$. The facultatively hetero-fermentative species *Lb. sake* and *Lb. curvatus* that proliferate in most 'anaerobic' meat storage systems are of major economic importance, causing eventual spoilage in VP and CO_2-flushed MAP meat packages. Lactic acid bacteria predominate and pseudomonads, Enterobacteria and *E.coli* are essentially absent when meat is stored in LP at 0.57 kPa (4.3 mm Hg) (Tables 11.3 and 11.10).

Lactic acid bacteria generally do not form malodorous substances, but tyramine is produced by *Carnobacterium*, and small concentrations of dimethylsulphide and methanethiol have been associated with the sour odour typical of VP and MAP (100% $[CO_2]$) meat stored for an extended duration. Flavour changes in meat occur when homo-fermentative *Lactobacillus* sp. producing acetic acid and D-lactate from L-lactate and glucose reach a maximum population of 10^7 organisms per gram (Borch and Agerheim, 1992), and a similar flavour change was detected before *Leuconostoc* sp. had attained maximum numbers. In vacuum-packaged beef, *Hafnia alvei* and/or *Serratia liquefaciens* produce malodorous diamines (putrescine and cadaverine), and also branched-chain esters, methanethiol and derivatives, dimethyldisulphide, dimethyltrisulphide, methylthioacetate, bis (methyl) methane and hydrogen sulphide. These sulphur compounds are associated with an odour ranging from 'boiled' to 'rotten' egg. *Shewenella putrefaciens* grown on high-pH vacuum-packed meat also produces sulphur compounds and a putrid/faecal odour, as well as 'greening' caused by sulphmyoglobin formation (Stanbridge and Davies, 1998).

Table 11.3. Inhibition of Pseudomonad growth at 0°C (Burg, 1990) and total bacterial growth at −1.1°C on lamb stored in LP at various pressures, or in NA (Burg, 1981).

		Relative generation time	
Pressure kPa (mm Hg)	% $[O_2]$	Pseudomonads	Total count
101.30 (760)	20.90	1.0	1.0
5.33 (40)	0.97	1.1	–
1.33 (10)	0.15	2.2	1.9
0.93 (7)	0.074	–	2.7
0.80 (6)	0.047	–	2.9
0.67 (5)	0.019	5.6	3.9
0.57 (4.3)	0.0	–	4.8

Extremely low O_2 partial pressures are needed to inhibit aerobic and microaerophilic bacterial growth during meat storage at a low temperature (Tables 11.3 and 11.4). The growth of almost all moulds is prevented in CAPTECH packages by scrubbers that reduce the $[O_2]$ to < 0.01%, multiplication of yeasts is greatly reduced under these conditions (Abe and Kondoh, 1989) and LP may be equally effective, since it lowers the $[O_2]$ to 0.0068%. Neither technical grade N_2 nor CA can decrease the $[O_2]$ sufficiently to prevent the growth of microaerophyllic bacteria, because technical N_2 typically contains at least 0.1–0.5% $[O_2]$, and current CA technology cannot produce intermodal container atmospheres containing < 0.6% $[O_2]$.

Table 11.4. Total plate counts on lamb after 40 days' storage at –1.1°C in LP at various pressures. Chambers were humidified (Jamieson, 1980c).

	Log_{10} of bacteria per cm^2 @ kPa (mm Hg)		
Treatment	0.67 (5)	0.8 (6)	0.93 (7)
Carcass, thighs	6.19	6.89	7.41
Carcass, ribs	7.80	4.67	6.11
Paper-wrapped ribs	5.82	6.71	6.85
Paper-wrapped leg	5.30	6.87	7.55
Saran-wrapped breast	6.30	7.11	7.72
EVA*-wrapped breast	7.02	7.60	7.47
Average	6.37	6.76	7.18

*EVA = ethylene vinyl acetate copolymer.

11.10 Effect of CO_2 on Microbial Spoilage

High [CO_2] inhibits bacterial respiration, is bacteriostatic, extends the lag phase of bacterial growth and increases the generation time of susceptible bacteria. The respiration and development of aerobic meat spoilage bacteria such as *Pseudomonas*, *Acinetobacter*, *Alteromonas*, *Yersinia*, *Enterobacter* and *Microbacterium*, is maximally inhibited by 20–30% [CO_2], the growth of gram negative aerobic pseudomonads and related psychrotrophs that produce off-flavours in meat is selectively inhibited by 20–100% CO_2 (Brody, 1989c)[6] and anaerobic bacterial growth decreases progressively as the [CO_2] is elevated to 80–100% (Baker *et al.*, 1986; Gill and Harrison, 1989; Hotchkiss, 1989; Stanbridge and Davies, 1998). At a lower temperature, the inhibitory effect of CO_2 on the growth of fungi and bacteria is greater, in direct relation to the increase in CO_2 solubility (English and Gerhardt, 1942; Tables 15.1 and 15.2). Vacuum packaging extends meat storage life by fourfold or more, and an atmosphere of pure CO_2 can more than double that result (Gill and Harrison, 1989).

High [CO_2] initially was considered unsuitable for meat storage at atmospheric pressure because > 20% can cause surface discoloration in the presence of even trace amounts of O_2 (Stanbridge and Davies, 1998). This difficulty is obviated in gas-tight packages flushed with 100% [CO_2], since this procedure reduces the [O_2] to less than 0.1% (Gill and Harrison, 1989; Stanbridge and Davies, 1998). High [CO_2] also causes an increase in the amount of exhudate released from meat. To ensure that the unsightly drip does not become obvious to the consumer, in CO_2-flushed MAP packs it is absorbed in pads or trapped in the patterned bottom of the MAP tray. In 100% [CO_2], offals may undergo chemical changes detrimental to quality (Garcia-López *et al.*, 1998), and fish quickly develop a powdery texture and carbonated bland taste due to acidification of the superficial tissue layers (Haard and Lee, 1982). High [CO_2] may also induce sweet off-odours in chicken.

Lactic acid bacteria such as facultatively anaerobic or anaerobic *Streptococci* and microaerophilic *Lactobacilli* are relatively insensitive to elevated [CO_2] (Jay *et al.*, 1962; Johnson, 1974; Gill and Harrison, 1989). Studies with vermillion rockfish, coho salmon, Gulf trout and Atlantic croaker indicate that while CO_2 is effective in inhibiting the growth of gram-negative bacteria such as pseudomonads that produce highly odoriferous trimethylamine and ammonia, 100% [CO_2] stimulates growth of gram-negative bacteria such as Lactobacilli (Brody, 1989b). CO_2 exerts its effect in part by lowering the pH of fish from a normal value of 6.3 to between 5.7 and 5.8, which tends to inhibit the growth of acid-sensitive organisms while allowing *Streptococci* and acid-tolerant Lactobacilli to develop (Brody, 1989b).

CO_2 is assimilated by most bacteria (Wood and Werkman, 1940), and when aerobic cultures are vigorously oxygenated with CO_2-free air, growth may be reduced (Werkman and Wood, 1942; Clifton, 1950; Stephenson, 1950; Thimann, 1955). Low CO_2 concentrations stimulate the growth of aerobic and facultatively anaerobic forms including *Str. haemolyticus* (Table 11.5), *Erwinia carotovora* and *E. atrospetica* (Wells, 1974; Fig. 7.2). The growth of *Bact. lactis aerogenes* is markedly inhibited in CO_2-free air and maximally stimulated by 0.08% [CO_2] (Fig. 11.3). Anaerobic bacterial growth also may require CO_2, especially among coliforms (Thimann, 1955). Development of *Pseudomonas aeruginosa*, a common cause of slime formation and meat spoilage, is stimulated and the bacterial generation time decreased by a factor of 2–3 in gas mixtures containing 50% air + 50% CO_2 (King, 1966 – referred to in Wells, 1974).

Table 11.5. Effect of CO_2 concentration on the growth of *Str. haemolyticus.* Conditions: pO_2 = 16 kPa (120 mm Hg); total pressure made up to 98.7 kPa (740 mm Hg) with N_2 (Pappenheimer and Hottle, 1940).

CO_2 pressure mm Hg (%)	Growth (mg bacterial N per 500 ml) 20 h	40 h
0.0 (0)	–	0.3
0.4 (0.05)	0	0.7
1.4 (0.18)	0.1	0.7
2.4 (0.32)	0.15	3.8
4.3 (0.57)	0.15	9.0
8.0 (1.05)	1.4	9.1
20.0 (2.63)	1.0	9.2
40.0 (5.26)	10.0	12.2

By removing all CO_2, LP restricts development of some bacterial types, including certain lactobacillus strains. Lactic acid bacteria are the predominant floras that eventually develop on meat stored in CO_2 packages, vacuum packages (VP) and in LP (Table 11.6). These organisms grow rapidly

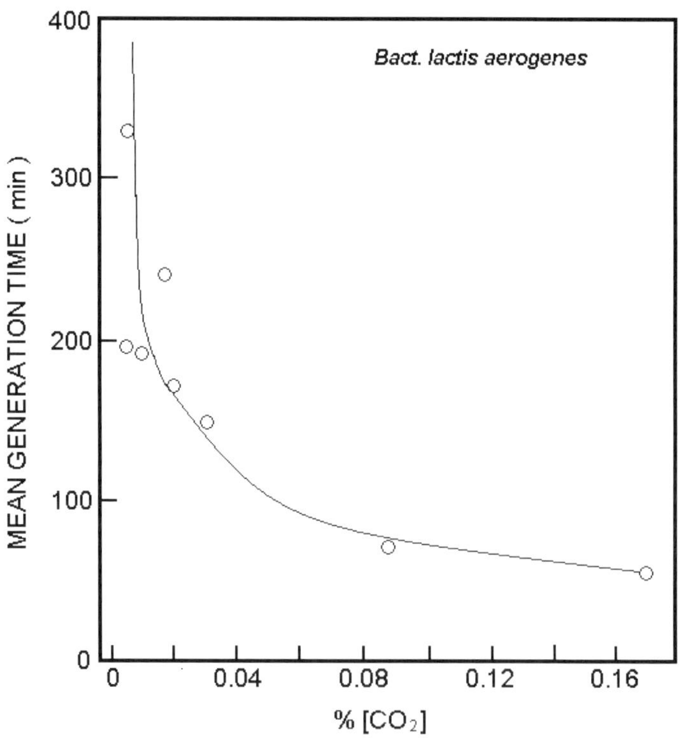

Fig. 11.3. Effect of [CO_2] concentration on the growth rate of *Bact. lactis aerogenes* (Monod, 1942).

Table 11.6. Microflora and weight loss of pork loins after 6 weeks' storage in LP at a pressure of 0.67 kPa (5 mm Hg) with water-ice frozen under the floor. De-boned, skinned and rolled pork loins, measuring approximately 60 cm long × 12 cm diameter (average weight = 4.8 kg/ea), were wrapped in 60 × 100 cm pieces of either 0.8 mil polyvinyl chloride (PVC) or 1 mil polyethylene (PE). Each basket containing four pork loins was shielded from radiation with a loosely fitting Mylar over-wrap. No differences were noted between the PVC and PE wraps, and therefore the results have been pooled. The thermostat was set at −1°C and the meat temperature was −1.5°C. Bacterial counts indicated in the table are an average of 33 samples, each from a 25 cm^2 surface swabbed ten times. Weight loss data are an average of 11 samples. The initial microflora were comprised of 2.4×10^4 total bacteria per cm^2, including 1.8×10^4 lactic acid bacteria per cm^2, 47 coliform per cm^2 and 8 *E. coli* per cm^2 (S.P. Burg, 1998, unpublished data).

	Numbers per cm^2	
Type of count	Average	Range
Total plate count at 35°C	1.8×10^6	$4.4–45.5 \times 10^5$
Total coliform (Petrifilm)	38.6	0.1–155
E. coli (Petrifilm)	4.2	< 0.1–12.9
Lactic acid bacteria at 30°C	1.4×10^6	$3.4–30.9 \times 10^5$
Weight loss (%)	2.9	0.3–6.5

in vacuum packages containing approximately 20% [CO_2] and more slowly in packages with 100% [CO_2] or in LP with no CO_2. *L. sake Lb706* inoculated into sterile beef slices reached maximum population in vacuum packages within 3 weeks at 2°C (Leisner *et al.*, 1995). The total bacterial count on pork cuts stored in vacuum packages for 2 weeks at −1.5°C was 5.2×10^6 per cm^2; after 4 weeks the total bacterial count on pork stored at the same temperature in packages containing 100% [CO_2] was 7.7×10^4 per cm^2; and in 6 weeks pork cuts stored at −1°C and a pressure of 0.57 kPa (4.3 mm Hg) developed a total plate count of 1.8×10^6 organisms per cm^2 (Gill and Harrison, 1989; S.P. Burg, 1995, unpublished). In each case, the flora was comprised largely of *Lactobacillus*, while coliform bacteria, including *E. coli*, did not develop. Sour aromatic odours characteristic of *Brochothrix thermosphacta* were detected within 4–5 weeks in vacuum-packed pork, and the meat had spoiled, whereas pork was odour-free at 6 weeks both in LP and 100% [CO_2] packages. LP at 0% [CO_2] and MAP containing 100% [CO_2] seem to be equally effective in preserving pork, and both methods improve storage life compared to vacuum packages. The LP data cannot be considered conclusive until longer storage periods are tested and organoleptic quality and the growth of enterobacteria and *B. thermosphacta* are evaluated.

11.11 Microbiological Safety

A meat storage method should not only inhibit the growth of the normal spoilage association, but in addition must protect the consumer against disease-causing microorganisms such as *Aeromonas hydrophila*, *Bacillus cereus*, *Campylobacter* sp., *Staphylococcus aureus*, *Clostridium botulinum*, *Cl. perfringens*, *Listeria monocytogenes*, *Versinia enterocolitica*, and enterobacteria including *Salmonella choleraesuis*, *S. enteritidis*, and toxigenic strains of *Escherichia coli* including 0157:H7 (Stanbridge and Davies, 1998). These disease organisms do not necessarily cause noticeable deterioration of the commodity before they become virulent. MAP and VP have not been found to increase the hazards from *Salmonella*, *S. aureus*, *Campylobacter* or *Vibrio parahaemolyticus*. (Brody, 1989a; Stanbridge and Davies, 1998) but theoretically food poisoning could be caused by enterotoxigenic *E. coli*, *A. hydrophila* and *V. enterocolitica*, and *L. monocytogenes* systematic infections, and *Cl. botulinum* type E neuromuscular paralysis is possible. Yet no *bona fide* outbreaks of food poisoning have ever been traced to CA/MA/vacuum packaging of fresh or precooked foods (Brody, 1989a; Stanbridge and Davies, 1998).

11.12 Botulism

Botulism is caused by the neurotoxin produced when *Cl. botulinum* grows on food. The exotoxin secreted by this organism is destroyed by boiling in water for 10 min, and therefore is hazardous mainly when uncooked food is eaten. The seven serologically distinct toxins and the respective types of organisms that produce them are designated A, B, C, D, E, F and G. Each toxin is neutralized by a complementary antitoxin of its type, but not by other specific antitoxins. Types A, B, E and F affect man and therefore are the primary concern. Toxins C, D and G are poisonous only to lower animals, such as chickens and water fowl (Bryan *et al.*, 1962).

Botulism is a rare but often fatal intoxication. Between 1963 and 1973, an average of 20 cases per year was reported in the USA (Myrvik *et al.*, 1974; Horowitz and Hughes, 1977). Botulism is not spread by fresh foods; all human cases have originated from foods that were improperly processed (Clifton, 1950; Horowitz and Hughes, 1977). The spores of *Cl. botulinum* are not killed in 3–5 h at 100°C, and therefore canning techniques using boiling water do not preclude the possibility of this intoxication. This accounts for the high incidence of botulism outbreaks following consumption of home-canned foods. Most cases of botulism in the USA result from home-canned vegetables, whereas in Europe the incidence is greatest from the consumption of smoked, salted or spiced fish and meats. *Cl. botulinum* is an anaerobic organism that normally cannot grow in fresh products because they typically are kept in O_2-containing air, but when hypobaric storage removes all O_2, a condition is established that theoretically might permit this organism to grow. This same condition is produced in canned goods, in vacuum- and CO_2-packaged meats and fish, and even in CA and MA.

The growth of *Cl. botulinum* and the production of toxin in a fresh, refrigerated product can only occur if the product has been exposed to and contaminated by a sufficient inoculum of the organism, the pH of the commodity is high enough to permit growth of the organism, there is too little O_2 present to kill the organism and the temperature is high enough to support growth and toxin production. Spores of *Cl. botulinum* are widespread in nature, especially in soil and mud, and since the organism is so ubiquitous, all raw foods must be considered to be potentially contaminated. The different types of *Cl. botulinum* have distinctive patterns of distribution, with Type A the most common form in the USA, and type B predominating in Europe. Types A and B occur in meat, fruit, but rarely in fish and shellfish. Type E occurs naturally and has caused outbreaks of botulism after consumption of processed fish and shellfish (Johannsen, 1965; Insalta *et al.*, 1967; Pace *et al.*, 1967; Craig *et al.*, 1968). It also occurs in canned chicken (Breed *et al.*, 1957), and can grow in vacuum-packaged meat (Taclindo *et al.*, 1967). Type F has been isolated from liver paste, venison jerky, marine sediments (Eklund *et al.*, 1967b) and crabs (Myrvik *et al.*, 1974). Most cases of human botulism are due to types A, B or E; those due to F are rare.

Cl. botulinum cannot grow and produce toxin at a pH lower than 4.5–5 (Ohye and Christian, 1967). The pH may be this low naturally in processed foods, including many canned fruits, but in fresh produce the pH is high enough to permit growth of the organism. The lethal effect of O_2 on anaerobic bacteria results from the production of peroxides during the aerobic metabolism of carbohydrates and other substrates (Morris, 1970). All cells tend to carry out these reactions, but aerobic cells contain enzymes such as catalase and superoxide dismutase, which destroy the peroxides. The relative ability of an organism to withstand exposure to O_2 is approximately proportional to its contents of these enzymes, and *Cl. botulinum* is particularly sensitive to the content of superoxide dismutase (Hewitt and Morris, 1975). The net rate of peroxide production has a positive temperature coefficient, and therefore the time required for O_2 to kill *Cl. botulinum* increases as the temperature is lowered. In one study, strict anaerobes were killed by atmospheric O_2 in 1.3–1.6 h at 37°C, whereas moderate anaerobes tolerated eight or more hours (Loesche, 1969). In another experiment, various anaerobes tolerated 2–10 h at 37°C, *Cl. perfringens* being the most tolerant (Walden and Hentges, 1975). Amongst the strict anaerobes related to *Cl. botulinum* (Breed *et al.*, 1957), *Cl. haemolyticum* grows equally well in the range 0–0.5% [O_2], is not killed by 0.7% [O_2] and grows sporadically even at 1% [O_2], while *Cl. novyi* grows even at 3% [O_2]. Although there apparently are no corresponding data for *Cl. botulinum*, these results suggest that the O_2 partial pressure during LP meat storage is low enough to support the growth of this organism, and that at 0°C the organism is not likely to be killed by intermittent exposure to O_2 during occasional ventings of the vacuum tank. Growth and toxin production by *Cl. botulinum* are not slowed by 100% [CO_2]. The gas has been reported to stimulate germination of *Cl. botulinum* spores, and in some studies to slightly accelerate both toxin production and growth (Brody, 1989a).

Psychrophylic types of *Cl. botulinum* (types B, E and F) can grow at and above 3.3°C (38°F), and mesophylic types above 10°C (50°F) (Michener and Elliott, 1964; Schmidt, 1964; Eklund *et al.*, 1967a,b; Ohye and Chrisian, 1967; Roberts and Hobbs, 1968; Rowley and Feeherry, 1970). At an optimal pH and growth medium composition, a heavy inoculum of types E and F does not result in toxin production for 1–4 months at 4.4°C (40°F), at 3.3°C (38°F) the safe period is approximately doubled and at 2.2°C (36°F) no toxin is produced (Schmidt *et al.*, 1962; Segner *et al.*, 1971; Lerke, 1973). Therefore, provided that the temperature is kept at or below 2.2°C during LP storage, there is no danger of toxin development. After a commodity is removed from an LP container, the temperature may not be so precisely controlled, but at that time meat is exposed to air and the organism should be killed. Theoretically, there is a possibility of toxin production in vacuum and CO_2 packages if their temperature becomes elevated during some stage of distribution, but this has never been reported, possibly because the psychrotrophic spoilage association swiftly outgrows *Cl. botulinum*, resulting in obvious odour, slime formation and spoilage long before a public health risk develops (Mossel *et al.*, 1975; Brody, 1989a). Although psychrotrophic, non-proteolytic strains of *Cl. botulinum* are able to grow on vacuum- and CO_2-packaged fresh meats, and consumption of raw unprocessed meats is common in Germany, Belgium and The Netherlands, the only outbreaks of meat-borne botulism worldwide (exclusive of fish) have been among the Inuit population in Northern Canada and Alaska, apparently because they eat raw, putrid meat regularly (Stanbridge and Davies, 1998). There is even less likelihood of botulism as a consequence of LP storage than from vacuum or CO_2 packaging (see 10.43, mushrooms).

11.13 Temperature

The optimum temperature for psychrophilic bacterial multiplication usually is above 20°C, but they can grow at 0°C or even sometimes at −10°C. The minimum temperature for growth of the more common spoilage association of fresh meats is approximately −7°C (Mossel *et al.*, 1975). It is important to store fresh meat as close as possible to its freezing point because the 'generation time' of these organisms is infinitely great at temperatures near the minimum for their growth (Potter, 1973).[7] The lowest commercially practicable temperature at which packaged meat can be held indefinitely without risk of freezing is considered to be −1.5 ± 0.5°C (Gill and Harrison, 1989).[8] Precise temperature control just above the meat's freezing point is especially helpful with Lactobacilli and pseudomonads because Lactocbacilli grow much more slowly at −1 or −1.5°C compared with 3.3°C, and the Q_{10} for the growth of *Pseudomonas* shifts progressively from 1.5 at 25°C to 10 at 1.6°C, rising extremely rapidly at still lower temperatures (Jay *et al.*, 1962; Gill and Harrison, 1989; Gill, 1998). In a 28-day LP lamb-storage test performed at 0.73 kPa (5.5 mm Hg), the bacterial generation time was 1.36-fold lower at

0.6°C compared to −1.1°C. After 2 weeks' storage, the total bacterial count on pork cuts stored in vacuum packages was 6.2×10^8 per cm^2 at 3°C, and 5.2×10^6 per cm^2 at −1.5°C; after 4 weeks the total bacterial count on pork stored in packages containing 100% [CO_2] was 4.7×10^8 per cm^2 at 3°C and only 7.7×10^4 per cm^2 at −1.5°C. In each case the flora was mainly comprised of *Lactobacillus* (Gill and Harrison, 1989).

Heat transmitted through a standard reverse-cycle refrigerated intermodal container's insulation makes it difficult to sustain a precise and uniform temperature throughout its interior. It is much easier to accurately maintain a stable, low meat temperature with LP's jacketed refrigeration system because it removes transmitted heat before it reaches the container's interior atmosphere (6.18; Alloca, 1980a; Sharp, 1985). This is an important advantage for meat storage.

A narrow differential ON–OFF thermostat was used to control the temperature in Grumman/Dormavac containers. It was assumed that the high specific heat of meat and lack of convection at a low pressure would prevent meat from changing temperature during a refrigeration cycle. Instead, in experiments in which the humidity was elevated with water-ice frozen under the floor and the return glycol temperature was sensed and controlled with a 1.8°C differential ON–OFF thermostat, the temperature of a small sample of polyethylene-wrapped pork loins rose or fell by as much as 0.9°C in 3 h, tracking the return and supply glycol temperatures (Fig. 11.4). During a typical refrigeration ON–OFF cycle, the average temperature difference between the meat and glycol was nearly 1°C, and the meat acquired or lost as much as 0.86 kcal/h (3.4 BTU/h) per 6.8 kg (15 lb) pork loin. The result was the same regardless of whether

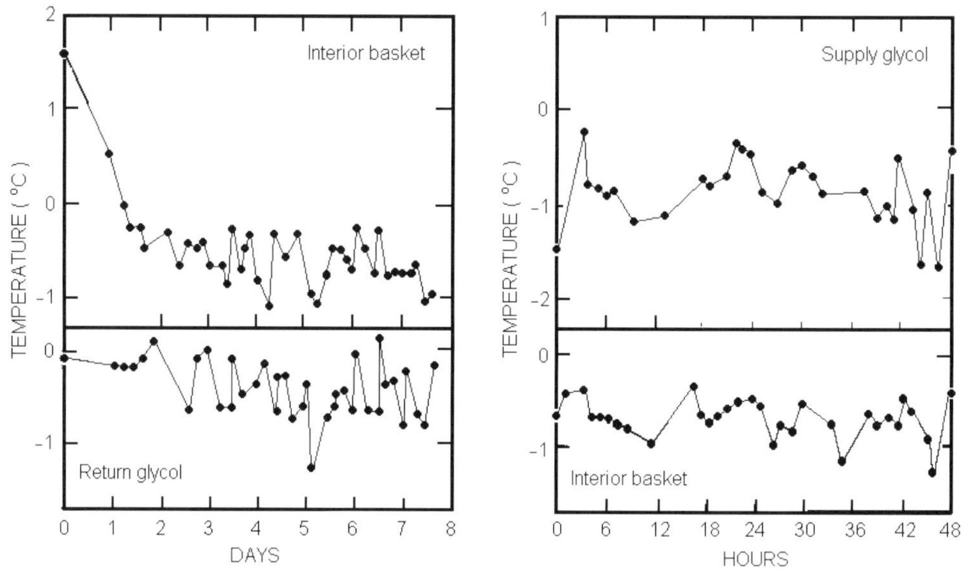

Fig. 11.4. (*left*) The temperature of an interior basket of polyethylene-wrapped pork loins tracks the return glycol temperature in a 12.2 cm (40 ft) Grumman/Dormavac container. Conditions: single basket containing four 6.8 kg polyethylene-wrapped pork loins (D = 12.7 cm, L = 50.8 cm) shielded from radiation by surrounding cardboard boxes; water-ice frozen under the floor; return glycol temperature controlled by an ON–OFF thermostat with a 1.8°C differential. (*right*) The temperature of an interior basket of polyethylene-wrapped pork loins tracks the supply glycol temperature in a 12.2 m (40 ft) Grumman/Dormavac container. Conditions: single basket containing four 6.8 kg polyethylene-wrapped pork loins (D = 12.7 cm, L = 50.8 cm) shielded from radiation by Mylar film; water-ice frozen under the floor; temperature controlled by an ON–OFF thermostat with a 1.8°C differential.

the meat was shielded from radiation by surrounding cardboard boxes (Fig. 11.4, *right*) or by a loosely fitting Mylar overwrap covering the baskets (Fig. 11.4, *left*). Without the radiation shield, if the entire surface area of each pork loin viewed the container wall, a 1.1°C temperature difference would transfer 0.86 kcal/h (3.4 BTU/h), but with the shield in place, radiation could not transfer a significant part of the heat that was gained or lost during each refrigeration ON–OFF cycle. Convection also cannot account for the cycling meat temperature, because even if the low heat capacity of the air–water mixture did not restrict convective heat transfer, with a 1°C temperature difference the low natural convective-film coefficient would only transfer 0.04 kcal/h (0.16 BTU/h) per pork loin. An evaporation–condensation cycle at the conditions existing in the container can transfer 0.91 kcal/h (3.6 BTU/h), and is the only heat transfer mode able to account for the fluctuating meat temperature. When the glycol temperature falls, the container dew point decreases and creates a vapour-pressure gradient between the meat and container atmosphere, which causes the meat to cool by evaporating water. When the glycol temperature increases, the container humidity rises and water condenses on the meat, releasing latent heat that elevates the meat temperature. Condensation on the meat surface is undesirable because it can promote mould and bacterial growth, and it increases weight loss, since part of the condensed water drips from the surface and may not be available to evaporate during the next refrigeration cycle. This problem was corrected in VacuFresh^{SM} by installing a proportional integral derivative thermostat[9] to control the meat temperature with a tolerance of ± 0.1°C (Fig. 11.5), making it feasible to keep meat at a constant temperature close to –1.5°C.

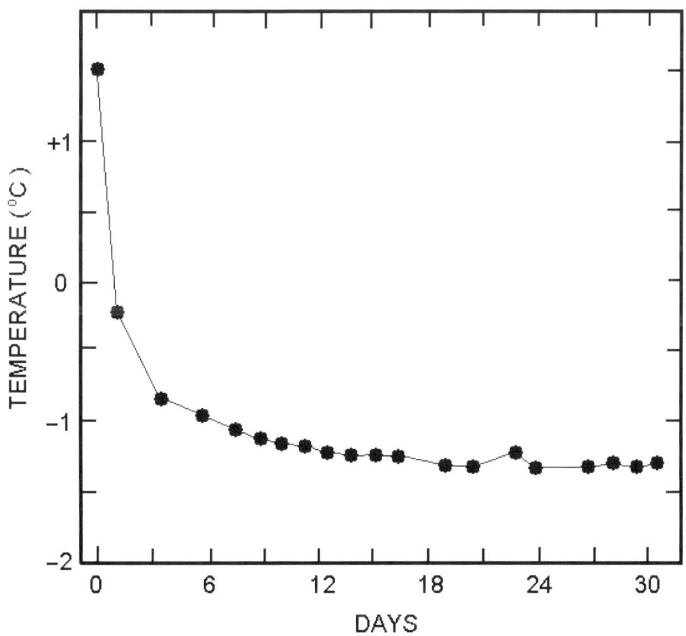

Fig. 11.5. Temperature of an interior basket of polyethylene-wrapped pork loins stored in a 6.1 m (20 ft) VacuFresh^{SM} container. Conditions: an interior basket containing four 6.8 kg polyethylene-wrapped pork lions (D = 12.7 cm; L = 50.8 cm) was shielded from radiation with a loosely fitting Mylar overwrap covering the basket; water-ice frozen under the floor; supply glycol temperature continuously controlled at –1.0 ± 0.1°C with a proportional integral derivative (PID) thermostat.[9]

11.14 Ice-crystal Formation

Ice crystals may shorten shelf life by puncturing membranes and causing fluid loss from muscle. They also lower metmyoglobin reducing power, and induce off-colour development and inferior bloom. Crust-freezing of surface water external to the cells ostensibly has none of these disadvantages, but is undesirable because the recipient of a shipment usually cannot distinguish between crust-freezing and ice-crystal formation within cells. He assumes that the meat has been frozen whenever ice is apparent on its surface.

Theoretically, the freezing point of water should be independent of the storage pressure (15.1), and yet it has been observed that during LP transport, meat sometimes forms superficial ice crystals at temperatures that would not be expected to cause such an occurrence at atmospheric pressure. LP and NA freezing-point data for pork and lamb are presented in Table 11.7. At atmospheric pressure, ice-crystal formation never occurs on lamb at –1.1°C; it occurs sporadically and superficially at –1.4°C; serious ice-crystal formation can be expected at –1.7°C and the surface usually will be crust-frozen; a temperature of –1.9°C invariably causes severe ice-crystal formation and crust-freezing; and the muscle freezes at or below –2.2°C (Burg, 1979b). Even though no significant difference between the temperature of various meat samples could be detected using ± 0.2°C-accuracy thermocouples imbedded in the muscle, significant ice-crystal formation occurred in lamb during a 12-day test at –0.8°C and a pressure of 0.67 kPa (5 mm Hg), but not at atmospheric pressure and a temperature of –1.4°C (Cicale, 1980; Tables 11.8 and 11.9).

Table 11.7. Freezing point of various commodities in LP and NA. Measurements were made by super-cooling the samples in a –4.4 to –6.7°C cabinet either in LP or NA, until ice-crystal formation began and the commodity's temperature abruptly rose to the freezing point (Burg, 1979b; Alloca 1980b). Data in parentheses from Gill and Harrison (1989) and Lutz and Hardenburg (1968).

	Freezing point (°C)	
Commodity	NA	LP
Pork	(–1.5 to –2.0)	–1.2
Lamb	(–1.4 to –1.7)	–0.8
Sweet cherries	–3.0 (–1.8)	–3.1 to –3.7
Mushrooms	(–0.9)	–0.3

The exact temperature at which ice-crystal formation begins is determined by the meat's moisture content. Drier meats, especially if they have a greater muscle pH, form ice crystals at lower temperatures and wetter meats with more free water form ice crystals more rapidly and at higher temperatures. This can cause the temperature for ice-crystal formation to vary from –1.7 to –3.3°C. The laboratory freezing experiments with lamb (Tables 11.8 and 11.9) were performed under conditions that caused a large weight loss in non-humidified LP chambers

Table 11.8. Ice-crystal formation in lamb cuts at –1.4°C storage (air) temperature (Cicale, 1980).

	Meat temp. (°C)			Weight loss (%)			Ice-crystal formation		
Sample	LP–dry	LP–wet	NA	LP–dry	LP–wet	NA	LP–dry	LP–wet	NA
Leg	–1.9	–1.6	–1.7	3.4	3.3	0.2	+++	+++	0
Leg	–1.2	–1.2	–1.4	–	–	–	++++	++	0
Loin	–1.7	–1.7	–1.8	3.5	3.1	0.3	+++	+++	0
Loin	–1.3	–1.2	–1.3	–	–	–	++++	++	0
Rib	–1.9	–1.9	–1.9	4.0	2.8	0.3	+++	+++	0
Rib	–1.2	–	–1.4	3.6	–	–	+++	–	0
Breast	–1.9	–1.9	–1.9	4.2	3.9	0.6	+++	+++	+
Breast	–1.2	–1.3	–1.4	6.0	4.2	1.1	+++	++	0
Shoulder	–1.2	–1.9	–1.9	3.9	3.5	0.9	+++	+++	++
Shoulder	–	–1.3	–1.4	–	4.2	0.9	–	++	0

preserved for 45 days, and at −1°C and 1.07 kPa (8 mm Hg) the result was even better, but at −2°C there was some loss of bloom and evidence of freezer burn. These results were confirmed in an experiment at Texas A & M University in which fore- and hindquarters of beef were cooled to 7.2°C, covered either with polyethylene bags or PVC self-tacking film, and stored in a 12.2 m (40 ft) Fruehauf LP intermodal container operated at −1°C and a pressure of 1.6 kPa (12 mm Hg). Evaluations at 3 and 6 weeks indicated preservation of at-harvest bloom and initial appearance, no slime or odour build-up, with 1% shrinkage in 3 weeks and 2% in 6 weeks. The beef was marketed locally with excellent acceptance. A few cartons of boxed ribs and rounds included during a 3-week test remained in good condition (S.P. Burg, 1977, unpublished data).

Test at 0.57–0.67 kPa (4.3–5 mm Hg)

Hindquarters of beef were stored by the Armour Meat Co. at −0.8°C either in a refrigerated room or in a humidified Grumman/Dormavac LP container operated at 0.67 ± 0.07 kPa (5 ± 0.5 mm Hg) (Shum, 1980; Table 11.13). The meat was fabricated from animals that had been slaughtered the previous day, and had a temperature of 4.4°C when they were received at the test site. Hindquarters kept in NA were unmarketable after 12 days, whereas after 41 days in LP storage samples had 8–9 days' shelf life remaining when they were transferred to air at 2.2°C (Table 11.13). The bacterial count was nearly 100-fold higher after 4 days in NA compared to 41 days in LP. In a companion experiment, bacteria proliferated more rapidly at −1°C and 1.33 kPa (10 mm Hg) than at 0.67 kPa (5 mm Hg). At the higher pressure, the total bacterial plate count on beef carcasses increased from an initial value of 500 per cm^2 to 6×10^6 organisms per cm^2 during a 21-day storage (Armour, 1978).

After 41 days in LP, beef butts and prime strip steaks stored at −1.7°C and a pressure of 0.57 kPa (4.3 mm Hg) had an average total bacterial count of 1.25×10^5 and 2.07×10^5 per cm^2, respectively. Lactobacilli comprised almost the entire bacterial population, with coliform and *E. coli* counts below 1 colony per cm^2 (Jamieson, 1984; Burg, 1987a).

The storage life of retail beef cuts packaged in various types of wraps was compared at −1°C and 95–98% RH in NA vs. LP at pressures of 0.57, 1.33 and 4.0 kPa (4.3, 10 and 30 mm Hg) (Cicale, 1979b,c). A beef round was fabricated into top round, bottom round, and knuckle and the pieces from each subprimal cut were sliced into 18 one-inch steaks and stored naked or wrapped in either ethylene vinyl acetate film[11] or aluminium foil. After 19 days, all steaks had excessive browning and were unmarketable, except those stored in LP at 0.57 kPa (4.3 mm Hg), which had remained in excellent, marketable condition with no off-odour. No off-flavour was noted by an informal tasting

Table 11.13. Average bacterial counts on beef hind quarters stored at −0.8°C either in humidified LP at a pressure of 0.67 kPa (5 mm Hg) or in NA (Shum, 1980).

Storage time (days @ −0.8°C)	Shelf life (days @ 2.2°C)	Bacteria per cm^2 (generations per day)	
		NA	LP
Initial	0	1.5×10^3	–
4	0	1.7×10^7 (3.38)	–
10	0	1.4×10^6	–
12	0	2.2×10^7 (2.00)	–
41	0	–	2.0×10^5 (0.17)
41	2	–	3.2×10^5 (0.35)
41	6	–	2.8×10^6 (0.78)
41	9	–	1.9×10^8 (2.03)

panel when these LP steaks were cooked following a 4-day case life test. The EVA-wrapped steaks still retained 2 days' case life after 24 days' storage in LP at 0.57 kPa (4.3 mm Hg), but after 27 days they began to brown.

Beef carcasses weighing 640 kg were loaded along with 6400 kg of lamb and veal carcasses into a 12.2 m (40 ft) Grumman/Dormavac LP intermodal container, and transported by road from San Angelo, Texas, to Baltimore, Maryland, then by ship to Bandar Shapur, Iran, and by truck to Teheran, arriving after 42 days. The container was operated without humidification at –1.7°C and a pressure of 0.60 kPa (4.5 mm Hg). At arrival, the beef was judged to be in excellent condition by government inspectors and officials of the Iranian Meat Organization (Mermelstein, 1979).

Even better LP microbiological results have been obtained with polyethylene-wrapped, Mylar-shielded beef rumps and strip steaks. After 6 weeks in LP storage at 0.6 kPa (4.5 mm Hg), the total bacterial count on strip steaks was only 2.1×10^4 per cm^2 when they were stored with water-ice under the floor and the thermostat set at –1°C. The meat temperature was –1.2°C, there was a 2.8% weight loss and per cm^2 the bacterial population consisted of 1.6×10^4 lactic acid bacteria, 1.0×10^{-1} coliform and 1.0×10^{-1} *E. coli*. After 6 weeks' LP storage, the total count was 2.8×10^5 per cm^2 on beef rumps, the weight loss 1.9% and per cm^2 there were 3.7×10^5 lactic acid bacteria, 1.0×10^{-1} coliform and $< 1.0 \times 10^{-1}$ *E. coli* (S.P. Burg, 1996, unpublished data).

11.19 Chicken

The development of undesirable microorganisms causing slime and odour formation is the primary cause of refrigerated fresh-poultry deterioration (Hotchkiss, 1989). Frozen chicken has not gained consumer acceptance where fresh, unfrozen chicken is available, mainly because of flavour loss and the cost of energy required for freezing. Icing non-frozen chicken is a relatively expensive means of bacterial control because of the cost of liquid ice and the fact that it constitutes from 30 to 50% of the shipping weight. Although less dry ice is required per load of chicken and there is little or no clean-up involved, the relatively high material cost, combined with a danger of inadvertently freezing the chicken, detracts from its usefulness, and CO_2 can impart an unpleasant odour. LP controls bacterial growth on chicken, eliminating the need for ice or freezing. It limits shrinkage, preserves flavour and appearance, and more than doubles the shelf life of chickens without the attendant mess of ice and the loss in useful weight shipped (Jamieson, 1976b).[12]

Tests at 1.33 kPa (10 mm Hg)

Fryer chickens developed an unpleasant odour, experienced excessive shrinkage and became covered with plaques of pseudomonad bacteria during 3–7 days' storage in NA at –1 to –2°C. Shrinkage and *Pseudomonas* development were almost completely prevented during 3 weeks at pressures ranging from 1.33 to 3.33 kPa (10–25 mm Hg), but chicken stored at 6.67–20.0 kPa (50–150 mm Hg) benefited only slightly. Use of a plastic wrap reduced shrinkage at the lower pressures (Burg, 1976a).

Drumsticks packaged six per shelf tray, with absorbent pads and plastic over-wrap, were stored at 1.7°C and 85–95% RH in LP at various pressures between 1.33 and 20.0 kPa (10 and 150 mm Hg). The storage life of controls kept in NA at the same temperature and humidity was 6 days, judged as the time to reach a surface bacterial count of 10^7 per cm^2. The drumsticks remained in saleable condition for 12 days at 1.33 kPa (10 mm Hg) and for progressively shorter times at higher pressures, equalling control air-storage at about 13.33 kPa (100 mm Hg = 2.6% $[O_2]$) (S.P. Burg, 1978, unpublished data). Whole eviscerated fryer chickens, legs and breasts packed in paper-lined waxed cartons, in polyethylene bags within waxed cartons or unwrapped in waxed cartons, were stored with humidification at 0 ± 1.1°C in a 6.1 m

(20 ft) Fruehauf intermodal hypobaric container operated at a pressure of 1.33 kPa (10 mm Hg). Iced controls were kept in waxed cartons at the same temperature. During 13 days in NA, the initial total plate count of 10^4 bacteria per cm^2 increased to 10^7 per cm^2, which is indicative of spoilage (Cox *et al.*, 1998), and the birds developed off-odours. Chickens stored in LP spoiled in 14–15 days if they were wrapped in butcher paper, 15–16 days in polyethylene and after 20 days the count was only 2×10^6 bacteria per cm^2 without a liner. The lower bacterial count in chicken stored without a liner was associated with a greater weight loss, suggesting that bacterial growth had been inhibited by reduced surface-water activity. During 9 days the weight loss was 2.3% in iced NA controls and 2.4, 3.3 and 5.9% in LP with plastic film, paper and no wrap, respectively. Broilers packed in ice and held in NA at 0 to −1°C spoiled in 18 days, judged as the time required for the total bacterial count to reach 10^6 per 6.45 cm^2 (McCall *et al.*, 1976). Broilers packed without ice, with polyethylene liners loosely folded over the birds, spoiled in 25–26 days when they were stored at 0 to −1°C in a 12.2 m Grumman/Dormavac intermodal hypobaric container operated at a pressure of 1.33 kPa (10 mm Hg). In another test at the same pressure, ice-packed controls spoiled in 12–14 days and birds stored in LP at −1°C lasted 25 days. Tasting panels found no differences in acceptability between hypobaric and ice-pack stored birds (Herring, 1976; Kline, 1976).

Tests at 0.57–0.67 kPa (4.3–5 mm Hg)

Freshly killed and cleaned chickens were weighed, swabbed and stored without top ice at −0.8°C in waxed boxes lined with butcher paper, either in NA or in LP at 0.67 kPa (5 mm Hg) and 96–98% RH (Lovizio, 1980; Jamieson, 1984). The initial surface bacterial count, 1.3×10^3 per cm^2, increased to a spoilage level of 10^7 per cm^2 after 15 days in NA. In LP, the chickens still had an acceptable appearance and their bacterial count hardly changed during storage for 21 days (3.1×10^2 per cm^2) and 28 days (1.4×10^4 per cm^2). The shelf life of the chickens at 3.3°C was 4–5 days when they were removed from LP storage after 28 days (Table 11.14).

11.20 Duck

Freshly killed and cleaned ducks were weighed, swabbed and stored at −0.8°C without top ice in waxed boxes lined with butcher paper, either in NA or in LP at 0.67 kPa (5 mm Hg) and 96–98% RH (Lovizio, 1980; Jamieson, 1984). After 21 days' LP storage, the shelf life at 3.3°C was 4 days; after 28 days' LP storage, the ducks retained an acceptable appearance and had a shelf life of 3 days (Table 11.15).

11.21 Fish (Atlantic Cod, *Gadus morhua*; Atlantic Herring, *Clupea harengus*; Spot Fish, *Leiostrous xanthurus*; Atlantic Salmon, *Salmo solar*; Caesar Grunt, *Haemulon carbonarium*; Gray Snapper, *Lutjanus griseus*; Atlantic mackerel, *Scomber scombrus*)

In the EEC Council Regulation 103/76 with later amendments, common marketing standards for certain fresh or chilled fish

Table 11.14. Total plate count on chickens stored in LP at 5 mm Hg, 96–98% RH, or in NA (Lovizio, 1980).

Storage time (days @ −0.8°C)	+Shelf life (days @ 3.3°C)	Bacterial count (organisms/cm^2)	
		NA	LP
0	–	1.3×10^3	1.3×10^3
6	–	2.9×10^3	–
8	–	5.7×10^4	–
12	–	1.3×10^6	–
15	–	9.1×10^6	–
21	–	–	3.1×10^2
21	2	–	8.1×10^4
28	–	–	1.4×10^4
28	2	–	2.9×10^4
28	5	–	7.1×10^6

Table 11.15. Total plate count on ducks stored in LP at 0.67 kPa (5 mm Hg) and 96–98% RH, or in NA (Lovizio, 1980).

Storage time (days @ −0.8°C)	+Shelf life (days @ 3.3°C)	Bacterial count (organisms/cm²)	
		NA	LP
0	–	8.0×10^3	8.0×10^3
6	–	1.7×10^4	–
8	–	1.6×10^5	–
12	–	3.3×10^6	–
21	–	–	1.0×10^4
21	2	–	7.8×10^4
28	–	–	2.5×10^4
28	2	–	4.0×10^5
28	5	–	4.2×10^7

are stipulated with rules for freshness and size categories (Mercantila, 1989a). Codex Alimentarius, the recommended code of practice for fresh fish (CAC/RCP 9–1976), states that this commodity should always be carried in ice, and that the temperature in the storage room must be a few degrees above 0°C in order to allow the ice to melt and continuously wash bacteria and undesirable metabolic products of natural decomposition and bacterial action from the fish surface. When modified-atmosphere packaging (MAP) is used, the temperature of the fish must be kept below 3.3°C to avoid type E botulism. The only international standard for the transport of preserved fish products is the Codex Recommended international code of practice for smoked fish (CAC/RCP 25–1979), which recommends that smoked fish, which is not treated in such a way as to prevent the outgrowth of *Cl. botulinum* type E, should always be kept at a temperature below 3.3°C.

The major causes of postharvest decline in fish quality are intrinsic chemical and physical alterations, referred to as 'autolysis', and changes resulting from bacterial growth and metabolism (Haard, 1992). Although bacteria cause gross spoilage, nevertheless sterile and non-sterile fish have the same shelf life because exclusion of bacteria does not alter the rate at which fish become unacceptable to sensory panellists. Most methods of preservation only affect microbial growth, whereas chilling also slows odour, flavour, colour, appearance and texture deterioration caused by autolytic chemical and biochemical processes such as proteolysis, lipid hydrolysis and lipid oxidation. Off-odours form from volatile bases, and sulphides and acids by bacterial action, but unpleasant odours and aromas also develop on sterile fish due to ammonia formation and the influence of lipolysis on oxidative rancidity. In both sterile and non-sterile fish, the rate of texture softening is the same. Bacterial TMAO reductase catalyses production of the 'fishy' aroma compound trimethylamine (TMA) from trimethylamine oxide (TMAO).

LP fish tests over seen by Grumman were performed at pressures of 1.33 kPa (10 mm Hg) or higher because of concerns about type E botulism (11.12). Even LP storage pressures higher than 1.33 kPa (10 mm Hg) conceivably could provide a low enough O_2 partial pressure to support *Cl. botulinum* growth and toxin production, and the organism is not likely to be killed by intermittent exposure to air during the time required for occasional ventings of the vacuum tank (Loesche, 1969; Walden and Hentges, 1975). The inclusion in modified-atmosphere packages of some O_2 with CO_2 or N_2 is believed to reduce the hazard from botulism in packaged fish, but this perception is probably erroneous because, regardless of the presence of added CO_2 and/or O_2, toxic growth of *C. botulinum* occurs within 24 h in fresh fish if it is abused sufficiently at 30–35°C (Brody, 1989b). For this reason legal regulations dictate that vacuum- or MA-packaged fresh or smoked fish must always be kept below 3.3°C. After fish are removed from an LP container, the temperature may not be so precisely controlled, but at that time the botulism organism will be killed because the commodity is exposed to atmospheric air. Theoretically, there is a much greater risk of toxin production in vacuum- or CO_2-packaged fish than there is in LP because the vacuum-packed and CO_2-flushed commodity remains in an anaerobic environment and might be exposed to an excessive temperature during distribution. Fatal accidents have resulted from improperly refrigerated vacuum-packaged smoked fish (Brody, 1989b).

Melting top-ice doubles or triples the storage life of fish by leaching out trimethylamine and ammonia, and rinsing away bacteria (Table 11.16). Usually the result of a hypobaric test has been judged by comparing the shelf life of top-iced fish preserved in NA vs. fish without top-ice stored in LP, both kept at the same temperature slightly above 0°C to ensure slow melting of the top-ice in NA. This tends to underestimate LP's effect because the optimal storage temperature in LP is several degrees lower. A better experimental protocol would be to store both treatments at their optimal temperature, or both with top-ice at > 0°C.

Freshly caught Caesar grunts (*Haemulon carbonarium*) and grey snappers (*Lutjanus griseus*) were stored naked or wrapped in plastic at +2°C either in NA or in LP at pressures ranging from 1.07 to 20 kPa (8–150 mm Hg). The gray snappers were gutted; the grunts were stored intact. In NA the fish developed slime and a foul odour in 4–6 days; in LP they were well preserved for up to several weeks by the optimal pressure range, 1.07–3.33 kPa (8–25 mm Hg). Higher pressures were ineffective, a PVC wrap was required to prevent drying and lower pressures were not tested (S.P. Burg, 1975, unpublished data).

Internal bacterial counts and the pH increased rapidly in iced spot fish (*Leiostrous xanthurus*) stored in NA at 1.1–2.2°C and more slowly in non-iced fish stored in LP at 1.33 kPa (10 mm Hg) and the same temperatures. Fish kept at 0.8 kPa (6 mm Hg) had the lowest bacterial count and best appearance, with clear eyes, firm flesh and almost no odour or surface slime (Tables 11.16 and 11.17). The hypobaric chamber had to be evacuated slowly to avoid swelling of the fish belly and bulging of the eyes, and vented gradually to prevent the eyes and belly from becoming depressed.

Table 11.16. Changes in the bacterial count, trimethylamine (TMA) and pH of non-gutted spot fish (*Leiostrous xanthurus*) stored with top-ice at 1.1–2.2°C in NA, and without ice at 1.1–2.2°C either in LP at 0.8 and 1.33 kPa (6 and 10 mm Hg) or in NA (S.P. Burg, 1975, unpublished data).

Treatment	Days stored	Bacteria/g	mg TMA/100 g	pH
At receipt	0	0	0.01	6.1
NA – no ice	15	1.8×10^5	6.7	6.6
NA – ice	15	7.9×10^3	2.4	6.2
10 mm Hg	15	1.3×10^4	5.0	6.2
6 mm Hg	15	2.0×10^3	3.4	6.0
NA – no ice	23	1.6×10^6	11.3	7.2
NA – ice	23	5.6×10^4	6.3	6.8
10 mm Hg	23	6.3×10^4	8.5	6.3
6 mm Hg	23	2.5×10^4	8.0	6.3

Table 11.17. Bacterial count on spot fish (*Leiostrous xanthurus*) stored at 1.1–2.2°C either in NA with top-ice, or in LP at 1.33 kPa (10 mm Hg) without ice (S.P. Burg, 1975, unpublished data).

	Bacteria/g			
Storage time (days)	Gutted fish		Non-gutted fish	
	NA – ice	10 mm Hg	NA – ice	10 mm Hg
0	1.8×10^3	1.8×10^3	1.8×10^3	1.8×10^3
4	1.0×10^3	1.7×10^3	–	–
5	–	–	5.0×10^4	3.4×10^4
9	4.0×10^5	1.7×10^5	–	–
12	–	–	2.0×10^6	1.8×10^5
13	2.4×10^7	2.0×10^5	–	–

Non-iced cod fish stored better at −1.1°C and a pressure of 1.33 kPa (10 mm Hg) compared to iced controls held at 0°C, but the LP fish were no longer acceptable after 3 weeks. The same pressure and temperature improved the storage life of fluke by approximately 10 days compared to iced controls. After 2 weeks, the LP-stored fish was of very good quality, and after 4 weeks it was acceptable but of poor quality (Jamieson, 1979a, 1984). Atlantic herring and cod fish held at 2–4°C and 100% RH at pressures ranging from 1.73 to 2.67 kPa (13–20 mm Hg) had acceptable odour and appearance for 3 days longer than iced controls, and retained borderline acceptability for an additional 5 days (Haard *et al.*, 1979; Martins, 1979). Samples of cod, herring (skin on) and mackerel (skin on) fillets, which had been stored at 0°C in NA with top-ice, and in LP without ice covered with a polyethylene film at 0 or −0.56°C at a pressure of 1.33 kPa (10 mm Hg), were analysed for eating quality by an eight-member experienced tasting panel, and for trimethylamine (TMA), total volatile base, moisture, fat content, peroxide, free fatty acid, thiobarbituric acid (TBA), total bacterial count at 25°C and *Ps. putrefaciens*, fluorescent pseudomonads and proteolytic bacterial counts (Haard *et al.*, 1979; Martins, 1979). Storage life was 12.3 days for cod and 13.7 days for herring in NA at 0°C; 14.1 days for cod and 14.9 days for herring in LP at 0°C; and 17.1 days for cod and 20.1 days for herring in LP at −0.56°C. In another study, the keeping time of cod and mackerel fillets at −1.1°C was 16.3 and 21.5 days in LP, respectively vs. 15.6 and 16.6 days for samples held on ice in NA at 0°C. LP slowed the growth of fluorescent pseudomonads and to a lesser extent *Ps. putrifaciens* and proteolytic bacteria, and lowered the peroxide and TBA values in stored herring fillets, indicating that the oxidation of fats had been slowed. This was reflected in an improved tasting-panel rating for both herring and mackerel fillets stored in LP (Varga *et al.*, 1979; Table 11.18). It was concluded that a pressure of 2.02 kPa (15 mm Hg) extended the storage life of these fish by 1.5–2.0-fold. During rapid venting, the eyes of the cod fish were sucked into the eye cavity.

Fresh eviscerated Atlantic herring and cod fish were stored at 2–4°C in NA, in CA flowing 0.2% [O_2], or in LP at 2.23–2.67 kPa (16.7–20.0 mm Hg = 0.33–0.42% [O_2]) (Figs 11.6 and 11.7). Herrings in NA were judged unacceptable after 8 days based on appearance, chemical and microbiological indices of fish freshness, and odour, while samples

Table 11.18. Peroxide and thiobarbituric acid (TBA) content and mean taste panel scores of herring and mackerel fillets stored with top-ice at 0°C in NA, or without ice at 1.33 kPa (10 mm Hg) either at 0°C or −0.56°C (Varga *et al.*, 1979).

Storage time (days)	Mean tasting panel score		H_2O_2 (m-equiv/1000 g)		TBA (mg-MA/1000 g)	
	Iced	LP	Iced	LP	Iced	LP
Herring fillets (skin on): LP at 0°C						
0	9.3	9.3	0.40	0.40	0.31	0.31
12	5.2	6.8	8.41	2.00	2.32	1.84
15	4.0	4.0	10.84	2.43	3.03	1.34
Herring fillets (skin on): LP at −0.56°C						
0	9.30	9.3	0.45	0.45	0.42	0.42
10	5.84	7.37	9.97	0.91	4.15	0.69
16	3.42	4.83	15.08	1.54	5.41	1.27
Mackerel fillets (skin on): LP at −1.1°C						
0	9.33	9.33	0.46	0.46	0.60	0.60
7	6.50	6.67	–	–	–	–
13	5.67	7.29	–	–	–	–
17	4.50	–	–	–	–	–
20	–	4.57	–	–	–	–

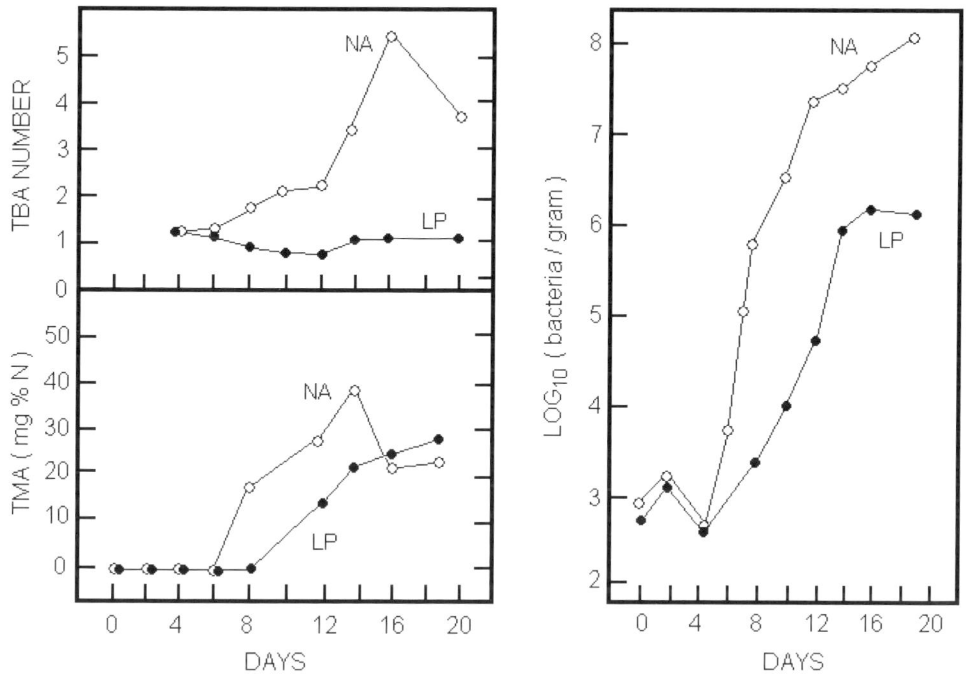

Fig. 11.6. (*upper left*) Thiobarbaturic acid (TBA) value, (lower left) trimethylamine (TMA) content and (*right*) total plate counts of gutted herring held at 2–4°C for various times. Storage atmosphere: air (□), or hypobaric at 2.67 kPa (20 mm Hg) (■). Data are from one experiment and are representative of two additional trials in which the hypobaric pressure was 2.23 kPa (16.7 mm Hg) (Haard *et al.*, 1979). A TBA value in the range 1–2 indicates that the fish will in all probability smell and taste rancid (Cornell, 1975).

stored in LP at 2.67 kPa (20 mm Hg) were acceptable until day 16. The primary bacterial flora which developed in both NA and LP were pseudomonads, with *Achromobacter* and *Acaligenes* species also present. Herring held at atmospheric and hypobaric pressure reached the fresh fish microbiological spoilage aerobic plate count guideline of 1.0×10^6 organisms per gram (Wehr, 1978) after 8 and 14 days, respectively. TMA, which is a by-product of bacterial metabolism, accumulated in parallel with the total plate count. LP suppressed TBA formation, indicating a reduced rate of lipid oxidation. LP also improved the storage life of cod, compared to fish held in NA or in CA with 0.2% $[O_2]$ (Fig. 11.7). According to the microbiological spoilage guideline, cod held in NA or CA was unacceptable in 6 days, while fish kept in LP was acceptable for at least 13 days. CA storage with flowing 0.2% $[O_2]$ did not improve storage life, bacterial counts, or TMA content, while in LP, although 0.35% $[O_2]$ was present, the lag phase for bacterial growth was extended, the log phase abbreviated and the fish retained a fresh odour. This indicates that factors in addition to low $[O_2]$ are involved in the effect which LP has on bacterial proliferation (also see 7.4; Tables 7.4 and 7.5). 'Belly-burn' rapidly developed after iced (1:1 w/v) intact herrings were removed from LP storage at 2.67 kPa (20 mm Hg). This probably was caused by damage to proteolytic enzyme-containing viscera brought about by rapid venting. Disruption of the swim bladder can be prevented or minimized by a gradual chamber evacuation and release (W. Jamieson, 1982, unpublished).

The shelf life of salmon was extended by hypobaric storage in a CO_2 atmosphere far more than by LP storage flowing air, or CA storage flowing 100% $[CO_2]$ (Haard and Lee, 1982; Haard, 1983; Figs 11.8 and 11.9).

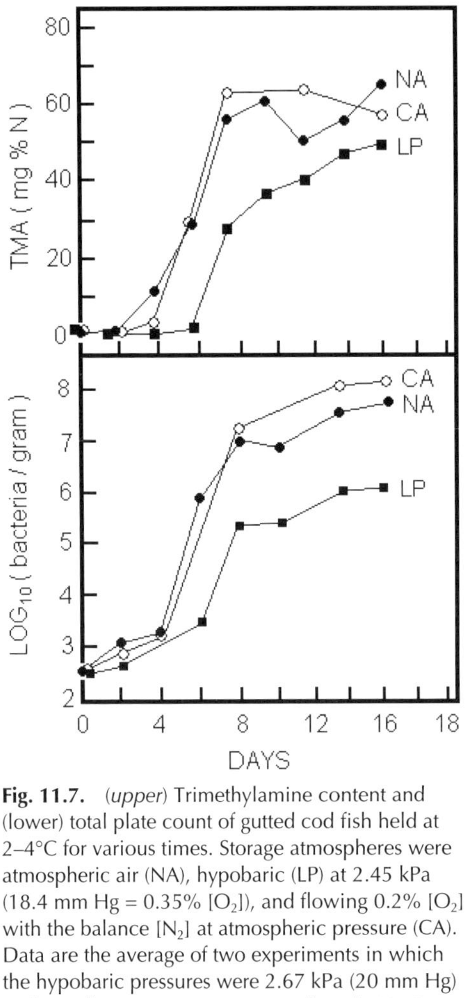

Fig. 11.7. (*upper*) Trimethylamine content and (lower) total plate count of gutted cod fish held at 2–4°C for various times. Storage atmospheres were atmospheric air (NA), hypobaric (LP) at 2.45 kPa (18.4 mm Hg = 0.35% [O_2]), and flowing 0.2% [O_2] with the balance [N_2] at atmospheric pressure (CA). Data are the average of two experiments in which the hypobaric pressures were 2.67 kPa (20 mm Hg) and 2.23 kPa (16.7 mm Hg) (Haard *et al.*, 1979).

Salmon steaks were cut 2 cm thick, perpendicular to the long axis of the fish, and individually packaged in plastic trays overwrapped with polyvinylchloride film (RMF 61, Bordan). Steaks had a shelf life of 6–8 days in NA at 3°C and developed an undesirable appearance and 'fishy' odour after 10 days. Steaks held in pure CO_2 at atmospheric pressure did not undergo rapid bacterial spoilage, but developed a powdery texture and a carbonated, bland taste due to acidification of the superficial tissue layers by the added CO_2 (Haard and Lee, 1982). Shelf life was extended in LP at a pressure of 2.02 kPa (15 mm Hg), provided that the hypobaric chambers were ventilated with pure CO_2 rather than air in order to sustain 1.3% [CO_2] in the saturated hypobaric atmosphere. The steaks did not develop an off-odour and had excellent appearance after 20 days. Both 100% [CO_2] at atmospheric pressure and CO_2 ventilation at 2.02 kPa (15 mm Hg) extended the lag phase and reduced the log phase of bacterial growth (Fig. 11.8, *left*), keeping the total bacterial count below the spoilage guideline of 10^6 organisms per gram, and even decreasing the total count after 16 days. Flowing CO_2 at a pressure of 2.02 kPa (15 mm Hg) was effective in slowing bacterial growth even when the salmon initially had a high initial plate count (Fig. 11.9, *left*). TMA formation was correlated with the total aerobic plate count (Figs 11.8 and 11.9, *right*), but the TMA values for samples stored with CO_2 at 2.02 kPa (15 mm Hg) never reached the level of samples held in CO_2 at atmospheric pressure.

Based on the experiments in which CO_2 was added during LP storage, Haard and Lee (1982) suggested that control of O_2 tension may not be the critical factor limiting bacterial growth or extending the storage life of fish in LP, and that other factors such as 'out-gassing' and low pressure stress may contribute to the retardation of bacterial growth that occurs under hypobaric conditions. However, there is a more plausible explanation for the beneficial effect that CO_2 had in preserving salmon fillets under hypobaric conditions. The 1.3% concentration of [CO_2] present in the hypobaric chambers should not have any significant direct effect on bacterial growth, since at least 20% and preferably 40–100% [CO_2] is required for CO_2 to have had a substantial anti-microbial action with seafood (Brody, 1989b). Instead, the improved storage in LP must be due to the elimination of O_2 by flushing with CO_2, rather than from any action of CO_2. This conclusion is supported by the test showing that 0.8 kPa (6 mm Hg = 0.04% [O_2]) is significantly better than 1.3 kPa (10 mm Hg = 0.15% [O_2]) for preserving spot fish (Table 11.16), by studies indicating that reduced O_2 packaging is an effective means to extend the shelf life of fish such as plaice and haddock

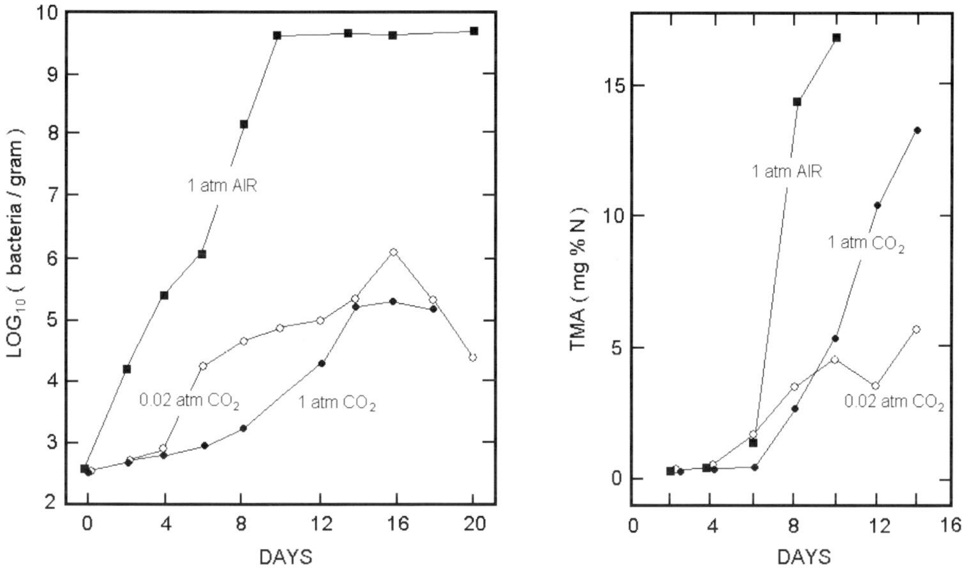

Fig. 11.8. Total aerobic plate counts (*left*) and trimethylamine content (*right*) of salmon steaks held at 3°C in air or 100% CO_2 at atmospheric pressure and in LP flushed with pure CO_2 at 2.02 kPa (15 m Hg = 0.02 atm) (Haard and Lee, 1982).

(Brody, 1989b), and by the fact that the tripling of storage life that resulted when CO_2 and LP were combined is similar in magnitude to the extension in storage life that occurs with various types of red meat when the pressure is lowered from 2 to 0.88 kPa (15 to 4.4 mm Hg) in order to eliminate all O_2 (Table 11.3). CO_2 added in MA packages reduces oxidative rancidity of fats by a similar mechanism, displacing O_2 from the package, rather than by exerting a direct action. If this is the mechanism by which CO_2 improved the storage of salmon fillets, the optimal LP pressure/temperature combination for fish storage must be –1°C at 0.57 kPa (4.3 mm Hg).

11.22 Horse Meat

Horse meat was successfully shipped from Texas to France, Belgium and Italy (Jamieson, 1984) in a 12.2 m (40 ft) Grumman/Dormavac hypobaric intermodal container operated at –1°C and a pressure of 1.33 kPa (10 mm Hg). Even better horse meat storage results when the pressure is lowered to 0.67 kPa (5 mm Hg; Table 11.19). A horse meat forequarter and one hindquarter, wrapped in cotton stockings, were hung in a 12.2 m Grumman/Dormavac intermodal LP container, and one piece each of the shoulder, outside round and rump were wrapped in butcher paper and placed in a cardboard box inside the container. The container was operated with humidification at a pressure of 0.67 kPa (5 mm Hg) and a temperature of –1 ± 0.5°C. Identical samples were stored in NA at the same temperature and humidity. The surface of the samples stored in NA gradually darkened, eventually turning brown-black, and by 13 days the colour of both quarters and boxed meat was unacceptable. An off-odour developed in the boxed meat after 5 days, and in the quarters after 8 days, but no sour or putrid odour was detected even after 13 days. During LP storage, the bacterial counts hardly changed from the initial values and after 20 days the horse meat was judged to be in excellent condition (Jamieson, 1980b). Colour was retained in the LP samples throughout the entire storage period, and with minimal trimming the inner muscle was cherry red and appeared

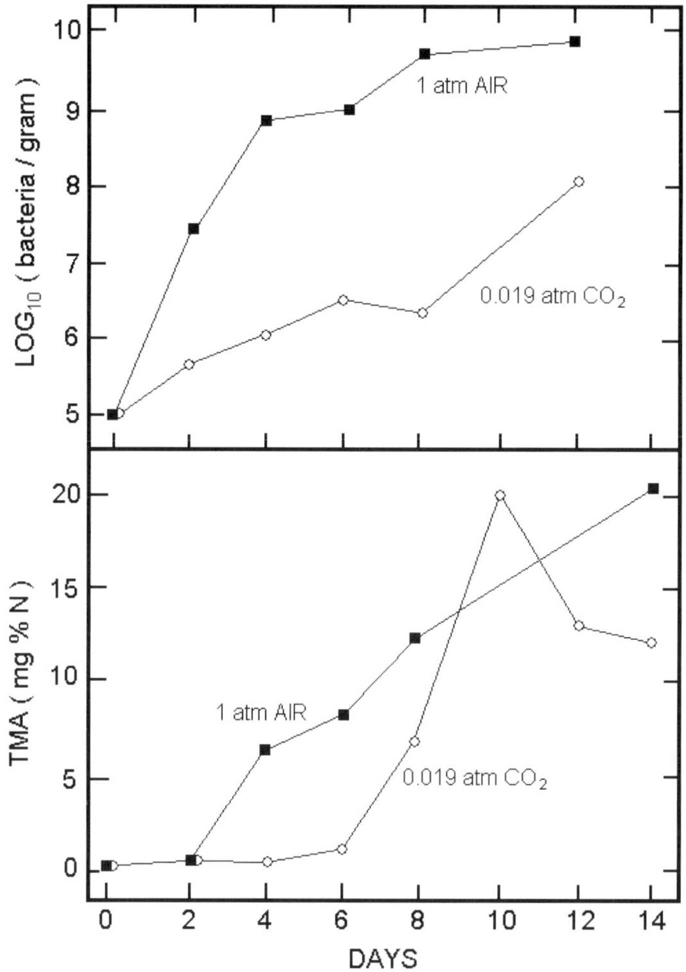

Fig. 11.9. Total aerobic plate counts (*upper*) and trimethylamine content (*lower*) of salmon steaks held at 3°C in air, or in LP flushed with pure CO_2 at 1.9 kPa (14 mm Hg = 0.19 atm). Note that the initial plate count was much higher in this experiment, compared to that described in Fig. 11.8 (Haard and Lee, 1982).

Table 11.19. Average total plate count of horse meat stored at −1 ± 0.5°C in LP with humidification at a pressure of 0.67 kPa (5 mm Hg) or in NA (Jamieson, 1980b).

	Microorganisms per cm^2	
Storage condition	Quarters	Boxed meat
Initial	2.5×10^2	2.2×10^2
7 days in NA	7.8×10^2	6.2×10^3
10 days in NA	6.8×10^3	1.4×10^4
13 days in NA	1.4×10^5	2.0×10^6
21 days in LP	6.4×10^2	3.3×10^3

moist (Burg, 1987a). The texture was judged extremely tender by a representative of Sumitomo.

After a relatively short storage period in vacuum bags, the surface of table-ready horse meat darkens, but upon removal from LP after 22 days at −1°C and a pressure of 0.57 kPa (4.3 mm Hg), a PVC-wrapped horse meat quarter and a polyethylene-wrapped horse meat pistol developed full bloom without the darkening caused by metmyoglobin formation (S.P. Burg, 1994, unpublished).

11.23 Lamb

Tests at 1.33 kPa (10 mm Hg)

Lamb carcasses weighing 9100 kg, with initial total plate counts of 155 bacteria per cm^2 for internal and 1550 bacteria per cm^2 for external surfaces, were stored naked, or wrapped with plastic or paper in a 12.2 m (40 ft) Grumman/Dormavac LP intermodal container operated with humidification at –1°C and a pressure of 1.33 kPa (10 mm Hg). During 22–30 days, the bacterial count increased by 2 log units on internal surfaces and 1 log unit on external surfaces, and in 43 days the average internal and external total plate counts increased to more than 1.5×10^6 bacteria per cm^2. No discoloration or off-odour of the lamb carcasses was detected during 43 days, and no differences were noted between the various wrapping methods. After various periods of LP storage, six lamb carcasses were removed and placed either in a refrigerated cooler, or they were fabricated into lamb chops, packaged and placed into a refrigerated showcase for a shelf-life study. Lamb held in LP storage for 30, 36 and 43 days had a refrigerated-cooler shelf life of 5, 4 and 2 days, respectively; in the showcase the shelf life of lamb chops was 4, 3 and 2 days, respectively (Restaino, 1978).

Fresh lamb carcasses weighing 9600 kg were wrapped in paper or PVC film, or hung naked in a 12.2 m (40 ft) Grumman/Dormavac LP container operated with humidification at –1°C and a pressure of 1.33 kPa (10 mm Hg). After 28 days in LP, 8200 kg were inspected and sold to a retail supermarket chain in Los Angeles. Selected cuts were fabricated and cooked for evaluation by a skilled tasting panel to compare the LP-stored lamb to fresh lamb purchased at a local supermarket. The LP lamb was found to be more tender by 28 of the 29 judges, 23 found it to be more juicy and 21 preferred its flavour (Grumman, 1978). The remaining carcasses were left in LP, and removed after 36 and 43 days. At these times they were not off-colour or odoriferous, and the bacteria levels were well within the USDA accepted limits. Their shelf life in a refrigerated cold room was 2–4 days. A sensory evaluation could not distinguish between lambs stored in LP for 28 and 43 days. The best general appearance was obtained with naked carcasses, although meat wrapped with PVC and paper was quite acceptable (Grumman Allied Industry, 1978, unpublished data).

A total of 622 lambs weighing 13,612 kg were cooled overnight to 0°C and then hung from straps in a 12.2 m (40 ft) Grumman/Dormavac LP intermodal container. Carcasses that could contact the aluminium walls were wrapped with cotton stockinettes; the remainder were left naked. The test was designed to simulate a 33-day shipment from Australia to Arab countries, Iran or the US West Coast, or a delayed shipment of up to 55 days maximum. The container was operated with humidification at –1°C and a pressure of 1.33 kPa (10 mm Hg). After 33 and 42 days' storage in LP, the average total plate counts were 5×10^4 and 1.2×10^7 organisms per cm^2, respectively. Retail cuts (chops, loins, legs) prepared immediately after 33–42 days' LP storage had a showcase life of 2–3 days; cuts prepared after 55 days' LP storage had a shelf life of only 1 day. Carcasses removed after 42 days had a shelf life of 5–6 days at 0–1.1°C and 4 days at 4.4°C. By 55 days in LP the shelf life of carcasses at 0–1.1°C was reduced to 2–4 days. A tasting panel comprised of 31 judges rated prepared lamb cut from carcasses after 33, 42 and 55 days' LP storage to be superior in juiciness, flavour and tenderness (Jamieson, 1980d).

Tests at 0.57–0.67 kPa (4.3–5 mm Hg)

Lamb storage at –1.7°C improved progressively when the pressure was reduced from 1.33 kPa (10 mm Hg) to 0.57 kPa (4.3 mm Hg) (Jamieson, 1980c; Tables 11.3 and 11.4; Burg, 1987a). Table 11.20 presents details of a laboratory test performed at –1.6°C and a pressure of 0.67 kPa (5 mm Hg).

A total of 622 chilled lamb carcasses were shipped from Australia to Miami in a 12.2 m (40 ft) Grumman/Dormavac

Table 11.20. Bacterial counts on chilled lamb stored for 40 days at –1.6°C and a pressure of 0.67 kPa (5 mm Hg) with humidification (Jamieson, 1980c).

	Log_{10} bacterial count/cm^2	
Location	Initial	40 days
Hind leg	3.72	4.03
Abdominal cavity	2.62	3.48
Abdominal flap	3.55	5.49
Brisket	3.38	5.51
Shoulder	3.53	4.40
Ribs	3.37	6.15
Mean	3.37	4.84

intermodal hypobaric container operated at 0.2 to –0.7°C and a pressure of 0.57 kPa (4.3 mm Hg). All of the carcasses were 'naked' except that those in contact with the aluminium bulkhead, sides and door were wrapped in disinfected cotton stockinettes (Jamieson, 1980d). At arrival after 42 days in transit, the lambs were rated by the Australian Veterinary attaché as follows:

> After 5 to 6 hours they showed good bloom and resembled 2 to 3 day-old chilled carcasses in appearance. Deep muscle temperature on unloading varied between –2.2 and –1.1°C. Bacterial swabs were taken from 25 lambs from three sites; namely the brisket, pleural cavity and crutch area. I am advised that standard plate counts indicated a range of 10^2 to 10^4 with one count each of 10^5 and 10^6. Paired samples correlated with the Grumman Industries' laboratory results. I am advised that these counts are rather similar to the plate counts one would expect from 48-hour carcasses.
> (Murray, 1980)

Unloading was also witnessed by a resident USDA inspector and the USDA Circuit Supervisor, and by the Director of the New York Office of the Australian Meat and Livestock Corporation. During the trip, the air temperature inside the container varied between –0.2 and –1.3°C, averaging –0.8°C. At receipt, the average total bacterial count was 3×10^4 and 1.6×10^4 per cm^2 according to analyses carried out by Grumman's Dormavac Laboratory and the American Bacteriological and Chemical Research Corp., respectively. The in-transit average weight loss per carcass was only 0.54%. A trained tasting panel judged the lamb to be superior in juiciness, flavour and tenderness compared to fresh lamb obtained from a local supermarket. Some of these carcasses were hung in a local supermarket's meat room at 0–1°C. They had 8–10 days' shelf life, and meat fabricated after 10 days was judged to have excellent flavour and juiciness (Burg, 1980b). The meat department manager at the supermarket found the bloom to be excellent and reported the meat to be very flavourful and juicy. After 2 days' storage in NA, a carcass was cut in half and both halves were placed in a single cardboard box with a polyethylene lower slip-sheet and kept at 1°C and 92% RH. The meat had at least 2–4 days' case life remaining, and even after 6 days was highly desirable with good bloom and surface condition, excellent flavour and juiciness. Immediately after unloading, two carcasses were hung at ambient conditions (21.1–26.7°C, 50–60% RH) to simulate an outdoor market in the Middle East. During the first 24 h, no objectionable odour or slime developed; a strong but not objectionable odour developed without slime in an additional 24 h; the carcass spoiled in 12 additional hours with slime and putrid odour. The test results indicate that lamb shipped for 42 days will tolerate the traditional Arab practice of hanging the carcasses for up to 24 h at outdoor ambient temperatures. During this interval, meat is periodically cut from the carcass by a butcher.

A shipment of 1130 stockinette-wrapped 2nd/3rd-grade lamb carcasses was sent from Australia to Dubai in a 12.2 m (40 ft) Grumman/Dormavac intermodal hypobaric container operated at 1°C and a pressure of 0.60 kPa (4.5 mm Hg). At receipt, after 55 days in transit, it was rated by the Australian Assistant Director of the Middle East Region as follows:

> Once stockinette was removed, the lambs revealed: No off-odour – in fact very little 'lamb' smell at all. Surface was dry but not desiccated with no feeling of 'stickiness'.

Estimated shelf life: 5–6 days. Overall the product was excellent.

(Conkey, 1981)

Details of a successful 40-day full container-load shipment of carcass lamb in a 12.2 m (40 ft) Grumman/Dormavac hypobaric container operated at −2°C and a pressure of 0.61 kPa (4.6 mm Hg) were described in Sharp (1985).

11.24 Pork

Tests at 1.33 kPa (10 mm Hg)

Pork loins and butts stored at −1°C and pressures ranging from 1.07 to 2.0 kPa (8–15 mm Hg) were still in excellent condition after 3 weeks with no odour; they retained their initial colour, there was no slime or desiccation, and very little weight loss. In NA, pork spoiled in 7 days (Burg, 1976a).

Boxed pork loins and spare ribs were stored with humidification in a 12.2 m (40 ft) Grumman/Dormavac LP intermodal container operated at −1°C and a pressure of 1.33 kPa (10 mm Hg). The loins were wrapped with paper or polyethylene; the spare ribs were stored in polyethylene bags. Sparerib swab samples taken after 13 days in NA indicated that the surface bacterial count had reached a spoilage level of 1.55×10^8 organisms per cm^2. The same count on spare ribs developed during 28 days in LP, at which time odour, discoloration and slime development became evident. Surface counts on both paper- and polyethylene-wrapped LP loins reached 1.55×10^7 per cm^2 during 20 days' storage. Freshly cut pork chops prepared from loins held in NA for 10 days had bacterial counts of 1.55×10^5 organisms per cm^2, whereas the same count did not result for 22–23 days when pork chops were prepared from LP-stored loins. The shrink rate was 3.5% in 2 weeks for paper-wrapped loins stored in LP and NA, but only 1.5% for polyethylene-wrapped LP loins (Herring, 1976; McClendon and Kammlah, 1976; Restaino *et al.*, 1976). LP loins had normal taste, cooking yield, evaporative loss, cooking time, tenderness and general acceptance. Spare ribs stored in polyethylene bags yielded similar results, except that they lost no weight during NA or LP storage. Their storage life was 13 days in NA and 28 days in LP, and their shelf life was a function of bacterial count at the time storage was terminated independent of the nature of storage. The bacterial flora that eventually developed were similar in LP and air (Jamieson, 1980c). Table 11.21 indicates the total plate count at 0–1°C after bacon, ham and pork loins were stored for 13 days in NA or in LP at a pressure of 1.33 kPa (10 mm Hg).

Pork loins obtained 17 h after slaughter were stored naked or wrapped in polyethylene or paper at 0°C, either in NA or in LP with humidification at a pressure 1.33 kPa (10 mm Hg). In NA, the initial plate count of 800 organisms per cm^2 increased to 10^7 organisms per cm^2 during 10 days, regardless of whether the meat was wrapped or left naked. After 28 days in LP, the total plate

Table 11.21. Total plate counts of various cuts of fresh pork after storage for 13 days at 0–1°C either in NA or in LP with humidification at a pressure of 1.33 kPa (10 mm Hg). Initial count ($\log_{10}$ number of bacteria per cm^2): bacon, 4.45; ham, 4.51; loins, 3.72 (Jamieson, 1980c).

	$\log_{10}$ bacterial count per cm^2					
	Bacon		Ham		Loins	
Test conditions	LP	NA	LP	NA	LP	NA
Paper-wrapped, pre-cooled	5.28	8.42	5.23	5.85	3.66	8.96
PE-wrapped, pre-cooled	5.91	8.34	3.89	7.66	3.20	8.08
Paper-wrapped	4.15	7.49	5.46	7.68	3.97	7.81
PE-wrapped	4.20	7.63	5.04	7.40	–	7.83

count was 5×10^6 bacteria per cm^2 for samples wrapped in polyethylene film, 5×10^5 bacteria per cm^2 for paper-wrapped pork and 2×10^5 bacteria per cm^2 for meat stored without a wrap. The lower count in the naked LP samples was associated with a slightly greater weight loss, indicating that bacterial growth had been inhibited by a decreased water activity at the meat's surface (Jamieson, 1976a, 1980a,e). In a subsequent laboratory test with paper-wrapped hams, after 4 weeks at 0 to –1°C, the surface bacterial count had reached 6.24×10^7 per cm^2 in NA and 1.55×10^4 per cm^2 in LP at a pressure of 1.33 kPa (10 mm Hg). After 5 weeks in LP, the count had only risen to 1.09×10^5 per cm^2 (Herring, 1976).

Fresh pork cuts weighing 9500 kg were shipped from Huron, South Dakota, to Honolulu, Hawaii, in a 12.2 m (40 ft) Grumman/Dormavac intermodal LP container operated with humidification at –1°C and a pressure of 1.33 kPa (10 mm Hg). Unloading after the 7-day voyage was witnessed by State meat inspectors and representatives of the USDA, Hormel, Armour, Foodland, US Cold Storage and Matson Navigation. The pork was judged to be in excellent condition and superior to that shipped conventionally (Mermelstein, 1979).

Tests at 0.57–0.67 kPa (4.3–5 mm Hg)

Pork butts were weighed, swabbed, wrapped in butcher paper and stored at –0.8°C in wax boxes, either in NA or in LP with humidification at a pressure of 0.67 kPa (5 mm Hg). In LP, the weight loss was 2.9% in 21 days and 3.5% in 28 days, and after 21–28 days the meat was moist with good colour and a shelf life of 4–5 days at 3.3°C. The pork lasted approximately four times longer in LP compared to NA (Lovizio, 1980; Jamieson, 1984; Table 11.22).

Pork cuts stored at –1°C and a pressure of 0.67 kPa (5 mm Hg) developed a total plate count of 1.8×10^6 organisms per cm^2 in 6 weeks (Table 11.6). At that temperature the same plate count developed in 4 weeks in pork cuts stored at 1.33 kPa (10 mm Hg), even though the initial plate count was 2.4×10^4 total bacteria per cm^2 for the 0.67 kPa (5 mm Hg) experiment and 8×10^2 bacteria per cm^2 for the 1.33 kPa (10 mm Hg) test (Burg, 1987a). Essentially, the entire microflora that developed at 0.67 kPa (5 mm Hg) was comprised of lactic acid bacteria.

Table 11.22. Total plate count on pork stored at –0.8°C in LP with humidification at a pressure of 0.67 kPa (5 mm Hg) or in NA (Lovizio, 1980).

Storage time (days @ –0.8°C)	Shelf life (days @ 3.3°C)	Bacterial count (organisms/cm²)	
		NA	LP
0	–	1.6×10^4	1.6×10^4
6	–	1.4×10^5	–
8	–	3.3×10^7	–
21	–	–	3.7×10^6
21	2	–	1.3×10^7
28	–	–	4.0×10^6
28	2	–	1.5×10^6
28	5	–	1.1×10^8

11.25 Shrimp (*Panaeus setiferus*)

Enzyme action in shrimp tissue and by contaminating microorganisms produces colour changes, off-flavours and off-odours in putrefactive and ammoniacal phases (Duggan and Strasburger, 1946). During the putrefactive phase, indole, which is never present in fresh shrimp, forms from tryptophan by bacterial action; inosine monophosphate (IMP), which contributes to the characteristic flavour of shrimp, decreases from approximately 3078 to 725 μg per 100 g of iced tissue, paralleling the loss in flavour during 10 days; and hypoxanthin and inosine accumulate (Flick and Lovell, 1972). Ammoniacal spoilage is a slower process characterized by the odour of ammonia, a rise in pH, an increased ratio of total volatile nitrogen to amino acid nitrogen and putrid odours arising from protein decomposition (Cobb *et al.*, 1973a,b). Shrimp tissue's proteolytic enzymes, especially cathepsin, contribute to ammoniacal breakdown by providing substrates for bacteria (De Almeida, 1955). Changes in shrimp tissue pH are a good indicator of spoilage: 7.5–8.25 is considered good quality, 8.16–8.40 is rated fair and > 8.4 is regarded as spoiled. Iced shrimp stored in NA for 10 days at 0°C increase in pH from 7.4 to 8.2 (Flick and Lovell, 1972). Other indicators of shrimp spoilage are trimethylamine (TMA) formation, which contributes to the 'bilgy' odour that develops (Fieger *et al.*, 1956), and 'black spot' or 'blotchy' appearance resulting from the oxidation of tyrosine to melanin by shrimp polyphenol oxidase and tyrosinase enzymes (Alford and Fieger, 1952; Odyll *et al.*, 1959).

The bacterial count of freshly caught shrimp is never higher than 10^4 bacteria per gram (Green, 1949; Lantz, 1951), but by the time fishing trawlers unload, approximately 90% of the shrimp have counts higher than 10^6 bacteria per gram (Carroll *et al.*, 1968; Cobb *et al.*, 1973b). The average count for market shrimp bought on the Gulf Coast of the US is 3.2×10^7 bacteria per gram (Green, 1949). *Achromobacter*, *Bacillus*, *Micrococcus* and *Pseudomonas* comprise 78% of the spoilage association, with *Pseudomonas* predominating and giving the shrimp a sticky touch and glossy appearance (Campbell and Williams, 1952; Williams and Rees, 1952).

The progressive changes in organoleptic quality of iced shrimp occur in three phases (Fieger and Bailey, 1954): phase I, from 0 to 7 days, orthophosphate disappears and there is a gradual loss of the characteristic sweet flavour (Flick and Lovell, 1972); phase II, from 8 to 14 days, the pH rises to 7.8–8.0 and when it surpasses 7.7, the shrimp become 'tasteless' (Fieger and Bailey, 1954; Fieger *et al.*, 1956); phase III, after 14 days, off-flavours develop. Even when shrimp are frozen, they develop off-flavours rather quickly, in part due to fat rancidity. The VacuFresh[SM] intermodal container has sufficient refrigeration capacity to permit frozen cargo to be carried under vacuum at high ambient temperatures. By excluding O_2, this should prevent the development of fat rancidity and off-flavours in frozen shrimp.

Melting ice benefits shrimp storage by controlling the temperature and washing bacteria from their surface, but it also removes leachate containing amino acids and phosphate (Velankar and Govidan, 1957). Since the flavour of shrimp is related to the amino nitrogen content and especially to the glycine content, icing reduces palatability (Simidu and Hujeta, 1954a,b). There is an increased bacterial count from the top to the bottom layers of shrimp packed in ice, probably due to percolation of bacteria (Green, 1949). The weight of stacked shrimp layered in ice increases the pressure on the lower shrimp layers, squeezing fluids from the tissue and increasing weight loss from the lower layer. As the weight of ice is very large compared to the weight of the shrimp, ice significantly increases the shipping cost for packed-out shrimp. LP eliminates these disadvantages by allowing fresh shrimp to be shipped without ice.

Tests at 1.06–1.33 kPa (8–10 mm Hg)

Freshly harvested shrimp were wrapped in PVC film and stored at −1 to +2°C in NA or at the same temperature in LP at pressures ranging from 1.07 to 16.67 kPa (8–125 mm Hg). In NA, shrimp became blotchy and developed severe bilgy and fishy odour within 4–6 days, whereas in LP they were preserved for 15 days. The optimal LP storage pressure was between 1.07 and 2.0 kPa (8 and 25 mm Hg). At a pressure of 1.07 kPa (8 mm Hg), the shrimp did not become bilgy, blotchy or develop off-odour during 15 days' storage (Burg, 1976a). In another laboratory study, shrimp was stored at −1.7°C either in NA with top-ice or in LP with humidification at a pressure of 1.33 kPa (10 mm Hg). After 13 days, bacterial growth was the same in the iced and hypobaric shrimp. LP inhibited melanosis (black spot), the hypobaric shrimp were firmer both before and after cooking, and flavour retention was superior in the hypobaric shrimp (Hartman and Jamieson, 1976).

Tests at 0.67 kPa (5 mm Hg)

Shrimp storage at 0 to –1.7°C improved when the pressure was lowered to 0.67 kPa (5.0 mm Hg). The white shrimp (*P. setiferus*) used in this study typically have an average bacterial count of 3.2×10^8 organisms per gram after 10 days' iced storage, at which time they are rated Grade 'A'. By 15 days, the average count rises to 8.0×10^8 bacteria per gram. After 42 days' storage at 0.67 kPa (5 mm Hg) the initial bacterial count of 2.1×10^6 organisms per gram had risen to 1.1×10^8 bacteria per gram, and at that time a trained tasting panel had equal preference for the LP shrimp and fresh (not frozen and thawed) samples bought from a local shop. LP also slowed the pH rise in shrimp (Hartman and Jamieson, 1977; Table 11.23). The optimal storage condition for shrimp undoubtedly is –1.7°C at a pressure of 0.57 kPa (4.3 mm Hg) (Burg, 1987a).

11.26 Veal

Tests at 0.60–0.67 kPa (4.5–5.0 mm Hg)

Hindquarters of veal prepared from animals slaughtered the previous day were procured from a local abattoir and transported in a refrigerated truck to the test site. At arrival, the meat's temperature was 4.4°C. Samples were stored at –0.8°C either in a refrigerated room or in a 12.2 m (40 ft) Grumman/ Dormavac LP intermodal container operated at 0.67 kPa (5 mm Hg) with humidification. In NA, the veal became unmarketable in 12 days, whereas after 41 days' storage in LP, the veal had a shelf life of 8–9 days at 2.2°C. Bacterial counts are summarized in Table 11.24.

Veal carcasses weighing a total of 410 kg were loaded into a 12.2 m (40 ft) Grumman/ Dormavac LP intermodal container as part of a mixed load, along with 6640 kg of lamb and beef carcasses. The container was operated with humidification at –1.7°C and a pressure of 0.6 kPa (4.5 mm Hg). After 42 days' transport by road from San Angelo, Texas, to Baltimore, Maryland, and then by ship to Bandar Shapur, Iran, and by truck to Teheran, Iran, at arrival the veal was judged to be in excellent condition by government inspectors and officials of the Iranian Meat Organization (Shum and Jamieson, 1981).

11.27 Examples

1. The atmospheric composition in a VacuFresh[SM] container was determined with water-ice frozen under the floor grating to simulate the humidity created by a full load. A single pork loin was suspended inside, shielded from radiation by Mylar. The conditions were:

Thermostat setting = –1°C
Vacuum pump (full capacity) = 71.4 m^3/h (42 cfm); corrected for efficiency at a suction pressure of 4.3 mm Hg
Container leak rate = 0.86 mm Hg/h (measured)
Internal container volume = 69.6 m^3 (750 ft^3)
Container steady-state pressure = 4.3 ± 0.01 mm Hg (measured)
Container steady-state air temperature = –0.9 ± 0.1°C (measured)
Pork loin steady-state temperature = –1.5 ± 0.1°C (measured)

Table 11.23. pH of shrimp samples throughout the test period (Hartman and Jamieson, 1977).

	pH at indicated days in storage					
Sample	0	8	22	30	37	42
NA (iced)	6.6	6.85	7.4	7.45	Spoiled	Spoiled
10 mm Hg	6.6	6.70	7.2	7.20	7.20	7.35
5 mm Hg	6.6	6.80	7.0	7.20	7.05	7.10

Table 11.24. Average bacterial counts on veal hindquarters stored at –0.8°C in LP with humidification at a pressure of 0.67 kPa (5 mm Hg) or in NA (Shum, 1980).

Storage time (days @ –0.8°C)	Shelf life (days @ 2.2°C)	Bacteria per cm^2 (generations per day)	
		NA	LP
Initial	0	8.0×10^2	8.0×10^2
4	0	1.1×10^7 (3.43)	–
10	0	3.7×10^6 (1.22)	–
12	0	2.2×10^7 (2.00)	–
41	0	–	4.8×10^4 (0.14)
41	2	–	9.8×10^4 (0.52)
41	6	–	8.7×10^6 (1.62)
41	9	–	5.9×10^8 (0.91)

Vapour pressure of ice at –1°C = 4.217 mm Hg
Vapour pressure of ice at –1.5°C = 4.046 mm Hg
Vapour pressure of water at –1.5°C = 4.105 mm Hg

Based on the container leak rate, container volume and pumping speed, the steady-state fixed gas partial pressures inside the container were:

O_2 = 0.000069 atm (0.052 mm Hg)
N_2 = 0.000267 atm (0.197 mm Hg)
CO_2 = $1.18 \times 10–^7$ atm (8.98×10^{-5} mm Hg)

The water-vapour pressure (4.051 mm Hg) is calculated as the difference between the 4.3 mm Hg total pressure and 0.249 mm Hg sum of the fixed gas pressures. As the saturated vapour pressure of water is 4.258 mm Hg at –1°C, therefore the RH is 95.1% [100 × (4.051/4.258)], and the water activity at the meat's surface is 4.051/4.105 = 0.986. Convection is ineffective at 4.3 mm Hg, and since the pork loin was shielded from radiation by Mylar, the meat cooled by evaporation until its temperature approached the container dew point and stabilized at –1.5°C. The vapour pressure of water, 4.105 mm Hg at –1.5°C, is decreased to approximately 4.031 mm Hg by dissolved solutes in the meat. Water-ice crust frozen on the meat's surface would have a vapour pressure of 4.046 mm Hg. Thus, the pork loin (dew point) temperature measurement indicates that the water vapour pressure in the container was in the range 4.031–4.046 mm Hg, in close agreement with the value based on total pressure and the partial pressures of the fixed gases.

2. (Kossons, 1994). Consider a 20 ft VacuFresh[SM] container stacked with wire boxes containing pork loins wrapped with polyethylene film having sufficient mass-transport resistance to restrict the meat's weight loss to 1% in 34 days (Fig. 11.1). The conditions are:

Container leak rate = 0.106 kPa/h (0.8 mm Hg/h)
Pumping speed = 24.3 m^3/h (14.3 cfm)
Thermostat set temperature = 0°C
Cargo weight = 8182 kg (18,000 lb)
Water-vapour partial pressure (p_{VO}) = 0.52 kPa (3.895 mm Hg)
Container air partial pressure (p_{AO}) = 0.093 kPa (0.699 mm Hg)
Container total pressure (p_C) = 0.61 kPa (4.594 mm Hg)
Molecular weight of water (MW_V) = 18.016
Molecular weight of air (MW_A) = 28.964

The properties of the gas/water-vapour mixture in the container (C) are:

Density (ρ_C) = 53.085 kg/m^3 (3.314 lb/ft^3)
Viscosity (μ) = 9.840 Pa·s (6.609×10^{-6} lb/ft·s)
Thermal conductivity (k) = 0.01659 W/m·°C (0.00974 BTU/h·ft·°F)

Specific heat (c_P) = 0.428 kcal/kg·°C (0.428 BTU/lb·°F)
Molecular weight (MW_C) = 19.68179

The polyethylene-wrapped pork loins are shaped like horizontal cylinders, diameter = 12.7 cm (5 in), length = 50.8 cm (20 in), weight = 6.818 kg (15 lb). Assume that the pork loins are 1°C colder than the 0°C air and wall, then for the cylindrical surface area (A) of $\pi DL = 0.2027$ m^2 (2.182 ft^2), and a weight loss of 0.000405 kg/h (0.0001838 lb/h), the heat-transfer rate into the cylinder required to continuously replace the latent heat consumed by evaporation is 0.04967 kcal/h (0.1971 BTU/h), and for a 1°C temperature difference the convective-film coefficient (h) required to transfer that amount of heat is:

$$h = 0.04967/0.2027 = 0.245 \text{ kcal/m}^2\cdot\text{h}\cdot°\text{C} \ (0.0502 \text{ BTU/h}\cdot\text{ft}^2\cdot°\text{F})$$

The actual value of h depends on the change in buoyancy caused by the evaporation of water from the meat. From data for air/water-vapour mixtures in humidifiers, the buoyancy effect is given as (Kreith and Bohn, 1997):

$$h/(\text{kg}\cdot \text{MW}\cdot p_O) = 0.26 \quad (11.4)$$

where p_O is the vapour pressure of water at the temperature of the meat. The kg of water evaporated is:

$$\text{kg} = 0.245/(0.26 \times 19.66 \times 4.594) = 0.0104 \text{ kg-moles/m}^2\cdot\text{h}\cdot\text{mm Hg} = 0.00214 \text{ lb-moles/ft}^2\cdot\text{h}\cdot\text{mm Hg}$$

and the average increase in vapour pressure at the polyethylene wrap's outside surface in the presence of natural convective flow is:

$$\Delta p_V = \text{d}M_V\,\text{d}t/(\text{kg}\cdot\text{MW}_V\cdot A) = 0.00219 \text{ mm Hg} \quad (11.5)$$

Since the total pressure is unchanged, this slight increase in vapour pressure offsets to some extent the increase in density caused by the mixture's temperature decrease as it passes over the cooler meat surface. The net change in density ($\Delta\rho$) is 1.1×10^{-6} and the Grashof and Prandtl numbers (6.12) are:

$$\text{Gr} = (g\cdot\rho\cdot\Delta\rho\cdot D^3)/\mu^2 = 20.07 \quad (11.6)$$
$$\text{Pr} = (c_P\cdot\mu)/k = 1.045$$
$$X = \text{Gr}\cdot\text{Pr} = 20.99$$

For this value of X, treating each individual box as a horizontal cylinder (Perry and Given, 1984):

$$\text{Nu} = (h\cdot D)/k = 1.09^{0.2} = 2.004 \quad (11.7)$$

and therefore $h = 0.2285$ kcal/m^2·h·°C (0.0468 BTU/h·ft·°F), which is slightly below the required value. For a Reynolds number in the range 4–40 (Perry and Given, 1984):

$$\text{Nu} = 0.911\ \text{Re}^{0.385}\text{Pr}^{0.33} \quad (11.8)$$

from which, if Nu = 2.004, then Re = 7.46. For this value of Re, the Nusselt number for forced convective flow would be the same as for natural convective flow on a single horizontal cylinder. For forced convective flow over staggered banks of tubes, a geometry resembling the pork loins (Perry and Given, 1984), with no leakage:

$$\text{Nu} = 1.086\ \text{Re}^{0.41}\text{Pr}^{0.33} \quad (11.9)$$

and for Re = 7.46, Nu = 2.512. This is a 25% increase, and assuming that a similar advantage would accrue for natural convective flow over staggered banks of cylinders, $h = 0.2866$ kcal/m^2·h·°C (0.0587 BTU/h·ft^2·°F). Therefore, a 1°C ΔT between the meat and container wall would be sufficient to compensate for the loss of latent heat caused by a 1% weight loss in 34 days, provided that the heat capacity of the gas-vapour mixture is sufficient to carry that amount of heat flowing the low-density mixture at the rate generated by natural convection.

If the pork loins are in wire baskets, stacked according to Fig. 11.1 with four loins per basket, in rows four baskets wide and eight tiers along the container length, then in the horizontal centre plane there would be an array of 128 pork loins occupying a total cross-sectional area of 8.259 m^2 (88.9 ft^2). The container cross-sectional area is approximately 10.414 m^2 (112.1 ft^2), leaving a flow area of approximately 2.155 m^2 (23.2 ft^2). The Reynolds number can be written as:

$$\text{Re} = (G{\cdot}D)/\mu = 7.6 \qquad (11.10)$$

where the mass velocity $(G) = 5.781 \times 10^{-4}$ kg/m²·s (1.183×10^{-4} lb/ft²·s), the total free convective flow $(G \times A)$ is 0.00125 kg/s (0.00275 lb/s), and for $c_P = 0.428$ kcal/kg·°C (0.428 BTU/lbm/°F), the heat capacity of the flow is 1.92 kcal/°C (4.24 BTU/°F). If the flow were to cool the entire 1°C temperature difference, the heat given up would be only 1.92 kcal/h (7.62 BTU/h), but the heat transfer required by the 128 loins in the centre plane layer alone is 6.35 kcal/h (25.2 BTU/h). From this it can be seen that the low heat capacity of the air limits the ability of natural convection so that at most it can provide only a part of the heat required by the first (external) layer (basket) of meat that it flows over.

The outer baskets receive heat both by natural convection from the container air-water mixture, and also by radiation from the container wall. As a simplification, the outer boxes will be separated into three groups, designated B, C and D (Fig. 11.1). With respect to the external surface area exposed to receive radiation, group B > group C > group D, and for the smaller boxes in groups B and D the heat and mass transfer is pro-rated from that of the larger boxes, based on the ratio of the amount of meat contained, 17 kg (37.5 lb) in the smaller baskets vs. 27.2 kg (60 lb) in the larger baskets.

The box mass-transport resistance (r_p) is inverse with both the log mean air pressure and the meat weight since the polyethylene wrap is assumed to provide most of the resistance. The value of r_p also varies directly with box external-surface area since that is the area used in making the mass-transfer calculations, i.e.:

$$dV/dt = \Delta\rho_V{\cdot}A/r_p \qquad (11.11)$$

The value of r_p, determined from weight-loss measurements at a higher pressure, is approximately 1.5 s/cm under the conditions cited for this example.

The method of calculation first assumes a total cargo weight-loss fraction that provided a container vapour pressure. Next, temperatures are assumed for each group of boxes, permitting calculations for the convective and radiant heat transfer into each box, the mean ambient pressure, the box mass-transfer resistance and the box evaporative-heat loss. Box temperatures were then adjusted until the combined radiant and convective heat input equalled the evaporative heat loss for each box. The weight loss for each group of boxes was then computed and summed to obtain the total weight loss and total weight loss fraction for one entire tier of boxes. (The interior boxes were assumed to have negligible weight loss). The total weight-loss fraction was then assumed to apply to the total cargo and compared with the initially assumed value. If the value did not agree, the initially assumed value was then adjusted and the calculations were repeated. The result is indicated in the Table 11.25 spreadsheet for the following container condition:

Leak rate, mm Hg/h	0.8
Pump rate, m³/h (cfm)	24.3 (14.3)
Container volume, m³ (ft³)	21.24 (750)
Container wall temperature, °C	0
Air leakage, m³/h (cfm)	0.02236 (0.01316)
Container air pressure, mm Hg	0.6993
Storage time, days	34
Cargo weight – total, kg (lb)	8172 (18,000)
Water-vapour loss rate – total, kg/min (lb/min)	0.002 (0.004408)
Container water-vapour partial pressure, mm Hg	4.322
Container total pressure, mm Hg	5.02
Radiant film coefficient, kcal/m²·h·°C (BTU/ft²·h·°F)	3.969 (0.813)
Natural convective film coefficient, kcal/·m²·h·°C (BTU/ft²·h·°F)	0.1665 (0.04)

Table 11.25. Spreadsheet (Kossons, 1994).

Quantity	B	C	D
Meat temperature, °C	–0.4273	–0.5043	0.5601
Saturated v.p. @ meat temperature, °C	4.446	4.422	4.405
Meat vapour pressure (mm Hg)*	4.381	4.357	4.340
Air-vapour temperature inside meat wrap, °C	–0.427	–0.504	–0.560
Box external surface area, m^2 (ft^2)	0.417 (4.484)	0.211 (2.227)	0.094 (1.012)
Air partial pressure in box, mm Hg	0.6407	0.6816	0.6903
Mean air pressure, mm Hg	0.6700	0.6816	0.6903
Box resistance, s/cm	1.4500	1.527	1.546
Box in–out vapour pressure gradient, mm Hg	0.0586	0.0350	0.0179
Vapour mass flux, kg/h	0.00124	0.000726	0.000395
(lbm/h)	(0.00273)	(0.00160)	(0.00087)
Evaporative heat loss, kcal/h	0.740	0.4339	0.2191
(BTU/h)	(2.9397)	(1.7238)	(0.86934)
Ratio (heat in)/(heat out)	1.0007	0.9999	1.0004
Cargo mass in category, kg (lbm)	177.1 (390)	109 (240)	320 (705)
Mass loss in category, kg (lbm)	6.585 (14.504)	2.376 (5.234)	17.079 (37.754)
% weight loss in 34 days	3.72	2.18	1.10

*Corrected for osmotic concentration (example 3).

While not shown on the spreadsheet, the water-vapour pressure of the interior boxes, corrected for dissolved solutes in the meat, equals the container vapour pressure; the temperature of the interior boxes is –0.619°C, and they experience negligible weight loss. The total cargo weight loss, including interior boxes, is 1.1% in 34 days, spread unevenly between box category B (3.72%) > C (2.18%) > D (1.10%).

3. Meat freezes at approximately –1.9°C. It must contain approximately 1 molar solute since the freezing point of water is lowered by 1.855°C per gram molecular weight of dissolved solute in 1000 ml. Raoult's Law states that in a dilute solution $P_A = X_A P^O_A$, where P_A is the partial pressure of A above a solution in which its mole fraction is X_A, and P^O_A is the vapour pressure of pure liquid A at the same temperature. In a 1 molar solution, the mole fraction of water is 1000/18 = 55.5556, and the vapour pressure of the solution (P_A) is (55.5556–1)/55.5556 = 0.982 P^O_A. The vapour pressure of ice at –1°C is 4.217 mm Hg. For meat to have the same vapour pressure as –1°C ice, its temperature would have to equal the water temperature that creates a vapour pressure of 4.217/0.982 = 4.294 mm Hg. The vapour pressure of water is 4.294 mm Hg at a temperature of –0.88°C, and the meat would be 0.12°C warmer than the ice.

Notes

1. With no special precautions during manufacture, the first prototype 6.1 m VacuFresh[SM] container had a leak rate of 0.7 mbar (0.5 mm Hg) per hour. The oil-sealed VacuFresh[SM] vacuum pump is far more efficient than the supercharged, water-sealed Dormavac pumping system. Even though the Dormavac pump has three times more capacity at atmospheric pressure, the VacuFresh[SM] pump has a greater capacity at 0.57–0.67 kPa (4.3–5 mm Hg).
2. The cause of these failed shipments was not an excessive pressure, but rather a malfunction of the refrigeration system caused by leaks created by shock and vibration.
3. This was a laborious process. A workman was sealed in the container and the pressure slowly lowered by 10 kPa (75 mm Hg) with suitable safeguards, creating a condition equivalent to that at an altitude of 1542 m (5000 ft). The 'man in the can' sprayed each weld seam and the door seal with soapy water and used a battery-operated light to check for bubbles indicative of leaks. Each leak was marked for repair.

4. Before meat freezes, the specific heat of veal = 0.71; lamb = 0.79; lean beef = 0.77; fatty beef = 0.60; pork = 0.68; poultry = 0.79 kcal/kg·°C (BTU/lb·°F). After freezing, the specific heat is much lower, veal = 0.39; lamb = 0.30; lean beef = 0.40, fatty beef = 0.35; pork = 0.38; poultry = 0.37 kcal/kg·°C (BTU/lb·°F). The latent heat of fusion is veal = 91; lamb = 83; pork = 86; lean beef = 100; fatty beef = 79; poultry = 106 kcal/kg·°C (BTU/lb·°F) (Potter, 1973).
5. Lamb carcasses are small enough to be broken, bent back on themselves, and packed in CO_2-flushed bags.
6. Pseudomonads grow on chicken breasts in 20% [O_2] + 80% [CO_2] (Kakopuri and Nychas, 1994).
7. For the Japanese trade, if the measured temperature is less than −2.2°C the meat is considered frozen, while at −1.7°C it is considered fresh.
8. At −1.5°C, pure water may innocuously crust-freeze on the surface of meat, while cells remain unfrozen due to their high concentration of dissolved solutes.
9. Within a proportionate 'band' around the set-point temperature, the PID thermostat varies the 'open' and 'closed' time of a solenoid valve configured to introduce hot compressor discharge gas into the refrigeration evaporator. If the glycol temperature is below set-point, the valve remains open longer during each fixed time interval; if it is above the set-point, it remains open for a shorter time. The controller adjusts the proportional band so that at the set-point (the mid-point of the proportional band) the ON/OFF ratio = 1. The integral and derivative functions help the controller automatically adjust to set-point changes and avoid large offsets when the heat load changes (13.14).
10. The relationship between water potential and RH is defined in equation 15.12.
11. Ethylene vinyl acetate (EVA) is an O_2-and-water-vapour-permeable film; aluminium foil is impermeable to both O_2 and water vapour.
12. The influence of LP and CA on the survival and multiplication of *Salmonella* sp. during the storage of poultry has been studied (Voigt and Haard, 1982), but results of this research could not be accessed.

12

Warehouse Design

The only hypobaric warehouse that has been placed into service is a 184 m^3 (6500 ft^3) metal structure built by Grumman Corporation. Concepts for concrete warehouses of larger size have been proposed, but it remains to be demonstrated that the structural and thermal problems associated with their design have been resolved.

12.1 Grumman's Hypobaric Warehouse

Grumman's hypobaric warehouse is a 19.8 m (65 ft) long × 3.35 m (11 ft) diameter polyurethane-insulated steel cylinder, serviced by an environmental control system capable of handling two identical cylinders (Fig. 12.1). The warehouse maintains a temperature of 0–18.3°C (32–65°F), a relative humidity of 95% and an absolute pressure of 1.33–13.3 ± 0.27 kPa (10–100 ± 2 mm Hg), flowing two air changes per hour. Refrigerated propylene glycol/water (50:50 wt/wt) is pumped through manifolded aluminium heat-exchange panels pinned with a slight offset over the entire inner surface of each cylindrical steel tank to create a 'jacketed' refrigeration system. Initially, the vacuum system consisted of a rotary liquid ring pump and a 19.1 m^3/min (320 cfm) gas-ballasted oil-seal trochoid pump combination, but the water-sealed stage was eliminated after it was found that gas ballast effectively prevented water from condensing in the sealing oil of the trochoid pump. Air withdrawn from the vacuum tank by a 34 m^3/min (1200 cfm) Roots-type blower is passed through a boiler equipped with a wattage-regulated immersion heater, automatic fill system and glycol-cooled heat exchanger. The boiler system saturates the low-pressure air at the tank temperature before the air re-enters the cylinder at the opposite end from which it had been withdrawn.

12.2 Alternative Steel-cylinder Design

A simple LP warehouse can be constructed by installing a saddle-mounted carbon steel or stainless steel non-insulated vacuum tank inside an insulated, temperature-controlled room. This arrangement has many advantages. The 'jacketed' refrigeration system prevents heat transmission through the cylinder's walls, and the air changes that are drawn into the warehouse are automatically preconditioned to the storage temperature because they originate from within the temperature-controlled room. The commodity and the interior of the cylinder are not exposed to a different temperature when the warehouse is vented and opened to remove or add inventory, and the temperature-controlled space outside the LP warehouse can be used to

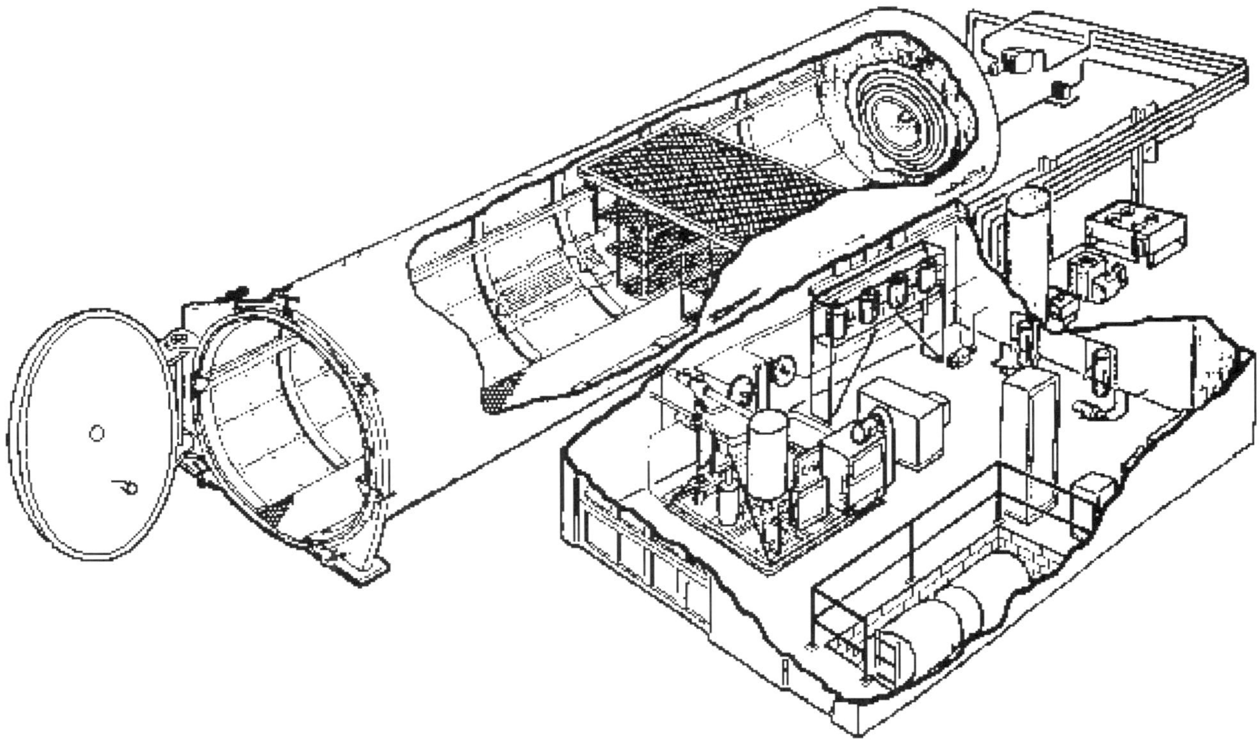

Fig. 12.1. Grumman's hypobaric warehouse. Diagrammatic view of a warehouse built to store rooted *Chrysanthemum* cuttings. The warehouse was situated with its door opening into a refrigerated room. To remove or add inventory, each day the storage racks were rolled out from the warehouse into the temperature-controlled space, moving on rails. One loading rack with shelves is shown inside the warehouse (Mermelstein, 1979).

precool commodity before it is stored, and keep it cool when it is removed. The vacuum system, pressure controls, pneumatic air mover and many of the other systems used in intermodal hypobaric transportation containers can be incorporated into this type of warehouse. This design concept uses a standard forced-air refrigeration system.

12.3 Concrete Warehouse Designs

Studies carried out for the Dormavac Corp. by Boeing Aircraft and the US Bridge Division of US Steel, and independently at the University of Guelph (Lougheed *et al.*, 1974), indicate that the cost of a sizable vacuum warehouse would be reduced by constructing the pressure vessel from concrete rather than metal.

12.4 Boeing/Preload Design

Because concrete structures are used as vacuum-containment vessels in nuclear power plants, the technology is well advanced (Paul *et al.*, 1970). A 1973 LP warehouse proposal submitted to Boeing by the Preload Co. (Spokane, Washington), a builder of nuclear containment vessels, contained the following specification:

TANK A: 3057 m^3 (108,000 ft^3)-capacity:

- Fixed-base cylindrical tank with a flat roof supported on nine interior columns.
- Inside diameter: 22.9 m (75 ft).
- Outside diameter: 23.2 m (76 ft) with 15.2 cm (6 in) wall thickness.
- Clear height = 7.6 m (25 ft).
- Roof slab thickness = 45.7 cm (18 in).
- 1.5 m (5 ft) × 2.1 m (7 ft) steel access door.
- Continuous plastic liner on concrete surfaces.

TANK B: 1019 m^3 (36,000 ft^3) capacity:

- Fixed-base cylindrical tank with a flat roof supported on four interior columns.
- Inside diameter = 14 m (46 ft).
- Outside diameter = 14.3 m (47 ft) with 15.2 cm (6 in) wall thickness.
- Clear height = 6.7 m (22 ft).
- Roof slab thickness = 45.7 cm (18 in).
- Floor slab thickness = 45.7 cm (18 in).
- 1.5 m (5 ft) × 2.1 m (7 ft) steel access door.
- Continuous plastic liner on concrete surfaces.

The general design criteria were:

- Vacuum: 10.1 kPa (76 mm Hg).[1]
- Maximum allowable leak rate: 5% of net volume per hour (STP). Based on experience at the Pickering Nuclear Reactor, a concrete tank without liner is sufficient in all other respects. With the proposed liner, expected leakage rates will be well below the maximum allowable rate.[1]
- Storage temperature range: 0–26.7°C (32–80°F).
- Earthquake zone: The soil strength requirements are minimal. The proposed tank structure is well suited for application in earthquake zones.
- Concrete, minimum 28-day cylinder strength: wall, roof, floor slabs and columns = 4.1×10^8 kg/m^2 (4000 psi).
- Applicable codes: ACI 318, Recommendations of ACI 344.

The study concluded that an effective metal door-to-concrete seal is feasible. The preliminary design envisioned a portal opening with tapered edges to effect a positive bearing for the steel door, with a rubber gasket along the edges of the steel door bearing against a studded steel plate lining the edges of the portal opening. Problems caused by entrapped air in the concrete under the influence of vacuum can be minimized and even totally eliminated by precautions taken in the design and placement of the concrete mix. Damage to the concrete from entrapped air can only result from an excessively high rate of pumping, far greater than the rates envisioned for the structure. The use of an impervious liner on the outside of the tank will by itself eliminate any possibility of damage to the concrete and obviate water leakage and moisture and air permeation problems. With an assumed life of 70 years and two cycles per month, the resulting number of cycles (approximately 1700) is well within the design limits of comparable concrete structural designs.

Although calculations of permeability show acceptable leakage rates through a concrete tank without a liner (Paul *et al.*, 1970), there is some uncertainty in the effect of shrinkage and temperature-induced cracks on leakage, and therefore a liner was incorporated into the design. A tank without a liner might well prove adequate, but experience in the industry is not sufficient to predict this with certainty. The initial unit could be constructed with a liner under the foundation only, with provisions for installing an external liner over the roof and wall later, if required. This could be accomplished at no increase in cost, since the same construction sequence would be employed in either case. This procedure allows the possibility of saving the cost of roof and wall liners for all units constructed.

The cost estimate (1972), exclusive of subsoil investigation or treatment, insulation, refrigeration, vacuum and humidification equipment, and a steel door was US$79.98 per m^3 (US$2.01 per ft^3) for a single 3057 m^3 (108,000 ft^3) capacity unit, and US$65.33 per m^3 (US$1.85 per ft^3) for 10 units; US$104.18 per m^3 (US$2.95) per ft^3 for a single 1019 m^3 (36,000 ft^3) capacity unit and US$97.82 per m^3 (US$2.77 per ft^3) for 10 units.

12.5 Crom Corporation Design

A circular wire-pre-stressed composite-concrete structure is ideal for a vacuum warehouse, and the pre-stressed tank is an attractive, durable, low-maintenance structure (Fig. 12.2). Using thin concrete-shell construction under vacuum loading results in compressive forces that develop the full strength capabilities of the concrete materials. The flexibility of shotcrete/concrete construction allows varying dimensions and incorporation of a variety of internal structural appurtenances to accommodate the hypobaric process. The Crom Corporation prepared designs for a 32.3 m (70 ft)

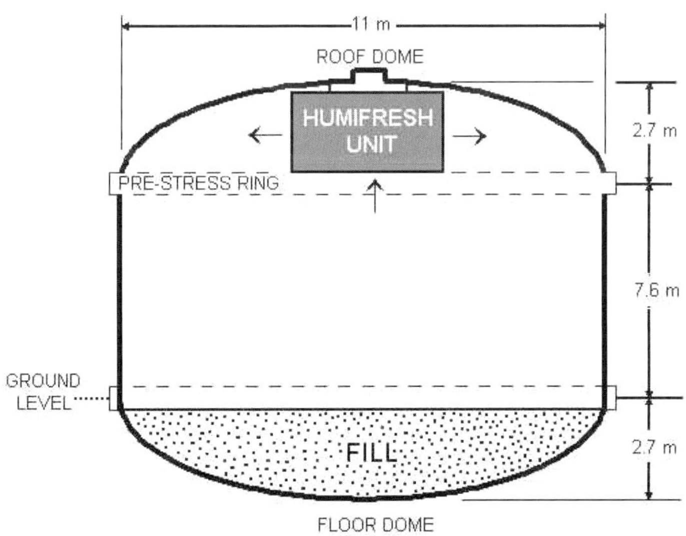

Fig. 12.2. Concept for a pre-stressed composite 10.7 m (35 ft) diameter × 7.6 m (25 ft) high concrete hypobaric warehouse with 706 m^3 (24,000 ft^3) of usable space (Courtesy of Crom Corporation, Gainesville, Florida).

diameter 2835 m^3 (100,135 ft^3) tank; a 23.1 m (50 ft) diameter 1446 m^3 (51,090 ft^3) tank; and a 10.7 m (36 ft) diameter (749.8 m^3) (26,485 ft3) tank, each with a 7.6 m (25 ft) wall height. The tank has shotcrete walls; a pre-stressed free-span dome ceiling requiring no supporting posts; a dome floor filled with compacted, clean sand under an 8.9 cm (3.5 in) thick concrete false floor over the concrete floor dome to provide a working floor inside the structure; a 0.91 m (3 ft) diameter concrete centre well with sump pump located at the centre of the floor dome; a 2.1 m (7 ft) wide × 3.2 m (10.5 ft) high air-tight door; a two-component, polyurethane, 40 mil hypolon-type membrane coating under the concrete dome floor; a 20 mil asphalt-neoprene coating on the exterior of the walls and dome roof; 7.6 cm (3 in) of 40 kg/m^3 (2.5 lb/ft^3)-density urethane-foam insulation covered with a foam-protective coating over the exterior wall and roof-dome surfaces; and an air-tight metal liner in the side wall.

The circular rooms can be efficiently stacked with conventional pallet bins up to ten high using a fork-lift. The usable capacity is 10,000 bushels of apples for the smaller tank, 20,000 bushels for the mid-sized tank and 40,000 bushels for the larger tank. In 1972, the cost for the complete structure (without equipment) was estimated at US$92.17–98.53 per m^3 (US$2.61–2.79 per ft^3) for the 749.8 m^3 (26,485 ft^3) warehouse; US$82.28–88.99 per m^3 (US$2.33–2.52 per ft^3) for the 1446 m^3 (51,090 ft^3) warehouse; and US$74.50–79.10 per m^3 (US$2.11–2.24 per ft^3) for the 2835 m^3 (100,136 ft^3) warehouse. By 1983, the price of the smaller tank had increased to US$211.88 per m^3 (US$6.00 per ft^3) and for the larger tank to US$134.19 per m^3 (US$3.80 per ft^3), both exclusive of the access door and equipment.

The predicted air-leakage rate, based on experience with the Ontario Hydro vacuum structure, is 1.3% of the tank volume per day, expressed at STP, where one saturated air change per hour at an operational pressure of 1.33 kPa (10 mm Hg) corresponds to a controlled leakage rate of 0.72% of the chamber volume per hour (STP). This calculation assumes that the diaphragm in the wall prevents significant leakage; a hot mopped or similar membrane under the floor results in negligible leakage through the lower dome; leakage past proper joints is negligible, as it was in the Ontario Hydro project; the leakage rate is 5.08 $m^3/s{\cdot}m^2{\cdot}atm$ per cm thickness (0.1 $in^3/ft^2{\cdot}min{\cdot}psi$ for a 1 in thickness); and the only significant leakage is through the upper 12.7 cm (5 in)-thick dome. It was concluded that the leakage would not be significant even if leakage at water stops was not negligible and an upper dome membrane was not used.

The Pressure Cool Co. proposed using a 20-ton 21C-type Humi-Fresh 'Filacell' high-humidity refrigeration system mounted in the roof dome to cool and humidify the smaller Crom warehouse. In that location the unit would provide an advantageous recirculation pattern of saturated, refrigerated air (Fig. 12.2; Krahn and Darby, 1971; Meredith, 1973). The Filacell system can be operated at atmospheric pressure initially to cool the commodity, and then at a low pressure after cool down is completed. A Humi-Fresh Cooling System makes use of direct-contact sensible heat exchange between chilled water and circulating air, utilizing 'Filacell' packing made of horizontally arranged polypropylene monofilaments. The cold water is uniformly distributed on the Filacell and trickles downward over the filaments, each of which is much like a pipe with the water on the outside, while the air flows upward in a counter-current pattern. Latent heat exchange is relatively insignificant in this type of heat transfer, other than to balance the moisture content of the air change, and as a result the air dew point temperature is the same as the water temperature. The system has a refrigeration coil incorporated into each air handler, with a shallow reservoir of water in the bottom and a pump circulating the water over the coil and Filacells. The refrigerant liquid and suction lines, as well as a condensate overflow line and make-up water-supply line, are piped directly to each air handler. The Filacell unit's 849 m^3/min (30,000 cfm) blower motor[2] must be cooled with a refrigerated jacket similar to that used in mines in order to prevent motor overheating due to the poor convective-film coefficient at a low pressure (6.21 and 6.22). The immersion pump providing water to the Filacell core would be installed in a water reservoir inside the vacuum warehouse in order to avoid seal problems. In 1973, the cost of the Filacell unit for the smaller warehouse was estimated as US$75,000, or US$99.94 per m^3 of space (US$2.83 per ft^3).

A 9.1 m^3/min (320 cfm) oil-sealed rotary-vein gas-ballasted pump is required to provide three-quarter air changes per hour for the smaller room. The vacuum-regulation system used on hypobaric intermodal containers would be adequate for this warehouse (Fig. 13.17). The cost of the larger 2836 m^3 (100,135 ft^3) warehouse, including electrical, insulation, refrigeration, the vacuum system, humidifying system and engineering work was estimated in 1973 as US$277,631 (US$97.82 per m^3 = US$2.77 per ft^3).

Table 12.1. Comparative cost (1973) of reinforced concrete and steel LP structures of different shapes with approximately 10,000 bu. capacity (Lougheed *et al.*, 1974). The rectangular steel building was 10.7 m (35 ft) W × 17.1 m (56 ft) L × 73.2 m (25 ft) H on a 1.1 m (3.5 ft) thick concrete floor; the circular steel structure of 13.7 m (45 ft) diameter and 7.3 m (24 ft) height had a flat roof without columns inside, and a 1.1 m (3.5 ft) thick concrete floor. The rectangular concrete building was 8.8 m (29 ft) W × 15.8 m (52 ft) L × 6.7 m (22 ft) H with an 2.4 m (8 ft) × 2.4 m (8 ft) access door, 7.6 cm (3 inch) wall thickness, and 8.9 cm (3.5 inch) roof and floor thickness; the cylindrical concrete building had an inside diameter of 13.4 m (44 ft), a spherical segment roof of 7.3 m (24 ft) radius, 30.5 cm (12 inch) thick walls and roof, and 1.1 m (3.5 ft) thick floor (Fig. 13.3).

	Rectangular	Cylindrical
Concrete	US$102,000	US$50,950 (domed)*
Steel	US$167,900	US$96,188 (flat ceiling)

*This is a lower estimate than that made by a builder of concrete cylindrical domed structures (12.5).

12.6 University of Guelph Design

A comparison of reinforced concrete and steel LP structures having different shapes and approximately 10,000-bushel capacity (Table 12.1) indicated that the best structural efficiency and highest ratio of volume to surface area are obtained with a spherically shaped building, but this shape was rejected because of its high cost and poor storage-space utilization (Lougheed *et al.*, 1974). The other extreme, a rectangular build, has a low ratio of volume to surface area, but it is very efficient in utilization of the storage space and well suited for loading and unloading. The construction of rectangular buildings for high external pressure has been well developed in building underground and underwater tunnels. The most cost-effective structure was judged to be a cylindrical domed concrete structure (Fig. 12.3), built in accord with the safety standards established for similarly shaped nuclear reactor containment vessels.

In addition to the cost of the basic concrete cylindrical structure, there is an additional (1974) cost of US$15,775 for internal insulation[3] and an external coating to reduce air leakage and provide weather protection, plus US$3500 for heavy electrically operated doors. The total cost, US$70,225 or approximately US$7.00 per bushel, was estimated to be 36.5% lower than for a CA storage of equal capacity. As an alternative to a single LP warehouse of 10,000-bushel capacity, an estimate was made of the cost for four cylindrical compartments having a total storage capacity of 9720 bushels. Each compartment is 8.5 m (28 ft) in diameter, has a 4.9 m (16 ft)-high vertical wall and spherical-segment roof with flat floor and accommodation for 2430 bushels. The 1974 cost of each compartment, including insulation, external coating and door, was US$22,800; the cost for four compartments was US$91,200, or 30% more than for a single building (Table 12.1).

The mechanical system (Fig. 12.4) was comprised of: (i) an air-conditioning system, to control the flow rate, temperature,

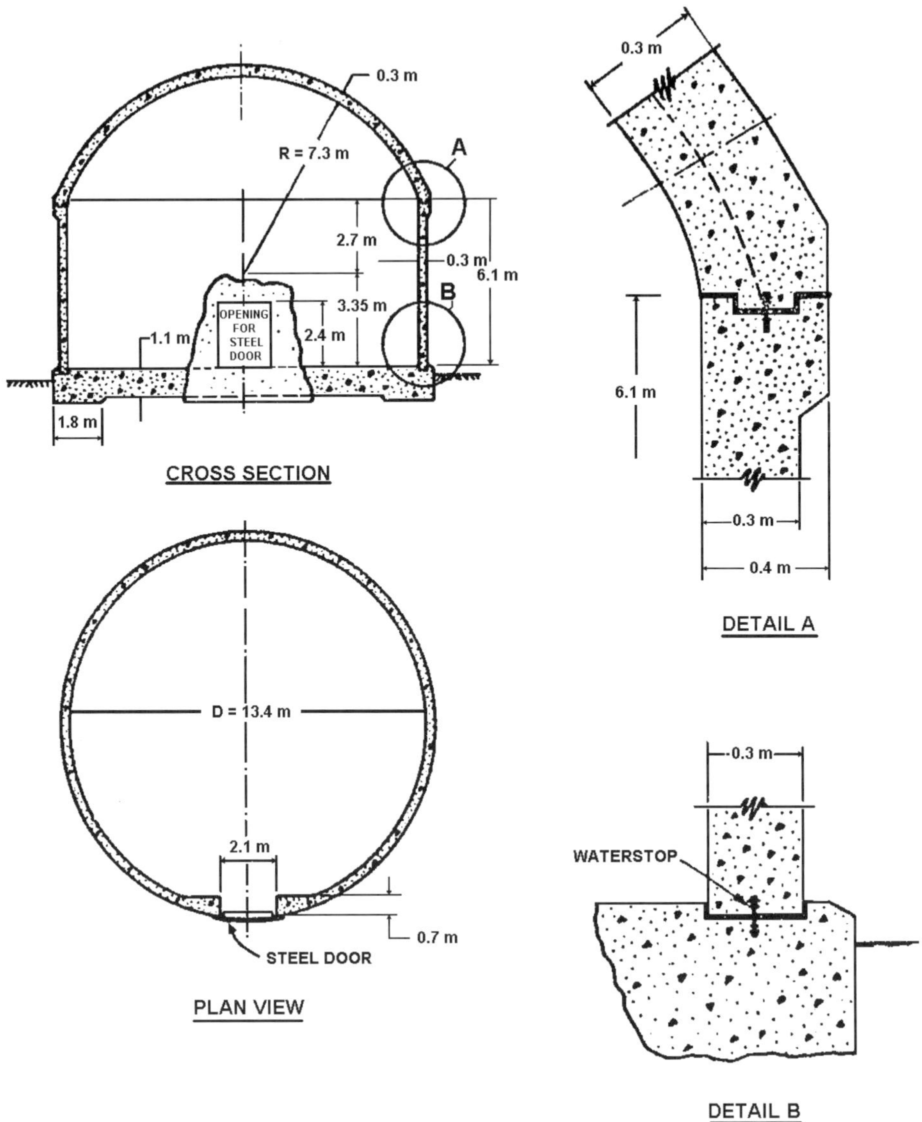

Fig. 12.3. University of Guelph reinforced-concrete warehouse design (Lougheed *et al.*, 1974).

pressure and humidity of the incoming air; (ii) a vacuum system to maintain the required low pressure in the storage and provide air flow; (iii) a temperature-control system to assist in maintaining constant temperature within the storage area by compensating for heat exchange through the enclosure's surface; (iv) an initial cooling system to reduce the time needed to create the required temperature inside the storage; and (v) an air-circulation system to provide adequate aeration inside the storage area. Each subsystem, except that for air circulation, is a closed loop, controlled by one or more sensors and controlling units (Fig. 12.4). The LP room is operated at atmospheric pressure while the commodity is cooled by a standard refrigeration unit, and then the vacuum pump and air-conditioning system are activated.

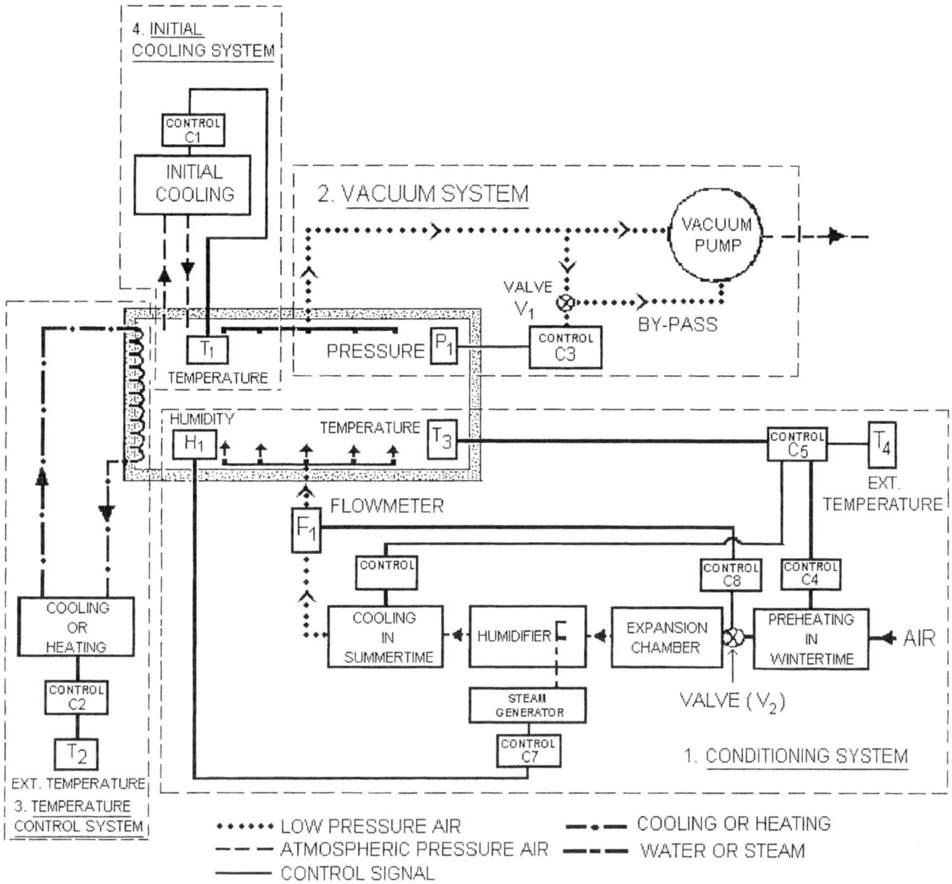

Fig. 12.4. Block diagram of the University of Guelph LP warehouse mechanical system (Lougheed *et al.*, 1974).

During low-pressure operation, before fresh air enters the storage area, it passes through a 198 W (376 BTU/h) preheater, an expansion chamber, a humidifier, and in summer through a 226 W (764 BTU/h) cooler. The estimated maximum amount of moisture to be provided by the humidifier is 0.91 kg (2 lb) of water per hour. The source of moisture is steam[4] generated at 110°C (230°F). The air-conditioning system is designed for an air change rate of half the warehouse volume per hour (196 m^3/h = 6700 ft^3/h). In the design study, two extreme conditions were considered: summer conditions with an atmospheric air temperature of 26.7°C (80°F); and winter conditions with incoming air at –23.3°C (–10°F). The integrated signal in the controlling unit (C_5), from the temperature inside the storage area (T_3) and the external temperature (T_4), automatically switches on the preheater by controlling unit (C_4) if it is cold, or switches off the preheater, activating the cooling unit located downstream of the expansion and humidification chambers. The airflow (rate of ventilation) is controlled by valve (V_2) in the expansion chamber in response to a signal from the flow meter (F_1) and control unit (C_8). Moisture is provided by the humidifier located downstream of the expansion chamber, the RH is sensed by the humidity sensor in the storage area (H_1)[5] and the amount of steam injected from the steam generator is regulated by controller C_7.

The vacuum system is comprised of a rotary vacuum pump liquid-sealed with glycol water. It was selected to avoid the problem of water condensation in an oil-seal pump.[6] The 3.2 m^3/min (112 cfm) pump will reduce the pressure to 10.13 kPa (76 mm Hg) in 4.5 h. Pressure sensor (P_1) operates by-pass valve (V_1) through control unit (C_3) to control the pressure.[7] Ventilation amounts to 0.32 m^3/min (11.67 cfm) of dry air at a pressure of 10.13 kPa (76 mm Hg).

A temperature-control subsystem compensates for heat transfer through the enclosure and eliminates temperature cycles and variation inside the storage space. This is a standard system used to operate air-conditioned storage buildings, and is capable of providing heating and cooling. The heat-transporting medium, glycol/water, is cooled or heated as required in response to the temperature sensor (T_2) and controlling unit (C_2), and circulated through a heat exchanger inside the storage located close to the walls.[8]

The air-circulation system uses jet-type venturi devices to provide air penetration into the boxes.[9] Similar devices are used in VacuFresh^{SM} containers (13.9 and Fig. 13.11).

12.7 Concrete Masonry Corporation (CMC) Precast Concept

The CMC structure (Fig. 12.5) requires 22 wall pieces weighing 9534 kg (21,000 lb) each, and 22 roof pieces weighing 10,442 kg (23,000 lb) each. The 0.91 m (3 ft) diameter supporting column weighs 15,436 kg (34,000 lb). A 2.7 m (9 ft) diameter × 0.91 m (3 ft) thick heavily reinforced ring-crown is required. The structure requires a custom metal-framed pressure-vessel door for a 3.0–3.4 m (10–11 ft) opening in the pressure vessel (manufactured by Babcox-Wilcoxin or Allis Chalmers). The door must seat like a keystone. In 1972, the cost was estimated at approximately US\$70.63/m^3 and US\$52.97/m^3 (US\$2.00 and US\$1.50 per ft^3) for 849 m^3 (30,000 ft^3) and 1982 m^3 (70,000 ft^3) warehouses, respectively. This

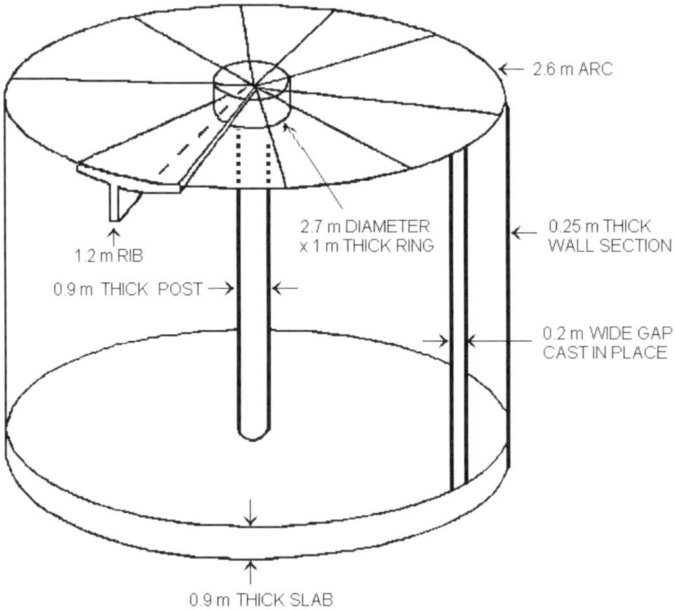

Fig. 12.5. Precast concept for a 1982 m^3 (70,000 ft^3) capacity, 18.3 m (60 ft) diameter × 7.6 m (25 ft) high hypobaric warehouse (courtesy of CMC Corporation, Cleveland, Ohio).

included amortization of the forms but not the cost of insulation, the door (estimated at US$10,000–12,000 if quick opening or US$3000–4000 if slow opening), a seal coat and the equipment. The CMC precast concept appears to be the most cost-effective large concrete structure. Two structures are 15–20% cheaper per unit than a single structure, and the cost decreases as the structure becomes larger.

12.8 Inherent Thermal Problem in a Concrete Warehouse Design

The thermal conductivity of concrete (1:2:4) and polyurethane insulation are approximately 1.4 W/m·K (10 BTU·in/ft²·h·°F) and 0.042 W/m·K (0.29 BTU·in/ft²·h·°F), respectively (Porges, 1982). In the Guelph design (Fig. 12.3), if the 15.2 cm (6 in) thick concrete walls were covered with 7.6 cm (3 in) of polyurethane, the thermal conductivity of the structure would be determined almost exclusively by the properties of the polyurethane insulation.

Assumed conditions:

storage temperature = 0°C (32°F)
storage pressure = 1.33 kPa (10 mm Hg)
convective heat-transfer coefficient at storage pressure = 0.34 W/m²·°C (0.06 BTU/h·ft²·°F)
greybody shape factor for radiation = 0.9
radiant heat transfer coefficient = 5.34 W/m²·°C (0.94 BTU/h·ft²·°F), see equation 6.34
ambient temperature = 22.2°C (72°F)
thermal conductivity of insulation = 0.042 W/m·°C (0.29 BTU.in/ft²·h·°F)
interior surface area of building = approximately 455.2 m² (4900 ft²)

To simplify the calculation, it will be assumed that both the stored commodity and the air inside the warehouse remain at 32°F (0°C). At steady-state, the transmitted heat equals the heat removed from the interior surface of the wall by radiation and convection (see bottom of page).

The temperature of the interior surface of the wall (T_{wall}) will be 35.9°F (7.02°C) when the ambient temperature is 72°F (22.2°C), and radiation will transfer heat into the warehouse at the rate of 19,110 BTU/h (560 W). Assuming that the building is loaded with 10,000 50 lb (22.7 kg) bushels of produce, radiation will cause an extra commodity weight loss of 2.6% each month. This will not be distributed evenly throughout the load, and instead will be largely confined to commodity with a line-of-sight exposure to the interior wall of the warehouse. Each 50 lb (22.7 kg) exterior bushel has an exposed surface area of approximately 1.22 ft² (0.113 m²), through which it will acquire sufficient heat by radiant transfer to lose an extra 5.5% of its weight per month. This serious problem was avoided in the Grumman hypobaric warehouse by lining the entire interior surface with cooling panels. It can to a considerable extent be alleviated in a concrete warehouse by installing perforated Mylar slip-sheets inside all exterior boxes to shield them from radiation (9.09). A single layer of Mylar would decrease the rate of 'extra' water loss from exterior boxes to 1.10% of the commodity weight per month, and a double thickness could lower this value to 0.22% each month at the stated conditions. Some advantage could be gained by increasing the insulation on the warehouse.

Notes

1. The warehouse should be able to operate at an internal pressure of 1.33 kPa (10 mm Hg), but this change is not significant because the force on the structure would only increase from 9.3 × 103 kg/m³ (13.23 psi) at a vacuum of 10.13 kPa (76 mm Hg) to 9.6 × 103 kg/m³ (13.62 psi) at 1.33 kPa (10 mm Hg). At an operational

$$4900 \times (0.29/3) \times (72 - T_{wall}) = 0.06 \times 4900\,(T_{wall} - 32) + 0.9 \times 0.94 \times 4900\,(T_{wall} - 32)$$

(transmitted heat) = (convection) + (radiation)

temperature of 0°C, the maximal leak-rate specification for one air change per hour of 1.33 kPa (10 mm Hg) saturated air is reduced to 0.7% of the net volume per hour = 16.8% per day (STP).

2. At atmospheric pressure, the 11.2 kW (15 hp) fan motor would introduce 1318.5 W (45,000 BTU/h) into the 849 m^3/min (30,000 cfm) of circulating air. About 85% of this heat is caused by compression, and only 15% is motor heat. This would warm the air by only 0.54°C (0.3°F). The cooling jacket will significantly decrease the amount of heat that is introduced into the air, but will not remove all of the compression heat. At a pressure of 10.13 kPa (76 mm Hg), the motor heat will be somewhat less, and the heat of compression should be reduced by nearly 90%. Even though the heat capacity of the air is lowered by approximately 90% at a pressure of 1.33 kPa (10 mm Hg), the air's temperature rise should be quite small using a jacketed pump motor.

3. It is not obvious why internal insulation was recommended. Closed-cell insulation 'explodes' when it is exposed to a vacuum. Open-cell insulation would have to be used inside the building, and it is less effective than closed-cell insulation, and prone to accumulating moisture.

4. Much less cooling would be required in summertime if the steam was generated in a vacuum boiler operated at the storage pressure, as it is in the Grumman warehouse design and Grumman/Dormavac intermodal container (Fig. 13.8, right).

5. A major difficulty with this arrangement is that the humidity sensor cannot distinguish whether the humidity originated from the incoming air or the stored commodity. When the same arrangement was used in Grumman intermodal containers, the commodity elevated the humidity sufficiently to cause the steam generator to turn off, causing an excess water loss from the commodity. The Grumman system had to be replaced by a modulating heater (9.02; 13.05; note 2 – chapter 13).

6. Water does not condense in the oil of a gas-ballasted oil-seal pump under the conditions used for LP storage. For the same pumping capacity, a liquid-sealed pump consumes much more energy than an oil-sealed pump, increasing the operational cost.

7. This type of OFF–ON pressure control causes the pressure to cycle and is less accurate and reliable than the modulating pressure control used in intermodal containers and the Grumman warehouse.

8. Unless the inner surfaces of the walls are completely covered with heat-exchange panels, as they were in the Grumman warehouse (12.01), almost all of the heat transmitted through the walls will be radiated into the commodity and cause an excessive water loss. On a warm day, the cooling coils (Fig. 12.4) will be colder than the air in the warehouse, lowering the air's dew point. This in turn will cause greater evaporation from the commodity, and the evaporative heat loss will be replaced by radiation from the walls. In cold weather, the walls will be colder than the set air temperature, and since the air-dry bulb is controlled at the set temperature, water will condense on the walls, and the air dew point temperature (and humidity) will be lowered.

9. A pneumatic air mover (13.09; Fig. 13.11) has a limited capacity to create flow across a pressure drop, especially in a vacuum. It suffices to flow air around the boxes, but not through them. Therefore most of the respiratory heat will be removed by evaporative cooling, creating a natural convective updraught through the pallet boxes.

13

Intermodal Container Design

13.1 Prototype No. 1

The first hypobaric intermodal container, built in 1969 by Fruehauf Corporation for the United Fruit Company, consisted of a 1.8 m (6 ft)-diameter cylindrical stainless-steel tank mounted inside a conventional 2.4 × 2.6 × 6.1 m (8 × 8.5 × 20 ft) insulated intermodal container equipped with a standard forced-air refrigeration unit. This prototype operated satisfactorily for many years and proved the feasibility of scaling-up the hypobaric concept to a commercial size (Fig. 13.1).

Heat infiltration was prevented and the tank's surface temperature was kept uniform by a 'cold jacket' created by circulating refrigerated air in the space between the tank and insulated intermodal container. Initially, a gas-ballasted, rotary-plunger vacuum pump was housed in an insulated box and kept warmer than 100°C, but the insulation was removed after testing proved that a 'hot pump' was not needed to prevent

Fig. 13.1. Prototype No. 1. View of the hypobaric tank located inside a 6.1 m (20 ft) insulated intermodal container. The door is drawn in by hold-down clamps with speed handles, and hinged from a vertical column mounted inside the cargo space. The door rotates on pillow block bearings and was sealed by means of a solid rubber O-ring mounted in a machined groove.

cargo water from condensing in the pump oil. Low-pressure air changes were humidified in a tower within which water was heated with a thermostatically controlled immersion heater. A snap-action ball cock regulated the tower's water level, using the low pressure in the humidifier to draw water from a vented reservoir. The highly reliable and accurate absolute pressure-regulation system developed by Fruehauf for this prototype has been used in all subsequent hypobaric intermodal containers and warehouses.

13.2 Prototype No. 2

Fruehauf Corporation's Prototype No. 2, manufactured in 1971 for the United Fruit Company, verified the concept of using a secondary coolant and 'cold plate' to refrigerate the evacuated space, and revealed numerous problems that needed to be addressed before commercial use would be possible. The door seal and humidification system were inadequate, leakage was excessive, the pumping system was unreliable and the container was grossly overweight and too costly (Figs 13.2 and 13.3).

A square aluminium vacuum tank was foamed in place inside a standard, non-insulated 2.44 × 2.44 × 12.2 m (8 × 8 × 40 ft) intermodal container. The tank's external reinforcing T-ribs were spaced to alternate with the conventional intermodal container's structural ribs, and the interior storage space was cooled by pumping refrigerated ethylene glycol: water (50:50 wt:wt) through a coil attached to the inner surface of the ceiling. The vacuum subsystem consisted of two oil-lubricated air compressors supercharged by a dry roots blower.

13.3 Grumman's Dormavac Container

The structural and mechanical design of Grumman's Dormavac container (Figs. 13.4 and 13.5) has been described by Byers (1977), Mermelstein (1979), Alloca (1980a), Jamieson (1980a,c), Sharp (1985) and in an article entitled 'Hypobarics' (Anon, 1997).

Fig. 13.2. Prototype No. 2. View of equipment end. Under-chassis equipment includes the vacuum-pump cabinet containing twin compressors (*left*), electric cabinet (*middle*), insulated water tank and a water purifier (*right*).

Fig. 13.3. Prototype No. 2. Interior view showing forklift floor, cooling coils attached to the underside of the roof and a rectangular humidifying tank mounted on the forward bulkhead.

Fig. 13.4. Grumman/Dormavac 12.2 m (40 ft) intermodal hypobaric container. View of door end, showing single door drawn in by four cam operators and suspended with slotted hinges to assist alignment (Alloca, 1980a).

Fig. 13.5. Grumman/Dormavac 12.2 m (40 ft) intermodal hypobaric container. View of equipment end. (A) Refrigeration system pumps coolant through lines on the container's inner walls causing them to act as a jacketed refrigeration cold plate. (B) Vacuum/water system draws humidified air from the container and recovers, stores and purifies up to 98% of the water for re-use. The unit can operate without water refill for at least 6 weeks. (C) Electrical system operates from a diesel generator or from ship or shore power. (D) Humidifier maintains relative humidity at 90–95%. (E) 38 hp diesel engine with 20 kW generator (Anon., 1977; Alloca, 1980a).

13.4 Grumman/Dormavac Structural Design

The basic structural element of Grumman's Dormavac container was a 22.9 cm (9 in) wide × 20.3 cm (8 in) high × 2.4 m (8 ft) long I-beam fabricated from two aluminium extrusions separated by a riveted inner fiberglass pultrusion web (Fig. 13.6). Fifty-four of these beams were welded together to form a 2.4 × 12 × 0.203 m (8 × 40 × 0.67 ft) double-walled panel with no metal-to-metal contact between the walls. The container was assembled from eight subassemblies consisting of two end frames, a bulkhead, door, and four longitudinal sections forming side walls, ceiling and the floor. These were fastened together mechanically with inner and outer longerons forming the corners, after which the open-wall assemblies were filled with foam-in-place polyurethane insulation. The inner skin provided the pressure seal. Two steel end frames with standard corner castings were formed as T-1 tool-steel weldments, and fastened to the basic container with fibreglass strips providing a thermal interface. The resultant overall coefficient of heat transfer per unit of wall area was approximately 0.0568 $kW/cm^2{\cdot}°C$ (0.1 $BTU/h{\cdot}ft^2{\cdot}°F$) for the complete assembly. Unequal expansion of the inner and outer walls caused by operational temperature differences as large as 78°C (140°F) was accommodated by a thermal splice located adjacent to the door end frame, exterior to the vacuum enclosure. The container had an interior usable volume of 42.45 m^3 (1500 ft^3), and including the equipment and fluids weighed 12,939 kg (28,500 lb).

13.5 Grumman/Dormavac Mechanical Design

Cooling fluid flowed through four aluminium extrusions that were welded along the entire length of the inner surface of each structural panel and manifolded in parallel at the equipment end to operate from a common refrigerated glycol supply. The heat-transfer coefficient was optimized by using 30 mm wide × 6 mm high (δ) rectangular cross-section glycol channels to create a large cooling-channel rectangular cross-section aspect ratio ($L/\delta = 30/6 = 5$; Fig.

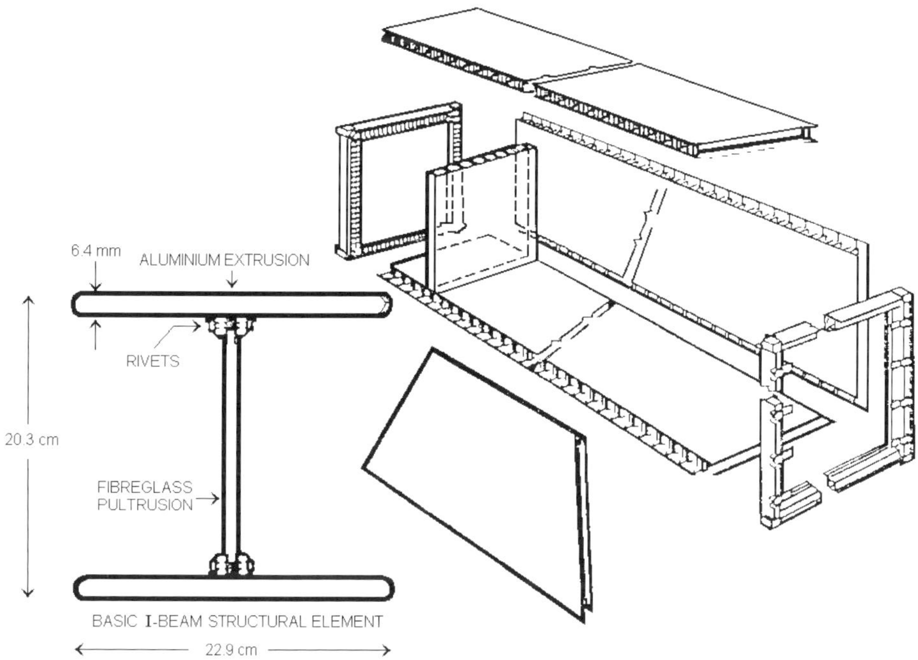

Fig. 13.6. Structure of the Grumman/Dormavac container. The basic structural element is a 22.9 cm (9 in) wide × 20.3 cm (8 in) high × 2.4 m (8 ft) long I-beam fabricated from two aluminium extrusions and a riveted inner web of fibreglass (pultrusion). Fifty-four of these beams were welded together to form a 2.4 × 12 × 0.203 m (8 × 40 × 0.67 ft) double-walled panel with no metal-to-metal contact between the inner and outer walls (Alloca, 1980a; Jamieson, 1980c).

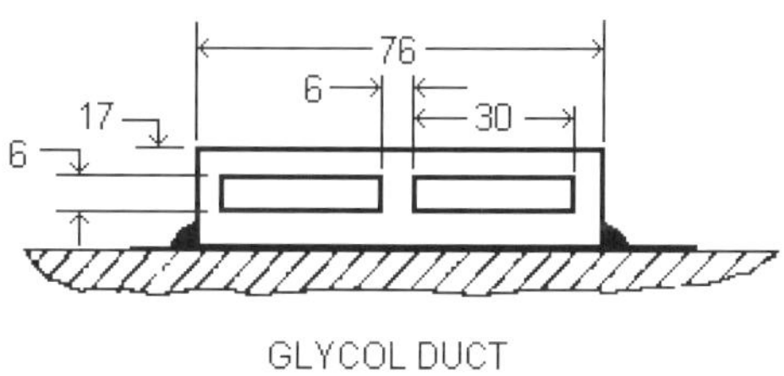

Fig. 13.7. Cross-section (mm) of glycol duct in a Grumman/Dormavac container (Sharp, 1985).

13.7).[1] By passing the outgoing and return streams through a common twin cavity extrusion, the cooling surface was maintained close to the mean temperature of the streams. The door was cooled by a coil welded at intervals to its surface and manifolded in parallel with the wall, roof and floor cooling extrusions.

The humidification subsystem (Fig. 13.8, *right*; Fig. 13.9) included a boiler with a 3 kW electrical immersion heating element, a water-supply tank, water-feed purifier and electrical controls. Steam produced in the insulated boiler flowed into the conduit carrying expanded in-leaking air from the pressure regulator to the hypobaric chamber. The wattage powering the electric immersion heater was preset to generate steam at a rate that saturated the controlled in-leaking air change in accordance with the operational temperature and vacuum-pump capacity at the set pressure, assuming that the incoming air contained no moisture and the insulated boiler lost no heat.[2] Minimum/maximum boiler water levels were maintained by high- and low-level sensors, which controlled the operation of a water-fill solenoid valve. When the solenoid valve opened, water entered the boiler through a water-purifying cartridge, drawn in from a vented water reservoir in response to the low pressure in the boiler. Steam was produced at a low temperature because the pressure in the humidifier was only slightly higher than the container's controlled pressure. All

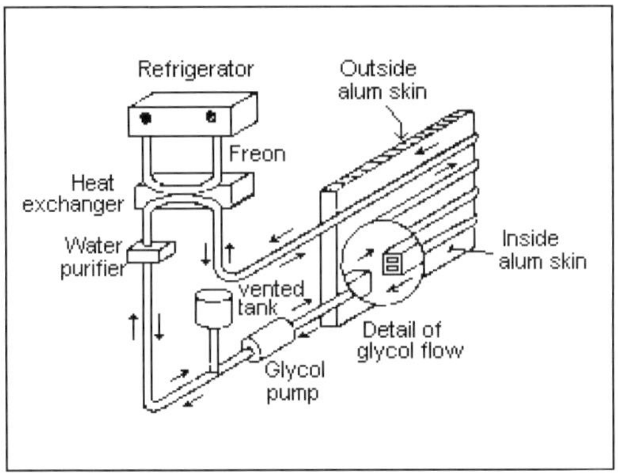

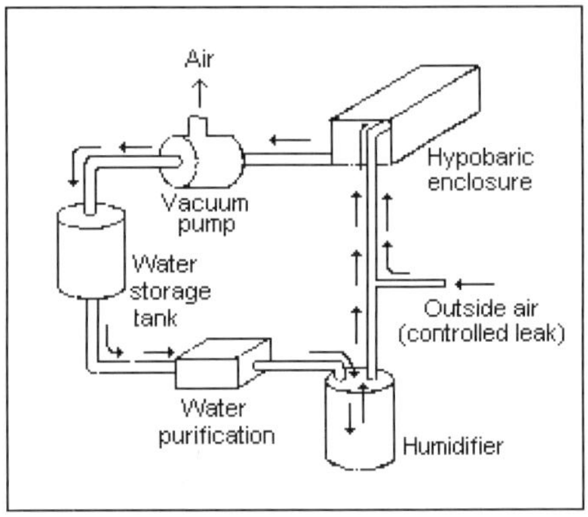

Fig. 13.8. (*left*) Grumman/Dormavac refrigeration subsystem draws propylene-glycol/water coolant from an expansion tank, and pumps it through cooling ducts welded to the interior walls of the container, converting the walls into cold plates. The vapour-pressure difference between the commodity and the air–vapour mixture in the cargo area causes water to evaporate from the commodity, removing respiratory heat. The evaporated moisture condenses on the wall panels, releasing the heat of vaporization into the coolant, which in turn transfers it through a heat exchanger to freon in a conventional refrigeration unit. (*right*) Vacuum/humidity subsystem uses a two-stage vacuum pump to reduce the pressure, change the air in the hypobaric enclosure, and remove commodity-generated gases and water vapour. The air and gases are vented to the atmosphere, and the water vapour is condensed for reuse. An absolute pressure regulation system mixes filtered outside air with cold steam generated in a boiler operated at the same pressure as the cargo area, and this mixture is distributed into the container to control the humidity and continuously supply O_2 to the commodity. The system changes the air in the cargo area several times each hour (Mermelstein, 1979).

water lines were traced with heating tape that automatically energized if the outside temperature fell below 4°C (40°F).

The pressure-control subsystem consisted of a pressure regulator, vacuum breaker and a vacuum-pump assembly comprised of a three-lobed rotary positive-displacement blower, liquid-ring second stage, water-cooling heat exchanger, two-gallon sealed water reservoir and a 50-gallon storage tank (Fig. 13.9). At suction (inlet) pressures between 3.33 and 20.0 kPa (25 and 150 mm Hg), the 5.6 kW (7.5 hp) vacuum pump's capacity was 11.15 m^3/min (120 cfm), but due to pump inefficiency this decreased to 3.72 m^3/min (40 cfm) at 0°C and an inlet pressure of 0.64 kPa (4.5 mm Hg). A vacuum breaker modulated the flow of air changes by sensing the container's pressure at the vacuum-pump inlet and comparing it to a set pressure signal from a much smaller pilot-operated, adjustable absolute-pressure regulator that provided a static, selected 'reference' pressure to the upper chamber of the vacuum breaker (an identical system is described in 13.21 and illustrated in Fig. 13.17).

The humidifying moisture had to be reclaimed in the vacuum pump because without water recovery a 700-gallon water tank would have been required for a design duration trip of 6 weeks, reducing the legally permissible over-the-road payload by 2724 kg (6000 lb). Liquid water was cooled by heat exchange vs. refrigerated glycol-water and injected into the interstage between the dry-lobed vacuum blower (supercharger) and water-sealed vacuum pump (Fig. 13.9). The cooling water kept the water-sealed vacuum pump's temperature close to the storage temperature, causing most of the humidification water to condense during isothermal compression in the pump. The recovered water overflowed from the pump's seal-water tank and gravity-drained into the storage tank.

The refrigeration subsystem (Fig. 13.8, *left*) was comprised of two independent 3.7 kW (5 hp) R-502 compressors, a single freon-glycol heat exchanger, dual

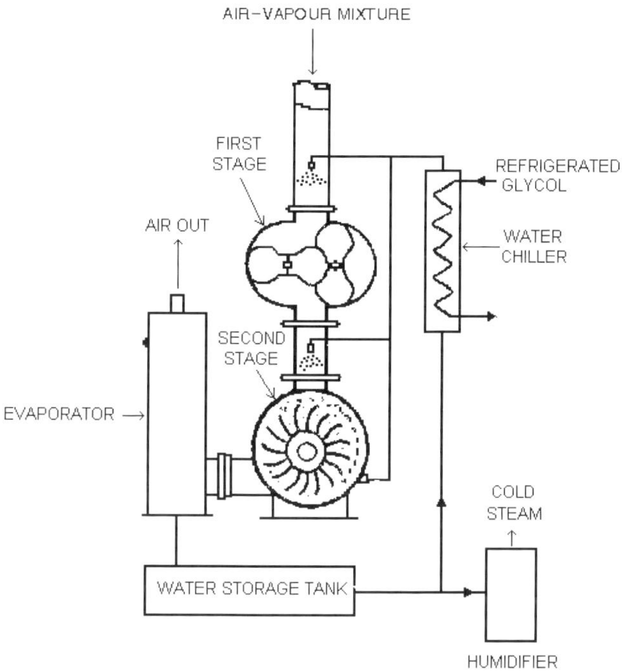

Fig. 13.9. Grumman/Dormavac two-stage vacuum system consisting of a rotary positive displacement Roots-type first-stage pump backed by a second-stage liquid-ring pump. Compression is nearly isothermal, pumping a saturated air–vapour mixture, and the combination of compression and cooling by introduction of chilled liquid into the pump results in vapour condensation, greatly improving the available displacement for the non-condensable gases being extracted by the system. The second stage is operated at a constant speed, but the speed of the first stage is regulated by a constant-torque fluid-drive coupling which automatically adjusts the displacement of the first stage to match the fixed displacement of the second stage (Huse, 1969, 1970).

condensers and a condensing fan. It produced 11.7 kW (40,000 BTU/h) of cooling at a container set-point of –1.1°C and a 49°C (120°F) ambient temperature. Inhibited 50:50 (wt:wt) propylene glycol:water cooling fluid was drawn from a vented expansion tank by a 1.3×10^{-3} m^3/s (20 gpm) centrifugal pump that provided 1.4×10^5 N/m^2 (20 psid) differential pressure to drive the fluid through parallel glycol loops in an R-502 heat exchanger (chiller) before the coolant flowed through the container's parallel cooling loops. The return glycol temperature was sensed by a thermistor and controlled after the coolant temperature decreased to within 0.8°C (1.5°F) of the set-point. In that temperature range, the primary refrigeration system was modulated using a combination of hot-gas bypass, switching off one or both compressors, and activating heaters, as necessary. Depending on the ambient temperature and set-point, the glycol stream increased in temperature by 0.5–1°C as it travelled along the container wall to the door end and returned to the equipment end. The glycol flow limited the temperature rise of the coolant within the container to less than 0.6°C (1°F) with a 49°C (88°F) temperature difference between the glycol and ambient temperatures. An independent study (Sharp, 1985) concluded that the temperature variance in the Grumman/Dormavac container was < 1°C and could be improved to < 0.5°C by a more efficient door cooling loop, whereas the spread is > 2°C in conventional containers and > 2.6°C in another type of 'jacketed' container. Approximately 35% of the return glycol flowed through the vacuum-pump heat exchanger (Fig. 13.9), removing 4.7 kW (16,000 BTU/h) of heat from the vacuum pump's cooling water.

13.6 Defects in Grumman's Dormavac Design

Grumman's Dormavac design had intrinsic difficulties for which there was no obvious remedy. The 655 linear metres of weld seams in the inner surface (vacuum barrier) of the wall panels and 1500 linear metres in the entire structure created a propensity for leakage and a high labour cost for fabrication. The large quantity of aluminium in the double-walled structure caused the container to have an excessive materials cost and weight. Water that condensed on the cold exterior of the inner wall panels became trapped within the double-walled structure and eventually impaired the insulation and corroded the 6061-series aluminium. Occasionally individual coolant lines, and especially the door loop, developed air embolism and were bypassed by coolant. The lines had to be repeatedly vented to flush out entrapped air, and because the door was not adequately cooled it tended to radiate heat into the load. The weight, cost and size of the equipment package were excessive. It required a massive built-in motor-generator set, a ponderous 230/460 VAC 3-phase transformer, and carried nearly 380 kg (840 lb) of water. Statutory on-board ship and over-the-road gross-weight restrictions limited the permissible payload to approximately 13,600 kg (30,000 lb) even though the container had sufficient space to load 18,200 kg (40,000 lb) of a dense cargo such as meat.

Grumman was prevented from developing an energy-efficient, reliable equipment package by the requirement for a humidification step and also by the excessive volumetric capacity of the water-sealed vacuum system needed to reclaim water and provide adequate pumping capacity at a very low pressure. Although otherwise preferable and more energy efficient, an oil-seal vacuum pump could not be used for this purpose because its exhaust is contaminated with oil mist, making water reclamation unfeasible. The refrigeration needed to cool the vacuum system's seal water consumed nearly 5 kW of electric power, the blower and water-sealed vacuum pump consumed 8.5 hp, an additional 3 kW was required to vaporize steam and at least 2 tons of refrigeration to remove respiratory and transmitted heat. A 38-hp diesel engine with a 20-kW generator had to be built into each equipment package, decreasing the length of available cargo space by 0.9 m, and increasing the total cost and weight. Freeze-protection heaters were required on the storage tank and pipes, and automatic drains had to be activated if the ambient temperature decreased to less than 4°C when the container was idle and not powered. The boiler's water-levelling devices frequently failed due to scale build-up even though the boiler was filled through a water purifier and flushed between trips. Because the humidification step saturated the incoming air changes, commodity water vaporized by respiratory heat condensed on the cold roof of the container, and dripped on to the boxes. The cartons eventually became soggy and collapsed unless they were protected by a water-impervious coating.

13.7 General Description of the VacuFresh^{SM} Container

VacuFresh^{SM} was designed to overcome the weight, space, price, excess energy and reliability deficiencies of Grumman's Dormavac container.[3] It provides the maximum possible space for cargo within the ISO and ANSI-specified external dimensions for a 6.1 m (20 ft) or 12.2 m (40 ft) cylindrical intermodal container, while at the same time the weight of the container and associated equipment has been reduced sufficiently to allow all of this space to be filled with dense perishable cargo without exceeding the gross over-the-road or on-board-ship weight limitations. The VacuFresh^{SM} cylindrical beam tank is self-supporting along its length from end-frame to end-frame.[4] A 12.2 m (40 ft) version will provide slightly more usable space (Fig. 6.14), weigh 50% less, and cost one-third as much for materials and labour compared to Grumman's Dormavac container.

When chilled meat, fish, poultry or shrimp are transported in VacuFresh^{SM}, the pressure regulation system is not used (Burg, 1987a). The storage tank is continuously flushed with cold steam released from the animal matter, the steam is circulated exclusively by natural convection and diffusion, and the pressure is controlled by the vapour pressure of the animal matter at the storage temperature. When a chilled plant commodity is shipped, atmospheric air is leaked into the tank at a controlled rate and passed through at a regulated pressure. Water loss from the plant matter is kept to a minimum by continuously evacuating the tank at a rate that is adjusted to saturate the incoming air changes with the amount of moisture evaporated from the weight and type of stored commodity in response to the small amount of respiratory heat produced at an optimal storage pressure (Burg, 1987b). A pneumatic air mover (Fig. 13.11), driven by the pressure difference created by the vacuum pump, circulates up to 28 m^3/min (2500 cfm) of low-pressure cargo air within the vacuum tank without producing heat or consuming additional power.

13.8 VacuFresh^{SM} Tank Container Specifications and Physical Parameters

The VacuFresh^{SM} 6.1 m (20 ft) intermodal container is designed to withstand a full internal vacuum and applicable handling/transportation load factors per ANSI MH5.1.2 M and ISO 1496–3. It meets all of the racking-test criteria for ABS and Lloyds' certification. The container and equipment package can operate in ambient conditions ranging between 0 and 100% relative humidity at −17.8 to +49°C (0–120°F), with in-transit loads of 2 *g* downward, 0.6 *g* lateral, 2 *g* longitudinal and 0.5 *g* upward. The safety factors utilized in the design are 1.15 yield safety factor (YSF) and 1.5 ultimate safety factor (USF) for depressurization; 1.15 USF for fail-safe depressurization with the complete loss of the structural integrity of one stiffening ring; 1.5 YSF and 2.0 USF for cargo, handling/transportation and any other operational induced forces, which are considered to act simultaneously or independently, whichever is critical. All material properties utilized in the design are minimum expected values.

The following exceptions apply:

- Container stacking is limited to six high below the deck during marine transport.
- The container meets the requirements for four-high stacking above deck in accordance with ISO 1493–3, and three-high stacking according to ANSI MH5.1.2 M.
- Load-transfer areas are not provided on the container.

The overall size and weight parameters are in accordance with ANSI MH5.1.2 M and ISO 1496–3 for a nominal 6.1 m (20 ft) container.

- Size = 6.1 m × 2.6 m × 2.4 m (20 ft × 8.5 ft × 8 ft).
- Volume (usable) = 21.2 m^3 (750 ft^3).
- Gross weight (ANSI MH5.1.2 M) = 20,345 kg (44,800 lb).
- Gross weight (ISO 1496–3) = 24,033 kg (52,920 lb).
- Tare weight (including equipment) = 3633 kg (8000 lb).
- Equipment weight = 681 kg (1500 lb).

Container heat loads:

- Thermal transmission – 2720 kcal/h @ 50°CTD (10,800 BTU/h @ 90°FTD).
- Equipment (incl. vacuum pump) – 2650 kcal/h (10,550 BTU/h) @ 60 Hz.

13.9 Special Features

Tank door

The neoprene door-seal extrusion, which is vulcanized into a 'circle', is snapped on to the door flange without adhesive (Fig. 13.10). During container pump-down, the extrusion's sealing bulb makes an initial contact with the triangular door sill and vents as it compresses. The self-aligning door, suspended with three degrees of freedom, is drawn inward as the vacuum increases, until the door extrusion's solid O-seal makes a final leak-tight closure with the door sill. After the operational pressure is reached, racking is prevented because the door is held in place by approximately 36,300 kg (80,000 lb) of external force. The low-cost, resilient rubber door-seal extrusion can be replaced in a matter of minutes if it is damaged.

Mounting clips

The beam tank[5] is attached to the container end frames by mounting clips. Aluminium mounting 'boxes' constructed at the top, bottom and both sides of the tank barrel on sections extending beyond the door and forward dish head 'bulkhead', are pinned by means of huck-bolts to steel mounting clips welded to the end frames (Fig. 13.12, *middle left*; the same system is used for Transvac; see Fig. 13.20, *upper left* and *upper middle*). Fabrication tolerances are accommodated by the positioning of the steel clips when they are welded to the end frames. Heat leakage through the mounting assembly is reduced by a fibreglass insulator sandwiched between each clip and aluminium box, and by fabricating the clips from T-1 tool steel, not for strength, but rather because T-1 has half the thermal conductivity of carbon steel.[6] T-1 steel decreases heat leakage by increasing the temperature gradient that develops across the clip. Heat leakage through the clips accounts for 41% of the total heat infiltration in a 6.1 m (20 ft) VacuFresh[SM] container. The overall heat-transfer coefficient per unit of wall area is approximately 0.0852 kW/cm^2·°C (0.15 BTU/h·ft^2·°F) for the 6.1 m (20 ft) container, and 0.0682 kW/cm^2·°C (0.12 BTU/h·ft^2·°F) for a 12.2 m (40 ft) container.

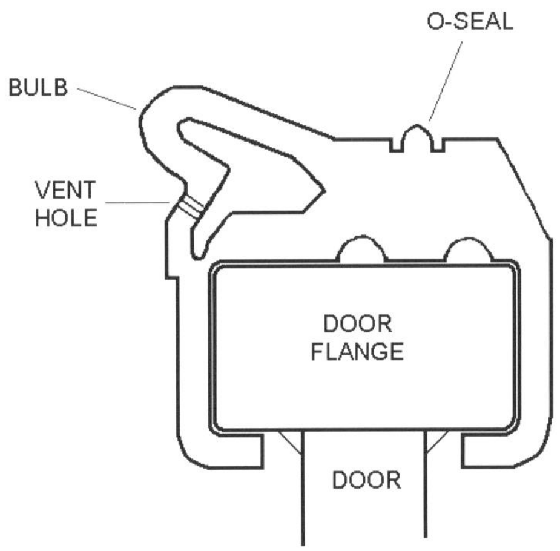

Fig. 13.10. Cross-section of the VacuFresh[SM] door seal attached to a flange welded to the door's edge.

Air mover

A pneumatic air mover is mounted beneath the port-side shelf close to the forward bulkhead (Fig. 13.12, *lower middle*; Fig. 13.11; Fig. 13.17, AH). The air mover operates on the venturi principle using small volumes of high-velocity fluid to accelerate large volumes of low-velocity fluid. The device accommodates the air changes introduced by the entire range of pumping speeds and pressures recommended for different commodity types (example 1). Air enters a concentric manifold chamber surrounding a venturi throat, where jets with a length/diameter ratio = 6, located symmetrically around the chamber, are positioned to expand the air into a reaction zone extending from 0.92 to 13.2 cm down-stream

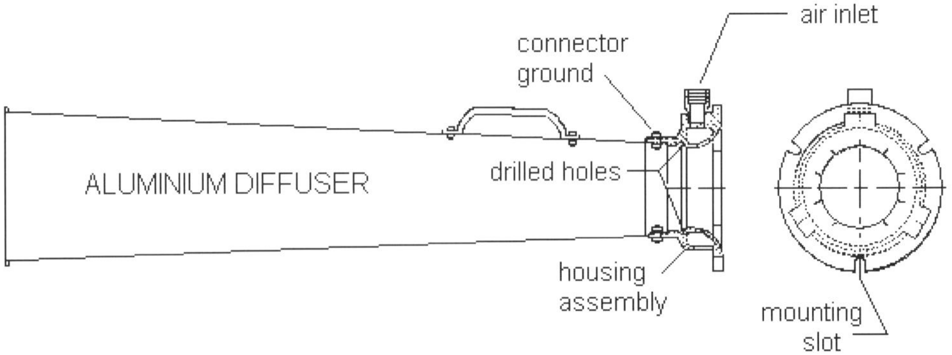

Fig. 13.11. Pneumatic air mover.

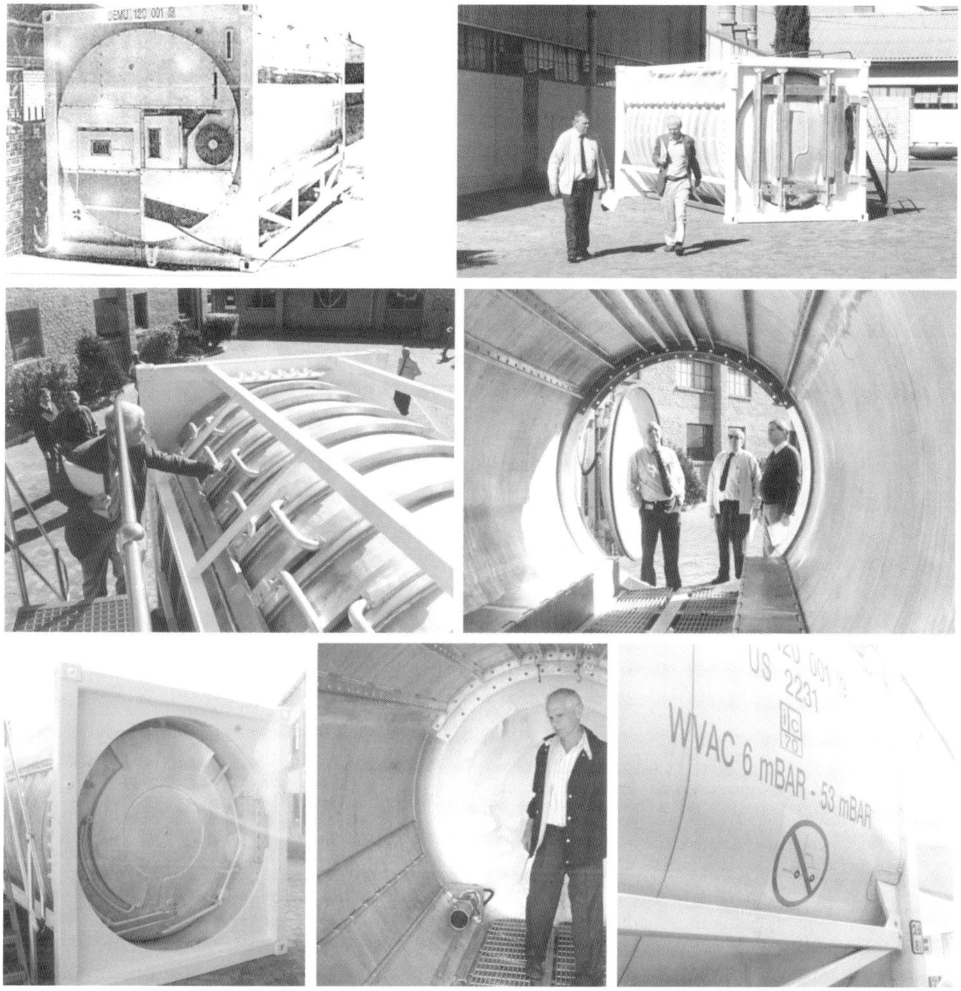

Fig. 13.12. VacuFresh^{SM} 6.1 m (20 ft) intermodal container. (*upper left*) Equipment end; (*upper right*) door view showing cooling coil and cam operators; (*middle left*) interconnected stiffening rings and ring splices; (*middle right*) interior view of shelves, meat rails, trolley, floor and duct; (*lower left*) expansion tank welded to forward head; (*lower middle*) air horn with shelf folded up; (*lower right*) insulation and cladding installed.

of the throat. It is here that the high-velocity air imparts its energy to slower-moving air and accelerates it, which in turn draws more air through the throat into the reaction zone. The jets must discharge air at supersonic velocity (turbulent flow) for the pneumatic air horn to operate at maximum efficiency. This critical flow occurs when the upstream pressure is at least 1.9 times larger than the pressure downstream of the jets. The VacuFresh^{SM} air horn has been specially designed to satisfy this requirement when the vacuum pump operates at and above 27% of its maximum capacity at 50 Hz, discharging 17.55 m^3/h (10.33 cfm; example 1).

The air mover mixes 23.9°C air circulating through the vacuum-control cabinet (Fig. 13.13, No. 12) with commodity--saturated low-pressure air drawn from inside the container. The air emitted by the jets of the air mover expands into an essentially 'infinite' volume, decreasing in humidity but hardly changing in temperature. Assuming that the storage temperature is 0°C, if the air mover mixes 40 parts of circulating high humidity container air with one part of incoming bone-dry 23.9°C air, the mixture

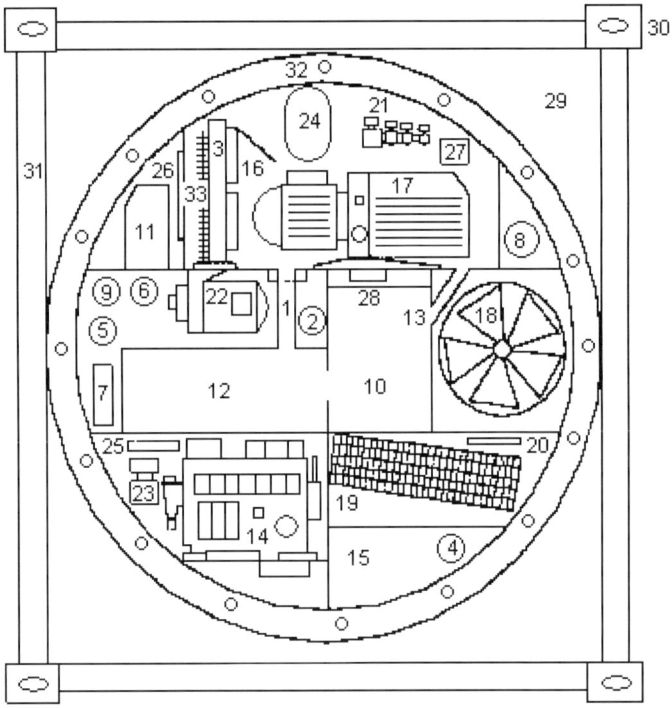

Fig. 13.13. Outline drawing of the VacuFresh^{SM} mechanical package. (1) Removable duct from vacuum-pump cabinet, through forklift slot, terminates in a 24-VDC air-circulating fan inside vacuum-control box No. 12; (2) tank vent valve; (3) vacuum pump heat exchanger (radiator) with Finbar® heating strips and drain pan; (4) air horn shut-off valve; (5) vacuum-sensing line to tank, strapped to (heated by) adjacent liquid line; (6) thermal expansion valve and hot gas inlet; (7) suction-line/liquid-line heat exchanger; (8) shut-off valve in suction line from tank to vacuum pump; (9) supply glycol line to tank, with shut-off valve and thermistor probe; (10) electric box; (11) brazed-plate heat exchanger (refrigerant to glycol); (12) vacuum-control box; (13) air-circulating duct from electric box to vacuum-pump cabinet; (14) compressor; (15) electric-cable storage bin; (16) fan-discharge air baffle, 24-VDC tube-axial fan and 21.1°C thermostatic valve regulating glycol flow to radiator; (17) vacuum pump; (18) condenser fan (5-bladed); (19) condenser coil; (20) discharge pressure regulator; (21) manifolded flow-control solenoid valves; (22) glycol pump; (23) suction-line solenoid valve; (24) vacuum pump oil-coalescing filter; (25) crankcase pressure regulator; (26) baffled air-inlet duct and filter screen; (27) vacuum pump vented reservoir for coalesced oil; (28) forklift slot; (29) reinforcing gusset plate; (30) corner casting; (31) end frame; (32) bolting flange; (33) Finbar® heater.

discharged from the air mover would have a temperature of 0.6°C and a relative humidity less than 6% below that of the mixture in the container.[7] Due to the incoming rarified air's low-heat capacity and close contact with the container walls and shelf, whilst flowing through the under-shelf duct, the mixture cools to 0°C before it is released at the door end. This increases the air's humidity by 4% before it is allowed to encounter the first commodity at the door end.

The air mover is mounted by three captivated bolts. It can be detached and the shelves folded down to provide additional cargo space when hanging-meat carcasses are transported.

13.10 Interior Structures

Floor supports

Crescent-shaped floor supports welded to the aluminium shell at the centreline of each stiffening ring add necessary strength at the point of maximum stress on the barrel. The bar-grill floor (Fig. 13.12 – *middle right* and *lower middle*) folds up to facilitate cleaning the tank shell below.

Shelves

Two full-length shelves (Fig. 13.12 – *middle right*), formed as continuous three-leaf aluminium hinges, are installed along the longitudinal sides of the tank by intermittent welding at each stiffening ring. This arrangement transmits load stresses exclusively into the stiffening rings, rather than into the tank shell. To reduce weight and cost, each shelf leaf has a different thickness, just sufficient to sustain the loading which it will encounter. The shelves fold down to increase the space available for hanging meat carcasses, and up to facilitate cleaning (Fig. 13.12 – *lower middle*).

Suction duct

The rectangular suction duct mounted along the floor's longitudinal centreline (Fig. 13.12 – *middle right*) has less than 10% variation in intake along its length. A 0.5-cm (3/16-in)-diameter hole is drilled in the duct's underside at the midpoint between each pair of floor supports, plus an additional hole in the capped door-end of the duct. Operating at full capacity, the vacuum pump's suction pressure is only 5% lower than the tank pressure. Essentially, the entire pressure drop is due to the small size of the holes in the suction duct.

Meat rails and trolleys

Seven meat rails and two trolleys are provided for hanging carcasses (Fig. 13.12 – *middle right* and *lower middle*).

Drain plug

An expandable drain plug's handle is accessible inside the tank through an opening in the floor grating at the floor centreline near the door. When the plug is loosened, it remains captivated by the handle. The plug creates a vacuum-tight seal.

13.11 VacuFresh^{SM} Mechanical Equipment

Mechanical equipment is housed in a circular package (Fig. 13.13, No. 32) to provide space for end-frame reinforcing gussets (Fig. 13.13, No. 29), which add sufficient strength to permit low-cost conventional extruded tubular-steel end frames (Fig. 13.13, No. 31) to be used in the VacuFresh^{SM} design, instead of the less reliable and more expensive T-1 tool steel end-frame weldments in the Grumman/Dormavac container.

13.12 Refrigeration System General Schematic

The aluminium tank shell, door and forward bulkhead of the vacuum tank serve as a cold-plate jacket. Ethylene-glycol/water coolant (50:50 wt:wt) is pumped from a vented expansion tank welded to the vacuum tank's forward bulkhead (Fig. 13.12, *lower left*), through service valves, and discharged from a centrifugal glycol pump into a glycol-R-404a heat exchanger. Most of the glycol cooled by the heat exchanger flows through a supply duct to the door end of the vacuum tank, and returns to the expansion tank either through the door cooling tubes and door return duct, or through the door-sill cooling duct and series-connected tank stiffening rings (Fig. 13.12, *upper right*, *middle left*; Fig. 13.14).

The stiffening rings are formed from rectangular extrusions rather than channels to eliminate the possibility of leakage through welded seams. Their pitch and

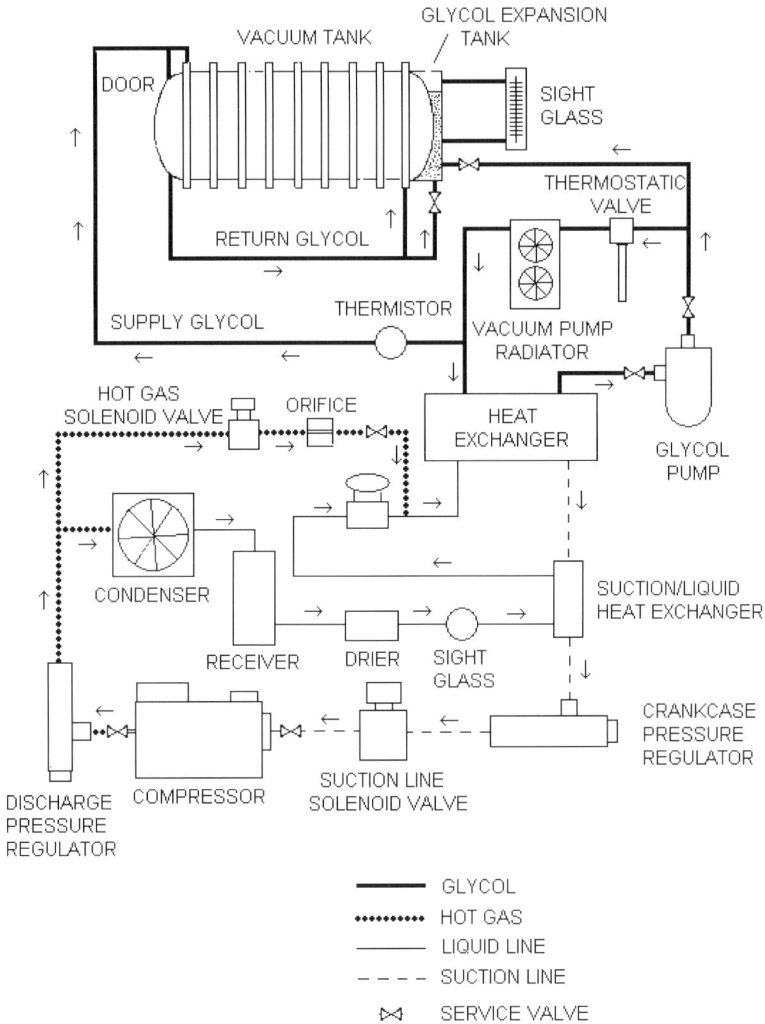

Fig. 13.14. General schematic of VacuFresh^{SM} refrigeration system.

dimensions (Fig. 13.15) and the thickness of the tank shell provide the required strength, minimize the tank's weight, maximize its volume and permit sufficient glycol flow through the series-connected rings to maintain the inner surface of the tank shell at a constant temperature.[8] The velocity of glycol flow through the stiffening rings, approximately 23 cm/s (0.75 ft/s), is sufficient to flush air bubbles through the system to the expansion tank where a baffle deflects them, after which they are vented to the atmosphere. The series-cooling system obviates the troublesome problem of air embolism experienced with the Grumman/Dormavac parallel-duct cooling system.

Low-profile fluid interconnections between the stiffening rings are located at the points of least stress around the tank's circumference, and the 'break' in each ring is bridged with a structural splice (Fig. 13.12 – *middle left*). Internal baffles cause the glycol to flow completely around one ring in a clockwise direction, through the interconnection, then around the next ring in a counter-clockwise direction, and the sequence is repeated the full length of the tank (Fig. 13.14). A separate parallel cooling

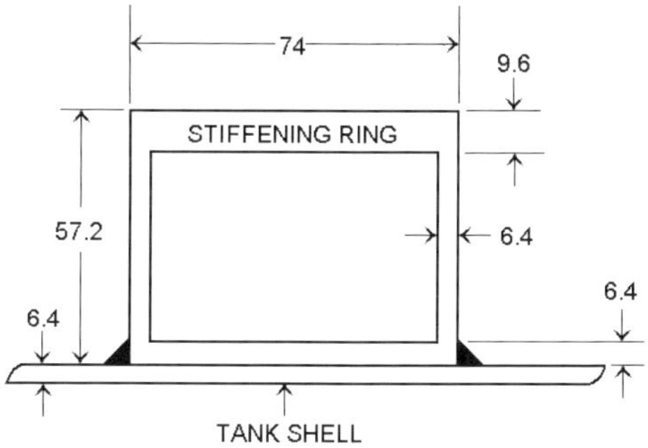

Fig. 13.15. Cross-sectional view of a stiffening ring welded to the exterior surface of the VacuFresh^{SM} tank shell. Dimensions in mm. The large hydraulic diameter [(4 × area)/(wetted perimeter = 49.2 mm)] makes it feasible to flow coolant through the series-connected stiffening rings at the volumetric rate needed to avoid a significant coolant temperature rise, without creating an excessive pressure drop.[9] The hydraulic diameter (D_h) of the cooling ducts in Grumman's Dormavac container was only 10 mm (Fig. 13.7).

coil is bonded with thermally conducting material to the door (Fig. 13.12 – *upper right*). The glycol expansion tank, which is welded to the aluminium head at the equipment bay end (Fig. 13.12 – *lower left*), cools the forward tank bulkhead. Coolant flowing through an extra stiffening ring added to the shell extension at the equipment end captures heat infiltrating through the end frames and supporting clips before it reaches the tank. Glycol flowing through the triangular door sill is used for that same purpose at the door end. The 1.1-kW (1.5-hp) centrifugal glycol pump (Fig. 13.13, No. 22) provides 2.3 kg/cm^2 (39 psid) differential pressure at 60 Hz to flow 2.3×10^{-3} m^3/s (35 gpm) of inhibited 50:50 (wt:wt) ethylene-glycol/water through the cooling system.

The jacketed refrigeration system maintains the tank shell within ± 0.2°C (± 0.36°F) throughout its length at set temperatures between –17.8 and +16°C (0–60.8°F) and ambient temperatures ranging from –17.8 to 49°C (0–120°F). With a 0°C (32°F) supply glycol temperature, the net cooling capacity at 60 Hz is 11,339 kcal/h (45,000 BTU/h). With the thermostat set at –17.8°C (0°F), the net cooling capacity is 5166 kcal/h (20,500 BTU/h). The net heating capacity at 60 Hz is 4248 kcal/h (16,860 BTU/h).

The refrigeration system has a 4.5 kW (6 hp) compressor (Fig. 13.13, No. 14); a condenser coil (Fig. 13.13, No. 19); and a 1.5 kW (2 hp) 5-bladed 40-cm (16-in)-diameter condenser fan (Fig. 13.13, No. 18) that draws 3.2 running amps at 50 Hz, and delivers 1.5 m^3/s (2100 cfm) at 2880 rpm, at a condenser static pressure of 3.8 cm (1.5 in) water column (wc). A receiver and the condenser serve as a reservoir capable of holding the full refrigerant charge. A crankcase pressure-regulating valve (Fig. 13.13, No. 25) limits compressor power consumption to 6.5 kW at start-up by preventing the suction pressure from exceeding 40 psig (272 kPa). A discharge pressure-regulating valve (13.13, No. 20) equipped with a dampening device to minimize the effects of pulsations, controls the discharge pressure at 285 psig (1895 kPa). This regulator prevents the head pressure from decreasing at a low ambient temperature, keeps the pressure differential across the expansion valve high enough to guarantee adequate refrigerant flow into the evaporator, and assures that the refrigeration system's capacity remains constant and the compressor discharge provides the same pressure of hot gas to the modulation system regardless of ambient conditions. The compact evaporator (28.7 cm high × 11.7 cm wide × 20.2 cm deep; Fig. 13.13, No. 11) is an 80-plate stainless steel R404a/glycol heat exchanger specially manufactured for VacuFresh^{SM} to avoid mal-distribution on the refrigerant side. Total heat of absorption is 11.8 kW (40,283 BTU/h) for a 7.8°C mean temperature difference at a glycol outlet temperature of 0°C, flowing 2.646 kg/s (39.4 gal/min) of 50:50 (wt:wt) ethylene-glycol/water at 157 kPa (8.1 psig) pressure drop. A bypass 'orifice' plate in the suction-line solenoid valve (Fig. 13.13, No. 23; Fig. 13.14) maintains the refrigeration capacity at 60% when the solenoid valve closes to modulate ('unload') the compressor. The hot-gas-pulsating solenoid valve (Fig. 13.14) has a life expectancy of 5,000,000 cycles, and should not need to be replaced for at least 5 years in normal use. It is located distant from the compressor discharge port in order to air-cool the hot gas sufficiently to avoid exceeding the solenoid valve's continuous operational temperature limit of 120°C (248°F). An upstream orifice (Fig. 13.14, orifice) limits the hot gas flow when the normally closed solenoid valve opens.

The refrigeration system is charged with Suva HP62 (R-404a) primary refrigerant and 50:50 (wt:wt) ethylene glycol:water secondary coolant. Suva HP62 (R-404a) is a near-azeotropic, non-flammable mixture of hydrofluorocarbon (HFC) refrigerants HFC-125, HFC-143a and HFC-134a. It is a non-ozone-depleting replacement for R-502, with capacity and efficiency values nearly equivalent to R-502, but with much lower global warming potential. The DuPont Acceptable Exposure Limit (AEL) to R-404a is 1000 ppm. This is the acceptable airborne exposure limit to which workers can be repeatedly exposed without adverse effects during a 40-h working week or 8-h day.

13.13 Heating Cycle

When heating is required, a three-element 2.7/3.9 kW, 3-phase, 50/60-Hz Finbar® heater (Fig. 13.13, No. 33) is energized. The heating element is mounted at the inlet side of a dual-fan radiator (Fig. 13.13, No. 3) located in the insulated vacuum-pump cabinet. Heat transferred into glycol passing through the radiator circulates throughout the vacuum-tank cooling/heating system. The Finbar® heaters cannot energize unless the glycol pump is ON and a pressure switch senses that there is adequate glycol flow. When the vacuum pump is operating, circuitry limits the heat output of the Finbar® heaters to 1.2/1.8 kW, single phase, 50/60 Hz. in order to lower power consumption and prevent the vacuum-pump cabinet temperature from increasing and overheating the pump oil if the thermostat requests heat with the vacuum pump ON. The vacuum pump's motor and compression heat (1.8/2.2 kW at 50/60 Hz) exceeds any heat demand that might be required to keep the tank at a controlled temperature, even when the ambient temperature is –18.9°C. Therefore, the Finbar® heaters do not normally energize when the vacuum pump is operating.

13.14 Temperature Modulation

A microprocessor proportional-integral-derivative (PID) thermostat (note 9, chapter 11) keeps the supply glycol temperature constant ± 0.1°C by cycling the hot-gas solenoid valve open/shut in response to a thermistor probe inserted in a well downstream of the glycol/R-404a heat exchanger. The vacuum-pump heat load improves the tank's temperature uniformity by forcing the compressor to continuously operate and the PID thermostat to modulate the refrigeration by means of hot-gas pulses vs. a constant heat input.

The thermostat measures the temperature twice each second during a 15-s interval with ± 0.055°C (± 0.1°F) accuracy, averages 30 readings, and presents the average on the digital display panel-meter, upgrading the result every 15 s. The operational sequence for a chilled mode is illustrated in Fig. 13.16. Heat cycle, hot-gas bypass and unload are locked out for frozen loads at set-point temperatures below –7°C (9.4°F).

13.15 Vacuum System General Schematic

The vacuum system general schematic is illustrated in Fig. 13.17. The 2.2 kW (3 hp) single-stage oil-sealed rotary-vane air-cooled vacuum pump has a nominal pumping speed of 78 m^3/h (45.9 cfm) at 60 Hz and 65 m^3/h (38.2 cfm) at 50 Hz (Table 13.1). Its efficiency remains high down to the lowest operational pressure, 0.6 kPa (4.5 mm Hg), which is typical for meat. With full gas ballast, the pump's ultimate total pressure is 0.05 kPa (0.19 mm Hg) at 60 Hz and 0.065 kPa (0.49 mm Hg) at 50 Hz. Universal mounts isolate the pump axially and radially from low-frequency vibration,

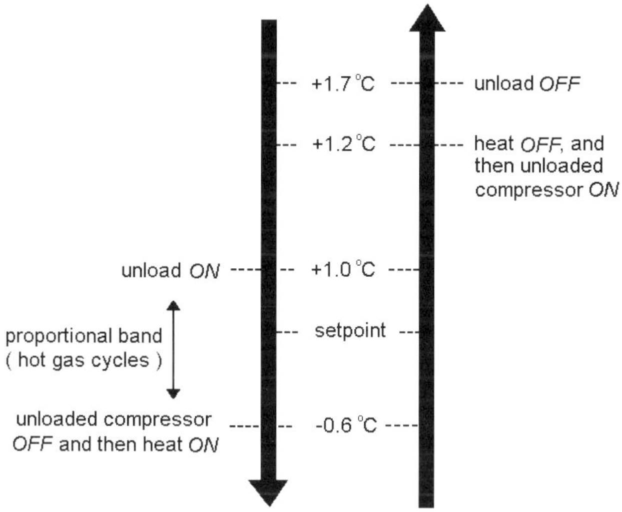

Fig. 13.16. Microprocessor thermostat operational sequence (chilled mode).

noise and severe shock. All hose connections to the pump are flexible and do not interfere with the functioning of the mounts. Because the pump is lubricated by a positive-displacement gear pump, it can be operated continuously at any inlet pressure.[9]

Before compression advances in the pumping chamber, 3.29 m^3/h (2.28 scfm) of 'gas ballast' air is admitted at 60 Hz, or 5/6 that amount at 50 Hz, in order to keep the compression ratio below 10:1. If the oil temperature is kept high enough, the exhaust valve of the gas-ballasted pump will open before the water condensation pressure is reached. The inlet vapour pressure (p_{VAP}), which can be continuously pumped without condensing water in the oil, is given by:

$$p_{VAP} = (B/S)p_{SUM}\frac{(p_{VAP,SAT} - p_{VAP,GB})}{(p_{SUM} - p_{VAP,SAT})} + p_{PERM}\frac{(p_{VAP,SAT})}{(p_{SUM} - p_{VAP,SAT})} \quad (13.2)$$

where $p_{SUM} = p_{EXHAUST} + \Delta p_{VALVE} + \Delta p_{EXHAUST\ FILTER}$ and $p_{EXHAUST}$, Δp_{VALVE}, and $\Delta p_{EXHAUST\ FILTER}$ are the exhaust pressure, pressure difference across the exhaust filter and pressure difference across the exhaust valve, respectively. B is the volume of gas-ballast air at 760 mm Hg which is admitted to the pump chamber per unit time, S is the nominal pumping speed (volume flow rate), $p_{VAP,\ SAT}$ is the saturation water vapour pressure at the working temperature of the pump, p_{PERM} is the pressure exerted by permanent gases at the pump inlet and $p_{VAP,GB}$ is the vapour pressure of the gas ballast air. For the VacuFresh[SM] pump, $B/S = 0.05$, $p_{SUM} < 998$ mm Hg, and since the gas-ballast air originates from the vacuum control cabinet where the maximum relative humidity is 80% at 20°C, $p_{VAP,GB} \leq 22$ mm Hg. Using these values, equation 13.2 can be simplified to the form:

$$p_{VAP} = 0.051\ (p_{VAP,\ SAT} - 22) + 0.001\ (p_{PERM})\ (p_{VAP,\ SAT}) \quad (13.3)$$

where the pressure units are mm Hg. When meat is shipped at –0.8.°C, $p_{VAP} = 4.32$ mm Hg, $p_{PERM} = 0$, $p_{VAP,SAT} = 107$ mm Hg (equation 13.3), and the saturation temperature (minimum operating oil temperature) is 53°C (127.4°F). A correction must be applied to account for the fact that the pump's oil-reclaiming system (Fig. 13.17) provides 5% of the gas-ballast air, and this additional air is saturated at a relatively high temperature, elevating the minimum permissible oil temperature to 54.6°C (130.3°F). This is well below both the pump's maximum permissible temperature, 104°C (230°F), and its nominal operating temperature in the vacuum-pump cabinet, 77°C (170°F). When produce is transported at various pressures and flow

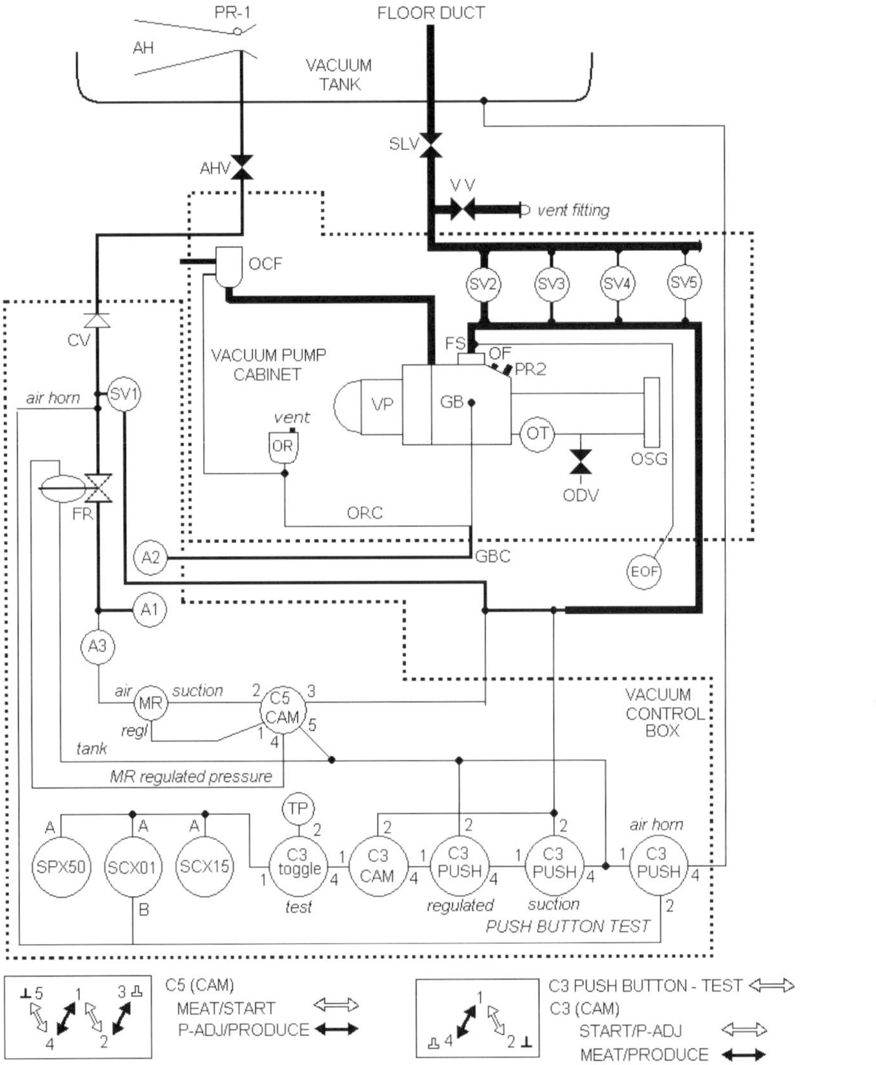

Fig. 13.17. VacuFresh[SM] vacuum system general schematic. The 4.8 mm (3/16″) ID polyurethane ether tubes used throughout the vacuum-control system are colour-coded to simplify identification.
A1, Fisher regulator air filter; A2, vacuum-pump gas-ballast air filter; A3, Moore regulator air filter; AH, air horn; AHV, air horn shut-off ball valve; C2, Clippard 2-way valve (push-button operated); C3, Clippard three-way valve (cam or push-button operated); C5, Clippard 5-way valve (cam operated); CV, check valve; EOF, emergency oil fill; FC, flow-control manifold; FR, Fisher regulator; FS, filter screen; GBC, gas-ballast flow-control hose; GB, gas-ballast inlet fitting; MR, Moore regulator; OCF, vacuum pump oil coalescing filter; ODV, vacuum pump oil drain valve; OF, oil fill plug; OR, oil reservoir; ORC, oil return capillary; OSG, oil level sight glass; OT, oil temperature sensor; PR-1, air-horn pressure-relief valve; PR-2, vacuum pump pressure relief valve; SCX01, differential pressure transducer (0–1 psid); SCX15, absolute pressure transducer (0–15 psia); SPX50, differential pressure transducer (0–8 psid); SLV, vacuum-pump suction-line shut-off ball valve; SV1, n/c bypass solenoid valve (Cv = 1.8)*; SV2, n/c rapid pump-down solenoid valve (Cv = 31.4)*; SV3, n/o flow-control solenoid valve (Cv = 1.6)*; SV4, n/o flow-control solenoid valve (Cv = 2.2)*; SV5, n/o flow-control solenoid valve (Cv = 1.6)*; TP, test port to check vacuum gauge accuracy; VP, vacuum pump; VV, vent valve.
*Cv is valve flow coefficient, gal/min, of water at 15.6°C (60°F) flowing under a valve pressure drop of 6.8947 kPa (1 psi); n/o = normally open; n/c = normally closed.

Table 13.1. Pumping speed vs. suction pressure.

Suction pressure kPa (mm Hg)	Pumping speed* m^3/h (cfm)
101.3 (760)	78.0 (45.9)
0.8 (6.0)	78.0 (45.9)
0.6 (4.5)	74.4 (43.8)
0.4 (3.0)	67.2 (39.5)
0.2 (1.5)	34.0 (24.0)

*Data for 60 Hz. For 50 Hz multiply by 5/6.

rates, the oil temperature required to prevent condensation in the pump oil is always somewhat lower because $p_{PERM} \neq 0$, and the additional air taken in at the pump inlet supplements the gas-ballast air (equation 13.2).

A small portion of the glycol emerging from the brazed-plate heat exchanger (Fig. 13.13, No. 11) is diverted to a copper dual-fan, fin-and-tube radiator (Fig. 13.13, No. 3) located inside the insulated vacuum-pump cabinet. A highly reliable reverse-acting melting-wax air-sensing thermostatic valve (Fig. 13.13, No. 16) modulates the flow of refrigerated glycol through the radiator at a rate that controls the temperature of air discharged from the upper radiator fan between 18.3 and 23.9°C (65 and 75°F) while at the same time the radiator removes the compression and motor heat generated by the vacuum pump. Warm glycol emerging from the radiator is recycled into the glycol pump's suction inlet (Fig. 13.14) to prevent the vacuum-pump heat from entering the glycol expansion tank and warming the forward tank bulkhead. Two 24/29 VAC fans are mounted on the radiator (Fig. 13.13, No. 16; Fig. 13.14). Each fan is rated for 259 l/s (550 cfm) free-air delivery, and provides 151 l/s (320 cfm) of air through the radiator at 3.8 mm (0.15 in) water-column static pressure (wc). The lower radiator fan continuously discharges air on to the vacuum pump, and a part of this air is drawn through the pump motor by its cooling fan; another part is drawn into the vacuum pump's cooling system by an integral vacuum pump cooling fan; the remainder and the air from the upper radiator fan is baffled (Fig. 13.13, No. 16) to flow over the cooling fins located along each side of the vacuum pump. Air that has been warmed in cooling the pump returns beneath the roof of the insulated vacuum-pump cabinet, to the suction side of the radiator.

The amount of oil vapour entrapped and lost in the vacuum pump exhaust is increased by gas-ballast airflow. A miniaturized high-efficiency coalescing system reclaims the emitted oil, and recycles it into the pump (Fig. 13.17). Air enters gas-ballast filter A2 and passes through gas-ballast flow-control hose GBC to vacuum pump VP. Oil mist entrapped in the humid vacuum-pump exhaust is coalesced in vacuum-pump coalescing filter OCF (Fig. 13.13, No. 24) and accumulates in the sump of OCF, from which it gravity-drains into vented oil reservoir OR. Hose GBC is sized to create a slight suction at the vacuum pump's gas-ballast inlet fitting GB, causing coalesced oil to be drawn from oil reservoir OR through oil-return hose ORC, to the pump. This oil is entrained in the gas-ballast airflow and continuously recycled into the pump along with a small amount of hot, water-saturated exhaust air from the coalescing filter. A length of copper refrigeration capillary tubing inserted in ORC limits the flow of saturated exhaust air to 5% of the total gas-ballast airflow. The gas-ballast air supply and air entering the tank through Fisher vacuum breaker FR is filtered through 3-micron hydraulic spin-on filters A2 and A1 to prevent accumulated matter from limiting the service life of the pump's internal oil demister and coalescing filter element OCF. The oil-coalescing element has a normal service life of many years.

The accumulated coalesced oil level can be viewed through the clear acrylic walls of oil reservoir OR, which serves as a sight glass. Due to the large air volume exhausted during the initial stages of tank evacuation, oil mist is rapidly lost in the vacuum pump's discharge. The suction developed in GBC may not be sufficient to return coalesced oil to the pump as rapidly as it accumulates in OCF. Since initially very little gas-ballast air is taken in because of the elevated pressure in the pumping chamber. Oil reservoir OR provides additional space to store this coalesced oil. As pump-down progresses, oil that has accumulated in OR is reclaimed because the rate at which oil mist is emitted from the vacuum pump declines, and the gas-ballast airflow and suction in GBC increase. The vacuum pump should not becontinuously operated with the pump's inlet at atmospheric pressure as this condition eventually might fill oil reservoir OR and coalescing filter OCF with liquid oil, causing them to overflow and expel oil into the pump exhaust. To prevent this occurrence, in the auto mode (13.17) the pump will stop operating if the suction pressure does not decrease by at least 14 kPa (105 mm Hg) during a 10-min period. If the oil level viewed in oil-sight glass OSG falls to a low level, oil can be added without opening the vacuum-pump cabinet, while the pump is running. Suction created at the pump inlet draws the added oil into the pump via a valved, small-diameter tube (Fig. 13.17, EOF).

13.16 Preventing Condensation in Cabinets, Conduits and Regulators

In previous hypobaric intermodal containers, when the ambient temperature was lower than the operational temperature, water sometimes condensed in the vacuum gauge, pipes, control lines, vacuum breakers and pressure regulators, and errors in regulation and measurement occurred due to the temperature coefficients of the controllers and vacuum gauge. These problems are avoided in VacuFresh^{SM} by balancing heat and cold sources.[10] Between 3.7 and 8.5 m^3/h (2.2–5 cfm) of air is continuously withdrawn from the vacuum-control cabinet to provide gas-ballast air to the vacuum pump, and air changes to the vacuum tank. An equivalent flow of ambient air is drawn through a filter screen (Fig. 13.13, No. 26) into the suction side of the radiator (Fig. 13.13, No. 3), replacing gas-ballast air and tank air changes drawn through filters A1 and A2 (Fig. 13.17) located in the vacuum-control cabinet (Fig. 13.13, No. 12). If the dew point of the incoming ambient air is higher than the temperature of the glycol passing through the radiator, moisture condenses on the radiator fins, drains into a trough located beneath the radiator and is discharged outside through a drain tube. This keeps the humidity in the vacuum-pump cabinet below 80% at 23.9°C (75°F).[11] Approximately 39 m^3/h (23 cfm) of radiator-fan discharge air is exhausted through an insulated duct (Fig. 13.13, No. 1) by a 29-VDC fan and released into the insulated vacuum-control box, from whence it flows to the insulated electric box (Fig. 13.13, No. 10), before returning to the vacuum-pump cabinet through two insulated ducts (Fig. 13.13, No. 13). The circulating air regulates the temperature of the insulated vacuum-control cabinet at approximately 23.9°C (75°F) and removes heat generated in the electric box. Condensation is prevented in tubing, piping, regulators and pressure transducers contained in these cabinets and the vacuum-pump cabinet because their temperature is maintained higher than the maximum container operating temperature that might be set for a horticultural commodity, 14.4°C (58°F). The temperature-controlled airflow also eliminates errors in regulation caused by the temperature coefficients of the electronic vacuum gauge, pressure regulators, and absolute voltage reference.[12]

During normal operation with meat or produce, the vacuum-pump exhaust is saturated with cargo water at close to 50°C. It is discharged into a separate cabinet (Fig. 13.13, No. 8) in order to warm the vacuum-pump suction hose and service valves that are located therein. This prevents cargo water from condensing and/or freezing inside the suction line and service valves during cold-weather operation. Condensation in the exposed vacuum-sensing line between the tank and vacuum-control box (Fig. 13.13, No. 5) is prevented by strapping it to the hot refrigerant liquid line and insulating both throughout their 'exposed' length.

13.17 Vacuum-control Circuitry

All circuits in the vacuum-control cabinet (Fig. 13.13, No. 12) are low voltage to prevent AC currents from inducing voltages in the control wires, and also to eliminate any high-voltage danger in servicing components. A 4PDT auto/manual toggle switch selects whether the vacuum system operates in a manual or automatic mode, allowing the operator to override and bypass the electronic logic circuits at any time and manage the equipment manually, independently of electronic control (Fig. 13.18). Automatic mode eliminates potential sources of operator error during start-up, and provides safeguards that are lacking in manual mode. Manual functions as a back-up fail-safe mode, increasing reliability by allowing the vacuum system to operate independently of electronic control. It is permissible to switch between manual and auto at any time regardless of whether the unit is OFF or operating.

13.18 Controlling the Pumping Speed

The flow-control system adjusts the pumping speed at a value that keeps the humidity in the container close to saturation with water evaporated from the commodity in response to its respiratory heat. The flow rate is controlled by a combination of four solenoid valves (normally closed SV2, and normally open SV3, SV4, SV5; Fig. 13.17), each of different capacity. The valves are operated by 8-position, 4-deck, rotary flow-control switch FCS (Fig. 13.18), which can select eight combinations of open and closed solenoid valves providing 27, 36, 44, 57, 63, 69, 79 and 100% pump capacity. In the auto mode, the desired pumping speed is set with FCS before the vacuum pump is turned ON, and solenoid valves SV2, SV3, SV4 and SV5, located in flow-control manifold FC (Fig. 13.13, No. 21), automatically open during evacuation of the tank to increase the pumping speed. The pump-down time (t, s) to reach pressure p (mbar) is calculated from the expression:

$$S_{EFF} = (V/t) \ln (1013/p) \qquad (13.4)$$

where S_{EFF} is the pumping speed, and the volume of the 6.1 m tank is $V = 21.24$ m^3 (750 ft^3). The pump-down time will be shorter than the value indicated in Table 13.2 if the starting atmospheric pressure is less than 1013 mbar. When differential-pressure switch SCX01 (Fig. 13.17) senses that the set operational pressure has been reached, the selected flow rate is automatically activated.

13.19 Detented Controller

A six-position detented controller (Fig. 13.19) coordinates the operation of the vacuum-control system during start-up and continuous operation by selecting combinations of four cam-operated electric switches (Fig. 13.18, ES-1,2,3,4) and cam-operated pneumatic valves (Fig. 13.17, C-3 CAM and C-5 CAM). The six positions of the controller sequence the following operations (Figs 13.17 and 13.18):

- Position 1 (OFF). When the power is turned ON, the refrigeration system starts.
- Position 2 (PREHEAT). The oil temperature must be higher than 12.8°C (55°F) before the vacuum pump can be started without overloading its motor. In the auto mode, vacuum-pump strip heaters VPH turn ON/OFF, warming the oil to between 12.8° and 15.6°C (55–60°F). In the manual mode, the heating strips turn ON regardless of the oil temperature, but with the strip heaters continuously ON and the pump OFF, the oil cannot overheat.
- Position 3 (TEST/ON). In the manual mode, the vacuum pump will immediately attempt to start regardless of the oil temperature. In the auto mode if the oil temperature is less than 12.8°C (55°F), the vacuum-pump heating strips turn ON/OFF, keeping the oil temperature between 12.8 and 15.6°C (55–60°F). Provided that the oil temperature has increased to 12.8°C (55°F),

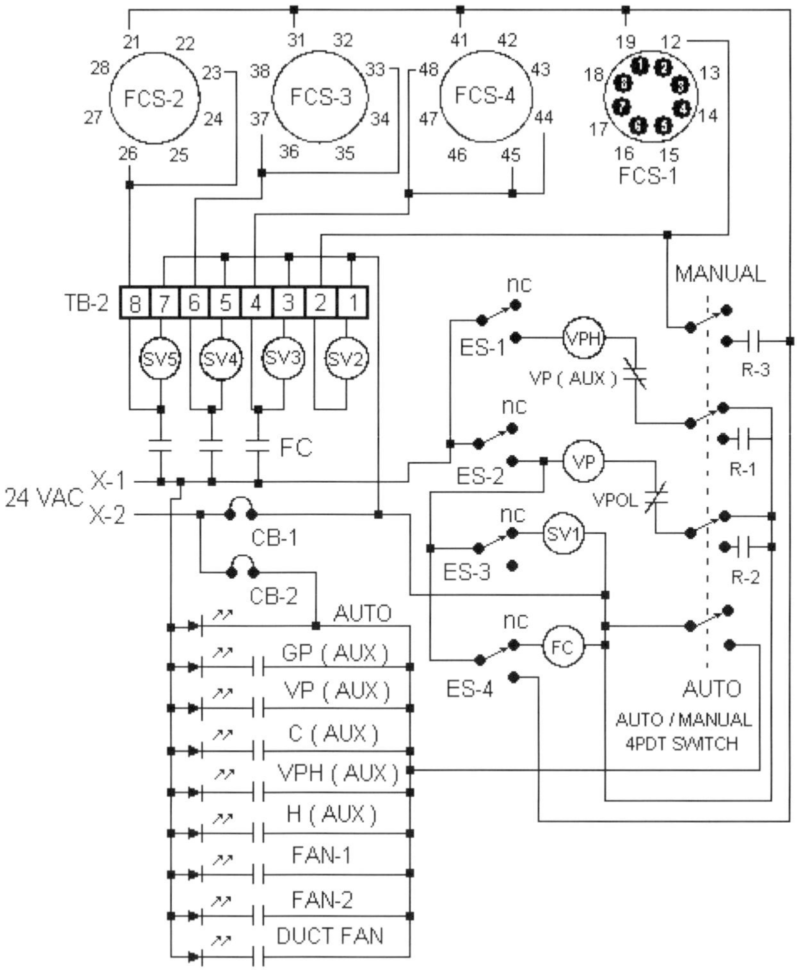

Fig. 13.18. Vacuum control circuit (20/24 VAC 50/60 Hz). C (AUX), compressor contactor auxiliary switch; CB-1,2, circuit breakers 1, 2; DUCT FAN, pressure sensor switch for duct fan (0.1 inch wc sensitivity); ES-1,2,3,4, cam-operated electric switches in detented controller; FAN-1,2, 0.1 inc wc pressure-sensor switches – vacuum-pump radiator fans 1,2; FC, flow-control contactor; FCS-1,2,3,4, 8-position flow-control rotary switch (decks 1, 2, 3, 4); GP (AUX), glycol pump contactor auxiliary switch; H (AUX), heater contactor auxiliary switch; R-1,2,3, 24-VAC, 5-amp, solid-state relays (5 VDC control); SV-1,2,3,4,5, solenoid valves 1, 2, 3, 4, 5; TB-2, terminal board 2; VP, vacuum pump-contactor coil; VP (AUX), n/o vacuum pump-contactor auxiliary switch; VPH, vacuum pump heater-contactor coil; VPH (AUX), n/o vacuum pump heater-contactor auxiliary switch; VPOL, vacuum pump contactor-overload switch.

as soon as the refrigeration system decreases the glycol temperature to 21.1°C (70°F), the vacuum pump automatically starts. Both in the auto and manual modes, cam-operated electric switches ES-1 and ES-2 cause all solenoid valves (SV2, SV3, SV4, SV5) in the flow-control manifold to close, cam-operated Clippard valve C-5 disables Moore regulator MR, and the vacuum gauge senses the vacuum-pump suction pressure through cam-operated Clippard valve C-3. If the pump is operating properly and the control system is not leaking, the suction pressure will decrease to 0.5–1.0 mbar after moisture has been evacuated from the piping.

- Position 4 (SET PRESSURE). Both in the auto and manual modes, cam-operated Clippard valve C-5 enables Moore regulator MR, and the digital vacuum gauge senses the pump suction pressure through 3-way, cam-operated valve C3. Cam-operated electric switch ES-2 opens bypass solenoid valve SV1, substituting a low-volume bypass pipe in place of the vacuum tank in the flow circuitry. This allows the operational pressure to be instantly preset since the bypass pipe is evacuated in a matter of seconds. Check valve CV isolates SV1 and FR from atmospheric air in the tank while the pressure is being adjusted. The pressure is set by varying the reference pressure that Moore absolute pressure regulator MR provides to Fisher vacuum-relief valve FR. The locking nut on the Moore regulator knob is loosened and the pressure is read on the digital vacuum gauge while turning the Moore regulator knob to decrease or increase the set pressure. The Moore regulator knob is relocked after the desired pressure reading has been displayed.[13]
- Position 5 (PRODUCE). Both in the auto and manual modes, Moore regulator MR is enabled by cam-operated Clippard 5-way valve C5, bypass solenoid valve SV1 is closed by cam-operated electric switch ES-2, and the tank pressure is sensed through cam-operated 3-way Clippard valve C3. In the auto mode, regardless of the flow setting, the logic system causes rapid pump-down solenoid valve SV2 to open while the tank is being evacuated. This provides 100% pump capacity to hasten the pressure reduction. If differential pressure switch SPX-50 does not sense that the pressure has decreased by 105 mm Hg in 10 min, the pump automatically stops.[14] Fisher vacuum regulator FR opens when the set pressure is

Table 13.2. Pump-down time from atmospheric pressure (101.3 kPa) for a 6.1 m (20 ft) container @ 60 Hz. Multiply by 5/6 for 50 Hz.

Pressure (kPa)	Time (min)	Pressure (kPa)	Time (min)
101.3	0	8.0	41.5
80.0	3.9	6.0	46.2
60.0	8.6	4.0	52.8
40.0	15.2	2.0	64.1
20.0	26.5	1.0	75.5
10.0	37.8	0.6	83.8

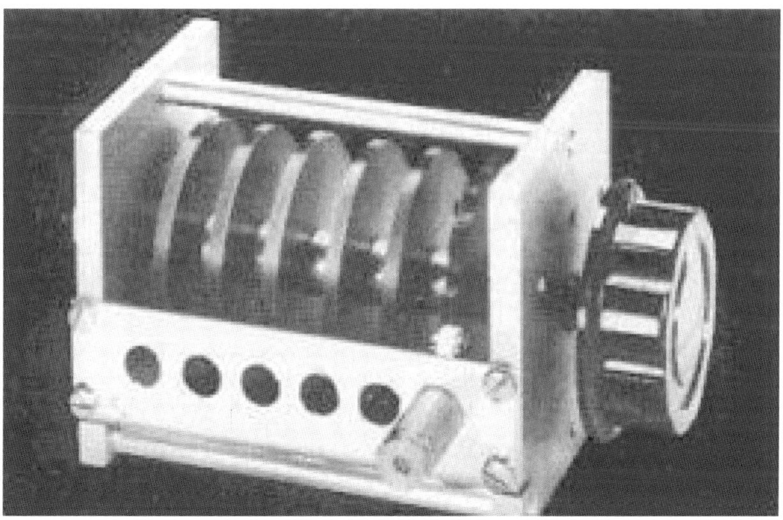

Fig. 13.19. Detented 6-position controller with one cam-operated pneumatic valve installed. Complete assembly has one 3-way valve, one 5-way valve and four cam-operated electric switches.

reached, pressurizing air horn AH. This causes differential pressure switch SCX-01 to sense that the air horn pressure exceeds the tank pressure, and respond by instructing the control system to activate the combination of open solenoid valves (SV2, SV3, SV4, SV5) preselected by the operator using eight-position rotary flow-control switch FCS. SV2 is automatically closed by a fail-safe timer after 150 minutes.

- Position 6 (MEAT). Both in the auto and manual modes, Moore regulator MR is disabled by Clippard 5-way cam-operated valve C-5, closing Fisher vacuum breaker FR and preventing the air supply to the Moore regulator from leaking into the vacuum-pump flow-control manifold. Bypass solenoid valve SV1 is closed by cam-operated electric switch ES-2, and the tank pressure is sensed through cam-operated Clippard 3-way-valve C3. The pump-down sequence in both manual and auto is the same as that which occurs in position 5, except that differential pressure switch SCX01 does not operate.

13.20 Electronic Vacuum Gauge

Absolute pressure gauge SCX15 is temperature-compensated and kept at a controlled temperature to ensure ± 0.5 mbar (± 0.38 mm Hg) accuracy (SenSym, 1998). The pressure is displayed on a miniature 2-VDC voltmeter, which has an automatic auto-zero, adjustable decimal point and bright LED digital display. The electronic gauge's output initially is amplified to 10.13 V per 1013 mbar, but when a comparator senses that the pressure is higher than 200 mbar, the output is reduced to 1.013 V per 1013 mbar by a resistance bridge, and the resulting voltage is displayed in whole units; i.e. 500 mbar = 500 mV. To improve accuracy below 200 mbar, the comparator causes the 10:1 resistance bridge to be bypassed and a decimal point to be inserted, allowing the display to read the pressure ± 0.1 mbar. The long-term stability of the *SCX15* gauge typically is ± 1.1 mbar per year or better. The gauge can be easily recalibrated, but this seldom is necessary. To record the pressure in the data logger, the *SCX15* output is reduced to 5 V per 1013 mbar, and compared to a precision 5 V reference located in the logger.

13.21 Pressure Regulation

Fisher vacuum-breaker FR senses the tank pressure above its diaphragm, and an absolute reference pressure provided by Moore regulator MR below its diaphragm (Fig. 13.17). FR regulates the controlled flow of in-leaking air to maintain the tank pressure 120 mbar lower than the reference pressure. The 120-mbar bias pressure, which is permanently applied above the FR diaphragm by an adjustable control spring, allows the regulating system to control the pressure at absolute values lower than 1.33 kPa (10 mm Hg) even though the Moore regulator cannot control the reference pressure below 2.67 kPa (20 mm Hg). Because the MR-sensing bellows contains a sealed vacuum, MR provides an absolute regulated pressure to FR, and therefore the pressure maintained within the tank container is insensitive to changes in altitude and barometric pressure. Although MR has a very low temperature coefficient between –40 and +180°C (–40 to +356°F), nevertheless it is mounted inside the temperature-regulated vacuum-control cabinet to limit the influence of bellows leakage and temperature on pressure stability.

MR is exhausted by the vacuum pump through C5 (cam) and continuously supplied with atmospheric air through filters A1 and A2. MR pilot-air consumption (SCFM) is given by:

$$\text{SCFM} = 0.003\ (\text{PSIA Setting})^{1/2} \quad (13.5)$$

This amount of air bleed is so small that it does not interfere with pressure regulation when horticultural commodities are transported, but it would adversely influence the storage life of meat by slightly elevating the [O_2] level in the vacuum tank. To prevent this occurrence, 5-way cam-operated Clippard valve C-5 (cam) blocks the suction connection to the Moore regulator (MR), disabling it when the detented controller is turned to position 6 (meat). This causes Fisher vacuum regulator FR to close and prevent air from entering the tank during meat shipments.[15] When plant produce is transported with the detented controller in position 5 (produce), incoming controlled air changes pass sequentially through air filter A1, Fisher vacuum breaker FR, check valve CV, and service valve V3 before pressurizing air horn AH and entering the tank through the orifices of the AH jets.

13.22 Momentary Test Buttons

Push-button-operated momentary three-way pneumatic slide valves (Fig. 13.17, C3, *push button test*) are provided to measure the air-horn pressure, vacuum-pump suction pressure and Moore regulator absolute reference pressure. When a button is held in, the test pressure registers on the vacuum meter's digital display, and as soon as the button is released the spring-loaded valve reverts to its original configuration and the gauge again indicates the vacuum tank's pressure.

Suction pressure test button

After the tank has stabilized at the set pressure with the detented controller in position No. 5 (produce), vacuum gauge readings taken before and after pushing the momentary suction-pressure test button indicate the tank-to-suction-pressure ratio. This is a measure of the vacuum pump flow capacity and should agree with the value selected by flow-control rotary switch FCS.

Air horn test button

When the air horn momentary test button is pushed with the detented controller in position 5 (produce), the digital vacuum gauge indicates the absolute pressure powering the air horn jets. This should agree with the value listed in a maintenance table for the air horn pressure corresponding to the selected flow at the operational pressure. The table identifies potential causes of deviation from the expected value.

Regulated pressure test button

The procedure used initially to set the pressure prior to pump-down cannot be used to reset the pressure after the container reaches its operational pressure because check valve CV (Fig. 13.17) downstream of Fisher regulator FR cannot reliably close under these conditions. To readjust the pressure after it has stabilized with the detented controller in position No. 5, the regulated pressure test button is held in and the adjustment knob on Moore regulator MR is turned to change the regulated pressure indicated on the vacuum gauge. The tank set-pressure will be elevated or decreased by the amount that the reference pressure is changed. The tank pressure will gradually adjust to the new setting during 15–30 min after the regulated pressure test button is released.

13.23 Data Logger

A 32-kB data log chip with clock calendar reference is included in the thermostat. The logger uses a precision 5-VDC reference for both ratiometric temperature and absolute-pressure measurements. Data are recorded at 1-h intervals for up to 60 days, after which the logger wraps around. The logger records tank pressure (mbar), supply glycol temperature (°C), vacuum-pump oil temperature (°C), vacuum-pump cabinet temperature (°C), tank wet-bulb temperature (°C), tank dry-bulb temperature (°C), ambient temperature (°C), the temperature of commodity inside the tank (°C), and the status (0/1 = off/on) of the vacuum pump, heater, unloader, compressor, condenser fan, duct fan and phase. The programme used to interface with the data logger is run on a personal computer connected through an RS-485 to RS-232 adapter cable. One end of the cable plugs into the computer parallel port; the other end into a 3-pin connector located on the electric box.

13.24 TransVac

TransVac was designed to increase the usable volume and simplify the loading of VacuFresh^{SM} containers without significantly increasing their weight and cost. A nearly square barrel design (Fig. 13.20, *lower left*), patented by Johann Wannenburg of BKS Engineering, Pretoria, S.A., is substituted for the circular barrel used in the VacuFresh^{SM} design. A circular-to-square transition is made at the door (Fig. 13.20, *upper middle*) and equipment ends (Fig. 13.20, *upper left*). The remainder of the design, including the end frames, mounting clips, door, expansion tank and equipment, remain essentially the same as in VacuFresh^{SM}. Glycol cooling is provided by a series system utilizing the stiffening rings (Fig. 13.20, *lower right*).

The square design has 19% greater above-floor volume, and 20% more usable volume when the non-usable space beneath the shelves in VacuFresh^{SM} is considered (Fig. 13.20, *upper right*). The square barrel weighs 234 kg (515 lb), which is a 22% increase but, the square design eliminates the requirement for shelves weighing 114 kg (250 lb) and therefore the net increase in weight is 120 kg (264 lb). The rest of the section masses remain more or less the same,

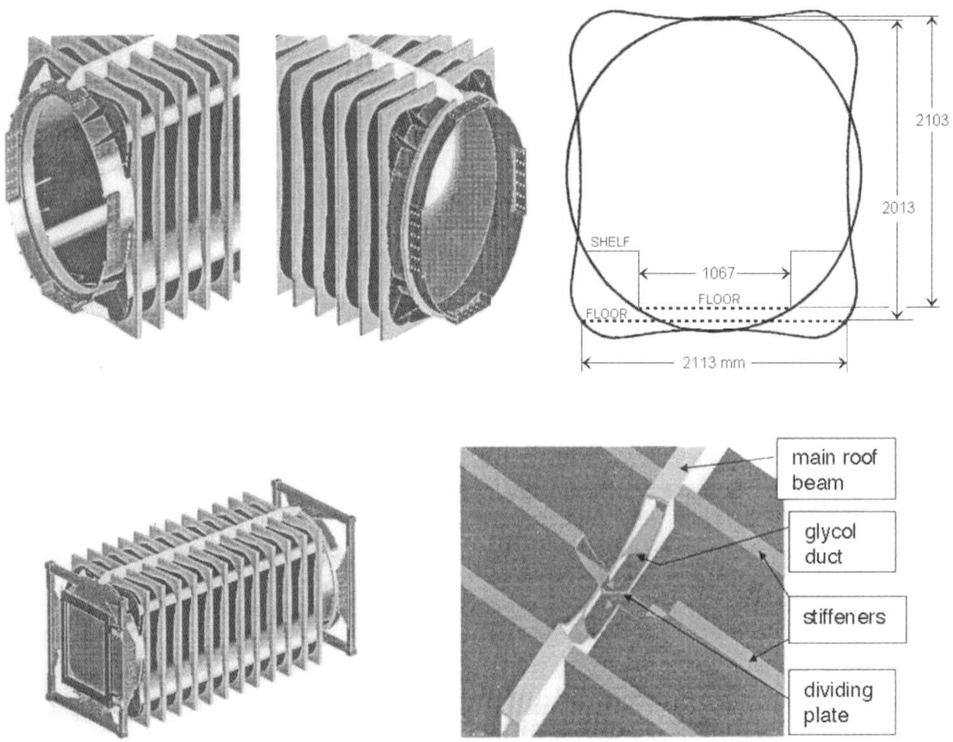

Fig. 13.20. (*lower left*) Vacuum tank with square barrel and circular end frames. (*lower right*) Cut-out detail of stiffeners and main roof beam. (*upper left*) Circular-to-square interface at equipment end. (*upper middle*) Circular-to-square interface at door end. (*upper right*) Diagrammatic layout difference between square and circular barrels (courtesy of BKS Engineering).

and therefore the total percentage increase in mass is only about 4%.

13.25 Example

1. For a 20° jet angle, when the flow is critical and $x/D_O < 100$, if both the jet fluid and entrained fluid are air, the entrainment of surrounding fluid at a distance x from the nozzle is given by:

$$q/q_O = 0.32x/D_O \text{ [for } 7 < x/D_O < 100] \quad (13.6)$$

where q is the volume flow rate at x, q_O the volume flow rate at $x = 0$ and D_O the nozzle diameter. The limiting induction ratio q/q_O (at $x/D_O = 100$) is approximately 32. Performance data at atmospheric pressure suggest that the ratio for the VacuFresh^SM air horn should approach 40 moving low-pressure saturated air, but this has not been confirmed because no method has yet been devised to measure this ratio under hypobaric conditions. The VacuFresh^SM flow-control system permits selection of pumping speeds ranging from a minimum of 17.6 m^3/h at 50 Hz (27%), to a maximum of 78 m^3/h at 60 Hz (100%). The flow through the jets is given by:

$$q_O = q_{VP}[1 - (p - p_V)]/p \quad (13.7)$$

where p is the storage pressure (atm), and p_V the partial pressure of water vapour (atm) in the storage air. At the optimal flow for all products, the number and diameter of the jets in the VacuFresh^SM air horn have been selected to create a supersonic exhaust and a pressure of less than 584 mbar upstream of the jets. The upstream pressure tends to be higher than 584 mbar when full flow is selected during the cool down period at storage pressures higher than 66.7–80.0 mbar (50–60 mm Hg), and also when limes are stored at a pressure of 20 kPa (150 mm Hg). This has the potential to create a control problem when the container is operated at a high altitude, where the barometric pressure may be as low as 75 kPa (563 mm Hg). To avoid this occurrence, when the pressure upstream of the jets increases to > 379 mbar (284.3 mm Hg = 5.5 psig) above the tank pressure, the excess is bypassed around the air horn and discharged into the air horn exhaust by means of 'pop-off' pressure relief valve PR-1 installed in the air horn collar (Fig. 13.17). PR-1 is factory preset and normally does not need to be reset.

Notes

1. In the range $2 < L/\delta < 10$, $\Pr < 10$ and $Ra_\delta < 10^{10}$:

$$Nu_\delta = 0.22(L/\delta)^{-1/4} [\Pr/(0.2 + \Pr)] Ra_\delta)^{0.28} \quad (13.1)$$

where $Ra_\delta = Gr_\delta \Pr$; $Gr_\delta = g\beta(T_1 - T_2)\delta^3/v^2$; $Nu_\delta = h\delta/k$; k is the thermal conductivity of the mixture and h the convective heat-transfer coefficient in a fluid layer of thickness δ (6.12; Kreith and Bohn, 1997).

2. Initially, the wattage was controlled by a dew point sensor installed at the forward end of the container and set to regulate the humidity close to saturation. This approach failed because invariably the humidity was brought close to saturation by moisture evaporating from the commodity. This caused the dew point sensor to turn off the humidification heater.

3. The VacuFresh^SM tank was designed by Frank Colangelo, a former Grumman aerospace structural engineer, assisted by Robert Alloca, a former Grumman aerospace mechanical engineer, and the author. The equipment package was designed by the author.

4. The longitudinal framing supports in the 6.1 m (20 ft) prototype (Fig. 13.11, *upper right, upper left, middle left, lower right*) were added to prevent damage, and are not structurally required.

5. 'Although some in the industry feel that the lack of longitudinal frames makes the barrel more vulnerable to damage, more than 20 years' experience in operating, as well as leasing, both beam and frame tanks, has shown this is not so', according to Stolt-Nielsen Leasing Managing Director Preben Hansen. 'The major regulatory bodies such as the DTI in the UK and the DOT in the United States agree that there is no disparity between the certification and approval of beam tanks', Hansen adds. 'While the cost of maintaining beam tanks may be slightly higher due to the insulation being more exposed, the greater payload and resultant freight cost savings far outstrip any extra maintenance and repair costs' (Cardin, 1994).

6. Carbon steel thermal conductivity = 0.108–0.124 cal/cm^2·cm·s·°C; T-1 steel = 0.058 cal/cm^2·cm·s·°C. Stainless steel will be substituted for the T-1 tool steel in subsequent containers because its thermal conductivity is only 0.036 cal/cm^2·cm·s·°C and it is available with sufficient yield and ultimate strength (American Society of Metals, 1983).
7. The humidity inside the storage area will be higher than the computed value because the incoming air is never bone-dry.
8. For equivalent volumetric flow rates, the pressure drop increases as a function of D_h.[4]
9. Most oil-seal rotary-vein vacuum pumps are lubricated using the pressure differential between the atmosphere and pump suction inlet to circulate oil. At a high inlet pressure they cannot be operated continuously because the pressure differential is insufficient to provide proper lubrication.
10. Temperature-sensitive components in Grumman's Dormavac container had to be wrapped with self-limiting heating tapes.
11. The radiator fan velocity is not high enough to entrain water droplets from the radiator coil.
12. In addition, all of these components are temperature-compensated.
13. Three large 2-way ball valves had to be readjusted manually and a small 3-way valve shifted initially to set the pressure in a Grumman/Dormavac container. After the pressure was set with the Moore regulator, the same valves had to be manually returned to their original position before the container could be pumped down. In VacuFresh[SM] these operations are automatically controlled by the detented sequencer.
14. This alarm condition will arise if tank pump-down is initiated with the door ajar or the door seal leaking, if the vent valve is open, or if the tank drain plug has not been installed. If any of these events has occurred, the vacuum pump automatically shuts OFF after the 10-min timing interval.
15. When meat is shipped, the air-horn service valve V3 (Fig. 13.17) can be closed to eliminate any possibility of air leaking in through the Fisher vacuum breaker.

14

Conclusions

During the author's lifetime, numerous attempts to develop commercial CA intermodal containers have been unsuccessful, but each failure has led to a renewed effort and technological advances. There has never been any doubt that CA 'works', and, undeterred by commercial setbacks, CA research has continued unabated in the UK and worldwide for more than 95 years (Thompson, 1998). Perhaps it is because CA has been 'in vogue' for so long that the attitude of postharvest physiologists toward hypobaric storage has been less generous. Independent LP research ceased and sceptics expressed concern about the merits of the 'new' technology as soon as Grumman terminated its Dormavac programme in April 1982, ending the first serious attempt to develop an LP intermodal container. It was conjectured that there must be something intrinsically wrong with the hypobaric process if Grumman could not make a commercial success of it after investing so much time, effort, money and highly skilled engineering talent. Even though LP had only been under development for a relatively short time, there was no expectation that the technology could be improved, or that new applications and better storage conditions might be discovered. A few researchers questioned whether LP worked as well as CA; others theorized that it was too expensive or complicated ever to be a commercial success; some claimed that LP failed satisfactorily to prevent fruit senescence (Theologis *et al.*, 1993a) and that it desiccates commodities (Hughes *et al.*, 1981); an adversarial lawyer insisted that Grumman abandoned the project after realizing that it would cost 'a fortune' to carry out the R&D that was still required; but all of these opinions had the common theme that they were expressed by persons lacking insight into the cause of Grumman's Dormavac failure, or of the unique advantages LP affords. The same pessimism remains fashionable 20 years later, and this academic bias tends to become a self-fulfilling prophecy as it is expressed by individuals with whom banks, technology developers, growers, shippers and venture capitalists consult for advice and recommendations.

After a few overseas demonstration shipments of limes, mangoes and papaya, the Grumman International sales force located another market, Australian lamb, to be sold in Iran. The first lamb shipment arrived in perfect condition after a long voyage, and the customer, Parvese Ltd, was so impressed they ordered 78 containers on the spot, but by the time Grumman was to make the first delivery of five production units to Iran, loaded with Australian lamb, things had changed. It was monsoon season, the temperature surpassed 49°C (120°F), and the rain and pounding, relentless seas did much damage, causing the five shipments to fail. Meanwhile the Shah of Iran had just been deposed, and when the containers arrived in Iran, Grumman did not allow its technicians to get off the boat to try to identify what had happened. Eventually, the buyer of the containers had difficulty re-entering Iran; Grumman had to repurchase 78 containers, and spent the next 18 months working out technical difficulties.

A view of these events from a different perspective suggests there was another lesson to be learned from the Iranian debacle. Much later, apprehensive when a licensee decided to send the first VacuFresh^{SM} container on its maiden demonstration voyage carrying horse meat from the USA to Belgium, the author sent a facsimile to the consulting stress analyst who had been instrumental in the design of the VacuFresh^{SM} tank. He had previously worked for Grumman in their Dormavac programme:

> We are considering a commercial full-scale shipment of horse meat carcasses from Houston to Belgium on about October 3. Do you have any recommendations about precautions we should observe or steps which should be taken to eliminate or decrease the possibility of structural or mechanical failure during the first commercial trial shipment?

The consulting engineer replied as follows:

> Back in the spring of 1996 [. . .] I visited the Aberdeen proving grounds to determine if their facilities were adequate to test our [VacuFresh^{SM}] container and to get a budgetary cost estimate. As a result of that meeting and subsequent correspondences, I had recommended that we conduct a series of tests at their Aberdeen proving grounds [. . .]. Let me further elaborate that my main concern is the survivability of the equipment package, keeping in mind that no analysis to determine the fragility level of the equipment components and associated piping was ever made. We are about to repeat the critical mistake that was made on the Grumman Dormavac program when containers were shipped out prematurely by Grumman Management/Marketing against the strong objections made by the Engineering Department. The rest is history: we had major failures in our equipment components and major leaks in our glycol and freon plumbing, causing a loss of approximately 80% of the product (lamb) being transported simultaneously from Australia to Iran. This premature introduction of the container into service and the subsequent failures led to cancellation of 100 units that were sold to Parvese Ltd. This further led to a total demise of the Grumman Dormavac program which was never able to overcome the disapprobation it earned in the market place by putting a product in service before it was validated successfully. [. . .] I cannot emphasize enough that I have serious concerns that this course of action will jeopardize the success of the VacuFresh^{SM} program.

The author expressed his strongest objections to the proposed shipment, recommended that structural and mechanical testing be carried out first, and refused to participate in the trial,[1] but nevertheless the shipment was sent. The horsemeat froze in transit due to an improper pre-trip adjustment of the refrigeration modulation system by an inexperienced mechanic, and heavy seas flooded the electric box, burning out contactors. After the container underwent water spray and high- and low-temperature testing in South Africa, the entire refrigeration system had to be redesigned, new water-tight electric and vacuum-control boxes were installed, the refrigerant and glycol piping were reconfigured because, as the stress analyst had warned, the pipes developed leaks due to expansion, contraction and vibration, just as the pipes did in Grumman's Dormavac container. Numerous other major changes had to be made before VacuFresh^{SM} was ready to be used for its first successful commodity test 18 months later.

The Grumman/Dormavac engineering specification was predisposed to failure for reasons that could not have been anticipated at the time it was drafted. The vacuum system was oversized as a precautionary measure because it was uncertain how much capacity was needed. This created a requirement for water reclamation and an inefficient water-seal vacuum pump, which increased the overall power requirement and equipment complexity. Nearly half of the available refrigeration was consumed cooling water for the vacuum system, leaving too little refrigeration capacity for a 49°C (120°F) day. An oversized, heavy, motor generator had to be built into each container, decreasing the space available for cargo and increasing the container weight and cost. There was no specification for cost or weight, and some have suggested that the philosophy 'whatever it costs is what they will have to pay' might have been a by-product born from fabricating military hardware under government contract.

Potential problems or disadvantages with LP were summarized in an article entitled 'LPS – Great expectations' (Lougheed *et al.*, 1977), which has remained the guideline by which LP is still judged more than 25 years later. All of the suggested difficulties, restated by Abeles *et al.* in 1992, have proved incorrect or been eliminated by advances in technology and a better understanding of the manner in which LP functions. The problems, according to Lougheed *et al.*, are indicated in the bulleted items that follow:

- Cost – initial cost and royalties; potential benefit/cost.

Whether or not the cost is greater for an LP container than it is for CA depends on what is meant by 'cost'. A grower or shipper wants to know his cost per day to rent the container. The 'useful life' of a standard intermodal container's insulated structure is seldom more than 7 years and it may be less for CA containers because racking destroys the leak integrity of the door seal required for CA usage. The longevity of intermodal refrigeration systems, CA control modules and LP equipment is approximately 5 years. For tax purposes, it is mandated that all intermodal containers must be depreciated over a 5–6 year period, and CA containers normally are financed over that term. However, most of the cost for VacuFresh^{SM} is in the structure, which is designed to last for more than 20 years.[2] Indeed, the industry's standard useful life expectancy for tank containers is 20 years with a refurbishment after 10 years at 20% of the container's initial cost. For this reason, tank containers sometimes are financed over a 15-year term. The same institutions that provide this service have offered long-term funding for VacuFresh^{SM}, making the daily cost of renting an LP container competitive with CA. Moreover, the manufacturing cost of VacuFresh^{SM} containers is expected to decrease after forms, special tooling and extrusion dies are amortized, and also due to improvements in design and manufacturing techniques, as well as cost efficiencies associated with an increased number of units fabricated in a production run.

- Insurance – implosions.

VacuFresh^{SM} qualifies under the 'non-hazardous' IMDG standard tank container classification because it does not carry cargo with a flashpoint of more than 61°C, there are no hazards from toxicity of corrosivity, it does not fall within the criteria listed by international authorities as requiring mandatory design safeguards and it has a pressure rating of < 20 psi, which is the lowest pressure category (Sea Containers Group, 1983). Lloyds of London has approved the VacuFresh^{SM} design, and the tank has been ISO tested and certified. There never was an insurance or implosion problem. The vacuum force only accounts for approximately 10% of the design load on a VacuFresh^{SM} container. The major design stresses are caused by shock (e.g. the impact during rail 'humping'), vibration, weight of the cargo and of containers stacked above, impact on the door of the cargo if the load should break free and mandated safety factors. These same stresses arise in all intermodal containers.

- Desiccation – if the humidification system fails or the chamber/building leaks and allows non-humidified air to enter the system.

VacuFresh^{SM} does not have a humidification system, its leak rate is extraordinarily low and desiccation has only been a problem in laboratory experiments using poorly designed or defective equipment (chapter 9). VacuFresh^{SM} containers maintain a saturated humidity by means of a jacketed refrigeration system, regulated pumping speed and a pneumatic air mover

(Table 14.1). The minimum weight loss during any type of storage depends mainly on the amount of respiratory heat available to evaporate water, but it can be increased by additional heat provided to the commodity from the environment (6.1). The inhibition of respiratory heat production may reach or exceed 90% in VacuFresh^{SM} (Fig. 4.2), whereas in CA it seldom exceeds 50% and usually is much less, creating the potential for more than a fivefold greater weight loss in CA, and a tenfold greater loss in NA, compared to LP. Heat transmitted through the insulation of CA and NA containers, and introduced in air changes or produced by evaporator fans, is transferred to the commodity by convection and radiation, causing an extra water loss. In addition, CA and NA containers lower the humidity by condensing water on the refrigeration evaporator coil, and CA atmosphere generators remove additional water (Table 14.1). The LP jacketed refrigeration system captures transmitted heat before it enters the tank's interior; low-pressure air changes introduce no latent heat and essentially no sensible heat; the LP pneumatic air mover does not produce motor heat; and LP containers do not have an evaporator coil to lower the humidity. A small amount of respiratory heat is the only latent energy available in LP to evaporate commodity water. If required, Mylar radiation shielding can be used as a liner within boxes to prevent radiant heat uptake by commodities stored in LP. Since convection is ineffective at a low pressure, the Mylar deprives the commodity of the only remaining heat-transfer mode capable of providing environmental heat to support commodity water evaporation (9.9).

- Power failures or breakdown of refrigeration system.

NA and CA containers and VacuFresh^{SM} use the same motor-generator set. The LP refrigeration system requires no defrost cycle or evaporator fans since it utilizes a secondary coolant, and the LP 'brazed-plate heat exchanger' is much simpler and more reliable than the forced-air evaporator coil and fans used in CA and NA containers. Temperature uniformity is more precise in VacuFresh^{SM}, and is not interrupted by the multiple daily defrost cycles required in CA and NA containers (Table 14.1).

- Heat transfer. Can the current system be scaled up to a 10,000-bushel room?

The LP heat exchange system is in use in a Yoder Brothers hypobaric warehouse, and has proved adequate.

- Energy costs. How do the energy costs compare with those of other methods of storage?

LP and NA containers use the same amount of energy, and are more energy-efficient than CA containers. A slight commodity water loss in LP transfers essentially all respiratory heat by evaporative cooling, and since the moisture is pumped out of the container, respiratory heat does not have to be removed by the refrigerator. LP eliminates the power requirement for evaporator fans.

- Response of produce to rapid pressure changes!

Pump-down and repressurization do not injure commodities when they are cooled in commercial vacuum coolers, even though the rates of evacuation and venting are considerably more rapid than they are in VacuFresh^{SM}. Amongst 100 commodities tested, no commodity was found to be harmed by evacuation in an LP container. Only grapefruit, fully coloured papaya fruits, and perhaps cherries are disturbed by rapid pressurization. In LP, the vacuum is released over a 20-min period to avoid damage even to these commodities.

- New disorders.

No new disorders have been observed to arise as a result of low-pressure storage. LP prevents internal breakdown and scald in apples, reduces the incidence of bitter pit and core flush, and prevents or controls low-temperature-induced flesh browning (3.25; 7.11).

- Pathogen control.

Table 14.1. Comparison of CA and LP intermodal container systems (Thompson, 1998; S.P. Burg, 2003, unpublished data).

	CA	LP
Regulation	Uses O_2, CO_2 and humidity sensors that are prone to failure and need constant recalibration.	Highly reliable, mechanical, absolute pressure regulator requires no calibration. No sensors.
Leakage	Leakage around the reefer insert, through the container 'skin' and at the door affects the composition of the atmosphere.	Vacuum-tight and leak-proof.
Altitude compensation	Rapid leakage occurs in transit when the container is subjected to altitude or barometric pressure changes.	Immune to changes in altitude and barometric pressure.
Bottled gas	Required for CO_2 (limited capacity).	None.
Water	Required for humidification.	Not required.
Ethylene scrubber	Mixture of activated alumina and Hisea material remains pressurized for several hours and then depressurizes via the O_2 venting lines. The O_2 flow is then routed through the bed for 20 min in order to scrub it, and the process is repeated.	None.
CO_2 + N_2 beds	Zeolite.	None.
Drier beds	Activated alumina.	None.
Humidification system	Water is removed by the N_2 generator then condenses and freezes on the refrigeration evaporator, decreasing the humidity. Water released during the defrost cycle and from the water trap is injected into the container. Once the water spray has equilibrated, the humidity is measured and the injector operated again if required.	Rotary flow-control switch is set to regulate the pumping speed.
Refrigeration	Forced-air cooling. End-to-end accuracy in the container $\pm$ 2°C. Usually two 30-min daily defrost cycles each day. 2-HP evaporator fan.	Jacketed system with secondary coolant to ensure high humidity. End-to-end accuracy in the container $\pm$ 0.2°C. No defrost cycle. 1.5-hp glycol pump (no evaporator fan).
Accuracy of control	90 $\pm$ 5% relative humidity; +1 to –0.5% CO_2 accuracy. O_2 can be reduced to 4% and sometimes to 2% or even 1.5% depending on air leakage. Accuracy of control is +1 to –0.5% between 4 and 21% O_2.	O_2 can be controlled between 0.000068% and 20 $\pm$ 0.008%. No CO_2 is present.
Power	2.5 kW air compressor + 8 kW for refrigeration.	2.5 kW vacuum pump + 7.5 kW for refrigeration.
Air change	Sealed system.	Up to 3.6 air changes per hour introduce essentially no heat because of their low density and water content after expansion.
Air circulation	2-hp electric evaporator fan provides 2000 cfm. Requires 0.5 ton of refrigeration.	2000 cfm using a pneumatic horn which has no moving parts and is powered by the pressure differential created by the vacuum pump. Requires no refrigeration.
Time to reach operational condition	1.5 days unless the container initially is flushed with bottled N_2. After door is opened the container must be ventilated before entering.	90–220 min depending on the size of the container. After releasing the vacuum the container can be entered immediately.

Lougheed *et al.* (1977) conjectured that 'the high relative humidity conditions possible in LP appear ideal for growth of fungi and resultant decay in storage'. To the contrary, humidities approaching saturation discourage mould growth on plant commodities provided that water does not condense on their surface (7.8). During LP storage, water condensation is avoided by ± 0.2°C temperature control (Table 14.1), and by using water-retentive wraps to keep the commodity's temperature slightly elevated above the storage air's temperature. The growth and sporulation of almost all moulds and bacteria of horticultural interest are prevented or severely retarded by the very low [O_2] and [CO_2] concentrations present around and within most commodities when they are kept in LP at their optimal storage pressure (7.1–7.4).

- Poor flavour of fruit and unsatisfactory ripening after LP.

LP cannot and does not remove fruit volatiles (3.25, 3.26). Tomatoes do not produce their normal volatiles if they ripen during LP storage, and eventually apples lose their ability to produce volatiles when they are kept at a low LP storage pressure, but these effects do not arise because of 'out-gassing', as first proposed. Instead, they are caused by low [O_2] and occur to the same extent when apples and tomatoes are stored in CA. Hypobaric storage has no deleterious influence on flavour or volatile production by bananas, tomatoes and other fruits when they ripen after hypobaric storage.

- Expensive to add 'functional' levels of CO_2.

The effects of abnormally low [CO_2] could not be critically tested until LP became available because prior to that time there was no practical way to continuously lower the atmospheric [CO_2] content and remove the gas from within a commodity's intercellular spaces. It is now apparent that LP's ability to drastically decrease both the ambient and intercellular [CO_2] levels is an important advantage, providing benefits that cannot be duplicated by elevating [CO_2]. Some of the most notable documented LP effects resulting from unusually low [CO_2] levels are stomatal opening in the dark (4.15), depressed growth of aerobic bacteria and fungi (7.2), retention of ascorbic acid (4.14) and inactivation of ethylene-forming enzyme (4.10).

Important advances in science occasionally are 'lost', only to be rediscovered many years later. Research on the plant hormone, ethylene, carried out prior to 1935, demonstrated that this gas is the fruit-ripening hormone; that it is produced by most plant tissues; that its formation is stimulated by auxin; and that it causes a wide variety of effects in flowers, seeds and vegetative tissues (Burg, 1962a; Abeles *et al.*, 1992). Perhaps because auxin had been in vogue for so long, by 1955 there was little interest in ethylene and the early literature drifted into obscurity. The gas was considered an oddity, a by-product of plants rather than a ripening hormone, and ethylene research practically ceased. It was only when the early findings were 'rediscovered' using the newly available technology of gas chromatography (Burg, 1962b) that plant physiologists again became aware of ethylene's significance and ethylene research blossomed into an important field of plant science (McGlasson, 1970).

Today postharvest physiologists seem unaware or unconcerned with literature published 20–40 years ago reporting that LP kills insects (Calderon *et al.*, 1966; Aharoni *et al.*, 1986), directly prevents mould and bacterial growth (Apelbaum and Barkai-Golan, 1977; Alvarez, 1979), utilizes the heat transfer properties of a vacuum to limit water loss (Burg and Kosson, 1982, 1983), opens stomates (Kirk and Andersen, 1986; Veierskov and Kirk, 1986), and in the presence of remarkably low [O_2] and the absence of [CO_2] provides a longer storage life than CA can achieve, preserving commodities that CA does not benefit (chapters 10 and 11). Most of the early LP studies were performed at pressures much higher than optimal, and there is an expectation that even better results are likely at or below the 1.33 to 2.67 kPa (10–20 mm Hg) pressure range.

Postharvest Physiology and Hypobaric Storage of Fresh Produce was written to reacquaint postharvest physiologists with the LP literature of the past, explain how LP functions and familiarize the reader with a new generation of LP intermodal containers capable of distributing highly perishable commodities.

Notes

1. The licensee responded by terminating the author's consulting activities for VacuFresh^{SM}, and entered into a protracted legal disagreement that ultimately led to abandonment of the LP patents.
2. VacuFresh^{SM} and TransVac containers will be outfitted with a strain-gauge system designed by Johann Vandenberg of BKS engineering. This system allows the container longevity to be predicted during a short testing period in order to rapidly confirm the engineering calculations and thus lower the cost of insurance.

15

Appendix – Influence of LP on Physical, Biological and Chemical Parameters

15.1 Physical and Chemical Properties

In vacuum technology, it is customary to divide the pressure region into somewhat arbitrary subdivisions within which the kinetic theory of gases predicts behavioural differences such as the type of gas flow that occurs. LP operates in a region of 'intermediate' pressure between 1 and approximately 200 mbar where flow is viscous, and in passing through a tube at 20°C it is characterized by the expressions $pd > 6 \times 10^{-1}$ mbar·cm and $\Lambda < d/100$, where p is the pressure (mbar), d is the diameter of the tube (cm) and Λ is the mean free path (cm) of air.

The following properties of gases are pressure-dependent: specific volume (v, m^3/kg) and density (ρ, kg/m^3); diffusion coefficient (D, cm^2/s); convective heat-transfer coefficient and convective coefficient for water-vapour condensation (h_m, kcal/$m^2 \cdot s \cdot K$); mean free path (Λ) and Reynolds number (Re). The boiling point of liquids is pressure-dependent, and their vapour pressure (e, atm) and chemical potential (μ, energy per mol) have a slight pressure dependency. Volumetric heat capacity (c_P, $J/m^3 \cdot K$), surface tension of water (σ_W, N/m), Prandtl number (Pr), Henry's Law coefficient (K_H, M/atm = [($mol_{aq}/dm^3{}_{aq}$)] per atm), thermal conductivity of gases and liquids (k, W/m·K), gas and liquid dynamic viscosity (μ, Pa·s), liquid density (ρ, kg/m^3), liquid diffusion coefficients (D, m^2/s), radiant heat transfer; and latent heats of vaporization (H_v, kcal/kg) and fusion (H_f, kcal/kg) are not significantly influenced by pressure in the 0.53 kPa (4 mm Hg) to 1 atm range. The effect of pressure on the freezing point of water is negligible.

15.2 Chemical Potential

The chemical potential of species j depends on its concentration, the pressure, electric potential and gravity (Noble, 1991):

$$\mu_j = \mu_j^* + RT \ln a_j + \overline{V}_j p + z_j FE + m_j gh \qquad (15.1)$$

where μ_j (energy per mol) is the chemical potential of species j; μ_j^* is an arbitrary reference state which cancels out when the chemical potential in one location is subtracted from that in another to obtain the chemical potential difference between two locations; a_j is the concentration-dependent chemical activity of j; $\overline{V}_j$ is the partial molal volume of species j; p is the pressure; z_j is an integer representing the charge number on species j, F is Faraday's constant; E the electric potential; m_j the mass per mol of species j; g is the gravitational acceleration term and h the height in the gravitational field. The gravitational term $m_j gh$ can be disregarded, since horticultural commodities and floral products have a small size

Table 15.1. The solubility of various gases and vapours in water illustrating Henry's Law, $K_H = C_a/p_{gas}$, where C_a is the molar concentration of a species in the aqueous phase (a), p_{gas} is the partial pressure of that species in the gas phase and K_H is the Henry's Law constant. Data from Edsall and Wyman, 1958; Hodgman *et al.*, 1962; Liley *et al.*, 1984; Sander, 1999; and *International Critical Tables* (Vol. 3), McGraw-Hill. A simple way to describe Henry's Law as a function of temperature is $K_H = K_H^o \times \exp[(-\Delta_{soln}H/R)(1/T - 1/T^o)]$, where $\Delta_{soln}H$ is the enthalpy of solution, K_H^o is the Henry's Law constant at T° (298.15°K), R is the gas constant, and $d \ln K_H/d(1/T) = \Delta_{soln}H/R$ (Sander, 1999). Values in parentheses are computed from the Henry Law constant at 25°C using the temperature function. All other values are measured. Henry's Law sometimes is expressed as $p_B = HX_B$, where p_B is the partial pressure (atm) of the species in the gas phase, X_B is its mole fraction in the aqueous phase, $1/K_H = 180H$ (atm/mol gas per mol water) and H represents the volatility instead of the solubility and is the inverse of K_H.

	Henry's Law coefficient K_H ($\times 10^3$) M/atm = [(mol$_{aq}$/dm$^3_{aq}$)]/atm						$-d \ln K_H$
Gas or vapour	0°C	10°C	20°C	25°C	30°C	40°C	d (1/T) (°K)
Ethane	4.41 (4.01)	(2.96)	2.11 (2.23)	1.95	(1.71)	(1.34)	2350
Ethylene	10.7 (8.52)	7.61 (6.74)	5.50 (5.43)	4.90	4.21 (4.44)	3.43 (3.67)	1800
Oxygen	2.20 (2.19)	1.70 (1.76)	1.41 (1.43)	1.30	1.20 (1.18)	1.02 (0.99)	1700
Carbon dioxide	76.1	53.4	39.4	34.0	30.4	24.4	**
Ethanol	–	–	–	1.9×10^5	–	–	6500
Ammonia	–	–	–	5.8×10^4	–	–	4100*

*See equation 3.20 and Table 3.2.
**See Table 3.2.

and there is no significant difference in the elevation of their various parts.

15.3 Electrochemical Potential

The electrochemical potential of ions (z_jFE) is virtually insensitive to physiological shifts in pressure, changing by approximately 0.03 mV per 0.1 MPa. Since the typical electric potential difference across a membrane in a biological system might be 100 mV, LP will change the electric potential by less than 0.03%. Therefore, the electrochemical potential term can be neglected and the chemical potential is (Noble, 1991):

$$\mu_j = \mu_{j*} + RT \ln a_j + \overline{V}_j p \qquad (15.2)$$

where $RT \ln a_j$ describes the chemical activity of j and the term $\overline{V}_j p$ represents the effect of pressure on the chemical potential of j.

15.4 Standard State

By convention, p in equation 15.2 is defined as *the pressure in excess of atmospheric pressure*. The additive constant μ_j^* in equation 15.2 is the chemical potential of species j for a specific reference state, attained when the activity of species j is unity ($RT \ln a_j = 0$), and the hydrostatic pressure equals atmospheric pressure ($\overline{V}_j p = 0$). Under these conditions $\mu_j = \mu_j^*$. When the pressure is less than 1 atm (0.1013 MPa), p is expressed as a (–) term, and for a total vacuum, $p = -0.1013$ MPa.

15.5 Hydrostatic Pressure, Water Activity and Osmotic Pressure

Plant cells develop significant hydrostatic (turgor) pressures because they have rigid walls and contain dissolved solutes. The term expressing the effect of hydrostatic pressure on the chemical activity of water is $\overline{V}_W p$, where $\overline{V}_W$ is the partial molal volume of water (18×10^{-6} m^3/mol), and p is the hydrostatic pressure *in excess of 1 atm* within the aqueous solution.

Solute added to an aqueous solution decreases the water's chemical potential by decreasing the water activity (a_W) in the

RT ln a_w term of equation 15.3, while at the same time increasing the solution's osmotic pressure, Π. Therefore Π and a_W are related, changing in opposite directions as a function of solute concentration:

$$RT \ln a_W = -\bar{V}_W \Pi \qquad (15.3)$$

where the subscript (W) refers to liquid water. As solutes are added, a_W decreases from its value of unity for pure water, causing ln a_W to be negative and Π positive, in accord with equation 15.3. The solute concentration in plant cells only reduces the vapour pressure of water by about 0.5 to 1% below the saturated water-vapour pressure at the same temperature (Burton, 1982; Noble, 1991), and since there typically is little if any dissolved solute present in water at the site from which it evaporates at the plant surface, the vapour pressure there is that of pure water unless a negative tension arises due to transpiration. In dilute solutions the osmotic pressure is described by the van't Hoff relationship:

$$\Pi = RTc \qquad (15.4)$$

where c is the molar concentration of dissolved solute. The osmotic pressure of a typical plant cell containing 0.3 osmolar solution is 0.67 MPa at 0°C. At 20°C it is 0.73 MPa.

15.6 Osmoregulation

Animal cells lack rigid walls and must osmoregulate in order to maintain their structural integrity, whereas the turgor pressure developed in plant cells due to the physical properties of the cell wall usually is sufficient to maintain an appropriate water potential and prevent excessive water uptake from damaging cells. Nevertheless, plant cells have pressure sensors that regulate solute content and turgor pressure (Gutknecht, 1968; Hastings and Gutknecht, 1974; Zimmerman and Steudle, 1974; Zimmerman *et al.*, 1976; Shagan and Bar-Zvi, 1993; Johansson *et al.*, 1996). The mechanism by which turgor is sensed may be a thickening or thinning of the cellular membrane caused by turgor-induced stretching (Alexandro and Lassalles, 1992). The approximately 1 atm lowering of turgor pressure under hypobaric conditions might be sensed and result in an osmotic adjustment.

15.7 Water Potential of Liquid Water

> Water potential is a short way to say the potential energy of water. For most applications, pure water is assigned zero potential energy, meaning that it has zero work capacity relative to water not pure or free. Water less than pure and free has a negative potential energy; water under pressure (e.g. turgor water) has positive potential energy. (Cook and Papendick, 1978)

The water potential, Ψ_W, of liquid cellular water, defined by equations 15.5 and 15.6, is the difference between the cellular hydrostatic pressure, p, and osmotic pressure, Π:

$$\Psi_W = (\mu_W - \mu_{W^*})/\bar{V}_W = p - \Pi \qquad (15.5)$$

Equation 15.5 indicates that an increase in hydrostatic pressure (turgor pressure) elevates a cell's water potential, and an increase in osmotic pressure lowers it. Equation 15.5 also can be written in the form:

$$\Psi = \Psi_P + \Psi_\Pi \qquad (15.6)$$

where Ψ_P (= p) is the hydrostatic or pressure potential and Ψ_Π (= −Π) is the osmotic potential. The hydrostatic pressure (turgor pressure) is reacted by the extracellular pressure (p_{EP}) exerted by surrounding cells on each other, by atmospheric pressure (p_{ATM}) and by the pressure exerted on the cell solution by the stretched cell wall (p_{WALL}). These parameters do not appear separately in equation 15.5 because they are included in the turgor pressure term:

$$\text{turgor pressure} = p = -(p_{WALL} + p_{ATM} + p_{EP}) \qquad (15.7)$$

When LP lowers the atmospheric pressure by nearly 0.1 MPa, the cellular turgor pressure (p) decreases by the same amount, reducing the cellular water potential by

0.1 MPa. Simultaneously, the hydrostatic pressure of liquid water external to the cells is lowered by 0.1 MPa, decreasing the liquid water potential Ψ_W by a corresponding amount, and therefore the water potential gradient between the cell contents and external liquid water and adjacent cells is unchanged. Since the hydrostatic pressure may be as high as 0.73 MPa when a typical plant cell is fully turgid at night, and as low as –3 MPa during daylight hours (Noble, 1991), a 0.1 MPa decrease under hypobaric conditions is well tolerated and may have no physiological significance. In summary, although LP lowers the turgor pressure by nearly 0.1 MPa, it does not change the water potential gradients in plant tissues.

Variations in turgor pressure (Δp) are related to a relative change in cellular volume ($\Delta V/V$) by the elastic modulus ($\in$) of the cell wall:

$$\in = \Delta p/(\Delta V/V) \quad (15.8)$$

Assuming a nominal value of $\in$ = 10 MPa, a 1% change in volume ($\Delta V/V = 0.01$) will give rise to a 0.1 MPa change in turgor pressure, which for a typical cell with a turgor pressure of 0.3 MPa represents a 33% change in hydrostatic pressure. The cellular osmotic pressure, which is proportional to volume, would be changed by only 1%.

Before horticultural commodities are harvested, they undergo daily fluctuations in turgor pressure in response to changes in the demand for and supply of water. After harvest, turgor only decreases since the commodity has been separated from its water source and continues losing water when it transfers respiratory and acquired heat by evaporative cooling. When the harvested commodity is precooled in preparation for transport or storage, depending on the method of cooling, the contribution of evaporation can vary from 0 to 100%. It is not unusual for evaporation to account for approximately 50% of the total heat transfer during conventional forced-air cooling. During a 25°C cool down, if half of the heat is removed by evaporation, the commodity will lose 2.3% of its water, and the volume of the remaining water contracts by 0.5% due to its cubical expansion coefficient.[1] Together, these effects cause a substantial reduction in turgor. Typically a plant cell decreases 10% in volume from full turgor to incipient plasmolysis (Noble, 1991), and stored fruits and vegetables cannot tolerate more than a 6% weight loss without a noticeable loss in quality (Burton, 1982).

15.8 Chemical Activity of Water

The relationship between the chemical activity of water, a_w, and water potential is given by:

$$\Psi_W = -RT \ln a_W/\overline{V}_W \quad (15.9)$$

from which it can be computed that when a fully turgid cell containing 0.5 M solute (Π = 1.12 MPa) is placed under hypobaric conditions, the chemical activity of the water in the cell is reduced by less than 0.07%. Equation 15.9 indicates that although water participates in many biochemical reactions, and the rates of these reactions are controlled in part by the cellular water activity, these reaction rates should not be significantly altered by a hypobaric condition.

15.9 Water Vapour Pressure and Relative Humidity

The effect of pressure on the vapour pressure of water is given by the Gibbs equation:

$$\ln(e^o/e) - \overline{V}_W(p - p^o)/RT \quad (15.10)$$

$$RH = 100\ (e/e^o) \quad (15.11)$$

where e^o and e are the vapour pressures at atmospheric pressure ($p^o = 0$) and the prevailing pressure (p), respectively, and $\overline{V}_W$ is the molar volume of liquid water (18 cm^3 per g·mole). Lowering the atmospheric pressure to 10 mm Hg ($p = -0.09997$ MPa) at 0°C only reduces the vapour pressure of water by 0.079%.

15.10 Water Potential of Water Vapour in the Gas Phase

The water potential of water vapour in atmospheric air (Ψ_{WV}) is formulated relative to $\overline{V}_w$ (rather than $\overline{V}_{WV}$) so that the proportionality factor between Ψ_{WV} and Ψ_W is the same in both cases, namely $\overline{V}_W$. This allows a direct comparison of Ψ_{WV} and Ψ_W:

$$\Psi_{WV} = -RT \ln (e^o/e/\overline{V}_W) \qquad (15.12)$$

Equation 15.12 indicates that at 90% RH and 0°C, the water potential of water vapour in air is –13.1 MPa. At 95% RH it is –6.38 MPa.

15.11 Humidity Control in an LP Container

During evacuation of a refrigerated hypobaric container, the gas mixture initially present in the vacuum tank is cooled and dried by expansion. The temperature decreases slowly, then more rapidly, until the set pressure is reached, after which the air temperature rapidly increases and returns to the value controlled by the refrigeration thermostat. Subsequently the air change, which continuously enters through the pressure regulator, expands into an essentially 'infinite' volume without changing temperature, and the water vapour that initially was present in each volume of atmospheric air is now present in a much larger volume of rarified air. This causes the relative humidity of the incoming air to be lowered in proportion to the expansion. When a hypobaric container is operated at 0°C and a pressure of 1.33 kPa (10 mm Hg), if the atmospheric air initially had a humidity of 90% at 27°C, after passage through the pressure regulator and cooling to the set temperature, the air's relative humidity will decrease to 6.9% at 0°C. The incoming rarified air's humidity will only be 2.2% at 0°C if the initial atmospheric relative humidity is 50% at 20°C.

A humidification step upstream of the pressure regulator was included in the original 'wet' LP patent (Burg, 1967) to prevent low-RH incoming air from desiccating the stored commodity. The 'dry' hypobaric method humidifies the rarified air by a different procedure (Burg, 1987a,b). It substitutes respiratory heat and product water for the electric heater and auxiliary water used in the 'wet' method's mechanical humidifier. Regardless of which humidification method is used, the relative humidity in the vacuum tank depends on the prevailing temperature, independent of pressure, and the mole fraction of water vapour increases as the pressure is lowered.

15.12 Influence of Pressure and Water Vapour on the Partial Pressures of O_2, CO_2 and Other Gases Present in the Vacuum Tank's Gas Mixture

The atmospheric partial pressures of O_2 (0.209 atm), CO_2 (0.00036 atm), ethylene (approximately 2×10^{-8} atm) and atmospheric contaminants such as carbon monoxide, sulphur dioxide, ammonia and industrial ethylene, are reduced in LP in proportion to the pressure reduction. To calculate the partial pressure of any atmospheric component present in the vacuum tank, a correction must be applied to account for the presence of saturated water vapour (Burg, 1977):

$$p_i = p_{atm,i}\, [(p - p_{wv})/p_{atm}] \qquad (15.13)$$

where p_i is the partial pressure of component i (atm) inside the vacuum tank, $p_{atm,i}$ is the partial pressure of component i in air at atmospheric pressure, p is the storage pressure, p_{atm} is the atmospheric pressure, p_{wv} is the saturated pressure of water vapour present in the vacuum tank and all values are at the prevailing storage temperature. In the presence of saturated water vapour at 0°C and a pressure of 1.33 kPa (10 mm Hg), the O_2 partial pressure is 0.00154 atm ($[O_2]$ = 0.154%), whereas in dry air at the same pressure and temperature the O_2 partial pressure is 0.00275 atm ($[O_2]$ = 0.275%).

Expansion at the pressure regulator automatically purifies the incoming air. At 0°C and the optimal storage pressure for many commodities (1.33 kPa = 10 mm Hg), the saturated water-vapour pressure is 4.579 mm Hg, and incoming air expands by 760/(10 – 4.579) = 140-fold when it passes through the pressure regulator. This causes the partial pressures of atmospheric CO_2 and contaminants such as carbon monoxide, sulphur dioxide, ammonia and industrial ethylene automatically to be reduced by 140-fold as these gases enter the evacuated space.

15.13 Diffusion in the Gas Phase

Chapter 3 considers gas and vapour exchange between cells and the adjacent intercellular air spaces, and mass transport through the intercellular system, boxes, wraps and the barrier air layers at the surface of the commodity and storage cartons. Diffusion through the plant surface occurs both in air-filled pores and through the liquid/solid/air phases of the cuticle. The resistance (s/cm) of the air-filled pores, r_p, and liquid/solid/air phases of the cuticle, r_c, act in parallel, so their combined resistance, $r_{p,c}$, is:

$$1/r_p + 1/r_c = 1/r_{p,c} \qquad (15.14)$$

The skin resistance, $r_{p,c}$, acts in series with the boundary air layer resistance, r_a. The total resistance of the plant surface ($r_{p,c,a}$) is given by:

$$r_{p,c,a} = r_{p,c} + r_a \qquad (15.15)$$

As a gas moves to or from the atmosphere through cells where it is consumed or produced, it encounters other resistances acting in series with $r_{p,c,a}$, including the resistance of the intercellular air spaces r_{ias}, boxes and wraps r_b, and the total liquid phase resistance Σr_{liq}, which is comprised of the combined liquid phase resistance of the individual cells (r_{cell}).

15.14 Definitions

In a mixture of two components, A and B, the molal concentration c_A is the number of component A molecules per unit volume of the mixture (g·mol/cm^3). The mole fraction x_A of component A is defined as $x_A = c_A/c$, where c is the total molal concentration of the mixture, and the mass concentration ρ_A of component A is the mass of component A per unit volume (g/cm^3). The molal fluxes N_A and N_B represent the number of moles of species A and B, respectively, that pass through a unit area perpendicular to the x axis per unit time. D_{AB} is the mass diffusivity (diffusion coefficient) of A in B, and D_{BA} is the mass diffusivity of B in A; they are equal to each other independently of the proportions of A and B:

$$D_{AB} = D_{BA} \equiv D \qquad (15.16)$$

If the mixture is considered to be a perfect gas, the molal concentrations c_A and c_B are related to the partial pressures p_A and p_B of the species A and B in the mixture by:

$$p_i = c_i\mathrm{RT} \qquad (15.17)$$

where c_i is the molal concentration of component i in the mixture (g·mol/cm^3), p_i is the partial pressure of component i in the mixture (atm), i = A or B, and R is the gas constant (80.205 cm^3·atm/g·mol·K).

15.15 Steady-state Equimolar Counter-diffusion in Gases

The exchange of respiratory CO_2 and O_2 with an RQ ≅ 1 is an example of steady-state equimolar counter-diffusion in gases. Component A (CO_2) diffuses through component B (O_2), and vice versa, at the same molal rate but in opposite directions. The molal fluxes of A and B are equal, and the total molal flux with respect to stationary coordinates is zero. This type of transport is described by the relationship:

$$N_A = (D/\mathrm{RT}) \cdot [(p_{A1} - p_{A2})/(x_2 - x_1)] \qquad (15.18)$$

where N_A is the molal flux of A (g·mol/s·cm^2), p_{A1} and p_{A2} are the partial pressures (atm) of A at $x = 1$ and $x = 2$, respectively, $(x_2 - x_1)$ is the length of the diffusion pathway, R is the gas constant

[80.205 cm^3·atm/g·mol·K], T the temperature (K) and D is the mass diffusivity (diffusion coefficient, cm^2/s) of A in B.

15.16 Steady-state Unidirectional Diffusion in Gases

Unidirectional diffusive escape of water vapour is the major mass-transport process occurring in stored commodities. Component A (water vapour) will flow through a plant's surface by steady-state unidirectional diffusion in component B (air), but component B will remain motionless relative to the stationary coordinates, and the molal flux N_B of component B is zero. The partial pressure of component A is highest toward the inner side of the surface and lowest at the outer side, and since the total pressure is the same at all locations, therefore the partial pressure of component B is higher at the outer side of the surface and lower at the inner side. B diffuses from the inner to the outer side of the surface due to the concentration gradient that arises, but the net flux of B remains zero because as rapidly as B diffuses toward the outer side of the surface it is supplied by the bulk movement occurring in the opposite direction. This type of transport is described by the relationship:

$$N_A = \left[\frac{pD}{RT(x_2 - x_1)}\right]\left[\frac{p_{A1} - p_{A2}}{p_{Bln}}\right] \quad (15.19)$$

where p is the total pressure of the mixture (atm), and the logarithmic mean partial pressure p_{Bln} of component B is defined as:

$$p_{B\,ln} = (p_{B2} - p_{B1})/\ln(p_{B2}/p_{B1}) \quad (15.20)$$

The logarithmic mean partial pressure term p_{Bln} approaches unity when the partial pressure of component A is very small compared to that of component B, and in that instance equation 15.19 reduces to the form of equation 15.18 for equimolar counter-diffusion of gases. At atmospheric pressure, the error introduced by ignoring the logarithmic mean partial pressure correction is insignificant because the partial pressures of gases and vapours, including water vapour, which are transported through the air phases in plants, are present in dilute amounts relative to air. For that reason this correction is not considered in biological calculations, and the mass flux for gas transport (g·mol/s) in air-filled pores through a commodity's entire surface area is given by:

$$\dot{m}_A = \Delta p_A\left[\frac{p_A}{r_p\,RTp_{Bln})}\right] \quad (15.21)$$

where A is the surface area (cm^3) through which the gas mass transfer is occurring, Δp_A is the partial pressure gradient of the diffusing gas across the length of the diffusing pathway, $\dot{m}_A$ is the mass transport flux (g·mol/s) and r_P (s/cm) is the resistance of the pores to gas exchange defined by the expression:

$$r_P = \Delta x/D \quad (15.22)$$

in which D is the binary diffusion coefficient of the diffusing gas in air, and Δx is the length of the diffusion path. Because the atmosphere is kept saturated in LP, the mole fraction of water vapour progressively increases at lower pressures, until below approximately 5.33–10.67 kPa (40–80 mm Hg), the mole fraction of water vapour accounts for a significant portion of the total mixture. Then the log mean partial pressure must be considered, and the resistance to water vapour flux through air-filled pores is given by (Burg and Kosson, 1983):

$$r_P = r_{P,R}\,\frac{\ln\left[\dfrac{p_R - p_{v,o}}{p_R - p_{v,i}}\right]}{\ln\left[\dfrac{p - p_{v,o}}{p - p_{v,i}}\right]} \quad (15.23)$$

where p_R is the reference pressure (atmospheric pressure); $r_{P,R}$ the pore resistance (s/cm) measured at atmospheric pressure and 273°K for the particular vapour pressure values $p_{v,i}$ and $p_{v,o}$ inside (i) and outside (o) the surface, respectively; and r_P is the pore resistance at pressure p. The log mean partial pressure correction does not need to be considered for a gas such as ethylene, because although the transport is unidirectional, the mole fraction of ethylene is dilute at all pressures.

15.17 Steady-state Equimolar Counter-diffusion in Liquids

The equations derived for gas diffusion, written in terms of partial pressures, can be rewritten for diffusion of molar concentrations c_A and c_B in liquids, where:

$$c_A + c_B = c = \text{total molal concentration (g·mol/cm}^3) \quad (15.24)$$

Steady-state equimolar counter-diffusion in liquids is described by the relationship:

$$N_A = D[(c_{A1} - c_{A2})/(x_2 - x_1)] \quad (15.25)$$

When gas A diffuses from an air/liquid interface (1), through a liquid phase, to another air/liquid interface (2), the molal concentration c_A (g·mol/cm³) of gas A present in the liquid phase at each air/liquid interface is:

$$c_A = K_H\, p_A \quad (15.26)$$

where p_A is the partial pressure of gas A in equilibrium with the liquid phase and K_H is the Henry's Law coefficient (M/atm) relating the partial pressure developed by dissolved solvent A in a liquid solvent B. Provided that the temperature is the same at interface 1 and 2, $K_{H,A1} = K_{H,A2}$, and by substitution into equation 15.25:

$$N_A = DK_H \frac{(p_{A1} - p_{A2})}{(x_2 - x_1)} \quad (15.27)$$

The resistance of the liquid phase in the cuticle is given by:

$$r_C = [(x_2 - x_1)]/AD \quad (15.28)$$

where the absorption coefficient A is the volume of gas, when reduced to 0°C and 760 mm Hg (1 atm), absorbed by one volume of water when the pressure of the gas itself, without the aqueous tension, amounts to 760 mm Hg (Table 15.2).

15.18 Steady-state Unidirectional Diffusion in Liquids

Steady-state unidirectional diffusion in liquids is described by the relationship:

$$N_A = \left[\frac{DK_H}{(x_2 - x_1)}\right]\left[\frac{(p_{A1} - p_{A2})}{p_{B\ln}}\right] \quad (15.29)$$

where the logarithmic mean mole fraction $p_{B\ln}$ of component B is defined by equation 15.20.

15.19 Binary Diffusion Coefficient of Gases

At pressures less than 20 atm, the diffusion coefficient ($D_{A,B}$) for a binary mixture of gases may be estimated from kinetic theory:

$$D_{AB} = [1.8583\ 10^{-3}\ T^{3/2}\ (1/M_A + 1/M_B)^{1/2}]/p\sigma^2_{AB}\Omega_{D,AB} \quad (15.30)$$

where M_A and M_B are the molecular weights of A and B, p is the total pressure (atm), σ_{AB} is the collision diameter (Å), and $\Omega_{D,AB}$ is the collision integral (a dimensionless

Table 15.2. Absorption coefficient (A), also known as partition coefficient (K); or bunsen coefficient B of various gases in pure water. Data from *Handbook of Chemistry and Physics*, CRC Publishing Co., 1962. The absorption coefficient (*A*) is the volume of gas, when reduced to 0°C and 760 mm Hg (1 atm), absorbed by one volume of water, when the pressure of the gas itself, without the aqueous tension, amounts to 760 mm Hg.

Temperature (°C)	Absorption coefficient			
	Ethylene	Oxygen	Carbon dioxide	Ammonia
0	0.256	0.0489	1.713	1299
10	0.172	0.0380	1.194	–
20	0.126	0.0310	0.878	690
30	0.095	0.0261	0.665	–
40	0.077	0.0231	0.530	–

function dependent on temperature and intermolecular forces). Values for σ_A, $(\varepsilon/k)_A$ and σ_B, $(\varepsilon/k)_B$ are presented in Table 15.3 (*upper*), and σ_{AB} is given by:

$$\sigma_{AB} = \tfrac{1}{2}\,(\sigma_A + \sigma_B) \qquad (15.31)$$

The p_{Bln} term approaches unity in very dilute solutions, and since in biological systems the amount of gas dissolved in a liquid phase always is very small, the correction for p_{Bln} is insignificant and can be ignored. This is equally true at atmospheric and hypobaric pressures. Equation 15.29 is identical to equation 15.27 when $p_{Bln} = 1$.

Table 15.3 (*lower*) lists values of $\Omega_{D,AB}$ as a function of kT/ε where:

$$kT/\varepsilon \cong T/(\varepsilon/k)_{AB} \qquad (15.32)$$

and the value of $(\varepsilon/k)_{AB}$ for the mixture is taken as:

$$(\varepsilon/k)_{AB} = [(\varepsilon/k)_A\,(\varepsilon/k)_B]^{1/2} \qquad (15.33)$$

For pressures less than about 10 atm, the diffusion coefficient for a binary mixture of gases A and B may also be estimated from the Fuller, Schettler and Giddings' relation:

$$D_{AB} = \frac{10^{-3}\,T^{1.75}[(M_A + M_B)(M_A M_B)]^{1/2}}{p[(\Sigma_v)_B^{1/3} + (\Sigma_v)_A^{1/3}]^2} \qquad (15.34)$$

where Σ_v is the atomic diffusion volume (Σ_v = 16.6, 20.1, 26.9, 12.7 and 40.9 for O_2, air, CO_2, water and ethylene, respectively).

Binary diffusion coefficients computed according to equations 15.30 and 15.34 are presented in Table 15.4 and compared with measured values when available. At 0°C, the average of the computed values [(equation 1 + equation 2)/2] is 0.224 (0.220) for water vapour, 0.139 (0.138) for CO_2 and 0.176 (0.178) cm²/s for O_2 (parentheses = measured values). Due to molecular interactions, neither the computed nor measured values conform precisely with Graham's Diffusion Law:

$$D_1/D_2 = M_2^{1/2}/M_1^{1/2} \qquad (15.35)$$

where D_1 and D_2 are the binary diffusion coefficients of the respective gases or vapours in air, and M_1 and M_2 are their molecular weights. If CO_2 diffusion in air is taken as a reference, according to Graham's Law the diffusion coefficients should be in the relationship CO_2 (1.0) < O_2 (1.17) <

Table 15.3. Parameter σ and ε/k (*upper*) and values of the collision integral $\Omega_{D,AB}$ (*lower*) for various gases (Özisik, 1985).

Substance	Molecular wt.	σ, Å	ε/k, K
Air	28.97	3.617	97.0
Carbon dioxide	44.01	3.996	190.0
Ethane	30.07	4.418	230.0
Ethylene	28.05	4.232	205.0
Oxygen	32.00	3.433	113.0
Water	18.00	2.641	809.1

kT/ε	$\Omega_{D,AB}$	KT/ε	$\Omega_{D,AB}$	kT/ε	$\Omega_{D,AB}$
0.3	2.662	1.2	1.320	2.1	1.057
0.4	2.318	1.3	1.273	2.2	1.041
0.5	2.066	1.4	1.233	2.3	1.026
0.6	1.877	1.5	1.198	2.4	1.012
0.7	1.729	1.6	1.167	2.5	0.9996
0.8	1.612	1.7	1.140	2.6	0.9878
0.9	1.517	1.8	1.116	2.7	0.9770
1.0	1.439	1.9	1.094	2.8	0.9672
1.1	1.375	2.0	1.075	2.9	0.9576

Table 15.4. Binary diffusion coefficients of gases and vapours in air at 0 and 25°C, 1 atm, computed from equations 15.30 (referred to as 1) and 15.34 (referred to as 2). Measured values are indicated in parentheses (Liley *et al.*, 1984; Özisik, 1985).

	Binary diffusion coefficient in air, cm^2/s					
	0°C			25°C		
Gas	Equation 1	Equation 2	Exptl.	Equation 1	Equation 2	Exptl.
Water vapour	0.233	0.216	(0.220)	0.267	0.252	(0.260)
Carbon dioxide	0.144	0.134	(0.138)	0.165	0.156	(0.161)
Oxygen	0.182	0.170	(0.178)	0.207	0.198	(0.207)
Ethylene	0.136	0.128	–	0.149	0.149	–
Ethane	0.122	0.121	–	0.139	0.141	–
Ammonia	–	–	(0.198)	–	–	–
Ethanol	–	–	(0.102)	–	–	(0.115)

ethylene (1.25) < water vapour (1.56). Instead, the computed and measured values are in the order ethylene (0.95) < CO_2 (1.0) < O_2 (1.27) < water vapour (1.60). This in part explains why the measured 'skin' resistance to ethylene mass transport in fruits always is greater than the resistance to CO_2 exchange (Table 2.3), whereas the opposite would be true according to Graham's Law. Estimates of a fruit's 'skin' resistance to CO_2, O_2 and ethylene typically are based on measurements of the fruit's surface area, each gas's production or consumption rate, and the ICC, IEC and internal O_2. Often it is tacitly assumed that the calculation is indicative of the resistance of the air-filled pores (stomates, lenticles and pedicel stem scar), but this fails to consider that gas exchange is occurring in parallel through the cuticle, where the resistance to CO_2 is small enough to influence the result, and the cuticular resistances to ethylene and O_2 are 10- and 25-fold larger, respectively (3.20; Table 3.12). This causes a commodity's 'skin' resistance to ethylene to exceed that for CO_2 by 10–50%.

15.20 Binary Diffusion Coefficient in Liquids

For a binary mixture containing a small concentration of gas (A) dissolved in liquid (B), the diffusion coefficient (D_{AB}) is given by:

$$D_{AB} = \frac{7.4 \times 10^{-8} (\varnothing M_B)^{1/2} T}{\mu_B \overline{V}_A^{\,0.6}} \quad (15.36)$$

where μ_B is the viscosity of solute B (cP) at temperature T (K), and ø is the association factor of solvent B (dimensionless). The recommended value for ø is 2.6 for gases dissolved in water. If a reliable value of the critical volume is known, the liquid molal volume $\overline{V}_A$ (cm^3/g·mol) of gas A calculated at its boiling point can be computed with approximately 3% accuracy from the expression:

$$\overline{V}_A = 0.285 \overline{V}_C^{\,1.048} \quad (15.37)$$

where $\overline{V}_C$ is the critical molar volume. Binary diffusion coefficients computed according to equations 15.36 and 15.37 for O_2, CO_2, ethylene and ethane diffusing in water are presented in Table 15.5, and compared with measured values when available.

15.21 Influence of Water Vapour on Physical and Chemical Parameters

The humidity in the LP vacuum tank is maintained close to saturation at the prevailing temperature, independent of pressure. Therefore the mole fraction of water vapour present in the atmosphere progressively increases as the pressure is lowered. The mole fraction of water vapour present in saturated air is 0.00609 at atmospheric pressure, whereas at 1.33 kPa (10 mm Hg) and 0°C, or at 2.67 kPa (20 mm Hg) and 13°C, it is 0.46 and 0.56, respectively. When meat is stored at −1°C and a pressure of 4.3 mm Hg, only water vapour is present.

The chemical and physical constants differ significantly for many properties of water vapour and air (Table 15.6), so new values must be computed for the hypobaric water vapour + air mixture.

Table 15.5. Binary diffusion coefficients of gases and vapours in water at 0 and 25°C, computed from equations 15.36 and 15.37. Measured values are indicated in parentheses (Liley *et al.*, 1984; Özisik, 1985).

Gas	$\bar{V}_A$ (cm³/g.mol)	Binary diffusion coefficient (cm²/s × 10⁵) 0°C	25°C
Carbon dioxide	33.3	1.09 (1.09)	2.05 (2.04)
Oxygen	27.4 (26.6)	1.24 (1.38)	2.34 (2.50)
Ethylene	44.1 (45.9)	0.84	1.58 (1.09)
Ethane	49.4	0.80	1.51
Ammonia	–	–	– (2.0)
Ethanol	–	–	– (1.26)

15.22 Viscosity of Water Vapour + Air Mixtures

The dynamic viscosity, μ, of binary gas mixtures is not a mole fraction average. It can be calculated from the expression (Herzfeld and Smallwood, 1952) (see equation 15.38 at bottom of page):

and μ_1 and μ_2 are the viscosity of components 1 and 2; m_1 and m_2 their masses (water = 18; air = 28.97); d_1 and d_2 their molecular diameters (air = 3.10 × 10–8 cm; water vapour = 2.80 × 10–8 cm); and c_1 and c_2 their concentrations (moles/cm³). The viscosity at 0°C decreases from 0.172×10^{-4} Pa·s in dry air, to 0.130×10^{-4} Pa·s in saturated air at 1.33 kPa (10 mm Hg), to 0.080×10^{-4} Pa·s in pure water vapour at 0.61 kPa (4.579 mm Hg). Pressure has a negligible effect on viscosity in the range from 0.32 kPa (2.4 mm Hg) to 1 atm (Table 15.7), but the apparent dynamic viscosity decreases at lower pressures due to slippage. Viscosity increases with temperature, somewhat more rapidly than predicted by the expression $\mu \sim T^{1/2}$.

Table 15.6. Physical and chemical constants of water vapour and air at 0°C, 1 atm. Binary diffusion coefficients at 1.33 kPa (10 mm Hg) are from Özisik (1985) or computed from equations 15.31–15.35.

	SI units	Water vapour	Air
Specific heat (c_p)	kJ/kg·K	1.8540	1.006
Viscosity (μ)	10^{-4} Pa·s	0.0802	0.172
Thermal conductivity (*k*)	W/m·K	0.0182	0.024
Molecular weight (*M*)	g·mol	18.0000	28.970
Prandtl number (Pr)	dimensionless	0.8200	0.720
Binary diffusion coefficient (*D*) at 1.33 kPa (10 mm Hg)			
Carbon dioxide	cm²/s	13.72	10.34
Ethylene	cm²/s	13.00	10.03
Oxygen	cm²/s	16.10	13.30

$$\mu = \frac{\mu_1[1+(c_1/c_2)\beta_2]+\mu_2[1+(c_2/c_1)\beta_1]+A}{[1+(c_1/c_2)\beta_2][1+(c_2/c_1)\beta_1]-A^2(4\mu_1/\mu_2)} \quad (15.38)$$

where:

$\beta_1 = (1/12)\,[(d_1 + d_2)/d_1]^2\,[2m_2/(m_1 + m_2)]^{1/2}\,[(5m_1 + 3m_2)/(m_1 + m_2)]$

$\beta_2 = (1/12)\,[(d_1 + d_2)/d_2]^2\,[2m_1/(m_1 + m_2)]^{1/2}\,[(3m_1 + 5m_2)/(m_1 + m_2)]$

$A^2 = 16\mu_1\mu_2\beta_1\beta_2\,[m_1m_2/(15m_1^2 + 34m_1m_2 + 15m_2^2)]$

Table 15.7. Mean free path, Λ, and apparent viscosity of air, μ, at different pressures and 15°C (Herzfeld and Smallwood, 1952). Viscosity at an atmospheric pressure of 750 mm Hg = μ_{750}.

p, mm Hg	μ/μ_{750}	Λ (mm × 10^4)
750.00	1.000	0.6
380.00	1.010	1.2
20.50	1.004	22.0
2.40	0.978	190.0
1.53	0.956	300.0
0.63	0.908	730.0

15.23 Thermal Conductivity of Water Vapour + Air Mixtures

The thermal conductivity (k) of a binary gas mixture is not a mole fraction average. It can be computed from the expression (Herzfeld and Smallwood, 1952) (see equation 15.39 at bottom of page):

and k_1, k_2 and $k_{1,2}$ are the thermal conductivity of components 1, 2 and their mixture (1 + 2), respectively. All other symbols are the same as for equation 15.38. At 0°C, the thermal conductivity decreases from 0.024 W/m·K in dry air at atmospheric pressure or at a pressure of 1.33 kPa (10 mm Hg), to 0.019 W/m·K in a saturated air/water-vapour mixture at 1.33 kPa (10 mm Hg), to 0.018 W/m·K in pure water vapour at 0.61 kPa (4.579 mm Hg). The thermal conductivity of a gas increases with pressure, but between 0.13 kPa (1 mm Hg) and 10 bar, the effect is negligible. The effect of temperature can be approximated by the expression:

$$k_1/k_2 = (T_2/T_1)^n \qquad (15.40)$$

where $n \cong 1.8$.

15.24 Heat Capacity of Water Vapour + Air Mixtures

At 0°C, the heat capacity at a constant pressure (c_P) increases from 1.006 kJ/kg·K in air, to 1.394 kJ/kg·K in air saturated with water vapour at 1.33 kPa (10 mm Hg), to 1.854 kJ/kg·K in pure water vapour at 0.61 kPa (4.579 mm Hg).

15.25 Prandtl (Pr) and Reynolds (Re) Numbers of Water Vapour + Air Mixtures

The Reynolds ($\mathrm{Re} = Dv\rho/\mu$) and Prandtl ($\mathrm{Pr} = c_P\mu/k$) numbers depend on the density (ρ), viscosity (μ), thermal conductivity (k) and heat capacity (c_P) of a gas, all of which change when the pressure is lowered and the proportion of water vapour to air increases in the storage mixture. These dimensionless numbers are important in calculations involving convective heat transfer, mass transfer and forced flow (6.12). At 0°C, the Prandtl number increases from 0.72 in dry atmospheric air, to 0.95 in air saturated with water vapour at 1.33 kPa (10 mm Hg), to 1.05 in 84.8% RH air at 0.61 kPa (4.579 mm Hg), but decreases to 0.82 in pure water vapour at 0.61 kPa (4.579 mm Hg).

15.26 Diffusion of a Dilute Gaseous Component in a Water Vapour + Air Mixtures

In the case of a dilute component i diffusing in a homogenous mixture of water vapour and air, where x_j is the mole fraction of j, and n is the number of components (n = 3

$$k_{1,2} = \frac{k_1[1+(c_1/c_2)\beta_2] + k_2[1+(c_2/c_1)\beta_1] + A}{[1+(c_1/c_2)\beta_2][1+(c_2/c_1)\beta_1] - A^2(4k_1/k_2)} \qquad (15.39)$$

where:

$$\beta_1 = [(d_1 + d_2)/2d_1]^2\,[2m_2/(m_1 + m_2)]^{1/2}\,[(30m_1^2 + 16\,m_1m_2 + 13m_2^2)/8\,(m_1 + m_2)^2]$$

$$\beta_2 = [(d_1 + d_2)/2d_2]^2\,[2m_1/(m_1 + m_2)]^{1/2}\,[(13m_1^2 + 16\,m_1m_2 + 30m_2^2)/2\,(m_1 + m_2)^2]$$

$$A^2 = 27(2k_1k_2)^{1/2}\,[(d_1 + d_2)^2/16\,d_1d_2]^2\,[(m_1m_2)^{5/4}/(m_1 + m_2)^{5/2}]$$

for a ternary mixture), the diffusion coefficient is given by Blanc's Law (Liley *et al.*, 1984):

$$D_{i,mix} = \left[\sum_{\substack{j=1 \\ \neq i}}^{n} (x_j / D_{i,j}) \right]^{-1} \qquad (15.41)$$

The binary diffusion coefficients of CO_2, O_2 and ethylene in water vapour, computed from equation 15.35 at 0°C and a pressure of 1 atm, are 0.181, 0.227 and 0.166 cm^2/s, respectively, compared to 0.138, 0.178 and 0.132 cm^2/s in air. The effects of water vapour and pressure on gas diffusion are additive (equation 15.41). Therefore at 1.33 kPa (10 mm Hg) and a temperature of 0°C, since the mole fraction of water vapour in the saturated air + water vapour mixture is 0.46, the ternary diffusion coefficients are 11.78, 15.02 and 11.08 cm^2/s for CO_2, O_2 and ethylene, respectively, compared to 10.48, 13.35 and 10.03 cm^2/s in dry air at that same pressure (Table 15.4). The large mole fraction of water vapour present in LP at 1.33 kPa (10 mm Hg) increases the diffusion coefficients by approximately 12%.

15.27 Effect of Air Partial Pressure on the Convective Film Coefficient for Condensation in a Water Vapour + Air Mixtures

In the absence of non-condensable gases, the heat-transfer coefficient for condensing pure water vapour is very large, 4020 $W/m^2 \cdot s$ on vertical surfaces and 8517 $W/m^2 \cdot s$ on horizontal surfaces (Knudsen *et al.*, 1984). If even a small amount of a non-condensable gas such as air is present, the coefficient is reduced many-fold. When a vapour containing non-condensable gas condenses, the non-condensable gas is left at the surface and the incoming condensable vapour must diffuse through this body of vapour–gas mixture collected in the vicinity of the condensate surface before the vapour reaches the cold surface to condense. The presence of non-condensable gas adjacent to the condensate surface acts as a thermal resistance to heat transfer (Özisik, 1985). The convective film coefficient is reduced twofold if 0.5% air is present, and 5% air can easily lower it by a factor of five. Under hypobaric conditions, since most of the air is evacuated from the commodity's intercellular system, the heat-transfer film coefficient for condensation is significantly increased. Non-condensable air will still comprise at least 50% of the gas mixture in the intercellular system, and therefore the film coefficient for condensation will be at least an order of magnitude less than that for pure water vapour.

Note

1. The cubical expansion coefficient of water is 0.207×10^{-3} per °C in the physiological temperature range.

Bibliography

Abdel-Rahman, M. (1977) Patterns of hormones, respiration and ripening enzymes during development, maturation and ripening of cherry tomato fruits. *Physiol. Plant.* 39, 115–118.

Abdul Raouf, U.M., Beuchat, L.R. and Ammar, M.S. (1993) Survival and growth of *Escherischia coli* 0157:H7 on salad vegetables. *Applied and Environmental Microbiology* 59, 1999–2006.

Abe, Y. and Kondoh, Y. (1989) Oxygen absorbers. In: Brody, A.L. (ed.) *Controlled/Modified Atmosphere/Vacuum Packaging of Foods.* Food & Nutrition Press, Trumbell, Connecticut, pp. 149–158.

Abeles, F.B. (1966) The effect of ethylene on auxin transport. *Plant Physiol.* 41, 946–948.

Abeles, F.B. (1967) Mechanism of action of abscission accelerators. *Physiol. Plant.* 20, 442–444.

Abeles, F.B. (1973) *Ethylene in Plant Growth.* Academic Press, New York.

Abeles, F.B. and Gahagan, H.E. (1968a) Abscission: the role of ethylene, ethylene analogues, carbon dioxide, and oxygen. *Plant Physiol.* 43, 1255–1258.

Abeles, F.B. and Gahagan, H.E. (1968b) Accelerated abscission of *Coleus* petioles by placing plants in a horizontal position. *Life Sciences* 7, 653–655.

Abeles, F.B. and Rubenstein, B. (1964) Regulation of ethylene evolution and leaf abscission by auxin. *Plant Physiol.* 39, 963–969.

Abeles, F.B. and Takeda, F. (1990) Cellulase activity and ethylene in ripening strawberry and apple fruits. *Scientia Horticulturae* 423, 269–275.

Abeles, F.B., Holm, R.E. and Gahagan, H.E. (1967) Abscission: the role of aging. *Plant Physiol.* 42, 1351–1356.

Abeles, F.B., Bosshart, R.P., Forrence, L.E. and Habig, W.H. (1970) Preparation and purification of glucanase and chitinase from bean leaves. *Plant Physiol.* 47, 129–134.

Abeles, F.B., Hershberger, W.L. and Dunn, L.J. (1989) Hormonal regulation, and intracellular localization of a 33-kD cationic peroxidase in excised cucumber cotyledons. *Plant Physiol.* 89, 664–668.

Abeles, F.B., Morgan, P.W. and Saltveit, M.E. (1992) *Ethylene in Plant Biology.* Academic Press, San Diego, California.

Abou Hadid, A.F., El-Beltagy, A.S., Smith, A.R. and Hall, M.A. (1986) Ethylene production by leaves of various plant species in response to stress imposition. *Acta Hort.* 190, 415–422.

Abrams, G.J.V. and Pratt, H.K. (1966) Effect of ethylene on the permeability of excised cantaloupe fruit tissue. *Plant Physiol.* 42, 299–301.

Achilea, O., Fuchs, Y., Chalutz, E. and Rot, I. (1985) The contribution of host and pathogen to ethylene biosynthesis in *Penicillium digitatum*-infected citrus fruit. *Physiol. Plant Pathol.* 27, 55.

Adachi, K. and Hamer, J.E. (1998) Divergent cAMP signaling pathways regulate growth and pathogenesis in the rice blast fungus *Magnaporthe grisea. Plant Cell* 10, 1361–1373.

Adachi, M., Nakabayashi, K., Azuma, R., Kurata, H., Takahashi, Y. and Shimokawa, K. (2000) The ethylene-induced chlorophyll catabolism of radish (*Raphanus sativus* L.) cotyledons: production of colorless fluorescent chlorophyll catabolite (FCC) *in vitro*. *Plant Growth Regl. Abstracts* 26, 136.

Adamicki, F. (1989) Przechowywanie warzyw w kontolowanej atmosferze. *Biuletyn Warzywniczy* Supplement 1, 107–113.

Adamicki, F. and Elkner, K. (1985) Effect of temperature and a controlled atmosphere on cauliflower storage and quality. *Biuletyn Warzywniczy* 28, 197–224.

Adams, D.O. and Yang, S.F. (1979) Ethylene biosynthesis: identification of 1-aminocyclopropane-1-carboxylic acid as an intermediate in the conversion of methionine to ethylene. *Proc. Natl Acad. Sci. USA* 76, 170–174.

Adams, K.B., Wu, M.T. and Salunkhe, D.K. (1976) Effects of subatmospheric pressures on patulin production and mycelial growth by *P. expansum* and *P. patulum*. *Lebensm.-Wiss. Rechnol.* 9, 153–155.

Adato, I. and Gazit, S. (1974) Water deficit stress, ethylene production and ripening in avocado fruits. *Plant Physiol.* 53, 45–46.

Adato, I. and Gazit, S. (1977) Role of ethylene in avocado fruit development and ripening. II. Ethylene production and respiration by harvested fruit. *J. Expt. Bot.* 28, 644–649.

Agar, I.T., Streif, J. and Bangerth, F. (1994) Effect of high CO_2 and controlled atmosphere (CA) storage on the keepability of blackberry cv. 'Thornfree'. *Commisions C2, D1, D2/3 of the International Institute of Refrigeration International Symposium 8–10 June, Istanbul, Turkey*, pp. 271–280.

Agar, I.T., Streif, J. and Bangerth, F. (1997) Effect of high CO_2 and controlled atmosphere (CA) on the ascorbic and dehydroascorbic acid contents of some berry fruits. *Postharvest Biol. Technol.* 11, 47–55.

Aharoni, N. (1978) Relationship between leaf water status and endogenous ethylene in detached leaves. *Plant Physiol.* 61, 658–662.

Aharoni, N. (1989) Inter-relationship between ethylene and growth regulators in the senescence of lettuce leaf discs. *J. Plant Growth Regl.* 8, 309–317.

Aharoni, N. and Lieberman, M. (1979) Ethylene as a regulator of senescence in tobacco leaf discs. *Plant Physiol.* 64, 801–804.

Aharoni, N., Ben-Yehoshua, S., Sinai, A., Aviram, C., Barkai-Golan, R. and Netzer, D. (1971) Delaying the deterioration of Romaine leaf lettuce. Experiments to delay the deterioration of lettuce, tomatoes, strawberries and celery for export (1969/1970). *Special Publication No. 2. Volcani Instit. Agr. Res.*, Division of Fruit and Vegetable Storage, Bet Dagan, Israel.

Aharoni, N., Blumenfeld, A. and Richmond, A.E. (1977) Hormonal activity in detached lettuce leaves as affected by leaf water content. *Plant Physiol.* 59, 1169–1173.

Aharoni, N., Lieberman, M. and Sisler, H.D. (1979a) Patterns of ethylene production in senescing leaves. *Plant Physiol.* 64, 796–800.

Aharoni, N., Anderson, J.D. and Lieberman, M. (1979b) Production and action of ethylene in senescing leaf discs: effect of indoleacetic acid, kinetin, silver ion, and carbon dioxide. *Plant Physiol.* 64, 805–809.

Aharoni, N., Philosoph-Hadas, S. and Barkai-Golan, R. (1985) Modified atmospheres to delay senescence and decay of broccoli. In: Blankenship, S.M. (ed.) *Controlled Atmospheres for Storage and Transport of Perishable Agricultural Commodities. Proceedings of the 4th National Controlled Atmosphere Research Conference*, North Carolina Horticulture Report No. 126. North Carolina State University, Raleigh, North Carolina, p. 169.

Aharoni, Y. (1968) Respiration of oranges and grapefruits harvested at different stages of development. *Plant Physiol.* 43, 99–102.

Aharoni, Y., Apelbaum, A. and Copel, A. (1986) Use of reduced atmospheric pressure for control of the green peach aphid on harvested head lettuce. *HortScience* 21, 469–470.

Aimi, R. (1963) Studies on the irritability of the pulvinus of *Mimosa pudica*. *Mag. Tokyo* 76, 374–380.

Akamine, E.K. (1957) Respiration and ethylene production in the purple passion fruit. *J. Amer. Soc. Hort. Sci.* 69, 221–225.

Akamine, E.K. (1976a) Postharvest handling of tropical ornamental cut crops in Hawaii. *HortScience* 11, 125–127.

Akamine, E.K. (1976b) Problems in shipping fresh Hawaiian tropical and subtropical fruits. *Acta Horticulturae* 57, 151–161.

Akamine, E.K., Goo, T. and Suehisa, R. (1979) Relationship between leaf darkening and chemical composition of leaves of species of *Protea*. *Flor. Rev.* 163, 62–63, 107–108.

Akapyan, T.N., Braunstein, A.E. and Goryachenkova, E.V. (1975) β-cyanoalanine synthase: purification and characterization. *Proc. Natl Acad. Sci. USA* 72, 1617–1621.

Albrigo, L.G. (1972) Distribution of stomata and epicuticular wax on oranges as related to stem end rind breakdown and water loss. *J. Amer. Soc. Hort. Sci.* 57, 220–223.

Albrigo, L.G. (1973) Ultrastructure of cuticular surfaces and stomata of developing leaves and fruit of the 'Valencia' orange. *J. Amer. Soc. Hort. Sci.* 97, 761.

Albrigo, L.G., Syvertsen, J.P. and Costa, J.T.A. (1982) Some stomatal characteristics of citrus and their implications to gas exchange. *Int. Hort. Congr.* Hamburg, 2, 1364 (abstract).

Alejar, A.A., Visser, R.D. and Spencer, M.S. (1988) Ethylene production by attached leaves or intact shoots of tobacco cultivars differing in their speed of yellowing during curing. *Plant Physiol.* 88, 329–332.

Alexandro, J. and Lassalles, J.P. (1992) Hydrostatic and osmotic pressure-activated channel in plant vacuole. *Biophys. J.* 60, 1326–1336.

Alford, J.A. and Fieger, E.A. (1952) The non-microbial nature of the black spots on ice-packed shrimp. *Food Technol.* 6, 217–219.

Alique, R., Rodrigo, M.E. and Canet, W. (1979) Experiments on the hypobaric storage of Picota cherries. *XV-th International Congress of Refrigeration*, 23–29 September 1979 C2, 106.

Alloca, R. (1979) Container operation at pressure below 10 mm Hg. Report for Grumman Allied Industries, Inc., Garden City, New York.

Alloca, R. (1980a) Dormavac – long-term storage of perishable commodities. Contributed by the Aerospace Division of the American Society of Mechanical Engineering at the *Intersociety Environmental Systems Conference*, San Diego, California, 14–17 July, 1980.

Alloca, R. (1980b) Dormavac mushroom transport: temperature gradients: a quick look at possible remedial action. Report for Grumman Allied Industries, Inc., Garden City, New York.

Alloca, R. (1981) Cherry freeze-point test. Report for Grumman Allied Industries, Inc., Garden City, New York.

Alonso, J.M. and Granell, A. (1995) A putative vacuolar processing protease is regulated by ethylene and also during fruit ripening in citrus fruit. *Plant Physiol.* 109, 541–547.

Alonso, J.M., Chamarro, J. and Granell, A. (1995a) A non-photosynthetic ferridoxin gene is induced by ethylene in citrus organs. *Plant Mol. Biol.* 29, 1211–1221.

Alonso, J.M., Chamarro, J. and Granell, A. (1995b) Evidence for the involvement of ethylene in the expression of specific RNAs during maturation of the orange, a non-climacteric fruit. *Plant Mol. Biol.* 29, 385–390.

Alonso, J.M., Hiramaya, T., Roman, G., Nourrizadeh, S. and Ecker, J.R. (1999) EIN2, a bifunctional transducer of ethylene and stress responses in *Arabidopsis*. *Science* 284, 2148–2152.

Aluja, M.W.M. and Liedo, P. (1990) *Fruit Flies, Biology and Management.* Springer, New York.

Alvarez, A.M. (1979) Effects of hypobaric storage on quality and disease incidence of papaya fruits. *Phytopathology* 69, 1020.

Alvarez, A.M. (1980) Improved marketability of fresh papaya by shipment in hypobaric containers. *HortScience* 15, 517–518.

American Society of Metals (1983) *ASM Metals Reference Book.*

Amrhein, N. and Schneebeck, D. (1980) Prevention of auxin-induced epinasty by α-aminooxyacetic acid. *Physiol. Plant.* 49, 62–64.

Amrhein, N., Schneebeck, D., Skorupka, H. and Tophof, S. (1981) Identification of a major metabolite of the ethylene precursor 1-aminocyclopropane-1-carboxylic acid in higher plants. *Naturwissenschaften* 68, 619–620.

Amrhein, N., Breuing, F., Eberle, J., Skorupka, H. and Tophof, S. (1982) The metabolism of 1-aminocyclopropane-1-carboxylic acid. In: Wareing, P.F. (ed.) *Plant Growth Substances.* Academic Press, London, pp. 249–258.

Anderson, J.D., Lieberman, M. and Stewart, R.N. (1979) Ethylene production by apple protoplasts. *Plant Physiol.* 63, 931–935.

Anderson, J.D., Bailey, B.A., Taylor, R., Sharon, A., Avni, A., Mattoo, A.K. and Fuchs, Y. (1993) Fungal xylanase elicits ethylene biosynthesis and other responses in tobacco. In: Pech, J.C., Latché, A. and Balagué, C. (eds) *Cellular and Molecular Aspects of the Plant Hormone Ethylene.* Kluwer Academic, Dordrecht, pp. 197–204.

Andreae, W.A. (1964) Auxin metabolism and root growth inhibition. In: Nitsch, J.P. (ed.) *Régulateurs Naturels de la Recherche Scientifique.* Éditions du Centre National de la Recherche Scientifique, Paris, pp. 559–573.

Andreae, W.A., Venis, M.A., Jursic, F. and Dumas, T. (1968) Does ethylene mediate root growth inhibition by indole-3-acetic acid? *Plant Physiol.* 43, 1375–1380.

Andres, J. (1995) The climacteric respiration rise in attached and detached tomato fruit. *Postharv. Biol. Technol.* 6, 287–289.

Andrich, G., Zinnai, A., Balzini, S., Silvestri, S. and Fiorentini, R. (1998) Aerobic respiration rate of Golden Delicious apples as a function of temperature and P_{O_2}. *Postharvest Biol. and Tech.* 14, 1–9.

Anet, E.F.L.J. (1972) Superficial scald, a functional disorder of stored apples. VIII. Volatile products from the autooxidation of α-farnescene. *J. Sci. Food Agric.* 23, 605–658.

Anon. (1920) Food Investigation Board. Department of Scientific and Industrial Research Report for the Year 1920, pp. 16–25.

Anon. (1961) *Banana Ripening Manual.* United Fruit Co.

Anon. (1966) *Matheson Gas Data Handbook.* Herst Litho Inc., New York.

Anon. (1977) Hypobarics. A new horizon for handling perishables. *Food Engineering* August, 65–68.

Anon. (1991) Dry aging of beef – the traditionalists' preference. *Meat Source* (September), 2.

Anon. (2001) Note on the biopathology and technique of storage and transport of fruit: the apple (part 6). *Plant Growth Regl. Abstracts* 27, 247.

Antunes, M.D.C. and Sfakiotakis, E.M. (2000) Effect of high temperature stress on ethylene biosynthesis, respiration and ripening of 'Hayward' kiwifruit. *Postharvest Biol. Technol.* 20, 251–259.

Apelbaum, A. (1979) Reducing sugar metabolism in potato tubers stored under low pressure. *Hassadeh* 59, 788–789.

Apelbaum, A. and Barkai-Golan, R. (1977) Spore-germination and mycelial growth and sporulation of some pathogenic fungi under hypobaric pressure. *Phytopathology* 67, 400–403.

Apelbaum, A. and Burg, S.P. (1971a) Altered cell wall microfibrillar orientation in ethylene-treated *Pisum sativum* stems. *Plant Physiol.* 48, 648–652.

Apelbaum, A. and Burg, S.P. (1971b) Effects of ethylene and 2,4-dichlorophenoxyacetic acid on cellular expansion in *Pisum sativum. Plant Physiol.* 50, 125–131.

Apelbaum, A. and Burg, S.P. (1972) Effect of ethylene on cell division and deoxyribonucleic acid synthesis in *Pisum sativum. Plant Physiol.* 50, 117–124.

Apelbaum, A. and Yang, S.F. (1981) Biosynthesis of stress ethylene induced by water deficit. *Plant Physiol.* 68, 594–596.

Apelbaum, A., Fisher, J.B. and Burg, S.P. (1972) Effect of ethylene on cellular differentiation in etiolated pea seedlings. *Amer. J. Bot.* 59, 697–705.

Apelbaum, A., Goldschmidt, E.E. and Ben-Yehoshua, S. (1976) Involvement of ethylene in the induction of color change in Shamouti oranges. *Plant Physiol.* 57, 836–838.

Apelbaum, A., Aharoni, Y. and Tempkin-Gorodeiski, N. (1977a) Effects of subatmospheric pressure on the ripening processes of banana fruits. *Trop. Agric. (Trinidad)* 54, 39–46.

Apelbaum, A., Zauberman, G. and Fuchs, Y. (1977b) Extending the storage life of avocado fruits by subatmospheric pressure. *HortScience* 12, 115–117.

Apelbaum, A., Zauberman, G. and Fuchs, Y. (1977c) Subatmospheric pressure storage of mango fruits. *Sci. Hortic.* 7, 153–160.

Apelbaum, A., Burgoon, A.C., Anderson, J.D., Solomos, T. and Lieberman, M. (1981) Some characteristics of the system converting 1-aminocyclopropane-1-carboxylic acid to ethylene. *Plant Physiol.* 67, 80–84.

Apelbaum, A., Goldhurst, A. and Icekson, I. (1985) Control by ethylene of arginine decarboxylase activity in pea seedlings and its implication for hormonal regulation of plant growth. *Plant Physiol.* 79, 635–640.

Archbold, H.K. (1932) Chemical studies in the physiology of apples. XII. Ripening processes in the apple and the relation of time of gathering to chemical changes in cold storage. *Ann. Bot.* 46, 407–459.

Armitage, M., Heins, R.H., Dean, S. and Carlson, W.H. (1980) Factors affecting petal abscission in the seed-propogated geranium. *J. Amer. Soc. Hort. Sci.* 105, 562–564.

Armour (1978) *Metmyoglobin formation in beef stored in Cryovac and Dormavac units.* Armour Food Company Technical Service Report, 3 March, 1978.

Aronoff, S. (1955) Translocation from soybean leaves. II. *Plant Physiol.* 30, 184–185.

Arteca, R.N. (1981) Non-permeant protein inhibitors. A possible tool in localization of ethylene synthesis. *Plant Physiol.* 67, 273.

Asen, S. and Lieberman, M. (1963) Ethylene oxide experimentation aimed at cut flower longevity. *Flor. Rev.* 131, 27–28.

Asen, S., Norris, K.H. and Stewart, R.N. (1971) Effect of pH and concentration of the anthocyanin-flavanol co-pigment complex on the color of 'Better Times' roses. *J. Amer. Soc. Hort. Sci.* 96, 770–773.

Ashby, B.H. (1970) *Protecting Perishable Foods During Transport by Motortruck.* Agricultural Handbook No. 105. ARS, USDA.

Atta-Aly, M.A., Saltveit, M.E. and Hobson, G.E. (1987) Effect of silver ions on ethylene biosynthesis by tomato fruit tissue. *Plant Physiol.* 83, 44–48.

Atta-Aly, M.A., Brecht, J.K. and Huber, D.J. (2000a) Ripening of tomato fruit locule gel tissue in response to ethylene. *Postharvest Biol. Technol.* 19, 239–244.

Atta-Aly, M.A., Brecht, J.K. and Huber, D.J. (2000b) Ethylene feedback mechanisms in tomato and strawberry fruit tissues in relation to fruit ripening and climacteric patterns. *Postharvest Biol. Technol.* 20, 151–162.

Audus, L.J. and Bakhsh, J.K. (1961) On the adaptation of pea roots to auxin and auxin homologues. In: Klein, R.M. (ed.) *Plant Growth Regulation.* 4th International Conference on Plant Growth Substances. Iowa State University Press, Ames, Iowa, pp. 109–126.

Autio, W.R. and Bramlage, W.J. (1982) Effect of AVG on maturation, ripening and storage of apples. *J. Amer. Soc. Hort. Sci.* 107, 1074–1077.

Autio, W.R. and Bramlage, W.J. (1986) Chilling sensitivity of tomato fruit in relation to ripening and senescence. *J. Amer. Soc. Hort. Sci.* 111, 201–204.

Avigad, G. and Dat, P.M. (1997) Carbohydrate metabolism: storage carbohydrates. In: Dey, P.M. and Harbone, J.B. (eds) *Plant Biochemistry.* Academic Press, San Diego, California, pp. 143–204.

Awad, M., Lamano, T.L.C. and Stradiotto, F.R. (1974) Effect of growth regulators and reduced pressure on the ripening of tomatoes (*Lycopersicon esculentum*). *Rev. Ceres* 21, 366–370.

Awad, M., Oliveira, A.I. and de Correa, D.L. (1975) The effect of ethephon, GA and partial vacuum on respiration in bananas (*Musa acuminata*). *Revista de Agricultura*, Piracicaba, Brazil 50, 109–113.

Ayub, R.A., Rombaldi, C., Petiprez, M., Latché, A., Pech, J.C. and Lelièvre, J.M. (1993) Biochemical and immunocytological characterization of ACC oxidase in transgenic grape cells. In: Pech, J.C., Latché, A. and Balagué, C. (eds) *Cellular and Molecular Aspects of the Plant Hormone Ethylene.* Kluwer Academic, Dordrecht, pp. 98–99.

Ayub, R.A, Guis, M., Ben Amor, M., Gillor, L., Roustan, J.P., Latché, A., Bouzayen, M. and Pech, J.C. (1996) Expression of ACC oxidase antisense gene inhibits ripening of cantaloupe melon fruits. *Nature Biotechnol.* 14, 862–866.

Back, E.A. and Cotton, R.T. (1925) The use of vacuum for insect control. *J. Agr. Res.* 31, 1035–1041.

Bain, J.M. (1958) Morphological, anatomical, and physiological changes in the developing fruit of Valencia orange, *Citrus senensis* (L.) Osbeck. *Austr. J. Botany* 6, 1–25.

Baker, J.E. (1975) Morphological changes during maturation and senescence. In: Pantastico, E.B. (ed.) *Postharvest Physiology, Handling and Utilization of Tropical and Subtropical Fruits and Vegetables.* AVI, Westport, Connecticut, pp. 128–147.

Baker, J.E. (1976) Superoxide dismutase in ripening fruits. *Plant Physiol.* 58, 644–647.

Baker, J.E., Wang, C.Y., Lieberman, M. and Hardenburg, R. (1977) Delay of senescence in carnations by a rhizobitoxine analog and sodium benzoate. *HortScience* 12, 38–39.

Baker, R.C., Qureshi, R.A. and Hotchkiss, J.H. (1986) Effect of an elevated level of carbon dioxide-containing atmosphere on the growth of spoilage and pathogenic bacteria at 2, 7 and 13°C. *Poultry Sci.* 65, 729.

Balatti, P.A. and Willemöes, J.G. (1989) Role of ethylene in the geotropic response of Bermudagrass (*Cynodon dactylon* L. Pers.) stolons. *Plant Physiol.* 91, 1251–1254.

Baldwin, E.A. (1993) Citrus fruits. In: Seymour, G.B., Taylor J.E. and Tucker, G.A. (eds) *Biochemistry of Fruit Ripening.* Chapman & Hall, London, pp. 107–149.

Baluska, F., Brailsford, R.W., Hauskrecht, M., Jackson, M.B. and Barlow, P.W. (1993) Cellular dimorphism in the maize root cortex: involvement of microtubules, ethylene, and gibberellin in the differentiation of cellular behavior in postmitotic growth zone. *Bot. Acta* 106, 394–403.

Baluska, F., Hauskrecht, M., Barlow, P.W. and Sievers, A. (1996) Gravitropism of the primary root of maize: a complex pattern of differential cellular growth in the cortex independent of the microtubule cytoskeleton. *Planta* 198, 310–318.

Bangerth, F. (1973) The effect of hypobaric storage on quality, physiology and storage life of fruits, vegetables and cut flowers. *Gartenbauwissenschaft* 38, 479–508.

Bangerth, F. (1974) Hypobaric storage of vegetables. *Acta Horticulturae* 1, 23–32.

Bangerth, F. (1975) The effect of ethylene on the physiology of ripening of apple fruits at hypobaric conditions. In: *Facteurs et Regulation de la Maturation des Fruits.* Colloques Internationaux de CNRS, Paris, 238, 183–188.

Bangerth, F. (1977) The effect of different partial pressures of carbon dioxide, ethylene and oxygen on the ascorbic acid content of vegetables. *Qual. Plant Foods Hum. Nutr.* 27, 125–134.

Bangerth, F. (1980) Function of abscisic-acid in the course of ripening of apple fruits. *Gartenbaumwissenschaft* 45, 224–228.

Bangerth, F. (1984) Changes in the sensitivity for ethylene during storage of apple and banana fruits under hypobaric conditions. *Scientia Hort.* (Amst.) 24, 151–163.

Bangerth, F. (1988) The effect of ethylene on the physiology of ripening of apple fruits at hypobaric conditions. In: *Facteurs et Régulation de la Maturation des Fruits.* Colloques Internationaux de CNRS, Paris, 238, 183–187.

Bangerth, F. and Streif, J. (1987) Effect of aminoethoxyvinylglycine and low-pressure storage on the post-storage production of aroma volatiles by Golden Delicious apples. *J. Sci. Food Agric.* 41, 351–360.

Bangga, M., Slaa, E.J., Blom, C.W.P.M. and Voesenek, L.A.C.J. (1996) Ethylene biosynthesis and accumulation under drained and submerged conditions. *Plant Physiol.* 112, 229–237.

Banks, H.J. and Annis, P.C. (1990) Comparative advantages of high CO_2 and low O_2 types of controlled atmospheres for grain storage. In: Calderon, M. and Barkai-Golan, R. (eds) *Food Preservation by Modified Atmospheres.* CRC Press, Boca Raton, Florida, pp. 93–122.

Banks, N.H. (1983) Evaluation of methods for determining internal gases in banana fruit. *J. Expt. Bot.* 34, 871–879.

Banks, N.H. (1985) The oxygen affinity of 1-aminocyclopropane-1-carboxylic acid oxidation in slices of banana fruit tissue. In: Roberts, J.A. and Tucker, G.A. (eds) *Ethylene and Plant Development.* Butterworths, London, pp. 29–36.

Banks, N.H., Elyatem, S. and Hammat, T. (1984) The oxygen affinity of ethylene production in apple slices. *Acta Horticulturae* 157, 257–260.

Bare, C.O. (1948) The effect of prolonged exposure to high vacuum on stored-product insects. *J. Econ. Ent.* 41, 109–110.

Barkai-Golan, R. (1990) Postharvest disease suppression by atmospheric modifications. In: Calderon, M. and Barkai-Golan, R. (eds) *Food Preservation by Modified Atmospheres.* CRC Press, Boca Raton, Florida, pp. 237–264.

Barker, A.V. and Corey, K.A. (1988) Ethylene evolution by tomato plants under nutrient stress. *HortScience* 23, 202–203.

Barmore, C.R. and Biggs, R.H. (1972) Ethylene diffusion through citrus leaf and fruit tissue. *J. Amer. Soc. Hort. Sci.* 97, 24–27.

Barnell, H.R. (1940) Studies on tropical fruits. VIII. Carbohydrate metabolism of the banana during ripening. *Ann. Bot. (N.S.)* 4, 39–71.

Barnell, H.R. (1941) Studies in tropical fruits. XI. Carbohydrate metabolism of the banana fruit during ripening under tropical condiitions. *Ann. Bot. (N.S.)* 5, 217–248.

Barr, C.E., Koh, M.S. and Ryan, T.E. (1974) NH_3 efflux as a means for measuring H^+ extrusion in *Nitella.* In: Zimmerman, U. and Dainty, J. (eds) *Membrane Transport in Plants.* Springer, New York, pp. 180–185.

Barry, C.S., Blume, B., Bouzayen, M., Cooper, W., Hamilton, A.J. and Grierson, D. (1996) Differential expression of the 1-aminocyclopropane-1-carboxylate oxidase gene family of tomato. *Plant J.* 9, 525–535.

Barry, C.S., Llop-Tous, M.I. and Grierson, D. (2000) The regulation of 1-aminocyclopropane-1-carboxylic acid synthase gene expression during the transition from system-1 to system-2 ethylene synthesis in tomato. *Plant Physiol.* 123, 979–986.

Bartlemew, E.T. (1926) Internal decline of lemons. III. Water deficit in lemon fruits caused by excessive leaf evaporation. *Amer. J. Bot.* 13, 102–117.

Bartley, J.M. (1984) Lipid metabolism in ripening apples. *Phytochemistry* 24, 2857–2859.

Bauchot, A.D., Mottram, D.S., Dodson, A.T. and John, P. (2000) Role of ethylene in aroma formation in cantaloupe Charentais melon. *Plant Growth Regl. Abstracts* 26, 136.

Baur, J.R. and Morgan, P.W. (1969) Effect of Picloram and ethylene on leaf movement in Huisache and Mesquite seedlings. *Plant Physiol.* 44, 831–838.

Baur, J.R. and Workman, M. (1964) Relationship between cell permeability and respiration in ripening banana fruit tissue. *Plant Physiol.* 39, 540–543.

Baur, P., Buchholz, A. and Schönherr, J. (1997) Diffusion in plant cuticles as affected by temperature and size of organic solutes: similarity and diversity among species. *Plant Cell Environ.* 20, 982–984.

Beament, J.W.S. (1946) The waterproofing process in eggs of *Rhodinius prolixus. Stähl. Proceedings of the Royal Society of London B,* 133, 407–418.

Bean, R.C. (1962) Photosynthesis in green fruits. *Calif. Agr.* 16, 13.

Beaudry, R.M. and Gran, C.D. (1993) Using a modified atmosphere packaging approach to answer some postharvest questions: factors influencing the lower oxygen limit. *Acta Horticulturae* 326, 203–212.

Beebe, S.D., Vanderzant, C., Hanna, M.O., Carpenter, Z.L. and Smith, G.C. (1976) Effect of initial internal temperature and storage temperature on the microbial flora of vacuum-packaged beef. *J. Milk Food Technol.* 39, 600–605.

Beevers, H. (1961) Respiratory Metabolism in Plants. *Row-Peterson Biological Monographs.*

Bekman, E.P., Saibo, N.J.M., De Cataldo, A., Regalado, A.P., Ricardo, C.P. and Rodrigues-Pousada, C. (2000) Differential expression of four genes encoding aminocyclopropane-1-carboxylate synthase in *Lupinus albus* during germination, and in response to indole-3-acetic acid and wounding. *Planta* 211, 663–672.

Ben-Amor, M., Flores, B., Latché, A., Bouzzayen, M., Pech, J.C. and Romojaro, F. (1999) Inhibition of ethylene biosynthesis in antisense ACC oxidase RNA prevents chilling injury in Charentais cantaloupe melons. *Plant Cell and Environment* 22, 1579–1586.

Bendall, D.S., Ransom, S.L. and Walker, D.A. (1960) Effects of CO_2 on the oxidation of succinate and reduced diphosphopyridine nucleotide by *Ricinus* mitochondria. *Biochem. J.* 76, 221.

Bender, R.J., Brecht, J.K., Sargent, S.A. and Huber, D.J. (2000a) Low-temperature controlled-atmosphere storage for tree ripe mangoes (*Mangifera indica* L.). *Acta Horticulturae* 509, 447–458.

Bender, R.J., Brecht, J.K., Sargent, S.A. and Huber, D.J. (2000b) Mango tolerance to reduced oxygen levels in controlled atmosphere. *J. Amer. Soc. Hort. Sci.* 125, 707–713.

Bengochea, T., Dodds, J.H., Evans, D.E., Jerie, P.H., Niepel, B., Shaari, A.R. and Hall, M.A. (1980) Studies on ethylene binding by cell-free preparations from cotyledons of *Phaseolus vulgaris* L. I: separation and characterization. *Planta* 148, 397–406.

Bennet-Clark, T.A., Younis, A.F. and Esnault, R. (1959) Geotropic behavior of roots. *J. Expt. Bot.* 10, 69–86.

Bennett, M.J., Marchant, A., Green, H.G., May, S.T., Ward, S.P., Millner, P.A., Walker, A.R., Schultz, B. and Feldman, K.A. (1996) *Arabidopsis AUX1* gene: a permease-like regulator of root gravitropism. *Science* 273, 948–950.

Ben-Yehoshua, S. (1964) Respiration and ripening of fruit discs of the avocado fruit. *Physiol. Plant.* 17, 71–80.

Ben-Yehoshua, S. (1969) Gas exchange, transpiration, and the commercial deterioration in storage of orange fruit. *J. Amer. Soc. Hort. Sci.* 94, 524–528.

Ben-Yehoshua, S. (1978) Delaying deterioration of individual citrus fruit by seal-packaging in film of high-density polyethylene. I. General effects. *Proceedings of the 3rd International Citrus Congress, Sydney, Australia*, pp. 110–115.

Ben-Yehoshua, S. (1986) Transpiration, water stress, and gas exchange of fruits and vegetables. In: Weichmann, J. (ed.) *Postharvest Physiology of Vegetables.* Marcel Dekker, New York, Chapter II F.

Ben-Yehoshua, S. (1989) Individual seal-packaging of fruit and vegetables in plastic film. In: Brody, A.L. (ed.) *Controlled/ Modified Atmosphere/Vacuum Packaging of Foods.* Food & Nutrition Press, Inc. Trumbell, Connecticut, pp. 101–119.

Ben-Yehoshua, S. and Aloni, B. (1974a) Effects of water stress upon ethylene production by detached leaves of Valencia oranges. *Plant Physiol.* 53, 863–865.

Ben-Yehoshua, S. and Aloni, B. (1974b) Endogenous ethylene and abscission of orange-leaf explants. *Bot. Gaz.* 135, 41–44.

Ben-Yehoshua, S. and Eaks, I.L. (1970) Ethylene production and abscission of fruit and leaves of orange. *Bot. Gaz.* 131, 144–150.

Ben-Yehoshua, S., Robertson, R.N. and Biale, J.B. (1963) Respiration & internal atmosphere of avocado fruit. *Plant Physiol.* 38, 194–201.

Ben-Yehoshua, S., Juven, B., Fruchter, M. and Halevy, A.H. (1966) Effects of ethylene oxide on opening and longevity of cut rose flowers. *J. Amer. Soc. Hort. Sci.* 89, 677–682.

Ben-Yehoshua, S., Kobiler, H. and Shapiro, B. (1979) Some physiological effects of delaying deterioration of citrus fruits by individual seal packaging in high-density polyethylene film. *J. Amer. Soc. Hort. Sci.* 104, 868–872.

Ben-Yehoshua, S., Burg, S.P. and Young, R. (1985) Resistance of citrus fruit to C_2H_4, O_2, CO_2, and H_2O mass transport. *Plant Physiol.* 79, 1048–1053.

Bérard, L.S. (1982) Crumbly breakdown an internal disorder of McIntosh apples held in low-pressure storage. *Phytoprotection* 63, 48–55.

Bérard, L.S. and Lougheed, E.C. (1978) Is low-pressure storage an alternative to controlled atmosphere for McIntosh apples? *HortScience* 13, 351.

Bérard, L.S. and Lougheed, E.C. (1982) Electrolyte leakage from daminozide-treated apples *Malus-Domestica* cultivar McIntosh held in air, low pressure, and controlled atmosphere storage. *J. Amer. Soc. Hort. Sci.* 107, 421–425.

Bérard, L.S., Lougheed, E.C., Murr, D.P. and Miller, S.R. (1976) Low-pressure storage of McIntosh apples. *Proceedings of the Can. Soc. Hort. Sci.* 97, 207–209.

Berrios, J.D.J., Swanson, B.G. and Cheong, W.A. (1998) Structural characteristics of stored black beans (*Phaseolus vulgaris* L.). *Scanning* 20, 410–417.

Berrios, J.D.J., Swanson, B.G. and Cheong, W.A. (1999) Physico-chemical characterization of stored black beans (*Phaseolus vulgaris* L.). *Food Research International* 32, 669–676.

Bertell, G., Bolander, E. and Eliasson, L. (1990) Factors increasing ethylene production enhance the sensitivity of root growth to auxins. *Physiol. Plant.* 79, 255–258.

Betts, G.D. (ed.) (1996) *A Code of Practice for the Manufacture of Vacuum and Modified Atmosphere Packaged Chilled Foods.* Campden and Chorleywood Food Research Association Guideline 11.

Beutelmann, P.H. and Kende, H. (1977) Membrane lipids in senescing flower tissue of *Opomoea tricolor. Plant Physiol.* 58, 888–893.

Beyer, E.M. (1975a) $^{14}C_2H_4$, its incorporation and metabolism in pea seedlings under aseptic conditions. *Plant Physiol.* 56, 273–278.

Beyer, E.M. (1975b) Abscission: support for a role of ethylene modification of auxin transport. *Plant Physiol.* 52, 1–5.

Beyer, E.M. (1975c) Abscission: the initial effect of ethylene is in the leaf blade. *Plant Physiol.* 55, 322–327.

Beyer, E.M. (1976) A potent inhibitor of ethylene action in plants. *Plant Physiol.* 58, 269–271.

Beyer, E.M. (1979) Effect of silver ion, carbon dioxide, and oxygen on ethylene action and metabolism. *Plant Physiol.* 63, 169–173.

Beyer, E.M. (1980) Recent advances in ethylene metabolism. *Dutch Plant Growth Regulator Group/British Plant Growth Regulator Group Monograph* 6, 35.

Beyer, E.M. (1985) Ethylene metabolism. In: Roberts, J.A. and Tucker, G. (eds) *Ethylene and Plant Development.* Butterworths, London, p. 125.

Beyer, E.M. and Morgan, P.W. (1970a) A method for determining the concentration of ethylene in the gas phase of plant tissues. *Plant Physiol.* 46, 352–354.

Beyer, E.M. and Morgan, P.W. (1970b) Effect of ethylene on the uptake, distribution, and metabolism of indoleacetic acid-1-^{14}C and -2-^{14}C and naphthaleneacetic acid 1-^{14}C. *Plant Physiol.* 63, 325–331.

Beyer, E.M. and Morgan, P.W. (1971) The role of ethylene modification of auxin transport. *Plant Physiol.* 48, 208–212.

Bhambhani, H.J. (1956) Responses of pests to fumigation – VI. Water loss and the mortality of *Calandra* sp. at reduced pressures. *Bull Ent. Res.* 47, 749–753.

Biale, J.B. (1946) Effect of oxygen concentration on respiration of the Fuerte avocado fruit. *Amer. J. Bot.* 33, 363–373.

Biale, J.B. (1954) Physiological requirement of citrus fruits. *Citrus Leaves* 34, 6–7, 31, 33.

Biale, J.B. (1960) The post-harvest biochemistry of tropical and subtropical fruits. *Advanc. Food Res.* 10, 293–354.

Biale, J.B. (1964) Growth, maturation and senescence in fruits. *Science* 146, 880–888.

Biale, J.B. and Young, R.E. (1971) The avocado pear. In: Hulme, A.C. (ed.) *The Biochemistry of Fruits and Their Products*, vol. 2. Academic Press, London, pp. 1–63.

Biale, J.B., Young, R.E. and Olmstead, A.J. (1954) Fruit respiration and ethylene production. *Plant Physiol.* 29, 168–174.

Biddulph, O. and Markle, J. (1944) Translocation of radioactive phosphorous in the phloem of the cotton plant. *Amer. J. Bot.* 31, 551–555.

Bieleski, R.L. (1993) Fructan hydrolysis drives petal expansion in the ephemeral day lily flower. *Plant Physiol.* 103, 213–219.

Biermann, B., Johnson, E.M. and Feldman, L.J. (1990) Characterization and distribution of a maize cDNA encoding a peptide similar to the catalytic region of second messenger dependent protein kinases. *Plant Physiol.* 94, 1609–1615.

Biggs, M.S., Woodson, W.R. and Handa, A.K. (1988) Biochemical basis of high-temperature inhibition of ethylene biosynthesis in ripening tomato fruits. *Physiol. Plant.* 72, 572–578.

Bischoff, J. (1984) Dry-aging beef: is it worth it? *Meat Ind.* 30, 12.

Bishop, D.W. (1952) Respiration and metabolism. In: Prosser, C.L. (ed.) *Comparative Animal Physiology.* W.B. Saunders, Philadelphia, Pennsylvania, pp. 209–289.

Bjorkman, T. (1988) Perception of gravity in plants. *Adv. Bot. Res.* 15, 1–41.

Blackman, F.F. (1953) Respiratory drifts. In: James, W.O. (ed.) *Plant Respiration.* Oxford at the Clarendon Press, London, pp. 40–62.

Blackman, F.F. (1954) *Analytic Studies in Plant Respiration.* Cambridge University Press, London.

Blake, T.J., Pharis, R.P. and Reid, D.M. (1980) Ethylene, gibberellins, auxin and the apical control of branch angle in a conifer, *Cupressus arizonica. Planta* 148, 64–68.

Blake, T.J., Reid, D.M. and Rood, S.B. (1983) Ethylene, indoleacetic acid and apical dominance in peas: a reappraisal. *Physiol. Plant.* 59, 481–487.

Blancaflor, E.B. and Hasenstein, K.H. (1993) Organization of microtubules in gravi-responding corn roots. *Planta* 191, 231–237.

Blancaflor, E.B. and Hasenstein, K.H. (1995) Time course and auxin sensitivity of cortical microtubule reorientation in maize roots. *Protoplasma* 185, 72–82.

Blanke, M.M. (1991) Respiration of apple and avocado fruits. *Postharv. News Inf.* 2, 429–436.

Blanke, M.M. and Bonn, E.L. (1985) Spaltöffnungen, fruchtoberflache und transpiration wachsender apelfrüchte der sorte 'Golden Delicious'. *Erwebsobstbau* 27, 139–143.

Blankenship, S.M. and Richardson, D.G. (1985) Development of ethylene biosynthesis and ethylene-induced ripening in d'Anjou pears during the cold requirement for ripening. *J. Amer. Soc. Hort. Sci.* 107, 807–812.

Blankenship, S.M. and Sisler, E.C. (1988) 2,5-Norbornadiene retards apple softening. *HortScience* 24, 313–314.

Blankenship, S.M. and Sisler, E.C. (1989) Ethylene binding changes in apple and morning glory during ripening and senescence. *J. Plant Growth Regul.* 8, 37–44.

Blankenship, S.M. and Sisler, E.C. (1993) Ethylene binding site affinity in ripening apples. *J. Amer. Soc. Hort. Sci.* 118, 609–612.

Blanpied, G.D. (1971) Apparatus for ethylene extraction from plant tissues. *Hort. Sci.* 6, 132–134.

Blanpied, G.D. (1972a) A study of ethylene in apple, red raspberry and cherry. *Plant Physiol.* 49, 627–630.

Blanpied, G.D. (1972b) Ethylene in apple and pear experimental CA chambers. *J. Amer. Soc. Hort. Sci.* 97, 207–209.

Blanpied, G.D. (1990a) Controlled atmosphere storage of apples and pears. In: Calderon, M. and Barkai-Golan, R. (eds) *Food Preservation by Modified Atmospheres.* CRC Press, Boca Raton, Florida, pp. 265–298.

Blanpied, G.D. (1990b) Low-ethylene CA storage for apples. *Postharvest News Inform.*1, 29–34.

Bleecker, A.B. and Patterson, S.E. (1997) Last exit: senescence, abscission and meristem arrest in *Arabidopsis. Plant Cell* 9, 1169–1179.

Bleecker, A.B., Estelle, M.A., Somerville, C. and Kende, H. (1988) Insensitivity to ethylene conferred by a dominant mutation in *Arabidopsis thaliana. Science* 241, 1086–1089.

Bollard, E.G. (1960) Transport in the xylem. *Annu. Rev. Plant Physiol.* 11, 141–166.

Bollard, E.G. (1970) The physiology and nutrition of developing fruits. In: Hulme, A.C. (ed.) *The Biochemistry of Fruits and Their Products*, vol. 1. Academic Press, London, pp. 387–425.

Boller, T.H. and Kende, H. (1980) Regulation of wound ethylene synthesis in plants. *Nature* 286, 259–260.

Bolwell, G.P. (1992) A role for phosphorylation in the down-regulation of phenylalanine ammonia lyase in suspension-cultured cells of French bean. *Phytochemistry* 31, 4081–4086.

Bolwell, G.P. (1995) Cyclic AMP, the reluctant messenger in plants. *Trends Biochem. Sci.* 20, 492–495.

Bommineni, V.R., Mathews, H., Clendennen, D.K., Wagoner, W., Dewey, V., Kellogg, J., Peters, S., Matsumura, W., Pieper, M., Kramer, M. and Wagner, D.R. (2000) Genetic engineering of fruits and vegetables with the ethylene control gene encoding S-adenosylmethionine hydrolase (SAMase). *Plant Growth Regl. Abstracts* 26, 311.

Bond, G. (1952) Some features of root growth in nodulated plants of *Myrica gale* L. *Ann. Bot.* 16, 467.

Bonghi, C., Tonutti, P. and Ramina, A. (2001) Fruit maturation: molecular aspects of the biosynthesis and the mechanism in ethylene production. *Plant Growth Regl. Abstracts* 27, 345.

Bonner, J. (1944) Accumulation of various substances in girdled stem of tomato plants. *Amer. J. Bot.* 31, 551–555.

Bonner, J. (1950) *Plant Biochemistry.* Academic Press, New York.

Bonner, W.D. (1957) Soluble oxidases and their functions. *Annu. Rev. Plant Physiol.* 8, 427–452.

Bonner, W.D. (1965) Mitochondria and electron transport. In: Bonner, W.D. and Varner, J.E. (eds) *Plant Biochemistry.* Academic Press, London, pp. 98–123.

Borch, E. and Agerheim, H. (1992) Chemical, microbial and sensory changes during the anaerobic cold storage of beef innoculated with a homofermentative *Lactobacillus* sp. or *Leuconostoc* sp. *Int. J. Food Microbiol.* 15, 99–108.

Borecka, H. (1986) The quality of sweet cherries stored under hypobaric conditions. *Acta Agrobotanica* 39, 195–200.

Borecka, H. and Olak, J. (1978) The effect of hypobaric storage conditions on the growth and sporulation of some pathogenic fungi. *Fruit Sci. Rep. Poland* 5, 39.

Borecka, H. and Parynow, H. (1985a) Effect of tomato postharvest fungicide treatment and storage conditions on the quality of fruits and biological value of tomato pulp and concentrated pulp. *Acta Agrobot* 35, 257–268.

Borecka, H. and Parynow, H. (1985b) Effect of tomato postharvest treatment and storage conditions on the biological activity of benomyl and methylthiophanate residues in fruits and their processed products. *Acta Agrobot* 35, 269–276.

Borgström, G. (1939a) *The Transverse Reactions of Plants.* Williams & Northgate Ltd, London.

Borgström, G. (1939b) Theoretical suggestions regarding the ethylene responses of plants and observations on the influence of apple-emanations. *Kungl. Fysiografiska Sallskapets* I *Lund Förhandlingar* 9, 135–174.

Borochov, A. and Woodson, W.R. (1989) Physiology and biochemistry of flower petal senescence. *Hortic. Rev.* 11, 15–43.

Borochov, A., Drori, A., Tirosh, T., Borochov-Neori, H. and Mayak, S. (1990) Quantitative and qualitative changes in membrane proteins during petal senescence. *J. Plant Physiol.* 136, 203–207.

Borochov, A., Spiegelstein, H. and Philosoph-Hadas, S. (1997) Ethylene and flower petal senescence: interrelationship with membrane lipid catabolism. *Physiol. Plant.* 100, 606–612.

Boron, W.F. and De Weer, P. (1976a) Intracellular pH transients in squid giant axons caused by CO_2, NH_3 and metabolic inhibitors. *J. Gen. Physiol.* 67, 91–112.

Boron, W.F. and De Weer, P. (1976b) Active proton transport stimulated by CO_2/HCO_3^-, blocked by cyanide. *Nature* (London) 259, 240–241.

Botjes, J.O. (1942) Aethyleen als stimulans bij hetauxineverbruik bij erwten en tomaten. *Proceedings of the Nederland Ajad. Wetensch* 45, 999–1002.

Botondi, R., Cardarelli, M.T. and Mencarelli, F. (2000) The role of ethylene in regulating cell-wall-degrading enzyme activity using antisense ACC-oxidase in cantaloupe melons. *Plant Growth Regl. Abstracts* 26, 228.

Bouzayen, M., Latché, A., Pech, J.C. and Marigo, G. (1989) Carrier-mediated uptake of 1-(malonylamino) cyclopropane 1-carboxylic acid in vacuoles isolated from *Chararanthus roseus* cells. *Plant Physiol.* 91, 1317–1322.

Bouzayen, M., Latché, A. and Pech, J.C. (1990) Subcellular localization of the sites of conversion of 1-aminocyclopropane-1-carboxylic acid into ethylene in plant cells. *Planta* 180, 175–180.

Bouzayen, M., Cooper, W., Barry, C., Zegzouti, H., Hamilton, A.J. and Grierson, D. (1993) EFE multigene family in tomato plants: expression and characterization. In: Pech, J.C., Latché, A. and Balagué, C. (eds) *Cellular and Molecular Aspects of the Plant Hormone Ethylene.* Kluwer Academic, Dordrecht, pp. 76–81.

Bowler, C. and Chua, N.H. (1994) Emerging themes of plant signal transduction. *Plant Cell* 6, 1529–1541.

Bowyer, J.R. and Leegood, R.C. (1997) Chapter 2. Photosynthesis. In: Dey, P.M. and Harbone, J.B. (eds) *Plant Biochemistry.* Academic Press, San Diego, California, pp. 49–110.

Boyer, J.S. (1976) Water deficits and photosynthesis. In: Kozlowski, T.T. (ed.) *Deficits and Plant Growth*, vol. 4. Academic Press, New York, pp. 153–190.

Boyer, J.S., Wong, S.C. and Farquhar, G.D. (1997) CO_2 and water vapor exchange across the leaf cuticle (epidermis) at various water potentials. *Plant Physiol.* 114, 185–191.

Bradford, K.J. (1983) Involvement of plant growth substances in the alteration of leaf gas exchange of flooded tomato plants. *Plant Physiol.* 73, 480–483.

Bradford, K.J. and Dilley, D.R. (1978) Effect of root anaerobiosis on ethylene production, epinasty and growth of tomato plants. *Plant Physiol.* 621, 506–509.

Bradford, K.J. and Hsiao, T.C. (1982) Stomatal behaviour and water relations of waterlogged tomato plants. *Plant Physiol.* 78, 1508–1513.

Bradford, K.J. and Yang, S.F. (1980) Xylem transport of 1-aminocyclopropane-1-carboxylic acid, an ethylene precursor, in waterlogged tomato plants. *Plant Physiol.* 65, 322–326.

Brady, C.J. (1987) Fruit ripening. *Annu. Rev. Plant Physiol.* 38, 155–178.

Brady, C.J., O'Connell, P.B.H., Smydzuk, J. and Wade, N.L. (1970) Permeability, sugar accumulation, and respiration rate in ripening banana fruits. *Austr. J. Biol. Sci.* 23, 1143–1152.

Brauner, L. and Zipperer, A. (1961) Über die Aufangsphasen der geotropischen Krümmungsbeweegung von *Avena*-Koleoptilen. *Planta* 57, 503–517.

Brecht, J.K., Chen, W.-X., Sargent, S.A., Cordasco, K. and Bartz, J.A. (2001) Exposure of green tomatoes to hot water affects ripening and reduced decay and chilling injury. *Plant Growth Regl. Abstracts* 27, 56.

Bredmose, N. (1979) The influence of subatmospheric pressure on storage life and keeping quality of cut flowers of 'Belinda' roses. *Acta Agriculturae Scandinavica* 29, 287–290.

Bredmose, N. (1980a) Afskårne roser kan lagres en måned ved lavt tryk. *Statens Planteavisforsøg*, Meddelelse nr. 1557, 82.

Bredmose, N. (1980b) Effects of low pressure on storage life and subsequent keeping quality of cut roses. *Acta Horticulturae* 113, 1–7.

Breed, R.S., Murray, E.G.D. and Smith, N.R. (1957) *Bergey's Manual of Determinative Bacteriology*, 7th edn. Williams and Wilkins, Baltimore, Maryland.

Brennan, E., Leone, I. and Daines, R.H. (1962) Ammonia injury to apples and peaches in storage. *Plant Dis. Rptr.* 46, 792–795.

Bridgen, M.P. and Staby, G.L. (1979) Low pressure and controlled atmosphere storage of tissue cultured *Nicotiana tabacum* 'Wisconsin 38'. *HortScience* 14, 47.

Bridgen, M.P. and Staby, G.L. (1981) Low pressure and low oxygen storage of *Nicotiana tabacum* and *Chrysanthemum* × *morifolium* tissue cultures. *Plant Science Letters* 22, 177–186.

Bridgen, M.P. and Staby, G.L. (1983) Protocols of low-pressure storage for plant tissue cultures. In: Evans, D.A., Sharp, W.R., Ammirato, P.V. and Yamada, Y. (eds) *Handbook of Plant Cell Culture*, volume I. Macmillan, New York, pp. 816–825.

Bridgen, M.P., Staby, G.L. and Sharp, W.R. (1978) Effects of low-pressure storage on the preservation of tissue cutlured *Chrysanthemum* × *morifolium*. *4th International Congress of Plant Tissue and Cell Culture*, Calgary, Canada, 20–25 August, 1978.

Briggs, G.E. and Robertson, R.N. (1957) Apparent free space. *Annu. Rev. Plant Physiol.* 8, 11–30.

Brink, J.A. and de Swart, G.H. (1986) The effect of sucrose in a vase solution on leaf browning of *Protea neriifolia*. *Acta Horticulturae* 185, 111–119.

Brisker, H.E. (1980) Delaying grapefruit harvest: hormonal and enzymatic control by endogenous and applied regulators. PhD dissertation, the Hebrew University of Jerusalem.

Brody, A.L. (1989a) Microbiological safety of modified/controlled atmosphere/vacuum packed foods. In: Brody, A.L. (ed.) *Controlled/Modified Atmosphere/Vacuum Packaging of Foods*. Food & Nutrition Press, Trumbell, Connecticut, pp. 159–174.

Brody, A.L. (1989b) Modified atmosphere packaging of seafoods. In: Brody, A.L. (ed.) *Controlled/Modified Atmosphere/Vacuum Packaging of Foods*. Food & Nutrition Press, Trumbell, Connecticut, pp. 59–65.

Brody, A.L. (1989c) Modified atmosphere/vacuum packaging of meat. In: Brody, A.L. (ed.) *Controlled/Modified Atmosphere/Vacuum Packaging of Foods*. Food & Nutrition Press, Trumbell, Connecticut, pp. 17–37.

Broglie, K.R., Biddle, P., Cressman, R. and Broglie, R. (1989) Functional analysis of DNA sequences responsible for ethylene regulation of a bean chitanase gene in transgenic tobacco. *Plant Cell* 1, 599–607.

Brooks, C. (1937) Some effects of waxing tomatoes. *Proceedings of the Am. Soc. Hort. Sci.* 35, 720.

Brooks, J. (1929) Post-mortem formation of metmyoglobin in red muscle. *Biochem. J.* 23, 1391.

Brooks, J. (1935) The oxidation of haemoglobin to methaemoglobin by oxygen. II. The relation between the rate of oxidation and the partial pressure of oxygen. *Proceedings of the Roy Soc. London* (Ser. B.) 118, 560.

Brooks, S.C. (1916) Studies in exosmosis. *Amer. J. Bot.* 3, 483–492.

Brown, A.E. and Swinburne, T.R. (1978) Stimulants of germination and appressorial formation by *Diaporthe permiciosa* in apple leachate. *Trans. Br. Mycol. Soc.* 71, 405.

Brown, A.W. (1985) CO_2 and intracellular pH. *Plant Cell and Environment* 8, 459–465.

Brown, G.E. (1978) Hypersensitive response of orange-colored Robinson tangerines to *Colletotrichum gloeosporioides* after ethylene treatment. *Phytopathology* 68, 700–706.

Brown, G.E. and Barmore, C.R. (1977) The effect of ethylene on the susceptibility of Robinson tangerines to anthracnose. *Phytopathology* 67, 120–123.

Brown, G.E. and Wilson, W.C. (1968) Mode of entry of *Diplodia natalensis* and *Phomopsis citri* into Florida oranges. *Phytopathology* 58, 736.

Brown, J.H., Legge, R.L., Sisler, E.C., Barker, J.E. and Thompson, J.E. (1986) Ethylene binding in senescing carnation petals. *J. Expt. Bot.* 37, 526–534.

Brown, K.M. (1997) Ethylene and abscission. *Physiol. Plant.* 100, 567–576.

Brown, W.D., Harris, L.S. and Olcott, H.S. (1963) Catalysis of unsaturated lipid oxidation by iron protoporphyrin derivatives. *Arch. Biochem. Biophys.* 101, 14.

Browning, G. (1974) 2-chloroethanephosphonic acid reduces transpiration and stomatal opening in *Coffea arabica* L. *Planta* 121, 175–179.

Brownleader, M.D., Harbon, J.B. and Dey, P.M. (1997) Carbohydrate metabolism: primary metabolism of monosaccharides. In: Dey, P.M. and Harbone, J.B. (eds) *Plant Biochemistry*. Academic Press, San Diego, California, pp. 111–141.

Bruinsma, J. (1983) Hormonal regulation of senescence, ageing, fading and ripening. In: Lieberman, M. (ed.) *Postharvest Physiology and Crop Preservation*. Plenum Press, New York, pp. 141–163.

Bryan, A.H., Bryan, C.A. and Bryan, C.G. (1962) *Bacteriology, Principles and Practice*, 6th edn. Barnes & Noble Books, New York.

Bubb, M. (1975a) Hypobaric storage. *Rpt. E. Malling Res. Sta.* 1974. Maidstone, Kent, p. 81.

Bubb, M. (1975b) Storage of soft fruits. Hypobaric storage. *Rpt. E. Malling Res. Sta.* 1974. Maidstone, Kent, p. 83.

Bucher, D. and Pilet, P.E. (1982) Ethylene effects on growing and gravireacting maize root segments. *Physiol. Plant.* 55, 1–4.

Bucher, D. and Pilet, P.E. (1983) Auxin effect on root growth and ethylene production. *Experientia* 39, 493–494.

Buchsbaum, R. (1948) *Animals Without Backbones*. University of Chicago Press.

Buck, J.B. (1962) Some physical aspects of insect respiration. *Annu. Rev. Entomology* 7, 27–56.

Buck, J.B. and Keisster, M.L. (1949) Tracheal filling in *Sciara* larvae. *Biol. Bull.* 97, 323–330.

Buddenbrock, W. and Rohr, G. (1922) Respiration of *Dixippus* and *Dytiscus* larvae. *Arch. Ges. Physiol.* 194, 218–236.

Buescher, R.W. and Adams, K. (1977) Reducing discoloration and quality deterioration in snap beans (*Phaseolus vulgaris*) by atmospheres enriched in CO_2. *Acta Horticulturae* 62, 55.

Buescher, R.W. and Adams, K. (1983) Basis for enhanced softening of snap beans (*Phaseolus vulgaris*) from postharvest exposure to carbon dioxide, *J. Amer. Soc. Hort. Sci.*108, 58.

Buescher, R.W., Sistrunk, W.A. and Brady, P.L. (1975) Effects of ethylene on metabolic and quality attributes of sweet potato. *J. Food Sci.* 40, 1018–1020.

Bufler, G. (1984) Ethylene-enhanced 1-aminocyclopropane-1-carboxylic acid synthase activity in ripening apples. *Plant Physiol.* 75, 192–195.

Bufler, G. (1986) Ethylene-promoted conversion of 1-aminocyclopropane-1-carboxylic acid to ethylene in peel of apple at various stages of fruit development. *Plant Physiol.* 80, 539–543.

Bufler, G. and Bangerth, F. (1983) Effects of propylene and oxygen on the ethylene-producing system of apples (*Malus sylvestris* cultivar Golden Delicious). *Plant Physiol.* 58, 486–492.

Bufler, G., Mor, Y., Reid, M.S. and Yang, S.F. (1980) Changes in 1-aminocyclopropane-1-carboxylic acid content of cut carnation flowers in relation to their senescence. *Planta* 150, 439–442.

Bui, A.Q. and O'Neil, S.D. (1998) Three 1-aminocyclopropane-1-carboxylate synthase genes regulated by primary and secondary pollination signals in orchid flowers. *Plant Physiol.* 116, 419–428.

Bulard, C. (1948) Inversion du georopisme par l'heteroauxine chez *Thuya orientalis*. *Comptes Rendus des Sciences*, Paris 227, 443.

Burg, S.P. (1958) Biogenesis of ethylene. PhD dissertation, Harvard University, Cambridge, Massachusetts.

Burg, S.P. (1962a) The physiology of ethylene formation. *Annu. Rev. Plant Physiol.* 13, 265–302.

Burg, S.P. (1962b) Vapour phase chromatography. In: *Modern Methods of Plant Analysis*, vol. 5. Springer, Berlin, pp. 97–158.

Burg, S.P. (1964) Studies on the formation and function of ethylene gas in plant tissues. *Régulateurs Naturels de la Croissance Végétale* 123, 719–725.

Burg, S.P. (1967) Method for storing fruit. US Patent 3,333,967 and US Patent reissue Re. 28,995 (1976).

Burg, S.P. (1968a) Ethylene formation in pea seedlings: its relation to the inhibition of bud growth caused by indole-3-acetic acid. *Plant Physiol.* 43, 1069–1074.

Burg, S.P. (1968b) Ethylene, plant senescence and abscission. *Plant Physiol.* 43, 1503–1511.

Burg, S.P. (1969) Progress Reports for United Fruit Co.

Burg, S.P. (1970) Progress Reports for United Fruit Co.

Burg, S.P. (1971) Progress Reports for United Fruit Co.

Burg, S.P. (1973a) Ethylene in plant growth. *Proc. Natl Acad. Sci. USA* 70, 591–597.

Burg, S.P. (1973b) Hypobaric storage of flowers. *HortScience* 8, 202–205.

Burg, S.P. (1975) Hypobaric storage and transportation of fresh fruits and vegetables. In: Haard, N.F. and Salunkhe, D.K. (eds) *Postharvest Biology and Handling of Fruits and Vegetables*. AVI, Westport, Connecticut, pp. 172–188.

Burg, S.P. (1976a) Extending the storage life of green sweet corn by means of hypobaric technology. *Low Pressure Storage Systems Laboratory Research Report.*

Burg, S.P. (1976b) Low temperature hypobaric storage of metabolically active matter. US patents 3,958,028 and 4,061,483.

Burg, S.P. (1977) Hypobaric preservation of fresh foods – principles of the system. *Second Annual World Fair for Technology Exchange*, 7–11 February, 1977. *First World Congress for Technology Transfer*, pp. 115–127.

Burg, S.P. (1978a) Hypobaric storage of green snap beans. Report for Grumman Allied Industries, Inc., Garden City, New York.

Burg, S.P. (1978b) Spoilage of yellow papayas during shipment from Hawaii. Report for Grumman Allied Industries, Inc., Garden City, New York.

Burg, S.P. (1979a) Apparatus and procedure for determining respiratory heat load, fermentation and the ability to cool by diffusive escape of water, and ability to vacuum cool individual products. Report for Grumman Allied Industries, Inc., Garden City, New York

Burg, S.P. (1979b) Ice crystal formation in lamb. Report for Grumman Allied Industries, Inc., Garden City, New York.

Burg, S.P. (1979c) Results of Continental farms carnation test. Report for Grumman Allied Industries, Inc., Garden City, New York.

Burg, S.P. (1980a) Mushroom storage. Report for Grumman Allied Industries, Inc., Garden City, New York.

Burg, S.P. (1980b) Shelf life of lamb – Australia/Miami shipment. Report for Grumman Allied Industries, Inc., Garden City, New York.

Burg, S.P. (1981) Storage of boxed lamb. Report for Grumman Allied Industries, Inc., Garden City, New York.

Burg, S.P. (1987a) Hypobaric storage of non-respiring animal matter without supplementary humidification. US Patent No. 4,655,048

Burg, S.P. (1987b) Hypobaric storage of respiring plant matter without supplementary humidification. US Patent No. 4,685,305

Burg, S.P. (1990) Theory and practice of hypobaric storage. In: Calderon, M. and Barkai-Golan, R. (eds) *Food Preservation by Modified Atmospheres.* CRC Press, Boca Raton, Florida, pp. 353–372.

Burg, S.P. (1992) Current status of hypobaric storage. *Acta Horticulturae* 326, 259–266.

Burg, S.P. (1997) Evaluation and analysis of the January 22 (1997) horse meat shipment and February (1997) tulip shipment. Progress Report for Welfit Oddy.

Burg, S.P. (2000) Hypobaric shipping system. *Florida Postharvest Horticultural Institute 'Innovations in Fresh Produce Transportation'*, 6 March, 2000. Universtity of Florida, Gainesville (abstract).

Burg, S.P. and Burg, E.A. (1960) Studies on the ethylene production of apple tissue. *Plant Physiol.* 35, 24–35.

Burg, S.P. and Burg, E.A. (1962a) Postharvest ripening of avocados. *Nature* 194, 398–399.

Burg, S.P. and Burg, E.A. (1962b) Role of ethylene in fruit ripening. *Plant Physiol.* 37, 179–189.

Burg, S.P. and Burg, E.A. (1965a) Ethylene action and the ripening of fruits. *Science* 148, 1190–1196.

Burg, S.P. and Burg, E.A. (1965b) Gas exchange in fruits. *Physiol. Plant.* 18, 870–884.

Burg, S.P. and Burg, E.A. (1965c) Relationship between ethylene production and ripening of bananas. *Bot. Gaz.* 126, 200–204.

Burg, S.P. and Burg, E.A. (1966a) The interaction between auxin and ethylene and its role in plant growth. *Proc. Natl Acad. Sci. USA* 55, 262–269.

Burg, S.P. and Burg, E.A. (1966b) Auxin-induced ethylene formation and its relation to flowering in the pineapple. *Science* 152, 1269.

Burg, S.P. and Burg, E.A. (1966c) Fruit storage at subatmospheric pressures. *Science* 153, 314–315.

Burg, S.P. and Burg, E.A. (1967a) Inhibition of polar auxin transport by ethylene. *Plant Physiol.* 42, 1224–1228.

Burg, S.P. and Burg, E.A. (1967b) Lateral auxin transport in stems and roots. *Plant Physiol.* 42, 891–893.

Burg, S.P. and Burg, E.A. (1967c) Molecular requirements for the biological activity of ethylene. *Plant Physiol.* 42, 144–152.

Burg, S.P. and Burg, E.A. (1968) Auxin-stimulated ethylene formation: its relationship to auxin inhibited growth, root geotropism and other plant processes. In: Wightman, F. and Setterfield, G. (eds) *Biochemistry and Physiology of Plant Growth Substances.* Runge Press, Ltd, Ottawa, pp. 1275–1294.

Burg, S.P. and Burg, E.A. (1969) Interaction of ethylene, oxygen and carbon dioxide in the control of fruit ripening. *Qualitas Plant Mater. Veg.* 19, 185–200.

Burg, S.P. and Burg, E.A. (1976) Prevention of microbial growth. Canadian Patent 997,532.

Burg, S.P. and Dijkman, M.J. (1967) Ethylene and auxin participation in pollen induced fading of *Vanda* orchid blossoms. *Plant Physiol.* 42, 1648–1650.

Burg, S.P. and Hentschel, W.J. (1974) Low-pressure storage of metabolically active material with open cycle refrigeration. US Patent 3,810,508.

Burg, S.P. and Kang, B.G. (1993) Gravity-dependent ethylene action. In: Pech, J.C., Latché, A. and Balagué, C. (eds) *Cellular and Molecular Aspects of the Plant Hormone Ethylene*. Kluwer Academic, Dordrecht, pp. 335–340.

Burg, S.P. and Kosson, R.L. (1982) Heat transfer under hypobaric conditions. *ASHRAE Transactions* 88.

Burg, S.P. and Kosson, R.L. (1983) Metabolism, heat transfer and water loss under hypobaric conditions. In: Lieberman, M. (ed.) *Post-harvest Physiology and Crop Preservation*. Plenum Press, New York, pp. 399–424.

Burg, S.P. and Stolwijk, A.J. (1959) A highly sensitive katharometer and its application to the measurement of ethylene and other gases of biological importance. *J. Biochem. Microbiol. Tech. Eng.* I, 245–249.

Burg, S.P. and Thimann, K.V. (1959) The physiology of ethylene formation in apples. *Proc. Natl Acad. Sci. USA* 45, 335–344.

Burg, S.P., Burg, E.A. and Marks, T. (1964) Relationship of solute leakage to solution tonicity in fruits and other plant tissues. *Plant Physiol.* 39, 185–195.

Burg, S.P., Apelbaum, A., Eisinger, W. and Kang, B.G. (1971) Physiology and mode of action of ethylene. *HortScience* 6, 359–364.

Burg, S.P., Apelbaum, A. and Kang, B.G. (1972) Control of cell division, expansion and differentiation by ethylene. In: Kaldewey, H. and Vardar, Y. (eds) *Hormonal Regulation of Plant Growth and Development*. Proceedings of the Adv. Study Inst. Izmir 1971. Verlag Chemie, Weinheim, Germany, pp. 263–279.

Burgheimer, F.T., McGill, J.N., Nelson, A.I. and Steinberg, M.P. (1967) Chemical changes in spinach stored in air and in controlled atmosphere. *Food Technol.* 21, 109.

Burmeister, D.M. and Dilley, D.R. (1995) A 'scald-like' controlled atmosphere storage disorder of Empire apples – a chilling injury induced by CO_2. *Postharvest Biol. and Technol.* 6, 1–7.

Burroughs, L.F. (1970) Amino acids. In: Hulme, A.C. (ed.) *The Biochemistry of Fruits and Their Products*, vol. 1. Academic Press, London, pp. 119–146.

Bursell, E. (1957) Spiracular control of water loss in the tsetse fly. *Proceedings of the Royal Ent. Soc. Lond. A* 32, 21–29.

Burström, H.G. (1957) On the adaptation of roots to β-indolyl-acetic acid. *Physiol. Plant.*10, 187–197.

Burström, H.G. (1971) Orientation and geotropic reaction of seminol roots of wheat. *Plant Physiol.* 25, 283–293.

Burton, W.G. (1950) Studies on the dormancy and sprouting of potatoes. I. The oxygen content of the potato tuber. *New Phytol.* 49, 121–134.

Burton, W.G. (1952) Studies on the dormancy and sprouting of potatoes. III. The effect upon sprouting of volatile metabolic products other than carbon dioxide. *New Phytol.* 51, 154–162.

Burton, W.G. (1974) The oxygen uptake in air and in 5% O_2, and the carbon dioxide output, of stored potato tubers. *Potato Res.* 17, 113–137.

Burton, W.G. (1982) *Postharvest Physiology of Food Crops*. Longman, London.

Burton, W.G. and Sprague, T. (1950) A note on the intercellular space of the potato tuber. *New Phytol.* 49, 8–10.

Buse, E.L. and Laties, G. (1993) Ethylene-mediated post-transcriptional regulation in ripening avocado (*Persea americana*) mesocarp discs. *Plant Physiol.* 102, 417–423.

Bush, D.R. (1993) Proton-coupled sugar and amino acid transporters in plants. *Annu. Rev. Plant Physiol.* 44, 513–542.

Bushnell, J. (1925) The relation of temperature to growth and respiration in the potato plant. *University Minn. Agr. Expt. Sta. Bul.* 34.

Butters, R.E. (1976) Promise of longer storage under low pressure confirmed in UK. *The Grower* 86, 195.

Butters, R.E. (1977) Low-pressure storage of flowers and cuttings can extend life. *Horticulture Industry* April, 5.

Byers, B. (1977) The Grumman Dormavac system. In: Dewey, D.H. (ed.) *Controlled Atmospheres for the Storage and Transport of Perishable Agricultural Commodities*. Proceedings of the 2nd National Controlled Atmosphere Research Conference, 5–7 April, 1977, Michigan State University, East Lansing, Michigan, pp. 82–85.

Byers, R.D., Baker, L.R., Sell, H.M., Herner, R.C. and Dilley, D.R. (1972) Ethylene: a natural regulator of sex expression of *Cucumis melo* L. *Proc. Natl Acad. Sci. USA* 69, 717–720.

Cabrera, R.M. and Saltveit, M.E. (1990) Physiological responses to chilling temperatures of intermittently warmed cucumber fruit. *J. Amer. Soc. Hort. Sci.* 115, 256–261.

Calderon, M. and Navarro, S. (1979) Increased toxicity of low-oxygen atmospheres supplemented with carbon dioxide on *Tribolium castaneum* adults. *Entomol. Exp. Appl.* 25, 39.

Calderon, M., Navarro, S. and Donahaye, E. (1966) The effect of low pressures on the mortality of six stored product insect species. *J. Stored Prod. Res.* 2, 135–140.

Cameron, A.C. (1982) Gas diffusion in bulky organs. PhD dissertation, University of California, Davis, California.

Cameron, A.C. and Reid, M.S. (1981) The use of silver thiosulfate anionic complex as a foliar spray to prevent flower abscission of *Zygocactus*. *HortScience* 16, 761–762.

Cameron, A.C. and Reid, M.S. (2001) 1-MCP blocks ethylene-induced petal abscission of *Pelargonium peltatum* but the effect is transient. *Postharvest Biol. Technol.* 22, 169–177.

Cameron, A.C. and Yang, S.F. (1982) A simple method for the determination of resistance to gas diffusion in plant organs. *Plant Physiol.* 70, 21–23.

Campbell, L.L. and Williams, O.B. (1952) The bacteriology of Gulf Coast shrimp and organoleptic changes with ice storage. *Food Technol.* 6, 125–126.

Camprubi, P. and Nichols, R. (1978) Effects of ethylene on carnation flowers (*Dianthus caryophyllus*) cut at different stages of development. *J. Hort. Sci.* 53, 17–22.

Canel, C., Bailey-Serres, J.N. and Roose, M.L. (1995) Pummelo fruit transcript homologous to ripening-induced genes. *Plant Physiol.* 108, 1323–1324.

Canny, M.J. (1995) Apoplastic water and solute movement: new rules for an old space. *Annu. Rev. Plant Physiol. Mol. Biol.* 46, 215–236.

Cao, X.-F., Linstead, P., Berger, F., Kieber, J.J. and Dolan, L. (1999) Differential ethylene sensitivity of epidermal cells is involved in the establishment of cell pattern in the *Arabidopsis* root. *Physiol. Plant.* 106, 311–317.

Cardin, T. (1994) *Intermodal Shipping*, November, 1994, p. 40.

Cardinale, F.C., Jennings, J.C. and Anderson, J.C. (1995) Use of the ethylene action inhibitor 1-methylcyclopropene to study the role of ethylene in elicitor-induced ethylene biosynthesis in tomato leaves. *Plant Physiol.* (Suppl.) 108, 140.

Carlton, B.C., Peterson, C.E. and Tolbert, N.E. (1961) Effect of ethylene and oxygen on production of bitter compound by carrot roots. *Plant Physiol.* 36, 550–552.

Carolus, R.L., Lipton, W.J. and Apple, S.B. (1953) Effects of packaging on quality and market acceptability of asparagus. *Mich. Agric. Exp. Stn. Q. Bull.* 35, 330.

Carpenter, A. and Potter, M. (1994) Controlled atmospheres. In: Sharp, J.L. and Hallman, G.J. (eds) *Quarantine Treatments for Pests of Foods.* Westview Press, Boulder, Colorado, pp. 171–198.

Carpenter, W.J. and Dilley, D.R. (1975) Investigations to extend cut flower longevity. Research Report 263. Agricultural Experiment Station, Michigan State University, East Lansing, Michigan, pp. 1–10.

Carroll, B.J., Reese, G.B. and Ward, B.Q. (1968) Microbial study of iced shrimp. *1965 Iced-Shrimp Symposium. Circular of the US Department of Interior, Bureau of Commercial Fisheries* 284, 1–17.

Catchpole, A.H. and Hillman, J.R. (1976) The involvement of ethylene in the coiled sprout disorder of potato. *Ann. Appl. Bot.* 83, 413–423.

Cazzonelli, C.I., Cavallaro, A.S. and Botella, J.R. (2000) Searching for the role of ethylene in non-climacteric fruits. *Plant Growth Regl. Abstracts* 26, 137.

Cebeci, T. (1974) Laminar-free-convective-heat transfer from the outer surface of a vertical slender circular cylinder. *Fifth Int. Heat Transfer Conf.*, vol. 3, pp. 15–19.

Chadwick, A.V. (1969) Regulation of root growth by auxin-ethylene interaction. PhD dissertation, University of Miami, Miami, Florida.

Chadwick, A.V. and Burg, S.P. (1967) An explanation of the inhibition of root growth caused by indole-3-acetic acid. *Plant Physiol.* 42, 415–420.

Chadwick, A.V. and Burg, S.P. (1970) Regulation of root growth by auxin-ethylene interaction. *Plant Physiol.* 45, 192–200.

Chahine, H., Gouble, B., Audergon, J.M., Souty, M., Albagnac, G., Jacquemin, G., Reich, M. and Hugues, M. (2000) Effect of ethylene on certain quality parameters of apricot fruit (*Prunus armeniaca* L.) during maturation and postharvest evolution. *Plant Growth Regl. Abstracts* 26, 56.

Chalutz, E. and Stahman, M.A. (1969) Induction of pisatin by ethylene. *Phytopathology* 59, 1972–1973.

Chalutz, D., DeVay, J.E. and Maxie, E.C. (1969) Ethylene-induced isocoumarin formation in carrot root tissue. *Plant Physiol.* 44, 235–241.

Chambers, T.C. and Possingham, J.V. (1963) Studies of the fine structure of the wax layer of Sultana grapes. *Austr. J. Biol. Sci.* 16, 818–825.

Chamel, A., Pineri, M. and Escoubes, M. (1991) Quantitative determination of water sorption by plant cuticles. *Plant Cell Environ.* 14, 87–95.

Chan, H.T. (1986a) Effects of heat treatments on the ethylene-forming enzyme in papayas. *J. Food Sci.* 51, 581–583.

Chan, H.T. (1986b) Heat inactivation of the ethylene-forming enzyme systems in cucumbers. *J. Food Sci.* 51, 1491–1493.

Chan, H.T., Sanxter, S. and Couey, H.M. (1985) Electrolyte leakage and ethylene production induced by chilling injury of papaya (*Carica papaya*). *HortScience* 10, 1070–1072.

Chan, K.W., Corley, R.H. and Seth, A.K. (1972) Effects of growth regulators on fruit abscission in the oil palm *Elaeis guineensis*. *Ann. Appl. Biol.* 71, 243–249.

Chang, C. (1996) The ethylene signal transduction pathway in *Arabidopsis*: an emerging paradigm? *Trends Biochem. Sci.* 21, 129–133.

Chang, C. and Shockey, J.A. (1999) The ethylene-response pathway: signal perception to gene regulation. *Current Opinions in Plant Biology* 2, 352–358.

Chang, C. and Stewart, R.C. (1998) Regulation of diverse signaling pathways in prokaryotes and eukaryotes. *Plant Physiol.* 117, 723–731.

Chang, C., Kwok, S.F., Bleeker, A.B. and Meyerowitz, E.M. (1993) *Arabidopsis* ethylene-response gene *ETR1*, similarity of product to two-component regulators. *Science* 262, 539–544.

Chang, C.W. (1975) Carbon dioxide and senescence in cotton plants. *Plant Physiol.* 55, 515–519.

Chang, D.C. and Han, S. (1975) The effect of ethephon and 2,4-D on the cellular differentiation of etiolated asparagus seedlings. *J. Chinese Soc. Hort. Sci.* 21, 5–13.

Chang, K.-J., Jacobs, S. and Cuatrecasas, P. (1975) Quantitative aspects of hormone-receptor interactions of high affinity: effect of receptor concentration and measurement of dissociation constants of labelled and unlabelled hormones. *Biochem. Biophys. Acta* 406, 294–303.

Chao, Q., Rothenberg, M., Solano, R., Roman, G., Terzaghi, W. and Ecker, J.R. (1997) Activation of the ethylene gas response pathway in *Arabidopsis* by the nuclear protein ETHYLENE-INSENSITIVE-3 and related proteins. *Cell* 89, 1133–1144.

Chapman, R.F. (1998) *The Insects. Structure and Function*, 4th edn. Cambridge University Press, Cambridge.

Chase, E.M. and Church, C.G. (1927) Effect of ethylene on the composition of fruits. *Indust. Eng. Chem.* 19, 1135–1138.

Chase, E.M. and Denny, F.E. (1924) Use of ethylene in the coloring of citrus fruit. *Indust. Eng. Chem.*16, 339–340.

Chau, K.F. and Alvarez, A.M. (1983) Effects of low-pressure storage on *Colletotrichum gloeosporioides* and postharvest infection of papaya. *HortScience* 18, 953.

Chaves, A.R. and Thomas, J.O. (1984) Effect of a brief CO_2 exposure on ethylene production. *Plant Physiol.* 76, 88–91.

Chaves-Franco, S.H. and Kader, A.A. (1993) Effects of CO_2 on ethylene biosynthesis in 'Bartlett' pears. *Postharvest Biol. and Technol.* 3, 183–190.

Chen, P.M. (2000) Promoting ripening of d'Anjou pears. *Good Fruit Grower* 51, 71–77.

Chen, P.M., Melleenthin, W.M., Kelly, S.B. and Facteau, T.U. (1981) Effects of low oxygen and temperature on quality retention of 'Bing' cherries during prolonged storage. *J. Amer. Soc. Hort. Sci.* 105, 533–535.

Chen, P.M., Olsen, K.L. and Maheruik, M. (1985) Effect of low-oxygen atmosphere on storage scald and quality preservation of 'Delicious' apples. *J. Amer. Soc. Hort. Sci.* 110, 16–20.

Chen, Y., Cann, M.J., Litvin, T.N., Iourgenko, V., Sinclair, M.L., Levin, L.R. and Buck, J. (2000) Soluble adenylyl cyclase as an evolutionarily conserved bicarbonate sensor. *Science* 289, 625–628.

Chen, Z. and Kiessig, D. (1991) Identification of a soluble salicylic acid-binding protein that may function in signal transduction in the plant disease resistance response. *Proc. Natl Acad. Sci. USA* 88, 8179–8183.

Chevillotte, P. (1973) Relationship between the reaction of cytochrome oxidase, oxygen and oxygen uptake in cells *in vivo*. *J. Theor. Biol.* 39, 277–295.

Chi, C.K. and Frenkel, C. (1977) Upsurge in respiration and peroxide formation in potato tubers as influenced by ethylene, propylene, and cyanide. *Plant Physiol.* 59, 515–518.

Cho, M.-S., Celikel, F., Dodge, L. and Reid, M.S. (2001) Sucrose enhances the postharvest quality of cut flowers of *Eustoma grandiflorum* (Raf.) *Shinn. Plant Growth Regl. Abstr.* 27, 348.

Choi, W. and Dean, R.A. (1997) The adenylate cyclase gene *MAC1* of *Magnaporthe grisea* controls appressorium formation and other aspects of growth and development. *Plant Cell* 9, 1973–1983.

Cholodny, N. (1927) Wuchshormone und Tropismen bei den Pflanzen. *Biol. Zentralbl.* 47, 604–626.

Chrispeels, M.J. and Maurel, C. (1994) Aquaporins: the molecular basis of facilitated water movement through living plant cells. *Plant Physiol.* 105, 9–13.

Christoffersen, R.E., McGarvey, D.J. and Savarese, P. (1993) Biochemical and molecular characterization of ethylene-forming enzyme in avocado. In: Pech, J.C., Latché, A. and Balagué, C. (eds) *Cellular and Molecular Aspects of the Plant Hormone Ethylene.* Kluwer Academic, Dordrecht, pp. 65–70.

Chu, C.L. (1984) Use of internal ethylene concentration as a maturity index of eleven apple cultivars. *Acta Horticulturae* 157, 129–134.

Chu, C.L., Lougheed, E.C. and Tiessen, H. (1976) Low-pressure storage of asparagus. *Proceedings of the Can. Soc. Hot. Sci.* 15 (abstract).

Church, I.J. and Parsons, A.L. (1995) Modified atmosphere packaging: a review. *J. Sci. Food Agric.* 67, 143–152.

Cicale, D. (1979a) Meat blank-off tests with Stan Burg. Report for Grumman Allied Industries, Inc., Garden City, New York.

Cicale, D. (1979b) Summary of beef retail cuts test. Report for Grumman Allied Industries, Inc., Garden City, New York.

Cicale, D. (1979c) Summary of meat pigment test. Report for Grumman Allied Industries, Inc., Garden City, New York.

Cicale, D. (1979d) Beef purge test plan. Report for Grumman Allied Industries, Inc., Garden City, New York.

Cicale, D. (1980) Summary of ice crystal test. Report for Grumman Allied Industries, Inc., Garden City, New York.

Cicale, D. and Jamieson, W. (1978) Preliminary investigation of the storage of Haas avocados under hypobaric conditions. Report for Grumman Allied Industries, Inc., Garden City, New York.

Clark, D.G., Klee, H.J., Barrett, J.E. and Nell, T.A. (2000) Horticultural performance of ethylene-insensitive petunias. *Plant Growth Regl. Abstracts* 26, 137.

Claypool, L.L. and Allen, F.W. (1948) Carbon dioxide production by deciduous fruits held at different oxygen levels during transit periods. *Proceedings of the J. Amer. Soc. Hort. Sci.* 51, 103–113.

Clements, H.F. (1935) Morphology and physiology of the pome lenticles of *Pyrus malus. Bot. Gaz.* 97, 101–117.

Clements, H.F. (1940) Movement of organic solutes in the sausage tree, *Kigelia africana. Plant Physiol.* 15, 689–700.

Clendennen, S.K., Kellogg, J.A., Wolff, K.A., Matsumara, W., Peters, S., Vanwinkle, J.E., Copes, B., Pieper, M. and Kramer, M.G. (2000) Genetic engineering of cantaloupe to reduce ethylene biosynthesis and control ripening. *Plant Growth Regl. Abstracts* 26, 138.

Clendenning, K.A. (1941) Studies of the tomato in relation to its storage. II. The effects of altered internal atmosphere upon the respiratory and ripening behavior of tomato fruits stored at 12.5°C. *Can. J. Res.* 19C, 500–518.

Clifford, P.E. and Oxlade, E.L. (1989) Ethylene production, georesponse, and extension growth in dandelion peduncles. *Can. J. Bot.* 67, 1927–1929.

Clifford, P.E., Reid, D.M. and Pharis, R.P. (1983) Endogenous ethylene does not initiate but may modify geobending – a role for ethylene in autotropism. *Plant Cell Environ.* 6, 433–436.

Clifford, P.E., Mousdale, M.A., Lund, S.J. and Oxlade, E.L. (1985) Differences in auxin level detected across geo-stimulated dandelion peduncles: evidence supporting a role for auxin in geotropism. *Ann. Bot.* 55, 293–296.

Clifton, C.E. (1950) *Introduction to the Bacteria.* McGraw Hill, New York.

Cline, M.G. (1991) Apical dominance. *Bot. Rev.* 57, 318–358.

Cliquet, J.B. and Morot-Gaudry, J.F. (1989) Ethephon and photosynthesis control in maize. *C.R. Acad. Sci. Paris* (Ser. III) 309, 317–322.

Cobb, B.F., Alaniz, I. and Thompson, C.A. (1973a) Biochemical and microbial studies on shrimp: volatile nitrogen and amino nitrogen analysis. *J. Food Sci.* 38, 431–436.

Cobb, B.F., Vanderzant, C., Thompson, C.A. and Custer, C.S. (1973b) Chemical characteristics, bacterial counts and potential shelf-life of shrimp from various locations on the Northwestern Gulf of Mexico. *J. Milk Food Tech.* 36, 463–468.

Cochrane, V.W. (1958) *Physiology of Fungi.* John Wiley & Sons, New York.

Collander, R. (1949) The permeability of plant protoplasts to small molecules. *Physiol. Plant.* 2, 300.

Collier, D.E. and Cummins, W.R. (1991) Respiratory shifts in developing petals of *Saxifraga cernua. Plant Physiol.* 95, 324–328.

Colman, B. and Espie, G.S. (1985) CO_2 uptake and transport in leaf mesophyll cells. *Plant Cell and Environment* 8, 449–457.

Condon, P. and Kuc, J. (1960) Isolation of a fungitoxic compound from carrot root tissue inoculated with *Ceratocystis fimbriata*. *Phytopathology* 50, 267–270.

Coniver, C.A. (1992) Foliage plants. In: Larson, R.A. (ed.) *Introduction to Floriculture*, 2nd edn. Academic Press, New York, pp. 569–601.

Conkey, R.L. (1981) Dormavac lambs. Out turn report, Dubai, 29 April, 1981.

Cook, H.T., Parsons, C.S. and McColloch, L.P. (1958) Methods to extend storage of fresh vegetables aboard ships of the U. S. Navy. *Food Tech.* 12, 548.

Cook, R.J. and Papendick, R.I. (1978) Role of water potential in microbial growth and development of plant disease, with special reference to postharvest pathology. *HortScience* 13, 11–16.

Coombe, B.G. (1976) Development of fleshy fruits. *Annu. Rev. Plant Physiol.* 27, 207–228.

Cooper, W.C. and Horanic, G. (1973) Induction of abscission at hypobaric pressures. *Plant Physiol.* 51, 1002–1004.

Cooper, W.C., Rasmussen, G.K. and Hutchison, D.J. (1969) Promotion of abscission of orange fruits by cycloheximide as related to site of treatment. *BioScience* 19, 443–444.

Coorts, G.D., Gartner, J.B. and McCollum, J.P. (1965) Effect of senescence and preservative on respiration in cut flowers of *Rosa hybrida*, 'Velvet Times'. *Proceedings of the J. Amer. Soc. Hort. Sci.* 86, 779–780.

Corey, K.A., Barker, A.V. and Cracker, L.E. (1987) Ethylene evolution by tomato plants under stress of ammonium toxicity. *HortScience* 22, 471–473.

Cornell, J.J. (1975) *Control of Fish Quality*. Fishing News Ltd, Surrey, UK, p. 129.

Cosgrove, D.J. and Cleland, R.E. (1983) Solutes in the free space of growing stem tissues. *Plant Physiol.* 72, 326–331.

Couey, H.M. and Wells, J.M. (1970) Low-oxygen or high carbon dioxide atmospheres to control postharvest decay of strawberries. *Phytopathology* 60, 47–49.

Couey, H.M., Follstad, M.N. and Uota, M. (1966) Low-oxygen atmospheres for control of postharvest decay of fresh strawberries. *Phytopathology* 56, 1339–1341.

Coursey, D.G. (1983) Postharvest losses in perishable foods of the developing world. In: Lieberman, M. (ed.) *Postharvest Physiology and Crop Preservation*. Plenum Press, New York, pp. 485–514.

Covey, R.P. (1970) Effect of oxygen tension on the growth of *Phytophthora cactorum*. *Phytopathology* 60, 358–359.

Cowan, I.R. and Milthorpe, F.L. (1968) Plant factors influencing the water status of plant tissues. In: Kozlowski, T.T. (ed.) *Water Deficits and Plant Structure*, vol. 1. Academic Press, New York, pp. 137–193.

Cox, N.A., Russell, S.M. and Bailey, J.S. (1998) The microbiology of stored poultry. In: Davies, A. and Board, R. (eds) *The Microbiology of Meat and Poultry*. Blackie Academic & Professional, London, pp. 266–287.

Crafts, A.S. (1961) *Translocation in Plants*. Holt, Rinehardt and Winston, New York.

Crafts, A.S. and Lorenz, O. (1944a) Composition of fruits and phloem exhudate of curcubites. *Plant Physiol.* 19, 131–138.

Crafts, A.S. and Lorenz, O. (1944b) Fruit growth and food transport in curcubits. *Plant Physiol.* 19, 326–337.

Craig, I.M., Hayes, S. and Pilcher, K.S. (1968) Incidence of *Clostridium botulinum* type E in salmon and other marine fish in the Pacific Northwest. *Appl. Microbiology* 16, 553–557.

Crocker, W. (1948) *Growth of Plants*. Reinhold, Norwood, Massachusetts.

Crocker, W. and Knight, L. (1908) Effect of illuminating gas and ethylene upon flowering carnations. *Bot. Gaz.* 46, 259–275.

Crocker, W., Zimmerman, P.W. and Hitchcock, A.E. (1932) Ethylene-induced epinasty in leaves and the relation of gravity to it. *Contrib. Boyce Thompson Inst.* 4, 177–218.

Crocker, W., Hitchcock, A.E. and Zimmerman, P.W. (1935) Similarities in the effects of ethylene and the plant auxins. *Contrib. Boyce Thompson Inst.* 7, 231–248.

Cronshaw, J. (1975) P-proteins. In: Aronoff, S., Dainty, J., Gorham, P.R., Sribvastava, S.M. and Swanson, C.A. (eds) *Phloem Transport*. Plenum, New York, pp. 79–115.

Crosset, R.N. and Campbell, D.J. (1975) The effects of ethylene in the root environment upon the development of barley. *Plant and Soil* 42, 453–464.

Cua, A.U. and Lizada, M.C.C. (1990) Ethylene production in the 'Caraboa' mango (*Magifera Indica* L.) fruit during maturation and ripening. *Acta Horticulturae* 269, 169–179.

Culpepper, C.W., Moon, H.H. and Lutz, J.M. (1936) The determination of the internal gases of plant tissue. *Science* 84, 398–400.

Curtis, O.F. (1929) Studies on the solute translocation in plants: experiments indicating that translocation is dependent on the activity of living cells. *Amer. J. Bot.* 16, 154–168.

Curtis, O.F. (1937) Vapor pressure gradients, water distribution in fruits, and so-called infrared injury. *Amer. J. Bot.* 24, 705–710.

Curtis, O.F. and Clark, D.G. (1950) *An Introduction to Plant Physiology*. McGraw Hill, New York.

Curtis, R.W. (1968) Mediation of a plant response to malformin by ethylene. *Plant Physiol.* 43, 76–80.

Curtis, R.W. (1987) Antiethylene properties of $AgNO_3$ and 2,5-norbornadiene in light and dark in *Vigna radiata*. *J. Plant Growth Regul.* 6, 41–56.

Cutaia, A.J. and Ordal, Z.J. (1964) Pigment changes in anaerobically packed ground beef. *Food Tech.* May, 163–166.

Dalling, M.J., Boland, G. and Wilson, J.H. (1976) Relation between acid proteinace activity and redistribution of nitrogen during grain development in wheat. *Austr. J. Plant Physiol.* 3, 721–730.

Dalrymple, G.D. (1967) *The Development of Controlled Atmosphere Storage of Fruits*. USDA, Division of Mktg. Utils. Sci.

Dalrymple, G.D. (1969) The development of an agricultural technology: controlled atmosphere storage of fruits. *Technol. Cult.* 10, 35.

Danish Research Service for Plant and Soil Science (1980) *LP-group report*. Research Centre for Horticulture, Institute for Glasshouse Crops, Kirstinebjergvej qo, DK-5792 Arsslev.

Darvill, A.G. and Albershein, P. (1984) Phytoalexins and their elicitors – a defense against microbial infections in plants. *Annu. Rev. Plant Physiol.* 35, 243–275.

Darwin, C.D. (1981) *The Power of Movement in Plants*. D. Appleton and Company, New York.

Darwin, F. (1898) Observations on stomata. *Phil. Trans. Roy. Soc. Lond. Ser. B* 190, 531–621.

d'Auzac, J., Bouteau, F., Chrestin, H., Clément, A., Jacob, J.L., Lacrotte, R., Prévot, J.C., Pulade-Renaud, V. and Rone, J.P. (1993) Stress ethylene in *Hevea brasiliensis*: physiological, cellular, molecular aspects. In: Pech, J.C., Latché, A. and Balagué, C. (eds) *Cellular and Molecular Aspects of the Plant Hormone Ethylene*. Kluwer Academic, Dordrecht, pp. 205–210.

Davenport, T.L. and Campbell, C.W. (1977a) Stylar-end breakdown: a pulp disorder in 'Tahiti lime'. *HortScience* 12, 246–248.

Davenport, T.L. and Campbell, C.W. (1977b) Stylar-end breakdown in 'Tahiti lime': aggravating effects of field heat and fruit maturity. *J. Amer. Soc. Hort. Sci.* 103, 484–486.

Davidson, O.W. (1949) Effects of ethylene on orchid flowers. *Proceedings of the J. Amer. Soc. Hort. Sci.* 54, 440–446.

Davies, D.D. (1973) Metabolic control in higher plants. In: Milborrow, B.V. (ed.) *Biosynthesis and its Control in Plants*. Academic Press, London, pp. 1–20.

Davies, K.M., Hobson, G.E. and Grierson, D. (1988) Silver ions inhibit the ethylene-stimulated production of ripening-related mRNAs in tomato. *Plant Cell and Environment* 11, 729–738.

Davies, K.M., Hobson, G.E. and Grierson, D. (1990) Differential effect of silver ions on the accumulation of ripening-related mRNAs in tomato fruit. *J. Plant Physiol.* 135, 708–713.

Davis, K.R., Lyon, G.D., Darvill, A.G. and Albersheim, P. (1984) Host-pathogen interactions. XXV. Endopolygalacturonic acid lyase from *Erwinia carotovora* elicits phytoalexin accumulation by releasing plant cell wall fragments. *Plant Physiol.* 74, 52–60.

Davis, P.L., Roe, B. and Bruemmer, J.H. (1973) Biochemical changes in citrus fruits during controlled atmosphere storage. *J. Food Sci.* 38, 225–229.

Davis, S.D. and McCree, K.J. (1978) Photosynthetic rate and diffusion conductance as a function of age in leaves of bush bean (*Phaseolus vulgaris* L.). *Crop Sci.* 18, 280–282.

Davson, H. (1960) *A Textbook of General Physiology*, 2nd edn. Little, Brown and Co., Boston, Massachusetts.

De Almeida, M.E.H. (1955) Amino acidos lwires em camaroes variacoes decorridas durante a decomposicao. *Rev. Inst. Adolfo Lutz* 15, 158–167.

De Laat, M.M., Brandenburg, D.C.C. and Van Loon, L.C. (1981) The modulation of conversion of 1-aminocyclopropane-1-carboxylic acid to ethylene by light. *Planta* 153, 193–200.

de Souza, L. and King, P.J. (1991) Mutants of *Nicotiana plumbaginifolia* with increased sensitivity to auxin. *Mol. Gen. Genet.* 231, 65–75.

De Swardt, G.H. (1977) Verbruikning van Protea-loofblare. *Landbouweekblad* 30, 30–33.

De Vries, H. (1872) Under einige Ursachen der Richtung bilateral-symmetrischer Pflanzentheile. *Arch. Bot. Inst. Wurzbur.* II, 223–277.

De Vries, H.S.M., Harren, F.J.M., Voessenek, L.A.C.J., Blom, C.W.P.M., Woltering, E.J., Van der Valk, H.C.P.M. and Reuss, J. (1995) Investigation of local ethylene emission from intact cherry tomatoes by means of photothermal deflection and photoacoustic detection. *Plant Physiol.* 107, 1371–1377.

Deikman, J. (1997) Molecular mechanisms of ethylene regulation of gene transcription. *Physiol. Plant.* 100, 561–566.

Deikmann, F. and Fisher, R.L. (1988) Organization of a DNA binding factor with the 5′-flanking region of an ethylene-responsive fruit ripening gene from tomato. *EMBO J.* 7, 3315–3320.

Deleano N.T. (1911) Ueber die Ableitung die plasmolysierten Blattstiele der Laubblatter. *Jahrb. Wiss. Bot.* 49, 129–186.

DellaPenna, D., Alexander, D.C. and Bennett, A.B. (1986) Molecular cloning of tomato fruit polygalacturonase: analysis of polygalacturonase levels during ripening. *Proc. Natl Acad. Sci. USA* 83, 6420–6424.

Denna, D.W. (1970) Transpiration and the waxy bloom in *Brassica oleracea L. Austr. J. Biol. Sci.* 23, 27–31.

Denny, F.E. (1923) Method of coloring citrus fruit. US Patent 1,475,938.

Denny, F.E. (1924a) Effect of ethylene on the respiration of lemons. *Bot. Gaz.* 77, 322–329.

Denny, F.E. (1924b) Hastening the coloration of lemons. *J. Agr. Res.* 27, 757–769.

Denny, F.E. (1935) Testing plant tissues for emanations causing leaf epinasty. *Contrib. Boyce Thompson Inst.* 7, 341–347.

Denny, F.E. (1936) Gravity position of tomato stems and their production of the emanation causing leaf epinasty. *Contrib. Boyce Thompson Inst.* 8, 88–104.

Denny, F.E. (1937) Leaf epinasty tests with volatile products from seedlings. *Contrib. Boyce Thompson Inst.* 9, 431–438.

Denny, F.E. (1946) Gas content of plant tissue and respiration measurements. *Contrib. Boyce Thompson Inst.* 14, 257–276.

Dervinis, C., Clark, D.G., Barrett, J.E. and Nell, T.A. (2000) Effect of pollination and exogenous ethylene on accumulation of ETR1 homologue transcripts during flower petal abscission in geranium (*Pelargonium* × *hortorum* L. H. Bailey). *Plant Molecular Biol.* 42, 847–856.

Desai, M.C. (1937) Effects of certain nutrient deficiencies on stomatal behavior. *Plant Physiol.* 12, 253–281.

DeVaux, H. (1891) Etude experumentale sur l'aeration des tissues massifs. Introduction a l'etude des mechanismes des exchages gazeux chez les plantes aeriennes. *Ann. Sci. Nat. Bot.* XIV, 279–396.

Dey, P.M., Brownleader, M.D. and Harbone, J.B. (1997) The plant, the cell and its molecular components. In: Dey, P.M. and Harbone, J.B. (eds) *Plant Biochemistry*. Academic Press, New York, pp. 1–47.

Dhawan, K.R., Bassi, P.K. and Spenser, M.S. (1981) Effects of carbon dioxide on ethylene production and action in intact sunflower plants. *Plant Physiol.* 68, 831.

Difco Manual (1953) Difco Laboratories.

Dilley, D.R. (1963) Apple scald induction by anaerobiosis. *Nature* 200, 1229–1230.

Dilley, D.R. (1971) Low-pressure storage. *Annu. Rept. Res. Grad. Educ. Hort. Dept. Mich. State. University*, p. 43.

Dilley, D.R. (1972) Hypobaric storage – a new concept for preservation of perishables. *102nd Annual Report of Michigan State Horticultural Society*, pp. 82–89.

Dilley, D.R. (1976) Hypobaric investigations with mushrooms (*Agaricus bisporus* cv. Golden Light). Progress Report to Grumman Allied Industries, Inc., Garden City, New York.

Dilley, D.R. (1977a) Application of the hypobaric system for storage and transportation of perishable agricultural commodities. *Second Annual World Fair for Technology Exchange*, 7–11 February, 1977. *First World Congress for Technology Transfer*, pp. 135–149.

Dilley, D.R. (1977b) The hypobaric concept for controlled atmosphere storage. In: Dewey, D.H. (ed.) *Controlled Atmospheres for the Storage and Transport of Perishable Agricultural Commodities. Proceedings of the 2nd National Controlled Atmosphere Research Conference*, 5–7 April, 1977, Michigan State University, East Lansing, Michigan, pp. 29–37.

Dilley, D.R. (1978) Approaches to maintenance of postharvest integrity. *J. Food Biochemistry* 2, 235–242.

Dilley, D.R. (1983) Manipulation of the postharvest atmosphere for preservation of food crops. In: Lieberman, M. (ed.) *Postharvest Physiology and Crop Preservation*. Plenum Press, New York, pp. 383–397.

Dilley, D.R. (1990) Historical aspects and perspectives on controlled atmosphere storage. In: Calderon, M. and Barkai-Golan, R. (eds) *Food Preservation by Modified Atmospheres*. CRC Press, Boca Raton, Florida, pp. 187–195.

Dilley, D.R. and Carpenter, W.J. (1973) Hypobaric storage of flowers – a physiological interpretation. *HortScience* 8, 273 (abstract).

Dilley, D.R., Carpenter, W.J. and Burg, S.P. (1975) Principles and application of hypobaric storage of cut flowers. *Acta Horticulturae* 41, 249–268.

Dilley, D.R., Kuai, J., Poneleit, L., Zhu, Y., Pekker, Y., Wilson, I.D., Burmeister, D.M., Gran, C. and Bowers, A. (1993) Purification and characterization of ACC oxidase and its expression during ripening in apple fruit. In: Pech, J.C., Latché, A. and Balagué, C. (eds) *Cellular and Molecular Aspects of the Plant Hormone Ethylene.* Kluwer Academic, Dordrecht, pp. 46–52.

Dillon, V.M. (1998) Yeasts and moulds associated with meat and meat products. In: Davies, A. and Board, R. (eds) *The Microbiology of Meat and Poultry.* Blackie Academic & Professional, London, pp. 85–117.

Dimock, A.W. and Tamen, J. (1969) Powdery mildew of roses. In: Mastalerz, J.W. and Langhans, R.W. (eds) *Roses.* Pennsylvania Rose Growers, New York State Flower Growers Association, and Roses Inc., pp. 163–184.

Dimock, A.W., Williamson, C.E. and Nelson, P.E. (1964) Diseases. In: Langhans, R.W. (ed.) *Chrysanthemums. A Manual of the Culture, Insects, Diseases and Economics of Chrysanthemums.* New York State and County Agricultural Extension Service, pp. 116–150.

Dixon, M. and Webb, E.C. (1959) *Enzymes.* Academic Press, New York.

Do, J.Y. and Salunkhe, D.K. (1975) Controlled Atmosphere Storage. Part 1 – Biochemical considerations. In: Pantastico, E.B. (ed.) *Postharvest Physiology, Handling and Utilization of Tropical and Subtropical Fruits and Vegetables.* AVI, Westport, Connecticut, pp. 175–185.

Dodds, J.H. and Hall, M.A. (1980) Plant hormone receptors. *Science Progress* (Oxford) 66, 513–535.

Dolk, H.E. and Postma, N. (1927) Frog respiration. *Ztschr. vergl. Physiol.* 10, 417–444.

Dong, J.G., Fernandez-Maculet, J.C. and Yang, S.F. (1992) Purification and characterization of 1-aminocyclopropane-1-carboxylic acid oxidase from ripe apple fruit. *Proc. Natl Acad. Sci. USA* 89, 9789–9793.

Doubt, S.L. (1917) The response of plants to illuminating gas. *Bot. Gaz.* 63, 209–224.

Dourtoglou, V., Koussissi, E. and Petritis, K. (2000) Evaluation of novel inhibitors of ACC oxidase possessing cyclopropane moiety. *Plant Growth Regl. Abstracts* 26, 153.

Dressler, B. (1978a) *Preliminary Investigation of the Storage of Florida Booth #8 Avocados Under Hypobaric Conditions.* Grumman Allied Industries, Inc., Garden City, New York.

Dressler, B. (1978b) *Storage of Kapoho Papaya for 28 Days Under Hypobaric Condtions.* Grumman Allied Industries, Inc., Garden City, New York.

Dressler, B. (1979a) Carnation foliage test status as-to-date. Report for Grumman Allied Industries, Inc., Garden City, New York.

Dressler, B. (1979b) Shelf life summary of California carnation static test conducted at Continental farms, Miami, Florida. Report for Grumman Allied Industries, Inc., Garden City, New York.

Dressler, B. (1980) Trip report, Ohio State University. November 17, 1980. Report for Grumman Allied Industries, Inc., Garden City, New York.

Dressler, B. and Jamieson, W. (1977) The effect of hypobaric storage on unrooted plant cuttings. Report for Grumman Allied Industries, Inc., Garden City, New York.

Driessche, T.V., Kevers, C., Collet, M. and Gaspar, T. (1988) *Acetabularia mediterranea* and ethylene: production in relation with development, circadian rhythms of emission, and response to external application. *Plant Physiol.* 133, 635–639.

Droby, S., Prusky, D., Jacoby, B. and Goldman, A. (1986) Presence of antifungal compounds in the peel of mango *Mangifera indica* fruits and their relation to latent infections of *Alternaria alternata. Physiol. Mol. Plant Pathol.* 29, 173–184.

Du Plessis, D.G.C. (1978) Leuko-antosianienbinders as beheermaatreël vir die voorkoming van bruinwording in protealooffbalre. MSc thesis, Rand Afrikaans University, Johannesburg.

Dubucq, M., Hofinger, M. and Gaspar, T. (1978) Auxin-controlled root growth and ethylene production. *Plant Cell Environ.* 1, 151–153.

Ducet, G. and Rosenberg, A.J. (1962) Leaf respiration. *Annu. Rev. Plant Physiol.* 13, 171–200.

Duggan, R.E. and Strasburger, L.W. (1946) Indole in shrimp. *Assoc. Offic. Agr. Chemists* 29, 177–188.

Dugger, W.M. (1952) The permeability of non-stomate leaf epidermis to carbon dioxide. *Plant Physiol.* 27, 489–499.

Dull, G.G. (1971) The pineapple: general. In: Hulme A.C. (ed.) *The Biochemistry of Fruits and their Products*, vol. 2. Academic Press, London, pp. 303–324.

Dull, G.G., Young, R.E. and Biale, J.B. (1967) Respiratory patterns in fruit of pineapple, *Anonas comosus*, detached at different stages of development. *Physiol. Plantarum.* 20, 1059–1065.

Dunlap, J.R. (1988) Regulation of ACC-dependent ethylene production by excised leaves from normal and albino *Zea mays* L. seedlings. *J. Expt. Bot.* 39, 1079–1089.

Duque, P. and Arrabara, J.D. (1999) Respiratory metabolism during cold storage of apple fruit.

II. Alternative oxidase is induced at the climacteric. *Physiol. Plant.* 107, 24–31.

Durkin, D. and Kuc, R. (1966) Vascular blockage and senescence of the cut rose flower. *Proceedings of the Amer. Soc. Hort. Sci.* 89, 683–688.

Durkin, D.J. (1992) Roses. In: Larson, R.A. (ed.) *Introduction to Floriculture*, 2nd edn. Academic Press, New York, pp. 67–92.

Eaks, I.L. (1956) Effects of modified atmosphere on cucumbers at chilling and non-chilling temperatures. *Proceedings of the J. Amer. Soc. Hort. Sci.* 67, 473.

Eaks, I.L. (1983) Effects of chilling on respiration and ethylene production of 'Haas' avocado fruit at 20°C. *HortScience* 18, 235.

Eames, A.J. and MacDaniels, L.H. (1947) *An Introduction to Plant Anatomy*, 2nd edn. McGraw-Hill, New York.

Eason, J.R. and Vre, L.D. (1995) Ethylene insensitive floral senescence in *Sandersonia aurantiaca* (Hook). *New Zealand J. Crop Hortic. Sci.* 23, 447–454.

Eckert, J.W. (1967) Application and use of postharvest fungicides. In: Torgeson, C. (ed.) *Fungicides*, vol. I. Academic Press, New York, pp. 287–378.

Eckert, J.W. (1975) Postharvest pathology. Part 1. General principles. In: Pantastico, E.B. (ed.) *Postharvest Physiology, Handling and Utilization of Tropical and Subtropical Fruits and Vegetables.* AVI, Westport, Connecticut, pp. 393–414.

Eckert, J.W. (1978a) Pathological diseases of fresh fruits and vegetables. In: Hutlin, H.O. and Milner, M. (eds) *Postharvest Biology and Biotechnology.* Food and Nutrition Press, Westport, Connecticut, pp. 121–145.

Eckert, J.W. (1978b) Postharvest diseases of citrus fruits. *Outlook on Agriculture* 9, 225–232.

Eckert, J.W. and Ratnayake, M. (1983) Host-pathogen interactions in postharvest diseases. In: Lieberman, M. (ed.) *Postharvest Physiology and Crop Preservation.* Plenum Press, New York, pp. 247–264.

Eckert, J.W., Rubio, P.P., Mattoo, A.K. and Thompson, A.K. (1975) Diseases of tropical crops and their control. In: Pantastico, E.B. (ed.) *Postharvest Physiology, Handling and Utilization of Tropical and Subtropical Fruits and Vegetables*, AVI, Westport, Connecticut, pp. 415–443.

Edmond, J.-P. and Nunes, M.C. (1999) Air transportation of horticultural products. Air Cargo Transportation Research Group, University of Laval, Québec, Canada.

Edsall, J.T. and Wyman, J. (1958) *Biophysical Chemistry*, vol. 1. Academic Press, New York.

Edwards, G.A. (1946a) Permeability of trachea. *Quart. J. Florida Acad. Sci.* 9, 1–5.

Edwards, G.A. (1946b) Sex differences in O_2 consumption: arthropods. *Cell & Comp. Physiol.* 27, 53–64.

Edwards, J.I., Saltveit, M.E. and Henderson, W.R. (1983) Inhibition of lycopene synthesis in tomato pericarp tissue by inhibitors of ethylene biosynthesis and reversal by applied ethylene. *J. Amer. Soc. Hort. Sci.* 108, 512–514.

Edwards, T.J., Maurer, G., Newman, J. and Prausnitz, J.M. (1978) Vapor-liquid equilibria in multi-component aqueous solutions of weak electrolytes. *American Inst. Chem. Eng. Jour.* 24, 966–976.

Ehsan, H., Reichheld, J.P., Roef, L., Witters, E., Lardon, F., Van Bockstaele, D., Van Montagu, M., Inze, D. and Van Onckelen, H. (1998) Effect of indomethacin on cell cycle dependent cyclic AMP fluxes in tobacco BY-2 cells. *FEBS Lett.* 422, 165–169.

Eilam, Y. (1965) Permeability changes in senescing tissue. *J. Expt. Bot.*16, 614–627.

Eisenberg, B.A. (1977) Low pressure and common cold storage of rooted and unrooted ornamental cuttings. PhD dissertation, Ohio State University, Ohio.

Eisenberg, B.A., Staby, G.L. and Fretz, T.A. (1977a) Low pressure and common cold storage of rooted and unrooted ornamental cuttings. *HortScience* 12, 396.

Eisenberg, B.A., Staby, G.L. and Fretz, T.A. (1977b) Low pressure vs. common cold storage of geraniums. *Ohio Agric. Res. and Development Center Research Circular* 226, 17–19.

Eisenberg, B.A., Staby, G.L. and Fretz, T.A. (1978) Low pressure and refrigerated storage of rooted and unrooted ornamental cuttings. *J. Amer. Soc. Hort. Sci.* 103, 732–737.

Eisinger, W.R. (1983) Regulation of pea internode expansion by ethylene. *Annu. Rev. Plant Physiol.* 34, 225–240.

Eisinger, W.R. and Burg, S.P. (1971) Ethylene-induced pea internode swelling; its relation to ribonucleic acid metabolism, wall protein synthesis, and cell wall structure. *Plant Physiol.* 50, 510–517.

Eisinger, W.R., Croner, L.J. and Tiaz, L. (1983) Ethylene-induced lateral expansion in etiolated pea stems. *Plant Physiol.* 34, 225–240.

Eklund, M.W., Poysky, F.T. and Wieler, D.I. (1967a) Characteristics of *Clostridium botulinum* type F isolated from the Pacific Coast of the United States. *Appl. Microbiology* 15, 1316–1323.

Eklund, M.W., Poysky, F.T. and Wieler, D.I. (1967b) Outgrowth and toxin production of non-proteolytic type B *Clostridium botulinum* at 3.3 to 5.6°C. *J. Bacteriol.* 93, 1461–1462.

El Nahal, A.K.M. (1953) Responses of pests to fumigation – IV. The responses of *Calandra* sp. to reduced pressures. *Bull Ent. Res.* 44, 651–656.

El-Beltagy, A.S. and Hall, M.A. (1974) Effect of water stress upon endogenous ethylene levels in *Vicia faba*. *New Phytol.* 73, 47–60.

El-Beltagy, A.S., Hewett, E.W. and Hall, M.A. (1976a) Effect of ethephon (2-chloroethyl phosphonic acid) on endogenous levels of auxin, inhibitors and cytokinins in relation to senescence and abscission in *Vica faba* L. *J. Hort Sci.* 51, 451–465.

El-Beltagy, A.S., Kapuya, J.A., Modkour, M.A. and Hall, M.A. (1976b) A possible endogenous rhythm in the internal ethylene levels in the leaves of *Lycopersicon esculentum* Mill. *Plant Sci. Lett.* 6, 175–180.

El-Goorani, M.A. and Sommer, N.F. (1979) Suppression of postharvest plant pathogenic fungi by carbon monoxide. *Phytopathology* 69, 834.

Eliasson, L., Bertell, G. and Bolander, E. (1989) Inhibitory action of auxin on root elongation not mediated by ethylene. *Plant Physiol.* 91, 310–314.

Elkashif, M.E. and Huber, D.J. (1988) Electrolyte leakage, firmness, and scanning electron microscopic studies of watermelon treated with ethylene. *J. Amer. Soc. Hort. Sci.* 113, 378–381.

El-Kazzaz, M.K., Sommer, N.F. and Kader, A.A. (1983) Ethylene effects on *in vitro* and *in vivo* growth of certain postharvest fruit-infecting fungi. *Phytopathology* 73, 998–1001.

Elliot, M.I. and Lougheed, E.C. (1977) Low-pressure storage of some vegetable seeds. *HortScience* 12, 397.

Ellmore, G.S. (1981) Root dimorphism in *Ludwigia peploides* (Onagraceae) structure and gas content of mature roots. *Amer. J. Bot.* 68, 557.

Else, M.A., Davies, W.J., Hall, K.C. and Jackson, M.B. (1993) Knowledge of xylem sap flow rate is a prerequisite for accurate estimates of hormone transport from roots to shoots. In: Pech, J.C., Latché, A. and Balagué, C. (eds) *Cellular and Molecular Aspects of the Plant Hormone Ethylene.* Kluwer Academic, Dordrecht, pp. 373–374.

Elsworth, J.F. and Martin, K.R. (1971) Flavinoids of the Proteaceae. Part 1. A chemical contribution to the studies on the evolutionary relationships in the South African Proteoidaceae. *J. S. Afr. Bot.* 37, 199–212.

Emery, R.J.N., Reid, D.M. and Chainnappa, C.C. (1994) Phenotypic plasticity of stem elongation in two ecotypes of *Stellaria longipes*: the role of ethylene and response to wind. *Plant Cell Environ.* 17, 691–700.

English, H. and Gerhardt, F. (1942) Effect of carbon dioxide and temperature on the decay of sweet cherries under simulated transit conditions. *Proceedings of the J. Amer. Soc. Hort. Sci.* 40, 172–176.

Erdmann, H., Griesbach, R.J., Lawson, R.H. and Mattoo, A.K. (1989) 1-Aminocyclopropane-1-carboxylic-acid-dependent ethylene production during reformation of vacuoles in evacuolated protoplasts of *Petunia hybrida*. *Planta* 179, 196–202.

Erkan, Z. and Bangerth, F. (1980) Untersuchungen uber den einfluss von phytohormonen und wachstumsregulatoren auf den wasserbrauch, das stomataverhalten und die photosynthesse von parika and tomaten pflanzen. *Angew. Botank.* 54, 207–220.

Ernest, L.C. and Valdovinos, J.G. (1971) Regulation of auxin levels in *Coleus blumei* by ethylene. *Plant Physiol.* 48, 402.

Esashi, Y. (1991) Ethylene and seed germination. In: Mattoo, A.K. and Suttle, J.C. (eds) *The Plant Hormone Ethylene.* CRC Press, Boca Raton, Florida, pp. 133–157.

Esashi, Y., Saijoh, Y., Saitoh, H., Ishida, S. and Satoh, S. (1985) CO_2 enhancement of anaerobic and hypoxic C_2H_4 production in the tissues of cocklebur seeds at different dormant states. *Austr. J. Plant Physiol.*12, 59–68.

Esashi, Y., Oshima, Y., Michiharu, A., Akiko, K. and Satoh, S. (1986) CO_2-enhancement of C_2H_4 production in tissues of imbided cocklebur seeds. *Austr. J. Plant Physiol.* 13, 417–429.

Esashi, Y., Fuwa, N., Kurota, A., Oota, H. and Abe, M. (1987) Interrelation between ethylene and carbon dioxide in relation to respiration and adenylate content in the pre-germination period of cocklebur seeds. *Plant Cell Physiol.* 28, 141–150.

Esquerre-Tugaye, M.T., Bottin, A., Rickauer, M., Sancan, J.P., Fournier, J. and Pouenat, M.L. (1993) Ethylene in early signalling phenomena at the plant-microorganism interface. In: Pech, J.C., Latché, A. and Balagué, C. (eds) *Cellular and Molecular Aspects of the*

Plant Hormone Ethylene. Kluwer Academic, Dordrecht, pp. 217–222.

Estelle, M. and Somerville, C. (1987) Auxin-resistant mutants of *Arabidopsis thaliana* with an altered morphology. *Mol. Gen. Genet.* 206, 200–206.

Evans, D.E., Bengochea, T., Cairns, A.J., Dodds, J.H. and Hall, M.A. (1982a) A study of the sub-cellular localization of an ethylene binding site in developing cotyledons of *Phaseolus vulgaris* L. by high resolution autoradiography. *Planta* 154, 48–52.

Evans, D.E., Bengochea, T., Cairns, A.J., Dodds, J.H. and Hall, M.A. (1982b) Studies of ethylene binding by cell-free preparations from cotylendons of *Phaseolus vulgaris* L. : subcellular localization. *Plant Cell Environ.* 5, 101–107.

Evans, D.E., Smith, A.R., Taylor, J.E. and Hall, M.A. (1987) Ethylene metablism in *Pisum sativum* L.: kinetic parameters, the effects of propylene, silver and carbon dioxide and the comparison with other systems. *Plant Growth Regul.* 3, 85.

Evans, M.L. (1991) Gravitropism: interaction of sensitivity modulation and effector redistribution. *Plant Physiol.* 95, 1–5.

Evensen, K.B. (1991) Ethylene responsiveness changes in *Pelargonium* × *domesticum* florets. *Physiol. Plant.* 82, 409–412.

Evensen, K.B., Clark, D.G. and Singh, A. (1993) Rapid ethylene-induced gene expression during petal senescence. In: Pech, J.C., Latché, A. and Balagué, C. (eds) *Cellular and Molecular Aspects of the Plant Hormone Ethylene*. Kluwer Academic, Dordrecht, pp. 278–283.

Eya, Y., Meller, Y., Lev-Yadun, S. and Fluhr, R. (1993) A basic-type PR-1 promoter directs ethylene responsiveness, vascular and abscission zone-specific expression. *Plant J.* 4, 225–234.

Eyles, M.J., Moir, C.J. and Davey, J.A. (1993) The effects of modified atmospheres on the growth of psychrotrophic pseudomonads on a surface in a model system. *Int. J. Food Microbiol.* 20, 97–107.

Fan, X.-T. and Mattheis, J.P. (2000) Yellowing of broccoli in storage is reduced by 1-methylcyclopropene. *HortScience* 35, 885–887.

Fan, X.-T., Blankenship, S.M. and Mattheis, J.P. (1999) 1-methylcyclopropene inhibits apple ripening. *J. Amer. Soc. Hort. Sci.* 124, 690–695.

Fan, X.-T., Mattheis, J.P. and Roberts, R.G. (2000) Biosynthesis of phytoalexin in carrot root requires ethylene action. *Physiol. Plant.* 110, 450–454.

Fan, X.-T., Argenta, L. and Mattheis, J.P. (2001) Inhibition of ethylene action by 1-methylcyclopropene prolongs storage life of apricots. *Plant Growth Regl. Abstracts* 27, 61.

Farquhar, G.D. and Sharkey, T.D. (1982) Stomatal conductance and photosynthesis. *Annu. Rev. Plant Physiol.* 33, 317–345.

Farquhar, G.D., Wetselaar, R. and Firth, P.M. (1979) Ammonia volatization from senescing leaves of maize. *Science* 203, 1257–1258.

Farquhar, G.D., Firth, P.M., Wetselaar, R. and Weir, B. (1980) On the gaseous exchange of ammonia between leaves and the environment: determination of the ammonia compensation point. *Plant Physiol.* 66, 710–714.

Feldman, L.J. (1984) Regulation of root development. *Annu. Rev. Plant Physiol.* 35, 223–242.

Feldman, L.J. (1985) Root gravitropism. *Physiol. Plant.* 65, 341–344.

Felix, G. and Meins, F., Jr (1987) Ethylene regulation of β-1,3-glucanase in tobacco. *Planta* 172, 386–392.

Fentem, P.A., Lea, P.J. and Stewart, G.R. (1983) Action of inhibitors of ammonia assimilation on amino acid metabolism in *Hordeum vulgare* L. (cv. Golden Promise). *Plant Physiol.* 71, 502–506.

Ferguson, W. (1942) Sleepiness in carnations. *Can. J. Agr.* 22, 509–518.

Fernández-Maculet, J.C., Dong, J.G. and Yang, S.F. (1993) Activation of 1-aminocyclopropane-1-carboxylate oxidase by carbon dioxide. *Biochem. Biophys. Res. Comm.* 193, 1168–1173.

Ferrarese, L., Trainotti, L., Moretto, P., Polverino de Laureto, P., Rascio, N. and Casaadoro, G. (1995) Differential ethylene-inducible expression of cellulase in pepper plants. *Plant Mol. Biol.* 29, 735–747.

Ferreira, D.I. (1983) Prevention of browning of leaves of *Protea neriifolia* R. Br. *Acta Horticulturae* 138, 273–276.

Ferreira, D.I. (1986) The influence of temperature on the respiration rate and browning of *Protea neriifolia* R. Br. inflorescences. *Acta Horticulturae* 185, 121–129.

Fidler, J.C. (1931) Studies in zymosis. IV. The accumulation of zymosis products in apples during senescence. *Biochem. J.* 27, 1614.

Fidler, J.C. (1950) Studies of the physiologically active volatile organic compounds produced by fruits. *J. Hort. Sci.* 25, 81–110.

Fidler, J.C. (1951) A comparison of aerobic and anaerobic respiration of apples. *J. Expt. Bot.* 2, 41.

Fidler, J.C. and North, C.J. (1962) Gas storage of apples in low concentrations of oxygen. *Bull. Int. Instit. Refrig.* 42, 25.

Fidler, J.C. and North, C.J. (1967) The effect of conditions of storage on the respiration of apples. I. The effects of temperature and concentrations of carbon dioxide and oxygen on the production of carbon dioxide and uptake of oxygen. *J. Hort. Sci.* 42, 189–206.

Fidler, J.C. and North, C.J. (1969a) Production of volatile organic compounds by apples. *J. Sci. Food Agric.* 20, 521–526.

Fidler, J.C. and North, C.J. (1969b) Production of volatile organic compounds by pears. *J. Sci. Food Agric.* 20, 518–520.

Fieger, E.A. and Bailey, M.E. (1954) Chemical prevention of black spot (melanogenesis) in ice stored shrimp. *Food Technol.* 8, 317–319.

Fieger, E.A., Bailey, M.E. and Novak, A.F. (1956) Chemical ices for shrimp preservation. *Food Technol.* 10, 578–583.

Field, R.J. (1985) The effect of temperature on ethylene production by plant tissues. In: Roberts, J.A. and Tucker, G.A. (eds) *Ethylene and Plant Development.* Butterworths, Boston, Massachusetts, pp. 47–69.

Fife, J.M. and Framptom, V.L. (1935) The effect of carbon dioxide upon the pH and certain nitrogen fractions of the sugar-beet plant. *J. Biol. Chem.* 109, 643–655.

Filner, B., Hertel, R., Steele, C. and Fan, V. (1970) Some aspects of geotropism in coleoptiles. *Planta (Berl.)* 94, 333–354.

Finger, F.L., Campanha, M.M., Barbosa, J.G. and Fontes, P.C.R. (2000) Influence of ethephon, silver thiosulfate and sucrose pulsing on bird-of-paradise vase life. *Plant Growth Regl. Abstracts* 26, 60.

Finlayson, S.A., Lee, I.-J. and Morgan, P.W. (1998) Phytochrome B and the regulation of circadian ethylene production in sorghum. *Plant Physiol.* 116, 17–25.

Finlayson, S.A., Lee, I.-J. and Morgan, P.W. (1999) The mechanism of rhythmic ethylene production in sorghum. The role of phytochrome B and simulated shading. *Plant Physiol.* 119, 1083–1089.

Finlayson, S.A., He, C.J., Lee, I.J., Drew, M.C., Mullet, J.E. and Morgan, P.W. (2000) Phytochrome B and ethylene rhythms in sorghum: biosynthetic mechanism and developmental effects. *Plant Growth Regl. Abstracts* 26, 138.

Fischer, A. (1885) Ueber den Inhalt der Siebrohren in der univerletzter Pflanze. *Ber. Deutsch. Bot. Ges.* 3, 230–239.

Fisher, C.W. (1950) Ethylene gas a problem in cut flower storage. *New York State Flower Growers Bul.* 61, 1–4.

Fisher, J.B. (1980) The vegetative and reproductive structure of papaya (*Carica papaya*) *Lyonia* 1, 191–208.

Fleurat-Lessard, F. (1990) Effect of modified atmospheres on insects and mites infesting stored products. In: Calderon, M. and Barkai-Golan, R. (eds) *Food Preservation by Modified Atmospheres.* CRC Press, Boca Raton, Florida, pp. 21–38.

Fleurat-Lessard, P. (1981) Ultrastructural features of the starch sheath of the primary pulvinus after gravistimulation of the sensitive plant (*Mimosa pudica* L.). *Protoplasma* 105, 177–184.

Flick, G.J. and Lovell, R.T. (1972) Post mortem biochemical changes in the muscle of Gulf shrimp, *Penaeus Aztecus. J. Food Sci.* 37, 609–611.

Flores, F. Ben, Amor, M., Jones, B., Pech, J.C., Bouzayen, M., Latché, A. and Romojaro, F. (2001a) The use of ethylene-suppressed lines to assess differential sensitivity to ethylene of various ripening pathways in Cantaloupe melons. *Physiol. Plant.* 113, 128–133.

Flores, F.B., Martínez-Madrid, M.C., Sánchez-Hidalgo, F.J. and Romojaro, F. (2001b) Differential rind and pulp ripening of transgenic antisense ACC oxidase melon. *Plant Physiol. Biochem.* 39, 37–43.

Fockens, F.H. and Meffert, H.F.T. (1972) Biophysical properties of horticultural products as related to loss of moisture during cooling down. *J. Sci. Food. Agric.* 23, 285–298.

Follstad, M.N. (1966) Mycelial growth rate and sporulation of *Alternaria tenuis, Botrytis cinerea, Cladosporium herbarum,* and *Rhizopus stolonifer* in low oxygen and high carbon dioxide atmospheres. *Phytopathology* 56, 1098–1099.

Fonteno, W.C. (1992) Geraniums. In: Larson, R.A. (ed.) *Introduction to Floriculture,* 2nd edn. Academic Press, New York, pp. 451–475.

Forsyth, F.R., Evans, C.A. and Lightfoot, H.J. (1969) Storage quality of McIntosh apples as affected by removal of ethylene from the storage atmosphere. *Can. J. Plant Sci.* 49, 567–572.

Forsyth, F.R., Hall, I.V. and Lightfoot, H.J. (1972) Diffusion of CO_2, O_2, and ethylene in cranberry fruit. *HortScience* 8, 45–46.

Fosket, D.E. and Morejohn, L.C. (1992) Structural and functional organization of tubulin. *Ann. Rev. Plant Physiol. Plant Mol. Biol.* 43, 201–240.

Fox, H.M., Wingfield, C.A. and Simmonds, B.G. (1937) Oxygen consumption in relation to habitat. *J. Expt. Biol.* 14, 210–218.

Fraenkel, G. and Herford, G.V.B. (1940) Effects of high temperatures on insects. *J. Expt. Bot.* 17, 386–395.

Freiberg, S.R. (1955) Effect of growth regulators on ripening, split peel, reducing sugars and diastatic activity of bananas. *Bot. Gaz.* 117, 113–119.

Frenkel, C. and Dyck, R. (1973) Auxin inhibition of ripening in Bartlett pears. *Plant Physiol.* 51, 6–9.

Frenkel, C. and Patterson, M.E. (1969) The effect of carbon dioxide on succinic dehydrogenase in pears during cold storage. *Hort. Sci.* 4, 165.

Frenkel, C., Klein, I. and Dilley, D.R. (1968) Protein synthesis in relation to ripening of pome fruits. *Plant Physiol.* 43, 1146–1153.

Frey-Wyssling, A. (1959) *Die pflanzliche Zellwand.* Springer, Berlin.

Frommhold, I. (1982) Effect of ethephon on stomatal opening in detached epidermal strips of tobacco leaves (*Nicotiana tabacum* L. cv. Samsun). *Biol. Plant. (Prague)* 24, 303–306.

Fuchs, Y. and Temkin-Gorodeiski, N. (1971) The course of ripening of banana fruits stored in sealed polyethylene bags. *J. Amer. Soc. Hort. Sci.* 96, 401–403.

Fujimoto, S.Y., Ohta, M., Usul, A., Shinshi, H. and Ohme-Takagi, M. (2000) *Arabidopsis* ethylene-responsive element-binding factors act as transcriptional activators or repressors of GCC box-mediated gene expression. *Plant Cell* 12, 393–394.

Fujino, D.W., Burger, D.W., Yang, S.F. and Bradford, K.J. (1988) Characterization of an ethylene overproducing mutant of tomato (*Lycopersicum esculentum* Mill. cultivar VFN8). *Plant Physiol.* 88, 774–779.

Fujino, D.W., Burger, D.W. and Bradford, K.J. (1989) Ineffectiveness of ethylene biosynthetic and action inhibitors in phenotypically reverting the *Epinastic* mutant of tomato (*Lycopersicum esculentum* Mill.). *J. Plant Growth Regul.* 8, 53–61.

Fukushima, T., Yarimizu, K., Kitamura, T. and Iwata, T. (1980) The relation between water stress and the climacteric in respiration of some fruits. *Scientia Horticulturae* 12, 259–264.

Furutani, S.C., Sakai, W.S., Hatta, T. and Nagao, M.A. (1987) NAA-induced leaf epinasty in *Crysanthemum. HortScience* 22, 100–101.

Gaastra, P. (1959) Photosynthesis of crop plants as influenced by light, carbon dioxide, temperature, and stomatal diffusion resistance. *Meded Landbouwhogesch Wageningen* 59, 1–68.

Gac, A. (1956) Contribution to the study of the influence of the relative humidity and the rate of circulation of air on the behavior of harvested fruit. *La Revue Generale du Froid* 33, 365–379; 505–531; 733–744; 833; 843; 969–979.

Gago, C.M.L., Montteiro, J.A. and Rodrigues, M.H. (2001) *Bougainvillea* post-production: NAA and STS control of bract abscission is subject to exogenous ethylene. *Plant Growth Regl. Abstr.* 27, 232.

Galaway, M.E., Masucci, J.D., Lloyd, A.M., Walbot, V., Davis, R.W. and Schiefelbein, J.W. (1994) The *TTG* gene is required to specify epidermal cell fate and cell patterning in the *Arabidopsis* root. *Dev. Biol.* 166, 740–754.

Gallardo, M.A., Melendo, J.M., Ureita, J.S. and Losa, C.G. (1987) Solubility of non-polar gases in cyclohexanone between 273.15 and 303.15 K at 101.32 kPa partial pressure of gas. *Can. J. Chem.* 65, 2198–2202.

Galliard, T. (1968) Aspects of lipid metabolism in higher plants. II. The identification and qualitative analysis of lipids from pulp of pre- and post-climacteric apples. *Phytochemsitry* 7, 1915–1922.

Galston, A.W., Tuttle, A.A. and Penny, P.J. (1961) A kinetic study of growth movements and photomorphogenesis in etiolated pea seedlings. *Amer. J. Bot.* 51, 853–858.

Gamble, R.L., Coonfield, M.L. and Schaller, G.E. (1996) Histidine kinase activity of the ETR1 ethylene receptor from *Arabidopsis. Proc. Natl Acad. Sci. USA* 95, 7825–7829.

Gane, R. (1934) The identification of ethylene among the volatile products of ripe apples. *Gt. Brit. Dept. Sci. Indus. Res., Food Invest. Bd. Rept.* 1934, 130.

Garcia-López, M.L., Prieto, M. and Otero, A. (1998) The physiological attributes of gram-negative bacteria associated with spoilage of meat and meat products. In: Davies, A. and Board, R. (eds) *The Microbiology of Meat and Poultry.* Blackie Academic & Professional, London, pp. 1–34.

Garrett, M. (1996) Controlled atmospheres provide an added value service: how is this achieved? *Food Trans Africa Seminar,* 12 March 1996, Durban, South Africa.

Gatehouse, J.A. (1997) Regulation of gene expression in plants. In: Dey, P.M. and Harbone, J.B. (eds) *Plant Biochemistry.* Academic Press, New York, pp. 353–385.

Gates, D.M. (1968) Transpiration and leaf temperature. *Annu. Rev. Plant Physiol.* 19, 211–238.
Gazit, S. and Blumenfeld, A. (1970) Response of mature avocado fruits to ethylene treatments before and after harvest. *J. Amer. Soc. Hort. Sci.* 95, 229–231.
Geeson, J.D., Browne, K.M. and Guaraldi, F. (1986) The effects of ethylene concentration in controlled atmosphere storage of tomatoes. *J. Appl. Biol.* 108, 605–610.
Gemma, H., Oogaki, C., Fukushima, M., Yamada, R. and Nose, Y. (1989) Preservation of some tropical fruits with an apparatus of low-pressure storage for practical use. Effect of low-pressure storage on fruits and vegetables. Part I. *J. Jpn. Soc. Food Sci. Technol.* 36, 509–518.
George, J.B., Marriott, J., Palmer, J.M. and Karikari, S.K. (1982) Sensitivity to water stress and ethylene of stored plantain fruits. *J. Expt. Bot.* 33, 1194–1201.
George, P. and Stratmann, C.J. (1952a) The oxidation of myoglobin to metmyoglobin by oxygen. *Biochem. J.* 51, 103.
George, P. and Stratmann, C.J. (1952b) The oxidation of myoglobin to metmyoglobin by oxygen. 2. The relation between the first order rate constant and the partial pressure of oxygen. *Biochem. J.* 51, 418–425.
Gepstein, S. and Thimann, K.V. (1980) The effect of light on the production of ethylene from 1-aminocyclopropane-1-carboxylic acid by leaves. *Planta* 149, 196–199.
Gepstein, S. and Thimann, K.V. (1981) The role of ethylene in the senescence of oat leaves. *Plant Physiol.* 68, 349–354.
Gerasopoulos, D. and Richardson, D.G. (1996) Effects of exogenous propylene and fruit calcium on ripening of nonchilled and chilled Anjou pears. *Postharv. Biol. Technol.* 8, 111–120.
Gerber, H. and Bussmann, A. (1958) Der Verlaaud des Vitamin-C-Gehaltes während der Lagerung. *Mitt. aus. dem Gebeit der Lebensmitteluntersuchung und Hygiene* 49, 192–200.
Gill, C.O. (1998) Microbial contamination of meat during slaughter and butchering of cattle, sheep and pigs. In: Davies, A. and Board, R. (eds) *The Microbiology of Meat and Poultry.* Blackie Academic & Professional, London, pp. 118–157.
Gill, C.O. and Harrison, J.C.L. (1989) The storage life of chilled pork packaged under carbon dioxide. *Meat Science* 26, 313–324.
Gillespie, B. and Thimann, K.V. (1961) The lateral transport of indole acetic acid C^{14} in geotropism. *Experientia* 17, 1–8.
Gillespie, B. and Thimann, K.V. (1963) Transport and distribution of auxin during tropistic response. I. The lateral migration of auxin in geotropism. *Plant Physiol.* 38, 214–225.
Gimenes, C.A. and Sommer, N.F. (1988) Oscillating pressure storage and gas exchange in apple fruit. *Pessquisa Agropecuaria Brasileira* 23, 145–152.
Gimmler, H., Weiss, C., Baier, M. and Hartung, W. (1990) The conductance of the plasmalemma for CO_2. *J. Expt. Bot.* 41, 785–795.
Giovannoni, J. (1997) Molecular genetic analysis of ethylene-regulated and developmental components of tomato fruit ripening. In: Kanellis, A.K., Chang, C., Kende, H. and Grierson, D. (eds) *Biology and Biotechnology of the Plant Hormone Ethylene.* Kluwer Academic, Dordrecht, pp. 133–140.
Giovannoni, J. (2001) Molecular biology of fruit maturation and ripening. *Annu. Rev. Plant Physiol. Plant Mol. Biol.* 52, 725–749.
Giovannoni, J., Fox, E., Kannan, P., Lee, S., Padmanabhan, V. and Vrebalov, J. (2000) Analysis of gene expression and mutants influencing ethylene responses and fruit development in tomato. *Plant Growth Regl. Abstracts* 26, 139.
Girsch, G.K. (1978) Susceptibility of full grown larvae of *Trogoderma granarium* Everts to varying concentrations of carbon dioxide at low oxygen tension. *Bull. Grain Technol.* 16, 199.
Glasziou, K.T., Sacher, J.A. and McCalla, D.R. (1960) On the effects of auxins on membrane permeability and pectic substances in bean endocarp. *Amer. J. Bot.* 47, 743–752.
Goddard, D.R. (1947) The respiration of cells and tissues. In: Höber, R. (ed.) *Physical Chemistry of Cells and Tissues.* Churchill, London, pp. 371–444.
Goeschl, J.D. and Kays, S.J. (1975) Concentration dependencies of some effects of ethylene on etiolated pea, peanut, bean and cotton seedlings. *Plant Physiol.* 55, 670–677.
Goeschl, J.D. and Pratt, H.K. (1968) Regulatory roles of ethylene in the etiolated growth habit of *Pisum sativum.* In: Wightman, F. and Setterfield, G. (eds) *Biochemistry and Physiology of Plant Growth Substances.* Runge Press Ltd, Ottawa, Canada, pp. 1129–1142.
Goeschl, J.D., Rappaport, D.L. and Pratt, H.K. (1966) Ethylene is a factor regulating the growth of pea epicotyls subjected to physical stress. *Plant Physiol.* 41, 877–884.
Goeschl, J.D., Pratt, H.K. and Bonner, B.A. (1967) An effect of light on the production of ethylene and the growth of the plumular portion of

the etiolated pea seedlings. *Plant Physiol.* 42, 1077–1080.

Goh, C.J., Halevy, A.H., Engel, R. and Kofranek, A.M. (1985) Ethylene evolution and sensitivity in cut orchid flowers. *Scientia Hortic.* 26, 57–67.

Gold, A.H. and Facciolli, G. (1972) Comparative effect of curly-top infection and ethylene treatment on tomatoes and other plants. *Phytopathol. Mediterr.* 11, 145.

Golding, J.B., McGlasson, W.B. and Wyllie, S.G. (2001) Relationship between production of ethylene and α-farnescene in apples, and how it is influenced by the timing of diphenylamine treatment. *Postharvest Biol. Technol.* 21, 225–233.

Goldschmidt, E.E., Huberman, M. and Goren, R. (1993) Probing the roles of endogenous ethylene in the degreening of citrus fruits with ethylene antagonists. *J. Plant Growth Regul.* 12, 325–329.

Goldschmidt, M.H.M., Cataldo, D.A., Karn, J., Brenneman, T. and Trip, P. (1974) The rapid non-polar transport of auxin in the phloem of intact *Coleus* plants. *Planta* 116, 301–317.

Good, X., Kellog, J.A., Wagoner, W., Langhoff, D., Matsumura, W. and Bestwick R.K. (1994) Reduced ethylene synthesis in transgenic tomatoes expressing *S*-adenosylmethionine hydrolase. *Plant Mol. Biol.* 26, 781–790.

Goodlass, G. and Smith, K.A. (1979) Effects of ethylene on root extension and nodulation of pea (*Pisum sativum* L.) and white clover (*Trifolium repens* L.). *Plant Soil* 51, 387–395.

Goodwin, R.H. (1972) Studies on roots. V. Effects of indoleacetic acid on the standard root growth pattern of *Phleum pretense*. *Bot. Gaz.* 133, 224–229.

Goos, R.D. and Tschirsch, M. (1962) Effect of environmental factors on spore germination, spore survival and growth of *Gloeosporium musarum*. *Mycologia* 34, 353.

Goren, R. and Sisler, E.C. (1986) Ethylene-binding characteristics in *Phaseolus*, *Citrus*, and *Ligustrum* plants. *J. Plant Growth Regul.* 4, 43–54.

Goren, R., Mattoo, A.K. and Anderson, J.D. (1984) Ethylene binding during leaf development and senescence and its inhibition by silver nitrate. *J. Plant Physiol.* 117, 243–248.

Goren, R., Winer, L. and Riov, J. (2000) Effect of ethylene on the oxidative decarboxylation pathway of indole-3-acetic acid. *Plant Growth Regl. Abstr.* 26, 139.

Gorney, J.R. and Kader, A. (1996) Controlled-atmosphere suppression of ACC synthase and ACC oxidase in 'Golden Delicious' apple during long-term cold storage. *J. Amer. Soc. Hort. Sci.* 121, 751–755.

Goss, S.P.A., Singh, R.J. and Kalyanaraman, B. (1999) Bicarbonate enhances the peroxidase activity of Cu,Zn-superoxide dismutase. Role of carbonate anion radical. *J. Biol. Chem.* 274, 28233–28239.

Goswami, K.K.A. and Audus, L.J. (1976) Distribution of calcium, potassium and phosphorus in *Helianthus annus* hypocotyls and *Zea mays* coleoptiles in relation to tropic stimulation and curvature. *Ann. Bot.* 40, 49–64.

Goszczynska, D. and Rudnicki, R.M. (1982) Long-term storage of carnations *Dianthus caryophyllus* cultivar Scania-3C cut at the green bud stage. *Sci. Hortic. (Amst.)* 17, 289–298.

Gougler, J.A. and Evans, M.L. (1981) Adaptation of corn roots to exogenously applied auxin. *Physiol. Plant.* 51, 394–398.

Govindarajan, S. (1973) Fresh meat color. In: *Critical Reviews in Food Technology*, vol. 4. CRC Press, Boca Raton, Florida, pp. 117–140.

Graell, J., Recasens, I., Salas, J. and Vendrell, M. (1993) Variability of internal ethylene concentrations as a parameter of maturity in apples. *Acta Horticulturae* 326, 277–284.

Graham, D. and Patterson, B.D. (1982) Responses of plants to low, nonfreezing temperatures: proteins, metabolism and acclimation. *Annu. Rev. Plant Physiol.* 33, 347–372.

Graham, D. and Reed, M.L. (1971) Carbonic anhydrase and the regulation of photosynthesis. *Nature New Biol.* 231, 81–82.

Gran, C.D. and Beaudry, R.M. (1993) Determination of the low-oxygen limit for several commercial apple cultivars by respiratory quotient breakpoint. *Postharvest Biol. and Tech.* 3, 259–267.

Gray, J.E., Picton, S., Fray, R., Hamilton, A.J., Smith, H., Barton, S. and Grierson, D. (1993) Altered gene expression, leaf senescence, and fruit ripening by inhibiting ethylene synthesis with EFE-antisense gene. In: Pech, J.C., Latché, A. and Balagué, C. (eds) *Cellular and Molecular Aspects of the Plant Hormone Ethylene*. Kluwer Academic, Dordrecht, pp. 82–89.

Green, B.E. (1969) Lipid oxidation and pigment changes in raw beef. *J. Food Sci.* 34, 110–113.

Green, E.P., Hukill, W.V. and Rose, D.H. (1941) Calorimetric measurements of the heat of respiration of fruits and vegetables. *USDA Tech. Bull.* 771.

Green, M. (1949) Bacteriology of shrimp, II. Quantitative studies of freshly caught and iced shrimp. *Food Research* 14, 372–383.

Green, P.B. (1963) Cell walls and the geometry of growth. In: *Meristems and Differentiation. Brookhaven Symp. Biol.* 16, 203–215.

Greenwood, D.J. (1967) *New Phytol.* 66, 337.

Gregg, K.E. (1973) Studies on the control of sex expression in the genera *Cychnoches* and *Catasetum*, Catasetinae, Orchidaceae. PhD dissertation. University of Miami, Miami, Florida.

Gregg, K.E. (1982) Sunlight-enhanced ethylene evolution by developing inflorescences of *Catasetum* and *Cycnoches* and its relation to female flower production. *Bot. Gaz.* 143, 466–475.

Grierson, D. (1987) Senescence in fruits. *HortScience* 22, 859–862.

Grierson, W. (1971) Chilling injury in tropical and subtropical fruits: IV. The role of packaging and waxing in minimizing chilling injury of grapefruit. *Proceedings of the Trop. Reg. Amer. Soc. Hort. Sci.* 15, 76–88.

Grierson, W. and Wardowski, W.F. (1978) Relative humidity effects on the postharvest life of fruits and vegetables. *HortScience* 13, 570–578.

Grignon, C. and Sentenac, H. (1991) pH and ionic conditions in the apoplast. *Annu. Rev. Plant Physiol. Mol. Biol.* 42, 103–128.

Grncarevic, M. and Radler, F. (1967) The effect of wax components on cuticular transpiration – model experiments. *Planta (Berl.)* 75, 23–27.

Grodzinski, B., Boesel, I. and Horton, R.F. (1981) The effect of light intensity on the release of ethylene from leaves. *J. Expt. Bot.* 33, 1185.

Groeschal, E.C., Nelson, A.E. and Steinberg, M.P. (1966) Changes in the color and other characteristics of green beans stored in controlled refrigerated atmosphere. *J. Food Sci.* 31, 488.

Grossmann, K. (1996) A role for cyanide, derived from ethylene biosynthesis, in the development of stress symptoms. *Physiol. Plant.* 97, 722–725.

Grumman Allied Industry (1978) Dormavac lamb gets high marks from Armour taste panel. Grumman Dormavac Progress Report, Vol. 1 No. 3, April 1978.

Grumman Allied Industry (1979) Lettuce test report.

Gu, Y.-Q., Yang, C.-M., Thara, V.K., Zhou, J.-M. and Martin, G.B. (2000) *Pti4* is induced by ethylene and salicylic acid, and its product is phosphorylated by the Pto kinase. *Plant Cell* 12, 771–785.

Guadagni, D.G., Bomben, J.L. and Hudson, J.S. (1971) Factors influencing the development of aroma in apple peels. *J. Sci. Food. Agric.* 22, 110.

Guern, J., Mathieu, Y., Péan, M., Pasquier, C., Beloeil, J.-C. and Lallemand, J.Y. (1986) Cytoplasmic pH regulation in *Acer psuedoplatanus* cells. I. A ^{31}P NMR description of acid-load effects. *Plant Physiol.* 82, 840–845.

Guis, M., Bouquin, T., Zegouti, H., Ayub, R.B., Amor, M., Lasserre, E., Botondi, R., Raynal, J., Latché, A., Bouzayes, M., Balagué, C. and Pech, J.C. (1997) Differential expression of ACC oxidase genes in melon and physiological characterization of fruit expressing the antisense ACC oxidase gene. In: Kanellis, A.K., Chang, C., Kende, H. and Grierson, D. (eds) *Biology and Biotechnology of the Plant Hormone Ethylene.* Kluwer Academic, Dordrecht, pp. 327–338.

Gunning, B.E.S. and Hardham, A.R. (1982) Microtubules. *Annu. Rev. Plant Physiol.* 33, 651–698.

Gutierrez Martínez, P., López Gómez R. and Gómez-Lim, M.A. (2001) Identification of an ETR-1 homologue from mango fruit expressed during fruit ripening and wounding. *J. Plant Physiol.* 158, 101–108.

Gutknecht, J. (1968) Salt transport in *Valonia*: inhibition of potassium uptake by small hydrostatic pressures. *Science* 160, 68.

Guttenberg, H. (1910) Üb. d. Zusammenwirken von Geotropismus und heliotropoismuss und d. tropisy. Empfindlichkeit in reiner u. unreiner Luft. *Jahrb. f. wss. Bot. Bd.* 47, S. 462.

Guttenberg, H. and Steinmetz, E. (1947) Effects of ethylene on growth hormone and growth. *Pharmazie* 2, 17–21.

Guy, M. and Kende, H. (1984) Conversion of 1-aminocyclopropane-1-carboxylic acid to ethylene by isolated vacuoles of *Pisum sativum* L. *Planta* 160, 281–287.

Guzman, P. and Ecker, J.R. (1990) Exploiting the triple response of *Arabidopsis* to identify ethylene-related mutants. *Plant Cell* 2, 513–523.

Ha, Y.-S. and Sohn, T.H. (1985) Effects of subatmospheric pressure storage on ultrastructural changes in the chloroplasts of apple fruits. *J. Korean Agric. Chem. Soc.* 28, 174–181.

Haard, N.F. (1983) Hypobaric storage of fresh fish. In: Sun Pan, B. (ed.) *Properties and Processing of Marine Foods.* Marine Food Science Series, National Taiwan College of Marine Science & Technology, Keelung, Taiwan, pp. 36–60.

Haard, N.F. (1992) Technological aspects of extending prime quality of seafood: a review. *J. Aquatic Food Product Technology* 1, 9–27.

Haard, N.F. and Lee, Y.Z. (1982) Hypobaric storage of Atlantic salmon in a carbon dioxide atmosphere. *Can. Inst. Food Sci. & Technol. J.* 15, 68–71.

Haard, N.F. and Salunkhe, D.K. (1975) *Symposium: Postharvest Biology and Handling of Fruits and Vegetables.* AVI, Westport, Connecticut.

Haard, N.F., Sharma, S.C., Wolfe, R. and Frenkel, C. (1974) Ethylene induced isoperoxidase changes during fiber formation in post-harvest asparagus. *Jour. Food Sci.* 39, 452–456.

Haard, N.F., Martins, I., Newbury, T. and Bota, R. (1979) Hypobaric storage of Atlantic herring and cod. *Can. Inst. Food Sci. & Technol. J.* 12, 84–87.

Haasbroek, F.J., Rousseau, G.G. and Villeirs, J.F. (1973) Effect of gamma rays on cut flower blooms of *Protea compacta* R. Br., *P. longiflora* Lamarck, and *Leucospermum cordifolium* Salisb. Ex Knight. *Agroplantae* 5, 42–53.

Habereder, H., Schrõder, G. and Ebel, J. (1989) Rapid induction of phenylalanine ammonia lyase and chalcone synthease mRNAs during fungus infection of soybean (*Glycine max* L.) roots or elicitor treatment of soybean cell cultures at the onset of phytoalexin synthesis. *Planta* 177, 58–65.

Haberlandt, G. (1927) *Physiological Plant Anatomy.* Macmillan, New York.

Hackett, M., Ho, C.H., Lin, Z., Foote, H.C.C., Fray, R.G. and Grierson, D. (2000) Nonsense inhibition of the *Nr* gene restores normal ripening to the tomato *never-ripe* mutant, consistent with the ethylene-receptor inhibition model. *Plant Physiol.* 124, 1079–1085.

Hadfield, K.A., Dang, T., Guis, M., Pech, J.C., Bouzayen, M. and Bennet, A.B. (2000) Characterization of ripening-regulated cDNAs and their expression in ethylene-suppressed Charentais melon fruit. *Plant Physiol.* 122, 977–983.

Haffner, K.E. (1992) Storage trials of 'Aroma' apples at the Agricultural University of Norway. *Acta Horticulturae* 326, 305–314.

Hahlbrock, K. and Scheel, D. (1989) Physiology and molecular biology of polypropanoid metabolism. *Annu. Rev. Plant Physiol. Plant Mol. Biol.* 40, 347–369.

Halder-Doll, H. and Bangerth, F. (1987) Inhibition of autocatalytic C_2H_4-biosynthesis by AVG applications and consequences on the physiological behavior and quality of apple fruits in cool storage. *Sci. Hortic.* 33, 87–96.

Halevy, A.H. and Mayak, S. (1974) Improvement of cut-flower quality opening and longevity by pre-shipment treatments. *Acta Horticulturae* 43, 335–347.

Halevy, A.H., Whitehead, C.S. and Kofranek, A.M. (1984) Does pollination induce corolla abscission of cyclamen flowers by promoting ethylene production? *Plant Physiol.* 75, 1090–1093.

Halevy, A.H., Porat, R., Spiegelstein, H., Borochov, A., Botha, L. and Whitehead, C.S. (1996) Short-chained saturated fatty acids in the regulation of pollination-induced ethylene sensitivity in *Phaleonopsis* flowers. *Physiol. Plant.* 97, 469–474.

Hall, A.E., Chen, Q.G., Findell, J.F., Schaller, E. and Bleecker, A.B. (1999) The relationship between ethylene binding and dominant insensitivity conferred by mutant forms of the ETR1 ethylene receptor. *Plant Physiol.* 121, 291–299.

Hall, C.W., Hardenburg, R.E. and Pantastico, E.B. (1975) Consumer packaging with plastics. In: Pantastico, E.B. (ed.) *Postharvest Physiology, Handling and Utilization of Tropical and Subtropical Fruits and Vegetables.* AVI, Westport, Connecticut, pp. 303–313.

Hall, I.V. and Forsyth, F.R. (1967) Production of ethylene by flowers following pollination and treatments with water and auxin. *Can. J. Bot.* 45, 1163–1166.

Hall, M.A. (1991) Ethylene metabolism. In: Mattoo, A.K. and Suttle, J.C. (eds) *The Plant Hormone Ethylene.* CRC Press, Boca Raton, Florida, pp. 65–80.

Hall, M.A., Kapuya, J.A., Sivakumaran, S. and John, A. (1977) The role of ethylene in the response of plants to stress. *Pestic. Sci.* 8, 217–223.

Hall, M.A., Howarth, C.J., Robertson, D., Sanders, I.O., Thomas, C. Jr. and Williams, R.A.N. (1987) Ethylene-binding proteins. In: Fox, J.E. and Jacobs, M. (eds) *U.C.L.A. Symposium on Molecular and Cellular Biology.* New Series V. 44. Alan R. Liss Inc., New York, pp. 335–344.

Hall, S.M. and Baker, D.A. (1972) The chemical composition of *Ricinus* phloem exudate. *Planta* 106, 131–140.

Hall, W.C. (1952) Evidence on the auxin-ethylene balance hypothesis of foliar abscission. *Bot. Gaz.* 113, 31–22.

Hall, W.C., Truchelut, G.B., Leinweber, C.L. and Herrero, F.R. (1957) Ethylene production by the cotton plant and its effects under experimental and field conditions. *Physiol. Plant.* 10, 306–317.

Hallaway, M. and Osborne, D.J. (1969) Ethylene: a factor in defoliation induced by auxins. *Science* 163, 1067–1068.

Hamilton, A.J., Lycett, G.W. and Grierson, D. (1990) Antisense gene that inhibits synthesis of the hormone ethylene in transgenic plants. *Nature* 346, 284–287.

Hamilton, A.J., Bouzayen, M. and Grierson, D. (1991) Identification of a tomato gene for the ethylene-forming enzyme by expression in yeast. *Proc. Natl Acad. Sci. USA* 88, 7434–7437.

Hamilton, A.J., Bouzayen, M. and Grierson, D. (1993) Identification of genes encoding EFE in tomato. In: Pech, J.C., Latché, A. and Balagué, C. (eds) *Cellular and Molecular Aspects of the Plant Hormone Ethylene*. Kluwer Academic, Dordrecht, pp. 71–75.

Hamm, R. (1975) Water-holding capacity of meat. In: Cole, D.J.A. and Lawrie, R.A. (eds) *Meat*. Avi, Westport, Connecticut, pp. 321–338.

Han, S.S., Halevy, A.H. and Reid, M.S. (1991) The role of ethylene and pollination in petal senescence and ovary growth of *Brodiaea*. *J. Amer. Soc. Hort. Sci.* 116, 68–72.

Hanan, J.J. (1966) Experiments with controlled atmosphere storage of carnations. *Proceedings of the J. Amer. Soc. Hort. Sci.* 90, 370–376.

Hanebuth, W. and Raschke, K. (1972) Is stomatal aperture controlled by light directly or through changes in the intercellular CO_2 concentration? *Plant Research '72*. MSU/AEC Plant Research Laboratory, Michigan State University, East Lansing, Michigan, pp. 139–144.

Haney, W.J. (1952) Snapdragon shattering. *Mich. Florist* 258, 24.

Hansen, A.D. and Kende, H. (1975) Ethylene-enhanced ion and sucrose efflux in morning glory flower tissue. *Plant Physiol.* 55, 663–669.

Hansen, D.J. and Bendixen, L.E. (1974) Ethylene-induced tropism of *Trifolium fragiferum* L. stolons. *Plant Physiol.* 53, 80–82.

Hansen, E. (1942) Quantitative study of ethylene production in relation to respiration of pears. *Bot. Gaz.* 103, 543–558.

Hansen, E. (1943) Relation of ethylene production to respiration and ripening of premature pears. *J. Amer. Soc. Hort. Soc.* 43, 69–72.

Hansen, E. and Blanpied, G.D. (1968) Ethylene-induced ripening of pears in relation to maturity and length of treatment. *Proceedings of the J. Amer. Soc. Hort. Sci.* 93, 807–812.

Hanstein, S. and Felle, H.H. (1999) The influence of atmospheric NH_3 on the apoplastic pH of green leaves; a noninvasive approach with pH-sensitive microelectrodes. *New Phytol.* 143, 333–338.

Harbaugh, B.K., Wilfret, G.J., Englehard, A.W., Walters, W.E. and Marousky, F.J. (1976) Evaluation of 40 ornamental plants for a mass-marketing system utilizing sealed polyethylene packages. *Proceedings of the Fla. State Hort. Soc.* 89, 320–323.

Hardenburg, R.E. (1973) Use of plastic films in maintaining quality of fresh fruits and vegetables during storage and marketing. *Amer. Soc. Heat. Refrig. Air Cond. Eng. Bul.* CH-73-77, 19–29.

Hardenburg, R.E., Watada, A.E. and Wang, C.L. (1986) *The Commercial Storage of Fruits and Vegetables, and Florist and Nursery Stocks*. USDA Agric. Handbook No. 66 (revised).

Harper, D.B. and Swinburne, T.R. (1979) 2,3-Dihydroxybenzoic acid and related compounds as stimulants of germinations of conidia of *Colletotrichum musae* (Berk. & Curt.) Arx. *Physiol. Plant Pathol.* 14, 363.

Harpham, N.V.J., Berry, A.W., Knee, E.M., Roveda-Hoyos, G., Raskin, L., Sanders, I.O., Smith, A.R., Wood, C.K. and Hall, M.A. (1991) The effect of ethylene on the growth and development of wild-type and mutant *Arabidopsis thaliana* (L.) Heynh. *Ann. Bot.* 68, 55–62.

Harpham, N.V.J., Berry, A.W., Holland, M.G., Moshkov, I.E., Smith, A.R. and Hall, M.A. (1996) Ethylene binding sites in higher plants. *Plant Growth Regul.* 18, 71–77.

Harpster, M.H., Lee, K.L. and Dunsmuir, P. (1997) Isolation and characterization of a gene encoding endo-*B*-1,4-glucanase from pepper (*Capsicum annuum* L.). *Plant Mol. Biol.* 33, 47–59.

Harris, D.R., Seberry, J.A., Wills, R.B.H. and Spohr, L.J. (2000) Effect of fruit maturity on efficiency of 1-methylcyclopropene to delay the ripening of bananas. *Postharvest Biol. Technol.* 20, 303–308.

Harrison, M.A. and Kaufman, P.N. (1982) Does ethylene play a role in the release of lateral buds (tillers) from apical dominance in oats? *Plant Physiol.* 70, 811–814.

Harrison, M.A. and Pickard, B.G. (1984) Burst of ethylene upon horizontal placement of tomato seedlings. *Plant Physiol.* 75, 1167–1169.

Harrison, M.A. and Pickard, B.G. (1986) Evaluation of ethylene as a mediator of gravitropism by tomato hypocotyls. *Plant Physiol.* 80, 592–595.

Hart, C.M., Nagy, F. and Meins, F. Jr (1993) A 61 bp enhancer element of the tobacco beta-1,3-glucanase B gene interacts with one or more regulated nuclear proteins. *Plant Mol. Biol.* 21, 121–131.

Hartman, S.A. and Jamieson, W. (1976) Preliminary study of the hypobaric storage of shrimp. Report for Grumman Allied Industries, Inc., Garden City, New York.

Hartman, S.A. and Jamieson, W. (1977) Summary of shrimp test conducted from August 25–Oct 6, 1977. Report for Grumman Allied Industries, Inc., Garden City, New York.

Hartmann, C., Drouet, A. and Morin, F. (1987) Ethylene and ripening of apple, pear and cherry fruit. *Plant Physiol. Bicohem.* 25, 505–512.

Haruenkit, R. and Thompson, A.K. (1996) Effect of O_2 and CO_2 levels on internal browning and composition of pineapples – Smooth Cayenne. *Proceedings of the International Conference on Tropical Fruits*, Kuala Lumpur, Malaysia, 23–26 July, 1966, pp. 343–350.

Harvey, E.M. and Rose, R.C. (1915) The effects of illuminating gas on root systems. *Bot. Gaz.* 60, 27–44.

Harvey, E.N. (1910) Studies on the permeability of cells. *J. Expt. Zool.* 10, 507–556.

Harvey, E.N. (1922) The permeability of cells for oxygen and its signficance for the theory of simulation. *J. Gen. Physiol.* 5, 215–222.

Harvey, E.N. (1928a) Photosynthesis in the absence of oxygen. *Plant Physiol.* 3, 85–89.

Harvey, J.M. and Pentzer, W.T. (1966) *Market Diseases of Grapes and Other Small Fruits.* Agricultural Handbook No. 189. USDA ARS, Market Quality Research Division.

Harvey, R.B. (1924) Sun scald of tomatoes. *Minn. Studies Plant Sci. Biol. Sci.* 5, 229–233.

Harvey, R.B. (1928b) Artificial ripening of fruits and vegetables. *Minn. Agr. State Research Bul.* 247.

Harwood, J.L. (1997) Plant lipid metabolism. In: Dey, P.M. and Harbone, J.B. (eds) *Plant Biochemistry.* Academic Press, New York, pp. 237–272.

Hasenstein, K.H., Blancaflor, E.B. and Lee, J.S. (1999) The microtubule cytoskeleton does not integrate auxin transport and gravitropism in maize roots. *Physiol. Plant.* 105, 729–738.

Hass, A.R.C. (1936) Growth and water relations of the avocado fruit. *Plant Physiol.* 11, 383–400.

Hassek, R.F. (1975) Hypobaric storage of cut flowers. Flower and Nursery Report for Commercial Growers. University of California Cooperative Extension Report, September–October, 1975, p. 9.

Hastings, D.F. and Gutknecht, J. (1974) Turgor pressure regulation: modulation of active potassium transport by hydrostatic pressure gradients. In: Zimmerman, U. and Dainty, J. (eds) *Membrane Transport in Plants.* Springer, Berlin, p. 79.

Hatfield, S.G.S. and Patterson, B.D. (1974) Abnormal volatile production by apples during ripening after controlled atmosphere storage. In: *Facteurs et Regulation de la Maturation des Fruits. Colloques Internationaux de CNRS, Paris*, 238, 57–64.

Hatton, T.T. and Cubbedge, R.H. (1982) Conditioning Florida grapefruit to reduce chilling injury during low-temperature storage. *J. Amer. Soc. Hort. Sci.* 107, 57–60.

Hatton, T.T. and Reeder, W.F. (1972) Quality of 'Lula' avocados stored in controlled atmospheres with or without ethylene. *J. Amer. Soc. Hort. Sci.* 97, 339.

Hatton, T.T. and Spalding, D.H. (1990) Controlled atmosphere storage of some tropical fruits. In: Calderon, M. and Barkai-Golan, R. (eds) *Food Preservation by Modified Atmospheres.* CRC Press, Boca Raton, Florida, pp. 303–315.

Hatton, T.T., Pantastico, E.B. and Akamine, E.K. (1975) Individual commodity requirements. In: Haard, N.F. and Salunkhe, D.K. (eds) *Symposium: Postharvest Biology and Handling of Fruits and Vegetables.* AVI, Westport, Connecticut, pp. 201–218.

Hauschild, A.H.W., Aris, B.J. and Hilsheimer, R. (1975) *Clostridium botulinum* in marinated products. *Can. Inst. Food Sci. Technol. Jour.* 8, 84–88.

Hayes, H.P. (1954) Vacuum cooling of produce. *Refrig. Engin.* 62, 47–49, 90, 92.

Hazelhoff, E.H. (1927) Function of spiracles. *Z. Vergl. Physiol.* 5, 179–190.

Heath, M.C. (1974) Chloroplast ultrastructure and ethylene production of senescing and rust-infected cowpea leaves. *Can J. Bot.* 52, 2591.

Heather, N.W. (1994) Pesticide quarantine treatments. In: Sharp, J.L. and Hallman, G.J. (eds) *Quarantine Treatments for Pests of Foods.* Westview Press, Boulder, Colorado, pp. 89–100.

Heinze, P.H. and Craft, C.C. (1953) Effectiveness of ethylene for ripening tomatoes. *Proceedings of the J. Amer. Soc. Hort. Sci.* 62, 397–404.

Hejnowicz, Z. (1961) The response of different parts of the cell elongation zone in root to external β-indolylacetic acid. *Acta Soc. Bot. Pol.* 30, 25–42.

Hejnowicz, Z. and Erickson, R.O. (1968) Growth inhibition and recovery in roots following

temporary treatment with auxin. *Physiol. Plant.* 21, 302–313.
Helder, R.J. (1956) The loss of substances by cells and tissues (salt glands). *Handb. d. Pflanzenphysiol.* 2, 468–488.
Hellickson, M.L. and Koca, R.W. (1992) Energy conservation and fruit quality preservation in fan-cycled controlled atmosphere apple storages. *8th Annual Washington State Tree Fruit Postharvest Conference*, Yakima, Washington. (Technical Paper No. 9854, Oregon State University Agr. Expt. Sta.), pp. 52–57.
Henchel, P.A. (1964) Physiology of plants under drought. *Annu. Rev. Plant Physiol.* 15, 336–386.
Henderson, J. and Osborne, D.J. (1994) Inter-tissue signalling during the 2-phase abscission of oil palm fruit. *J. Expt. Bot.* 41, 203–210.
Hennessey, T.L., Freeden, A.L. and Field, C.B. (1993) Environmental effects on circadian rhythms in photosynthesis and stomatal opening. *Planta* 189, 369–376.
Henry, M.F., de Troostembergh, J.C. and Nyns, E.J. (1977) Biogenesis and properties of the mitochondrial cyanide-insensitive alternative respiratory pathway in the yeast *Saccharomycopsis lipolytica*. In: Degn, H., Lloyd, D. and Hill, G.C. (eds) *Federation of European Biochemical Societies*, vol. 49. Pergamon Press, Oxford, UK, pp. 55–65.
Hensel, W. and Iversen, T.H. (1980) Ethylene production during clinostat rotation and effect on root geotropism. *Z. Pflanzenphysiol.* 97, 343–352.
Henskens, H., Somhors, D. and Woltering, E.J. (1993) Expression of two ACC synthease mRNAs in carnation flower parts during ageing and following treatment with ethylene. In: Pech, J.C., Latché, A. and Balagué, C. (eds) *Cellular and Molecular Aspects of the Plant Hormone Ethylene.* Kluwer Academic, Dordrecht, pp. 323–334.
Henze, J. (1969) Beziehungen zwischen respiration und innerer atmosphäre bei CO_2-empfindlichen kernobstsorten. *Qual. Plant. Mater. Veg.* 19, 229–242.
Herreid, C.F. (1980) Hypoxia in invertebrates. *Comp. Biochem. Physiol.* A67, 311.
Herreid, C.F., Hull, R.J. and Prawel, D.A. (1981) Energetics of cockroach locomotion. *J. Exptl. Biol.* 94, 189–202.
Herring, H.K. (1976) Summary of evaluations conducted to date with the Grumman/Dormavac hypobaric container. Armour Research Center Inter-Office Correspondence, 1 June, 1976.
Herzfeld, K.F. and Smallwood, H.M. (1952) The kinetic theory of ideal gases. In: Taylor, H.S. and Glasstone, S. (eds) *Treatise on Physical Chemistry*, 3rd edn. D.Van Nostrand, New York.
Hewitt, J. and Morris, J.G. (1975) Superoxide dismutase in some obligatory anaerobic bacteria. *FEBS Lett.* 50, 315–318.
Hewitt, S.P. and Curtis, O.F. (1948) The effect of temperature on loss of dry matter and carbohydrate from leaves by respiration and translocation. *Amer. J. Bot.* 35, 746–755.
Hiestand, W.A. (1931) Oxygen tension and respiration; Arthropoda. *Physiol. Zool.* 4, 246–270.
Hild, V. and Hertel, R. (1972) Initial phases of gravity-induced lateral auxin transport and geotropic curvature in corn coleoptiles. *Planta* 108, 245–248.
Hill, P.W., Raven, J.A., Loubet, B., Fowler, D. and Sutton, M.A. (2001) Comparison of gas exchange and bioassay determinations of the ammonia compensation point in *Luzula sylvatica* (Huds.) Gaud. *Plant Physiol.* 125, 477–487.
Hill, R.B.H., McGlasson, W.B., Graham, D., Lee, T.H. and Hall, E.G. (1989) *Postharvest – An Introduction to the Physiology and Handling of Fruit and Vegetables.* AVI, New York.
Himelblau, E., Mirsa, H., Lin, S.-J., Culotta, V.C., Peñarrubia, L. and Amasino, R.M. (1998) Identification of a functional homolog of the yeast copper homeostasis gene *ATX1* from *Arabidopsis*. *Plant Physiol.* 117, 1227–1231.
Hine, J. and Weimar, R.D. (1965) Carbon basicity. *J. Am. Chem. Soc.* 87, 3387–3396.
Hirayama, T. and Alonoso, J.M. (2000) Ethylene captures a metal! Metal ions are involved in ethylene perception and signal transduction. *Plant Cell Physiol.* 41, 548–555.
Hirayama, T., Kieber, J.J., Hirayama, N., Kogan, M., Guzman, P., Nourrrizadeh, S., Alonso, J.M., Dailey, W.P., Dancis, A. and Ecker, J.R. (1999) RESPONSIVE-TO-ANTAGONIST1, a Menkes/Wilson disease-related copper transporter, is required for ethylene signaling in *Arabidopsis*. *Cell* 97, 383–393.
Hitchcock, A.E. (1935) Tobacco as a test plant for comparing the effectiveness of preparations containing growth substances. *Contrib. Boyce Thompson Inst.* 7, 349–364.
Hitchcock, A.E., Crocker, W. and Zimmerman, P.W. (1932) The effect of illuminating gas on the lily, narcissus, tulip, and hyacinth. *Contrib. Boyce Thompson Inst.* 4, 155–176.
Hoad, S.P., Grace, J. and Jeffree, C.E. (1996) A leaf disc method for measuring cuticular conductance. *J. Expt. Bot.* 47, 431–437.

Hoad, S.P., Grace, J. and Jeffree, C.E. (1997) Humidity response of cuticular conductance of beech (*Fagus sylvatica*) leaf discs maintained at high relative water content. *J. Expt. Bot.* 48, 1969–1975.

Hobson, G.E. (1967) The effects of alleles at the '*never ripe*' locus on the ripening of tomato fruit. *Phytochemistry* 6, 1337–1341.

Hobson, G.E. and Grierson, D. (1993) Tomato. In: Seymour, G.B., Taylor, J.E. and Tucker, G.A. (eds) *Biochemistry of Fruit Ripening.* Chapman & Hall, London, pp. 405–442.

Hobson, G.E., Nichols, R., Davies, J.N. and Atkey, P.T. (1984) The inhibition of tomato fruit ripening by silver. *J. Plant Physiol.* 116, 21–29.

Hodges, C.H. and Coleman, L.W. (1984) Ethylene-induced chlorosis in the pathogenesis of *Bipolaris sorokiniana* leaf spore of *Poa pratensis. Plant Physiol.* 75, 462–465.

Hodgman, C.D., Weast, R.C., Shankland, R.S. and Selby, S.M. (eds) (1962) *Handbook of Chemistry and Physics.* CRC, Cleveland, Ohio.

Hoerl, A.E., Nashed, M.Z., McKetta, J.J. and Silberberg, I.H. (1984) Mathamatics. In: Perry, R.H., Green, D.W. and Maloney, J.O. (eds) *Perry's Chemical Engineers Handbook.* McGraw-Hill, New York, Section 2, 1–116.

Hoffman, N.E. and Yang, S.F. (1982) Enhancement of wound-induced ethylene synthesis by ethylene in preclimacteric cantaloupe. *Plant Physiol.* 69, 317–322.

Hoffman, N.E., Yang, S.F. and MecKeon, T. (1982) Identification of 1-(malonylamino)-cyclopropane-1-carboxylic acid as a major conjugate of 1-aminocyclopropane-1-carboxylic acid, an ethylene precursor in higher plants. *Biochem. Biophys. Res. Commun.* 104, 765–770.

Hoffman, R. and Heale, J.B. (1987) Cell death, 6-methoxymellein accumulation and induced resistance to *Botrytis cinerea* in carrot root slices. *Physiol. Mol. Plant Path.* 30, 67–75.

Hoffman, R., Roebroeck, E. and Heale, J.B. (1988) Effects of ethylene biosynthesis in carrot root slices on 6-methoxymellein accumulation and resistance to *Botrytis cinerea. Physiol. Plant.* 42, 1094–1102.

Hofman, P.J., Jopbin-Décor, M., Meiburg, G.F., Macnish, A.J. and Joyce, D.C. (2001) Ripening and quality responses of avocado, custard apple, mango and papaya fruit to 1-methylcyclopropene. *Austr. J. Expt. Agr.* 41, 567–572.

Hollaway, P.J. (1982) Structure and histochemistry of plant cuticular membranes: an overview. In: Cutter, E.G., Alvin, K.L. and Price, C.E. (eds) *The Plant Cuticle.* Academic Press, London, p. 144.

Holmgren, P., Jarvis, P.G. and Jarvis, M.S. (1965) Resistance to carbon dioxide and water vapour transfer in leaves of different plant species. *Physiol. Plantarum* 18, 557–573.

Holzapfel, W.H. (1998) The gram-positive bacteria associated with meat and meat products. In: Davies, A. and Board, R. (eds) *The Microbiology of Meat and Poultry.* Blackie Academic & Professional, London, pp. 35–84.

Hong, J.-H. and Gross, K.C. (2000) Involvement of ethylene in development of chilling injury in fresh-cut tomato slices during cold storage. *J. Amer. Soc. Hort. Sci.* 125, 736–741.

Hong, J.-H. and Lere, S.-K. (2000) Comparison characteristics of vine- and room-ripened tomato fruit. *Plant Growth Regl. Abstracts* 26, 57.

Hong, S.-B., Kuo, J.M., Mullins, L.S. and Raushel, F.M. (1995) CO_2 is required for the assembly of the binuclear metal center of phosphotriesterase. *J. Am. Chem. Soc.* 117, 7581–7582.

Hopkins, W.G. (1995) *Introduction to Plant Physiology.* John Wiley & Sons, New York.

Horiuchi, Y. and Imaseki, H. (1986) High temperature sensitivity of auxin-induced ethylene production in mung bean hypocotyl sections. *Plant Cell Physiol.* 27, 453–461.

Horowitz, M.A. and Hughes, J.M. (1977) Food-borne botulism in the United States. *J. Infect. Disease* 136, 153–159.

Horrocks, R.L. (1964) Wax and the water permeability of apple cuticles. *Nature* 203, 547.

Horton, R.F. (1991) Methyl jasmonate and transpiration in barley. *Plant Physiol.* 96, 583–585.

Hotchkiss, J.H. (1989) Modified atmosphere packaging of poultry and related products. In: Brody, A.L. (ed.) *Controlled/Modified Atmosphere/Vacuum Packing of Foods.* Food & Nutrition Press, Trumbull, Connecticut, pp. 39–58.

Høyer, L. (1985) Bud and flower drop in *Begonia eliator* 'Sirene' caused by ethylene and darkness. *Acta Hort.* 167, 387–394.

Hrazdina, G., Kiss, E., Rosenfield, C., Norelli, J.L. and Aldwinckle, H.S. (2001) Down regulation of ethylene production in apples. *Plant Growth Regl. Abstr.* 27, 347.

Hruschka, H.W. (1977) Postharvest weight loss and shrivel in five fruits and five vegetables. USDA Agr. Mkt. Res. Rpt. 1059.

Hsiao, T.C. (1973) Plant responses to water stress. *Annu. Rev. Plant Physiol.* 24, 519–570.

Hu, X.-Q., Yu, X. and Chen, L.-G. (2001) Studies of Chinese bayberry fruits on some physiological characters during storage. *Plant Growth Regl. Abstracts* 27, 249.

Hua, J. and Meyerrowitz, E. (1998) Ethylene responses are negatively regulated by a receptor gene family in *Arabidopsis thaliana. Cell* 94, 261–271.

Hua, J., Chang, C., Sun, Q. and Meyerowitz, E.M. (1995) Ethylene insensitivity conferred by *Arabidopsis ERS* gene. *Science* 269, 1712–1714.

Hua, J., Sakai, S., Nourizadeh, S., Chen, Q.G., Bleecker, A.B., Ecker, J.R. and Meyerowitz, E.M. (1998) *EIN4* and *ERS2* are members of the putative ethylene receptor gene family in *Arabidopsis. Plant Cell* 10, 1321–1332.

Huang, P.-L., Do, Y.-Y., Huang, F.-C., Thay, T.-S. and Chang, T.-W. (1997) Characterization and expression analysis of a banana gene encoding 1-aminocyclopropane-1-carboxylate oxidase. *Biochem. Mol. Biol. Int.* 41, 941–950.

Huber, B. (1941) Gesichertes und Problematisches in der Wanderung der Assimilate. *Ber. Deutsch Bot. Ges.* 59, 181–194.

Huber, B., Schmidt, E. and Jahnel, H. (1937) Untersuchungen über den Assimilatstrom. I. *Tharandter Forstl. Jahrb.* 88, 1017–1050.

Huelin, F.E. and Coggiola, I.M. (1970) Superficial scald, a functional disorder of apples. V. Oxidation of farnescene and its inhibition by diphenylamine. *J. Sci. Food Agric.* 21, 44.

Huelin, F.E. and Kennett, B.H. (1958) Superficial scald, a functional disorder of stored apples. I. The role of volatile substances. *J. Sci. Food Agric.* 9, 657–666.

Huelin, F.E. and Kennett, B.H. (1959) Nature of the olefins produced by apples. *Nature* 184, 996.

Huffman, D.L. (1974) Effect of gas atmospheres on microbial quality of pork. *J. Food Sci.* 39, 723.

Huffman, D.L., Davis, K.A., Marple, D.N. and McGuire, J.A. (1975) Effect of gas atmospheres on microbial growth, color and pH of beef. *J. Food Sci.* 40, 1229–1231.

Hughes, P.A., Thompson, A.K., Plumbley, R.A. and Seymour, G.B. (1981) Storage of Capsicums (*Capsicum annum* cultivar Bellboy) under controlled atmosphere, modified atmosphere and hypobaric conditions. *J. Hortic. Sci.* 56, 261–266.

Hulme, A.C. (1951) Apparatus for the measurement of gaseous conditions inside an apple fruit. *J. Expt. Bot.* 2, 65–85.

Hulme, A.C. (1956) Carbon dioxide injury and the presence of succinic acid in apples. *Nature* 178, 218–219.

Hulme, A.C. (1971) The mango. In: Hulme, A.C. (ed.) *The Biochemistry of Fruits and Their Products*, vol. 1. Academic Press, London, pp. 233–254.

Hulme, A.C. and Rhodes, M.J.C. (1971) Pome fruits. In: Hulme, A.C. (ed.) *The Biochemistry of Fruits and Their Products*, vol. 1. Academic Press, London, pp. 333–373.

Hulme, A.C., Rhodes, J.C., Wooltorton, L.S.C. and Galliard, T. (1968) Biochemical changes associated with ripening of apples. *Qual. Plant Matter. Veg.* 19, 1–3, 18.

Hummel, C.E. and Stoddard, E.S. (1957) Methods of improving food preservation in home refrigerators. *Refrig. Eng.* 65, 33–39.

Huse, H. (1969) Design and performance characteristics of staged positive displacement and liquid ring vacuum pumps. ASME paper 69-WA/PID-17, November, 1969.

Huse, H. (1970) Thermodynamic characteristics of staged mechanical vacuum pumps on condenser service. ASME paper 70-WA/PID-10, July, 1970.

Husted, S. and Schjoerring, J.K. (1995) Apoplastic pH and ammonium concentration in leaves of *Brassica napus* L. *Plant Physiol.* 109, 1453–1460.

Husted, S. and Schjoerring, J.K. (1996) Ammonia flux between oilseed rape plants and the atmosphere in response to changes in leaf temperature, light intensity, and air humidity. *Plant Physiol.* 112, 67–74.

Husted, S., Mattsson, M. and Schjoerring, J.K. (1997) Ammonia compensation points in two cultivars of *Hordeum vulgare* L. during vegetative and generative growth. *Plant Cell Environ.* 19, 1299–1306.

Hyodo, H. (1991) Stress/wound ethylene. In: Mattoo, A.K. and Suttle, J.C. (eds) *The Plant Hormone Ethylene*. CRC Press, Boca Raton, Florida, pp. 43–63.

Hyodo, H. and Fujinami, H. (1989) The effects of 2,5-norbornadiene on the induction of the activity of phenylalanine ammonia-lyase in wounded mesocarp tissue of *Curcubita maxima. Plant Cell. Physiol.* 30, 857–860.

Iannetta, P.P.M., Wyman, M., Neelam, A., Jones, C., Taylor, M.A., Davies, H.V. and Sexton, R. (2000) A casual role for ethylene and endo-β-1,4-glucanase in the abscission of red raspberry (*Rubus idaeus*) drupelets. *Physiol. Plant.* 110, 535–543.

Ievinish, G. and Kreicbergs, O. (1992) Endogenous rhythmicity of ethylene production in growing intact cereal seedlings. *Plant Physiol.* 100, 1389–1391.

Iglesias, D.J., Tadeo, F.R., Legaz, F., Primo-Nillo, E. and Talon, M. (2001) *In vivo* sucrose stimulation of colour change in citrus fruit epicarps: interactions between nutritional and hormonal signals. *Physiol. Plant.* 112, 244–250.

Ilangantileke, S.G. and Salokhe, V. (1989) Low-pressure atmosphere storage of Thai mango. *Proceedings of the 5th International Controlled Atmosphere Research Conference*, Wenatchee Washington, USA, 14–16 June, pp. 103–117.

Ilangantileke, S.G., Turla, L. and Chen, R. (1989) Pretreatment and hypobaric storage life of mango fruit. Canadian/American Society of Agricultural Engineering. Paper No. 896,036. Quebec, Canada (25–28 June).

Iljin, W.S. (1927) Die Durchlassigkeit des *Protoplasmas*; ihre quantitive Bestimmung und ihre Beeinflussung durch salze and durch die wasser stofffenkonzentration. *Protoplasma* 3, 558–602.

Ilker, Y. and Morris, L.L. (1975) Alleviation of chilling injury of okra. *HortScience* 10, 324.

Imaseki, H. (1970) Induction of peroxidase activity by ethylene in sweet potato. *Plant Physiol.* 46, 172–174.

Imaseki, H. (1991) The biochemistry of ethylene synthesis. In: Mattoo, A.K. and Suttle, J.C. (eds) *The Plant Hormone Ethylene.* CRC Press, Boca Raton, Florida, pp. 1–41.

Imaseki, H., Teranishi, T. and Uritani, I. (1968) Production of ethylene by sweet potato roots infected by black rot fungus. *Plant Cell Physiol.* 9, 761–789.

Imms, A.D. (1949) *Outlines of Entomology.* Methuen & Co. Ltd, London.

Imoolehin, E.D. and Grogan, R.G. (1980) Effects of oxygen, carbon dioxide and ethylene on growth, sclerotial production, germination, and infection by *Sclerotinia minor. Phytopathology* 70, 1158.

Ingraham, M. and Dainty, R.H. (1971) Changes caused by microbes in spoilage of meats. *J. Appl. Bacteriol.* 34, 21–39.

Insalta, N.F., Fredericks, G.J., Berman, J.H. and Borker, E. (1967) *Clostridium botulinum* type E in frozen vacuum-packed fish. *Food Technol.* 21, 296–298.

Ireland, J.P., Meynhardt, J.T. and Strauss, J.M. (1967) When *Proteas* become sailors. *Fmg. S. Afr.* 43, 33–35.

Irwin, P.L. and Dilley, D.R. (1980) The effect of hypobaric and low oxygen storage on the ripening of Empire and Idared apple fruits. *HortScience* 15, 424.

Isenberg, F.M. and Hartman, J. (1958) Vacuum cooling vegetables. *New York State College of Agriculture, Cornell Extension Bulletin* 1012.

Ishizawa, K. and Easashi, Y. (1984) Gaseous factors involved in the enhanced elongation of rice coleoptiles under water. *Plant Cell Environ.* 7, 239–245.

Itzhaki, H., Maxson, J.M. and Woodson, W.R. (1994) An ethylene-responsive enhancer element is involved in the senescence-related expression of the carnation glutathione-*S*-transferase (*GST1*) gene. *Proc. Natl Acad. Sci. USA* 91, 8925–8929.

Izumi, H., Nakatani, T. and Ogikuba, H. (1999) Controlled-atmosphere storage of fresh-cut spinach at various temperatures. *HortScience* 34, 507.

Jackson, M.B. (1979) Is the diageotropic tomato mutant ethylene deficient? *Plant Physiol.* 46, 347–351.

Jackson, M.B. (1983) Regulation of root growth and morphology by ethylene and other externally applied growth substances. In: Jackson, M.B. and Stead, A.D. (eds) *Growth Regulators in Root Development.* Proceedings of the Brit. Plant Growth Regulator Group. Monograph 10, 103–116.

Jackson, M.B. (1991) Ethylene in root growth and development. In: Mattoo, A.K. and Suttle, J.C. (eds) *The Plant Hormone Ethylene.* CRC Press, Boca Raton, Florida, pp. 159–181.

Jackson, M.B. and Campbell, D.J. (1975) Movement of ethylene from roots to shoots, a factor in the responses of tomato plants to waterlogged soil conditions. *New Phytol.* 74, 397.

Jackson, M.B., Drew, M.C. and Giffard, S.C. (1981) Effects of applying ethylene to the root system of *Zea mays* on growth and nutrient concentration in relation to flooding. *Physiol. Plant.* 46, 347–351.

Jacobs, M.H. (1920) To what extent are the physiological effects of carbon dioxide due to hydrogen ions? *Amer. J. Physiol.* 51, 321–331.

Jacobs, M.H. (1922) The influence of ammonium salts on cell reaction. *J. Gen Phsyiol.* 5, 181–188.

Jacob-Wilk, D., Holland, D., Goldschmidt, E.E., Riov, J. and Eyal, Y. (1999) Chlorophyll breakdown by chlorophyllase: isolation and functional expression of the *Chlase1* gene from ethylene-treated *Citrus* fruit and its regulation during development. *Plant Journal* 20, 653–661.

Jadhav, S.J., Patil, B.C. and Salunkhe, D.K. (1973) Control of potato greening under hypobaric storage: in hypobaric storage, reduced pressure inhibits light-induced greening of potato tubers. *Food Engineering* 45, 111–112.

Jaffe, M.J., Huberman, M., Johnson, J. and Telewski, F.W. (1985) Thigmomorphogenesis: the induction of callose formation and ethylene evolution by mechanical perturbation in bean stems. *Physiol. Plant.* 64, 271–279.

Jahn, O.L. (1975) Ripening response of green tomatoes to ethylene concentration and temperature. *Citrus and Vegetable Magazine*, 18–24.

Jahn, O.L., Chace, W.G. and Cubbedge, R.H. (1973) Degreening response of 'Hamlin' oranges in relation to temperature, ethylene concentration, and fruit maturity. *J. Amer. Soc. Hort. Sci.* 98, 177–181.

James, W.O. (1953) *Plant Respiration.* Oxford University Press, Oxford.

Jamieson, W. (1976a) Hypobaric storage of pork loins. Report for Grumman Allied Ind. Inc., Garden City, New York.

Jamieson, W. (1976b) Hypobaric storage of fresh chicken. Report for Grumman Allied Ind. Inc., Garden City, New York.

Jamieson, W. (1979a) Fish test summary. Report for Grumman Allied Industries, Inc., Garden City, New York.

Jamieson, W. (1979b) Use of hypobaric conditions for refrigerated storage of meat, fruits and vegetables. *Food Technol.* 34, 64–71.

Jamieson, W. (1980a) Dormavac low pressure storage of perishable commodities. *Prog. Food Nutr. Sci.* 4, 61–72.

Jamieson, W. (1980b) Horsemeat test. Report for Grumman Allied Industries, Inc., Garden City, New York.

Jamieson, W. (1980c) Hypobaric intermodal containers. *CSIRO Food Research Quarterly* 40, 129–137.

Jamieson, W. (1980d) Test shipment of lamb – Australia to United States (April 22–June 2, 1980). Report to Grumman Allied Industries, Inc., Garden City, New York.

Jamieson, W. (1980e) Use of hypobaric conditions for refrigerated storage of meats, fruits and vegetables. *Food Technology* March, 64–72.

Jamieson, W. (1984) *Commodity Information Bulletin.* Dormavac Systems Corp.

Jamieson, W. and Dressler, B. (1977) Storage of Long Island Jerseybelle strawberries under hypobaric conditions. Report for Grumman Allied Industries, Inc., Garden City, New York.

Jana, S. and Choudhuri, M.A. (1982) Ethylene prodution and senescence in submerged aquatic angiosperms. *Aquatic Bot.* 13, 359–365.

Janes, B.S. (1962) Leaf-clearing techniques to assist fungal spore germination counts. *Nature* 4820, 1099–1100.

Jardine, J.D., Dilley, D.R. and Price, H.C. (1984) Hypobaric storage of celery transplants. *HortScience* 19, 869.

Jay, E.G. and Pearman, G.C. (1971) Susceptibility of two species of *Tribolium* (Coleoptera: Tenebrionidae) to alterations of atmospheric gas concentrations. *J. Stored Prod. Res.* 7, 181–186.

Jay, E.G., Arbogast, R.T. and Pearman, G.C. (1971) Relative humidity: its importance in the control of stored-product insects with modified atmospheric gas concentrations. *J. Stored Prod. Res.* 6, 325.

Jay, M., Kittaka, R.S. and Ordal, Z.J. (1962) The effect of temperature and packaging material on the storage life and bacterial flora of ground beef. *Food Technol.* 16, 95–98.

Jeffery, D., Smith, C., Goodenough, P., Prosser, I. and Grierson, D. (1984) Ethylene-independent and ethylene-dependent biochemical changes in ripening tomatoes. *Plant Physiol.* 74, 32–38.

Jensen, H.E.K. and Bredmose, N. (1979) Planter under lavtryk. *Gartner Tidende* 95, 512–513.

Jensen, H.E.K. and Rasmussen, P.M. (1978) Lavtryksopbevaring af stiklinger. I. Principper og indledende forsøg. *Tidsskrift for Planteavl* 82, 623–632.

Jerie, P.H. and Hall, M.A. (1978) The identification of ethylene oxide as a major metabolite of ethylene in *Vicia faba* L. *Proceedings of the Royal Soc., London Ser. B* 200, 87.

Jerie, P.H., Shaari, A.R. and Hall, M.A. (1979) The compartmentation of ethylene in developing cotyledons of *Phaseolus vulgaris* L. *Planta* 144, 503–507.

Jiao, X., Philosoph-Hada, S., Su, L. and Yang, S.F. (1986) The conversion of 1-(malonylamino) cyclopropane-1-carboxylic acid to 1-aminocyclopropane-1-carboxylic acid. *Plant Physiol.* 81, 637–641.

Jobbling, J., McGlasson, W.B. and Dilley, D.R. (1991) Induction of ethylene synthesizing competency in Granny Smith apples by exposure to low temperature in air. *Postharv. Biol. Technol.* 1, 111–118.

Johannsen, A. (1965) *Clostridium botulinum* type E in foods and the environment generally. *J. Appl. Bacteriology* 28, 90–94.

Johansson, I., Larsson, C. and Kjellbom, P. (1996) The major integral proteins of spinach leaf plasma membranes are putative aquaporins and are phosphorylated in response to Ca^{2+}

and apoplastic water potential. *Plant Cell* 8, 1181–1191.

John, P., Reynolds, E.A., Prescott, A.G. and Bauchot, A.D. (2000) ACC oxidase in the biosynthesis of ethylene. *Plant Growth Regl. Abstracts* 26, 141.

Johni, B.M., Guha, S. and Maheshwari, S.C. (1965) Auxin-induced reversal of geotropic response in onion seedlings. *Nature* 205, 1026–1027.

Johnson, B.E. and Brun, W.A. (1966) Stomatal density and responsiveness of banana fruit stomates. *Plant Physiol.* 41, 99–101.

Johnson, B.Y. (1974) Chilled vacuum-packed beef. *CSRIO Food Res. Quart.* 34, 14.

Johnson, D.S. (1976) Influence of water loss on the storage quality of apples. *Chemstry and Industry* 24, 1044–1046.

Johnson, D.S. and Ertan, U. (1983) Interaction of temperatures amd oxygen level on the respiration rate and storage quality of Idared apples. *J. Hort. Sci.* 58, 527.

Johnson, D.S., Prinja, J. and Smith, S.M. (1990) The use of controlled atmosphere (CA) condition for control of bitter pit and superficial scald in Bramley's Seedling apples. *Proceedings of the 5th Controlled Atmosphere Research Conference*, Wenatchee, Washington, vol. 1, pp. 157–165.

Johnson, D.S., Dover, C.J. and Pearson, K. (1993) Very low oxygen storage in relation to ethanol production and control of superficial scald in Bramley's seeling apples. *Acta Horticulturae* 326, 175–182.

Johnson, H.B. and Buford, W.R. (1963) Heat-evolution rates of some Texas-grown fruits and vegetables. *Rio Grande Valley Hort. Soc. Jour.* 17, 170–175.

Jones, H.G. (1981) Carbon dioxide exchange of developing apple (*Malus pumila* Mill.) fruits. *J. Expt. Bot.* 32, 1203–1210.

Jones, H.G. and Higgs, K.H. (1982) Surface conductance and water balance of developing apple (*Malus pumila* Mill.) fruits. *J. Expt. Bot.* 33, 67–77.

Jones, M.L. and Woodson, W.R. (1999) Interorgan signaling following pollination in carnation. *J. Amer. Soc. Hort. Sci.* 124, 598–604.

Jones, M.L., Woodson, W.R. and Lindstrom, J.T. (2000) Regulation and function of pollination-induced ethylene in carnation and petunia flowers. *Plant Growth Regl. Abstracts* 26, 142.

Jones, W.W. and Kubpta, H. (1940) Some chemical and respirational changes in the papaya fruit during ripening, and the effects of cold storage on these changes. *Plant Physiol.* 15, 711–717.

Jordan, W.R., Morgan, P.W. and Davenport, T.L. (1972) Water stress enhances ethylene-mediated leaf abscission in cotton. *Plant Physiol.* 50, 756–764.

Jorgensen, E.G. (1968) Jacketed food storage for high humidity environments. *Can. Refrig. & Air Cond.* 34, 16–19.

Joyce, D.C., Macnish, A.J., Hofman, P.J., Simons, D. and Reid, M.S. (2000) Use of 1-methylcyclopropene to modulate banana ripening. *Plant Growth Regl. Abstracts* 26, 154.

Ju, Z.-G. and Bramlage, W.J. (2001) Developmental changes of cuticular constituents and their association with ethylene during fruit ripening in 'Delicious' apples. *Postharvest Biol. Technol.* 21, 257–263.

Ju, Z.-G. and Curry, E.A. (2000) Evidence that α-farnescene biosynthesis during fruit ripening is mediated by ethylene regulated gene expression in apples. *Postharvest Biol. Technol.* 19, 9–16.

Jung, H.-W. and Hwang, B.-K. (2000) Isolation, partial sequencing, and expression of pathogenesis-related cDNA genes from pepper leaves infected by *Xanthomonas campestris* cv. vesicatoria. *Molecular Plant-Microbe Interactions* 13, 136–142.

Kader, A.A. (1983) Postharvest quality maintenance of fruits and vegetables in developing countries. In: Lieberman, M. (ed.) *Post Harvest Physiology and Crop Preservation.* Plenum Press, New York, pp. 455–470.

Kader, A.A. and Morris, L.L. (1974) The effects of reduced oxygen tension attained by hypobaric conditions vs. modified atmospheres on the ripening behavior of tomato fruits. *HortScience* 9, 274 (abstract).

Kader, A.A. and Morris, L.L. (1975a) Regulation of tomato fruit ripening. *Plant Physiol.* 55 (annual supplement – abstract).

Kader, A.A. and Morris, L.L. (1975b) Amelioration of chilling injury symptoms on tomato fruits. *HortScience* 10, 324.

Kader, A.A., Klaustermeyer, J.A., Morris, L.L. and Kasmire, R.F. (1975) Extending the shelf-life of mature-green chili peppers by modified atmospheres. *HortScience* 10, 335.

Kajiura, I. (1973a) Hypobaric storage of fruits. *Agric. Hort. (Tokyo)* 48, 1031–1036.

Kajiura, I. (1973b) Low pressure storage of fruits. Simple apparatus for low pressure storage, and its application to white peaches and Jonathan apples. *J. Food Sci. Technol. (Tokyo)* 20, 331–334.

Kajiura, I. (1975) CA storage and hypobaric storage of white peach 'Okubo'. *Scientia Horticulture* 3, 179–187.

Kajiura, I. and Iwata, M. (1971) Effects of gas concentration on fruits. III. Effects of oxygen concentrations and the delayed storage on white peach 'Okubo' fruits. *J. Japanese Soc. Hort. Sci.* 40, 94–100.

Kakopuri, A. and Nychas, G.J. (1994) Storage of poultry meat under modified atmospheres or vacuum packs: possible role of microbial metabolites as indictors of spoilage. *J. Appl. Bacteriol.* 76, 163–172.

Kalinin, F.L. and Kurchii, B.A. (1986) Effect of ethephon on the content of endogenous phytohormones in winter rye plants. *Fizil. Biokhim. Kultur Rasr.* 18, 182–185.

Kalra, S.K. and Brooks, J.L. (1973) Lipids of ripening tomato fruit and its mitochondrial fraction. *Phytochemistry* 12, 487–492.

Kaltaler, R.E.L. and Boodley, J.W. (1970) The effect of various concentrations of ethylene on the respiration rate, water uptake, fresh weight changes, pH changes and visual appearance of 'Red American Beauty' roses during senescence. *HortScience* 5 (abstract), 356.

Kamerbeek, G.A. and deMunk, W.J. (1976) A review of ethylene effects in bulbous plants. *Sci. Hort. (Amst.)* 4, 101–115.

Kanellis, A.K. and Roubelakis-Angelakis, K.A. (1993) Grape. In: Seymour, G.B., Taylor, J.E. and Tucker, G.A. (eds) *Biochemistry of Fruit Ripening.* Chapman & Hall, London, pp. 189–234.

Kanellis, A.K., Solomos, T. and Mattoo, A.K. (1989a) Hydrolytic enzyme activities and protein pattern of avocado fruit ripened in air and low oxygen with and without ethylene. *Plant Physiol.* 90, 259–266.

Kanellis, A.K., Solomos, T., Meehta, A.M. and Mattoo, A.K. (1989b) Decreased cellulose activity in avocado fruit subjected to 2.5% oxygen correlates with lower cellulose protein and gene transcript levels. *Plant Cell Physiol.* 30, 817–823.

Kanellis, A.K., Solomos, T. and Roubelakis-Angelakis, K.A. (1991) Suppression of cellulose and polygalacturonase and induction of alcohol dehydrogenase isoenzymes in avocado fruit mesocarp subjected to low oxygen stress. *Plant Physiol.* 96, 269–274.

Kanellis, A.K., Loulakakis, K.A., Hassan, M.M. and Roubelakis-Angelakis, K.A. (1993) Biochemical and molecular aspects of low-oxygen action on fruit ripening. In: Pech, J.C., Latché, A. and Balagué, C. (eds) *Cellular and Molecular Aspects of the Plant Hormone Ethylene.* Kluwer Academic, Dordrecht, pp. 117–122.

Kang, B.G. (1979) Epinasty. In: Haupt, W. and Feinlab, M.E. (eds) *Physiology of Movements.* Springer, Berlin, pp. 647–667.

Kang, B.G. and Burg, S.P. (1972a) Relation of phytochrome-enhanced geotropic sensitivity to ethylene production. *Plant Physiol.* 50, 132–135.

Kang, B.G. and Burg, S.P. (1972b) Ethylene as a natural agent inducing plumular hook formation in pea seedlings. *Planta* 104, 275–281.

Kang, B.G. and Burg, S.P. (1972c) Involvement of ethylene in phytochrome-mediated carotenoid synthesis. *Plant Physiol.* 49, 631–633.

Kang, B.G. and Burg, S.P. (1973) Role of ethylene in phytochrome-induced anthocyanin synthesis. *Planta* 110, 227–235.

Kang, B.G. and Burg, S.P. (1974a) Ethylene action on lateral auxin transport in tropistic responses, leaf epinasty, and horizontal nutation. In: *Plant Growth Substances,* Hirokawa, Tokyo, pp. 1090–1094.

Kang, B.G. and Burg, S.P. (1974b) Red light enhancement of the phototropic response of etiolated pea stems. *Plant Physiol.* 53, 445–448.

Kang, B.G. and Ray, P.M. (1969a) Ethylene and carbon dioxide as mediators in the response of the bean hypocotyl hook to light and auxins. *Planta (Berl.)* 87, 206–216.

Kang, B.G. and Ray, P.M. (1969b) Role of growth regulators in the bean hypocotyl hook opening response. *Planta* 87, 193–205.

Kang, B.G., Yocum, C.S., Burg, S.P. and Ray, P.M. (1967) Ethylene and carbon dioxide mediation of hypocotyl hook-opening response. *Science* 156, 958–959.

Kang, B.G., Newcomb, W. and Burg, S.P. (1971) Mechanism of auxin-induced ethylene production. *Plant Physiol.* 47, 504–509.

Kao, C.H. and Yang, S.F. (1982) Light inhibition of the conversion of ACC to ethylene in leaves is mediated through carbon dioxide. *Planta* 155, 261–266.

Kao, C.H. and Yang, S.F. (1983) Role of ethylene in the senescence of rice leaves. *Plant Physiol.* 73, 881–885.

Kapuya, J.A. and Hall, M.A. (1977) Diurnal variations in endogenous ethylene levels in plants. *New Phytol.* 79, 233–237.

Karaoulanis, G. (1983) The formation of enzymatic products in the fruits during growth and storage. In: Lieberman, M. (ed.) *Post Harvest*

Physiology and Crop Preservation. Plenum Press, New York, pp. 231–245.

Karvouni, Z., John, I., Taylor, J.E., Watson, C.F., Turner, A.J. and Grierson, D. (1995) Isolation and characterization of a melon cDNA clone encoding phytoene synthase. *Plant Mol. Biol.* 44, 283–307.

Kasahara, M., Yashiro, K., Sakamoto, T. and Ohmori, M. (1997) The *Spirulina platensis* adenylate cyclase gene, cyaC, encodes a novel signal transduction protein. *Plant Cell Physiol.* 38, 828–836.

Kasai, Y., Kato, M., Aoyama, J. and Hyodo, H. (2000) Ethylene production and increase in 1-aminocyclopropane-1-carboxylate oxidase activity during senescence of broccoli florets. *Plant Growth Regl. Abstracts* 26, 57.

Katayama, M. and Ohmori, M. (1997) Isolation and characterization of multiple adenylate cyclase genes from the cyanobacterium *Anabaena* sp. strain PCC 7120. *J. Bacteriol.* 179, 3588–3593.

Katayama, M., Wada, Y. and Ohmori, M. (1995) Molecular cloning of the cyanobacterial adenylate cyclase gene from the filamentous cyanobacterium *Anabaena cylindrica. J. Bacteriol.* 177, 3873–3878.

Kathiresan, A., Reid, D.M. and Chinnappa, C.C. (1996) Light and temperature entrained circadian regulation of activity and mRNA accumulation of 1-aminocyclopropane-1-carboxyolic acid oxidase in *Stellaria longipes. Planta* 199, 329–335.

Kaufman, J., Hardenburg, R.E. and Lutz, J.M. (1956) Weight losses and decay of Florida and California oranges in mesh and perforated polyethylene consumer bags. *Proceedings of the J. Amer. Soc. Hort. Sci.* 67, 701–705.

Kaufman, P., Pharis, R.P., Reid, D.M. and Beall, F.D. (1985) Investigations into the possible regulation of negative gravitropic curvature in intact *Avena sativa* plants and in isolated stem segments by ethylene and gibberellins. *Plant Physiol.* 65, 237–244.

Kawai, M., Aotsuka, S. and Uchimiya, H. (1998) Isolation of a cotton *CAP* gene: a homologue of adenylyl cyclase-associated protein highly expressed during fiber elongation. *Plant Cell Physiol.* 39, 1380–1383.

Kays, S.J. and Pallas, J.E. (1980) Inhibition of photosynthesis by ethylene. *Nature* 285, 51–52.

Kays, S.J., Nicklow, C.W. and Simons, D.H. (1974) Ethylene in relation to the response of roots to physical impedance. *Plant Soil* 40, 565–571.

Kazemi, S. and Kefford, N.P. (1974) Apical correlative effects in leaf epinasty of tomato. *Plant Physiol.* 54, 512–519.

Ke, D.Y. and Kader, A.A. (1989) Tolerances and responses of fresh fruits to O_2 levels at or below 1%. *Proceedings of the 5th International Controlled Atmosphere Research Conference*, Wenatchee, Washington, 14–16 June, pp. 209–216.

Ke, D.Y. and Kader, A.A. (1992) Potential of controlled atmospheres for postharvest insect disinfestation of fruits and vegetables. *Postharvest News and Information* 3, 31–37N.

Keele, C.A. and Neil, E. (1961) Structure and functions of the kidney. In: *Sampson Wright's Applied Physiology.* Oxford University Press, London.

Kefeli, V.I., Rakitinia, Y.A., Vlasov, P.V., Jalilova, F. and Kalevich, A.E. (1993) Various conditions of illumination and ethylene evolution. In: Pech, J.C., Latché, A. and Balagué, C. (eds) *Cellular and Molecular Aspects of the Plant Hormone Ethylene.* Kluwer Academic, Dordrecht, pp. 347–358.

Kelley, M.O. and Bradford, K.J. (1986) Insensitivity of the *diageotropica* tomato mutant to auxin. *Plant Physiol.* 82, 713–717.

Kelley, M.O. and Saltveit, M.E. (1988) Effect of endogenously synthesized and exogenously applied ethanol on tomato fruit ripening. *Plant Physiol.* 88, 143–147.

Kende, H. and Baumgartner, B. (1974) Regulation of aging in flowers of *Ipomoea tricolor* by ethylene. *Planta (Berl.)* 116, 279–289.

Kende, H. and Boller, T. (1981) Wound ethylene and 1-aminocyclopropane-1-carboxylate synthase in ripening tomato fruit. *Planta* 151, 476–481.

Kenza, M., Umiel, N. and Borochov, A. (2000) The involvement of ethylene in the senescence of *Ranunculus* cut flowers. *Postharvest Biol Technol.* 19, 287–290.

Ketsa, S. and Herner, R.C. (1987) Effect of carbon dioxide on wound ethylene production by fruit pericarp of *rin* mutant tomato. *Thai J. Agric. Sci.* 20, 57–64.

Ketsa, S. and Herner, R.C. (1989) A negative feedback of wound ethylene production by fruit pericarp tissue of *rin* mutant tomato. *Thai J. Agric. Sci.* 22, 223–232.

Ketsa, S. and Rugkong, A. (2000) Ethylene production, senescence and ethylene sensitivity of *Dendrobium* 'Pompadour' flowers following pollination. *J. Hort. Sci. Biotech.* 75, 149–153.

Keys, A.J., Bird, I.F., Cornelius, M., Lea, P.J., Wallsgrove, R.M. and Miflin, B.J. (1978)

Photorespiratory nitrogen cycle. *Nature (London)* 275, 741–743.

KiCheol, S. and JungNam, S. (1999) Ethylene production in *Lilium* Oriental Hybrid 'Casa Blanca' florets during vase life. *J. Korean Soc. Hort. Sci.* 40, 467–469.

Kidd, F. (1915) The controlling influence of carbon dioxide. III. The retarding effect of carbon dioxide on respiration. *Proceedings of the Roy. Soc. (London) B* 89, 136–156.

Kidd, F. and West, C. (1925) The course of respiratory activity throughout the life of an apple. Food Invest. Bd. Rept. London for 1924, pp. 27–34.

Kidd, F. and West, C. (1927a) Gas storage of fruit. Special Rept. Food Invest. Bd. Dept. Sci. Ind. Res. 30.

Kidd, F. and West, C. (1927b) A relation between the concentration of O_2 and CO_2 in the atmosphere, rate of respiration and the length of storage of apples. Food Inv. Bd. Rept. London for 1925, 1926, pp. 41–42.

Kidd, F. and West, C. (1928) Atmosphere control in fruit storage. Food Invest. Bd. Rept. London for 1927, p. 32.

Kidd, F. and West, C. (1934) The influence of composition of the atmosphere upon incidence of the climacteric. Food Invest. Bd. Rept. London for 1933.

Kidd, F. and West, C. (1938a) The action of carbon dioxide on the respiratory activity of apples. Food Invest. Bd. Rept. London for 1937, p. 101.

Kidd, F. and West, C. (1938b) Individual variation in apples. Food Invest. Bd. Rept. London for 1937, pp. 114–115.

Kidd, F. and West, C. (1945) Respiratory activity and duration of life of apples gathered at different stages of development and subsequently maintained at a constant temperature. *Plant Physiol.* 20, 467–504.

Kidd, F. and West, C. (1949a) Carbon dioxide injury in relation to the maturity of apples and pears. Food. Invest. Bd. Rept. London for 1939, pp. 64–68.

Kidd, F. and West, C. (1949b) Resistance of the skin of the apple fruit to gaseous exchange. Food Invest. Bd. Rept. London for 1939, pp. 59–64.

Kieber, J.J. (1997) The ethylene response pathway in *Arabidopsis*. *Annu. Rev. Plant Physiol. Plant Mol. Biol.* 48, 277–296.

Kieber, J.J. and Ecker, J.R. (1993) Ethylene gas: it's not just for ripening any more! *Trends Genet.* 9, 356–362.

Kieber, J.J., Rothenberg, M., Roman, G., Feldmann, K.A. and Ecker, J.R. (1993) *CTR1*, a negative regulator of ethylene response pathway in *Arabidopsis*, encodes a member of the Raf family of protein kinases. *Cell* 72, 427–441.

Kim, H.O., Hewett, E.W. and Lallu, N. (2000) The role of ethylene in kiwifruit softening. *Plant Growth Regl. Abstracts* 26, 142.

Kim, J.H., Kim, W.T., Kang, B.G. and Yang, S.F. (1997) Induction of 1-aminocyclopropane-1-carboxylate oxidase mRNA by ethylene in mung bean hypocotyls: involvement of both protein phosphorylation and dephosphorylation in ethylene signaling. *Plant J.* 11, 399–405.

Kim, J.-K., Solomos, T. and Gross, K.C. (1999) Changes in cell-wall galactosyl and soluble galactose content in tomato fruit stored in low-oxygen atmospheres. *Postharvest Biol. Technol.* 17, 33–38.

Kim, K.S., Lee, K.L., Hong, S.Y. and Sohn, T.H. (1969a) Studies on the reduced pressure storage of fruits. II. Preservation of Jonathan under various storage chamber pressures. *J. Korean Agr. Chem. Soc.* 11, 77–82 (Hort. Abst. 42, 387, 1972).

Kim, K.S., Park, Y.T., Hong, S.Y. and Sohn, T.H. (1969b) Studies on the reduced pressure storage of fruits. I. Comparison of reduced pressure storage with other storages for American Summer Pearmain. *J. Korea Agr. Chem. Soc.* 11, 67–76. (Hort. Abstr. 42, 386, 1972).

Kim, M. and Oogaki, C. (1986) Characteristics of respiration and ethylene production in fruits transferred from low-pressure storage to ambient atmosphere. *J. Jpn. Soc. Hortic. Sci.* 55, 339–347.

Kim, Y.-K., Kawano, T., Li, D. and Kolattukudy, P.E. (2000) A mitogen-activated protein kinase required for induction of cytokinesis and appresorium formation by host signals in the conidia of *Colletotrichum gloeosporioides*. *Plant Cell* 12, 1331–1344.

Kinane, J., Dalvin, S., Binddslev, L., Hall, A., Gurr, S. and Oliver, R. (2000) Evidence that the cAMP pathway controls emergence of both primary and appressorial germ tubes of barley powdery mildew. *Mol. Plant Microbe Interact.* 13, 494–502.

King, J.J., Stimart, D.P., Fisher, R.H. and Bleecker, A.B. (1995) A mutation altering auxin homeostasis and plant morphology in *Arabidopsis*. *Plant Cell* 7, 2023–2027.

Kirk, H.G. and Andersen, A.S. (1986) Influence of low-pressure storage on stomatal opening and rooting of cuttings. *Acta Horticulturae* 181, 393–397.

Kirk, H.G., Skytt Andersen, A., Veierskov, B., Johansen, E. and Aabrandt, Z. (1986) Low-pressure storage of hibiscus cuttings. Effect on stomatal opening and rooting. *Ann. Bot.* 58, 389–396.

Kirkham, M.B. (1983) Effect of ethephon on the water status of a drought-sensitive cultivar of winter wheat. *Z. Pflanzenphysiol.* 112, 103–112.

Kirrschbaum, M.U.F. and Pearcy, R.W. (1988) Gas exchange analysis of the relative importance of stomatal and biochemical factors in photosynthetic induction in *Alocasia macrorrhiza. Plant Physiol.* 86, 782–785.

Kitagawa, H., Kawada, K. and Tarutani, T. (1977) Degreening of 'Satsuma' mandarin in Japan. *Proceedings of the Int. Soc. Citriculture* 1, 219–223.

Klaustermeyer, J.A., Kader, A.A. and Morris, L.L. (1977) Effect of controlled atmospheres on insect control in harvested lettuce. In: Dewey, D.H. (ed.) *Controlled Atmospheres for the Storage and Transport of Perishable Agricultural Commodities.* Proceedings of the 2nd National Controlled Atmosphere Research Conference, 5–7 April, 1977, Michigan State University, East Lansing, Michigan, pp. 302–304.

Klee, H.J. (1993) Ripening physiology of fruit from transgenic tomato (*Lycopersicum esculentum*) plants with reduced ethylene synthesis. *Plant Physiol.* 102, 911–916.

Klee, H.J., Horsch, R. and Rogers, S. (1987a) *Agrobacterium*-mediated plant transformation and its further application to plant biology. *Annu. Rev. Plant Physiol.* 38, 467–486.

Klee, H.J., Horsch, R.B., Hinchee, M.A., Hein, M.B. and Hoffman, N.L. (1987b) The effects of overproduction of two *Agrobacterium tumefaciens* T-DNA auxin biosynthetic gene products in transgenic petunia plants. *Genes Dev.* 1, 86–96.

Klee, H.J., Hayford, M.B., Kretzmer, K.A., Barry, G.F. and Kishore, G.M. (1991) Control of ethylene synthesis by expression of a bacterial enzyme in transgenic tomato plants. *Plant Cell* 3, 1187–1193.

Klein, J.D. and Lurie, S. (1991) Postharvest heat treatment and fruit quality. *Postharvest News and Information* 2, 15–19.

Kleiner, D. (1981) The transport of NH_3 and NH_4^+ across biological membranes. *Biochim. Biophys. Acta* 639, 41–52.

Kline, R.W. (1976) Storage of broilers in Grumman hypobaric container. Progress Report. Armour Food Research, Scottsdale, Arizona, 20 May, 1976.

Knapp, J., Moureau, P., Schuch, W. and Grierson, D. (1989) Organization and expression of polygalacturonase and other ripening-related genes in Ailsa Craig 'Never ripe' and 'ripening inhibitor' tomato mutants. *Plant Mol. Biol.* 12, 105–116.

Knee, E.M., Hangarter, R.P. and Knee, M. (2000) Interactions of light and ethylene in hypocotyl hook maintenance in *Arabidopsis thaliana* seedlings. *Physiol. Plant.* 108, 208–215.

Knee, M. (1973) The effect of controlled atmosphere storage on respiratory metabolism of apple fruit tissue. *J. Sci. Food Agric.* 24, 1289.

Knee, M. (1987) Development of ethylene biosynthesis in pear fruits at -1°C. *J. Expt. Bot.* 38, 1724–1733.

Knee, M. (1990) Ethylene effects in controlled atmosphere storage of horticultural crops. In: Calderon, M. and Barkai-Golan, R. (eds) *Food Preservation by Modified Atmospheres.* CRC Press, Boca Raton, Florida, pp. 225–235.

Knee, M. (1993) Pome fruit. In: Seymour, G.B., Taylor, J.E. and Tucker, G.A. (eds) *Biochemistry of Fruit Ripening.* Chapman & Hall, London, pp. 325–346.

Knee, M. (1995) Do tomatoes on plants behave as climacteric fruits? *Physiol. Plant.* 95, 211–216.

Knee, M. and Hatfield, S.G.S. (1981a) The metabolism of alcohols by apple fruit tissue. *J. Sci. Food Agric.* 593, 1981.

Knee, M. and Hatfield, S.G.S. (1981b) Benefits of ethylene removal during apple storage. *Ann. Appl. Biol.* 98, 157–165.

Knee, M., Looney, N.E., Hatfield, S.G.S. and Smith, S.M. (1983) Initiation of a rapid ethylene synthesis by apple and pear fruits in relation to storage temperature. *J. Expt. Bot.* 34, 1207–1212.

Knee, M., Hatfield, S.G.S. and Bramlage, W.J. (1987) Response of developing apple fruits to ethylene treatment. *J. Expt. Bot.* 38, 972–979.

Knegt, E., Kramer, S.J. and Bruuinsma, J. (1974) Pectin changes and internal ethylene concentrations in ripening tomato fruit. *Colleques Internationaux de CNRS, Paris* 238, 355–358.

Kneissl, M.L. and Deikman, J. (1996) The tomato *E8* gene influences ethylene biosynthesis in fruit but not in flowers. *Plant Physiol.* 112, 537–547.

Knudsen, J.G., Bell, K.J., Holt, A.D., Hottel, H.C., Sarofim, A.F., Standiford, F.C., Stuhlbarg, D. and Uhl, V.W. (1984) Heat transmission. In: Crawford, H.B. and Eckes, B.E. (eds) *Perry's Chemical Engineers' Handbook.* McGraw-Hill, New York, Section 10, pp. 1–68.

Kodama, H., Hamada, T., Horiguchi, G., Nishimura, M. and Iba, K. (1994) Genetic enhancement of cold tolerance by expression of a gene for chloroplast ω-3 fatty acid desaturase in transgenic tobacco. *Plant Physiol.* 105, 601–605.

Kodama, H., Horiguchi, G., Nishiuchi, T., Nishimura, M. and Iba, K. (1995) Fatty acid desaturation during chilling acclimation is one of the factors involved in conferring low-temperature tolerance to young tobacco leaves. *Plant Physiol.* 107, 1177–1185.

Kofranek, A.M. (1992) Cut chrysanthemums. In: Larson, R.A. (ed.) *Introduction to Floriculture*, 2nd edn. Academic Press, New York, pp. 3–42.

Kok, B. (1965) Photosynthesis: the path of energy. In: Bonner, W.D. and Varner, J.E. (eds) *Plant Biochemistry*. Academic Press, New York, pp. 903–960.

Kolattukudy, P.E. (1980) Bipolyester membranes of plants: cutins and suberin. *Science* 208, 990–1000.

Kolattukudy, P.E. (1981) Structure, biosynthesis, and biodegredation of cutin and suberin. *Annu. Rev. Plant Physiol.* 32, 539–567.

Koning, R.E. (1986) The role of ethylene in corolla unfolding in *Ipomoea nil* (Convulaceae). *Amer. J. Bot.* 74, 510–516.

Konings, H. and Jackson, M.B. (1979) A relationship between the rates of ethylene production by roots and the promoting or inhibiting effects of exogenous ethylene and water on root elongation. *Z. Pflanzenphysiol.* 92, 385–397.

Konze, J.R., Jones, J.F., Boller, T. and Kende, H. (1980) Effect of 1-aminocyclopropane-1-carboxylic acid on the production of ethylene in senescing flowers of *Ipomoea tricolor* Cav. *Plant Physiol.* 30, 533–541.

Kossons, R. (1994) Storage of boxed pork in a VacuFresh[SM] container. Memorandum to VacuFresh[SM].

Kosugi, S. and Ohasgi, Y. (2000) Cloning and DNA-binding properties of tobacco ethylene-insensitive3 (EIN3) homolog. *Nucleic Acid Research* 28, 960–967.

Kraemer, E.O. (1952) The colloidal state and surface chemistry. In: Taylor, H.S. and Glasstone, S. (eds) *Treatise on Physical Chemistry*. D. Van Nostrand, New York, pp. 509–692.

Krahn, T.R. and Darby, D.E. (1971) The 'Filacell' system for refrigerated vegetable storage. Presented at *Annual Meeting Canadian Society for Horticultural Science*, Lethbridge, Alberta, Canada, 4–7 July, 1971.

Krebs, H.A. (1943) Carbon dioxide assimilation in heterotrophic organisms. *Annu. Rev. Biochem.* 12, 529–550.

Kreith, F. and Bohn, M.S. (1997) *Principles of Heat Transfer*, 5th edn. PWS, Boston, Massachusetts.

Krejci, M., Teddarik, K. and Janak, J. (1959) Rapid chromatographic separation and determination of helium–neon–hydrogen and nitrogen–krytpon–methane mixtures and determination of argon in the presence of oxygen at room temperature. *ISA Proceedings of the Anal. Instr. Div. 2nd International Gas Chromatography Symposium, 1959.* Preprints 2, pp. 105–110.

Kressin Muday, G., Lomax, T.L. and Rayle, D.L. (1995) Characterization of the growth and auxin physiology of roots of the tomato mutant, *diageotropica*. *Planta* 195, 548–553.

Krogh, A. (1920) Diffusion theory of tracheal respiration. *Arch. Ges. Physiol.* 179, 95–120.

Krogh, A. (1941) *The Comparative Physiology of Respiratory Mechanisms.* University of Pennsylvania Press, Pittsburgh, Pennsylvania.

Krotkov, G. (1941) The respiratory metabolism of McIntosh apples during ontogeny, as determined at 22°C. *Plant Physiol.* 16, 799–812.

Kuai, J. and Dilley, D.R. (1992) Extraction, partial purification and characterization of 1-aminocyclopropane-1-carboxylic acid oxidase from apple fruit. *Postharvest Biol. and Tech.* 1, 203–211.

Kubo, Y., Inaba, A. and Nakamura, R. (1989) Effects of high CO_2 on respiration in various horticultural crops. *Scientific Reports of the Faculty of Agriculture, Okayama University* 73, 27–33.

Kuc, J. (1982) Induced immunity to plant diseases. *BioScience* 32, 854–860.

Kuraishi, S. (1976) Ineffectiveness of cytokinin-induced chlorophyll retention in hypostomatous leaf discs. *Plant and Cell Physiol.* 17, 875–885.

Kuraishi, S. and Ishikawa, F. (1977) Relationship between transpiration and amino acid accumulation in *Brassica* leaf discs treated with cytokinins and fusiococcin. *Plant and Cell Physiol.* 18, 1273–1279.

Kurkdjian, A. and Guern, J. (1989) Intracellular pH: measurement and importance in cell activity. *Annu. Rev. Plant Physiol. Plant Mol. Biol.* 40, 271–303.

Kurkdjian, A., Leguay, J. and Guern, J. (1978) Measurement of intracellular pH and aspects of its control in higher plant cells cultivated in liquid medium. *Respiration Physiology* 33, 75–89.

Laan, P.A. (1934) Effects of ethylene on growth hormone formation in *Avena* and *Vicia*. *Rec. Trav. Botan. Neerl.* 31, 691–742.

LaFuente, M.T., Lopez-Galvez, G., Cantwell, M. and Yang, S.F. (1996) Factors influencing ethylene-induced isocoumarin formation and increased respiration in carrots. *J. Amer. Soc. Hort. Sci.* 121, 537–542.

Lakshminarayana, S. (1973) Respiration and ripening patterns in the life cycle of the mango fruit. *J. Hort. Sci.* 48, 227–233.

Lakshminarayana, S. and Subramanyan, H. (1970) CO_2 injury and fermentative decarboxylation in mango fruit at low-temperature storage. *J. Food Sci. Technol.* 7, 148–152.

Lambers, H. (1985) Respiration in intact plants and tissues. Its regulation and dependence on environmental factors, metabolism and invaded organisms. In: Douce, R. and Day, D.A. (eds) *Higher Plant and Cell Respiration Encyclopedia of Plant Physiology. New Series* 18, 418–473.

Lanahan, M.B., Yen, H.H.C., Giovannoni, J.J. and Klee, H.J. (1994) The *never-ripe* mutation blocks ethylene perception in tomato. *Plant Cell* 6, 521–530.

Landrock, A.H. and Ayres, J.C. (1955) Discoloration of fresh red meat and its relationship to oxygen permeability. *Food Technol.* 9, 194.

Lang, J.M., Eisinger, W.R. and Green, P.B. (1982) Effects of ethylene on the orientation of microtubules and cellulose microfibrils in pea epicotyl cells with polylamellate walls. *Protoplasma* 110, 5–14.

Langridge, I.W. and Sharples, R.O. (1972) Storage under reduced atmospheric pressure. *East Mall. Res. Sta. Annu. Rept. for 1971*, p. 76.

Lantz, A.W. (1951) Shrimp processing. *Fisheries Research Board Can. Prog. Repts.* Pacific Coast Stas. 89, 82–83.

Larrigaudiére, C. and Vendrell, M. (1993) Short-term activation of the conversion of 1-aminocyclopropane-1-carboxylic acid to ethylene in rewarmed Granny Smith apples. *Plant Physiol. Biochem.* 31, 585–591.

Larsen, P. (1956) Geotropic responses in roots. Some theoretical and technical problems. In: Wain, R.L. and Wightman, F. (eds) *The Chemistry and Mode of Action of Plant Growth Substances.* Butterworths, London, pp. 76–90.

Larsen, P. (1973) Gravity sensing in plants. In: Sneath, P.H.A. (ed.) *Life Sciences and Space Research XI.* Akademie, Berlin, pp. 141–154.

Larson, R.A. (1992) Azaleas. In: Larson, R.A. (ed.) *Introduction to Floriculture*, 2nd edn. Academic Press, New York, pp. 223–248.

Lashbrook, C.C. and Bennett, A.B. (1993) Functional analysis of CX-Cellulase (endo-β-1-4-glucanase) gene expression in transgenic tomato fruit. In: Pech, J.C., Latché, A. and Balagué, C. (eds) *Cellular and Molecular Aspects of the Plant Hormone Ethylene.* Kluwer Academic, Dordrecht, pp. 123–128.

Lashbrook, C.C., Gonzalez-Bosch, C. and Bennett, A.B. (1994) Two divergent endo-β-1,4-glucanase genes exhibit overlapping expression in ripening fruit and abscissing flowers. *Plant Cell* 6, 1485–1493.

Lashbrook, C.C., Tiernan, D.M. and Klee, H.J. (1998) Differential regulation of the tomato *ETR* gene family throughout plant development. *Plant J.* 15, 243–252.

Lasserre, E., Bouquin, T., Hernandez, J.A., Bull, J., Pech, J.C. and Balagué, C. (1996) Structure and expression of three genes encoding ACC oxidase homologs from melon (*Cucumis melo* L.). *Mol. Gen. Genet.* 251, 81–90.

Latché, A., Dupille, E., Rombaldi, C., Cleyet-Marel, J.C., Lelièvre, J.M. and Pech JC (1993) Purification, characterization and subcellular localization of ACC oxidase from fruits. In: Pech, J.C., Latché, A. and Balagué, C. (eds) *Cellular and Molecular Aspects of the Plant Hormone Ethylene.* Kluwer Academic, Dordrecht, pp. 39–45.

Laties, G.G. (1964) Physiological aspects of membrane function in plant cells during development. In: *Cellular Membranes in Development.* Academic Press, New York, pp. 299–320.

Laties, G.G. (1982) The cyanide-resistant alternative path in higher plant respiration. *Annu. Rev. Plant Physiol.* 33, 519–555.

Lau, O.L. (1985) Storage responses of four apple cultivars to low-O_2 atmosphere. In: Blankenship, S.M. (ed.) *Controlled Atmospheres for Storage and Transport of Perishable Agricultural Commodities.* Proceedings of the 4th National Controlled Atmosphere Research Conference, North Carolina Horticultural Report No. 126. North Carolina State University, Raleigh, North Carolina, p. 43.

Lau, O.L. and Yang, S.F. (1976) Inhibition of ethylene production by cobaltous ion. *Plant Physiol.* 57, 88–92.

Lau, O.L. and Yastremski, R. (1993) The use of 0.7% storage oxygen to attenuate scald symptoms in 'Delicious' apples: effect of apple strain and harvest maturity. *Acta Horticulturae* 326, 183–189.

Lau, O.L., Liu, Y. and Yang, S.F. (1986) Effects of fruit detachment on ethylene biosynthesis and loss of flesh firmness, skin color, and starch in ripening Golden Delicious apples. *J. Amer. Soc. Hort. Sci.* 111, 731–734.

Lawton, K.A., Raghothama, K.G., Goldsbrough, P.B. and Woodson, W.R. (1990) Regulation of senescence-related gene expression in carnation flower petals by ethylene. *Plant Physiol.* 93, 1370–1375.

Lea, P.J. (1997) Primary nitrogen metabolism. In: Dey, P.M. and Harbone, J.B. (eds) *Plant Biochemistry*. Academic Press, New York, pp. 273–313.

Leather, G.R., Forrence, L.E. and Abeles, F.B. (1972) Increased ethylene production during clinostat experiments may cause leaf epinasty. *Plant Physiol.* 49, 183–186.

Lebermann, K.W., Nelson, A.I. and Steinberg, M.P. (1968) Postharvest changes in broccoli stored in modified atmospheres. *Food Technol.* 22, 490–493.

Ledward, D.A. (1970) Metmyoglobin formation in beef stored in carbon dioxide-enriched and oxygen-depleted atmospheres. *J. Food Sci.* 35, 33.

Lee, J.S., Chang, W.-K. and Evans, M.L. (1990) Effects of ethylene on the kinetics of curvature and auxin redistribution in gravistimulated roots of *Zea mays*. *Plant Physiol.* 94, 1770–1775.

Lee, S., Seo, W. and Ota, Y. (1981) Studies on physiological activity of ethylene in rice plants. III. Diurnal variation in ethylene evolution in leaf and panicle. *Japan J. Crop Sci.* 50, 396–400.

Lee, Y.H. and Dean, R.A. (1993) cAMP regulates infection structure formation in the plant pathogenic fungus *Magnaporthe grisea*. *Plant Cell* 5, 693–700.

Lehman, A., Black, R. and Ecker, J.R. (1996) *HOOKLESS1*, an ethylene response gene, is required for differential cell elongation in the *Arabidopsis* hypocotyl. *Cell* 85, 183–194.

Leisner, J.J., Greer, C.G., Dilts, B.D. and Stiles, M.E. (1995) Effect of growth of selected lactic acid bacteria on storage life of beef stored under vacuum and in air. *Int. J. Food Microbiol.* 26, 231–243.

Lelièvre, J.M., Tichit, L., Larrigaudière, C., Vendrell, M., Pech, J.C. and Latché, A. (1995) Cold-induced accumulation of 1-aminocyclopropane-1-carboxylate oxidase protein in Granny Smith apples. *Postharv. Biol. Technol.* 5, 11–17.

Lelièvre, J.M., Latché, A., Jones, B., Bouzayen, M. and Pech, J.C. (1997a) Ethylene and fruit ripening. *Physiol. Plant.* 101, 727–739.

Lelièvre, J.M., Tichit, L., Dao, P., Fillion, L., Nam, Y.W., Pech, J.C. and Latché, A. (1997b) Effect of chilling on the expression of ethylene biosynthesis genes in Passe-Crassane pear (*Pyrus communis* L.) fruits. *Plant Mol. Biol.* 33, 847–855.

Lelièvre, J.M., Amor, M.B., Flores, B., Gomez, M.C., El-Yahyaqui, F., Chatenet, C.D., Périn, C., Hernandez, J.A., Romojaro, F., Latché, A., Bouzzayen, Z.M., Pitrat, M., Dogimont, C. and Pech, J.C. (2000) Ethylene-regulated genes and clarification of the role of ethylene in the regulation of ripening and quality in cantaloupe melon fruit. *Plant Growth Regl. Abstracts* 26, 231.

Lemon, E. and van Houtte, R. (1980) Ammonia exchange at the land surface. *Argon J.* 72, 876–883.

Leng, Q., Mercier, R.W., Yao, W. and Berkowitz, G.A. (1999) Cloning and first functional characterization of a plant cyclic nucleotide-gated cation channel. *Plant Physiol.* 121, 753–61

Lengeler, K.B., Davidson, R.C., D'souza, C., Harashima, T.S., Dhen, W.-C., Eang, P., Pan, X., Waugh, M. and Heitman, J. (2000) Signal transduction cascades regulating fungal development and virulence. *Microbiol. and Mol. Biol. Reviews* 64, 746 (abstract).

Lentz, C.P. and Rooke, E.A. (1964) Rates of moisture loss of apples under refrigerated storage conditions. *Food Technol.* 18, 119–121.

Leonard, E.R. (1941) Studies in tropical fruits, X. Preliminary observations on transpiration during ripening. *Ann. Bot. (N.S.)* 5, 89–119.

Leonard, E.R. and Wardlaw, C.W. (1941) Studies in tropical fruits. XII. The respiration of bananas during storage at 53°F, and ripening at controlled temperatures. *Ann. Bot. (N.S.)* 5, 378–391.

Lerke, P. (1973) Evaluation of potential risk of botulism from seafood cocktails. *Appl. Microbiology* 25, 807–810.

Leshuk, J.A. and Saltveit, M.E. (1990) Controlled atmosphere storage requirements and recommendations for vegetables. In: Calderon, M. and Barkai-Golan, R. (eds) *Food Preservation by Modified Atmospheres*. CRC Press, Boca Raton, Florida, pp. 315–352.

Lessler, M.A. (1947) Effect of a temperature gradient on distribution of water in apples. *Bot. Gaz.* 109, 90–94.

Lev, S., Sharon, A., Hadar, R., Ma, H. and Horwitz, B.A. (1999) A mitogen-activated protein kinase of the corn leaf pathogen *Cochliobolus heterstrophus* is involved in conidiation, appressorium formation and pathogenicity:

diverse roles of mitogen-activated protein kinase homologs in foliar pathogens. *Proc. Natl Acad. Sci. USA* 96, 13542 (abstract).

Levin, A., Sonego, L., Zutkhi, Y. and Ben Arie, R. (1993) Effects of CO_2 on ethylene production by apples at high and low O_2. In: Pech, J.C., Latché, A. and Balagué, C. (eds) *Cellular and Molecular Aspects of the Plant Hormone Ethylene.* Kluwer Academic, Dordrecht, pp. 150–151.

Levitt, K.L., Stein, D.B. and Rubenstein, B. (1987) Promotion of stomatal opening by indoleacetic acid and ethrel in epidermal strips of *Vicia faba* L. *Plant Physiol.* 85, 318–321.

Leybold-Heraeus (1991) *Vacuum Technology: Its Foundations, Formulae, and Tables.* Leybold-Heraeus Vacuum Products, Inc.

Leyser, H.M., Lincoln, C., Timpte, C., Lammer, D., Turner, J. and Estelle, M. (1993) *Arabidopsis* auxin-resistance gene *AXR1* encodes a protein related to ubiquitin-activating enzyme E1. *Nature* 364, 161–164.

Leyser, H.M., Pockett, F.B., Dharmasiri, S. and Estelle, M. (1996) Mutations in the *AXR3* gene of *Arabidopsis* result in altered auxin response including ectopic expression from the SAUR-AC1 promoter. *Plant J.* 10, 403–413.

Li, S., Andrews, P.K. and Patterson, M.E. (1994) Effects of ethephon on the respiration and ethylene evolution of sweet cherry (*Prunus avium* L.) fruit at different developmental stages. *Postharv. Biol. Technol.* 4, 235–243.

Li, Z.-G., El-Sharkawy, I. and Lelièvre, J.M. (2001) Effects of temperature, propylene and 1-MCP on ethylene biosynthesis and expression of ethylene receptor homolog gene *ETR1* during ripening of pear fruit. *Plant Growth Regl. Abstracts* 27, 250.

Liao, L.-J., Lin, Y.-H., Huang, K.-L., Chen, W.-S. and Cheng Y.-M. (2001) Postharvest life of cut rose flowers as affected by silver thiosulfate and sucrose. *Plant Growth Regl. Abstracts* 27, 155.

Licciaedello, J.J., Ronsivalli, L.J. and Slanin, J.W. (1967) Effect of oxygen tension on the spoilage microflora of irradiated and non-irradiated haddock (*Melanogrammus aeglefinus*) fillets. *J. Appl. Bact.* 30, 239.

Lidster, P.D., Lightfoot, H.J. and McRae, K.B. (1983) Production and regeneration of principal volatiles in apples stored in modified atmospheres and air. *J. Food Sci.* 48, 400.

Lieberman, M. (1979) Biosynthesis and action of ethylene. *Annu. Rev. Plant Physiol.* 30, 533–591.

Lieberman, M. and Knegt, E. (1977) Influence of ethylene on indole-3-acetic acid concentration in etiolated pea epicotyl tissue. *Plant Physiol.* 60, 475–477.

Lieberman, M. and Mapson, L.W. (1962) Inhibition of the evolution of ethylene and the ripening of fruit by ethylene oxide. *Nature* 196, 660–661.

Lieberman, M., Asen, S. and Mapson, L.W. (1964) Ethylene oxide an antagonist of ethylene in metabolism. *Nature* 204, 756–758.

Lieberman, M., Kunishi, A., Mapson, L.W. and Wardale, D.A. (1966) Stimulation of ethylene production in apple tissue slices by methionine. *Plant Physiol.* 41, 376–382.

Lieke, H. and von Guttenberg, H. (1961) Über die Bedeutung des Auxins für die epinastiche Bewegung plagiotroper Seitensprosse. *Naturwissenschaften* 48, 604–605.

Liley, P.E., Reid, R.C. and Buck, E. (1984) Physical and chemical data. In: Crawford, H.B. and Eckes, B.E. (eds) *Perry's Chemical Engineers' Handbook.* McGraw-Hill, New York, Section 3, pp. 1–291.

Lincoln, C., Britton, J.H. and Estelle, M. (1990) Growth and development of *axr1* mutants of *Arabidopsis. Plant Cell* 2, 1071–1080.

Lincoln, J.E. and Fischer, R.L. (1988) Diverse mechanisms for the regulation of ethylene-inducible gene expression. *Mol. Gen. Genet.* 212, 71–75.

Lincoln, J.E., Cordes, S., Read, E. and Fischer, R.L. (1987) Regulation of gene expression by ethylene during *Lycopersicum esculentum* (tomato) fruit development. *Proc. Natl Acad. Sci. USA* 84, 2793–2797.

Lincoln, J.E., Campbell, A.D., Oetiker, J., Rottmann, W.H., Oeller, P., Shen, W. and Theologis, A. (1993) *LE-ACS4*, a fruit ripening and wound-induced 1-aminocyclopropane-1-carboxylate synthase gene of tomato (*Lycopersicum escultentum*). *J. Biol. Chem.* 269, 19422–19430.

Liochev, S.I. and Fridovich, I. (2002) Copper, zinc superoxide dismutase and H_2O_2. Effects of bicarbonate on inactivation and oxidations of NADPH and urate, and on consumption of H_2O_2. *J. Biol. Chem.* 277, 34674–34678.

Lipe, J.A. and Morgan, P.W. (1972a) Ethylene: role in fruit abscission and dehiscence processes. *Plant Physiol.* 50, 759–764.

Lipe, J.A. and Morgan, P.W. (1972b) Ethylene: response of fruit dehiscence to CO_2 and reduced pressure. *Plant Physiol.* 50, 765–768.

Lipe, J.A. and Morgan, P.W. (1973) Ethylene, a regulator of young fruit abscission. *Plant Physiol.* 51, 949–953.

Lipton, W.J. (1965) Postharvest responses of asparagus spears to high-carbon dioxide and low-oxygen atmospheres. *Proceedings of the J. Amer. Soc. Hort. Sci.* 86, 347–356.

Lipton, W.J. (1971) *Controlled Atmosphere Effects on Lettuce Quality in Simulated Export Shipments.* ARS 51–55. USDA, ARS, Hyattsville, Maryland.

Lipton, W.J. (1975) Controlled atmospheres for fresh vegetables and fruits – why and when. In: Haard, N.F. and Salunkhe, D.K. (eds) *Symposium: Postharvest Biology and Handling of Fruits and Vegetables.* AVI, Westport, Connecticut, pp. 130–142.

Lipton, W.J. (1977a) Recommendations for CA storage of broccoli, Brussels sprouts, cabbage, cauliflower, asparagus and potatoes. In: Dewey, D.H. (ed.) *Controlled Atmospheres for the Storage and Transport of Perishable Agricultural Commodities. Proceedings of the 2nd National Controlled Atmosphere Research Conference,* 5–7 April 1977, Michigan State University, East Lansing, Michigan, p. 277.

Lipton, W.J. (1977b) Toward an explanation of disorders of vegetables induced by high CO_2. In: Dewey, D.H. (ed.) *Controlled Atmospheres for the Storage and Transport of Perishable Agricultural Commodities. Proceedings of the 2nd National Controlled Atmosphere Research Conference,* 5–7 April, 1977, Michigan State University, East Lansing, Michigan, pp. 137–141.

List, A. (1969) Transient growth responses of the primary roots of *Zea mays. Planta* 87, 1–19.

Little, C.R., Taylor, H.J. and McFarlane, F. (1985) Postharvest and storage factors affecting superficial scald and core flush of 'Granny Smith' apples. *HortScience* 20, 1080–1082.

Littlefield, N.A., Wankier, B.A., Salunkhe, D.K. and McGill, J.N. (1966) Fungistatic effects of controlled atmospheres. *Appl. Microbiol.* 14, 579–581.

Littmann, M.D. (1971) Effect of water loss on the ripening of climacteric fruits. *Queensl. J. Agric. Anim. Sci.* 29, 103–113.

Litvin, T.N., Kamenetsky, M., Zarifyan, A., Bucks, J. and Levin, L.R. (2003) Kinetic properties of 'soluble' adenylyl cyclase. Synergism between calcium and bicarbonate. *J. Biol. Chem.* 278(18), 15,922–15,926.

Liu, F.W. (1976) Storing ethylene-pretreated bananas in controlled atmosphere and hypobaric air. *J. Amer. Soc. Hort. Sci.* 101, 198–201.

Liu, X.-J., Shiomi, S., Nakatsuka, A., Kubo, Y., Nakamura, R. and Inaba, A. (1999) Characterization of ethylene biosynthesis associated with ripening in banana fruit. *Plant Physiol.* 121, 1257–1265.

Liu, Y., Hoffman, N.E. and Yang, S.F. (1985a) Ethylene promotes the capability to malonate 1-aminocyclopropane-1-carboxylic acid and D-amino acids in preclimacteric tomato fruits. *Plant Physiol.* 77, 891–895.

Liu, Y., Hoffman, N.E. and Yang, S.F. (1985b) Ethylene-promotion of 1-aminocyclopropane-1-carboxylic acid participates in autoinhibition of ethylene synthesis in grapefruit flavedo discs. *Planta* 164, 565–568.

Liu, Y., Hoffman, N.E. and Yang, S.F. (1985c) Promotion by ethylene of the capability to convert 1-aminocyclopropane-1-carboxylic acid to ethylene in preclimateric tomato and cantaloupe fruit. *Plant Physiol.* 77, 407–411.

Lizada, C. (1993) Mango. In: Seymour, G.B., Taylor, J.E. and Tucker, G.A. (eds) *Biochemistry of Fruit Ripening.* Chapman & Hall, New York, pp. 255–271.

Lloyd, F.E. (1908) The physiology of stomata. *Publ. Carnegie Inst. Wash,* p. 314.

Lockhardt, C.L. (1967) Influence of controlled atmospheres on the growth of *Gloeosporium album in vitro. Can. J. Plant Sci.* 47, 649–651.

Lockhardt, C.L. (1968) Influence of various carbon dioxide and oxygen concentrations on the growth of *Fusarium oxysporum in vitro. Can. J. Plant Sci.* 48, 451.

Lockhardt, C.L. (1969) Effect of CA on storage rot pathogens. In: Dewey, D.H., Herner, R.C. and Dilley, D.R. (eds) *Controlled Atmospheres for the Storage and Transport of Horticultural Crops. Proceedings of the National Controlled Atmosphere Research Conference,* 27 and 28 January, 1969, Michigan State University, East Lansing, Michigan, pp. 113–121.

Loesche, W.J. (1969) Oxygen sensitivity of various anaerobic bacteria. *Appl. Microbiol.* 18, 723–727.

Lomax, T.L., Coenen, C., Gaiser, J.C., Hopkins, R., Rayle, D.L. and Rice, M.S. (1993) Auxin perception and the regulation of tomato growth and development. In: Yoder, J. (ed.) *Molecular Biology of Tomato: Fundamental Advances and Crop Improvement.* Technomic, Lancaster, Pennsylvania, pp. 129–138.

Looney, N.E., McGlasson, W.B. and Coombe, B.G. (1974) Control of fruit ripening in peach, *Prunus persica*: action of succinic acid-2-2-dimethylhydrazide and (2-chloroethyl) phosphonic acid. *Austr. J. Plant Physiol.* 1, 77–86.

Lorimer, G.H. (1981) The carboxylation and oxygenation of ribulose 1,5-biphosphate: the primary events in photosynthesis and photorespiration. *Annu. Rev. Plant Physiol.* 32, 349–383.

Lorimer, G.H. (1983) Carbon dioxide and carbamate formation: the makings of a biochemical control system. *Trends Biochem. Sci.* 8, 65–68.

Lougheed, E.C. and Franklin, E.W. (1965) Influence of carbon dioxide on the uptake of water by asparagus. *Nature* 207, 1313.

Lougheed, E.C. and Franklin, E.W. (1974) Ethylene production increased by bruising of apples. *HortScience* 9, 192–193.

Lougheed, E.C., Franklin, E.W., Papple, D.J., Pattie, D.R., Malinowski, H.K. and Wenneker, A. (1974) *A feasibility study of low pressure storage.* Horticultural Science Department and School of Engineering. University of Guelph, Guelph, Ontario, Canada.

Lougheed, E.C., Murr, D.P., Harney, P.M. and Sykes, J.T. (1976) Low-pressure storage of seeds. *Experientia* 32, 1159–1161.

Lougheed, E.C., Murr, D.P. and Bérard, L. (1977) LPS – Great expectations. In: Dewey, D.H. (ed.) *Horticultural Report: Controlled Atmospheres for the Storage and Transport of Perishable Agricultural Commodities. Proceedings of the 2nd National Controlled Atmosphere Research Conference*, 5–7 April, 1977, Michigan State University, East Lansing, Michigan, pp. 3–44.

Lougheed, E.C., Murr, D.P. and Bérard, L. (1978) Low-pressure storage of horticultural crops. *HortScience* 13, 21–27.

Love, J.W. (1980) Kalanchoe. In: Larson, R.A. (ed.) *Introduction to Floriculture*, 2nd edn. Academic Press, New York, pp. 409–434.

Lovell, P.J., Lovell, P.H. and Nichols, R. (1987) The control of flower senescence in petunia (*Petunia hybrida*). *Ann. Bot.* 60, 49–59.

Loveridge, J.P. (1968) The control of water loss in *Locusta migratoria migratorioides* R. & F. I. Cuticular water loss. *J. Expt. Biology* 49, 1–13.

Loveys, B.R. and Kriederman, P.E. (1974) Internal control of stomatal physiology and photosynthesis. I. Stomatal regulation and associated changes of endogenous levels of abscisic and phaseic acid. *Austr. J. Plant Physiol.* 1, 407–415.

Lovizio, S. (1980) Summary of pork, chickens, and ducks container test. Report for Grumman Allied Industries, Inc., Garden City, New York.

Ludlow, M.M. and Wilson, G.L. (1971) Photosynthesis of tropical pasture plants. III: Leaf age. *Austr. J. Biol. Sci.* 24, 1077–1087.

Lukaszewska, A.J. and Gorin, N. (1988) Effect of ethylene treatment on changes in the weight and carbohydrate content of corollas from cut 'Sonia' roses. *Bull. Polish Acad. Sci.* 36, 113–119.

Lumsden, D.V., Wright, R.C., Whiteman, T.M. and Byrnes, J.W. (1940) Ethylene injury to cut flowers in cold-storage rooms. *Science* 92, 243–244.

Lund, S.T., Stall, R.E. and Klee, H.J. (1998) Ethylene regulates the susceptible response to pathogen infection in tomato. *Plant Cell* 10, 371–382.

Luo, Y.-B., Hao, S.-P. and Sheng, J.-P. (2000) Characteristics of postharvest physiology of antisense ACS transgenic tomato fruits. *Plant Growth Regl. Abstracts* 27, 62.

Lurie, S., Handros, A., Fallik, E. and Shapira, R. (1996) Reversible inhibition of tomato fruit gene expression at high temperature. *Plant Physiol.* 110, 1207–1214.

Lutz, J.M. and Hardenburg, R.E. (1968) *The Commercial Storage of Fruits, Vegetables, and Florist and Nursery Stocks.* USDA Agricultural Handbook No. 66.

Lyon, C.J. (1962) Gravity factor for auxin transport. *Science* 137, 432–433.

Lyon, C.J. (1963a) Auxin factor in branch epinasty. *Plant Physiol.* 38, 145–152.

Lyon, C.J. (1963b) Auxin transport in leaf epinasty. *Plant Physiol.* 38, 567–574.

Lyon, C.J. (1965) Action of gravity on basipetal transport of auxin. *Plant Physiol.* 40, 953–961.

Lyon, C.J. (1970) Ethylene inhibition of auxin transport by gravity in leaves. *Plant Physiol.* 45, 644–646.

Lyon, C.J. (1972) Auxin control for the orientation of pea roots grown on a clinostat or exposed to ethylene. *Plant Physiol.* 50, 417–420.

Lyons, J.M. (1973) Chilling injury in plants. *Annu. Rev. Plant Physiol.* 24, 445–466.

Lyons, J.M. and Pratt, H.K. (1964) Effect of stage of maturity and ethylene treatment on respiration and ripening of tomato fruits. *J. Amer. Soc. Hort. Sci.* 84, 491–500.

Lyons, J.M., McGlasson, W.B. and Pratt, H.K. (1962) Ethylene production, respiration, and internal gas concentration in cantaloupe fruits at various stages of maturity. *Plant Physiol.* 37, 31–36.

Macey, R.I. (1984) Transport of water and urea in red blood cells. *Am. J. Physiol.* 246, C195–C203.

Macháchová, I., Chauvaaux, N., Dewitte, W. and Van Onckelen, H. (1997) Diurnal fluctuations in ethylene formation in *Chenopodium rubrum*. *Plant Physiol.* 113, 981–985.
Mack, W.B. (1927) The action of ethylene in accelerating the blanching of celery. *Plant Physiol.* 2, 103.
MacLeod, R.F., Kader, A.A. and Morris, L.L. (1976) Stimulation of ethylene and CO_2 production by mature green tomatoes by impact bruising. *HortScience* 11, 604–606.
Macnish, A.J., Joyce, D.C., Hofman, P.J. and Simons, D.H. (1999) Involvement of ethylene in postharvest senescence of *Boronia heterophylla* flowers. *Austr. J. Exptl. Agric.* 39, 911–913.
Macnish, A.J., Joyce, D.C., Faragher, J.D. and Reid, M.S. (2000a) Use of 1-methylcyclopropene to prevent floral organ abscission from ethylene-sensitive Proteaceae. *Plant Growth Regl. Abstracts* 26, 154.
Macnish, A.J., Joyce, D.C., Hofman, P.J., Simons, D.H. and Reid, M.S. (2000b) 1-Methylcyclopropene treatment efficacy in preventing ethylene perception in banana fruit and *Grevillea* and waxflower flowers. *Austr. J. Expt. Agri.* 40, 471–481.
Macnish, A.J., Simons, D.H., Joyce, D.C., Faragher, J.D. and Hofman, P.Y. (2000c) Responses of native Australian cut flowers to treatment with 1-methylcyclopropene and ethylene. *HortScience* 35, 254–255.
Madhavan, S., Chrominski, A. and Smith, B.N. (1983) Effect of ethylene on stomatal opening in tomato and carnation leaves. *Plant Cell Physiol.* 24, 569–572.
Magness, J.R. (1920) Composition of gases in intercellular spaces of apples and potatoes. *Bot. Gaz.* 70, 308–316.
Magness, J.R. and Ballard, W.S. (1926) The respiration of Bartlett pears. *J. Agric. Res.* 32, 801–832
Maher, E.P. and Martingale, S.J.B. (1980) Mutants of *A. thaliana* with altered responses to auxin and gravity. *Biochem. Genet.* 18, 1041–1053.
Maier, V.P., Brewster, L.C. and Hsu, A.C. (1973) Ethylene-accelerated liminoid metabolism in citrus fruits: a process for reducing juice bitterness. *J. Agr. Food Chem.* 21, 490–495.
Makhlouf, J., Willemont, C., Arul, J.J., Castaigne, F. and Emond, J.P. (1989) Regulation of ethylene biosynthesis in broccoli flower buds in controlled atmospheres. *J. Amer. Soc. Hort. Sci.*114, 955–958.
Malamy, J., Carr, J.P., Klessig, D.F. and Raskin, I. (1990) Salicylic acid: a likely endogenous signal in the resistance response of tobacco to viral infection. *Science* 250, 1002–1004.
Mangin, L. (1896) Sur la végétation dans une atmosphére viciée par le respiration. *Comptes Rendues de l'Académie des Sciences (Paris)* 122, 747–749.
Mansell, R.L., Rose, G.W., Richardson, B. and Miller, R.L. (1968) Effects of prolonged reduced pressure on the growth and nitrogen content of turnip (*Brassica rapa* L.). Technical Report, US Air Force School of Aerospace Medicine, Brooks AFB, Texas, pp. 68–100
Mansfield, E.S. and Meidner, H. (1971) Endogenous stomatal movements in *Tradescantia virginia*. *New Phytol.* 70, 923–928.
Mansfield, T.A. and Heath, O.V.S. (1963) Studies in stomatal behavior IX. Photoperiodic effects on rhythmic phenomena in *Xanthium pennsylvanicum*. *J. Expt. Bot.* 15, 334–352.
Mansour, R., Latché, A., Vaillant, V., Pech, J.C. and Reid, M. (1986) Metabolism of 1-aminocyclopropane-1-carboxylic acid in ripening apple fruits. *Physiol. Plant.* 66, 495–502.
Maotani, T., Kawase, K., Kamuro, Y. and Hirai, K. (1983) Effects of ethylene on peel puffing of *Satsuma* mandarin. *J. Japan Soc. Hort. Sci.* 52, 238–242.
Mapson, L.W. (1970a) Biosynthesis of ethylene and the ripening of fruit. *Endeavor* 29, 29–33.
Mapson, L.W. (1970b) Vitamins in fruits. In: Hulme, A.C. (ed.) *The Biochemistry of Fruits and their Products.* Academic Press, New York, pp. 369–384.
Mapson, L.W. and Burton, W.G. (1962) The terminal oxidases of potato tuber. *Biochem. J.* 82, 19–25.
Marcellin, P. (1963) Mesure de la diffusion des gaz dans les organes vegetaux. *Bull. Soc. Fr. Physiol. Veg.* 9, 29–45.
Marcellin, P. (1974) Conditions physiques de la circulation des gaz respiratoires a travers la masse des fruits et maturation. *Facteurs et Regulation de la Maturation des Fruits* 238, 241–251.
Marcellin, P. and Chaves, A. (1983) Effects of intermittent high CO_2 on storage life of avocado fruits in relation to respiration and ethylene production. *Acta Horticulturae* 138, 155.
Marousky, F.J. (1977a) Commodity requirements and recommendations for flowers and nursery stock. In: Dewey, D.H. (ed.) *Controlled Atmospheres for the Storage and Transport of Perishable Agricultural Commodities. Proceedings of the 2nd National Controlled Atmosphere Research*

Conference, 5–7 April, 1977, Michigan State University, East Lansing, Michigan, pp. 287–291.

Marousky, F.J. (1977b) Controlled atmosphere and low-pressure storage of gladiolus, chrysanthemums and snapdragons. In: Dewey, D.H. (ed.). *Proceedings of the 2nd National Controlled Atmosphere Research Conference*, Michigan State University, East Lansing, Michigan, Hort. Rept. 28, 116–121.

Martin, E.S. and Meidner, H. (1971) Endogenous stomatal movements in *Tradescantia virginiana. New Phytol.* 70, 923–928.

Martin, F., Winspear, M.J., MacFarlane, J.D. and Oaks, A. (1983) Effect of methionine sulphoximine on the accumulation of ammonia in C_3 and C_4 leaves. *Plant Physiol.* 71, 177–181.

Martin, J.T. and Juniper, B.E. (1970) *The Cuticle of Plants.* St Martin's Press, Edward Arnold, London.

Martinez, G., Serrano, M., Petel, M.T., Amoros, A., Riquelme, F. and Romojaro, F. (1993) Ethylene production during the ripening of Cherimoya (*Anono chermiola*, Mill.). In: Pech, J.C., Latché, A. and Balagué, C. (eds) *Cellular and Molecular Aspects of the Plant Hormone Ethylene.* Kluwer Academic, Dordrecht, pp. 148–149.

Martins, I. (1979) Microbiological changes occurring during hypobaric and conventional storage of Atlantic cod. Honors thesis, Memorial University of Newfoundland.

Mason, T.G. and Lewin, C.J. (1926) On the rate of carbohydrate transport in the Greater Yam, *Discorea alta. Linn. Sci. Proceedings of the Royal Dublin Society* 18, 203–205.

Massia, A. (1998) Superoxide dismutase and catalase activities in apple fruit during ripening and post-harvest and with special reference to ethylene. *Physiol. Plant.* 104, 668–672.

Mastalerz, J.W. (1969) Chapter 20. Low temperature dry storage. In: Masterlerz, J.W. and Langhans, R.W. (eds) *Roses. A Manual on the Culture, Management, Diseases, Insects, Economics and Breeding of Greenhouse Roses.* Pennsylvania Flower Growers, New York State Flower Growers Association, Inc., and Roses Incorporated.

Masucci, J.D. and Schiefelbein, J.W. (1994) The *rhd6* mutation of *Arabidopsis thaliana* alters root-hair initiation through an auxin- and ethylene-associated process. *Plant Physiol.* 106, 1335–1346.

Masucci, J.D. and Schiefelbein, J.W. (1996) Hormones act downstream of *TTG* and *GL2* to promote root-hair outgrowth during epidermis development in the *Arabidopsis* root. *Plant Cell* 8, 1505–1517.

Mathieu, Y., Guern, J., Péan, M., Pasquier, C., Beloeil, J.-C. and Lallemand, J.Y. (1986) Cytoplasmic pH regulation in *Acer pseudoplatanus* cells. II. Possible mechanisms involved in pH regulation during acid-load. *Plant Physiol.* 82, 846–852.

Mathooko, F.M. (1996) Regulation of ethylene biosynthesis in higher plants by carbon dioxide. *Postharvest Biol. and Technol.* 7, 1–26.

Mathooko, F.M., Kubo, Y., Inaba, A. and Nakamura, R. (1995) Characterization of the regulation of ethylene biosynthesis in tomato fruit by carbon dioxide and diazocyclopentadiene. *Postharvest Biol. and Technol.* 5, 221–233.

Mathooko, F.M., Tsunashima, Y., Owino, W.Z.O., Kuboi, Y. and Inaba, A. (2001) Regulation of genes encoding ethylene biosynthetic enzymes in peach (*Prunus persica* L.) fruit by carbon dioxide and 1-methylcyclopropene. *Postharvest Biol. Technol.* 21, 265–281.

Mattei, A. (1972) L'emission organique volatile de banane et ses variations en fonction des conditions de maturation, particulièrement de la temperature. PhD dissertation, University of Paris.

Mattoo, A.K. and Lieberman, M. (1977) Localization of the ethylene-synthesizing system in apple tissue. *Plant Physiol.* 60, 794–799.

Mattoo, A.K. and Lieberman, M. (1982) Role of silver ions in controlling senescence and conversion of 1-aminocyclopropane-1-carboxylic acid to ethylene. *Plant Physiol.* 69, S-18.

Mattoo, A.K. and White, W.B. (1991) Regulation of ethylene biosynthesis. In: Mattoo, A.K. and Suttle, J.C. (eds) *The Plant Hormone Ethylene.* CRC Press, Boca Raton, Florida, pp. 21–42.

Mattoo, A.K., Baker, J.E., Chalutz, E. and Lieberman, M. (1977) Effect of temperature on the ethylene-synthesizing systems of apple, tomato, and *Penicillium digitatum. Plant Cell Physiol.* 18, 715–719.

Mattoo, A.K., Kanellis, A.K., Solomos, T. and Mehta, A.M. (1988) Low-oxygen atmospheres and gene expression in relation to ethylene action. In: Jona, R. (ed.) *Proceedings of the Symposium on the Physiology of Fruit Drop, Ripening, Storage, and Postharvest Processing of Fruits*, Turin, Italy, 3–4 October.

Mattsson, M. and Schjoerring, J.K. (1996) Ammonia emission from young barley plants: influence of N source, light/dark cycles and

inhibition of glutamine synthetase. *J. Expt. Bot.* 47, 477–484.
Mattsson, M., Häusler, R.E., Leegood, R.C., Lea, P.J. and Schjoerring, J.K. (1997) Leaf atmosphere NH_3 exchange in barley mutants with reduced activities of glutamine synthetase. *Plant Physiol.* 114, 1307–1312.
Mauch, F., Mauch-Mani, B. and Boller, T. (1988) Antifungal hydrolases in pea tissue. II. Inhibition of fungal growth by combinations of chitinase and β-1,3-glucanase. *Plant Physiol.* 88, 936–942.
Maunders, M.J., Holdsworth, M.J., Slater, A., Knapp, J.E., Bird, C.R., Schuch, W. and Grierson, D. (1987) Ethylene stimulates the accumulation of ripening-related mRNAs in tomato fruits. *Proceedings of the J. Amer. Soc. Hort. Sci.* 84, 491–500.
Maurel, C. (1997) Aquaporins and water permeability of plant membranes. *Annu. Rev. Plant Physiol. Plant Mol. Biol.* 48, 399–429.
Maxie, E.C., Farnham, D.S., Mitchell, F.G. and Sommer, N.F. (1974) Temperature management and quality of carnation flowers and rose buds. Calif. Flower and Nursery Rept. July, pp. 5–8.
Maxson, J.M. and Woodson, W.R. (2000) Ethylene-responsive gene expression during carnation flower senescence. *Plant Growth Regl. Abstracts* 26, 58.
Mayak, S. and Dilley, D.R. (1976) Regulation of senescence in carnation (*Dianthus caryophyllus*). Effect of abscisic acid and carbon dioxide on ethylene production. *Plant Physiol.* 58, 663–665.
Mayak, S. and Halevy, A.H. (1971) Water stress as the cause for failure of bud opening in iris. *J. Amer. Soc. Hort. Sci.* 96, 482–483.
Mayak, S. and Halevy, A.H. (1972) Interrelationships of ethylene and abscisic acid in the control of rose petal senescence. *Plant Physiol.* 50, 341–346.
Mayak, S., Vadia, Y. and Dilley, D.R. (1977) Regulation of senescence in carnation (*Dianthus caryophyllus*) by ethylene. Mode of action. *Plant Physiol.* 59, 591–593.
Mayak, S., Legge, R.L. and Thompson, J.E. (1982) Ethylene formation from 1-aminocyclopropane-1-carboxylic acid (ACC) by microsomal membranes from senescing carnation flowers. *Plant Physiol.* 67, S-279.
Mayak, S., Legge, R.L. and Thompson, J.E. (1983) Superoxide radical production by microsomal membranes from senescing carnation flowers. *Phytochemistry* 22, 1375–1380.
Mayne, R.G. and Kende, H. (1986) Ethylene biosynthesis in isolated protoplasts of *Vicia faba* L. – requirement for membrane integrity. *Planta* 151, 159–165.
Mazliak, P. (1965) The lipids of the avocado (*Persea americana* var. Fuerte); II. Variation of the composition in fatty acids of lipids of the pericarp according to the atmosphere surrounding the ripening fruit. *Fruits* 20, 117–120, 122.
Mazliak, P. (1983) Plant membrane lipids: changes and alterations during aging and senescence. In: Lieberman, M. (ed.) *Postharvest Physiology and Crop Preservation.* Plenum Press, New York, pp. 123–140.
McAdams, W.H. (1954) *Heat Transmission*, 3rd edn. McGraw-Hill, New York.
McAfee, J.A. and Morgan, P.W. (1971) Rates of production and internal levels of ethylene in vegetative cotton plant. *Plant Cell Physiol.* 12, 839–847.
McCall, W.A., Grauman, G.S., Restaino, L. and Hill, W.M. (1976) Broiler chickens stored in Grumman hypobaric container. Progress Report for Armour Research Center, Scottsdale, Arizona, 29 March.
McClendon, E. and Kammlah, H.W. (1976) Hypobaric storage of fresh pork. Progress Report for Armour Research Center, Scottsdale, Arizona, 8 June.
McClung, C.R. (2001) Circadian rhythms in plants. *Annu. Rev. Plant Physiol. Mol. Biol.* 52, 139–162.
McClung, L.S. and Lindberg, R.B. (1957) The study of obligately anaerobic bacteria. In: *Manual of Microbiological Methods.* McGraw-Hill, New York, pp. 120–139.
McColloch, L.P. (1962) Chilling injury and alternaria rot of bell peppers. USDA Market Res. Rpt. 536.
McColloch, L.P. and Worthington, J.T. (1952) Low temperature as a factor in the susceptibility of mature-green tomatoes to alternaria rot. *Phytopathology* 42, 425–427.
McColloch, L.P., Cook, H.T. and Wright, W.R. (1966) *Market Diseases of Tomatoes, Peppers and Eggplants.* USDA Agr. Handbook 28.
McConchie, R., Lang, N.S. and Gross, K.C. (1991) Carbohydrate depletion and leaf blackening in *Protea neriifolia. J. Amer. Soc. Hort. Sci.* 116, 1019–1024.
McCormack, A.A. (1972) Effect of ethylene degreening on decay of Florida citrus fruits. *Proceedings of the Florida State Hort. Soc.* 84, 270.
McDonald, R.E. and Milhjer, W.R. (1994) Quality and condition maintenance. In: Sharp, J.L. and Hallman, G.J. (eds) *Quarantine*

Treatments for Pests of Foods. Westview Press, Boulder, Colorado, pp. 249–277.

McGill, J.N., Nelson, A.J. and Steinberg, M.P. (1966) Effects of modified storage atmospheres on ascorbic acid and other quality characteristics of spinach. *J. Food Sci.* 31, 510–517.

McGlasson, W.B. (1970) The ethylene factor. In: Hulme, A.C. (ed.) *The Biochemistry of Fruits and Their Products*, vol. 1. Academic Press, London, pp. 475–519.

McGlasson, W.B. (1985) Ethylene and fruit ripening. *HortScience* 20, 51–54.

McGlasson, W.B. and Pratt, H.K. (1964a) Effects of ethylene on cantaloupe fruits harvested at various ages. *Plant Physiol.* 39, 120–127.

McGlasson, W.B. and Pratt, H.K. (1964b) Effects of wounding on respiration and ethylene production by cantaloupe fruit tissues. *Plant Physiol.* 39, 128–132.

McGlasson, W.B. and Wills, R.B.H. (1972) Effects of oxygen and carbon dioxide on respiration, storage life, and organic acids of green bananas. *Austr. J. Biol. Sci.* 25, 35–42.

McGlasson, W.B., Palmer, J.K., Vendrell, M. and Brady, C.J. (1971) Metabolic studies with banana fruit slices. II. Effects of inhitors on respiration, ethylene synthesis and ripening. *Austr. J. Biol. Sci.* 24, 1103–1114.

McGlasson, W.B., Abdi, N. and Holford, P. (2000) Does inhibition of ACO activity in Japanese-type plums account for the suppression of ethylene production in attached fruit by the tree factor and the suppressed climacteric? *Plant Growth Regl. Abstr.* 26, 144.

McKenzie, R.J. and Lovell, P.H. (1992) Perianth abscission in montbretia (*Crocosmia* × *Crocosmiiflora*). *Ann. Bot.* 69, 199–207.

McKeown, A.W. and Lougheed, E.C. (1980) Low-pressure storage of some vegetables. *Acta Horticulturae* 116, 83–100.

McKeown, A.W., Lougheed, E.C. and Murr, D.P. (1978) Compatibility of cabbage, carrots and apples in low-pressure storage. *J. Amer. Soc. Hort. Sci.* 103, 749–752.

McManus, M.T., Hunter, D.A., Yoo, S.D. and Gong D (2000) ACC oxidase expression and leaf ontogeny in white clover. *Plant Growth Regl. Abstr.* 26, 144.

McMichael, B.L., Jordan, W.R. and Powell, R.D. (1972) An effect of water stress on ethylene production by intact cotton petioles. *Plant Physiol.* 49, 658–660.

McMullen, L.M. and Stiles, M.E. (1993) Microbial ecology of fresh pork stored under modified atmosphere at −1, 4.4 and 10 degrees C. *Int. J. Food Microbiol.* 18, 1–14.

McMurchie, E.J., McGlasson, W.B. and Eaks, I.L. (1972) Treatment of fruit with propylene gives information about the biogenesis of ethylene. *Nature* 237, 235–236.

McRae, D.G., Coker, J.A., Legge, R.L. and Thompson, J.E. (1983) Bicarbonate/CO_2-facilitated conversion of 1-aminocyclopropane-1-carboxylic acid to ethylene in a model system and intact tissues. *Plant Physiol.* 73, 784–790.

Meidner, H. (1986) Cuticular conductance and the humidity response of stomata. *J. Expt. Bot.* 37, 517–525.

Meigh, D.F. (1959) Nature of the olefins produced by apples. *Nature* 184, 1072–1073.

Meigh, D.F. (1969) Production of farnescene and incidence of scald in stored apples. *Qual. Plant Matter. Veg.* 19, 243–254.

Meigh, D.F., Jones, T.D. and Hulme, A.C. (1967) The respiration climacteric in the apple. Production of ethylene and fatty acids in fruit attached to and detached from the tree. *Phytochemistry* 6, 1507–1515.

Melampy, R.M. and Willis, E.R. (1939) Respiration of larvae and pupae of honeybee. *Physiol. Zool.* 12, 302–311.

Mellanby, K. (1934) The site of loss of water from insects. *Proceedings of the Royal Ent. Soc. Lond. (B)* 116, 130–149.

Mercantila Publishers (1989a) *Guide to Food Transport; Fish, Meat and Dairy Products.* Mercantila Publishers, Copenhagen.

Mercantila Publishers (1989b) *Guide to Food Transport; Fruit and Vegetables.* Mercantila Publishers, Copenhagen.

Meredith, D. (1973) The Humifresh system design and operating experience. *Ashrae Semi-Annual Meeting*, 28 January–1 February, Chicago, Illinois.

Mermelstein, N.H. (1979) Hypobaric transport and storage of fresh meats and produce earns IFT Food Technology Industrial Achievement Award. *Food Tech.* 33, 32–35; 38–40.

Métraux, J.-P. and Kende, H. (1984) The cellular basis of the elongation response in submerged deep-water rice. *Plant Physiol.* 72, 441–446.

Métraux, J.-P, Signer, H., Ryais, J., Ward, E., Wyss-Benz, M., Gaudin, J., Raschdorf, K., Schmid, E., Blum, W. and Inverrardi, B. (1990) Increase in salicylic acid at the onset of systemic acquired resistance in cucumber. *Science* 250, 1004–1006.

Metzidakis, J. and Sfakiotakis, E. (1993) Inhibition of ethylene biosynthesis and suppression of cellulase and polygalacturonase in avocado fruit subjected to low oxygen storage. In:

Pech, J.C., Latché, A. and Balagué, C. (eds) *Cellular and Molecular Aspects of the Plant Hormone Ethylene.* Kluwer Academic, Dordrecht, pp. 129–135.

Miccolis, V. and Saltveit, M.E. (1991) Morphological and physiological changes during fruit growth and maturation of seven melon cultivars. *J. Amer. Soc. Hort. Sci.* 116, 1025–1029.

Michael, M.Z., Savin, K.W., Baudinette, S.C., Graham, M.W., Chandler, S.F., Lu, C.-Y., Caeser, C., Gautrais, I., Young, R., Nugent, G.D., Stevenson, K.R., O'Conner, E.L-J., Cobbett, C.S. and Cornish, E.C. (1993) Cloning of ethylene biosynthetic genes involved in petal senescence of carnation and petunia, and their antisense expression in transgenic plants. In: Pech, J.C., Latché, A. and Balagué, C. (eds) *Cellular and Molecular Aspects of the Plant Hormone Ethylene.* Kluwer Academic, Dordrecht, pp. 298–303.

Michaeli, R., Philosoph-Hadas, S., Riov, J. and Meir, S. (1999) Chilling-induced leaf abscission of *Ixora coccinea* plants. I. Induction of oxidative stress via increased sensitivity to ethylene. *Physiol. Plant.* 107, 166–173.

Michaeli, R., Philosoph-Hadas, S., Riov, J. and Meir, S. (2000) Role of ethylene in mediating the chilling-induced leaf abscission of *Ixora* plants. *Plant Growth Regl. Abstr.* 26, 145.

Michener, H.D. (1938) The action of ethylene on plant growth. *Amer. J. Bot.* 25, 711–720.

Michener, H.D. and Elliott, R.P. (1964) Minimum growth temperature for food poisoning, fecal-indicator, and psychotrophilic microorganisms. *Advan. Food. Res.* 13, 349–396.

Michiyama, H. and Saka, H. (1988) Endogenous ethylene levels in leaf sheath and panicle of rice plant. *Japan J. Crop Sci.* 57, 366–370.

Miller, C.O. (1951) Promoting effect of cobaltous and nickelous ions on expansion of etiolated bean leaf discs. *Arch. Biochem. Biophys.* 32, 216–218.

Miller, G.W. and Hsu, W.J. (1965) Effects of carbon dioxide-bicarbonate mixture on oxidative phosphorylation by cauliflower mitochondria. *Biochem. J.* 97, 615.

Miller, J.M. and Conn, E.E. (1980) Metabolism of hydrogen cyanide by higher plants. *Plant Physiol.* 65, 1199–1202.

Miller, L.P. (1933) Effect of various chemicals on the sugar content, respiratory rate, and dormancy of potato tubers. *Contr. Boyce Thomp. Inst.* 5, 213–234.

Miller, L.P. (1934) Time relations in effect of ethylene chlorohydrin in increasing and of ethyl alcohol in decreasing the respiration of potato tubers. *Contr. Boyce Thomp. Inst.* 6, 123–128.

Miller, L.P. (1935) Further experiments on the effect of halogenated aliphatic compounds on the respiration of potato tubers. *Contr. Boyce Thomp. Inst.* 7, 1–17.

Miller, P.L. (1964) Respiration – aerial gas transport. In: *The Physiology of Insects*, vol. 3. Academic Press, New York, chapter 4.

Mingo-Castel, A.N., Negm, F.B. and Smith, O.E. (1964) Effect of carbon dioxide and ethylene on tuberization of isolated potato stolons cultured in vitro. *Plant Physiol.* 53, 798–801.

Miquel, M., James, D., Dooner, H. and Browse, J. (1993) *Arabidopsis* requires polyunsaturated lipids for low-temperature survival. *Proc. Natl Acad. Sci. USA* 90, 6208–6212.

Mirza, J.I., Olsen, G.M., Iversen, T.-H. and Maher, E.P. (1984) The growth and gravitropic responses of wild-type and auxin-resistant mutants of *Arabidopsis thaliana. Physiol. Plant.* 60, 516–522.

Mita, S., Kirita, C., Kato, M. and Hyodo, H. (1999) Expression of ACC synthase is enhanced earlier than that of ACC oxidase during fruit ripening of mem (*Prunus mume*). *Physiol. Plant.* 107, 319–328.

Mitchell, D.J. and Zentmyer, G.A. (1971) Effects of oxygen and carbon dioxide tensions on growth of several species of *Phytophthora. Phytopathology* 61, 787–791.

Mitchell, J.W. and Marth, P.C. (1944) Effects of 2,4-dichlorophenoxyacetic acid on the ripening of detached fruit. *Bot. Gaz.* 106, 199–207.

Moe, R. (1975) The effect of growing temperature on keeping quality of cut roses. *Acta Horticulturae* 41, 77–88.

Moe, R. and Fjeld, T. (1985) Keeping quality of pot plants as influenced by ethylene. *Gartner Tidende* 48, 1580–1583.

Molisch, H. (1884) Über die Ablenkung der Wurzeln voniher normalen Wachstumsrichting durch Gase (Äärotropismus) Kaissserl. *Akad. Wiss. Wien.* 90, 111–196.

Molz, F.J. and Boyer, J.S. (1978) Growth-induced water potentials in plant cells and tissues. *Plant Physiol.* 62, 423–429.

Monod, J. (1942) *La Croissance des Cultures Bactériennes.* Hermann et Coe, Paris.

Montaldi, E.R. (1969) Gibberellin–sugar interaction regulating the growth habit of bermudagrass (*Cynodon dactylon* L. Pers.). *Experientia* 25, 91–92.

Moore, R. and Evans, M.L. (1986) How roots perceive and respond to gravity. *Amer. J. Bot.* 73, 574–587.

Moreshet, S. (1970) Effect of environmental factors on cuticular transpiration resistance inversely related to rh. *Plant Physiol.* 46, 815–818.

Moreshet, S. and Green, G.C. (1980) Photosynthesis and diffusion conductance of the Valencia orange fruit under field conditions. *J. Expt. Bot.* 31, 15–27.

Morgan, J.A. and Parton, W.J. (1989) Characteristics of ammonia volatilization from spring wheat. *Crop Sci.* 29, 726–731.

Morgan, P.W. and Baur, J.I. (1970) Involvement of ethylene in picloram-induced leaf movement response. *Plant Physiol.* 46, 655–659.

Morgan, P.W. and Durham, J.I. (1980) Ethylene production and leaflet abscission in *Melia azedarach* L. *Plant Physiol.* 66, 88–92.

Morgan, P.W. and Gausman, H.W. (1966) Effects of ethylene on auxin transport. *Plant Physiol.* 41, 445–452.

Morgan, P.W. and Hall, W.C. (1962) Effect of 2,4-dichlorophenoxyacetic acid on the production of ethylene by cotton and grain sorghum. *Physiol. Plant.* 15, 420–427.

Morgan, P.W. and Hall, W.C. (1964) Accelerated release of ethylene by cotton following application of indolyl-3-acetic acid. *Nature* 201, 99.

Morgan, P.W., Beyer, E. and Gausman, H.W. (1968) Ethylene effects on auxin physiology. In: Wightman, F. and Setterfield, G. (eds) *Biochemistry and Physiology of Plant Growth Substances.* Runge Press Ltd, Ottawa, pp. 1255–1273.

Morgan, P.W., Meyer, R.E. and Merkle, M.G. (1969) Chemical stimulation of ethylene evolution and bud growth. *Weed Sci.* 17, 353–355.

Morgan, P.W., He, C.-J., Greef, J.A. and Proft, M.P.D. (1990) Does water deficit stress promote ethylene synthesis by intact plants? *Plant Physiol.* 94, 1616–1624.

Morgan, P.W., Sarquis, J.I., Jordan, W.R. and Drew, M.C. (1993) Regulation of ethylene synthesis in maize root by responses to stress. In: Pech, J.C., Latché, A. and Balagué, C. (eds) *Cellular and Molecular Aspects of the Plant Hormone Ethylene.* Kluwer Academic, Dordrecht, pp. 232–237.

Mori, H., Nakagawa, N., Ono, T., Yamagishi, N. and Imaseki, H. (1993) Structural characteristics of ACC synthase isozymes and differential expression of their genes. In: Pech, J.C., Latché, A. and Balagué, C. (eds) *Cellular and Molecular Aspects of the Plant Hormone Ethylene.* Kluwer Academic, Dordrecht, pp. 1–6.

Morin, F., Riggault, C. and Hartmann, C. (1985) Conséquences d'un séjour au froid sur le métabolisme de l'éthylène au cours de la maturation de la poire Passe-Crassane après récolte. *Physio. Vég.* 23, 353–359.

Morris, J.G. (1970) The biochemical basis for oxygen sensitivity. *J. Gen. Microbiology* 60, iii.

Morris, L.L. and Kader, A.A. (1975) Postharvest physiology of tomato fruits. In: *Fresh Market Tomato Research – 1974,* Vegetable Crops series 171, University of California, Davis, California, pp. 36–48.

Morris, L.L. and Platenius, H. (1938) Low-temperature injury to certain vegetables after harvest. *Proceedings of the Am. Soc. Hot. Sci.* 36, 609–613.

Morris, L.L., Strand, L.L. and Adamick, F. (1981) Factors affecting ripening and quality of mature-green tomatoes. *Acta Horticulturae* 116, 133–139.

Morrow, L. and Lorenz, K. (1974) Some physiochemical properties of starches as affected by changes in atmospheric pressure. *J. Food Sci.* 39, 467–470.

Moshkov, I.E., Novikova, G.V., Smith, A.R. and Hall, M.A. (1993) In vitro study of ethylene binding sites in pea seedlings. In: Pech, J.C., Latché, A. and Balagué, C. (eds) *Cellular and Molecular Aspects of the Plant Hormone Ethylene.* Kluwer Academic, Dordrecht, pp. 195–196.

Mossel, D.A., Dijkmann, K.E. and Snijders, J.M.A. (1975) Microbial problems in handling and storage of fresh meats. In: Cole, D.J.A. and Lawrie, R.A. (eds) *Meat.* AVI, Westport, Connecticut, pp. 223–246.

Mott, K.A. and Berry, J.A. (1986) Effects of pH on activity and activation of ribulose-1,5-biphosphate carboxylase at air level CO_2. *Plant Physiol.* 82, 77–82.

Moutinho, A., Hussey, P.J., Trewavas, A.J. and Malho, R. (2001) cAMP acts as a second messenger in pollen tube growth and reorientation. *Proc. Natl Acad. Sci. USA* 98, 10481–10486.

Muday, G.K., Lomax, T.L. and Eayle, D.L. (1995) Characterization of the growth and auxin physiology of roots of the tomato mutant, *diageotropica. Planta* 195, 548–553.

Mueller, S.C. and Brown, R.M. (1982a) The control of cellulose microfibril deposition in higher plants. I. Can directed membrane flow orient cellulose microfibrils? Indirect evidence from freeze-fractured plasma membranes of maize and pine seedlings. *Planta* 154, 489–500.

Mueller, S.C. and Brown, R.M. (1982b) The control of cellulose microfibril deposition in the cell wall of higher plants. II. Freeze-fracture microfibril patterns in maize seedling tissues following experimental alteration with colchicine and ethylene. *Planta* 154, 501–515.

Muir, R.M. and Richter, E.W. (1970) The measurement of ethylene from plant tissues and its relation to auxin effect. In: Carr, D.J. (ed.) *Plant Growth Substances. Proceedings of the 7th International Conference on Plant Growth Substances*, Canberra, Australia, 7–11 December, 1970, pp. 518–525.

Mulder, P.W.A. (1977) *Primêre meganismes betrokke by die bruinwording van verhandeling.* R.A.U., Johannesburg.

Mulkey, T.J., Kuzmanoff, K.M. and Evans, M.L. (1981) Promotion of growth and shift in the auxin dose/response relationship in maize roots treated with the ethylene biosynthesis inhibitors aminoethoxyvinylglycine and cobalt. *Plant Sci. Letter* 25, 43–48.

Mulkey, T.J., Kuzmanoff, K.M. and Evans, M.L. (1982) Promotion of growth and hydrogen ion efflux by auxin in roots of maize pretreated with ethylene biosynthesis inhibitors. *Plant Physiol.* 70, 186–188.

Müller, R., Serek, M., Sisler, E.C. and Andersen, A.S. (1997) Post-storage quality and rooting ability of *Epipremnum pinnatum* L. cutttings after treatment with ethylene inhibitors. *J. Hort. Sci.* 72, 445–452.

Müller, R., Lind-Iversen, S., Stummann, B.M. and Serek, M. (2000a) Expression of genes for ethylene biosynthetic enzymes and an ethylene receptor in senescing flowers of potted roses. *J. Hort. Sci. Biotech.* 75, 12–15.

Müller, R., Sisler, E.C. and Serek, M. (2000b) Stress-induced ethylene production, ethylene binding, and the response to the ethylene action inhibitor 1-MCP in miniature roses. *Scientia Horticulturae* 83, 51–9.

Müller, R., Andersen, A.S., Serek, M. and Stummann, B.M. (2001) Physiological and genetic reasons for different postharvest characteristics in miniature potted rose. *Plant Growth Regl. Abstracts* 27, 253.

Mullins, E.D., McCollum, T.G. and McDonald, R.E. (1999) Ethylene: a regulator of stress-induced ACC synthase activity in non-climacteric fruit. *Physiol. Plant.* 107, 1–7.

Mullins, E.D., McCollum, T.G. and McDonald, R.E. (2000) Consequences on ethylene metabolism of inactivating the ethylene receptor sites in diseased non-climacteric fruit. *Postharvest Biol. Technol.* 19, 155–164.

Munk, W.J. (1971) Bud necrosis, a storage disease of tulips. II. Analysis of disease promoting storage conditions. *Neth. J. Pl. Path.* 77, 177–186.

Munk, W.J. (1972) Bud necrosis, a storage disease of tulips. III. The influence of ethylene and mites. *Neth. J. Pl. Path.* 78, 168–178.

Munk, W.J. (1973a) Bud necrosis, a storage disease of tulips. IV. The influence of ethylene concentration and storage temperature on bud development. *Neth. J. Pl. Path.* 79, 13–22.

Munk, W.J. (1973b) Flower bud blasting in tulips caused by ethylene. *Neth. J. Pl. Path.* 79, 41–53.

Munk, W.J. and Rooy, M. (1971) The influence of ethylene on the development of 5°C-precooled 'Apeldoorn' tulips during forcing. *HortScience* 6, 40–41.

Muñoz, M.T., Escribano, M.I. and Merodio, C. (2000) Ethylene and polyamine synthesis in cherimoya fruit under high CO_2 levels. Adaptive mechanism to chilling damage. *Plant Growth Regl. Abstracts* 26, 146.

Munro, H.A.U. (1959) The response of *Tenebroides mauritanicus* (L.) and *Tenebrio molitor* (L.) to methyl bromide at reduced pressures. *J. Sci. Food. Agric.* 7, 366–379.

Murata, T. and Ueda, Y. (1967) Studies on the storage of spinach. *J. Jpn. Soc. Hort. Sci.* 36, 449.

Murray, A.J., Hobson, G.E., Schuch, W. and Bird, C.R. (1993a) Isolation and characterization of a melon cDNA clone encoding phytoene synthase. *Plant Mol. Biol.* 44, 283–307.

Murray, A.J., Hobson, G.E., Schuch, W. and Bird, C.R. (1993b) Reduced ethylene synthesis in EFE antisense tomatoes has differential effects on fruit ripening processes. *Postharv. Biol. Technol.* 2, 301–303.

Murray, A.J., Hobson, G.E., Schuch, W. and Bird, C.R. (1993c) Expression of EFE antisense RNA in tomato causes retardation of leaf senescence and most fruit-ripening characteristics. In: Pech, J.C., Latché, A. and Balagué, C. (eds) *Cellular and Molecular Aspects of the Plant Hormone Ethylene.* Kluwer Academic, Dordrecht, pp. 327–328.

Murray, D.R. (1997) *Carbon Dioxide and Plant Responses.* John Wiley & Sons Inc., New York.

Murray, G. (1980) Trial shipment of chilled lambs – Dormavac container. Report to Australian Meat and Livestock Commission, 26 June.

Myer, B.S. and Anderson, D.B. (1952) *Plant Physiology.* D. Van Nostrand Company, Canada.

Myrvik, Q.N., Pearsall, N.N. and Weiser, R.S. (1974) *Fundamentals of Medical Bacteriology and Mycology.* Lea & Febiger, Philadelphia, Pennsylvania.

Nadeeau, J.A., Bui, A.Q., Zhang, X. and O'Neil, S.D. (1993) Interorgan regulation of post-pollination events in orchid flowers. In: Pech, J.C., Latché, A. and Balagué, C. (eds) *Cellular and Molecular Aspects of the Plant Hormone Ethylene.* Kluwer Academic, Dordrecht, pp. 304–309.

Nakagawa, N., Mori, H., Yamasaki, K. and Imaseki, H. (1991) Cloning of complementary DNA for auxin-induced 1-aminocyclopropane-1-carboxylate synthase and differential expression of the gene by auxin and wounding. *Plant Cell Physiol.* 32, 1153–1163.

Nakajima, N. and Imaseki, H. (1986) Purification and properties of 1-aminocyclopropane-1-carboxylate synthase of mesocarp of *Curcubita maxima* Duch. fruits. *Plant Cell Physiol.* 27, 969.

Nakajima, N., Mori, H., Yamazaki, K. and Imaseki, H. (1990) Molecular cloning and sequence of a complementary DNA encoding 1-aminocyclopropane-1-carboxylate synthase induced by tissue wounding. *Plant Cell Physiol.* 31, 1021–1029.

Nakatsuka, A., Shiomi, S., Kubo, Y. and Inaba, A. (1997) Expression and internal feedback regulation of ACC synthase and ACC oxidase genes in ripening tomato fruit. *Plant Cell Physiol.* 38, 1103–1110.

Nakatsuka, A., Murachi, S., Okunishi, H., Shiomi, S., Nakano, R., Kubo, Y. and Inaba, A. (1998) Differential expression and internal feedback regulation of 1-aminocyclopropane-1-carboxylate synthase, 1-aminocyclopropane-1-carboxylate oxidase, and ethylene receptor genes in tomato fruit during development and ripening. *Plant Physiol.* 118, 1295–1305.

Nakayama, M. and Oata, Y. (1983) Physiological function of ethylene and low pressure storage of apple. *Japan Agr. Res. Quart.* 17, 30–33.

Nakhoul, N.L., Davis, B.A., Romero, M.F. and Boron, W.F. (1998) Effect of expressing the water channel aquaporin-1 on the CO_2 permeability of *Xenopus* oocytes. *Amer. J. Physiol.* 274 (*Cell. Physiol.* 43), C543–C548.

Narayanan, E.S. and Bhambhani, H.J. (1956) Effect of reduced pressures on *Tribolium castaneum* Hbst. (Tenebrionidae, Coleoptera) and *Trogoderma granarium* Everts (Dermestidae, Coleoptera). *Indian J. Ent.* 18, 196–198.

Navarro, S. (1975) Progress Report for the Year 1974/1975. Agric. Res. Organization, Division of Stored Products, Tel-Aviv-Yafo, Israel, p. 13.

Navarro, S. and Calderon, M. (1979) Mode of action of low atmospheric pressures on *Ephestia cautella* (Wlk.) pupae. *Experientia* 35, 620–1.

Navarro, S. and Donahaye, E. (1972) An apparatus for studying the effects of controlled low pressures and compositions of atmospheric gases on insects. *J. Stored Prod. Res.* 8, 223–226.

Neljubow, D. (1901) Über die horizontale Nutation der Stengel von *Pisum sativum* und einiger anderen Pflanzen. *Beih. Bot. Centralbl.* 10, 128–129.

Neljubow, D. (1910) Geotropismus in der laboratoriumsluft. *Ber. Deutsch. Bot. Ges.* 29, 97–112.

Newmann, J.P., van Doorn, W. and Reid, M.S. (1989) Carbohydrate stress causes leaf blackening. *J. Intl. Protea Assn.* 18, 44–46.

Nichols, R. (1966) Ethylene production during senescence of flowers. *J. Hort. Sci.* 41, 279–290.

Nichols, R. (1968) The response of carnations to ethylene. *HortScience* 43, 335–349.

Nichols, R. (1973) Senescence of cut carnation flower: respiration and sugar status. *J. Hort. Sci.* 48, 111–121.

Nichols, R. (1975) Senescence and sugar status of the cut flower. *Acta Horticulturae* 41, 21–30.

Nichols, R. and Hammond, J.B. (1973) Storage of mushrooms in prepacks: the effects of changes in carbon dioxide and oxygen on quality. *J. Sci. Food Agric.* 24, 1371–1381.

Nichols, R. and Kofranek, A.M. (1982) Reversal of ethylene inhibition of tulip stem elongation by silver thiosulphate. *Sci. Hortic.* 17, 71–79.

Nick, P., Bergfeld, R., Schäfer, E. and Schopfer, P. (1990) Unilateral reorientation of microtubules at the outer epidermal wall during photo- and gravitropic curvature of maize coleoptiles and sunflower hypocotyls. *Planta* 181, 162–168.

Nick, P., Schäfer, E., Hertel, R. and Furuya, M. (1991) On the putative role of microtubules in gravitropism of maize coleoptiles. *Plant Cell Physiol.* 32, 873–880.

Niederl, S., Kirsch, T., Riederer, M. and Schreiber, L. (1998) Co-permeability of ^{3}H-labeled water and ^{14}C-labeled organic acids across isolated plant cuticles; investigating cuticular paths of diffusion and predicting cuticular transpiratation. *Plant Physiol.* 116, 117–123.

Nielsen, K.H. and Schjoerring, J.K. (1998) Regulation of apoplastic NH_4^+ concentration in leaves of oilseed rape. *Plant Physiol.* 118, 1361–1368.

Noble, P.S. (1974) *Introduction to Biophysical Plant Physiology.* W. H. Freeman, San Francisco, California.

Noble, P.S. (1991) *Physiochemical and Environmental Plant Physiology.* Academic Press, San Diego, California.

Noichinda, S. (1997) Respiration rate and ethylene production of lime fruit. *Jour. King Mongkut's Inst. Tech. North Bangkok* 7, 27–31.

Novikova, G.V., Moshkov, I.E., Smith, A.R. and Hall, M.A. (1993) Ethylene and phosphorylation of pea epicotyl proteins. In: Pech, J.C., Latché, A. and Balagué, C. (eds) *Cellular and Molecular Aspects of the Plant Hormone Ethylene.* Kluwer Academic, Dordrecht, pp. 371–372.

Nowak, J. (1981) Chemical pretreatment of snapdragon spikes to increase cut-flower longevity. *Scientia Hort.* 15, 255–262.

Nychas, G.-J.E., Drosinos, E.H. and Board, R.G. (1998) Chemical changes in meat. In: Davies, A. and Board, R. (eds) *The Microbiology of Meat and Poultry.* Blackie Academic & Professional, London, pp. 288–326.

Odawara, S., Watanabe, A. and Imaseki, H. (1977) Involvement of cellular membrane in regulation of ethylene production. *Plant Cell Physiol.* 18, 569–575.

Odyll, C.P., Alexander, J. and Lee, C.F. (1959) Influence of vessel-handling practices on formation of black spot in shrimp. *Com. Fisheries Rev.* 21, 10–16.

Oeller, P.W., Min-Wong, L., Taylor, L.P., Pike, D.A. and Theologis, A. (1991) Reversible inhibition of tomato fruit senescence by antisense RNA. *Science* 254, 437–439.

Oetiker, J.H. and Yang, S.F. (1995) The role of ethylene in fruit ripening. *Acta Horticulturae* 398, 167–178.

Ogata, K., Yamauchi, N. and Minamide, T. (1975) Physiological and chemical studies on ascorbic acid in fruits and vegetables. I. Changes in ascorbic acid content during maturation and storage of okra. *J. Jpn. Soc. Hort. Sci.* 44, 192–195.

Ohme-Takagi, M. and Shinshi, H. (1995) Ethylene-inducible DNA binding proteins that interact with an ethylene-responsive element. *Plant Cell* 7, 173–182.

Ohmura, T. and Howell, R.W. (1960) Inhibitory effect of water on oxygen consumption by plant materials. *Plant Physiol.* 35, 184–188.

Ohye, D.F. and Christian, J.H.B. (1967) Combined effects of temperature, pH and water activity on growth and toxin production of *Cl. botulinum* types A, B and E. In: Ingram, M. and Toberts, T.A. (eds) *Botulism 1966, Proceedings of the 5th International Symposium on Food Microbiology, Moscow.* Chapman and Hall, London, pp. 217–221.

O'Leary, M.H., Jaworski, R.J. and Hartman, F.L. (1979) ^{13}C nuclear magnetic resonance study of the CO_2 activation of ribulosebiphosphate carboxylase from *Rhodospirillium rubrum. Proc. Natl Acad. Sci. USA* 76, 673–675.

Oliver, B.G., Panepinto, J.C., Askew, D.S. and Rhodes, J.C. (2002) cAMP alteration of growth rate of *Aspergillus fumigatus* and *Aspergillus niger* is carbon-source dependent. *Microbiol.* 148, 2627–2633.

Olorunda, A.O., Meheriuk, M. and Looney, N.E. (1978) Some postharvest factors associated with the occurrence of chilling injury in banana. *J. Sci. Food Agric.* 29, 213–218.

Olsen, G.M. and Iversen, T.H. (1980) Growth and curvature in seedlings of *Pisum sativum* and an ageotropic mutant. *Physiol. Plant.* 50, 269–274.

Olsen, K.L. (1985) Controlled atmosphere storage of apples. *Wash. State Coop. Ext. Postharvest Pomology Newsl.* 3, 25.

Olson, D.C., White, J.A., Edeman, L., Harkins, R.N. and Kende, H. (1991) Differential expression of two genes for 1-aminocyclopropane-1-carboxylate synthase in tomato fruits. *Proc. Natl Acad. Sci. USA* 88, 5340–5344.

Onoda, A., Koizumi, T., Yamamoto, K., Furruya, T., Yamakawa, H. and Ogawa, K. (1989a) A study on variable low-pressure storage of cabbage and turnip. *Nippon Shokuhin Kogyo Gakkaishi* 36, 369–374.

Onoda, A., Ohmori, S., Shiina, T. and Hoshino, K. (1989b) Measurement of gas components in the low-pressure storage chambers. *J. Jpn. Soc. Food Sci. Technol.* 36, 403–408

Osborne, D.J. (1982) The ethylene regulation of cell growth in specific target tissues in plants. In: Wareing, P.F. (ed.) *Plant Growth Substances 1982.* Academic Press, New York, pp. 279–290.

Osborne, D.J. and Mullins, M.G. (1969) Auxin, ethylene and kinetin in a carrier-protein model system for the polar transport of auxins in petiole segments of *Phaseolus vulgaris. New Phytol.* 68, 977–991.

Oxley, T.A. and Wickenden, G. (1963) The effect of restricted air supply on some insects which infest grain. *Ann. Appl. Biol.* 51, 313–324.

Özisik, M.N. (1985) *Heat Transfer. A Basic Approach.* McGraw-Hill, New York.

Ozuna, R., Year, R., Ortega, K. and Tallman, G. (1985) Analysis of guard cell viability and action in senescing leaves of *Nicotiana glauca* (Graham), tree tobacco. *Plant Physiol.* 79, 7–10.

Pace, P.J., Krumbiegel, E.R., Angelotti, R. and Wisniewski, H.J. (1967) Demonstration and isolation of *Clostridium botulinum* types from whitefish chubs collected at fish smoking plants of the Milwaukee area. *Appl. Microbiology* 15, 877–884.

Pal, R.K. and Buescher, R.W. (1993) Respiration and ethylene evolution of certain fruits and vegetables in response to CO_2 in controlled atmosphere storage. *J. Food Science and Technology Mysore* 30, 29–32.

Pallaghy, C.K. and Raschke, K. (1972) No stomatal response to ethylene. *Plant Physiol.* 49, 275–276.

Pallaghy, C.K., Lorimer, F., Bierer, D. and Raschke, K. (1970) No stomatal response to ethylene and CO_2 analogues in leaf sections of *Zea mays*. *Plant Research '70.* MSU/AEC Plant Research Laboratory, Michigan State University, East Lansing, Michigan, pp. 53–54.

Pallas, J.E. and Kays, S.J. (1982) Inhibition of photosynthesis by ethylene – a stomatal effect. *Plant Physiol.* 70, 598–601.

Palmer, J.H. (1964) Comparative study of the effects of applied indoleacetic acid and horizontal orientation of the primary shoot upon internode extension and petiole epinasty in *Helianthus annuus*, and the modifying influence of gibberellic acid. *Planta* 61, 283–297.

Palmer, J.H. (1972) Roles of ethylene and indolyl-yl-acetic acid in petiole epinasty in *Helianthus annuus* and the modifying influence of gibberellic acid. *J. Expt. Bot.* 23, 733–743.

Palmer, J.H. (1973) Ethylene as a cause of transient petiole epinasty in *Helianthus annuus* during clinostat experiments. *Physiol. Plant.* 28, 188–193.

Palmer, J.H. (1976) Failure of ethylene to change the distribution of indoleacetic acid in the petiole of *Coleus blumei* × *frederici* during epinasty. *Plant Physiol.* 58, 513–515.

Palmer, J.H. (1985) Epinasty, hyponasty, and related topics. In: Pharis, R.P. and Reid, D.M. (eds) *Encyclopedia of Plant Physiology.* Springer, Berlin, pp. 139–168.

Palmer, J.H. and Phillips, I.D.J. (1963) The effect of terminal bud, indoleacetic acid, and nitrogen supply on the growth and orientation of the petiole of *Helianthus annuus. Physiol. Plant. 16*, 572–584.

Palmer, J.K. (1971) The banana. In: Hulme, A.C. (ed.) *The Biochemistry of Fruits and Their Products*, vol. 2. Academic Press, New York, pp. 65–105.

Palmer, O.H. and Halsall, D.M. (1969) Effect of transverse gravity stimulation, gibberellin and indoleacetic acid upon polar transport of IAA-C^{14} in the stem of *Helianthus annuus. Physiol. Plant.* 22, 59–67.

Pantastico, E.B. (1968) Postharvest physiology of fruits. I. Chilling injury. *Philipp. Agr.* 11, 697–730.

Pantastico, E.B. (1975) Structure of fruits and vegetables. In: Pantastico, E.B. (ed.) *Postharvest Physiology, Handling and Utilization of Tropical and Subtropical Fruits and Vegetables.* AVI, Westport, Connecticut, pp. 1–24.

Pantastico, E.B., Grierson, W. and Soule, J. (1967) Chilling injury in tropical fruits: I. Bananas. *J. Amer. Soc. Hort. Sci. Proceedings of the Trop. Reg.* 11, 82–91.

Pantastico, E.B., Soule, E. and Grierson, W. (1968) Chilling injury in tropical fruits: II. Limes and grapefruit. *J. Amer. Soc. Hort. Sci. Proc. Trop. Reg.* 12, 171–183.

Pantastico, E.B., Chattopadhyay, T.K. and Subramanyam, H. (1975a) Storage and commercial storage operations. In: Pantastico, E.B. (ed.) *Postharvest Physiology, Handling and Utilization of Tropical and Subtropical Fruits and Vegetables.* AVI, Westport, Connecticut, pp. 314–318.

Pantastico, E.B., Mattoo, A.K., Muata, T. and Ogata, K. (1975b) Chilling injury. In: Pantastico, E.B. (ed.) *Postharvest Physiology, Handling and Utilization of Tropical and Subtropical Fruits and Vegetables.* AVI, Westport, Connecticut, pp. 339–362.

Pantastico, E.B., DeFossard, R.A., Safran, H., Venter, F., Akamine, E.K. and Subramanyam, H. (1975c) Physiological disorders other than chilling injury. In: Pantastico, E.B. (ed.) *Postharvest Physiology, Handling and Utilization of Tropical and Subtropical Fruits and Vegetables.* AVI, Westport, Connecticut, pp. 363–392.

Pappenheimer, A.M. and Hottle, G.A. (1940) Effect of certain purines and CO_2 on growth of strain of group A haemolytic *Streptococcus. Proceedings of the Soc. Expt. Biol. Med.* 44, 645.

Park, I.S. and Hausinger, R.P. (1995) Requirement of carbon dioxide for *in vitro* assembly of the urease nickel metallocenter. *Science* 267, 1156–1158.

Parrish, F.C. (1986) Effect of dry and wet aging on the palatability of loin and rib steaks. *44th*

NAMP Annual Convention, 6 November, 1986, Rancho Mirage, California.

Parrish, F.C., Boles, J.A., Rust, R.E. and Olson, D.G. (1991) Dry and wet aging effects on palatability attributes of beef loin and rib steaks from three quality grades. *J. Food Sci.* 56, 601–603.

Parsons, C.S. and Spalding, D.H. (1971) Influence of controlled atmospheres, storage temperatures, and state of ripeness on the development of four fungal rots in tomatoes. *HortScience* 6, 294.

Parsons, C.S., Gates, J.E. and Spalding, D.H. (1964) Quality of some fruits and vegetables after holding in nitrogen atmospheres. *J. Amer. Soc. Hort. Sci.* 84, 549–556.

Parsons, C.S., Asen, S. and Stuart, N.W. (1967) Controlled atmosphere storage of daffodil flowers. *J. Amer. Soc. Hort. Sci. Proc.* 90, 506–514.

Parsons, C.S., Anderson, R.E. and Penny, R.W. (1970) Storage of mature green tomatoes in controlled atmospheres. *J. Amer. Soc. Hort. Sci.* 95, 791–793.

Parton, W.J., Morgan, J.A., Aletenhofen, J.M. and Harper, L.A. (1988) Ammonia volatization from spring wheat plants. *Agronomy Journal* 80, 419–425.

Parups, E.V. (1973) Control of ethylene-induced responses in plants by a substituted benzothiadiazole. *Physiol. Plant.* 29, 365–370.

Parups, E.V. (1977) Control of ethylene-induced permeability of plant cells by substituted benzothiadiazole. *Physiol. Plant.* 39, 290–294.

Paster, N. (1990) Modified atmospheres for preventing molds and mucotoxin production in stored grain. In: Calderon, M. and Barkai-Golan, R. (eds) *Food Preservation by Modified Atmospheres.* CRC Press, Boca Raton, Florida, pp. 39–55.

Patterson, B.D., Hatfield, S.G.S. and Knee, M. (1974) Residual effects of controlled atmosphere storage on the production of volatile compounds by two apple varieties. *J. Sci. Food Agric.* 25, 843.

Patterson, D.R., Lyle, E.W. and Earhart, D.R. (1975) Crown gall and ethylene content of rose plants. *HortScience* 10, 381.

Patterson, D.S. (1975a) Hypobaric storage: a means of extending chrysanthemum cut flower longevity. *Washington Florists' Association 'Washline'* 3, 4–5.

Patterson, D.S. (1975b) The effects of preservative solution and hypobaric storage on chrysanthemum flower longevity. MSc thesis, Washington State University, Pullman, Washington.

Patterson, M.E. and Melsted, J.L. (1977) Sweet cherry handling and storage alternatives. In: Dewey, D.H. (ed.) *Controlled Atmospheres for the Storage and Transport of Perishable Agricultural Commodities. Proceedings of the 2nd National Controlled Atmosphere Research Conference*, Michigan State University, East Lansing, Michigan, p. 29.

Patterson, M.E. and Melstad, J.L. (1978) Improvement of prune quality and condition by hypobaric storage. *HortScience* 13, 351.

Paul, R.C., Watt, D.G., Haines, N.S. and Havard, D.G. (1970) Design features and concrete requirements for a vacuum building at Pickering nuclear generating station. *The American Concrete Institute Seminar on Concrete for Nuclear Reactors*, West Berlin, October, 1970.

Paull, R.E. and Chen, N.J. (2000) Heat treatment and fruit ripening. *Postharvest Biol. Technol.* 21, 37.

Paull, R.E., Criley, R.A., Parvin, P.E. and Goo, T. (1980) Leaf blackening in cut *Protea exima*: the importance of water relations. *Acta Hort.* 113, 159–166.

Paxton, J.D. (1981) Phytoalexins – a working redefinition. *Phytopathol. Z.* 101, 106–109.

Payton, S., Fray, R.G., Brown, S. and Grierson, D. (1996) Ethylene receptor expression is regulated during fruit ripening, flower senescence, and abscission. *Plant Mol. Biol.* 31, 1227–1231.

Peacock, B.C. (1972) Role of ethylene in the initiation of fruit ripening. *Queensl. J. Agric. Anim. Sci.* 29, 137–145.

Pearman, G.C. and Jay, E.G. (1970) The effect of relative humidity on the toxicity of carbon dioxide to *Trilobium casteneum* in peanuts. *J. Georgia Ent. Soc.* 5, 61–64.

Pearson, J., Clough, E.C.M., Woodall, J., Havill, D.C. and Zhang, X.H. (1998) Ammonia emissions to the atmosphere from leaves of wild plants and *Hordeum vulgare* treated with methionine sulphoximine. *New Phytol.* 138, 37–48.

Pech, J.C., Guis, M., Botondi, R., Ayub, R., Bouzayen, M., Leliévre, J.M., El-Yahyaoui, F. and Latché, A. (2000) Ethylene-dependent and ethylene-independent pathways in a climacteric fruit, the melon. *Plant Growth Regl. Abstracts* 26, 146.

Peiser, G.D., Wang, T.-T., Hoffman, N.E., Yang, S.F. and Walsh, C.T. (1984) Formation of cyanide from carbon 1 of 1-aminocyclopropane-1-carboxylic acid during its conversion to ethylene. *Proc. Natl Acad. Sci. USA* 81, 3059–3063.

Pelter, G.Q. (1975) The effects of subatmospheric pressure upon the breakdown, color, respiration, and ethylene evolution of cranberry fruit. MSc thesis, Washington State University, Pullman, Washington.

Peñarrubia, L., Aguilar, M., Margossian, L. and Fischer, R.L. (1993) Effect of E8 protein on ethylene biosynthesis during tomato fruit ripening. In: Pech, J.C., Latché, A. and Balagué, C. (eds) *Cellular and Molecular Aspects of the Plant Hormone Ethylene.* Kluwer Academic, Dordrecht, pp. 100–105.

Penmetsa, R.V. and Cook, D.R. (1997) A legume ethylene-insensitive mutant hyperinfected by its rhizobial symbiont. *Science* 275, 527–530.

Peoples, M.B., Beilharz, V.C., Waters, S.P., Simpson, R.J. and Dalling, D.J. (1980) Nitrogen redistribution during grain growth in wheat (*Triticum aestivum* L.). II. Chloroplast senescence and the degradation of ribulose-1-5-biphosphate carboxylase. *Planta* 149, 241–251.

Pérez-Rodríguez, J. and Valpuesta, V. (1996) Expression of glutamine synthetase genes during natural senescence of tomato leaves. *Physiol. Plant.* 97, 576–582.

Perez-Trejo, M.S., Janes, H.W. and Frenkel, C. (1981) Mobilization of respiratory metabolism in potato tubers by carbon dioxide. *Plant Physiol.* 67, 514–517.

Perkins-Veazie, P., Nonnecke, G.R. and Gladon, R.J. (1982) Ethylene involvement in raspberry ripening. In: Pech, J.C., Latché, A. and Balagué, C. (eds) *Cellular and Molecular Aspects of the Plant Hormone Ethylene.* Kluwer Academic, Dordrecht, pp. 144–145.

Perkins-Veazie, P., Clark, J.R., Huber, D.J. and Baldwin, E.A. (2000) Ripening physiology in 'Navaho' thornless blackberries: color, respiration, ethylene production, softening and compositional changes. *J. Amer. Soc. Hort. Sci.* 125, 357–363.

Perry, R.H. and Given, D. (1984) *Perry's Chemical Engineers' Handbook*, 6th edn. McGraw-Hill, New York.

Pesis, E., Marinansky, R., Zauberman, G. and Fuchs, Y. (1993) Reduction of chilling injury symptoms of stored avocado fruit by prestorage treatment with high nitrogen atmosphere. *Acta Horticulturae* 343, 251–255.

Pfleiderer, H. (1933) Kritische unterrsuchungen zur methodic der transpiration bestimmung an abgesechnittenen sprossen. *Z. Bot.* 26, 305–327.

Phan, C.-T. (1963) Production d'éthylène par les fleurs. *Compt. Rend. Acad. Agric. (France)* 49, 53–59.

Phan, C.-T. (1965) Une technique pour le dosage de l'éthylène émis en très petites quantités. *Phytochemistry* 4, 353–355.

Phan, C.-T. (1970) Conversion of various substrates to ethylene by flowers. *Physiol. Plant.* 23, 981–984.

Philosoph-Hadas, S., Meir, S. and Aharoni, N. (1985a) Autoinhibition of ethylene production in tobacco leaf discs: enhancement of 1-aminocyclopropane-carboxylic acid conjugation. *Physiol. Plant.* 63, 431–437.

Philosoph-Hadas, S., Meir, S. and Aharoni, N. (1985b) Carbohydrates stimulate ethylene production in tobacco leaves. II. Sites of stimulation in the ethylene biosynthesis pathway. *Plant Physiol.* 78, 139–143.

Philosoph-Hadas, S., Aharoni, N. and Yang, S.F. (1986) Carbon dioxide enhances the development of the ethylene-forming enzyme in tobacco leaf discs. *Plant Physiol.* 82, 925–929.

Philosoph-Hadas, S., Meir, S. and Aharoni, N. (1994) Role of ethylene in the senescence of water cress leaves. *Physiol. Plant.* 90, 553–559.

Philosoph-Hadas, S., Meir, S., Rosenberger, I. and Halevy, A.H. (1996) Regulation of the gravitropic response and ethylene biosynthesis in gravistimulated snapdragon spikes by calcium chelators and ethylene inhibitors. *Plant Physiol.* 110, 301–310.

Philosoph-Hadas, S., Friedman, H., Berkovitz-Simantov, R., Rosenberger, I., Woltering, E.J., Halevy, A.H. and Meir, S. (2000) Involvement of ethylene biosynthesis and action in regulation of the gravitropic response of cut flowers. *Plant Growth Regl. Abstracts* 26, 146.

Pickard, B.G. (1985) Early events in geotropism of seedling shoots. *Annu. Rev. Plant Physiol.* 36, 55–75.

Pickett, F.B., Wilson, A.K. and Estelle, M. (1990) The *aux1* mutation of *Arabidopsis* confers both auxin and ethylene resistance. *Plant Physiol.* 94, 1462–1466.

Picton, S. and Grierson, D. (1988) Inhibition of expression of tomato-ripening genes at high temperature. *Plant Cell Environ.* 11, 265–272.

Picton, S., Barton, S.L., Bouzayen, M., Hamilton, A.J. and Grierson, D. (1993) Altered fruit ripening and leaf senescence in tomatoes expressing an antisense ethylene-forming enzyme transgene. *Plant Journal* 3, 469–481.

Pieniazek, S.A. (1942) External factors affecting water loss from apples in cold storage. *Refrig. Eng.* 44, 171–173.

Pierik, R., Verkerke, W., Voesenek, R.L.A.C.J., Blom, K.C.W.P.M. and Visser, E.J.W. (2000) Thick root syndrome in cucumber (*Cucumis sativus* L.), a description of the phenomenon and an investigation of the role of ethylene. *Ann. Bot.* 84, 755–762.

Pierson, M.D., Collins-Thompson, D.L. and Ordal, Z.J. (1970) Microbiological, sensory and pigment changes of aerobically and anaerobically packaged beef. *Food Technol.* 24, 129–133.

Plata, M.C., Sanchez de Medina, L., Martinez-Cayuela, M., Faus, M.J. and Gil, A. (1987) Changes in texture, protein content and polyphenoloxidase and peroxidase activities in cherimoya induced by ripening in hypobaric atmospheres or in presence of sulfite. *Revista de Agroquimica y Tecnologia de Alimentos* 27, 215–224.

Platenius, H. (1942) Effect of temperature on the respiration rate and the respiratory quotient of some vegetables. *Plant Physiol.* 17, 179–197.

Pleasants, A.L. (1930) The effect of nitrate fertilizers on stomatal behavior. *J. Elisha Mitchell Sci. Soc.* 46, 95–116.

Plich, H. (1987) Regulating factors of ethylene production in apple fruits. *Fruit Sci. Rpts.* 14, 57–63.

Poenicke, E.F., Kays, S.J., Smittle, D.A. and Williamson, R.E. (1977) Ethylene in relation to postharvest deterioration in processing cucumbers. *J. Amer. Soc. Hort. Sci.* 102, 303–306.

Poland, G.L. and Wilson, R.M. (1933) Banana chilling studies. Effect of low temperatures on appearance and edibility. United Fruit Co., Research Department Bulletin No. 46.

Poneleit, L.S. and Dilley, D.R. (1993) Carbon dioxide activation of 1-amino-cyclopropane-1-carboxylate (ACC) oxidase in ethylene biosynthesis. *Postharvest Biol. and Technol.* 3, 191–199.

Poole, R.T., Conover, C.A. and Staby, G.L. (1985) Low-oxygen storage allows foliage plants two weeks of rest. *Greenhouse Mgr.* 3, 33–35.

Porat, R., Halevy, A.H., Serek, M. and Borochov, A. (1995a) An increase in ethylene sensitivity following pollination is the initial event triggering an increase in ethylene production and enhanced senescence of *Phaleonopsis* orchid flowers. *Physiol. Plant.* 93, 778–784.

Porat, R., Shlomo, E., Serek, M., Sisler, E.C. and Borochov, A. (1995b) 1-Methylcyclopropene inhibits ethylene action in cut phlox flowers. *Postharvest Biol. Technol.* 6, 313–319.

Porges, F. (1982) *Handbook of Heating, Ventilation and Air Conditioning*, 8th edn. Butterworths, London.

Porritt, S.W., Lister, P.D. and Meheriuk, M. (1975) Postharvest factors associated with the occurrence of breakdown in Spartan apple. *Can. J. Plant Sci.* 55, 743–747.

Porter, M.A. and Grodzinski, B. (1984) Acclimation to high CO_2 in bean carbonic anhydrase and ribulose biphosphate carboxylase. *Plant Physiol.* 74, 413–416.

Portis, A.R. (1992) Regulation of ribulose biphosphate carboxylase/oxygenase activity. *Ann. Rev. Plant Physiol. Plant Mol. Biol.* 43, 415–437.

Possingham, J.V., Chambers, T.C., Radler, F. and Grncarevic, M. (1967) Cuticular transpiration and wax structure and composition of leaves and fruit of *Vitis vinifera*. *Austr. J. Biol. Sci.* 20, 1149–1153.

Potter, N.A. and Griffiths, D.G. (1947) The effect of temperature and gas mixtures on the production of volatile substances by apples during storage. *J. Pomol. Hort. Sci.* 23, 171–177.

Potter, N.N. (1973) *Food Science.* AVI, Westport, Connecticut.

Poulssen, K.P., Guleria, S.S. and Møller, D. (1982) Experiments with hypobaric storage of fruits and vegetables. *21st Int. Hort. Cong.*, Hamburg, Germany. 1234a (abstract)

Prasad, T.K. and Cline, M.G. (1985) Gravistimulation direction, ethylene production and shoot elongation in the release of apical dominance in *Pharbitis nil*. *J. Expt. Bot.* 36, 1969–1975.

Prasad, T.K. and Cline, M.G. (1987) Shoot inversion inhibition of stem elongation. A posssible role for ethylene-induced glycoprotein and lignin. *Plant Physiol.* 85, 104–108.

Prasad, T.K., Hosokawa, Z. and Cline, M.G. (1989) Shoot inversion-induced ethylene production: a general phenomenon? *J. Plant Growth Regul.* 8, 71–77.

Pratt, H.K. (1971) Melons. In: Hulme, A.C. (ed.) *The Biochemistry of Fruits and Their Products*, vol. 2. Academic Press, London, pp. 207–232.

Pratt, H.K. and Goeschl, J.D. (1968) The role of ethylene in fruit ripening. In: Wightman, F. and Setterfield, G. (eds) *Biochemistry and Physiology of Plant Growth Substances.* Runge Press, Ottawa, pp. 1295–1302.

Pratt, H.K. and Reid, M.S. (1974) Chinese gooseberry: seasonal patterns in fruit growth and maturation, ripening, respiration and the role of ethylene. *J. Sci. Food Agr.* 25, 747–757.

Pratt, H.K. and Workman, M. (1962) Studies on the physiology of tomato fruits. III. The effect of ethylene on respiration and ripening behavior of fruits stored at 20°C after harvest. *Proceedings of the J. Amer. Soc. Hort. Sci.* 81, 467–478.

Preger, T. and Gepstein, S. (1984) Carbon-dioxide-independent and -dependent components of light inhibition of the conversion of 1-aminocyclopropane-1-carboxylic acid to ethylene in oat leaves. *Physiol. Plant.* 60, 187–191.

Pressman, E., Huberman, M., Aloni, B. and Jaffe, M.J. (1984) I. The effect of mechanical perturbation and the involvement of ethylene in petiole pithiness in celery. *Plant Cell Physiol.* 25, 891–897.

Price, N.R. and Walter, C.M. (1987) A comparison of some effects of phosphine, hydrogen cyanide and anoxia in the lesser grain borer, *Rhyaopertha dominica* (F.) (Coleoptera: Bostrychidae). *Comp. Biochem. Physiol.* 33, 86C.

Prince, T.A. (1989) Modified atmosphere packaging of horticultural commodities. In: Brody, A.L. (ed.) *Controlled/Modified Atmosphere/Vacuum Packing of Foods.* Food & Nutrition Press, Trumbull, Connecticut, pp. 67–100.

Prince, T.A., Cunningham, M.S. and Peary, J.S. (1987) Floral and foliar quality of potted Easter lilies under STS or phenidone application, refrigerated storage, and simulated shipment. *J. Amer. Soc. Hort. Sci.* 112, 469–473.

Prosser, C.L., Bishop, D.W., Brown, F.A., Jahn, T.L. and Wulff, V.J. (1952) *Comparative Animal Physiology.* W. B. Sauders & Co., Philadelphia, Pennsylvania.

Prusky, D., Keen, N.T., Sims, J.J. and Midland, S.L. (1982) Possible involvement of an antifungal diene in the latency of *Colletotrichum gloeosporioides* on unripe avocado fruits. *Phytopathology* 72, 1578.

Prusky, D., Keen, N.T. and Eaks, I. (1983) Further evidence for the involvement of a preformed antifungal compound in the latency of *Colletotrichum gloeosporioides* on unripe avocado fruits. *Physiol. Plant Pathol.* 22, 189.

Prusky, D., Kobiler, I., Jacoby, B., Sims, J.J. and Midland, S.L. (1985) Inhibitors of avocado lipoxygenase: their possible relationship with the latency of *Colletotrichum gloeosporioides. Physiol. Plant Pathol.* 29, 173.

Prusky, D., Leikin-Frenkel, A. and Madi, L. (2000) Ethylene enhances the antifungal diene content in idioblasts from avocado mesocarp. *Plant Growth Regl. Abstracts* 26, 147.

Purvis, A.C. (1979) Differential inhibition of ethylene by $AgNO_3$ and α, α[1] dipyridyl in the degreening of calamondin fruit. *Plant Physiol.* 63, S-91.

Purvis, A.C. (1981) Sequence of chloroplast degreening in calamondin fruit as influenced by ethylene and $AgNO_3$. *Plant Physiol.* 66, 624–627.

Purvis, A.C. and Barmore, C.R. (1981) Involvement of ethylene in chlorophyll degradation in the peel of citrus fruits. *Plant Physiol.* 68, 854–856.

Pushka, H.M.C. and Srivastava, H.C. (1963) Studies on the effect of protective coating on the refrigerated storage of bell peppers. *Food Sci. (Mysore)* 12, 192–196.

Quazi, M.H. and Freebairn, H.T. (1970) The influence of ethylene, oxygen and carbon dioxide on the ripening of bananas. *Bot. Gaz.* 131, 5–14.

Radin, J.W. and Loomis, R.S. (1969) Ethylene and carbon dioxide in the growth and development of cultured radish roots. *Plant Physiol.* 44, 1584–1589.

Radin, J.W., Parker, L.I. and Guinn, G. (1982) Water relations of cotton plants under nitrogen deficiency. V. Environmental control of abscisic acid accumulation and stomatal sensitivity to abscisic acid. *Plant Physiology* 79, 1066–1077.

Ragg, H., Kuhn, D.N. and Hahlbrock, K. (1981) Coordinated regulation of 4-coumarate: CoA ligase and phenylalanine ammonia lyase mRNAs in cultured plant cells. *J. Biol. Chem.* 256, 10061–10065.

Rahman, A., Tsurumi, S., Amakawa, T., Soga, K., Hoson, T., Goto, N. and Kamisaka, S. (2000) Involvement of ethylene and gibberellin signalings in chromosaponin I-induced cell division and cell elongation in the roots of *Arabidopsis* seedlings. *Plant Cell Physiol.* 41, 1–9.

Rahman, A., Amakawa, T., Goto, N. and Tsurumi, S. (2001) Auxin is a positive regulator for ethylene-mediated response in the growth of *Arabidopsis* roots. *Plant Cell Physiol.* 42, 301–307.

Rahn, O. (1941) Notes on the CO_2 requirement of bacteria. *Growth* 5, 197–199.

Raison, J.K. and Orr, G.O. (1986) Phase transitions in liposomes formed from the polar lipids of mitochondria from chilling-sensitive plants. *Plant Physiol.* 81, 807–811.

Rakitin, V.Y. (1988) Ethylene activation of cyanide-resistant respiration in lemon fruits. *Soviet Plant Physiol.* 34, 559–566.

Ramsey, G.B. and Butler, L.F. (1927) Injury to onions and fruits caused by exposure to ammonia. *J. Agr. Res.* 37, 339–348.

Ramsey, G.B., Friedman, B.A. and Smith, M.A. (1967) *Market Diseases of Beets, Chicory, Endive, Escarole, Globe Artichokes, Lettuce, Rhubarb, Spinach and Sweet Potatoes.* USDA Agricultural Handbook No. 155.

Ranawaka, R.A.K. and Wijeratnam, R.S.W. (2001) Ethylene-regulated ripening under tropical conditions and the development of crown rot in banana. *Plant Growth Regl. Abstracts* 27, 346.

Ransom, S.L. (1953) Zymasis and acid metabolism in higher plants. *Nature (London)* 172, 252.

Ransom, S.L., Walker, D.A. and Clarke, I.D. (1957) The inhibition of succinic oxidase by high CO_2 concentrations. *Biochem J.* 66, 57.

Ransom, S.L., Walker, D.A. and Clarke, I.D. (1960) Effects of carbon dioxide on mitochondrial enzymes from *Ricinus. Biochem. J.* 76, 221–225.

Rao, D.N., Nair, K.K.S. and Sakkhare, P.Z. (1998) Meat microbiology and spoilage in tropical countries. In: Davies, A. and Board, R. (eds) *The Microbiology of Meat and Poultry.* Blackie Academic & Professional, London, pp. 220–265.

Raschke, K. (1960) Heat transfer between the plant and the environment. *Annu. Rev. Plant Physiol.* 11, 111–126.

Raschke, K. (1972) Saturation kinetics of the velocity of stomatal closing in response to CO_2. *Plant Physiol.* 49, 229–234.

Raschke, K. (1975) Stomatal action. *Annu. Rev. Plant Physiol.* 26, 309–340.

Raschke, K. and Pierce, M. (1973) Stomatal regulation of gas exchange between the plant and environment. *Plant Research '73.* MSU/AEC Plant Research Laboratory, Michigan State University, East Lansing, Michigan, pp. 25–28.

Raschke, K., Dunn, W. and Lorimer, F. (1970) Saturation kinetics of the velocity of stomatal closing in response to CO_2. *Plant Research '70.* MSU/AEC Plant Research Laboratory, Michigan State University, East Lansing, Michigan, pp. 41–43.

Raskin, I. (1992) Ethylene in vegetative growth. In: Mattoo, A.K. and Suttle, J.C. (eds) *The Plant Hormone Ethylene.* CRC Press, Boca Raton, Florida, pp. 183–192.

Rauser, W.E. and Horton, R.F. (1975) Rapid effects of indoleacetic acid and ethylene on the growth of intact pea roots. *Plant Physiol.* 55, 443–447.

Raven, J.A. and Smith, F.A. (1976) Cytoplasmic pH regulation and electrogenic H^+ extrusion. *Curr. Adv. Plant Sci.* 8, 649–660.

Raven, J.A., Wollenweber, B. and Handley, L.L. (1992) Ammonia and ammonium fluxes between photolithotrophs and the environment in relation to the global nitrogen cycle. *New Phytol.* 121, 5–18.

Raz, V. and Fluhr, R. (1992) Calcium requirement for ethylene-dependent responses. *Plant Cell* 4, 1123–1130.

Raz, V. and Fluhr, R. (1993) Ethylene signal is transduced via protein phosphorylation events in plants. *Plant Cell* 5, 523–530.

Rea, P.A. and Poole, R.J. (1993) Vacuolar H^+-translocating pyrophosphatases. *Annu. Rev. Plant Physiol.* 44, 157–180.

Reeder, W.F. and Hatton, T.T. (1971) Storage of Lula avocados in controlled atmosphere – 1970 test. *The Citrus Industry* February, 5–6.

Regeimbal, L.O., Vacha, G.A. and Harvey, R.B. (1927) The effect of ethylene on the respiration of bananas during ripening. *Plant Physiol.* 2, 357–359.

Reid, J.B. (1986) Internode length in *Pisum.* Three further loci, *lh, ls* and *lk. Ann. Bot.* 57, 577–592.

Reid, M.S. and Pratt, H.K. (1970) Ethylene and the respiration climacteric. *Nature* 226, 976–977.

Reid, M.S. and Pratt, H.K. (1972) Effects of ethylene on potato tuber respiration. *Plant Physiol.* 49, 252–255.

Reid, M.S. and Wu, M. (1992) Ethylene and flower senescence. *Plant Growth Regl.* 11, 37–43.

Reid, M.S., Rhodes, M.J.C. and Hulme, A.C. (1973) Changes in ethylene and carbon dioxide during ripening of apples. *J. Sci. Food Agric.* 24, 971–979.

Reid, M.S., Mor, Y. and Kofranek, A.M. (1981) Epinasty in poinsettias. The role of auxin and ethylene. *Plant Physiol.* 67, 950–952.

Rennel, L. and Thompson, A.K. (1994) Carambola in controlled atmosphere. *Inter-American Institute for Cooperation on Agriculture, Tropical Fruits Newsletter* 11, 7.

Restaino, L. (1978) Hypobaric storage of lamb carcasses. Progress report for Armour Food Research, Scottsdale, Arizona, 6 March.

Restaino, L., Grauman, G.S., McCall, W.A. and Hill, W.A. (1976) Hypobaric storage of fresh pork. Progress Report for Armour Research Center, Scottsdale, Arizona, 14 May.

Rhoads, A., Trocano, J. and Brennan, E. (1973) Ethylene gas as a cause of injury to Easter lilies. *Plant Disease Reporter* 57, 1023–1024.

Rhodes, M.J.C. (1970) The climacteric and ripening of fruits. In: Hulme, A.C. (ed.) *The Biochemistry of Fruits and Their Products*, vol. 1. Academic Press, New York, pp. 521–533.

Ribereau-Gayon, P. (1972) *Plant Phenolics.* Oliver and Boyd, Edinburgh, UK.

Richter, O. (1906) Über den Einfluss verunreinigter Luft auf Heliotropismus and geotripsmus. *Sitz. Ber. K.u. K. Akad. Wein* 115, I. 1.

Richter, O. (1910) Die horizontale Nutation. *Ebenda* 119, I. 1.

Ridge, I. (1973) The control of cell shape and rate of cell expansion by ethylene and cell wall extensibility in etiolated peas. *Acta Bot. Neerl.* 22, 144–158.

Ridge, I. and Osborne, D. (1973) Hydroxyproline and peroxidases in cell walls of *Pisum sativum*: regulation by ethylene. *J. Expt. Bot.* 21, 843–856.

Riederer, M. and Schönherr, J. (1990) Effects of surfactants on water permeability of isolated cuticles and on the composition of their cuticular waxes. *Pestic. Sci.* 29, 85–94.

Rikin, A., Chalutz, E. and Anderson, J.D. (1984) Rhythmicity in ethylene production in cotton seedlings. *Plant Physiol.* 75, 493–495.

Riov, J. and Yang, S.F. (1982a) Autoinhibition of ethylene production in citrus peel discs. Suppression of 1-aminocyclopropane-1-carboxylic acid synthesis. *Plant Physiol.* 69, 687–690.

Riov, J. and Yang, S.F. (1982b) Effects of exogenous ethylene on ethylene production in citrus leaf tissue. *Plant Physiol.* 70, 136–141.

Risse, L.A. and Hatton, T.T. (1982) The sensitivity of watermelons to ethylene during storage. *HortScience* 17, 946–948.

Ritchie, R.J. (1987) The permeability of ammonia, methylamine and ethylamine in the Charophyte *Chara colallina* (*C. australis*). *J. Expt. Bot.* 38, 67–76.

Roberts, E.A., Wills, R.B.H. and Scott, K.J. (1965) The effects of change in concentration of carbon dioxide and oxygen on storage behavior of 'Jonathan' apples. *Austr. J. Expt. Agric. Anim. Husb.* 5, 161.

Roberts, I.N., Lloyd, C.W. and Roberts, K. (1985) Ethylene-induced microtubule reorientations: mediation by helical arrays. *Planta* 164, 439–447.

Roberts, J.K.M., Ray, P.M., Wade-Jardetsky, N. and Jardetsky, O. (1980) Estimation of cytoplasmic and vacuolar pH in higher plant cells by ^{31}P NMR. *Nature* 283, 870–872.

Roberts, J.K.M., Wemmer, D., Ray, P.M. and Jardetsky, O. (1982) Regulation of cytoplasmic and vacuolar pH in maize root tips under different experimental conditions. *Plant Physiol.* 69, 1344–1347.

Roberts, T.A. and Hobbs, G. (1968) Low-temperature growth characteristics of *Clostridia. J. Appl. Microbiol.* 31, 75–88.

Roberts, W.K. and Selitrennikoff, C.P. (1988) Plant and bacterial chitanases differ in antifungal activity. *J. Gen. Microbiol.* 134, 169–176.

Roberts, W.K. and Selitrennikoff, C.P. (1990) Zeamatin, an antifungal protein from maize with membrane-permeabilizing activity. *J. Gen. Microbiol.* 136, 1771–1778.

Robinson, J.E., Brown, K.M. and Burton, W.G. (1975) Storage characteristics of some vegetables and soft fruits. *Ann. Appl. Biol.* 81, 339–408.

Robinson, R.L., Anderson, J.M., Barick, M.W., Bufkin, B.A. and Ross, C.H. (1987) Phase behavior of coal fluids. Data for correlation development. Report DOE/PC/60039-12 Order No. DE87007781. *Energy Res. Abstract* 12 (13).

Robinson, S.A., Stewart, G.R. and Phillips, R. (1992) Regultation of glutamate dehydrogenase activity in relation to carbon limitation and protein catabolism in carrot cell suspension cultures. *Plant Physiol.* 98, 1190–1195.

Roblin, G. (1976) Inhibition of the pulvinar geotropic movements of *Mimosa pudica* by triiodobenzoic acid. *Die Naturwissenschaften* 63, 42.

Roblin, G. (1979) Les mouvements révolutifs des feuilles de *Mimosa pudica* L. *Biol. Plant.* 21, 57–65.

Roblin, G. and Pérault, J.-M. (1985) Effects of ethephon, 1-aminocyclopropane-1-carboxylic acid, and inhibitors of ethylene synthesis on the gravitropically induced movement of *Mimosa pudica* pulvinus. *Plant Physiol.* 77, 922–925.

Rocha, C.R.C., Schröppel, K., Hareus, D., Marrcil, A., Dignard, D., Taylor, B.N., Thomas, D.Y., Whiteway, M. and Leberrer, E. (2001) Signaling through adenylyl cyclase is essential for hyphal growth and virulence in the pathogenic fungus *Candida albicans. MBS Abstracts* 12, 3631–3634.

Rochwell, E. and Highberger, J.H. (1927) The necessity of CO_2 for the growth of bacteria, yeasts, and molds. *J. Infect. Dis.* 40, 438–446.

Rodde, K. (2000) Postharvest treatments for transport. 'Innovations in Fresh Produce Transportation', 6 March, Florida Postharvest Horticultural Institute, University of Florida, Gainesville, Florida.

Rodriguez, F.I., Esch, J.J., Hall, A.E., Binder, B.M., Schaller, G.E. and Bleecker, A.B. (1999) A copper cofactor for the ethylene receptor *ETR1* from *Arabidopsis*. *Science* 283, 996–998.

Rogers, M.N. (1973) An historical and critical review of postharvest physiology research on cut flowers. *HortScience* 8, 189–194.

Rogers, M.N. (1992) Snapdragons. In: Larson, R.A. (ed.) *Introduction to Floriculture*, 2nd edn. Academic Press, New York, pp. 93–112.

Rogiers, S.Y. and Knowles, N.R. (1999) A comparison of preharvest and postharverst ethylene production and respiration rates of Saskatoon (*Amelanchier alnifolia* Nutt.) fruit during development. *Can. J. Bot.* 77, 323–332.

Roh, K., Son, K.-C., In, B.-G. and Sisler, E.C. (2001) Effect of 1-methylcyclopropene and ethylene on the action mechanism and ripening of banana. *Plant Growth Regl. Abstracts* 27, 151.

Roios, A. and Boron, W.F. (1981) Intracellular pH. *Physiol. Rev.* 61, 296–434.

Roman, G., Lubarsky, B., Kieber, J.J., Rothenberg, M. and Ecker, J.R. (1995) Genetic analysis of ethylene signal transduction in *Arabidopsis thaliana*: five novel mutant loci integrated into a stress response pathway. *Genetics* 139, 1393–1409.

Romani, R. and French, K. (1977) Temperature-dependent changes in the polysomal population of senescent (ripening) pear fruit. *Plant Physiol.* 60, 930–932.

Romano, C.P., Hein, M.B. and Klee, H.J. (1991) Inactivation of auxin in tobacco transformed with the indoleacetic acid-lysine synthetase gene of *Pseudomonas savastonoi*. *Genes Dev.* 5, 438–446.

Romano, C.P., Cooper, M.L. and Klee, H.J. (1993) Uncoupling auxin and ethylene effects in transgenic tobacco and *Arabidopsis* plants. *Plant Cell* 5, 181–189.

Rombaldi, C., Petitprez, M., Cleyet-Marel, J.C., Rougé, P., Latché, A., Pech, J.C. and Leliévre, J.M. (1993) Immunocytolocalization of ACC oxidase in tomato fruits. In: Pech, J.C., Latché, A. and Balagué, C. (eds) *Cellular and Molecular Aspects of the Plant Hormone Ethylene*. Kluwer Academic, Dordrecht, pp. 96–97.

Rose, D.H. (1939) Effect of ammonia on nuts in storage. *Ice and Refrig.* 96, 147–148.

Ross, A.F. (1961) Systemic acquired resistance induced by localized virus infections in plants. *Virology* 14, 340–358.

Ross, G.S., Knighton, M.L. and Lay-Lee, M. (1992) An ethylene-related cDNA from ripening apples. *Plant Mol. Biol.* 19, 231–238.

Ross, J.J. and Reid, J.B. (1986) Internode length in *Pisum*. The involvement of ethylene with the gibberellin-insensitive *erectoides* phenotype. *Physiol. Plant.* 67, 673–679.

Rottmann, W.H., Peter, G.F., Oeller, P.W., Keller, J.A., Shen, N.F., Nagy, B.P., Taylor, L.P., Campbell, A.D. and Theologis, A. (1991) 1-Aminocyclopropane-1-carboxylate synthase in tomato is encoded by a multigene family whose transcription is induced during fruit and floral senescence. *J. Mol. Biol.* 22, 937–961.

Roux, E.R. (1940) Respiration and maturity in peaches and plums. *Ann. Bot.* 4, 317–327.

Rowley, D.B. and Feeherry, F. (1970) Conditions affecting germination of *Clostridium botulinum* 62A spores in a chemically defined medium. *J. Bacteriol.* 104, 1151–1157.

Roy, S., Anantheswaran, R.C. and Beelman, R.B. (1995). Fresh mushroom quality as affected by modified atmosphere packaging. *J. Food Sci.* 60, 334–340.

Rubenstein, B. and Abeles, F.B. (1965) Relationship between ethylene evolution and leaf abscission. *Bot. Gaz.* 126, 255–259.

Ruddnicki, R.M., Braum, J.W. and Khan, A.A. (1978) Low pressure and ethylene seed germination. *Physiol. Plant.* 43, 189.

Rudich, J., Halevy, A.H. and Kedar, N. (1972) The level of phytohormones in monoecious and gynoecious cucumbers as affected by photoperiod and ethephon. *Plant Physiol.* 50, 585–590.

RueySong, L. (1999) Flower senescence and quality preservation of cut *Oncidium*. *J. Agr. Forestry* 48, 63–83.

Rufelt, H. (1957) Influence of growth substances on the geotropic reactions of wheat roots. *Physiol. Plant.* 10, 509–520.

Rufelt, H. (1961) Geotropism in roots and shoots. *Annu. Rev. Plant Physiol.* 10, 409–430.

Rupasinghe, H.P.V., Murr, D.P., Paliyath, G. and Skoog, L. (2000) Inhibitory effect of 1-MCP on ripening and superficial scald development in 'McIntosh' and 'Delicious' apples. *J. Hort. Sci. and Biotech.* 75, 271–276.

Russell, W. and Thimann, K.V. (1990) The second messenger in apical dominance controlled by auxin. In: Pharis, R.P. and Rood, S.B. (eds) *Plant Growth Substances 1988*. Springer, Berlin, pp. 419–427.

Ryall, A.L. (1963) *Proceedings of the Seventeenth National Conference on Handling Perishables*, Purdue University, Indiana, 11–14 March.

Ryall, A.L. and Lipton, W.J. (1972) *Handling, Transportation and Storage of Fruits and Vegetables*, vol. 1, 2nd edn. AVI, Westport, Connecticut

Sacalis, J.N. (1978) Ethylene evolution by petioles of sleeved poinsettia plants. *HortScience* 13, 594.

Sacher, J.A. (1959) Studies on auxin-membrane permeability relations in fruit and leaf tissues. *Plant Physiol.* 34, 365–372.

Sacher, J.A. (1962) Relations between changes in membrane permeability and the climacteric in banana and avocado. *Nature* 195, 577–578.

Sacher, J.A. (1966) Permeability characteristics and amino acid incorporation during senescence (ripening) of banana tissue. *Plant Physiol.* 41, 701–708.

Sacher, J.A. (1973) Senescence and postharvest physiology. *Annu. Rev. Plant Physiol.* 24, 197–224.

Sachs, T. and Thimann, K.V. (1964) Release of lateral buds from apical dominance. *Nature* 201, 939–940.

Sakai, H., Hua, J., Chen, G.Q., Chang, C., Bleecker, A.B. and Myerowitz, E.M. (1988) *ETR2* is an *ETR1*-like gene involved in ethylene signal transduction in *Arabidopsis*. *Proc. Natl Acad. Sci. USA* 95, 5812–5817.

Saleh, B. and Watts, B.M. (1968) Substrates and intermediates in the enzymatic reduction of metmyoglobin in ground beef. *Food Sci.* 33, 353–357.

Salisbury, F.B. and Wheeler, R.M. (1981) Interpreting plant responses to clinostating. I. Mechanical stresses and ethylene. *Plant Physiol.* 67, 677–685.

Saltveit, M.E. (1977) Carbon dioxide, ethylene and color development in ripening mature green bell peppers. *J. Amer. Soc. Hort. Sci.* 102, 523–525.

Saltveit, M.E. (1993) Internal carbon dioxide and ethylene levels in ripening tomato fruit attached to or detached from the plant. *Physiol. Plant.* 89, 204–210.

Saltveit, M.E. and Dilley, D.R. (1978a) Rapidly induced wound ethylene from excised segments of etiolated *Pisum sativum* L., cv. Alaska. II. Oxygen and temperature dependency. *Plant Physiol.* 61, 675–679.

Saltveit, M.E. and Dilley, D.R. (1978b) Rapidly induced wound ethylene from excised segments of etiolated *Pisum sativum* L. cv. Alaska. I. Characterization of the response. *Plant Physiol.* 61, 447–450.

Saltveit, M.E. and Larson, R.A. (1979) Mechanical stress induces ethylene production and epinasty in poinsettia cultivars. *J. Am. Soc. Hortic. Sci.* 104, 452.

Saltveit, M.E. and Larson, R.A. (1981) Reducing leaf epinasty in mechanically stressed poinsettia plants. *J. Amer. Soc. Hort. Sci.* 106, 156–159.

Saltveit, M.E. and Larson, R.A. (1983) Effect of mechanical stress and inhibitors of protein synthesis on leaf epinasty in mechanically stressed poinsettia plants. *J. Amer. Soc. Hort. Sci.* 108, 253–257.

Saltveit, M.E., Bradford, K.J. and Dilley, D.R. (1978) Silver ion inhibits ethylene synthesis and action in ripening fruits. *J. Amer. Soc. Hort. Sci.* 103, 472–475.

Salunkhe, D.K. and Wu, M.T. (1973) Effects of subatmospheric-pressure storage on ripening and associated chemical changes of certain deciduous fruits. *J. Amer. Soc. Hort. Sci.* 98, 113–116.

Salunkhe, D.K. and Wu, M.T. (1974) Subatmospheric storage of fruits and vegetables. *Lebensmitt. Wiss. Technol.* 7, 261–267.

Salunkhe, D.K. and Wu, M.T. (1975) Subatmospheric storage of fruits and vegetables. In: Haard, N.F. and Salunkhe, D.K. (eds) *Symposium: Postharvest Biology and Handling of Fruits and Vegetables.* AVI, Westport, Connecticut, pp. 153–188.

Sampson, A.W. and Allen, L.M. (1909) Influence of physical factors on transpiration. *Miss. Bot. Studies* 4, 33–59.

Sander, R. (1999) *Compilation of Henry's Law Constants for Inorganic and Organic Species of Potential Importance in Environmental Chemistry*, vol. 3, www.mpch-mainz.mpg.de/~sander/res/henry/html (accessed 10 November, 2002).

Sanders, I.O., Smith, A.R., Taylor, J.E. and Hall, M.A. (1986) Ethylene metabolism and action. *Physiol. Plant.* 65, 723–726.

Sanders, I.O., Smith, A.R. and Hall, M.A. (1991) Ethylene binding in epicotyls of *Pisum sativum* L. cv. 'Alaska'. *Planta* 183, 209–217.

Sanders, S.A., Williams, C.H. and Massey, V. (1999) The roles of two amino acid residues in the active site of L-lactate monoxygenase. *J. Biol. Chem.* 274, 22289–22295.

Sankarapandi, S. and Zweier, J.L. (1999) Bicarbonate is required for the peroxidase function of Cu, Zn-superoxide dismutase at physiological pH. *J. Biol. Chem.* 274, 1226–1295.

Sankhla, N., Mackay, W.A. and Davis, T.D. (2001) Extension of vase life and prevention of ethylene-induced flower shattering in *Lupinus havardii* by 1-methylcyclopropene. *Plant Growth Regl. Abstr.* 27, 253.

Santrucek, J. and Slavik, B. (1990) A method for separating cuticular and stomatal components of gas exchange by amphistomatous leaves. *J. Expt. Bot.* 41, 803–813.

Sargent, J.A., Osborne, D.J. and Dunford, S.M. (1984) Cell separation and its hormonal control during fruit abscission in the Gramineae. *J Expt. Bot.* 35, 1663–1674.

Sarkar, S.K. and Phan, C.T. (1979) Naturally occurring and ethylene-induced phenolic compounds in carrot root. *J. Food Protection* 42, 526–534.

Sarquis, J.I., Jordan, W.R. and Morgan, P.W. (1991) Ethylene evolution from maize (*Zea mays* L.) seedling roots and shoots in response to mechanical impedance. *Plant Physiol.* 96, 1171–1177.

Sastry, S.K., Baird, C.D. and Buffington, D.E. (1978) Transpiration rates of certain vegetables and fruits. *Ashrae Transactions* 84, 237–255.

Satler, S.O. and Thimann, K.V. (1983) Metabolism of oat leaves during senescence. VII. The interaction of carbon dioxide and other atmospheric gases with light in controlling chlorophyll loss and senescence. *Plant Physiol.* 71, 67–70.

Satoh, S. and Esashi, Y. (1986) Inactivation of 1-aminocyclopropane-1-carboxylic acid synthase of etiolated mung bean hypocotyl segments by its substrate, *S*-adenosyl methionine. *Plant Cell Physiol.* 27, 285–291.

Satoh, S. and Yang, S.F. (1988) *S*-adenosylmethionine-dependent inactivation and radiolabeling of 1-aminocyclopropane-1-carboxylic acid synthase isolated from tomato fruits. *Plant Physiol.* 88, 109–114.

Sawamura, M. (1981) Levels of endogenous ethylene in attached citrus fruits. *Agric. Biol. Chem.* 45, 2935–2937.

Sawamura, M. and Miyazaki, T. (1989) Effects of exogenous ethylene on ethylene production in discs of mume fruits. *J. Hort. Sci.* 64, 633–638.

Sawamura, M., Knegt, E. and Bruinsma, J. (1978) Levels of endogenous ethylene, carbon dioxide, and soluble pectin, and activities of pectin methylesterase and polygalacturonase in ripening tomato fruit. *Plant Cell Physiol.* 19, 1061–1069.

Schaller, G.E. and Bleecker, A.B. (1995) Ethylene binding sites generated in yeast expressing the *Arabidopsis ETR1* gene. *Science* 270, 1809–1811.

Schaller, G.E., Ladd, A.N., Lanahan, M.B., Spanbauer, J.M. and Bleeker, A.B. (1995) The ethylene response mediator ETR1 from *Arabidopsis* forms a disulfide-linked dimmer. *J. Biol. Chem.* 270, 12526–12530.

Schierle, J. and Schwark, A. (1988) Asymetric synthesis and concentration of ethylene in the hypocotyl hook of *Phaseolus vulgaris*. *J. Plant Physiol.* 133, 325–331.

Schierle, J., Rohwer, F. and Bopp, M. (1989) Distribution of ethylene synthesis along the etiolated pea shoot and its regulation by ethylene. *J. Plant Physiol.* 134, 331–337.

Schjoerring, J.K., Kyllingsbaek, A., Mortensen, J.V. and Byskov-Nielsen, S. (1993) Field investigations of ammonia exchange between barley plants and the atmosphere. I. Concentration profiles and flux densities of ammonia. *Plant Cell and Environment* 16, 161–167.

Schjoerring, J.K., Husted, S. and Mattson, M. (1998) Physiological parameters controlling plant–atmosphere ammonia exchange. *Atmospheric Environment* 32, 491–498.

Schmidt, C.F. (1964) Spores of *Cl. botulinum*: formation, resistance, germination. In: Lewis, K.H. and Cassel, K. (eds) *Botulism*. Proceedings of a Symposium. Public Health Service Publication No. 999. FP-1. USPHS, Washington, pp. 79–82.

Schmidt, C.F., Lechowich, R.V. and Folinazzo, J.F. (1962) Growth and toxin production of type E *Clostridium botulinum* below 40°F. *J. Food Sci.* 26, 626–630.

Schönherr, J. (1976a) Water permeability of isolated cuticular membranes: the effect of pH and cations on diffusion, hydrodynamic permeability and size of polar pores in the cutin matrix. *Planta* 128, 113–126.

Schönherr, J. (1976b) Water permeability of isolated cuticular membranes: the effect of cuticular waxes on diffusion of water. *Planta* 131, 159–164

Schönherr, J. (1982) Resistances of plant surfaces to water loss: transport properties of cutin, suberin and associated lipids. In: Lange, O.L., Nobel, P.S., Osmond, P. and Ziegler, H. (eds). *Encyclopedia of Plant Physiology*, vol. 12B. Springer, Berlin, pp. 153–179.

Schönherr, J. (2000) Calcium chloride ions penetrate plant cuticles via aqueous pores. *Planta* 212, 112–118.

Schönherr, J. and Bukovac, M. (1973) Ion exchange properties of isolated tomato cuticular membrane: exchange capacity, nature of fixed charges and cation selectivity. *Planta* 109, 73–93.

Schönherr, J. and Merida, T. (1981) Water permeability of plant cuticular membranes: the effects of humidity and temperature on the permeability of non-isolated cuticles of onion bulb scales. *Plant Cell Environ.* 4, 349–354.

Schönherr, J. and Schmidt, H.W. (1979) Water permeability of plant cuticles. Dependence of permeability coefficients of cuticular transpiration on vapor pressure saturation deficit. *Planta* 144, 391–400.

Schönherr, J., Eckl, K. and Gruler, H. (1979) Water permeability of plant cuticles: the effect of temperature on diffusion of water. *Planta* 147, 21–26.

Schreiber, L. and Riederer, M. (1996a) Ecophysiology of cuticular transpiration: comparative investigation of cuticular water permeability of plant species from different habitats. *Oceologia* 107, 426–432.

Schreiber, L. and Riederer, M. (1996b) Determination of diffusion coefficients of octadecanoic acid in isolated cuticular waxes and their relationship to cuticular water permeabilities. *Plant Cell Environ.* 19, 1075–1082.

Schreiber, L., Skrabs, M., Hartman, K.D., Diammantopoulos, P., Simanova, E. and Santrucek, J. (2001) Effect of humidity on the cuticular water permeability of isolated cuticular membranes and leaf discs. *Planta* 214, 274–282.

Schwark, A. and Bopp, M. (1993) Interaction of ethylene and auxin in the regulation of hook growth. II. The role of ethylene in different growing regions of the hypocotyl hook of *Phaseolus vulgaris*. *J. Plant Physiol.* 142, 585–592.

Schwark, A. and Schierle, J. (1992) Interaction of ethylene and auxin in the regulation of hook growth. I. The role of auxin in different growing regions of the hypocotyl hook of *Phaseolus vulgaris*. *J. Plant Physiol.* 140, 562–570.

Scott, E.M. and Baker, K.C. (1947) Anatomy of Washington navel orange rind in relation to water spot. *Bot. Gaz.* 108, 459–475.

Scott, W.J. (1957) Water relations of food spoilage microorganisms. *Advan. Food Res.* 7, 83–127.

Sea Containers Group (1983) *General Guide to Tank Container Operation.*

Seberry, J.A. and Hall, E.G. (1970) Controlled atmosphere storage of citrus fruits. *Agr. Gaz. of New South Wales* October, 564–567.

Segall, R.H. (1967) Bacterial soft rot, bacterial necrosis, and alternaria rot of tomatoes as influenced by field washing and postharvest chilling. *Plant Disease Reptr.* 51, 151.

Segner, W.P., Schmidt, C.F. and Boltz, J.K. (1971) Minimal growth temperature, sodium chloride tolerance, pH sensitivity, and toxin production of marine and terrestrial strains of *Clostridium botulinum* type C. *Appl. Microbiology* 22, 1025–1029.

Sena Gomes, A.R. and Kozlowski, T.T. (1980) Responses of *Melaleuca quinquenervia* seedlings to flooding. *Physiol. Plant.* 49, 473.

SenSym (1998) *Solid-state Pressure Sensors Handbook.* Vol. 16, 2-3–2-15.

Serek, M. and Reid, M.S. (2000) Ethylene and postharvest performance of potted *Kalanchoë*. *Postharvest Biol. Technol.* 18, 43–48.

Serek, M., Jones, R.B. and Reid, M.S. (1994a) Role of ethylene in opening and senescence of *Gladiolus* sp. flowers. *J. Amer. Soc. Hort. Sci.* 119, 1014–1019.

Serek, M., Reid, M.S. and Sisler, E.C. (1994b) A volatile inhibitor improves the postharvest life of potted roses. *J. Amer. Soc. Hort. Sci.*119, 572–577.

Serek, M., Sisler, E.C. and Reid, M.S. (1994c) Effects of 1-MCP on the vase life and ethylene response of cut flowers. *Plant Growth Regul.* 16, 93–97.

Serek, M., Sisler, E.C. and Reid, M.S. (1994d) Novel gaseous inhibitor of ethylene binding prevents ethylene effects in potted flowering plants. *J. Amer. Soc. Hort. Sci.* 119, 1230–1233.

Serek, M., Sisler, E.C. and Reid, M.S. (1995a) Effects of 1-MCP on the vase life and ethylene response of cut flowers. *Plant Growth Regul.* 16, 93–97.

Serek, M., Sisler, E.C., Tirosh, T. and Mayak, S. (1995b) 1-methylcyclopropene prevents bud, flower and leaf abscission of geraton waxflower. *HortScience* 30, 1310.

Serek, M., Tamari, G., Sisler, E.C. and Borochov, A. (1995c) Inhibition of ethylene-induced cellular senescence symptoms by 1-methylcyclopropene, a new inhibitor of ethylene action. *Physiol. Plant.* 94, 229–232.

Serek, M., Sisler, E.C. and Reid, M.S. (1996) Ethylene and the postharvest performance of miniature roses. *Acta Horticulturae* 424, 145–149.

Serrano, M., Pretel, M.T., Botella, M.A. and Amorós, A. (2001) Physiochemical changes during date ripening related to ethylene production. *Food Sci. Technol. Int.* 7, 31–36.

Sessa, G., Raz, V., Savaldi, S. and Fluhr, R. (1996) PK12, a plant dual-specificity protein kinase

Streif, J. and Bangerth, F. (1976) The effect of different partial pressures of oxygen and ethylene on the ripening of tomato fruits. *Sci. Hort.* 5, 227–237.

Stritmatter, C.F. and Ball, E.G. (1954) The intracellular distribution of cytochrome components and of oxidative enzyme activity in rat liver. *J. Cell. Comp. Physiol.* 43, 57–78.

Stutte, C.A. and Weiland, R.T. (1978) Gaseous nitrogen loss and transpiration of several crop and weed species. *Crop Sci.* 18, 887–889.

Stutte, C.A., Weiland, R.T. and Blem, R.R. (1979) Gaseous nitrogen loss from soybean foliage. *Argon J.* 71, 95–97.

Suge, H. (1971) Stimulation of oat and rice mesocotyl growth by ethylene. *Plant and Cell Physiol.* 12, 831–837.

Suge, H. and Kusanagi, T. (1975) Ethylene and carbon dioxide: regulation of growth and nodal root development in floating rice. *Plant Cell Physiol.* 16, 65–72.

Sugiyama, H. and Yang, K.H. (1975) Growth potential of *Clostridium botulinum* in fresh mushrooms packaged in semipermeable plastic film. *Appl. Microbiology* 15, 964–969.

Sun, J.-A., Jameson, P.E. and Clemens, J. (2001) Water relations and stamen abscission in cut flowers of selected Myrtaceae. *Plant Growth Regl. Abstr.* 27, 254.

Sun, X.-C., Zhai, C.B., Cui, M., Chen, Y., Levin, L.R., Buck, J. and Bonanno, J.A. (2003) HCO_3-dependent soluble adenylyl cyclase activates the cystic fibrosis transmembrane conductance regulator in corneal endothelium. *A. J. Physiol. Cell Physiol.* 284(5), C1114–C1122.

Supek, F., Superkova, L., Nelson, H. and Nelson, N. (1996) A yeast manganese transporter related to the macrophage protein involved in conferring resistance to mycobacteria. *Proc. Natl Acad. Sci. USA* 93, 5105–5110.

Suttle, J.C. (1988) Effect of ethylene treatment on polar IAA transport, net IAA uptake, and specific binding of *N*-1-naphthylphthalamic acid in tissues and microsomes isolated from etiolated pea epicotyls. *Plant Physiol.* 88, 795–799.

Suttle, J.C. (1991) Ethylene interactions with other endogenous growth substances. In: Mattoo, A.K. and Suttle, J.C. (eds) *The Plant Hormone Ethylene.* CRC Press, Boca Raton, Florida, pp. 115–131.

Suttle, J.C. and Hulstrand, J.F. (1991) Ethylene-induced leaf abscission in cotton seedlings. The physiological bases for age-dependent differences in sensitivity. *Plant Physiol.* 95, 29–33.

Suttle, J.C. and Kende, H. (1980a) Ethylene action and loss of membrane integrity during petal senescence in *Tradescantia. Plant Physiol.* 65, 1067–1072.

Suttle, J.C. and Kende, H. (1980b) Ethylene and senescence in petals of *Tradescantia. Plant Physiol.* 62, 267–271.

Sutton, M., Asman, W.A.H. and Schjoerring, J.K. (1994) Dry deposition of reduced nitrogen. *Tellus* 46B, 255–273.

Sutton, M., Schjoerring, J.K. and Wyers, P. (1995) Plant–atmosphere exchange of ammonia. *Philos. Trans. R. Soc. London – Biol. Sci.* 351, 261–278.

Swanson, C.A. and El-Shishiny, E.D. (1958) Translocation of sugars in the Concord grape. *Plant Physiol.* 33, 33–37.

Tabak, H.H. and Cooke, W.B. (1968) The effects of gaseous environments on the growth and metabolism of fungi. *Bot. Rev.* 34, 126–252.

Taclindo, C.J., Midura, T., Nygaard, G.S. and Bodily, H.L. (1967) Examination of prepared foods in plastic packages for *Clostridium botulinum. Appl. Microbiology* 15, 426–430.

Takahashi, H., Suge, H. and Jaffe, M.J. (1991) Agravitropic growth and its relation to the formation of the plumular hook in etiolated shoots of the pea mutant, *ageotropicum. J. Plant Physiol.* 138, 216–222.

Takata, M. (1975) The effects of ethylene on the respiration and ripening of fruits. *J. Jpn. Soc. Hort. Sci.* 44, 82–88.

Takata, M. (1981) Effects of silver ion on the ripening of Japanese persimmon fruits. *J. Japan. Soc. Hort. Sci.* 53, 314–319.

Tammen, J. (1961) Diseases. In: Langhans, R.W. (ed.) *Carnations. A Manual of the Culture, Insects, and Diseases and Economics of Carnations.* New York State and County Agricultural Extension Service, pp. 75–89.

Tan, Z.-Y. and Thimann, K.V. (1989) The roles of carbon dioxide and abscisic acid in the production of ethylene. *Physiol. Plant.* 75, 13–19.

Tanimoto, M., Roberts, K. and Dolan, L. (1995) Ethylene is a positive regulator of root hair dvelopment in *Arabidopsis thaliana. Plant J.* 8, 943–948.

Tatsuki, M. and Mori, H. (1999) Rapid and transient expression of 1-aminocyclopropane-1-carboxylate synthase isogenes by touch and wound stimuli in tomato. *Plant Cell Physiol.* 40, 709–715.

Taylor, G.E. and Gunderson, C.A. (1986) The response of foliar gas exchange to exogenously applied ethylene. *Plant Physiol.* 82, 653–657.

Taylor, G.E. and Gunderson, C.A. (1988) Physiological site of ethylene effects on carbon dioxide assimilation in *Glycine max* L. Merr. *Plant Physiol.* 86, 85–92.

Taylor, M.N., Joyce, D.C., Wearing, A.H. and Simons, D.H. (2001) Influence of fungal pathogens and environmental conditions on disease severity, flower fall, and desiccation of harvested *Geraldton* waxflower. 2. Studies with commercial packages. *Austr. J. Exptl. Agric.* 41, 105–115.

Tepper, B. (1978) Maximum storage density for chrysanthemum cuttings in the Dormavac container. Report to Grumman Allied Industries, Inc., Garden City, New York.

Terai, H. (1993) Behavior of 1-aminocyclopropane-1-carboxylic acid (ACC) and ACC synthase responsible for ethylene production in normal and mutant (*nor* and *rin*) tomato fruits at various stages of ripeness. *J. Jpn. Hortic. Soc.* 61, 805–812.

Tetlow, I.J. and Farrar, J.F. (1993) Apoplastic sugar concentration and pH in barley leaves infected with brown rust. *J. Expt. Bot.* 44, 929–936.

Theologis, A. (1992) One rotten apple spoils the whole bushel: the role of ethylene in fruit ripening. *Cell* 70, 181–184.

Theologis, A. (1995) Ethylene sensors: how perceptive! *Science* 270, 1774.

Theologis, A. and Laties, G.G. (1978) Respiratory contribution of the alternative path during various stages of ripening in avocado and banana fruits. *Plant Physiol.* 62, 249–255.

Theologis, A. and Laties, G.G. (1982a) Potentiating effect of pure oxygen on the enhancement of respiration by ethylene in plant storage organs: a comparative study. *Plant Physiol.* 69, 1031–1035.

Theologis, A. and Laties, G.G. (1982b) Selective enhancement of alternative-path capacity in plant storage organs in response to ethylene plus oxygen: a comparative study. *Plant Physiol.* 69, 1036–1039.

Theologis, A., Oeller, P.W. and Min-Wong, L. (1993a) Modifying fruit ripening by suppressing gene expression. In: Pech, J.C., Latché, A. and Balagué, C. (eds) *Cellular and Molecular Aspects of the Plant Hormone Ethylene.* Kluwer Academic, Dordrecht, pp. 19–23.

Theologis, A., Oeller, P.W., Min-Wong, L., Rottmann, W.H. and Gantz, D. (1993b) Use of a tomato mutant constructed with reverse genetics to study ripening, a complex developmental process. *Dev. Genet.* 14, 281–295.

Thimann, K.V. (1952) Plant growth hormones. In: Thimann, K.V. (ed.) *The Action of Hormones in Plants and Invertebrates.* Academic Press, New York.

Thimann, K.V. (1955) *The Life of Bacteria.* Macmillan, New York.

Thimann, K.V. (1972) The natural plant hormones. In: Stewart, F.C. (ed.). *Plant Physiology*, vol. VIB, *Physiology of Development: The Hormones.* Academic Press, New York.

Thimann, K.V. and Satler, S.O. (1979a) Relation between leaf senescence and stomatal closure: senescence in the light. *Proc. Natl Acad. Sci. USA* 76, 2295–2298

Thimann, K.V. and Satler, S.O. (1979b) Relation between leaf senescence and stomatal opening: senescence in darkness. *Proc. Natl Acad. Sci. USA* 76, 2770–2773.

Thimann, K.V., Yokum, C.S. and Hacket, D.P. (1954) Terminal oxidases and growth in plant tissues. III. Terrminal oxidation in potato tuber tissue. *Arch. Biochem. Biophys.* 53, 239–257.

Thimann, K.V., Tetley, R.M. and Krivak, B.M. (1977) Metabolism of oat leaves during senescence. V. Senescense in light. *Plant Physiol.* 59, 448–454.

Thines, E., Weber, R.W.S. and Talbot, N.J. (2000) MAP kinase and protein kinase A-dependent mobilization of triacetylglycerol and glycogen during appressorium turgor generation by *Magnaporthe grisea. Plant Cell* 12, 1703–1718.

Thomas, M. (1925) A quantitative study of the production of ethyl alcohol and acetaldehyde by cells of higher plants in relation to concentration of oxygen and carbon dioxide. *Biochem. J.* 19, 927.

Thomas, M. (1929) The production of ethyl alcohol and acetaldehyde by fruits in relation to the injuries occurring in storage. II. Injuries to apples and pears occurring in the presence of oxygen and in the absence of accumulations of carbon dioxide in the storage atmosphere. *Ann. Appl. Biol.* 18, 60–74.

Thomas, R.C. (1974) Intracellular pH of snail neurons measured with a new pH-sensitive glass microelectrode. *J. Physiol. (London)* 238, 159–180.

Thomma, B.P.H.J., Eggermont, K., Tierens, K.F.M.J. and Broekaert, W.F. (1999) Requirement of functional *ethylene-insensitive 2* gene for efficient resistance of *Arabidopsis* to infection by *Botrytis cinerea. Plant Physiol.* 121, 1093–1101.

Thompson, A.K. (1998) *Controlled Atmosphere Storage of Fruits and Vegetables.* CAB International, Wallingford, UK.

Thompson, J.E., Mayak, S., Shinitzky, M. and Halevy, A.H. (1981) Modulation of membrane senescence in cut carnation flowers by treatment with ethylene. *Plant Physiol.* 67, S-278.

Thompson, J.E., Mayak, S., Shinitzky, M. and Halevy, A.H. (1982) Acceleration of membrane senescence in cut carnation flowers by treatment with ethylene. *Plant Physiol.* 69, 859–863.

Thompson, J.S., Harlow, R.L. and Whitney, J.F. (1983) Copper(I)-olefin complexes. Support for the proposed role of copper in the ethylene effect in plants. *J. Am. Chem. Soc.* 105, 3522–3527.

Thornton, B.C. and Sullivan, W.N. (1964) Effects of a high vacuum on insect mortality. *J. Econ. Ent.* 57, 852–854.

Thornton, N.C. (1930) The use of carbon dioxide for prolonging the life of cut flowers with special reference to roses. *Amer. J. Bot.* 17, 614–626.

Thornton, N.C. (1933a) Carbon dioxide storage. III. The influence of carbon dioxide on oxygen uptake by fruits and vegetables. *Contr. Boyce Thomp. Inst.* 5, 371–402.

Thornton, N.C. (1933b) Carbon dioxide storage. IV. The influence of carbon dioxide on the acidity of plant tissue. *Contr. Boyce Thomp. Inst.* 5, 403–418.

Thornton, N.C. (1934a) Carbon dioxide storage. VII. Changes in flower color as evidence of the effectiveness of carbon dioxide in reducing the acidity of plant tissue. *Contr. Boyce Thomp. Inst.* 6, 403–405.

Thornton, N.C. (1934b) Carbon dioxide storage. VI. Lowering the acidity of fungal hyphae by treatment with carbonic acid. *Contr. Boyce Thomp. Inst.* 6, 395–402.

Thornton, N.C. (1935) Carbon dioxide storage. VIII. Chemical changes in potato tubers resulting from exposure to carbon dioxide. *Contr. Boyce Thomp. Inst.* 7, 113–118.

Thornton, N.C. (1937a) Carbon dioxide storage. X. The effect of carbon dioxide on the ascorbic acid content, respiration, and pH of asparagus tips. *Contr. Boyce Thomp. Inst.* 9, 137–148.

Thornton, N.C. (1937b) Carbon dioxide storage. XI. The effect of carbon dioxide on the ascorbic acid (vitamin C) content of some fruits and vegetables. *Proceedings of the J. Amer. Soc. Hort. Sci.* 35, 200–201.

Thornton, N.C. (1940) The facts about 'artificially' ripened fruit. *Food Ind.* 12, 48–50; 51–52.

Thornton, N.C. (1943) Carbon dioxide storage. XIV. Influence of carbon dioxide, oxygen and ethylene on the vitamin C content of ripening bananas. *Contr. Boyce Thomp. Inst.* 13, 201–220.

Thornton, N.C. (1946) Factors influencing vitamin C content of asparagus, banana and seedlings of garden pea during growth or in storage. *Contr. Boyce Thomp. Inst.*14, 295–304.

Thornton, N.C. and Stetterstrom, C. (1940) Toxicity of ammonia, chlorine, hydrogen cyanide, hydrogen sulphide and sulphur dioxide gases. III. Green plants. *Contr. Boyce Thomp. Inst.* 11, 343–356.

Thornton, N.C., Woodruff, R., Burg, S. and Dilley, D. (1971) Low-pressure system: a new approach to controlled atmosphere storage. *HortScience* 6, 279 (abstract).

Tian, M.S., Prakash, S., Elgar, H.J., Young, H., Burmeister, D.M. and Ross, G.S. (2000) Response of strawberry fruit to 1-methylcyclopropene (1-MCP) and ethylene. *Plant Growth Regl.* 32, 83–90.

Tian, S.-P., Fan, Q., Xu, Y., Wang, Y. and Jiang, A.-L. (2001) Evaluation of the use of high CO_2 concentrations and cold storage to control *Monilinia fruticola* on sweet cherries. *Postharvest Biol. Technol.* 22, 53–60.

Tieman, D.M., Taylor, M.G., Ciardi, J.A. and Klee, H.J. (2000) The tomato ethylene receptors NR and LeETR4 are negative regulators of ethylene response and exhibit functional compensation within a multigene family. *Proc. Natl Acad. Sci. USA* 97, 5663–5668.

Tigchelaar, E.C. and McGlasson, W.B. (1977) Tomato ripening mutants: a key role for polygalacturonase in fruit ripening? *Plant Physiol.* 59, S-121.

Tigchelaar, E.C., McGlasson, W.B. and Buescher, R.W. (1978) Genetic regulation of tomato fruit ripening. *HortScience* 13, 508–513.

Ting, S.V. and Attaway, J.A. (1971) Citrus fruits. In: Hulme, A.C. (ed.) *The Biochemistry of Fruits and Their Products*, vol. 1. Academic Press, London, pp. 107–169.

Tissera, P. and Ayres, P.G. (1986) Endogenous ethylene affects the behavior of stomata in epidermis isolated from rust-infected faba bean (*Vicia faba* L.). *New Phytol.* 104, 53–61.

Tjosvold, S.A., Wu, M.-J. and Reid, M.S. (1994) Reduction of post-production quality loss in potted miniature roses. *HortScience.* 29, 293–294.

Toledo, R., Steinberg, M.P. and Nelson, A.I. (1969) Heat of respiration of fresh produce as affected by controlled atmosphere. *J. Food Sci.* 34, 261.

Toler, R.W., Dukes, P.D. and Jenkins, S.F. (1966) Growth responses of *Fusarium oxysporum f.*

tracheiphilum in vitro to varying oxygen and carbon dioxide tensions. *Phytopathology* 56, 183–186.

Tolle, W.E. (1962) Film permeability requirements for storage of apples. USDA Tech. Bul. 1257.

Tolle, W.E. (1969) Hypobaric storage of mature green tomatoes. USDA Agr. Research Rept. 842, pp. 1–9.

Tolle, W.E. (1972) Hypobaric storage of fresh produce. *1972 Yearbook*. United Fresh Fruit & Vegetable Association, pp. 27, 28, 30, 34, 36, 43.

Toole, V.K., Bailey, W.K. and Toole, W.H. (1964) Factors influencing dormancy of peanut seeds. *Plant Physiol.* 39, 822–832.

Tovar, B., García, H.S. and Mata, M. (2001) Physiology of precut mango. I. ACC and ACC oxidase activity of slices subjected to osmotic dehydration. *Food Research Int.* 34, 207–215.

Trewavas, A.J. and Jones, A.M. (1981) Consequences of hormone-binding studies for plant growth substance research. *What's New in Plant Physiology* 12, 5–8.

Trout, S.A., Tindale, G.B. and Huelin, F.E. (1938) The storage of oranges with special reference to locality, maturity, respiration and chemical composition. Austr. Comm. Sci. Ind. Res. Pamphlet No. 80.

Trout, S.A., Hall, E.G., Robertson, R.N., Hackney, E.M.V. and Sykes, S.M. (1942) Studies in the metabolism of apples. *Austr. J. Exptl. Biol. Med. Sci.* 20, 219–231.

Trout, S.A., Heulin, F.E. and Tindale, G.B. (1960) The respiration of Washington navel and Valencia oranges. Austr. Comm. Sci. Ind. Research Organization, Div. Food Pres. and Transport, Tech. Paper 14, 5–11.

Tucker, G.A. and Brady, C.J. (1987) Silver ions interrupt tomato fruit ripening. *J. Plant Physiol.* 127, 163–169.

Tucker, M.L. and Laties, G.G. (1984) Interrelationships of polysome prevalence, gene expression, and respiration during ripening of ethylene and/or cyanide-treated avocado fruit. *Plant Physiol.* 74, 307–315.

Tucker, M.L. and Laties, G.G. (1985) The dual role of oxygen in avocado fruit respiration: kinetic analysis in computer modeling of diffusion-affected respiratory oxygen isotherms. *Plant Cell and Environ.* 8, 117–127.

Turnbull, C.G.N., Sinclair, E.R., Anderson, K.L., Nissen, R.J., Shorter, A.J. and Lanham, T.E. (1999) Routes of ethephon uptake in pineapple (*Ananas comosus*) and reasons for failure of flower induction. *J. Plant Growth Regul.* 18, 145–152.

Turrell, F.M. (1936) The area of the internal exposed surface of dicotyledon leaves. *Amer. J. Bot.* 23, 255–264.

Turrell, F.M. (1942) A quantitative morphological analysis of large and small leaves of alfalfa, with special reference to internal surface area. *Amer. J. Bot.* 29, 400–415.

Turrell, F.M. and Klotz, L.J. (1940) Density of stomata and oil glands and incidence of water spot in the rind of Washington navel orange. *Bot. Gaz.* 101, 862.

Twigg, M.C. and Link, C.B. (1951) Nutrient deficiency symptoms and leaf analysis of azaleas grown in sand culture. *Proceedings of the J. Amer. Soc. Hort. Sci.* 57, 369–375.

Uknes, S., Dincher, S., Friedrich, L., Negrotto, D., Williams, S., Thompson-Taylor, H., Potter, S., Ward, E. and Rylas, J. (1993) Regulation of pathogenesis-related protein-1a gene expression in tobacco. *Plant Cell* 5, 159–169.

Ulrich, R. (1970) Organic acids. In: Hulme, A.C. (ed.) *The Biochemistry of Fruits and Their Products*, vol. 1. Academic Press, New York, pp. 89–118.

Ulrich, R. (1975) Controlled atmosphere storage. Part 2. Physiological and practical considerations. In: Pantastico, E.B. (ed.) *Postharvest Physiology, Handling and Utilization of Tropical and Subtropical Fruits and Vegetables.* AVI, Westport, Connecticut, pp. 186–200.

Ulrich, R. and Marcellin, P. (1955) Voies et modalitiés des échanges de gaz carbonique et d'oxygéne des fruits avec l'atmosphère ambiante. *J. Recherches Centre Nat. Recherche Sci. Labs. Bellevue (Paris)* 6, 241–251.

Ulrich, R. and Paulin, A. (1954) Sur la complexité des conditions thermiques de la maturation des poires Passe-Crassane. *Academie D'Agriculture de France.* 7 April, 1954, pp. 1–3.

Uota, M. (1969a) Carbon dioxide suppression of ethylene-induced sleepiness of carnation blooms. *J. Amer. Soc. Hort. Sci.* 94, 598–601.

Uota, M. (1969b) Commodity requirements and recommendations for flowers and nursery stock. *Proceedings of the 1969 National Controlled Atmosphere Research Conference*, 27/28 January, Michigan State University, East Lansing, Michigan, Hort. Rept. 9, pp. 109–122.

Uota, M. (1970) Sleepiness of carnation blooms – how much ethylene does it take? *Flor. Rev.* 146, 35, 65–67.

Uota, M. and Garazsi, M. (1967) Quality and display life of carnation blooms after storage in controlled atmospheres. USDA Marketing Research Rept. No. 796, pp. 1–9.

Ursin, V.M. and Bradford, K.J. (1989) Auxin and ethylene regulation of petiole epinasty in two developmental mutants of tomato, *diageotropica* and *Epinastic*. *Plant Physiol.* 90, 1341–1346.

Uys, A.E. (1980) Die invloed van waterverlies op loofblaarverbruining by *Proteas neriifolia* R.Br. MSc thesis, Rand Africaans University, Johannesburg.

Vaclavik, J. (1973) Effect of different leaf ages on the relationship between the CO_2 uptake and water vapour efflux of tobacco plants. *Biol. Plant.* 233–236.

Valdovinos, J.G., Ernest, L.C. and Henry, E.W. (1967) Effect of ethylene and gibberellic acid on auxin synthesis in plant tissue. *Plant Physiol.* 42, 1803–1806.

Valdovinos, J.G., Ernest, L.C. and Jensen, T.E. (1972) Studies on the action of ethylene in physiological processes of plant cells. In: Carr, D.J. (ed.) *Plant Growth Substances 1970.* Springer, Berlin, pp. 493–501.

Valley, G. (1927) The effect of carbon dioxide on bacteria. *Quart. Rev. Biol.* 3, 209–224.

Valley, G. and Rettger, L.F. (1927) The influence of carbon dioxide on bacteria. *Jour. Bact.* 14, 101–137.

van den Berg, L. (1981) The role of humidity, temperature and atmospheric composition in maintaining vegetables of North America. In: Teranishi, R. and Bararere-Benitez, H. (eds) *Quality of Selected Fruits and Vegetables of North America.* ACS Symposium Series No. 170, 95–107.

van den Berg, L. and Lentz, C.P. (1978) High-humidity storage of vegetables and fruits. *HortScience* 13, 565–569.

Van der Straeten, D., Van Wiemeersch, L., Goodman, H.M. and Van Montagu, M. (1990) Cloning and sequence of two different cDNAs encoding 1-aminocyclopropane-1-carboxylase synthase in tomato. *Proc. Natl Acad. Sci. USA* 87, 4859–4863.

Van der Straeten, D., Djudzman, A., Van Caeneghem, W., Smalle, J. and Van Montagu, M. (1993) Genetic and physiological analysis of a new locus in *Arabidopsis* that confers resistance to 1-aminocyclopropane-1-carboxylic acid and ethylene and specifically affects the ethylene signal transduction pathway. *Plant Physiol.* 102, 401–408.

van Gardingen, P.R. and Grace, J. (1992) Vapour pressure deficit response of cuticular conductance in intact *Fagus sylvatica L. J. Expt. Bot.* 43, 1293–1299.

Van Steveninck, R.F.M. (1962) Potassium fluxes in red beet tissue during its 'lag' phase. *Physiol. Plantarum* 15, 211–215.

van Uffelen, J.A.M. (1974) Ethylene promotes rapid yellowing of cucumbers. *Tuindeerij* 14, 16–17.

Varder, Y. and Kaldewey, H. (1972) Auxin transport and apical dominance in *Pisum sativum.* In: Kaldewey, H. and Varder, Y. (eds) *Hormonal Regulation in Plant Growth and Development.* Verlag Chemie GmbH, Weinheim, Germany, pp. 401–411.

Varga, M., Rusznak, A. and Niki, K. (1982) Effect of ethrel on IAA content and on its distribution in cucumber seedlings. *Biochem. Physiol. Pflanzen* 177, 659–669.

Varga, S., Michalik, P., Keith, B., Sims, G.G. and Regier, L.W. (1979) Keeping time of lean and fatty fish in hypobaric storage. New Series Circular 71. Fisheries and Oceans, Canada, Technology Branch, Halifax, Nova Scotia.

Veen, B.W. (1970a) Control of plant cell shape by cell wall structure. *Koninkl. Nederi. Akad. van Wetensschappen–Amsterdam* 73, 118–121.

Veen, B.W. (1970b) Orientation of microfibrils in parenchyma cells of pea stem before and after longitudinal growth. *Koninkl. Nederi. Akad. van Wetensschappen–Amsterdam* 73, 113–117.

Veen, H. (1979) Effects of silver on ethylene synthesis and action in cut carnations. *Planta* 145, 467–470.

Veen, H. (1983) Silver thiosulfate: an experimental tool in plant science. *Sci. Hort.* 20, 211–224.

Veen, H. (1987) Use of inhibitors of ethylene action. In: Reid, M.S. (ed.) *Manipulation of Ethylene Responses in Horticulture. Acta Horticulturae* 201, 213–222.

Veierskov, B. and Kirk, H.G. (1986) Senescence in oat leaf segments under hypobaric conditions. *Physiol. Plant.* 66, 283–287.

Velankar, N.K. and Govidan, T.K. (1957) The free amino acid nitrogen of the skeletal muscle of some marine fishes and invertebrates. *Current Sci. (India)* 26, 285–286.

Velkov, L. (1974) Chemical changes in Jonathan, Golden Delicious and Red Delicious apples stored in controlled atmosphere. *Hortic. Viticult. Sci.* 11, 21.

Vendrell, M. and McGlasson, W.B. (1971) Inhibition of ethylene production in banana fruit tissue by ethylene treatment. *Austr. J. Biol. Sci.* 24, 885–895.

Ververidis, P. and Dilley, D.R. (1994) Mechanism studies of CO_2 activation of 1-aminocyclopropane-1-carboxylate (ACC) oxidase: evidence for a lysyl residue involvement. *Plant Physiol.* 105, S-33.

Vigers, A.J., Roberts, W.K. and Selitrennikoff, C.P. (1991) A new family of antifungal proteins. *Mol. Plant–Microbe Interact.* 4, 315–323.

Vitagliano, C. and Hoad, G.V. (1978) Leaf stomatal resistance, ethylene evolution and ABA levels as influenced by (2-chloroethyl) phosphonic acid. *Sci. Hort.* 8, 101–106.

Voigt, M.V. and Haard, N.F. (1982) Hypobaric storage of poultry. Influence of reduced pressure and controlled atmospheres on the survival and enumeration of Salmonella species. DSS Contract 01531-9-0350, Agriculture, Canada.

Volotovski, I.D., Sokolovsky, S.G., Molchan, O.V. and Knight, M.R. (1998) Second messengers mediate increases in cytosolic calcium in tobacco protoplasts. *Plant Physiol.* 117, 1023–1030.

Von Caemmerer, S. and Farquhar, G.D. (1981) Some relationships between the biochemistry of photosynthesis and the gas exchange of leaves. *Planta* 153, 376–387.

von Loesecke, H.W. (1950) *Bananas: Chemistry, Physiology, Technology.* Interscience, New York.

Vyas, S.P., Bohra, S.P. and Sankhla, N. (1973) Antagonism between morphactin and ethylene in root-coiling of *Ipomoea pentaphylla. Z. Pflanzenphysiol.* 69, 185–188.

Wachter, W. (1905) Untersuchungen uber den austritt von zucheraus den zellen der speicherorgane von *Allium cepa* and *Beta vulgaris. Prings. Jahrb. fur wiss. Bot.* 41, 165–187.

Wade, N.L. (1975) Physiology of cold-storage disorders of fruit and vegetables. In: Lyons, J.M., Graham, D. and Raison, J.K. (eds) *Low Temperature Stress in Crop Plants.* Academic Press, New York, p. 81.

Wade, N.L. and Bishop, D.G. (1978) Changes in the lipid composition of ripening banana fruits and evidence for an associated increase in cell membrane permeability. *Biochim. Biophys. Acta* 529, 454–464.

Waelti, H., Cavalieri, R.P. and Johnson, M. (1992) Humidification and weight loss of apples in CA storage. *8th Annual Washington Tree Fruit Postharvest Conference*, Yakima, Washington, pp. 45–47.

Wager, H.G. (1973) The effect of subjecting peas to air enriched with CO_2. *J. Expt. Bot.* 25, 338–351.

Waithaka, K., Reid, M.S. and Dodge, L.L. (2001) Cold storage and flower keeping quality of cut tuberose (*Polianthes tuberose* L.). *J. Hort. Sci. Biochem.* 76, 271–275.

Walden, W.C. and Hentges, D.J. (1975) Differential effects of oxygen and oxidation-reduction potential on the multiplication of three species of anaerobic intestinal bacteria. *Appl. Microbiology* 30, 781–785.

Walker, N.A., Beilby, M.J. and Smith, F.A. (1979a) Amine uniport at the plasmalemma of charophyte cells. I. Current–voltage curves, saturation kinetics, and effects of unstirred layers. *J. Membrane Biol.* 49, 21–55.

Walker, N.A., Mith, F.A. and Beilby, M.J. (1979b) Amine uniport at the plasma membrane of charophyte cells. II. Ratio of matter to charge transported and permeability of free base. *J. Membrane Biol.* 49, 283–296.

Wallach, A. (1937) Beiträge zur kenntnis der wasseraufnahme durch die luftswurzeln tropischer Orchideen. *Z. Bot.* 33, 433–468.

Walter, M.H. (1989) The induction of phenylpropanoid biosynthetic enzymes by ultraviolet light or fungal elicitor in cultured parsley cells is overridden by heat-shock treatment. *Planta* 177, 1–8.

Wang, C.Y. (1977) Effect of CO_2 treatment on storage and shelf life of peppers. *J. Amer. Soc. Hortic. Sci.* 102, 808.

Wang, C.Y. (1979) Effect of a short-term high-CO_2 treatment on the market quality of stored broccoli. *J. Food Sci.* 44, 1478.

Wang, C.Y. (1983) Postharvest responses of Chinese cabbage to high-CO_2 treatment or low-O_2 storage. *J. Amer. Soc. Hort. Sci.* 108, 125–129.

Wang, C.Y. (1990) Physiological and biochemical effects of controlled atmosphere on fruits and vegetables. In: Calderon, M. and Barkai-Golan, R. (eds) *Food Preservation by Modified Atmospheres.* CRC Press, Boca Raton, Florida, pp. 197–223.

Wang, C.Y. and Adams, D.O. (1982) Chilling-induced ethylene production in cucumbers (*Cucumis sativus* L.). *Plant Physiol.* 69, 424–427.

Wang, C.Y., Mellenthin, W.M. and Hansen, E. (1971) Effect of temperature on development of premature ripening in Bartlett pears. *J. Amer. Soc. Hort. Sci.* 96, 122–126.

Wang, C.Y., Mellenthin, W.M. and Hansen, E. (1972) Maturation of 'Anjou' pears in relation to chemical composition and reaction to ethylene. *J. Amer. Soc. Hort. Sci.* 97, 9–12.

Wang, C.Y., Baker, J.E., Hardenburg, R.E. and Lieberman, M. (1977) Effects of two analogs of rhizobitoxin and sodium benzoate on senescence of snapdragons. *J. Amer. Soc. Hort. Sci.* 102, 517–520.

Wang, H. and Woodson, W.R. (1989) Reversible inhibition of ethylene action and interruption of petal senescence in carnation flowers by norbornadiene. *Plant Physiol.* 89, 434–438.

Wang, S.F., Haard, N.F. and Dimarco, G.R. (1971) Chlorophyll degradation during controlled atmosphere storage of asparagus. *J. Food Sci.* 36, 657.

Wang, Z. and Dilley, D.R. (1997) The relationship of α-farnescene production and its oxidation product 6-methyl-5-heptene-2-one to superficial scald of Granny Smith, Law Rome, Red Delicious and Idared apples during controlled atmosphere and air storage. *Proceedings of the 7th International Controlled Atmosphere Conference* 2, 98–104.

Wang, Z. and Dilley, D.R. (2000) Hypobaric storage removes scald-related volatiles during the low-temperature induction of superficial scald of apples. *Postharvest Biol. Tech.* 18, 191–199.

Wang, Z., Kosittrakun, M. and Dilley, D.R. (2000a) Temperature and atmosphere regimens to control a CO_2-linked disorder of 'Empire' apples. *Postharvest Biol. Tech.* 18, 183–189.

Wang, Z.-Y., MacRae, E.A., Wright, M.A., Bolitho, K.M., Ross, G.S. and Atkinson, R.G. (2000b) Polygalacturonase gene expression in kiwi fruit: relationship to fruit softening and ethylene production. *Plant Molecular Biology* 42, 317–328.

Wankier, D.N., Salunkhe, D.K. and Campbell, W.F. (1970) Effects of controlled atmosphere storage on biochemical changes in apricots and peaches. *J. Amer. Soc. Hort. Sci.* 95, 604–609.

Ward, C.M. (1975) Hypobaric storage. 1974 Rept. National. Veg. Res. Sta., Wellesbourne, UK, p. 85.

Ward, E.R., Uknes, S.J., Williams, S.C., Dincher, S.S., Wiederhold, D.L., Alexander, D.C., Ahi-Goy, P., Métraux, J.-P. and Rylas, J.A. (1991) Coordinate gene activity in response to agents that induce systemic acquired resistance. *Plant Cell* 3, 1085–1094.

Wardlaw, C.W. (1961) *Banana Diseases.* Longmans, Green and Co., London.

Wardlaw, C.W. and Leonard, E.R. (1936a) Studies in tropical fruits. I. Preliminary observations on some aspects of development, ripening and senescence, with special reference to respiration. *Ann. Bot. (London)* 50, 621–653.

Wardlaw, C.W. and Leonard, E.R. (1936b) Studies in tropical fruits. II. Observations on internal gas concentrations in fruit. *Ann. Bot. (London)* 50, 655–676.

Wardlaw, C.W. and Leonard, E.R. (1940) Studies in tropical fruits. IX. The respiration of bananas during ripening at tropical temperatures. *Ann. Bot. (N.S.)* 4, 269–315.

Wardlaw, C.W., Leonard, E.R. and Barnell, H.R. (1939) Studies in tropical fruits, VII. *Ann. Bot. (N.S.)* 3, 845–860.

Warner, H.L. and Leopold, A.C. (1971) Timing of growth regulator responses in peas. *Biochem. Biophys. Res. Comm.* 44, 989–994.

Watkins, C.B., Bowen, J.H. and Walker, V.J. (1989) Assessment of ethylene production by apple cultivars in relation to commercial harvest dates. *New Zealand J. Crop Hort. Sci.* 17, 327–331.

Watkins, C.B., Nock, J.F. and Whitaker, B.S. (2000) Responses of early-, mid- and late-season apple cultivars to postharvest application of 1-methylcyclopropene (1-MCP) under air and controlled atmosphere storage conditions. *Postharvest Biol. Technol.* 19, 17–32.

Wattanabe, K., Kamo, T., Nishikawa, F. and Hyodo, H. (2001) Effect of methyl jasmonate on senescence of broccoli florets. *Plant Growth Regl. Abstracts* 63.

Wawrzynczak, A. and Goszczynska, D.M. (1999) Effects of some inhibitors of ethylene biosynthesis and action on cut carnation (*Dianthus caryophyllus* L. cv. Charlotte) flower longevity. *J. Fruit and Ornamental Plant Research* 7, 105–117.

Wehr, M.H. (1978) Attitudes and policies of state governments. Overview. Microbiological standards for food. An Update. *Food Tech.* 32, 63.

Weiler, T.C. (1993) Easter lilies. In: Larson, R.A. (ed.) *Introduction to Floriculture*, 2nd edn. Academic Press, New York, pp. 333–364.

Weinstein, L.H. (1957) Senescence of roses. I. Chemical changes associated with senescence of cut Better Times roses. *Contrib. Boyce Thomp. Inst.* 19, 33–48.

Weis-Fogh, T. (1964) Diffusion in insect wing muscle, the most active tissue known. *J. Expt. Biol.* 41, 229–256.

Weisner, J. (1878) Die undulierende Nutation der Internodien. *Sitz. Ber. K. u. K. Ak. Wien* 77, 15.

Welby, E.M. and McGregor, B. (1997) *Agricultural Export Transportation Handbook.* USDA Agricultural Handbook 700.

Wells, A.W. (1962) Effects of storage temperature and humidity on the loss of weight by fruit. USDA Market Res. Rept. 539.

Wells, J.M. (1970) Modified atmosphere, chemical, and heat treatments to control postharvest decay of California strawberries. *Plant Disease Reporter* 54, 431–434.

Wells, J.M. (1974) Growth of *Erwinia carotovora*, *E. atroseptica*, and *Pseudomonas fluorescens* in low-oxygen and high-carbon dioxide atmospheres. *Phytopathology* 64, 1012–1015.

Wells, J.M. and Spalding, D.H. (1975) Stimulation of *Geotrichum candidum* by low-oxygen and high-carbon dioxide atmospheres. *Phytopathology* 65, 1229.

Wells, J.M. and Uota, M. (1970) Germination and growth of five fungi in low-oxygen and high-carbon dioxide atmospheres. *Phytopathology* 60, 50–53.

Went, F.W. (1927) Wuchsstoff und Wacstum. *Rec. Trav. Bot. Neerland.* 25, 1–116.

Werkman, C.H. and Wood, H.G. (1942) Heterotrophic assimilation of CO_2. *Adv. in Enzymol.* 2, 135–182.

Whalen, M.C. and Feldman, L.J. (1988) The effect of ethylene on the root growth of *Zea mays* seedlings. *Can. J. Bot.* 66, 719–723.

Whealy, C.A. (1992) Carnations. In: Larson, R.A. (ed.) *Introduction to Floriculture*, 2nd edn. Academic Press, New York, pp. 43–65.

Wheeler, R.M. and Salisbury, F.B. (1981) Gravitropism in higher plants. I. A role for ethylene. *Plant Physiol.* 67, 686–690.

Wheeler, R.M., White, R.G. and Salisbury, F.B. (1986) Gravitropism in higher plant shoots. IV. Further studies on participation of ethylene. *Plant Physiol.* 82, 534–542.

Whitehead, C.S. and Bossè, C.A. (1991) The effect of ethylene and short-chain saturated fatty acids on ethylene sensitivity and binding in ripening bananas. *J. Plant Physiol.* 137, 358–362.

Whitehead, C.S. and Halevy, A.H. (1989) Ethylene sensitivity: the role of short-chain saturated fatty acids in pollination-induced senescence of *Petunia hybrida* flowers. *Plant Growth Regul.* 8, 41–54.

Whitehead, C.S., Halevy, A.H. and Reid, M.S. (1982) Roles of ethylene and 1-aminocyclopropane-1-carboxylic acid in pollination and wound-induced senescence of *Petunia hybrida* flowers. *Physiol. Plant.* 61, 643.

Whitehead, C.S., Halevy, A.H. and Reid, M.S. (1984) Control of ethylene synthesis during development and senescence of carnation petals. *J. Amer. Soc. Hort. Sci.* 109, 473–475.

Whittaker, D.J., Smith, G.S. and Gardner, R.C. (1997) Expression of ethylene biosynthetic genes in *Actinidia chinensis* fruit. *Plant Mol. Biol.* 34, 45–55.

Wickson, M. and Thimann, K.V. (1958) The antagonism of auxin and kinetin in apical dominance. *Physiol. Plant.* 11, 62–74.

Wigglesworth, V.B. (1930) Trachael respiration. *Proceedings of the Roy. Soc. Lond. B* 106, 229–250.

Wigglesworth, V.B. (1931) Insect respiration. *Biol. Rev.* 6, 181–220.

Wigglesworth, V.B. (1935) The regulation of respiration in the flea, *Xenopsylla cheopsis* Roths. (Pulicidae). *Proceedings of the Roy. Ent. Soc. Lond. B* 118, 397–419.

Wigglesworth, V.B. (1950) *Insect Physiology.* John Wiley & Sons, Inc., New York.

Wigglesworth, V.B. (1972) *The Principles of Insect Physiology.* Methuen, London.

Wigglesworth, V.B. (1983) The physiology of insect tracheoles. *Adv. Insect Physiol.* 17, 85–148.

Wigglesworth, V.B. and Gillett, J.D. (1936) The loss of water during ecdysis in *Rhodnius proxlixus* (Hemiptera). *Proceedings of the Roy. Ent. Soc. Lond. A* 11, 104–107.

Wilfret, G.J. (1992) Gladiolus. In: Larson, R.A. (ed.) *Introduction to Floriculture*, 2nd edn. Academic Press, New York, pp. 143–157.

Wilkins, H.F. (1965) Factors affecting carbon dioxide and ethylene gas production in flowers of the carnation (*Dianthus caryophyllus* Linn.) cultivar Red Gayety. PhD dissertation, University of Illinois, Urbana, Illinois.

Wilkins, H.F. and Swanson, B.T. (1975) The relationship of ethylene to senescence. *Acta Horticulturae* 41, 133–142.

Wilkinson, B.G. (1963) Effect of time of picking on ethylene production by apples. *Nature* 199, 715–716.

Wilkinson, B.G. (1970) Physiological disorders of fruit after harvesting. In: Hulme, A.C. (ed.) *The Biochemistry of Fruits and Their Products*, vol. 1. Academic Press, New York, pp. 537–554.

Wilkinson, J.Q., Lanahan, M.B., Yen, H.C., Giovannoni, J.J. and Klee, H.J. (1995) An ethylene-inducible component of signal transduction encoded by *Never-ripe*. *Science* 270, 1807–1809.

Wilkinson, J.Q., Lanahan, M.B., Clark, D.G., Bleeker, A.B., Chang, C., Meyerowitz, E.M. and Klee, H.J. (1997) A dominant mutant receptor from *Arabidopsis* confers ethylene insensitivity in heterologous plants. *Nature Biotechnol.* 15, 444–447.

Willaman, J.J. and Beaumont, J.H. (1927) The effect of accumulated carbon dioxide on plant respiration. *Plant Physiol.* 3, 45–59.

Williams, J.P., Kahn, M.U., Mitchell, K. and Johnson, G. (1988) The effect of temperature on the level and biosynthesis of unsaturated fatty acid diacetylglycerols of *Brassica napus* leaves. *Plant Physiol.* 87, 904–910.

Williams, J.P., Khan, M.U. and Wong, S. (1996) Fatty acid desaturation in monogalactosyldiacylglycerol of *Brassica napus* leaves during low temperature acclimation. *Physiol. Plant.* 96, 258–262.

Williams, M.W. and Patterson, M.E. (1964) Nonvolatile organic acids and core breakdown of Bartlett pears. *J. Agric. Food. Chem.* 12, 80–83.

Williams, O.B. and Rees, F.B. (1952) The bacteriology of Gulf Coast shrimp II. The intestinal flora. *Texas J. Sci.* 4, 55–58.

Williamson, C.E. (1950) Ethylene, a metabolic product of diseased or injured plants. *Phytopathology* 40, 205–208.

Willmer, C.M. (1988) Stomatal sensing of the environment. *Biol. J. Linn. Soc.* 34, 204–217.

Willmer, C.M. and Fricker, M. (1996) *Stomata*, 2nd edn. Chapman & Hall, London.

Wills, R.B.H. and Kim, G.H. (1995) Effect of ethylene on postharvest life of strawberries. *Postharv. Biol. Technol.* 6, 249–255.

Wills, R.B.H., McGlasson, W.B., Graham, D., Lee, T.H. and Hall, E.G. (1989) *Postharvest. An Introduction to the Physiology and Handling of Fruit and Vegetables.* AVI, Van Nostrand Reinhold, New York.

Wills, R.B.H., Harris, D.R. and Seberry, J.A. (1999) Delayed ripening of bananas through minimization of ethylene. *Tropical Agric.* 76, 279–282.

Wills, R.B.H., Warton, M.A. and Ku, V.V.V. (2000) Ethylene levels associated with fruit and vegetables during marketing. *Austr. J. Expt. Agric.* 40, 465–470.

Wills, R.B.H., Warton, M.A., Mussa, D.M.D.N. and Chew, L.P. (2001) Ripening of climacteric fruits initiated at low ethylene levels. *Austr. J. Expt. Agric.* 41, 89–92.

Wilmer, C.M., Wilson, A.B. and Jones, H.G. (1988) Changing sensitivities of stomata to abscisic acid and CO_2 as leaves and plants age. *J. Expt. Bot.* 39, 401–410.

Wilson, A.K., Pickett, F.B., Turner, J.C. and Estelle, M. (1990) A dominant mutation in *Arabidopsis* confers resistance to auxin, ethylene and abscisic acid. *Mol. Gen. Genet.* 222, 377–383.

Wink, M. (1997) Special nitrogen metabolism. In: Dey, P.M. and Harbone, J.B. (eds) *Plant Biochemistry*. Academic Press, New York, pp. 439–486.

Wittenbach, V.A. and Bukovac, M.J. (1973) Cherry fruit abscission: effect of growth substances, metabolic inhibitors and environmental factors. *J. Amer. Soc. Hort. Sci.* 98, 348–351.

Woeste, K.E. and Kieber, J.J. (2000a) A strong loss-of-function mutation in *RAN1* results in constitutive activation of the ethylene response pathway as well as a rosette-lethal phenotype. *Plant Cell* 12, 443–445.

Woeste, K.E. and Kieber, J.J. (2000b) Characterization of *Arabidopsis* ethylene-overproducing mutants. *Plant Growth Regl. Abstracts* 26, 150.

Wolley, J.T. (1962) Potato tuber respiration and ventilation. *Plant Physiol.* 37, 793.

Woltering, E.J. (1984) Ethyleengevoeligheid van zomerbloemen. Voorbehandeling voorkomt schade. *Vadblad voor de Bloemisterij* 17, 34–37.

Woltering, E.J. (1987) Effects of ethylene on ornamental pot plants: a classification. *Scientia Hortic.* 31, 283–294.

Woltering, E.J. (1990) Interorgan translocation of 1-aminocyclopropane-1-carboxylic acid coordinates senescence in emasculated *Cymbidium* flowers. *Plant Physiol.* 92, 837–845.

Woltering, E.J. (1991) Regulation of ethylene biosynthesis in gravi-stimulated *Kniphofia* (hybrid) flower stalks. *J. Plant Physiol.* 138, 443–449.

Woltering, E.J. (1993) Roles of ethylene, ACC and short-chain saturated fatty acids in inter-organ communication during senescence of *Cymbidium* flowers. In: Pech, J.C., Latché, A. and Balagué, C. (eds) *Cellular and Molecular Aspects of the Plant Hormone Ethylene.* Kluwer Academic, Dordrecht, pp. 310–316.

Woltering, E.J. and van Doorn, W.G. (1988) Role of ethylene in senescence of petals: morphological and taxonomical relationships. *J. Expt. Bot.* 39, 1605–1616.

Wood, H.G. and Werkman, C.H. (1940) The relationship of bacterial utilization of CO_2 to succinic acid formation. *Biochem. J.* 34, 129.

Woodrow, I.E. and Berry, J.A. (1988) Enzymatic regulation of photosynthetic CO_2 fixation in C3 plants. *Annu. Rev. Plant Physiol. Plant Mol. Biol.* 39, 533–594.

Woodrow, I.E. and Mott, K.A. (1992) Biphasic activation of ribulosebiphosphate carboxylase in spinach leaves as determined by non-steady state CO_2 exchange. *Plant Physiol.* 99, 298–303.

Woodrow, I.E., Kelly, M.E. and Mott, K.A. (1996) Limitation of the rate of ribulosebiphosphate carboxylase activation by carbamylation and ribulose-biphosphate carboxylase activity: development and tests of a mechanistic model. *Austr. J. Plant Physiol.* 23, 141–149.

Woodrow, L. and Grodzinski, B. (1987) Ethylene evolution from bracts and leaves of poinsettia, *Euphorbia pulcherrima* Willd. *J. Expt. Bot.* 38, 24–32.

Woodruff, R. (1971) LPS Studies – Strawberries var. Shasta. Progress Report to United Fruit Co., 19 July, 1971.

Woods, S.L., Roberts, J.A. and Taylor, I.B. (1984) Ethylene-induced root coiling in tomato seedlings. *Plant Growth Regul.* 2, 217–225.

Woodson, W.R. and Lawton, K.A. (1988) Ethylene-induced gene expression in carnation petals. Relation to autocatalytic ethylene production and senescence. *Plant Physiol.* 87, 498–503.

Woolhouse, H.W. (1981) The general biology of plant senescence and the role of nucleic acid in protein turnover in the control of senescence processes which are genetically programmed. In: Lieberman, M. (ed.) *Post-harvest Physiology and Crop Preservation.* Plenum Press, New York, pp. 1–43.

Workman, M. and Pratt, H.K. (1957) Studies on the physiology of tomato fruits. II. Ethylene production at 20°C as related to respiration, ripening, and date of harvest. *Plant Physiol.* 32, 330–334.

Workman, M., Pratt, H.K. and Morris, L.L. (1957) Studies on the physiology of tomato fruits. I: Respiration and ripening behavior at 20°C as related to date of harvest. *J. Amer. Soc. Hort. Sci. Proc.* 69, 352–365.

Wray, V., Schiel, O., Berlin, J. and Witte, L. (1985) Phosphorous-31 nuclear magnetic resonance investigation of the *in vivo* regulation of intracellular pH in cell suspension cultures of *Nicotiana tabacum*: the effects of oxygen supply, nitrogen and external pH change. *Arch. Biochem. Biophys.* 236, 731–740.

Wright, M. (1974) The effect of chilling on ethylene production, membrane permeability and water loss of leaves of *Phaseolus vulgaris*. *Planta (Berl.)* 120, 63–69.

Wright, M. and Simon, E.W. (1973) Chilling injury in cucumber leaves. *J. Expt. Bot.* 24, 400–411.

Wright, M., Mousdale, D.M.A. and Osborne, D.J. (1978) Evidence for a gravity-regulated level of endogenous auxin controlling cell elongation and ethylene production during geotropic bending in grass nodes. *Biochem. Physiol. Pflanzen* 172, 581–596.

Wright, S.T.C. (1969) An increase in the 'inhibitor-B' content of detached wheat leaves following a period of wilting. *Planta* 86, 10–20.

Wright, S.T.C. (1977) The relationship between leaf water potential and the levels of abscisic acid and ethylene in excised wheat leaves. *Planta* 34, 183–189.

Wright, S.T.C. (1980) The effect of plant growth regulator treatment on the levels of ethylene emanating from excised turgid and wilted wheat leaves. *Planta* 148, 381–388.

Wright, S.T.C. and Hiron, R.W.P. (1969) (+)-abscisic acid, the growth inhibitor induced in wheat leaves by a period of wilting. *Nature* 224, 719–720.

Wu, M.T. and Salunkhe, D.K. (1972a) Fungistatic effects of subatmospheric pressures. *Experientia* 28, 866–867.

Wu, M.T. and Salunkhe, D.K. (1972b) Subatmospheric pressure storage of fruits and vegetables. *Utah Sci.* 33, 29–31.

Wu, M.T., Jadhav, S.J. and Salunkhe, D.K. (1972) Effects of subatmospheric pressure storage on ripening of tomato fruits. *J. Food Sci.* 37, 952–956.

Wuttke, M.S., Buck, J. and Levin, L.R. (2001) Bicarbonate-regulated soluble adenylyl cyclase. *JOP* 2, 154–158.

Xu, J.R. and Hamer, J.E. (1996) MAP kinase and cAMP signaling regulate infection structure formation and pathogenic growth in the rice blast fungus *Magnaporthe grisea*. *Genes & Development* 10, 2696–2706.

Xue, C., Park, G., Choi, W., Zheng, L., Dean, R.A. and Xu, J.-R. (2002) Two novel fungal virulence genes specifically expressed in appressoria of the rice blast fungus. *Plant Cell* 14, 2107–2109.

Yahia, E.M. and Hernandez, T. (1993) Tolerance and response of harvested mango to insecticidal low-O_2 atmospheres. *HortScience* 28, 1031–1033.

Yahia, E.M., Medina, F. and Rivera, M. (1989) The tolerance of mango and papaya to atmospheres containing very high levels of CO_2 and/or very low levels of O_2 as a possible insect control treatment. *Proceedings of the 5th International Controlled Atmosphere Conference*, 14–16 June, Wenatchee, Washington. *Other commodities and storage recomendations* 2, 77–89.

Yahia, E.M., Rivera, M. and Hernandez, O. (1992) Responses of papaya to short-term insecticidal O_2 atmosphere. *J. Amer. Soc. Hort. Sci.* 117, 96–99.

Yalpani, N., Silverman, P., Wilson, T.M.A., Kleier, D.A. and Raskin, I. (1991) Salicylic acid is a systemic signal and an inducer of pathogenesis-related proteins in virus-infected tobacco. *Plant Cell* 3, 809–818.

Yamamotto, M., Miki, T., Ishiki, Y., Fujinami, K., Yanagisawa, Y., Nakagawa, H., Ogura, N., Hirabayashi, T. and Sato, T. (1995) The synthesis of ethylene in melon fruit during early stage of ripening. *Plant Cell Physiol.* 326, 591–596.

Yamauchi, N. and Watada, A.E. (1991) Regulated chlorophyll degradation in spinach leaves during storage. *J. Amer. Soc. Hort. Sci.* 116, 58–62.

Yan, Y.-L. and Fan, Y.-R. (2001) Changes of respiratory rate, the contents of ethylene and peroxidase of cut *Rosa hybrida* Hort. during the period inserted in water. *Plant Growth Regl. Abstracts* 27, 155.

Yang, R.F., Cheng, T.S. and Shewfelt, R.L. (1990) The effect of high temperature and ethylene treatment on the ripening of tomatoes. *J. Plant Physiol.* 136, 368–372.

Yang, S.F. (1980) Regulation of ethylene biosynthesis. *HortScience* 15, 238–243.

Yang, S.F. and Hoffman, N.E. (1984) Ethylene biosynthesis and its regulation by higher plants. *Annu. Rev. Plant Physiol.* 35, 155–189.

Yang, S.F., Dong, J.G., Fernandez-Maculet, J.C. and Olson, D.C. (1993) Apple ACC oxidase: purification and characterization of the enzyme and cloning of its cDNA. In: Pech, J.C., Latché, A. and Balagué, C. (eds) *Cellular and Molecular Aspects of the Plant Hormone Ethylene.* Kluwer Academic, Dordrecht, pp. 59–64.

Yao, R.-L., Li, Y.-R., Yang, L.-T. and Ye, Y.-P. (2000) Effects of ethephon on the IAA oxidase activity in sugarcane stems. *Plant Growth Regl. Abstracts* 26, 318.

Yashiro, K., Sakamoto, T. and Ohmori, M. (1996) Molecular characterization of an adenylate cyclase gene of the cyanobacterium *Spirulina platensis. Plant Mol. Biol.* 31, 175–181.

Yeager, A.F., Richards, M.C., Phillips, T.G. and others (1945) *The Storage of Hubbard Squash.* N.H. Agr. Expt. Sta. Bull. 356.

Yeang, H.Y. and Hillman, J.R. (1981a) Control of lateral bud growth in *Phaseolus vulgaris* L. by ethylene in the apical shoot. *J. Expt. Bot.* 32, 395–404.

Yeang, H.Y. and Hillman, J.R. (1981b) Extraction of endogenous ethylene and ethane from leaves of *Phaseolus vulgaris* L. *J. Expt. Bot.* 32, 381–394.

Yeang, H.Y. and Hillman, J.R. (1984) Ethylene and apical dominance. *Physiol. Plant.* 60, 275–282.

Yemm, E.W. (1935) The respiration of barley plants. II. Carbohydrate concentration, and carbon dioxide production in starving leaves. *Proceedings of the Roy. Soc. (London) C* 117, 504–525.

Yen, H.C., Lee, S., Tanksley, S.D., Lanahan, M.B., Klee, H.J. and Giovannoni, J.J. (1995) The tomato *never-ripe* locus regulates ethylene-inducible gene expression and is linked to a homolog of the *Arabidopsis ETR1* gene. *Plant Physiol.* 107, 1343–1353.

Ying, T.-J., Zeng, G.-W., Barry, C.S. and Grierson, D. (1999) Differential expressions of ACS synthase and ACC oxidase gene families in fruit of *Epinastics*, a tomato mutant over-expressing ethylene. *Plant Growth Regl. Abstracts* 26, 151.

Yip, K.-C. and Hew, C.S. (1989) Ethylene-induced cyanide-resistant respiration in orchid petal cells. *Plant Growth Regul.* 8, 365–373.

Yip, W.-K. (1993) Characterization of the 1-aminocyclopropane-1-carboxylate (ACC) synthase isozyme (genes) in tomato. In: Pech, J.C., Latché, A. and Balagué, C. (eds) *Cellular and Molecular Aspects of the Plant Hormone Ethylene.* Kluwer Academic, Dordrecht, pp. 13–18.

Yip, W.-K. and Yang, S.F. (1988) Cyanide metbolism in relation to ethylene production in plant tissues. *Plant Physiol.* 88, 473–476.

Yip, W.-K., Jiao, X.-Z. and Yang, S.F. (1988) Dependence of in vivo ethylene production rate on 1-aminocyclopropane-1-carboxylic acid content and oxygen concentration. *Plant Physiol.* 87, 553–558.

Yip, W.-K., Dong, J., Kenny, J.W., Thompson, G.A. and Yang, S.F. (1990) Characterization and sequencing of the active site of 1-aminocyclopropane-1-carboxylate synthase in tomato. *Proc. Natl Acad. Sci. USA* 87, 7930–7934.

Yoon, I.S., Mori, H., Kim, J.H., Kang, B.G. and Imaseki, H. (1997) *VR-ACS6* is an auxin-inducible 1-aminocyclopropane-1-carboxylate synthase gene in mungbean (*Vigna radiata*). *Plant Cell Physiol.* 38, 217–224.

Yoshi, H. and Imaseki, H. (1982) Regulation of auxin-induced ethylene biosynthesis. Repression of inductive formation of ACC synthase by ethylene. *Plant Cell Physiol.* 23, 639–649.

Yoshida, T. (1975) Nitrogen atmosphere and pest insects. Reports on pest insects in milk powder. *J. Food Hygenic Society Japan* 16, 1–11.

Young, L.M., Evans, M.L. and Hertel, R. (1990) Correlations between gravitropic curvature and auxin movement across gravistimulated roots of *Zea mays*. *Plant Physiol.* 92, 792–796.

Young, R.E. and Biale, J.B. (1962) Carbon dioxide effects on fruit respiration. Measurement of oxygen uptake in continuous gas flow. *Plant Physiol.* 37, 409–415.

Yu, Y.B. and Yang, S.F. (1979) Auto-induced ethylene production and its inhibition by amninoethoxyvinylglycine and cobalt ion. *Plant Physiol.* 64, 1974–1977.

Yu, Y.B., Adams, D.O. and Yang, S.F. (1980) Inhibition of ethylene production by 2,4-dinitrophenol and high temperature. *Plant Physiol.* 66, 286–290.

Zacarias, L. and Reid, M.S. (1972) Inhibition of ethylene action prevents root penetration through compressed media in tomato (*Lycopersicum esculentum*) seedlings. *Physiol. Plant.* 86, 301–307.

Zacarias, L. and Reid, M.S. (1990) Role of growth regulators in the senescence of *Arabidopsis thaliana* leaves. *Physiol. Plant.* 890, 549–554.

Zacarias, L., Tadeo, F.R., Bono, R. and Primo-Millo, E. (1993) Abscission studies. I. A new mutant of navel oranges. In: Pech, J.C., Latché, A. and Balagué, C. (eds) *Cellular and Molecular Aspects of the Plant Hormone Ethylene*. Kluwer Academic, Dordrecht, pp. 284–290.

Zagory, D. (1992) Choosing the right film; no 'crap shoot'. *Packer/Shipper* May/June, 24–25.

Zanella, A. (2001) Scald control and quality improvement of Granny Smith apples by postharvest treatment with 1-MCP. *Plant Growth Regl. Abstracts* 27, 348.

Zarembinski, T.I. and Theologis, A. (1994) Ethylene biosynthesis and action: a case in conservation. *Plant Mol. Biol.* 26, 1579–1597.

Zauberman, G. and Fuchs, Y. (1973) Ripening process in avocado stored in ethylene atmosphere in cold storage. *J. Amer. Soc. Hort. Sci.* 98, 477.

Zelitch, I. (1969) Stomatal control. *Annu. Rev. Plant Physiol.* 20, 329–350.

Zenk, M.H. (1964) Isolation, biosynthesis, and function of indoleacetic acid conjugates. In: Nitsch, J.P. (ed.) *Régulateurs Naturels de la Recherche Scientifique, Paris*, pp. 241–249.

Zeroni, M. (1972) Ethylene regime during ripening of figs of *Ficus sycamorus* L. PhD dissertation, Tel-Aviv University, Israel.

Zeroni, M., Galil, J. and Ben-Yehoshua, S. (1976) Autoinhibition of ethylene formation in nonripening stages of fruit of the Sycamore fig (*Ficus sycamorus* L.). *Plant Physiol.* 57, 647–650.

Zeroni, M., Jerie, P.H. and Hall, M.A. (1977) Studies on the movement and distribution of ethylene in *Vicia faba* L. *Planta* 134, 119–125.

Zhang, H., Joseph, J., Felix, C. and Kalyanaraman, B. (2000) Bicarbonate enhances the hydroxylation, nitration, and peroxidation reactions catalyzed by copper, zinc superoxide dismutase and hydrogen peroxide. *J. Biol. Chem.* 274, 34576–34583.

Zhang, H., Joseph, J., Gurney, M., Becker, D. and Kalyanaraman, B. (2002) Bicarbonate enhances peroxidase activity of Cu,Zn-superoxide dismutase. Role of carbonate anion radical and scavenging of carbonate anion radical by metalloporphyrin antioxidant enzyme mimetics. *J. Biol. Chem.* 277, 1013–1020.

Zhang, Z.-H., Wang, W.-J., Gao, Y., Fang, Z. and Zhang, Y.-G. (2001) Changes of respiration and endohormones during the fruit ripening of walnut. *Plant Growth Regl. Abstracts* 27, 345.

Zhao, H. and Murata, T. (1988) A study on the storage of muskmelon 'Earl's Favourite'. *Bull. Faculty Agri., Shizouki University* 38, 713 (abstract).

Zhou, D., Kalaitzis, P., Mattoo, A.K. and Tucker, M. (1996) The mRNA for an ETR1 homolog in tomato is consititutively expressed in vegetative and reproductive tissues. *Plant Mol. Biol.* 30, 1331–1338.

Zhu-Salzman, K., Salzman, R.A., Koiwa, H., Murdock, L.L., Bressan, R.A. and Hasegawa, P.M. (1998) Ethylene negatively regulates local expression of plant defense lectin genes. *Physiol. Plant.* 104, 365–372.

Zieslin, N. and Halevy, A.H. (1969) Petal blackening in 'Baccara' roses. *J. Amer. Soc. Hort. Sci.* 95, 629–631.

Zieslin, N., Halevy, A.H., Mor, Y., Bachrach, A. and Sapir, I. (1972) Promotion of renewal canes in roses by ethephon. *HortScience* 7, 75–76.

Zieslin, N., Kohl, H.C., Kofranek, A.M. and Halevy, A.H. (1978) Changes in the water status of cut roses and its relationship to bent-neck phenomenon. *J. Amer. Soc. Hort. Sci.* 103, 176–179.

Zimmerman, M.H. (1960) Transport in the phloem. *Annu. Rev. Plant Physiol.* 11, 167–190.

Zimmerman, P.W. and Hitchcock, A.E. (1933) Initiation and stimulation of adventitious roots caused by unsaturated hydrocarbon gases. *Contrib. Boyce Thomp. Inst.* 5, 351–369.

Zimmerman, P.W. and Wilcoxon, F. (1935) Several chemical growth substances which cause initiation of roots and other growth responses in plants. *Contrib. Boyce Thompson Inst.* 7, 209–229.

Zimmerman, P.W., Hitchcock, A.E. and Crocker, W. (1931) The movement of gases into and through plants. *Contrib. Boyce Thomp. Inst.* 3, 313–320.

Zimmerman, P.W., Crocker, W. and Hitchcock, A.E. (1933) Initiation and stimulation of roots from exposure of plants to carbon monoxide gas. *Contrib. Boyce Thomp. Inst.* 5, 1–17.

Zimmerman, U. and Steudle, E. (1974) The pressure dependence of the hydraulic conductivity, the membrane resistance and membrane potential during turgor pressure regulation in *Valonia utricularis*. *J. Membrane Biol.* 16, 331.

Zimmerman, U., Steudle, E. and Lelkes, P.I. (1976) Turgor pressure regulation in *Valonia utricularis*. Effects of cell wall plasticity and auxin. *Plant Physiol.* 58, 605.

Zippin, J.H., Levin, L.R. and Buck, J. (2001) CO_2/HCO_3^-–responsive soluble adenylyl cyclase as a putative metabolic sensor. *Trends Endocrinol. Metab.* 12, 366–370.

Ziska, L.H. and Bunce, J.A. (1993) The influence of elevated CO_2 and temperature on seed germination and emergence from soil. *Field Crops Res.* 34, 147–157.

Ziv, M., Koller, D. and Halevy, A.H. (1976) Ethylene and the geotropic response of lateral branchers in peanuts (*Arachis hypogaea* L.). *Plant Cell Physiol.* 17, 333–339.

Zobel, R.W. (1973) Some physiological characteristics of the ethylene-requiring tomato mutant *diageotropica*. *Plant Physiol.* 52, 385–389.

Zobel, R.W. (1974) Control of morphogenesis in ethylene-requiring tomato mutant, *diageotropica*. *Can. J. Bot.* 52, 735–741.

Zuzunaga, M., Serrano, M., Martínez-Romero, D., Valero, D. and Riquelme, F. (2001) Comparative study of two plum (*Prunus salicina* Lindl.) cultivars during growth and ripening. *Food Sci. Technol. Int.* 7, 123–130.

Index